Civil War Soldiers
Of
Bedford County
Pennsylvania

The front cover is a colorized picture of an original tin type photograph of Andrew John Himes.

Andrew John Himes died from dysentery at Fort Monroe, Virginia on October 7th, 1864, while serving in the 3rd Pennsylvania Heavy Artillery. Andrew was 29 years old when he died and was buried at the Blackheart Farm Cemetery outside of Everett. His wife, Susan Blackhart Himes, was at home with 2 small children when she received the news of Andrew's death. Their son John Alexander would have been 5 years old and daughter Eutoka Phoebe was born months earlier on June 11th, 1864. It is not known if Andrew had the opportunity to see his newborn daughter before his passing. Susan passed on February 23, 1868, at 28 and was buried beside her husband. Their orphaned children were raised by two different relatives.

Susan Blackhart Himes - Andrew's Wife

Martha Dolbough Himes - Andrew's Mother

John Himes - Andrew's Father

John Himes is pictured at a "Fathers for Fallen Soldiers of the Civil War" event in 1870.

John and Martha's three sons, Samuel, John A. and Andrew, fought for the Union Army in the Civil War.

A special thank you to the late Mildred Himes Lawson, Hope Creighton & Shaun Creighton for sharing the picture on the front cover and of the Himes family on this page. The bibliography contains information on the generous contributors of other pictures in this book.

Civil War Soldiers of Bedford County Pennsylvania by Kevin Mearkle

First Edition

Copyright © 2021 Kevin Mearkle

All rights reserved. No portion of this book may be reproduced in any form without permission from the author, except as permitted by U.S. copyright law. For permissions contact: kevinmearklehistory@gmail.com

*This book is dedicated to the Civil War Soldiers of
Bedford County who gave their lives for our country*

Contents

	Introduction	vi
	Authors Notes on Civil War Research	vii
1.	**Overview**	1
2.	**1861**	6
	1st Bull Run	7
3.	**1862**	9
	1st Kernstown	10
	Edisto Island & Pocotaligo	12
	Fair Oaks	14
	Seven Days Battle	16
	2nd Bull Run	20
	South Mountain	22
	Harpers Ferry	24
	Antietam	26
	Fredericksburg	30
4.	**1863**	35
	Chancellorsville	36
	Gettysburg	38
	Ft. Wagner, SC	44
	Mine Run	46
5.	**1864**	49
	Plymouth, NC	50
	Wilderness	54
	Spotsylvania	58
	Bermuda Hundred & Drewry's Bluff	62
	Cold Harbor	68
	Petersburg Siege Overview	74
	Initial Petersburg Assault	76
	Jerusalem Plank Road	80
	Battle of the Crater	82
	Globe Tavern	84
	Boydton Plank Road	86
	Monocacy	88
	Opequon	90
	Cedar Creek	94
6.	**1865**	99
	Fort Stedman	100
	White Oak Road & Five Forks	104
	Petersburg Final Assault	106
	Sailor's Creek	110
	Appomattox	112
	Andersonville POW Camp	116
7.	**Those who made the Ultimate Sacrifice**	125
8.	**Father & Sons**	149
9.	**Families with Four or More Enlistments**	161
10.	**Youngest Soldiers to Enlist**	183
11.	**Oldest Soldiers to Enlist**	195
12.	**Veterans Living in the 1930's & 1940's**	201
13.	**Medal of Honor Recipients**	213
14.	**Confederate Veterans**	217

Contents

15. Regiments

Regiments Overview	221
8th Pennsylvania Reserves	222
11th Pennsylvania Infantry	226
55th Pennsylvania Infantry	228
76th Pennsylvania Infantry	243
77th Pennsylvania Infantry	246
82nd Pennsylvania Infantry	248
84th Pennsylvania Infantry	250
91st Pennsylvania Infantry	252
99th Pennsylvania Infantry	256
101st Pennsylvania Infantry	261
107th Pennsylvania Infantry	265
110th Pennsylvania Infantry	269
125th Pennsylvania Infantry	273
133rd Pennsylvania Infantry	275
138th Pennsylvania Infantry	284
149th Pennsylvania Infantry	296
171st Pennsylvania Infantry	298
184th Pennsylvania Infantry	302
194th Pennsylvania Infantry	305
205th Pennsylvania Infantry	307
208th Pennsylvania Infantry	309
2nd Pennsylvania Cavalry	316
18th Pennsylvania Cavalry	317
21st Pennsylvania Cavalry	319
22nd Pennsylvania Cavalry	322
McKeage's Pennsylvania Militia	327
1st Maryland PHB Cavalry	330
2nd Maryland PHB Infantry	331
3rd Maryland PHB Infantry	333
U.S.C.T. Regiments	335

16. Gallery of Bedford County Soldiers

Individual & Family Photographs	337
Veterans Group Photographs	355
Last surviving family member	363

17. Unconfirmed Listing of Deaths   364

18. Detailed Alphabetical Listing   365

Photograph Index   544

Acknowledgments   549

Bibliography and Photograph Courtesy Listing   550

Introduction

Bedford County men and boys were casualties on many battlefields during the Civil War. They were citizen soldiers, who months earlier were civilians living with their families in houses and on property, many of us call home today. Some fought and died on the hallowed grounds of the Cornfield at Antietam. Others were cut down during senseless charges on Marye's Heights at Fredericksburg. They were among the badly outnumbered troops, whose heroics on the first day of Gettysburg, enabled the Union Army to seize the crucial high ground surrounding the town. Bedford County citizens were on the extreme right flank of the Union Army during the horrific battle of the Wilderness. They suffered heavy casualties during an early morning assault at Cold Harbor, that Ulysses Grant later admitted was his biggest regret of the war. Large numbers of county soldiers were on the front lines during the nine-month siege of Petersburg. Some languished in hellish prisoner of war camps and many were eyewitnesses to the Confederate surrender at Appomattox. Three county soldiers received our nation's highest honor, the Medal of Honor for actions taken during the war. Some never returned home. Others returned with ruined health or with wounds they would suffer from the remainder of their lives.

The Civil War may appear to be a long-ago event, but as recently as 1930, 94 county Civil War veterans were still living. Martin Myers and Winfield Scott Conrad lived to see a new generation of young men depart their homes during the Second World War. The 1942 obituary of Martin Myers mentioned, he recently "volunteered his services for the national defense." It appears Martin was supporting war bond efforts during the immediate aftermath of Pearl Harbor. As of the publishing of this book, there are many seniors in our communities who were Bedford County citizens at the same time as Civil War veterans. Remarkably, the last known immediate family member of a Bedford County Civil War soldier passed in 2008.

As I was finishing up the research for this book, I read an article in the Bedford Gazette about the last Civil War widow dying on December 16th, 2020. Stories exist about younger women marrying aging veterans, during the depression years of the 1930s, to qualify for widow's pensions, but it was assumed the last Civil War widow died some time ago. Helen Viola Jackson of Missouri kept secret her marriage to a Civil War veteran, including her own family, until 3 years before her death at age 101. The Sons of Union Veterans of the Civil War and other historical organizations reviewed her claims, and all recognized her as the last surviving Civil War widow. She never made a request for a widow's pension and stated at the end of her life, "This is the only man who ever loved me." In some ways, the Civil War did not happen all that long ago.

Ralph Waldo Emerson once asserted, "There is properly no history; only biography." This book provides a sampling of brief biographical glimpses on some of our Civil War soldiers. Around the turn of the 20th Century, the phrase "A picture is worth a thousand words", became part of the American lexicon. It is regretful photographs are not available of each of our Civil War Veterans, and especially for those who never returned home to joyful reunions with loved ones. But we are blessed to have significant numbers of photographs of our Civil War veterans and their families. These available pictures help bring to life the compelling stories of our Civil War generation.

Historian David McCullough once stated, "History is who we are and why we are the way we are." Carl Jung, a founder of analytical psychology, posed the question, "Who has fully realized that history is not contained in thick books but lives in our very blood?" Simply put…our history is part of who we are…

Authors Notes on Civil War Research

The following is an overview of the research resources used in this book.

Civil War Era Records and Information on Bedford County Soldiers

Samuel Bates was appointed by Governor Andrew Curtin and the Pennsylvania Legislature to create a historical record of Pennsylvania's involvement in the Civil War in 1864. Bates took 7 years to complete the monumental task of listing information on over 300,000 Civil War soldiers in the Commonwealth. The Bates 5-volume series titled "History of the Pennsylvania Volunteers, 1861-1865" is a standard reference book for Pennsylvania's Civil War regimental rosters and histories; but the source materials available to Samuel Bates were not error free or complete. The Pennsylvania Historical & Museum Commission website simply states, "information in the Bates series is incomplete." Many Civil War buffs believe Bates may have only captured information on 80% of Pennsylvania Civil War soldiers, according to Ronn Palm, a well-known curator of the Museum of Civil War Images in Gettysburg.

Civil War era record keeping on individual soldiers was lacking for a variety of reasons. Some records were lost, not kept or were incomplete. If a soldier was captured or hospitalized, the fate of the soldier may not have been known to regimental officers maintaining muster rolls. Some officers perished during the war, taking personal knowledge of the fate of some soldiers with them to their graves. Sometimes officers had to rely on memory and uncorroborated information to complete war-end muster reports.

By far, the biggest research challenge in identifying Bedford County soldiers was the lack of residential information of individual soldiers on Civil War documents. The Bates 5-volume books did not list residential information of soldiers. The following is a picture of a typical regimental listing in the book.

The Pennsylvania Civil War Archives contains index Card Records of individual soldiers initially prepared for Samuel Bates to help compile information in his book. Today these Index Card Records are available on the Pennsylvania State Archives website. Index Cards of soldiers are listed in alphabetical order of last and first names.

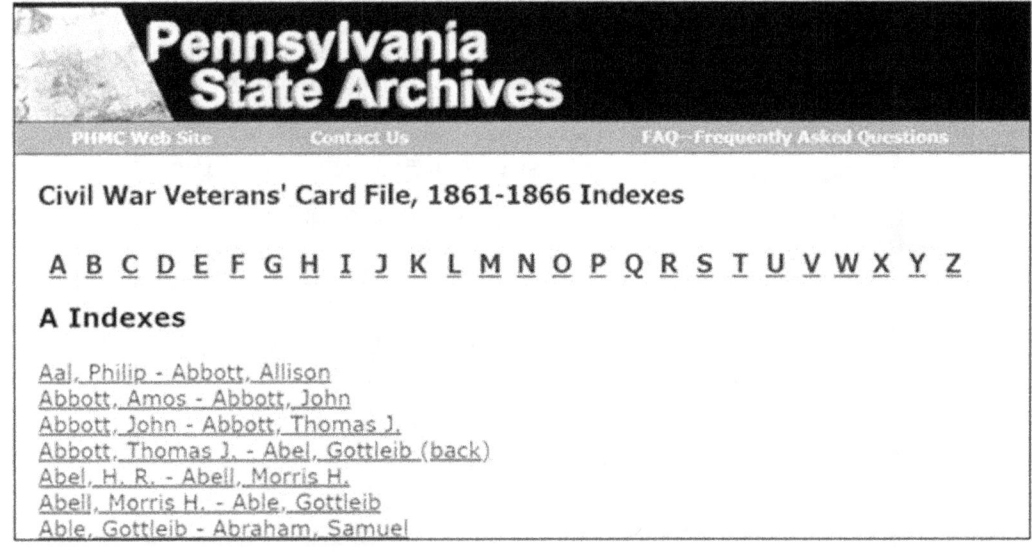

viii *Authors Notes on Civil War Research*

Many Pennsylvania soldiers enlisted at locations outside of the county where they lived. Enlistment locations of soldiers were routinely recorded on muster documents, but residential locations were not. The following is an example of a typical Index Card Record of an individual soldier in the Pennsylvania State Archives. In this example, Abbey, George G of Company H of the 83rd PA Infantry Enrolled (enlisted) on 8-18-61 at Girard, PA. Please note no Residence location is listed.

```
Abbey, George G.           H - 83 I              2 - 1290
Enrolled: 8-18-61          At: Girard, Pa.
M. I: 8-27-61  As: pvt.    At: Erie, Pa.
M. O: 9-20-64
Discharged:
Age at enrollment: 26      Complexion:
Height:                    Eyes:
Hair:                      Occupation:
Residence:
Remarks: Pro. to Sgt. date unknown.
```

Later the Office of the Adjutant General added additional information to some Index Card Records from muster rolls and other Civil War documents, including age at enrollment, physical description, residence, and birthplace. The following is an example of an Index Card Record with more complete information on an individual soldier, including a residence listed as Clarion County.

```
Aaler, Martin              D - 62 I              2 - 467
Enrolled 7-24-61                      at Armstrong County, Pa.
M.I.   7-24-64   As Pvt.              at Pittsburgh, Pa.
M.O.   (7-13-64 Bates)
Discharged:
Age at enrollment 18       Complexion Fair
Height 5-9½                Eyes Brown
Hair Brown                 Occup. Wagoner
Residence: Clarion County
Remarks:
```

Unfortunately, Index Card Records with residential information was more of an exception than the rule. To illustrate this issue, I reviewed the first 500 names listed on Civil War Veterans Index Cards in the Pennsylvania State Archives to determine how many of the Index Cards listed residences. Only 162 out of the first 500 Index Card Records or approximately 1 out of every 3 Civil War soldiers had residential information listed. This book lists 4209 Civil War soldiers who lived in Bedford County or an immediate surrounding community during their lifetimes. Approximately half of these soldiers enlisted for the war outside of Bedford County. Those who enlisted in other areas of Pennsylvania and in other states were much more difficult to identify.

In addition, some information on Civil War records including Pennsylvania Index Cards is not always accurate. The following Index Card Record of Cornelius Rice does not list a residence but states he was born in Huntingdon, PA. Cornelius is my great-great-grandfather and my family continues to live on the same property Cornelius farmed during most of his life. His parents grew up and are buried in Monroe Township near Clearville. I am quite sure neither Cornelius nor anyone in his immediate family ever lived in Huntingdon.

```
Rice, Cornelius            2d K - 78 I           2 - 1073
Enrolled: 2-28-65          At: Hollidaysburg, Pa.
M.I. 3-1-65  As: Pvt.      At:    "         "
M.O.
Discharged
Age at enrollment: 28      Complexion: Fair
Height: 5'8"               Eyes: Grey
Hair: Fair                 Occupation: Farmer
Residence:                 (Born) Huntingdon, Pa.
Remarks: Absent. Sick since 8-17-65 at Bloody Run, Pa.
Sick at MO. (B to)
```

Authors Notes on Civil War Research

Various muster roll documents are available for Civil War soldiers. The following is a picture of a portion of the original Muster-Out Roll of the 101st Pennsylvania Infantry, Company G, completed when the regiment was disbanded at the conclusion of the war. Muster Rolls records were filled out by officers in the regimental company. Muster-Out Rolls include the name, rank, age, muster dates, muster locations and some casualty information.

The following are four examples of individual muster roll documents for John B Amos of the 55th Pennsylvania Infantry. From left to right is a Special Muster Roll showing John was present on April 18th, 1863; a Company Muster Roll for May & June 1864 listing John as being wounded at Cold Harbor on June 3rd, 1864 and was absent when the roll was taken; a Hospital Muster Roll with information on the wound and location where he was being treated; and an Individual Muster Roll listing a promotion to 2nd Lieutenant. Copies of Individual Muster Roll documents and other Compiled Military Service Records (CMSR) are available upon request from the National Archives.

Period newspapers are substantial sources of information. Civil War soldiers periodically provided letters with casualty information to the Bedford Gazette and Bedford Inquirer for publication. Some correspondence from the warfront provided casualty information not listed in other Civil War resources. The following is a portion of a letter sent from Samuel B Schwartz of the 110th Pennsylvania Infantry to the Bedford Inquirer, published on August 12th, 1864.

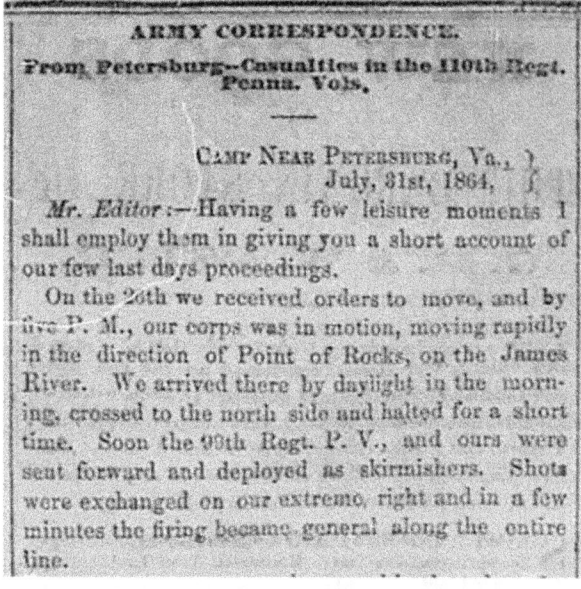

Research Resources generated from the end of the Civil War through the early years of the 20th Century

Civil War Soldiers who were severely disabled during the war were initially eligible for pensions at the end of their enlistment. In 1879 the Arrears Act was passed which provided veterans, who had not previously requested pensions for wartime disabilities, a lump sum payment to cover the time between when they left the military and when they applied for a pension. Veterans still needed to prove they were disabled as a direct result of their time in the service.

Pension request documents submitted to the Bureau of Pensions often contain details on what a Civil War soldier experienced during the war, including wounds suffered at specific battles. Some Pension documents provide details not found anywhere else. Copies of Civil War pension records are available upon request from the National Archives.

Authors Notes on Civil War Research xi

In 1890, a special census was conducted on Union Civil War veterans and their widows to help with the pension claims process. The 1890 Veterans census records were compiled in each township and borough. Census workers routinely asked veterans to provide their Civil War discharge paperwork. Information listed on these census records includes regiment, company, muster dates, and current disability. Some also mention battlefield wounds. The following is a picture of a portion of the Monroe Township 1890 Veterans Census form.

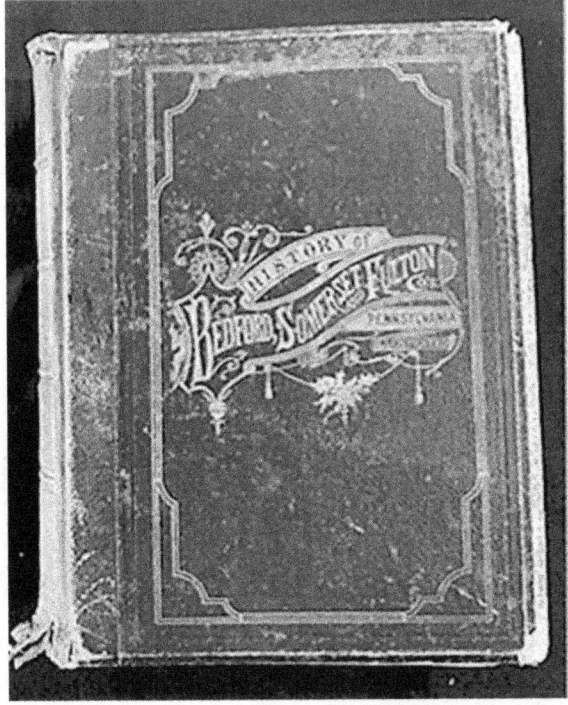

The "History of Bedford, Somerset and Fulton Counties", by Waterman, Watkins & Co. was published in 1884. A 54-page chapter, "War of the Rebellion" provides a brief history, muster, and some casualty information on regimental companies organized in Bedford, Somerset, and Fulton counties. Some county soldiers who enlisted in other counties in Pennsylvania are listed on the last two pages. The book includes drafted soldiers who mustered in regimental companies organized in Bedford County. Census records show some of these soldiers appear to be residents of other counties in Pennsylvania.

The "Biographical review containing life sketches of leading citizens of Bedford and Somerset Counties, Pennsylvania" book was published in 1899. This book provided information on a limited number of Bedford County Civil War veterans.

Floyd Hoenstine's "Soldiers of Blair County" book published in 1940 contains information on some Civil War soldiers who lived near Claysburg and County residents in the southern part of Morrisons Cove. The book also included an overview of some locally recruited regiments.

Some Veterans wrote Regimental History books after the war. These books are substantial sources of firsthand information that include some information on individual soldiers. The following is an example of a book written on the 138th Pennsylvania Infantry Regiment. Most of these books are out of copyright and are available to view at no cost on archive.com.

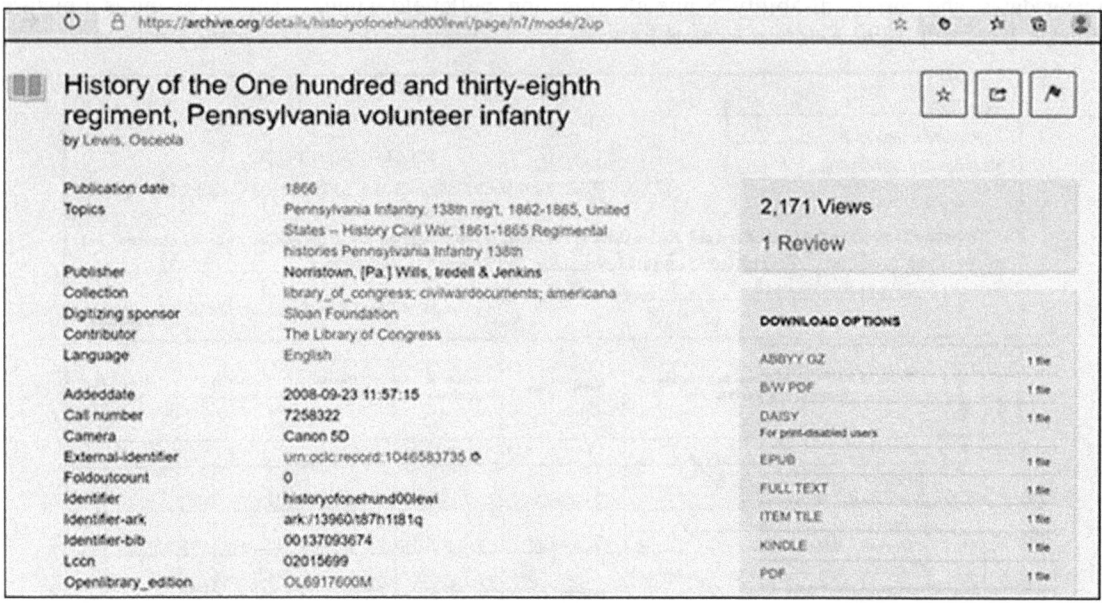

A limited number of books have been published in recent years based on correspondence, and memoirs of local Civil War soldiers are excellent sources of firsthand information.

The Grand Army of the Republic, commonly referred to as the GAR, was the fraternal organization of Civil War Veterans. There were several local chapters in Bedford County. The GAR was officially dissolved in 1956, with the death of its last surviving member. Some GAR documents, including Personal War Sketches of individual soldiers, are available at historical societies and libraries. These documents provide interesting firsthand testimony of wartime experiences.

Local veteran Frank McCoy complied a listing of Bedford County Civil War soldiers buried in Bedford County cemeteries in 1912. He also provided a listing of county soldiers who died during the war, which was published in the Bedford Gazette on February 13th, 1914. Additional listings were published in the Bedford Inquirer on May 22nd, 1908 and Bedford Gazette on May 21st, 1909, that included veterans from other wars. A listing was published in the Bedford Gazette on July 2nd, 1915, of the Civil War soldiers still living in Bedford County. In the early part of the 20th Century, the Bedford Gazette published listings of Veterans who had passed since the previous Memorial Day. These lists were helpful in identifying county Civil War veterans.

Current resources of Civil War Records, Information and Pictures of Bedford County Soldiers

Ancestry.com enabled the identification of many county soldiers. Census record links were especially helpful in determining if a soldier lived in Bedford County or an immediate surrounding community during their lifetime. Ancestry.com provides family tree history details and links to many Civil War records.

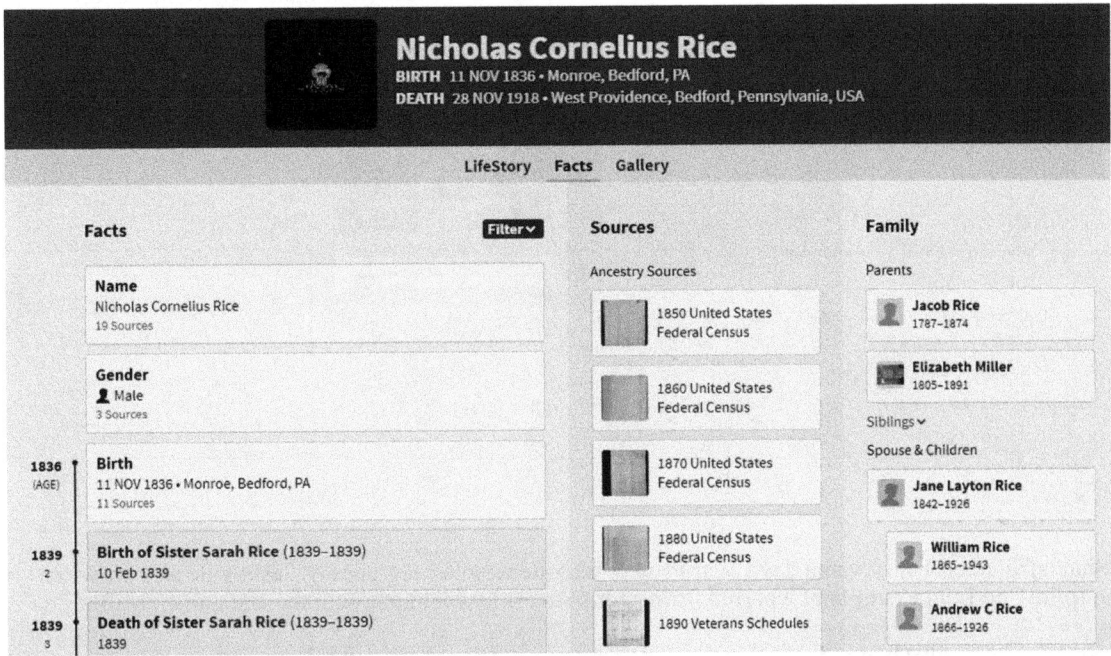

The following is an example of a Muster record on Ancestry linked to Cornelius Rice. Information on this document includes rank, age, enlistment date and location. The remarks section on the roll states Cornelius was sick at Bloody Run (hospital) since August 17th, 1865. Cornelius was deathly sick from Typhoid Fever contracted in Nashville, TN, and was sent back to Bedford County to recuperate in a local hospital.

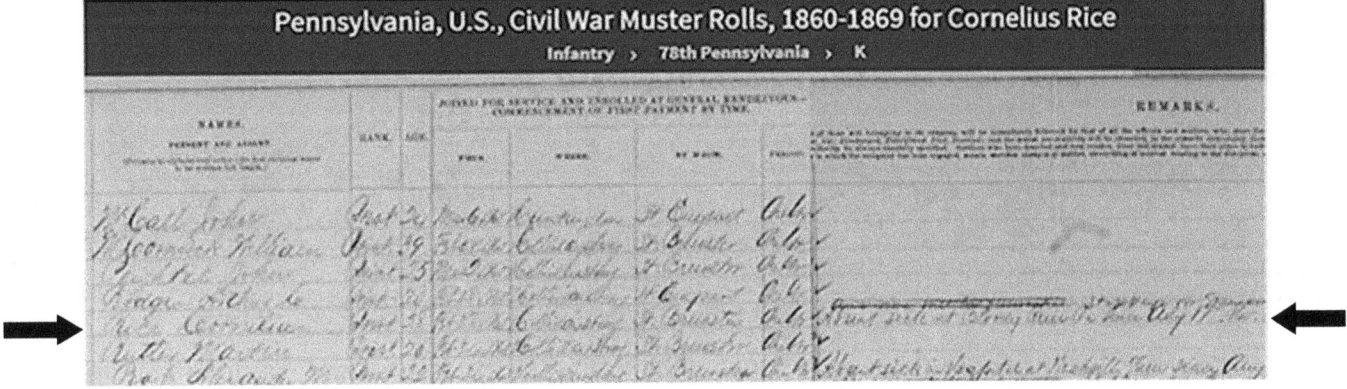

Ancestry includes pension request records on some soldiers. There are other PA soldiers with the name Cornelius Rice listed on Civil War records. Widow's names on these types of records help match a soldier with the correct regiment.

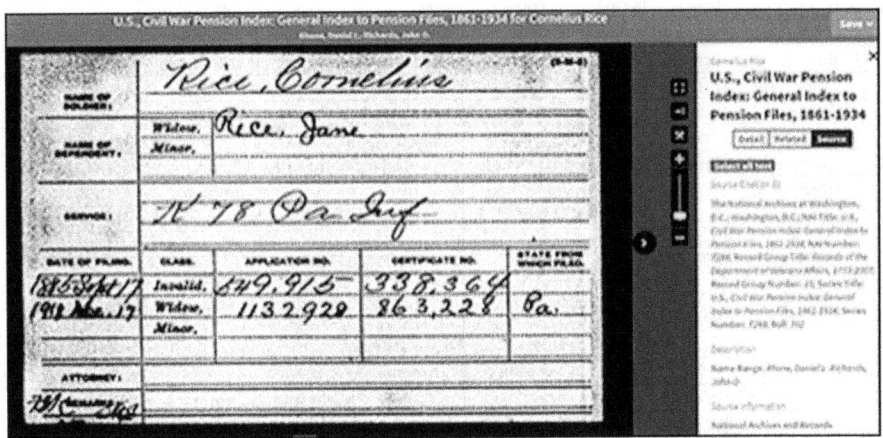

xiv   *Authors Notes on Civil War Research*

The Find A Grave website was useful in determining burial locations, family members, and Civil War service information. Many veterans listed Civil War regiments on their gravestones. Since there were 6 county Civil War soldiers who shared the name William H Miller, photographs of headstones helped match the soldier with the correct cemetery. The structure of the Find A Grave website also enabled visual searches of gravestone photographs at most county cemeteries. Photographs of gravesites with Civil War markers and American flags enabled the identification of many additional soldiers.

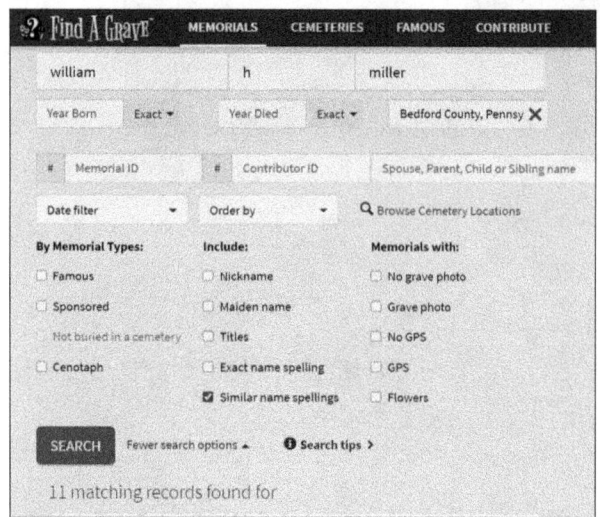
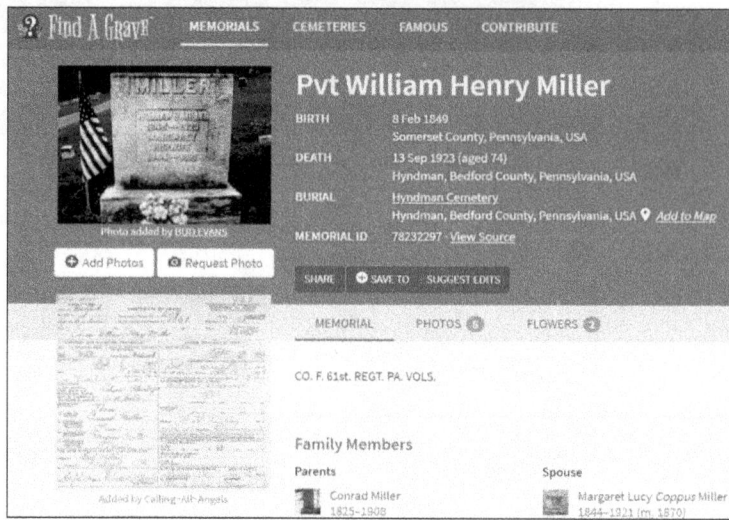

The Penn State affiliated Pennsylvania Civil War Project website enabled searches by last name and provided service record details. The following is an example of a partial listing for soldiers by regiment with the last name "Miller."

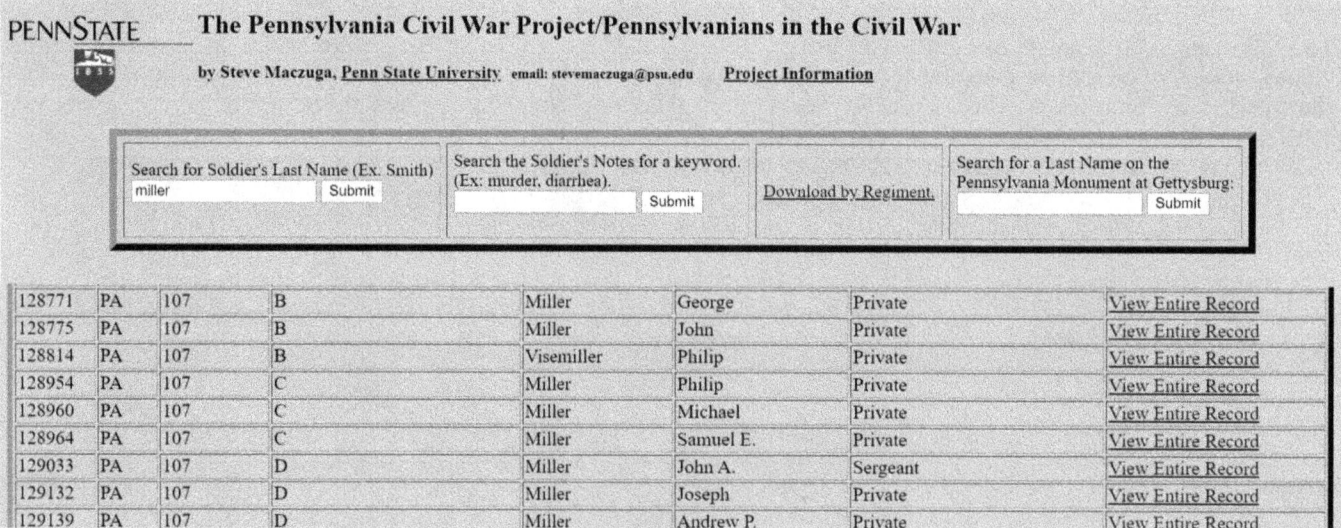

The regimental listings on this website allowed soldiers' names to be sorted in alphabetical order. This report was especially helpful when searching for Civil War records on soldiers with different spellings of last names on various documents. The following is an example of a partial report for the 107th PA regiment.

Reg't	Comp.	Last Name	First Name	Rank	Muster in Date	Notes
107	K	Ackerman	William	Sergeant	January 9, 1862	Promoted to Commissary Sergeant on February 18, 1865.
107	H	Adams	James	Private	August 10, 1864	Substitute - never joined company.
107	I	Adams	Simon	Private	September 20, 1864	Substitute - never joined company.
107	A	Adams	William T.	Private	February 10, 1862	Discharged on Surgeons certificate on August 19, 1862.
107	K	Ague	Alexander	Private	August 26, 1864	Substitute - discharged by General Order on June 7, 1865.
107	A	Albert	Joseph	Private	February 18, 1862	Captured at Weldon R. R., Va. Prisoner - August 19, 1864 to February 27, 1865
107	B	Albic	Joseph	Private	August 3, 1864	Substitute - never joined company.
107	F	Allen	George	Private	March 8, 1862	Wounded at Bull Run, Va. on August 30, 1862, and at Gettysburg, Pa.
107	F	Allen	Jared	Private	March 8, 1862	Deserted on July 13, 1863.
107	E	Allison	George	Private	March 3, 1862	Not on muster-out roll.
107	G	Allison	Joseph	Private	March 12, 1862	Deserted on March 18, 1862.
107	A	Alliton	Thomas	Private	August 17, 1864	Substitute - never joined company.
107	B	Altmyer	Peter	Corporal	June 3, 1864	Missing in action at Dabneys Mills, Va. on February 6, 1865.
107	B	Amey	Henry	Private	September 3, 1864	Substitute - discharged by General Order on June 7, 1865.
107	F	Amos	John	Private	June 28, 1864	Substitute - never joined company.
107	H	Amsley	James	Private	January 9, 1862	Died on July 9, of wounds received at Gettysburg, Pa. on July 1, 1863

Authors Notes on Civil War Research xv

The Historical Data Systems website is an invaluable resource for identification of soldiers and their service records. Historical Data System's data base includes soldiers who enlisted in all states in the Union and Confederate armies. This website and the Pennsylvania Civil War Project website enabled the identification and listed service records of large numbers of county soldiers. Interestingly, both websites contained information on some soldiers not listed on the other's system. The following screenshot shows available search options on the Historical Data Systems website including name, state served, residence (if known), and if a soldier survived the war. In this example, a search was made for William Miller in a Pennsylvania regiment. The search yielded 346 soldiers with the name William Miller who enlisted in Pennsylvania Regiments, including 58 soldiers listed as William H Miller.

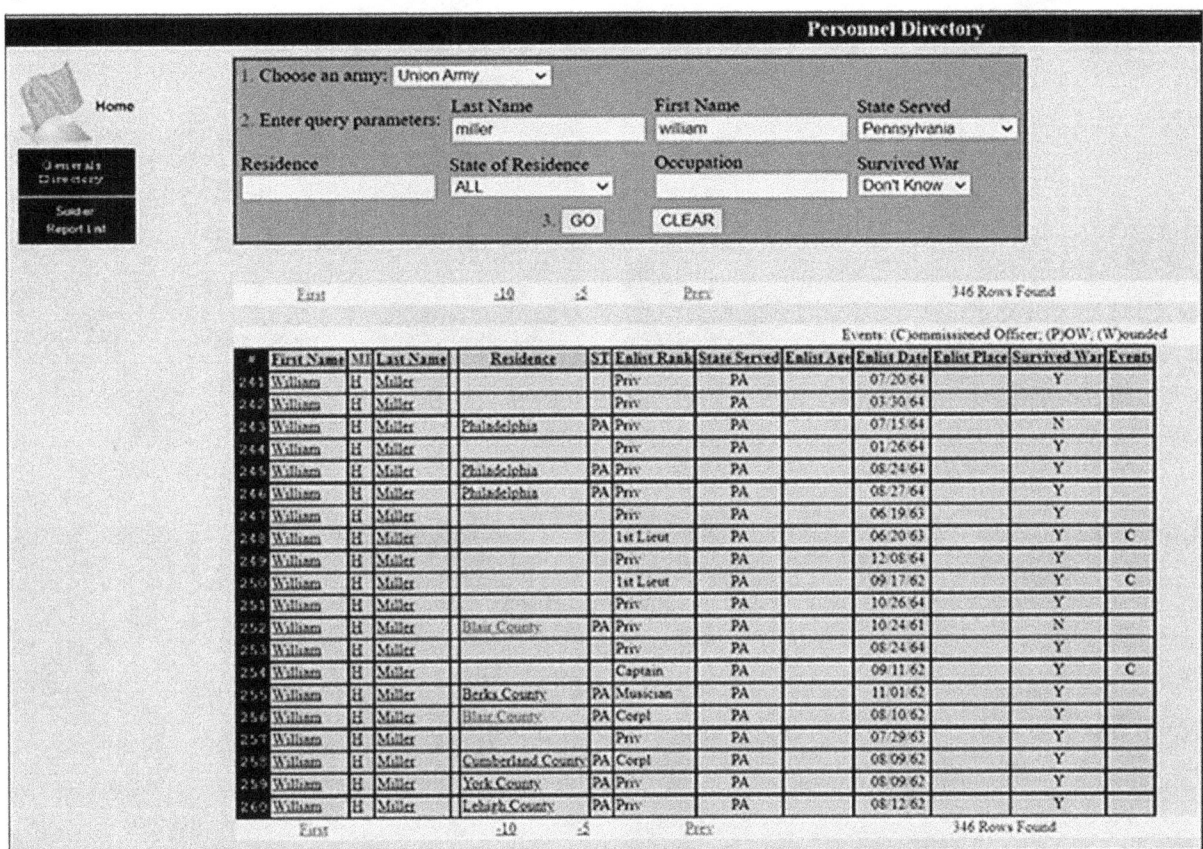

The following is an example of a relatively complete service record of a county soldier, including muster, promotion, and casualty information available on the Historical Data Systems website.

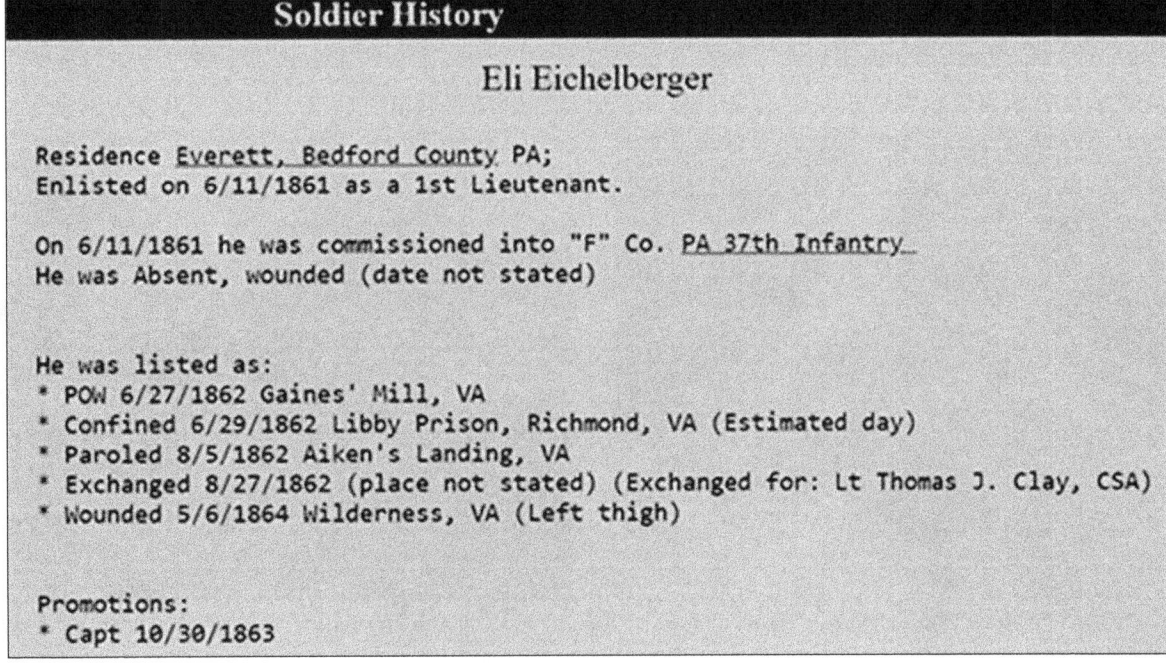

Our Historical Societies, Museums and many smaller Historical Groups & Organizations conserve artifacts, pictures and documents on Bedford County history not available anywhere else. Without local support, this book would have lacked large numbers of pictures and much information. Many excellent books have been published in recent years on local history with the support of our many organizations. The following are some of our historical entities who exhibit local treasures and are open to visitors during the year. Please continue to support those who are preserving our heritage.

Bedford County Historical Society

Bloody Run Historical Society

Fort Bedford Museum

Old Bedford Village

Broad Top Area Coal Miners Museum

Lincoln Highway Heritage Corridor in Latrobe

Authors Notes on Civil War Research   xvii

Please note, locating a soldier's service record from a single Civil War research resource would have been the exception, and not the rule. Generally, it was necessary to search for details on each soldier in multiple resources to compile a more complete service record. Information on each soldier listed in this book was cross-referenced and compiled from the following non-exhaustive listing of resources.

- Bedford County Historical Society - website: http://bedfordpahistory.com/
- Bloody Run Historical Society - website: http://bloodyrunhistory.org/
- Fort Bedford Museum - website: www.fortbedfordmuseum.org
- 1890 Veterans Census records listed by Township and Borough
- Ancestry - website: https://www.Ancestry.com/
- Bedford County Census Records
- Bedford Gazette & Bedford Inquirer published many Civil War Veterans listings in the early 1900s
- Bedford Gazette & Bedford Inquirer published many wartime Civil War articles
- Bedford Gazette and Bedford County Press published many Civil War articles over the last 50 years
- Biographical Review: containing life sketches of leading citizens of Bedford and Somerset Counties, Pennsylvania. Boston: Biographical Review Pub. Co., 1899
- Census Records by Township and Borough
- Civil War Muster Rolls
- Civil War Pension Records
- Civil War Pension Request Records
- Civil War Prisoner of War Records - website: www.civilwarprisoners.com
- Civil War Regimental History books
- Civil Works Administration January 1934 listing on the Bedford Historical Society website
- Find A Grave - website: https://www.findagrave.com/
- Fold3 Military Records - website: https://www.fold3.com
- Frank McCoy 1912 Listing - Soldiers Dead: List of those buried in the cemeteries and graveyards of Bedford County
- GAR - Civil War veteran fraternal organization documents including Personal War Sketches
- Historical Data Systems (listing of Civil War soldiers from all states) - website: http://www.civilwardata.com/
- History of Bedford, Somerset and Fulton Counties, PA by Waterman, Watkins & Co. - 1884
- History of Pennsylvania Volunteers, 1861-1865 by Samuel Bates
- National Park Service - Soldiers and Sailors database - website: https://www.nps.gov/civilwar/soldiers-and-sailors-database.htm
- Newspapers.com - Civil War Veterans Obituaries and Articles - website: https://www.newspapers.com/archive
- Pennsylvania Civil War Project/Pennsylvanians in the Civil War by Steve Maczuga, Penn State University - website: http://personal.psu.edu/~sam21/cw1.html
- Pennsylvania Historical & Museum Commission - website: https://www.phmc.pa.gov/archives/pages/default.aspx
- Pennsylvania State Archives
- Pennsylvania Civil War Archives -Veterans Index Cards- website: http://www.digitalarchives.state.pa.us/archive.asp
- Pennsylvania Volunteers of the Civil War - website: http://www.pacivilwar.com/
- Regimental Muster Records
- Soldiers of Blair County Pennsylvania by Floyd Hoenstine (veterans in Woodbury & Claysburg) - 1940
- U.S Department of Veterans Affairs - Nationwide Gravesite Locator - website: https://gravelocator.cem.va.gov/

The lion's share of photographs in this book were the compliments of the many individual members of the Ancestry & Find A Grave websites, the Bedford Historical Society, and Barbara Sponsler Miller of the Bloody Run Historical Society. Ronn Palm, the curator of the Museum of Civil War Images in Gettysburg, graciously provided many images of individual soldiers from his vast collection. The sources of the photographs in this book are listed in the Bibliography and Photograph Courtesy section of the book.

Civil War Research

Occasionally names are mentioned on a website, a document or a listing as possibly being a Bedford County Civil War soldier with little to no information associated with a name. Please note, there are surprisingly high numbers of Civil War soldiers who share the same name. As previously mentioned, there were 58 Pennsylvania solders with the name William H Miller including 7 who were listed on various resources as possibly being from Bedford County.

Initial Source	Name	Rank	Company	Regiment	State	Unit
Cumberland Valley - 1890 Veterans Census	Miller, William H (1)	Pvt	E	50th	PA	Infantry
History of Bedford, Somerset & Fulton Counties Book - 1884	Miller, William H (2)	Pvt	H	55th	PA	Infantry
Frank McCoy 1912 Listing	Miller, William H (3)	Pvt	A	84th	PA	Infantry
History of Bedford, Somerset & Fulton Counties Book - 1884	Miller, William H (4)	Pvt	G	99th	PA	Infantry
History of Bedford, Somerset & Fulton Counties Book - 1884	Miller, William H (5)*	Pvt	G	93rd	PA	Infantry
Hyndman - 1890 Veterans Census	Miller, William H (6)	Pvt	F	61st	PA	Infantry
Findagrave.com	Miller, William H (7)	Pvt	D	100th	PA	Infantry

Only the 5th highlighted soldier listed above was determined not to have been a Bedford County resident. Even less common names often yielded more than one soldier in Pennsylvania Civil War records and documents. The lack of residential information on Civil War era documents was a fundamental challenge in compiling the listing of Bedford County Civil War soldiers for this book and can also present challenges for descendants who are trying to research information on their Civil War ancestors.

A note of caution for those doing Civil War research, examples were found on Ancestry.com and Findagrave.com where service records of another soldier with the same or similar name was listed. Resources that can help confirm correct information include pension records and pension requests listing the names of family members, newspaper obituaries with Civil War service information, and gravestones engraved with Civil War regiments.

Different spellings of last names can also present challenges. It is not unusual to see different spellings of last names in various Civil War documents and census records. Different spellings of last names made some records difficult to locate and easy to miss entirely.

Confirmation of the battles a Civil War ancestor took part in will require some additional research. The following is the verbiage from the National Archives website on this topic, "Do not assume that a particular individual participated in a battle if (1) his unit was at the battle and (2) the person appears likely to have been with that unit. In the War Department's view, and from a strict adherence to objective information in existing evidence, such an assumption cannot ordinarily be made. No roll call was recorded just before a unit entered battle. As noted above, there are a variety of reasons why a particular individual may not have been present at that time: different companies in the regiment may have had different assignments, or an individual soldier may have been absent due to sickness, desertion, temporary assignment to other duties, or other causes."

According to the National Archives the following is a listing of records which provide very strong evidence that someone was at a battle include the following.

- Records showing death, wounds, or capture at battle.

- Postcards or testimony, found in pension files, wherein the veteran names the battles in which he participated.

- Some Compiled Military Service Records (CMSRs) that specifically record presence at a battle.

- Mention of a person's presence at a battle in the Official Records.

- Mention of participation in battle in a regimental history book.

- Other records, such as a receipt for a horse killed in action.

After 4 years of research, I consider this book to be more of a departure document than a finished work. The unfortunate reality is the identification of some Bedford County Civil War soldiers will probably remain lost to history. It is my hope we can continue to add information to the Civil War records for the known county soldiers. My assumption is much of the additional information to be gained may come from descendants who possess letters, diaries, pictures, pension and CMSR documents of their ancestors. Pension and Compiled Military Service Records (CMSR) often contain personal information, including casualty details not listed anywhere else. Pension & CMSR records can be ordered from the National Archives for a fee. There are 4209 county soldiers identified in this book. Unfortunately, the costs involved in ordering Pension & CMSR records for each identified soldier would be prohibitive. The following is a website link to the National Archives Civil War Records: Basic Research Sources.... https://www.archives.gov/research/military/civil-war/resources

Walking in the Footsteps of Bedford County Civil War Soldiers

Residents of Bedford County live within a relatively short distance of where many county soldiers fought and died during the Civil War. The battlefield of Antietam, just south of Hagerstown, MD is a closer driving distance than Gettysburg. So is Winchester, VA. It would take less than 3 hours to reach Fredericksburg from Breezewood. Richmond is a 4-hour drive during a non-rush hour commute.

Visiting Civil War battlefields offers an opportunity for a memorable family trip and is a wonderful way for parents and grandparents to help pass along the rich heritage of Bedford County to our next generation. Many Civil War sites are along some of the most scenic drives in America. The Shenandoah Valley region surrounding Winchester certainly needs no buildup. Fredericksburg is a historic town and a wonderful place to spend a night or a weekend. George Washington's boyhood home is also located just across the Rappahannock River from Fredericksburg. The historic downtown area of Richmond is well worth a visit and is central to many Civil War battlefields.

Primary Cluster Areas of Major Civil War Battles

The map on the right lists some larger scale battles with Bedford County casualties. There are 4 primary cluster areas of major Civil War battles.

1. Hagerstown/Frederick area battles: Antietam, South Mountain, Harpers Ferry and Monocacy
2. Winchester area battles: 1st Kernstown, Opequon and Cedar Creek
3. Fredericksburg area battles: Fredericksburg, Wilderness, Chancellorsville, Spotsylvania and Mine Run
4. Richmond area battles: Petersburg, Cold Harbor, Fair Oaks, 7 Days Battle and Drewry's Bluff

National Park Service rangers are very knowledgeable and can help pinpoint where regiments of Bedford County soldiers were positioned during a battle. A surprising number of resources also exist on the internet, including regimental maps that show the location and troop movements on Civil War battlefields. On the bottom right is an example of a partial regimental map of troop movements in the Cornfield area of the Antietam battlefield from 6am to 7am.

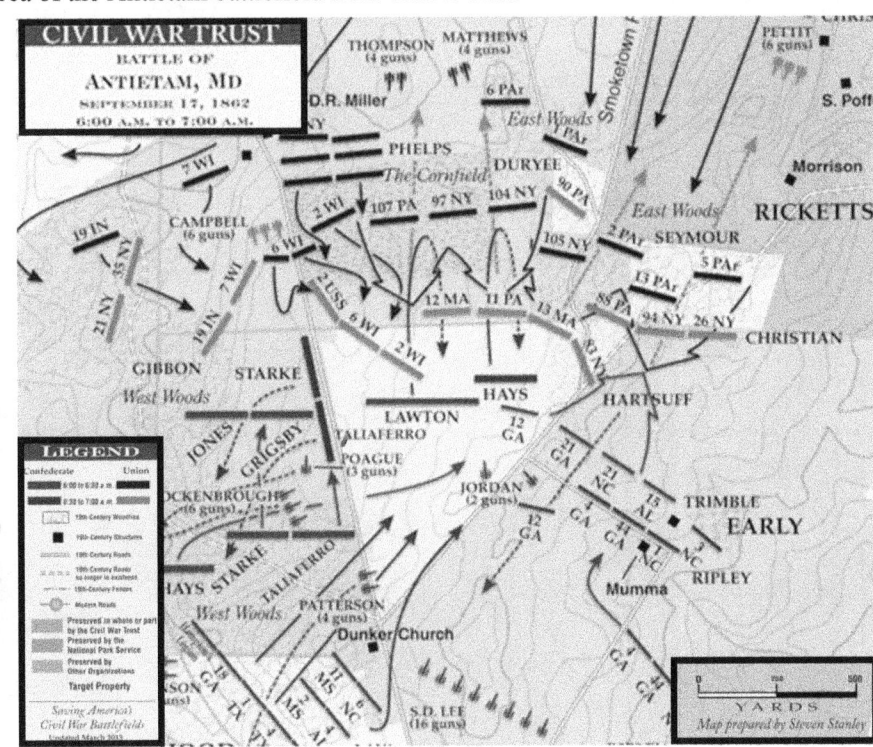

"A soldier dies twice... once whenever they take their last breath and a second time when their name is thought of for the last time"
  - *Wreaths Across America*

1. Overview

"Freedom in not Free" is engraved on the Korean War Veterans Memorial in Washington, DC. Bedford County Civil War Veterans would have needed no convincing of this truth.

The Civil War was by far the deadliest war in American history. Exact numbers of Union and Confederate deaths will never be known because of deficiencies in record keeping. A widely cited historical estimate has been 620,000 soldiers died in the Union and Confederate Armies. Historian David Hacker completed a recent study and calculated a range of between 650,000 to 850,000 likely lost their lives during the war. For a historical comparison, let's use the lower 620,000 estimate. The next highest number of American deaths during a war was 405,399 during World War II. In fact, more Americans lost their lives during the Civil War than in all other military conflicts combined from the Revolutionary War through the Korean War. On a percentage basis, 3% of all American soldiers lost their lives during World War II. During the Civil War, 16% of all Union Army soldiers lost their lives. A detailed comparison of the number of American deaths suffered during the Revolutionary War, Civil War, World War I and World War II and information on the death rates of Bedford County soldiers in some Civil War regiments are listed on page 221.

Another perspective on the loss of life during the Civil War would be in percentage of population terms. The total American population in 1860 was around 31 million people. The U.S. population in 2010 was 308 million, which is about 10 times larger than during the Civil War. As a percentage of population, 620,000 deaths in the 1860s would equate to over 6 million Americas today.

A high number of Bedford County soldiers lost their lives during the Civil War. The total number of identified deaths of soldiers who were living in Bedford County from 1861 through 1866 was 530. The Civil War ended 1865, but some soldiers returned home with severe wounds, ruined health and disease that they died from the year after the war ended. A great western migration had taken place during the years preceding the Civil War. This book lists an additional 26 soldiers who died that were born in Bedford County but fought in Union regiments of other states, including one soldier enlisted in a U.S. Regular Army regiment. Also listed are soldiers who lived in communities just outside of the Bedford County line. There were 29 deaths of soldiers living in immediate border areas including Claysburg, Broad Top, and Crystal Spring. Combined, 585 perished among the 4,209 county soldiers identified in this book equating to 1 in every 7 county soldiers who enlisted in the Civil War lost their lives. There were 43 families who lost more than one immediate family member, including 7 families who lost three of their loved ones and two families who lost a father and a son. Tragedy touched many in Bedford County during the war. Families who did not lose a loved one almost certainly would have known others who were not as fortunate.

Total casualty figures refer to soldiers killed in action, missing in action, died from wounds after the battle, died of disease, wounded, captured or injured during a battle. There were 1,491 known casualties among the 4,209 county soldiers identified in this book. This known Bedford County Civil War soldier casualty rate of 35% is extraordinarily high by any historical measure in American history.

There were many reasons for the shockingly high casualty numbers, including significant technological advances in weaponry in the years leading up to the Civil War. Smooth-bore muskets fired round lead balls and had an effective range of 50 to 100

Unidentified Union soldiers along the west bank of the Rappahannock River at Fredericksburg, Virginia in 1863

yards. Smooth-bore muskets were still being used early in the Civil War, but both armies transitioned to the more deadly .58 caliber rifled muskets with an effective range of over 300 yards.

The standard rifle munition during the war was the recently invented Minie Ball. This soft lead conical shaped bullet flattened out upon hitting the human body causing horrendous gaping wounds. When a Minie ball shattered an arm or leg bone, amputation was often necessary. Estimations were as high as 1 in every 13 Civil War soldiers returned home without one or more limbs. Unfortunately, for Civil War soldiers, weapons had changed, but military tactics had not. Charges were much more deadly when facing enemy rifle fire, with over three times the effective killing range of the muskets of previous wars. Other advances in weaponry also proved devastating. Many massive frontal assaults made by both armies ended as horrific vortexes of death during the war.

Other technological advances in the 19th century likely led to higher casualties. Railroads were relatively new when the war broke out, over 75% of the 30,000 miles of rail lines used during the Civil War did not exist before 1850. Railroads enabled both armies to rush men and supplies over long distances. More battles were being fought with larger numbers of soldiers resulting in greater numbers of casualties. Some Bedford County soldiers were rushed to the warfront by train and became battlefield casualties less than a month after mustering in the army.

Mobile telegraph wagons provided near instantaneous communications from the warfront. Over 15,000 miles of telegraph lines generated 6.5 million messages during the Civil War, increasing the velocity of warfare. For the first-time, the instruments of war included hot air balloons, submarines, and ironclad ships. The actions of one Confederate ironclad ship in North Carolina led to 52 Bedford County soldiers becoming prisoners of war in one day in 1864, 14 of the captured soldiers later perished in Confederate Prisoner of War camps.

Photography was another recent invention that had a profound impact on the war. For the first time, battlefield photographs provided a visual record of the unprecedented carnage at the battle of Antietam. When the American public viewed the shocking images of bloated, lifeless bodies of dead soldiers, civilian perceptions of the war changed.

Conversely, some lifesaving advances in medicine and a basic understanding of the importance of hygiene were a generation or two away. These unfortunate realities, combined with a lack of clean drinking water in massive, overcrowded camps, led to tragic consequences for many Civil War soldiers. Disease proved to be more deadly than bullets. Nearly 400,000 soldiers in both armies died from disease, almost twice the number of those who lost their lives from battlefield wounds. Over 150,000 soldiers died from Dysentery and Typhoid from drinking contaminated water. Other major causes of death were Pneumonia, Measles, Malaria and Tuberculosis. Hygiene was all but non-existent. Doctors and nurses rarely cleaned their hands or medical equipment between patients. Penicillin did not exist until after World War I, so any infection could be deadly. When a soldier contracted a disease, doctors often had no effective way to treat it. The total known battlefield deaths for Bedford County soldiers were 237. Many more died from disease, including the men and boys who perished in hellish Confederate POW camps.

The 1912 reunion of Bedford County Veterans of the 55th PA Infantry Regiment. The 496 men who enlisted in the 55th from Bedford County suffered 272 casualties, the highest number of county casualties of any regiment during the Civil War.

Western Theater

Eastern Theater

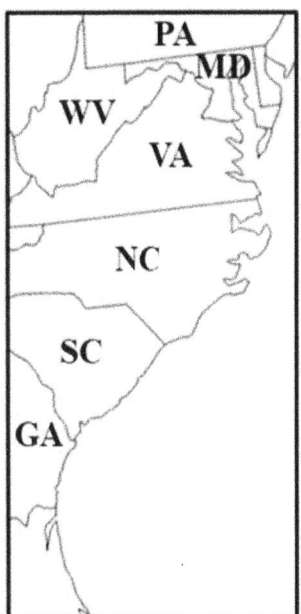

The American Civil War took place in two primary regions, referred to as the Western and Eastern Theaters. With few exceptions, Bedford County Soldiers enlisted in regiments who fought in the Eastern Theater, because of geographical proximity reasons.

Most Union soldiers in the Western Theater were from Ohio, Indiana, Illinois, Wisconsin, Michigan, Minnesota, Iowa and the border states of Kentucky and Missouri. In the years prior to the Civil War, families from Bedford County migrated to these states. When the Civil War broke out, some men who were born and raised in Bedford County enlisted in the regiments of their new home states.

The 77th Pennsylvania Infantry Regiment enlisted recruits from Huntingdon, Fulton, Blair Counties, and some men who lived in the border areas of Bedford County. The 77th Pennsylvania Infantry was organized in Pittsburgh on October 15th, 1861 and left 3 days later for Louisville, KY, becoming one of the few regiments with local recruits to fight in the Western Theater. In early 1865, many 78th Pennsylvania Infantry recruits from the county traveled to Nashville, TN to replace other 3-year enlistment soldiers who mustered out of the Union Army.

Men and boys who had lived in Bedford County prior to the war were participants in the most famous battles in Western Theater. Some became casualties at Chickamauga (GA), Corinth (MS), Franklin (TN), Ft. Donelson (TN), Kennesaw Mountain (GA), Shiloh (TN) and Vicksburg (MS). Union troops in the Western Theater experienced great success during the first two years of the Civil War, culminating with the Confederate surrender of Vicksburg on July 4th, 1863. This book lists the identified Bedford County soldiers in the Western Theater.

Over 95 percent of the battlefield casualties of Bedford County soldiers took place in the Eastern Theater. The lion's share of these battles took place in Virginia, Maryland, and Pennsylvania. Bedford County is located close to two of the most consequential battles of the Civil War. Robert E Lee planned two invasions of Pennsylvania hoping to win a decisive battle on Northern soil and force Abraham Lincoln to end the war. Lee was turned away at Antietam, near Hagerstown, which enabled Lincoln to issue the Emancipation Proclamation freeing all slaves in the Confederacy. More Americans died at Antietam on September 17th, 1862 than on any other day, in any war, in our nation's history. The greatest battle ever fought on American soil is just down the Lincoln Highway from Bedford County. Robert E Lee was defeated at Gettysburg and never again threatened the North with a major invasion. There were many other epic battles fought within a short distance of Bedford County. This book summarizes many of the major battles in the Eastern Theater. Bedford County soldiers suffered significant numbers of casualties during most of these battles. The primary Union Army in the Eastern Theater was the Army of the Potomac. It was heavily engaged with the primary Confederate Army in the East, known as the Army of Northern Virginia. The focus of this chapter is solely on the battles in the Eastern Theater of the war. For simplicity in the narrative of this book, I shortened the name of the "Army of the Potomac" to the "Union Army," and the "Army of Northern Virginia" to the "Confederate Army."

Civil War battles were often referred to by different names in the North and South. In the North, battles were named after a close-by body of water or another prominent physical feature near the battlefield. The South generally referred to the same battle by the name of the closest town. For example, the battle of "Bull Run" was the common name used in the North, while the South referred to the same battle as "Manassas". This chapter lists the additional names referring to the same battle.

Bedford County Township & Borough Map

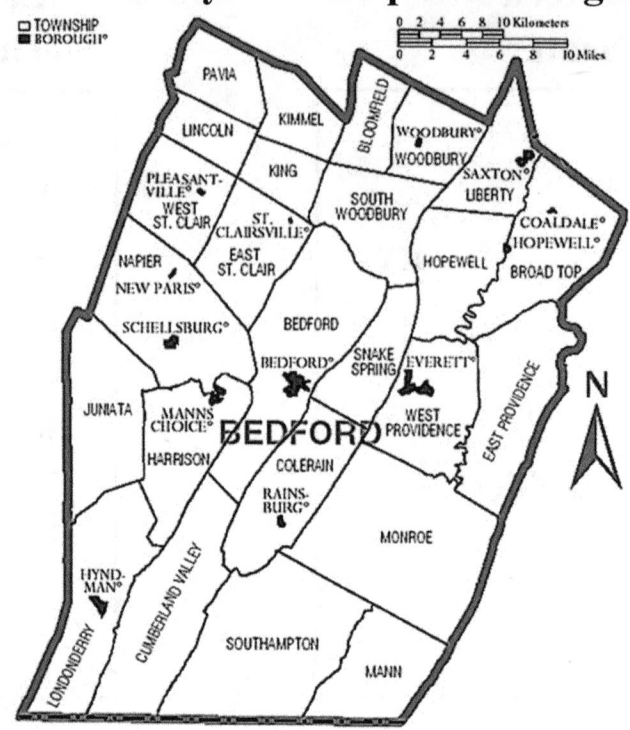

Residential information of individual soldiers and their families listed in this book is often sourced from Township and Borough Census records. Please note the following name changes from the Civil War era. Bloody Run officially changed its name to Everett in 1873. Union Township changed to Pavia Township in the 1990s.

The following is the basic Union Army organizational structure and number of soldiers in each unit at full strength. On a battlefield, a Company or Regiment may have had half or less the number of men listed below because of disease and prior battlefield casualties.

Company - 100 soldiers

Regiment - 10 Companies or 1000 soldiers

Brigade - 2 to 5 Regiments

Division - 2 to 3 Brigades

Corps - 2 to 3 Divisions

Army - 2 to 3 Corps

Many Regimental Companies recruited soldiers who lived in the same city, township or county, but Civil War soldiers most often identified with their Regiment. During the Civil War, Regiments were the basic maneuver unit in the army. This book lists Regiments as the primary fighting unit for individual soldiers.

Infantry regiments, cavalry regiments and artillery batteries were the 3 primary army groups in the Civil War. Some Pennsylvania infantry, cavalry, and artillery regiments can be referred to by two names. The following is a brief overview of the regimental numbering system, which can be confusing.

The numbering system for Pennsylvania regiments was the chronological order of being organized, regardless of whether the units were infantry, cavalry, or artillery. The first 29 regiments organized in Pennsylvania were for 90-day enlistment periods and were numbered the 1st through the 29th Pennsylvania regiments. There were plans for additional non-activated regiments referred to as Reserves. When the Southern insurrection was not quickly quelled, the Reserve units were activated, mostly on 3-year enlistment periods. There were 15 Pennsylvania Reserve Infantry Regiments numbered the 1st through the 15th Reserves. These regiments were also referred to as the 30th through 44th Pennsylvania regiments organized during the Civil War. For example, the 1st Pennsylvania Reserve Regiment could also be referred to as the 30th Pennsylvania Regiment, the 2nd Pennsylvania Reserve Regiment as the 31st Pennsylvania Regiment, and so forth for all 15 Reserve units. The only Reserve Company recruited directly out of Bedford County was Company F of the 8th Pennsylvania Reserves. This unit is the well-known "Hopewell Rifles" Company. The army also referred to the 8th Pennsylvania Reserves as the 37th Pennsylvania Regiment. There are examples of county soldiers enlisted in different infantry regiments with the same numerical name. The 11th Pennsylvania Infantry Regiment enlisted 36 Bedford County soldiers and the 11th Pennsylvania Reserves Regiment enlisted 7 county soldiers. Reserve regimental number names are more commonly used for those units. For simplicity purposes, in the narrative of this book, all Reserve infantry regiments are referred to as "Reserves" and non-Reserve infantry regiments as "Infantry."

There were 22 Pennsylvania Cavalry Regiments organized during the Civil War, along with some Independent units. After the surrender at Appomattox, some additional short-term Provisional Calvary units formed. Pennsylvania calvary units also have both a calvary regiment number and a regiment number in the chronological order of organization. For example, the 22nd Pennsylvania Cavalry could also be referred to as the 185th Pennsylvania Regiment. This book will refer to only the Cavalry Regiment names. Please note early in the war, some cavalry and infantry units were in the same regiment before being separated.

Most Artillery units also had an artillery regiment number and a regiment number in chronological order of organization. The following are two examples. The 1st Pennsylvania Light Artillery was initially part of the 14th PA Reserve Regiment and was the 43rd PA Regiment organized, and the 2nd Pennsylvania Heavy Artillery was also the 112th Regiment organized. This book will only reference the Artillery Regiment and Battery number names. Information on some Independent Artillery units are also listed.

Union Army enlistments periods mostly ranged from 90 days to 3 years. When the Rebel army threatened Pennsylvania during the Antietam and Gettysburg campaigns, civilian volunteers joined emergency Pennsylvania militias with shorter enlistment periods. The Enlistment and Muster date were often the same, but it was not unusual for a soldier to Muster in the army a few weeks after enlisting. A soldiers Muster dates are primarily listed in this book.

The following is a partial list of abbreviations in this book. Casualty: "KIA" - Killed in Action, "Wound.-Died" - Wounded and Died after the battle, "POW" - Prisoner of War, "Inj." - Injured, "POW-Died" - Died after being taken prisoner, and "MIA" - Missing in Action (presumed to have been killed in battle or died in a POW camp). Rank: "Pvt." - Private, "Corp." - Corporal, "Sgt." - Sergeant, "Lt." - Lieutenant, "Capt." - Captain. During the Gettysburg campaign, many county soldiers enlisted in McKeage's Militia. The abbreviation for this Militia in the book is "Mil.". Many African American soldiers from the county volunteered in United States Colored Troops regiments. "USCT" was the common abbreviation used during the Civil War and is still the abbreviation commonly used today. Some Independent Cavalry and Artillery regiments were organized during the war. The abbreviation used in the book for these units is "Ind." regiment. The following is a clarification on Age at Muster: Some soldiers enlisted in multiple regiments during the war. Most of the Muster Ages listed is the age of a soldier when they first joined the army. I also listed approximate ages at first muster when only a birth year record is available. Rank in this book is listed as the highest rank a soldier achieved up to the Confederate surrender at Appomattox on April 9th, 1865. Some Civil War Soldiers from Bedford County shared the same name. Soldiers who share the exact same name are differentiated by a number within parentheses added to the end of each name.

There are likely many more casualties suffered by Bedford County soldiers than listed in this book. The Bedford Historical Society and some descendants on Ancestry.com provided copies of Civil War pension documents listing battlefield wounds and prisoner of war references not found in other available Civil War records. It appears many non-debilitating battlefield wounds treated at regimental field hospitals are not listed on widely available records. Prisoner of War records also appear to be incomplete. The following is an example. There were over 90 Bedford County soldiers enlisted in the Hopewell Rifles Company of the 8th Pennsylvania Reserves in the spring of 1862. Lieutenant James Cleaver stated two-thirds of the Hopewell Rifles were captured during the Union Army morning withdrawal to Gaines Mill on June 27th, 1862. The widely available Civil War records list only 38 Hopewell Rifles Company soldiers being captured. It is possible some soldiers were being treated for sickness in a hospital during the battle, but 38 Prisoners of War on that day appears to be a low number.

James Cleaver is pictured on the right after the Civil War. He was born in England in 1838. His family immigrated to America in 1845. His father, Reverend Charles Cleaver, was the minister at the Barndollar Methodist Church in Bloody Run (Everett) for 2 years, starting in 1860. James mustered as a 1st Sergeant in the Hopewell Rifles company of the 8th Pennsylvania Reserves on June 11th, 1861. He was captured, along with many Hopewell Rifles soldiers during the withdrawal to Gaines Mill in June 1862. James spent 42 days in the Libby prisoner of war camp in Richmond. He suffered a leg wound at the battle of Fredericksburg on December 13th, 1862. James was still recovering in a hospital in Washington during the battle of Gettysburg in early July 1863. He received a promotion to a 2nd Lieutenant in October 1863. James was wounded a second time on May 10th, 1864, during the battle of Spotsylvania Court House. The 3-year enlistment for the 8th Reserves Regiment ended a few weeks later and James returned home to Bloody Run. James married 1865. After his first wife died, he remarried in 1885. James was elected Register and Recorder of Bedford County in 1872 and served as a Prothonotary of Bedford County from 1896 to 1902. He passed at 72 and was buried at the Bedford Cemetery in 1911. His obituary stated he received 4 wounds during the Civil War. Widely available records for James only listed being wounded twice. This is a good example of wounds not always being recorded.

2. 1861

Rebel forces bombarded a federal fortification in Charleston Harbor on April 12th and forced the surrender of Fort Sumter two days later. Abraham Lincoln made an urgent request for 75,000 troops to put down the Southern Rebellion on April 16th, 1861. The first volunteer company organized in Bedford County to respond to Lincoln's request was Company G of the 13th Pennsylvania Infantry Regiment. This unit was referred to as the Taylor Guards, named after Zachary Taylor, the 12th President of the United States and hero of the Mexican American War. The Bedford Inquirer published a "Patriotic Meetings" article on April 26th describing community leaders making eloquent and patriotic speeches at several impromptu meetings in Bedford, Bloody Run, East Providence Township, Schellsburg and Rainsburg. Many volunteers enlisted with the Taylor Guards during these rallies. This article ended with the following proclamation, "Never before in the history of our County was there so much enthusiasm on any subject as there is on our present difficulties, and the determination to uphold the Government in this war forced upon it by the Southern traitors." Most Pennsylvania regiments organized in early 1861 were for 90-day enlistment periods because no one expected the war to last long. Many soldiers on the following listing volunteered a second time in other regiments, after their initial enlistments. Some would later become casualties in a war that lasted far longer and was much bloodier than anyone in Bedford County imagined in April 1861. In 1861, 16 county soldiers died from disease and there were no known battlefield casualties.

1st Bedford County Volunteers of the Civil War and the PA regiment joined

Name	Regiment
Armstrong, Thomas	13th
Baker, George W	2nd
Barndollar, John W	13th
Barndollar, William P	13th
Barr, John	3rd
Barr, Reuben	14th
Bartholomew, Borchiel	13th
Bartlebaugh, M	14th
Bartlebaugh, Philip	14th
Bartlebaugh, Silas M	14th
Beck, Henry	7th
Boehm, John W	13th
Boor, William A	13th
Boring, Henry J	14th
Borland, Zachariah	13th
Bowers, John	13th
Bowman, William	13th
Bradley, Alexander	13th
Brown, Jacob	14th
Brown, Jeremiah	13th
Burket, George S	14th
Cook, John F	13th
Copenhaver, David A	27th
Cypher, George W	5th
Dasher, James H	10th
Davis, James W	13th
Davis, Richard	13th
Deffenbaugh, Samuel S	5th
Defibaugh, Harrison	13th
Elder, Samuel	13th
Elliott, David S	13th
Engle, Henry	14th
Filler, John H	13th
Filler, William T	13th
Gates, Martin V B	14th
Gates, Theophilus R	13th
Gollipher, Justice	13th
Grimes, John C	14th

Name	Regiment
Guy, Robert	13th
Hafer, Alexander H	13th
Hafer, William H	13th
Hartman, Frederick	13th
Helm, John B	13th
Helsel, Henry S	3rd
Hildebrand, Alexander	13th
Hill, Aaron	13th
Hissong, Josiah	13th
Hornig, Frederick	13th
Ickes, Adam	14th
Jacoby, Edward	13th
Karchner, David	13th
Karder, William	13th
Kay, Ezra P	13th
Kay, Harry H C	13th
Kegg, Jacob	13th
Kelly, John T	13th
King, Hezekiah	3rd
Kreiger, John	13th
Lightmigator, Ernestus	13th
Lightningstar, Augustus	13th
Little, James	14th
Lowery, Samuel	13th
Malone, John	14th
Manges, Abraham	14th
Mardis, Samuel	14th
Mauk, Joseph W	14th
May, Marcus	13th
McIlnay, John F	14th
McGirr, Matthias	13th
McKee, David	14th
McMurtrie, James	5th
McQuillen, Hiram	13th
Medley, William	13th
Mellen, Thomas	13th
Miller, Clement R	13th
Miller, John H	13th

Name	Regiment
Mohn, Frederick	13th
Moore, John B	13th
Mopps, Edward S	13th
Mordus, Samuel	14th
Mower, Alexander C	13th
Mullin, David W	13th
Munshower, George W	13th
Nulton, William H	13th
Pearson, Edward P	25th
Peck, Jacob B	13th
Penn, William J	13th
Pilkington, James H	13th
Pilkington, Richard P	13th
Poorman, Franklin H	13th
Price, William	3rd
Prosser, Alexander	2nd
Rabe, Frederick W	13th
Ramsey, Eli B	13th
Ramsey, Oliver C	13th
Rinard, Samuel	5th
Saupp, James	13th
Shellar, William	13th
Shoeman, David	14th
Slack, Francis M	13th
Statler, William	13th
Steel, David S	13th
Steinman, Mathew C	15th
Tate, Samuel B	13th
Thomas, Joseph A	13th
Tobias, Calvin	5th
Tobias, Samuel H G	13th
Washabaugh, William H	13th
Wentling, George	13th
Williams, Richard	13th
Wilson, Hugh	13th
Wilson, William	13th
Wolf, William H	25th
Wonech, Michael	13th

1st Bull Run - July 21, 1861

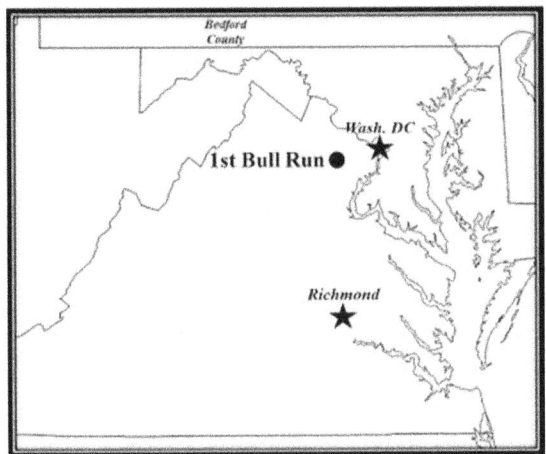

Battle also referred to as
First Manassas

No Known Bedford Co. Casualties

Two soldiers kneeling along Bull Run. The barely visible structure in the background is the Sudley Church, which served as a field hospital during the battle.

1st Bull Run was the first major battle of the Civil War. The two major railroad lines intersecting at Manassas Junction, 30 miles southwest of Washington, had great strategic significance for both the Union and Confederate Armies. Control of railroad lines would remain hotly contested during all four years of the war. On the morning of the battle, hundreds of Washington civilians famously jumped in their buggies and headed to Manassas to picnic and experience the spectacle of the Union Army putting down the Southern insurrection. The battle ended with Confederates charging down a hill, projecting a full-throated version of the soon to be famous "Rebel Yell" as Union Army lines crumbled. 1st Bull Run was as a complete fiasco for the Union Army. Union troops fled the battlefield in a full, disorganized retreat the entire way back to Washington, intermixed with civilians, horrified at the bloodshed they had just witnessed. Union Army casualties totaled 460 KIA, 1,124 wounded, and over 1,300 missing or POW. The Confederates lost 387 KIA, 1,582 wounded with 13 missing or POW. The only Pennsylvania regiment on the battlefield at 1st Bull Run was the 27th Infantry. David A Copenhaver was enlisted in the 27th PA Infantry during the battle, but no documentation is available to confirm he was on the battlefield that day. Little else is known about David other than being listed as a Saxton resident on the 1890 Liberty Township Veterans Census. 1st Bull Run would be the last larger scale battle in the Eastern Theater in which loved one's in Bedford County did not receive news of a father, a brother or a son being a casualty.

3. 1862

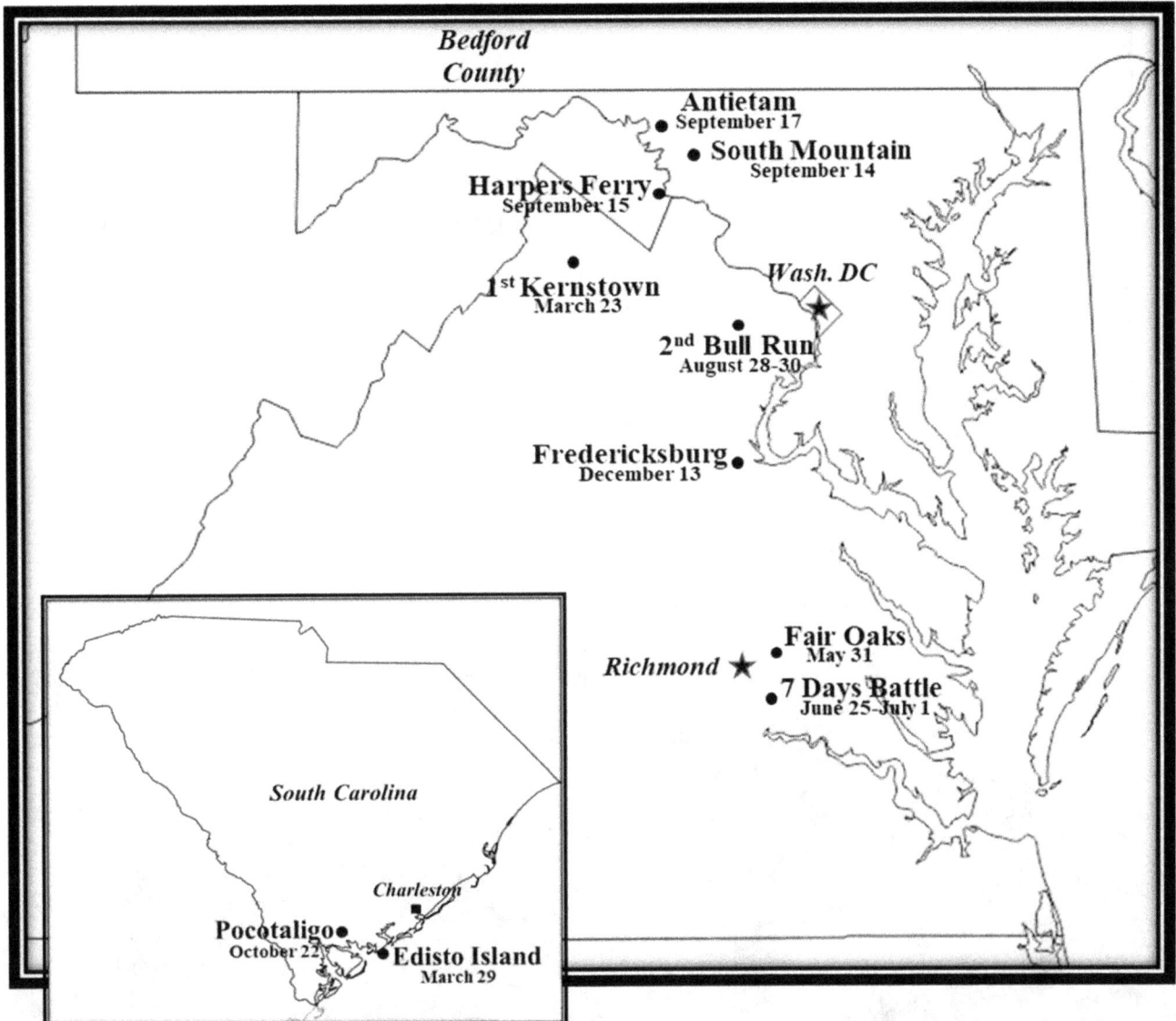

The Civil War took a deadly turn in 1862. Epic clashes between the Union and Confederate armies resulted in unprecedented numbers of dead and wounded lying on battlefields. As the size of their armies dramatically increased, shocking numbers of soldiers began dying from disease in large unsanitary camps. 1862 was also a terrible year for Bedford County. In 1862, there were at least 231 battlefield casualties, including 50 soldiers who died from their wounds. Another 70 county soldiers died from disease. The dates of deaths and casualties for some soldiers are not listed on available records; therefore, it is likely the total number of deaths and casualties were higher in both 1861 and 1862 than is listed.

After the disastrous defeat at 1st Bull Run, Abraham Lincoln installed George McClellan as Commander of the Union Army in the Eastern Theater. McClellan showed a skill set for organizing and training an army and was popular with his troops. In November 1861, Lincoln appointed McClellan as general-in-chief of all Union armies in the Eastern and Western theaters. Over time Lincoln became increasingly frustrated with McClellan's lack of action in prosecuting a war against the Rebel army. McClellan habitually overestimated the strength of the Confederate forces and made repeated requests for more troops before advancing on the Rebel army. In March 1862, Lincoln relieved McClellan of his general-in-chief responsibilities and ordered McClellan to focus on mounting an offensive on Richmond, the capitol city of the Confederacy. McClellan transported over 100,000 soldiers by ship to a Union controlled installation on the Chesapeake Bay east of Richmond, bypassing much of the Confederate army concentrated in the northern part of Virginia. This offensive became known as the Peninsula Campaign.

Later, in 1862, two of the most famous battles of the Civil War took place at Antietam and Fredericksburg. One battle was a major setback for the Confederacy and the other a disaster for the Union Army.

In 1862, a separate Union Army group in South Carolina faced off with Confederate at Edisto Island and Pocotaligo. Another Union Army group in the Shenandoah Valley battled the troops of Stonewall Jackson during the spring of 1862. Inexperienced soldiers suffered the first battlefield casualties from a unit recruited in Bedford County during the 1st Battle of Kernstown.

1st Kernstown - March 23, 1862

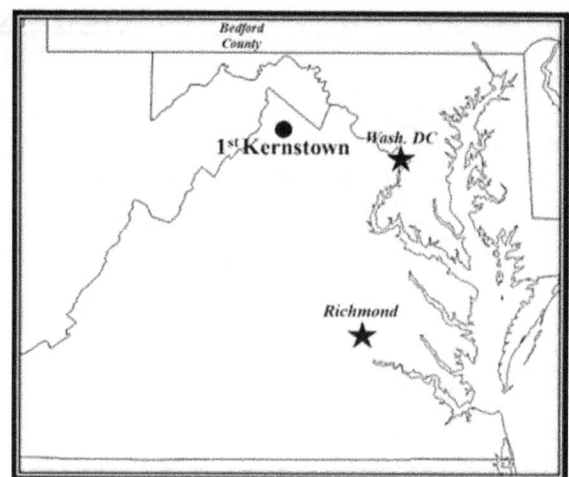

Known Bedford Co. Casualties - 9

KIA or Died from Wounds - 5

Wounded - 4

A drawing of the charge of 5 Union regiments including the 110th Pennsylvania Infantry on the famed Stonewall Brigade.

On a Sunday morning, over one hundred men and boys from Bedford County who had recently volunteered to join the 110th and 84th Pennsylvania Infantry regiments were preparing for combat for the first time. Just six months earlier, they would have been attending church services with their loved ones back home. On this day they would face off against Stonewall Jackson and his famous Stonewall brigade at the 1st Battle of Kernstown on the outskirts of Winchester, VA. That afternoon, the 110th PA Infantry and four other regiments were in a desperate struggle for control of a stone wall on the left flank of the Confederate lines. Initial Union attacks were repulsed as furious fighting continued. Late in the afternoon, the 84th PA Infantry and four other regiments launched an attack on the right flank of the Confederate lines while the Union cavalry swept around the left flank. An hour later, with ammunition running low, Confederate lines began to crumble. Within minutes, the rout turned into a panicked Rebel retreat. Over 200 Confederates had been captured by 7:00 and Jackson was withdrawing his troops from the field. The citizen soldiers of Bedford County who were working as farmers, laborers, and carpenters just months earlier were among the front-line troops that handed Stonewall Jackson, one of history's legendary generals, his only tactical battlefield defeat of the entire Civil War. Kernstown would be the first of many costly Civil War battles for Bedford County families. It was an especially tragic for the College family. James and his younger brother John William were both wounded at Kernstown. John William succumbed to his wounds four days after the battle and older brother James died at his home in Yellow Creek on May 11th, 1862.

1st Kernstown - March 23, 1862

Known Bedford Co. Casualty Listing

Name	Muster Age	Rank	State	Regiment	Company	Casualty
Baker, David N	18	Pvt.	PA	110th Infantry	D	Wounded
Burke, Samuel	36	Pvt.	PA	84th Infantry	A	Wound.-Died
College, James	28	Pvt.	PA	110th Infantry	C	Wound.-Died
College, John W	16	Pvt.	PA	110th Infantry	C	Wound.-Died
Croft, Philip P	23	Pvt.	PA	110th Infantry	C	KIA
Davis, William A	25	Pvt.	PA	84th Infantry	A	Wounded
Ferguson, John	19	Pvt.	PA	110th Infantry	C	KIA
Grimes, Henry W	20	Pvt.	PA	84th Infantry	C	Wounded
Price, David J	19	Corp.	PA	110th Infantry	C	Wounded

Thomas Livingston was born in Hopewell Township on March 17th, 1837. He mustered as a 1st Sergeant in the 110th Pennsylvania Infantry on October 24th, 1861. Thomas suffered a wound during the Battle of Spotsylvania on May 12th, 1864. He survived the war and married in 1865. He moved to Iowa after his 1st wife died in 1868. Later in his life, he moved back to Bedford County. Thomas passed at age 71 and is buried in the Everett Cemetery.

Ezra Brisbin was a 32-year-old resident of Bedford when he mustered in as a Captain in the 110th Pennsylvania Infantry on October 24th, 1861. His wife was at home with two toddlers during the war. Ezra resigned his commission in the 110th Infantry on June 16th, 1862. He passed at age 65 and is buried at the Grandview Cemetery in Tyrone.

Edisto Island & Pocotaligo - 1862

Known Bedford Co. Casualties - 18
KIA or Died from Wounds - 4
Wounded & POW - 1
Wounded - 7
POW - 6

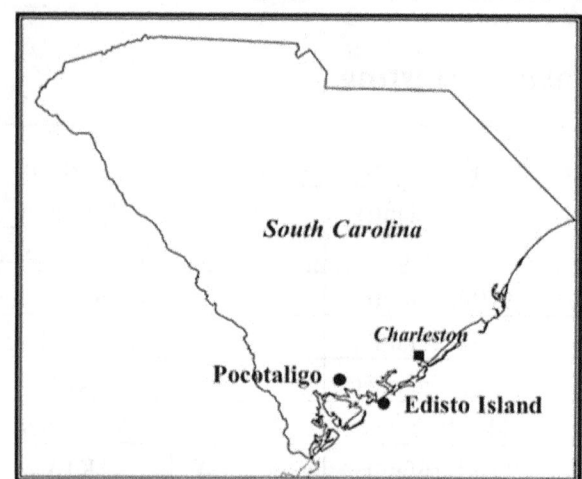

Recently freed slaves are pictured at the Edisto Island, SC plantation of James Hopkinson in 1862. Being witnesses to similar scenes would have reminded Bedford County soldiers on why they were there.

The 55th PA Infantry arrived on Edisto Island in February 1862. Southern planters had abandoned Edisto Island in late 1861, and thousands of escaped slaves were seeking refuge there on former plantation land. The mission of the 55th was to initiate a campaign against Confederate forces around Charleston and to protect the African Americans who were on Edisto Island. On March 29th, the 55th PA Infantry repulsed a detachment of dismounted Confederate cavalry who crossed a river and launched an attack on Union troops at Edisto Island. On October 22nd, the Union Army attempted to destroy the Charleston and Savannah Railroad line during the Battle of Pocotaligo. Approximately 5,000 Union soldiers initially drove back Rebel troops guarding the rail line. When Confederate reinforcements arrived, the Union Army realized a targeted rail bridge was out of their reach and withdrew from the battlefield. In addition to the battlefield casualties listed on the next page, the following 13 Bedford County soldiers in the 55th PA Infantry died from disease during the South Carolina campaign: Solomon Adams, Andrew Butler, John Coffee, Charles Engle, Eli Harbaugh, James M Holler, Jacob Kinley, Irving Little, Levi Long, Philip S Miller, Malachi B Mock, John Moyer and Philip Murphy.

Edisto Island & Pocotaligo - 1862

Known Bedford Co. Casualty Listing

Name	Muster Age	Rank	State	Regiment	Casualty	Battle	Date
Garlinger, Walter E	22	Corp.	PA	55th Inf.	POW	Edisto Island, SC	3/29/62
Gollipher, Silas	19	Sgt.	PA	55th Inf.	Wound.-POW	Edisto Island, SC	3/29/62
Gollipher, Silas					Wounded	Edisto Island, SC	4/17/62
Lockard, Thomas R	19	Pvt.	PA	55th Inf.	POW	Edisto Island, SC	3/29/62
Mars, John	21	Pvt.	PA	55th Inf.	POW	Edisto Island, SC	3/29/62
Ream, Isaac	23	Pvt.	PA	55th Inf.	POW	Edisto Island, SC	3/29/62
Ritchey, Jonas	18	Pvt.	PA	55th Inf.	Wound.-Died	Edisto Island, SC	3/29/62
Saupp, John	17	Pvt.	PA	55th Inf.	KIA	Edisto Island, SC	3/29/62
Werning, John	39	Pvt.	PA	55th Inf.	POW	Edisto Island, SC	3/29/62
Whitaker, Christian		Pvt.	PA	55th Inf.	POW	Edisto Island, SC	3/29/62
Claycomb, Frederick	40	Pvt.	PA	55th Inf.	Wounded	Pocotaligo, SC	10/22/62
Claycomb, John	33	Pvt.	PA	55th Inf.	Wounded	Pocotaligo, SC	10/22/62
Gephart, John	21	Pvt.	PA	76th Inf.	Wounded	Pocotaligo, SC	10/22/62
Klahre, Theodore M	20	Corp.	PA	76th Inf.	Wounded	Pocotaligo, SC	10/22/62
Leech, Wlliam	18	Pvt.	PA	55th Inf.	KIA	Pocotaligo, SC	10/22/62
Martin, William L	23	Sgt.	PA	55th Inf.	KIA	Pocotaligo, SC	10/22/62
Miller, John W	24	Pvt.	PA	55th Inf.	Wounded	Pocotaligo, SC	10/22/62
Fleegle, Isaac S	23	Pvt.	PA	55th Inf.	Wounded		

Theodore Klahre was born in Germany on February 5th, 1841 and immigrated with his parents to the Bedford County in 1858. Theodore suffered a wound during the battle of Pocotaligo, SC. At the battle of Cold Harbor in June 1864, he was struck by a bullet in front of his right ear; the bullet went through his jaw and lodged in his throat. Reportedly, fourteen days later, he spat out the bullet. Theodore survived the war, lived until the age of 69, and was buried in the Everett Cemetery in 1910.

Silas Gollipher was born in Schellsburg in April 1842. He mustered as a Sergeant in the 55th Pennsylvania Infantry on October 11th, 1861. Silas was wounded and taken prisoner on Edisto Island on March 29th, 1862, and was exchanged for a confederate prisoner. Silas suffered a second wound three weeks later on April 17th and received a disability discharge for his wounds on January 15, 1863. He passed at the home of his daughter when he was 77 years old and is buried in the Schellsburg Cemetery.

John Claycomb was born in Blair County and moved to St. Clair Township in 1850. He mustered into the 55th PA Infantry on November 5th, 1861 with a wife and 5 children at home. John suffered a severe leg wound which required amputation in October 1862 at the Battle of Pocotaligo. He received a disability discharge on March 14th, 1863. John passed on October 15th, 1873, and is buried at the Fairview Cemetery in Altoona.

Fair Oaks - May 31, 1862

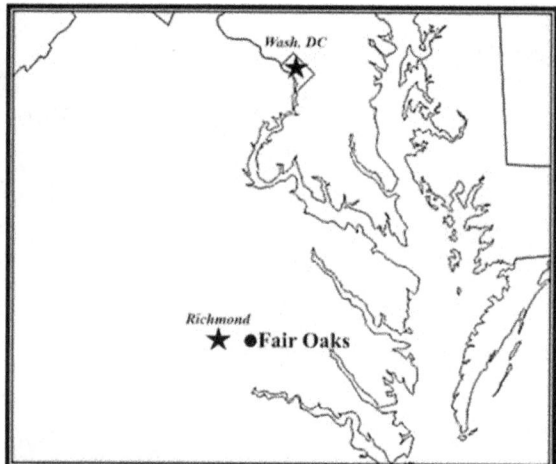

Battle also referred to as Seven Pines

Known Bedford Co. Casualties - 11
KIA- 1
Died while POW - 1
Wounded - 9

Union Troops at Fair Oaks minutes before a Confederate attack.

The Peninsula Campaign began in early April 1862. McClellan transported over 100,000 men by ship to Fort Monroe, VA about 75 miles southeast of Richmond. The ever-cautious McClellan took over a month to proceed toward Richmond after initially being bluffed by greatly outnumbered Confederate troops at Yorktown. By late May, the Union Army reached Fair Oaks just 6 miles east of Richmond and were close enough to see the church spires in the Confederate capitol. Confederate commander Joseph Johnston grew concerned the Union Army would surround Richmond and launched an attack on May 31st. Fair Oaks ended as a technical draw with about 5,000 casualties suffered by each army. The most consequential event of the battle was General Johnston being severely wounded. Jefferson Davis' decision to replace Johnston with Robert E Lee would profoundly change the direction of the Civil War. Lee was a more audacious risk-taker than his predecessor. Lee immediately recalled Stonewall Jackson from the Shenandoah Valley and planned an offensive against the Union forces threatening Richmond. Conversely, Union commander McClellan was shaken by the carnage he witnessed at Fair Oaks and stated after the battle, "Victory has no charms for me when purchased at such a cost." Lee would exploit McClellan's cautious, risk adverse personality over the next six months which eventually led to McClellan being fired by Abraham Lincoln after the Battle of Antietam.

Fair Oaks - May 31, 1862

Known Bedford Co. Casualty Listing

Name	Muster Age	Rank	State	Regiment	Company	Casualty
Bessor, John		Corp.	PA	101st Infantry	D	Wounded
Boerkamp, Henry	30	Pvt.	PA	101st Infantry	G	Wounded
Clingerman, Peter	21	Pvt.	PA	101st Infantry	D	Wounded
Geller, Solomon	35	Pvt.	PA	101st Infantry	G	POW-Died
Huffman, William B	33	Pvt.	PA	101st Infantry	G	Wounded
Keagy, John T	20	Corp.	PA	101st Infantry	D	Wounded
Knipple, William H	18	Corp.	PA	101st Infantry	G	Wounded
Miller, Martin D	20	Pvt.	PA	101st Infantry	D	KIA
Mills, Andrew J	21	Pvt.	PA	101st Infantry	D	Wounded
Shinn, Job	24	Pvt.	PA	53rd Infantry	I	Wounded
Uperaft, John		Pvt.	PA	61st Infantry	F	Wounded

William H Stuckey was born in Colerain Township on April 6th, 1841. He mustered in the 101st Pennsylvania Infantry on November 1st, 1861. William survived 3 years of combat and mustered out on November 1st, 1864. His brother David enlisted in the 184th Pennsylvania Infantry and was taken prisoner during the battle of Jerusalem Plank Road near Petersburg on June 22nd, 1864. David died of Scurvy in the Andersonville POW camp on November 18th, 1864. William married after the war. He passed at age 76 and is buried at the Everett Cemetery.

Simon P Kegg was born on November 6th, 1838 in Colerain Township. He was working on a farm when he mustered in the 101st Pennsylvania Infantry on November 1st, 1861. Muster Roll records show Simon was at the Foster Hospital in Bern, NC on April 20th, 1864, when the entire Union garrison was captured at Plymouth, NC, including many Bedford County solders in the 101st PA Infantry. Almost all the enlisted men in the 101st were taken to Andersonville. Simon mustered out on November 23rd, 1864. He married in 1867 and moved his family to Mansfield, Ohio in the 1880s. He passed at age 72 and is buried at the Mansfield Cemetery.

Seven Days Battle - June 25 - July 1, 1862

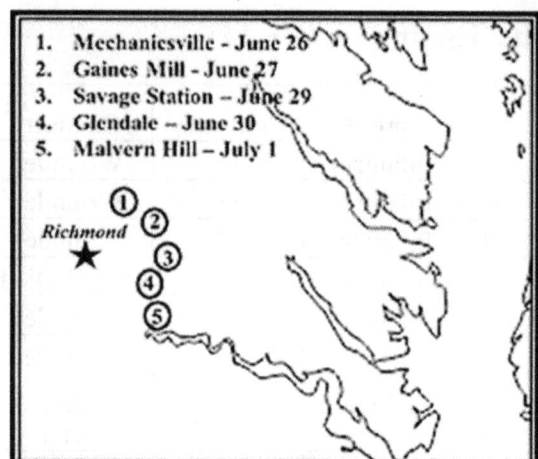

Series of Battles within 10 miles of Richmond

Known Bedford Co. Casualties – 50
KIA or Died from Wounds - 4
Wounded & POW - 2
POW - 37
Wounded or Injured - 7

Union field hospital during the 7 Days Battle.

New Confederate commander Robert E Lee seized the offensive initiative during a series of battles collectively referred to as the 7 Days Battle. Union forces won four of the five battles and inflicted heavy casualties on the attacking Confederates. Gaines Mill was the only decisive Confederate victory, but Lee achieved his strategic objectives. The initial battle of Mechanicsville took place less than 5 miles from Richmond. A week later Lee had pushed a retreating McClellan to Harrison's Landing on the James River over 15 miles southeast of Richmond. Lee had saved Richmond, rebuilt the moral of his army, and installed a Southern confidence in his leadership that would last the rest of the war. The Peninsula Campaign, which had started with much promise for the Union Army, ended with McClellan's failure to capture Richmond. The Hopewell Rifles company of the 8th Pennsylvania Reserves were on the front lines at Mechanicsville and did not heed the order to withdrawal to Gaines Mill. Two-thirds of the Company were captured and taken to Richmond. Most of the Hopewell Rifles company were exchanged for Confederate prisoners within 6 weeks. During the 7 Days battle, Confederates forces took 20,000 casualties compared to 16,000 for the Union army.

Known Bedford Co. Casualty Listing

Name	Muster Age	Rank	State	Regiment	Comp.	Casualty	Battle
Amick, George W	23	Pvt.	PA	8th Reserves	F	POW	Gaines Mill
Armstrong, David B	25	1st Sgt.	PA	8th Reserves	F	POW	Gaines Mill
Barber, James	27	Pvt.	PA	8th Reserves	F	POW	Gaines Mill
Bollinger, David S	23	Pvt.	PA	8th Reserves	F	POW	Gaines Mill
Brown, George	27	Pvt.	PA	8th Reserves	F	POW	Gaines Mill

Seven Days Battle - June 25 – July 1, 1862

Name	Muster Age	Rank	State	Regiment	Comp.	Casualty	Battle
Brumbaugh, Levi	15	Pvt.	PA	8th Reserves	F	POW	Gaines Mill
Carnell, John	20	Pvt.	PA	8th Reserves	F	POW	Gaines Mill
Cleaver, James	23	2nd Lt.	PA	8th Reserves	F	POW	Gaines Mill
Cook, Joseph S	19	Pvt.	PA	8th Reserves	F	POW	Gaines Mill
Dasher, William H	21	Corp.	PA	8th Reserves	F	POW	Gaines Mill
Eastright, Christian	24	Pvt.	PA	8th Reserves	F	POW	Gaines Mill
Edwards, Allison	27	Pvt.	PA	8th Reserves	F	POW	Gaines Mill
Eichelberger, Eli	24	Capt.	PA	8th Reserves	F	POW	Gaines Mill
Eichelberger, John S	32	Capt.	PA	8th Reserves	F	POW	Gaines Mill
Evans, Johnson	22	Pvt.	PA	8th Reserves	F	POW	Gaines Mill
Figart, Henry J	21	Pvt.	PA	8th Reserves	F	POW	Gaines Mill
Foor, Samuel S	23	Pvt.	PA	8th Reserves	F	Wounded	Mechanicsville
Foor, William H	24	Pvt.	PA	8th Reserves	F	KIA	Glendale
Foster, Aaron	25	Pvt.	PA	8th Reserves	F	POW	Gaines Mill
Gamble, Robert	35	Pvt.	PA	8th Reserves	F	POW	Gaines Mill
Garlick, Christian C	23	Pvt.	PA	8th Reserves	F	POW	Gaines Mill
Garrett, Alexander A	21	Pvt.	PA	8th Reserves	F	KIA	Glendale
Gates, George W	18	Pvt.	PA	13th Reserves	D	Wounded	7 Days Battle
Gates, James	23	Pvt.	PA	8th Reserves	F	POW	Gaines Mill
Grubb, Wilson	25	Pvt.	PA	8th Reserves	F	POW	Gaines Mill
Hainsey, Josiah	28	1st Sgt.	PA	1st Light Art.	A	Wounded	Gaines Mill
Headrick, David	21	Pvt.	PA	8th Reserves	F	Wound.-POW	Gaines Mill
Heffner, George	20	Corp.	PA	8th Reserves	F	POW	Gaines Mill
Holdcraft, William	38	Pvt.	PA	8th Reserves	F	Wound.-POW	Glendale
Horton, David	24	Sgt.	PA	8th Reserves	F	POW	Gaines Mill
Juda, George V A	20	Corp.	PA	8th Reserves	F	POW	Gaines Mill
Kerns, Mark C	25	Capt.	PA	1st Light Art.	G	Wounded	Gaines Mill
Layton, John	20	Pvt.	PA	101th Infantry	D	KIA	Savage Station
Leader, George W	22	Pvt.	PA	8th Reserves	F	POW	Gaines Mill
Line, William	27	Sgt.	PA	138th Infantry	E	Wounded	Malvern Hill
Manges, Abraham	18	Pvt.	US	12th Infantry		Wounded	Gaines Mill
Martin, David	22	Pvt.	PA	8th Reserves	F	KIA	Mechanicsville
Penrod, John B Jr	19	Pvt.	PA	8th Reserves	F	POW	Gaines Mill
Piper, Lewis M	30	Pvt.	PA	8th Reserves	F	POW	Gaines Mill
Piper, Luther R	26	Corp.	PA	8th Reserves	F	POW	Gaines Mill
Ritchey, William D	21	Corp.	PA	8th Reserves	F	Injured	Mechanicsville
Robb, Conrad		Pvt.	PA	8th Reserves	F	POW	Gaines Mill
Shaw, Matthew P	22	Pvt.	PA	8th Reserves	F	POW	Gaines Mill
Showalters, Henry		Pvt.	PA	8th Reserves	F	POW	Gaines Mill
Taylor, Thomas A	20	Pvt.	PA	8th Reserves	F	POW	Gaines Mill
Tobias, John B	20	Pvt.	PA	8th Reserves	F	POW	Gaines Mill
Waltz, Lewis	37	Capt.	PA	8th Reserves	F	POW	Gaines Mill
Warsing, Alexander	33	Pvt.	PA	8th Reserves	F	POW	Gaines Mill
Whisel, William H	20	Pvt.	PA	8th Reserves	F	POW	Gaines Mill
White, Edmund H	25	Pvt.	PA	8th Reserves	F	POW	Gaines Mill

Seven Days Battle - June 25 - July 1, 1862

Lewis B Waltz was born around 1827. Little is known about the early part of his life. Lewis was mustered as a 2nd Lieutenant into the Hopewell Rifles Company of the 8th Pennsylvania Reserves on June 11th, 1861. He was captured along with most of the Hopewell Rifles early in the morning of June 27th during the Union withdrawal to Gaines Mills from the Mechanicsburg battlefield. Lewis was exchanged on August 27th, 1862 for Confederate Lewis M Slaughter of the 17th Virginia Infantry. He received a promotion to Captain on October 30, 1863. Lewis mustered out with the rest of his regiment on May 26th, 1864. He returned to Bedford County and married Margaret Long on August 9th, 1864. Lewis passed on June 26th, 1881, and is buried in the Saint Luke's Cemetery in Saxton.

Photograph of Ellerson's Mill. The Hopewell Rifles company of the 8th Pennsylvania Reserves were positioned near Ellerson's Mill on the Mechanicsville battlefield. Confederate troops advanced across the open ground in the background of this photograph toward the Bedford County soldiers in the 8th Reserves.

Seven Days Battle - June 25 – July 1, 1862

One of the most well-known veterans in Bedford County history was born in Hopewell on January 4th, 1840. Eli Eichelberger mustered in the Hopewell Rifles company of the 8th Pennsylvania Reserves as a 1st Lieutenant on June 11th, 1861. He was taken prisoner during the Union withdrawal to Gaines Mills on June 27th, 1862, and confined at Libby Prison in Richmond. Eli was paroled on August 5th, 1862 at Aiken's Landing in Virginia, and exchanged for a captured Confederate officer before rejoining the Hopewell Rifles on August 27th. He received a promotion to Captain on October 30th, 1863. Eli suffered a wound in the left thigh on May 6th, 1864 during the battle of the Wilderness. He recovered from his wound and returned home to Bedford County after the 8th Reserves completed their enlistment. Eli married Helen Wishart on December 18th, 1866, and was a father to two sons. Eli was the proprietor of the E. Eichelberger and Son store in Saxton. He passed at age 75 on May 18th, 1915 and is buried in the Everett Cemetery.

Christian C Garlick was 23 years old when he left his home in Monroe Township to muster in the 8th Pennsylvania Reserves on June 11th, 1861. He was captured along with most of the Hopewell Rifles company on June 27th, 1862. The 8th Reserve Infantry Regiment mustered out on May 15th after almost 3 years of combat. Christian immediately reenlisted in the 191st Pennsylvania Infantry and took part in many battles during the siege of Petersburg. The 191st was at Appomattox when Robert E Lee surrendered to Ulysses S Grant. Christian mustered out after the war ended on July 18th, 1865. He married Henrietta Stuckey on September 7th, 1865 and was a father to 6 children. Christian passed when he was 38 years old and was buried at Rockhill Cemetery in Monroe Township in 1876.

2nd Bull Run - August 29-30, 1862

Battle also referred to as Second Manassas

Known Bedford Co. Casualties - 15
KIA - 4
Wounded & POW - 2
POW - 2
Wounded - 7

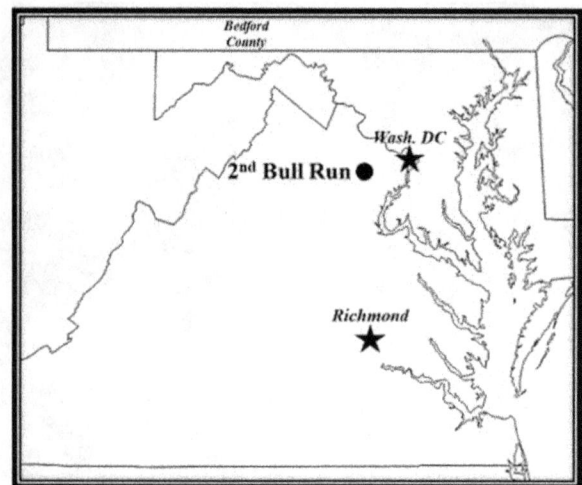

Children playing across from Calvary Troops at Sudley's Ford on the Bull Run battlefield.

After failing to take Richmond, Abraham Lincoln recalled McClellan's army from Harrison's Landing with orders to unite with a second Union Army group commanded by John Pope in Northern Virginia. Robert E Lee moved his army north from Richmond to strike Pope's Army before McClellan arrived. For a second time in as many years, Union and Confederate forces faced off against each other at Manassas. McClellan resisted orders to join up with Pope and did not arrive at Manassas in time for the 2nd Battle of Bull Run. The 1st day of the battle ended in a stalemate, with the troops of Pope and Stonewall Jackson taking heavy losses. That evening, Pope misread the maneuvers of Jackson's troops as preparations for a retreat. Pope was also unaware Confederate General James Longstreet had moved his 28,000-man division through the Thoroughfare Gap and had arrived on the Bull Run battlefield. The next day, Pope renewed the attack, and was driven back by Confederate artillery. A coordinated counterattack by Longstreet and Jackson's divisions forced a Union Army retreat across Bull Run. That night Pope withdrew his army toward Washington DC. The 2nd Battle of Bull Run ended as a decisive Confederate victory. The defeated Union Army suffered almost 14,000 casualties compared to around 9,000 for the Confederates. Lee sensed an opportunity and made plans to invade Pennsylvania. With Pope's demoralized army in disarray, Lincoln reluctantly re-appointed McClellan as Commander of all Union forces in the Eastern theater.

2nd Bull Run - August 29-30, 1862

Known Bedford Co. Casualty Listing

Name	Muster Age	Rank	State	Regiment	Comp.	Casualty	Casualty Date
Figart, Henry J	21	Pvt.	PA	8th Reserves	F	KIA	8/29/62
Focht, George W	20	1st Lt.	PA	107th Infantry	I	POW	8/30/62
Gracey, James A		Pvt.	PA	107th Infantry	H	Wound.-POW	8/30/62
Heffner, George	20	Corp.	PA	8th Reserves	F	KIA	8/29/62
Kerns, Mark C		Capt.	PA	1st Light Art.	G	KIA	8/30/62
Lysinger, George W		1st Sgt.	PA	107th Infantry	H	POW	8/30/62
Mellott, Cornelius	20	Pvt.	PA	11th Infantry	A	KIA	8/30/62
Riley, George W	41	Sgt.	PA	107th Infantry	H	Wounded	8/30/62
Rollins, Thomas J	28	Pvt.	GA	60th Infantry	K	Wounded	8/28/62
Roush, James Levi	23	Corp.	PA	6th Reserves	D	Wounded	
Showalters, Henry		Pvt.	PA	8th Reserves	F	Wound.-POW	
Sigel, Stephen	36	Pvt.	PA	11th Infantry	A	Wounded	8/28/62
Swartz, Abraham	17	Pvt.	PA	11th Infantry	A	Wounded	8/30/62
Swartz, William B	44	Pvt.	PA	107th Infantry	I	Wounded	8/28/62
Weaverling, Adam	42	2nd Lt.	PA	11th Infantry	A	Wounded	8/30/62

George and Susan Foor Riley married in 1849 and were raising a family in East Providence Township when the Civil War broke out. Susan was at home with 8 children under the age of 12 when George volunteered in the 107th Pennsylvania Infantry on January 9th, 1862. Susan received word later that year, George had been wounded during the 2nd Battle of Bull Run and he later mustered out of the army on March 1st, 1863.

1864 was a tragic year for the Riley and Foor families. Susan passed on June 30th, 1864, at age 29. Less than two weeks later her brother, Martin T Foor, was taken prisoner during the battle of Monocacy. Martin died in a Confederate prison on December 6th, 1864. George reenlisted in the 208th Pennsylvania Infantry less than 3 months after his wife's death on September 7th, 1864. He survived the war but died on June 18th, 1868, at age 46, leaving behind orphan children to be raised by relatives. Both are buried at the Ray's Cove Christian Church Cemetery outside of Breezewood.

South Mountain - September 14, 1862

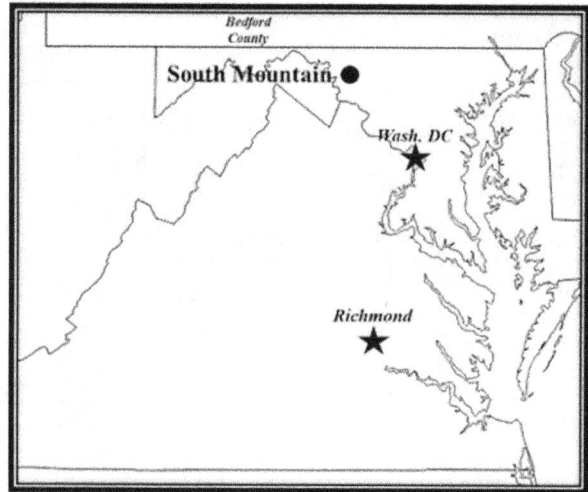

Also referred to as the Battle of Boonsboro Gap

Known Bedford Co. Casualties - 9
KIA or Died from Wounds - 3
Wounded - 6

Andrew J Foor was 31 years old when he left his East Providence Township home to muster in the 107th Pennsylvania Infantry on March 11th, 1862. The 107th took casualties at 2nd Bull Run, South Mountain, and Antietam. Andrew later reenlisted in the 199th PA Infantry on October 1st, 1864, and mustered out at the end of war. He lived until the age of 78 and is buried at the Rays Cove Christian Cemetery.

Thomas L Salkeld was 31 years old when he mustered in the 107th Pennsylvania Infantry on Jan. 20th, 1862 leaving behind a pregnant wife and 4 small children at home. Thomas was one of five brothers who volunteered in the Union Army. All five brothers survived the war. Thomas passed around the age of 67.

Robert E Lee's army marched North and crossed the Potomac river near Leesburg on September 4th. When Lee reached Frederick, MD, he made a fateful decision to split his army. Lee did not expect the cautious McClellan would launch an attack on his divided army. Stonewall Jackson was sent to clear out Union forces at Harpers Ferry and Martinsburg, WV and rejoin Lee's main Army group near the village of Boonsboro on the west side of South Mountain. Some Union soldiers found a copy of Lee's invasion plans at an abandoned Confederate camp, near Frederick on September 13th. McClellan, realizing Lee's army was vulnerable to attack while split in two, advanced 60,000 Union soldiers to South Mountain. The Battle of South Mountain was fought on September 14th on the most rugged terrain of any battlefield during the Civil War. All known Bedford County casualties on that day were at the Turner's Gap area of the battlefield. By nightfall Union troops had driven the Confederates from their entrenched positions on higher ground. The Rebel army was handed a significant defeat at South Mountain. Lee's offensive initiative of the invasion was lost, and Union troops were in position to deliver a decisive blow on his divided army. Casualties during the battle of South Mountain were 2,300 for the Union Army and 2,600 for the retreating Confederates.

South Mountain - September 14, 1862

Known Bedford Co. Casualty Listing

Name	Muster Age	Rank	State	Regiment	Company	Casualty
Eidenbaugh, John	37	Pvt.	PA	107th Infantry	H	Wounded
Figart, Levi H	17	Pvt.	PA	107th Infantry	H	Wounded
Foor, Jonathan S	22	Pvt.	PA	107th Infantry	H	Wounded
Gates, George W	18	Pvt.	PA	13th Reserves	D	Wounded
Horton, George	21	Corp.	PA	8th Reserves	F	Wounded
Kay, William H	20	Corp.	PA	8th Reserves	F	Wound.-Died
Mellott, Frederick	21	Pvt.	PA	12th Reserves	K	KIA
Riley, Andrew J	16	Pvt.	PA	107th Infantry	H	Wounded
Shaw, Matthew P	22	Pvt.	PA	8th Reserves	F	KIA

Drawing of Union Troops charging Confederates holding the higher ground at Turners Gap during the Battle of South Mountain.

Harpers Ferry - September 15, 1862

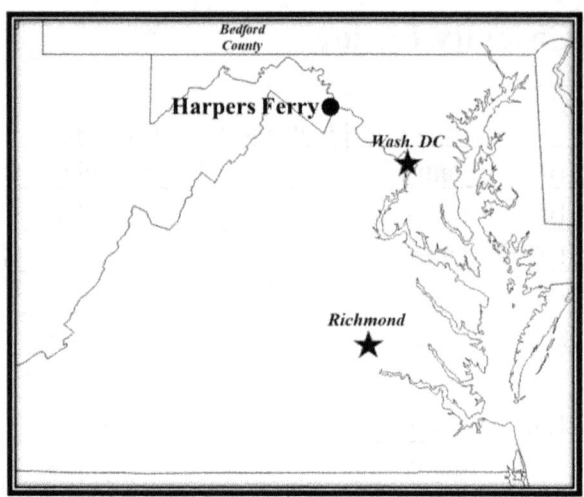

Known Bedford Co. Casualties

POW - 7

Harper's Ferry shown after its evacuation by the rebels in 1861. The Railroad bridge was left in ruins.

Lee expected the isolated Union troops at Harpers Ferry would evacuate their garrison and cede control of the northern Shenandoah Valley to the Confederates during the Rebel invasion. When the Union garrison did not evacuate, Stonewall Jackson intended to capture Harpers Ferry quickly and rejoin the rest of the invading Confederate army. The Confederates surrounded Harpers Ferry on September 13th, but Union troops refused to surrender. The following day the Confederates were defeated at South Mountain, putting Lee's divided army in peril. On September 15th, the Confederate Army opened a heavy artillery bombardment from the higher ground surrounding Harpers Ferry and finally forced the Union garrison to surrender. The 3rd Maryland PHB Infantry was one of the Union regiments defending Harpers Ferry. Over 30 Bedford County soldiers were enlisted in the regiment during the Battle. All soldiers who were at Harpers Ferry on September 15th were captured and exchanged for Confederate POWs. It is highly likely many more than the 7 Bedford County soldiers listed on the next page were taken prisoner on that day, but additional POW records have not been located.

Harpers Ferry - September 15, 1862

Known Bedford Co. Casualty Listing

Name	Muster Age	Rank	State	Regiment	Company	Casualty
Hendershot, Samuel C	27	Pvt.	MD	3rd PHB Infantry	C	POW
Karns, Jabez	28	Pvt.	MD	3rd PHB Infantry	C	POW
Linn, Hugh	52	Pvt.	MD	3rd PHB Infantry	B	POW
Linn, Riley	17	Pvt.	MD	3rd PHB Infantry	B	POW
Linn, William	14	Pvt.	MD	3rd PHB Infantry	B	POW
Smith, Nathan P R	18	Pvt.	MD	3rd PHB Infantry	B	POW
Sponsler, George W	18	Pvt.	MD	3rd PHB Infantry	C	POW

John Lowery was 46 years old when he volunteered in the 3rd Maryland PHB Infantry on September 28th, 1861. His wife and 7 children waited for his return during the war. John mustered out of the army on May 29th, 1865. He was likely captured with the rest of his regiment at Harpers Ferry and paroled a few days later. John died at age 68 and was buried at the Everett Cemetery in 1883.

George W Sponsler was 18 years old when left his home in Everett with his 16-year-old brother Solomon to join the Union Army. Both mustered in the 3rd Maryland PHB Infantry on November 5th, 1861. George was captured with all other Union soldiers who were defending Harpers Ferry on September 15th, 1862. Two other brothers also volunteered. John mustered in the 22nd PA Cavalry when he was 17 in 1863. William substituted for another man in the 208th PA Infantry when he was 16 years old. William mustered in on March 4th, 1865 and suffered a serious wound 3 weeks later during the battle of Ft. Stedman that required the amputation of his leg. All four brothers survived the war. George married after mustering out in 1865 and was a father to 13 children. He passed at age 41 and was buried in the Everett Cemetery in 1885.

Antietam - September 17, 1862

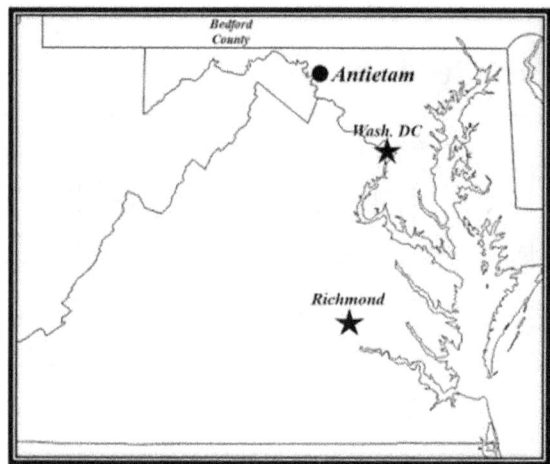

Also referred to as the Battle of Sharpsburg

Known Bedford Co. Casualties - 31
KIA or Died from Wounds - 13
Wounded - 18

This photograph of Confederate dead lying near the Dunker Church is one of the most iconic images of the Civil War. Bedford County soldiers in the 125th Pennsylvania infantry would have reached the wooded area just to the right of the Dunker Church before being counter-attacked by an overwhelming Confederate force and driven back.

Lee's main army group crossed Antietam Creek and assumed defensive positions on the rolling hills just outside of Sharpsburg on September 15th to wait for the rest of the Confederate army to arrive from Harpers Ferry. The 60,000 troops in McClellan's army advanced cautiously from South Mountain. The first Union troops did not cross Antietam Creek until the afternoon on September 16th. That night 14,000 Confederate troops arrived from Harpers Ferry, increasing Lee's forces to 35,000 men at Sharpsburg. The bloodiest day in American military history began at daybreak on September 17th, 1862, when Union General Joseph Hooker advanced his Corp across the Miller farm and into a cornfield. The 11th Pennsylvania Infantry were among the first Union regiments to face heavy fire as they emerged from the 20-acre Cornfield. This Cornfield area was the site of the most savage fighting at Antietam. One vicious Union attack after another was met with an equally vicious counterattack by Confederates during the first three hours of the battle. By around 9am, Union forces advanced from the Cornfield to the Dunker Church area of the battlefield before being driven back.

Antietam - September 17, 1862

Known Bedford Co. Casualty Listing

Name	Muster Age	Rank	State	Regiment	Company	Casualty
Baker, Franklin S	18	Pvt.	PA	125th Infantry	E	KIA
Black, George W Z	19	Capt.	PA	107th Infantry	H	Wounded
Bradley, James A	30	Pvt.	PA	8th Reserves	F	Wounded
Breneman, Michael B	24	Pvt.	PA	125th Infantry	C	Wounded
Bryant, James	25	Pvt.	PA	125th Infantry	F	Wounded
Burkholder, George		Pvt.	PA	125th Infantry	H	Wounded
Chaney, Levi	19	Pvt.	PA	107th Infantry	H	Wounded
Davis, James W	30	Pvt.	PA	28th Infantry	O	Wounded
Dean, Franklin		Pvt.	PA	8th Reserves	F	Wounded
Dell, Moses	21	Pvt.	PA	1st Light Art.	F	Wounded
Fessler, Samuel		Pvt.	PA	107th Infantry	H	KIA
Foor, George W	43	Pvt.	PA	107th Infantry	H	KIA
Frazey, Frederick L	21	Pvt.	PA	11th Infantry	A	Wounded
Gaster, James H	18	Sgt.	PA	107th Infantry	H	Wound.-Died
Gates, James	23	Pvt.	PA	8th Reserves	F	Wound.-Died
Harclerode, David	20	Pvt.	PA	125th Infantry	E	Wounded
Horton, George	21	Corp.	PA	8th Reserves	F	Wound.-Died
Jamison, Benjamin	19	Pvt.	PA	125th Infantry	B	Wounded
Kelley, John A	23	Corp.	PA	125th Infantry	D	KIA
Lear, John	26	Pvt.	PA	125th Infantry	E	KIA
Leighty, John Q	23	Corp.	PA	8th Reserves	F	Wound.-Died
Malone, William	24	Pvt.	PA	8th Reserves	F	Wound.-Died
Maugle, Joseph	22	Pvt.	PA	8th Reserves	F	Wounded
Morse, David	16	Pvt.	PA	11th Infantry	A	Wound.-Died
Pee, Frances W	21	Sgt.	PA	11th Infantry	A	Wounded
Riley, William (1)		Pvt.	PA	11th Infantry	A	Wound.-Died
Sigel, Stephen	36	Pvt.	PA	11th Infantry	A	Wounded
VanOrmer, William W	20	Capt.	PA	53rd Infantry	I	Wounded
Weaverling, David	22	2nd Lt.	PA	11th Infantry	A	Wounded
Weaverling, Jacob P	33	Pvt.	PA	11th Infantry	A	KIA
White, Edmund H	25	Corp.	PA	8th Reserves	F	Wounded

Almost all the Bedford County casualties at Antietam took place in the Cornfield and Dunker Church areas of the battlefield. The only exception is Captain William Van Ormer of the 53rd PA infantry who was wounded during the Sunken Road assault. The Cornfield and nearby areas accounted for about 6,500 of the 10,300 Confederate casualties and almost 7,300 of the 12,400 Union casualties at the battle of Antietam. General Hooker stated in his official report of the battle, "most of the Cornfield's stalks had been cut as closely as could have been done with a knife". The battle shifted to the Sunken Road area of Antietam during the middle of the day. Later in the afternoon, the battle shifted a third time to the Burnside Bridge area of the battlefield. Union troops were about to overtake the Confederate lines near the Burnside Bridge when the last of Stonewall Jackson's troops arrived from Harpers Ferry just in time to halt the Union advance. The battle ended in a tactical draw, with both sides taking shockingly high casualties. Union General McClellan failed to press an outnumbered Confederate Army during the battle and a lost another chance to destroy Lee's army after the battle by allowing the Confederates to retreat across the Potomac river unmolested. Lincoln removed McClellan as Commander of the Union Army on November 7, 1862 because of his lack of initiative in prosecuting the war. Lee's first invasion of the North had failed. This costly battle enabled Abraham Lincoln to issue the Emancipation Proclamation that would end slavery in America if the Union Army prevailed in the war. The exact number of casualties at Antietam will never be known, but estimations are 3,650 Americans in the Union and Confederate Armies lost their lives on September 17th, 1862. For a historical perspective, approximately 2,500 Americans lost their lives on D-Day during World War II.

Antietam - September 17, 1862

Knap's Pennsylvania Light Artillery Battery immediately after the battle of Antietam. Americus Enfield, a well-known Bedford County doctor after the war, was enlisted in this unit during the battle of Antietam and is possibly in this photograph.

Benjamin F Jamison was 19 when he left Snake Spring Valley to volunteer in the 125th PA Infantry on August 13th, 1862. He was one of over thirty known Bedford County casualties at Antietam. He received a second wound and was taken prisoner at Cold Harbor in 1864. Benjamin was taken to Andersonville and survived 9 months in Confederate POW camps. After the war, Benjamin was a Teacher and a Justice of the Peace. He passed at age 77 and was buried at St. John's Cemetery in Loysburg in 1920.

Michael B Breneman grew up in Huntington County. He was 24 years old when he volunteered in the 125th Pennsylvania Infantry on August 11th, 1862 with his brother George. Both brothers were wounded at Antietam. Michael was hit by a Minie ball in the left leg, but survived. George succumbed to his wounds on November 10th, 1862. Michael became a well-known Bedford County Doctor after the war. He lived to be 82 years old and was buried in the Fockler Cemetery in Saxton in 1921.

Antietam - September 17, 1862

Close-up view of the photograph on the left. Note the U.S. flag and the burial party visible in the background. The Cornfield area of the battlefield is in the background.

Frederick L Frazey left his home in East Providence Township when he was 21 years old to enlist in the 11th Pennsylvania Infantry on September 30th, 1861. Frederick was wounded at Antietam. He mustered out of the Union Army on Mar. 18th, 1863. Frederick married the following year and moved west. He lived to be 84 and is buried in Urbana, Illinois.

David Harclerode was born in Friends Cove on Nov. 13th, 1841. He mustered in the 125th Pennsylvania Infantry on August 13th, 1862. A month later, he suffered a wound at Antietam. David mustered out of the 125th Infantry on May 18th, 1863. He married in 1864 and was a father to 6 children. David passed at age 69 in 1911 and is buried in the Everett Cemetery.

Fredericksburg - December 13, 1862

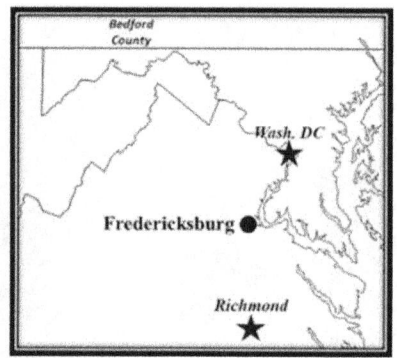

Known Bedford Co. Casualties - 64
KIA or Died from Wounds -16
Wounded or Injured - 47
POW - 1

The men on the next page charged over the 600 yards of open ground in this photo to reach Marye's Heights in the background. Confederate artillery was on the high ground, with infantry lined behind a stone wall at the foot of the hill.

Confederate dead behind the heavily defended stone wall at the foot of Marye's Heights. Bedford County soldiers in the 133rd made it to within 50 yards of this wall. No other Union troops are believed to have gotten any closer on that day.

Fredericksburg - December 13, 1862

Known Bedford Co. Casualty List - Marye's Heights Charge

Name	Muster Age	Rank	State	Regiment	Company	Casualty
Armstrong, Joseph M	21	Pvt.	PA	133rd Infantry	C	Wounded
Ashcom, George Jr	23	2nd Lt.	PA	133rd Infantry	C	Wounded
Barkman, Thomas C	19	Pvt.	PA	133rd Infantry	K	Wounded
Barndollar, Jacob W	21	Pvt.	PA	133rd Infantry	C	Wounded
Barndollar, James E	18	Pvt.	PA	133rd Infantry	C	KIA
Border, Henry	17	Pvt.	PA	133rd Infantry	K	Wounded
Cambell, Robert	19	Pvt.	PA	133rd Infantry	K	Wounded
Cooper, Joshua H	20	Pvt.	PA	133rd Infantry	C	KIA
Crissey, John C	16	Pvt.	PA	133rd Infantry	D	Wounded
Daugherty, James	16	Pvt.	PA	110th Infantry	C	KIA
Everhart, David	21	Pvt.	PA	110th Infantry	C	KIA
Gallagher, Edward	32	Pvt.	PA	133rd Infantry	C	KIA
Gogley, James H	21	Pvt.	PA	133rd Infantry	C	Wounded
Johnston, John W	19	Pvt.	PA	133rd Infantry	C	Wounded
Justice, Edward	24	Pvt.	PA	133rd Infantry	C	Wounded
Klahre, Herman T	20	Pvt.	PA	133rd Infantry	K	Wounded
Klotz, Jocob B	25	Pvt.	NY	59th Infantry	H	Wounded
Kochendarfer, John Z	22	Pvt.	PA	133rd Infantry	C	Wounded
Lambert, Joseph C	18	Corp.	PA	133rd Infantry	D	Wounded
McCleary, George B	20	Pvt.	PA	133rd Infantry	K	Wound.-Died
McClellan, John	25	Pvt.	PA	133rd Infantry	K	Wounded
McClellan, Josiah	22	Pvt.	PA	133rd Infantry	K	Wounded
McCullip, Alexander	22	Pvt.	PA	133rd Infantry	C	Wounded
Mentzer, Jacob M	31	Pvt.	PA	133rd Infantry	C	KIA
Miller, Jacob W	23	Pvt.	PA	133rd Infantry	C	Wounded
Mumper, Henry	22	Pvt.	PA	133rd Infantry	K	Wounded
Nycum, William H	21	Corp.	PA	133rd Infantry	C	Wounded
Over, Benjamin	22	Pvt.	PA	133rd Infantry	C	Wounded
Perrin, John	35	Pvt.	PA	133rd Infantry	C	KIA
Riley, James H	22	Pvt.	PA	133rd Infantry	K	Wounded
Roy, James	24	Pvt.	PA	133rd Infantry	C	Wounded
Scutchall, Samuel	18	Pvt.	PA	133rd Infantry	C	KIA
Sheeder, Henry F	25	Pvt.	PA	53rd Infantry	C	Wounded
Shroyer, Andrew G	20	Pvt.	PA	133rd Infantry	K	Wounded
Stailey, George E	21	Pvt.	PA	133rd Infantry	C	Wounded
Steele, David F	21	Pvt.	PA	133rd Infantry	K	KIA
Swank, George W	24	Pvt.	PA	133rd Infantry	C	Wounded
Wagner, Joseph H	31	Pvt.	PA	131st Infantry	D	Wounded
Weaverling, James T	21	Corp.	PA	133rd Infantry	K	Wounded
Woy, James H	19	Pvt.	PA	126th Infantry	B	Wounded

Fredericksburg - December 13, 1862

Known Bedford Co. Casualties on the Union Left at Fredericksburg

Name	Muster Age	Rank	State	Regiment	Company	Casualty
Hixon, Henry H	21	Pvt.	PA	11th Infantry	A	Wounded
Manspeaker, George	18	Pvt.	PA	11th Infantry	A	KIA
Baker, Andrew C	20	Pvt.	PA	13th Cavalry	B	Injured
Imes, John	22	Pvt.	PA	13th Reserves	G	Wound.-Died
Humbert, Wesley C	19	Corp.	PA	142nd Infantry	C	Wounded
Miller, Josiah C	18	Pvt.	PA	142nd Infantry	D	Wounded
Hainsey, Josiah	28	1st Sgt.	PA	1st Light Art.	A	Wounded
Bessor, Philip	21	Sgt.	PA	6th Reserves	D	Wounded
Mellen, Thomas	35	Pvt.	PA	6th Reserves	D	KIA
Griffin, Andrew H	24	Pvt.	PA	7th Cavalry	B	Wounded
Bowser, Emanuel	18	Pvt.	PA	8th Reserves	F	KIA
Cleaver, James	23	2nd Lt.	PA	8th Reserves	F	Wounded
Edwards, Allison	27	Pvt.	PA	8th Reserves	F	Wounded
Eichelberger, John S	32	Capt.	PA	8th Reserves	F	Wounded
Evans, Johnson	22	Pvt.	PA	8th Reserves	F	Wounded
Foor, Samuel S	23	Pvt.	PA	8th Reserves	F	POW
Foster, Aaron	25	Pvt.	PA	8th Reserves	F	Wounded
Holsinger, Frank	25	Pvt.	PA	8th Reserves	F	Wounded
Imes, Aaron	21	Pvt.	PA	8th Reserves	F	Wounded
Manspeaker, Barclay	23	Pvt.	PA	8th Reserves	F	KIA
Piper, Lewis M		Pvt.	PA	8th Reserves	F	Wounded
Piper, Luther R	26	Corp.	PA	8th Reserves	F	Wound.-Died
Tobias, John B		Corp.	PA	8th Reserves	F	Wounded
Whisel, William H		Pvt.	PA	8th Reserves	F	Wounded

There were many bad days for the North during the Civil War, perhaps none worse than December 13th, 1862. It was also a tragic day for Bedford County. The 64 known casualties at Fredericksburg were the most of any single day during the Civil War, and possibly the most casualties on any day, of any war, in Bedford County history. Just about everything that could go wrong, went wrong at the battle of Fredericksburg. Ambrose Burnside was appointed the new commander of the Union Army in the Eastern Theater on November 7th. His predecessor had just been fired for not aggressively pursuing the Rebel army. Burnside immediately planned an offensive to position the Union Army between Lee's army and Richmond. When Burnside reached the banks of the Rappahannock River across from Fredericksburg on November 17th, the requested pontoon boats needed to the build portable bridges had not arrived because of bureaucratic delays. A Union engineering regiment did not start assembling the portable bridges until December 11th. This delay provided time for Lee to concentrate the Confederate Army around the town of Fredericksburg to prepare for the Union Army attack he knew was coming.

Burnside's revised plan was for the 120,000-man Union Army to cross the Rappahannock River and launch a two-prong attack against the 80,000 Confederates positioned around Fredericksburg. Union General George Meade would lead a thrust on General Stonewall Jackson's lines, positioned 3 to 4 miles south of Fredericksburg. A second Union attack was planned on General James Longstreet's forces positioned at Marye's Heights, just outside of the town of Fredericksburg. This second Union attack was intended to be a diversion to keep Longstreet from reinforcing Stonewall Jackson troops during Meade's main assault. During the battle, Meade's forces temporarily broke through Confederate lines, but needed Union reinforcements were not provided, and Jackson's troops launched a successful counterattack. After Meade's assault failed, wave after wave of Union soldiers continued charging the impregnable Confederate positions at Marye's Heights, suffering horrific casualties. The Confederate Artillery commander on Marye's Heights prophetically stated before the battle, "A chicken could not live on that field when we open fire". The Union Army suffered around 12,500 killed and wounded at Fredericksburg, over two-thirds of the casualties were at Marye's Heights. Most Bedford County casualties were during these charges. The Confederates lost around 4,200 men on that day. Robert E Lee commented to James Longstreet while observing the ill-fated Union charges on Marye's Heights from a nearby hill, "It is well war is so terrible. We should grow too fond of it."

Fredericksburg - December 13, 1862

(Above) Joseph C Lambert suffered a wound during the Marye's Heights charge by the 133rd Pennsylvania Infantry. He later enlisted in the 21st Cavalry and suffered a Sabre wound to the head at Amelia Springs during the Appomattox campaign. Joseph lived to be 91 and was buried in Cambria Co. in 1935.

(Left) Jacob W Barndollar was born in Monroe Township on February 6th, 1841. Jacob was 20 years old when he volunteered in the 133rd Pennsylvania Infantry on August 13th, 1862. Four months later, Jacob and his brother James were among the Bedford County soldiers in the 133rd PA Infantry that charged Marye's Heights in a near suicidal frontal assault. Confederate Infantry poured rifle fire from behind a stone wall while Confederate Artillery Units on top of the hill rained cannon fire on the attacking Union soldiers. Jacob was standing beside his brother when James was mortally wounded. Before Jacob could kneel to attend to his brother, he was also hit in the chest. The photograph on the left is an exceedingly rare picture of a Civil War veteran recovering from a wound shortly after the battle. Jacob recovered and mustered out of the 133rd on May 26th, 1863 at the end of the 9-month enlistment. In 1864, Jacob married Nancy H Gogley and reenlisted in the Civil War. He mustered in the 186th PA Infantry Regiment on August 5th, 1864. Jacob survived the war and became a father to seven children. Jacob passed on April 15th, 1921 at age 80 and is buried in the Everett Cemetery.

... But in a larger sense, we cannot dedicate, we cannot consecrate, we cannot hallow this ground. The brave men, living and dead, who struggled here, have consecrated it, far above our poor power to add or to detract. The world will little note, nor long remember what we say here, but it can never forget what they did here.

Abraham Lincoln
portion of the Gettysburg Address

4. 1863

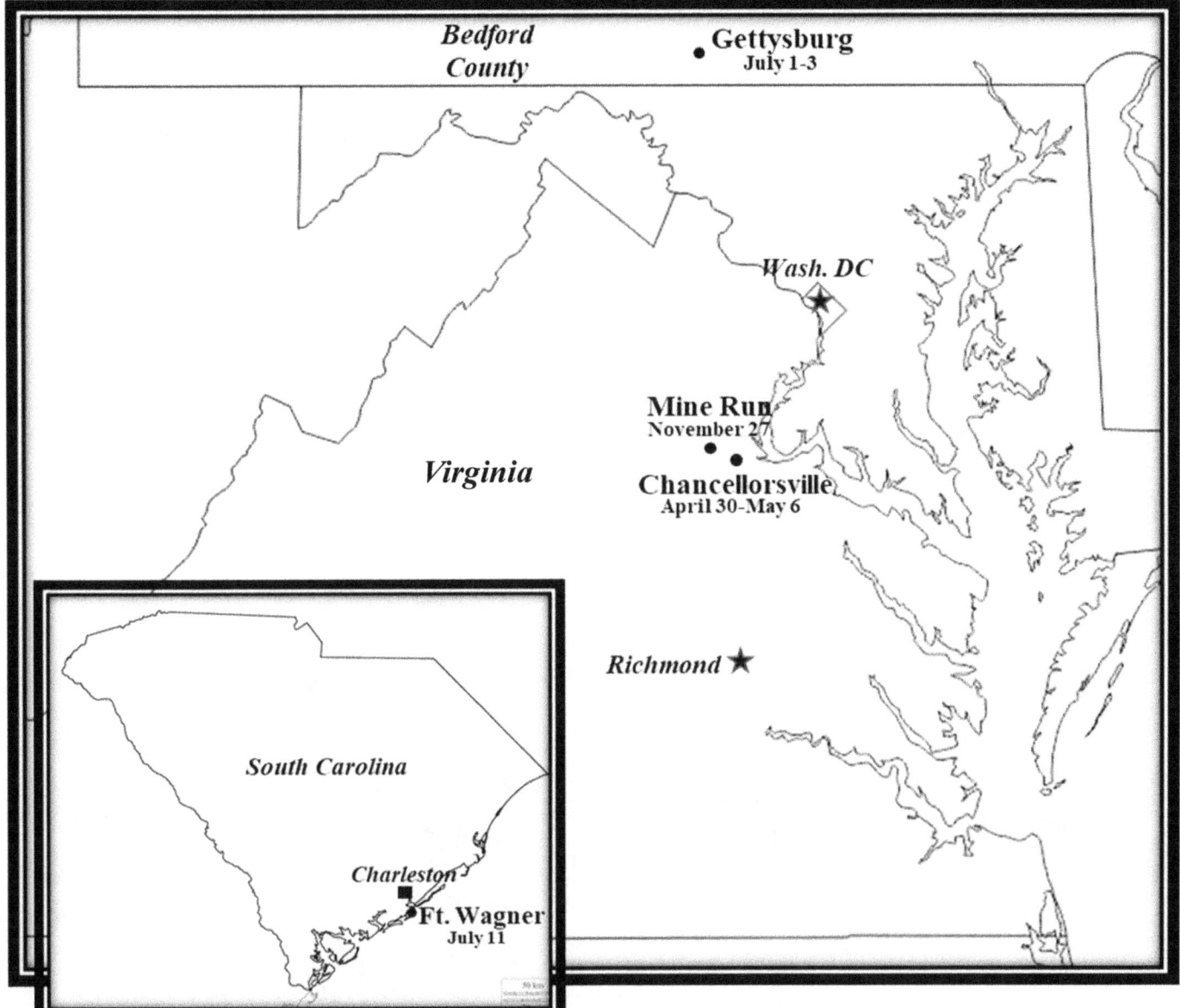

Gettysburg was the central battle of 1863 in the Eastern Theater and possibly the pivotal battle of the entire war. The stakes at Gettysburg could not have been much higher. Lee launched an invasion of Pennsylvania hoping to win a decisive battle on northern soil that would force Lincoln to end the war and divide America into separate countries forever. The war was not universally popular in the North. Some Peace Democrats were in favor of negotiating a settlement to end the costly war Lincoln was waging. On the 3rd day of the epic battle, Confederate troops were decimated during Picket's Charge. The Confederacy's best chance to prevail in the war evaporated as surviving stragglers of this charge streamed back from Cemetery Ridge. The 51,000 casualties suffered by both armies were by far the most of any single battle of the Civil War. A couple months after Gettysburg, the second most costly battle of the war took place at Chickamauga, GA in the Western Theater. Both armies suffered 34,624 casualties during this battle.

The fifth largest number of casualties of the war was in a major battle leading up to Gettysburg. A stunning Confederate victory at Chancellorsville resulted in both armies taking a combined 24,000 casualties. Weeks later the Confederate army routed a Union garrison in Winchester Virginia, clearing the path for the Rebel invasion of Pennsylvania. Many of the Union troops not captured at Winchester fled to Bloody Run. This Gettysburg Campaign story is well-detailed in the book "From Winchester to Bloody Run".

Further south, Bedford County troops in South Carolina took casualties during a July assault on Fort Wagner, which guarded the port city of Charleston. The battle of Fort Wagner was featured in the 1989 movie "Glory". Bedford County soldiers also suffered casualties during the last campaign of 1863 at Mine Run.

During 1863, there were 151 identified Bedford County casualties, including at least 34 soldiers who died from battlefield wounds and 47 more perished from disease. Please note all available Civil War records do not provide dates of known casualties, including deaths, therefore it is likely the casualty numbers are higher than identified in this book in 1863.

Chancellorsville - Apr. 30 - May 6, 1863

Known Bedford Co. Casualties - 18

KIA or MIA - 4

Wounded & POW - 1

Wounded or Injured - 10

POW - 3

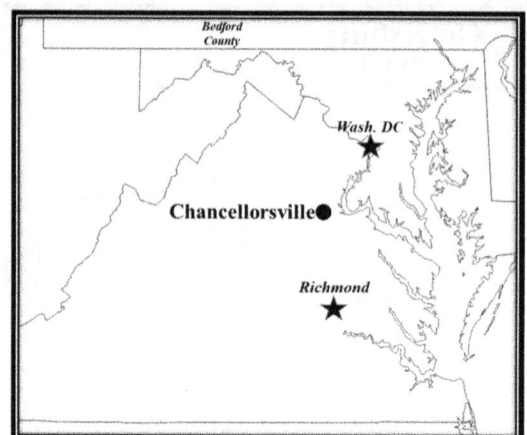

110th Pennsylvania Infantry is pictured on April 24th, 1863 in Falmouth, VA just prior to the Battle of Chancellorsville.

Joseph Hooker replaced Ambrose Burnside as commander of the Union Army in January 1863. In April 1863, the Union and Confederate armies were still facing each other from across the river at Fredericksburg. Hooker hoped to surprise Lee by sending one third of his army around Fredericksburg to strike the rear of Confederate lines. Lee recognized what Hooker was attempting and split his army to meet the Union forces sweeping around his flank. Despite having superior numbers of men, Hooker fell back into a defensive position. Lee and Stonewall Jackson met on the evening of May 1st, and developed an audacious plan to split their smaller Confederate army a second time and launch a surprise attack on the right flank of Hooker's lines. Jackson's 14,000 troops moved stealthily through woods and launched a devastating attack on the Union Army. After forcing a Union Army retreat, Stonewall Jackson was accidentally shot by his own troops during a nighttime reconnaissance ride outside of Confederate lines. The wound required the amputation of his left arm. Chancellorsville was possibly Robert E Lee's most spectacular victory of the war, but it came at a steep price. Upon hearing that Jackson had been shot, Lee lamented, "Jackson has lost his left arm, and I have lost my right". Stonewall Jackson died 8 days later. Robert E Lee followed up on the Confederate success at Chancellorsville with his second plan to invade Pennsylvania.

Chancellorsville - Apr. 30 – May 6, 1863

Known Bedford Co. Casualty Listing

Name	Muster Age	Rank	State	Regiment	Company	Casualty
Blake, William B	21	Pvt.	PA	125th Infantry	B	Wounded
Border, John S	21	Pvt.	PA	110th Infantry	C	INJ
Bruner, George	24	Pvt.	PA	126th Infantry	B	Wounded
Burket, Elias S	27	Pvt.	PA	84th Infantry	D	Wounded
Giffin, Peter	22	Pvt.	MD	3rd Reg Infantry	A	Wounded
Harbaugh, William H	22	Pvt.	PA	84th Infantry	A	POW
Hartman, George L	21	Corp.	PA	110th Infantry	C	MIA
Ickes, Adam	21	Pvt.	US	12th Infantry	A	Wounded
Lambright, William	21	Pvt.	PA	84th Infantry	A	POW
Lewis, John D		Pvt.	PA	125th Infantry	F	MIA
McCullip, Alexander	22	Pvt.	PA	133rd Infantry	C	Wounded
McLaughlin, James A	18	Corp.	PA	102nd Infantry	I	Wounded
Miller, Wyrman S	19	Pvt.	PA	148th Infantry	H	KIA
Peterson, William A	20	Pvt.	PA	84th Infantry	A	Wound.-POW
Ralston, David E	23	Pvt.	PA	110th Infantry	C	KIA
Salkeld, Samuel W	23	Pvt.	PA	126th Infantry	B	Wounded
Smith, John W	22	Corp.	PA	110th Infantry	C	POW
Woy, James H	19	Pvt.	PA	126th Infantry	B	Wounded

Elias S Burket was 26 years old when he left his Pavia Township home to muster in the 84th Pennsylvania Infantry on September 15th, 1862. Elias was wounded on May 3rd, 1863 during the battle of Chancellorsville. He mustered out of the 84th Infantry on January 13, 1865, and returned to Bedford County. Elias married in 1869 and was a father to 6 children. He passed at age 64 and is buried in the Nicodemus Cemetery in Henrietta.

William B Blake was born on September 12th, 1841. He volunteered in the 125th Pennsylvania Infantry on August 10th, 1862. William suffered a wound on May 3rd, 1863 at Chancellorsville and mustered out of the 125th two weeks later. William enlisted in the 208th Pennsylvania Infantry on August 18th, 1864. He married after the war and was a father to 3 children who survived to adulthood. William passed at age 67 and was buried at the Hopewell Cemetery in 1909.

John W Smith grew up in Hopewell and volunteered in the 110th PA Infantry in 1861 when he was 22. He suffered a hip wound and was taken prisoner at Port Republic on June 9th, 1862. John was taken prisoner a second time at Chancellorsville. At the Battle of the Wilderness on May 6th, 1864, a Minie ball passed through his elbow. John survived the war and passed in 1905 at age 65. He is buried at St. Paul's Cemetery in Yellow Creek.

Gettysburg - July 1-3, 1863

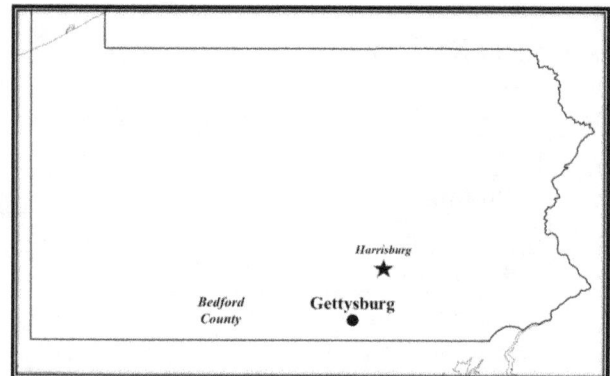

Known Bedford Co. Casualties - 25
KIA or Died from Wounds - 3
POW-Died - 1
Wounded & POW - 3
Wounded - 13
POW - 5

Union soldiers near McPherson's Woods on the 1st Day of Gettysburg

The tide of war was favoring the Confederates as they marched North in June 1863. Robert E Lee had racked up impressive battlefield victories over the previous 12 months against four different Union Commanders. For the second time in as many years, Pennsylvania was being threatened by a major invasion of the Rebel army. Governor Curtin made an urgent call for 50,000 volunteers to join emergency militias to help defend Pennsylvania. Fear spread throughout the commonwealth when Lee's army crossed into Pennsylvania during the last week of June. On July 1st, a division of Confederate Infantry ran into Union Cavalry at a small crossroads town in south central Pennsylvania and both armies converged. The epic battle of Gettysburg had begun. By late afternoon, 30,000 Confederates had overwhelmed the 20,000 Union troops positioned on the outskirts of the town. But the heroic efforts of these outnumbered soldiers enabled the Union Army to seize defensible positions on high ground around Gettysburg. The strategically important positions of Little Round Top, Cemetery Ridge, and Culp's Hill were successfully defended during the next 2 days. On the afternoon of the second day, 90,000 Union troops successfully defended their 5-mile-long line against 71,000 Confederates during fierce fighting. Early in the afternoon of the third day, the Confederates unleashed a tremendous artillery bombardment on the center of the Union Army lines before over 11,000 Confederates emerged from the woods and marched toward Cemetery Ridge. A few Confederate soldiers briefly breached the Union line at the Bloody Angle before the attack collapsed. Pickets Charge was a disaster for the Confederates, who lost approximately half their attacking force during the assault. The following evening, a demoralized Southern army retreated toward Virginia in a driving rainstorm, officially ending the bloodiest battle of the Civil War. Northern troops suffered 23,000 casualties including 3,100 soldiers who lost their lives. The Confederate Army suffered 28,000 casualties, including 3,900 who died during the 3-day battle.

Gettysburg - July 1-3, 1863
Known Bedford Co. Casualty Listing

Name	Muster Age	Rank	State	Regiment	Company	Casualty	Casualty Date
Enfield, Americus	14		PA	Knap's Light Art.	E	Wounded	7/2/63
Fessler, John B		Pvt.	PA	151st Infantry	H	Wounded	7/1/63
Foor, Jonathan S	22	Pvt.	PA	107th Infantry	H	POW	7/1/63
Gates, Martin	22	Pvt.	PA	110th Infantry	C	Wounded	7/2/63
Gracey, Alfred	18	Sgt.	PA	107th Infantry	H	POW	7/1/63
Gracey, James A	21	Pvt.	PA	107th Infantry	H	POW	7/1/63
Gracey, William C	42	1st Lt.	PA	107th Infantry	H	Wounded	7/1/63
Hammer, Joseph D		Pvt.	PA	142nd Infantry	D	Wound.-Died	7/1/63
Hays, Alexander Y	17	Pvt.	PA	110th Infantry	C	Wounded	7/2/63
Hixon, Henry H	21	Pvt.	PA	11th Infantry	A	Wound.-POW	7/1/63
Holsinger, Josiah	19	Pvt.	PA	110th Infantry	C	Wounded	7/2/63
Humbert, Wesley C	19	Corp.	PA	142nd Infantry	C	Wounded	7/1/63
Lamison, George W	23	Pvt.	PA	110th Infantry	C	Wound.-Died	7/2/63
Lohr, Benjamin F	20	Pvt.	PA	142nd Infantry	D	POW	7/1/63
Miller, John I	14	Pvt.	PA	110th Infantry	C	Wounded	7/2/63
Moore, John B	18	Sgt.	PA	110th Infantry	C	Wound.-POW	7/2/63
Moses, Emanuel	17	Pvt.	PA	18th Cavalry	K	POW-Died	7/6/63
Pee, Frances W	21	Sgt.	PA	11th Infantry	A	Wound.-POW	7/1/63
Potter, Levi	23	Pvt.	PA	56th Infantry	A	Wounded	7/1/63
Querry, Matthias	26	Pvt.	PA	53rd Infantry	C	Wounded	
Rollins, Thomas J	29	Pvt.	GA	60th Infantry	K	POW	7/3/63
Sparks, Uriah	24	Sgt.	PA	107th Infantry	H	Wounded	7/1/63
Steinman, Mathew C	23	Pvt.	PA	62nd Infantry	M	Wounded	7/2/63
Tobias, Samuel H G	19	1st Sgt.	PA	110th Infantry	C	KIA	7/2/63
VanOrmer, William W	20	Capt.	PA	53rd Infantry	I	Wounded	7/2/63

Uriah Sparks is pictured on the left with his brother John. Uriah was 19 years old when he left his home in Hopewell to enlist as a sergeant in the 107th Pennsylvania Infantry on March 12th, 1862. He was wounded during desperate fighting on the 1st Day of the battle of Gettysburg.

Uriah's two younger brothers also joined Pennsylvania Infantry Regiments. Both John and William were 15 years old when they volunteered. John mustered in the 194th Pennsylvania Infantry on July 22nd, 1864 and William mustered in the 101st Pennsylvania Infantry on November 1st, 1861. William was captured at Plymouth, NC on April 20th, 1864 and was taken to Andersonville. He survived over 10 months in Confederate POW camps.

All three brothers survived the war. Uriah passed in 1893 when he was 50 years old. John also passed at age 50 in 1900. Both are buried at the Providence Union Cemetery in Everett. William passed in 1915 when he was 68-year-old and is buried in the Wilmore United Brethren Church Cemetery in Cambria County.

Gettysburg Casualties July 1, 1863

Jonathan S Foor was born on December 20th, 1839, in East Providence Township. He was 22 years old when he mustered in the 107th Pennsylvania Infantry on January 9th, 1862. His wife gave birth to their third child 9 days later. Jonathan suffered a wound on September 14th, 1862 at South Mountain during the Antietam campaign. The following year, he was taken prisoner during the 1st day of the Battle of Gettysburg. Jonathan spent the next 17 months in several Confederate POW camps, including Andersonville, before being released on December 11th, 1864. Jonathan was a father to 10 children. He passed at age 68 is buried in the Rays Cove Christian Church Cemetery outside of Breezewood.

Alfred Gracey was 19 years old when left his East Providence Township home with his father William and two brothers to volunteer in the 107th Pennsylvania Infantry on February 10th, 1862. During the first day of fighting at Gettysburg, William suffered a neck wound and both Alfred and James were taken prisoner. Only their youngest brother George was not a casualty on that day. Alfred and James survived 20 months of captivity at several confederate POW camps. Both were released on February 27th, 1865. Albert was so weak, he needed to be carried from the Andersonville POW camp to board the train home. It took a year of rest and medical care to recover. All four Gracey family members survived the war. Alfred passed around the age of 74 and is buried in the Everett Cemetery.

Frances W Pee was born on January 27th, 1840. He left his East Providence Township home to volunteer in the 11th Pennsylvania Infantry on Sep. 30th, 1861. His wife gave birth to their first child on March 28th, 1862. Six months later, Frances suffered a shoulder wound at Antietam. The following year, he received a second wound above the elbow at Gettysburg on July 1st, 1863. Francis survived the war and was a father to 13 children. He passed at age 71 and is buried at the Mt. Pleasant Cemetery in Mattie.

Benjamin Lohr was born on August 27th, 1842. He mustered into the 142nd Pennsylvania Infantry on August 22, 1862. Benjamin was captured during the 1st day of Gettysburg. There is no record of how long Benjamin was held as a prisoner of war. He was judged physically unable to continue his enlistment in the 142nd PA Regiment and was transferred to the Veterans Reserve Corp on November 28th, 1863. Benjamin married after the war and raised 10 children and was listed as a farmer in the 1880 East St. Clair Twp. Census. He lived to be 80 years old and is buried in the Grandview Cemetery in Cambria Co.

Gettysburg - July 2, 1863

James Levi Roush was one of only 64 Union soldiers awarded the Medal of Honor at Gettysburg. James was born in Bedford County on February 11th, 1838. He was living near Sarah Furnace when he mustered in the 6th Pennsylvania Reserve Infantry on July 27th, 1861. James was shot in the face during the 2nd Battle of Bull Run, but recovered and rejoined his regiment. On the 2nd Day of Gettysburg, deadly accurate Confederate sharpshooters were targeting Union soldiers from a log house near the foot of Little Round Top. While under heavy fire, James Levi Roush and 5 other men charged the log house and captured all 12 Confederate sharpshooters. James mustered out of the army on June 11th, 1864, with the rest of the 6th PA Reserve Infantry and returned home. He married the following year and was a father to 8 children. James passed at age 68 and is buried at the Saint Patrick's Cemetery in Newry.

James Levi Roush

Little Round Top is pictured in the middle background. A small section of the Big Round Top is visible on the right. James Levi Roush was awarded the Medal of Honor for his bravery near the foot of Little Round Top, close to where the photographer would have been standing when the picture was taken.

Gettysburg - July 1-3, 1863

110th Pennsylvania Infantry, Company C in early 1863. The 110th was in heavy combat in the Rose Woods area of the battlefield near the Wheatfield during the 2nd day at Gettysburg. Many of the Soldiers in this Company were from Woodbury Township. The identities of the soldiers in this picture are not known.

Confederate soldiers gathered for burial at the edge of the Rose Woods, July 5, 1863.

Gettysburg Casualties - July 2, 1863

(Left) John I Miller was 14 years old when he left his Morrisons Cove home to volunteer in the 110th Pennsylvania Infantry on December 19, 1861. He suffered a face wound during the 2nd day of Gettysburg in the Rose Woods area of the battlefield near the Wheatfield. John recovered from his wound and mustered out of the army on April 10th, 1864. He married after the war and raised a family. John passed at age 48 on March 6th, 1896. He is buried in the Bedford Cemetery.

(Right) William W VanOrmer was 20 years old when he mustered as a 1st Sergeant in the 53rd Pennsylvania Infantry on October 10, 1861. He was wounded three times during the Civil War. William suffered a wound at Antietam on September 17th, 1862 near the Sunken Road, at Gettysburg on July 2nd, 1863 in the Wheatfield and at Spotsylvania in May 1864. He received three promotions and mustered out as a Captain on June 30th, 1865. He married in 1868 and was a father of 6 children. William lived to be 81 years old and is buried in the Schellsburg Cemetery.

(Left) Americus Enfield was born on April 7th, 1847. Americus was 14 years old when he mustered into the Knap's Independent Light Artillery, Battery E on June 13th, 1861. He was wounded on the 2nd Day of Gettysburg near Culp's Hill. Americus re-enlisted as a Surgeon in the 22nd Pennsylvania Cavalry on February 24th, 1864. He married in 1870 and was a father to 6 children. Americus was a well-known Doctor in Bedford County after the war and can often be identified in Civil War veterans group pictures by his tall stature and prominent sideburns. He lived to be 83 years old and was buried in the Bedford Cemetery in 1931. Inscribed on his gravestone is "He lived for Humanity."

Ft. Wagner, SC - July 11 & 18, 1863

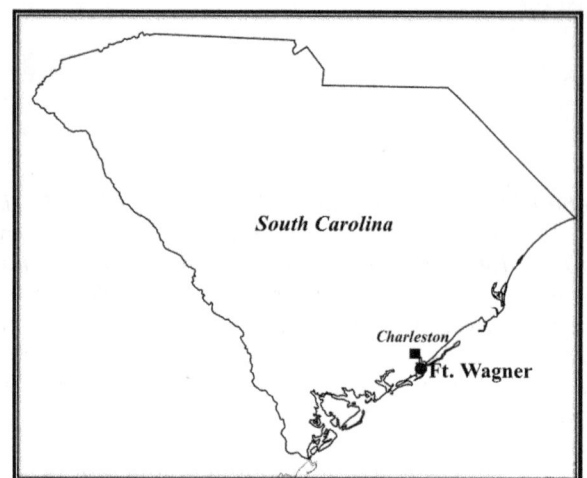

Known Bedford Co. Casualties - 7
KIA - 5
POW-Died – 1
POW - 1

View of Ft. Wagner after being evacuated by the Confederates on September 7, 1863.

Fort Wagner was a heavily defended Confederate beach fortification on Morris Island that guarded the southern approach to Charleston Harbor. The imposing fort had 30-foot-high walls, a water-filled moat studded with spikes, land mines buried in the sand, and was defended by 14 cannons. The narrow beach terrain on Morris Island also made it difficult for Union troops to attack the fort en masse. Bedford County troops in the 76th Pennsylvania Infantry took part in two Union assaults on Fort Wagner in July 1863. On the July 11th charge, no Union soldiers came close to breaching the walls. All the Bedford County casualties in the 76th Pennsylvania Infantry at Fort Wager were during this initial assault. A week later, a larger Union force charged the fort with heavy support of artillery units and ironclad vessels. One of the Union regiments charging Fort Wagner on July 18th was the 54th Massachusetts Infantry, the 2nd Black regiment organized during the Civil War. Troops on that day captured the outer rifle pits surrounding the fortification, but this assault was also repelled. The soldiers of the 54th Massachusetts showed great courage during the attack, which led to additional Black regiments being organized during the Civil War. The Confederates withstood the siege of Fort Wagner for another 6 weeks before evacuating on September 7, 1863. Fort Wagner had historical significance because it was one of the first times Black troops took part in combat during the Civil War and the first African American was awarded the Medal of Honor for heroism during this battle.

Ft. Wagner, SC - July 11 & 18, 1863

Known Bedford Co. Casualty Listing

Name	Muster Age	Rank	State	Regiment	Company	Casualty
Fetter, Joseph	18	Pvt.	PA	76th Infantry	E	KIA
Foor, Daniel V	19	Pvt.	PA	76th Infantry	E	KIA
Kagarice, Ebenezer	20	Pvt.	PA	76th Infantry	C	KIA
Riceling, William	28	Pvt.	PA	76th Infantry	E	POW-Died
Roush, Ernest	32	Pvt.	PA	76th Infantry	E	POW
Steckman, Daniel H	19	Pvt.	PA	76th Infantry	E	KIA
Washabaugh, William H	25	Pvt.	PA	76th Infantry	E	KIA

William P Barndollar was born in Bloody Run (Everett) on April 27th, 1840. He was two days shy of his 21st birthday when he mustered in the 13th Pennsylvania Infantry as a second Lieutenant on April 25th, 1861. Company G of the 13th Regiment was the first unit recruited out of Bedford County during the Civil War. The enlistment was for only 90-days because no one expected the war to last long. William mustered into the 76th Pennsylvania Regiment as a 1st Lieutenant on October 9th, 1861. William led troops during the battles of Pocotaligo, South Carolina in 1862 and Ft. Wagner, South Carolina in 1863 before mustering out on March 10th, 1864. He married in 1865 and was a father to 5 children. William passed when he was 65 years old and is buried in the Everett Cemetery.

(Left) The Storming of Fort Wagner is illustrated in an 1890 lithograph by Kurz and Allison. Two frontal assaults took place within a week on a narrow strip of beach on Morris Island. Troops charging Fort Wagener were under heavy fire from entrenched confederate troops and artillery. Once Union soldiers reached the outer gun pit defenses, fierce hand to hand combat ensued with Confederate troops.

Mine Run - November 27, 1863

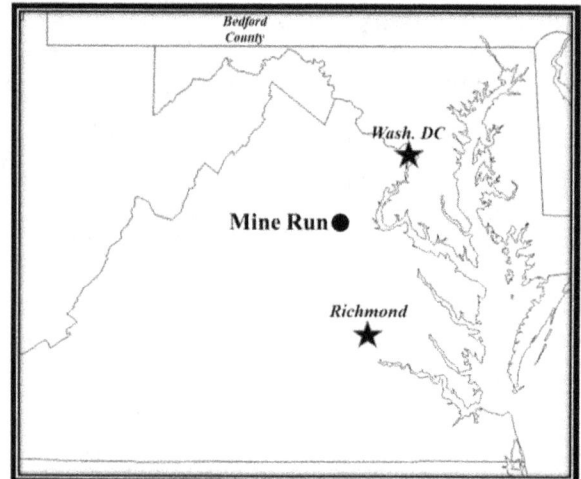

Battle also referred to as Payne's Farm

Known Bedford Co. Casualties - 17
KIA, Wounded & Died – 7
POW & Died - 2
Wounded – 8

Union Signal Corps on a hill above the Rapidan River overlooking the Mine Run Battlefield.

Union commander George Meade made one last attempt to destroy Lee's army before winter. Poor weather, difficult terrain, and shortages of river crossing materials needed to move troops across the rain swollen Rapidan River, led to many delays. On November 27th, the battle of Mine Run surged back and forth with both armies attempting flanking maneuverers. The day ended with neither side gaining an advantage. The Union Army had suffered 952 casualties compared to 545 for the Confederates. That evening, Lee ordered his troops to fall back across Mine Run Creek to a defensive position on higher ground after recognizing the superior number of Union troops on the battlefield. Meade decided against a crossing of Mine Run Creek to avoid another potential Fredericksburg type defeat. On December 2nd, Meade withdrew the Union Army from the banks of Mine Run Creek, ending the campaign. Part of the Mine Run battlefield was in a thick, densely wooded area referred to as the Wilderness. Six months later, Bedford County troops would suffer some of their highest casualties of the war on this same terrain. The following spring, Lincoln replaced Meade with the more aggressive Ulysses S Grant, who had brilliant success against the Confederate army in the Western Theater the previous two years.

Mine Run - November 27, 1863

Known Bedford Co. Casualty Listing

Name	Muster Age	Rank	State	Regiment	Company	Casualty
Burket, David	21	Pvt.	PA	138th Infantry	E	Wounded
Carrell, Daniel	18	Pvt.	PA	138th Infantry	E	Wounded
Dicken, John	21	Pvt.	PA	138th Infantry	D	POW-Died
Hellman, Daniel	18	Pvt.	PA	138th Infantry	D	KIA
Hochard, John A	19	Pvt.	PA	138th Infantry	D	Wounded
Kennard, John H	21	Pvt.	PA	138th Infantry	D	Wounded
Lowery, John E	25	Pvt.	PA	138th Infantry	D	Wound.-Died
Mock, Aaron	21	Pvt.	PA	138th Infantry	D	Wounded
Robb, George W	24	Pvt.	PA	138th Infantry	F	KIA
Smith, Jacob F	32	Pvt.	PA	138th Infantry	F	Wounded
Speck, Henry	39	Pvt.	PA	138th Infantry	E	Wounded
Stiffler, John H	18	Pvt.	PA	138th Infantry	E	Wound.-Died
Stuckey, Simon C	22	1st Sgt.	PA	138th Infantry	D	KIA
Taylor, Matthew P	31	Pvt.	PA	138th Infantry	D	Wounded
Ward, Samuel	32	Pvt.	PA	138th Infantry	E	POW-Died
Wentz, Philip	25	Pvt.	PA	138th Infantry	D	KIA
Whip, Jacob	20	Sgt.	PA	138th Infantry	F	Wound.-Died

Aaron Mock was born on April 7th, 1841 in St. Clair Township. He volunteered in the 138th Pennsylvania Infantry on August 29, 1862. Aaron suffered wounds at the battle of Mine Run but recovered and rejoined his regiment. He was captured during the battle of the Wilderness on May 6, 1864 and taken to Andersonville. Sergeant John B Hammer of the 138th provided the following testimony on Aaron's pension affidavit, "he did not see Aaron from the day of his capture until after the war had ended in July 1865 and the claimant was pretty well used up, being dull of hearing and looked very badly." Emanuel Harbaugh provided the following testimony "I was captured at the same time and place as Aaron and was with him in Andersonville Prison. I watched Aaron crawl from place to place on hands and knees while suffering from Piles and Scurvy. I was also with him in the Florence SC prison camp, and from the treatment received there we had little chance of recovery. We were getting worse all the time. I was finally paroled and sent home; he also came home about the same time in December 1864." Aaron eventually recovered and married after the war. He was a father of 14 children. Aaron lived to be 79 years old and is buried in the Mt. Union Cemetery in Lovely.

5. 1864

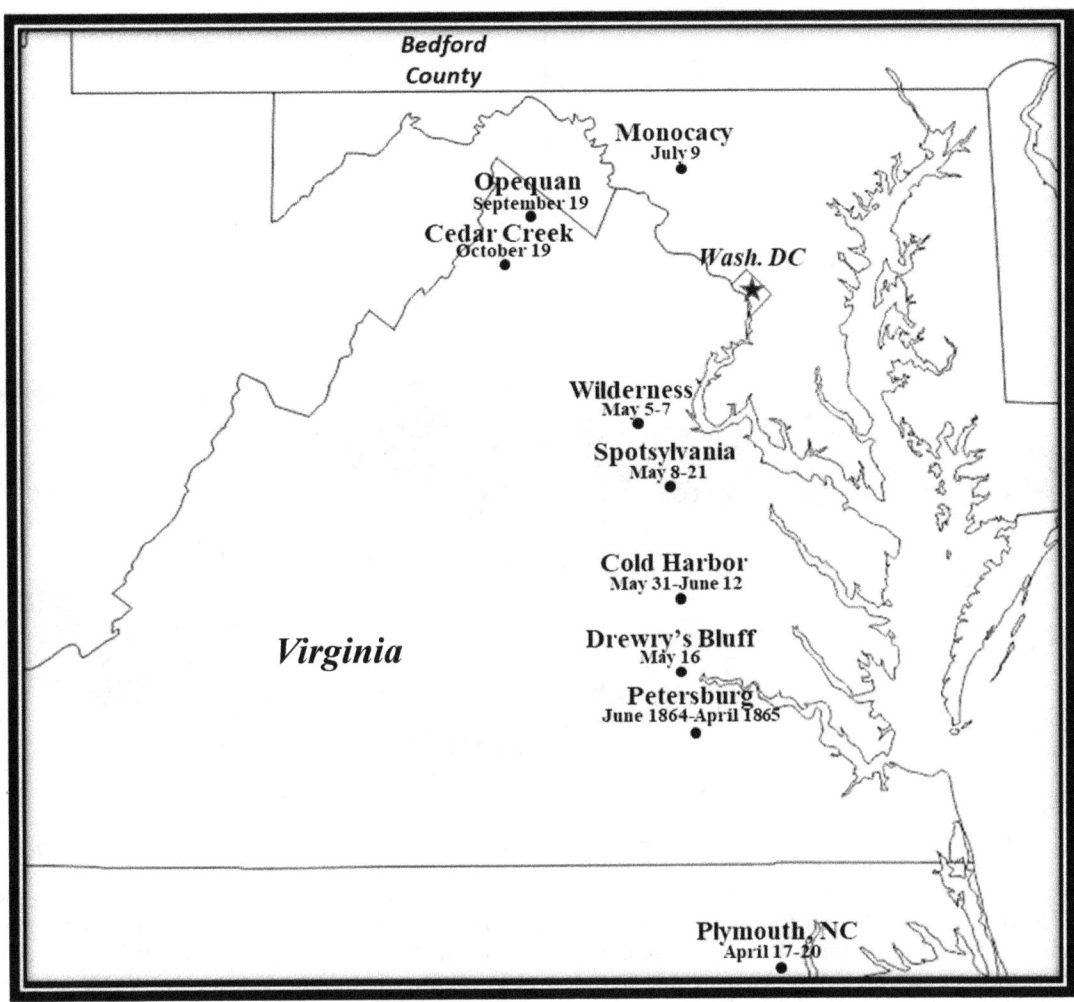

An advisor to Abraham Lincoln commented on the rumors of Ulysses S Grant's drinking habits. Lincoln replied, "Well, I wish some of you would tell me the brand of whiskey Grant drinks. I would like to send a barrel of it to my other generals." Lincoln was often frustrated by the failures of his generals in the Eastern Theater during the first three years of the war, including squandered opportunities to destroy a retreating Confederate army after the battles of Antietam and Gettysburg. Grant had outstanding success in the Western Theater, including a stunning victory at Vicksburg in July 1863. In March 1864, Lincoln promoted Grant to lieutenant general of the entire Union Army. The last person to hold this high rank was George Washington.

Grant lost little time in mounting an offensive against Robert E Lee in the Eastern Theater. Grant and the Union Army relentlessly drove toward Richmond during the Overland Campaign, fighting a series of battles that resulted in massive casualties for both armies. During the first couple weeks of the Overland Campaign, Spotsylvania and the Wilderness became the 3rd and 4th most costly battles of the entire war. Less than 6 weeks after the Overland Campaign began, the Union Army had crossed the James River and were in front of Petersburg where vital railroad lines supplying Richmond and the Confederate army intersected. If Petersburg fell to the Union Army, the immediate evacuation of the capitol city of the Confederacy would be necessary. A series of bloody battles took place during the 9-month Siege of Petersburg.

Grant appointed Philip Sheridan, a young and aggressive cavalry general, Commander of the Union Army in the Shenandoah Valley in August 1864. Sheridan's instructions were to defeat the Rebel army and to lay waste to the rich Shenandoah farming region by burning crops and storage facilities. Sheridan accomplished both during the successful Shenandoah Campaign of 1864.

1864 was the deadliest year of the Civil War for Bedford County. The 670 known casualties were more than twice the number of any other year. Most striking is the number of deaths; 265 soldiers lost their lives, 123 from wounds suffered on the battlefield, and another 142 died from disease, including 81 soldiers who never returned home from prisoner of war camps. Please note some available Civil War records do not provide dates of known casualties, including deaths, therefore it is likely the casualty numbers are higher than identified here in 1864.

The first significant battle for Bedford County soldiers in 1864 was highly tragic. Fifty-two soldiers were taken prisoner on one day during the battle of Plymouth in North Carolina. Fourteen of these soldiers suffered greatly before perishing in southern POW camps.

Plymouth, NC - April 17-20, 1864

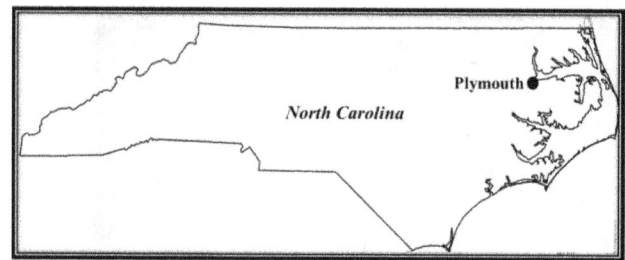

Known Bedford Co. Casualties - 52
POW-Died – 14
Wounded & POW – 2
POW - 36

Photograph of the ironclad ship, the CSS Albemarle taken from the waterfront in Plymouth.

Union soldiers captured the port town of Plymouth in 1862. It was strategically located near the mouth of the Roanoke River in coastal North Carolina. Union troops based in Plymouth launched frequent raids in North Carolina and threatened a vital railroad line that helped supply Robert E Lee's army. The Confederates planned a joint land and river assault in the spring of 1864 to dislodge the 3,000 troop Union garrison at Plymouth. On April 17th, over 10,000 Confederate troops were repulsed during multiple attacks on Union fortifications around Plymouth. Early on the morning on April 19th, the newly built Confederate Ironclad, the CSS Albemarle, attacked the two most powerful Union ships on the Roanoke. The CSS Albermarle sunk one ship and severely damaged the other. That same afternoon, the Confederates renewed land attacks on the Union fortifications at Plymouth, supported by the rifled cannon fire from the CSS Albermarle. The Union garrison was surrounded on land and being bombarded from the river and left with no other options but surrender on April 20th. With few exceptions, the enlisted men captured at Plymouth were taken to Andersonville and the officers sent to other POW camps in South Carolina and Georgia.

Known Bedford Co. Casualty Listing

Name	Muster Age	Rank	State	Regiment	Company	Casualty
Adams, David	22	Pvt.	PA	101th Infantry	K	POW
Anderson, James	21	Pvt.	PA	101th Infantry	G	POW-Died
Banner, Adam	26	Pvt.	PA	103rd Infantry	E	POW
Beam, Daniel	59	Pvt.	PA	101th Infantry	D	POW-Died
Beegle, David F	26	1st Lt.	PA	101th Infantry	D	POW
Beltz, Abraham	32	Pvt.	PA	101th Infantry	G	POW-Died
Bequeth, William H	17	Pvt.	PA	101th Infantry	D	POW
Bessor, John		Corp.	PA	101th Infantry	D	POW-Died
Brown, Jacob D	24	Corp.	PA	101th Infantry	D	POW
Brown, John W	25	Pvt.	PA	101th Infantry	D	POW
Brown, Joseph L	33	Pvt.	PA	101th Infantry	G	Wound.-POW
Brown, Samuel D	19	Pvt.	PA	101th Infantry	D	POW

Plymouth, NC - April 17-20, 1864

Known Bedford Co. Casualty Listing

Name	Muster Age	Rank	State	Regiment	Company	Casualty
Carnell, Samuel	24	Corp.	PA	101th Infantry	D	POW
Clingerman, Peter	21	Pvt.	PA	101th Infantry	D	POW
Compher, Alexander	41	Capt.	PA	101th Infantry	D	POW
Conley, Isaiah	31	1st Lt.	PA	101th Infantry	G	POW
Defibaugh, John	34	Pvt.	PA	101th Infantry	G	POW-Died
England, Jacob	21	Pvt.	PA	101th Infantry	D	POW
Foor, Francis L	19	Pvt.	PA	101th Infantry	D	POW-Died
Gollipher, Justice	23	Pvt.	PA	101th Infantry	G	POW-Died
Hanks, Benjamin A	28	Sgt.	PA	101th Infantry	D	POW-Died
Hanks, Caleb	19	Pvt.	PA	101th Infantry	D	POW-Died
Hanks, David F	18	Pvt.	PA	101th Infantry	D	POW-Died
Hanks, Nelson	28	Pvt.	PA	101th Infantry	D	POW-Died
Hazelett, Moses	19	Pvt.	PA	101th Infantry	B	POW
Helm, John B	27	2nd Lt.	PA	101th Infantry	G	POW
Hetrick, Daniel L	16	Pvt.	PA	101th Infantry	D	POW
King, Thomas	50	1st Lt.	PA	101th Infantry	G	POW
Knipple, William H	18	Corp.	PA	101th Infantry	G	POW
Lightningstar, Augustus	25	Pvt.	PA	101th Infantry	G	POW
Linn, Henry	22	Sgt.	PA	101th Infantry	D	POW
Longenecker, Jacob H	22	1st Lt.	PA	101th Infantry	D	POW
Martin, James P	18	Pvt.	PA	101th Infantry	F	POW
McEldowney, George E	27	Pvt.	PA	101th Infantry	D	POW
McEldowney, Samuel J	34	1st Sgt.	PA	101th Infantry	D	POW
Miller, John (1)	33	Pvt.	PA	101th Infantry	G	POW-Died
Mills, Andrew J	21	Pvt.	PA	101th Infantry	D	POW
Mills, Franklin G	18	Musician	PA	101th Infantry	D	Wound.-POW
Mullin, David W	33	Capt.	PA	101th Infantry	G	POW
Oler, James W	20	Pvt.	PA	101th Infantry	D	POW
Potter, Martin L	18	Pvt.	PA	101th Infantry	D	POW
Rice, Abraham	27	Sgt.	PA	101th Infantry	D	POW
Rice, Isaac	27	Corp.	PA	101th Infantry	D	POW-Died
Shoemaker, George F	20	Corp.	PA	101th Infantry	D	POW
Siler, James P	20	Pvt.	PA	101th Infantry	D	POW
Slick, Thomas W	23	Pvt.	PA	101th Infantry	G	POW
Smith, Amos	18	Corp.	PA	101th Infantry	D	POW
Smith, Joseph A	19	Pvt.	PA	101th Infantry	D	POW
Sparks, William	15	Pvt.	PA	101th Infantry	D	POW
Stone, Reuben M	26	Sgt.	PA	101th Infantry	D	POW
Vaughan, Ephraim	18	Musician	PA	101th Infantry	D	POW
Wilson, George W		Pvt.	PA	101th Infantry	D	POW-Died

Plymouth, NC - April 17-20, 1864

Alexander Compher was born in Friends Cove on October 19th, 1820. He married Mary Ann Whetstone in January 1844. Tragically, Mary Ann and their two young children died at separate times within a 5-year period. Alexander remarried in November 1849. Alexander and Barbara Ann Mills Compher were living on a farm in Colerain Township with their 4 children when he mustered as a Captain in the 101st Pennsylvania Infantry in February 1862. On April 17th, 1864, over 10,000 Confederate troops surrounded the garrison defending Plymouth, North Carolina, and around 3,000 men from several Union regiments were forced to surrender. Alexander was taken to a Columbia, South Carolina, prisoner of war camp. He survived Confederate captivity and returned home to his wife and children on March 12th, 1865. Alexander Compher was listed as a hotel keeper in Altoona on the 1870 census. In 1875, he was appointed as Postmaster of Stuckeysville in Bedford County. He later moved to Nebraska and died at age 81 in 1902 in Colorado. Alexander is buried at the Fairmont Cemetery in Fillmore County, Nebraska.

The "Field of Honor" is a memorial park in remembrance of those who were engaged at the Battle of Plymouth, NC in April 1864.

A fort at Plymouth is named in Alexander Compher's honor. The Field of Honor began April 2014 with the erection of the Battle of Plymouth monument on the 150th Anniversary of the battle. This monument was erected by the Civil War Plymouth Pilgrims Descendants Society in conjunction with the Washington County, NC Historical Society.

According to the Civil War Plymouth Pilgrims Descendants Society website, there are plans for this field to evolve into a beautiful park that will give a peaceful and respectful place to come and honor ancestors and those with whom they served. A Unity Walkway was unveiled on the 155th Anniversary in April 2019.

Field of Honor

Located at the site of Fort Compher, Plymouth, NC

Plymouth, NC - April 17-20, 1864

(Left) Jacob H Longenecker was raised on a farm near Woodbury. He was a 22-year-old school principal when he mustered as a private in the 101st Pennsylvania Infantry on June 20th, 1862. He received a promotion to 1st Lieutenant and Adjutant in 1863. Jacob was captured with the rest of the garrison at Plymouth, NC, and spent time in several POW camps. In early 1865, he escaped from a prison in Charlotte, NC, but was recaptured two weeks later. He was exchanged for a Confederate officer on March 2, 1865, and mustered out of the army after the war ended. Jacob married in 1869 and was a father to 3 sons. He served as a Bedford County judge from 1890 until 1901. His picture is on display at the Courthouse. Jacob passed when he was 77 and was buried at the Bedford Cemetery in 1916.

(Right) William Sparks was 15 years old when he left his home in West Providence Township to muster in the 101st Pennsylvania Infantry on November 1st, 1861. He was captured at Plymouth, NC, and was taken to Andersonville. On October 2nd, 1864, he was transferred to a POW camp in Florence, South Carolina. William was released on March 3rd, 1865 and mustered out of the army on June 25th, 1865. He married in 1869 and was a father to two sons. William passed when he was 68 and is buried at the Wilmore Brethren Church Cemetery in Cambria County.

(Left) David F Beegle was born on June 2nd, 1835. He mustered as a 1st Lieutenant in the 101st Pennsylvania Infantry on Nov. 1st, 1861 leaving behind a wife, a toddler, and a newborn at their home in Waterside. He was captured along with the entire garrison at Plymouth, NC in 1864. David survived captivity and returned home after the war. He passed at age 64 and is buried in the Oak Ridge Cemetery in Altoona.

Wilderness - May 5 - 7, 1864

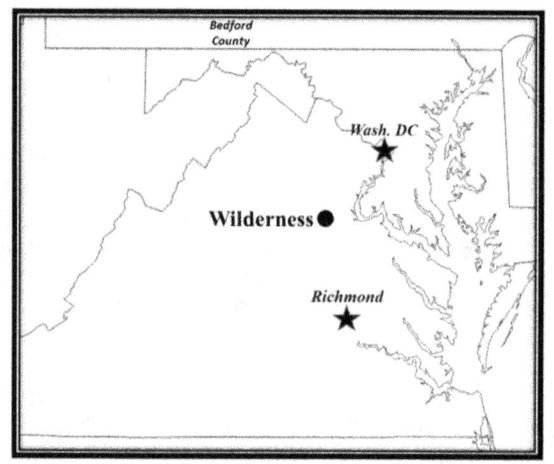

Known Bedford Co. Casualties - 65
KIA, MIA, Wounded & Died – 16
POW & Died - 6
Wounded & POW – 2
POW - 2
Wounded - 39

Abraham Lincoln promoted Ulysses S Grant to lieutenant general of the army in the Spring of 1864 after demonstrating success in the Western Theater of the War the previous two years. Grant faced off against Robert E Lee for the first time at the battle of the Wilderness. Lee positioned his troops in and around a heavily wooded area with dense undergrowth to negate the numerical advantages of the Union Army. The three-day battle erupted into a series of chaotic skirmishes in difficult to maneuver terrain. Gunsmoke and fires created by exploding shells made it difficult for both armies to identify enemy forces. On May 7th, the battle of the Wilderness was at a stalemate. The Union Army had suffered around 17,000 casualties compared to 10,000 for the Confederates. Grant withdrew from the Wilderness battlefield and swung around the right flank of Lee's lines, repositioning the Union Army between the Confederate army and Richmond. Union troops famously cheered when they realized Grant was continuing their march south instead of retreating. The Wilderness was the 1st battle of the Overland Campaign. Grant had promised Lincoln the Union Army would relentlessly press the Rebel Army and continue the drive toward the Confederate capital, regardless of the outcome of individual battles. Grant was determined to keep this promise.

Known Bedford Co. Casualty Listing

Name	Muster Age	Rank	State	Regiment	Company	Casualty	Casualty Date
Ake, John G	20	Pvt.	PA	138th Infantry	E	Wounded	5/6/64
Ake, William H	25	Corp.	PA	138th Infantry	E	KIA	5/6/64
Amick, William B	26	Sgt.	PA	138th Infantry	E	KIA	5/6/64
Armstrong, Albert	19	Pvt.	PA	138th Infantry	F	Wounded	5/6/64
Baner, Franklin	17	Pvt.	PA	138th Infantry	F	Wounded	5/6/64
Baughman, George	22	Sgt.	PA	138th Infantry	D	POW-Died	5/6/64
Beals, Nicholas H	22	Pvt.	PA	138th Infantry	D	Wounded	5/6/64
Bessor, Philip	21	Sgt.	PA	6th Reserves	D	Wounded	
Biddle, Andrew B	16	Pvt.	PA	138th Infantry	E	Wounded	5/5/64
Brant, Andrew J	37	Pvt.	PA	149th Infantry	B	Wounded	5/5/64
Brant, Benjimen F	32	Pvt.	PA	149th Infantry	B	Wounded	5/5/64
Brumbaugh, Francis M		Pvt.	PA	110th Infantry	C	Wound.-Died	5/5/64
Carrell, Joseph	20	Pvt.	PA	138th Infantry	E	Wounded	5/6/64
Claycomb, Conrad	26	Pvt.	PA	138th Infantry	E	Wounded	5/6/64
Cook, John H	17	Pvt.	PA	138th Infantry	E	Wound.-POW	5/6/64
Cook, Levi	18	1st Sgt.	PA	138th Infantry	F	Wounded	5/6/64
Corle, Abraham	22	Pvt.	PA	138th Infantry	E	Wound.-Died	5/6/64
Corle, William C	21	Pvt.	PA	138th Infantry	D	Wounded	5/6/64
Defibaugh, William H	18	Pvt.	PA	138th Infantry	E	MIA	5/6/64
Eichelberger, Eli	24	Capt.	PA	8th Reserves	F	Wounded	5/6/64
Evans, Johnson	22	Pvt.	PA	8th Reserves	F	Wounded	5/6/64
Fritz, Daniel E	25	Corp.	PA	18th Cavalry	K	POW-Died	5/5/64

Wilderness - May 5 - 7, 1864

Known Bedford Co. Casualty Listing

Name	Muster Age	Rank	State	Regiment	Company	Casualty	Casualty Date
Gaudig, Herman	39	Pvt.	MA	15th Infantry	D	Wounded	
Grim, William	30	Pvt.	PA	149th Infantry	K	POW-Died	5/5/64
Grimes, Henry W	20	Pvt.	PA	84th Infantry	C	Wound.-Died	5/6/64
Hammer, John B	21	Sgt.	PA	138th Infantry	D	Wounded	5/6/64
Harbaugh, Emanuel	19	Pvt.	PA	138th Infantry	D	POW	5/6/64
Hardinger, George W	28	Pvt.	PA	149th Infantry	B	POW-Died	5/5/64
Hays, Alexander Y	17	Pvt.	PA	110th Infantry	C	Wounded	5/6/64
Hellman, George	21	Pvt.	PA	138th Infantry	D	Wounded	5/6/64
Heltzel, Jonathan D	19	Pvt.	PA	110th Infantry	C	KIA	5/6/64
Hixon, Erastus J	21	Corp.	PA	138th Infantry	D	KIA	5/6/64
Hochard, John A	19	Pvt.	PA	138th Infantry	D	Wounded	5/6/64
Holler, George W	36	Pvt.	PA	138th Infantry	F	Wounded	5/6/64
Horton, David	24	Sgt.	PA	8th Reserves	F	Wounded	
Huffman, Josiah	22	Corp.	PA	138th Infantry	D	Wounded	5/6/64
Kegg, Levi	40	Corp.	PA	149th Infantry	B	Wounded	5/5/64
Kegg, Nathaniel	21	Pvt.	PA	138th Infantry	E	Wounded	
King, Harrison H	23	Corp.	PA	138th Infantry	E	Wounded	5/5/64
Lay, Joseph	28	Pvt.	PA	138th Infantry	E	MIA	5/6/64
Leighty, George	24	Pvt.	PA	8th Reserves	F	Wounded	5/6/64
Leonard, John D	21	Pvt.	PA	138th Infantry	E	Wounded	5/6/64
Maugle, Joseph	22	Pvt.	PA	8th Reserves	F	Wounded	5/8/64
May, Jacob	26	Pvt.	PA	149th Infantry	I	MIA	5/5/64
May, John W	23	Corp.	PA	138th Infantry	F	Wounded	5/6/64
McCleary, Henry	22	Sgt.	PA	138th Infantry	D	Wounded	5/6/64
Meloy, Biven D	22	Pvt.	PA	138th Infantry	E	Wounded	5/6/64
Miller, Henry	33	Pvt.	PA	138th Infantry	F	Wound.-Died	5/6/64
Miller, Jackson	22	Sgt.	PA	138th Infantry	F	KIA	5/6/64
Miller, Thomas J	23	Pvt.	PA	138th Infantry	D	POW-Died	5/6/64
Mock, Aaron	21	Pvt.	PA	138th Infantry	D	Wound.-POW	5/6/64
Morgart, William	23	Pvt.	PA	18th Cavalry	K	POW-Died	5/5/64
Mountain, George R	30	Pvt.	PA	84th Infantry	E	Wounded	5/6/64
Price, Joseph J	20	Corp.	PA	138th Infantry	D	KIA	5/6/64
Robinson, Thomas S		Pvt.	VT	2nd Infantry	H	Wounded	5/6/64
Robinson, William J	23	Pvt.	PA	138th Infantry	E	Wounded	5/6/64
Shaffer, Joseph H		Pvt.	MD	6th Infantry	D	POW	5/5/64
Smith, Jacob F	32	Pvt.	PA	138th Infantry	F	Wound.-Died	5/6/64
Smith, John W	22	Corp.	PA	110th Infantry	C	Wounded	5/6/64
Smith, Miles N	18	Pvt.	PA	138th Infantry	E	Wounded	5/6/64
Snyder, Jonathan	30	1st Sgt.	PA	138th Infantry	D	Wounded	5/6/64
Steuby, Conrad G	37	Pvt.	PA	138th Infantry	F	Wounded	5/6/64
Summerville, Charles	21	Pvt.	PA	138th Infantry	D	MIA	5/6/64
Swaney, William S	20	Pvt.	PA	110th Infantry	C	Wounded	
Wolford, William	30	Pvt.	PA	149th Infantry	B	KIA	5/5/64

Wilderness - May 5 - 7, 1864

"Fire in the Wilderness" is an illustration of Union soldiers carrying comrades to safety from nearby fires in the background. Countless numbers of wounded Union and Confederate soldiers perished in raging infernos on the night of May 6th.

Skulls and bones of unburied soldiers remained on the Wilderness Battlefield a year after the battle in 1865.

Wilderness - May 5 - 7, 1864

(Left) Lewis A May was born in Rainsburg on December 23rd, 1824. He married Anna Margaret Weisel in 1847. Lewis volunteered as a Captain in the 138th Pennsylvania Infantry on August 29th, 1862 while his wife and several children waited for his return at their home in Cumberland Valley. He led Bedford County soldiers in combat at some of the most pivotal battles in the Civil War. Lewis received a promotion to Lt. Colonel on February 12th, 1865 and was present at the Confederate surrender at Appomattox. After the war he worked as a blacksmith and made chairs in the Bedford area. Lewis lived to be 82 years old and is buried in the Woods Church Cemetery in Colerain Township. His obituary stated he was one of the bravest and most gallant soldiers who entered the service from this county.

(Right) David Horton was born in Hopewell on March 7th, 1837. He mustered as a Corporal in the Hopewell Rifles Company of the 8th Pennsylvania Reserves on June 11th, 1861. David was one of the tallest soldiers in his regiment at 6'4". He was captured during the Union withdrawal to Gaines Mills with many other soldiers in the Hopewell Rifles on June 27th, 1862. David suffered a head and knee wound at the battle of the Wilderness in early May 1864. He mustered out as a sergeant on May 26th, 1864. David married in 1866 and was a father to six children. He passed at age 76 and is buried at the Maplewood Cemetery in Elkins, West Virginia.

(Left) Andrew B Biddle was born in Bedford Township on November 4th, 1842. He left his Bedford home to volunteer in the 138th Pennsylvania Infantry on August 29th, 1862. Andrew suffered a wound on May 5th, 1864, during the Battle of the Wilderness. He recovered from his wounds and mustered out at the end of the war on April 22nd, 1865. Andrew married in 1868 and was a father of 5 sons. He passed at age 75 and is buried at the Trinity United Church Cemetery in Colerain Township.

Spotsylvania - May 8 - 21, 1864

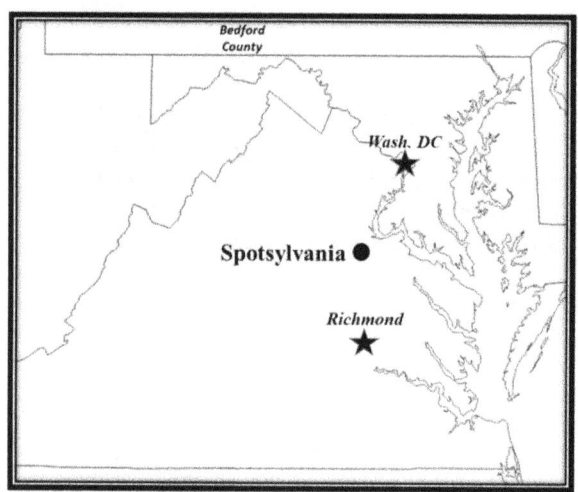

Battle also referred to as Spotsylvania Court House

Known Bedford Co. Casualties - 35
KIA, MIA, Wounded & Died – 10
POW - 3
Wounded - 22

Above are Confederate entrenchments at the Muleshoe Salient. Confederate troops feverishly constructed these earthworks on May 8th, 1864, to prepare for a Union Army attack they knew was coming.

Spotsylvania - May 8 - 21, 1864

Known Bedford Co. Casualty Listing

Name	Muster Age	Rank	State	Regiment	Company	Casualty	Casualty Date
Bessor, Philip	21	Sgt.	PA	6th Reserves	D	Wounded	
Carn, Jeremiah	24	Pvt.	PA	149th Infantry	D	Wounded	5/8/64
Claybaugh, James	26	Pvt.	PA	3rd Cavalry	G	POW	5/15/64
Cleaver, James	23	2nd Lt.	PA	8th Reserves	F	Wounded	5/10/64
Eastright, Christian	24	Pvt.	PA	8th Reserves	F	Wound.-Died	5/8/64
Evans, Isaiah	27	Pvt.	PA	149th Infantry	I	KIA	5/8/64
Feather, Simon	17	Pvt.	PA	138th Infantry	E	Wounded	5/19/64
Fleegle, Daniel	18	Pvt.	PA	93rd Infantry	I	KIA	5/12/64
Foster, Aaron	25	Pvt.	PA	8th Reserves	F	Wounded	5/8/64
Gates, George W	18	Pvt.	PA	13th Reserves	D	Wounded	5/10/64
Glessner, Philip	34	Pvt.	PA	148th Infantry	G	Wounded	5/13/64
Holdcraft, William	38	Pvt.	PA	8th Reserves	F	Wounded	
Hurley, William	32	Pvt.	PA	84th Infantry	E	Wounded	5/12/64
Juda, George V A	20	Corp.	PA	8th Reserves	F	Wound.-Died	5/8/64
Lashley, Isaac W	21	Pvt.	MD	8th Infantry	C	Wound.-Died	5/8/64
Leighty, Joseph	22	Pvt.	PA	8th Reserves	F	Wounded	5/12/64
Livingston, Thomas G	24	2nd Lt.	PA	110th Infantry	C	Wounded	5/12/64
Malone, Charles		Pvt.	PA	8th Reserves	F	Wounded	5/8/64
Manspeaker, David	24	Pvt.	PA	8th Reserves	F	KIA	5/13/64
Marshall, Henry	23	Pvt.	PA	8th Reserves	F	POW	5/8/64
McDaniel, Hiram	19	Pvt.	PA	149th Infantry	I	Wound.-Died	5/10/64
McFarland, Joseph	22	Pvt.	PA	8th Reserves	F	Wounded	5/8/64
Miller, George W	31	Pvt.	PA	149th Infantry	I	Wounded	5/10/64
O'Neal, John E	19	Corp.	PA	138th Infantry	D	Wounded	5/12/64
Price, Daniel J	21	Pvt.	PA	138th Infantry	E	Wounded	5/12/64
Reed, Isaac	23	Pvt.	PA	143rd Infantry	E	Wound.-Died	5/10/64
Ritchey, William D	21	Corp.	PA	8th Reserves	F	POW	5/8/64
Shoeman, Peter	30	Pvt.	PA	49th Infantry	H	Wound.-Died	5/10/64
Speer, William H	18	Pvt.	PA	110th Infantry	C	Wounded	5/12/64
Steele, Thomas	23	Pvt.	PA	149th Infantry	K	Wounded	5/8/64
Stonerook, Simon B	20	Sgt.	PA	110th Infantry	C	Wounded	
Swaney, William S	20	Pvt.	PA	110th Infantry	C	Wounded	
VanOrmer, William W	20	Capt.	PA	53rd Infantry	I	Wounded	
Williams, John P	26	Pvt.	PA	8th Reserves	F	Wounded	5/8/64
Woodcock, George W	18	Pvt.	PA	53rd Infantry	C	Wound.-Died	5/12/64

Spotsylvania was the 2nd battle of the Overland Campaign. The Confederates quickly built earthen work fortifications along a mile-long front called the Mule Shoe Salient on May 8th. During the next 12 days, the Union Army launched a series of attacks attempting to break through the entrenched Confederate lines. The most vicious hand to hand fighting of the entire war took place over a 20-hour period near the "Bloody Angle" section of the Mule Shoe Salient during a rainstorm on May 12th. The following is a description of the scene from a Union Officer, "Our own killed were scattered over a large space near the Angle while in front of the captured breastworks the enemy's dead, vastly more numerous than our own, were piled upon each other in some places four layers deep, exhibiting every ghastly phase of mutilation. Below the mass of fast-decaying corpses, the convulsive twitching of limbs and the writhing of bodies showed that there were wounded men still alive and struggling to extricate themselves from the horrid entombment." A Confederate soldier wrote the following description of the repeated attacks at the Bloody Angle, "Many of them (Union soldiers) were shot dead and sank down on the breastworks without pulling their feet out of the mud. Many others plunged forward when they were shot and fell headlong into the trench among us. All the time a drizzling rain was falling. The blood shed of the dead and wounded in the trench mixed with the mud and water. It became more than shoe deep, and soon it was smeared all over our clothes. We could hardly tell one another apart." On May 21st, Grant again disengaged his army from a bloody stalemate and swung around the right flank of the Confederate lines to continue the march toward Richmond. Spotsylvania ended as an inconclusive battle with 18,000 Union and 11,000 Confederate casualties.

Spotsylvania - May 8 - 21, 1864

Aaron Foster was born in the Round Knob area of Broad Top Township on May 17th, 1836. He married Martha Ellen Dachenbach in 1859. Aaron mustered in the Hopewell Rifles Company of the 8th Pennsylvania Reserves on June 11th, 1861, leaving behind a wife and young daughter on their Broad Top farm. He was taken prisoner during the withdrawal to Gaines Mill on June 27th, 1862, and later released. Later that year, Aaron injured his shoulder at the battle of Fredericksburg. He suffered wounds to his lung and foot on May 8th, 1864 during the battle of Spotsylvania. Aaron remained a cripple for the rest of his life from these wounds. He mustered out of the Union Army with the 8th PA Reserve Regiment on May 26th, 1864. Aaron was a father to 8 more children after the war. He passed at age 62 and is buried at Duvalls Cemetery in Coaldale.

Above is a painting of the "Attack at the Bloody Angle" by Thure de Thulstrup. The Bloody Angle was close to the apex of the Muleshoe Salient and site of horrific hand to hand fighting.

Spotsylvania - May 8 - 21, 1864

Thomas G Livingston was born in Hopewell on March 17th, 1837. He mustered in the 110th Pennsylvania Infantry on October 24th, 1861. A Minie ball passed through his hip and lower back during hand-to-hand combat at the Mule Shoe Salient on May 12th. Thomas recovered from these wounds and was promoted to 1st Sergeant on March 18, 1865. He received another promotion to 2nd Lieutenant just before the war ended. He returned home after the war and was a father to 6 children. Thomas passed at age 70 and was buried at the Everett Cemetery in 1909.

Daniel J Price left his home in Bedford Township to volunteer in the 138th Pennsylvania Infantry on Aug 29th, 1862. The following year the first of his 12 children was born. Daniel suffered a wound during the Assault on the Mule Shoe Salient on May 12, 1864, and was transferred to the Veterans Reserve Corp the next day. During the Civil War, it was common for soldiers to be assigned to the VRC when they were no longer physically fit for combat. He mustered out of the army on July 3, 1865. Daniel lived to be 88 years old and was buried at the Trinity United Church of Christ Cemetery in Osterburg in 1929.

Bermuda Hundred Campaign - May 1864

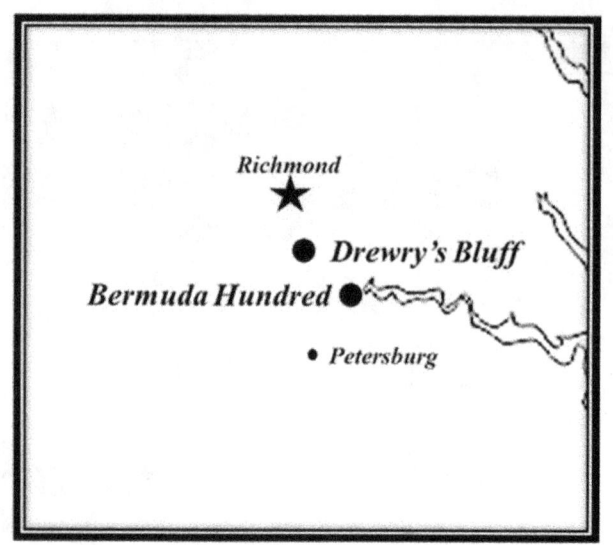

Drewry's Bluff - Bedford Co. Casualties - 52
KIA, MIA, Wounded & Died – 7
POW & Died - 15
Wounded & POW – 2
POW - 13
Wounded - 15

Bermuda Hundred - Casualties - 17
KIA, MIA, Wounded & Died – 5
Wounded - 12

During the Overland Campaign, a separate Union offensive was taking place closer to Richmond. Union troops disembarked from a transport ship at the Bermuda Hundred Landing about 15 miles from Richmond on May 5th. The Bermuda Hundred Campaign began with about 25,000 Union soldiers marching toward the Confederates garrison defending Fort Darling on Drewry's Bluff that guarded the James River approach to Richmond. The primary objectives of the Bermuda Hundred Campaign were to interrupt Confederate supply lines, disrupt communications, and occupy the Rebel troops around Richmond to deny Robert E Lee the opportunity for reinforcements during Ulysses Grant's Overland Campaign. Bedford County soldiers took heavy casualties when 18,000 Confederates counter-attacked Union forces near Drewry's Bluff on May 16th. The Union Army accomplished some of their primary objectives, but the Bermuda Hundred Campaign ended as a modest victory for the Confederates. Losses were high as a percentage of troops on the battlefield. The Union Army suffered around 4,000 casualties compared to 3,000 for the Confederate troops.

Confederate Fort Darling (upper left) was located 90 feet above the James River on Drewry's Bluff.

Drewry's Bluff - May 15 & 16, 1864

Known Bedford Co. Casualty Listing

Name	Muster Age	Rank	State	Regiment	Company	Casualty
Adams, Philip	18	Pvt.	PA	55th Infantry	H	POW-Died
Allison, Edward	18	Pvt.	PA	55th Infantry	K	POW-Died
Amos, John B	28	1st Sgt.	PA	55th Infantry	D	Wounded
Anderson, Henry	19	Pvt.	PA	55th Infantry	H	POW
Barnhart, John H	27	2nd Lt.	PA	55th Infantry	D	KIA
Bartlebaugh, Silas M	20	1st Sgt.	PA	55th Infantry	I	POW
Beaver, Simon J	16	Pvt.	PA	55th Infantry	H	POW
Boor, William A	21	Sgt.	PA	55th Infantry	D	POW
Bowser, David	22	Pvt.	PA	55th Infantry	K	Wounded
Bridaham, Henry W	29	Pvt.	PA	55th Infantry	H	POW-Died
Burket, Frederick	33	Pvt.	PA	55th Infantry	K	POW-Died
Corle, Alexander B	18	Pvt.	PA	55th Infantry	K	POW
Dasher, James H	23	Pvt.	PA	55th Infantry	A	POW
Detwiler, Joseph	43	Pvt.	PA	55th Infantry	K	Wounded
Ealy, John C	18	Sgt.	PA	55th Infantry	H	Wounded

View from within Fort Darling of the James River taken during the Civil War.

Drewry's Bluff - May 16, 1864

Known Bedford Co. Casualty Listing

Name	Muster Age	Rank	State	Regiment	Company	Casualty
Exline, Jacob	22	Pvt.	PA	55th Infantry	K	POW-Died
Feather, William	22	Pvt.	PA	55th Infantry	K	Wounded
Filler, Joseph	49	Lt. Col	PA	55th Infantry	K	Wounded
Hainsey, Frederick	22	Sgt.	PA	55th Infantry	H	Wound.-POW
Hand, Henry		Pvt.	PA	55th Infantry	H	POW-Died
Hissong, Josiah	22	2nd Lt.	PA	55th Infantry	H	Wounded
Hunt, John T	25	Corp.	PA	55th Infantry	K	POW-Died
James, John W	21	Pvt.	PA	55th Infantry	I	Wound.-POW
Kennedy, Samuel	20	Corp.	PA	55th Infantry	D	KIA
King, Samuel T	15	Pvt.	PA	55th Infantry	H	Wounded
Lemon, Henry	18	Corp.	PA	55th Infantry	H	Wounded
Lingenfelter, David	42	Pvt.	PA	55th Infantry	I	POW
Lybarger, Henry G	22	Pvt.	PA	55th Infantry	D	Wounded
Mars, John	21	Pvt.	PA	55th Infantry	H	POW-Died
McGee, James	22	Pvt.	PA	55th Infantry	I	POW-Died
Metzger, James	25	Capt.	PA	55th Infantry	D	POW
Meyers, Levi	19	Pvt.	PA	55th Infantry	H	POW-Died
Miller, Matthew	36	Pvt.	PA	55th Infantry	D	MIA
Miller, Solomon H	29	Sgt.	PA	55th Infantry	H	POW-Died
Mitchell, James P	18	Pvt.	PA	55th Infantry	H	POW-Died
Noffsker, Martin	23	Pvt.	PA	55th Infantry	I	POW
Nottingham, William	31	Pvt.	PA	55th Infantry	D	POW
Prosser, David W	15	Corp.	PA	55th Infantry	D	POW
Radebaugh, Jacob L	19	Sgt.	PA	55th Infantry	K	MIA-Died
Shaffer, George W	15	Pvt.	PA	55th Infantry	K	Wounded
Shaffer, Jacob J	19	Pvt.	PA	55th Infantry	H	POW-Died
Sholl, Isaac	20	Pvt.	PA	55th Infantry	H	Wounded
Shrader, William O	18	Pvt.	PA	55th Infantry	H	POW-Died
Smith, Jesse	40	Pvt.	PA	55th Infantry	D	Wound.-Died
Smith, Samuel		Pvt.	PA	55th Infantry	I	KIA
Steckman, Levi	19	Pvt.	PA	55th Infantry	D	Wounded
Stickler, Samuel	23	Pvt.	PA	55th Infantry	D	POW-Died
Thompson, Jeremiah H	32	Pvt.	PA	55th Infantry	D	POW
Wentz, Isaac	21	Pvt.	PA	55th Infantry	K	Wounded
White, James S	41	Pvt.	PA	55th Infantry	D	Wound.-Died
Whysong, Samuel	29	Pvt.	PA	55th Infantry	K	Wounded
Wonderly, William H	19	Pvt.	PA	55th Infantry	I	POW

Drewry's Bluff - May 16, 1864

John B Amos was born in Bedford on April 13th, 1833 and his wife Sarah on March 5th, 1835. They married in 1852. John mustered in the 55th Pennsylvania Infantry on October 12th, 1861, while Sarah and four small children waited for his return at their home in Bedford. He received two promotions to Corporal and Sergeant in 1862. John suffered a wound at Drewry's Bluff on May 16th, 1864. Two weeks later, he suffered wounds to his jaw and neck during the early morning charge at Cold Harbor on June 3rd, 1864. The photograph on the right may have been taken prior to this wound. John received another promotion to 1st Sergeant on September 9th, 1864, 2nd Lieutenant on June 8th, 1865, before mustering out of the army in August 1865. Sarah and John were parents of eight more children after the war. John was an outdoor enthusiast and frequently hired as a guide for fishing excursions by guests staying at local resorts. He lived to be 67 years old and was buried at the Bedford Cemetery in 1901. His obituary stated he was an honest, kind-hearted man and had many friends who sincerely mourned his death.

William A Boor was a 21-year-old Silversmith living in Cumberland Valley when he volunteered in the 13th Pennsylvania Infantry on April 25th, 1861. This unit was the first Civil War Company organized in Bedford County. He mustered in the 55th Pennsylvania Infantry on October 12th, 1861 as a corporal. William received a promotion to Sergeant on July 1st, 1863. During the attack on Drewry's Bluff on May 16th, 1864, William was captured and taken to Andersonville. He was exchanged in Savannah on November 14th, 1864. William received a Discharge on a Surgeons Certificate in 1865, presumably for being in poor health. Like countless other Civil War veterans, wounds and ruined health likely led to early graves. He died at age 41 and is buried in the Bedford Cemetery.

Bermuda Hundred Campaign – May 1864
Known Bedford Co. Casualty Listing

Name	Muster	Rank	State	Regiment	Comp.	Casualty	Battle	Casualty
Agnew, Levi J	20	Corp.	PA	76th Infantry	E	Wound.-Died	Chester Station	5/7/64
Allenbarger, George	32	Pvt.	PA	55th Infantry	K	Wounded	Foster's Plantation	5/20/64
Allison, John M	21	Pvt.	PA	1st Light Art.	M	Wounded	Bermuda Hundred	5/1/64
Burket, Baltzer	20	Pvt.	PA	55th Infantry	K	KIA	Foster's Plantation	5/20/64
Cobler, John A	18	Pvt.	PA	55th Infantry	K	Wounded	Bermuda Hundred	5/20/64
Crist, Solomon	49	Pvt.	PA	55th Infantry	K	MIA	Foster's Plantation*	
Ellenberger, George	32	Pvt.	PA	55th Infantry	K	Wounded	Foster's Plantation	5/20/64
Jackson, Charles	27	Pvt.	PA	55th Infantry	H	Wounded	Half Way Station	5/13/64
Kessler, John	27	Pvt.	PA	55th Infantry	H	Wounded	Half Way Station	5/13/64
Kline, James S	19	Pvt.	PA	55th Infantry	I	Wound.-Died	Bermuda Hundred	5/19/64
Mock, Emanuel A	22	Pvt.	PA	55th Infantry	K	Wounded	Bermuda Hundred	5/21/64
Mock, Paul S	38	Sgt.	PA	55th Infantry	I	Wounded	Bermuda Hundred	5/18/64
Mock, Wlliam A	33	1st Sgt.	PA	55th Infantry	K	KIA	Foster's Plantation	5/20/64
Ornst, John		Pvt.	PA	55th Infantry	H	Wounded	Chesterfield	5/9/64
Risling, John H	26	Pvt.	PA	55th Infantry	H	Wounded	Half Way Station	5/13/64
Statler, Samuel F	13	Pvt.	PA	55th Infantry	H	Wounded	Foster's Plantation	5/19/64
Trott, Benjamin	36	Pvt.	PA	55th Infantry	H	Wounded	Chesterfield	5/9/64

Union earthworks at the Bermuda Hundred near the Point of Rocks battlefield.

Bermuda Hundred Campaign - May 1864

Emanuel A Mock was born in Union (Pavia) Township on January 15th, 1842. His first enlistment was in McKeage's Militia during the Gettysburg campaign. Emanuel mustered in the 55th Pennsylvania Infantry on February 19th, 1864. He suffered a severe wound during the Bermuda Hundred Campaign on May 20, 1864. The 1890 Pavia Veterans census stated Emanuel's right arm was shot off below the elbow. Emanuel married in 1865 and was a father to 6 children. He lived to be 70 years old and is buried in the Mt. Union Cemetery in Lovely.

Samuel F Statler was 13 years old when he left his home in Schellsburg to muster in the 55th PA Infantry on February 2nd, 1862. He is believed to be one of the youngest soldiers to carry a rifle in the Union Army during the Civil War. Samuel suffered a wound during the Bermuda Hundred Campaign on May 19th, 1864. Samuel was one of the oldest surviving county veterans of the Civil War when he passed at age 87 in 1936. He is buried in the Everett Cemetery.

The 55th PA Infantry disembarked from a transport ship at the above wharf at Bermuda Hundred Landing, Virginia, before marching toward Fort Darling on Drewry's Bluff.

Cold Harbor – May 31 - June 12, 1864

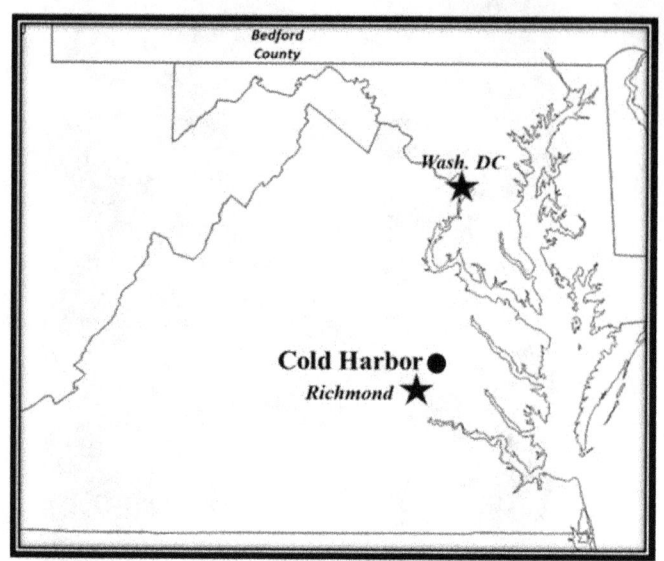

Known Bedford Co. Casualties - 83
KIA, MIA, Wounded & Died – 24
POW & Died - 1
POW - 6
Wounded, Injured - 52

Known Bedford Co. Casualty Listing

Name	Muster Age	Rank	State	Regiment	Company	Casualty	Casualty Date
Allison, Nathaniel	18	Pvt.	PA	55th Infantry	K	Wounded	6/3/64
Allison, Noah	19	Pvt.	PA	138th Infantry	D	KIA	6/5/64
Amick, William M	19	Sgt.	PA	55th Infantry	H	Wound	6/4/64
Amos, John B	28	2nd Lt.	PA	55th Infantry	D	Wounded	6/3/64
Bagley, Moses G	25	Pvt.	PA	138th Infantry	E	Wounded	6/1/64
Baner, Franklin	17	Pvt.	PA	138th Infantry	F	Wounded	6/1/64
Barkman, Hezekiah	33	Corp.	PA	138th Infantry	D	Wounded	6/1/64
Barnett, David	31	Pvt.	PA	184th Infantry	A	Wounded	
Beals, Nicholas H	22	Pvt.	PA	138th Infantry	D	Wounded	6/1/64
Beltz, Adam	24	Pvt.	PA	138th Infantry	E	KIA	6/5/64
Berkhimer, Levi	17	Pvt.	PA	184th Infantry	A	Wounded	
Boose, Isaac	15	Pvt.	PA	102nd Infantry	I	Wounded	6/3/64
Bowers, George	19	Pvt.	PA	184th Infantry	A	Wound.-Died	
Bowser, Daniel L	30	Pvt.	PA	55th Infantry	K	INJ	6/3/64
Brown, George D	20	Pvt.	PA	184th Infantry	A	Wounded	
Cain, John	43	Pvt.	PA	55th Infantry	D	Wounded	6/3/64
Cobler, Allen	37	Pvt.	PA	138th Infantry	E	Wounded	6/1/64
Conley, Martin L	27	Corp.	PA	138th Infantry	E	MIA	6/1/64
Cook, John F	25	Corp.	PA	184th Infantry	A	Wounded	
Cox, Samuel	51	Pvt.	PA	149th Infantry	D	INJ	6/3/64
Darr, Abraham	20	Sgt.	PA	55th Infantry	H	KIA	6/3/64
Davis, Charles M	24	Pvt.	PA	55th Infantry	H	KIA	6/5/64
Doyle, Martin P		1st Lt.	PA	21st Cavalry	E	Wounded	6/3/64
Drenning, Henry G	38	Corp.	PA	55th Infantry	K	KIA	6/3/64
Fink, Abraham	29	Pvt.	PA	148th Infantry	C	Wounded	6/3/64

Cold Harbor – May 31 – June 12, 1864

Known Bedford Co. Casualty Listing

Name	Muster Age	Rank	State	Regiment	Company	Casualty	Casualty Date
Fleegle, George W	21	Pvt.	PA	138th Infantry	E	Wounded	6/1/64
Foreman, William	34	Pvt.	PA	2nd Heavy Art.	L	POW-Died	6/2/64
Frauenfelter, Adam		Pvt.	PA	55th Infantry	H	Wounded	6/3/64
Fry, Solomon W	31	2nd Lt.	PA	55th Infantry	I	Wounded	6/3/64
Garrett, John C	20	Pvt.	PA	110th Infantry	C	POW	6/1/64
Gates, Theophilus R	31	Corp.	PA	55th Infantry	K	Wounded	6/3/64
Gilliam, Michael	21	Pvt.	PA	55th Infantry	D	Wounded	6/3/64
Hagan, John	18	Pvt.	PA	184th Infantry	A	Wounded	6/3/64
Hainsey, Valentine	14	Pvt.	PA	55th Infantry	H	Wound.-Died	6/3/64
Hartman, John P	21	Pvt.	PA	110th Infantry	C	POW	6/1/64
Hess, Daniel A	24	2nd Lt.	PA	55th Infantry	H	Wounded	6/3/64
Holsinger, Josiah	18	Pvt.	PA	110th Infantry	C	POW	6/1/64
Jamison, Benjamin	21	Pvt.	PA	110th Infantry	B	POW	6/2/64
Johnston, Charles W	18	Pvt.	PA	184th Infantry	A	Wounded	6/3/64
Kelly, Henry	20	Pvt.	PA	138th Infantry	F	MIA	6/1/64
Kinton, David		Pvt.	PA	55th Infantry	K	KIA	6/3/64
Klahre, Theodore M	20	Corp.	PA	76th Infantry	E	Wounded	6/1/64
Leader, Henry H	20	Pvt.	PA	2nd Heavy Art.	H	POW	6/2/64
Leasure, Josiah G	20	Pvt.	PA	138th Infantry	D	Wounded	6/1/64
Livingston, John A	22	Capt.	PA	55th Infantry	H	Wounded	6/3/64
Lowery, Emanuel	30	Pvt.	PA	138th Infantry	D	Wounded	6/5/64
Lutz, Simon S	24	Pvt.	PA	184th Infantry	A	Wound.-Died	6/3/64
May, Francis M	36	Pvt.	PA	148th Infantry	G	Wounded	6/3/64
May, Hiram	17	Pvt.	PA	138th Infantry	F	Wounded	6/1/64
McChesney, John	26	Corp.	PA	55th Infantry	H	Wounded	6/3/64
McCoy, Shannon E	22	Corp.	PA	138th Infantry	F	MIA	6/1/64
McDaniel, Daniel	21	Pvt.	PA	2nd Heavy Art.	K	POW	6/2/64
McGregor, John	20	Pvt.	PA	55th Infantry	I	Wound.-Died	6/3/64
McLaughlin, Charles P	19	1st Lt.	PA	138th Infantry	F	KIA	6/1/64
Metzger, Solomon S	23	Capt.	PA	55th Infantry	D	Wounded	6/3/64
Miller, James H	23	1st Lt.	PA	55th Infantry	H	Wounded	6/3/64
Mock, Andrew	33	Pvt.	PA	55th Infantry	K	Wounded	6/3/64
Mock, Emanuel	27	Pvt.	PA	138th Infantry	D	Wounded	6/1/64
Mock, Mathias	24	Pvt.	PA	184th Infantry	A	KIA	6/3/64
Moser, Jeremiah	21	Corp.	PA	138th Infantry	F	Wounded	6/1/64
Nycum, John	28	Pvt.	PA	138th Infantry	D	Wound.-Died	6/1/64
Potter, James	18	Pvt.	PA	184th Infantry	A	Wounded	6/3/64
Price, Michael H	43	Corp.	PA	184th Infantry	A	Wounded	
Reighard, George W	17	Pvt.	PA	184th Infantry	A	Wounded	
Riley, William	53	Pvt.	PA	55th Infantry	D	Wound.-Died	6/3/64
Rininger, Eli	21	Pvt.	PA	55th Infantry	H	Wounded	6/3/64

Cold Harbor – May 31 - June 12, 1864
Known Bedford Co. Casualty Listing

Name	Muster Age	Rank	State	Regiment	Company	Casualty	Casualty Date
Rush, David	29	Pvt.	PA	138th Infantry	F	Wounded	6/1/64
Semler, Reuben J	18	Pvt.	PA	55th Infantry	D	Wound.-Died	6/3/64
Shoop, Joseph L	22	Pvt.	PA	55th Infantry	I	Wound.-Died	6/3/64
Skillington, Robert M	23	Pvt.	PA	184th Infantry	A	Wounded	6/3/64
Slick, Allen	48	Pvt.	PA	55th Infantry	H	KIA	6/3/64
Smith, Josiah N	22	Sgt.	PA	184th Infantry	A	Wounded	
Snowberger, Theodore	20	Pvt.	PA	184th Infantry	A	Wounded	
Snowden, David	24	Pvt.	PA	184th Infantry	A	KIA	6/3/64
Steckman, Francis	18	Corp.	PA	138th Infantry	E	Wound.-Died	6/1/64
Stephens, John G	19	Pvt.	PA	184th Infantry	A	Wounded	
Stull, William	27	Pvt.	PA	188th Infantry	A	Wounded	6/1/64
Tewell, Joseph	19	Corp.	PA	55th Infantry	K	Wound.-Died	6/2/64
Treese, Adie Bell	17	Pvt.	PA	55th Infantry	I	Wounded	6/3/64
Wagoner, August		Pvt.	PA	184th Infantry	B	KIA	6/4/64
Witman, John	30	Corp.	PA	184th Infantry	A	Wounded	
Wolf, Richard	18	Pvt.	PA	55th Infantry	H	Wounded	6/1/64
Wright, Charles C	22	Pvt.	PA	184th Infantry	A	Wounded	6/3/64

Photograph of the extreme left of Confederate lines at Cold Harbor. Rebel soldiers behind these crude breastworks were defending a position very close to where Bedford County Soldiers in the 55th PA Infantry were charging on the morning of June 3rd, 1864. Barely visible on the far right, behind the breastworks, is a house whose fireplace and chimney are still standing.

Cold Harbor – May 31 - June 12, 1864

The last battle of the Overland Campaign took place at Cold Harbor, northeast of Richmond. Grant's relentless offensive had brought the Union Army to within 10 miles of the Confederate capitol in less than a month. Bedford County soldiers suffered more casualties at Cold Harbor than in any other battle during the Civil War. On June 1st, County troops in the 138th Pennsylvania Infantry took heavy casualties during an early evening assault. The 138th took part in a charge through a gap in Confederate lines before taking fire from three sides and being driven back. Two days later, Grant ordered a massive early morning frontal attack on the entrenched Confederate lines. Over 7,000 Union soldiers were killed or wounded within an hour before Grant halted the attack. Nearly half of the Bedford County casualties at Cold Harbor were during this assault. Bedford County soldiers in the 55th Pennsylvania Infantry, who had just taken heavy casualties during the Bermuda Hundred campaign, were cut down by artillery and rifle fire as they crossed a cornfield on the extreme left of Confederate lines. The 55th suffered over 20 casualties within minutes during this near suicidal charge. An opposing Confederate soldier described the shotgun type munitions of double-canister artillery shells, being fired at Union soldiers as "deadly, bloody work." In the no-man's-land between the lines of the two armies were thousands of wounded Union soldiers lying on the battlefield from the ill-fated charge. No truce was agreed upon to allow the removal of the wounded until three days after the June 3rd attack. By then, many of these wounded men had suffered greatly before dying. Grant later wrote in his memoirs, "I have always regretted the last assault at Cold Harbor was ever made."

Grant ordered no further large-scale attacks at Cold Harbor. For the next 9 days after the early morning assault on the 3rd, both armies adopted trench warfare type tactics that would become familiar to all during the siege of Petersburg over the next 9 months. The Union Army took around 13,000 casualties at Cold Harbor compared to 2,500 for the Confederates. It was Robert E Lee's last major battlefield success of the Civil War. With both armies deadlocked at Cold Harbor, Grant planned yet another sweeping maneuver around the right flank of the Confederate army. This time he would catch Lee off-guard by stealthily shuttling his army across the James River.

The Overland Campaign ended with the Union Army taking over 50,000 casualties in little more than a month. The Confederates lost 33,000 men during the campaign, a higher number of casualties on a percentage basis than the Union Army and troops Robert E Lee could ill-afford to lose. Bedford County soldiers suffered at least 190 casualties during all battles of the Overland Campaign and an additional 69 casualties during the separate Bermuda Hundred Campaign.

James H Miller was born on August 4th, 1838. He mustered as a Sergeant in the 55th Pennsylvania Infantry on December 4th, 1861 and received a promotion to 2nd Lieutenant on May 3rd, 1863. James suffered a wound at Cold Harbor during the early morning charge on Confederate lines on June 3rd. He received a second wound two weeks later during the Initial Assault of Petersburg on June 18th. James was discharged from the army on October 11th, 1864, presumably for the debilitating wounds suffered earlier in the year. When he died on July 11th, 1872, at age 33, James left behind a wife and a 4-year-old son. He is buried at the Schellsburg Cemetery.

Cold Harbor – May 31 - June 12, 1864

George D Brown was born in Woodbury Township on January 16th, 1844. He mustered in the 184th Pennsylvania Infantry on May 12th, 1864. George received a wound at Cold Harbor, less than a month after joining the Union Army. During the siege of Petersburg on November 24th, 1864, George suffered a severe wound that required the amputation of his left arm. He spent the next 6 months recovering at the Armory Square Hospital in Washington DC and was only three blocks from Ford Theater on the night Abraham Lincoln was assassinated. Three of his brothers enlisted in the 101st Pennsylvania Infantry. All three were captured during the battle of Plymouth, NC in April 1864 and taken to the Andersonville prisoner of war camp. George, Jacob, John, and Samuel Brown all survived the war and returned to Bedford County. George married in 1874 and was a father to 6 children. Despite his disability, he was a watch and clock maker during much of his life. George lived to be 83 years old and is buried at the Potter Creek Cemetery in Woodbury.

The killing fields of the Cold Harbor battlefield after the Civil War had ended in 1865.

Cold Harbor – May 31 - June 12, 1864

(Right) Richard Wolf was born in Fishertown on February 12th, 1846. He left his home in St. Clair Township to muster in the 55th Pennsylvania Infantry on February 29th, 1864. Richard suffered a wound on June 1st, 1864 at Cold Harbor. He recovered from his wounds and mustered out of the army on August 30th, 1865. Richard married in 1867 and was a father to 6 children. He passed at age 72 and is buried in the Fishertown Cemetery. His obituary stated he was a kind husband, father and neighbor and will be greatly missed at home.

(Left) Allen Cobler was a 38-year-old shoemaker living in Shellsburg when mustered in the 138th Infantry on August 29th, 1862. Allen suffered a wound on June 1st, 1864 at Cold Harbor. He recovered from the wounds and mustered out of the army after the war ended on June 23rd, 1865. Allen is buried at the Schellsburg Cemetery. Little else is known about Allen other than being listed as residing in Westmoreland County on the 1890 Veterans Census.

(Right) Gabriel Burket was born in Union (Pavia) Township on January 10th, 1838. He married Leah Mock in 1860. Gabriel was a 26-year-old constable when he mustered in the 55th Pennsylvania Infantry on February 19th, 1864. He was wounded during the Overland Campaign in 1864. Gabriel recovered from the wound and mustered out of the army on August 20th, 1865. Gabriel was a father to 9 children after the war. He passed at age 73 and is buried in the Mount Zion Cemetery in Pavia.

Petersburg Siege June 1864 - April 1865

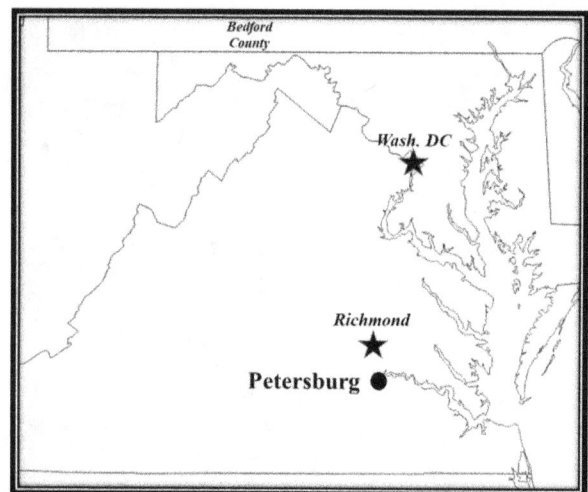

In one of the brilliant maneuvers of the war, Ulysses S Grant surprised Robert E Lee by quickly building a pontoon bridge (pictured below) on June 14th, 1864 that spanned the two thousand foot wide James River which enabled Union forces to attack unsuspecting Confederate troops during the initial assault of Petersburg on June 16th to the 18th.

During the Overland Campaign, Lee prophetically stated to one of his generals, "We must destroy Grant's Army before it reaches the James River. If he gets there, the war will become a siege and then it will be a mere matter of time." Petersburg was a strategically important railroad hub in the South. Both Grant and Lee understood, if Union troops ever captured Petersburg, Richmond would be cut off from needed supplies and the capitol city of the Confederacy would fall. Bedford County troops took part in many battles around Petersburg during the 9-month siege before Confederate lines were finally broken in April 1865. Overviews are provided on five battles during Siege of Petersburg in 1864 that resulted in at least 103 Bedford County casualties. An additional 47 identified casualties were suffered by county troops during many smaller scale battles and skirmishes around Petersburg in 1864.

Petersburg Siege - 1864

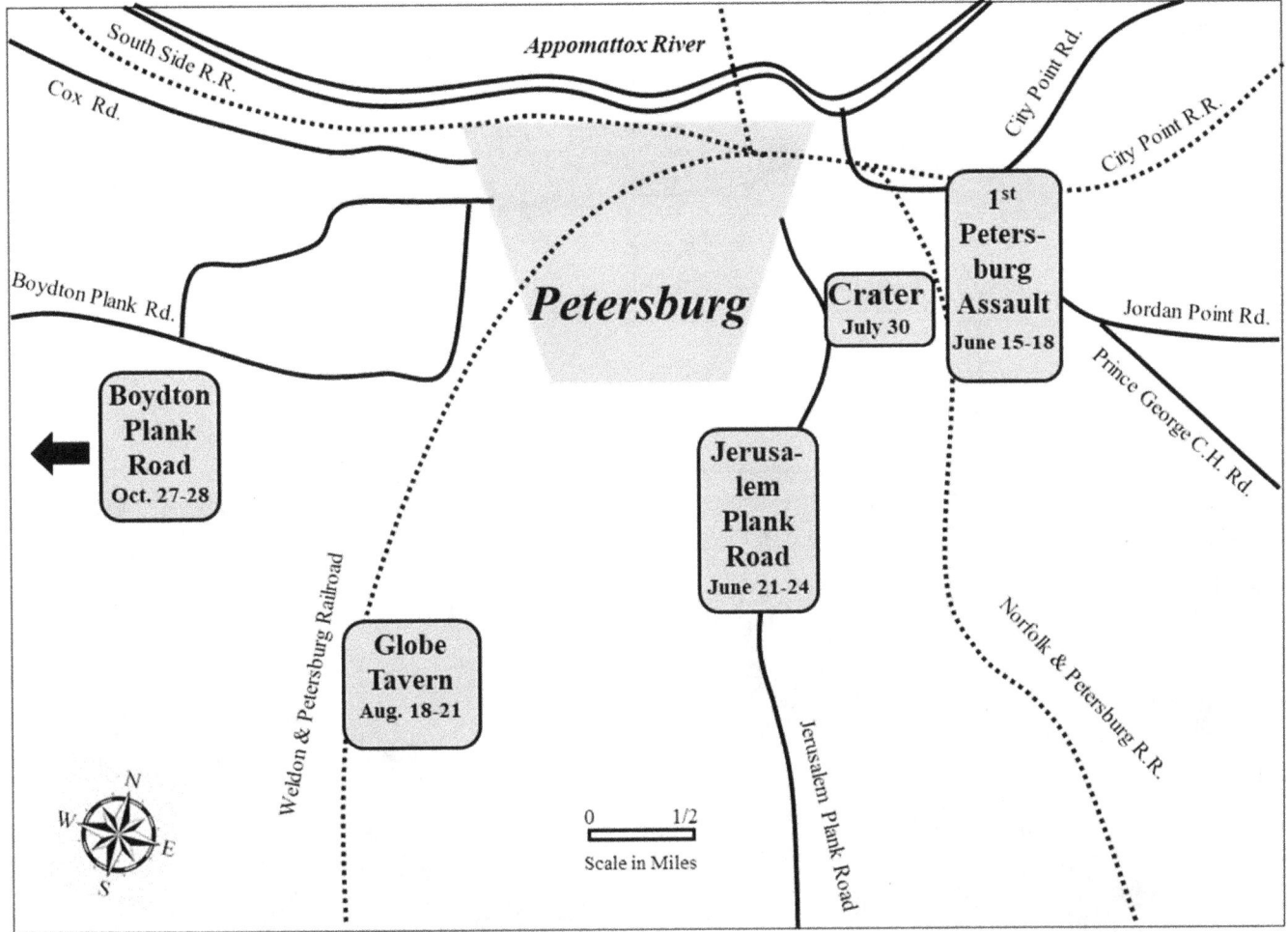

Major Union assaults during the Siege of Petersburg in 1864.

Ulysses S Grant was a brilliant innovator of warfare strategies. Prior to Grant, Robert E Lee was afforded the time and space necessary to maneuver men and shift resources from one front to another. This enabled Lee to achieve success against a larger and better equipped foe. Grant relentlessly pressured Lee during the Overland Campaign, often simultaneously engaging all Confederate troops on the battlefield. This denied Lee the opportunity to reinforce threatened areas in his lines and kept the Confederate general from initiating offensive actions of his own. During the Siege of Petersburg, Grant made multiple assaults on the flanks of Lee's army, stretching out and thinning Confederate lines and disrupting supplies and communications. These assaults would eventually extend the trench lines over 30 miles around Petersburg.

Grant also understood that civilians supporting war efforts would need to suffer significant hardship to reduce the political will to continue fighting the war. Grant's brilliant campaign during the Siege of Vicksburg in 1863 enabled the Union Army to control the Mississippi River, reducing the flow of supplies to the southern army and civilians. The Union navy had already been blockading southern port cities during the previous two years, and the Siege of Petersburg would place additional hardships on civilians living in the capitol city of Richmond.

Grant brought his concept of "Total War" to the Shenandoah Valley in 1864. Grant instructed Philip Sheridan to engage and defeat the Confederate Army to deny Lee reinforcements in Petersburg, and to destroy the agricultural industry in the Shenandoah to deny the Confederate Army and civilians needed food supplies for the following year. An overview of the battle of Monocacy in Maryland; and the battles of Opequon and Cedar Creek in the Shenandoah Valley will follow the overview of battles during the siege of Petersburg in 1864.

After crossing the James River on June 15th, surprised and outnumbered Confederate troops positioned outside of Petersburg mounted a spirited defense. Union generals on the battlefield did not press the Rebel defenders during the next two days, squandering an opportunity to capture Petersburg before Lee's main army group arrived on June 18th. The bloody Union assaults on June 18th were unsuccessful in dislodging the Confederates from their earthwork defenses in front of Petersburg. Both armies dug in and the Civil War entered a new type of siege warfare which became the precursor to the Trench Warfare of World War I fifty years later.

Initial Petersburg Assault - June 15-18, 1864

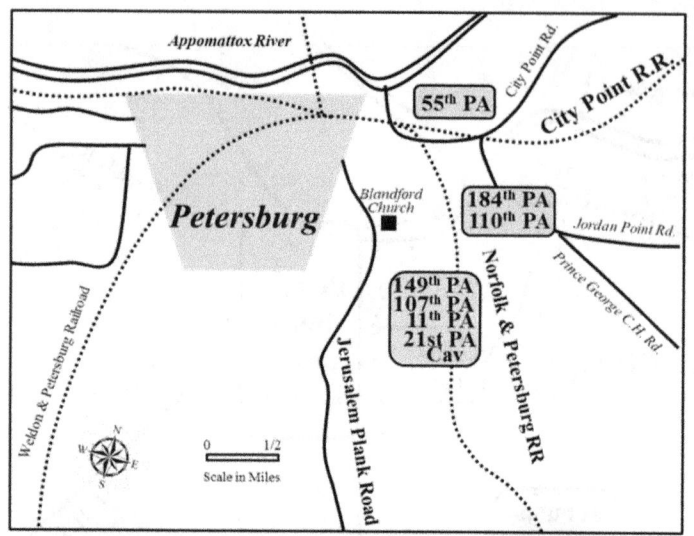

Known Bedford Co. Casualties - 42

KIA, Wounded & Died – 14

Wounded - 28

Map on left shows approximate regiment positions on the battlefield

John A Livingston was born in Bedford on June 21st, 1840. He left his home in Colerain Township to muster as a 1st Lieutenant in the 55th Pennsylvania Infantry on October 11th, 1861. John received a promotion to Captain on May 3rd, 1863. He suffered a wound at Cold Harbor during the charge on Confederate lines on June 3rd, 1864. John was wounded a second time two weeks later during the Initial Assault on Petersburg on June 18th. He mustered out of the army on October 11th, 1864. John married after the war, moved to Missouri and was a father to 5 children. John lived to be 80 years old and is buried in the Oak Hill Cemetery in Monett, Missouri. His obituary stated, "his efforts to better mankind were evident through life and his goodness of character is left as a heritage to his relatives and friends."

Christian Harr was born in Fulton County on January 21st, 1844. He volunteered as a substitute for another man in the 149th Pennsylvania Infantry on October 2nd, 1863. During the Civil War, a draftee could pay another man to take their place. The 149th was one of the most distinguished regiments in the Union Army. They wore bucktails on the back of their caps as a symbol of their marksmanship. Christian suffered wounds during the Initial Assault on Petersburg on June 18th, 1864. He survived his wounds and mustered out of the army on December 13th, 1864. He married Margaret Hammer, a sister of Captain Hezekiah Hammer of the 55th Pennsylvania Infantry in 1868. Christian and Margaret were parents to 13 children. Christian lived to 87 years old and was buried in 1931 at the Mount Union Cemetery in Lovely.

Initial Petersburg Assault - June 15-18, 1864

Known Bedford Co. Casualty Listing

Name	Muster Age	Rank	State	Regiment	Company	Casualty	Casualty Date
Amick, William M	19	Sgt.	PA	55th Infantry	H	Wound.-Died	6/18/64
Chamberlain, Joseph	29	Pvt.	PA	107th Infantry	H	Wounded	6/18/64
Conner, Jonas	19	Pvt.	PA	107th Infantry	H	Wounded	6/18/64
Croyle, William H	18	Pvt.	PA	55th Infantry	H	Wound.-Died	6/16/64
Darr, Henry H	27	1st Sgt.	PA	55th Infantry	H	Wounded	6/18/64
Daugherty, David L	26	Pvt.	PA	55th Infantry	H	Wound.-Died	6/16/64
Dibert, John J	26	3rd Sgt.	PA	55th Infantry	K	KIA	6/16/64
Gardiner, Francis L	37	Pvt.	PA	55th Infantry	K	Wounded	6/18/64
Garlinger, Walter E	22	Corp.	PA	55th Infantry	H	Wounded	6/18/64
Gates, Jeremiah E	17	Pvt.	PA	149th Infantry	I	Wounded	6/18/64
Geller, Jesse	30	Pvt.	PA	55th Infantry	H	Wounded	6/18/64
Harbaugh, George W	20	Pvt.	PA	55th Infantry	H	Wound.-Died	6/18/64
Harbaugh, Robert	23	Pvt.	PA	55th Infantry	K	KIA	6/18/64
Harr, Christian	19	Pvt.	PA	149th Infantry	D	Wounded	6/18/64
Hillegass, Henry P	30	Pvt.	PA	55th Infantry	H	Wounded	6/15/64
Hillegass, John C	22	Pvt.	PA	55th Infantry	H	Wounded	6/18/64
Hixon, Henry H	21	Pvt.	PA	11th Infantry	A	Wounded	6/18/64
Houck, Ezekiel J	28	Pvt.	PA	84th Infantry	E	KIA	6/17/64
Imler, Isaac M	36	Sgt.	PA	55th Infantry	K	KIA	6/18/64
Jackson, Charles	27	Pvt.	PA	55th Infantry	H	Wounded	6/16/64
James, Edward V	19	Pvt.	PA	55th Infantry	K	KIA	6/18/64
Justice, Edward	24	Pvt.	PA	110th Infantry	C	Wound.-Died	6/18/64
Keller, Jacob		Sgt.	PA	21st Cavalry	E	Wounded	6/18/64
Leonard, Jerome	25	Sgt.	PA	55th Infantry	D	Wound.-Died	6/18/64
Licher, John S		Pvt.	PA	55th Infantry	H	Wounded	6/18/64
Livingston, John A	22	Capt.	PA	55th Infantry	H	Wounded	6/18/64
McChesney, John	26	Corp.	PA	55th Infantry	H	Wounded	6/18/64
Miller, David	56	Pvt.	PA	55th Infantry	H	Wounded	6/18/64
Miller, James H	23	1st Lt.	PA	55th Infantry	H	Wounded	6/18/64
Mock, John	34	Pvt.	PA	55th Infantry	K	Wounded	6/16/64
Mock, Tobias B	22	Pvt.	PA	55th Infantry	K	Wound.-Died	6/18/64
Nichols, Charles	32	Pvt.	PA	21st Cavalry	E	Wounded	6/18/64
Ream, Isaac	23	Pvt.	PA	55th Infantry	H	Wounded	6/16/64
Sholl, Isaac	20	Pvt.	PA	55th Infantry	H	Wounded	6/18/64
Skillington, Robert M	23	Pvt.	PA	184th Infantry	A	Wounded	6/16/64
Smith, William F	19	Pvt.	PA	21st Cavalry	K	Wounded	6/18/64
Snowberger, Theodore	20	Pvt.	PA	184th Infantry	A	Wound.-Died	6/18/64
Spidle, Wilson	22	Pvt.	PA	55th Infantry	D	Wound.-Died	6/18/64
Summerland, Peter J	18	Pvt.	PA	55th Infantry	I	Wounded	6/18/64
Wentz, Henry	26	Pvt.	PA	55th Infantry	K	Wounded	6/18/64
Whiteneck, George W		Pvt.	PA	21st Cavalry	I	Wounded	6/18/64
Williams, Harrison P	26	Pvt.	PA	149th Infantry	I	Wounded	6/18/64

Initial Petersburg Assault - June 15-18, 1864

John and Mary Elizabeth McChesney are pictured on their wedding day on July 4th, 1867 in Woodbury. John volunteered in the 55th PA Infantry on September 20th, 1861. He suffered a wound during the early morning charge at Cold Harbor on June 3rd, 1864. John suffered a second wound two weeks later during the Initial Assault on Petersburg. He received a discharge from the army that same day because of the severity of his wounds. Pension records show a request on February 17th, 1868 for debilitating wounds suffered during the war. John died on June 22nd, 1870 when he was 31 years old, leaving behind a widow and a 20-month-old son. Mary was also expecting a second child. He is buried in the Hopewell Cemetery. Mary remarried after John's death, moved to Illinois, and passed in 1923.

Union troops standing on the captured ground of outer Confederate lines in front of Petersburg during the Initial Assault.

Initial Petersburg Assault - June 15-18, 1864

Zophar P and Mary Horton are pictured on their wedding day on October 10th, 1865. Zophar was born in April 1842 and mustered in the Hopewell Rifles Company of the 8th Pennsylvania Reserves on June 11th, 1861. After mustering out of the 8th Reserves, he immediately reenlisted in the 191st Infantry. Zophar suffered a fractured leg during the siege of Petersburg. He mustered out of the army after the war ended on June 29th, 1865.

He was one of 4 brothers who enlisted in the Civil War. Jonathan, Reuben, and Milton Horton enlisted in the 77th Pennsylvania Infantry. Reuben died from wounds during the battle of Chattahoochee, GA, in July 1864. Jonathan died the following December. The cause of death is not listed on available records. Milton survived the war.

Zophar and Mary moved to Everett after they married and were parents to 9 children. Zophar passed at age 67 in 1909. Mary was 84 years old when she passed in 1931. They are buried in the Everett Cemetery.

Harrison P Williams is pictured with his wife Margaret and son Frank in the early 1870s. Harrison was born in Monroe Township in 1837. He married Margaret Dunkle on February 20th, 1862. Their 1st child was born later that year. Harrison volunteered in McKeage's emergency militia during the Gettysburg campaign in July 1863 and enlisted in the 149th PA Infantry on August 26th, 1863, after the Confederate Army retreated from Pennsylvania. During the heavy combat of the Overland Campaign in 1864, Harrison wrote in his diary "as they advanced toward the James River, the fighting became more terrible. The dead and wounded were being carried back, and the scene was one which tried the souls of men." Harrison was himself wounded on June 18th during the Initial Assault on Petersburg. A Minie ball passed through his right instep and required a lengthy hospital stay. He mustered out of the army after the war ended on June 2nd, 1865. Harrison and Margaret were parents to 12 children. Harrison passed in 1899 and is buried at the Rockhill Cemetery in Monroe Township. Son Frank died in 1907 and wife Margaret passed in 1918.

Jerusalem Plank Road - June 21-24, 1864

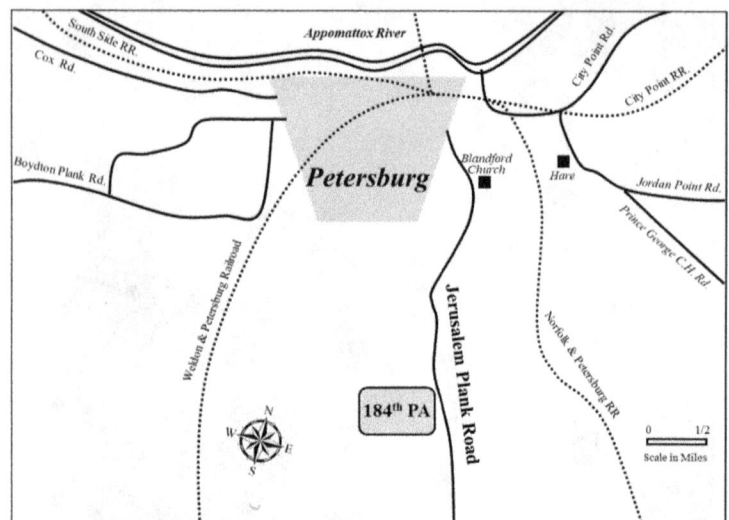

Also referred to as the 1st Battle of Weldon Railroad

Known Bedford Co. Casualties - 32

KIA, Wounded & Died – 4

POW & Died - 16

Wounded & POW – 1

POW - 9

Wounded - 2

The above map shows the approximate location of the 184th Pennsylvania Infantry Regiment during the battle of Jerusalem Plank Road. Ulysses S Grant launched the offensive to destroy the Weldon and Petersburg Railroad line after the failed attempt to overrun Confederate defenses during the initial assault on Petersburg. The 184th PA Infantry were among the Union troops that marched west through unfamiliar, heavily wooded terrain on June 22nd. The lack of major East-West roads around Petersburg made movements difficult for the Union Army. A gap developed in the Union lines and the Confederates launched an attack through the gap and drove behind the rear of several Union regiments, capturing around 1,700 soldiers. Bedford County soldiers in the 184th PA Infantry were among those taken prisoner. It was a terrible day for county soldiers in the 184th. Three lost their lives during the battle, and 1 of the 2 wounded soldiers later died. Tragically, 16 of the 25 county soldiers in the 184th PA Infantry taken prisoner on that day never returned home to Bedford County. Jerusalem Plank Road was an inconclusive battle. The Confederates maintained control of the Weldon and Petersburg railroad, but the Union Army damaged a short segment of the rail line and moved their siege lines further west. The Union forces took over 2,000 casualties compared to less than 1,000 for the Confederates.

Charles C Wright was born on April 16th, 1842 in Schellsburg. Charles and his younger brother Edmund left their St. Clair Township home to muster in the 184th Pennsylvania Infantry on May 12th, 1864. Three weeks after joining the army, Charles suffered a chest wound at Cold Harbor during the early morning charge on Confederate lines on June 3rd. Nineteen days later Charles, Edmund and many other Bedford County soldiers enlisted in the 184th PA Infantry were captured during the battle of Jerusalem Plank Road. Most of the enlisted men were taken to the notorious Andersonville prisoner of war camp, including Charles and Edmund. The brothers were later transferred to the Lawton, Georgia POW camp on October 31st, 1864. Charles and Edmund were moved a third time, but this time they were separated and sent to different POW camps. Charles was taken to Baldwin, Florida, and Edmund to Florence, South Carolina. Edmund was released on February 26th, 1865 but died less than a week later from starvation and exhaustion in Wilmington, NC. Charles was released two months later from the Baldwin POW camp and arrived at his home in St. Clair Township in poor health. After recovering, he went into the mercantile business in Altoona. Charles never married and passed at age 73 in 1916. He is buried in the Pleasantville Cemetery in Alum Bank.

Jerusalem Plank Road - June 21-24, 1864

Known Bedford Co. Casualty Listing

Name	Muster Age	Rank	State	Regiment	Company	Casualty	Casualty Date
Barnett, Samuel	19	Pvt.	PA	184th Infantry	A	POW-Died	6/22/64
Bechtel, Isaac S	20	Pvt.	PA	184th Infantry	A	KIA	6/22/64
Blackburn, Martin	40	Pvt.	PA	184th Infantry	A	POW-Died	6/22/64
Davidson, Samuel	20	Pvt.	PA	184th Infantry	A	POW-Died	6/22/64
Defibaugh, John W	21	Sgt.	PA	184th Infantry	A	POW	6/22/64
Devore, Lewis	18	Pvt.	PA	184th Infantry	B	KIA	6/22/64
Ensley, Christopher	34	Corp.	PA	184th Infantry	A	POW-Died	6/22/64
Evans, Nathan C	30	Capt.	PA	184th Infantry	A	Wound.-POW	6/22/64
Irvine, Wilson	19	Pvt.	PA	184th Infantry	A	POW-Died	6/22/64
Klahre, Herman T	20	Corp.	PA	184th Infantry	A	Wound.-Died	6/22/64
Knox, James H	26	1st Sgt.	PA	184th Infantry	A	POW-Died	6/22/64
Layton, Samuel	32	Pvt.	PA	184th Infantry	A	POW-Died	6/22/64
Lee, John	18	Corp.	PA	184th Infantry	A	Wounded	6/22/64
Line, Jacob	33	Corp.	PA	184th Infantry	A	KIA	6/22/64
Marshall, Henry L	31	Pvt.	PA	184th Infantry	A	POW-Died	6/22/64
Orris, Jacob	35	Pvt.	PA	184th Infantry	A	POW-Died	6/22/64
Otto, Henry S	16	Pvt.	PA	184th Infantry	A	POW-Died	6/22/64
Over, David S	18	Pvt.	PA	184th Infantry	A	POW	6/22/64
Over, Jacob Z	22	Sgt.	PA	184th Infantry	A	POW	6/24/64
Rhodes, George	19	Pvt.	PA	184th Infantry	A	POW	6/22/64
Shoemaker, Austin	20	Pvt.	PA	110th Infantry	C	POW	6/23/64
Smith, Charles S	18	Pvt.	PA	191st Infantry	D	Wounded	6/23/64
Stuckey, David H	28	Corp.	PA	184th Infantry	A	POW-Died	6/22/64
Swoveland, William	31	Pvt.	PA	184th Infantry	A	POW-Died	6/22/64
Teeter, Christian	22	Pvt.	PA	184th Infantry	A	POW-Died	6/22/64
Trout, Sylvester	19	Pvt.	PA	184th Infantry	A	POW	6/22/64
Watson, Henry S	18	Pvt.	PA	184th Infantry	A	POW-Died	6/22/64
Watson, John R	45	Pvt.	PA	184th Infantry	E	POW	6/22/64
Wolfhope, John	23	Pvt.	PA	184th Infantry	A	POW-Died	6/22/64
Wright, Charles C	22	Pvt.	PA	184th Infantry	A	POW	6/22/64
Wright, Edmund S	20	Pvt.	PA	184th Infantry	A	POW-Died	6/22/64
Zembower, Josiah A	21	Pvt.	PA	184th Infantry	G	POW	6/22/64

Battle of the Crater - July 30, 1864

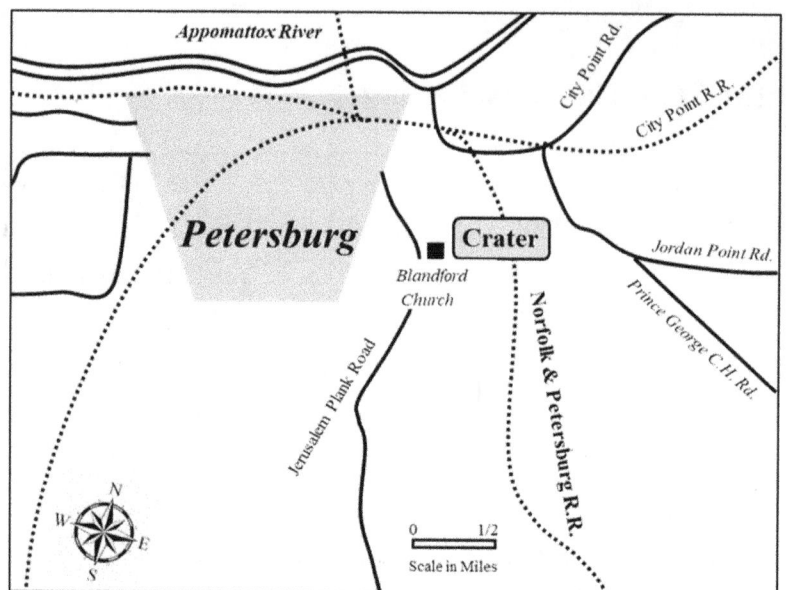

USCT Combat Regiments at the Battle of the Crater

19th, 23rd, 27th, 29th, 30th, 31st, 39th and 43rd

The Crater.

Both armies were facing each other from heavily entrenched earthworks in front of Petersburg. A Union officer who was a mining engineer as a civilian proposed ending the stalemate by packing the end of a tunnel with explosives to blow a gap in the Confederate lines. In late June, experienced miners in the 48th Pennsylvania Regiment began digging a 500-foot-long shaft. Early on the morning of July 30th, a fuse was lit on 4 tons of gunpowder that created a 200-foot-high explosion that instantly killed 278 Confederate soldiers. When the dust cleared, a crater measuring 170 feet long, 60 feet wide and 30 feet deep had been created. Unfortunately for Union soldiers, the day before the explosion, the plan of attack had changed. General Meade decided to not use USCT regiments specifically trained to lead the assault and substituted an unprepared regiment made up of white men. Meade was concerned about being blamed for needlessly sacrificing black soldiers if the attack failed. After the explosion, the unprepared lead regiment was late in launching the attack and confusion reigned. Black troops followed the lead regiment, but the element of surprise had been lost. During the counterattack, Confederates poured heavy fire on Union troops in and around the crater. Some Union troops trapped within the crater were little more than sitting ducks. By early afternoon, the Battle of the Crater had ended in disaster. The Union Army had lost almost 4,000 men to around 1,500 for the Confederates. Grant stated the attack was "the saddest affair I have witnessed in this war."

Battle of the Crater - July 30, 1864

Known Bedford Co. Casualty Listing

Name	Muster Age	Rank	Unit	Regiment	Company	Casualty
Hollinger, Stephen	17	Pvt.	USCT	43rd Infantry	A	Wound.- Died
Strathers, James	27	Pvt.	USCT	43rd Infantry	B	Wounded
Strathers, Willis	24	Pvt.	USCT	43rd Infantry	C	KIA*

The Bedford County Historical Society website lists information on county USCT soldiers who served during the Civil War. Willis Strathers is listed as killed in action. Neither the Frank McCoy 1912 Civil War listing nor the Civil Works Administration list compiled in 1934 referenced the battle where Willis lost his life. Jeremiah M Mickley compiled a history of 43rd USCT regiment in 1866. Mickley recorded the 43rd took significant numbers of casualties in two battles. At the Battle of the Crater, 28 enlisted men were killed, 94 wounded, and 12 were missing. The loses at the battle of Boydton Plank Road were 7 enlisted men killed, 18 wounded and 1 taken prisoner. The Historical Data Systems Civil War data base lists James Strawthers in company B of the 43rd USCT regiment as being wounded at the Battle of the Crater. It is not uncommon for Civil War listings to have different spellings of last names of soldiers. James and Willis were brothers. It appears likely Willis lost his life during the Battle of the Crater.

USCT Troops at Petersburg

Globe Tavern - Aug. 18-21, 1864

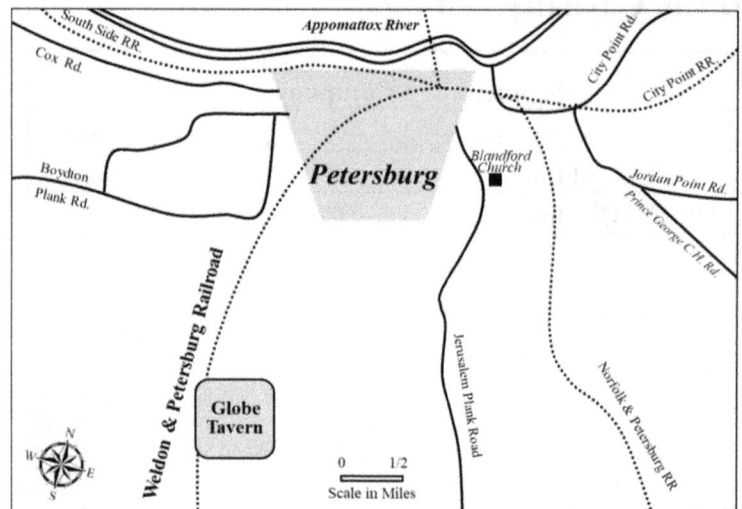

Also referred to as the 2nd Battle of Weldon Railroad

Known Bedford Co. Casualties - 11
MIA – 1
POW & Died - 5
POW - 5

All Bedford County casualties during the Battle of Globe Tavern were in the vicinity of the Blick House shown in the background.

After a disastrous offensive at the Battle of the Crater, Ulysses S Grant again attempted to sever the Weldon & Petersburg Railroad line. Bedford County soldiers in the 11th, 91st, 149th, and 191st Pennsylvania Infantry regiments were among the troops that reached the Weldon & Petersburg Railroad line on the morning of August 18th. Sections of railroad tracks were being destroyed when Confederate forces launched an attack in the afternoon. The Confederates renewed the attack the next day and drove through a gap in Union lines. Desperate hand to hand fighting took place before Union troops mounted a successful counterattack. Bedford County casualties at the Battle of Globe Tavern were likely during this attack. Further Confederate assaults on the following days also failed. Grant had won a significant victory. The Union Army had severed the Confederate railroad supply line from Wilmington, NC, and extended trench lines several miles west. Four thousand casualties were suffered by the Union Army, including around 3,000 soldiers captured during the Confederate assault on August 19th. The Confederates took around 2,000 casualties during the battle.

Globe Tavern - Aug. 18-21, 1864

Known Bedford Co. Casualty Listing

Name	Muster Age	Rank	State	Regiment	Company	Casualty	Casualty Date
Barmond, Nathaniel		Pvt.	PA	191st Infantry	H	POW-Died	8/19/64
Brower, George		Corp.	PA	191st Infantry	E	POW	8/19/64
Carothers, William H	22	Pvt.	PA	190th Infantry	D	POW	8/19/64
Chamberlain, Joseph		Pvt.	PA	107th Infantry	H	MIA	8/19/64
Griffith, Abel	19	Pvt.	PA	191st Infantry	H	POW	8/19/64
Grubb, George	23	Pvt.	PA	149th Infantry	I	POW-Died	8/19/64
Lysinger, George W	28	1st Sgt.	PA	107th Infantry	H	POW-Died	8/19/64
Malone, Charles		Pvt.	PA	191st Infantry	H	POW-Died	8/19/64
Manspeaker, John	17	Pvt.	PA	11th Infantry	A	POW	8/19/64
Newcomer, Joseph	21	Pvt.	PA	191st Infantry	C	POW	8/19/64
Thomas, Joseph A	19	Pvt.	PA	12th Reserves	I	POW-Died	8/19/64

Abel Griffith was born in Broad Top Township January 23, 1845. He mustered in the Hopewell Rifles Company of the 8th Pennsylvania Reserves on March 11th, 1864. The 8th PA Reserves regiment mustered out of the Union Army on May 15th, 1864. Abel reenlisted in the 191st Pennsylvania Infantry that same day. He was taken prisoner on August 19th, 1864, during the battle of Globe Tavern and released on October 7th, 1864. Abel survived the war and married in 1871. He was a father to 3 children. Abel passed when he was 76 and is buried in the Everett Cemetery.

James A Grove was born in Ray's Cove (Breezewood) in 1846 and was 17 years old when he enlisted in the 107th Pennsylvania Infantry on Apr 26th, 1864. Less than 2 months after joining the Union Army, he suffered a wound during the siege of Petersburg on June 20th, 1864. James returned home after the war ended and was a father to six children. His last surviving child lived until 1985. James lived to be 81 years old and was buried at the Everett Cemetery in 1927. His obituary stated, "he made lasting friendships in Everett, was well known throughout the county, and his loss will be sorely felt by his family, relatives, and friends."

Boydton Plank Road - Oct. 27-28, 1864

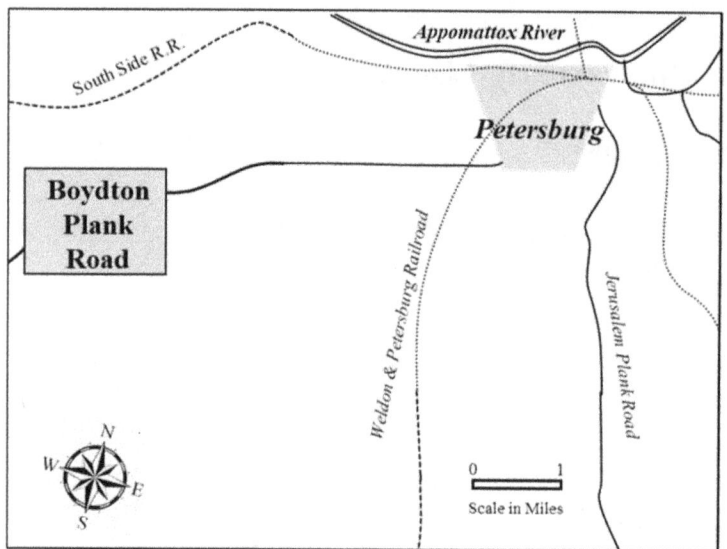

Also referred to as the 1st Battle of Hatcher's Run and Burgess Mill

Known Bedford Co. Casualties - 15
KIA & Wound.-Died - 4
POW & Died - 1
POW - 1
Wounded - 9

A Union soldier on picket duty in front of Petersburg.

The primary objective of Boydton Plank Road offensive was to capture the Southside Railroad, the last remaining rail line supplying the Confederate troops defending Petersburg. Approximately 30,000 Union troops marched west and attacked Confederate positions close to the intersection of Boydton Plank Road and Hatcher's Run, 3 miles south of the Southside rail line. On October 27th, the 91st PA Infantry led an assault on the extreme right of the Confederate lines just east of Hatcher's Run. The commanding Union general reported "the 91st PA Infantry was in a lively firefight with the enemy and advanced to within 100 to 200 yards of entrenched Confederate lines behind fallen timber. When further advancement was not possible, temporary entrenchments were thrown up establishing a Union line." All Bedford County casualties in the 91st PA Infantry at the Battle of Boydton Plank Road were during this assault. 21st Pennsylvania Cavalry casualties during the battle were likely during dismounted exchanges with Confederate cavalry units. The battle ended the following day with Union forces unable to dislodge the Confederates from entrenched positions below the Southside Railroad. The Union Army suffered 1800 casualties to around half that amount for the Confederates. Lee kept the last Petersburg rail supply line open. The Confederate victory at Boydton Plank Road was a needed morale booster for a besieged rebel army.

Boydton Plank Road - Oct. 27-28, 1864

Known Bedford Co. Casualty Listing

Name	Muster Age	Rank	State	Regiment	Company	Casualty
Corle, Francis B	37	Pvt.	PA	91st Infantry	G	Wound.-Died
Corle, Jonathan	37	Pvt.	PA	91st Infantry	G	KIA
Doyle, Martin P		2nd Lt.	PA	21st Cavalry	E	Wounded
Dull, John	19	Pvt.	PA	184th Infantry	A	POW-Died
Imler, William H	21	Pvt.	PA	91st Infantry	B	Wounded
Lingenfelter, David R	35	Pvt.	PA	91st Infantry	B	Wounded
Mock, George	42	Pvt.	PA	110th Infantry	D	Wounded
Oster, Samuel C	20	Pvt.	PA	91st Infantry	B	Wounded
Pearson, Henry C		1st Lt.	PA	21st Cavalry	H	Wounded
Pennel, Andrew J	22	Pvt.	PA	91st Infantry	F	Wounded
Shehan, James		Pvt.	PA	21st Cavalry	E	POW
Stephens, John G	19	Pvt.	PA	184th Infantry	A	KIA
Stineman, John	26	Pvt.	PA	91st Infantry	B	KIA
Stineman, Thomas B	23	Pvt.	PA	91st Infantry	B	Wounded
Stoner, Daniel F	44	Pvt.	PA	91st Infantry	B	Wounded

Isaiah Collins was born and raised in Southampton Township. He mustered in the 91st Pennsylvania Infantry on September 21st, 1864, leaving behind a wife and one-year-old son at home. Isaiah suffered a wound during the siege of Petersburg. There is no information on the battle or date of his wounds. He survived the war but died on February 8th, 1867. Isaiah was only 28 years old. There is no information on the cause of death. Countless Civil War soldiers returned home with debilitating wounds and ruined health, which led to early graves. Isaiah is buried in the Mt. Zion Cemetery in Chaneysville.

John G Feaster was born in St. Clair Township on October 11th, 1843. He mustered in the 91st Pennsylvania Infantry on September 21st, 1864. John carried a pocket Hymnal book during the war. This small hymnal, which appears to have bloodstains on the front cover, is in the possession of his descendants. John survived the war and mustered out on May 30th, 1865. He married in 1866 and was a father of 4 children. John passed at age 71 in 1915 and is buried in the Fishertown Cemetery.

Monocacy - July 9, 1864

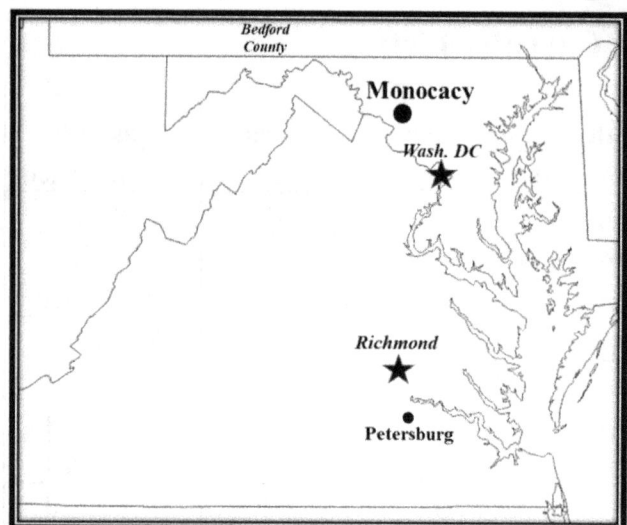

Known Bedford Co. Casualties - 18
MIA, Wounded & Died - 3
POW & Died - 1
POW - 6
Wounded & POW - 1
Wounded - 7

Monocacy Junction. Many Civil War battles were fought in close proximity to railroad lines.

Monocacy was a major battle prior to the Shenandoah Campaign in the fall of 1864. In June 1864, Lee instructed General Jubal Early to march his 15,000-man army north from Petersburg to threaten Washington, DC. Lee hoped this action would divert Union troops away from Petersburg. Early's forces crossed the Potomac River at Shepardstown on July 6th and veered southeast toward Washington, DC. Grant immediately dispatched troops from Petersburg to help defend the panicked city of Washington. The outnumbered Union forces of General Lew Wallace collided with Early's army, east of Frederick, along the Monocacy River at a Baltimore and Ohio Railroad junction. Wallace's approximate 7,500 Union troops included 2,500 inexperienced 100-day enlistment soldiers. The invading Confederates also had 40 artillery pieces compared to only 7 for the Union. Wallace was hoping to delay the advancing Rebel army long enough for Union reinforcements to reach Washington. The outmanned and outgunned Union troops mounted a spirited defense, pushing back multiple Confederate assaults. Late in the afternoon, Early ordered an attack with some of his best troops who had been held in reserve. The rested Confederate troops included veteran remnants of the famed Stonewall Brigade and the Louisiana Tigers. Union soldiers were running low on ammunition and exhausted from fighting all day when a Union retreat was finally ordered. Most Bedford County casualties at Monocacy likely happened during this last Confederate assault. The Union forces lost almost 2,000 men compared to 1,100 for the Confederates. Early's rebel army was so exhausted from the fighting, they needed to regroup the following day before continuing the march on Washington. By the time the rebel army reached the outskirts of Washington, Union reinforcements had arrived from Petersburg. Early's army was turned away two days later and returned to the Shenandoah Valley. Union General Lew Wallace made the following observation on the Battle of Monocacy, "From every point of view, it was heroism."

Monocacy - July 9, 1864
Known Bedford Co. Casualty Listing

Name	Muster Age	Rank	State	Regiment	Company	Casualty
Bailey, William H	19	Pvt.	PA	138th Infantry	E	POW
Ball, Daniel M	24	Pvt.	PA	138th Infantry	F	Wounded
Blackburn, Joseph	30	Pvt.	PA	138th Infantry	E	Wound.-Died
Carrell, Daniel	18	Pvt.	PA	138th Infantry	E	MIA
Earnest, William	32	Pvt.	PA	138th Infantry	F	Wounded
Ferguson, William	24	Sgt.	PA	138th Infantry	D	POW
Foor, Martin T	37	Corp.	PA	138th Infantry	F	POW-Died
Gardner, Charles	18	Pvt.	PA	138th Infantry	E	POW
Geller, George	16	Pvt.	PA	138th Infantry	F	Wounded
Gillam, George	24	Corp.	PA	138th Infantry	D	Wounded
Gordon, Isaac	18	Pvt.	PA	138th Infantry	E	POW
Imler, George R	21	Pvt.	PA	138th Infantry	E	Wound.-POW
Kellerman, James L	28	Pvt.	PA	138th Infantry	F	Wounded
Ling, William H	21	Pvt.	PA	138th Infantry	D	POW
Owens, Chauncey	22	Pvt.	PA	138th Infantry	F	Wounded
Shaffer, Harvey E	21	Sgt.	PA	138th Infantry	F	Wounded
Stineman, William	18	Pvt.	PA	138th Infantry	E	POW
Yarnell, Jesse	43	Pvt.	PA	138th Infantry	D	Wound.-Died

George R Imler was born in St. Clairsville on September 16th, 1842. He mustered in the 138th Pennsylvania Infantry on August 29th, 1862. George suffered a wound and was taken prisoner during the Battle of Monocacy. No information is available on how long he was held at the Danville POW camp in Virginia before being released. George survived the war and mustered out on June 23rd, 1865. His first wife died in 1877 with two small children at home. George married again in 1880 and was a father to 11 children with his second wife. He lived to be 85 years old and is buried in the Lutheran Cemetery in Woodbury.

Chauncey Owens was born in 1836 and was listed with his family on the 1850 Cumberland Valley Census. Chauncey volunteered in the 138th Pennsylvania Infantry on August 29th, 1862. He was wounded on July 9th, 1864 during the Battle of Mococacy. Chauncey recovered from the wound and mustered out of the army on June 23rd, 1865. He moved to Ohio after the war. Chauncey married in 1866 and was a father of 3 children. He passed around the age of 92 in Tulsa, Oklahoma in 1928.

Opequon - September 19, 1864

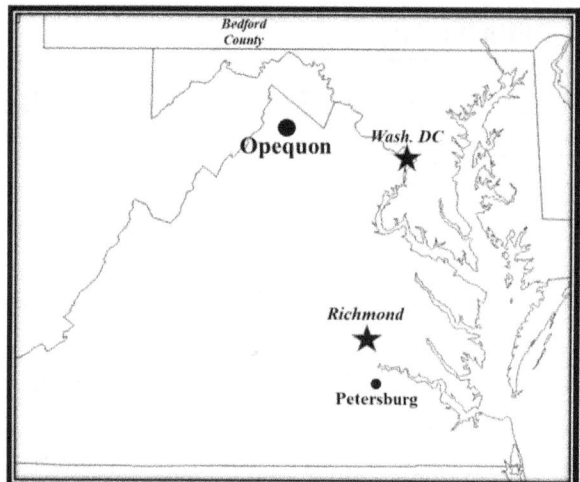

Also referred to as the 3rd battle of Winchester

Known Bedford Co. Casualties - 22
KIA, MIA, Wounded & Died - 3
POW - 1
Wounded & Injured - 18

Image of the Largest Cavalry Charge of the Civil War at the Battle of Opequon. Col. James Schoonmaker was awarded the Medal of Honor for leading the 22nd PA, 14th PA and 8th Ohio Cavalry regiments on the charge on Fort Star on the extreme right flank of the Union lines. Over 100 Bedford County soldiers were enlisted in the 22nd PA Calvary on that day.

On August 1st, Grant appointed Philip Sheridan, Commander of the Union Army in the Shenandoah. After taking command, Sheridan was initially cautious but moved to strike the Confederates when General Jubal Early spread out his army between Winchester and Martinsburg, WV. The Confederates slowed Sheridan's advance at a narrow canyon on the morning on September 19th to allow Early time to gather his dispersed troops. Just before noon, the armies clashed east of Winchester in the bloodiest battle ever fought in the Shenandoah Valley. The Confederates held their ground during initial Union assaults, but were gradually driven back toward the town of Winchester. With the Confederate army on its heels, Sheridan ordered the largest cavalry charge of the Civil War. Late in the afternoon, 8,000 cavalry troops thundered around the Confederate left flank and rebel lines crumbled. Casualties were exceedingly high. The Union Army lost over 5,000 men and the Confederates nearly 4,000. The resounding Union victory at the Battle of Opequon marked the beginning of the end for the Confederacy in the Shenandoah.

Opequon - September 19, 1864

John A Felton was born in Monroe Township on April 15th, 1843, He left his home in East Providence Township to volunteer in McKeage's emergency militia when the Confederates invaded Pennsylvania during the battle of Gettysburg. He later enlisted in the 22nd Pennsylvania Cavalry on February 26th, 1864. John injured his leg when his horse was shot out from underneath him during the charge on Fort Star during the battle of Opequon. Pension documents state he suffered the rest of his life from a lame leg. John married in 1867 and was a father of 13 children. He lived to be 80 years old and is buried in the Mt. Zion Lutheran Cemetery in Breezewood.

Known Bedford Co. Casualty Listing

Name	Muster Age	Rank	State	Regiment	Company	Casualty
Ball, Andrew M	22	Pvt.	PA	87th Infantry	B	Wounded
Clark, Samuel M	18	Pvt.	PA	138th Infantry	E	KIA
Cooper, Nathan	39	Pvt.	PA	49th Infantry	E	Wounded
Feight, John W	25	Capt.	PA	138th Infantry	F	Wounded
Felton, John A	20	Pvt.	PA	22nd Cavalry	H	Injured
Hammer, John B	20	Sgt.	PA	138th Infantry	D	Wounded
Holler, George W	36	Pvt.	PA	138th Infantry	F	Wounded
James, Nathaniel	20	Pvt.	PA	138th Infantry	D	MIA
King, Harrison H	23	Corp.	PA	138th Infantry	E	Wounded
Lowry, Oliver	24	Pvt.	PA	138th Infantry	F	Wounded
Lucas, William	21	Pvt.	PA	138th Infantry	D	Wounded
May, Hiram	17	Pvt.	PA	138th Infantry	F	Wounded
McVicker, Jesse	29	Pvt.	PA	102nd Infantry	K	Wounded
Miller, William O	18	Pvt.	IA	24th Infantry	C	Wound.-Died
Over, James E	27	Corp.	PA	138th Infantry	E	Wounded
Riffle, William	17	Pvt.	PA	138th Infantry	E	Wounded
Ritchey, Frederick G	21	Corp.	PA	138th Infantry	F	Wounded
Shroyer, Joseph	19	Pvt.	PA	138th Infantry	F	Wounded
Shroyer, Moses	17	Pvt.	PA	138th Infantry	D	Wounded
Sipes, Charles W	24	Pvt.	IA	28th Infantry	F	POW
Stuckey, John S	27	Capt.	PA	138th Infantry	D	Wounded
Valentine, John	42	Pvt.	PA	138th Infantry	F	Wounded

Opequon - September 19, 1864

(Right) Marcus May was born on October 8th, 1842. He mustered in the 138th Pennsylvania Infantry on August 29th, 1862 with his brother Samuel. Another brother, Hiram joined the 138th PA Infantry in June 1863. Two other brothers also enlisted in the Union Army during the war. Hiram suffered a wound at the battle of Opequon on September 19th, 1864. All five brothers survived the war. Marcus married in 1870 and was a father to 4 children. He lived to be 87 years old and was buried in the Lybarger Cemetery in Madley in 1930.

(Left) John S Stuckey was born in Schellsburg on April 24th, 1834. He left his farm in Napier Township to muster as a Captain in the 138th Pennsylvania Infantry on September 2nd, 1862. John suffered a severe wound during the battle of Opequon that required amputation of his right leg. He was discharged on February 8th, 1865. John married in 1866, moved to Nebraska in the 1870s and was a father to 4 children. John passed at age 62 in 1897 and is buried in the Evergreen Cemetery, in Lexington, Nebraska.

(Right) Joseph Mowry was born around 1838. There is little available information on Joseph. He mustered in the 22nd Pennsylvania Cavalry on July 11th, 1863. After the war ended, he left the Cavalry on June 18th, 1865 prior to his Company being mustered out. Joseph passed around the age of 64 on August 19th, 1902, and is buried in the Bedford Cemetery.

Opequon - September 19, 1864

John Border was born on August 8th, 1837. He left his farm in Yellow Creek to volunteer in the 110th Pennsylvania Infantry on September 2nd, 1861. John's physical description at enlistment was he stood 5'6" tall, with brown hair, brown eyes, and a fair complexion. The 1890 Hopewell Veterans Census stated he suffered a hip wound but does not list the battle or date. He mustered out of the 110th on December 27th, 1862. John mustered in the 22nd Pennsylvania Cavalry on July 11th, 1863. John survived the war and married in 1866. He was a father to 7 children. John passed at age 65 and is buried in the Bethel Brethren Cemetery in Tatesville.

William B Filler was born in Rainsburg on September 15th, 1845. He was 16 when he left his family farm in Colerain Township to muster in the 101st Pennsylvania Infantry on November 11th, 1861. William received a disability discharge on July 11th, 1862. He recovered and reenlisted in the 22nd Pennsylvania Cavalry on July 11th, 1863. William received a promotion to Sergeant on January 4th, 1864. He was captured during the battle of Brown's Gap, VA on September 26th, 1864 and held in the Libby, Belle Island & Salisbury POW camps before being released on March 8th, 1865. William married in 1870 and was a father to 10 children. He was a skilled carpenter and a postmaster in Rainsburg. William passed at age 74 in 1920 and is buried in the Woods Church Cemetery in Colerain Twp.

Cedar Creek - October 19, 1864

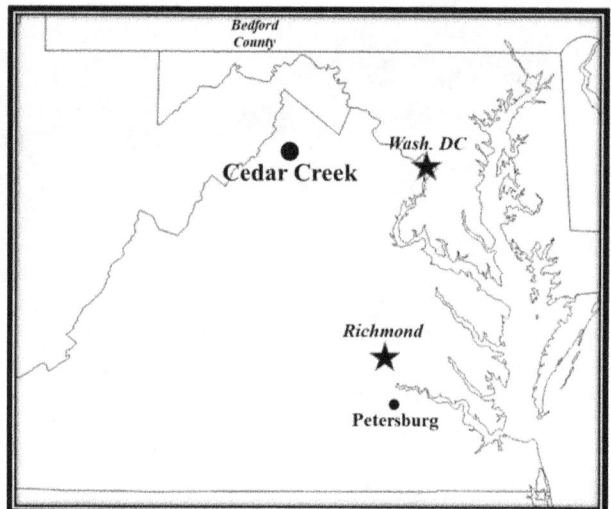

Also referred to as Belle Grove

Known Bedford Co. Casualties - 24
KIA, Wounded & Died - 4
POW - 1
Wounded - 19

Following the victory at the Battle of Opequon, Union troops began burning the farming harvest of the rich Shenandoah Valley to deny the Confederates needed food supplies to continue fighting the war. Robert E Lee sent 3,000 reinforcements to General Jubal Early's depleted army for an offensive to regain control of the "Breadbasket of the Confederacy". General Sheridan was traveling back from meetings in Washington as Early was advancing his army toward Union forces concentrated along Cedar Creek. After a daring night-time march, the Confederates launched a pre-dawn surprise attack on October 19th. Dense fog aided the attacking rebel army as they quickly routed unsuspecting Union troops. By the time Sheridan reached his army after his famous 12-mile gallop from Winchester, Union lines had been driven 3 miles from its encampments. Fortunately for Sheridan, Early had halted his attack. Many of his famished soldiers fell out of line to pillage food supplies left behind by retreating Union soldiers. Sheridan immediately reorganized his army for a counterattack. In the middle of the afternoon, Union General George Custer's Cavalry units swept around the left flank of the Confederate army, causing their lines to break. A full retreat quickly turned into a rout. Twenty Bedford County soldiers were enlisted in the 18th PA Cavalry during the battle of Cedar Creek and may have taken part in Custer's charge. The Union Army suffered over 5,000 casualties and the Confederates lost almost 3,000 men. Cedar Creek was a resounding Union victory and marked the end of Sheridan's highly successful Shenandoah Valley Campaign.

Sheridan's Ride was painted by Thure de Thulstrup. Sheridan arrived at Cedar Creek at a full gallop to rally his army. Generations of school children memorized a poem written by Thomas Buchanan immortalizing "Sheridan's Ride".

Cedar Creek - October 19, 1864

Known Bedford Co. Casualty Listing

Name	Muster Age	Rank	State	Regiment	Company	Casualty
Ball, Daniel M	24	Pvt.	PA	138th Infantry	F	Wounded
Barclay, Joseph	28	Sgt.	PA	138th Infantry	F	Wounded
Beard, Daniel	25	Sgt.	PA	138th Infantry	E	Wounded
Beltz, George W		Pvt.	PA	138th Infantry	D	Wounded
Beltz, John A	33	Pvt.	PA	138th Infantry	D	Wounded
Bortz, Martin S	23	Capt.	PA	138th Infantry	F	Wounded
Carpenter, Abraham	20	Corp.	PA	138th Infantry	E	Wounded
Clark, George W	40	Capt.	IA	22nd Infantry	K	Wounded
Craine, David D	18	Pvt.	PA	138th Infantry	E	KIA
Crawford, Joseph	15	Pvt.	WV	15th Infantry	B	POW
Feight, William F	17	Pvt.	PA	138th Infantry	F	Wounded
Geller, John	34	Sgt.	PA	138th Infantry	F	Wounded
Gump, John A	19	1st Lt.	PA	138th Infantry	D	Wound.-Died
Hiner, John E		Pvt.	PA	67th Infantry	C	Wound.-Died
Horton, Oliver	32	Capt.	PA	138th Infantry	D	Wounded
Kellerman, James L	28	Pvt.	PA	138th Infantry	F	Wounded
Kelly, William	31	Pvt.	PA	138th Infantry	F	Wounded
Miller, Abraham	18	Pvt.	PA	138th Infantry	F	Wounded
Miller, Ephraim B	22	Pvt.	PA	138th Infantry	F	Wounded
Ridenbaugh, Samuel	23	Corp.	PA	138th Infantry	E	Wounded
Ritchey, Frederick G	21	Corp.	PA	138th Infantry	F	Wounded
Snyder, Jonathan	30	1st Sgt.	PA	138th Infantry	D	Wound.-Died
Speck, Henry	39	Pvt.	PA	138th Infantry	E	Wounded
Yarnell, John	17	Pvt.	PA	138th Infantry	D	Wounded

Members of the Sheridan's Veterans Association pose on the Valley Pike in 1883 and point to the location where General Sheridan joined the Army of the Shenandoah at Cedar Creek.

Cedar Creek - October 19, 1864

Abraham L Carpenter was born and raised in Londonderry Township. He was 20 years old when he left his family farm to volunteer in the 138th PA Infantry on August 29th, 1862. He received a promotion to Corporal prior to the battle of Cedar Creek. Abraham suffered a wound at Cedar Creek on October 19th, 1864 and mustered out that same day from the debilitating wound. He survived the war and married in 1868. Abraham, his wife and two children moved to Iowa in the 1870s. He lived to be 91 and was buried at the Highland Cemetery in Wichita, KS in 1933.

Jonathan Snyder left his home in Monroe Township to volunteer in the 138th Pennsylvania Infantry in August 1862. Jonathan was wounded during the battle of the Wilderness on May 6th, 1864. He recovered from the wound and rejoined his regiment. Five months later, Jonathan suffered severe wounds in both legs at Cedar Creek and died 3 days later. Jonathan was initially buried in what is now the National Cemetery at Winchester. Emanuel O'Neal traveled to Winchester in early November 1864 to recover the body of his cousin for burial at the Ash-Snyder Cemetery in Monroe Township. Jonathan was 32 years old.

Cedar Creek - October 19, 1864

George Clark was born in Schellsburg in 1821. George moved west and was working as a lawyer in Iowa when he mustered as a Captain in the 22nd Iowa Infantry on August 14th, 1862. His wife and 3 small children waited at home for his return during the war. He suffered a severe wounded at Cedar Creek but survived and returned to Iowa. George passed when he was 63 years old and was buried at the Oakland Cemetery in Iowa City, IA in 1885.

Martin S Bortz was 22 when he left his Cumberland Valley home to enlist as a 1st Sergeant in the 138th Pennsylvania Infantry on August 29th, 1862. Martin suffered a wound at Cedar Creek on October 19th, 1864. He received a promotion to 1st Lieutenant on December 1st, 1864 and Captain on February 21st, 1865. Martin married after mustering out in 1865 and was a father of five children. He passed at age 79 in 1918 and is buried at the Union Cemetery in Centerville.

Photograph of the battlefield at Cedar Creek taken in the 1800s, looking Northwest from the vantage point of Confederate lines toward where the Union Army would launch a counterattack in the afternoon of October 19, 1864.

6. 1865

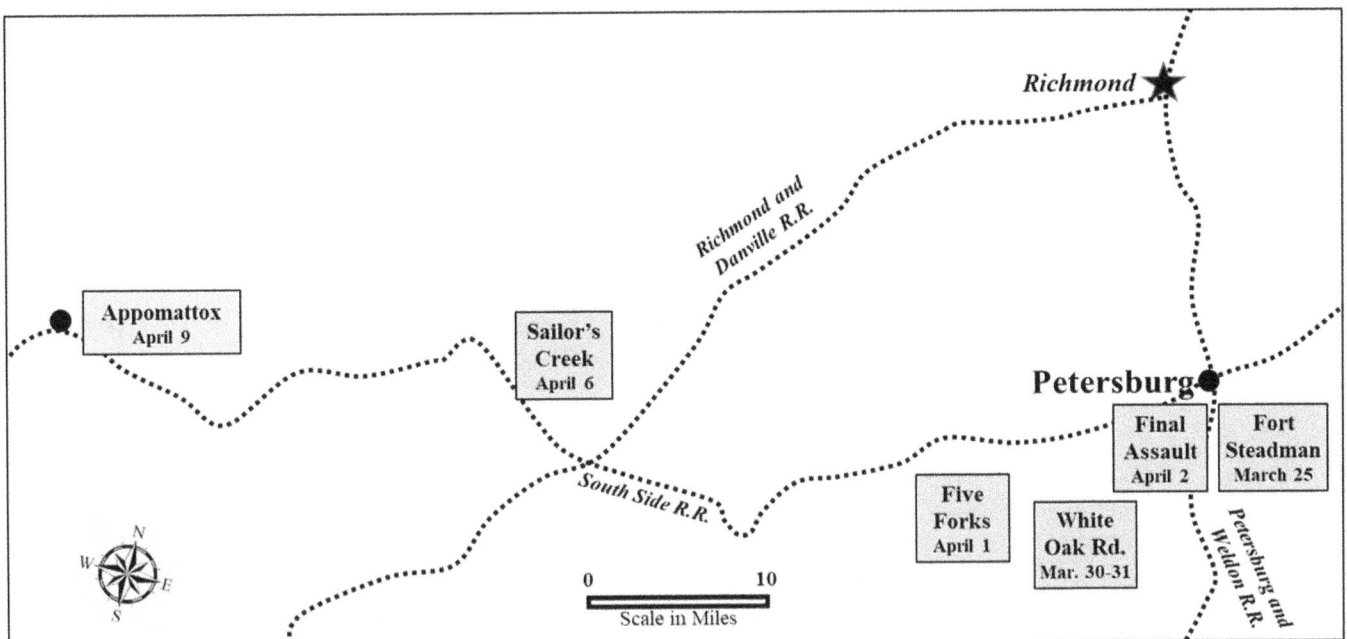

In the Fall and Winter of 1864, the war had taken a turn for the worse for the Confederacy. The Union Army had burned a large swath of the crops and barns in the rich Shenandoah farming region. William Tecumseh Sherman's famous March to the Sea cut a wide path of destruction across Georgia, including the dismantling of railroad tracks, burning of crops, and seizure of livestock. Everything that would enable the South's ability to continue fighting the war was being destroyed or confiscated. Ulysses S Grant's concept of "Total War" was exhausting the Confederate Army and taking a heavy toll on Southern civilians. Grant's strategy of applying constant pressure on the Confederate army had also prevented Lee from mounting an offensive during the Siege of Petersburg prior to one last desperate assault on March 25th, 1865.

There were 164 known casualties in 1865, including 24 soldiers who died of battlefield wounds and 62 from disease. Please note all available Civil War records do not provide dates of known casualties, including deaths, therefore it is likely the casualty numbers are higher than identified for 1865.

Photograph of earthwork fortifications created by both the Union and Confederate armies at Petersburg.

Fort Stedman - March 25, 1865

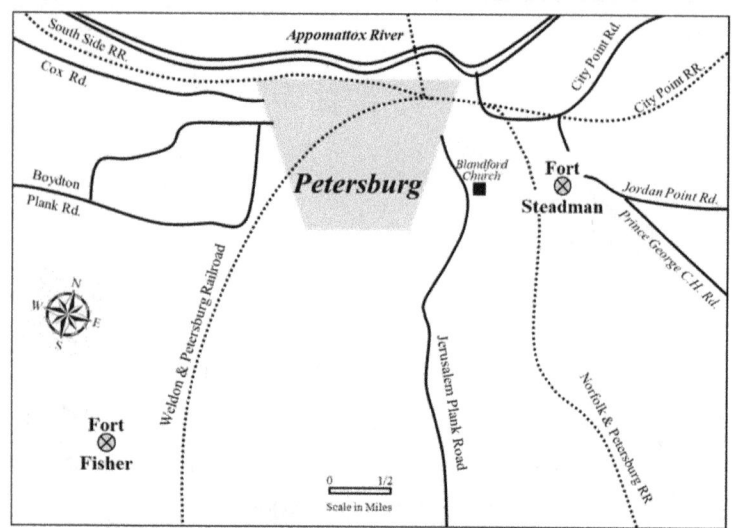

Combined Confederate Assaults on Fort Stedman & Fort Fisher

Known Bedford Co. Casualties - 20

KIA - 1

Wounded - 19

Photograph was taken after the fall of Petersburg. The Union picket line is shown in the foreground at Ft. Stedman. The men in the background are standing at the Confederate picket line, illustrating how close the two were to each other.

The besieged Confederate troops defending Petersburg were suffering from chronic disease and a lack of food. Desertion had become a serious problem, further thinning Confederate lines. Robert E Lee desperately needed to breakout from the 9-month siege and unite his troops in Petersburg with the Confederate army of Joseph Johnston in North Carolina. An assault was devised to sever the 30-mile-long Union line in two and force Grant to redeploy troops around Petersburg. Lee calculated these actions would provide the best chance for his army to slip out of Petersburg. The focus of the attack would be Fort Stedman. Three other Union fortifications along the Union line were targeted, including Ft. Fisher, southwest of Petersburg. Eleven thousand Confederates launched a surprise attack at 4:00 in the morning on March 25th. Fort Stedman fell quickly. The Confederate plan included the seizure of other redoubt fortifications behind Union lines near Fort Stedman. The assault stalled when the Confederates could not locate these fortifications and the attacking Rebel force began taking fire from three directions. Fort Stedman was recaptured by the Union Army by 8:15 that morning. The Confederates suffered the loss of over 4,000 soldiers during the attack, more than twice the number of Union casualties. These were men the Confederates could not afford to lose. Fort Stedman was the last major offensive taken by Robert E Lee in the Civil War.

Fort Stedman - March 25, 1865

Known Bedford Co. Casualty Listing

Name	Muster Age	Rank	State	Regiment	Company	Casualty	Assault
Brant, Henry	16	Pvt.	PA	93rd Infantry	H	Wounded	Ft. Fisher
Brown, Benjamin F	18	Pvt.	PA	208th Infantry	H	Wounded	Ft. Stedman
Chamberlain, Eli G	19	Pvt.	PA	208th Infantry	K	KIA	Ft. Stedman
Clevenger, Franklin	18	1st Sgt.	MD	3rd Infantry	B	Wounded	Petersburg
Fickes, Cyrus W	21	Pvt.	PA	200th Infantry	C	Wounded	Ft. Stedman
Gibson, Henry F	18	Corp.	PA	208th Infantry	H	Wounded	Ft. Stedman
Hollar, Philip V	40	Pvt.	PA	208th Infantry	K	Wounded	Ft. Stedman
May, William E	22	Pvt.	PA	93rd Infantry	D	Wounded	Ft. Fisher
Miller, Abraham M	30	Pvt.	PA	82nd Infantry	C	Wounded	Ft. Fisher
Miller, Armstrong	20	Pvt.	PA	200th Infantry	C	Wounded	Ft. Stedman
Mock, Henry	20	Pvt.	PA	200th Infantry	C	Wounded	Ft. Stedman
Mortimer, John L	38	Pvt.	PA	67th Infantry	C	Wounded	Ft. Fisher
O'Neal, James R	32	1st Sgt.	PA	208th Infantry	K	Wounded	Ft. Stedman
Perrin, Jonathan	37	Pvt.	PA	93rd Infantry	A	Wounded	Ft. Fisher
Phillipi, Franklin		1st Lt.	PA	93rd Infantry	E	Wounded	Ft. Fisher
Ritchey, Daniel S	23	Pvt.	PA	208th Infantry	K	Wounded	Ft. Stedman
Snow, William H	19	Pvt.	PA	207th Infantry	C	Wounded	Ft. Stedman
Sponsler, William	16	Pvt.	PA	100th Infantry	B	Wounded	Ft. Stedman
Williams, Samuel W		Corp.	PA	208th Infantry	K	Wounded	Ft. Stedman
Wilt, Daniel H	42	Pvt.	PA	208th Infantry	K	Wounded	Ft. Stedman

View from outside of Ft. Stedman. Confederate forces crossed the field in the foreground during the assault on March 25th.

Fort Stedman - March 25, 1865

(Right) Philip V Hollar was born on January 11th, 1824. His wife passed in 1860 with 3 small children at home on their family farm in West Providence Township. Philip remarried in 1861. He enlisted in the 208th Pennsylvania Infantry on September 7th, 1864, when he was 40 years old. Philip suffered a wound during the Ft. Stedman assault, but survived. He mustered out at the end of the war and returned to his farm in West Providence Township. Philip passed in 1888 and is buried in the Everett Cemetery.

(Left) Cyrus W Fickes was born on January 26th, 1844. He enlisted in the 21st Pennsylvania Cavalry on June 26th, 1863, while a wife and a young son waited for his return to their home in St. Clair Township. Cyrus mustered out of the Calvary on February 20th, 1864. His second enlistment was in the 200th Pennsylvania Infantry in August 1864. Cyrus suffered a wound during the Ft. Stedman attack. He survived and mustered out on June 6th, 1865. His brother was not as fortunate. James M Fickes mustered in the 101st PA Infantry when he was 16 years old and died of disease on December 8th, 1862. Cyrus was a father to 6 more children after the war. He passed at age 64. Both brothers are buried in the Fishertown Cemetery.

Fort Stedman - March 25, 1865

Two views from inside Fort Stedman

White Oak Road & Five Forks - March 30-April 1, 1865

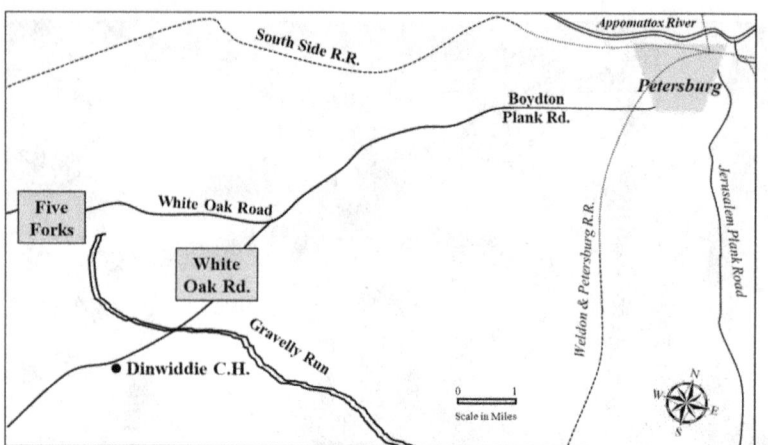

White Oak Road also referred to as Gravelly Run & Hatchers Run

Gravelly Run and Five Forks Known Bed. Co. Casualties - 15
KIA, Wounded & Died - 4
Wounded - 10
POW - 1

Captured Confederate Soldiers at Five Forks

General Philip Sheridan and two divisions of Union cavalry were recalled from the Shenandoah Valley. On March 29th, Grant instructed the aggressive Sheridan to swing around the right flank of the Confederate lines and attempt to sever the Southside Railroad line. Grant also dispatched the 5th Infantry Corps to attack the extreme right of Confederate lines. Lee shifted troops to meet this threat, further thinning his defense lines. On March 31st, General George Pickett moved 4 miles beyond the end of Confederate lines to meet Sheridan's cavalry. Pickett halted Sheridan's advance near Dinwiddie Court House, then withdrew his troops to Five Forks. That same day, the Union 5th Infantry Corps approached Confederate positions at White Oak Road before being driven back across Gravely Run. In the middle of the afternoon, the 5th Infantry Corps successfully counterattacked and pushed the Confederates back across White Oak Road. Union forces lost 1400 men on that day to about half that number for the Confederates. The next afternoon at Five Forks, the Union 5th Infantry Corps attacked the left side of Confederate lines while Sheridan led a cavalry charge around their right flank. Rebel lines quickly caved, and over 2400 soldiers were captured. Five Forks was a resounding victory for the Union and a crushing loss for the Confederates. The last railroad supply line into Petersburg would soon be severed. Grant sensed the dire situation facing the Confederate Army defending Petersburg and planned to launch a major assault the next day.

White Oak Road & Five Forks - March 30-April 1, 1865

Known Bedford Co. Casualty Listing

Name	Muster Age	Rank	State	Regiment	Company	Casualty	Date	Battle
Cashman, Jacob	23	Pvt.	PA	107th Infantry	E	Wound.-Died	3/30/65	White Oak Rd.
Chamberlain, David		Sgt.	PA	21st Cavalry	D	Wounded	4/1/65	Five Forks
Corle, James L	20	Pvt.	PA	55th Infantry	I	KIA	3/30/65	White Oak Rd.
Dull, Lewis	19	Pvt.	PA	55th Infantry	K	Wounded	3/30/65	White Oak Rd.
Gates, George W	18	Pvt.	PA	190th Infantry	D	POW	3/31/65	White Oak Rd.
Gordon, Joseph	16	Pvt.	PA	55th Infantry	K	Wounded	3/30/65	White Oak Rd.
Hare, Henry	16	Pvt.	PA	210th Infantry	K	Wounded	3/31/65	White Oak Rd.
Hissong, Josiah	22	2nd Lt	PA	55th Infantry	H	Wounded	3/30/65	White Oak Rd.
Kagarice, John	29	Pvt.	PA	198th Infantry	F	Wound.-Died	3/31/65	White Oak Rd.
Martin, Josiah	16	Pvt.	PA	107th Infantry	A	Wounded	3/31/65	White Oak Rd.
Mock, Andrew	33	Pvt.	PA	55th Infantry	K	Wound.-Died	3/31/65	White Oak Rd.
Ritchey, John	20	Pvt.	PA	55th Infantry	K	Wounded	3/30/65	White Oak Rd.
Smith, Peter	21	Pvt.	PA	11th Cavalry	A	Wounded	4/1/65	Five Forks
Whitaker, John H	16	Pvt.	PA	91st Infantry	I	Wounded	3/31/65	Five Forks
Young, Alexander	19	Pvt.	PA	191st Infantry	H	Wounded	4/1/65	Five Forks

John H Whitaker was born in Union (Pavia) Township on June 21st, 1848. He was 16 years old when he left his home in St. Clair Township to muster in the 91st Pennsylvania Infantry. John suffered a leg wound below his knee on March 31st, 1865, during the Battle of Five Forks. He was discharged from the army on June 19th, 1865, at Chester Hospital in Philadelphia. John married after the war and was a father to 3 children. He died at the young age of 31 in 1879 and is buried at Pleasantville Cemetery in Alum Bank.

Josiah Hissong was 21 when left Bedford Township to muster in the 55th Pennsylvania Infantry in October 1861. Josiah married Elizabeth Amick while on leave in February 1864. He was first wounded at Drewry's Bluff in May 1864 and a second time at the battle of Chaffin's Farm in September 1864. Two months later, their first child was born. Josiah suffered a serious hip wound at White Oak Road in March 1865. He recovered and returned home to Bedford County. Josiah was a father to 7 more children after the war. He lived to be 84 and was buried at the Schellsburg Cemetery in 1924.

Petersburg Final Assault - April 2nd, 1865

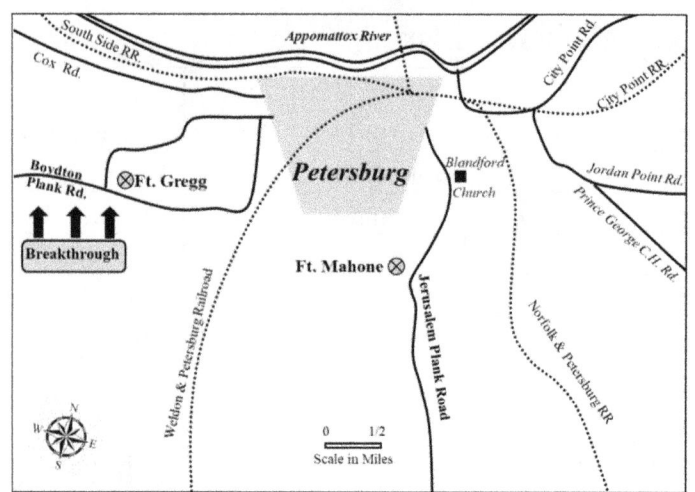

Known Bedford Co. Casualties - 33
KIA, Wound.-Died - 8
Wounded - 25

Ulysses S Grant ordered a pre-dawn frontal assault on Confederate lines at Petersburg on April 2nd. Union artillery launched a massive bombardment of the Confederate fortifications at midnight. Advance teams of Union soldiers hacked through wooden spiked obstacles to clear a path for fixed bayonet infantry charges at 4:00am. Initial waves of Union assaults on Fort Mahone were met with double cannister cannon shot and volleys of rifle fire from entrenched Confederates. Desperate hand to hand fighting took place as Union troops scrambled over the earthwork fortifications. Union soldiers reached the inner defenses of Fort Mahone, but a counterattack stopped any further advance.

The "Breakthrough" assault took place southwest of Petersburg. Union troops took over 2,000 casualties during initial infantry charges to reach Confederate entrenchments near Boydton Plank Road. Hand to hand fighting broke out all along the trench line. Within 20 minutes, rebel soldiers began surrendering and thousands of Union troops poured through the broken Confederate lines.

Just north of Boydton Plank Road, two Confederates on horseback were rapidly approaching two Bedford County soldiers who became separated from the rest of 138th Pennsylvania Infantry during the chaos of the Breakthrough. The Confederates shouted for John Mauk of Cumberland Valley and Daniel Wolford of Londonderry Township to surrender. One of the Bedford County soldiers shouted, "I can't see it!" and both fired a volley. A bullet struck Confederate Major General A. P. Hill just above the heart, killing one of the most famous figures of the Civil War.

Lee urgently notified Jefferson Davis he could "hold his position no longer" and Richmond needed to be evacuated immediately. The 9-month Siege of Petersburg was over, Richmond would soon be in Union hands, and Grant began a relentless pursuit of the fleeing Southern Army during the brief Appomattox Campaign.

Known Bedford Co. Casualty Listing

Name	Muster Age	Rank	State	Regiment	Company	Casualty	Assault
Avey, William H	28	Pvt.	PA	54th Infantry	H	Wounded	Breakthrough
Cartwright, Franklin J	16	Pvt.	PA	205th Infantry	C	Wounded	Fort Mahone
Clark, Ferdinand	19	Pvt.	PA	208th Infantry	H	Wounded	Fort Mahone
Felton, Peter S	20	Pvt.	PA	208th Infantry	K	KIA	Fort Mahone
Fluck, Porter	20	Pvt.	PA	87th Infantry	C	KIA	Breakthrough
Foor, Simon P	32	Pvt.	PA	208th Infantry	K	KIA	Fort Mahone
Funk, John D		Pvt.	PA	208th Infantry	K	Wounded	Fort Mahone
Garretson, Benjamin H	21	Pvt.	PA	205th Infantry	C	Wound.-Died	Fort Mahone
Gillam, Thomas	34	Pvt.	PA	93rd Infantry	A	Wounded	Breakthrough
Gray, George W (2)	41	Sgt.	PA	138th Infantry	E	Wounded	Breakthrough
Growden, Joseph	31	Pvt.	PA	45th Infantry	C	Wounded	Breakthrough
Hammer, Hezekiah	21	Capt.	PA	55th Infantry	K	Wounded	Breakthrough
Hellman, George	21	Pvt.	PA	138th Infantry	D	Wounded	Breakthrough
Hite, John	34	Pvt.	PA	205th Infantry	I	KIA	Fort Mahone
Lehn, Philip		Pvt.	PA	208th Infantry	H	Wounded	Fort Mahone
Leonard, John B	27	Pvt.	PA	208th Infantry	K	Wounded	Fort Mahone
Long, Joseph C	25	1st Sgt.	PA	208th Infantry	H	Wounded	Fort Mahone

Petersburg Final Assault - April 2nd, 1865

Josiah Garretson (pictured on the left) and Benjamin Garretson (on the right) grew up on a family farm in Napier Township. Josiah was born on March 22nd, 1839 and Benjamin on September 17th, 1841. Benjamin mustered in the 21st Cavalry on July 8th, 1863, and enlisted a second time in the 205th PA Infantry on August 27th, 1864. Josiah mustered in the 55th PA Infantry on February 29th, 1864. Both brothers were in heavy combat on April 2nd, 1865, during the final assault on Petersburg. Josiah survived fierce fighting against determined Confederate troops at Fort Gregg. After the capture of Fort Gregg, Josiah went to meet up with his younger brother and discovered Benjamin was severely wounded during the Fort Mahone assault. Benjamin had been shot in the left frontal lobe of his head. He was transported to McClellan General Hospital in Philadelphia, where he died on May 27th, 1865. Josiah returned home after the war and married in 1868. He was a father of 8 children. Josiah passed when he was 49 in 1888. Both brothers are buried at the Friends Cemetery in Spring Meadow.

Known Bedford Co. Casualty Listing

Name	Muster Age	Rank	State	Regiment	Company	Casualty	Assault
Morrow, B Moritimer	27	Lt. Col	PA	205th Infantry	F&S	Wounded	Fort Mahone
Osborn, Peter	37	Pvt.	PA	208th Infantry	K	Wound.-Died	Fort Mahone
Ott, Nicholas	26	Pvt.	PA	208th Infantry	H	Wounded	Fort Mahone
Price, Daniel M	21	Pvt.	PA	205th Infantry	C	Wounded	Fort Mahone
Slack, Francis M	23	1st Sgt.	PA	138th Infantry	E	Wounded	Breakthrough
Slick, John A	33	Pvt.	PA	208th Infantry	H	Wound.-Died	Fort Mahone
Slick, William S	18	Sgt.	PA	138th Infantry	D	Wounded	Breakthrough
Smeltzer, John B	18	Pvt.	PA	205th Infantry	C	Wounded	Fort Mahone
Snow, William H	19	Pvt.	PA	207th Infantry	C	Wounded	Fort Mahone
Snyder, John W	19	Pvt.	PA	49th Infantry	B	Wounded	Breakthrough
Sparks, James	33	Pvt.	PA	208th Infantry	K	Wounded	Fort Mahone
Strayer, Nicholas	30	Pvt.	PA	205th Infantry	C	Wound.-Died	Fort Mahone
Stuckey, Elias B	19	Corp.	PA	138th Infantry	D	Wounded	Breakthrough
Walter, Moses	17	Pvt.	PA	205th Infantry	C	Wounded	Fort Mahone
Williams, Alvah R	33	Pvt.	PA	208th Infantry	K	Wounded	Fort Mahone
Woy, Ezekiel C	16	Pvt.	PA	208th Infantry	K	Wounded	Fort Mahone

Petersburg Final Assault - April 2nd, 1865

Most Bedford County Casualties on April 2nd were during the attacks on Ft. Mahone, known as Fort Damnation. Advance teams of Union axe-men hacked apart the X-shaped wood obstacles pictured above to clear a path for charging infantry troops.

John W Snyder of 49th PA Infantry is seated on the left. He died on June 5th, 1871 and is buried in Duvall's Cemetery in Coaldale. John suffered a wound during the Breakthrough on April 2nd. Samuel Cupp (standing) was from Center County. Thomas Reader (right) from Chester County was killed at Spotsylvania in 1864.

William S Slick was born on October 6th, 1843 and grew up on a farm in St. Clair Township. William was an 18-year-old surveyor when he mustered as a Corporal in the 138th Pennsylvania Infantry on September 2nd, 1862. He received a promotion to Sergeant on February 24th, 1865. William was wounded during the Final Assault of Petersburg on April 2nd. He recovered and mustered out on June 23rd, 1865. William married in 1866 and was a father to 11 children. He passed at age 66 and is buried in the Pleasantville Cemetery in Alum Bank.

Alva R Williams is pictured on June 24th, 1865 after the amputation of his right arm from wounds received during the Fort Mahone assault on April 2nd. Alva mustered in the 208th PA Infantry on September 7th, 1864, while his wife and three children under the age of six remained on their family farm in Monroe Township. He returned home to continue farming after the war and was a father to six more children. Alva passed in 1897 at age 66 and is buried at the Rock Hill Cemetery in Monroe Township.

Petersburg Final Assault - April 2nd, 1865

Ft. Gregg pictured on the right has been referred to as the "Confederate Alamo". Nearly every Confederate soldier defending Fort Gregg fought until killed, wounded, or captured. Fourteen Union soldiers received the Medal of Honor for their bravery in taking Fort Gregg. Many veterans recalled this assault as being the fiercest of the war. Confederates were desperately holding on to Fort Gregg to allow time for the rest of Robert E Lee's troops to retreat from Petersburg. Hezekiah Hammer of the 55th and William Avey of the 54th Pennsylvania Regiments suffered wounds here, possibly from cannister shot. Cannister shot was a shotgun type munition being fired from cannons inside the fort.

Hezekiah Hammer amassed a remarkable record of bravery during the Civil War. Hezekiah was born in Union (Pavia) Township in 1840. He mustered as a private in the 55th Pennsylvania Infantry on November 5th, 1861. Hezekiah received 6 promotions during the war. He was promoted to Corporal in November 1863, Sergeant in June 1864, and 1st Sergeant in September 1864. The following is a verbatim excerpt of the official army record of the battle of Chaffin's Farm during the Siege of Petersburg. "Sergt. Hezekiah Hammer, Company K, Fifty-fifth Pennsylvania, with great fearlessness rushed forward, seized the colors from the wounded color bearer, and brought them off the field, and is recommended for promotion another grade for his gallantry (being already a lieutenant) to His Excellency the Governor of Pennsylvania. Would that his whole regiment had emulated his example." Hezekiah received a promotion to 2nd Lieutenant in October 1864, 1st Lieutenant in November 1864, and Captain in December 1864. While on leave, he married Sarah Ann Taylor on March 19th, 1865, and immediately returned to Petersburg to rejoin his regiment. Two weeks later, Hezekiah suffered severe wounds during the assault on Fort Gregg. His left arm needed to be amputated and his right arm was broken. Hezekiah recovered from his wounds and returned home. He worked as a storekeeper after the war and was a father to 9 children who survived to adulthood. Hezekiah passed on February 8th, 1910, at age 69 and is buried at the Pleasantville Cemetery.

Sailor's Creek - April 6th, 1865

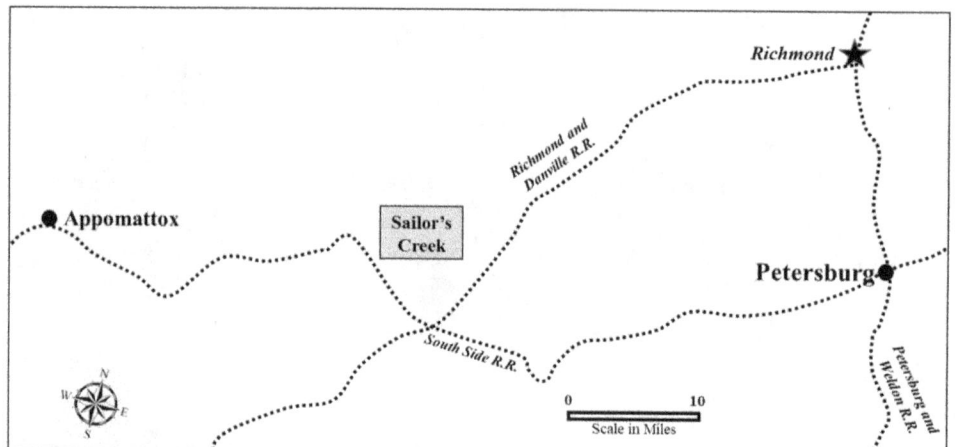

Known Bedford Co. Casualties - 11

KIA, Wound. & Died - 4

Wounded - 6

POW - 1

Photograph taken from the Confederate position on the Sailor's Creek battlefield. The 138th Pennsylvania Infantry advanced across the farthest left open field in the background toward Sailor's Creek at the bottom of the hill.

Approximately fifty thousand beleaguered Confederate troops retreated west from Petersburg. On April 4th, a hungry Confederate Army expected to receive food rations at the Amelia Court House railroad depot. A logistical mix-up resulted in plenty of ordinance being found, but no food supplies in the rail cars. This delayed their march until the next day and enabled the pursuing Union Army time to close in on the fleeing Confederates. On April 6th, Union cavalry units were nipping at the heels of retreating Confederate columns. A gap developed between columns and 11,000 Confederates were cut off from the rest of Lee's army. A joint attack by Union cavalry and infantry units routed the separated rebel troops, resulting in almost 9,000 casualties, including around 7,700 prisoners of war. The captured men included 6 Confederate generals. Lee observed the 2,000 stragglers who avoided being captured streaming back toward Confederate lines and remarked, "My God! Has the army dissolved?" Close to a quarter of Lee's remaining troops were lost. April 6th became known as "Black Tuesday" in the South. The next day, the Confederates failed to destroy a bridge that would have delayed the pursuit of the Union Army. That same day, Grant sent a note to Lee opening a dialog on surrender.

Sailor's Creek - April 6th, 1865

Known Bedford Co. Casualty Listing

Name	Muster Age	Rank	State	Regiment	Company	Casualty
Bridenstine, Jacob	34	Pvt.	PA	99th Infantry	B	Wounded
Burns, Joseph H	28	Pvt.	PA	82nd Infantry	K	KIA
Dickerhoof, Simon	38	Major	PA	138th Infantry	E	POW
Fetters, John		Pvt.	PA	99th Infantry		KIA
Hess, Daniel A	24	2nd Lt.	PA	55th Infantry	H	Wound.-Died
King, Harrison H	23	Corp.	PA	138th Infantry	E	Wounded
Lease, Robert H	21	Pvt.	PA	138th Infantry	D	Wounded
Little, John Pius	34	Pvt.	PA	99th Infantry	B	KIA
May, John L	38	Pvt.	PA	67th Infantry	C	Wounded
Smith, William H	18	Pvt.	PA	99th Infantry	I	Wounded
Young, Edward H	25	Corp.	PA	119th Infantry	D	Wounded

William H Smith was 18 years old when he left his home in Southampton Township to volunteer as a substitute for another man in the 99th PA Infantry on March 15th, 1865. During the Civil War, it was not uncommon to take part in combat soon after enlisting. Three weeks after joining the army, William received a wound at Sailor's Creek, the last major battle before Lee's surrender at Appomattox. He married after the war, was a father to 10 children, and enjoyed the companionship of a favorite pet in his later years. William passed at age 77 and was buried at the Rockhill Cemetery in Monroe Township in 1923.

Oliver Horton pictured with his wife Louise. Oliver was 32 years old when he mustered as a 1st Sergeant in the 138th Pennsylvania Infantry on August 29th, 1862. Louise remained at their home in Monroe Township with 6 children during the war. Oliver suffered a wound at the battle of Cedar Creek on October 19th, 1864. He received a promotion to Captain on February 20th, 1865 and led troops during the battle of Sailor's Creek. Oliver and Louise moved to Iowa to farm after the war and were parents to 3 more children. Louise passed when she was 83 in 1912. Oliver lived to be 89 years old. At his funeral in 1919, a sword taken from a Confederate Officer at Spotsylvania 55 years earlier was placed in his casket and was buried with him at his request.

Appomattox - April 9th, 1865

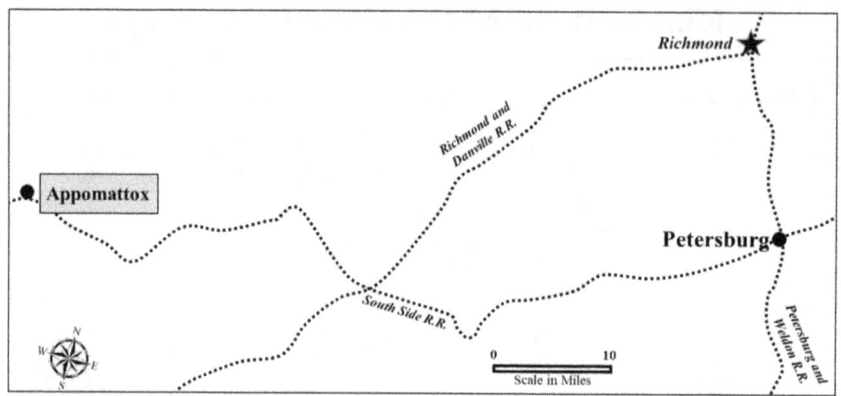

The McLean's House at Appomattox in 1865 where Robert E Lee surrendered to Ulysses S Grant.

Late in the afternoon of April 8th, Union cavalry units destroyed a Confederate supply train at Appomattox Station and attacked an Artillery unit capturing 25 cannons. On the morning of April 9th, Robert E Lee made one last attempt to punch through the Union lines surrounding his army. Confederate cavalry initially drove dismounted Union cavalry troops from their position on a nearby hill, but additional Union infantry regiments were rapidly approaching the battlefield. Lee realized the situation was hopeless and sent a note to Grant, requesting to negotiate the terms of surrender. Grant and Lee met that afternoon at the home of Wilmer McLean at Appomattox Court House. Grant offered more generous terms than Lee likely expected. Confederate soldiers received pardons and were allowed to return home with private property, including horses, and officers could keep side arms. Lee replied to Grant, "This will have the best possible effect upon the men. It will be very gratifying and will do much toward conciliating our people," Famished Confederate soldiers were also provided food rations. The victorious Union Army conducted a sober formal surrender ceremony on April 12th. Union troops were ordered to attention and to "carry arms" as a sign of respect. The defeated Southern army reciprocated with their own respectful gestures as they marched between the lines of their former enemies. After Lee's surrender at Appomattox, over 100,000 additional Confederate troops throughout the South would lie down their arms over the next couple months. The Civil War Soldiers from Bedford County would soon be coming home.

Appomattox - April 9th, 1865

Partial Listing of Regiments at Appomattox during the Surrender

11th Pennsylvania Infantry
53rd Pennsylvania Infantry
55th Pennsylvania Infantry
56th Pennsylvania Infantry
82nd Pennsylvania Infantry
84th Pennsylvania Infantry
91st Pennsylvania Infantry
93rd Pennsylvania Infantry
99th Pennsylvania Infantry
107th Pennsylvania Infantry

110th Pennsylvania Infantry
138th Pennsylvania Infantry
184th Pennsylvania Infantry
191st Pennsylvania Infantry
8th USCT Infantry
41st USCT Infantry
45th UCST Infantry
127th UCST Infantry
2nd Pennsylvania Cavalry
21st Pennsylvania Cavalry

Unidentified Union Soldiers during the Confederate Surrender at Appomattox.

Appomattox - April 9th, 1865

(Left) Isaac Nicodemus was born on May 18th, 1833. He married Catherine Carroll in Schellsburg in 1856. Isaac volunteered in the 138th Pennsylvania Infantry on August 29th, 1862, leaving behind a wife and 3 small children at their Napier Township home. The 138th PA infantry took part in heavy combat during the pivotal battles of the Wilderness, Cold Harbor, Monocacy, Opequon, and Cedar Creek before witnessing the Confederate surrender at Appomattox. Isaac survived the war and returned home to his family. Isaac lived to be 77 and was buried in the Schellsburg Cemetery in 1910. His obituary stated his comrades in the 138th Pennsylvania Infantry were his pallbearers.

(Right) Henry I Claar was born in Queen on August 2nd, 1842. He was working on a farm when he volunteered in the 55th Pennsylvania Infantry on December 4th, 1861. Henry received a wound during the Siege of Petersburg in the Summer of 1864. The photograph on the right is somewhat unusual. A bandage on the right ankle and foot covers a wound. A thick ink line was added across the top of the bandage, possibly showing the path of a bullet or shell fragment. More ink streamed to his toes, suggesting a trail of blood from the wound. Henry married Susan Stiffler on September 17th, 1864 in Claysburg while recuperating. After recovering from his wound, Henry rejoined the 55th and mustered out with the regiment on August 30th, 1865. Henry and Susan were parents to 16 children. Henry passed when he was 76 in 1919. Susan lived to the age of 80 and passed in 1826. They are buried in the Upper Claar Cemetery in Queen.

Appomattox - April 9th, 1865

(Left) Thomas H Farber was born on August 19th, 1837. Thomas was working as a Carpenter in St. Clair Township when he mustered in the 55th Pennsylvania Infantry in October 1861. He received promotions to Corporal on October 4th, 1863, and to Sergeant on September 26th, 1864. The 55th took part in heavy combat during the battles of Drewry's Bluff, Cold Harbor, and the Initial Assault of Petersburg. Fittingly, the 55th which suffered the highest number of casualties of any Bedford County unit during the war, was present to witness the Confederate Surrender at Appomattox. Thomas mustered out on August 30th, 1865. He married in 1866 and was a father to two sons. Thomas died in 1874 at the young age of 37 and is buried at the Bedford Cemetery.

(Right) Frederick Feight was born on April 22nd, 1842. The 1860 Cumberland Valley Township Census listed Frederick as a farm worker. He mustered in the 2nd Pennsylvania Cavalry on November 2nd, 1861. The 2nd PA Cavalry was on the battlefield at Antietam, Gettysburg, and was present during the Confederate surrender at Appomattox. Frederick mustered out on June 17th, 1865. The 2nd PA Cavalry spent much of the war in the Shenandoah Valley. After the war, Frederick moved to Winchester and worked as a teamster. He met his wife in Winchester and was a father to 6 daughters. Frederick passed at age 71 in 1914 and is buried at the Mount Pleasant Meeting House Cemetery in Frederick County, Virginia.

Andersonville Prisoner of War Camp

Known Bedford Co. POW's at Andersonville - 132

Died while a POW – 68

Wounded & POW – 5

POW - 59

Largest Number of Bedford Co. Andersonville POW's by Regiment

101st PA Infantry – 44

184th PA Infantry – 25

55th PA Infantry - 24

The first prisoners began arriving in Andersonville in February 1864 while the stockade was being built. During the war, almost 50,000 Union soldiers were prisoners of war at Andersonville. Nearly 13,000 died from disease, malnutrition, and inadequate shelter. Percentage wise, the Bedford County numbers were much worse. Over half of the Bedford County men and boys who walked through the gates of Andersonville perished there, or shortly after being moved to another Confederate POW camp. The horrors of Andersonville are well documented. A fifteen-foot-high timber wall enclosed a 26-acre rectangular enclosure originally built to hold 10,000 prisoners. By August 1864, Andersonville was overflowing with three times that many men. Planned barracks type housing never materialized because of the price and scarcity of lumber. This left Union soldiers with only small makeshift tents made of scraps of wood and cloth to try to find some relief from an unforgiving Georgia mid-summer sun. A stream ran thru the middle of the camp was the only source for water, but it quickly became fouled from human waste. The swampy conditions in Andersonville were rife with disease. Dysentery, scurvy, and diarrhea were the major causes of death. Malaria, cholera, typhoid, and smallpox were also common. Rations were completely inadequate. Malnourished prisoners received small amounts of cornmeal, bean soup and salt pork. Some developed scurvy from a lack of vegetables or fruit. The "deadline" was a small wooden fence inside the prison to keep inmates away from the outer walls of the stockade. Guards often shot prisoners if they touched the deadline fence or ventured too close.

When Union troops occupied Atlanta on September 2nd, 1864, within striking distance of Andersonville, Confederate authorities moved many prisoners at Andersonville to other POW camps in South Carolina and coastal Georgia. Andersonville remained a POW camp with smaller numbers of prisoners until the war ended in April 1865. Generations of people in the Mattie area of Bedford County have retold the story of a local man who survived Andersonville and was not recognized by his family when he returned to his home. Unfortunately, the identity of this soldier is unknown as of the publishing of this book, and may be lost to history. There were many other deplorable POW camps in both the North and South with horrific living conditions. Andersonville was just the worst of the worst. Approximately 30,000 Union and 26,000 Confederate soldiers who walked through the gates of all Civil War prisoner of war camps never returned home alive.

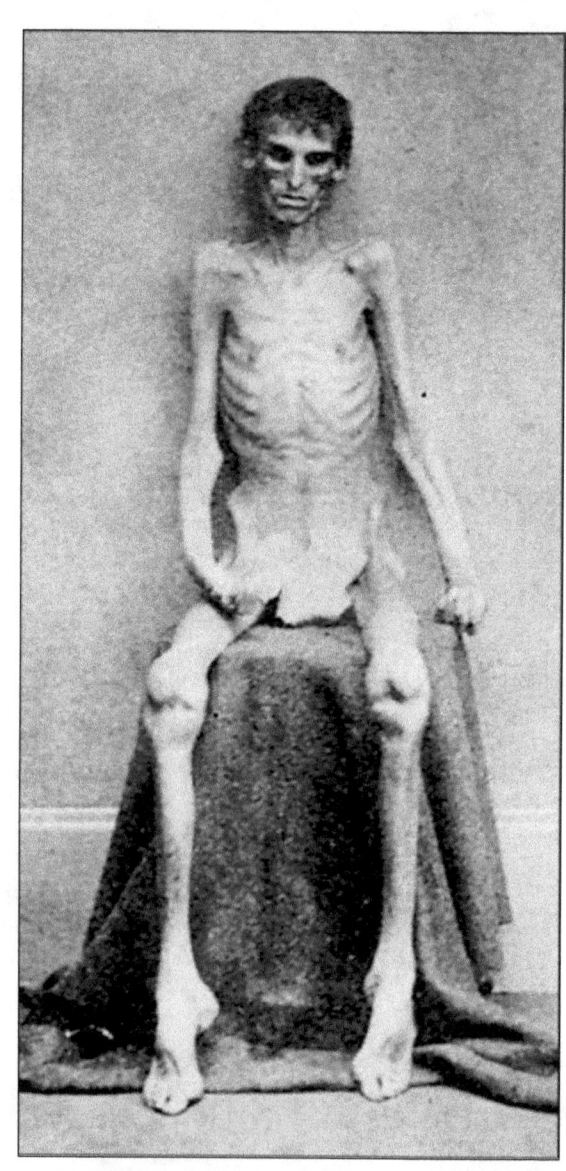

Unidentified emaciated Andersonville POW after his release in 1865

Andersonville Prisoner of War Camp

Known Bedford Co. POW's at Andersonville

Name	Muster Age	Rank	State	Regiment	Casualty	Battle
Adams, David	22	Pvt.	PA	101th Infantry	POW	Plymouth, NC
Adams, Philip	18	Pvt.	PA	55th Infantry	POW-Died	Drewry's Bluff
Agnew, George H		Pvt.	PA	55th Infantry	POW-Died	
Allison, David	20	Pvt.	PA	55th Infantry	POW-Died	2nd Drewry's Bluff*
Allison, Edward	18	Pvt.	PA	55th Infantry	POW-Died	2nd Drewry's Bluff
Anderson, James	21	Pvt.	PA	101th Infantry	POW-Died	Plymouth, NC
Banner, Adam	26	Pvt.	PA	103rd Infantry	POW	Plymouth, NC
Barnett, Ephraim	29	Pvt.	PA	149th Infantry	POW-Died	North Anna River
Barnett, Samuel	19	Pvt.	PA	184th Infantry	POW-Died	Jerusalem Plank Road
Baughman, George	22	Sgt.	PA	138th Infantry	POW-Died	
Beam, Daniel	59	Pvt.	PA	101th Infantry	POW-Died	Plymouth, NC
Beaver, Simon J	16	Pvt.	PA	55th Infantry	POW	2nd Drewry's Bluff
Beltz, Abraham	32	Pvt.	PA	101th Infantry	POW-Died	Plymouth, NC
Bennett, Jacob	23	Pvt.	PA	55th Infantry	POW-Died	
Bequeth, William H	17	Pvt.	PA	101th Infantry	POW	Plymouth, NC
Bessor, John	26	Corp.	PA	101th Infantry	POW-Died	Plymouth, NC
Blackburn, Martin	40	Pvt.	PA	184th Infantry	POW-Died	Jerusalem Plank Road
Blymyer, William		Pvt.	PA	20th Cavalry	POW-Died	
Boor, William A	21	Sgt.	PA	55th Infantry	POW	2nd Drewry's Bluff
Bridaham, Henry W	29	Pvt.	PA	55th Infantry	POW-Died	
Brown, Jacob D	24	Corp.	PA	101th Infantry	POW	Plymouth, NC

Samuel J McEldowney was born in Rainsburg on October 11th, 1827. Samuel enlisted in the Mexican War when he was 17 and took part in combat during the decisive battles around Mexico City. He returned to Friends Cove after the war and married Sarah Mary Oliver in 1850. Samuel mustered as a 1st Sergeant in the 101st Pennsylvania Infantry on Jan 13th, 1862, while Sarah and their 5 children tended to their family farm. Three days after their 6th child was born on April 17th, 1864, the entire Union garrison at Plymouth, NC was captured. Samuel was taken to Andersonville with the other enlisted men. He escaped, but bloodhounds hunted him down and he was brought back to the inhumane living conditions at Andersonville. He was released on April 1st, 1865, after almost a year in captivity. Samuel returned home and lived the rest of his life in Bedford County. He passed when he was 75 and is buried in the Everett Cemetery.

Andersonville Prisoner of War Camp
Known Bedford Co. POW's at Andersonville

Name	Muster Age	Rank	State	Regiment	Casualty	Battle
Brown, John W	25	Pvt.	PA	101th Infantry	POW	Plymonth, NC
Brown, Joseph L	33	Pvt.	PA	101th Infantry	Wound.-POW	Plymonth, NC
Brown, Samuel D	19	Pvt.	PA	101th Infantry	POW	Plymonth, NC
Burket, Frederick	33	Pvt.	PA	55th Infantry	POW-Died	2nd Drewry's Bluff
Carnell, Samuel	24	Corp.	PA	101th Infantry	POW	Plymonth, NC
Clingerman, Peter	21	Pvt.	PA	101th Infantry	POW	Plymonth, NC
Corle, Alexander B	18	Pvt.	PA	55th Infantry	POW	Drewry's Bluff
Dasher, James H	23	Pvt.	PA	55th Infantry	POW	2nd Drewry's Bluff
Davidson, Samuel	20	Pvt.	PA	184th Infantry	POW-Died	Jerusalem Plank Road
Defibaugh, John	34	Pvt.	PA	101th Infantry	POW-Died	Plymonth, NC
Diehl, Espy	27	Pvt.	PA	55th Infantry	POW-Died	
Dively, John	17	Pvt.	PA	110th Infantry	POW-Died	
Dull, John	19	Pvt.	PA	184th Infantry	POW-Died	Boydton Plank Road
Edwards, John R	25	Pvt.	PA	13th Cavalry	POW	Sulpher Springs
England, Jacob	21	Pvt.	PA	101th Infantry	POW	Plymonth, NC
Ensley, Christopher	34	Corp.	PA	184th Infantry	POW-Died	Petersburg

Rations being issued in Andersonville Prison on August 17, 1864. The view is from the main gate.

Andersonville Prisoner of War Camp

Known Bedford Co. POW's at Andersonville

Name	Muster Age	Rank	State	Regiment	Casualty	Battle
Exline, Jacob	22	Pvt.	PA	55th Infantry	POW-Died	2nd Drewry's Bluff
Foor, Francis L	19	Pvt.	PA	101th Infantry	POW-Died	Plymonth, NC
Foor, Jonathan S	22	Pvt.	PA	107th Infantry	POW	Gettysburg
Foreman, William	34	Pvt.	PA	3rd Heavy Art.	POW-Died	Cold Harbor
Fritz, Daniel E	25	Corp.	PA	18th Cavalry	POW-Died	Wilderness
Gollipher, Justice	23	Pvt.	PA	101th Infantry	POW-Died	Plymonth, NC
Gracey, Alfred	18	Sgt.	PA	107th Infantry	POW	Gettysburg
Griffin, Andrew H		Pvt.	PA	7th Cavalry	POW	Dallas, GA
Hand, Henry		Pvt.	PA	55th Infantry	POW-Died	
Hanks, Benjamin A	28	Sgt.	PA	101th Infantry	POW-Died	Plymonth, NC
Hanks, Caleb	19	Pvt.	PA	101th Infantry	POW-Died	Plymonth, NC
Hanks, David F	18	Pvt.	PA	101th Infantry	POW-Died	Plymonth, NC
Hanks, Nelson	28	Pvt.	PA	101th Infantry	POW-Died	Plymonth, NC
Harbaugh, Emanuel	19	Pvt.	PA	138th Infantry	POW	Wilderness
Hardinger, George W	28	Pvt.	PA	149th Infantry	POW-Died	Wilderness
Hazelett, Moses	19	Pvt.	PA	101th Infantry	POW	Plymonth, NC
Henry, James W	17	Sgt. Maj.	IA	15th Infantry	Wound.-POW	Atlanta, GA

Thomas W Slick was born and raised near St. Clairsville. He was 23 years old when he married Mary Jane Smith on October 16th, 1861. Two months later he volunteered in the 101st PA Infantry. Their 1st son was born the following year. Thomas was captured on April 20th, 1864 and taken to Andersonville. Two months after his capture, his two-year-old son died. Thomas was released from Andersonville on February 26, 1865 and was a father to 6 sons born after the war. He passed at age 50 in 1889 and is buried in the Old Union Cemetery in Osterburg.

Jacob England was a 21-year-old man living in Colerain Township when he mustered in the 101st Pennsylvania Infantry on November 1st, 1861. Jacob was one of the many Bedford County soldiers captured at Plymouth, NC and taken to Andersonville. He was later sent to a POW camp in Florence, SC and released on April 2nd, 1865. Jacob married after the war and was the father of 8 children. He moved his family to Illinois in 1870. Jacob passed at age 46 and is buried in the Westfall Cemetery in Victoria, IL. The ruined health of some POW survivors led to early graves.

Andersonville Prisoner of War Camp
Known Bedford Co. POW's at Andersonville

Name	Muster Age	Rank	State	Regiment	Casualty	Battle
Hetrick, Daniel L	16	Pvt.	PA	101th Infantry	POW	Plymonth, NC
Hite, Samuel C	29	Pvt.	PA	55th Infantry	POW-Died	
Holsinger, Josiah	18	Pvt.	PA	110th Infantry	POW	Cold Harbor
Irvine, Wilson	19	Pvt.	PA	184th Infantry	POW-Died	Jerusalem Plank Road
Jamison, Benjamin	19	Pvt.	PA	110th Infantry	POW	Cold Harbor
Knipple, William H	18	Corp.	PA	101th Infantry	POW	Plymonth, NC
Knox, James H	26	1st Sgt.	PA	184th Infantry	POW-Died	Jerusalem Plank Road
Layton, Samuel	32	Pvt.	PA	184th Infantry	POW-Died	Jerusalem Plank Road
Leader, Henry H	20	Pvt.	PA	2nd Heavy Art.	POW	Cold Harbor
Lightmigator, Ernestus		Corp.	PA	111th Infantry	POW	
Lightningstar, Augustus	24	Pvt.	PA	101th Infantry	POW	Plymonth, NC
Linn, Henry	22	Sgt.	PA	101th Infantry	POW	Plymonth, NC
Lowery, Daniel	18	Pvt.	IN	2nd Cavalry	POW-Died	Newnan, GA
Malone, John	20	Pvt.	PA	191st Infantry	POW	
Manges, Abraham	18	Pvt.	US	14th Infantry	POW	
Mars, John	21	Pvt.	PA	55th Infantry	POW-Died	Edisto Island, SC
Marshall, Henry L	31	Pvt.	PA	184th Infantry	POW-Died	Jerusalem Plank Road
Martin, James P	18	Pvt.	PA	101th Infantry	POW	Plymonth, NC
Mayers, John		Pvt.	PA	2nd Cavalry	POW-Died	

Bird's-eye view photograph of Andersonville Prison taken on August 17th, 1864.

Andersonville Prisoner of War Camp

Known Bedford Co. POW's at Andersonville

Name	Muster Age	Rank	State	Regiment	Casualty	Battle
McDaniel, Daniel	21	Pvt.	PA	2nd Heavy Art.	POW	Cold Harbor
McEldowney, George E	27	Pvt.	PA	101th Infantry	POW	Plymonth, NC
McEldowney, Samuel J	34	1st Sgt.	PA	101th Infantry	POW	Plymonth, NC
Miller, John	33	Pvt.	PA	101th Infantry	POW-Died	Plymonth, NC
Miller, Thomas J	23	Pvt.	PA	138th Infantry	POW-Died	Wilderness
Mills, Andrew J	21	Pvt.	PA	101th Infantry	POW	Plymonth, NC
Mills, Franklin G	18	Musician	PA	101th Infantry	Wound.-POW	Plymonth, NC
Mitchell, James P	18	Pvt.	PA	55th Infantry	POW-Died	
Mock, Aaron	21	Pvt.	PA	138th Infantry	Wound.-POW	Wilderness
Morgart, William	23	Pvt.	PA	18th Cavalry	POW-Died	Wilderness
Moyer, John	30	Pvt.	PA	2nd Cavalry	POW-Died	
Oler, James W	20	Pvt.	PA	101th Infantry	POW	Plymonth, NC
Otto, Henry S	16	Pvt.	PA	184th Infantry	POW-Died	Jerusalem Plank Road
Over, David S	18	Pvt.	PA	184th Infantry	POW	Jerusalem Plank Road
Over, Jacob Z	22	Sgt.	PA	184th Infantry	POW	Jerusalem Plank Road
Potter, Martin L	18	Pvt.	PA	101th Infantry	POW	Plymonth, NC
Prosser, David W	15	Corp.	PA	55th Infantry	POW	2nd Drewry's Bluff
Rhodes, George	19	Pvt.	PA	184th Infantry	POW	Jerusalem Plank Road

Josiah A Zembower was born in Cumberland Valley on August 16th, 1842. Josiah was married with a newborn son when he mustered in the 184th Pennsylvania Infantry on May 17th, 1864. A month later, he was captured at the Battle of Jerusalem Plank Road during the siege of Petersburg and sent to Andersonville. Josiah survived the next 10 months of captivity and was released on April 28th, 1865. He was the father of eleven more children born after the war. Josiah died in 1916 at age 73 and is buried at the Centenary Methodist Church Cemetery in Cumberland, MD.

George F Shoemaker was 19 years old when he volunteered in the 101st Pennsylvania Infantry. He was captured at Plymouth, NC on April 20th, 1864 and taken to Andersonville. George was transferred to the Florence, SC, POW camp in October 1864 before being released on March 1st, 1865. He married in 1866 and was the father of seven children. George had a farm near Imler and lived to be 96 years old. He was buried at St. Mark's Cemetery in King in 1937. George stated in a Bedford Gazette article in 1933, he contracted scurvy while a POW which he suffered from his entire life.

Andersonville Prisoner of War Camp
Known Bedford Co. POW's at Andersonville

Name	Muster Age	Rank	State	Regiment	Casualty	Battle
Rice, Abraham	27	Sgt.	PA	101th Infantry	POW	Plymonth, NC
Rice, Isaac	27	Corp.	PA	101th Infantry	POW-Died	Plymonth, NC
Ritchey, John C	18	Pvt.	PA	138th Infantry	Wound.-POW	
Roudabush, Benjamin	16	Pvt.	PA	55th Infantry	POW-Died	
Shaffer, Jacob J	19	Pvt.	PA	55th Infantry	POW-Died	
Shoemaker, Austin	20	Pvt.	PA	110th Infantry	POW	Jerusalem Plank Road
Shoemaker, George F	20	Corp.	PA	101th Infantry	POW	Plymonth, NC
Shrader, Auterbine	18	Pvt.	PA	55th Infantry	POW-Died	
Shrader, William O	18	Pvt.	PA	55th Infantry	POW-Died	
Siler, James P	20	Pvt.	PA	101th Infantry	POW	Plymonth, NC
Slick, Philip	18	Pvt.	IL	9th Cavalry	POW-Died	Tupelo, Miss.
Slick, Thomas W	23	Pvt.	PA	101th Infantry	POW	Plymonth, NC
Smith, Amos	18	Corp.	PA	101th Infantry	POW	Plymonth, NC
Smith, Joseph A	19	Pvt.	PA	101th Infantry	POW	Plymonth, NC
Snowden, John W	19	Pvt.	PA	2nd Cavalry	POW	
Snyder, John W	19	Pvt.	MD	3rd PHB Inf.	POW-Died	Jonesville
Sparks, William	15	Pvt.	PA	101th Infantry	POW	Plymonth, NC
Steckler, Charles	22	Pvt.	PA	55th Infantry	POW-Died	
Stephens, Jacob	51	Pvt.	PA	11th Cavalry	POW-Died	Reams Station
Stine, David	30	Pvt.	PA	11th Cavalry	POW-Died	Suffolk

Andersonville Prison Camp in 1864

Andersonville Prisoner of War Camp
Known Bedford Co. POW's at Andersonville

Name	Muster Age	Rank	State	Regiment	Casualty	Battle
Stone, Reuben M	26	Sgt.	PA	101th Infantry	POW	Plymonth, NC
Stuckey, David H	28	Corp.	PA	184th Infantry	POW-Died	Jerusalem Plank Road
Summerville, Abner	35	Pvt.	PA	55th Infantry	POW-Died	
Swoveland, William	31	Pvt.	PA	184th Infantry	POW-Died	Jerusalem Plank Road
Teeter, Christian	22	Pvt.	PA	184th Infantry	POW-Died	Jerusalem Plank Road
Tewell, Daniel	34	Corp.	PA	4th Cavalry	POW-Died	Sulpher Springs
Trout, Sylvester	19	Pvt.	PA	184th Infantry	POW	Jerusalem Plank Road
Turner, Thomas		Pvt.	PA	184th Infantry	POW	Jerusalem Plank Road
Vaughan, Ephraim	18	Musician	PA	101th Infantry	POW	Plymonth, NC
Waltman, William H	20	Pvt.	PA	184th Infantry	POW	Petersburg
Ward, Samuel	32	Pvt.	PA	138th Infantry	POW-Died	Mine Run
Watson, Henry S	18	Pvt.	PA	184th Infantry	POW-Died	Jerusalem Plank Road
Watson, John R	45	Pvt.	PA	184th Infantry	POW	Jerusalem Plank Road
Wilson, George W	50	Pvt.	PA	101th Infantry	POW-Died	Plymonth, NC
Wolfhope, John	23	Pvt.	PA	184th Infantry	POW-Died	Jerusalem Plank Road
Wonderly, William H	19	Pvt.	PA	55th Infantry	POW	Drewry's Bluff
Wright, Charles C	22	Pvt.	PA	184th Infantry	POW	Jerusalem Plank Road
Wright, Edmund S	20	Pvt.	PA	184th Infantry	POW-Died	Jerusalem Plank Road
Wright, Thomas	20	Pvt.	PA	13th Cavalry	POW	2nd Deep Bottom
Zembower, Josiah A	21	Pvt.	PA	184th Infantry	POW	Jerusalem Plank Road

Photograph of the Burial of Prisoners at Andersonville taken on August 17th, 1864

President Warren G Harding dedicated the Tomb of the Unknown Soldier at Arlington National Cemetery on November 11th, 1921. During his address, he echoed the thoughts of Abraham Lincoln on memorializing the deaths of soldiers by stating "Our part is to atone for the losses of the heroic dead by making a better Republic for the Living."

7. Those who made the Ultimate Sacrifice

Luther R Piper was born on September 28th, 1834. His twin brother Willie died in 1839. Luther was the grandson of John Piper, who was a Lieutenant Colonel during the Revolutionary War. John Piper also constructed Fort Piper in Cypher to help protect local settlers from Indian attacks. His father William Piper was a General during the War of 1812 and a member of the United States Congress. Luther mustered as a corporal in the Hopewell Rifles company of the 8th Pennsylvania Reserves on June 11th, 1861. Luther was captured during the withdrawal from the Mechanicsville battlefield to Gaines Mill on June 27th, 1862. There is no available record of when he was released from a prisoner of war camp. He suffered severe wounds at the battle of Fredericksburg on December 13th, 1862. Luther died from his wounds on January 1st, 1863 at Finley General Hospital in Washington, DC. He was 28 years old. Luther is buried at the historic Piper Cemetery in Cypher.

There were 585 Civil War soldiers who perished from 1861 to 1866, who were either living in Bedford County, an immediate border community, or were previous residents of Bedford County and enlisted in the regiments of other states. Five hundred and Thirty of the fallen soldiers were residents of Bedford County. Twenty-nine soldiers had lived just across the county border, including 14 in Blair County, 9 in Fulton County, 5 in Huntingdon County and 1 in Somerset County. Twenty-six soldiers were born or grew up in Bedford County before moving west. One soldier died while a member of a United States Army regiment. The following are the residence abbreviations in the last column (Res.) of this listing: B–Bedford County, BL–Blair County, HC–Huntington County, FC–Fulton County, SC–Somerset County, IA - Iowa, IL - Illinois, IN–Indiana, KS–Kansas, MI–Michigan, MO–Missouri, OH–Ohio, US–US Army regiment. Four sets of soldiers shared the same name on this listing. Please note a (1) and a (2) are added to the names of these soldiers on the following pages and in the Detailed Alphabetical Listing chapter.

The Muster Ages listed on the following pages are when soldiers first joined the army during their initial enlistment. Some Bedford County soldiers were especially young when they died. Nineteen boys were 16 years old or younger when they first mustered in the army, including 4 who were 14 years old when they left their homes. One hundred and forty-four were teenagers when they entered the war. Three hundred and sixty-two men were in their 20s and 30s and thirty-three died in their 40s and 50s. There are forty-six others without available documentation of their age. Many never experienced some of the joys in life, including marriage and the birth of their children. Those with young families were denied the opportunity to watch their children grow and witness their children become parents themselves. Some older soldiers left behind grandchildren. All would have missed future family gatherings during holidays and celebrations of landmark events.

Widows with children and some disabled parents who relied on fallen soldiers to put food on the table sometimes found themselves in dire financial straits. The number of new applications overwhelmed the Pension Bureau, and the approval process sometimes took years. Some delays resulted from requests for records that simply did not exist. Other times signed affidavits were necessary from doctors and fellow soldiers who could be difficult to find. Some widows remarried, many did not, and all were denied the opportunity to grow old with the father of their children.

The "Unconfirmed Listing of Deaths" section of the book on page 364 lists an additional 170 Civil War soldiers, who lost their lives. These soldiers were mentioned in various books, listings and records as possibly being from Bedford County. After reviewing census records and other available information, it appears some of these fallen soldiers likely did not live in Bedford County during their lifetimes. Others on this listing have little to no available documentation to confirm they were from Bedford County or information confirming their deaths occurred during the years of 1861-1866.

Tragedy touched many during the Civil War. US Census data from 1850 show families with six to nine children were common. All the soldiers listed on the following pages were sons, brothers, husbands, and fathers to others.

William M Amick was born in 1843. He was listed on the 1850 Bedford Township Census with his parents, John and Mary Amick and six sisters. William mustered as a corporal in the 55th Pennsylvania Infantry in October 1861. His family later read his name on a casualty report published in the Bedford Inquirer in 1864. William was first wounded during the battle of Cold Harbor on June 4th, 1864. He suffered a more severe wound two weeks later during the Initial Assault on Petersburg that resulted in his leg being amputated. William never recovered and died on August 12th, 1864. William is buried in the Hampton National Cemetery in Virginia.

Joseph Armstrong was born on February 23rd, 1827 and married Anna Rebecca Shaffer on May 27th, 1851. Joseph volunteered in the 11th PA Infantry in September 1861. Anna was at home with their 4 small children in Hopewell Township when Joseph left for the war. A fifth child was born in 1862. In a letter to his wife dated July 11th, 1862, Joseph wrote "Keep my little ones well… for I never expect to see them again in this world." Joseph received a disability discharged on July 1st, 1865 but was able to see his wife and children before his passing. He died at home when he was 38 years old on July 12th, 1865 and is buried in the Everett Cemetery.

Name	Muster Age	Rank	State	Regiment	Muster In Date	Casualty	Date of Death	Battle	Res.
Adams, Philip	18	Pvt.	PA	55th Infantry	10/11/61	POW-Died	6/19/65	Drewry's Bluff	B
Adams, Solomon	17	Pvt.	PA	55th Infantry	10/12/61	Died	6/25/62		B
Agnew, George H	29	Pvt.	PA	55th Infantry	2/19/64	POW-Died	11/27/64		B
Agnew, Levi J	20	Corp.	PA	76th Infantry	10/9/61	Wound.-Died	3/26/65	Bermuda Hundred	B
Ake, Thomas	17	Pvt.	PA	11th Cavalry	9/5/61	Wound.-Died	10/16/62	Carrsville	B
Ake, William H	25	Corp.	PA	138th Infantry	8/29/62	KIA	5/6/64	Wilderness	B
Akers, Amos	22	Pvt.	OH	94th Infantry	8/24/62	POW-Died	6/10/64	Chickamauga, GA	FC
Allison, David	20	Pvt.	PA	55th Infantry	11/5/61	POW-Died	9/27/64	Drewry's Bluff*	B
Allison, Edward	18	Pvt.	PA	55th Infantry	2/10/64	POW-Died	6/24/64	Drewry's Bluff	B
Allison, Jacob W	32	Pvt.	PA	55th Infantry	11/4/62	Died	12/14/63		B
Allison, John	19	Pvt.	PA	55th Infantry	11/4/61	Died	1/23/66		B
Allison, Joseph D	39	Pvt.	PA	138th Infantry	8/29/62	Died	1/26/64		B
Allison, Noah	19	Pvt.	PA	138th Infantry	8/29/62	KIA	6/5/64	Cold Harbor	B
Alloway, Simon P	28	Pvt.	IA	28th Infantry	9/3/62	Died	8/4/63		HC
Amick, Josiah	16	Pvt.	PA	101st Infantry	11/1/61	Died	11/15/62		B
Amick, William B	26	Sgt.	PA	138th Infantry	8/29/62	KIA	5/6/64	Wilderness	B
Amick, William M	19	Sgt.	PA	55th Infantry	10/11/61	Wound.-Died	8/12/64	Petersburg - Initial	B
Anderson, James	21	Pvt.	PA	101st Infantry	12/28/61	POW-Died	12/1/64	Plymonth, NC	B
Anderson, John (1)	24	Pvt.	IL	55 Infantry	10/17/61	Died	9/6/62		IL
Anderson, John (2)	28	Pvt.	PA	63rd Infantry	7/11/63	KIA	5/24/64	North Anna River	B
Anderson, William W	22	Major	PA	20th Cavalry	3/2/64	KIA	1/17/65		B

William J Baughman was born in 1837. His parents, George and Mary, were listed on the 1850 West Providence Township Census with their four sons - William, Josiah, George, and Jeremiah and five daughters. Tragically, William, Josiah and George died during the Civil War. William died at Douglas General Hospital in Washington, DC, on April 26th, 1864 while enlisted in the 2nd PA Cavalry. The cause of death is not listed on available records. He is buried in the Baughman Union Cemetery. A single reference was found stating William was married to Julia A Baughman. No other information on his wife is known.

George Baughman and two of his brothers died in the Civil War. George married Elizabeth Mortimer in 1862. They were expecting their first child when George left their West Providence Township home to volunteer in the 138th PA Infantry in August 1862. Three months later, his brother Josiah was killed near Chaneysville while trying to apprehend a deserter. His other brother William died in late April 1864. Two weeks after William's death, George was captured during the battle of the Wilderness on May 6th, 1864, and taken to the Andersonville POW camp. George died on September 13th, 1864 of Scurvy at age 24. He is buried in the Andersonville National Cemetery.

Name	Muster Age	Rank	State	Regiment	Muster In Date	Casualty	Date of Death	Battle	Res.
Andrews, Hiram			PA	Infantry		Died	1/5/65		B
Arenze, Charles		Pvt.	PA	11th Infantry	11/1/61	Died	11/29/62		B
Armstrong, Joseph	34	1st Lt.	PA	11th Infantry	9/30/61	Died	7/12/65		B
Ayres, John	40	Pvt.	PA	84th Infantry	9/16/62	Died	Unknown		BL
Baker, David N	18	Pvt.	PA	13th Cavalry	2/27/64	Died	8/31/64		B
Baker, Franklin S	18	Pvt.	PA	125th Infantry	8/13/62	KIA	9/18/62	Antietam	B
Barber, William A	18	1st Lt.	PA	11th Cavalry	8/18/61	POW-Died	Unknown	Darbytown Road	B
Barmond, Nathaniel		Pvt.	PA	191st Infantry	5/15/64	POW-Died	3/4/65	Globe Tavern	B
Barndollar, James E	18	Pvt.	PA	133rd Infantry	8/29/62	KIA	12/13/62	Fredericksburg	B
Barnes, David		Pvt.	MD	1st PHB Inf.	4/23/64	Died	8/5/64		B
Barnett, Ephraim	29	Pvt.	PA	149th Infantry	8/14/63	POW-Died	8/23/64	North Anna River	B
Barnett, Samuel	19	Pvt.	PA	184th Infantry	5/12/64	POW-Died	10/15/64	Jerusalem Plank Rd.	B
Barnhart, John H	27	2nd Lt.	PA	55th Infantry	10/12/61	KIA	5/16/64	Drewry's Bluff	B
Bartholow, George F	22	Pvt.	PA	45th Infantry	7/5/64	Died	5/31/65		B
Bartlebaugh, John	27	Corp.	PA	55th Infantry	9/20/61	MIA	9/29/64	Chaffin's Farm	B
Barton, Asa	20	Pvt.	PA	77th Infantry	10/9/61	Died	10/10/62		FC
Baughman, George	22	Sgt.	PA	138th Infantry	8/29/62	POW-Died	9/13/64	Wilderness	B
Baughman, Josiah	29	1st Lt.	PA	138th Infantry	8/30/62	KIA	11/13/62	Chaneysville PA	B
Baughman, William	24	Pvt.	PA	2nd Cavalry	11/4/61	Died	4/26/64		B

William Beaver was born in 1827. He married Hannah Egolf in 1846. They lived near Schellsburg prior to moving to Iowa in 1855. William volunteered in the 5th Iowa Infantry in July 1861 while Hannah remained at home with their 5 children. He suffered a chest wound near New Madrid, MO on March 4th, 1862, and died the next day. William was the 1st known Bedford County battlefield death of the war. Soon after her husband's death, Hannah and the children returned to Bedford County. Their 16-year-old son, Simon, volunteered in the 55th PA Infantry in February 1864. Simon was captured at the battle of Drewry's Bluff in May 1864 and taken to Andersonville. He returned home in April 1865 with ruined health from the terrible living conditions at the POW camp. This likely led to his early grave at age 30.

William W Beltz was born on May 3rd, 1835. He married Christina Beneigh in Schellsburg in 1856. Three months after their 3rd child was born, William volunteered in the 2nd Maryland PHB Infantry in September 1861. He was taken prisoner on January 3rd, 1864 at Ridgeville, VA. William was the 1st known Bedford County soldier to perish at the Andersonville prison camp. He died at age 28 from chronic diarrhea on April 10th, 1864, and was buried in the Andersonville National Cemetery. Christina never remarried and raised 3 young children on her own.

Name	Muster Age	Rank	State	Regiment	Muster In Date	Casualty	Date of Death	Battle	Res.
Beals, George		Pvt.	MD	2nd PHB Inf.	3/25/62	Died	6/12/62		B
Beam, Daniel	59	Pvt.	PA	101st Infantry	11/1/61	POW-Died	8/8/64	Plymouth, NC	B
Beaver, William	34	Pvt.	IA	5th Infantry	7/15/61	POW-Died	3/5/62	New Madrid, MO	IA
Bechtel, Isaac S	20	Pvt.	PA	184th Infantry	5/12/64	KIA	6/22/64	Jerusalem Plank Rd.	B
Beeny, Frederick	22	Pvt.	IL	77th Infantry	9/2/62	Died	6/18/64		IL
Beichtel, Lewis	20	Pvt.	PA	4th Cavalry	8/16/61	MIA	Unknown		BL
Beisel, George M	29	Pvt.	PA	55th Infantry	3/2/64	Wound.-Died	10/27/64		B
Beltz, Abraham	32	Pvt.	PA	101st Infantry	12/2/61	POW-Died	12/3/64	Plymouth, NC	B
Beltz, Adam	24	Pvt.	PA	138th Infantry	8/29/62	KIA	6/5/64	Cold Harbor	B
Beltz, William W	26	Pvt.	MD	2nd PHB Inf.	9/10/61	POW-Died	4/10/64	Ridgeville, VA	B
Bennett, Jacob	23	Pvt.	PA	55th Infantry	10/12/61	POW-Died	9/1/64		B
Berkheimer, John	18	Pvt.	PA	200th Infantry	9/4/64	Died	9/4/65		B
Bessor, John	26	Corp.	PA	101st Infantry	11/1/61	POW-Died	4/20/64	Plymouth, NC	B
Blackburn, Joseph	30	Pvt.	PA	138th Infantry	8/17/62	Wound.-Died	9/1/64	Monocacy, MD	B
Blackburn, Levi	26	Pvt.	PA	138th Infantry	8/29/62	Died	11/19/62		B
Blackburn, Martin	40	Pvt.	PA	184th Infantry	5/12/64	POW-Died	10/11/64	Jerusalem Plank Rd.	B
Blymyer, William		Pvt.	PA	20th Cavalry	1/18/64	POW-Died	7/28/64		B
Bollinger, William H	24	Pvt.	PA	21st Cavalry	7/8/63	Died	11/14/64		B
Boor, John A	21	Pvt.	PA	138th Infantry	9/1/62	Died	8/17/63		B
Bousch, Christian	25	Pvt.	PA	149th Infantry	8/20/63	Wound.-Died	6/23/64	North Anna River	B

John Berkheimer was born on June 3rd, 1846. He was listed on the 1860 Union (Pavia) Township Census with his parents, John Sr. and Sarah, and nine siblings. John mustered in the 200th Pennsylvania Infantry in September 1864. Three months later a sergeant wrote the following in a letter addressed to John's parents (in verbatim) "With trembling hand I wright to lette you know that John layes here very sick". John was fighting a terrible fever when the letter was written, but rejoined his regiment in early 1865. He mustered out of the army in May 1865 and returned home. John died 4 months later at age 19 from the long-term effects of the fever. He is buried in the Old Union Cemetery in Osterburg.

Frederick Burket was born in Pavia on August 2nd, 1831. He married Sarah Claar in 1856. Frederick mustered in the 55th Pennsylvania Infantry in March 1864. Sarah and their 6 children waited for his return to their Union (Pavia) Township home. Two months after joining the 55th, he was captured during the battle of Drewry's Bluff on May 16th and taken to Andersonville. Frederick died of Scurvy on October 16th, 1864, and was buried in the Andersonville National Cemetery. Sarah remarried in 1873 and lived to be 86 years old.

Name	Muster Age	Rank	State	Regiment	Muster In Date	Casualty	Date of Death	Battle	Res.
Bowen, Philip B	33	Pvt.	PA	61st Infantry	2/25/64	KIA	7/12/64	Ft. Stevens, DC	B
Bowers, George	19	Pvt.	PA	184th Infantry	5/12/64	Wound.-Died	9/9/64	Cold Harbor	B
Bowman, Daniel H	21	Pvt.	PA	110th Infantry	9/24/61	Wound.-Died	10/15/64	1st Deep Bottom	B
Bowman, George	24	Pvt.	PA	110th Infantry	10/24/61	Died	4/23/62		B
Bowser, Emanuel	18	Pvt.	PA	8th Reserves	6/19/61	KIA	12/13/62	Fredericksburg	B
Bowser, Job	24	Pvt.	PA	205th Infantry	8/27/64	Died	12/12/64		B
Brallier, Rueben	18	Pvt.	PA	11th Infantry	9/30/61	Died	2/8/62		B
Brallier, William	39	Pvt.	PA	107th Infantry	8/21/64	Died	12/30/64		B
Bridaham, Henry W	29	Pvt.	PA	55th Infantry	2/15/64	POW-Died	7/22/64	Drewry's Bluff	B
Broad, Isaac N	27	Pvt.	PA	55th Infantry	9/23/61	Died	3/8/64		B
Brown, Jeremiah	35	Pvt.	US	5th Light Art.	9/1/61	Died	Unknown		B
Brumbaugh, Francis		Pvt.	PA	110th Infantry	2/24/64	Wound.-POW-Died	Unknown	Wilderness	B
Bryant, John	26	Pvt.	PA	107th Infantry	2/21/62	Wound.-Died	8/2/63	Culpeper C.H.	B
Bryant, William			MD			POW-Died	Unknown		B
Bulger, Andrew	19	Pvt.	PA	110th Infantry	10/24/61	Died	2/15/63		B
Burke, Samuel	36	Pvt.	PA	84th Infantry	10/1/61	Wound.-Died	2/22/63	1st Kernstown	B
Burket, Baltzer	19	Pvt.	PA	55th Infantry	3/2/64	KIA	5/20/64	Bermuda Hundred	BL
Burket, Frederick	33	Pvt.	PA	55th Infantry	3/2/64	POW-Died	10/16/64	Drewry's Bluff	B
Burket, James	20	Pvt.	PA	99th Infantry	2/25/65	Died	Unknown		B
Burket, Philip	17	Pvt.	PA	55th Infantry	2/27/64	Died	4/18/64		B

Joel Clark was born in Bedford on February 4th, 1824. A younger brother was also born in Bedford in 1830. His family moved west prior to Joel marrying Susannah Lesley in Darke County, Ohio in 1843. Joel mustered in the 142nd Indiana Infantry in October 1864, while Susannah and their 8 children tended to their family farm. Joel died at age 41 from chronic diarrhea at the Pest House Hospital in Tennessee on June 24th, 1865. He is buried at the Nashville National Cemetery. Sarah remarried in 1866 and lived to be 86 years old.

Chauncey Corle was born on February 13th, 1837 in Union (Pavia) Township. He grew up in a large family of 11 children. Chauncey and Barbara Roudabush married in November 1859. Ten months later, she died at age 20. The following year Chauncey and his brother Eli mustered in the 55th Pennsylvania Infantry in November 1861. Eli died soon after leaving Bedford County on November 21st, 1861, at Camp Curtin in Harrisburg. Chauncey was wounded during the siege of Petersburg and died on August 23rd, 1864, at McDougal Hospital in Fort Schuyler, NY. He was 27 years old. Chauncey is buried at the Mt. Zion Cemetery in Pavia. Two other brothers, Isaac and William, survived the war and returned home.

Name	Muster Age	Rank	State	Regiment	Muster In Date	Casualty	Date of Death	Battle	Res.
Burket, William			PA	99th Infantry		MIA	Unknown		B
Burns, Joseph H	27	Pvt.	PA	82nd Infantry	9/21/64	KIA	4/6/65	Sailor's Creek	B
Butler, Andrew	43	Pvt.	PA	55th Infantry	11/5/61	Died	10/11/62		B
Carmack, Daniel	25	Pvt.	OH	65th Infantry	10/10/61	KIA	12/62-1/63	Stones River, TN	OH
Carmack, John A	35	Pvt.	PA	55th Infantry	2/10/64	Wound.-Died	10/25/64	Petersburg	B
Carrell, Daniel		Pvt.	PA	138th Infantry	8/29/62	MIA	7/9/64	Monocacy, MD	B
Cashman, Jacob	23	Pvt.	PA	107th Infantry	8/27/64	Wound.-Died	5/11/65	White Oak Road	B
Cessna, Charles	42	Pvt.	IL	75th Infantry	8/25/62	Died	1865		IL
Cessna, Martin	24	Pvt.	PA	190th Infantry	6/1/64	Died	4/11/65		B
Chamberlain, Eli G	19	Pvt.	PA	208th Infantry	9/7/64	KIA	3/25/65	Ft. Stedman	B
Chamberlain, Joseph	29	Pvt.	PA	107th Infantry	4/7/64	MIA	8/19/64	Globe Tavern	B
Clark, Jameson	19	Pvt.	PA	22nd Cavalry	2/26/64	KIA	8/22/64	Charlestown, WV	B
Clark, Joel	38	Pvt.	IN	142nd Infantry	10/13/64	Died	6/24/65		IN
Clark, Samuel M	18	Pvt.	PA	138th Infantry	8/29/62	KIA	9/19/64	Opequon	B
Clark, Zachariah	31	Pvt.	PA	107th Infantry	10/21/64	Died	5/15/65		B
Clingerman, Jeremiah	26	Pvt.	PA	171st Infantry	11/10/62	Died	6/29/63		B
Coffee, John	35	Pvt.	PA	55th Infantry	11/5/61	Died	11/10/62		B
College, James	28	Pvt.	PA	110th Infantry	10/24/61	Wound.-Died	5/11/62	1st Kernstown	B
College, John W	16	Pvt.	PA	110th Infantry	10/24/61	Wound.-Died	3/27/62	1st Kernstown	B
Collins, John T			PA	99th Infantry		Died	12/15/65		B

Those who made the Ultimate Sacrifice 131

Francis "Frank" Corle was born in Pavia in 1827. Frank married Elizabeth Whysong in 1847. Tragically, 4 of their children died of Scarlet Fever within a month in 1859. Francis and Elizabeth were raising their five surviving children on a family farm when Francis was drafted. He enlisted in the 91st PA Infantry on September 21st, 1864 with his cousin Jonathan Corle. Thirty-six days later, Jonathan was killed and Frank was mortally wounded at the battle of Boydton Plank Road. Frank died two days later at a Regimental Hospital on October 29th. Both are buried at the Mt. Zion Cemetery in Pavia. Elizabeth never remarried and passed at age 64 in 1892.

Abraham Darr was born in 1842. Abraham and his brother Henry left their home near Schellsburg to muster in the 55th Pennsylvania Infantry in October 1861. Both were promoted to Sergeant during the war. Abraham was killed during a near suicidal charge on entrenched Confederate lines at the battle of Cold Harbor on June 3rd, 1864. Two weeks later Henry suffered a wound during the Initial Assault of Petersburg on June 18th and received a disability discharge a year later. Henry moved his family to Ohio after the war and passed when he was 44 in 1879. He is buried in the Wooster Cemetery in Wayne County, Ohio. Abraham's burial site remains unknown.

Name	Muster Age	Rank	State	Regiment	Muster In Date	Casualty	Date of Death	Battle	Res.
Conley, Martin L	27	Corp.	PA	138th Infantry	8/29/62	MIA	6/1/64	Cold Harbor	B
Conner, David	23	Pvt.	PA	133rd Infantry	8/15/62	Died	1/7/63	Fredericksburg*	B
Cook, Dennis		Pvt.	IL	51st Infantry	10/12/64	MIA	11/30/64	Franklin, TN	IL
Cook, Edward H	27	Pvt.	PA	5th Heavy Art.	9/1/64	Died	10/5/64		B
Cooper, Jesse V	18	Pvt.	PA	101st Infantry	11/1/61	Died	7/30/62		B
Cooper, Joshua H	20	Pvt.	PA	133rd Infantry	8/29/62	KIA	12/13/62	Fredericksburg	B
Corle, Abraham	22	Pvt.	PA	138th Infantry	8/29/62	Wound.-Died	5/6/64	Wilderness	B
Corle, Anthony	19	Pvt.	PA	84th Infantry	10/24/61	MIA	6/9/62	Port Republic	B
Corle, Chauncey	24	Corp.	PA	55th Infantry	11/5/61	Wound.-Died	8/23/64	Petersburg	B
Corle, Eli	22	Pvt.	PA	55th Infantry	11/5/61	Died	11/21/61		B
Corle, Francis B	37	Pvt.	PA	91st Infantry	9/21/64	Wound.-Died	10/29/64	Boydton Plank Rd.	B
Corle, Henry W	22	Pvt.	PA	46th Infantry	7/14/63	Wound.-Died	5/18/64	Resaca, GA	B
Corle, James L	20	Pvt.	PA	55th Infantry	8/28/61	KIA	3/30/65	White Oak Road	B
Corle, Jonathan	37	Pvt.	PA	91st Infantry	9/21/64	KIA	10/27/64	Boydton Plank Rd.	B
Coughenour, Samuel	18	Pvt.	IA	22nd Infantry	8/27/62	Wound.-Died	10/15/63	Vicksburg, MS	IA
Craine, David D	18	Pvt.	PA	138th Infantry	8/29/62	KIA	10/19/64	Cedar Creek	B
Cramer, Jacob	20	Pvt.	PA	110th Infantry	10/24/61	Died	10/25/62		B
Crist, Solomon	49	Pvt.	PA	55th Infantry	3/2/64	MIA	8/26/64	Bermuda Hundred	B
Croft, Alexander	36	Sgt.	PA	110th Infantry	10/24/61	Died	2/5/62		B
Croft, Philip P	23	Pvt.	PA	110th Infantry	10/24/61	KIA	3/23/62	1st Kernstown	B

John Defibaugh was born in 1820. He was listed on the Bedford Census in 1840 with his parents, John Sr. and Isabella. John moved to Marion, Ohio, and married Sarah Hallowell in 1847. Sarah passed in 1859, leaving behind 5 children at home. John enlisted in the 101st Pennsylvania Infantry in 1861 while the children remained in Ohio with his late wife's parents. He was wounded and taken prisoner at the battle of Plymouth, NC, on April 20th, 1864. John perished at the Andersonville POW camp from Anasarca on August 5th, 1864. He is buried at the Andersonville National Cemetery. The 5 orphaned children were raised by his wife's family in Ohio.

Jacob J Dibert was born near Imlertown on September 29th, 1822. He married Sarah Whysong around 1850. Jacob is forever linked to one of the most famous stories in Bedford County history. Jacob is the man who located "The Lost Children of the Alleghenies." The bodies of the young Cox brothers were found at the foot of a birch tree with a shattered top, just as Jacob had dreamed. Jacob mustered in the 55th PA Infantry in November 1861 while Sarah and their 5 children waited for his return on their family farm. He died of Chronic Dysentery at age 29 on October 26th, 1864 near the Point of Rocks battlefield in Virginia, and is buried at the City Point National Cemetery. Sarah later remarried and passed at age 64 in 1895.

Name	Muster Age	Rank	State	Regiment	Muster In Date	Casualty	Date of Death	Battle	Res.
Crouse, Harry C	21	2nd Lt.	PA	55th Infantry	9/20/61	KIA	8/20/64	Petersburg	B
Croyle, William H	18	Pvt.	PA	55th Infantry	10/11/61	Wound.-Died	9/5/64	Petersburg - Initial	B
Darr, Abraham	20	Sgt.	PA	55th Infantry	10/11/61	KIA	6/3/64	Cold Harbor	B
Darr, Charles S	21	Pvt.	PA	11th Cavalry	3/26/63	Died	8/18/63		B
Darr, Christian J		Pvt.	PA	171st Infantry	11/12/62	Died	7/12/63		B
Daugherty, David L	26	Pvt.	PA	55th Infantry	8/28/61	Wound.-Died	6/16/64	Petersburg - Initial	B
Daugherty, James	16	Pvt.	PA	110th Infantry	10/24/61	KIA	12/13/62	Fredericksburg	B
Davidson, Samuel	20	Pvt.	PA	184th Infantry	5/12/64	POW-Died	10/28/64	Jerusalem Plank Rd.	B
Davis, Charles M	24	Pvt.	PA	55th Infantry	8/28/61	KIA	6/5/64	Cold Harbor	B
Davis, Isaiah M	24	Pvt.	PA	8th Reserves	6/11/61	Died	11/28/61		B
Decker, James M	18	Pvt.	PA	53rd Infantry	1/4/64	Died	7/22/64		B
Defibaugh, Jacob F	20	Pvt.	PA	101st Infantry	12/6/61	Died	1/3/62		B
Defibaugh, John	34	Pvt.	PA	101st Infantry	1861	POW-Died	8/5/64	Plymonth, NC	B
Defibaugh, William H	18	Pvt.	PA	138th Infantry	8/29/62	MIA	5/6/64	Wilderness	B
Devore, Lewis	18	Pvt.	PA	184th Infantry	5/12/64	KIA	6/22/64	Jerusalem Plank Rd.	B
Dibert, Jacob J	39	Corp.	PA	55th Infantry	11/5/61	Died	10/26/64		B
Dibert, John J	26	3rd Sgt.	PA	55th Infantry	8/11/61	KIA	6/16/64	Petersburg - Initial	B
Dibert, John Jackson	29	Pvt.	PA	99th Infantry	2/25/65	Died	6/24/65		B
Dicken, James H	16		PA	2nd Cavalry	9/1/61	POW-Died	Unknown		B
Dicken, John	21	Pvt.	PA	138th Infantry	8/29/62	POW-Died	12/2/63	Mine Run	B
Diehl, Adam	20	Pvt.	PA	91st Infantry	9/21/64	Died	8/2/65		B

Benjamin Eshelman was born on a farm in East Providence Township in 1838 and was listed as a carpenter on the 1860 East Providence Township Census. Benjamin mustered in the 92nd Illinois Infantry in September 1862. The enlistment rolls described Benjamin as being single, 5'7 ½" tall with brown hair, grey eyes, and a fair complexion. Benjamin died from Typhoid Fever on January 17th, 1863 in Danville, Kentucky and is buried in the Danville National Cemetery. His brothers George and John joined PA regiments during the war. John enlisted in the 88th PA Infantry and died at age 35 on June 5th, 1865, after returning home from the war. He is buried at the Old Rays Hill Cemetery in Breezewood. John left behind a wife and five children.

James M Fickes was born in 1846. He left the family farm in St. Clair Township when he was 15 years old to volunteer in the 101st Pennsylvania Infantry in February 1862. James was listed as 5'5 ¾" tall with dark hair and dark eyes. He received a disability discharge at a hospital at Fort Monroe, Virginia, on November 8th, 1862. He died a month later at home and was buried at the Fishertown Cemetery. James was among the youngest soldiers from Bedford County to die during the war. His brother, Cyrus, suffered a wound while enlisted in the 200th PA Infantry at the battle of Fort Stedman during the siege of Petersburg on March 25th, 1865. Cyrus survived the war but died young at age 33.

Name	Muster Age	Rank	State	Regiment	Muster In Date	Casualty	Date of Death	Battle	Res.
Diehl, Daniel	18	Pvt.	PA	55th Infantry	2/27/64	Died	7/30/64		B
Diehl, Espy	27	Pvt.	PA	55th Infantry	2/27/64	POW-Died	10/23/64		B
Diehl, Joshua	20	Saddler	OH	4th Cavalry	11/11/61	Died	4/15/62		OH
Dively, John	17	Pvt.	PA	110th Infantry	2/23/64	POW-Died	8/31/64		BL
Dodson, Albert	22	Pvt.	PA	19th Cavalry	9/17/63	Died	4/27/64		BL
Dodson, Joseph			IN			Died	1865		B
Drenning, Henry G	38	Corp.	PA	55th Infantry	11/5/61	KIA	6/3/64	Cold Harbor	B
Dull, John	19	Pvt.	PA	184th Infantry	5/12/64	POW-Died	10/28/64	Boydton Plank Rd.	B
Dull, Samuel A	19	Pvt.	MD	2nd PHB Inf.	4/9/62	POW-Died	2/22/64	Moorefield Junction	B
Duvall, Thomas	32	Pvt.	PA	77th Infantry	2/29/64	Died	8/14/64		FC
Eamick, Josiah	18	Pvt.	PA	101st Infantry	11/1/61	Died	11/15/62		B
Eastright, Christian	24	Pvt.	PA	8th Reserves	7/11/61	Wound.-Died	5/8/64	Spotsylvania	B
Eckels, Francis S	17	Pvt.	PA	76th Infantry	2/14/62	POW-Died	12/22/63		B
Eckels, John F	20	Pvt.	PA	76th Infantry	10/9/61	POW-Died	1/21/64		B
Edwards, Hiram	21	Pvt.	PA	8th Reserves	6/11/61	Died	8/12/61		B
Elliott, John	22	Pvt.	PA	2nd Cavalry	11/4/61	Wound.-Died	5/5/63		B
Engle, Charles	23	Pvt.	PA	55th Infantry	10/12/61	Died	11/7/62		B
Engle, Henry	40	Pvt.	PA	121st Infantry	8/4/62	POW-Died	1/16/65	Peebles Farm	B
Ensley, Christopher	34	Corp.	PA	184th Infantry	5/12/64	POW-Died	8/26/64	Jerusalem Plank Rd.	B
Ernst, John	41	Pvt.	OH	42nd Infantry	11/27/61	Died	3/22/63		OH
Eshelman, Benjamin	28	Pvt.	IL	92nd Infantry	9/4/62	Died	1/17/63		IL

Francis L Foor was born in East Providence Township in 1842. His parents, David and Henrietta Foor, had four sons who were casualties in the Civil War. Jonathan Foor was wounded at South Mountain, taken prisoner at Gettysburg, and spent 17 months in POW camps, including Andersonville. Samuel was wounded at the battle of Mechanicsville and captured at Fredericksburg. William H H suffered a wound, but the battle is not listed on available records. All survived the war except for Francis. Francis mustered in the 101st PA Infantry in November 1861. He was captured at Plymouth, NC in April 1864 and taken to Andersonville. Francis was later transferred to a Florence, SC. This is where he likely died on November 15th, 1864. His burial site remains unknown.

Benjamin Garretson was born on a family farm on September 17th, 1841. His parents, Nathan and Mary Garretson, are listed on the 1850 Napier Township Census with their children - Ann, Margaret, Josiah, Benjamin, Sophia, and Sarah. The tragic story of Benjamin and Josiah during the Petersburg Final Assault on April 2nd, 1865 is detailed on page 107. 1865 was a tragic year for the Garretson family. Father Nathan died on January 1st. Five months later, the family buried Benjamin alongside his father at the Friends Cemetery in Spring Meadow. He died at age 24. His brother Josiah survived heavy combat while a member of the 55th PA Infantry and mustered out on May 11th, 1865 after the war ended.

Name	Muster Age	Rank	State	Regiment	Muster In Date	Casualty	Date of Death	Battle	Res.
Eshelman, John W		Pvt.	PA	88th Infantry		Died	6/5/65		B
Evans, Isaiah	24	Pvt.	PA	149th Infantry	9/29/63	KIA	5/8/64	Spotsylvania	B
Evans, William H	18	Pvt.	PA	101st Infantry	12/28/61	Died	6/27/62		B
Everhart, David	21	Pvt.	PA	110th Infantry	10/24/61	KIA	12/13/62	Fredericksburg	B
Exline, Jacob	22	Pvt.	PA	55th Infantry	10/11/61	POW-Died	9/7/64	Drewry's Bluff	B
Feather, George W	18	Pvt.	PA	138th Infantry	8/29/62	Died	10/25/62		B
Feight, Abraham	18	Pvt.	PA	138th Infantry	8/29/62	Died	11/13/62		B
Feight, Levi		Pvt.	PA	57th Infantry	9/21/64	Died	2/21/65		B
Felton, Peter S	18	Pvt.	PA	208th Infantry	9/7/64	KIA	4/2/65	Petersburg - Final	B
Ferguson, John	19	Pvt.	PA	110th Infantry	10/24/61	KIA	3/23/62	1st Kernstown	B
Fessler, Samuel		Pvt.	PA	107th Infantry	1/9/62	KIA	9/17/62	Antietam	B
Fetter, Joseph	18	Pvt.	PA	76th Infantry	10/9/61	KIA	7/11/63	Ft. Wagner, SC	B
Fetter, Joseph J	20	Pvt.	PA	76th Infantry	10/9/61	Died	6/9/62		B
Fetters, Job	27	Pvt.	PA	171st Infantry	11/2/62	Died	11/27/62		B
Fetters, John		Pvt.	PA	99th Infantry		KIA	4/6/65	Sailor's Creek	B
Fickes, James M	16	Pvt.	PA	101st Infantry	2/18/62	Died	12/8/62		B
Figart, Henry J	21	Pvt.	PA	8th Reserves	6/11/61	KIA	9/17/62	2nd Bull Run	B
Finnegan, Daniel	43	Pvt.	PA	55th Infantry	9/4/61	Died	10/6/64		B
Fleegle, Daniel	18	Pvt.	PA	93rd Infantry	3/31/64	KIA	5/12/64	Spotsylvania	B
Fleegle, Jacob G	41	Pvt.	PA	76th Infantry	10/9/61	Died	7/9/62		B
Fletcher, Henry C	22	Pvt.	PA	22nd Cavalry	2/27/64	Died	4/20/64		B
Fluck, Porter	20	Pvt.	PA	87th Infantry	6/4/64	KIA	4/2/65	Petersburg - Final	B
Fluke, John W	16	Pvt.	PA	22nd Cavalry	2/27/64	Died	7/3/64		B

Daniel G Heltzel was born in Woodbury on August 6th, 1841. He married Julia Ann Exline in 1860. Daniel volunteered in the 138th Pennsylvania Infantry in August 1862, leaving behind Julia Ann and a young son at their St. Clairsville Twp home. Daniel died at the Relay House Hospital of Typhoid Fever in November 1862. He was 21. Daniel is believed to have been buried close to the hospital in an unmarked grave. Five months after his death, a daughter was born. Julia Ann remarried in 1865 to Simon Feather. Simon and her first husband, Daniel, mustered in the 138th PA Infantry on the same day in 1862. Simon suffered a wound during the war at the battle of Spotsylvania in 1864. Julia Ann and Simon were parents to six children.

Andrew J Himes was born on May 7th, 1835. Andrew's photograph is on the cover of this book. Information on Andrew and his family are inside the front cover. His father, John Himes Sr was a veteran of the War of 1812. Two of Andrew's brothers were also in the Civil War. John Jr volunteered in the 133rd Pennsylvania Infantry and later reenlisted in the 208th Pennsylvania Infantry. Samuel Himes enlisted in the 186th Pennsylvania. Both brothers survived the war. Andrew died when he was 29 years old of Dysentery at Fort Monroe, Virginia in October 1864 while enlisted in the 3rd PA Heavy Artillery. He is buried at the Blackheart Cemetery outside of Everett.

Name	Muster Age	Rank	State	Regiment	Muster In Date	Casualty	Date of Death	Battle	Res.
Foor, Daniel V	19	Pvt.	PA	76th Infantry	11/21/61	KIA	7/11/63	Ft. Wagner, SC	B
Foor, Francis L	19	Pvt.	PA	101st Infantry	11/1/61	POW-Died	11/15/64	Plymonth, NC	B
Foor, George W	43	Pvt.	PA	107th Infantry	2/10/62	KIA	9/17/62	Antietam	B
Foor, Mark W	19	Pvt.	PA	8th Reserves	6/11/61	Died	12/4/61		B
Foor, Martin T	37	Corp.	PA	138th Infantry	8/29/62	POW-Died	12/6/64	Monocacy, MD	B
Foor, Simon P	32	Pvt.	PA	208th Infantry	9/7/64	KIA	4/2/65	Petersburg - Final	B
Foor, William H	24	Pvt.	PA	8th Reserves	6/11/61	KIA	6/30/62	Seven Days Battle	B
Foreman, William	28	Pvt.	PA	2nd Heavy Art.	2/29/64	POW-Died	4/5/65	Cold Harbor	B
Foster, William A	19	Pvt.	PA	55th Infantry	10/11/61	Died	8/4/64		B
Frazier, William	21	Pvt.	PA	55th Infantry	11/5/61	Wound.-Died	6/9/64		B
Fritz, Daniel E	25	Corp.	PA	18th Cavalry	10/27/62	POW-Died	9/4/64	Wilderness	B
Fry, John		Pvt.	USCT	43rd Infantry	3/26/64	KIA	7/3/64	Petersburg	B
Fry, Michael	38	Pvt.	PA	84th Infantry	10/24/61	Died	3/16/66		B
Funk, Enoch	21	Pvt.	PA	11th Infantry	9/30/61	KIA	5/23/64	North Anna River	B
Furney, Nelson	31	Pvt.	PA	148th Infantry	8/26/63	Died	11/24/64		B
Gallagher, Edward	32	Pvt.	PA	133rd Infantry	8/29/62	KIA	12/13/62	Fredericksburg	B
Gamble, Robert	35	Pvt.	PA	8th Reserves	6/11/61	Died	9/2/63		B
Garretson, Benjamin H	21	Pvt.	PA	205th Infantry	8/27/64	Wound.-Died	5/27/65	Petersburg - Final	B
Garretson, Moses R	19	Pvt.	PA	55th Infantry	10/11/61	Died	10/15/64		B
Garrett, Alexander A	21	Pvt.	PA	8th Reserves	6/11/61	KIA	6/30/62	Seven Days Battle	B
Gaster, James H	18	Sgt.	PA	107th Infantry	1/9/62	Wound.-Died	2/6/63	Antietam	B
Gates, James	23	Pvt.	PA	8th Reserves	6/11/61	Wound.-Died	10/16/62	Antietam	B

Chapter 7

Name	Muster Age	Rank	State	Regiment	Muster In Date	Casualty	Date of Death	Battle	Res.
Gates, Samuel	18	Pvt.	PA	110th Infantry	10/24/61	Died	3/12/62		B
Geller, Solomon	35	Pvt.	PA	101st Infantry	12/28/61	POW-Died	6/16/62	Fair Oaks	B
Gephart, William			PA	76th Infantry		KIA	1864	Mill Springs Gap*	B
Gilliam, Wilson E	19	Pvt.	PA	101st Infantry	11/1/61	Died	5/15/62		B
Gladwell, George W	28	Pvt.	PA	55th Infantry	2/27/64	Died	6/20/65		B
Gogley, James H	21	Pvt.	PA	186th Infantry	2/24/64	Died	1/6/65		B
Gollipher, Justice	22	Pvt.	PA	101st Infantry	12/28/61	POW-Died	10/15/64	Plymonth, NC	B
Gordon, William	22	Pvt.	PA	55th Infantry	11/5/61	POW-Died	5/22/64	Drewry's Bluff*	B
Grim, William	30	Pvt.	PA	149th Infantry	8/26/62	POW-Died	9/20/64	Wilderness	B
Grimes, Henry W	20	Pvt.	PA	84th Infantry	10/21/61	Wound.-Died	5/16/64	Wilderness	B
Grubb, George	23	Pvt.	PA	149th Infantry	8/26/63	POW-Died	10/16/64	Globe Tavern	B
Gump, John A	19	1st Lt.	PA	138th Infantry	9/12/62	Wound.-Died	10/20/64	Cedar Creek	B
Hainsey, Valentine	14	Pvt.	PA	55th Infantry	8/28/61	Wound.-Died	6/17/64	Cold Harbor	B
Hamilton, John						Died	10/8/65		B
Hammel, Peter S	20	Pvt.	PA	54th Infantry	1861	Died	10/9/61		BL
Hammer, Joseph D		Pvt.	PA	142nd Infantry	8/22/62	Wound.-Died	9/9/63	Gettysburg	B
Hand, Henry		Pvt.	PA	55th Infantry	10/11/61	POW-Died	9/29/64	Drewry's Bluff	B
Hanks, Benjamin A	28	Sgt.	PA	101st Infantry	11/1/61	POW-Died	9/27/64	Plymonth, NC	B
Hanks, Caleb	19	Pvt.	PA	101st Infantry	11/1/61	POW-Died	10/26/64	Plymonth, NC	B
Hanks, David F	18	Pvt.	PA	101st Infantry	11/1/61	POW-Died	11/1/64	Plymonth, NC	B
Hanks, Nelson	28	Pvt.	PA	101st Infantry	11/1/61	POW-Died	9/15/64	Plymonth, NC	FC
Harbaugh, Eli	16	Pvt.	PA	55th Infantry	11/5/61	Died	1/27/62		B
Harbaugh, George W	20	Pvt.	PA	55th Infantry	10/11/61	Wound.-Died	7/11/64	Petersburg - Initial	B
Harbaugh, John	32	Pvt.	PA	55th Infantry	10/12/61	Died	9/27/63		B
Harbaugh, Robert	23	Pvt.	PA	55th Infantry	2/19/64	KIA	6/18/64	Petersburg - Initial	B
Harbaugh, Wilson	18	Pvt.	PA	55th Infantry	3/2/64	Died	3/28/64		B
Hardinger, George W	28	Pvt.	PA	149th Infantry	8/26/63	POW-Died	7/31/64	Wilderness	B
Harris, Joshua	23	Pvt.	MA	55th Infantry	6/15/63	Died	9/8/63		B
Hartle, John B	21	Pvt.	PA	99th Infantry	2/25/65	Died	5/21/65		B
Hartman, George L	21	Corp.	PA	110th Infantry	9/24/61	MIA	5/3/63	Chancellorsville	B
Heavener, George W	24	Corp.	PA	208th Infantry	9/7/64	Died	5/9/65	Petersburg	B
Hebner, Frederick	23	Pvt.	MD	3rd PHB Inf.	11/4/61	Died	8/6/65		FC
Heffner, George	20	Corp.	PA	8th Reserves	6/11/61	KIA	8/29/62	2nd Bull Run	B
Heffner, John F	38	Pvt.	PA	125th Infantry	8/8/62	Died	3/5/63		B
Hellman, Daniel	18	Pvt.	PA	138th Infantry	8/29/62	KIA	11/27/63	Mine Run	B
Heltzel, Daniel G	21	Pvt.	PA	138th Infantry	8/29/62	Died	11/1/62		B
Heltzel, Jonathan D	19	Pvt.	PA	110th Infantry	10/24/61	KIA	5/6/64	Wilderness	B
Heltzel, Simon		Pvt.	PA	138th Infantry	9/29/62	Died	11/9/62		B
Hemming, Augustus	14	Pvt.	PA	2nd Cavalry	12/20/61	Died	11/20/63		B
Herring, George W	44	Corp.	PA	55th Infantry	11/5/61	Died	8/25/63		B
Hess, Daniel A	24	2nd Lt.	PA	55th Infantry	10/11/61	Wound.-Died	4/20/65	Sailor's Creek	B
Hileman, John	27	Pvt.	PA	55th Infantry	11/5/61	Died	12/1/61		B
Himes, Andrew J	29	Pvt.	PA	3rd Heavy Art.	3/8/64	Died	10/7/64		B
Hiner, John E		Pvt.	PA	67th Infantry	1864	Wound.-Died	4/23/65	Cedar Creek	B
Hite, John	34	Pvt.	PA	205th Infantry	8/17/64	KIA	4/2/65	Petersburg - Final	BL

Isaac M Imler was born on February 26th, 1825. He married Catherine Whysong around 1845. Isaac mustered in the 55th Pennsylvania Infantry in November 1861 while Catherine and their 5 children tended to their family farm in Union (Pavia) Township. He received a promotion to Sergeant in February 1864. Isaac was killed during the Initial Assault on Petersburg on June 18th, 1864. He was 39 years old. Catherine remarried in 1882 after raising her children. Her new husband, Ferdinand Ritchey, had also lost a spouse. His 1st wife, Martha, passed in 1865. Ferdinand served alongside Isaac Imler as a corporal in the same regimental company during the Civil War. Isaac, Catherine, Ferdinand, and Martha are buried at the Mt. Zion Cemetery in Pavia.

George V A Juda was born around 1841. He mustered in the Hopewell Rifles company of the 8th Pennsylvania Reserves in June 1861. George was taken prisoner during the withdrawal from the Mechanicsville battlefield on June 27th, 1862. There is no record of the date of his release from a prisoner of war camp. He suffered a severe wound on May 8th, 1864, during the battle of Spotsylvania just days before the end of his enlistment in the 8th Reserves. George died at a hospital in Washington, DC, on June 25th, 1864. Little else is known about George other than being listed as a Bedford County resident in the Pennsylvania Civil War Archives and in other records. He is buried at the Arlington National Cemetery.

Name	Muster Age	Rank	State	Regiment	Muster In Date	Casualty	Date of Death	Battle	Res.
Hite, Joseph	19	Pvt.	MD	2nd PHB Inf.	10/31/61	Died	7/7/64		B
Hite, Perry	17	Pvt.	PA	2nd Cavalry	6/5/62	Died	11/2/62		B
Hite, Samuel C	29	Pvt.	PA	55th Infantry	2/23/64	POW-Died	7/16/64		BL
Hixon, Akers J	23	Sgt.	PA	101st Infantry	11/1/61	Died	7/21/62		FC
Hixon, Erastus J	21	Corp.	PA	138th Infantry	8/29/62	KIA	5/6/64	Wilderness	B
Hixon, Joel B	18	Pvt.	PA	101st Infantry	11/1/61	Died	10/15/62		FC
Hockenberry, Jonathan	18	Pvt.	PA	107th Infantry	10/6/64	Died	12/31/64		B
Hoenstine, David	18	Pvt.	PA	138th Infantry	8/29/62	Died	11/4/62		B
Hoffman, Jacob	18	Pvt.	PA	76th Infantry	11/21/61	Died	6/28/62		B
Hoffman, John	23	Pvt.	PA	101st Infantry	12/28/61	Died	5/19/62		B
Holler, James M	18	Pvt.	PA	55th Infantry	11/11/61	Died	9/2/62		B
Hollinger, Stephen	17	Pvt.	USCT	43rd Infantry	2/27/64	Wound.-Died	11/10/66	Battle of the Crater	B
Hook, William	23	Pvt.	PA	171st Infantry	11/2/62	Died	12/16/62		B
Hoover, Martin	24	Pvt.	PA	171st Infantry	11/2/62	Died	5/24/63		B
Hoover, Nathaniel	19	Pvt.	PA	55th Infantry	2/6/64	Died	3/30/64		B
Horn, Levi		Pvt.	PA	22nd Cavalry	9/6/62	MIA	Unknown		B
Horton, Alexander	36	Pvt.	PA	77th Infantry	2/29/64	Died	11/9/64		HC
Horton, George	21	Corp.	PA	8th Reserves	9/4/61	Wound.-Died	9/17/62	Antietam	B
Horton, Jonathan	22	Corp.	PA	77th Infantry	10/9/61	Died	12/12/64		B

Name	Muster Age	Rank	State	Regiment	Muster In Date	Casualty	Date of Death	Battle	Res
Horton, Reuben H	20	Pvt.	PA	77th Infantry	2/29/64	Wound.-Died	8/13/64	Chattahoochee, GA	B
Houck, Ezekiel J	28	Pvt.	PA	84th Infantry	4/8/64	KIA	6/17/64	Petersburg - Initial	HC
Hughes, Bailey	40	Pvt.	PA	149th Infantry	9/22/63	Died	12/18/63		B
Hunt, John T	25	Corp.	PA	55th Infantry	11/4/62	POW-Died	5/16/64	Drewry's Bluff	B
Hurley, Daniel W	18	Sgt.	PA	200th Infantry	8/31/64	Died	1/30/65		B
Hurley, John	18	Pvt.	PA	76th Infantry	10/28/61	Died	11/30/61		B
Hyde, Abraham	21	Pvt.	PA	55th Infantry	11/5/61	Died	11/20/61		B
Ickes, George	22	Pvt.	PA	138th Infantry	9/2/62	Died	11/14/62		B
Imes, John	22	Pvt.	PA	13th Reserves	7/4/61	Wound.-Died	Unknown	Fredericksburg	B
Imler, Daniel	22	Pvt.	PA	138th Infantry	9/19/64	Died	1/4/65		BL
Imler, Isaac M	36	Sgt.	PA	55th Infantry	11/5/61	KIA	6/18/64	Petersburg - Initial	B
Irvine, Wilson	19	Pvt.	PA	184th Infantry	5/12/64	POW-Died	10/27/64	Jerusalem Plank Rd.	B
James, Edward V	19	Pvt.	PA	55th Infantry	3/2/64	KIA	6/18/64	Petersburg - Initial	B
James, Jesse T	19	Pvt.	PA	84th Infantry	10/24/61	Died	9/28/63		B
James, John A	14	Pvt.	PA	55th Infantry	10/11/61	Wound.-Died	6/20/64		B
James, Nathaniel	20	Pvt.	PA	138th Infantry	8/29/62	MIA	9/19/64	Opequon	B
Johnson, Cromwell O	26	Surg	PA	5th Reserves	3/9/63	Died	12/21/64		B
Juda, George V A	20	Corp.	PA	8th Reserves	6/11/61	Wound.-Died	6/25/64	Spotsylvania	B
Justice, Edward	24	Pvt.	PA	110th Infantry	2/1/64	Wound.-Died	6/21/64	Petersburg - Initial	B
Kagarice, Ebenezer	20	Pvt.	PA	76th Infantry	7/11/61	KIA	7/11/63	Ft. Wagner, SC	B
Kagarice, John	29	Pvt.	PA	198th Infantry	8/31/64	Wound.-Died	5/17/65	White Oak Road	B
Karns, Jacob	30	Pvt.	PA	22nd Cavalry	2/26/64	Died	9/17/64		B
Kay, William H	20	Pvt.	PA	8th Reserves	6/11/61	Wound.-Died	9/18/62	South Mountain	B
Kegg, Emanuel	37	Pvt.	PA	18th Cavalry	11/14/62	Died	6/26/63		B
Kelley, John A	23	Corp.	PA	125th Infantry	8/13/62	KIA	9/17/62	Antietam	B
Kelly, Henry	20	Pvt.	PA	138th Infantry	8/29/62	MIA	6/1/64	Cold Harbor	B
Kennedy, Samuel	20	Corp.	PA	55th Infantry	10/12/61	KIA	5/16/64	Drewry's Bluff	B
Kerns, Mark C		Capt.	PA	1st Light Art.	7/26/61	KIA	8/30/62	2nd Bull Run	B
King, Watson	20	Pvt.	PA	76th Infantry	10/9/61	Died	6/18/62		B
Kinley, Jacob	43	Pvt.	PA	55th Infantry	11/5/61	Died	10/13/62		B
Kinton, David		Pvt.	PA	55th Infantry	10/11/61	KIA	6/3/64	Cold Harbor	B
Klahre, Herman T	20	Corp.	PA	184th Infantry	5/12/64	Wound.-Died	7/20/64	Jerusalem Plank Rd.	B
Kline, James S	19	Pvt.	PA	55th Infantry	10/11/61	Wound.-Died	5/20/64	Bermuda Hundred	B
Knee, Augustus D	21	Pvt.	IL	53rd Infantry	2/18/62	Died	6/21/62		IL
Knox, James H	24	1st Sgt.	PA	184th Infantry	5/12/64	POW-Died	2/23/65	Jerusalem Plank Rd.	B
Koontz, James	23	Pvt.	IA	5th Cavalry	11/1/62	KIA	7/18/64	Chehaw, AL	IA
Lamison, George W	23	Pvt.	PA	110th Infantry	10/24/61	Wound.-Died	8/3/63	Gettysburg	B
Lamison, Thomas	22	Pvt.	PA	110th Infantry	10/24/61	Died	6/23/63		B
Langdon, Samuel	36	Sgt.	PA	208th Infantry	9/7/64	Died	6/14/65		B
Lashley, Isaac W	21	Pvt.	MD	8th Infantry	8/8/62	Wound.-Died	6/1/64	Spotsylvania	B
Lay, Joseph		Pvt.	PA	138th Infantry	8/29/62	MIA	5/6/64	Wilderness	B
Layton, David	25	Pvt.	PA	101st Infantry	11/1/61	Died	4/1/63		B
Layton, Henry	20	Pvt.	PA	78th Infantry	2/28/65	Died	6/21/65		B
Layton, John (1)	20	Pvt.	PA	101st Infantry	2/3/62	KIA	6/29/62	Seven Days Battle	B
Layton, John (2)	37	Pvt.	PA	138th Infantry	8/29/62	Wound.-Died	10/6/64	Opequon*	B
Layton, Samuel	32	Pvt.	PA	184th Infantry	5/12/64	POW-Died	8/18/64	Jerusalem Plank Rd.	B
Leach, George E	31	1st Sgt.	PA	55th Infantry	11/5/61	Died	2/26/64		B

Those who made the Ultimate Sacrifice 139

Henry Layton was born on July 13th, 1844. He was raised on a family farm near Mattie with four sisters. Henry mustered in the 78th Pennsylvania Infantry in February 1865 with Cornelius Rice, his sister Jane's husband. They were among the few Bedford County soldiers sent to the Western Theater during the war. Henry died of Typhoid Fever on June 21st, 1865 at a Nashville, TN hospital. He was 20 years old. Cornelius was also deathly sick from Typhoid Fever and was sent back to a hospital in Bloody Run (Everett) to recover. The following year, Cornelius and Jane began farming on land that was originally intended for Henry. Henry is buried at the Stevens Chapel Cemetery near Mattie.

Malachi B Mock was born on a farm near Pavia on April 21st, 1838. Malachi and his two brothers, Josiah and Tobias, volunteered together in the 55th Pennsylvania Infantry in 1861. Malachi died of Typhoid Fever on November 7th, 1862, in Beaufort, South Carolina. Tobias was wounded during the Initial Assault of Petersburg on June 18th, 1864. He was taken to a hospital at Fort Schuyler, NY, where he died on August 7th, 1864. Josiah was captured in 1864 during the siege of Petersburg and held in the Salisbury and Libby POW camps before being paroled in Annapolis. Josiah died on March 22nd, 1865, from ruined health caused by the inhumane living conditions in Confederate prison camps. All three brothers are buried at the Mt. Zion Cemetery in Pavia.

Name	Muster Age	Rank	State	Regiment	Muster In Date	Casualty	Date of Death	Battle	Res.
Leader, William W	31	Sgt.	MI	12th Infantry	12/25/63	KIA	9/4/64	Wihite River, AK	MI
Lear, John	26	Pvt.	PA	125th Infantry	8/13/62	KIA	9/17/62	Antietam	B
Leasure, Amos	17	Pvt.	MD	1st PHB Cav.	4/23/64	KIA	8/30/64	Summit Point	B
Leasure, Solomon	32	Pvt.	PA	149th Infantry	9/29/63	Died	1/29/64		B
Leech, Wlliam	18	Pvt.	PA	55th Infantry	11/5/61	KIA	10/22/62	Pocotaligo, SC	B
Leighty, John Q	23	Corp.	PA	8th Reserves	6/11/61	Wound.-Died	9/21/62	Antietam	B
Leonard, Jerome	25	Sgt.	PA	55th Infantry	10/12/61	Wound.-Died	8/11/64	Petersburg - Initial	B
Lewis, John D		Pvt.	PA	125th Infantry	8/12/62	MIA	5/2/63	Chancellorsville	HC
Line, Jacob	33	Corp.	PA	184th Infantry	5/12/64	KIA	6/22/64	Jerusalem Plank Rd.	B
Lines, Jacob		Pvt.	PA	8th Reserves	6/7/61	Died	5/3/62		B
Lingenfelter, John	15	Pvt.	IL	103rd Infantry	10/2/62	Died	3/22/64		IL
Little, Irving	24	Pvt.	PA	55th Infantry	10/15/61	Died	10/12/62		BL
Little, John Pius	34	Pvt.	PA	99th Infantry	2/24/64	KIA	4/6/65	Sailor's Creek	B
Logue, James	23	Pvt.	PA	171st Infantry	11/2/62	Died	6/26/63		B
Logue, James F	21	Pvt.	PA	53rd Infantry	9/21/64	Died	1/26/65	Petersburg	B
Long, Joseph E	26	Pvt.	PA	101st Infantry	3/10/65	Died	8/23/65		B
Long, Levi	30	Pvt.	PA	55th Infantry	10/12/61	Died	7/27/62		B
Lowery, Daniel	18	Pvt.	IN	2nd Cavalry	10/7/61	POW-Died	8/29/64	Newnan, GA	IN
Lowery, John E	25	Pvt.	PA	138th Infantry	8/29/62	Wound.-Died	11/28/63	Mine Run	B
Lowery, William H	21	Corp.	PA	138th Infantry	8/29/62	Died	4/15/64	Brandy Station	B

Name	Muster Age	Rank	State	Regiment	Muster In Date	Casualty	Date of Death	Battle	Res.
Luman, John	20	Pvt.	US	16th Infantry	1/24/63	Died	8/28/63		US
Lutz, Simon S	24	Pvt.	PA	184th Infantry	5/12/64	Wound.-Died	6/3/64	Cold Harbor	B
Lysinger, George W	28	1st Sgt.	PA	107th Infantry	1/11/62	POW-Died	12/19/64	Globe Tavern	B
Malone, Charles		Pvt.	PA	191st Infantry	5/15/64	POW-Died	12/10/64	Globe Tavern	B
Malone, William	24	Pvt.	PA	8th Reserves	6/11/61	Wound.-Died	10/24/62	Antietam	B
Manspeaker, Barclay	23	Pvt.	PA	8th Reserves	6/19/61	KIA	12/13/62	Fredericksburg	B
Manspeaker, David	24	Pvt.	PA	8th Reserves	6/11/61	KIA	5/13/64	Spotsylvania	B
Manspeaker, George	18	Pvt.	PA	11th Infantry	9/30/61	KIA	12/13/62	Fredericksburg	B
Mars, John	21	Pvt.	PA	55th Infantry	10/11/61	POW-Died	9/30/64	Drewry's Bluff	B
Marshall, Henry L	31	Pvt.	PA	184th Infantry	5/12/64	POW-Died	10/23/64	Jerusalem Plank Rd.	B
Marshall, Moses F	42	Corp.	PA	55th Infantry	11/5/61	Died	12/5/61		B
Martin, David	22	Pvt.	PA	8th Reserves	6/11/61	KIA	6/26/62	Seven Days Battle	B
Martin, Samuel		Pvt.	PA	22nd Cavalry	7/9/63	KIA	9/26/64	Browns Gap	B
Martin, William L	23	Sgt.	PA	55th Infantry	11/5/61	KIA	10/22/62	Pocotaligo, SC	B
Mauk, Joseph W	25	Sgt.	PA	14th Infantry	4/24/61	Died	8/3/61		B
May, Jacob	25	Pvt.	PA	149th Infantry	8/26/63	MIA	5/5/64	Wilderness	B
May, Jacob L	41	Pvt.	PA	99th Infantry	9/20/64	Died	4/12/65	Appomattox	B
Mayers, John		Pvt.	PA	2nd Cavalry	10/15/61	POW-Died	1864		B
McCleary, George B	20	Pvt.	PA	133rd Infantry	8/15/62	Wound.-Died	4/10/63	Fredericksburg	B
McCoy, Shannon E	22	Corp.	PA	138th Infantry	8/29/62	MIA	6/1/64	Cold Harbor	B
McDaniel, Hiram	19	Pvt.	PA	149th Infantry	8/27/63	Wound.-Died	5/14/64	Spotsylvania	B
McDaniel, Lewis		Sgt.	PA	22nd Cavalry	2/27/64	KIA	8/21/64	Berryville	B
McDonald, James	24	Pvt.	PA	13th Reserves	5/29/61	Died	6/25/63		B
McDonald, Samuel		Pvt.	PA	97th Infantry	3/3/65	Died	9/23/65		B
McDonald, William, Jr	18	Pvt.	PA	101st Infantry	11/1/61	Died	7/3/62		B
McEnespy, Samuel	15	Pvt.	PA	13th Cavalry	10/5/63	Died	9/16/64	Coggin's Point, VA	B
McGee, James	22	Pvt.	PA	55th Infantry	10/15/61	POW-Died	5/27/64	Drewry's Bluff	B
McGee, John H	45	Pvt.	PA	147th Infantry	11/1/61	Died	5/5/63		B
McGee, William	23	Pvt.	PA	55th Infantry	10/15/61	MIA	9/29/64	Chaffin's Farm	B
McGregor, John	20	Pvt.	PA	55th Infantry	9/20/61	Wound.-Died	6/6/64	Cold Harbor	B
McIlnay, James	18	Pvt.	PA	110th Infantry	10/24/61	Died	6/15/62		B
McIlnay, John						KIA	3/23/63	Winchester	B
McLaughlin, Charles P	19	1st Lt.	PA	138th Infantry	8/29/62	KIA	6/1/64	Cold Harbor	B
McMullen, Charles		Pvt.	PA	55th Infantry	9/26/62	Wound.-Died	6/20/64		B
McPherson, John		Pvt.	USCT	3rd Infantry	7/23/63	Died	8/26/65		B
McVicker, J Clay	19	Pvt.	PA	55th Infantry	2/27/64	Died	4/18/64		B
Mearkle, Henry		Pvt.	PA	22nd Cavalry	7/23/63	Died	7/24/64	Leetown, MD	B
Mellen, Thomas	35	Pvt.	PA	6th Reserves	1/1/62	KIA	12/13/62	Fredericksburg	B
Mellott, Caleb	28	Pvt.	PA	82nd Infantry	11/28/64	POW-Died	7/6/65		FC
Mellott, Cornelius	20	Pvt.	PA	11th Infantry	9/30/61	KIA	8/30/62	2nd Bull Run	B
Mellott, Frederick	21	Pvt.	PA	12th Reserves	8/10/61	KIA	9/14/62	South Mountain	B
Mellott, John L	18	Pvt.	PA	11th Infantry	9/30/61	Died	12/31/64		B
Mentzer, Jacob M	31	Pvt.	PA	133rd Infantry	8/13/62	KIA	12/13/62	Fredericksburg	B
Meyers, Levi	19	Pvt.	PA	55th Infantry	3/2/64	POW-Died	5/20/64	Drewry's Bluff	B
Miller, Henry	33	Pvt.	PA	138th Infantry	8/29/62	Wound.-Died	5/20/64	Wilderness	B
Miller, Hezekiah H	18	Pvt.	PA	110th Infantry	1861	Died	8/6/64		B

Abraham Morgart was born on April 20th, 1823, in East Providence Township. He married Margaret Morgret in 1847. Abraham mustered in the 79th Pennsylvania Infantry in February 1865 while his wife and 7 children waited for his return at their East Providence Township home. He died in Washington, DC, on June 21st, 1865. Abraham was 42 years old. He is buried at the Morgart and Morgret Cemetery outside of Everett. Margaret never remarried and passed when she was 54 years old in 1880.

John Moyer was born in 1831. He married Margaret Harrier in 1852. John volunteered in the 2nd Pennsylvania Cavalry in November 1861, leaving behind a wife and 6 children at their Londonderry Township home. He died of disease on May 10th, 1864, at the Andersonville POW camp. John is buried at the Andersonville National Cemetery. Mowyer is the spelling of his last name on the headstone. Margaret raised her children on her own and never remarried. She passed at age 62 in 1899 and is buried at the Hyndman Cemetery.

Name	Muster Age	Rank	State	Regiment	Muster In Date	Casualty	Date of Death	Battle	Res.
Miller, Jackson	22	Sgt.	PA	138th Infantry	8/29/62	KIA	5/6/64	Wilderness	B
Miller, John	33	Pvt.	PA	101st Infantry	12/28/61	POW-Died	8/15/64	Plymonth, NC	B
Miller, Martin D	20	Pvt.	PA	101st Infantry	2/26/62	KIA	5/31/62	Fair Oaks	B
Miller, Matthew	36	Pvt.	PA	55th Infantry	9/26/62	MIA	5/16/64	Drewry's Bluff	B
Miller, Peter S	29	Pvt.	PA	99th Infantry	2/25/65	Died	8/24/65		B
Miller, Philip S	29	Sgt.	PA	55th Infantry	10/11/61	Died	9/28/62		B
Miller, Solomon H	29	Sgt.	PA	55th Infantry	10/11/61	POW-Died	6/8/64	Drewry's Bluff	B
Miller, Thomas J	23	Pvt.	PA	138th Infantry	8/29/62	POW-Died	9/15/64	Wilderness	B
Miller, William H	21	Pvt.	PA	84th Infantry	10/24/61	Died	2/24/62		B
Miller, William O	18	Pvt.	IA	24th Infantry	8/29/62	Wound.-Died	2/13/65	Opequon	IA
Miller, Wyrman S	19	Pvt.	PA	148th Infantry	8/16/62	KIA	5/3/63	Chancellorsville	B
Mitchell, James P	18	Pvt.	PA	55th Infantry	2/29/64	POW-Died	10/17/64	Drewry's Bluff	B
Mock, Andrew	33	Pvt.	PA	55th Infantry	3/2/64	Wound.-Died	3/31/65	White Oak Road	B
Mock, George W	34	Pvt.	PA	125th Infantry	8/13/62	Died	1/23/63		B
Mock, Josiah B	26	Sgt.	PA	55th Infantry	11/5/61	POW-Died	3/22/65		B
Mock, Malachi B	24	Pvt.	PA	55th Infantry	11/5/61	Died	11/7/62		B
Mock, Mathias	24	Pvt.	PA	184th Infantry	5/12/64	KIA	6/3/64	Cold Harbor	B
Mock, Tobias B	22	Pvt.	PA	55th Infantry	10/11/61	Wound.-Died	8/7/64	Petersburg - Initial	B
Mock, Wliam A	33	1st Sgt.	PA	55th Infantry	11/5/61	KIA	5/22/64	Bermuda Hundred	B
Morgart, Abraham	41	Pvt.	PA	79th Infantry	2/23/65	Died	6/21/65		B
Morgart, William	23	Pvt.	PA	18th Cavalry	9/29/62	POW-Died	11/1/64	Wilderness	B

142   Chapter 7

Name	Muster Age	Rank	State	Regiment	Muster In Date	Casualty	Date of Death	Battle	Res.
Morse, David	16	Pvt.	PA	11th Infantry	9/30/61	Wound.-Died	1/3/63	Antietam	B
Morse, Jesse W	22	Pvt.	PA	99th Infantry	3/23/65	Died	5/15/65		B
Morse, Samuel	19	Pvt.	PA	101st Infantry	2/8/62	Died	6/25/62		B
Mortimore, David	21	Pvt.	PA	22nd Cavalry	7/11/63	Died	9/18/63		B
Mosell, William	24	Corp.	PA	55th Infantry	9/20/61	Died	4/15/64		B
Moser, William	26	Pvt.	OH	55th Infantry	9/30/61	Wound.-Died	7/25/65	Cross Keys	OH
Moser, William S	31	Corp.	PA	55th Infantry	2/29/64	Died	7/15/64		B
Moses, Emanuel	17	Pvt.	PA	18th Cavalry	10/29/62	POW-Died	11/18/63	Gettysburg	B
Mountain, Richard D	16	Pvt.	PA	77th Infantry	10/9/61	Wound.-Died		Chickamauga, GA	HC
Mower, Alexander C	27	Musician	PA	55th Infantry	10/12/61	Died	1/28/65		B
Moyer, John (1)	30	Pvt.	PA	2nd Cavalry	11/4/61	POW-Died	5/10/64		B
Moyer, John (2)	18	Pvt.	PA	55th Infantry	10/11/61	Died	12/11/62		B
Mull, William	32	Pvt.	PA	82nd Infantry	7/5/64	Died	7/10/65		B
Murphy, George	29	Pvt.	OH	14th Infantry	8/21/61	Died	6/13/64		OH
Murphy, Philip	21	Pvt.	PA	55th Infantry	12/20/61	Died	7/12/62		B
Nichols, Charles	32	Pvt.	PA	21st Cavalry	7/21/63	KIA	3/21/65	Petersburg	B
Norton, Franklin G	14	Musician	PA	101st Infantry	12/28/61	Died	2/23/62		B
Nycum, John	28	Pvt.	PA	138th Infantry	8/29/62	Wound.-Died	6/28/64	Cold Harbor	B
Nycum, Josiah	23	Pvt.	PA	2nd Cavalry	10/18/61	Died	8/24/64		B
Nycum, Upton	22	Sgt.	PA	2nd Cavalry	11/4/61	Wound.-Died	12/12/63		B
Oldham, Michael	17	Pvt.	PA	55th Infantry	2/15/64	Died	4/6/64		B
Oler, John W	16	Musician	PA	101st Infantry	11/1/61	Died	1/2/62		B
Oliver, James M	29	Pvt.	OH	121st Infantry	8/18/62	KIA	6/27/64	Kennesaw Mtn.	OH
O'Neal, Hezekiah	18	Pvt.	PA	138th Infantry	8/29/62	Wound.-Died	12/4/63		B
Orris, Jacob	35	Pvt.	PA	184th Infantry	5/12/64	POW-Died	8/1/64	Jerusalem Plank Rd.	B
Osborn, Peter	37	Pvt.	PA	208th Infantry	9/5/64	Wound.-Died	4/3/65	Petersburg - Final	B
Otto, Henry	22	Pvt.	PA	101st Infantry	12/28/61	Died	9/26/62		B
Otto, Henry S	16	Pvt.	PA	184th Infantry	5/12/64	POW-Died	9/18/64	Jerusalem Plank Rd.	B
Palmer, Casper	22	Corp.	KS	10th Infantry	8/12/61	Died	12/30/61		KS
Palmer, John	17	Pvt.	PA	55th Infantry	2/19/64	POW-Died	3/8/65		B
Pearson, Francis	17	Pvt.	PA	110th Infantry	10/24/61	Died	12/31/62		B
Pennell, Henry C	19	Pvt.	PA	76th Infantry	10/9/61	Died	6/29/62		B
Perdew, Nathan	22		PA	16th Cavalry	9/6/62	Wound.-Died	12/1/64	2nd Deep Bottom	B
Perrin, John	35	Pvt.	PA	133rd Infantry	8/13/62	KIA	12/13/62	Fredericksburg	B
Piper, Luther R	26	Corp.	PA	8th Reserves	6/11/61	Wound.-Died	1/1/63	Fredericksburg	B
Price, Abraham	20	Pvt.	PA	138th Infantry	8/29/62	Died	10/19/62		B
Price, John	33	Pvt.	PA	184th Infantry	5/12/64	Died	3/22/65		B
Price, Joseph J	20	Corp.	PA	138th Infantry	8/29/62	KIA	5/6/64	Wilderness	B
Radcliff, James	18	Pvt.	PA	138th Infantry	8/29/62	Died	7/26/64		B
Radebaugh, Jacob L	19	Sgt.	PA	55th Infantry	11/5/61	MIA	Unknown	Drewry's Bluff	B
Ralston, David E	23	Pvt.	PA	110th Infantry	12/19/61	KIA	5/3/63	Chancellorsville	B
Rea, William H	22	Pvt.	PA	138th Infantry	2/1/62	Died	9/23/63		B
Reed, Andrew J	17	Pvt.	PA	55th Infantry	2/27/64	Died	11/17/64		B
Reed, Isaac	23	Pvt.	PA	143rd Infantry	6/15/63	Wound.-Died	6/1/64	Spotsylvania	B
Renninger, Frederick	17	Pvt.	PA	84th Infantry	10/24/61	Died	3/4/62		B
Ressler, Abraham	31	Pvt.	PA	101st Infantry	2/13/62	Died	6/6/62		B

Twin brothers Isaac (left) & Abraham Rice (right) were born on September 14th, 1834. They left their family farm in Monroe Township to volunteer in the 101st Pennsylvania Infantry in November 1861. Three other brothers also enlisted in the Union Army. Solomon Rice enlisted in the 149th PA Infantry, Cornelius Rice in the 78th PA Infantry, and Jonathan Rice in the 99th PA Infantry. Isaac & Abraham were captured with the entire Union garrison at Plymouth, NC on April 20th, 1864. The brothers were taken to Andersonville. They were later transferred to a POW camp in Charleston, SC. Prisoners at this POW camp lived in horrible conditions in the interior of a horse track. At least 257 men died from exposure and disease within weeks. Isaac died on September 21st, 1864, and was buried in a mass grave near the track with over 200 other prisoners. His body was later reburied in an unmarked grave at Beaufort National Cemetery in Beaufort, SC. Abraham remained a prisoner until February 27th, 1865. He mustered out after his release and returned to his home near Robinsonville. Three other brothers also survived the war. Abraham married in 1866 and was the father of seven children. He moved his family to Kansas in the 1880s. Abraham passed at age 60 and was buried at the Pleasant Ridge Cemetery in Radium, Kansas in 1895.

Name	Muster Age	Rank	State	Regiment	Muster In Date	Casualty	Date of Death	Battle	Res.
Rice, Isaac	27	Corp.	PA	101st Infantry	11/1/61	POW-Died	9/21/64	Plymonth, NC	B
Riceling, William	28	Pvt.	PA	76th Infantry	11/21/61	POW-Died	11/2/63	Ft. Wagner, SC	B
Riffle, Thomas	41	Pvt.	MO	35th Infantry	9/1/62	Died	10/6/63		MO
Riley, William (1)		Pvt.	PA	11th Infantry	9/30/61	Wound.-Died	4/14/63	Antietam	B
Riley, William (2)	53	Pvt.	PA	55th Infantry	3/2/64	Wound.-Died	6/4/64	Cold Harbor	B
Rinard, Emanuel	22	Corp.	PA	77th Infantry	10/9/61	Died	7/30/64		B
Rininger, Frederick		Pvt.	PA	84th Infantry	10/24/61	Died	3/4/62		B
Ritchey, David	28	Pvt.	PA	208th Infantry	9/7/64	Died	1/21/65		B
Ritchey, Henry S	21	Corp.	PA	101st Infantry	11/1/61	Died	6/2/62		B
Ritchey, Jacob M	37	Pvt.	PA	199th Infantry	10/1/64	Died	12/23/64		B
Ritchey, Jonas	18	Pvt.	PA	55th Infantry	11/5/61	Wound.-Died	3/29/62	Edisto Island, SC	BL
Ritchey, Samuel Y	26	Pvt.	PA	8th Reserves	7/29/61	Died	1864		B
Robb, George W	24	Pvt.	PA	138th Infantry	9/4/62	KIA	11/27/63	Mine Run	B
Robinson, William	35	Pvt.	PA	99th Infantry	2/24/65	Died	5/13/65		B
Robison, Jonas	23	Pvt.	PA	101st Infantry	11/1/61	Died	6/1/62		B
Roudabush, Benjamin	16	Pvt.	PA	55th Infantry	9/25/63	POW-Died	9/29/64		B
Rowland, John	34	Pvt.	PA	88th Infantry	9/24/64	Died	8/4/65		B
Rowser, Philip	20	Pvt.	PA	55th Infantry	10/11/61	Died	12/30/61		B
Ruby, John (1)	21	Pvt.	PA	101st Infantry	11/1/61	Died	10/30/61		B
Ruby, John (2)	23	Pvt.	PA	55th Infantry	2/27/64	Died	4/29/64		B
Sanno, Frederick	43	Corp.	PA	55th Infantry	11/5/61	Died	3/6/63		B
Saupp, John	17	Pvt.	PA	55th Infantry	11/5/61	KIA	3/29/62	Edisto Island, SC	B

Jacob M Ritchey was born in Rays Cove in East Providence Township on March 27th, 1827. He married Eliza Jane Manspeaker on April 3rd, 1851. Jacob mustered in the 199th Pennsylvania Infantry in October 1864, while Eliza and their 6 children tended to their family farm. He died in a Hampton Virginia Hospital of Typhoid Fever on December 23rd, 1864. Jacob was 37 years old when he died. Eliza raised their children on her own and never remarried. She passed at age 65 in 1892 and is buried beside her husband at the Asbury Methodist Episcopal Church Cemetery in Graceville.

William Saylor was born in 1832. He married Rachel Harbaugh in 1854. Their first child was born in Lovely the following year. William left behind a wife and 3 children when he mustered in the 82nd Pennsylvania Infantry in November 1864. The 82nd PA Infantry was in combat during the siege of Petersburg and witnessed the surrender at Appomattox. Three months after the war ended, William died of Typhoid Fever on July 13th, 1865 at Wilson's Station, VA. William is buried at the Poplar Grove National Cemetery in VA. His wife Rachel also lost a brother during the war. Robert Harbaugh was killed during the Initial Assault on Petersburg on June 18th, 1865. Rachel never remarried and passed in 1890.

Name	Muster Age	Rank	State	Regiment	Muster In Date	Casualty	Date of Death	Battle	Res.
Saylor, William	37	Pvt.	PA	82nd Infantry	11/18/64	Died	5/31/65		B
Scutchall, David	22	Pvt.	PA	8th Reserves	6/11/61	Died	1/5/63		B
Scutchall, Samuel	18	Pvt.	PA	133rd Infantry	8/13/62	KIA	12/13/62	Fredericksburg	B
Semler, Reuben J	18	Pvt.	PA	55th Infantry	12/20/61	Wound.-Died	6/9/64	Cold Harbor	B
Shaffer, Jacob J	19	Pvt.	PA	55th Infantry	2/19/64	POW-Died	9/30/64	Drewry's Bluff	B
Sharp, George	39	Corp.	PA	61st Infantry	9/26/64	Died	2/19/65		B
Sharp, James	28	Pvt.	PA	171st Infantry	11/10/62	Died	7/3/63		B
Shauf, Cornelius	18	Pvt.	PA	8th Reserves	6/11/61	Died	12/17/61		B
Shaw, Matthew P	22	Pvt.	PA	8th Reserves	6/19/61	KIA	9/14/62	South Mountain	B
Shippley, Lorenzo D	40	Pvt.	PA	171st Infantry	11/2/62	Died	7/21/63		B
Shoemaker, Isaac F	26	Corp.	PA	101st Infantry	12/6/61	Died	11/10/64		B
Shoeman, Peter	30	Pvt.	PA	49th Infantry	9/26/63	Wound.-Died	5/18/64	Spotsylvania	B
Shoop, Joseph L	22	Pvt.	PA	55th Infantry	10/11/61	Wound.-Died	8/7/64	Cold Harbor	BL
Showman, William	47	Pvt.	PA	101st Infantry	12/2/61	Died	7/9/62		B
Shrader, Auterbine	18	Pvt.	PA	55th Infantry	2/19/64	POW-Died	9/29/64		B
Shrader, William O	18	Pvt.	PA	55th Infantry	2/29/64	POW-Died	8/27/64	Drewry's Bluff	B
Shull, Isaac		Pvt.	IN	84th Infantry	8/11/62	KIA	9/19/63	Chickamauga, GA	B
Skillington, John	24	Pvt.	PA	49th Infantry	6/14/64	Died	9/14/64		B
Sleek, Jesse W		Pvt.	PA	184th Infantry	5/15/64	Died	1866		B
Slick, Allen	48	Pvt.	PA	55th Infantry	2/19/64	KIA	6/3/64	Cold Harbor	B
Slick, Hezekiah B	22	Pvt.	PA	55th Infantry	2/19/64	POW-Died	2/6/65		B
Slick, John A	33	Pvt.	PA	208th Infantry	9/8/64	Wound.-Died	4/5/65	Petersburg - Final	B
Slick, Philip	18	Pvt.	IL	9th Cavalry	9/19/61	POW-Died	10/11/64	Tupelo, Miss.	IL

Those who made the Ultimate Sacrifice 145

Information on Jonathan Snyder is also included in the overview of the battle of Cedar Creek on page 96. His parents, John and Sophia Snyder, raised four sons and three daughters on a family farm in Monroe Township. All four sons enlisted in the Union Army. Jonathan died from wounds suffered at the battle of Cedar Creek in October 1864. John was enlisted in the 2nd Maryland PHB Cavalry when he was captured on January 3rd, 1864, and taken to Andersonville. He perished there as a prisoner of war in April 1864. Joseph enlisted in the 4th Pennsylvania Cavalry and Leonard in the 26th Illinois Infantry. Both Joseph and Leonard survived the war.

David Stine was born in Claysburg in 1831. He married Catherine Dively in 1852. David volunteered in the 11th Pennsylvania Cavalry in September 1861 while Catherine and their six children waited at home for his return. He was captured in Suffolk, VA on November 10th, 1863 and was taken to Andersonville in 1864. David is believed to have perished on October 8th, 1864, and buried in an unmarked grave in the Andersonville National Cemetery. His brother Jacob enlisted in the 77th PA Infantry and a half-brother Abraham Burket mustered in the 11th PA Cavalry with David. Both survived the war. Catherine raised their children on her own and never remarried. She passed at age 78 in 1913.

Name	Muster Age	Rank	State	Regiment	Muster In Date	Casualty	Date of Death	Battle	Res.
Slick, Samuel K	20	Pvt.	PA	101st Infantry	12/28/61	Wound.-Died	12/19/62		B
Smith, David	19	Pvt.	PA	138th Infantry	8/29/62	Wound.-Died	9/20/62		B
Smith, George W	21	Pvt.	PA	101st Infantry	2/8/62	Died	Unknown		B
Smith, Jacob	32	Pvt.	PA	28th Infantry	9/21/64	Died	5/19/65		B
Smith, Jasper W	31	Pvt.	PA	55th Infantry	9/26/62	POW-Died	5/20/64		B
Smith, Jesse	40	Pvt.	PA	55th Infantry	2/17/64	Wound.-Died	5/27/64	Drewry's Bluff	B
Smith, Joseph	24	Pvt.	PA	28th Infantry	11/4/64	Died	1865		B
Smith, Nathan	21	Pvt.	PA	2nd Cavalry	11/4/61	KIA	11/11/63	Rappahannock	B
Smith, Samuel		Pvt.	PA	55th Infantry	9/20/63	KIA	5/16/64	Drewry's Bluff	BL
Snook, Emanuel	21	Corp.	PA	55th Infantry	12/4/61	Wound.-Died	7/6/65	Petersburg	B
Snowberger, David	31	Pvt.	PA	55th Infantry	10/12/61	Died	1/25/64		B
Snowberger, Theodore	20	Pvt.	PA	184th Infantry	5/12/64	Wound.-Died	9/1/64	Petersburg - Initial	B
Snowden, David	24	Pvt.	PA	184th Infantry	5/12/64	KIA	6/3/64	Cold Harbor	B
Snyder, John W	19	Pvt.	MD	2nd PHB Inf.	9/11/61	POW-Died	4/24/64	Jonesville	B
Snyder, Jonathan	30	1st Sgt.	PA	138th Infantry	8/29/62	Wound.-Died	10/22/64	Cedar Creek	B
Sparks, Jacob	21	Pvt.	PA	133rd Infantry	8/29/62	Died	3/16/63		B
Spidle, Wilson	22	Pvt.	PA	55th Infantry	10/12/61	Wound.-Died	7/11/64	Petersburg - Initial	B
Steckler, Charles	22	Pvt.	PA	55th Infantry	7/22/63	POW-Died	9/29/64		B
Steckman, Daniel H	19	Pvt.	PA	76th Infantry	10/9/61	KIA	7/11/63	Ft. Wagner, SC	B
Steckman, Francis	18	Corp.	PA	138th Infantry	8/29/62	Wound.-Died	6/5/64	Cold Harbor	B
Steckman, John B	18	Corp.	PA	138th Infantry	8/29/62	Died	12/23/62		B
Steele, David F	21	Pvt.	PA	133rd Infantry	8/15/62	KIA	12/13/62	Fredericksburg	B

Abner Summerville was born in 1827 and raised in a large family in Southampton Township. He volunteered in the 55th PA Infantry in October 1862 with a wife and 6 children at home. Abner was captured in 1864. He was likely taken to Andersonville and later to Millen, GA, where he died in October 1864. Abner is buried in the Lawton National Cemetery. Two of his brothers also perished in the war. Sylvanus was enlisted in the 55th PA Infantry and suffered a wound at Chaffin's Farm in September 1864. He died 11 days later. Charles was reported as MIA during the battle of the Wilderness in May 1864 while enlisted in the 138th PA Infantry. Only one brother survived the war. Robert mustered out of the 208th PA Infantry in June 1865.

Noah Tipton was born around 1827 in Fairhope Township in Somerset County. He married Elizabeth May in 1848 and was listed as a farmer on the 1850 Harrison Township Census. Noah enlisted in the 61st Pennsylvania Infantry in September 1864, while Elizabeth and their 8 children tended to their family farm. Noah mustered out after the war ended on June 20th, 1865. While in the army, Noah contracted malaria. He died in Buffalo Mills from the effects of this disease in October 1865, the same month his 9th child, a daughter, was born. Noah is buried in the Burkhart Burial Ground in Fairhope Township. Elizabeth remarried after the war. She died at age 50 in 1880.

Name	Muster Age	Rank	State	Regiment	Muster In Date	Casualty	Date of Death	Battle	Res.
Stephens, Jacob	51	Pvt.	PA	11th Cavalry	12/24/63	POW-Died	10/17/64	Reams Station	B
Stephens, John G	19	Pvt.	PA	184th Infantry	5/12/64	KIA	10/27/64	Boydton Plank Rd.	B
Stickler, Samuel	23	Pvt.	PA	55th Infantry	10/12/61	POW-Died	6/18/64	Drewry's Bluff	B
Stiffler, John H	18	Pvt.	PA	138th Infantry	8/29/62	Wound.-Died	1/3/64	Mine Run	B
Stine, David	30	Pvt.	PA	11th Cavalry	9/5/61	POW-Died	10/8/64	Suffolk, VA	BL
Stineman, John	26	Pvt.	PA	91st Infantry	9/21/64	KIA	10/27/64	Boydton Plank Rd.	B
Strathers, Willis	24	Pvt.	USCT	43rd Infantry	3/19/64	KIA	7/30/64*	Battle of the Crater*	B
Strayer, Nicholas	30	Pvt.	PA	205th Infantry	8/27/64	Wound.-Died	5/12/65	Petersburg - Final	B
Stuckey, David H	28	Corp.	PA	184th Infantry	5/12/64	POW-Died	11/18/64	Jerusalem Plank Rd.	B
Stuckey, Simon C	22	1st Sgt.	PA	138th Infantry	9/2/62	KIA	11/27/63	Mine Run	B
Stufft, Michael	41	Pvt.	PA	91st Infantry	9/21/64	Died	6/22/65		B
Stufft, William S	32	Pvt.	PA	171st Infantry	11/2/62	Died	6/13/63		B
Summerville, Abner	35	Pvt.	PA	55th Infantry	10/26/62	POW-Died	10/31/64		B
Summerville, Charles	21	Pvt.	PA	138th Infantry	8/29/62	MIA	5/6/64	Wilderness	B
Summerville, Sylvanus	22	Pvt.	PA	55th Infantry	10/12/61	Wound.-Died	10/10/64	Chaffin's Farm	B
Swoveland, William	31	Pvt.	PA	184th Infantry	5/12/64	POW-Died	8/10/64	Jerusalem Plank Rd.	B
Tate, George H	18	Pvt.	PA	101st Infantry	2/8/62	Died	7/24/63		B
Taylor, Ambrose K	19	Sgt.	PA	110th Infantry	10/24/61	KIA	7/27/64	1st Deep Bottom	B
Teeter, Christian	22	Pvt.	PA	184th Infantry	5/12/64	POW-Died	9/13/64	Jerusalem Plank Rd.	B
Tetwiler, Peter	24	Pvt.	PA	53rd Infantry	10/17/61	Died	4/13/65		B
Tewell, Daniel	34	Corp.	PA	4th Cavalry	9/16/61	POW-Died	5/16/64	Sulpher Springs	B

William Wagerman was born in Wolf Creek Run near Madley on February 20th, 1838. William was raised in a large family of 7 brothers and 5 sisters. He married Catharine Smith on March 16th, 1862, and volunteered in the 138th Pennsylvania Infantry 6 months later. William died of Typhoid Fever in Alexandria on December 8th, 1863. He was 25 years old. After his brother Samuel mustered out of the 3rd Maryland PHB Infantry on May 29th, 1865, he was sent to recover William's body for reinterment in the Wagerman Family Cemetery in Madley. Samuel married William's widow Catherine in 1865, and they were parents of two daughters.

John Wolfhope was born in New Baltimore on February 21st, 1839 and raised on a farm with 4 brothers and 6 sisters. John was working on the family farm when he mustered in the 184th Pennsylvania Infantry in May 1864. Two months after enlisting, he was captured during the battle of Jerusalem Plank Road on June 22nd, 1864. John was taken to Andersonville, where he died of dysentery on October 4th, 1864. He was 25 years old. John is buried in the Andersonville National Cemetery. Another photograph of John was donated to the museum at Andersonville by a descendant.

Name	Muster Age	Rank	State	Regiment	Muster In Date	Casualty	Date of Death	Battle	Res.
Tewell, Joseph	19	Corp.	PA	55th Infantry	11/5/61	Wound.-Died	6/17/64	Cold Harbor	B
Thomas, Joseph A	19	Pvt.	PA	12th Reserves	3/29/62	POW-Died	4/24/65	Globe Tavern	B
Thomas, Samuel	40		MO	Home Brigade		Died	8/18/61		MO
Thompson, David	42	Pvt.	PA	110th Infantry	10/24/61	POW-Died	7/23/64	Petersburg	B
Thorpe, John W	21	Musician	PA	138th Infantry	8/29/62	Died	8/2/64		B
Tipton, Noah	37	Pvt.	PA	61st Infantry	9/26/64	Died	10/1/65		B
Tobias, Samuel H G	19	1st Sgt.	PA	110th Infantry	10/24/61	KIA	7/2/63	Gettysburg	B
Trail, Nathan B	26	Pvt.	MD	1st PHB Cav.	2/27/64	Died	4/9/64		B
Trott, Benjamin	36	Pvt.	PA	55th Infantry	3/2/64	Died	6/28/65		B
Troutman, George W	16	Pvt.	PA	138th Infantry	8/29/62	Died	12/20/62		B
Truax, George M	19	Pvt.	PA	101st Infantry	11/1/61	Died	11/30/62		B
Turner, Andrew J	20	Corp.	PA	55th Infantry	11/5/61	Died	12/3/61		B
Veatch, Samuel	26	Pvt.	PA	101st Infantry	12/6/61	Wound.-Died	8/1/62		B
Wagerman, William	24	Pvt.	PA	138th Infantry	9/29/62	Died	12/8/63		B
Wagoner, August		Pvt.	PA	184th Infantry	5/12/64	KIA	6/4/64	Cold Harbor	B
Walker, William A	24	Pvt.	PA	21st Cavalry	7/8/63	Died	1/4/64		B
Walter, Joseph H	33	Pvt.	PA	19th Cavalry	9/17/63	Died	9/3/64		B
Ward, Samuel	32	Pvt.	PA	138th Infantry	8/29/62	POW-Died	7/30/64	Mine Run	B
Warsing, Alexander	33	Pvt.	PA	8th Reserves	6/19/61	Died	8/14/64		B
Washabaugh, William H	25	Pvt.	PA	76th Infantry	10/9/61	KIA	7/11/63	Ft. Wagner, SC	B
Waters, David Y	31	Sgt.	IL	20th Infantry	6/13/61	Died	3/8/65		IL

George Woy was born on June 24th, 1839 in East Providence Township. His parents John and Elizabeth raised 8 children on a family farm. George mustered in the 77th Pennsylvania Infantry in November 1861. The 77th was one of the few regiments with recruits from Bedford County who fought in the Western Theater of the Civil War. George was killed during the battle of Liberty Gap, TN, the day after his 24th birthday on June 25th, 1863. He is believed to have been initially buried near the Liberty Gap battlefield before being re-interred in the Stones River National Cemetery in Tennessee after the war ended. His brother Joseph was also in the Union Army. Joseph mustered in the 22nd Pennsylvania Cavalry in February 1864 and suffered a wound during the battle of Snicker's Gap, Virginia, on July 17th, 1864. Joseph survived the war and returned home to his wife and children.

Name	Muster Age	Rank	State	Regiment	Muster In Date	Casualty	Date of Death	Battle	Res.
Watkins, Jesse	33	Pvt.	PA	55th Infantry	10/15/61	Died	4/26/63		B
Watson, Henry S	18	Pvt.	PA	184th Infantry	5/12/64	POW-Died	1/2/65	Jerusalem Plank Rd.	B
Weaverling, Jacob P	33	Pvt.	PA	11th Infantry	3/21/62	KIA	9/17/62	Antietam	B
Wentz, John	32	Pvt.	PA	55th Infantry	11/5/61	Died	10/15/65		B
Wentz, Philip	25	Pvt.	PA	138th Infantry	8/29/62	KIA	11/27/63	Mine Run	B
Weyandt, James	27	Pvt.	PA	99th Infantry	2/25/65	Died	1865		B
Whip, Jacob	20	Sgt.	PA	138th Infantry	9/29/62	Wound.-Died	11/27/63	Mine Run	B
Whitaker, Christian	20	Pvt.	US	1st Artillery	2/22/63	Died	1865		B
White, James S	41	Pvt.	PA	55th Infantry	9/2/62	Wound.-Died	5/29/64	Drewry's Bluff	B
Wilkins, William	19	Pvt.	PA	11th Infantry	9/30/61	Died	7/12/62		B
Williamson, Gideon	39	Pvt.	PA	21st Cavalry	7/21/63	Died	12/31/64		B
Wilson, George W	50	Pvt.	PA	101st Infantry	3/7/62	POW-Died	10/26/64	Plymonth, NC	B
Wilson, John	40	Pvt.	PA	55th Infantry	10/11/61	Wound.-Died	6/24/64		B
Wisegarver, Daniel	19	Pvt.	PA	13th Cavalry	9/30/63	Died	2/29/64		B
Wisegarver, William V	21	Pvt.	PA	18th Cavalry	2/29/64	KIA	4/1/64		B
Witt, John	22	Pvt.	PA	171st Infantry	11/1/62	Died	2/17/63		B
Wolfhope, John	25	Pvt.	PA	184th Infantry	5/12/64	POW-Died	10/4/64	Jerusalem Plank Rd.	SC
Wolford, William	30	Pvt.	PA	149th Infantry	9/24/63	KIA	5/5/64	Wilderness	B
Woodcock, George W	18	Pvt.	PA	53rd Infantry	1/29/64	Wound.-Died	5/23/64	Spotsylvania	FC
Woy, David M	21	Pvt.	PA	12th Cavalry	2/14/62	Died	2/1/63		B
Woy, George	22	Pvt.	PA	77th Infantry	11/22/61	KIA	6/25/63	Liberty Gap, TN	B
Wright, Edmund S	20	Pvt.	PA	184th Infantry	5/12/64	POW-Died	6/22/64	Jerusalem Plank Rd.	B
Wright, James	21	Pvt.	PA	99th Infantry	2/25/65	Died	5/27/65		B
Yantz, Henry	36	Pvt.	PA	67th Infantry	11/28/64	Died	6/15/65		B
Yarnell, Jesse	43	Pvt.	PA	138th Infantry	8/29/62	Wound.-Died	7/22/64	Monocacy, MD	B

8. Father & Sons

Adam Weaverling on the left and his son Thomas H Weaverling on the right are pictured in their Civil War uniforms. Adam and Thomas Weaverling are emblematic examples of the sacrifices of a father and his son during the Civil War. Adam Weaverling was born on February 6th, 1819, in East Providence Township, and married Susan Hollar in 1842. Their oldest son, Thomas, was born on January 16th, 1848. Adam mustered as a 2nd Lieutenant in the 11th Pennsylvania Infantry on September 30th, 1861, leaving behind his wife Susan and 4 children to tend to their family farm in West Providence Township. He suffered a wound during the 2nd battle of Bull Run on August 30th, 1862, and received a discharge for his wounds on January 23rd, 1863. Six months later, Governor Curtin of Pennsylvania made an urgent call for volunteers to serve in emergency militias to help defend Pennsylvania from the invading Confederate army. Adam volunteered as a 1st Lieutenant in McKeage's Militia. He mustered out in early August after Robert E Lee was turned away at Gettysburg. Adam may have thought his service in the Union Army had ended before his 16-year-old son Thomas volunteered in the 208th Pennsylvania Infantry on September 7th, 1864. Adam and his wife Susan likely had deep concerns about their young son entering the war. Three days later, Adam mustered as a Captain in the same regimental company as Thomas. Both experienced combat during the siege of Petersburg but survived the war and returned home to the family farm. Thomas married in 1874 and was a father of 5 children. He moved his family to Kansas in the late 1870s but returned to Bedford County to continue farming in Colerain Township in the early 1880s. His mother Susan passed in 1896 when she was 72. Adam Weaverling passed the following year at age 78. Thomas passed in 1915 when he was 67 years old. Thomas' last surviving child lived until 1969.

There is little information on the total numbers of fathers and sons who served during the Civil War. There were 67 identified families from Bedford County with a father and son(s) in the Union Army. Twenty of these 67 families mourned the death of a loved one during the war. Two families lost both a father and a son. George W Foor was killed in action at Antietam in September 1862, and his son Daniel V Foor lost his life during the attack on Fort Wagner in South Carolina in July 1863. Jacob Stephens died as a POW in Andersonville in October 1864. His son, John G Stephens, was killed during the battle of Boydton Plank Road the same month his father died. Additional information on both families is on the next page.

The 67 Bedford County families with a father and son(s) in the war suffered a combined 75 casualties. Twenty-seven of the fathers and sons served together in the same regiment. Nineteen fathers in the 67 families were over the age of 50 during their enlistment. Nathaniel Gates was serving as a chaplain at age 71 when he picked up a rifle and fought alongside his sons in the 13th Pennsylvania Reserves. Twenty-four of the 67 Bedford County families had sons 16 years old and younger when they enlisted. Alexander Ake, Jacob Castner, William Linn, and Franklin Norton were 14 years old when they left their homes to enter the war. There are an extraordinary number of compelling stories about Bedford County's Civil War generation. The following listing of the fathers, sons and their families belong in their own special category.

(left) Leonard J Foor grew up on a family farm in East Providence Township with 6 siblings. Leonard was 5 years old when his father and brother left their home to join the Union Army. His father George was born around 1818. He married Anna Cooper in December 1841. Their oldest son, Daniel, was born in 1843. Daniel volunteered in the 76th Pennsylvania Infantry in November 1861. Three months later, his father George volunteered in the 107th Pennsylvania Infantry. Bedford County soldiers in the 107th suffered casualties at the 2nd Battle of Bull Run in late August 1862 and the battle of South Mountain two weeks later. Heroic Union Army fighting at South Mountain forced Robert E Lee to halt his planned invasion of Pennsylvania to reunite his divided army near Sharpsburg Maryland. 3 days later, Bedford County soldiers took heavy casualties at Antietam. George was killed near the Cornfield area of the battlefield on September 17th. Less than a year later, Daniel was killed during a charge at Fort Wagner in South Carolina on July 11th, 1863. There is no record of where Daniel is buried. His mother Anna passed when she was 60 and was buried beside her husband George at the Rays Cove Christian Church Cemetery in 1884. Leonard passed in 1924 at age 75 and was buried in the same cemetery as his parents.

Cyrus Riffle is pictured on the left with his half-siblings Eve Stevens Custer and Jacob Stevens Jr. Cyrus was born in 1842. His father died when he was a toddler. His mother Sarah remarried Jacob Stevens, and he grew up with 5 half-siblings in South Woodbury Township. Cyrus volunteered in the 133rd Pennsylvania Infantry in August 1862 and took part in the tragic Marye's Heights charge at Fredericksburg four months later. He mustered out with the regiment on May 26th, 1863. Three days after mustering out, his mother Sarah died, leaving behind a husband and six children. Their father, Jacob Sr mustered in the 11th Pennsylvania Cavalry in December 1863. Jacob Sr was captured at the battle of Reams Station in June 1864 and taken to Andersonville. He perished there on October 17th, 1864. His son John enlisted in the 184th Pennsylvania Infantry in May 1864. John was killed during the battle of Boydton Plank Road on October 27th, 1864. Cyrus had reenlisted in the 194th Infantry in July 1864, but mustered out of the army in November 1864 to return to Bedford County to take care of his four orphaned siblings still living at home including 14-year-old Jacob Jr and 8-year-old Eve.

David Carpenter is pictured later in life on the left. His father Curtis and mother Delilah are on the right prior to 1871. Curtis Carpenter was born in April 1819 in Frederick County, Maryland, and married Delilah Bender around 1840. Curtis and Delilah are listed on the 1860 South Woodbury Township Census with eight children. Their oldest son, David, volunteered in the 110th Pennsylvania Infantry in November 1861 when he was 15 years old. His description at enlistment was he stood 5'6" tall with gray eyes, brown hair, fair completion and claimed he was 18. Like many underage boys, he misrepresented his age. Available records show David was a POW with no dates or other information listed. David's second enlistment was in the 6th US Cavalry in October 1862. Their father Curtis left his job as a tanner to muster as a Sergeant in the 22nd Pennsylvania Cavalry in July 1863. A second son, Samuel, volunteered in McKeage's Militia that same month when he was 15 years old. After the Militia disbanded, Samuel mustered in the 13th Pennsylvania Cavalry in August 1863. All three family members survived combat in Union Cavalry units and returned home to South Woodbury Township. Curtis was listed on the 1870 census as a farmer. He was also a Sunday school superintendent in Loysburg. Delilah died when she was 50 years old in 1871. Curtis lived another 9 years and passed at age 70. Samuel died when he was 54 years old in 1902. David lived to be 82 and passed in 1928.

Name	Muster Age	Rank	State	Regiment	Muster In	Muster Out	Casualty	Cas. Year	Battle
Ake, Castleton	41	Sgt.	PA	McKeage's Mil.	7/2/63	8/8/63			
Ake, Alexander	14	Musician	PA	McKeage'sMil.	7/2/63	8/8/63			
Allison, Joseph D	39	Pvt.	PA	138th Infantry	8/29/62		Died	1864	
Allison, Nathaniel	18	Pvt.	PA	55th Infantry	2/19/64	8/30/65	Wounded	1864	Cold Harbor
Beaver, William	34	Pvt.	IA	5th Infantry	7/15/61		POW-Died	1862	New Madrid, MO
Beaver, Simon J	16	Pvt.	PA	55th Infantry	2/19/64	6/15/64	POW	1864	Drewry's Bluff
Blackburn, John M	45	Sgt., QM	PA	21st Cavalry	7/8/63	2/20/64	Wounded		
Blackburn, Cyrus E	21	Pvt.	PA	22nd Cavalry	2/25/64	6/24/65			
Bollman, David F	43	Pvt.	PA	107th Infantry	9/21/64	6/7/65			
Bollman, David R	21	Pvt.	PA	55th Infantry	2/27/64	8/30/65	Wounded	1864	Drewry's Bluff*
Bollman, George F	18	Pvt.	PA	McKeage's Mil.	7/2/63	8/8/63			
		Pvt.	PA	22nd Cavalry	2/16/64	6/24/65			
Brant, William	41	Pvt.	PA	107th Infantry	9/21/64	5/15/65			
Brant, Henry	16	Pvt.	PA	93rd Infantry	11/26/64	6/27/65	Wounded	1865	Ft. Fisher
Brant, Shannon	19	Pvt.	PA	55th Infantry	2/27/64	7/9/65	Wounded		
Bridenthal, Henry B	46	Pvt.	PA	55th Infantry	10/11/61	6/26/63			
Bridenthal, Thomas	17	Pvt.	PA	13th Cavalry	9/30/62	7/14/65			
Carpenter, Curtis J	44	Sgt.	PA	22nd Cavalry	7/11/63	2/5/64			
Carpenter, David B	15	Pvt.	PA	110th Infantry	11/17/61	10/27/62	POW	1862	
		Pvt.	US	6th Cavalry	10/27/62	4/4/65			
Carpenter, Samuel	15	Pvt.	PA	13th Cavalry	8/27/63	7/14/65			
		Pvt.	PA	McKeage's Mil.	7/2/63	8/8/63			

Chapter 8

Josiah and Louisa Aker Edwards are pictured on the left circa 1890. Their son Daniel is on the right circa 1910. Josiah was born in Union (Pavia) Township in September 1825 and married Louisa Aker in 1846. Daniel was born on August 16th, 1848. Josiah, Louisa are listed on the 1860 Union Township Census with 5 children. Fifteen-year-old Daniel volunteered in the 55th Pennsylvania Infantry in February 1864. Two weeks later, his father left his job as a farm laborer to join his son in the same infantry regiment while Louisa remained home with their other children. Both would take part in heavy combat during the Overland Campaign in May and June 1864, and the 9-month siege of Petersburg. Josiah received a Discharge on a Surgeons Certificate on March 1865. No record of the reason for the discharge is on available records. Daniel mustered out in August 1865 and rejoined his family in Bedford County. Josiah, Louisa, and Daniel moved to Illinois in 1865 after the war and settled on farmland in Iowa in the 1870s. Daniel married in 1866 and was a father to 6 children who survived to adulthood. Louisa passed at age 78 in 1903. Josiah lived to the age of 86 and was buried beside his wife in 1912. Daniel passed in 1926 at age 77.

Name	Muster Age	Rank	State	Regiment	Muster In	Muster Out	Casualty	Cas. Year	Battle
Castner, John B	42	1st Lt.	PA	133rd Infantry	8/25/62	2/12/63			
Castner, Jacob H	14	Pvt.	PA	133rd Infantry	8/13/62	5/26/63			
		Pvt.	PA	194th Infantry	7/22/64	9/6/64			
		Pvt.		22nd Cavalry	9/6/64				
Chamberlain, Philip	50	Pvt.	PA	208th Infantry	9/7/64	6/1/65			
Chamberlain, Fernando C	16	Pvt.	PA	194th Infantry	7/21/64	11/5/64			
Clites, Soloman (2)	36	Pvt.	PA	28th Infantry	9/21/64	7/18/65			
Clites, Levi	18	Pvt.	PA	McKeage's Mil.	7/2/63	8/8/63			
		Pvt.	PA	211th Infantry	9/9/64	Jun-65			
Cooper, Barton A	42	Pvt.	PA	107th Infantry	9/21/64	6/7/65			
Cooper, David M	15	Pvt.	PA	22nd Cavalry	7/11/63	6/15/65			
Cooper, Joshua H	20	Pvt.	PA	133rd Infantry	8/29/62		KIA	1862	Fredricksburg
Corle, Martin	51	Pvt.	PA	55th Infantry	11/4/62	2/16/63			
Corle, Alexander B	18	Pvt.	PA	55th Infantry	2/19/64	5/22/65	POW	1864	Drewry's Bluff
Croft, George	57	Pvt.	PA	22nd Cavalry	7/11/63	2/5/64			
Croft, Philip P	23	Pvt.	PA	110th Infantry	10/24/61		KIA	1862	1st Kernstown
Defibaugh, Daniel Sr	61	Sgt.	OH	43rd Infantry	7/12/61	7/13/65			
Defibaugh, Anderson	33		OH	43rd Infantry	12/7/61	7/13/65			
Defibaugh, Daniel Jr	18	Pvt.	OH	43rd Infantry	2/10/64	7/13/65			
Defibaugh, Harrison	21	Pvt.	PA	13th Infantry	4/25/61	8/6/61			
		Pvt.	PA	McKeage's Mil.	7/2/63	8/8/63			
Dodson, William	55	Pvt.	PA	13th Cavalry	2/29/64	7/17/65			
Dodson, Andrew J	17	Pvt.	PA	125th Infantry	8/15/62	5/18/63			
Dodson, Samuel B	21	Pvt.	PA	19th Cavalry	9/30/63	5/14/66			

(Above) Louisa Metzger Holler pictured later in life. (Right) John M Holler on the left with his father George W Holler in the only known photograph of a Bedford County father and son pictured together in their uniforms during the Civil War.

George W Holler was born in the Dry Ridge area of Juniata Township in February 1826. He married Louisa Metzger in September 1845. Their first son, John M Holler, was born the following year. George left Dry Ridge with his 15-year-old son John to volunteer in the 138th Pennsylvania Infantry in August 1862, while Louisa and their other 6 children tended to the family farm. Over the next 3 years, George and John served side by side in heavy combat during some of the most consequential battles of the Civil War. George was first wounded during the battle of the Wilderness on May 6th, 1864. He recovered from his wounds and rejoined his son in the 55th. George received a second wound during the battle of Opequon on September 19th, 1864. Both father and son survived the war and returned to their Dry Ridge farm in June 1865. John married in 1868 and was a father to 12 children. George lived to be 75 and passed in 1901. Louisa passed two years later at age 70. John passed when he was 74 in 1920. All are buried at the Trinity United Church of Christ Cemetery in Juniata Township.

Name	Muster Age	Rank	State	Regiment	Muster In	Muster Out	Casualty	Cas. Year	Battle
Dull, Valentine	45	Pvt.	PA	138th Infantry	8/29/62	6/23/65			
Dull, John	19	Pvt.	PA	184th Infantry	5/12/64		POW-Died	1864	Boydton Plank Rd.
Dull, Lewis	19	Pvt.	PA	55th Infantry	2/19/64		Wounded	1865	White Oak Rd.
		Pvt.	PA	McKeage's Mil.	7/2/63	8/8/63			
Edwards, Josiah V	38	Pvt.	PA	55th Infantry	3/2/64	3/6/65			
Edwards, Daniel L	15	Pvt.	PA	55th Infantry	2/19/64	8/30/65			
Engle, Henry	40	Pvt.	PA	14th Infantry	4/24/61	8/7/61			
		Pvt.	PA	121st Infantry	8/4/62		POW-Died	1865	Peebles Farm
Engle, Barney	17	Pvt.	PA	12th Cavalry	1/8/62	7/20/65			
Filler, Joseph	49	Lt. Col	PA	55th Infantry	11/5/61	11/13/64	Wounded	1864	Drewry's Bluff
		Lt. Col	PA	55th Infantry			POW	1864	Petersburg
Filler, William T	19	Pvt.	PA	13th Infantry	4/25/61	8/6/61			
		Pvt.	PA	138th Infantry	8/29/62	6/23/65			
Finnegan, Daniel	43	Pvt.	PA	55th Infantry	9/4/61		Died	1864	
Finnegan, Henry	19	Pvt.	PA	55th Infantry	2/14/64	10/21/64			
Fleegle, Jacob G	41	Pvt.	PA	76th Infantry	10/9/61		Died	1862	
Fleegle, George W	21	Pvt.	PA	138th Infantry	6/10/63	6/23/65	Wounded	1864	Cold Harbor
Fleegle, Simon S	16	Pvt.	PA	76th Infantry	10/9/61	5/18/63			

Name	Muster Age	Rank	State	Regiment	Muster In	Muster Out	Casualty	Cas. Year	Battle
Fletcher, Jacob	42	Pvt.	PA	22nd Cavalry	7/11/63	2/5/64			
		Pvt.	PA	208th Infantry	9/7/64	6/1/65			
Fletcher, Henry C	22	Pvt.	PA	22nd Cavalry	2/27/64		Died	1864	
Fletcher, John L	42	Sgt.	PA	133rd Infantry	8/13/62	5/26/63			
Fletcher, Winfield S	15	Pvt.	PA	22nd Cavalry	7/11/63	6/15/65			
Foor, Abraham T	40	Pvt.	PA	107th Infantry	1/9/62	11/21/62	Wounded		
Foor, Brazella	20	Pvt.	PA	McKeage's Mil.	7/2/63	8/8/63			
		Pvt.	PA	208th Infantry	9/7/64	6/1/65			
Foor, Samuel H	16	Pvt.	OH	25th Infantry	9/5/64	7/13/65			
Foor, George W	43	Pvt.	PA	107th Infantry	2/10/62		KIA	1862	Antietam
Foor, Daniel V	19	Pvt.	PA	76th Infantry	11/21/61		KIA	1863	Ft. Wagner, SC
Gates, Nathaniel	70	Chaplain	PA	13th Reserves	1862	1863			
Gates, Andrew G		Pvt.	IL	56th Infantry			Wounded		
Gates, Charles W			OH				Wounded		
Gates, George W	18	Pvt.	PA	13th Reserves	6/26/61	5/31/64	Wounded	1862	Seven Days Battle
		Pvt.	PA	13th Reserves			Wounded	1862	South Mountain
		Pvt.	PA	13th Reserves			Wounded	1864	Spotsylvania
		Pvt.	PA	190th Infantry	5/31/64	6/28/65	POW	1865	White Oak Rd.
Gates, Jacob C	20	Pvt.	PA	13th Reserves	6/26/61	9/29/62			
	22	Capt.	PA	13th Cavalry	12/9/63	7/14/65	POW	1864	Beefsteak Raid
Gates, Jeremiah E	17	Pvt.	PA	149th Infantry	9/29/63	5/17/65	Wounded	1864	Petersburg - Initial
Gates, Samuel K			OH				Wounded		
Gates, Theophilus R	31	Pvt.	PA	13th Infantry	4/25/61	8/6/61			
		Corp.	PA	55th Infantry	2/3/62	8/30/65	Wounded	1864	Cold Harbor
Gates, William	38	Pvt.	PA	192nd Infantry	2/14/65	8/24/65			
Gates, Thomas	15	Pvt.	PA	192nd Infantry	2/14/65	8/24/65			
Geller, John	34	Sgt.	PA	138th Infantry	8/29/62	4/6/65	Wounded	1864	Cedar Creek
Geller, George	16	Pvt.	PA	138th Infantry	6/15/63	6/23/65	Wounded	1864	Monocacy, MD
Gordon, Jeremiah	50	Pvt.	PA	55th Infantry	10/12/61	8/2/62			
Gordon, James S	24	Sgt.	PA	171st Infantry	11/2/62	8/8/63			
Gracey, William C	42	1st Lt.	PA	107th Infantry	2/10/62	3/4/65	Wounded	1863	Gettysburg
Gracey, Alfred	18	Sgt.	PA	107th Infantry	1/9/62	4/1/65	POW	1863	Gettysburg
Gracey, George E	16	Pvt.	PA	107th Infantry	2/10/62	7/13/65	POW		
Gracey, James A	21	Pvt.	PA	107th Infantry	4/4/62	4/29/65	Wound.-POW	1862	2nd Bull Run
		Pvt.	PA	107th Infantry			POW	1863	Gettysburg
Grimes, John C	49	Corp.	PA	14th Infantry	5/2/61	8/7/61			
		Pvt.	PA	84th Infantry	1861				
Grimes, Henry W	20	Pvt.	PA	84th Infantry	10/21/61		Wounded	1862	1st Kernstown
		Pvt.	PA	84th Infantry			Wound.-Died	1864	Wilderness
Grimes, Jacob R	18	Pvt.	PA	84th Infantry	10/12/61	11/6/62			
		Pvt.	PA	205th Infantry	8/27/64	6/2/65			

Kinsey, Dewalt

Kinsey, Peter Sr

Kinsey, Benjamin F

Peter Kinsey Sr was born in February 1811 in Napier Township. He married Jane Frazier in 1829. Jane was the granddaughter of John Frazier. John Frazier was a guide and interpreter for George Washington during the French and Indian War. Peter Sr and his 4 sons - Benjamin born in 1830, John in 1839, Dewalt in 1840 and Peter Jr in 1842 enlisted in the Union Army. Fifty-year-old Peter Sr and his youngest son Peter Jr left their home in Napier Township to volunteer in the 55th Pennsylvania Infantry in November 1861. John volunteered in the 138th Pennsylvania Infantry in 1862 and transferred to the 2nd Maryland PHB Infantry the following year. Brothers Benjamin and Dewalt mustered together in the 206th Pennsylvania Infantry in 1864. There are no records of any casualties among the 5 members of the Kinsey family, despite being in heavy combat. Peter Sr and his wife Jane died within a month of each other in 1888. Benjamin passed in 1912, Dewalt in 1914, John in 1916, and Peter in 1922.

Kinsey, Peter JR

Kinsey, John B

Alexander Messersmith is pictured on the left circa 1865 and son George is photographed later in life on the right. Alexander was born near Clearville in February 1823. He married Elizabeth Feight in February 1845. Their oldest son George was 17 years old when he left the family farm in Monroe Township to volunteer in McKeage's Militia in July 1863 during the Gettysburg campaign. After the Militia disbanded, he enlisted in the 22nd Pennsylvania Cavalry in February 1864. Alexander was 41 years old when he mustered in the 208th Pennsylvania Infantry in September 1864. Less than two months later his wife Elizabeth died at age 46, leaving behind a teenage daughter and son to tend to the family farm. Both Alexander and George survived the war and returned home in June 1865. Alexander remarried and was a father to 8 more children. George married in 1865 and was a father to 8 children who survived to adulthood. Alexander passed when he was 70 years old in 1893 and George lived to be 82 and passed in 1928.

Name	Muster Age	Rank	State	Regiment	Muster In	Muster Out	Casualty	Cas. Year	Battle
Hafer, Wilson	49	Pvt.	PA	133rd Infantry	8/14/62	1/5/63			
Hafer, Frank M	21	1st Lt.	PA	2nd Cavalry	11/5/61	6/17/65			
Hafer, George W		Pvt.	PA	2nd Cavalry	3/16/64	6/17/65			
Hafer, William H	21	Pvt.	PA	13th Infantry	4/25/61	8/6/61			
		Pvt.	PA	2nd Cavalry	11/4/61	4/10/65	Wounded	1863	
Hedding, Ephraim G	42	2nd Lt.	MD	3rd PHB Inf.	6/28/61	5/29/65			
Hedding, James E	19	Pvt.	PA	22nd Cavalry	7/16/63	2/5/64			
		Pvt.	MD	3rd PHB Inf.	2/11/65	5/29/65			
Hedding, Noah	20	Pvt.	MD	3rd PHB Inf.	10/12/61	5/29/65			
Hedding, Samuel E	18	Pvt.	MD	3rd PHB Inf.	2/27/64	5/29/65			
Holler, George W	36	Pvt.	PA	138th Infantry	8/29/62	6/13/65	Wounded	1864	Wilderness
							Wounded	1864	Opequon
Holler, John M	15	Pvt.	PA	138th Infantry	8/29/62	6/23/65			
Householder, James	44	Pvt.	PA	208th Infantry	9/7/64	6/1/65			
Householder, John	15	Pvt.	PA	208th Infantry	9/7/64	6/1/65			
Kinsey, Peter Sr	50	Pvt.	PA	55th Infantry	11/5/61	4/22/63			
Kinsey, Benjamin F	33	Pvt.	PA	206th Infantry	9/2/64	6/26/65			
Kinsey, Dewalt	24	Pvt.	PA	206th Infantry	9/2/64	6/26/65			
Kinsey, John B	22	Pvt.	PA	138th Infantry	8/29/62	4/5/63			
		Pvt.	MD	2nd PHB Inf.	4/5/63	5/29/65			
Kinsey, Peter JR	19	Corp.	PA	55th Infantry	11/5/61	8/30/65			
Lingerfelt, Abram	47	Pvt.	PA	55th Infantry	8/28/61	8/30/65			
Lingerfelt, Aaron	16	Pvt.	PA	55th Infantry	8/28/61	8/30/65			

William Nelson pictured later in life on left, his father John Nelson pictured in his Civil War uniform in the center and stepmother Susan Cypher Nelson is on the right. John Nelson was born in June 1821. He married Elizabeth Heffner in 1846. This first son, William, was born in April 1847. Elizabeth died in 1851. John remarried Susan Cypher in 1853 and she helped raise 6-year-old William and his 4-year-old brother John. William was 15 years old when he mustered as a corporal in the 18th Pennsylvania Cavalry in November 1862. His father John mustered as a 1st Lieutenant in the same cavalry unit a month later, leaving behind Susan and 7 children at home. John was shot in the shoulder and hip while skirmishing with Mosby's Raiders, an infamous Confederate guerilla unit near Chantilly, Virginia, on February 26th, 1863. The wound required amputation of his right leg at the hip. He mustered out of the Cavalry on May 14th, 1864 and bought a farm and mill in Cessna. William remained in the 18th PA Cavalry until mustering out after the war ended in October 1865. He returned home to help his father run the flour mill. John was later elected associate Judge in Bedford County and lived to be 83 years old. His obituary stated he suffered almost constantly from the wounds suffered during the war. Susan passed two years after her husband in 1906 at age 73. William moved to Southern California after retiring and passed in 1917 when he was 70.

Name	Muster Age	Rank	State	Regiment	Muster In	Muster Out	Casualty	Cas. Year	Battle
Linn, Hugh Sr	52	Pvt.	MD	3rd PHB Inf.	9/28/61	9/28/64	POW	1862	Harpers Ferry
Linn, Hugh Jr	23	Pvt.	PA	171st Infantry	11/2/62	8/8/63			
Linn, Riley	17	Pvt.	MD	3rd PHB Inf.	9/28/61	9/28/64	POW	1862	Harpers Ferry
Linn, William	14	Pvt.	MD	3rd PHB Inf.	9/28/61	9/28/64	POW	1862	Harpers Ferry
Madara, James W	49	Col	US						
Madara, David W	21	Capt.	PA	55th Infantry	9/20/61	4/20/62			
May, Jacob L	41	Pvt.	PA	99th Infantry	9/20/64		Died	1865	Appomattox
May, Samuel S	18	Pvt.	PA	208th Infantry	9/7/64	6/1/65			
McGregor, Robert	47	Pvt.	PA	55th Infantry	10/15/61	10/15/64			
McGregor, John	20	Pvt.	PA	55th Infantry	9/20/61		Wound.-Died	1864	Cold Harbor
Messersmith, Alexander	41	Pvt.	PA	208th Infantry	9/7/64	6/1/65			
Messersmith, George	17	Pvt.	PA	McKeage's Mil.	7/2/63	8/8/63			
		Pvt.	PA	22nd Cavalry	2/26/64	6/24/65			
Mills, Jacob H	36	Pvt.	PA	101st Infantry	1/23/62	5/24/62			
Mills, Franklin G	18	Musician	PA	101st Infantry	12/6/61	6/21/65	Wound.-POW	1864	Plymonth, NC
Mobley, Denton	42	Sgt.	PA	18th Cavalry	10/29/62				
Mobley, Ezekiel	18	Pvt.	PA	133rd Infantry	8/14/62	5/23/63			
		Pvt.	PA	205th Infantry	8/27/64	6/2/65			
Myers, Henry Sr	59	Pvt.	PA	194th Infantry	7/22/64	11/5/64			
Myers, Henry Jr									
Nelson, John	41	1st Lt.	PA	18th Cavalry	12/16/62	5/14/64	Wounded	1863	Chantilly
Nelson, William N	15	Corp.	PA	18th Cavalry	10/29/62	11/6/65			
Norton, James	43	Pvt.	PA	55th Infantry	10/12/61	1/17/63			
Norton, Franklin G	14	Musician	PA	101st Infantry	12/28/61		Died	1862	

Elizabeth Swartz Salkeld pictured with her son James W Salkeld on the left and her husband Bernard on the right. Bernard Salkeld was born in June 1815. He married Elizabeth Swartz in 1844. Their family was listed on the 1850 Colerain Township census with 3 sons, who later enlisted in the Union Army. Bernard volunteered in the 158th Pennsylvania Infantry in November 1862, while his wife Elizabeth remained home with 5 of their younger children. Son John was 15 years old when he volunteered in the 107th Pennsylvania Infantry in April 1862. James volunteered in the 13th Pennsylvania Cavalry in October 1863 and was captured during Confederate Beefsteak Raid at Coggin's Point Virginia on September 16th, 1864. There is no record of his release date. A third son, David is believed to have been killed during the Civil War according to Salkeld descendants. Unfortunately, no military records have been located on David. David may have died in 1865. Bernard, James, and John survived the war and returned home. James married in 1869 and was a father to 6 children. Elizabeth passed at age 83 in 1907. Bernard lived to be 93 years old and passed two years after his wife.

Name	Muster Age	Rank	State	Regiment	Muster In	Muster Out	Casualty	Cas. Year	Battle
Penrod, John B Sr	47	Pvt.	PA	8th Reserves	11/12/61	11/26/62			
Penrod, Henry C	18	Pvt.	PA	8th Reserves	6/11/61	10/27/62			
		Pvt.	PA	194th Infantry	7/22/64	9/6/64			
				6th Cavalry	10/27/62	4/29/64	Wounded		
				97th Infantry	9/6/64	7/17/65			
Penrod, John B Jr	19	Pvt.	PA	8th Reserves	6/11/61	10/27/62	POW	1862	Seven Days Battle
	23	Corp.	PA	194th Infantry	7/22/64	9/6/64			
		Pvt.	PA	6th Cavalry	10/27/62	4/29/64			
				97th Infantry	9/6/64	7/17/65			
Potter, John (1)	51	Pvt.	PA	22nd Cavalry	2/15/64	6/24/65			
Potter, David R	22	Corp.	IN	47th Infantry	10/16/61	12/6/65			
Potter, Martin L	18	Pvt.	PA	101st Infantry	11/1/61	6/25/65	POW	1864	Plymonth, NC
Riffle, Thomas	41	Pvt.	MO	35th Infantry	9/1/62		Died	1863	
		Pvt.	MO	1st SM Cavalry					
Riffle, Albert J	16	Pvt.	PA	55th Infantry	2/19/64	8/30/65			
Riffle, William	17	Pvt.	PA	138th Infantry	8/29/62	6/23/65	Wounded	1864	Opequan
Riley, William (2)	53	Pvt.	PA	55th Infantry	3/2/64		Wound.-Died	1864	Cold Harbor
Riley, Benjamin F	24	Corp.	MD	1st PHB Cav.	4/23/64	6/28/65			
Riley, George W (2)	19	Pvt.	PA	107th Infantry	1/9/62	7/13/65			
Riley, John (1)	28	Pvt.	MD	1st PHB Cav.	4/23/64				
		Corp.	PA	184th Infantry	6/9/64				
Risling, John	45	Pvt.	PA	55th Infantry	10/12/61	10/29/62			
Risling, John H	26	Pvt.	PA	55th Infantry	10/11/61	8/30/65	Wounded	1864	Bermuda Hundred

Joseph Shroyer pictured on the left during the Civil War. His father Jacob pictured on the right circa 1890. Jacob was born in January 1826 in Londonderry Township. Jacob mustered in the 19th Pennsylvania Cavalry in August 1863 while his wife Rachael and 5 younger children tended to their family farm in Londonderry Township. Joseph was 20 years old and working as a farm hand in Juniata Township when he volunteered in the 138th PA Infantry in August 1862. Joseph suffered a wound during the battle of Opequon on September 19th, 1864. GAR documents state Joseph left the hospital against surgeon's orders to rejoin his regiment. Both father and son survived the war and returned to Bedford County in June 1865. Joseph married after the war and was a father to 11 children. His father Jacob passed in 1899 at age 73. Jacob's obituary stated he was an honest, upright man and highly esteemed by all who knew him. Joseph died 9 years later at age 65. During his life, Joseph suffered from rheumatism caused by contracting typhoid fever while in the army.

Name	Muster Age	Rank	State	Regiment	Muster In	Muster Out	Casualty	Cas. Year	Battle
Ritchey, Ferdinand	44	Corp.	PA	55th Infantry	11/5/61	8/30/65			
Ritchey, Jacob K	16	Pvt.	PA	138th Infantry	9/19/64	6/23/65			
Ritchey, John C	18	Pvt.	PA	138th Infantry	8/29/62	6/16/65	Wound.-POW	1864	
Salkeld, Bernard	47	Pvt.	PA	158th Infantry	11/4/62	8/12/63			
Salkeld, David E*	17						KIA		
Salkeld, James W	18	Pvt.	PA	13th Cavalry	10/3/63	6/19/65	POW	1864	Coggin's Point
Salkeld, John F	15	Musician	PA	107th Infantry	1/20/62	7/13/65			
Shauf, John	52	Pvt.	PA	107th Infantry	2/10/62	5/8/62			
		Pvt.	PA	McKeage's Mil.	7/3/63	8/8/63			
Shauf, Cornelius	18	Pvt.	PA	8th Reserves	6/11/61		Died	1861	
Shauf, John J	17	Pvt.	MD	1st PHB Cav.	3/19/64	6/28/65			
Shroyer, Jacob	37	Pvt.	PA	19th Cavalry	8/11/63	6/1/65			
Shroyer, Joseph	19	Pvt.	PA	138th Infantry	8/29/62	6/23/65	Wounded	1864	Opequan
Snively, Andrew J	55	Pvt.	PA	22nd Cavalry	7/11/63	2/5/64			
Snively, John A		Pvt.	PA	22nd Cavalry	7/11/63	2/5/64			
Snyder, George W Sr.	50	Pvt.	PA	125th Infantry	8/13/62	4/2/63			
		Pvt.	PA	205th Infantry	8/17/64	7/2/65			
Snyder, George W Jr.	23	Pvt.	PA	55th Infantry	2/19/64	8/30/65			
		Pvt.	PA	125th Infantry	8/13/62	5/18/63			
Snyder, James	23	Musician	PA	76th Infantry	10/17/61	7/18/65			
Snyder, Christopher	35	Blacksmith	PA	13th Cavalry	3/21/64				
Snyder, William B	16	Pvt.	PA	125th Infantry	8/12/62	5/18/63			

Christopher Snyder is pictured on the left with his wife Catherine Bookhamer Snyder. Their oldest son William is on the right. Christopher was born in 1829. William was born in August 1846. William volunteered in the 125th Pennsylvania Infantry on August 12th, 1862, just prior to his 16th birthday. A month later, the 125th was in heavy combat near the Dunkard Church at the battle of Antietam. The nine-month enlistment of the 125th Infantry ended after the battle of Chancellorsville and William returned home in May 1863. Christopher Snyder enlisted in the 13th Pennsylvania Cavalry in March 1864, leaving behind Catherine and 6 younger children in their Middle Woodbury Township home. He survived the war and continued to work in his trade as a blacksmith. William married in 1870 and was a father to 11 children. Christopher passed when he was 74 in 1903. Catherine passed 4 years later at age 77. William lived until 1912 and passed at age 65.

Name	Muster Age	Rank	State	Regiment	Muster In	Muster Out	Casualty	Cas. Year	Battle
Sparks, Mahlon	42	1st Sgt.	OH	64th Infantry	10/2/61	5/21/65			
Sparks, George W	18	Pvt.	OH	163rd Infantry	5/12/64	9/19/64			
Sparks, Henry H	20	Pvt.	OH	64th Infantry	10/2/61	12/3/65			
Sparks, John M	18	Pvt.	OH	25th Infantry	6/8/61	3/17/62			
		Pvt.	OH	12th Light Art.	3/17/62	6/25/64			
Sparks, Wesley W	24	Sgt.	OH	163rd Infantry	5/12/64	9/10/64			
Stephens, Jacob	51	Pvt.	PA	11th Cavalry	12/24/63		POW-Died	1864	Reams Station
Stephens, John G	19	Pvt.	PA	184th Infantry	5/12/64		Wounded	1864	Cold Harbor
		Pvt.	PA	184th Infantry			KIA	1864	Boydton Plank Rd.
Swartz, William B	44	Pvt.	PA	107th Infantry	3/8/62	12/10/62	Wounded	1862	2nd Bull Run
Swartz, John W	17	Pvt.	PA	194th Infantry	7/22/64	9/6/64			
		Pvt.	PA	97th Infantry	9/6/64	7/17/65			
Trail, John	43	Pvt.	PA	171st Infantry	11/2/62	8/8/63			
Trail, George T	19	Corp.	PA	199th Infantry	9/29/64	6/28/65			
Trail, Nathan	17	Pvt.	MD	1st PHB Cav.	9/11/64	10/23/64			
Walter, Herman	37	Pvt.	PA	28th Infantry	9/21/64				
Walter, James A	18	Pvt.	MD	2nd PHB Inf.	10/17/63	5/19/65			
Watson, John R	45	Pvt.	PA	184th Infantry	5/12/64		POW	1864	Jerusalem Plank Rd.
Watson, Henry S	18	Pvt.	PA	184th Infantry	5/12/64		POW-Died	1865	Jerusalem Plank Rd.
Weaverling, Adam	42	2nd Lt.	PA	11th Infantry	9/30/61	1/20/63	Wounded	1862	2nd Bull Run
		1st Lt.	PA	McKeage's Mil.	7/2/63	8/8/63			
		Capt.	PA	208th Infantry	9/10/64	6/1/65			
Weaverling, Thomas H	16	Pvt.	PA	208th Infantry	9/7/64	6/1/65			
Yarnell, Jesse	43	Pvt.	PA	138th Infantry	8/29/62		Wound.-Died	1864	Monocacy, MD
Yarnell, John	17	Pvt.	PA	138th Infantry	8/29/62	6/23/65	Wounded	1864	Cedar Creek
Yarnell, William H	23	Pvt.	PA	21st Cavalry	2/28/65	7/8/65			

9. Families with Four or More Enlistments

During the Civil War, thousands of people in Bedford County experienced the emotional trepidation of saying goodbye to a loved one, not knowing if they would see each other again. Over 600 Bedford County families said goodbye to more than one family member. At least 81 families witnessed four or more members of their family leave home to enter the war. Forty-six of these 81 families ended up grieving the death of an immediate family member, including 16 families who suffered the loss of more than one of their loved ones. All the photographs in the following pages of "Families with Four or More Enlistments" are pictures of people who experienced the loss of sons, brothers, or husbands.

The striking photograph on the right is of Ann Croft McDonald with a grandson standing over her left shoulder. The years following the Civil War would have been a time of profound sadness for Ann because of the loss of her 3 sons.

Ann Croft married William McDonald in 1835. Four sons and a daughter were born over the following 10 years, while William worked as a stonemason in Woodbury Township. Three of their sons, James, William Jr and Samuel, volunteered in the Union Army in 1861. Oldest son James mustered in the 13th Pennsylvania Reserves on May 29th, leaving behind a wife and two young daughters. William Jr volunteered in the 101st Pennsylvania Infantry on November 1st while his wife remained at home with their young son. Samuel volunteered in the 110th Pennsylvania Infantry on December 19th, when he was 16 years old.

The following year, William Jr contacted Typhoid Fever during the Peninsula Campaign and died on July 3rd, 1862 in Portsmouth, VA. In May 1863, James McDonald sent his wife Mary Ann a letter from the Acquia Landing Hospital in Virginia stating he was suffering severe pain from a bacterial infection. This was the last correspondence Mary Ann would ever receive from her husband. Mary Ann repeatedly wrote unanswered letters to James. Officers in the 13th PA Cavalry responded to her letters addressed to them, but they were unaware of the fate of her husband. In August 1864, Mary Ann provided testimony for a pension request that she had not heard from her husband in over a year and was destitute. She stated James was always kind to her, provided money to support their family, and believed he must have died. A letter was finally received in November 1864 stating James had died a year and a half earlier on June 25th, 1863, at the Judiciary Square Hospital in Washington, DC.

A fourth son John mustered in the 208th Pennsylvania Infantry on September 8th, 1864, while his wife and two young children eagerly waited for his return to their South Woodbury Township home. Samuel mustered out of a second enlistment in August 1864 and volunteered a 3rd time as a substitute for another man in the 97th Pennsylvania Infantry in March 1865. The Confederates surrendered at Appomattox a month later. John returned home to his wife and children in June 1865. Samuel mustered out in August 1865 but returned to Bedford County stricken with Typhoid Fever. He died at home of the same disease that took his brother William on September 23rd, 1865.

Life became exceedingly difficult for the parents of these fallen soldiers after the war. Besides the emotional trauma of losing 3 sons, they struggled with health and money issues. William Sr and his sons had worked together as stonemasons prior to the war. He became afflicted with a spinal disease that left him crippled and unable to work after 1862. Affidavits stated Ann supported herself and her husband by washing clothes, husking corn, and picking up stones out of the neighboring fields while going through the difficult, years long process of gaining approval for a Civil War pension. William's condition worsened until he could no longer move around on crutches because of paralysis of his lower extremities. Ann was eventually forced to take him to an Alms House when she could no longer lift him up. William died in 1872 in poverty at the Alms House. Ann died in 1895 at when she was 80. William Sr and Ann are buried in the Pottery Cemetery in Woodbury. Additional family information is on the next page.

John McDonald is pictured above with wife Sarah Miller McDonald and their children George, baby Elsie and Clara Annie on the right. Sarah Miller McDonald was born in Everett in 1837. John was born the following year in Loysburg. They married in 1860 and daughter Clara Annie was born in 1861. John's parents and his wife must have felt a great deal of apprehension when he enlisted in the 208th Infantry in September 1864. Ann and William Sr had already lost one son, and a second son had been missing for over a year. To their great relief, John survived combat during the siege of Petersburg and returned home in June 1865. Tragically, three months later, the family suffered the loss of Samuel, another of John's brothers.

William and Sarah's daughter, Clara Annie wrote a memoir year later recalling the story told to her of her aunt Margaret running through the house crying when she realized her 16-year-old baby brother, Samuel had runoff in the middle of the night to join the Union army. In August 1865, Samuel returned home from the war stricken with Typhoid Fever and died a month later in the same house he left to join the army.

After his father passed in 1872, John and Sarah moved to Nebraska and eventually homesteaded in Colorado. John became a County Commissioner and used the masonry skills learned from his father to build a stone house in Fleming, Colorado that is still standing today. Sarah passed in 1907 and John in 1913. Both are buried at the Haxtun Cemetery in Phillips County, Co.

Margaret McDonald Shimer pictured on the near right was born in 1840. Her sister-in-law, Hannah Spotts McDonald, is on the far right. Margaret suffered the loss of three brothers, and Hannah lost her husband. Margaret was also married to a Civil War soldier. Her husband, William Harry Shimer, was a Sergeant in the 110th Pennsylvania Infantry. His regiment took part in heavy combat during the 2nd day of Gettysburg near the Wheatfield. William Harry survived the war and returned home to the great relief and joy of his wife. In the memoir written by Clara Annie McDonald, she recalls William Harry's return as "a bright and shinning time in those otherwise dark months whose depressing circumstances were felt even by the small children." Margaret's mother Ann Croft McDonald is listed as living with Margaret and William Harry in the 1880 Census. Margaret passed in 1912 at when she was 72. Hannah Spotts was born in Clinton County in 1840. She married Margaret's brother, William McDonald, in 1856 and their son Harrison was born in 1859. After William's death in 1862, she lived with her mother in Centre County. She remarried and was a mother to two more children. Hannah passed in 1917 at age 77.

Nancy Anna Schrader was born in St. Clair Township in 1817. She married Solomon Adams in 1846. They raised 10 children on their farm near Ryot. Four of their sons enlisted in the Union Army. Philip mustered in the 55th Pennsylvania Infantry in October 1861, John Q volunteered in McKeage's Militia during the Gettysburg Campaign in 1863. George joined his brother in the 55th Pennsylvania Infantry in February 1864, and William mustered in the 100th Pennsylvania Infantry in November 1864. William was the only brother who was married and his wife and young daughter waited for his return at their St. Clair Township home.

Philip suffered a severe wound on May 16th, 1864, at the battle of Drewry's Bluff. He died two months later at the Annapolis General Hospital on June 19th, 1864, and was buried in the Hospital Cemetery. Philip was 23 years old when he died. The other brothers survived the war and returned to Bedford County. Nancy survived her husband by 10 years and passed in 1906 when she was 88. Solomon and Nancy are buried in the Stone Church Cemetery in Fishertown.

Name	Muster Age	Rank	State	Regiment	Muster In	Muster Out	Casualty	Cas. Year	Battle
Adams, George W (2)	17	Pvt.	PA	55th Infantry	2/29/64				
Adams, Philip	18	Pvt.	PA	55th Infantry	10/11/61		POW-Died	1864	Drewry's Bluff
Adams, William (2)	27	Pvt.	PA	100th Infantry	11/14/64	7/24/65			
Adams, John Q	19	Pvt.	PA	McKeage's Mil.	7/2/63	8/8/63			
Anderson, James	21	Pvt.	PA	101st Infantry	12/28/61	2/5/63	POW-Died	1864	Plymonth, NC
Anderson, Henry	18	Pvt.	PA	McKeage's Mil.	7/2/63	8/8/63			
		Pvt.	PA	55th Infantry	2/19/64	8/30/65	POW	1864	Drewry's Bluff
Anderson, John (1)		Pvt.	IL	55th Infantry	10/17/61		Died	1862	
Anderson, William F	27	Pvt.	IL	102nd Infantry	9/2/62	3/16/63			
Barkley, David T	18	Pvt.	PA	79th Infantry	9/1/61	9/24/64			
Barkley, George W	21	Corp.	PA	138th Infantry	8/29/62	6/13/65			
Barkley, Jacob T	27	Pvt.	PA	91st Infantry	9/21/64	5/30/65	Wounded		
Barkley, Josiah T	34	Pvt.	PA	208th Infantry	9/7/64	6/1/65			
Barkley, Samuel	23	Corp.	PA	138th Infantry	8/29/62	6/23/65			
Barks, Alfred	30	Sgt.	USCT	41st Infantry	9/21/64				
Barks, John R	26	Corp.	USCT	32nd Infantry	2/12/64	8/22/65			
Barks, Moore	20	Pvt.	USCT	32nd Infantry	2/7/64	8/22/65			
Barks, William T	23	Pvt.	USCT	54th Infantry	3/21/63	8/8/65			
Barney, Jacob	29	Pvt.	MD	3rd PHB Inf.	2/22/64	5/29/65			
Barney, John H	22	Pvt.	MD	3rd PHB Inf.	11/19/61	5/29/65			
Barney, Dennis A	18	Pvt.	MD	3rd PHB Inf.	3/28/64	5/29/65			
Barney, Isaac	31	Pvt.	MD	3rd PHB Inf.	9/28/61	5/29/65			
Beltz, Adam	24	Pvt.	PA	138th Infantry	8/29/62		KIA	1864	Cold Harbor
Beltz, William W	26	Pvt.	MD	2nd PHB Inf.	9/10/61		POW-Died	1864	Ridgeville
Beltz, Oliver	21	Corp.	MD	2nd PHB Inf.	7/25/64	7/1/65			
Beltz, Samuel G	28	Pvt.	PA	88th Infantry	3/31/65	7/6/65			

Elizabeth Ferguson Anderson is seated beside her daughter, Susan. Elizabeth Ferguson married John Anderson Jr in Bedford County when she was 18. John died when he was 51 years old in 1852. Four of their 8 children - John, James, William, and Henry enlisted in the Union Army. Two sons died in the Civil War. John enlisted in the 55th Illinois Infantry in October 1861 and died on September 6th, 1862, at age 25 in Henderson, IL. He is buried at the Red-Blue Cemetery in Henderson. James left his home in St. Clair Township to volunteer in the 101st Pennsylvania Infantry in December 1861. He was captured on April 20th, 1864, with the entire 101st PA Infantry Regiment garrisoned at Plymouth, NC, and taken to Andersonville. James was transferred from Andersonville to a prison in Savannah where he died on December 1st, 1864. He was 22 years old.

William volunteered in the 102nd Illinois Infantry in September 1862. Six months later, William received a Discharge on a Surgeon's Certificate for being physically unable to continue serving in the Union Army. Henry volunteered in McKeage's Militia during the Gettysburg Campaign in 1863 and enlisted a second time in the 55th Pennsylvania Infantry in February 1864. Henry was captured during the battle of Drewry's Bluff on May 16th and taken to the Libby prison in Richmond. Henry was released after 3 months in captivity and survived the war. Their sister, Susan, married a Civil War Veteran from Ohio after the war and moved to Nebraska. Toward the end of her life, Elizabeth moved to Nebraska to live with Susan. Elizabeth died in 1890 on her 83rd birthday. Susan passed in 1926 when she was 86.

Name	Muster Age	Rank	State	Regiment	Muster In	Muster Out	Casualty	Cas. Year	Battle
Beltz, Abraham	32	Pvt.	PA	101st Infantry	12/2/61		POW-Died	1864	Plymonth, NC
Beltz, Frederick B	20	Pvt.	PA	135th Infantry	8/16/62	5/24/63			
Beltz, George W		Pvt.	PA	138th Infantry	8/29/62	6/23/65	Wounded	1864	Cedar Creek
Beltz, William Y	25	Pvt.	PA	135th Infantry	8/16/62	5/24/63			
Bennett, Enos	21	Pvt.	MD	1st PHB Cav.	9/7/64	6/28/65	POW		
Bennett, Abraham	33	Pvt.	PA	171st Infantry	11/10/62				
Bennett, David	27	Pvt.	PA	171st Infantry	11/10/62	8/8/63			
Bennett, Joseph	37	Pvt.	PA	91st Infantry	9/21/64	6/1/65			
Bennett, Jacob	23	Pvt.	PA	55th Infantry	10/12/61		POW-Died	1864	
Bowser, Moses	32	Pvt.	PA	49th Infantry	8/22/63	12/5/64	Wounded		
Bowser, Isaac B	34	Pvt.	PA	133rd Infantry	8/5/62	5/24/63			
		Sgt.	PA	187th Infantry	5/4/64				
Bowser, Daniel L	30	Pvt.	PA	55th Infantry	3/2/64	8/30/65	Injured	1864	Cold Harbor
Bowser, David	22	Pvt.	PA	55th Infantry	3/2/64	8/30/65	Wounded	1864	Drewry's Bluff
Bowser, George L	40	Pvt.	PA	133rd Infantry	8/5/62	5/24/63			
		Sgt.	PA	18th Cavalry	2/29/64	10/31/65			
Bowser, John J	25	Pvt.	PA	12th Cavalry	5/18/64	7/20/65			
Brown, Jacob D	24	Corp.	PA	101st Infantry	2/8/62	6/12/65	POW	1864	Plymouth, NC
Brown, Samuel D	19	Pvt.	PA	101st Infantry	11/1/61	6/13/65	POW	1864	Plymouth, NC
Brown, John W	25	Pvt.	PA	101st Infantry	2/8/62	6/20/65	POW	1864	Plymouth, NC
Brown, George D	20	Pvt.	PA	184th Infantry	5/12/64	6/9/65	Wounded	1864	Cold Harbor
		Pvt.	PA	184th Infantry			Wounded	1864	Petersburg

Anna Beltz Kipp is pictured on the left, and her brother William W Beltz is on the right. Anna and William grew up on a family farm in Harrison Township with seven other siblings. Two of the four brothers who enlisted in the Civil War never returned home. William left Juniata Township to muster in the 2nd Maryland PHB Infantry in September 1861, leaving behind a wife and 3 young children at their Dry Ridge home. He was captured at Ridgeville, VA on January 3rd, 1864 and taken to a prison in Richmond. William was transferred to Andersonville, where he died on April 10th, 1864 of Chronic Diarrhea. He was 28 years old. William is buried at the Andersonville National Cemetery. Adam Beltz volunteered in the 138th Pennsylvania Infantry in August 1862, while his wife and two small children waited for his return at their home in Bedford. Adam was killed during the battle of Cold Harbor on June 5th, 1864. He is buried in the Cold Harbor National Cemetery.

Their younger brother Oliver volunteered in the 2nd MD PHB Infantry, the same infantry regiment as his older brother William, in July 1864. A fourth brother, Samuel, enlisted in the 88th Pennsylvania in March 1865, the month before the Confederates surrendered at Appomattox. Oliver and Samuel returned to Bedford County in July 1865. Anna married Civil War veteran Lewis Kipp in 1869. Lewis served in the 2nd Maryland PHB Infantry with her brothers William and Oliver. Anna and Lewis were parents to 13 children. Anna passed in 1905 when she was 56 years old.

Name	Muster Age	Rank	State	Regiment	Muster In	Muster Out	Casualty	Cas. Year	Battle
Burns, Francis P	17	Pvt.	PA	104th Infantry	2/18/65	8/25/65			
Burns, Michael	24	Pvt.	PA	205th Infantry	8/27/64	6/2/65			
Burns, Sylvester	18	Pvt.	PA	22nd Cavalry	2/27/64	6/24/65			
Burns, William H (1)	24	Saddler	PA	22nd Cavalry	2/23/64	10/31/65			
Bussard, Andrew	21	Pvt.	PA	78th Infantry	3/7/65	7/18/65			
Bussard, Daniel S	25	Pvt.	PA	77th Infantry	10/9/61	8/1/63			
Bussard, Emanuel S	31	Pvt.	PA	208th Infantry	9/7/64	6/26/65			
Bussard, John S	34	Pvt.	PA	99th Infantry	2/24/65	5/31/65			
Bussard, Joseph S	23	Pvt.	PA	133rd Infantry	8/7/62	2/14/63			
		Corp.	PA	208th Infantry	9/7/64	6/26/65			
Bussard, Simon S	22	Pvt.	PA	99th Infantry	2/25/65	5/30/65			
Clark, Jameson	19	Pvt.	PA	McKeage's Mil.	7/2/63	8/8/63			
	19	Pvt.	PA	22nd Cavalry	2/26/64		KIA	1864	Charlestown, WV
Clark, John (1)	25	Pvt.	PA	208th Infantry	9/7/64	6/1/65			
Clark, Simon	22	Pvt.	PA	McKeage's Mil.	7/2/63	8/8/63			
		Pvt.	PA	208th Infantry	9/7/64	6/1/65			
Clark, Zachariah		Pvt.	PA	107th Infantry	10/21/64		Died	1865	

Enos Bennett is pictured in his later years. He grew up with 10 siblings on a farm in Southampton Township. Enos was 1 of 5 brothers who enlisted in the Union Army. Jacob was the first to volunteer. He mustered in the 55th Pennsylvania Infantry in October 1861. In 1864, Jacob was taken prisoner. No available record lists the battle or date of his capture. Jacob died of Scurvy in Andersonville on September 1st, 1864. He was 26 years old and was buried at the Andersonville National Cemetery.

Abraham and David enlisted in the 171st Pennsylvania Infantry in November 1862 and mustered out in August 1863. The same month Jacob died in September 1864, two other brothers enlisted in the Union Army. Joseph mustered in the 91st Pennsylvania Infantry and Enos volunteered in the 1st Maryland PHB Cavalry. Pension records show Enos was a prisoner of war during his enlistment, but there is no available information on the location or date. Enos and Joseph survived the war and returned home in June 1865. Enos married in 1873 and was a father of 9 children. He passed when he was 70 in 1914.

Name	Muster Age	Rank	State	Regiment	Muster In	Muster Out	Casualty	Cas. Year	Battle
Clevenger, George W	26	Pvt.	PA	97th Infantry	2/21/65	7/23/65			
Clevenger, Jacob A	15	Pvt.	PA	184th Infantry	9/3/64	6/2/65			
Clevenger, Adam	21	Pvt.	PA	126th Infantry	8/12/62	5/20/63			
			PA	20th Cavalry	1863	1864			
		Pvt.	PA	22nd Cavalry	2/18/64	6/24/65			
Clevenger, Franklin	18	1st Sgt.	MD	3rd Infantry	12/1/61	7/26/65	Wounded	1865	Ft. Steadman
Clevenger, Harrison	19	Pvt.	MD	3rd PHB Inf.	2/21/62	5/29/65			
Clingerman, Peter	21	Pvt.	PA	101st Infantry	12/6/61	6/25/65	POW	1864	Plymonth, NC
Clingerman, Harrison	21	Pvt.	PA	149th Infantry	8/26/63	6/24/65			
Clingerman, Jeremiah	26	Pvt.	PA	171st Infantry	11/10/62		Died	1863	
Clingerman, Joseph	34	Pvt.	PA	171st Infantry	11/5/62	8/8/63			
College, James (2)	28	Pvt.	PA	110th Infantry	10/24/61		Wound.-Died	1862	1st Kernstown
College, David	22	Pvt.	PA	110th Infantry	10/24/61				
		Pvt.	PA	9th Cavalry	2/4/64	7/18/65			
College, John W	16	Pvt.	PA	110th Infantry	10/24/61		Wound.-Died	1862	1st Kernstown
College, Simon	27	Pvt.	PA	22nd Cavalry	7/28/63	2/5/64			
		Pvt.	PA	208th Infantry	9/8/64	6/1/65			
Conner, Lewis	19	Pvt.	PA	133rd Infantry	8/15/62	5/26/63			
		Sgt.	PA	22nd Cavalry	2/26/64	6/24/65			
Conner, Adam	30	Pvt.	PA	208th Infantry	9/6/64	6/1/65			
Conner, Isaac	34	Corp.	PA	22nd Cavalry	2/26/64	6/24/65	Injured	1864	Fisher's Hill
Conner, David (2)	23	Pvt.	PA	133rd Infantry	8/15/62		Died	1863	Fredricksburg*
Conner, Jonas	19	Pvt.	PA	107th Infantry	4/29/64	7/23/65	Wounded	1864	Petersburg - Initial
Corle, Chauncey	24	Corp.	PA	55th Infantry	11/5/61		Died	1864	
Corle, Eli	22	Pvt.	PA	55th Infantry	11/5/61		Died	1861	
Corle, William C	21	Pvt.	PA	138th Infantry	8/29/62	6/23/65	Wounded	1864	Wilderness
Corle, Isaac	26	Pvt.	PA	172nd Infantry	10/28/62	8/1/63			

Isaac Bowser is pictured later in life on the left, and his brother Moses is on the right. Isaac and Moses grew up on a family farm near Blue Knob with seven other siblings. Six Bowser brothers enlisted in the Union Army. All 6 left behind wives and children when they enlisted in the Union Army. Isaac and George volunteered in the 133rd Pennsylvania Infantry in August 1862. Four months later, Bedford County soldiers in the 133rd PA Infantry took heavy casualties during the ill-fated charge on Marye's Heights at the battle of Fredericksburg. Moses Bowser mustered in the 49th Pennsylvania Infantry on August 22nd, 1863. Moses suffered a wound in 1864 and lost the ability to use his right arm. Daniel and David mustered in the 55th Pennsylvania Infantry in March 1864. David suffered a wound in the lung and shoulder in May 1864 during the battle of Drewry's Bluff. The following month, Daniel suffered a broken collar at Cold Harbor. John was the last brother to enlist. He mustered in the 12th Pennsylvania Cavalry in May 1864.

All six brothers returned home to their wives and children, but one brother died after his return. There is no information on the cause of John's death, and available records differ on whether he died in 1866 or in 1869. Moses passed in 1897 at age 66. Isaac passed when he was 77 years old in 1905.

Name	Muster Age	Rank	State	Regiment	Muster In	Muster Out	Casualty	Cas. Year	Battle
Cypher, George W	23	Corp.	PA	5th Infantry	4/19/61	7/21/61			
		Sgt.	PA	1st Cavalry	8/28/61	9/9/64			
Cypher, Henry S	18	Pvt.	PA	76th Infantry	10/9/61	11/28/64			
Cypher, Jacob F	19	Corp.	PA	76th Infantry	10/9/61	11/28/64			
Cypher, Thomas F	27	Pvt.	PA	125th Infantry	8/12/62	5/18/63			
		Pvt.	VT		3/10/65	3/29/67			
Darr, Charles S	21	Pvt.	PA	12th Infantry	12/2/61				
	24	Pvt.	PA	11th Cavalry	3/26/63		Died	1863	
Darr, Christian J		Pvt.	PA	171st Infantry	11/12/62		Died	1863	
Darr, George W	29	Pvt.	PA	100th Infantry	11/28/64	7/24/65			
Darr, Joseph A	31	Pvt.	PA	46th Infantry	7/1/63	8/19/63			
Darr, Michael S			PA	55th Infantry					
Davis, James P	20	Sgt., QM	PA	22nd Cavalry	7/11/63	6/24/65			
Davis, David	26	Pvt.	PA	49th Infantry	8/22/63	8/9/65			
Davis, Isaiah M	24	Pvt.	PA	8th Reserves	6/11/61		Died	1861	
Davis, John M (2)	20	Pvt.	PA	110th Infantry	10/24/61	10/24/64	Wounded	1864	1st Deep Bottom
Dell, James	18	Pvt.	PA	12th Cavalry	3/6/62	7/20/65			
Dell, John	23	Pvt.	PA	11th Cavalry	9/9/61	9/4/64			
Dell, Moses	21	Pvt.	PA	1st Light Art.	7/8/61	7/11/64	Wounded	1862	Antietam
Dell, Peter	20	Pvt.	PA	125th Infantry	8/13/62	5/18/63	Wounded		
Dell, Samuel	14	Pvt.	PA	76th Infantry	2/22/64	7/18/65			

Lewis Conner is pictured on the left during the Civil War. His brother Isaac is on the right. Five Conner brothers, who grew up together on a family farm in East Providence Township, enlisted in the Civil War. Lewis and David volunteered in the 133rd Pennsylvania Infantry in August 1862. Four months later, the 133rd PA charged Marye's Heights during the battle of Fredericksburg on December 13th. David died less than 4 weeks later, on January 7th. He was 23 years old. There is no available record of whether David was wounded during the charge. Lewis mustered out in May 1863 with the rest of the 133rd Regiment.

Four Conner brothers enlisted in the Union Army in 1864. Lewis and Isaac volunteered in the 22nd Cavalry in February. Isaac left behind a wife and two small children. Isaac was injured when his horse was shot out from underneath him during the battle of Fisher's Hill on September 21st, 1864. Jonas mustered in the 107th Pennsylvania Infantry on April 29th and suffered a wound two months later during the Initial Assault on Petersburg on June 18th. Adam mustered in the 208th Pennsylvania Infantry in September 1864. Lewis, Adam, Isaac, and Jonas survived the war and mustered out in June and July 1865. Isaac was a father to 13 more children born after the war. He passed at age 77 in 1907. Lewis married after the war and was a father of 8 children. He passed in 1915 at age 72.

Name	Muster Age	Rank	State	Regiment	Muster In	Muster Out	Casualty	Cas. Year	Battle
Fletcher, Samuel	18	Pvt.	OH	15th Infantry	9/7/61	9/18/64	POW	1862	Stones River, TN
Fletcher, David	33	Pvt.	PA	211th Infantry	9/5/64	6/2/65			
Fletcher, John (2)	21	Pvt.	OH	121st Infantry	9/11/62	6/8/65			
Fletcher, Henry	18	Pvt.	OH	15th Infantry	10/16/63	11/21/65			
Foor, Francis L	19	Pvt.	PA	101st Infantry	11/1/61		POW-Died	1864	Plymouth, NC
Foor, Jonathan S	22	Pvt.	PA	107th Infantry	1/9/62	3/20/65	Wounded	1862	South Mountain
		Pvt.	PA	107th Infantry			POW	1863	Gettysburg
Foor, Samuel S	23	Pvt.	PA	8th Reserves	6/11/61	5/26/64	Wounded	1862	7 Days Battle
		Pvt.	PA	8th Reserves			POW	1862	Fredricksburg
Foor, William H H	25	Pvt.	PA	107th Infantry	1/9/62	2/11/64	Wounded		
Foor, Abraham T	40	Pvt.	PA	107th Infantry	1/9/62	11/21/62	Wounded		
Foor, Lucius	33	Pvt.	PA	186th Infantry	3/24/64	8/15/65			
Foor, Martin T	37	Corp.	PA	138th Infantry	8/29/62		POW-Died	1864	Monocacy, MD
Foor, Richard T	31	Pvt.	PA	186th Infantry	3/24/64	8/15/65			
Gardner, Adam	22	Pvt.	PA	55th Infantry	2/27/64	5/26/65	Wounded	1864	
Gardner, George	28	Pvt.	PA	14th Cavalry	11/23/62	5/31/65			
Gardner, John	22	Pvt.	PA	55th Infantry	10/12/61	8/30/65	Wounded		
Gardner, Samuel	37	Corp.	PA	55th Infantry	10/12/61	8/30/65	Wounded		

Joseph Corle is pictured with his wife, Mary Crist Corle. Joseph and Mary married in 1830 and raised a large family of 11 children on their farm in Union (Pavia) Township. Four of their sons enlisted in the Civil War. Chauncy and Eli volunteered in the 55th Pennsylvania Infantry on November 5th, 1861. Eli died at Camp Curtin in Harrisburg less than 3 weeks later of Cholera. He was 22 years old. It was the second time in a little over a year that Chauncy had suffered the loss of a loved one. His wife Barbara passed in September 1860.

In 1862, two more brothers enlisted in the Union Army. William volunteered in the 138th Pennsylvania Infantry in August, and Isaac enlisted in the 172nd Pennsylvania Infantry in October. Isaac returned home in August 1863 at the end of his enlistment. William suffered a wound during the battle of the Wilderness on May 6th, 1864, but recovered.

Chauncey was wounded weeks later during the siege of Petersburg and taken to a hospital at Fort Schuyler in New York City. He died there on August 23rd, 1864. Chauncy was 27 years old. There is a picture of Chauncey in "Those who made the Ultimate Sacrifice" section of the book on page 130. William survived the war, married in 1865, and was a father to 11 children. Isaac married after the war and was a father to one son. Their father Joseph passed in 1896 at age 84 and their mother Mary lived to be 90 years old and passed in 1901.

Name	Muster Age	Rank	State	Regiment	Muster In	Muster Out	Casualty	Cas. Year	Battle
Garretson, Moses R	19	Pvt.	PA	55th Infantry	10/11/61		Died	1864	
Garretson, Edwin	24	Corp.	PA	21st Cavalry	7/8/63	2/20/64			
Garretson, George R	17	Musician	PA	101st Infantry	12/28/61	11/7/62			
		Pvt.	PA	55th Infantry	2/15/64	12/30/64			
Garretson, Thomas	39	Pvt.	PA	84th Infantry	10/24/61				
Gates, Theophilus R	31	Pvt.	PA	13th Infantry	4/25/61	8/6/61			
		Corp.	PA	55th Infantry	2/3/62	8/30/65	Wounded	1864	Cold Harbor
Gates, Jeremiah E	17	Pvt.	PA	149th Infantry	9/29/63	5/17/65	Wounded	1864	Petersburg - Initial
Gates, George W	18	Pvt.	PA	13th Reserves	6/26/61	5/31/64	Wounded	1862	7 Days Battle
		Pvt.	PA	13th Reserves			Wounded	1862	South Mountain
		Pvt.	PA	13th Reserves			Wounded	1864	Spotsylvania
		Pvt.	PA	190th Infantry	5/31/64	6/28/65	POW	1865	White Oak Road
Gates, Jacob C	20	Pvt.	PA	13th Reserves	6/26/61	9/29/62			
	22	Capt.	PA	13th Cavalry	12/9/63	7/14/65	POW	1864	Beefsteak Raid
Gates, Andrew G		Pvt.	IL	56th Infantry			Wounded		
Gates, Charles W			OH				Wounded		
Gates, Samuel K			OH				Wounded		

James P Davis grew up on a farm with 10 siblings in Hopewell Township. Three of his brothers - Isaiah, John and David also enlisted in the Union Army. Isaiah volunteered in the Hopewell Rifles company of the 8th Pennsylvania Reserves in June 1861. Isaiah died of disease at Camp Pierpont in Langley, Virginia, on November 28th, 1861. He was 24 years old.

John mustered in the 110th Pennsylvania Infantry in October 1861. Three years later, John suffered a wound during the siege of Petersburg at the battle of Deep Bottom on July 27th, 1864. He mustered out of the 110th PA Infantry at the end of his 3-year enlistment and returned home in October 1864. A third brother, David, mustered in the 49th Pennsylvania in August 1863.

James enlisted as a corporal in the 22nd Pennsylvania Cavalry a week after the Confederates retreated from Gettysburg in July 1863. He was promoted to Quarter Master Sergeant in January 1864. Later that year, the 22nd PA Cavalry took part in the largest Cavalry charge of the Civil War during the battle of Opequon in Winchester.

David and James survived the war and returned to Bedford County. James died in 1872 at the young age of 29 and is buried in the Everett Cemetery. There is no available information on the cause of death.

Name	Muster Age	Rank	State	Regiment	Muster In	Muster Out	Casualty	Cas. Year	Battle
Gibson, George G	32	Pvt.	PA	208th Infantry	9/7/64	6/1/65			
Gibson, Henry F	18	Pvt.	PA	133rd Infantry	8/15/62	5/26/63			
		Corp.	PA	208th Infantry	9/5/64	6/1/65	Wounded	1865	Ft. Steadman
Gibson, John (2)		Pvt.	PA	11th Infantry	3/17/64	1/1/65			
Gibson, William Y	34	Pvt.	PA	133rd Infantry	8/15/62	5/26/63			
Gordon, William	22	Pvt.	PA	55th Infantry	11/5/61		POW-Died	1864	Drewry's Bluff*
Gordon, George G	21	Pvt.	PA	158th Infantry	11/4/62	8/12/63			
Gordon, Isaac	18	Pvt.	PA	138th Infantry	8/29/62	6/23/65	POW	1864	Monocacy, MD
Gordon, Joseph	16	Pvt.	PA	McKeage's Mil.	7/2/63	8/8/63			
		Pvt.	PA	55th Infantry	2/19/64	7/22/65	Wounded	1865	White Oak Road
Hainsey, Valentine	14	Pvt.	PA	55th Infantry	8/28/61		Wound.-Died	1864	Cold Harbor
Hainsey, Adam R	15	Corp.	PA	76th Infantry	10/17/61	7/18/65			
Hainsey, Josiah	28	1st Sgt.	PA	1st Light Art.	5/28/61	5/28/64	Wounded	1862	7 Days Battle
		1st Sgt.	PA	1st Light Art.			Wounded	1862	Fredricksburg
		Sgt.	PA	45th Infantry	12/15/64	7/17/65			
Hainsey, Henry	28	Pvt.	OH	181st Infantry	10/7/64	7/14/65			
Hainsey, John	23	Pvt.	PA	76th Infantry	9/4/61	11/28/64			
Heffner, Samuel	28	Pvt.	PA	McKeage's Mil.	7/2/63	8/8/63			
		Pvt.	PA	101st Infantry	3/10/65	6/25/65			
Heffner, Daniel	19	Pvt.	PA	1st Light Art.	10/4/64	6/30/65			
Heffner, James	29	Pvt.	PA	3rd Heavy Art.	9/26/64	10/14/65			
Heffner, John	22	Pvt.	PA	125th Infantry	8/12/62	5/18/63			
Henry, George A	20	Capt.	IA	4th Infantry	8/15/61	7/24/65	Wounded	1862	Pea Ridge, AK
Henry, James W	17	Sgt. Maj	IA	15th Infantry	12/1/61	7/24/65	Wounded	1862	Corinth, MS
		Sgt. Maj	IA	15th Infantry			Wound.-POW	1864	Atlanta, GA
Henry, John B	20	Pvt.	IA	2nd Infantry	5/27/61	7/12/65			
Henry, Porter W	23	Pvt.	IA	1st Cavalry	7/31/61	11/1/62			

Hannah Miller Garretson pictured on the left was born in 1802. She married Aaron Garretson in 1820 and they raised a large family in St. Clair Township. Aaron passed in 1851 at age 56. Four of their sons enlisted in the Union Army. In October 1861, Moses volunteered in the 55th Pennsylvania Infantry and Thomas volunteered in the 84th Pennsylvania Infantry. George was 17 when he volunteered in the 101st Pennsylvania Infantry in December 1861. Edwin was the last brother to enlist in the war. He served in the 21st Cavalry from July 1863 until February 1864.

Moses suffered a sunstroke in South Carolina in 1863 and developed tuberculosis, also referred to as consumption in the Civil War era. He received treatment at two hospitals in Baltimore and Philadelphia. After his brother Edwin mustered out of the army, he moved west to Iowa and took Moses, who was still recovering from tuberculosis, with him. Moses died on October 15th, 1864, in Linn County, Iowa. He was 21 years old. The other Garretson brothers survived the war and migrated west to Iowa with their mother. Hanna lived to be 90 years old and was buried in the Dunkard Cemetery in Linn County, Iowa in 1892.

Name	Muster Age	Rank	State	Regiment	Muster In	Muster Out	Casualty	Cas. Year	Battle
Hite, David H	21	Pvt.	PA	148th Infantry	6/10/63	2/14/64			
Hite, George	40	Pvt.	PA	87th Infantry	6/18/64	6/29/65			
Hite, John	34	Pvt.	PA	205th Infantry	8/17/64		KIA	1865	Petersburg - Final
Hite, Samuel C	29	Pvt.	PA	55th Infantry	2/23/64		POW-Died	1864	
Hixon, Henry H	21	Pvt.	PA	11th Infantry	9/30/61	7/1/65	Wounded	1862	Fredricksburg
		Pvt.	PA	11th Infantry			Wound.-POW	1863	Gettysburg
		Pvt.	PA	11th Infantry			Wounded	1864	Petersburg - Initial
Hixon, Perry	21	Pvt.	PA	22nd Cavalry	2/26/64	6/24/65			
Hixson, Aquilla	25	Pvt.	PA	158th Infantry	11/1/62				
Hixson, Lewis B	28	Pvt.	PA	186th Infantry	3/3/64	8/15/65			
Holler, George W	36	Pvt.	PA	138th Infantry	8/29/62	6/13/65	Wounded	1864	Wilderness
		Pvt.	PA	138th Infantry			Wounded	1864	Opequan
Holler, Alexander	31	Pvt.	MD	2nd PHB Inf.	2/24/65	5/29/65			
Holler, James M	18	Pvt.	PA	55th Infantry	11/11/61		Died	1862	
Holler, Joseph M	34	Pvt.	PA	171st Infantry	11/1/62	8/1/63			
		Pvt.	MD	2nd PHB Inf.	3/21/65	5/29/65			
Holler, Samuel	32	Pvt.	PA	49th Infantry	9/25/63				
Holler, William H	36	Pvt.	PA	82nd Infantry	11/14/64	7/13/65			
Hook, Elias	22	Pvt.	PA	171st Infantry	11/10/62	8/8/63			
Hook, George	24	Pvt.	PA	171st Infantry	11/10/62	8/8/63			
			PA	138th Infantry					
Hook, James	27	Pvt.	PA	171st Infantry	11/10/62	8/8/63			
		Pvt.	PA	79th Infantry	9/21/64	7/12/65			
Hook, William	23	Pvt.	PA	171st Infantry	11/2/62		Died	1862	

Rachel and Uriah Gordon pictured above were married in 1834. They raised a large family on their farm in Union (Pavia) Township. Four of their sons enlisted in the Civil War. Oldest son William volunteered in the 55th Pennsylvania Infantry in November 1861. Isaac volunteered in the 138th Pennsylvania Infantry in August 1862. George joined the 158th Pennsylvania Infantry on November 4th, 1862. Their youngest son Joseph volunteered in McKeage's Militia during the Gettysburg Campaign when he was 16.

Joseph joined his brother William in the 55th Pennsylvania Infantry in February 1864. William was taken prisoner and died in a Richmond prison on May 22nd, 1864. There is no available record of the battle, but he most likely suffered a wound and was captured on May 16th during the battle of Drewry's Bluff, when over 20 other Bedford County soldiers in the 55th PA Infantry were captured. Isaac was taken prisoner during the battle of Monocacy on July 9th, 1864, and held at a POW camp in Danville, VA until February 21st, 1865. Joseph suffered a wound at the battle of White Oak Road on March 30th, 1865. Joseph, Isaac, and George survived the war and returned home to Pavia. Uriah passed when he was 64 in 1874 and Rachel also passed at age 64 in 1879.

Name	Muster Age	Rank	State	Regiment	Muster In	Muster Out	Casualty	Cas. Year	Battle
Horton, Zophar P	19	Pvt.	PA	8th Reserves	6/11/61	5/15/64			
		Pvt.	PA	191st Infantry	6/9/64	6/29/65	Injury		Petersburg
Horton, Jonathan	22	Corp.	PA	77th Infantry	10/9/61		Died	1864	
Horton, Milton	22	Corp.	PA	77th Infantry	10/9/61	9/2/64			
Horton, Reuben H	20	Pvt.	PA	77th Infantry	2/29/64		Wound.-Died	1864	Chattahoochee, GA
Johnston, John W	19	Pvt.	PA	133rd Infantry	8/31/62	5/26/63	Wounded	1862	Fredricksburg
Johnson, John J	44	Pvt.	PA	91st Infantry	9/21/64	6/1/65			
Johnson, Joshua	30	Pvt.	PA	149th Infantry	8/26/63				
Johnston, Charles W	18	Pvt.	PA	184th Infantry	5/12/64	7/14/65	Wounded	1864	Cold Harbor
King, Samuel T	15	Pvt.	PA	55th Infantry	2/19/64	8/30/65	Wounded	1864	Drewry's Bluff
King, Erastus	33	Pvt.	PA	148th Infantry	5/10/63	6/1/65	Wounded	1864	Po River
King, Hezekiah	23	Pvt.	PA	3rd Infantry	4/20/61	7/29/61			
		Pvt.	PA	16th Cavalry	2/16/65	8/11/65			
King, John T	18	Pvt.	PA	76th Infantry	10/9/61	2/18/63			
Kinsey, Peter JR	19	Corp.	PA	55th Infantry	11/5/61	8/30/65			
Kinsey, John B	22	Pvt.	PA	138th Infantry	8/29/62	4/5/63			
		Pvt.	MD	2nd PHB Inf.	4/5/63	5/29/65			
Kinsey, Benjamin F	33	Pvt.	PA	206th Infantry	9/2/64	6/26/65			
Kinsey, Dewalt	24	Pvt.	PA	206th Infantry	9/2/64	6/26/65			

Pictured above are Elizabeth Treaster Hainsey and her husband Adam F Hainsey. Adam is wearing his fallen son's medal. Adam and his first wife Ester married in 1832 and were the parents of 8 children. Their youngest child was born a month before Ester died in 1849. Adam married Elizabeth in 1852 and they raised the children in Greenfield and Union (Pavia) Township. Five of their sons enlisted in the Union Army.

Valentine was 14 years old when he volunteered in the 55th Pennsylvania Infantry in August 1861. Three years later, Valentine suffered a severe wound during the early morning frontal assault on Confederate lines at Cold Harbor on June 3rd, 1864. He died two weeks later on June 17th.

Josiah volunteered in the 1st Pennsylvania Light Artillery in May 1861. He suffered a left thigh and right foot wounds during the battle of Gaines Mill on June 27th, 1862. Josiah recovered from the wounds but was injured by a Confederate artillery shell that hit the carriage of the cannon he was sighting during the battle of Fredericksburg in December 1862. He later testified the impact of this artillery shell was the cause of his heart trouble the rest of his life. After mustering out of the 1st PA Artillery, Josiah reenlisted in the 45th Pennsylvania Infantry in December 1864. John mustered in the 76th Pennsylvania Infantry in September 1861. 15-year-old Adam joined his older brother John in the 76th Pennsylvania Infantry a month later. A 5th brother Henry enlisted in the 181st Ohio Infantry in October 1864. Josiah, John, Adam, and Henry survived the war. Adam lived to be 83 and passed in 1896. Elizabeth passed when she was 79 years old in 1909.

Name	Muster Age	Rank	State	Regiment	Muster In	Muster Out	Casualty	Cas. Year	Battle
Knipple, John A	21	Pvt.	PA	84th Infantry	10/24/61	2/28/63			
		Corp.	PA	3rd Heavy Art.	2/23/64	11/9/65			
Knipple, Andrew J	29	Pvt.	PA	101st Infantry	2/18/62	1/12/63	POW		
		Pvt.	PA	19th Cavalry	9/17/63	5/14/66			
Knipple, William H	18	Corp.	PA	101st Infantry	12/28/61	6/25/65	Wounded	1862	Fair Oaks
		Corp.	PA	101st Infantry			POW	1864	Plymonth, NC
Knipple, Frederick L	30	Pvt.	PA	13th Cavalry	2/26/64	7/14/65			
Knipple, George W	29	Pvt.	PA	13th Cavalry	2/26/64	7/14/65			
Koontz, James	23	Pvt.	IA	5th Cavalry	11/1/62		KIA	1864	Chehaw, AL
Koontz, John	33	Pvt.	IA	5th Cavalry	11/1/62	6/17/65			
Koontz, Peter	26	Pvt.	IA	5th Cavalry	11/1/62	6/17/65			
Koontz, William	38	Pvt.	IA	5th Cavalry	11/1/62	6/13/65			
Lamison, David M	22	Pvt.	PA	93rd Infantry	11/14/64	6/10/65			
Lamison, George W	23	Pvt.	PA	110th Infantry	10/24/61		Wound.-Died	1863	Gettysburg
Lamison, James H	16	Pvt.	PA	22nd Cavalry	7/11/63	10/31/65			
Lamison, Thomas	22	Pvt.	PA	110th Infantry	10/24/61		Died	1863	

Samuel McGee and his wife Margarette Lear are pictured above after the Civil War ended. Samuel and Margaret married in the early 1850s. Both grew up in large families in Middle Woodbury Township. Four McGee brothers enlisted in the Union Army.

William and James McGee volunteered in the 55th Pennsylvania Infantry in October 1861. James was captured on May 16th, 1864 during the battle of Drewry's Bluff and taken to Libby Prison in Richmond. James died on May 27th at the Libby prison hospital. He was 25 years old. There is no available record to confirm James was wounded prior to being captured. William was listed as missing in action on September 29th, 1864, at the battle of Chaffin's Farm. He is presumed to have been killed. William was 24 years old.

Samuel mustered in the 138th Pennsylvania Infantry in August 1862. Four days later, Margarette gave birth to their 5th child. A fourth brother, David mustered in the 3rd Pennsylvania Artillery in November 1862. David and Samuel survived the war and returned to Bedford County. Samuel and Margarette were the parents of 8 children who survived into adulthood. Samuel passed when he was 67 in 1893. Margarette died 8 years later, at age 68.

Name	Muster Age	Rank	State	Regiment	Muster In	Muster Out	Casualty	Cas. Year	Battle
Lashley, Daniel	23	Pvt.	PA	55th Infantry	10/12/61	10/26/64			
Lashley, Henry C	23	Pvt.	PA	55th Infantry	10/12/61	10/27/64	Wounded		
		Pvt.	PA	99th Infantry	3/16/65	5/15/65			
Lashley, John W	29	Pvt.	MD	1st PHB Cav.	4/23/64	6/28/65			
Lashley, Robert	17	Corp.	PA	12th Cavalry	9/3/64	6/1/65			
Lear, Daniel J	17	Pvt.	PA	55th Infantry	9/20/61	8/30/65			
Lear, Franklin	15	Pvt.	PA	77th Infantry	2/27/65	12/6/65			
Lear, John	26	Pvt.	PA	125th Infantry	8/13/62	9/17/62	KIA	1862	Antietam
Lear, Thomas	27	Pvt.	PA	5th Reserves	8/25/61	8/25/64			
Leonard, Jerome	25	Sgt.	PA	55th Infantry	10/12/61		Wound.-Died	1864	Petersburg - Initial
Leonard, Adam P	26	Pvt.	PA	107th Infantry	9/26/64	6/6/65			
Leonard, Henry N	17	Pvt.	PA	138th Infantry	8/29/62	6/23/65			
Leonard, John D	21	Pvt.	PA	138th Infantry	8/29/62	6/23/65	Wounded	1864	Wilderness
Manspeaker, Daniel*	27	Pvt.	PA	77th Infantry	10/9/61				
Manspeaker, David	24	Pvt.	PA	8th Reserves	6/11/61		KIA	1864	Spotsylvania
Manspeaker, Jacob	35	Pvt.	PA	56th Infantry	9/21/64	5/31/65			
Manspeaker, John (1)	34	Pvt.	PA	208th Infantry	9/7/64	6/1/65			
Manspeaker, George	18	Pvt.	PA	11th Infantry	9/30/61		KIA	1862	Fredricksburg
Manspeaker, James									
Manspeaker, John (2)	17	Pvt.	PA	11th Infantry	9/30/61	7/24/65	POW	1864	Globe Tavern
Manspeaker, Samuel	17	Pvt.	PA	17th Cavalry	9/2/64	6/16/65			

John and Mary Anne Lauderbaugh Mellott are pictured during happier times in 1859. Within 5 years after this photograph was taken, 3 of their sons would have died in the Civil War. They married in the early 1830s and raised a family of 6 sons and 4 daughters on a farm near Mattie. All six sons enlisted in the Union Army.

Frederick volunteered in the 12th Pennsylvania Reserves in August 1861. John and Cornelius volunteered in the 11th Pennsylvania Infantry the following month. Cornelius was killed on August 30th, 1862 during the 2nd Battle of Bull Run when he was 21. He is buried at the US Soldiers & Airman's National Cemetery in Washington, DC. Two weeks later, Frederick was killed at Turner's Gap during the battle of South Mountain on September 14th, 1862. He was buried at the Antietam National Cemetery. Frederick was 23 years old. John died of Remittent Fever on December 31st, 1864. He is buried at the Mount Pleasant Cemetery in Mattie. John was 21 years old when he died.

Henry mustered in the 158th Pennsylvania Infantry in November 1862. Henry completed his enlistment and returned home the following August. Thomas was 16 years old when he volunteered in the 11th PA Cavalry in February 1864. A sixth brother, Jacob mustered in the 208th Pennsylvania Infantry in September 1864. Thomas and Jacob survived the war and returned home during the summer of 1865. Their mother Mary Anne passed in 1885 when she was 69. Her husband John died 12 years later at age 87. Both are buried at the Mount Pleasant Cemetery in Mattie.

Name	Muster Age	Rank	State	Regiment	Muster In	Muster Out	Casualty	Cas. Year	Battle
May, John L	38	Pvt.	PA	67th Infantry	11/28/64	6/28/65	Wounded	1865	Sailor's Creek
May, Daniel H	31	Corp.	PA	82nd Infantry	11/14/64	7/13/65			
May, Marcus	19		PA	13th Infantry	1861	1861			
		Corp.	PA	138th Infantry	8/29/62	6/23/65	Wounded		
May, Hiram	17	Pvt.	PA	138th Infantry	6/12/63	6/23/65	Wounded	1864	Cold Harbor
		Pvt.	PA	138th Infantry			Wounded	1864	Opequan
May, Samuel M	22		MD	2nd PHB Inf.	8/31/61	1/1/62			
		Sgt.	PA	138th Infantry	8/29/62	3/30/63			
			MD	2nd PHB Inf.	3/30/63	2/27/65	POW	1864	Moorefield
McDonald, William, Jr	18	Pvt.	PA	101st Infantry	11/1/61		Died	1862	
McDonald, John	25	Pvt.	PA	208th Infantry	9/8/64	6/6/65			
McDonald, James	24	Pvt.	PA	13th Reserves	5/29/61	12/1/62	Died	1863	
McDonald, Samuel	16	Pvt.	PA	110th Infantry	12/19/61				
		Pvt.	PA	97th Infantry	3/3/65	8/28/65	Died	1865	
McGee, Samuel	37	Pvt.	PA	138th Infantry	8/26/62	6/23/65			
McGee, David	35	Pvt.	PA	3rd Art.	11/8/62	12/3/64			
McGee, James	25	Pvt.	PA	55th Infantry	10/15/61		POW-Died	1864	Drewry's Bluff
McGee, William	23	Pvt.	PA	55th Infantry	10/15/61		MIA	1864	Chaffin's Farm
Mellott, Cornelius	20	Pvt.	PA	11th Infantry	9/30/61		KIA	1862	2nd Bull Run
Mellott, Jacob L	19	Pvt.	PA	208th Infantry	9/7/64	6/1/65			
Mellott, Frederick	21	Pvt.	PA	12th Reserves	8/10/61		KIA	1862	South Mountain
Mellott, Henry T	30	Pvt.	PA	158th Infantry	11/4/62	8/12/63			
Mellott, John L	18	Pvt.	PA	11th Infantry	9/30/61		Died	1864	
Mellott, Thomas S	16	Pvt.	PA	11th Cavalry	2/27/64	8/13/65			

Pictured from the left are son Jeremiah Miller, mother Elizabeth Ryder Miller, father Martin G Miller and son Wiremen Miller. Elizabeth and Martin married in 1834. They were raising their children on a farm in Napier Township prior to moving their family to Iowa in the fall of 1851. Four of their sons enlisted in Iowa regiments during the war.

Jeremiah and George volunteered in the 11th Iowa Infantry in October 1861. George suffered a wound during the battle of Corinth in Mississippi on October 4th, 1862. He received a discharged because of the severity of his wounds and returned home to Iowa. William mustered in the 24th Iowa Infantry in August 1862. William was wounded during the Battle of Opequon in Winchester, VA on September 19th, 1864. He died from his wounds on February 13th, 1865 and is buried at the Winchester National Cemetery. William was 21 years old.

Wireman joined his brother Jeremiah in the 11th Iowa Infantry in August 1864. Jeremiah and Wireman survived the war and returned to Iowa in the summer of 1865. Wireman married the following year and raised 6 children on a family farm. Jeremiah married in 1867 and also raised his 4 children on a farm. Elizabeth passed when she was 64 in 1876. Martin lived to be 82 before passing in 1894. Jeremiah passed when he was 71 in 1913. Wiremen passed at age 72 in 1917.

Name	Muster Age	Rank	State	Regiment	Muster In	Muster Out	Casualty	Cas. Year	Battle
Miller, Joseph	28	Sgt.	PA	55th Infantry	1/1/64	8/28/65			
Miller, Armstrong	20	Pvt.	PA	21st Cavalry	6/16/63	2/20/64			
		Pvt.	PA	200th Infantry	8/26/64	5/30/65	Wounded	1865	Ft. Steadman
Miller, Charles	19	Pvt.	PA	200th Infantry	8/26/64	6/4/65			
Miller, James	23	Pvt.	OH	79th Infantry	8/13/62	2/23/63			
Miller, Thomas J (1)	26	Pvt.	PA	100th Infantry	11/14/64	7/24/65			
Miller, Abraham M	30	Pvt.	PA	82nd Infantry	11/14/64	7/13/65	Wounded	1865	Ft. Fisher
Miller, Elijah	27	Pvt.	PA	50th Infantry	2/24/65	5/10/65			
Miller, Ephraim B	22	Pvt.	PA	138th Infantry	8/29/62	6/23/65	Wounded	1864	Cedar Creek
Miller, Michael C	33	Pvt.	PA	149th Infantry	2/24/65	5/5/65			
Miller, Jeremiah J	19	1st Sgt.	IA	11th Infantry	10/3/61	7/15/65			
Miller, Wireman	17	Pvt.	IA	11th Infantry	8/25/64	6/2/65			
Miller, William O	18	Pvt.	IA	24th Infantry	8/29/62		Wound.-Died	1865	Opequan
Miller, George W (2)	21	Pvt.	IA	11th Infantry	10/3/61		Wounded	1862	Corinth, MS
Mock, George W	34	Pvt.	PA	125th Infantry	8/13/62		Died	1863	
Mock, Harrison	19	Pvt.	PA	133rd Infantry	8/13/62	5/26/63			
Mock, Josiah	38	Pvt.	PA	79th Infantry	2/23/65	5/29/65			
Mock, Mathias	24	Pvt.	PA	133rd Infantry	8/13/62	5/26/63			
		Pvt.	PA	184th Infantry	5/12/64		KIA	1864	Cold Harbor
Nevitt, James M	20	Pvt.	PA	133rd Infantry	8/13/62	5/26/63	Wounded		Fredricksburg*
		Pvt.	PA	3rd Heavy Art.	3/9/64	11/9/65			
Nevitt, Joseph H	26	Pvt.	IA	35th Infantry	9/4/62	8/10/65			
Nevitt, William E	33	Wagoner	OH	67th Infantry	11/4/61	12/7/65			
Nevitt, Thomas									

Wilson Nycum pictured on the right grew up on a farm in Monroe Township with 6 brothers. Four of the Nycum brothers enlisted in the Union Army.

Upton volunteered in the 2nd Pennsylvania Cavalry in September 1861. Upton suffered a wound in 1863. There is no available record of the battle or date of the wound. Upton later died at a hospital in Arlington, Virginia on December 12th, 1863. He was 24 years old. Upton is buried at the Bethal Frame Church Cemetery in Monroe Township. John and Bernard volunteered in the 138th Pennsylvania Infantry in August 1862. John suffered severe wounds on June 1st, 1864, during the battle of Cold Harbor. He died in Washington DC on June 28th, 1864 at age 29. John is buried at the Mt. Pleasant Cem. in Mattie.

Wilson enlisted in the 22nd PA Cavalry in July 1863. Bernard and Wilson survived the war and returned to Bedford County. Wilson moved out west in 1866, married in 1870. He was a father of 5 children. Wilson passed at age 58 in 1891 and is buried in Kansas.

Name	Muster Age	Rank	State	Regiment	Muster In	Muster Out	Casualty	Cas. Year	Battle
Nycum, Wilson	20	Pvt.	PA	22nd Cavalry	7/11/63	2/5/64			
Nycum, Bernard (2)	26	Pvt.	PA	138th Infantry	8/29/62	6/23/65			
Nycum, John (1)	28	Pvt.	PA	138th Infantry	8/29/62		Wound.-Died	1864	Cold Harbor
Nycum, Upton	22	Sgt.	PA	2nd Cavalry	11/4/61		Wound.-Died	1863	
Rice, Abraham	27	Sgt.	PA	101st Infantry	11/1/61	6/22/65	POW	1864	Plymonth, NC
Rice, Isaac	27	Corp.	PA	101st Infantry	11/1/61		POW-Died	1864	Plymonth, NC
Rice, Cornelius	28	Pvt.	PA	78th Infantry	2/28/65	9/11/65			
Rice, Jonathan	37	Pvt.	PA	99th Infantry	2/24/65	5/29/65			
Rice, Soloman	42	Pvt.	PA	149th Infantry	8/26/63	5/29/65	Wounded	1864	North Anna River
Ritchey, Adam S	30	Pvt.	PA	133rd Infantry	8/13/62	5/26/63			
		Pvt.	PA	194th Infantry	7/22/64	9/6/64			
		Pvt.	PA	97th Infantry	9/6/64	7/17/65			
Ritchey, Daniel S	23	Pvt.	PA	McKeage's Mil.	7/2/63	8/8/63	Wounded	1865	Ft. Steadman
		Pvt.	PA	208th Infantry	9/7/64	7/29/65			
Ritchey, David (2)	28	Pvt.	PA	208th Infantry	9/7/64		Died	1865	
Ritchey, Joseph	22	Pvt.	PA	8th Reserves	6/11/61	2/26/62			
		Pvt.	PA	186th Infantry	3/24/64	8/15/65			
Salkeld, Samuel W	23	Pvt.	PA	126th Infantry	8/12/62	5/20/63	Wounded	1863	Chancellorsville
		Pvt.	PA	49th Infantry	6/4/64	7/15/65			
Salkeld, Thomas L	31	Pvt.	PA	107th Infantry	1/20/62	2/11/64			
Salkeld, John N	21	Pvt.	PA	28th Infantry	8/17/61				
		Pvt.	PA	147th Infantry	4/1/62	4/3/65			
Salkeld, Bernard	47	Pvt.	PA	158th Infantry	11/4/62	8/12/63			
Salkeld, Jacob F	25	Pvt.	PA	158th Infantry	11/4/62	8/12/63			

Jane Layton Rice and Cornelius Rice are pictured on the left were married in 1864. Five Rice brothers enlisted in the Union Army. Twin brothers Abraham and Isaac volunteered in the 101st Pennsylvania Infantry in November 1861. Abraham and Isaac were captured along with the entire Union garrison at Plymouth, NC on April 20th, 1864, and taken to Andersonville. They were transferred from Andersonville to a POW camp in Charleston, SC where Isaac died of Chronic Diarrhea on September 21st, 1864. He was 30 years old. Isaac is buried at the Beaufort National Cemetery in an unmarked grave. Abraham was released from captivity on February 27th, 1865, and returned home.

Solomon Rice mustered in the 149th Pennsylvania Infantry in August 1863. He suffered a wound during the battle of North Anna River on May 23rd, 1864. Jonathan Rice enlisted in the 99th PA Infantry on February 24th, 1865. Four days later, Cornelius Rice and his wife's brother, Henry Layton, enlisted together in the 78th Pennsylvania Infantry and were sent to Nashville, TN. The following month, Henry's parents, William and Mary Layton, bought a neighboring property for Henry to farm when he returned from the war. Henry Layton died of disease in Nashville, TN in June 1865. Cornelius also became deathly ill in Nashville and was sent home to a hospital in Bloody Run (Everett) to recover. Cornelius and three brothers survived the war.

In 1866, the land originally bought for Henry was passed on to Jane and Cornelius. They raised a large family on this farm near Mattie. Cornelius lived to be 82 and passed in 1918. Jane lived to be 83 and was buried with her husband in 1926.

Name	Muster Age	Rank	State	Regiment	Muster In	Muster Out	Casualty	Cas. Year	Battle
Shull, Henry R	27	Pvt.	PA	55th Infantry	11/5/61	2/21/63			
		Pvt.	US	1st Light Art.	2/21/63				
Shull, Isaac		Pvt.	IN	84th Infantry	8/11/62		KIA	1863	Chickamauga, GA
Shull, Jacob	17	Pvt.	PA	Ind Light Art.	2/25/64	6/30/65			
Shull, Joseph	21	Corp.	IN	84th Infantry	7/26/62	6/14/65			
Slick, Josiah	15	Corp.	PA	55th Infantry	10/11/61	12/6/64	Wounded		
		2nd Lt.	USCT	107th Infantry	12/6/64				
Slick, Thomas W	23	Pvt.	PA	101st Infantry	12/28/61	6/25/65	POW	1864	Plymonth, NC
Slick, Abner W	29	Pvt.	PA	171st Infantry	11/2/62	11/10/62			
Slick, Samuel K	20	Pvt.	PA	101st Infantry	12/28/61		Wound.-Died	1862	
Smith, Philip E	23	Pvt.	PA	55th Infantry	10/12/61	11/23/64	Wounded		
Smith, Aquilla	21	Pvt.	PA	45th Infantry	9/21/64	6/7/65			
Smith, Gideon	25	Pvt.	PA	45th Infantry	6/1/64	5/30/65	Wounded		
Smith, James S	35	Pvt.	PA	45th Infantry	8/4/64	7/17/65	Wounded		
Snyder, Jonathan	30	1st Sgt.	PA	138th Infantry	8/29/62		Wounded	1864	Wilderness
		1st Sgt.	PA	138th Infantry			Wound.-Died	1864	Cedar Creek
Snyder, Leonard N	21	Pvt.	IL	26th Infantry	8/17/61	7/20/65			
Snyder, Joseph N		Pvt.	PA	4th Cavalry					
Snyder, John W (1)	19	Pvt.	MD	2nd PHB Inf.	9/11/61		POW-Died	1864	Jonesville
Sparks, Solomon C	41	Wagoner	IL	93rd Infantry	10/13/62	6/23/65			
Sparks, Abraham J		Capt.	IL	146th Infantry	9/17/64	7/8/65			
Sparks, John E	22	Musician	IL	57th Infantry	12/26/61	3/20/62			
		1st Lt.	IL	151st Infantry	2/21/65	1/24/66			
Sparks, David W	27	Pvt.	IL	12th Infantry	5/2/61	8/1/61			
		1st Lt.	IL	93rd Infantry	10/13/62	11/15/62	Injured		
Sparks, Joseph R			IL	151st Infantry					

Leonard Snyder is on the left. His brother Joseph Snyder is pictured with his family in the 1890s on the right. Five Snyder brothers and 3 sisters grew up on a farm in Monroe Township.

Four of the brothers enlisted in the Union Army. Leonard was the 1st to volunteer in the 26th Illinois Infantry in August 1861. John volunteered in the 2nd Maryland PHB Infantry the following month and was captured during the battle of Jonesville on January 3rd, 1864. He was taken to the Andersonville POW camp and was one of the first soldiers to perish there. John died of Chronic Diarrhea on April 24th, 1864. He is buried in the Andersonville National Cemetery. John was 22 years old when he died. Jonathan mustered in the 138th Pennsylvania Infantry in August 1862. He was wounded during the battle of the Wilderness on May 6th, 1864. Jonathan recovered from the wound and rejoined his regiment. He suffered severe wounds in both legs at the battle of Cedar Creek on October 19th, 1864. He died at the General Sheridan Hospital in Winchester three days later. Jonathan was 32 years old. He is buried at the Snider-Ash Cemetery in Monroe Township. A 4th brother, Joseph enlisted in the 4th PA Cavalry. Leonard and Joseph survived the war and moved west. Leonard passed in 1900 in Colorado when he was 60. Joseph married in 1873 in Illinois. He passed at age 75 in 1911 in Washington.

Name	Muster Age	Rank	State	Regiment	Muster In	Muster Out	Casualty	Cas. Year	Battle
Sparks, George W	18	Pvt.	OH	163rd Infantry	5/12/64	9/19/64			
Sparks, Henry H	20	Pvt.	OH	64th Infantry	10/2/61	12/3/65			
Sparks, John M	18	Pvt.	OH	25th Infantry	6/8/61	3/17/62			
		Pvt.	OH	12th Light Art.	3/17/62	6/25/64			
Sparks, Wesley W	24	Sgt.	OH	163rd Infantry	5/12/64	9/10/64			
Sponsler, George W	18	Pvt.	MD	3rd PHB Inf.	11/5/61	5/29/65	POW	1862	Harpers Ferry
Sponsler, John W	17	Pvt.	PA	22nd Cavalry	7/22/63	6/15/65			
Sponsler, Solomon	16	Pvt.	MD	3rd PHB Inf.	11/5/61	5/19/65			
Sponsler, William	16	Pvt.	PA	100th Infantry	3/4/65	7/4/65	Wounded	1865	Ft. Steadman
Sproat, William A	23	Pvt.	PA	6th Cavalry	3/2/65	6/16/65			
Sproat, George R			US	2nd Cavalry					
Sproat, James R	26	Pvt.	PA	88th Infantry	3/11/65	6/20/65			
Sproat, Joseph R	26	Corp.	PA	133rd Infantry	8/13/62	5/26/63			
Stailey, George E	21	Pvt.	PA	133rd Infantry	8/13/62	2/7/63	Wounded	1862	Fredricksburg
		Corp.	PA	208th Infantry	9/7/64	6/1/65			
Stailey, Henry C	21	Musician	PA	208th Infantry	9/7/64	6/1/65			
Stailey, Thomas	17		IN	147th Infantry	1863	1864			
			US	18th Infantry	1864	9/8/68			
Stailey, William A	26	Corp.	PA	22nd Cavalry	2/26/64	6/24/65			

Fannie Metzgar Steele and her husband Solomon Steele are pictured holding bibles in an undated drawing. They married in 1826 and raised 7 sons and 2 daughters on a farm in Yellow Creek. Five of their sons enlisted in Pennsylvania regiments.

David was 21 and Levi was 18 when they volunteered in the 133rd Pennsylvania Infantry on August 15th, 1862. Their 16-year-old brother Edward joined them 3 days later. Edward mustered out of the 133rd after only two months. It appears the army may have determined he was underage. It was a fortunate turn of events for Edward and his family. Two months later, Bedford County soldiers in the 133rd took part in a near suicidal charge on Marye's Heights during the battle of Fredericksburg on December 13th, 1862. David was killed during that charge. He was 21 years old and is likely buried in the unknown soldiers' section of the Fredericksburg National Cemetery on Marye's Heights. His brother Levi received a Discharge on a Surgeons Certificate on January 31st, 1863, possibly from a wound he suffered on December 13th. Levi later reenlisted in the 208th Pennsylvania Infantry in September 1864.

Thomas enlisted in the 149th Pennsylvania Infantry in August 1863. Thomas suffered a wound at Laurel Hill on May 8th, 1864 during the Spotsylvania Campaign, but recovered to rejoin his regiment. John enlisted in the 22nd Pennsylvania Cavalry in February 1864. Levi, Thomas, and John survived the war and returned home in June 1865. Solomon passed at age 74 in 1875 and Fannie lived another 20 years and passed when she was 81.

Name	Muster Age	Rank	State	Regiment	Muster In	Muster Out	Casualty	Cas. Year	Battle
Steele, John W	27	Pvt.	PA	22nd Cavalry	2/26/64	6/24/65			
Steele, David F	21	Pvt.	PA	133rd Infantry	8/15/62		KIA	1862	Fredricksburg
Steele, Edward	16	Pvt.	PA	133rd Infantry	8/18/62	10/9/62			
Steele, Levi H	18	Pvt.	PA	133rd Infantry	8/15/62	1/31/63			
		Pvt.	PA	208th Infantry	9/7/64	6/1/65			
Steele, Thomas	23	Pvt.	PA	149th Infantry	8/26/63	6/24/65	Wounded	1864	Spotsylvania
Summerville, Abner	35	Pvt.	PA	55th Infantry	10/26/62		POW-Died	1864	
Summerville, Robert	27	Pvt.	PA	208th Infantry	9/7/64	6/1/65			
Summerville, Charles	21	Pvt.	PA	138th Infantry	8/29/62		MIA	1864	Wilderness
Summerville, John B	35	Pvt.	PA	138th Infantry	2/24/64	6/23/65			
		Pvt.	PA	McKeage's Mil.	7/2/63	8/8/63			
Summerville, Sylvanus B	22	Pvt.	PA	55th Infantry	10/12/61		Wound.-Died	1864	Chaffin's Farm
Walker, Thomas G	31	Pvt.	PA	171st Infantry	11/2/62	8/8/63			
		Pvt.	PA	91st Infantry	9/21/64	5/30/65			
Walker, Isaac	19	Pvt.	PA	205th Infantry	8/26/64	6/2/65			
		Pvt.	PA	21st Cavalry					
Walker, Asahel	22	Pvt.	PA	84th Infantry	10/24/61	10/24/64			
Walker, Benjamin H	30	Corp.	PA	84th Infantry	10/24/61	1/13/65			
		Corp.	PA	57th Infantry	1/13/65	6/29/65			
Walker, Morris	25	Pvt.	PA	84th Infantry	10/24/61				
Walker, William A (1)	24	Pvt.	PA	21st Cavalry	7/8/63		Died	1864	

Thomas G Walker is pictured during the Civil War. He was born in Pleasantville in 1832 and was raised on a farm with 8 brothers and 2 sisters. Six Walker brothers enlisted in Union Army.

Asahel, Benjamin, and Morris volunteered in the 84th Pennsylvania Infantry in October 1861. William mustered in the 21st PA Cavalry in July 1863 with a wife and young son at home. Later that year, William contracted Typhoid Fever and was sent home to recover. He died shortly after returning to Bedford County on January 4th, 1864. After his brother died, Isaac enlisted in the 205th Pennsylvania Infantry in August 1864.

Thomas mustered in the 171st Pennsylvania Infantry in November 1862 while his wife Margaret and 4 small children waited at home for his return. Thomas reenlisted in the 91st Pennsylvania Infantry in September 1864. Margaret gave birth to their fifth child two months later. The 91st PA Infantry witnessed the Confederate surrender at Appomattox, and Thomas returned home in May 1865. Four of his brothers also survived and returned to Bedford County. Thomas and Margaret were parents to three more children born after the war. Margaret passed in 1910 at age 78. Thomas lived to be 85 years old and passed in 1917. They are buried at the Pleasantville Cemetery in Alum Bank.

Name	Muster Age	Rank	State	Regiment	Muster In	Muster Out	Casualty	Cas. Year	Battle
Walter, George I	27	Pvt.	PA	13th Cavalry	2/26/64	7/14/65			
Walter, Joseph H	33	Pvt.	PA	19th Cavalry	9/17/63	9/16/64	Died	1864	
Walter, Michael H	23	Pvt.	MD	1st Reg Cavalry	9/4/61	9/4/64	Wounded	1864	Deep Run
Walter, Samuel H	20	Pvt.	PA	19th Cavalry	9/17/63	5/14/66	Wounded		
Waters, David Y	31	Sgt.	IL	20th Infantry	6/13/61		Died	1865	
Waters, Isaac O	24	Corp.	IL	25th Infantry	8/9/61	8/30/64			
Waters, James B	20	Pvt.	IL	76th Infantry	11/17/63	6/13/65			
Waters, John F		Pvt.	IL	2nd Cavalry	4/28/64	11/22/65			
		Sgt.	IL	20th Infantry	6/13/61	11/1/61			
Waters, William	16	Pvt.	IL	25th Infantry	8/9/61	9/5/64			
Wentz, Isaac	21	Pvt.	PA	55th Infantry	11/5/61	8/30/65	Wounded	1864	Drewry's Bluff
Wentz, Adam	21	Pvt.	PA	55th Infantry	11/5/61	2/21/63			
		Pvt.	US	1st Light Art.	2/21/63				
Wentz, Henry	26	Pvt.	PA	55th Infantry	11/5/61	1865	Wounded	1864	Petersburg - Initial
Wentz, John	32	Pvt.	PA	55th Infantry	11/5/61	6/3/62	Died	1865	
Williams, Harrison P	26	Sgt.	PA	McKeage's Mil.	7/2/63	8/8/63			
		Pvt.	PA	149th Infantry	8/26/63	6/2/65	Wounded	1864	Petersburg - Initial
Williams, Alvah R	33	Pvt.	PA	208th Infantry	9/7/64	6/8/65	Wounded	1865	Petersburg - Final
Williams, David F	35	Pvt.	PA	171st Infantry	11/2/62	8/8/63			
Williams, Jonas	23	Musician	PA	McKeage's Mil.	7/2/63	8/8/63			
Williams, Wilson M	31	Sgt.	PA	208th Infantry	9/7/64	6/1/65			

John W Woy pictured on the right, grew up on a farm in East Providence Township with 4 brothers and 3 sisters. All four Woy brothers enlisted in the Union Army.

David Woy volunteered in the 12th Pennsylvania Cavalry in February 1862. David died on February 1st, 1863, at age 21. There is no record of the location or cause of death. David is buried at the Mount Zion Lutheran Church Cemetery near Breezewood.

James Woy mustered in the 126th Pennsylvania Infantry in August 1862. James was wounded during the charge on Marye's Heights at the battle of Fredericksburg, but recovered and rejoined his regiment. Four months later, James suffered a severe arm wound during the battle of Chancellorsville on May 3rd, 1863. He was still being treated in a hospital when the 126th Regiment mustered out at the end of the nine-month enlistment. James recovered from the wounds at Chancellorsville and reenlisted in the 208th PA Infantry with his brother Ezekiel in September 1864. Ezekiel was wounded during the Final Assault on Petersburg on April 2nd, 1865. John Woy mustered in the 22nd Pennsylvania Cavalry in February 1864. The 22nd took part in the largest Cavalry charge of the war in September 1864, when 8000 mounted soldiers thundered toward Confederate lines at the battle of Opequon.

Ezekiel, James, and John survived the war and returned to Bedford County in June 1865. John married in 1880 and was a father to 10 children. He lived to be 86 years old and passed in 1931.

Name	Muster Age	Rank	State	Regiment	Muster In	Muster Out	Casualty	Cas. Year	Battle
Wishart, James	25	2nd Lt.	PA	77th Infantry	10/9/61	4/24/62			
Wishart, Harvey S	24	1st Sgt.	PA	126th Infantry	8/12/62	5/20/63			
		Capt.	PA	208th Infantry	9/11/64	6/1/65			
Wishart, Samuel	28	2nd Lt.	PA	20th Cavalry	7/26/63	10/3/63			
Wishart, Henry	29	Capt.	PA	77th Infantry	10/9/61	2/2/63			
Witt, George	25	Pvt.	MD	2nd PHB Inf.	10/31/61	1/31/62			
		Pvt.	MD	5th Infantry	11/10/64	9/1/65			
Witt, Dennis	30	Pvt.	PA	61st Infantry	9/26/64	6/20/65			
Witt, Jacob	21	Pvt.	PA	138th Infantry	9/2/62	1/5/65			
Witt, John	22	Pvt.	PA	171st Infantry	11/1/62		Died	1863	
Witt, Jonathan	20	Pvt.	MD	2nd PHB Inf.	3/30/65	5/29/65			
Woy, John W	19	Pvt.	PA	22nd Cavalry	2/22/64	6/24/65			
Woy, David M	21	Pvt.	PA	12th Cavalry	2/14/62		Died	1863	
Woy, Ezekiel C	16	Pvt.	PA	208th Infantry	9/7/64	6/6/65	Wounded	1865	Petersburg - Final
Woy, James H	19	Pvt.	PA	126th Infantry	8/12/62	5/20/63	Wounded	1862	Fredricksburg
		Pvt.	PA	126th Infantry			Wounded	1863	Chancellorsville
		Pvt.	PA	208th Infantry	9/5/64	6/1/65			
Young, Aaron	21	Corp.	USCT	24th Infantry	2/16/65	10/1/65			
Young, Daniel D	29	Sgt.	USCT	24th Infantry	2/15/65	10/1/65			
Young, Jacob P	33	Pvt.	USCT	24th Infantry	2/28/65	10/1/65			
Young, Peter	28	Pvt.	USCT	127th Infantry	8/27/64	10/20/65			

10. Youngest Soldiers to Enlist

Samuel Statler pictured as a young man on the left and on the right as the Bedford Volunteer Fire Chief in 1885. Samuel was born on September 4th, 1848 in Berlin, Somerset County. His father passed when he was 4 years old. Samuel was adopted by an uncle and attended school in Shellsburg for 9 years. Samuel ran away when he was 13 to volunteer in the 55th Pennsylvania Infantry in October 1861. He was wounded during the Bermuda Hundred Campaign on May 19th, 1864, and suffered a second wound during the siege of Petersburg. His brother Nelson was severely wounded during the Final Assault on Petersburg on April 2nd, 1865 that required the amputation of his right arm. Nelson died in a hospital in Alexandria two months later. Samuel and Nelson were separated after their father's death, and there is no record of Nelson living in Bedford County during his brief life. After the war Samuel practiced dentistry for 35 years, was the Bedford Volunteer Fire Chief for 18 years, served on the Bedford Borough Council, was General Manager of the Bedford County Fair Association for 27 years and a prominent member of the Presbyterian Church. He married Margaret Schell in 1881 and was a father to 3 daughters. Samuel lived to be 87 years old. His 1836 obituary stated he was "actively interested in many things tending to improve Bedford."

When war broke out in 1861, President Lincoln mandated boys under the age of 18 needed the permission of their parents to enlist in the Union Army. The following year, when it became clear the Civil War would be far bloodier and last much longer than anyone initially expected, he prohibited anyone under the age of 18 from fighting in the war. This did little to stem the number of underage boys who left home to join the Union Army. In an era long before official identification was available, many boys simply lied about their age to recruiters. There are stories of boys writing the number 18 on the bottom of one of their shoes so they could "truthfully" state they were over 18. Other underage boys signed up for non-combat positions. But enlisting as a drummer or bugler did little to keep them out of harm's way. Many carried canteens, attended to wounded, and relayed orders on battlefields with bullets buzzing and artillery shells exploding all around them. Some picked up rifles and took part in the fighting. An estimated 200,000 boys were 16 years old and younger when they enlisted in the Union Army, including 300 boys 13 years old and younger.

At least 181 boys from Bedford County were 16 years old and younger when they volunteered, including two boys who were 13 and one boy who was 12 years old. Christian C Eicher was born on November 28th, 1848 and was growing up on a farm in Liberty Township when he enlisted as a Musician in the 12th Pennsylvania Reserves in June 1861. He had yet to reach his 16th birthday when his enlistment ended in June 1864, after 3 years of heavy combat. Christian is pictured in a veterans' group photograph in front of Eichelberger's store in 1890 on page 357. Both 13-year-old boys suffered wounds on a battlefield. Nicholas Bowser was born in July 1848 and grew up on a farm in Yellow Creek. He enlisted as a private in the 55th Pennsylvania Infantry in October 1861. His enlistment papers described Nicholas as being 5'4" tall, with fair complexion, blue eyes, and light hair. His parents would have later read his name on a casualty report published in the Bedford Inquirer on June 3rd, 1864. There is no available record of the battle or date, but it is likely he was wounded during the Drewry's Bluff / Bermuda Hundred Campaign. Nicholas survived almost 4 years of heavy combat and returned home after the war ended. The other 13-year-old soldier was Samuel Statler pictured above. Casualties were high among the 181 identified Bedford County soldiers who entered the war when they were 16 years old and younger. 19 lost their lives, 24 more suffered battlefield wounds, and an additional 15 were captured and held in hellish Confederate prisoner of war camps. The large numbers of boys who volunteered to enter the deadliest conflict in American history are among the most compelling stories of the Civil War. The following pages list the Bedford County soldiers, 16 years old and younger when they mustered in the Union Army.

184   Chapter 10

Jacob R Callahan pictured above circa 1900. He was born in Pleasantville on October 27th, 1844, and was the youngest of 11 children. Jacob volunteered in the 54th Pennsylvania Infantry on October 21st, 1861. He was described in his enlistment papers as being 5'6" with light complexion, grey eyes, and light hair. Jacob mustered out after the war ended on May 31st, 1865. He married in 1868 and was a father to 10 children who survived to adulthood. Jacob passed when he was 70 and was buried near his home in Allegheny County in 1915.

Thomas Drenning pictured with his wife Julianne Calhoun Drenning on the left and a sister on the right. Thomas was born on May 8th, 1847 in Bedford Township. He mustered in the 2nd Pennsylvania Cavalry in March 1864. Thomas took part in the siege of Petersburg and was enlisted during the surrender at Appomattox. He married Julianne Calhoun in 1866 and they were parents to 3 children. Julianne passed in 1904. Thomas remarried in 1908 and was a father to 4 more children. His last surviving child lived until 2005. Thomas passed in 1927 at age 79.

Name	Muster	Rank	State	Regiment	Muster In	Muster	Casualty	Cas.	Battle
Ake, Alexander	14	Musician	PA	McKeage's Mil.	7/2/63	8/8/63			
Akers, John T	16	Pvt.	PA	101st Infantry	11/1/61				
Amick, Josiah	16	Pvt.	PA	101st Infantry	11/1/61		Died	1862	
Arnold, Humphrey Y	15	Musician	PA	55th Infantry	12/30/61	12/30/64			
Barnett, Henry C	16	Pvt.	PA	9th Reserves	5/4/61	12/1/62			
Beaver, Simon J	16	Pvt.	PA	55th Infantry	2/19/64	6/15/64	POW	1864	Drewry's Bluff
Biddle, Andrew B	16	Pvt.	PA	138th Infantry	8/29/62	4/22/65	Wounded	1864	Wilderness
Bishop, David P	16	Pvt.	PA	82nd Infantry	11/22/64	5/31/65	Wounded		
Bollinger, Jacob	15		PA	76th Infantry	1861	1865			
Boose, Isaac	15	Pvt.	PA	102nd Infantry	8/18/61	5/16/65	Wounded	1864	Cold Harbor
Bowser, John B	15	Pvt.	PA	192nd Infantry	2/11/65	8/24/65			
Bowser, Nicholas	13	Pvt.	PA	55th Infantry	10/10/61	6/6/65	Wounded	1864	Drewry's Bluff*
Brant, Henry	16	Pvt.	PA	93rd Infantry	11/26/64	6/27/65	Wounded	1865	Ft. Fisher
Bretz, Calton L	16	Pvt.	PA	7th Infantry	7/4/63	8/11/63			
Brumbaugh, Levi	15	Pvt.	PA	8th Reserves	6/19/61	5/26/64	POW	1862	Seven Days Battle
Burket, Albert L	15	Pvt.	MD	2nd PHB Inf.	2/21/65	6/29/65			
Callahan, Jacob R	16	Pvt.	PA	54th Infantry	6/28/61	5/15/64			
Carmack, Thomas J	15	Corp.	OH	64th Infantry	11/10/62	5/22/65	POW		Franklin, TN
Carpenter, David B	15	Pvt.	PA	110th Infantry	11/17/61	10/27/62	POW		

Jacob Clevenger was born on September 28th, 1848. Jacob was one of 5 brothers who enlisted in the Union Army. Franklin suffered a wound at Ft. Stedman on March 25th, 1865 while enlisted in the 3rd MD Infantry. Harrison enlisted in 3rd MD PHB Infantry. George enlisted in the 97th PA Infantry and Adam enlisted in the 126th PA Infantry and later in the 20th & 22nd PA Cavalry units. Jacob volunteered in the 184th PA Infantry on September 3rd, 1864. After the war, Jacob worked as a tanner in Everett. Jacob married in 1868 and was a father of 2 children. He passed when he was 67 in 1915.

Daniel Edwards pictured with his family circa the middle 1880s. He was born on August 16th, 1848 and grew up in Union (Pavia) Township. Daniel volunteered in the 55th Pennsylvania Infantry in February 1864. Two weeks later, his father joined the same regiment. Both survived heavy combat in 1864. Photographs of Daniel and his father Josiah are in the "Fathers and Sons" chapter of this book. Daniel married Mary Virginia Jones in 1866 and raised a family on a farm in Iowa. Mary passed in 1920, Daniel in 1926, the older daughter in the picture died in 1947 and younger daughter in 1959.

Name	Muster Age	Rank	State	Regiment	Muster In	Muster Out	Casualty	Cas. Year	Battle
Carpenter, Samuel	15	Pvt.	PA	McKeage's Mil.	7/2/63	8/8/63			
		Pvt.	PA	13th Cavalry	8/27/63	7/14/65			
Carson, John (1)	16	Pvt.	PA	22nd Infantry	9/16/62	9/29/62			
Cartwright, Austin	15	Pvt.	PA	McKeage's Mil.	7/2/63	8/8/63			
Cartwright, Franklin J	16	Pvt.	PA	205th Infantry	8/26/64	6/5/65	Wounded	1865	Petersburg - Final
Castner, Jacob H	14	Pvt.	PA	133rd Infantry	8/13/62	5/26/63			
Chamberlain, Fernando C	16	Pvt.	PA	194th Infantry	7/21/64	11/5/64			
Clevenger, Jacob A	15	Pvt.	PA	184th Infantry	9/3/64	6/2/65			
Cobler, Francis C	16	Pvt.	PA	McKeage's Mil.	7/2/63	8/8/63			
College, John W	16	Pvt.	PA	110th Infantry	10/24/61		Wound.-Died	1862	1st Kernstown
Conrad, Jacob	15	Pvt.	PA	205th Infantry	9/10/64	6/2/65			
Conrad, Winfield S	16	Pvt.	PA	55th Infantry	2/18/64	8/30/65			
Cooper, David M	15	Pvt.	PA	22nd Cavalry	7/11/63	6/15/65			
Crawford, Joseph	15	Pvt.	WV	15th Infantry	2/29/64	1865	POW	1864	Cedar Creek

Americus Enfield pictured circa 1900. Americus was born on April 7th, 1847. He mustered in the Knap's Independent Light Artillery unit on June 13th, 1861. Knap's Light Artillery took part in heavy combat near the Cornfield on the Antietam battlefield in September 1862. The following year, Americas suffered a wound during the 2nd Day of Gettysburg at Culp's Hill. Americus enlisted as a Surgeon in the 22nd Pennsylvania Cavalry in February 1864. He became a well-known Doctor in Bedford County after the war. Americus married in 1870 and was a father to 6 children. He was 83 years old when he passed in 1931.

Henry Estep pictured during the Civil War. He was born on February 5th, 1845. Henry volunteered in the 49th Pennsylvania Infantry on August 30th, 1861. His enlistment papers described Henry as being 5'2½" with dark hair and eyes. The 49th PA regiment was on the battlefield at Antietam, Fredericksburg, Chancellorsville, Gettysburg, and the Overland Campaign. Henry mustered out in October 1864 and returned home. He married in 1874 and lived most of his life in the Six Mile Run area. Henry passed when he was 57 in 1902.

Name	Muster Age	Rank	State	Regiment	Muster In	Muster Out	Casualty	Cas. Year	Battle
Crissey, John C	16	Pvt.	PA	133rd Infantry	8/14/62	4/22/63	Wounded	1862	Fredricksburg
Croil, John T	16	Pvt.	PA	149th Infantry	8/25/63	5/3/65			
Cumpson, Benjamin	15	Pvt.	PA	99th Infantry	3/17/65	7/1/65			
Daugherty, James	16	Pvt.	PA	110th Infantry	10/29/61		KIA	1862	Fredricksburg
Davis, R DeCharmes	16	Corp.	USCT	32nd Infantry	2/12/64	9/2/65			
Dell, Samuel	14	Pvt.	PA	76th Infantry	2/22/64	7/18/65			
Dibert, Isaac	16	Musician	PA	142nd Infantry	8/25/62	5/29/65			
Dicken, James H	16		PA	2nd Cavalry	9/1/61		POW-Died		
Drenning, Thomas	16	Pvt.	PA	2nd Cavalry	3/11/64	6/17/65			
Edwards, Daniel L	15	Pvt.	PA	55th Infantry	2/19/64	8/30/65			
Eichelberger, Winfield S	15	Pvt.	PA	McKeage's Mil.	7/3/63	8/8/63			
Eicher, Christian C	12	Musician	PA	12th Reserves	6/15/61	6/11/64			
Enfield, Americus	14		PA	Knap's Light Art.	6/13/61	2/24/64	Wounded	1863	Gettysburg
Estep, Henry C	16	Pvt.	PA	49th Infantry	8/30/61	10/31/64			
Feight, Henry H	16	Pvt.	PA	138th Infantry	8/29/62	6/23/65			
Felix, John	16	Sgt.	PA	21st Cavalry	7/8/63	2/20/64			
Fickes, James M	16	Pvt.	PA	101st Infantry	2/18/62	11/8/62	Died	1862	
Filler, William B	16	Pvt.	PA	101st Infantry	11/1/61	7/11/62			
Fishel, George W	16	Pvt.	PA	110th Infantry	10/24/61	10/24/64			

Youngest Soldiers to Enlist

George Gracey is pictured outside the Altoona post office where he worked in 1883. George was 16 when left his East Providence Township home with his father William and two brothers to volunteer in the 107th Pennsylvania Infantry in February 1862. All four took part in desperate fighting on the 1st Day of Gettysburg. His brothers, Alfred and James, were taken prisoner and spent the next 20 months in Confederate POW camps. George and his father narrowly missed capture, but William suffered a wound in his neck. All four members of the family survived the war. George's first wife died in 1873. He remarried 3 years later and was a father to 3 sons. George passed in 1911 at age 65.

Joseph Gordon pictured later in life. Joseph was born on May 23rd, 1847 in Pavia and was the youngest of 4 brothers to enlist in the Civil War. Joseph left the family farm near Pavia to join his brother William in the 55th PA Infantry in February 1864. Three months later, William was wounded and taken prisoner during the Bermuda Hundred Campaign and died at a Confederate prison on May 22nd, 1864. Joseph suffered a wound at Gravely Run on March 30th, 1865. He survived the war and returned home. Joseph married in 1869 and was a father of 10 children. He named his first-born son, Grant Ulysses Gordon. Joseph lived to be 85 years old and passed 1932.

Name	Muster Age	Rank	State	Regiment	Muster In	Muster Out	Casualty	Cas. Year	Battle
Fleegle, Simon S	16	Pvt.	PA	76th Infantry	10/9/61	5/18/63			
Fletcher, Winfield S	15	Pvt.	PA	22nd Cavalry	7/11/63	6/15/65			
Fluke, John W	16	Pvt.	PA	22nd Cavalry	2/27/64		Died	1864	
Foor, Samuel H	16	Pvt.	OH	25th Infantry	9/5/64	7/13/65			
Gates, Thomas	15	Pvt.	PA	192nd Infantry	2/14/65	8/24/65			
Geller, George	16	Pvt.	PA	138th Infantry	6/15/63	6/23/65	Wounded	1864	Monocacy, MD
Gollipher, Espy	15	Musician	PA	55th Infantry	10/11/61	8/30/65			
Gordon, John F	16	Pvt.	MD	1st PHB Cav.	4/23/64	6/28/65			
Gordon, Joseph	16	Pvt.	PA	McKeage's Mil.	7/2/63	8/8/63			
		Pvt.	PA	55th Infantry	2/19/64	7/22/65	Wounded	1865	White Oak Road
Gracey, George E	16	Pvt.	PA	107th Infantry	2/10/62	7/13/65	POW		
Grass, Cephas	16	Pvt.	PA	McKeage's Mil.	7/2/63	8/8/63			
Grove, William H	16	Pvt.	IN	6th Infantry	8/28/62				
Hahn, Isaac	16	Pvt.	PA	55th Infantry	9/18/63	8/30/65			
Hainsey, Adam R	15	Corp.	PA	76th Infantry	10/17/61	7/18/65			
Hainsey, Valentine	14	Pvt.	PA	55th Infantry	8/28/61		Wound.-Died	1864	Cold Harbor
Harbaugh, Allen	15	Pvt.	PA	205th Infantry	9/1/64	6/2/65			
Harbaugh, Eli	16	Pvt.	PA	55th Infantry	11/5/61		Died	1862	
Harbaugh, Joseph	16	Pvt.	PA	79th Infantry	3/7/65	6/23/65			

Daniel Hetrick pictured later in life. Daniel was born in Morrison's Cove on July 5th, 1845. He volunteered in the 101st Pennsylvania Infantry in November 1861 when he was 16 years old. Daniel was captured along with the entire Union garrison at Plymouth, NC, on April 20th, 1864. He languished in terrible conditions in the Andersonville and Florence POW camps until his release on December 11th, 1864. Daniel returned home and was a well-known doctor in Pleasantville for 40 years. He married after the war and was a father to 3 sons. Daniel attended the Pennsylvania Memorial dedication at Andersonville in 1905, two years before he passed at age 61.

Oliver Hollingshead pictured during the Civil War. Oliver was born on March 19th, 1847, in Huntington County. He mustered in the 22nd Pennsylvania Cavalry in February 1864 when he was 16. Two weeks later, his father Samuel joined Oliver in the same Cavalry unit. Oliver and his father were enlisted together in the 22nd during in the largest Cavalry charge of the Civil War at the battle of Opequon at Winchester in September 1864. Both mustered out of the cavalry after the war ended. Oliver spent much of his life in Saxton after he returned home. He married in 1878 and was a father to 3 children. Oliver passed in 1916 when he was 69 years old.

Name	Muster Age	Rank	State	Regiment	Muster In	Muster Out	Casualty	Cas. Year	Battle
Hare, Henry	16	Pvt.	PA	210th Infantry	9/14/64	7/5/65	Wounded	1865	White Oak Rd.
Harlow, Charles E	16	Pvt.	PA	2nd Cavalry	12/17/63	6/28/65			
Heckman, William	15	Pvt.	PA	107th Infantry	2/24/62				
Hemming, Augustus	14	Pvt.	PA	2nd Cavalry	12/20/61		Died	1863	
Hemminger, Abraham O	16	Pvt.	PA	18th Cavalry	2/25/64	10/31/65			
Hess, Jacob	15	Pvt.	MD	3rd PHB Inf.	2/12/62	2/12/65			
Hess, John	16	Pvt.	IL	50th Infantry	9/12/61	7/13/65			
Hetrick, Daniel L	16	Pvt.	PA	101st Infantry	11/1/61	6/13/65	POW	1864	Plymonth, NC
Hite, Benjimen F	16	Pvt.	MD	2nd PHB Inf.	5/18/62	5/1/65			
Hite, David	15	Pvt.	PA	101st Infantry	5/3/64	6/25/65			
Hite, George W	16	Pvt.	PA	210th Infantry	9/8/65	6/27/65			
Holler, John M	15	Pvt.	PA	138th Infantry	8/29/62	6/23/65			
Hollingshead, Oliver S	16	Pvt.	PA	22nd Cavalry	2/5/64	6/24/65			
Householder, John	15	Pvt.	PA	208th Infantry	9/7/64	6/1/65			
Humbert, Moses	16	Pvt.	PA	5th Heavy Art.	9/1/64	6/30/65			
Ickes, William M	16	Pvt.	PA	91st Infantry	10/8/64	7/10/65			
James, John A	14	Pvt.	PA	55th Infantry	10/11/61		Wound.-Died	1864	
Kauffman, Isaac	15	Pvt.	PA	93rd Infantry	11/26/64	6/27/65			
Kegg, James P	16	Pvt.	PA	55th Infantry	2/29/64	8/30/65			

Youngest Soldiers to Enlist

Moses Humbert was born on April 5th, 1848 in Harrison Township. His older brother Wesley volunteered in the 142nd Pennsylvania Infantry in August 1862. Wesley was wounded at the battle of Fredericksburg on December 13th, 1862 and at Gettysburg on July 1st, 1863. Moses mustered in the 5th Pennsylvania Heavy Artillery in September 1864. Both brothers survived the war. Moses married in 1871 and was a father to 11 children. He passed in 1919 when he was 71.

James and Elizabeth Knupp Kegg are pictured on their wedding day on January 26th, 1871. James Kegg was born in Schellsburg on May 28th, 1847. He volunteered in the 55th Pennsylvania Infantry in February 1864 and survived heavy combat during the following 14 months. The 55th PA Infantry witnessed the Confederate surrender at Appomattox in April 1865. After the war, James worked as a stonemason and moved to Johnstown in 1871. Tragically, James and Elizabeth lost their 17-year-old son William in the Great Johnstown Flood in 1889. Elizabeth passed in 1918 at age 67. James lived to be 81 years old and passed in 1929.

Name	Muster Age	Rank	State	Regiment	Muster In	Muster Out	Casualty	Cas. Year	Battle
Kegg, Levi R	14	Pvt.	MD	2nd PHB Inf.	10/29/61	10/31/64			
Keyser, Samuel	15	Pvt.	PA	50th Infantry	9/28/64	6/2/65			
King, Samuel T	15	Pvt.	PA	55th Infantry	2/19/64	8/30/65	Wounded	1864	Drewry's Bluff
Knox, Otho S	16	Pvt.	PA	55th Infantry	10/12/61	12/30/64	Wounded	1864	
Lamison, James H	16	Pvt.	PA	22nd Cavalry	7/11/63	10/31/65			
Latta, Abraham	16	Pvt.	PA	208th Infantry	9/12/64	6/1/65			
Layton, George W	15	Pvt.	PA	99th Infantry	2/16/65	7/15/65			
Lear, Franklin	15	Pvt.	PA	77th Infantry	2/27/65	12/6/65			
Lehman, William H	16	Pvt.	PA	184th Infantry	5/12/64				
Leighty, George (1)	15	Pvt.	MD	3rd PHB Inf.	2/26/64	5/29/65			
Lesh, John A	16	Pvt.	PA	7th Reserves	2/9/64	5/31/64			
Lingenfelter, John	15	Pvt.	IL	103rd Infantry	10/2/62		Died	1865	
Lingenfelter, Thaddeus	16	Pvt.	PA	77th Infantry	2/24/65	12/6/65			
Lingerfelt, Aaron	16	Pvt.	PA	55th Infantry	8/28/61	8/30/65			
Linn, William	14	Pvt.	MD	3rd PHB Inf.	9/28/61	9/28/64	POW	1862	Harpers Ferry
Luman, David	15	Pvt.	OH	64th Infantry	10/7/64	6/5/65			
Lyons, George W	16	Sgt.	USCT	41st Infantry	9/29/64				
Lysinger, Joseph H	16	Pvt.	IL	146th Infantry	9/8/64	7/8/65			
Martin, Josiah	16	Pvt.	PA	107th Infantry	10/6/64	6/5/65	Wounded	1865	White Oak Road

William Lehman was born in Harrison Township on May 28th, 1847. He was one of 3 brothers to join the Union Army. William enlisted in the 184th Pennsylvania Infantry in May 1864 and was in combat during the siege of Petersburg. His brother Henry enlisted in the 93rd Pennsylvania Infantry in September 1864. Espy joined the 93rd PA Infantry two months later. All three brothers survived the war. William married in 1866 and was a father to 9 children. He passed in 1917 when he was 69 years old.

Thaddeus Lingenfelter was born on August 15th, 1848. He grew up on a farm in Greenfield Township with 11 siblings. Thaddeus mustered in the 77th Pennsylvania Infantry on February 24th, 1865 and was sent to Tennessee to replace soldiers who had fulfilled their enlistment. He mustered out in December 1865 and returned home. Thaddeus married in 1873 and raised a large family of 9 children on a farm in Greenfield Township. He passed at age 60 in 1909.

Name	Muster Age	Rank	State	Regiment	Muster In	Muster Out	Casualty	Cas. Year	Battle
McCoy, Francis P	15	Pvt.	PA	81st Infantry	2/24/64	6/25/65	Wounded		
McCray, James	16	Pvt.	PA	77th Infantry	3/2/62	7/2/65			
McDonald, Samuel	16	Pvt.	PA	110th Infantry	12/19/61				
McEnespy, James B	16	Pvt.	PA	55th Infantry	10/12/61				
McEnespy, Samuel	15	Pvt.	PA	McKeage's Mil.	7/2/63	8/8/63			
		Pvt.	PA	13th Cavalry	10/5/63		Died	1864	Coggin's Point
Mellott, Peter	16	Pvt.	MD	3rd PHB Inf.	11/9/61	5/29/65			
Mellott, Thomas S	16	Pvt.	PA	11th Cavalry	2/27/64	8/13/65			
Metzler, Henry Clay	16	Pvt.	MD	3rd PHB Inf.	2/17/62	5/29/65			
Mickey, Rankins	16	Pvt.	PA	55th Infantry	10/12/61	8/30/65			
Miller, John I	14	Pvt.	PA	110th Infantry	12/19/61	4/10/64	Wounded	1863	Gettysburg
Morse, David	16	Pvt.	PA	11th Infantry	9/30/61		Wound.-Died	1863	Antietam
Mountain, Richard	16	Pvt.	PA	77th Infantry	10/9/61		Wound.-Died	1863	Chickamauga
Murphy, Elias	16	Pvt.	PA	55th Infantry	9/26/62	1/12/65	Wounded	1864	Chaffin's Farm
Murrie, David	14	Pvt.	MD	2nd PHB Inf.	9/20/61	9/28/64			
Musselman, George	16	Pvt.	PA	205th Infantry	8/23/64	6/2/65			
Nelson, William N	15	Corp.	PA	18th Cavalry	10/29/62	11/6/65			
Neville, Henry	15	Pvt.	PA	46th Infantry	7/1/63	8/19/63			
Neville, J Richard	15	Pvt.	VA	23rd Cavalry					
Norton, Franklin G	14	Musician	PA	101st Infantry	12/28/61		Died	1862	
Oler, John W	16	Musician	PA	101st Infantry	11/1/61		Died	1862	

Joseph H Lysinger was born on May 7th, 1848 in Ray's Hill (Breezewood). His parents moved their family to Illinois in 1852. He mustered in the 146th Illinois Infantry Regiment in September 1864 and returned home after the war ended. Joseph homesteaded on 160 acres in Nebraska in 1873 and married in 1891. He passed in 1919 in Nebraska when he was 70 years old.

Francis McCoy was born on February 29th, 1848. He mustered in the 81st Pennsylvania Infantry on February 24th, 1864. The 1890 Londonderry Township Veterans Census listed Francis as suffering a wound during the war, but there is no available information on the date or battle. Francis was a well-known veteran after the war. He compiled a listing titled "Dead: List of those buried in the cemeteries and graveyards of Bedford County 1912" and provided a listing of county soldiers who died during the war for a Bedford Gazette article in 1914. Francis married in 1878 and was a father of 5 children. He passed at age 78 in 1924.

Name	Muster Age	Rank	State	Regiment	Muster In	Muster Out	Casualty	Cas. Year	Battle
Otto, Henry S	16	Pvt.	PA	184th Infantry	5/12/64		POW-Died	1864	Jerusalem Plank Rd.
Pearson, Josiah	15	Pvt.	PA	McKeage's Mil.	7/3/63	8/8/63			
Pennington, James F	15	Pvt.	PA	171st Infantry	11/4/62	8/7/63			
Porter, Willam H	15	Pvt.	PA	76th Infantry	2/22/65	7/18/65			
Price, George W	16	Pvt.	PA	99th Infantry	3/4/65	7/1/65			
Prince, Edwin S	16	Pvt.	OH	162nd Infantry	5/20/64	9/4/64			
Prosser, David W	15	Corp.	PA	55th Infantry	10/12/61	6/15/65	POW	1864	Drewry's Bluff
Replogle, Simon L	15	Pvt.	PA	194th Infantry	7/22/64	11/5/64			
Riffle, Albert J	16	Pvt.	PA	55th Infantry	2/19/64	8/30/65			
Riley, Andrew J	16	Pvt.	PA	107th Infantry	2/10/62	5/3/63	Wounded	1862	South Mountain
Ritchey, Jacob K	16	Pvt.	PA	138th Infantry	9/19/64	6/23/65			
Ritchey, James T	15	Pvt.	PA	107th Infantry	4/26/62	7/13/65			
Rohm, David F	16	Pvt.	IL	58th Infantry	3/17/65	3/10/66			
Rohm, John S	16	Pvt.	MD	3rd PHB Inf.	4/15/64	5/29/65			
Roudabush, Benjamin	16	Pvt.	PA	55th Infantry	9/25/63		POW-Died	1864	
Salkeld, John F	15	Musician	PA	107th Infantry	1/20/62	7/13/65			
Sanderson, Samuel K	16	Pvt.	PA	7th Reserves	7/4/63	8/11/63			
Sanderson, Theodore C	15	Pvt.	PA	149th Infantry	2/25/65	6/26/65			
Shaffer, George W	15	Pvt.	PA	55th Infantry	2/19/64	7/7/65	Wounded	1864	Drewry's Bluff
Shimer, John	14	Pvt.	PA	200th Infantry	8/25/64	4/21/65			

David Murrie was born in Scotland on July 29th, 1847. His parents immigrated to America in 1854 and the family settled in Allegany County, MD. David was 14 years old when he mustered in the 2nd Maryland PHB Infantry with his father in September 1861. They served together in Maryland and the Shenandoah Valley before mustering out of the army in September 1864. David married a girl from Schellsburg in 1871 and they raised a family of 7 children in Bedford County. He passed in 1910 at age 62.

Josiah Pearson was born in South Woodbury Township on August 25th, 1847. His 17-year-old brother Francis volunteered in the 110th Pennsylvania Infantry in October 1861. Francis died of Pernicious Fever at a regimental hospital in Falmouth, VA on December 31st, 1862. Josiah volunteered in McKeage's Militia during the Gettysburg Campaign in 1863 when he was 15 years old. He enlisted as a bugler in the 1st US Cavalry after mustering out of the militia. Josiah lived in California after the war before returning to Bedford County. He married in 1882 and was a father of 6 children. Josiah passed in 1906 and is buried at the Pleasantville Cemetery.

Name	Muster Age	Rank	State	Regiment	Muster In	Muster Out	Casualty	Cas. Year	Battle
Shoop, John	16	Pvt.	PA	McKeage's Mil.	7/2/63	8/8/63			
Showalter, Simon P	15	Pvt.	PA	8th Reserves	4/23/61	10/27/62			
Shroyer, John	16	Pvt.	PA	208th Infantry	8/18/64	6/1/65			
Simmons, Thomas H	14	Sgt.	MD	3rd PHB Inf.	11/11/61	6/29/65			
Slick, Josiah	15	Corp.	PA	55th Infantry	10/11/61	12/6/64	Wounded		
Snyder, Jacob H	15	Pvt.	PA	McKeage's Mil.	7/2/63	8/8/63			
Snyder, William H	16	Pvt.	PA	125th Infantry	8/12/62	5/18/63			
Sparks, John	15	Pvt.	PA	194th Infantry	7/22/64	11/5/64			
Sparks, William	15	Pvt.	PA	101st Infantry	11/1/61	6/25/65	POW	1864	Plymonth, NC
Sponsler, Solomon	16	Pvt.	MD	3rd PHB Inf.	11/5/61	5/19/65			
Sponsler, William	16	Pvt.	PA	100th Infantry	3/4/65	7/4/65	Wounded	1865	Ft. Steadman
Statler, Samuel F	13	Pvt.	PA	55th Infantry	2/2/62	8/8/65	Wounded	1864	Drewry's Bluff
Steele, Edward	16	Pvt.	PA	133rd Infantry	8/18/62	10/9/63			
Stoudnour, Jacob	16	Pvt.	PA	76th Infantry	10/9/61	11/28/64			
Stroud, George W	16	Pvt.	NY	13th Light Art.	9/15/64	8/30/65			
Swaney, Samuel J	16	Pvt.	PA	110th Infantry	2/28/64	6/28/65			
Swartz, Espy	15	Pvt.	PA	2nd Cavalry	12/10/61	12/13/64			
Swartz, Henry	16	Pvt.	PA	194th Infantry	7/22/64	11/5/64			
Taylor, William Y	14	Sgt.	PA	110th Infantry	6/28/62	5/31/65			
Tewell, Moses	14	Pvt.	PA	4th Cavalry	9/16/61	9/18/64			
Trail, Hugh	15	Pvt.	MD	1st PHB Inf.	2/13/65	6/28/65			
Trail, George T	16	Pvt.	MD	3rd PHB Inf	12/31/61	9/30/62	POW	1862	Moorfield
Troutman, George W	16	Pvt.	PA	138th Infantry	8/29/62		Died	1862	

James Ritchey is pictured with a grandson. James Ritchey was born on February 21st, 1847. He was raised on a farm with 9 siblings in West Providence Township. James volunteered in the 107th Pennsylvania Infantry in April 1862 when he was 15. Months later, the Bedford County soldiers in the 107th suffered many casualties during the battles of the Second Bull Run, South Mountain, and Antietam. The 107th also took casualties during heroic fighting on the 1st day at Gettysburg. James survived to witness the Confederate surrender at Appomattox and returned home after 3 years of combat. He later married and raised a family in California. James passed in 1910 when he was 63 years old.

Thomas Simmons is pictured with his granddaughter, who appears to be around the same age as Thomas was during the Civil War. Thomas was born on January 31st, 1847. He was 14 years old when he volunteered in the 3rd Maryland PHB Infantry in November 1861. The 3rd Maryland PHB Infantry surrendered with the entire garrison at Harpers Ferry to Stonewall Jackson's troops during the Antietam Campaign. Thomas was likely captured and paroled days after the battle. The 3rd Maryland PHB later distinguished themselves during the battle of Monocacy in July 1863. Thomas married in 1865 after the war ended and was a father to 9 children. He lived to be 81 years old and was buried in the Providence Union Cemetery in Everett in 1929.

Name	Muster Age	Rank	State	Regiment	Muster In	Muster Out	Casualty	Cas. Year	Battle
Tyson, Samuel H	14	Musician	PA	110th Infantry	10/24/61				
VanOrman, John W	16	Pvt.	PA	79th Infantry	1/14/65	8/1/65			
Waters, William	16	Pvt.	IL	25th Infantry	8/9/61	9/5/64			
Weaverling, Thomas H	16	Pvt.	PA	208th Infantry	9/7/64	6/1/65			
Weaverling, William T (1)	16	Corp.	PA	133rd Infantry	8/15/62	5/26/63			
Weyandt, Jeremiah	16	Pvt.	PA	McKeage's Mil.	7/2/63	8/8/63			
Weyant, Lafayette	15	Pvt.	PA	13th Cavalry	2/26/63	8/14/65	POW	1863	Winchester
Whitaker, John H	16	Pvt.	PA	91st Infantry	10/7/64	6/19/65	Wounded	1865	Five Forks
Wilkins, John H	16	Pvt.	PA	186th Infantry	2/24/64	7/24/65			
Wolf, Edmund	15	Pvt.	PA	55th Infantry	2/19/64	8/30/65			
Woy, Ezekiel C	16	Pvt.	PA	208th Infantry	9/7/64	6/6/65	Wounded	1865	Petersburg - Final

11. Oldest Soldiers to Enlist

The photograph on the right is likely Civil War veteran Theophilus Gates. One descendant believes this picture may actually be his father, Nathaniel Gates. Nathaniel was born on February 12th, 1791. Nathaniel fought in the War of 1812 and was captured by Indians fighting for the British and held in Canada until the end of the war. The 1860 Bedford Township Census listed Nathaniel as a Clergyman. Nathaniel had 7 sons who enlisted in the Union Army.

Around the age of 71, Nathaniel volunteered as a Chaplain in the 13th Pennsylvania Reserves. Two of his seven sons had already volunteered in this regiment. An Altoona Tribune article on August 8th, 1900, noted the following, "Nathaniel's fighting blood got something the better of his theology and he joined the ranks of active service with his sons. The commanding officer, however, noticing his advanced age, refused to permit him to share the hardships of that arduous campaign and while commending his courage and patriotism, made him an active member of the Christian Commission at Washington, DC." The Christian Commission supported Union troops with supplies, medical services, and religious literature. Records and newspaper articles note his 7 sons suffered a combined eight battlefield wounds and two sons were taken prisoner. All 7 of his sons survived the war. Later, when his eyesight failed, Nathaniel became known locally as the "Blind Preacher". Toward the end of their lives, Nathaniel and his wife Sarah were kept in separate rooms at the County Alms House for the destitute until their son Theophilus took them in at his East St. Clair Township home. Nathaniel and Sarah both passed in 1880.

The average life expectancy in the United States was 78.9 years in 2020. In 1860, the year before the beginning of the Civil War, life expectancy was 39.4 years. Army regulations initially set the maximum enlistment age at 35. The enlistment age was raised to 45 in the Enrolment Act of 1863. A study by E.B. Long identified only 7012 Union soldiers who enlisted at age 45 during the Civil War. Enlistment numbers dropped to 2366 for all men 50 and older.

There were valid reasons for the Army regulations setting the maximum age of 45 for enlistments. The daily life of a Civil War soldier was often physically grueling. Union soldiers wore heavy wool uniforms during the unforgiving mid-summer heat in the South. Soldiers often became fatigued during marches, exhausted by being weighted down with 50 to 60 pounds of equipment. Marches sometimes covered 15 miles and forced marches occasionally extended longer distances. Soldiers were often thirsty, and good water was hard to find. Marches during driving rainstorms and walking in mud were difficult. Disease and sickness were chronic problems because of unsanitary conditions in large camps. Life in the Union army could be exceedingly difficult, even for younger men in their physical prime.

There were at least 76 Bedford County soldiers 45 and older when they enlisted in the Civil War, including 37 men 50 and older. Remarkably there were two soldiers over the age of 60, Theophilus Gates listed above, and Daniel Defibaugh Sr. born in Waynesburg (Everett) in 1800. Daniel was listed on the 1860 Hopewell Township Census before moving west and volunteering with his son Anderson in the 43rd Ohio Infantry in December 1861. Another son, Harrison, was a member of the Taylor Guards, the first volunteer company recruited out of Bedford County. Harrison later volunteered in McKeage's Militia during the Gettysburg Campaign and in the 2nd Pennsylvania Cavalry. Daniel Jr joined his father and brother in the 43rd Ohio infantry in February 1864. All four members of the Defibaugh family survived the war. Daniel Sr lived to be 87 and was buried at the Harvey Cemetery in Urbana, IL, in 1887.

The number of Casualties and Discharges on Surgeon's Certificates were high among the 76 older Bedford County soldiers identified in this book. Eight men lost their lives, including 3 who languished in the inhumane living conditions of Andersonville before dying. Five more survived being held in prisoner of war camps. Five suffered wounds or were injured on a battlefield. An additional 13 soldiers received Discharges on Surgeon's Certificates (abbreviated to "DSC" in this book). Discharges on Surgeon's Certificates were disability discharges for soldiers judged physically unable to continue serving in the army.

Samuel Cox was born on May 19th, 1812. Susanna Slonaker had just turned 20 when she married Samuel in December 1847. They were the parents of the "Lost Children of the Alleghenies." Their tragic story is among the most widely known in the history of Bedford County. Two sons, George, age 7 and Joseph age 5, wondered off from their cabin in Spruce Hollow near Pavia while their father was hunting on the morning of April 24th, 1856. Susanna thought Samuel had taken the boys with him. When Samuel returned later that morning, they realized the boys were missing. Panic set in when the boys were not quickly located, and by afternoon there were 200 people searching the area. Dusk came, and the boys had not been found. Snow was still on the mountains, storm clouds were approaching, and the boys were only clad in thin jackets. Samuel pleaded for the searchers to continue looking for the boys, which many did by torch light. By the next morning, the weather had turned windy and cold. A thousand people were estimated to have joined the search by midday. Many others would stream in from surrounding communities over the next two weeks to continue the desperate search over rugged terrain. Some unfairly suspected Samuel had something to do with their disappearance. There was speculation the boys had being devoured by wild beasts. Wild rumors spread about the boys being taken by gypsies or were being held in a secret catholic hideaway. A sorcerer from Morrisons Cove and a witch from Somerset County were brought in to see if they could help locate the boys. Mystic beliefs were not uncommon in this era.

Jacob Dibert and his wife lived 15 miles away from Spruce Hollow and heard about the boy's disappearance four days later. They had children about the same age and found the story distressing. Ten days after the boys went missing, Jacob had a dream of the location where the boys could be found. He described the dream to his wife Sarah, and she thought it resembled a narrow valley on the farm she grew up on near Blue Knob. That night Jacob had the same dream. Jacob's dream returned a third straight night, and this time the vision of the location of the boys was more vivid than before. The next day, Jacob and his wife went to her childhood home and told her brother Harrison Whysong about Jacob's Dream. Harrison thought Jacob was crazy, but agreed to accompany him. One by one, the exact details of the dream were realized. A dead deer was passed, a little shoe was found, a beech log laying over a stream was crossed. They went up a ridge, down a ravine, and came upon the birch tree with a broken top from the dream. Beneath this tree, the bodies of the two boys were found. Jacob paid for the boy's gravestone from the reward money.

Five years later, Jacob Dibert volunteered in the 55th PA Infantry while Sarah and their 5 children tended to the family farm. He died of Chronic Dysentery in October 1864. He was 42 years old. Photographs of Jacob are in "Those who made the Ultimate Sacrifice" chapter on page 132 and in the 55th PA Infantry overview on page 236. Samuel Cox mustered in the 149th PA Infantry when he was 51 years old, while his wife Susanna and their 6 surviving children waited for his return. Samuel suffered a severe injury when a hickory tree fell on him after being struck by an artillery shell during the Battle of Cold Harbor. Jacob survived the war and lived to be 86 years old. Susanna passed 3 years after Samuel in 1899 at age 71. Samuel, Susanna, George, and Joseph are buried at the Mt. Union Cemetery in Lovely.

Name	Muster Age	Rank	State	Regiment	Muster In	Muster Out	Casualty & DSC listing	Cas. Year	Battle
Barr, John	45	Sgt.	PA	3rd Infantry	4/20/61	7/29/61			
Beam, Daniel	59	Pvt.	PA	101st Infantry	11/1/61	6/25/65	POW-Died	1864	Plymonth, NC
Blackburn, John M	45	Sgt., QM	PA	21st Cavalry	7/8/63	2/20/64	Wounded		
Blackburn, Samuel S	45	Pvt.	PA	99th Infantry	9/21/64	6/31/65			
Bowman, George W	53	Pvt.	PA	208th Infantry	9/6/64	5/11/65	DSC		
Bridenthal, Henry B	46	Pvt.	PA	55th Infantry	10/11/61	6/26/63	DSC		

Oldest Soldiers to Enlist 197

James Madara was born on January 20th, 1813, in Franklin County. According to the History of Bedford, Somerset and Fulton County book published in 1884, his great grandfather was John Ray, the Scottish trader who first settled in Raystown. James moved to Woodbury Township with his family when he was a boy. James married Jane Wishart in 1833 and was a father to 7 children. Though receiving a limited formal education, he eventually became superintendent and manager of the Sarah and Bloomfield Furnace operations for Peter Shoenberger. Dr. Shoenberger was considered the King of Pennsylvania Ironmasters in the pre-Civil War era. James was also responsible for erecting the Rodman furnace that provided iron for many of the large artillery pieces used by the Union army during the Civil War. Secretary of War Edwin Stanton appointed James Madara as Government Inspector of Iron and he held the rank of Colonel until the close of the Civil War. His son, David, was a Captain in the 55th Pennsylvania Infantry from September 1861 until April 1862.

James died of pneumonia on May 2nd, 1879, when he was 66 years old. His obituary stated he was a "careful businessman, with profound judgement and sterling integrity. He was honored and respected at home and abroad; loved by his employees and will be remembered for the charities and deeds of kindness during his life". His biography in the History of Bedford, Somerset, and Fulton counties book in 1884 paints a picture of an extraordinary man.

Name	Muster Age	Rank	State	Regiment	Muster In	Muster Out	Casualty & DSC listing	Cas. Year	Battle
Bulger, Daniel B	47	Pvt.	PA	208th Infantry	9/8/64	6/1/65			
Carson, John (2)	48	Pvt.	USCT	22nd Inf.	12/29/63	10/16/65			
Chamberlain, Philip	50	Pvt.	PA	208th Infantry	9/7/64	6/1/65			
Corle, Martin	51	Pvt.	PA	55th Infantry	11/4/62	2/16/63	DSC		
Cox, Samuel	51	Pvt.	PA	149th Infantry	10/2/63	6/24/65	INJ	1864	Cold Harbor
Crist, Solomon	49	Pvt.	PA	McKeage's Mil.	7/2/63	8/8/63			
		Pvt.	PA	55th Infantry	3/2/64		MIA	1864	Bermuda Hundred
Croft, George	57	Pvt.	PA	22nd Cavalry	7/11/63	2/5/64			
Croft, George W	46	Pvt.	PA	51st Infantry	7/3/63	9/2/63			
Defibaugh, Daniel Sr	61	Sgt.	OH	43rd Infantry	7/12/61	7/13/65			
Defibaugh, James C	57	Pvt.	PA	91st Infantry	10/8/64	7/10/65			
Dodson, William	55	Pvt.	PA	13th Cavalry	2/29/64	7/17/65			
Dull, Valentine	45	Pvt.	PA	138th Infantry	8/29/62	6/23/65			
Filler, Joseph	49	Lt. Col.	PA	55th Infantry	11/5/61	11/13/64	POW	1864	Petersburg
		Lt. Col.	PA	55th Infantry			Wounded	1864	Drewry's Bluff
Gates, Nathaniel	71	Chap.	PA	13th Reserves	1862	1863	DSC		
Good, George	47	Pvt.	PA	55th Infantry	2/9/64	8/30/65			
Gordon, Jeremiah	50	Pvt.	PA	55th Infantry	10/12/61	8/2/62	DSC		
Griffith, Michael	46	Pvt.	PA	8th Reserves	3/25/64	5/15/64			
		Pvt.	PA	91st	5/15/64	7/18/65			
Grimes, John C	49	Corp.	PA	14th Infantry	5/2/61	8/7/61			
		Pvt.	PA	84th Infantry	1861				
Growden, John S	50	Pvt.	PA	133rd Infantry	8/5/62	5/26/63			
Hafer, Wilson	49	Pvt.	PA	133rd Infantry	8/14/62	1/5/63	DSC		Fredricksburg*
Hazlett, Richard	45	Pvt.	PA	56th Infantry	10/1/61	5/31/65			
Himes, William	54	Pvt.	PA	McKeage's Mil.	7/2/63	8/8/63			
		Pvt.	PA	7th Cavalry	9/15/64	6/23/65			

Susan and John McFerren are pictured above. John was born in Cumberland Valley on January 26th, 1815. He married Susan Heming in 1840 and they raised 9 children on a family farm. John volunteered in the 210th Pennsylvania Infantry in September 1864, when he was 49 years old. The 210th was in combat during the siege of Petersburg and took heavy casualties at Lewis Farm near Gravelly Run on March 29th, 1865. The 210th Pennsylvania Infantry were eyewitnesses to the Confederate Surrender at Appomattox on April 9th and marched in the Grand Review in Washington DC in early May. After the war ended, John returned home to Cumberland Valley to continue working on the family farm with his wife. John lived to be 80 years old. Susan passed 6 years later when she was 83 and was buried beside her husband in the McFerren Farm Cemetery in 1901.

Name	Muster Age	Rank	State	Regiment	Muster In	Muster Out	Casualty & DSC listing	Cas. Year	Battle
Hiner, William	47	Pvt.	IA	33rd Infantry	9/6/62	5/9/63			
Hoover, Daniel	55	Pvt.	KY	8th Infantry	1/30/62				
Howser, Lewis	56	Sgt.	PA	172nd Infantry	10/28/62				
King, Daniel	45	Pvt.	PA	92nd Infantry	1/11/64	6/21/65			
King, John V	50	Major			1863				
King, Thomas	50	1st Lt.	PA	101st Infantry	12/28/61	3/11/65	POW	1864	Plymonth, NC
Kinsey, Peter Sr	50	Pvt.	PA	55th Infantry	11/5/61	4/22/63			
Lape, Abraham	48	Pvt.	PA	171st Infantry	11/1/62	8/8/63	Wounded	1863	New Berne, NC
			PA	97th Infantry					
Lingerfelt, Abram	47	Pvt.	PA	55th Infantry	8/28/61	8/30/65			
Linn, Hugh Sr	52	Pvt.	MD	3rd PHB Inf.	9/28/61	9/28/64	POW	1862	Harpers Ferry
Love, John R	53	Pvt.	USCT	41th Infantry	9/21/64				
Madara, James W	49	Col.							
McClure, Thomas	45	Chap.	PA	151st Infantry	2/3/63	7/27/63			
McFerren, John	49	Pvt.	PA	210th Infantry	9/16/64	5/30/65			
McGee, John H	45	Pvt.	PA	147th Infantry	11/1/61	5/5/63	Died	1863	
McGregor, Robert	47	Pvt.	PA	55th Infantry	10/15/61	10/15/64			
McGregor, William	51	Pvt.	PA	55th Infantry	10/15/61	5/14/62			
		Pvt.	PA	13th Cavalry	9/26/63	6/23/65			
McVicker, James A	45	Saddler	PA	21st Cavalry	7/18/63	2/20/64			
Mench, John	48	Pvt.	PA	99th Infantry	2/25/65	5/29/65			
Metz, Felty	47	Pvt.	PA	13th Infantry	9/12/62	9/26/62			
Miller, David	56	Pvt.	PA	55th Infantry	3/2/64	6/8/65	Wounded	1864	Petersburg - Initial
Mock, George (2)	55	Pvt.	IN	129th Infantry	3/6/64	4/26/65			
Mopps, Edward S	55	1st Lt.	PA	13th Infantry	4/25/61	8/6/61			
Morris, John	45	Musician	PA	194th Infantry	7/22/64	11/5/64			
Mullin, John	47	Pvt.	PA	138th Infantry	8/29/62	3/5/64			

Mary and John Shauf are pictured above. John Shauf was born in Germany on April 23rd, 1809. His family immigrated to America in 1837. Two years later he married Mary Feight in Schellsburg. They were living on a family farm in East Providence Township with their 9 children when the Civil War broke out. Their oldest son, Cornelius joined the Hopewell Rifles Company of the 8th Pennsylvania Reserves on June 11th, 1861. Cornelius died of disease at Camp Pierpont in Langley, VA on December 17th, 1861. He was 19 years old. John was 52 years old when he volunteered in the 107th Pennsylvania Infantry less than two months after his son had died. He received a Discharge on a Surgeons Certificate on May 5th, 1862, for being judged too old for the rigors of an infantry soldier. John volunteered a second time in McKeage's Militia during the Gettysburg Campaign and returned home after the Confederates retreated from Pennsylvania. A second son volunteered in the 1st Maryland PHB Cavalry in March 1864 and mustered out after the war ended. Mary passed when she was 59 in 1874. John passed 7 years later at age 72.

Name	Muster Age	Rank	State	Regiment	Muster In	Muster Out	Casualty & DSC listing	Cas. Year	Battle
Myers, Henry Sr	59	Pvt.	PA	194th Infantry	7/22/64	11/5/64			
Neff, Frederick	51	Pvt.	PA	138th Infantry	8/29/62	1/15/65			
Penrod, John B Sr	47	Pvt.	PA	8th Reserves	11/12/61	11/26/62	DSC		
Perry, Wythe	51	Pvt.	USCT	3rd Infantry	7/23/63	10/31/65			
Potter, John (1)	49	Pvt.	PA	101st Infantry	11/1/61	2/6/62	DSC		
		Pvt.	PA	22nd Cavalry	2/15/64	6/24/65			
Riley, William (2)	53	Pvt.	PA	55th Infantry	3/2/64		Wound.-Died	1864	Cold Harbor
Risling, John	45	Pvt.	PA	55th Infantry	10/12/61	10/29/62	DSC		
Salkeld, Bernard	47	Pvt.	PA	158th Infantry	11/4/62	8/12/63			
Sams, Wilson	45	Pvt.	PA	208th Infantry	9/7/64	6/1/65			
Shauf, John	52	Pvt.	PA	107th Infantry	2/10/62	5/8/62	DSC		
		Pvt.	PA	McKeage's Mil.	7/3/63	8/8/63			
Showman, William	47	Pvt.	PA	101st Infantry	12/2/61		Died	1862	
Slick, Allen	48	Pvt.	PA	55th Infantry	2/19/64		KIA	1864	Cold Harbor
Slick, William W (1)	45	Pvt.	PA	101st Infantry	2/18/62	8/26/62	DSC		
Snively, Andrew J	55	Pvt.	PA	22nd Cavalry	7/11/63	2/5/64			
Snyder, George W Sr.	50	Pvt.	PA	125th Infantry	8/13/62	4/2/63			
		Pvt.	PA	205th Infantry	8/17/64	7/2/65	DSC		
Stephens, Jacob	51	Pvt.	PA	11th Cavalry	12/24/63		POW-Died	1864	Reams Station
Stevens, Denton	57	Pvt.	PA	149th Infantry	9/29/63	4/21/64	DSC		
Vickroy, James R	52	Pvt.	PA	138th Infantry	8/30/62	11/28/63			
Ward, Jeremiah	46	Pvt.	PA	107th Infantry	10/21/64	3/26/65			
Watson, John R	45	Pvt.	PA	184th Infantry	5/12/64		POW	1864	Jerusalem Plank Rd.
Williams, Richard (1)	45	Pvt.	PA	194th Infantry	7/18/64	11/6/64			
Wilson, George W	50	Pvt.	PA	101st Infantry	3/7/62		POW-Died	1864	Plymonth, NC
Wolford, Samuel	50		PA	49th Infantry			POW	1865	

12. Veterans Living in the 1930's & 1940's

Civil War Veteran Jonas Imler is pictured above with his wife Elizabeth in the late 1920s. Jonas was born on a farm near Imler on October 13th, 1844. He was the youngest of 7 children. Jonas attempted to volunteer in the Union Army in Loysburg but was turned away because of his small stature. He persuaded a young man from Roaring Springs to enlist with a recruiter in Hollidaysburg using his name. Jonas used those papers to muster in the 205th Pennsylvania Infantry on August 28th, 1864. He survived combat during the siege of Petersburg and returned home in June 1865. Jonas married Elizabeth Smith in January 1867 and they raised 5 children on a 250-acre farm in King Township. Jonas also served two terms as a Bedford County Commissioner. Elizabeth passed at age 84 in 1930. Jonas lived to be 94 years old and passed in 1938. They are buried at the Imler Valley Cemetery.

The lifetimes of many Civil War Veterans spanned a remarkable era of significant change. In the decades leading up to the Civil War, railroad lines proliferated across America, dramatically reducing the time it took for people and cargo to reach destinations. Telegraph machines quickly relayed information over long distances and photography had recently been invented. Each had a profound impact on the Civil War.

The velocity of innovation increased in the later part of the 19th century. The inventions of the telephone, phonograph, electric light bulb, portable motion-picture cameras, alternating current electricity and the first automobile powered by an internal-combustion engine would change everyday life in America. Civil War veterans living in the early decades of the 20th Century would have read about the Wright Brothers first flight at Kitty Hawk, witnessed motor vehicles become common on our roads and listened to music on the radio from the comfort of their homes. Many Civil War Veterans witnessed a new generation of Bedford County men and boys depart for Europe during the 1st World War. Some experienced Prohibition during the Roaring 20s, and a surprising number of aging veterans lived during the hard times of the Great Depression. There were at least 95 Civil War Veterans from Bedford County who were still living in 1930, including 4 who were alive in 1940. Two Bedford Civil War veterans would have heard the shocking news of the Japanese surprise attack on Pearl Harbor on December 7th, 1941.

Albert Woolson was the last known surviving Union Army soldier from the Civil War. He was born in New York in 1850 and died on August 2nd, 1956, in Minnesota. Albert lived to see televisions become prized possessions in many homes in America. To add some historical perspective, if Albert had lived another month, he could have joined other Americans on being "All Shook Up" after watching the gyrations of an up-and-coming rock and roller from Memphis Tennessee on the Ed Sullivan Show. Life Magazine on August 20th, 1956 wrote the following on the passing of the last Civil War Veteran, "This chapter in our history has been closed. Something deeply and fundamentally American is gone."

Both photographs on this page were taken in 1907. Veteran Samuel Statler is pictured above on horseback in Bedford. He lived until 1936. Below pictured in the back seat of a Cadillac are veterans Americus Enfield on the left and William B Filler on the right. The driver is Art Fletcher, who owned the first Cadillac in Bedford County. Enfield and Filler where part of a group who made an automobile trip to the Gettysburg battlefield. Devil's Den is visible in the background. In this early era of automobile transportation, the rough roads on the Lincoln Highway had deep ruts until closer to Gettysburg. They departed Bedford County at 8 a.m. and did not arrive in Gettysburg until 6 p.m. Gasoline was purchased at hardware stores for 13 cents a gallon because no gas stations existed on route 30. Americus lived until 1931, and William passed in 1920.

Martin L Myers was born on February 24th, 1847. Martin left his Woodbury home to muster in the 99th Pennsylvania Infantry on March 3rd, 1865. He enlisted as a substitute for a man who begged Martin to go in his place. Martin witnessed the Confederate surrender at Appomattox the following month. After the war, Martin learned the wagon-making trade from his father. Martin married in 1873 and raised 9 children at his West Main Street home in Everett. He owned a carriage and wheel making business in Everett from 1874 until 1896. Martin closed the doors of his business when his handmade products could no longer compete with lower cost mass produced parts. Afterward, he built houses for many years. Martin passed when he was 94 on January 12th, 1942. His obituary stated he had recently "volunteered his services for the national defense." He died two months after Pearl Harbor, and it appears he was helping to sell War Bonds during the 2nd World War.

Winfield Scott Conrad was the last known surviving Civil War Veteran from Bedford County. He passed on March 11th, 1942, at age 94. Winfield was born on May 16th, 1847 and went to work at a young age. The 1860 Hopewell Township census listed Winfield as forgeman when he was 13 years old. Winfield was 16 years old when he volunteered in the 55th Pennsylvania Infantry on February 18th, 1864. The 55th experienced heavy combat in 1864 during the battles of Drewry's Bluff, Cold Harbor, and the siege of Petersburg. Winfield witnessed the Confederate surrender at Appomattox in April 1865 before returning home. He married in 1869 and was a father of 2 daughters. Winfield worked as a miner for 24 years in the Broad Top area before becoming a constable in Huntingdon County for 20 years. He retired as a clerk at the county courthouse when he was 89 years old.

Name	Muster Age	Lived Until	Rank	State	Regiment	Muster In	Muster Out	Casualty	Battle
Bailey, James	21	1935	Pvt.	PA	138th Infantry	8/29/62			
Baker, Samuel N	17	1934	Pvt.	PA	186th Infantry	7/2/63	8/8/63		
			Pvt.	PA	187th Infantry	2/10/65	8/15/66		
Barndollar, William G	18	1936	Pvt.	PA	194th Infantry	7/21/64	11/5/64		
Bennett, John E	19	1930	Pvt.	PA	184th Infantry	5/12/64	7/14/65	Wounded	
Bond, George M	23	1930	Sgt.	PA	12th Cavalry	1/27/62	7/20/65		
Bookhamer, John	22	1936	Pvt.	PA	77th Infantry	2/28/65	12/6/65		
Bookhamer, Thomas	22	1930	Pvt.	PA	22nd Cavalry	7/29/63	2/5/64		
			Pvt.	PA	77th Infantry	2/28/65	12/6/65		
Boylan, John	22	1930	Pvt.	PA	76th Infantry	1/4/65	7/8/65		
Brant, Henry	16	1932	Pvt.	PA	93rd Infantry	11/26/64	6/27/65	Wounded	Ft. Fisher
Bridenthal, Thomas	17	1930	Pvt.	PA	13th Cavalry	9/30/62	7/14/65		
Bruner, George	24	1932	Pvt.	PA	126th Infantry	8/12/62	5/20/63	Wounded	Chancellorsville
Calhoun, Christopher P	20	1932	1st Lt.	PA	138th Infantry	8/29/62	6/23/65		
Cambell, Samuel L		1938	Sgt.	PA	6th Heavy Art.	9/2/64	6/13/65		
Carpenter, Abraham	20	1933	Corp.	PA	138th Infantry	8/29/62	10/19/64	Wounded	Cedar Creek

George Bond was born on November 28th, 1838. He volunteered as a private in the 12th Pennsylvania Cavalry in January 1862. His enlistment description stated he was 5'5" tall, with a dark complexion, blue eyes, and dark hair. George received a promotion to Sergeant and spent much of the war in the Shenandoah Valley. He mustered out in July 1865. George married after the war and was a father to two daughters and a son. He lived to be 91 years old and was buried at the Pleasant Ridge Church Cemetery in Buffalo Mills in 1930.

Henry Brant pictured with his wife Hester. Henry was born on March 12th, 1848 and grew up on a farm in Harrison Township. He volunteered as a substitute in the 93rd Pennsylvania Infantry in November 1864, when he was 16. Henry received a wound at Fort Fisher during the siege of Petersburg on March 25th, 1865. Henry mustered out with his regiment in June 1865 and returned home. He married Hester Holler in 1871 and they raised 7 children on their farm in Juniata Township. Hester passed in 1915 when she was 67 years old. Henry lived to be 84 and passed in 1932.

Name	Muster Age	Lived Until	Rank	State	Regiment	Muster In	Muster Out	Casualty	Battle
Castner, Jacob H	14	1931	Pvt.	PA	133rd Infantry	8/13/62	5/26/63		
Chamberlain, Fernando	16	1930	Pvt.	PA	194th Infantry	7/21/64	11/5/64		
Chaney, Levi	19	1933	Pvt.	PA	107th Infantry	2/24/62	2/11/64	Wounded	Antietam
Comp, Adam A	21	1935	Pvt.	PA	82nd Infantry	11/14/64	7/15/65		
Conner, Jonas	19	1931	Pvt.	PA	107th Infantry	4/29/64	7/23/65	Wounded	Petersburg - Initial
Conrad, Winfield S	16	1942	Pvt.	PA	55th Infantry	2/18/64	8/30/65		
Crawford, Joseph	15	1933	Pvt.	WV	15th Infantry	Feb-64	1865	POW	Cedar Creek
Crist, David T	19	1934	Pvt.	PA	99th Infantry	3/17/65	7/1/65		
Croft, Levi G	17	1937	Pvt.	OH	11th Infantry	9/2/64	6/11/65	Wounded	
Custer, Joseph	19	1935	Pvt.	PA	6th Heavy Art.	9/3/64	6/15/65		
Davis, R DeCharmes	16	1930	Corp.	USCT	32nd Inf.	2/12/64	9/2/65		
Dicken, Thomas W	19	1940	Pvt.	PA	133rd Infantry	8/14/62	5/26/63		
Dodson, John (1)	21	1931	Pvt.	PA	97th Infantry	9/21/64	6/28/65		
Duvall, John N	21	1931	Pvt.	PA	97th Infantry	9/19/64	6/28/65		
Edmonson, Joseph	17	1932	Pvt.	VA	3rd Cavalry	8/14/61	Aug-64	Wounded	Boonsboro, MD
Enfield, Americus	14	1931		PA	Knap's Light Art.	6/13/61	2/24/64		
			Surgeon	PA	22nd Cavalry	2/24/64	10/31/65		
Feather, William	22	1933	Pvt.	PA	McKeage's Mil.	7/2/63	8/8/63		
			Pvt.	PA	55th Infantry	3/2/64	8/30/65		
Fletcher, Winfield S	15	1934	Pvt.	PA	22nd Cavalry	7/11/63	6/15/65		
Fluke, Samuel B	24	1935	Musician	PA	205th Infantry	8/26/64	6/2/65		
Gates, George	20	1931	Pvt.	PA	9th Cavalry	5/31/64	7/18/65	POW	
Gordon, Joseph	16	1932	Pvt.	PA	McKeage's Mil.	7/2/63	8/8/63		
			Pvt.	PA	55th Infantry	2/19/64	7/22/65	Wounded	White Oak Road

David Crist was born on July 22nd, 1845 and grew up on a farm in Union (Pavia) Township. He left home to enlist in the 99th Pennsylvania Infantry in March 1865. The following month the 99th was present at Appomattox during the Confederate surrender. He married Isabella Hammer in July 1865, the same month he returned from the war. They raised 8 children on their family farm in Pavia Township. Isabella passed in 1923 at age 76. David was 89 years old when he passed in 1934.

Samuel Fluke was born on August 2nd, 1840 in South Woodbury Township. He mustered in the 205th PA Infantry on August 26th, 1864, and received a promotion to Fife Major a week later. During the Civil War, musicians were in the thick of battles and considered essential to each regiment. He married in 1867 and was a father of 9 children. Samuel opened a jewelry and photography store in Woodbury in 1868. Samuel was also a successful surveyor and was credited with developing the street layout plan for the town of Roaring Springs. He was one month shy of his 95th birthday when he passed in 1935.

Thomas Dicken is pictured in the buggy on the left, circa 1915. His wife Melitta is in the front seat of the buggy on the right, peeking between two men. Surrounding them are members of their large family. Thomas Dicken was born on a farm in Cumberland Valley Township on January 20th, 1843. He volunteered in the 133rd Pennsylvania Infantry in August 1862. Four months later, Bedford County soldiers in the 133rd suffered 34 casualties during a charge on Marye's Heights at the battle of Fredericksburg. Thomas moved to Kansas after the war and married Melitta Lane in 1874. They raised 12 children on a farm in Crowley County, Kansas. Melitta passed at age 64 in 1916. Thomas lived to be 97 years old and passed on January 24th, 1940.

Allen Harbaugh was born in Lovely on November 23rd, 1848. He was raised on a farm with 2 brothers who also enlisted in the Union Army. Eli Harbaugh was 16 years old when he volunteered in the 55th PA Infantry on November 5th, 1861. He died of Smallpox in South Carolina less than 3 months later. Emanuel volunteered in the 138th PA Infantry in 1862. Emanuel was captured during the battle of the Wilderness in May 1864 and taken to Andersonville. Allen mustered in the 205th PA Infantry in September 1864. Both Allen and Emanuel survived the war. Allen married in 1866 and was a father to 7 children. He lived to be 85 and passed in 1934.

William R Hasenpat was born in Philadelphia on August 15th, 1842. He volunteered in the 7th PA Infantry in September 1862 during the Antietam campaign and mustered out of this emergency unit after the Confederates retreated from Maryland. William volunteered a second time in an emergency militia in June 1863 during the Gettysburg campaign. He worked as an accountant for most of his life in Philadelphia before living in the home of his daughter in Bedford toward the end of his life. William lived to be 89 years old and was buried in the Bedford Cemetery in 1832.

George Henry was born on June 29th, 1841, into a prominent family in Bedford. His uncle James Henry was a well-known doctor in the county. His father Alexander moved the family to Iowa in 1852. George and his three brothers Porter, James and John volunteered in Iowa regiments during the Civil War. George was a Captain in the 4th Iowa Infantry and was wounded at the battle of Pea Ridge, Arkansas in 1862. All four brothers survived the war. George passed when he was 88 in 1930.

Jonathan Horton was born on July 11th, 1843 in Broad Top Township. Jonathan volunteered in the 133rd PA Infantry in August 1862. The 133rd took part in a near suicidal charge of Marye's Heights at Fredericksburg in December 1862. There is no available record of Jonathan suffering a wound during the charge, but he received a Discharge on a Surgeons Certificate three months later for being physically unable to continue serving in the army. He moved west after the war, married, and raised a family on a farm in Nebraska. Jonathan lived to be 88 and passed in 1931.

Name	Muster Age	Lived Until	Rank	State	Regiment	Muster In	Muster Out	Casualty	Battle
Grove, Henry C	18	1931	Pvt.	PA	McKeage's Mil.	7/2/63	8/8/63		
			Pvt.	OH	191st Infantry	3/8/65	8/27/65		
Harbaugh, Allen	15	1934	Pvt.	PA	205th Infantry	9/1/64	6/2/65		
Harr, Christian	19	1931	Pvt.	PA	149th Infantry	10/2/63	12/13/64	Wounded	Petersburg - Initial

Samuel King is pictured wearing a Civil War ribbon. He was born in Schellsburg on March 2nd, 1848, and volunteered in Bedford when he was 15 years old. Samuel mustered in the 55th Pennsylvania Infantry in February 1864. Three months later, Samuel suffered a wound during the battle of Drewry's Bluff on May 16th, 1864. He recovered from the wound and returned home to his family. Samuel married in 1872 and was a father of 5 children. He worked as a plaster for much of his life. Samuel lived to be 84 years old and passed in 1932.

Jacob Manges was born on a Shade Township farm in Somerset County on February 14th, 1846. He enlisted in the 12th Pennsylvania Cavalry in March 1864 and was in the Shenandoah Valley during much of the war. He mustered out in July 1865 and returned home. Jacob married after the war and was listed as a farmer on the 1880 Napier Township Census. He was a father to 12 children and named his first son - Ulysses Grant Manges. Jacob lived to be 86 years old and passed in 1932.

Name	Muster Age	Lived Until	Rank	State	Regiment	Muster In	Muster Out	Casualty	Battle
Hasenpat, William R	20	1932	Pvt.	PA	7th Infantry	9/12/62	9/26/62		
			Pvt.	PA	32nd Infantry	6/26/63	8/1/63		
Heffner, David H	22	1932	Pvt.	PA	167th Infantry	11/12/62	8/12/65		
Henry, George A	20	1930	Capt.	IA	4th Infantry	8/15/61	7/24/65	Wounded	Pea Ridge, AK
Henry, Porter W	23	1932	Pvt.	IA	1st Cavalry	7/31/61	11/1/62		
Hess, John	16	1936	Pvt.	IL	50th Infantry	9/12/61	7/13/65		
Hixon, Henry H	21	1932	Pvt.	PA	11th Infantry	9/30/61	7/1/65	Wounded	Fredericksburg
			Pvt.	PA	11th Infantry			Wound.-POW	Gettysburg
			Pvt.	PA	11th Infantry			Wounded	Petersburg - Initial
Hixon, Jared H	17	1937	Pvt.	PA	22nd Infantry	8/9/64	5/24/65		
Horton, Jonathan A	21	1931	Pvt.	PA	133rd Infantry	8/29/62	3/14/63		Fredericksburg*
Hunt, John H	22	1932	Musician	PA	13th Infantry	9/12/62	9/26/62		
			Musician	PA	178th Infantry	11/5/62	7/27/63		
			Pvt.	PA	187th Infantry	1/16/64	8/3/65		
Husler, Thomas J	18	1933	Pvt.	PA	13th Cavalry	2/11/64	6/13/65		
Ickes, Joseph H	24	1930	Pvt.	PA	84th Infantry	9/24/61			
Imler, Jonas C	19	1938	Pvt.	PA	205th Infantry	8/26/64	6/2/65		
Isett, James M	19	1930	Pvt.	PA	194th Infantry	7/22/64	11/5/64		
Karns, Jacob Jr	30	1934	Pvt.	PA	209th Infantry	9/2/64	6/19/65		
King, Samuel T	15	1932	Pvt.	PA	55th Infantry	2/19/64	8/30/65	Wounded	Drewry's Bluff
Lambert, Joseph C	18	1935	Corp.	PA	133rd Infantry	8/14/62	5/27/63	Wounded	Fredericksburg
			Sgt.	PA	21st Cavalry	7/9/63	7/8/65		
Leighty, George (1)	15	1939	Pvt.	MD	3rd PHB Inf.	2/26/64	5/29/65		
Lysinger, Martin G	18	1935	Corp.	PA	McKeage's Mil.	7/2/63	8/8/63		
Manges, Jacob A	19	1932	Pvt.	PA	12th Cavalry	3/30/64	7/20/65		

The photograph on the left is a five-generation picture of the William A McGregor family taken in 1936. William was born on a farm in Napier Township on June 24th, 1845. He enlisted in the 99th Pennsylvania Infantry in March 1865. William married Mary Ann Harbaugh two months after returning home from the war. They raised 5 children in Pleasantville. William served as a Justice of the Peace for 45 years and retired as the local Registrar when he was 89 years old. William is holding his great-great-grandson John Earl Statler. Standing from the left is Mildred Hartman, Calvin McGregor, and Elsie Hartman. William passed on February 11th, 1937, at age 91. William and his wife Mary Ann are buried in the Pleasantville Cemetery.

Jacob Mellott was born on a farm near Mattie on July 31st, 1845. Three of his brothers died in the Civil War. John and Cornelius Mellott volunteered in the 11th PA Infantry in 1861. Cornelius was killed during the 2nd Battle of Bull Run in August 1862, and John died of Remittent Fever in 1864. Frederick volunteered in the 12th PA Reserves and was killed at the battle of South Mountain in September 1862. Jacob mustered in the 208th PA Infantry in 1864. Jacob and two other brothers who also enlisted in the Union Army survived the war. He married in 1871 and was a father to 6 children. Jacob lived to be 91 and was buried at the Everett Cemetery in 1936.

Joseph Penrose was born on a family farm in East St. Clair Township on June 6th, 1845. In a 1904 memoir, he recalled watching his grandfather providing meals to runaway slaves. One day a runaway kissed him and cried as if her heart would break because she had to leave behind a little boy about his age. Joseph wrote it was an experience he never forgot. He mustered in the 21st Pennsylvania Cavalry in 1863 and witnessed the Confederate surrender at Appomattox. Joseph married in 1874 and was a father to two daughters. He served as a Justice of the Peace for 27 years. Joseph lived to be 84 and passed in 1930.

Name	Muster Age	Lived Until	Rank	State	Regiment	Muster In	Muster Out	Casualty	Battle
Masters, Frank M	18	1941	Pvt.	PA	194th Infantry	7/22/64	11/5/64		
May, Daniel H	31	1931	Corp.	PA	82nd Infantry	11/14/64	7/13/65		
May, Marcus	19	1930	Corp.	PA	13th Infantry	1861			
				PA	138th Infantry	8/29/62	6/23/65	Wounded	

Robert M Skillington was 21 years old when he left his home in Snake Spring Valley to volunteer in the 133rd Pennsylvania Infantry in August 1862. He took part in the Union charge at Marye's Heights during the battle of Fredericksburg in December 1862. Robert returned home in May 1863 after the 133rd Regiment completed their 9-month enlistment. He reenlisted in the 184th Pennsylvania Infantry on May 12th, 1864. Three weeks later, Robert suffered a wound at Cold Harbor during the early morning assault of Confederate lines on June 3rd. He was wounded a second time during the Initial Assault on Petersburg two weeks later. That same week, his brother John enlisted in the 49th Pennsylvania Infantry. John died of Typhoid Fever on September 14th, 1864. Robert survived the war and returned to Bedford County. He married in 1876 and raised 6 children in his East Providence Township home. He lived to be 90 years old and was buried at the Mt. Zion Cemetery in Breezewood in 1931.

George Rowzer was born on June 8th, 1842 in Napier Township. He enlisted in the 107th Pennsylvania Infantry in September 1864. George suffered a wound to his right arm on February 6th, 1865 at the battle of Dabney's Mills. He recovered from the wound and returned home in June 1865. George worked as a shoemaker and a farmer after the war. He married in 1870 and was a father to 6 children. George lived to be 94 years old and was buried in the Pleasantville Cemetery in 1936.

Absolom Showalter was born in West Providence Township in May 1848. He enlisted in the 48th Pennsylvania Infantry in January 1865. The 48th took part in heavy combat at Fort Mahone during the Final Assault on Petersburg on April 2nd, 1865. Absolom mustered out in July 1865. After the war ended, Absolom worked on a farm and was as a postmaster in East Providence Township. He married in 1883 and was a father to one daughter. Absolom lived to be 82 years old and passed in 1931.

Name	Muster Age	Lived Until	Rank	State	Regiment	Muster In	Muster Out	Casualty	Battle
McClain, Jesse O	21	1930		PA	3rd Heavy Art.	7/29/64	11/7/65		
McGregor, William A	19	1937	Pvt.	PA	99th Infantry	3/17/65	7/1/65		
Mellott, Jacob L	19	1936	Pvt.	PA	208th Infantry	9/7/64	6/1/65		
Metzger, John S	17	1935	Pvt.	MD	2nd PHB Inf.	3/28/65	6/28/65		

George Sparks was born on November 6th, 1945. His father Mahlon was born in Bloody Run (Everett) and listed in the 1840 West Providence Township Census before moving their family to Ohio after George was born. Mahlon and four of his sons joined Ohio regiments during the Civil War. Mahlon and his son Henry enlisted in the 64th Ohio Infantry. George and Wesley enlisted in the 163rd Ohio Infantry. John enlisted in the 12th and 25th Ohio regiments. Mahlon and all four sons survived the war. George married in 1869 and raised two daughters in Missouri. He worked as a carpenter during most of his life. George lived to be 87 years old and passed in 1933.

Samuel Weyandt and his wife Mary Warsing Weyandt pictured on their wedding day in the early 1870s. Samuel was 18 when he volunteered in the 125th Pennsylvania Infantry in August 1862. The following month, the 125th was in heavy combat near the Dunker Church at the battle of Antietam. He mustered out of the regiment in May 1863. Samuel re-enlisted in the 3rd Heavy Artillery in February 1864 and returned home in November 1865. Samuel and Mary were parents of 4 children. Mary passed when she was 74 in Broad Top Township in 1923. Samuel lived to be 95 years old and was buried at the Old Claysburg Cemetery in 1938.

Name	Muster Age	Lived Until	Rank	State	Regiment	Muster In	Muster Out	Casualty	Battle
Metzler, Henry Clay	16	1935	Pvt.	MD	3rd PHB Inf.	2/17/62	5/29/65		
Murphy, Philip (2)	18	1937	Pvt.	OH	189th Infantry	3/3/65	9/28/65		
Myers, Martin L	18	1942	Pvt.	PA	99th Infantry	3/20/65	7/1/65		
O'Neal, Samuel	18	1936	Pvt.	PA	195th Infantry	4/13/65	1/31/66		
Penrose, Joseph	18	1930	Pvt.	PA	21st Cavalry	7/8/63	2/20/64		
			Pvt.	PA	205th Infantry	8/26/64	6/2/65		
Prince, Edwin S	16	1934	Pvt.	OH	162nd Infantry	5/20/64	9/4/64		
			Pvt.	OH	184th Infantry	2/19/65	9/20/65		
Ralston, William H	24	1932	Sgt.	PA	110th Infantry	10/24/61	2/11/63		
			1st Sgt.	PA	184th Infantry	5/12/64	7/14/65		
Reighard, George W	17	1936	Pvt.	PA	184th Infantry	5/12/64	5/15/65	Wounded	Cold Harbor
Robinette, Henry C	20	1932	Pvt.	MD	1st PHB Cav.	4/23/64	6/28/65		
Robison, Henry C	17	1935	Pvt.	PA	49th Infantry	10/24/61	10/24/64		
Rowzer, George C	22	1936	Pvt.	PA	107th Infantry	9/21/64	6/2/65	Wounded	Dabney's Mills
Schetrompf, Peter C	17	1930	Pvt.	MD	3rd PHB Inf.	10/7/63	5/29/65		
Shoemaker, George F	19	1937	Corp.	PA	101st Infantry	11/1/61	6/3/65	POW	Plymonth, NC
Showalter, Absolom	17	1931	Pvt.	PA	48th Infantry	1/20/65	7/17/65		
Skillington, Robert M	21	1931	Pvt.	PA	133rd Infantry	8/13/62	5/26/63		
			Pvt	PA	184th Infantry	5/12/64	7/14/65	Wounded	Cold Harbor
			Pvt	PA	184th Infantry			Wounded	Petersburg - Initial

James Walter is pictured with his wife Christie and granddaughter Mabel Walter circa 1930. James Walter was born near New Buena Vista on August 2nd, 1845. He volunteered in the 2nd Maryland PHB Infantry in October 1863. His father Herman enlisted in the 28th Pennsylvania Infantry the following year. Both survived the war and returned home to Juniata Township. James married Christie Otto on Christmas day 1865 and they raised 9 children who survived to adulthood. They moved their family to Ohio in the late 1880s. James lived to be 88 and passed in 1934, Christie was 87 when she passed in 1938. Mabel passed in 2012 at age 96.

William Woodcock was born in Wells Valley, Fulton County in October 1843. He mustered as a private in the 77th Pennsylvania Infantry on October 9th, 1861. William received a promotion to 1st Lieutenant of the Signal Corps before mustering out in August 1863. His half-brother George died of disease in June 1863 while enlisted in the 76th Ohio Infantry. After the war, he taught school, was a high school principal and a successful lawyer in Blair County. William married for the 1st time in 1901 and was a father to 2 daughters. He was a trustee of the American University in Washington, DC, and of Dickinson College in Carlisle. He lived to be 91 and passed in 1935.

Name	Muster Age	Lived Until	Rank	State	Regiment	Muster In	Muster Out	Casualty	Battle
Smith, Daniel A		1931	Pvt.	OH	18th Infantry	9/23/61	1/30/64		
Smith, William R (1)	19	1932	Pvt.	PA	32nd Infantry	6/26/63			
Sparks, George W	18	1933	Pvt.	OH	163rd Infantry	5/12/64	9/19/64		
Stailey, Thomas	17	1930		IN	147th Infantry	1863	1864		
				US	18th Infantry	1864	9/8/68		
Statler, Samuel F	13	1936	Pvt.	PA	55th Infantry	10/10/61	8/8/65	Wounded	Bermuda Hundred
Swartz, John W	17	1935	Pvt.	PA	194th Infantry	7/22/64	9/6/64		
			Pvt.	PA	97th Infantry	9/6/64	7/17/65		
Tipton, Noah (2)	21	1932	Pvt.	PA	138th Infantry	9/29/62			
Walter, James A	18	1934	Pvt.	MD	2nd PHB Infa.	10/17/63	5/19/65		
Wentling, Samuel J	18	1930	Pvt.	PA	210th Infantry	9/5/64	5/30/65	Wounded	
Wertz, Thomas	18	1930	Pvt.	PA	194th Infantry	7/22/64	11/5/64		
Weyandt, Samual S	18	1938	Pvt.	PA	125th Infantry	8/13/62	5/18/63		
			Pvt.	PA	3rd Heavy Art.	2/23/64	11/9/65		
Weyant, Lafayette	15	1931	Pvt.	PA	13th Cavalry	2/26/63	8/14/65	POW	Winchester
Wigfield, James W	26	1936	Pvt.	PA	57th Infantry	7/5/64	6/29/65		
Witt, Jonathan	20	1930	Pvt.	MD	2nd PHB Inf.	3/30/65	5/29/65		
Woodcock, William L		1935	Lt.	PA	77th Infantry	10/9/61	Aug-63		
Woy, John W	19	1931	Pvt.	PA	22nd Cavalry	2/22/64	6/24/65		

*"No Person was ever honored for what he received.
Honor has been the reward for what he gave.*

- Calvin Coolidge

13. Medal of Honor Recipients

The Medal of Honor has been awarded approximately 3500 times in our nation's history. There were 1523 recipients during the Civil War, including 210 soldiers from Pennsylvania. Three soldiers from Bedford County were awarded the Medal of Honor for actions taken during the Civil War.

James Levi Roush is pictured with his medals. On the left is a photograph of the 6th Pennsylvania Reserves Monument, which stands about 25 yards north of Wheatfield Road across the intersection with Ayers Avenue at Gettysburg. The location of the monument is near where James Levi Roush and 5 other soldiers captured 12 Confederate sharpshooters on July 2nd, 1863.

In the 1890s veterans began returning to Gettysburg to attend monument dedication ceremonies for their regiments. In August 1896, 40 regimental alumni of the 6th Pennsylvania Reserves sent a petition to the Secretary of War for consideration of the 6 men in their regiment who saved the lives of many near the foot of Little Round Top during the 2nd day of Gettysburg. A year later, James Levi Roush received the Medal of Honor on August 3rd, 1897.

After days of hard marching from Frederick, MD, the 6th Pennsylvania Reserves reached the northern slope of Little Round Top in the middle of the afternoon on July 2nd, 1863. The 6th observed Union lines deteriorating in the Devils Den and Wheatfield areas of the battlefield. James Levi Roush and the 6th Pennsylvania Reserves were soon rushing toward the Wheatfield. Confederate sharpshooters were taking deadly aim at Union troops from the John Weikert house, a short distance to the right of their lines. Sergeant George Mears asked for 5 volunteers to help clear the house. Roush was among the volunteers who charged over 80 feet of open ground and crawled over a fence to reach the house. The 6 men burst through the door and forced the 12 Confederates inside to surrender. This act of heroism was largely forgotten until surviving veterans of the 6th Pennsylvania Reserves recalled the action during their monument dedication ceremony in 1890, near where the Weikert house once stood.

On the left is an early illustration by Franklin Briscoe of the Pennsylvania Reserves charging toward the Wheatfield. The smoke-covered Little Round Top is in the near background and the Big Round Top is shown in the distant background. The Weikert house would have stood a short distance to the right of the closest standing soldier in this picture.

Abraham Kerns Arnold was born on March 24th, 1837 and raised in Bedford Borough. He graduated from West Point in 1859 and received a commission as a 1st Lieutenant in the US 2nd Cavalry in July 1859. Abraham transferred to the US 5th Cavalry in 1861 and was brevetted to Captain after being cited "for gallant and meritorious services" at Gaines Mill in June 1862. Within a single week during the Overland Campaign, Abraham was twice recognized for valor. He was brevetted to major after the Wilderness battle of Todd's Tavern on May 7th & 8th, 1864. Two days later, Abraham was awarded the Medal of Honor for heroic actions taken at Davenport Bridge, VA. The following is the citation, "The President of the United States of America, in the name of Congress, takes pleasure in presenting the Medal of Honor to Captain Abraham Kerns Arnold, United States Army, for extraordinary heroism on the 10th of May 1864, while serving with 5th U.S. Cavalry, in action at Davenport Bridge, Virginia. By a gallant charge against a superior force of the enemy, Captain Arnold extricated his command from a perilous position in which it had been ordered." Abraham married in 1866 and was a father to 3 children. He was a career officer and became a Brigadier General during the Spanish-American War. Abraham passed at age 64 and was buried in the Saint Philip's Church Cemetery in Garrison, NY in 1901.

Kurz & Allison illustration of the desperate Cavalry fight during the battle of the Wilderness, near Todd's Tavern in 1864. On May 7, the two opposing cavalries met at Todd's Tavern at 4:00pm. They engaged in a slashing cavalry battle until after dark. The battle resumed the next morning with both sides taking heavy losses.

Photograph taken in the late 1800s of Shy's Hill outside of Nashville. The site of Andrew's heroism.

Andrew Jackson Sloan was born on May 9th, 1835 in St. Clair Township. He was the oldest of 9 children. His family moved west in 1855 and bought an 160 acres farm in northeast Iowa. Andrew and his younger brother Samuel left the family farm to muster in the 12th Iowa Infantry on September 23rd, 1861. Andrew's physical description in his military records stated he was 5'9" tall, with hazel eyes, dark hair, and a dark complexion. The 12th Iowa experienced heavy combat in the western theater of the war. But disease took the highest toll on the 981 men who originally enlisted in the regiment. Two hundred and fifty men from the 12th Iowa died from disease, including those who perished in prisoner of war camps. Hospital records show Andrew was treated for disease in 1862 and a second time in the fall of 1864. His brother Samuel mustered out of the 12th Iowa Infantry on December 13th, 1863, and transferred to the Veterans Reserve Corp. During the Civil War, it was common for soldiers who suffered debilitating wounds or sickness to transfer to non-combat support regiments. There are no available records on the reason for Samuel's transfer.

Andrew re-enlisted in the 12th Iowa Infantry in early 1864. While on a 30-day leave, he married Martha Sterling on April 15th, 1864, and returned to his regiment soon after the wedding. During the battle of Nashville, on December 16th, 1864, the 12th Iowa Infantry charged up a steep hill while under heavy fire. The regiment captured over 200 prisoners, four 12-pound Napoleon cannons and two battle flags. Officers in the regiment recommended Andrew for the Medal of Honor for being in front of the charge and placing his life in great peril by seizing the Rebel flag of the 1st Louisiana Battery. During the Civil War, any soldier holding a battle flag was an immediate target of enemy fire. Secretary of War Edwin Stanton awarded the Medal of Honor to Andrew Sloan in Washington, D.C. on February 24th, 1865. He mustered out of the army in January 1866 and returned home a hero. Tragically, Andrew died in 1875 from a freak hunting accident. He climbed a tree for a better vantage point, lost his grip, and fell on some rocks below. He left behind a young widow and three children. Andrew was 39 years old and is buried in the Platt Cemetery in Colesburg, Iowa.

There are no known pictures of Andrew Jackson Sloan. Andrew's brother Samuel Sloan is on the left. Their mother Mary Ann "Polly" Cuppett Sloan is in the center. Andrew's wife, Martha Sterling, is on the right, the year after she remarried 1879. Sadly, she spent the last 14 years of her life in a mental hospital in Iowa. Martha died at age 57 and was buried in the hospital cemetery in 1904. Mary Ann passed in 1901. Samuel lived to be 89 and passed in 1929.

14. Confederate Enlistments

Little documentation exists on the total numbers of brothers who fought on opposite sides of the Civil War. Two families with brothers who fought in the Union and Confederate armies are known to have lived in Bedford County during their lifetimes.

Reuben A Riley volunteered in the Union Army, and John S Riley enlisted as a Confederate. Their parents, Andrew Riley and Margaret Sleychk, married in 1812. The spelling of Sleychk eventually changed to Slick or Sleek. John was born on December 12th, 1817, and Reuben on June 7th, 1819. Their father, Andrew was listed on the 1820 St. Clair Township Census before moving the family West in the early 1820s, eventually settling in Randolph County, Indiana.

Reuben Riley volunteered in the 8th Indiana Infantry from April to August 1861 while his wife and 5 children waited for his return at their home in Hancock County, IN. He reenlisted as a Captain in the 5th Indiana Cavalry in October 1862 and mustered out on Christmas Day 1863.

His brother John was a Mexican War veteran who went to California during the Gold Rush before returning to Texas in the 1850s. He enlisted in Waul's Confederate Cavalry in 1862, leaving behind a wife and 6 children at home. He was captured during the battle of Vicksburg and taken to a Union POW camp in Alton, Illinois. John escaped from the camp and lived for a time with his mother in Indiana under the name of John Sleychk. Three more brothers born in either Ohio or Indiana also fought on opposite sides of the Civil War. Two of the brothers were in Union Army regiments. George was a Lt. Colonel in the 134th Indiana Infantry and Benjamin enlisted in the 79th Indiana Infantry. Andrew enlisted in the 8th Missouri Confederate Infantry. It does not appear any of the brothers would have opposed each other on a battlefield. All 5 brothers survived the war. Reuben passed when he was 74 in 1893. John was still a practicing doctor in Denton, TX in 1914, the year before he passed at age 97.

Reuben Riley

John Riley circa 1910

Philip Knee

David H Knee circa 1880

The Knee brothers grew up together in Woodbury Township. Philip was born in 1825, David in 1832 and Augustus in 1841. Philip and Augustus volunteered in the Union Army and David joined the Confederate Army. Please note the picture of Philip is in question. This photograph of the unnamed Union soldier was originally a possession of David. One descendant believes the picture was of his brother Philip. Another descendant questioned whether the picture could be Philip because the soldier appears to be younger than 35, which was Philip's age at enlistment. The uniform in the photograph was compared with that of another Bedford County soldier in the MD 2nd PHB Infantry, and it appears the uniforms are the same. Another possibility is the picture could have been of his other brother Augustus, who died during the war.

In 1850, David was living in the home of a non-relative apprenticing to be a tailor. After his apprenticeship, David moved south to Wardensville, VA (now West Virginia) about 60 miles south of Cumberland, MD and opened a tailor shop. David's muster dates are unavailable. According to a family historian, David first joined the 14th Virginia Militia. The Militia mustered in during July 1861 on a 6-month enlistment. Many of the 14th VA Militia soldiers re-enlisted in the 18th Virginia Cavalry in December 1861. Records confirm David was a Quarter-Master Sergeant in the 18th VA Cavalry. The 18th VA Cavalry was at Gettysburg, arriving on the battlefield at noon on July 3rd, and was tasked with guarding ammunition and supply trains during the Confederate retreat.

Unlike the Riley brothers, it was entirely possible David and his brother Philip were combatants on the same battlefield. Philip Knee mustered in the 2nd Maryland PHB Infantry on February 18th, 1862. Both Philip's regiment and David's regiment spent much of the war in the Shenandoah Valley region. The regimental histories of the 18th VA Cavalry, and the 2nd MD PHB Infantry state both were on the same battlefield during the Lynchburg Campaign in June 1864. At the end of the war, David was paroled in Winchester, VA on April 25th, 1865. Philip mustered out of the Union Army the following month. Their brother Augustus died of Typhoid Fever while a member of the Illinois 53rd Infantry on June 21st, 1862 in Corinth, MS. After the war, David returned to Wardensville, served two terms as mayor of the town, and continued working in his tailor shop until his death in 1896. Philip passed in 1892 and is buried in the Chaneysville Methodist Cemetery.

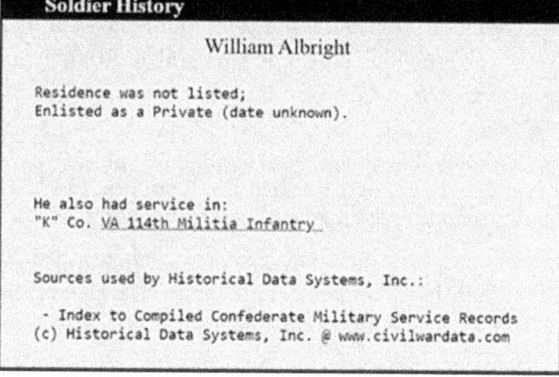
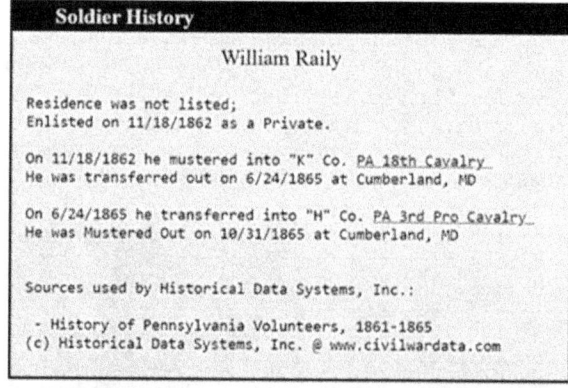

William Raley Albright pictured above was born in 1843 on a farm in Londonderry Township. William had a highly unusual wartime experience. He enlisted in both the Confederate and Union armies during the Civil War. Confirmation on two of his enlistments are shown on the above right. The 114th Virginia Militia was organized in Hampshire County south of Cumberland, Maryland. The Militia was activated on July 14, 1861 and disbanded on April 8, 1862. Many soldiers in the 114th Militia volunteered to serve in the 33rd Virginia Infantry. No documented muster dates were located for William in the 114th Virginia Militia or the 33rd Virginia Infantry. William deserted the Confederate army and volunteered in the 18th Pennsylvania Cavalry on November 18, 1862. The Pennsylvania State Archives on Civil War Veterans described William at enlistment as a 19-year-old farmer standing 6 foot tall with dark hair, dark eyes, and a fair complexion. William enlisted under his mother's maiden name. The spelling of his last name on various Union Army records was Raily. There was a good reason for the name change. If William were ever captured and identified as a Confederate deserter, he likely would have been hung on the spot. Seven months after mustering in the Union Cavalry, the former Confederate was in combat during the battle of Gettysburg. A statue of the 18th Pennsylvania Cavalry is near Big Round Top, where William would have taken part in a charge at the conclusion of Picket's Charge. Above on the lower right, William Raily's name is engraved on the 18th PA Cavalry plaque on the Pennsylvania State Monument at Gettysburg, confirming he was a participant in the battle. The 18th PA Cavalry was also on the battlefield at the Wilderness, Cold Harbor, Opequon and Cedar Creek. After the war ended, the 18th and the 22nd PA Cavalry regiments merged to form the 3rd PA Provisional Cavalry in June 1865. William mustered out in Cumberland, Maryland, in October 1865, near his home in Bedford County. He married in 1868 and raised 7 children who survived to adulthood on a family farm in Londonderry Township. William passed at age 69 in 1913.

How unusual would it have been for a soldier born and raised North of the Mason-Dixon line to have fought in both the Confederate and Union armies during the war? There is scant documentation on this topic, but it appears to be quite unusual. There were approximately 5,600 Confederate prisoners of war who agreed to take an Oath of Allegiance to the United States and join Union regiments. These converts were referred to as "Galvanized Yankees" and most were assigned to non-combat duty in Western frontier states. Interestingly, over 250 of these repatriated "Galvanized Yankees" had originally enlisted in the Union Army before being captured by the Confederates. They later agreed to join the Confederate army to gain their release from hellish southern POW camps. These 250 hapless soldiers had the dubious distinction of being captured by both the Confederate and Union armies during the war.

More documentation exists of soldiers born in Southern states who became disillusioned with the Confederacy and fought in the Union Army. There were some isolated areas within the Confederacy with pro Unionist sentiments. One example was Winston County, Alabama. Winston County was a predominantly poor area of subsistence farmers who did not own slaves and viewed succession from the Union as an illegal act. Twice as many men from Winston County joined Union Regiments than enlisted in the Confederate Army. Many joined the 1st Alabama Union Cavalry regiment. Of the 2,678 identified white soldiers from Alabama who fought in Union Regiments, over 2000 enlisted in this unit. County native John C Latta transferred as an officer in the 1st Alabama Union Cavalry.

John C Latta was born on January 18th, 1838. He grew up on an East Providence Township farm with two brothers who also served in the Union Army. John mustered as a corporal in the 57th Illinois Infantry on December 26th, 1861. His enlistment record noted he was a farmer from Manlius Illinois, stood 5'9 ½" tall with brown hair, blue eyes, and a light complexion. John received a commission as a 1st Lieutenant in the 1st Alabama Union Army Cavalry on December 22nd, 1862. This Union regiment was filled primarily with recruits from northern Alabama that opposed secession. General William Tecumseh Sherman chose the 1st Alabama Cavalry to be his escort during his famous "March to the Sea" campaign in November and December 1864. John received a promotion to Captain during the war and suffered an arm wound in March 1865. His younger brothers, William H Latta, mustered in the 148th Pennsylvania Infantry in August 1863, and 16-year-old Abraham volunteered in the 208th Pennsylvania Infantry in September 1864. All three brothers mustered out of the Union Army after the war ended. John returned to Illinois, married in 1866 and was a father to 7 children. He moved to South Dakota after the mother of his children passed in 1892. John passed in 1915 at age 77. He is buried at the Medicine Hill Cemetery in Hughes County, SD.

The life story of William Hinson, also known as Oliver Perry Niley, and how he came to Bedford County is one of the most fascinating local stories of the Civil War. Oliver Perry Niley was born in Yazoo City, Mississippi on December 7th, 1842. Oliver first volunteered in the Mississippi 10th Infantry in March 1861 on a 1-year enlistment. He re-enlisted in the 1st Mississippi Light Artillery on April 30th, 1862. He was captured on July 4th, 1863 during the Confederate surrender at Vicksburg. Vicksburg is one of the most important battles of the Civil War resulting in the Union Army gaining full control of the Mississippi river thus denying the Confederacy a primary transportation route to move troops and supplies. Victorious general, Ulysses S Grant, received a promotion to lieutenant general and commander of the entire Union Army the following March. After his capture, Oliver was sent to a Northern Prisoner of War camp in Alton, Illinois, in November 1863. In February 1864, Oliver was being transferred by train to a POW camp in Delaware. During a train stop near Gallitzin in Cambria County, he made an escape into the nearby heavily wooded mountains. Oliver immediately changed his name to his maternal Grandfather's name, William Hinson. He was given shelter by a Quaker family in Somerset County before finding work making barrels near Pleasantville. Within weeks, William married his boss's daughter, Catherine Miller, and the first of their 11 children was born in February 1865, a year after William escaped from the train. William was a well-respected man in the community during his lifetime, eventually becoming a Justice of the Peace. Just before he passed away in 1925, William called his family together to tell them for the first time the truth about his past as a Confederate Soldier who escaped from a POW train during the Civil War.

John C Latta

William Hinson aka Oliver Perry Niley

Chapter 14

There are 11 identified former Confederate soldiers who lived in Bedford County during their lifetimes. Several have documentation confirming they were born and raised in Southern states before moving to Bedford County after the Civil War ended. A Broad Top Township Veterans GAR Voter Roll listed Daniel and Fred Rice as Confederates living in Kearney. No information is available to determine whether both men were Bedford County residents prior to the war or moved from the South afterwards to work in Coal Mines. Rice is a common last name in Bedford County and the South. 1890 Broad Top Township Veterans Census listed James Woods and Joseph Edmonson next to each other, possibly indicating they lived in the same household. The only information on this Census was they were former Confederates living in Kearney. Records exist of Edmonson being born and raised in Virginia prior to moving to Bedford County after the war. There is no other census or other information on a James Woods living in Bedford County prior to or after the Civil War ended. Woods is a common name in the South. There are 102 Confederate soldiers listed in the Historical Systems database with the name of James Woods. The Southern economy remained in ruin for decades after the war ended. Whether Daniel Rice, Fred Rice and James Woods were Southerners who moved to the Broad Top area seeking work after the war may never be known.

Confederates	Muster Age	Rank	State	Reg't	Unit	Muster In	Muster Out	Information
Albright, William	18	Pvt	VA	114th	Inf			Refer to the biography on a previous page
Edmonson, Joseph	17	Pvt	VA	3rd	Cav	8/14/61	8/1/64	Born in Mecklinburg, VA; Wounded in right leg near Boonsboro, MD on 7/8/63 during the Confederate retreat from Gettysburg; Listed on 1880 Dinwiddie, VA Census. Listed on the 1890 Broad Top Veterans Census as a Confederate currently living in Kearney.
Gubernater, Charles	37	Pvt	LA	3rd	Cav	9/1/64		Born in Virginia and listed on a county census for the 1st time in 1870 as working as a blacksmith in Bedford Township; Listed on the 1890 Bedford Twp Veterans Census.
Hinson, William	19	Pvt	MS	1st	LA	4/30/62		Refer to the biography on a previous page
Knee, David H	32	Sgt	VA	18th	Cav			Refer to the biography on a previous page.
Neville, J Richard	15	Pvt	VA	23rd	Cav	3/14/64		Born in Virginia; Historical Data Systems lists Richard Neville as enlisting in Winchester, VA; He appears to have moved to Hyndman sometime after being listed on the 1870 Winchester, VA Census.
Rice, Daniel								Bedford Historical Society record - On GAR Voter rolls listing for Kearney in Broad Top Township
Rice, Fred								Bedford Historical Society record - On GAR Voter rolls listing for Kearney in Broad Top Township
Riley, John S		Surg	TX	Waul's	Cav			Refer to the biography on a previous page.
Rollins, Thomas J	28	Pvt	GA	60th	Inf	5/10/62		Born Spartanburg, SC; Listed on the 1850 Spartanburg Census; Wounded at 2nd Bull Run 8/28/62; POW at Gettysburg 7/3/63; Deserted on 6/6/64 and released by the Union Army on 6/8/64 on the condition he remained North of the Ohio River.
Woods, James C								Listed in 1890 Broad Top Veterans Census as a Confederate living in Kearney in Broad Top Township.

15. Regimental Listings

Regiments were the basic maneuver unit in the army during the Civil War and were the military unit Civil War soldiers most identified with. Bedford County soldiers enlisted in over 500 different State, US Army and USCT Regiments during the Civil War. Overviews are provided on 29 regiments with larger numbers of Bedford County Soldiers in this chapter. A listing of county African American soldiers who enlisted in various USCT regiments is also provided. Casualty figures for each regiment are categorized by Killed in Action, Died from Wounds, Missing in Action (presumed killed in battle or died as POW), Died (includes from disease & unidentified Battle wounds), Wounded-POW-Died (wounded, taken prisoner and later died), Died while POW, Wounded & POW (wounded and taken prisoner), POW, Wounded, and Injured (in battle). Casualty Percentages are calculated by dividing the known Bedford County casualties by the number of Bedford County enlistments in each regiment. Total casualty figures include multiple casualties suffered by a soldier during the war. For example, if a soldier was wounded in two separate battles, it is counted as two casualties for that regiment. Multiple casualties being suffered by a soldier is the reason county soldiers in the 8th Pennsylvania Reserves had a casualty rate percentage of over 100%. Please note the second column on the listing of soldiers in each regiment includes the pages where photographs of soldiers can be located. Photographs of some gold-star family members are included in this book and are designated with a dash after the page number with a family member abbreviation: -B (Brother), -D (Daughter), -M (Mother), -P (Parents), -S (Sister), -So (Son) and -W (Wife).

As previously mentioned, there were likely many more battlefield casualties than are listed on available records. Casualty rates and total numbers of deaths during the Civil War were extraordinarily high by any historical measure in American history. Below is an estimated number of American deaths and percentage of deaths during the Revolutionary War, Civil War (Union Army only), World War I and World War II according to the Department of Veterans Affairs in Washington, DC. Top right is a listing of Bedford County known deaths and percentage of deaths by regiment during the Civil War. Listed are the regiments with a percentage of death rates higher than 10%.

Regiment	Total County Soldiers	Total Deaths	Percentage of Deaths
184th PA Infantry	82	30	38%
101st PA Infantry	140	45	32%
11th PA Infantry	35	11	31%
8th PA Reserves	98	29	30%
55th PA Infantry	496	127	26%
2nd PA Cavalry	42	10	24%
110th PA Infantry	111	26	23%
18th PA Cavalry	23	5	22%
84th PA Infantry	45	10	22%
138th PA Infantry	279	60	22%
149th PA Infantry	53	11	21%
77th PA Infantry	55	8	15%
107th PA Infantry	78	10	13%
125th PA Infantry	47	6	13%
99th PA Infantry	112	13	12%

American War	Total Soldiers	Total Deaths	Percentage of Deaths
Revolutionary War	217,000	4,435	2%
Civil War (Union)	2,213,363	364,511	16%
World War I	4,734,991	116,516	2%
World War II	16,112,566	405,399	3%

An unidentified soldier proudly holds the 8th Pennsylvania Reserves battle flag.

8th Pennsylvania Reserves Listing

			Year	Major Battles with Casualties	Known Casualties
Bedford County Enlistments	98				
Known Casualties	99		1862	Seven Days Battle, Jun. 25 - Jul. 1	44
Casualty Rate Percentage	101%			2nd Bull Run, Aug. 28-30	3
Killed in Action	9			South Mountain MD, Sep. 14	3
Died From Wounds	8			Antietam, Sept. 17	8
Missing in Action	9			Fredericksburg, Dec. 13	14
Died	3		1864	Wilderness, May 5-7	5
Wounded & POW	3			Spotsylvania, May 8-21	12
POW	39				
Wounded	30				
Injured	1				

The 8th Pennsylvania Reserves suffered the highest percentage of casualties of any Bedford County unit during the Civil War. The regiment recruited Bedford County volunteers for Company F in June 1861. This unit became known as the Hopewell Rifles. They were organized at Camp Wilkins near Pittsburgh before being sent to Washington, DC, at the end of July. The Hopewell Rifles were front-line troops in many pivotal battles during the first 3 years of the war. Their initial combat experience was during the Seven Days Battle near Richmond in June 1862. The Hopewell Rifles were positioned on the front lines of the Mechanicsville battlefield and were not informed of the Union Army's retreat to Gaines Mill. Available records show at least 38 county soldiers were captured. Most were taken to Richmond prisoner of war camps before being exchanged for Confederate soldiers held by the Union Army within a couple months. The Hopewell Rifles were among the troops that defeated entrenched Confederates on steep terrain at Turners Gap during the battle of South Mountain that halted Robert E Lee's planned invasion of Pennsylvania. Several days later, the Hopewell Rifles took part in some of the most horrific fighting of the war in the Cornfield at Antietam. They suffered heavy casualties at Fredericksburg when Pennsylvania Reserve troops initially broke through Stonewall Jackson's lines before being driven back. The Hopewell Rifles took part in heavy combat at the battle of the Wilderness and Spotsylvania prior to the regiment completing their 3-year enlistment on May 24th, 1864.

Above is a rare wartime group photograph of Bedford County soldiers. Soldiers from the Hopewell Rifles Company pose in front of a Virginia residence in 1861. Eli Eichelberger is the only identified soldier. He is standing behind one of the chairs with his arms folded. The seated boy reportedly followed the Hopewell Rifles around for a while during the war.

8th Pennsylvania Reserve Listing

Name	Pict. Page	Known Casualties	Battle
Adams, Daniel			
Amick, George W	225	POW	7 Days Battle
Armstrong, David B	225	POW	7 Days Battle
Barber, James		POW	7 Days Battle
Barmond, John			
Barmond, Nathaniel			
Bollinger, David S		POW	7 Days Battle
Bowser, Emanuel		KIA	Fredericksburg
Bradley, James A		Wounded	Antietam
Brown, George		POW	7 Days Battle
Brumbaugh, Levi		POW	7 Days Battle
Callahan, Jacob R	225		
Carnell, John		POW	7 Days Battle
Cleaver, James	5	POW	7 Days Battle
		Wounded	Fredericksburg
		Wounded	Spotsylvania
Cook, Joseph S		POW	7 Days Battle
Dasher, William H		POW	7 Days Battle
Davis, Isaiah M	170-B	Died	
Dean, Franklin		Wounded	Antietam
Eastright, Christian		POW	7 Days Battle
		Wound.-Died	Spotsylvania
Edwards, Allison		POW	7 Days Battle
		Wounded	Fredericksburg
Edwards, Hiram		Died	
Eichelberger, Eli	19, 222, 225, 357, 360	POW	7 Days Battle
		Wounded	Wilderness
Eichelberger, John S		POW	7 Days Battle
		Wounded	Fredericksburg
Eichelberger, William H		Wounded	
Evans, Johnson		POW	7 Days Battle
		Wounded	Fredericksburg
		Wounded	Wilderness
Figart, Henry J		KIA	2nd Bull Run
		POW	7 Days Battle
Foor, Mark W		Died	
Foor, Samuel S		POW	Fredericksburg
		Wounded	7 Days Battle
Foor, William H		KIA	7 Days Battle
Foster, Aaron	60	POW	7 Days Battle
		Wounded	Fredericksburg
		Wounded	Spotsylvania
Gamble, Robert		Died	
		POW	7 Days Battle
Garlick, Christian C	19	POW	7 Days Battle
Garrett, Alexander A		KIA	7 Days Battle
Gates, James		POW	7 Days Battle
		Wound.-Died	Antietam

Name	Pict. Page	Known Casualties	Battle
Girdan, Isaac			
Griffith, Abel	85, 362		
Griffith, Michael			
Grubb, Wilson		POW	7 Days Battle
Headrick, David		Wound.-POW	7 Days Battle
Heffner, George		KIA	2nd Bull Run
		POW	7 Days Battle
Holdcraft, William		Wound.-POW	7 Days Battle
		Wounded	Spotsylvania
Holsinger, Frank		Wounded	Fredericksburg
Horton, David	57, 360	POW	7 Days Battle
		Wounded	Wilderness
Horton, George		Wound.-Died	Antietam
		Wounded	South Mountain
Horton, Zophar P	79		
Iams, David			
Imes, Aaron		Wounded	Fredericksburg
Jordan, Daniel			
Juda, George V A	137	POW	7 Days Battle
		Wound.-Died	Spotsylvania
Kay, William H		Wound.-Died	South Mountain
Leader, George W		POW	7 Days Battle
Leighty, George		Wounded	Wilderness
Leighty, John Q		Wound.-Died	Antietam
Leighty, Joseph		Wounded	Spotsylvania
Lines, Jacob		Died	
Linn, Jacob B			
Malone, Charles		Wounded	Spotsylvania
Malone, John			
Malone, William		Wound.-Died	Antietam
Manspeaker, Barclay		KIA	Fredericksburg
Manspeaker, David		KIA	Spotsylvania
Marshall, Henry		POW	Spotsylvania
Martin, David		KIA	7 Days Battle
Maugle, Joseph		Wounded	Antietam
		Wounded	Wilderness
May, P V			
McFarland, Daniel M			
McFarland, Joseph		Wounded	Spotsylvania
McKee, Alexander H			
Paul, John			

8th Pennsylvania Reserves Listing

Name	Pict. Page	Known Casualties	Battle
Paul, John			
Penrod, Henry C			
Penrod, John B Jr		POW	7 Days Battle
Penrod, John B Sr			
Piper, Lewis M	361	POW	7 Days Battle
		Wounded	Fredericksburg
Piper, Luther R	125	POW	7 Days Battle
		Wound.-Died	Fredericksburg
Reed, William B	306		
Ritchey, Joseph			
Ritchey, Samuel Y		Died	
Ritchey, William D	224, 359	Injured	7 Days Battle
		POW	Spotsylvania
Robb, Conrad		POW	7 Days Battle
Ross, Oliver P			
Scutchall, David		Died	
Shauf, Cornelius	199-P	Died	
Shaw, Matthew P		KIA	South Mountain
		POW	7 Days Battle
Shields, James			

Name	Pict. Page	Known Casualties	Battle
Showalter, Simon P			
Showalters, Henry		Wound.-POW	2nd Bull Run
		Wounded	7 Days Battle
Smith, Charles S			
Taylor, Thomas A		POW	7 Days Battle
Tobias, John B		POW	7 Days Battle
		Wounded	Fredericksburg
Tricker, George			
Waltz, Lewis B	18, 225	POW	7 Days Battle
Warsing, Alexander		Died	
		POW	7 Days Battle
Whisel, William H	225, 360	POW	7 Days Battle
		Wounded	Fredericksburg
White, Edmund H		POW	7 Days Battle
		Wounded	Antietam
Williams, John			
Williams, John P		Wounded	Spotsylvania
Young, Alexander			
Young, Joel T			

William D Ritchey (right) pictured with his brother James H (left) who enlisted in the 22nd PA Cavalry. William was born in West Providence Township on March 14th, 1840. He was living in Hopewell when he joined the 8th Pennsylvania Reserves on June 11th, 1861. He suffered an injury leaping over a fence and a ditch during an infantry charge at the battle of Mechanicsville. After recovering from surgery, William returned to his regiment on March 1st, 1863. William was captured at Spotsylvania on May 8th, 1864, but escaped four days later. He mustered out with the 8th PA Reserves regiment on May 24th, 1864. William immediately re-enlisted in the PA 191st Infantry. He was later wounded during the Siege of Petersburg. He survived the war and was a father to 7 children. His last surviving child passed in 1993. William passed at age 66 and is buried at the Bethel Brethren Cemetery in Tatesville. James died in 1868 and is buried in the Hinish Family Cemetery. His cause of death is not listed on available records.

8th Pennsylvania Reserve Listing

Eli Eichelberger, Captain

David B Armstrong, 1st Sergeant

Jacob R Callahan, Sergeant Major

George W Amick, Private

Lewis Waltz, Captain

William H Whisel, Private

11th Pennsylvania Infantry Listing

Bedford County Enlistments	35
Known Casualties	27
Casualty Rate Percentage	77%
Killed in Action	4
Died from Wounds	2
Died	5
Wounded & POW	2
POW	1
Wounded	13

Year	Major Battles with Casualties	Known Casualties
1862	2nd Bull Run, Aug. 28-30	4
	Antietam, Sep. 17	7
	Fredericksburg, Dec. 13	2
1863	Gettysburg, Jul. 1-3	2
1864	North Anna River, May 23-26	3
	Petersburg - Initial Assault, Jun. 15-18	1
	Globe Tavern, Aug. 18-21	1
1865	Appomattox, April 9 (at Surrender)	

Bedford County volunteers in the 11th Pennsylvania Infantry were front line combatants who helped turn back both of Robert E Lee's planned invasions of Pennsylvania. The 11th PA Infantry regiment was organized in Harrisburg and Westmoreland County in August 1861. Bedford County soldiers volunteered for 3-year enlistments and received initial training at Camp Curtin in Harrisburg before being sent to Maryland on November 27th. Some county soldiers served in the 11th PA Infantry until the war ended and experienced heavy combat in many of the most well-known battles of the Civil War. County soldiers in the 11th were among the first Union troops to face a murderous fire as they emerged from the Cornfield at first light during the battle of Antietam. The 11th Pennsylvania Infantry was positioned on the Union left in the Slaughter Pen area of the Fredericksburg battlefield. County soldiers in the 11th were among the outnumbered Union troops on the 1st day of Gettysburg, whose determined heroics enabled the Union Army to seize the crucial high ground surrounding the town. The 11th Pennsylvania Infantry took casualties during the Overland Campaign in late spring 1864 and during the siege of Petersburg before witnessing the Confederate surrender at Appomattox. The 78% casualty rate is the 3rd highest of any regiment with significant numbers of county soldiers.

Name	Pict. Page	Known Casualties	Battle
Arenze, Charles		Died	
Armstrong, Joseph	126	Died	
Brallier, Rueben		Died	
Corle, Leonard			
Cornell, William		Wounded	Hatcher's Run
Frazey, Frederick L	29	Wounded	Antietam
Frazey, Henry P		Wounded	Dabney's Mills
Funk, Enoch		KIA	North Anna River
Gibson, John			
Gibson, Joseph			
Grimes, George W			
Hixon, Henry H		Wound.-POW	Gettysburg
		Wounded	Fredericksburg
		Wounded	Petersburg - Initial Assault
Livingston, Thomas	11, 61		
Mahoney, John			
Manspeaker, George		KIA	Fredericksburg
Manspeaker, John		POW	Globe Tavern
McLaughlin, Collin	227		
Mellott, Cornelius	175-P	KIA	2nd Bull Run
Mellott, John L	175-P	Died	
Morse, David		Wound.-Died	Antietam
Nabona, John			
Pee, Frances W	40, 227, 363	Wound.-POW	Gettysburg
		Wounded	Antietam
Riley, William (1)		Wound.-Died	Antietam
Sigel, Stephen		Wounded	2nd Bull Run
		Wounded	Antietam
Stoutnour, Samuel R		Wounded	North Anna River
Streight, John			
Swartz, Abraham		Wounded	2nd Bull Run
Veatch, John G	227		
Weaverling, Adam	149	Wounded	2nd Bull Run
Weaverling, David		Wounded	Antietam
		Wounded	North Anna River
Weaverling, Jacob P		KIA	Antietam
Wilkins, Josephus			
Wilkins, William		Died	
Wilson, Joseph B			
Woodcock, Oliver E			

11th Pennsylvania Infantry Listing

(left) Members of the 11th PA Infantry on Dedication Day of their monument in Gettysburg in 1890. In the center of the monument on a bottom pedestal lays a likeness of their beloved mascot "Sallie." Below left is a closeup photograph of Sallie. This statue is one of the most well-known and beloved monuments at Gettysburg. The following is a verbatim excerpt from the Gettysburgsculptures.com website. "Sallie would be given to the members of the regiment when only a puppy. She would serve with the men in the ranks and suffer the hardships of long campaigns as well as the victories won on many fields. It was reported Sallie would be seen in the ranks and in line of battle barking as bullets whizzed by her. At Gettysburg she became separated from her friends of the 11th during their retreat through town. Not sure of where to go, she would return to the battle line of the first day's battle and would be found several days later resting among the dead and wounded until she rejoined her comrades of the 11th. Sallie would serve with the 11th until she was killed during the battle of Hatcher's Run, Virginia, on February 6, 1865. Today she is remembered on the western face of the monument, as she calmly looks across the fields of Gettysburg in search of her old friends."

Colin McLaughlin, Private

Frances & Sarah Pee. Sarah was at home with a son born 6 months before Frances was wounded at Antietam.

Adaline & Pvt. John Veach married in 1872 after John's first wife and forth child died during childbirth.

55th Pennsylvania Infantry Listing

Bedford County Enlistments	496
Known Casualties	272
Casualty Rate Percentage	55%
Killed in Action	19
Died from Wounds	28
Missing in Action	5
Died	46
Died while POW	29
Wounded & POW	4
POW	22
Wounded	118
Injured	1

Year	Major Battles with Casualties	Known Casualties
1862	Edisto Island SC, Mar. 29	10
	Pocotaligo, SC, Oct. 22	5
1864	Drewry's Bluff, May 16	52
	Bermuda Hundred, May	15
	Cold Harbor, May 31 - Jun. 12	29
	Petersburg - Initial Assault, Jun. 15-18	28
	Chaffin's Farm, Sep. 28-30	9
1865	White Oak Road, Mar. 30-31	6
	Petersburg - Final Assault, April 2	1
	Sailor's Creek, April 6	1
	Appomattox, April 9 (at Surrender)	

The 132 deaths and 276 total known casualties of the 55th Pennsylvania Infantry were by far the most suffered by any unit from Bedford County during the Civil War. The 55th PA Infantry regiment was organized during the summer and fall of 1861. Volunteers in Companies D, H, and K were recruited in Bedford County and Company I was recruited in Blair and Bedford. County troops were initially trained at Camp Curtin in Harrisburg before leaving for South Carolina in November 1861. Bedford County soldiers suffered at least 15 battlefield casualties and 13 deaths from disease during the South Carolina campaign in 1862. Late spring in 1864, the 55th Pennsylvania Infantry regiment was sent on a mission southeast of Richmond to disrupt Confederate supply lines, interrupt communications, and occupy Confederate forces to deny Robert E Lee troop reinforcements during Ulysses Grant's relentless Overland Campaign. The Bermuda Hundred Campaign culminated with the 55th Pennsylvania Infantry being driven back from a Confederate fortress overlooking the James River during the battle of Drewry's Bluff on May 16th. Bedford County soldiers in the 55th suffered their heaviest losses of the war, with 68 battlefield casualties within a two-week period in May. On June 3rd, the 55th took heavy casualties during a tragic early morning charge on entrenched Confederate lines at Cold Harbor. Two weeks later, county soldiers suffered 28 battlefield casualties during the initial assault on Petersburg. The 55th took part in several more offensives during the 9-month siege of Petersburg. Fittingly, many Bedford County Soldiers who had survived much combat over the previous 3 years were eyewitnesses to the Confederate surrender at Appomattox.

John A Livingston, Captain

Daniel M Wonders, Sergeant - Quarter Master

55th Pennsylvania Infantry Listing

Name	Pict. Page	Known Casualties	Battle	Name	Pict. Page	Known Casualties	Battle
Adams, George W				Boor, William A	65	POW	Drewry's Bluff
Adams, John				Bowman, John			
Adams, Philip	163-M	POW-Died	Drewry's Bluff	Bowman, William			
Adams, Samuel	234			Bowser, Daniel L		Injured	Cold Harbor
Adams, Solomon		Died		Bowser, David		Wounded	Drewry's Bluff
Agnew, George H		POW-Died		Bowser, Nicholas		Wounded	Drewry's Bluff*
Agnew, William K				Bradley, Francis P			
Allen, William				Brady, Peter			
Allenbarger, George		Wounded	Bermuda Hundred	Brant, Shannon		Wounded	
Allison, David		POW-Died	Drewry's Bluff*	Bridaham, Henry W		POW-Died	Drewry's Bluff
Allison, Edward		POW-Died	Drewry's Bluff	Bridenthal, Henry B			
Allison, Jacob W		Died		Broad, Isaac N		Died	
Allison, John		Died		Brookins, John			
Allison, John M				Burket, Baltzer		KIA	Bermuda Hundred
Allison, Nathaniel		Wounded	Cold Harbor	Burket, Frederick	129	POW-Died	Drewry's Bluff
Amick, William M	126	Wound.-Died	Petersburg - Initial Assault	Burket, Gabriel	73, 356	Wounded	
		Wounded	Cold Harbor	Burket, Jacob	235		
Amos, John B	65, 234	Wounded	Cold Harbor	Burket, Philip		Died	
		Wounded	Drewry's Bluff	Bush, Charles A			
Anderson, Henry		POW	Drewry's Bluff	Butler, Andrew		Died	
Andrews, John				Buxton, George W			
Arnold, Albin C	235			Byerly, James F			
Arnold, Almon				Cain, John		Wounded	Cold Harbor
Arnold, Humphrey Y				Carley, Peter A			
Ayers, Charles				Carmack, John A		Wound.-Died	Petersburg
Baker, John C		Wounded		Carson, Samuel R			
Baker, William				Cessna, William	235	Wounded	
Barkheimer, John		Wounded		Claar, Henry I	114, 362	Wounded	
Barnhart, John H		KIA	Drewry's Bluff	Claycomb, Frederick		Wounded	Pocotaligo, SC
Bartlebaugh, John		MIA	Chaffin's Farm	Claycomb, John	13	Wounded	Pocotaligo, SC
Bartlebaugh, Silas M		POW	Drewry's Bluff	Cobler, Francis C			
Beaver, Simon J		POW	Drewry's Bluff	Cobler, John A		Wounded	Bermuda Hundred
Bedell, Edmund				Coffee, John		Died	
Beisel, George M		Wound.-Died		Cole, John			
Beltz, William H				Cole, Samuel			
Beneigh, John				Conrad, Winfield S	203		
Bennett, Daniel		Wounded		Corle, Alexander B		POW	Drewry's Bluff
Bennett, Jacob	166-B	POW-Died		Corle, Chauncey	130, 169-P	Wound.-Died	Petersburg
Berkett, Harvey				Corle, Eli	169-P	Died	
Berkhimer, John				Corle, James L		KIA	White Oak Road
Bingaman, David				Corle, Martin			
Bird, William				Corle, Michael S			
Birkhimer, Samuel				Corley, Peter A		Wounded	
Blackburn, Nathan F				Crist, Francis T		Wounded	
Bloom, Jacob				Crist, John T			
Bloom, John				Crist, Solomon		MIA	Bermuda Hundred
Bollman, David R		Wounded	Drewry's Bluff*				

55th Pennsylvania Infantry Listing

Name	Pict. Page	Known Casualties	Battle
Croft, Jeremiah			
Crossan, John M			
Crouse, Harry C		KIA	Petersburg
Crouse, Henry			
Crouse, John H			
Croyle, James A			
Croyle, Martin			
Croyle, William H		Wound.-Died	Petersburg - Initial Assault
Dannaker, John			
Dannaker, William			
Darr, Abraham	131	KIA	Cold Harbor
Darr, David H	235	Wounded	
Darr, Henry H		Wounded	Petersburg - Initial Assault
Darr, Jacob			
Darr, Michael S			
Dasher, James H		POW	Drewry's Bluff
Daugherty, David L		Wound.-Died	Petersburg - Initial Assault
Davis, Charles M		KIA	Cold Harbor
Davis, Ephraim W			
Davis, Wilson			
Deremer, Henry			
Detwiler, Joseph		Wounded	Drewry's Bluff
Dibert, David		Wounded	
Dibert, Jacob J	132, 236	Died	
Dibert, John J		KIA	Petersburg - Initial Assault
Diehl, B James	236	Wounded	
Diehl, Daniel		Died	
Diehl, Espy		POW-Died	
Diehl, Henry			
Diehl, John			
Diehl, Samuel J	236		
Dorsey, Wlliam C			
Drenning, Henry G		KIA	Cold Harbor
Drips, Thomas			
Dull, Lewis		Wounded	White Oak Road
Ealy, John C		Wounded	Drewry's Bluff
Earnest, Alexander			
Eckhard, Jacob			
Edenbo, Daniel H			
Edwards, Daniel L	152, 185		
Edwards, Josiah V	152, 236		
Egan, Charles			
Ellenberger, George		Wounded	Bermuda Hundred
Engle, Charles		Died	
Evans, George W			
Exline, Jacob		POW-Died	Drewry's Bluff
Fagans, James			
Farber, Thomas H	115		

Name	Pict. Page	Known Casualties	Battle
Feather, William		Wounded	Drewry's Bluff
Feight, William W		Wounded	
Filler, John H			
Filler, Joseph		POW	Petersburg
		Wounded	Drewry's Bluff
Finnegan, Daniel		Died	
Finnegan, Henry			
Fisher, Andrew			
Fleegle, Isaac S	237, 361	Wounded	
Foster, William A		Died	
Fox, Henry W			
Francis, William			
Frauenfelter, Adam		Wounded	Cold Harbor
Frazier, William		Wound.-Died	
Freeburn, Richard H			
Fry, Solomon W		Wounded	Cold Harbor
Furlong, Edward			
Gardiner, Francis L		Wounded	Petersburg - Initial Assault
Gardner, Adam	238, 356	Wounded	
Gardner, John		Wounded	
Gardner, Samuel		Wounded	
Garland, Matthew		Wounded	
Garlinger, Walter E		POW	Edisto Island, SC
Garretson, George R		Wounded	Petersburg - Initial Assault
Garretson, Josiah P	107, 237		
Garretson, Moses R	171-M	Died	
Gates, Theophilus R	195, 237	Wounded	Cold Harbor
Geller, Jesse		Wounded	Petersburg - Initial Assault
Geyer, John C			
Gibson, James			
Gilliam, Michael		Wounded	Cold Harbor
Gladwell, George W		Died	
Gollipher, Espy			
Gollipher, Silas	13	Wound.-POW	Edisto Island, SC
		Wounded	Edisto Island, SC
Gonden, John W			
Good, George			
Gordon, Jeremiah			
Gordon, Joseph	187, 238, 356	Wounded	White Oak Road
Gordon, William	172-P	POW-Died	Drewry's Bluff*
Gray, George W			
Hackard, John A			
Hagerty, Daniel			

55th Pennsylvania Infantry Listing

Name	Pict. Page	Known Casualties	Battle	Name	Pict. Page	Known Casualties	Battle
Hahn, Isaac				Imler, Isaac M	137	KIA	Petersburg - Initial Assault
Hainsey, Frederick		Wound.-POW	Drewry's Bluff	Imler, John		Wounded	
Hainsey, Valentine	173-P	Wound.-Died	Cold Harbor	Jackson, Charles		Wounded	Bermuda Hundred
Hale, William						Wounded	Petersburg - Initial Assault
Haley, Josiah	238						
Hammer, Daniel R				James, Edward V		KIA	Petersburg - Initial Assault
Hammer, Hezekiah	109, 237, 358	Wounded	Petersburg - Final Assault	James, John A		Wound.-Died	
Hammer, Samuel	358			James, John W		Wound.-POW	Drewry's Bluff
Hand, Henry		POW-Died	Drewry's Bluff	Johnson, Edward			
Hand, James				Johnson, Samuel J			
Harbaugh, Eli		Died		Keeffe, Joseph			
Harbaugh, George W		Wound.-Died	Petersburg - Initial Assault	Keely, Thomas			
Harbaugh, John		Died		Kegg, Jacob			
Harbaugh, Robert		KIA	Petersburg - Initial Assault	Kegg, James P	189, 239		
				Kennedy, Samuel		KIA	Drewry's Bluff
Harbaugh, Wilson		Died		Kessler, John		Wounded	Bermuda Hundred
Hartley, William				King, Samuel T	207	Wounded	Drewry's Bluff
Hartzel, Francis B				Kinley, Jacob		Died	
Henry, Daniel B				Kinsey, Peter Jr	155, 239		
Herring, George W		Died					
Hess, Benjamin				Kinsey, Peter Sr	155		
Hess, Daniel A		Wound.-Died	Sailor's Creek	Kinton, David		KIA	Cold Harbor
		Wounded	Cold Harbor	Kipp, Jonas			
Hileman, John		Died		Kline, James S		Wound.-Died	Bermuda Hundred
Hillebrandt, Henry				Knapp, James M			
Hillegass, Henry P		Wounded	Petersburg - Initial Assault	Knox, Otho S		Wounded	
				Koontz, George			
Hillegass, John C		Wounded	Petersburg - Initial Assault	Kreiger, John			
				Kromer, George			
Hissong, Josiah	105, 238	Wounded	Chaffin's Farm	Lair, Moses			
		Wounded	Drewry's Bluff	Lashley, Daniel			
		Wounded	White Oak Road	Lashley, Henry C		Wounded	
Hite, Samuel C		POW-Died		Leach, George E		Died	
Hockenberry, John				Lear, Daniel J			
Hogan, James				Lee, James			
Hogan, John				Lee, Winfield S			
Holler, James M		Died		Leech, Thomas			
Holt, William				Leech, Wlliam		KIA	Pocotaligo, SC
Hoover, Nathaniel		Died		Lehman, Josiah M			
Hoover, Thomas G				Lemon, Henry		Wounded	Drewry's Bluff
Horne, John D				Leonard, Jerome		Wound.-Died	Petersburg - Initial Assault
Hughes, Edwin		Wound.-POW	Chaffin's Farm	Leonard, Philip	239		
Hunt, John T		POW-Died	Drewry's Bluff	Leppert, Gustavus			
Hunt, Samuel B	238			Lewis, Franklin			
Hyde, Abraham		Died		Lewis, William			
Hyde, John				Licher, John S		Wounded	Petersburg - Initial Assault
Ickes, Henry		Wounded		Ling, David C			

55th Pennsylvania Infantry Listing

Name	Pict. Page	Known Casualties	Battle
Ling, Isaac	235, 358	Wounded	Petersburg
Lingenfelter, David		POW	Drewry's Bluff
Lingerfelt, Aaron			
Lingerfelt, Abram			
Lingerfelt, Josiah			
Little, David			
Little, Irving		Died	
Little, James			
Livingston, John A	76, 228, 240	Wounded	Cold Harbor
Lockard, Thomas R		Wounded	Petersburg - Initial Assault
Lockard, Thomas R		POW	Edisto Island, SC
Long, John A			
Long, Levi		Died	
Lowry, William N			
Luther, Frederick H			
Lybarger, Henry G	240	Wounded	Drewry's Bluff
Lyons, Thomas H	239		
Madara, David W			
Maloney, William A			
Manges, George W	240	Wounded	
Mars, John		POW	Edisto Island, SC
Marshall, Henry L		POW-Died	Drewry's Bluff
Marshall, Moses F		Died	
Martin, William L		KIA	Pocotaligo, SC
Mathews, Hiram			
May, Daniel S			
May, Joseph			
May, Joseph C	241		
May, William A			
McChesney, John	78	Wounded	Cold Harbor
McCormick, William		Wounded	Petersburg - Initial Assault
McEnespy, James B			
McFarland, James			
McGee, James	174-B	POW-Died	Drewry's Bluff
McGee, William	174-B	MIA	Chaffin's Farm
McGregor, John		Wound.-Died	Cold Harbor
McGregor, Robert			
McGregor, William			
McKee, David			
McMullen, Charles		Wound.-Died	
McVicker, J Clay		Died	
Metzger, James		POW	Drewry's Bluff
Metzger, Solomon S		Wounded	Cold Harbor
Meyers, Levi		POW-Died	Drewry's Bluff
Mickey, Rankins			

Name	Pict. Page	Known Casualties	Battle
Millburn, William			
Miller, David		Wounded	Petersburg - Initial Assault
Miller, James H	71	Wounded	Cold Harbor
Miller, John		Wounded	Petersburg - Initial Assault
Miller, John D		Wounded	
Miller, John W		Wounded	Pocotaligo, SC
Miller, Joseph	240		
Miller, Matthew		MIA	Drewry's Bluff
Miller, Nelson B			
Miller, Philip S		Died	
Miller, Solomon H		POW-Died	Drewry's Bluff
Miller, William H		Wounded	Petersburg
Mitchell, James P		POW-Died	Drewry's Bluff
Mock, Andrew		Wound.-Died	White Oak Road
Mock, Anthony		Wounded	Cold Harbor
Mock, Emanuel A	67	Wounded	Bermuda Hundred
Mock, John		Wounded	Petersburg - Initial Assault
Mock, Josiah B		POW-Died	
Mock, Malachi B	139	Died	
Mock, Paul S		POW	Chaffin's Farm
Mock, Tobias B		Wounded	Bermuda Hundred
Mock, Tobias B		Wound.-Died	Petersburg - Initial Assault
Mock, Wlliam A		KIA	Bermuda Hundred
Moore, James E			
Moore, William G			
Moran, Thomas			
Mosell, William		Died	
Moser, William S		Died	
Mower, Abraham C			
Mower, Alexander		Died	
Mower, Edward E	241		
Mower, John H			
Mowry, Richard S			
Moyer, John		Died	
Mullin, Alexander S			
Mullin, George S			
Murphy, Elias		Wounded	Chaffin's Farm
Murphy, James S		Wounded	
Murphy, Philip		Died	
Mushbaum, John			
Musselman, Simon			
Myer, John			
Myers, Joseph		Wounded	
Newman, John			
Noffsker, Martin		POW	Drewry's Bluff

55th Pennsylvania Infantry Listing

Name	Pict. Page	Known Casualties	Battle
Norton, James			
Nottingham, William		POW	Drewry's Bluff
Oldham, Michael		Died	
Ornst, John		Wounded	Bermuda Hundred
Oyler, Abraham		Wounded	Chaffin's Farm
Oyler, William			
Palmer, John		POW-Died	
Parsons, William			
Peck, Jacob B	241		
Penrose, Andrew J			
Phillips, Daniel W			
Phillips, Scott			
Pleacher, Andrew		POW	
Polta, Augustus			
Porter, Andrew J	239		
Pote, Jacob B			
Presser, Philip			
Prosser, David W		POW	Drewry's Bluff
Radebaugh, Daniel W			
Radebaugh, Jacob L		MIA	Drewry's Bluff
Rahn, Edwin L			
Ream, Isaac		POW	Edisto Island, SC
		Wounded	Petersburg - Initial Assault
Reed, Andrew J		Died	
Reese, George L			
Reily, Michael			
Ressler, William			
Riffle, Albert J			
Riley, William		Wound.-Died	Cold Harbor
Rininger, Eli		Wounded	Cold Harbor
Rislenbatt, August			
Risling, John			
Risling, John H		Wounded	Bermuda Hundred
Ritchey, Adam			
Ritchey, Daniel B		Wounded	
Ritchey, David		Wounded	Drewry's Bluff*
Ritchey, Ferdinand			
Ritchey, John		Wounded	White Oak Road
Ritchey, Jonas		Wound.-Died	Edisto Island, SC
Roach, Thomas			
Robb, John M		Wounded	
Robinson, Tobias			
Roby, George W			
Rollins, Andrew			
Roudabush, Benjamin		POW-Died	
Rowser, Philip		Died	
Rowzer, John S			
Ruben, Joseph			
Ruby, John		Died	
Sanno, Frederick		Died	

Name	Pict. Page	Known Casualties	Battle
Saupp, Frank D	242	Wounded	
Saupp, John		KIA	Edisto Island, SC
Semler, Reuben J		Wound.-Died	Cold Harbor
Shaeffer, Sebastian			
Shaffer, George W	356	Wounded	Drewry's Bluff
Shaffer, Jacob J		POW-Died	Drewry's Bluff
Shine, James			
Shoenfelt, Henry			
Sholl, Isaac		Wounded	Drewry's Bluff
		Wounded	Petersburg - Initial Assault
Shoop, Joseph L		Wound.-Died	Cold Harbor
Shrader, Auterbine		POW-Died	
Shrader, William O		POW-Died	Drewry's Bluff
Shull, Henry R			
Shultz, Frederick			
Sleek, Andrew J			
Slick, Allen		KIA	Cold Harbor
Slick, Hezekiah B		POW-Died	
Slick, Josiah	336	Wounded	
Slick, Nicholas			
Slick, William W		Wounded	
Smith, Benjamin S			
Smith, Daniel			
Smith, Jasper W		POW-Died	
Smith, Jeremiah			
Smith, Jesse		Wound.-Died	Drewry's Bluff
Smith, Philip E		Wounded	
Smith, Robert C	241	Wounded	Chaffin's Farm
Smith, Samuel		KIA	Drewry's Bluff
Snellrider, David			
Snook, Emanuel		Wound.-Died	Petersburg
Snowberger, David		Died	
Snyder, George W Jr.			
Sorber, Martin V			
Spidle, Wilson		Wound.-Died	Petersburg - Initial Assault
Stahley, Henry			
Stambaugh, Joseph		Wounded	
Statler, Samuel F	67, 183, 202, 355, 359	Wounded	Bermuda Hundred
		Wounded	Petersburg
Steckler, Charles		POW-Died	
Steckman, Levi		Wounded	Drewry's Bluff
Stengal, Jacob			
Stickler, Samuel		POW-Died	Drewry's Bluff
Stiffler, George C			
Stineman, Daniel			

55th Pennsylvania Infantry Listing

Name	Pict. Page	Known Casualties	Battle
Stingle, Jacob			
Stoutnour, James H	242		
Straney, Edward			
Stratton, Jeremiah		Wounded	
Struckman, Charles		Wounded	
Summerland, John			
Summerland, Peter J		Wounded	Petersburg - Initial Assault
Summerville, Abner	146	POW-Died	
Summerville, Sylvanus		Wound.-Died	Chaffin's Farm
Swartz, Francis	242	Wounded	
Swartz, John			
Taslinger, Walter E			
Tewell, Joseph		Wound.-Died	Cold Harbor
Thompson, Jeremiah H		POW	Drewry's Bluff
Travis, John L			
Treese, Adie Bell		Wounded	Cold Harbor
Trott, Benjamin		Died	
		Wounded	Bermuda Hundred
Turner, Andrew J		Died	
VanHorn, James F			
Vickroy, Orrin G			
Walker, William M			
Walters, David			
Warner, Pius			
Waters, David		Wounded	
Watkins, Jesse		Died	

Name	Pict. Page	Known Casualties	Battle
Weisel, William W	242		
Welsh, John F			
Wentz, Adam			
Wentz, Henry		Wounded	Petersburg - Initial Assault
Wentz, Isaac	234	Wounded	Drewry's Bluff
Wentz, John		Died	
Werning, John		POW	Edisto Island, SC
Wheeler, Daniel A			
Whitaker, Christian		POW	Edisto Island, SC
White, Andrew J		Wounded	
White, James S		Wound.-Died	Drewry's Bluff
Whysong, John			
Whysong, Samuel	242	Wounded	Drewry's Bluff
Wilson, John		Wound.-Died	
Wisegarver, David			
Wisel, George C			
Wogan, James P			
Wolf, Edmund	235		
Wolf, Richard	73	Wounded	Cold Harbor
Wonderly, John B	357		
Wonderly, William		POW	Drewry's Bluff
Wonders, Daniel M	228		
Wonders, Henry			
Woodward, George			
Yeck, Frederick			
Yost, Francis F			

Samuel Adams, Private

John B Amos, 2nd Lieutenant

Isaac Wentz, Private

55th Pennsylvania Infantry Listing

Edmund Wolf, Private

David H Darr, Private

Isaac Ling, Private

Albin C. Arnold, Private

Jacob Burket, Private

William Cessna, Private (seated in the middle). The other soldiers have not been identified.

55th Pennsylvania Infantry Listing

Josiah Edwards, Private

James B Diehl, Private

Samuel J Diehl, Private (Left). The other soldier has not been identified.

Jacob J Dibert, Corporal

55th Pennsylvania Infantry Listing

Isaac S Fleegle, Private

Theophilus R Gates, Corporal

Hezekiah Hammer, Captain

Josiah P Garretson, Private

55th Pennsylvania Infantry Listing

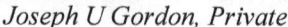

Joseph U Gordon, Private

Adam Gardner, Private

Samuel B Hunt, Private

Josiah Haley, Musician

Elizabeth Amick married 2nd Lt. Josiah Hissong while he was on leave in February 1864. In May 1864, Josiah was wounded during the battle of Drewry's Bluff. He suffered a second wound at Chaffin's Farm in September 1864. Two months later, their first child was born. Josiah was wounded a third time in March 1865 at White Oak Road before returning home to his wife and newborn child.

55th Pennsylvania Infantry Listing

Peter Kinsey Jr, Corporal

James Polk Kegg, Private

Thomas H Lyons, Captain

Andrew Porter, 2nd Lieutenant mustered out of the 55th on October 11th, 1864, at the end of his 3-year enlistment. He married Matilda Stuckey on December 8th, 1864. The photograph may be a wedding picture.

Philip Leonard, Private

55th Pennsylvania Infantry Listing

Captain John Livingston suffered a wound during the Initial Assault on Petersburg on June 18th, 1864. Later that year he married Mary Louise Statler in Schellsburg.

George Manges married Sarah Wilson in Schellsburg in 1862. He enlisted as a private in February 1864, the same month their 2nd child was born. George suffered a wound later in 1864 but survived and returned home after the war ended.

Henry G Lybarger, Private

Joesph Miller, Sergeant

55th Pennsylvania Infantry Listing

Joseph C May, Private

Edward E Mower, Musician

Jacob B Peck, Corporal

Robert C Smith, Sergeant

55th Pennsylvania Infantry Listing

Frank D Saupp, 1st Lieutenant

Frances Swartz, Private

James H Stoutnour, Musician

Samuel Whysong, Private

William W Weisel, Private

76th Pennsylvania Infantry Listing

				Known
Bedford County Enlistments	127	Year	Major Battles	Casualties
Known Casualties	29	1862	Pocotaligo, SC, Oct. 22	2
Casualty Rate Percentage	23%	1863	Ft. Wagner SC, Jul. 11	7
Killed in Action	6	1864	Bermuda Hundred, May	1
Died from Wounds	1		Cold Harbor, May 31 - Jun. 12	1
Died while POW	3		2nd Deep Bottom, Aug. 14-20	2
Died	6			
POW	3			
Wounded	10			

The 76th Pennsylvania Infantry regiment recruited Bedford County soldiers for Company E in October 1861 and the volunteers were sent to South Carolina that same month. Additional county soldiers were assigned to other companies in the 76th PA Infantry during the war. The battle of Pocotaligo in October 1862 was the regiment's initial combat experience. In July 1863, County soldiers took part in two assaults on the near impregnable Fort Wagner that guarded the southern approach to Charleston Harbor. Entrenched Confederates repelled both charges, resulting in Union Army troops taking heavy casualties. Bedford County casualties were taken during the first charge on July 11th. "Glory" a 1989 full-length motion picture focused on the story of the assault of Fort Wagner by another Union Army regiment, the 54th Massachusetts, one of the first African American regiments to experience combat during the war. The 76th PA Infantry took part in the Bermuda Hundred Campaign, Cold Harbor, and the siege of Petersburg in 1864. The 76th was routed to North Carolina in 1865, where Bedford County soldiers witnessed the surrender of a second rebel army in the Eastern Theater. Confederate General Joseph E. Johnston surrendered to General William T. Sherman at Bennett Place in Durham, North Carolina, on April 26th. The regiment mustered out on July 18th, 1865 after the war had ended.

Richard P Pilkington, Captain (Seated on Left). The other soldiers have not been identified.

George W Cessna, Private

76th Pennsylvania Infantry Listing

Name	Pict. Page	Known Casualties	Battle	Name	Pict. Page	Known Casualties	Battle
Adams, William				Hickok, Edwin H			
Agnew, Levi J		Wound.-Died	Bermuda Hundred	Hoffman, Jacob		Died	
Alcorn, George				Hoyman, Samuel			
Baker, Abraham				Humbert, Daniel			
Barndollar, William	45, 361	POW		Hurley, John		Died	
Basore, George				Jones, Samuel			
Bennett, Artemas S				Kagarice, Ebenezer		KIA	Ft. Wagner, SC
Bloom, John				Keiser, David O			
Boehm, John W				Kennedy, James			
Bolinger, Alexander				Kiester, Levi			
Bollinger, Jacob				King, John T			
Boylan, John				King, Watson		Died	
Brown, John				Klahre, Theodore M	13	Wounded	Cold Harbor
Burket, Henry						Wounded	Pocotaligo, SC
Burket, John B				Knabb, Albert		Wounded	2nd Fair Oaks
Burns, Charles S				Leader, John		Wounded	
Cessna, George W	243			Leary, James M		Wounded	
Charleston, John				Lewis, James A			
Chester, Edward				Lindsay, Charles B			
Crick, Andrew				Long, Joseph			
Cutler, Jonathan				Lucas, Henry			
Cypher, Henry S				Lyon, Alexander			
Cypher, Jacob F				Lyon, William			
Dell, Samuel				Lyon, William M			
Donahoe, Patrick				Martin, Thomas			
Duffy, James		POW		Meredith, Charles B			
Eckels, Francis S		POW-Died		Middleton, James M			
Eckels, John F		POW-Died		Miller, Andrew			
Elliott, David S				Miller, Clement R			
Faust, John J				Mills, Samuel			
Feese, Edward				Mills, Uriah			
Feidler, Michael				Mittong, John W			
Fetter, John				Moore, John			
Fetter, Joseph		KIA	Ft. Wagner, SC	Moore, Wlliam			
Fetter, Joseph J		Died		Morris, Henry			
Fleegle, Jacob G		Died		Mortimer, John			
Fleegle, Simon S				Moyer, Alexander			
Foor, Daniel V	150-B	KIA	Ft. Wagner, SC	Myers, Samuel			
Furcht, Frederick				Negley, David F			
Gates, Martin V B		Wounded	Petersburg	Null, George W			
Gephart, John	245	Wounded	Pocotaligo, SC	Nulton, Henry H			
Gephart, William		KIA	Mill Springs Gap*	Parsons, George W			
Glidenell, Thomas				Patton, Abraham			
Hainsey, Adam R	245			Pennell, Henry C		Died	
Hainsey, George				Pierrant, Joseph			
Hainsey, John				Pilkington, Richard	243	Wounded	2nd Deep Bottom
Hazlett, George M				Porter, Willam H			
Hazzard, Philip				Prilles, Joseph			
Helsel, Edward				Ramage, Thomas R			
Helsel, Henry S				Ray, William H			
Hershey, James				Reilley, George W			
				Riceling, William		POW-Died	Ft. Wagner, SC

76th Pennsylvania Infantry Listing

Name	Pict. Page	Known Casualties	Battle
Rough, John			
Roush, Ernest		POW	Ft. Wagner, SC
Royal, Clark			
Shunk, Jacob			
Slott, Samuel			
Smith, Levi		Wounded	2nd Deep Bottom
Smith, Seth S		Wounded	
Snave, Joseph W			
Snyder, James	245		
Sohn, Calvin			
Spidel, Matthew			
Steckman, Daniel H		KIA	Ft. Wagner, SC

Name	Pict. Page	Known Casualties	Battle
Stoudenour, John			
Stoudnour, Jacob			
Sutton, Joseph			
Swope, Thomas J			
Taylor, James			
Walker, Charles W			
Wall, Albert			
Waumbaugh, Lewis			
Washabaugh, William H		KIA	Ft. Wagner, SC
Wertz, Jacob A			
Wiltner, James			
Wray, William H			
Young, Thomas J			

John Gephart, Private

Adam R Hainsey, Corporal

James Snyder, Musician

77th Pennsylvania Infantry Listing

Bedford County Enlistments	55			Known
Known Casualties	16	Year	Major Battles with Casualties	Casualties
Casualty Rate Percentage	29%	1862	Stones River, TN, Dec. 31 - Jan. 2	1
Killed in Action	1	1863	Liberty Gap, TN, Jun. 25	2
Died from Wounds	2		Chickamauga, GA, Sep. 19	3
Died	5	1864	Chattahoochee, GA, Jul. 7	1
POW	1		Franklin, TN, Nov. 30	1
Wounded	7			

The 77th Pennsylvania Infantry was one of the very few locally recruited units to fight in the Western Theater of the war. The 77th PA Infantry regiment recruited soldiers in Huntingdon, Fulton, and Blair Counties. Over 50 soldiers from Bedford County also volunteered in this regiment and were sent to Louisville, KY on October 18th, 1861. County soldiers in the 77th were combatants in many of the well-known battles of the Civil War including Shiloh (Tennessee) in April 1862 which was the 6th bloodiest battle of the war with 23,746 total casualties and Stones River (Tennessee) in late December 1862/early January 1863, the 7th costliest battle with 23,515 casualties. The 77th was also on the battlefield at Chickamauga (Georgia) in September 1863 where 34,624 total casualties were suffered in both armies. Only the 51,000 casualties during the battle of Gettysburg were higher during the entire war. County soldiers in the 77th Pennsylvania Infantry took part in many significant Union Army victories before the regiment mustered out on December 6th, 1865 after the war had ended. Fifteen percent of the Bedford County soldiers who volunteered in the 77th Pennsylvania Infantry lost their lives during the Civil War.

Name	Pict. Page	Known Casualties	Battle
Barton, Asa		Died	
Berkstresser, David S			
Bivens, William		Wounded	Stones River, TN
Black, Erastus	247		
Blackburn, Harmon		Wounded	
Bookhamer, John			
Bookhamer, Thomas			
Burger, Joseph S			
Burket, David			
Bussard, Daniel S			
Childers, Randall			
Clites, Solomon			
Cooper, George M		Wounded	Liberty Gap, TN
Crall, John		Wounded	
Diggins, Jesse			
Dively, George M	247		
Dively, Martin	247		
Dively, Morgan			
Duvall, Thomas		Died	
Geisler, Lewis H			
Groomer, Anthony			
Herald, Jacob			
Horton, Alexander		Died	
Horton, Jonathan		Died	
Horton, Milton			
Horton, Reuben H		Wound.-Died	Chattahoochee, GA
Houck, McKenzie		Wounded	Chickamauga, GA
Kegg, Joseph			
Lear, Franklin			
Lingenfelter, Thaddeus	190		
Machtley, William H			
McCray, James			
McCue, William			
McHugh, William			
Mountain, Richard D		Wound.-Died	Chickamauga, GA
Norris, Harrison			
Rinard, Emanuel		Died	
Ritchey, George S			
Shock, Daniel	306		
Snow, William J			
Speece, Harry	357		
Stine, Jacob			
Stoner, David E		Wounded	Franklin, TN
Tipton, Levi			
Walter, Jacob			
Walters, Jacob D			
Walters, Jacob W			
Willett, William			
Wishart, Henry			
Wishart, James	247		
Wolf, John D		POW	Chickamauga, GA
Woodcock, William L	211		
Woy, George	148	KIA	Liberty Gap, TN
Zimmerman, Samuel			
Zimmerman, William		Wounded	

77th Pennsylvania Infantry Listing

James Wishart, 2nd Lieutenant

George M Dively, Private

Erastus Black, Private

Martin Dively, Private

82nd Pennsylvania Infantry Listing

Bedford County Enlistments	45
Known Casualties	10
Casualty Rate Percentage	22%
Killed in Action	1
Died	2
Died while POW	1
Wounded	6

Year	Major Battles with Casualties	Known Casualties
1865	Ft. Fisher, Mar. 25	1
	Sailor's Creek, Apr. 6	1

The 82nd Pennsylvania Infantry regiment initially recruited soldiers from Philadelphia and Allegheny County in 1861. Most of the Bedford County soldiers in the 82nd mustered in as replacement troops during the second half of 1864. The 82nd PA Infantry was in combat during the Overland Campaign in the spring of 1864 and in the Shenandoah Valley Campaign during the fall of 1864. County soldiers in the 82nd Infantry would have taken part in combat in several battles during the siege of Petersburg in 1865, including Dabney's Mills, Fort Fisher, and the Final Assault on Petersburg. The 82nd were among the Union troops that pursued the Confederate Army after their retreat from Petersburg in early April 1865. County soldiers in the 82nd Pennsylvania Infantry were eyewitnesses to the Confederate surrender in Appomattox on April 9th. The regiment was disbanded on July 13th, 1865 after the war ended.

Name	Pict. Page	Known Casualties	Battle
Andrews, Joseph	358	Wounded	
Ball, John		Wounded	
Bishop, David P		Wounded	
Bondebust, John M			
Bowers, John C			
Bridenstein, John E			
Burns, Joseph H		KIA	Sailor's Creek
Calhoun, David C			
Claycomb, Andrew			
Comp, Adam A			
Corley, David	249		
Corley, Jacob	249		
Cutchall, Dutton			
Dibert, Jacob			
Diehl, Adam			
Edwards, George W			
Fryberg, John L			
Harrier, Adam			
Hartzell, William B		Wounded	
Heltsel, John D			
Hill, William M			
Hixon, Amos			
Holler, William H			

Name	Pict. Page	Known Casualties	Battle
Hoopingardner, George			
Imler, John R			
Lake, John			
Machtley, John F	358		
May, Daniel H	249, 291		
May, George F			
Mechtley, John E			
Mellott, Caleb		POW-Died	
Mellott, Daniel B	249	Wounded	
Miller, Abraham M		Wounded	Ft. Fisher
Mull, William		Died	
Oldham, Thomas			
Palmer, Elijah N			
Rice, John			
Richards, William			
Roudabush, John M	249		
Saylor, William	144	Died	
Schinich, William H			
Sexton, William S			
Sparks, John C	283, 359		
Suter, William S			
Wertz, John A			

82nd Pennsylvania Infantry Listing

Jacob Corley, Private

Daniel H May, Corporal

Private John & Elizabeth Roudabush married in 1857. Elizabeth was at home in Union Township (Pavia) with 3 small children while John was enlisted in the war.

Private Daniel, Mary & son Riley Mellott circa 1860. Daniel mustered in the 82nd in November 1864 and was wounded in the face by a Minie Ball, likely during a battle in 1865. An 1866 pension request indicates the wound was debilitating.

David Corley, Private

84th Pennsylvania Infantry Listing

Bedford County Enlistments	46			**Known**
Known Casualties	21	**Year**	**Major Battles with Casualties**	**Casualties**
Casualty Rate Percentage	46%	1862	1st Kernstown, Mar. 23	3
Died from Wounds	2		Port Republic, Jun. 9	1
Missing in Action	1	1863	Chancellorsville, Apr. 30 - May 6	4
Died	7	1864	Wilderness, May 5-7	2
Wounded & POW	1		Spotsylvania, May 8-21	1
POW	3		Petersburg Initial Assault Jun. 17	1
Wounded	7		2nd Deep Bottom, Aug. 14-20	1

Twenty-two percent of the Bedford County soldiers who volunteered in the 84th Pennsylvania Infantry lost their lives in the Civil War. The 84th PA Infantry regiment recruited soldiers in Blair County for some companies. Many Bedford County volunteers in this regiment lived in border areas to Blair County. The regiment was organized at Huntingdon and Camp Curtin in Harrisburg from August to October 1861. Bedford County soldiers in the 84th PA Infantry experienced their first significant combat during the 1st Battle of Kernstown, where they took part in dealing legendary general Stonewall Jackson with his only tactical battlefield defeat of the entire war. The largest number of identified casualties for county soldiers in the 84th was at the battle of Chancellorsville in 1863. During this battle, Stonewall Jackson stealthily moved his troops through a wooded area around the right flank of the Union Army and launched a devastating attack, resulting in one of the most significant victories of the war for the Confederates. In May 1864, county soldiers in the 84th took casualties during the battle of the Wilderness, very near to the Chancellorsville battlefield the previous year. The 84th also took part in the siege of Petersburg. On January 13th, 1865, the 84th merged into the 57th Pennsylvania Infantry. This consolidated regiment mustered out of the army on June 29th, 1865, after the conclusion of the war.

Name	Pict. Page	Known Casualties	Battle
Ayres, John		Died	
Benton, David H			
Burke, Samuel		Wound.-Died	1st Kernstown
Burket, Elias S	37	Wounded	Chancellorsville
Claar, Martin			
Claar, Thomas	257		
Corle, Anthony		MIA	Port Republic
Davis, Nathan H			
Davis, William A	358	Wounded	1st Kernstown
Evans, John J			
Fry, Michael		Died	
Garretson, Thomas			
Gates, William H			
Grimes, Henry W		Wound.-Died	Wilderness
		Wounded	1st Kernstown
Grimes, Jacob R			
Grimes, John C			
Harbaugh, Jason	251		
Harbaugh, William H		POW	Chancellorsville
Houck, Ezekiel J		Died	Petersburg-Initial
Hurley, William		Wounded	Spotsylvania
Ickes, Joseph H			
James, Jesse T		Died	
Knipple, John A	251		

Name	Pict. Page	Known Casualties	Battle
Lambright, William		POW	Chancellorsville
Miller, William H		Died	
Mock, Josiah D			
Morningstar, Peter		POW	2nd Deep Bottom
Morrison, John			
Morrow, B Moritimer			
Mountain, George R		Wounded	Wilderness
Peterson, William A		Wound.-POW	Chancellorsville
Piper, Thompson F		Wounded	
Reed, Lewis S			
Renninger, Frederick		Died	
Rinard, Jacob			
Rininger, Frederick		Died	
Smith, John B			
Walker, Asahel	358		
Walker, Benjamin H			
Walker, Morris			
Walter, John H		Wounded	
Weisel, John H			
Weyandt, Jacob			
Weyant, Joseph	251		
White, Silas			
Witters, Jacob M			

84th Pennsylvania Infantry Listing

Jason Harbaugh, Private

Brothers: On left Private John A Knipple - 84th PA Infantry & Corporal William Knipple - 101st PA Infantry

Joseph Weyant, Private

91st Pennsylvania Infantry Listing

Bedford County Enlistments	72			**Known**
Known Casualties	17	**Year**	**Major Battles with Casualties**	**Casualties**
Casualty Rate Percentage	24%	1864	Boydton Plank Road, Oct. 27-28	9
Killed in Action	2	1865	Five Forks, April 1	1
Died from Wounds	1		Appomattox, April 9 (at Surrender)	
Died	2			
Wounded	12			

The 91st Pennsylvania Infantry regiment was organized in December 1861. Soldiers in this regiment were initially recruited out of Philadelphia. Only 3 Bedford County soldiers mustered in the regiment prior to 1864. Most county soldiers in the 91st enlisted en masse on September 21st, 1864. A month later these soldiers took heavy casualties at the battle of Boydton Plank Road during an assault on entrenched Confederates positions in an attempt to cut the lone remaining train line supplying Rebel troops around Petersburg. The 91st PA Infantry took part in several more battles during the siege of Petersburg, including White Oak Road and Five Forks during late March and early April 1865. County soldiers in the 91st were among the Union Army troops that pursued Robert E Lee's army after the Confederates retreated from Petersburg and were eyewitnesses to the Confederate surrender at Appomattox on April 9th. The regiment mustered out of the Union Army on July 10th after the war had ended.

Name	Pict. Page	Known Casualties	Battle
Barkley, Jacob T		Wounded	
Bennett, George S			
Bennett, Henry	255		
Bennett, Joseph			
Berkheimer, William	253		
Bowen, Alva			
Burket, Elias			
Burns, James			
Clark, John			
Clark, Philip	255		
Collins, Isaiah	87	Wounded	
Corl, Michael			
Corle, Francis B	131	Wound.-Died	Boydton Plank Rd
Corle, Jonathan		KIA	Boydton Plank Rd
Defibaugh, James C			
Dickens, William			
Diehl, Adam		Died	
Elbin, Otho			
Feaster, John G	87,		
Feather, Josiah			
Fordan, George G			
Himes, Thomas	255		
Ickes, Adam	253		
Ickes, Alexander			
Ickes, Daniel W			
Ickes, William M	253		

Name	Pict. Page	Known Casualties	Battle
Imler, Adam H		Wounded	Weldon Railroad
Imler, Martin			
Imler, William H		Wounded	Boydton Plank Rd
Jay, Thomas	300		
Johnson, Abel			
Johnson, Asa			
Johnson, Emanuel			
Johnson, John J			
Johnson, William P			
Leasure, John G			
Lecrone, William K			
Lingenfelter, David R		Wounded	Boydton Plank Rd
Martin, George W			
May, George			
McGregor, Elijah			
Miller, George	358		
Mock, Samuel A	358		
Mock, Samuel S			
Mock, William M			
Oster, Samuel C	255	Wounded	Boydton Plank Rd
Pennel, Andrew J	254	Wounded	Boydton Plank Rd
Russell, Abraham			
Russell, David			
Russell, Isaac			
Russell, John			
Shaffer, Zachariah A			
Simmons, William	254		

91st Pennsylvania Infantry Listing

Name	Pict. Page	Known Casualties	Battle
Smith, Bartley			
Smith, Isaac W			
Smith, Lewis M			
Smith, Samuel			
Smouse, Abner G		Wounded	
Smouse, Samuel			
Snyder, Joseph			
Steckman, David B			
Stineman, John		KIA	Boydton Plank Rd
Stineman, Thomas B		Wounded	Boydton Plank Rd
Stoner, Daniel F		Wounded	Boydton Plank Rd
Stufft, Michael		Died	
Tewell, George			
Treese, Willaim			
Troutman, James W		Wounded	
Walker, Thomas G	181, 254, 358		
Weaver, Samuel			
Whitaker, John H	105	Wounded	Five Forks
Whitaker, Joseph	254		

John G Feaster, Private

William M Ickes, Private

William S Berkheimer, Private

Adam Ickes, Private

91st Pennsylvania Infantry Listing

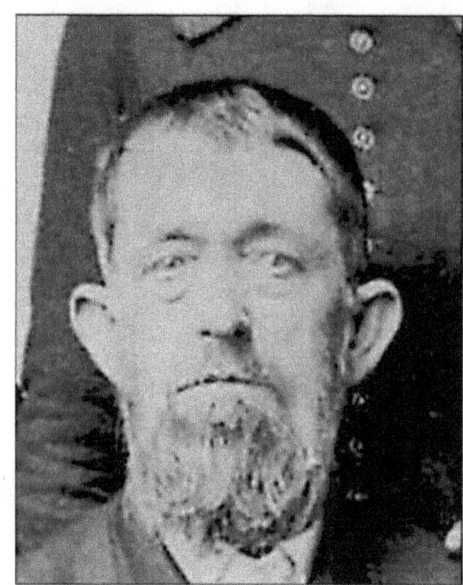

William Simmons, Private

Thomas G Walker, Private

Joseph Whitaker, Private

Andrew J Pennel, Private

91st Pennsylvania Infantry Listing

Thomas Himes, Private

Samuel C Oster, Private

Private Philip & Sarah Clark. Sarah was at home with 7 children when Philip enlisted in the 91st PA Infantry on September 21st, 1864. An 8th child was born on March 7th, 1865.

Private Henry Bennett and wife Ann Perdew Bennett. Ann was at home with 3 small children while her husband was fighting in the war.

99th Pennsylvania Infantry Listing

Bedford County Enlistments	112
Known Casualties	19
Casualty Rate Percentage	17%
Killed in Action	2
Missing in Action	1
Died	10
Wounded	6

Year	Major Battles with Casualties	Known Casualties
1865	Sailor's Creek, April 6	4
	Appomattox, April 9	1

The 99th Pennsylvania Infantry regiment was organized in 1861. Soldiers in this regiment were initially recruited from the city of Philadelphia, Schuylkill and Lancaster counties. Only one Bedford County soldier is known to have enlisted in the 99th prior to 1864. Most were mustered in the regiment as replacement soldiers in early 1865. County soldiers took part in the Final Assault of Petersburg on April 2nd. The 99th PA Infantry pursued the Confederate Army during their retreat from Petersburg and took casualties at Sailor's Creek, the last major battle prior to Robert E Lee surrendering to Ulysses S Grant. County soldiers in the 99th were witnesses to the Confederate surrender at Appomattox on April 9th. The 99th Pennsylvania Regiment mustered out of the army on July 1st, 1865.

Name	Pict. Page	Known Casualties	Battle
Allison, Andrew			
Amick, George W	255		
Baker, Jacob S			
Baker, William			
Barefoot, George W			
Barkman, Philip			
Bartholow, William E			
Baughman, Abraham A	258		
Bennett, Archer			
Bennett, Henry R			
Bennett, Israel M			
Benton, Emanuel			
Blackburn, Samuel S			
Bowers, Jacob			
Bowser, Valentine	257		
Bridenstine, Jacob		Wounded	Sailor's Creek
Bridenthal, David S			
Briggle, John M			
Burket, James		Died	
Burket, William		MIA	
Bush, John		Wounded	
Bussard, John S			
Bussard, Simon S			
Claar, Daniel			
Claar, Lewis			
Claar, Thomas	257		
Claycomb, William L			
Collins, John T		Died	
Cook, John	259		
Corle, Frederick	258		

Name	Pict. Page	Known Casualties	Battle
Crist, David T	205, 258		
Crouse, Samuel			
Cumpson, Benjamin			
Cushman, Adam			
Deaner, Michael			
Dibert, John Jackson		Died	
Edwards, Benjamin			
Eichelberger, John S			
Eicher, Samuel			
Ellenberger, John			
Everhart, Alexander		Wounded	Petersburg
Feather, Michael	259, 356		
Feathers, Henry			
Fetter, Joseph B			
Fetters, John		KIA	Sailor's Creek
Fletcher, John			
Fluke, George	258		
Hartle, John B		Died	
Herring, John B	361		
Hetrick, Jacob T			
Himes, Oliver			
Hiner, John M			
Hymes, Oliver			
Iames, David			
Iames, John			
Ickes, William			
Kelly, Thomas			
Kichinann, Adam			

99th Pennsylvania Infantry Listing

Name	Pict. Page	Known Casualties	Battle	Name	Pict. Page	Known Casualties	Battle
Krielman, Johan A				Poorman, Franklin H			
Kuchman, J Adam				Potts, John A			
Lashley, Henry C				Price, George W			
Layton, George W				Raley, Daniel			
Leader, Simon H				Redinger, Peter			
Little, John Pius		KIA	Sailor's Creek	Rice, Jonathan			
Long, Abraham B				Robinson, William		Died	
Luman, Aaron				Rowzer, Joseph O			
May, Jacob L		Died	Appomattox Campaign	Savits, Henry			
May, Josiah				Slack, George	260		
McGregor, William A	208, 358			Slonaker, John G			
Mearkle, David S	259	Wounded		Smith, Isaac M			
Mellen, William S				Smith, Jacob			
Meloy, John L				Smith, Samuel H	358		
Mench, John				Smith, William H	111, 260	Wounded	Sailor's Creek
Miller, Isaac C				Spangler, William H	259		
Miller, Peter S		Died		Stiffler, Thomas H			
Miller, William H				Taylor, George W			
Millin, William S	260			Teeter, Samuel			
Morse, James		Wounded		Watson, John N			
Morse, Jesse W		Died		Weimer, John S			
Myers, Martin L	203			Weyandt, James		Died	
Nunemaker, Peter	259, 358			Whitaker, William P			
Oaks, Jacob R				Whited, David L	282		
Over, David H				Wigfield, Wesley			
Park, Lounzo F				Wilkins, Ephraim	260		
Penrose, Mahlon				Wright, James		Died	

Valentine Bowser, Private

Thomas Claar, Private

99th Pennsylvania Infantry Listing

George Fluke, Private

Private Fred Corle & wife Sophia Shull Corle. Sophia was at home with 5 children while Fred was at the war-front in Petersburg.

David T Crist, Private

Abraham A Baughman, Private

99th Pennsylvania Infantry Listing

David S Mearkle, Private

Peter Nunemaker, Private

William H Spangler, Private

John Cook, Private

Michael Feather, Private

George Slack, Private

Pvt. William Smith & wife Maria Northcraft possibly posing for a Wedding photo in 1868

Private William S Millin (on right) is pictured with son Harold Millin in his WWI uniform

Ephraim Wilkins, Private

101st Pennsylvania Infantry Listing

Bedford County Enlistments	140
Known Casualties	91
Casualty Rate Percentage	65%
Killed in Action	2
Died from Wounds	2
Died	26
Died while POW	15
Wounded & POW	2
POW	36
Wounded	8

Year	Major Battles with Casualties	Known Casualties
1862	Fair Oaks, May 31	9
	Seven Days Battle, Jun. 25 - Jul. 1	1
1864	Plymonth NC, Apr. 20	51

Bedford County volunteers in the 101st Pennsylvania Infantry suffered the 2nd highest percentage of deaths of any unit during the Civil War. The regiment was organized in late 1861 and early 1862. Volunteers in Company D were recruited out of Bedford County, and Company G was composed of volunteers from Allegheny and Bedford counties. Initial drilling and instruction took place at Camp Curtin in Harrisburg. The first combat experience for the 101st was during the Peninsula Campaign in the spring and summer of 1862. The 101st Infantry took heavy casualties at the battle of Fair Oaks on May 31st. During the battle, county soldiers were positioned on the extreme right flank of the Union Army lines and held their ground for over an hour against a much larger Confederate force before withdrawing. The following year the 101st PA Regiment was transported by ship to Plymouth, NC, to conduct offensive operations against Confederate forces in North Carolina. On April 17th, 1864, a Confederate force of over 10,000 soldiers launched an attack on the 3,000 troop Union garrison in Plymouth. Two days later, a Confederate ironclad ship, the C.S.S. Albermarle, cleared out all Union Navy vessels in the vicinity. The garrison was surrounded and taking fire from a superior force on land and an ironclad ship on the Roanoke River. The following day, the entire garrison surrendered. Fifty-two Bedford County soldiers were taken prisoner. The enlisted men were taken to Andersonville and officers were sent to POW camps in South Carolina and Georgia. Fourteen soldiers captured on that day never returned home to Bedford County.

Name	Pict. Page	Known Casualties	Battle
Adams, David		POW	Plymouth, NC
Akers, John T			
Alhborn, Augustus			
Amick, Josiah		Died	
Anderson, James	164-M,S	POW-Died	Plymouth, NC
Bannon, Joseph J			
Barkman, Daniel			
Beam, Daniel		POW-Died	Plymouth, NC
Beegle, David F	53	POW	Plymouth, NC
Beltz, Abraham		POW-Died	Plymouth, NC
Bequeth, William H		POW	Plymouth, NC
Bessor, John		POW-Died	Plymouth, NC
		Wounded	Fair Oaks
Boerkamp, Henry		Wounded	Fair Oaks
Booty, William P			
Brown, Jacob D		POW	Plymouth, NC
Brown, John W		POW	Plymouth, NC
Brown, Joseph L		Wound.-POW	Plymouth, NC
Brown, Samuel D		POW	Plymouth, NC
Carnell, George W			
Carnell, Samuel		POW	Plymouth, NC
Clark, Robert A			
Clingerman, Peter		POW	Plymouth, NC
		Wounded	Fair Oaks
Compher, Alexander	52, 264	POW	Plymouth, NC

Name	Pict. Page	Known Casualties	Battle
Conley, Isaiah	263	POW	Plymouth, NC
Cooper, Jesse V		Died	
Croft, Daniel S			
Defibaugh, Jacob F		Died	
Defibaugh, John	132, 263	POW-Died	Plymouth, NC
Dibert, David		Wounded	
Eamick, Josiah		Died	
England, Jacob	119	POW	Plymouth, NC
Evans, Isaiah			
Evans, Nathan C	303		
Evans, William H		Died	
Fickes, James M	133	Died	
Filler, William B	93, 202, 326		
Filler, William C			
Fockler, Jacob L			
Foor, Francis L	134	POW-Died	Plymouth, NC
Garretson, George R			
Geller, Solomon		POW-Died	Fair Oaks
Gilliam, Michael			
Gilliam, Wilson E		Died	
Gollipher, Justice		POW-Died	Plymouth, NC
Hageman, Alexander B			
Hanks, Benjamin A		POW-Died	Plymouth, NC
Hanks, Caleb		POW-Died	Plymouth, NC
Hanks, David F		POW-Died	Plymouth, NC
Hanks, Jacob C			

101st Pennsylvania Infantry Listing

Name	Pict. Page	Known Casualties	Battle
Hanks, Nelson		POW-Died	Plymouth, NC
Hanks, Thompson			
Hazelett, Moses		POW	Plymouth, NC
Heffner, Samuel	264		
Helm, John B	263	POW	Plymouth, NC
Hetrick, Daniel L	188, 358	POW	Plymouth, NC
Hite, David			
Hite, Jacob A			
Hixon, Akers J		Died	
Hixon, Joel B		Died	
Hoffman, John		Died	
Howard, William			
Huffman, William B		Wounded	Fair Oaks
Hull, Abraham			
Iams, Daniel			
Keagy, John T		Wounded	Fair Oaks
Kegg, Levi			
Kegg, Simon P	15		
Kennard, Wliam B			
King, Thomas		POW	Plymouth, NC
Knipple, Andrew J		POW	
Knipple, William H	251	POW	Plymouth, NC
		Wounded	Fair Oaks
Layton, David		Died	
Layton, John		KIA	7 Days Battle
Lightningstar, Augustus		POW	Plymouth, NC
Linn, Henry		POW	Plymouth, NC
Long, Joseph E		Died	
Longenecker, Jacob H	53, 264, 356	POW	Plymouth, NC
Martin, James P		POW	Plymouth, NC
May, Harvey			
McDonald, William, Jr	162-W	Died	
McEldowney, George E		POW	Plymouth, NC
McEldowney, Samuel J	117	POW	Plymouth, NC
Miller, John		POW-Died	Plymouth, NC
Miller, John H			
Miller, Martin D		KIA	Fair Oaks
Miller, Matson J			
Miller, Watson J			
Mills, Andrew J	264	POW	Plymouth, NC
		Wounded	Fair Oaks
Mills, Franklin G		Wound.-POW	Plymouth, NC
Mills, Jacob H			
Morse, Samuel		Died	
Mortimer, John			
Mortimore, David			

Name	Pict. Page	Known Casualties	Battle
Moss, Jacob			
Mower, John H			
Mullin, David W		POW	Plymouth, NC
Norton, Franklin G		Died	
Oler, James W		POW	Plymouth, NC
Oler, John W		Died	
Otto, Henry		Died	
Over, Jacob Z			
Pfeifier, John M			
Pittman, Daniel H			
Pittman, John			
Potter, John			
Potter, Martin L		POW	Plymouth, NC
Ressler, Abraham		Died	
Rice, Abraham	143	POW	Plymouth, NC
Rice, Isaac	143	POW-Died	Plymouth, NC
Ritchey, Henry S		Died	
Roberts, John			
Robison, Jonas		Died	
Rock, George J	264		
Ruby, John		Died	
Sheaffer, Anthony			
Shoemaker, George F	121	POW	Plymouth, NC
Shoemaker, Isaac F		Died	
Shoemaker, Jacob W			
Showman, William		Died	
Siler, James P		POW	Plymouth, NC
Slick, Samuel K		Wound.-Died	
Slick, Thomas W	119	POW	Plymouth, NC
Slick, William W			
Smith, Amos		POW	Plymouth, NC
Smith, Andrew J			
Smith, George W		Died	
Smith, Joseph A		POW	Plymouth, NC
Smith, Joseph L			
Sparks, William	53	POW	Plymouth, NC
Stone, Reuben M		POW	Plymouth, NC
Strong, William			
Stuckey, William H	15, 263		
Sweitzer, Daniel F			
Tate, George H		Died	
Truax, George M		Died	
Vaughan, Ephraim		POW	Plymouth, NC
Vaughan, John W			
Veatch, Samuel		Wound.-Died	
Wilson, George W		POW-Died	Plymouth, NC
Wolford, George W			

101st Pennsylvania Infantry Listing

John Defibaugh, Private

William H Stuckey, Private

Officers of 101st PA Infantry, Company G. 1st Lieutenant Isaiah Conley is seated on the left. 2nd Lieutenant John B Helm is likely pictured standing on the left. John was identified as being one of the two soldiers in the back row and was 27 years old when he mustered in the army. The soldier standing on the right appears to be older.

101st Pennsylvania Infantry Listing

Jacob H Longenecker, 1st Lieutenant

George J Rock, Private

Andrew J Mills, Private

Captain Alexander & Barbara Compher. Barbara was at home with 4 children while Alexander was leading county troops in the war.

Samuel Heffner, Private

107th Pennsylvania Infantry Listing

				Known
Bedford County Enlistments	78	Year	Major Battles with Casualties	Casualties
Known Casualties	39	1862	2nd Bull Run, Aug. 28-30	5
Casualty Rate Percentage	50%		South Mountain MD, Sep. 14	4
Killed in Action	2		Antietam, Sep. 17	5
Died from Wounds	3	1863	Gettysburg, Jul. 1-3	5
Missing in Action	1	1864	Petersburg - Initial Assault, Jun. 15-18	2
Died	3		Globe Tavern, Aug. 18-21	2
Died while POW	1	1865	Dabney's Mills, Feb. 5-7	3
Wounded & POW	1		White Oak Road, Mar. 30-31	2
POW	7		Appomattox, April 9 (at Surrender)	
Wounded	21			

Some Bedford County volunteers in the 107th Pennsylvania Infantry took part in turning back both Confederate Army attempts to invade Pennsylvania during the Civil War. Soldiers in the 107th PA Infantry were recruited throughout Pennsylvania, including Bedford County, during the winter of 1861-1862 and organized in Harrisburg. Approximately half of the county soldiers who enlisted in the 107th did so in the early part of 1862 with the remainder mustering in the regiment in the summer and fall of 1864. The 107th suffered the highest number of casualties of any Bedford County unit at Turners Gap during the battle of South Mountain. The Union Army victory at this crucial battle halted the Confederate invasion of Pennsylvania. Robert E Lee realized his divided army was vulnerable, and chose to wait near Sharpsburg, MD for the troops of Stonewall Jackson to return from Harpers Ferry to reunite with his main army group. Three days later the 107th took casualties during horrific fighting in the Cornfield at the battle of Antietam. The 107th suffered the highest number of county casualties during the 1st day of Gettysburg when outnumbered Union troops held back the invading Confederate Army on the western edge of the town of Gettysburg. This enabled the Union army to seize the crucial high ground surrounding the town, which was successfully defended during the next two days. The entire 107th PA Infantry Regiment suffered 165 casualties out of the 255 men on the battlefield at Gettysburg. County soldiers in the regiment also took part in the Overland Campaign in the spring of 1864, the siege of Petersburg, and were eyewitnesses to the Confederate surrender at Appomattox on April 9th, 1865. The regiment mustered out on July 13th after the war had ended. Thirteen percent of the Bedford County soldiers who enlisted in the 107th Pennsylvania Infantry lost their lives during the Civil War.

John Wareham, Private

James A Grove, Private

107th Pennsylvania Infantry Listing

Name	Pict. Page	Known Casualties	Battle
Black, George W Z		Wounded	Antietam
Bollman, David F			
Bowen, Silas J			
Brallier, William		Died	
Brant, William	267		
Bryant, John		Wound.-Died	Culpeper Courthouse
Buck, John			
Burket, Samuel			
Cashman, Jacob		Wound.-Died	White Oak Road
Chamberlain, Joseph		MIA	Globe Tavern
		Wounded	Petersburg - Initial Assault
Chaney, Levi		Wounded	Antietam
Christ, John			
Clark, Zachariah		Died	
Conner, Jonas		Wounded	Petersburg - Initial Assault
Cooper, Barton A	267		
Corle, Aaron			
Davis, Andrew J			
Davis, Samuel			
Eidenbaugh, John		Wounded	South Mountain
Ellis, Enos			
Fessler, Samuel		KIA	Antietam
Fetter, Samuel		POW	Dabney's Mills
Figart, Levi H		Wounded	South Mountain
Focht, George W		POW	2nd Bull Run
Fockler, George	246		
Foor, Abraham T		Wounded	
Foor, Andrew J	22		
Foor, George W	150-S	KIA	Antietam
Foor, Jeremiah	268		
Foor, John T			
Foor, Jonathan S	40	POW	Gettysburg
		Wounded	South Mountain
Foor, William H H		Wounded	
Gaster, James H		Wound.-Died	Antietam
Gracey, Alfred	40	POW	Gettysburg
Gracey, George E	187	POW	
Gracey, James A		POW	Gettysburg
		Wound.-POW	2nd Bull Run
Gracey, William C		Wounded	Gettysburg

Name	Pict. Page	Known Casualties	Battle
Grove, James A	85, 265	Wounded	Petersburg
Heckman, William			
Heinish, James			
Hillegass, Frederick			
Hockenberry, Jonathan		Died	Petersburg
Kay, Ezra P			
Kelley, John		Wounded	Petersburg
Leonard, Adam P	266		
Lyon, Samual			
Lysinger, George W		POW	2nd Bull Run
		POW-Died	Globe Tavern
Lysinger, John			
Martin, Josiah		Wounded	White Oak Rd
McDaniel, Jason			
Miller, Andrew P			
Miller, David P			
Mullenix, George			
Nycum, George			
Nycum, John W	267		
Pote, Andrew B			
Riley, Andrew J		Wounded	South Mountain
Riley, George W (1)	21	Wounded	2nd Bull Run
Riley, George W (2)			
Riley, Jacob			
Rinard, George W		Wounded	Dabney's Mills
Ritchey, James T	193		
Rohm, William H			
Rowzer, George C	209, 358	Wounded	Dabney's Mills
Salkeld, John F			
Salkeld, Thomas L	22, 267		
Shauf, John	199		
Smith, John			
Smith, John F			
Snowberger, Daniel			
Sparks, Uriah	39	Wounded	Gettysburg
Streightif, Samuel	268		
Stuckey, George W			
Swartz, William B		Wounded	2nd Bull Run
Ward, Jeremiah			
Wareham, John	265	Wounded	
Woodcock, Walter W			
Wright, Edwin V		Wounded	

107th Pennsylvania Infantry Listing

Thomas L Salkeld, Private

John W Nycum, Private

William Brant, Private on left pictured with his brother

Barton A Cooper, Private

107th Pennsylvania Infantry Listing

Private Samuel Streightif with his granddaughter

Private Adam & Emma Leonard. Emma was at home with a toddler when Adam enlisted in the 107th.

Private Jeremiah Foor with his wife Hannah and their daughter Margaret Foor Waltman.

110th Pennsylvania Infantry Listing

Bedford County Enlistments	111
Known Casualties	59
Casualty Rate Percentage	53%
Killed in Action	8
Died from Wounds	5
Missing in Action	1
Died	9
Wounded, POW & Died	1
Died while POW	2
Wounded & POW	2
POW	7
Wounded	23
Injured	1

Year	Major Battles with Casualties	Known Casualties
1862	1st Kernstown, Mar. 23	6
	Port Republic, Jun. 9	2
	Fredericksburg, Dec. 13	2
	Chancellorsville, Apr. 30 - May 6	4
1863	Gettysburg, Jul. 1-3	7
1864	Wilderness, May 5-7	5
	Spotsylvania, May 8-21	4
	Cold Harbor, May 31 - Jun. 12	4
	Petersburg - Initial Assault, Jun. 15-18	1
	Jerusalem Plank Road, Jun. 21-23	1
	1st Deep Bottom, Jul. 27	5
	Boydton Plank Road, Oct. 27-28	1
1865	Appomattox, April 9 (at Surrender)	

The 110th PA Infantry recruited soldiers in Bedford County for Company C in the fall of 1861. County volunteers in the 110th took casualties in many of the well-known battles in the Eastern Theater of the war. The 110th PA Infantry were among the regiments that handed Stonewall Jackson his only tactical battlefield defeat of the Civil War during their initial combat experience at the battle of 1st Kernstown. Casualties were taken during two of the worst Union Army defeats of the war at Fredericksburg and Chancellorsville. The 110th Pennsylvania Infantry Regiment took part in heavy combat near the Wheatfield on the 2nd day of fighting at Gettysburg and suffered the most casualties of any Bedford County unit during the three-day battle. County soldiers in the 110th suffered 13 known casualties during the Overland Campaign in the spring of 1864 at the battles of the Wilderness, Spotsylvania, and Cold Harbor. Additional casualties were taken during the siege of Petersburg in the summer and fall of 1864. After over 4 years of heavy combat, the soldiers in the 110th Pennsylvania Infantry regiment were eyewitness participants to the Confederate surrender at Appomattox on April 9th, 1865. The regiment also took part in the Grand Review parade in Washington, D.C. on May 23rd before mustering out on June 28th, 1865.

110th PA Infantry Monument at Gettysburg

The Pennsylvania State Memorial commemorates the 34,530 Pennsylvania soldiers who fought at Gettysburg by listing the name of each soldier confirmed to have been on the field during the battle.

110th Pennsylvania Infantry Listing

Name	Pict. Page	Known Casualties	Battle
Allen, William			
Baker, David N		Wounded	1st Kernstown
Banks, John			
Beegle, John A			
Blake, Samuel			
Blake, Simon			
Blake, Thomas			
Border, Andrew			
Border, John	93	Wounded	
Border, John S	272	Injured	Chancellorsville
		Wounded	Kelly's Ford
Bowman, Daniel H		Wound.-Died	1st Deep Bottom
Bowman, George		Died	
Brisbin, Ezra D	11, 271		
Brumbaugh, Francis		Wound.-POW-Died	Wilderness
Bulger, Andrew		Died	
Bulger, Levi M			
Carpenter, David B	151	POW	
Castner, John W			
Chilcoat, Isaac			
College, David			
College, James		Wound.-Died	1st Kernstown
College, John W		Wound.-Died	1st Kernstown
Cramer, Jacob		Died	
Criswell, Joseph		Wounded	
Croft, Alexander		Died	
Croft, Philip P		KIA	1st Kernstown
Daugherty, James		KIA	Fredericksburg
Davis, John M		Wounded	1st Deep Bottom
Davis, Martin L			
Davis, Porter R			
Defibaugh, David	272		
Detwiler, William			
Dively, John		POW-Died	
Everhart, David		KIA	Fredericksburg
Fackler, Samuel O			
Ferguson, John		KIA	1st Kernstown
Fishel, George W			
Fluke, Oliver P	355		
Fockler, Samuel	272		
Garrett, Albert T			
Garrett, John C		POW	Cold Harbor

Name	Pict. Page	Known Casualties	Battle
Gates, Joseph	271		
Gates, Martin		Wounded	Gettysburg
Gates, Samuel		Died	
Gates, William H			
Hamilton, John C			
Hartman, George L		MIA	Chancellorsville
Hartman, John P		POW	Cold Harbor
Harwood, Richard			
Hays, Alexander Y		Wounded	Gettysburg
			Wilderness
Heltzel, Jonathan D		KIA	Wilderness
Holsinger, Josiah		POW	Cold Harbor
		Wounded	Gettysburg
Jamison, Benjamin	28, 272	POW	Cold Harbor
Justice, Edward		Wound.-Died	Petersburg - Initial Assault
Kay, Ezra P			
Kay, Harry H C			
Kay, Isaac			
Kelly, David			
Lamison, George W		Wound.-Died	Gettysburg
Lamison, Thomas		Died	
Lane, David C			
Livingston, Thomas G	11, 61	Wounded	Spotsylvania
McDonald, Samuel			
McIlnay, James		Died	
Miller, Andrew			
Miller, Hezekiah H		Died	
Miller, John I	43	Wounded	Gettysburg
Mock, George		Wounded	Boydton Plank Road
Moore, James B			
Moore, John B		Wound.-POW	Gettysburg
Morgan, Dennis			
Newton, James			
Pearson, Francis		Died	
Price, David J		Wounded	1st Kernstown
Ralston, David E		KIA	Chancellorsville
Ralston, William H			
Roberts, William			
Schwartz, Samuel B			
Scott, Cornelius W		Wounded	
Seabrooks, George			
Shimer, William H H			

110th Pennsylvania Infantry Listing

Name	Pict. Page	Known Casualties	Battle
Shoemaker, Austin		POW	Jerusalem Plank Road
Shoemaker, Benjamin	272		
Smith, David S			
Smith, John W	37	POW	Chancellorsville
		Wound.-POW	Port Republic
		Wounded	Wilderness
Smith, Samuel H		Wounded	1st Deep Bottom
Speer, William H		Wounded	Spotsylvania
Stonerook, Aaron B			
Stonerook, Simon B		Wounded	Spotsylvania
Stout, Richard F			
Straley, James			
Sutton, Jonathan A		Wounded	1st Deep Bottom
Swaney, David R P			
Swaney, Samuel J			
Swaney, William S		Wounded	Spotsylvania
			Wilderness

Name	Pict. Page	Known Casualties	Battle
Tasker, George			
Taylor, Ambrose K		KIA	1st Deep Bottom
Taylor, William Y			
Tetwiler, Jacob D		Wounded	
Tetwiler, William			
Thompson, David		POW-Died	Petersburg
Tobias, John			
Tobias, Samuel H G		KIA	Gettysburg
		Wounded	Port Republic
Tyson, Samuel H			
Wallace, Samuel G			
Williamson, Gideon			
Wilson, James A			
Woodward, James A			
Woolett, Sylvester B			
Young, Edwin			
Young, George N			

Joseph Gates, Sergeant

Ezra D Brisbin, Captain, circa 1870

110th Pennsylvania Infantry Listing

(left) John S Border, Private

(right) Benjamin Shoemaker, Sgt.

(left) David Defibaugh, Private circa 1860

(right) Samuel Fockler, Private

Private Benjamin & Caroline Whetstone Jamison pictured with their grandchildren

125th Pennsylvania Infantry Listing

Bedford County Enlistments	47			**Known**
Known Casualties	13	**Year**	**Major Battles with Casualties**	**Casualties**
Casualty Rate Percentage	28%	1862	Antietam, Sep. 17	8
Killed in Action	3	1863	Chancellorsville, Apr. 30 - May 6	1
Missing in Action	1			
Died	2			
Wounded	7			

Soldiers in the 125th Pennsylvania Infantry regiment were recruited in Blair, Huntingdon, and Cambria Counties for a nine-month enlistment in August 1862. Most of the Bedford County volunteers in this regiment mustered in on August 8th, 1862. Five weeks removed from being civilians, the 47 men and boys from Bedford County in the 125th were front-line combatants on the deadliest day in American military history. Heavy fighting had already been raging for 3 hours at Antietam on the morning of September 17th when the 125th advanced over 500 yards from the Cornfield area of the battlefield to the West Woods near the Dunker Church. Colonel Jacob Higgins stated on his official report, "I ordered my skirmishers to rally and gave the command to commence firing. A most destructive fire caused the enemy to halt. I held him here for some time, until I discovered two regiments of them moving around my right, while a brigade charged on my front. On looking around and finding no support in sight, I was compelled to retire. Had I remained in my position two minutes longer, I would have lost my whole command." The monument below, near the Dunker Church states, "the 125th PA Infantry was the first to reach the Woods. Being far advanced and without sufficient support, it was outflanked by the enemy and retired behind batteries in field in rear and subsequently saved the guns of Monroe's Battery from capture." Within 20 minutes around 9am, the green troops of the 125th PA Infantry had suffered 229 casualties, including 4 color bearers who were shot down. All 8 of the known Bedford County casualties were likely during this action. During the battle of Antietam, there were approximately 22,000 casualties on both sides, including 3,650 soldiers who lost their lives. This costly battle forced Robert E Lee to abort his invasion plans of Pennsylvania and retreat back across the Potomac River. The 125th PA Infantry also took casualties during the battle of Chancellorsville before the regiment completed their enlistment on May 18th, 1863.

William C Kean, Corporal

125th PA Monument at Antietam

David Harclerode, Private

125th Pennsylvania Infantry Listing

Name	Pict. Page	Known Casualties	Battle
Baker, Franklin S		KIA	Antietam
Barr, Thomas M	274, 357		
Beegle, John A			
Benton, Emanuel M			
Blake, William B	37	Wounded	Chancellorsville
Boyes, George			
Bradley, Thomas			
Breneman, Michael B	28, 357	Wounded	Antietam
Brown, Jacob			
Brumbaugh, Jacob			
Bryant, James		Wounded	Antietam
Burkholder, George		Wounded	Antietam
Cypher, Thomas F			
Dasher, James H			
Dell, Peter		Wounded	
Dively, Gabriel			
Dodson, Andrew J	274		
Fagans, James			
Fockler, Jacob L			
Geisler, Lewis H			
Green, William B			
Harclerode, David	29, 273	Wounded	Antietam
Hazzard, David			
Heffner, John			
Heffner, John F		Died	
Homan, William			
Jamison, Benjamin	28, 272	Wounded	Antietam
Kean, William C	273		
Kelley, John A		KIA	Antietam
Lear, John		KIA	Antietam
Lewis, John D		MIA	Chancellorsville
Long, Samuel L			
McLaughlin, William H			
McMurtrie, J R			
Mock, George W		Died	
Moore, Joseph			
Owens, Richard			
Saxton, Henry C			
Snyder, George W Jr.			
Snyder, George W Sr.			
Snyder, William H	160		
Spangler, Jeremiah			
Suder, Charles H			
Swisher, Daniel			
Trout, Alexander			
Weyandt, Samual S	210		
Wright, Thomas			

Andrew J Dodson, Private

Thomas M Barr, Private

133rd Pennsylvania Infantry Listing

Bedford County Enlistments	198	**Year**	**Major Battles with Casualties**	**Known Casualties**
Known Casualties	42	1862	Fredericksburg, Dec. 13	34
Casualty Rate Percentage	21%	1863	Chancellorsville, Apr. 30 - May 6	1
Killed in Action	7			
Died from Wounds	1			
Died	2			
POW	1			
Wounded	31			

The 133rd Pennsylvania Infantry regiment recruited volunteers in Companies C & K in Bedford County during the summer of 1862 for a 9-month enlistment. Recruits from Somerset and Cambria Counties volunteered in other companies in the 133rd. The regiment was organized at Camp Curtin in Harrisburg in August 1862 before being sent to Washington, D.C. later that same month. When Robert E Lee threatened an invasion of Pennsylvania, the 133rd departed Washington but arrived at Antietam the day after the battle. Two months later the 133rd PA Infantry arrived in Falmouth, VA across the river from Fredericksburg, a month before they would take part in one of the most horrific assaults of the Civil War. Humphrey's Charge was the last in a series of Union assaults on Marye's Heights during the battle of Fredericksburg. Marye's Heights was heavily defended by Confederate artillery, raining cannon fire from the top of the hill, on charging Union soldiers while infantry units poured rifle fire from behind a stone wall at the foot of the slope. Colonel Speakman of the 133rd PA infantry wrote the following in his official report. "On December 13th, 1862, between 2:00 & 3:00, the 133rd crossed the Rappahannock river while being fired upon by enemy artillery. After passing through the town of Fredericksburg, the regiment fixed their bayonets and marched toward Marye's Heights. After advancing 250 yards, the 133rd came across Union infantry soldiers lying on the ground, neither advancing nor retreating from their position. The 133rd was ordered to charge over the prostrate troops and advanced to within 50 yards of the stonewall at the bottom of Marye's Heights. This position was held for almost an hour under terrific fire from enemy infantry and artillery before being ordered to withdraw." No other regiment is believed to have made it any closer to the stonewall at Marye's Heights on that day. Bedford County soldiers in the 133rd suffered 34 casualties, including 8 who lost their lives. The 133rd PA also took part in the battle of Chancellorsville before the regiment completed their enlistment on May 24th, 1863.

Martin Moser, Private

John W Fisher, Private

133rd Pennsylvania Infantry Listing

Name	Pict. Page	Known Casualties	Battle	Name	Pict. Page	Known Casualties	Battle
Amick, William	278			Edwards, Jonathan B	279		
Armstrong, Joseph M		Wounded	Fredericksburg	Elder, Daniel S			
Ashcom, George Jr		Wounded	Fredericksburg	Evans, George W			
Bagley, Samuel				Evans, William			
Baker, George W				Figart, David			
Barkman, Thomas C		Wounded	Fredericksburg	Fink, Valentine	279		
Barndollar, Jacob W	33	Wounded	Fredericksburg	Fisher, Henry H	279		
Barndollar, James E		KIA	Fredericksburg	Fisher, John W	275		
Barndollar, James J	278			Fleegle, William			
Barndollar, Martin D	278			Fletcher, John L			
Bayer, George M				Fluck, Porter			
Benner, Samuel B				Foor, James H	279	Wounded	
Blake, Simon				Fore, James F			
Blankley, Job				Foreman, William			
Boore, Jocob C				Foster, Joseph E			
Border, Henry		Wounded	Fredericksburg	Fulton, Adam			
Bowser, George L				Gallagher, Edward		KIA	Fredericksburg
Bowser, Isaac B	167			Gaster, Ezekiel W	279		
Brechbiel, Abraham				Gates, John W			
Brown, William P				Gibson, Henry F			
Burch, Thomas H				Gibson, William Y			
Burget, Isaac				Gogley, Jacob			
Bussard, Joseph S				Gogley, James H		Wounded	Fredericksburg
Butterfield, G W				Gray, Ellis J			
Butts, James B	278			Grove, Robert C			
Cambell, Robert		Wounded	Fredericksburg	Growden, John S			
Carnell, James				Grubb, Harvey	279, 326		
Carson, Daniel				Hafer, Wilson			
Castner, Jacob H				Hanks, Albert B			
Castner, John B				Hanks, William H	322		
Chamberlain, Joseph				Hann, Philip			
College, James				Hartman, Frederick			
Conner, David	168-B	Died	Fredericksburg*	Hawman, John C	322		
Conner, Lewis	168, 279			Hayes, William			
				Heltzel, David S			
Cooper, Joshua H		KIA	Fredericksburg	Himes, John	314		
Cottle, Jacob				Hoffman, John O	357		
Crissey, John C		Wounded	Fredericksburg	Horner, Henry			
Croyle, James A				Horton, Jonathan A	206		
Daugherty, Jacob W				Imler, Adam H			
Deremer, William				Johnston, John W	283	Wounded	Fredericksburg
Dicken, Thomas W	205			Jones, David W			
Downey, Michael				Justice, Edward		Wounded	Fredericksburg
Dunkle, David	283			Kauffman, David			
Dunkle, Simon				Keagy, Samuel			
Durno, William Jr	278			King, Philip V			

133rd Pennsylvania Infantry Listing

Name	Pict. Page	Known Casualties	Battle
Klahre, Herman T		Wounded	Fredericksburg
Kochendarfer, John Z		Wounded	Fredericksburg
Lamberson, David			
Lambert, Joseph C	33	Wounded	Fredericksburg
Langdon, Samuel			
Leader, David F			
Lee, Henry W			
Lewis, Simon P			
Longenecker, John S			
Lucas, Joshua T			
Lysinger, John			
Mack, Joseph			
Madden, Cyrus			
Malone, John			
Maugle, Solomon			
McCleary, George B		Wound.-Died	Fredericksburg
McClellan, John		Wounded	Fredericksburg
McClellan, Josiah	281	Wounded	Fredericksburg
McCullip, Alexander		Wounded	Chancellorsville
		Wounded	Fredericksburg
McDaniel, Daniel	281		
McDaniel, George W	279		
McDaniel, Lewis			
Meloy, John L			
Mentzer, Jacob M		KIA	Fredericksburg
Miller, Jacob			
Miller, Jacob B			
Miller, Jacob W		Wounded	Fredericksburg
Mills, Jacob			
Mixel, Samuel			
Mobley, Ezekiel			
Mock, Harrison			
Mock, Mathias			
Mock, Tobias			
Morse, Morgan			
Moser, Martin	275		
Mumper, Henry		Wounded	Fredericksburg
Nevitt, James M	280	Wounded	Fredericksburg*
Newcomer, Joseph		POW	
Nycum, William H	280	Wounded	Fredericksburg
Osborn, William			
Ott, Michael			
Over, Benjamin	282	Wounded	Fredericksburg
Peck, Jesse			
Penrod, George			

Name	Pict. Page	Known Casualties	Battle
Perrin, John		KIA	Fredericksburg
Pilkington, James H			
Potter, John			
Price, Daniel M		Wounded	
Protheroe, David			
Raley, Daniel			
Ramsey, Alexander	282		
Reed, Thomas			
Refley, William			
Richter, Adam	281		
Riffle, Cyrus	150		
Riley, Jacob E			
Riley, James H	280	Wounded	Fredericksburg
Ritchey, Adam S			
Roy, James		Wounded	Fredericksburg
Scritchfield, Hezekiah			
Scutchall, John			
Scutchall, Samuel		KIA	Fredericksburg
Shade, James A			
Shaffer, Abraham	283		
Shaffer, Samuel			
Shroyer, Andrew G		Wounded	Fredericksburg
Singleton, Samuel		Wounded	Fredericksburg*
Skillington, Robert M	209		
Smith, Albert			
Smith, Jacob			
Smith, Jacob N			
Smith, John P			
Snyder, William			
Souser, Henderson	282, 361		
Sparks, Jacob		Died	
Sparks, James H	309		
Sparks, John C	283, 359		
Sparks, Joseph H			
Sparks, Silas H	283		
Speice, Louis D	307		
Spielman, Martin V			
Sproat, Joseph R			
Stailey, George E		Wounded	Fredericksburg
Steele, David F	180-P	KIA	Fredericksburg
Steele, Edward			
Steele, Levi H	357		
Stoner, Joshua			
Stoudnour, William			
Stoutnour, Samuel R			

133rd Pennsylvania Infantry Listing

Name	Pict. Page	Known Casualties	Battle
Swaney, David R P			
Swank, George W		Wounded	Fredericksburg
Tate, Samuel B			
Taylor, John W			
Thompson, William			
VanHorn, John M			
Weaverling, James T		Wounded	Fredericksburg
Weaverling, William T			
Welsh, William F			

Name	Pict. Page	Known Casualties	Battle
Wertz, Henry			
Whited, David L	282		
Whittaker, Jonathan			
Wilkinson, William			
Williams, Charles			
Williams, Jacob	282		
Williams, Samuel D			
Winters, John J			
Yeagle, Simon B			

James J Barndollar, 1st Sergeant

Martin D Barndollar, Private

William Durno, Corporal

James B Butts, Musician

War-time photograph of Private William & Sophia Amick. William appears to be holding a bible. They were married in 1853. Sophia was at home with three children when William volunteered in the 133rd PA Infantry. She gave birth to their 4th child in December 1862, the same month as William took part in Humphrey's Charge during the battle of Fredericksburg.

133rd Pennsylvania Infantry Listing

Ezekiel W Gaster, Musician

Harvey Grubb, Private

Jonathan B Edwards, Corp

Valentine Fink, Private

George W McDaniel, Private

James H Foor, Private

Private Henry & Catherine Fisher married in 1860. Catherine was at home in South Woodbury Township with 3 small children while Henry was away in the war.

Lewis Conner, Private

133rd Pennsylvania Infantry Listing

Left to right:: Pamelia Nycum, John Q. Adams Nycum, and William Henry Harrison Nycum in a rare circa 1850 photograph of the children who were growing up in Ray's Hill (Breezewood). William was a corporal in the 133rd PA. Both brothers mustered in the 186th Pennsylvania Regiment in March 1864.

James M Nevitt, Private

Private James & Sarah Riley wedding picture in 1867

133rd Pennsylvania Infantry Listing

Josiah McClellan, Private

Adam Richter, Private

(left to right) Susan, Daniel (Private), Elizabeth and Ida McDaniel pictured circa 1871-1872

133rd Pennsylvania Infantry Listing

Rebecca & Private David Whited

Jacob Williams

Alexander Ramsey, Private

Henderson Souser, Corporal

Benjamin Over, Private

133rd Pennsylvania Infantry Listing

Private John W Johnston, he is possibly pictured with his wife Barbara and one of their daughters.

Lucy and Private David Dunkle

Humphrey's Division memorial at the Fredericksburg National Cemetery on Marye's Heights

Silas H Sparks, Private

Abraham Shaffer, Corporal

John C Sparks, Private

138th Pennsylvania Infantry Listing

Bedford County Enlistments	279			**Known**
Known Casualties	176	**Year**	**Major Battles with Casualties**	**Casualties**
Casualty Rate Percentage	63%	1863	Mine Run, Nov. 27	17
Killed in Action	15	1864	Wilderness, May 5-7	40
Died from Wounds	14		Spotsylvania, May 8-21	3
Missing in Action	8		Cold Harbor, May 31 - Jun. 12	20
Died	18		Monocacy MD, Jul. 9	18
Died while POW	5		Opequan, Sep. 19	16
Wounded & POW	4		Fisher's Hill, Sep. 21-22	2
POW	8		Cedar Creek, Oct. 19	21
Wounded	104	1865	Petersburg - Final Assault, April 2	5
			Sailor's Creek , April 6	3
			Appomattox, April 9 (at Surrender)	

The 138th Pennsylvania Infantry suffered the second highest number of casualties of any Bedford County unit during the war. The 138th PA Infantry recruited volunteers for Companies D, E & F in Bedford County during the summer of 1862. County volunteers reported for initial training at Camp Curtin in Harrisburg on August 16th, 1862. Most of the 176 Bedford County casualties in the regiment were suffered in 1864 during the late spring Overland Campaign and the following fall during the triumphant Shenandoah Valley campaign. The 138th was on the extreme right flank of the Union Army lines during the battle of the Wilderness, fighting in some of the thickest underbrush and heavily wooded terrain of the war. Heavy losses were suffered on May 6th when Confederate forces mounted a surprise flanking attack that rolled back Union Army lines. County troops took many casualties the following month during an attempt to drive through a gap in Confederate lines at the battle of Cold Harbor. The heroic efforts of the 138th and other vastly outnumbered Union Army troops during the battle of Monocacy bought desperately needed time for reinforcements to be deployed to defend Washington, D.C. That fall, the 138th took part in the successful Shenandoah Campaign that helped to break the back of the Confederacy. The following spring, the 138th were among the troops that first broke through the Confederate lines during the Final Assault on Petersburg on April 2nd, 1865. Fittingly, after much combat and casualties, Bedford County troops in the 138th PA Infantry witnessed the Confederate surrender at Appomattox. The 138th Regiment mustered out of the army on June 23rd, 1865 after suffering the 5th highest percentage of casualties of any Bedford County unit during the war.

1st Lt. Christopher P Calhoun prior to his promotion

Lewis A May, Lt. Colonel

138th Pennsylvania Infantry Listing

Name	Pict. Page	Known Casualties	Battle
Ake, John G		Wounded	Wilderness
Ake, William H		KIA	Wilderness
Allison, Joseph D		Died	
Allison, Noah		KIA	Cold Harbor
Amick, William B		KIA	Wilderness
Armstrong, Albert		Wounded	Wilderness
Bagley, Moses G		Wounded	Cold Harbor
Bailey, James			
Bailey, John W			
Bailey, William H		POW	Monocacy, MD
Ball, Daniel M		Wounded	Cedar Creek
		Wounded	Monocacy, MD
Baner, Franklin		Wounded	Cold Harbor
		Wounded	Wilderness
Barclay, Joseph		Wounded	Cedar Creek
Barkley, George W			
Barkley, Samuel			
Barkman, David			
Barkman, Hezekiah		Wounded	Cold Harbor
Baughman, George	127	POW-Died	Wilderness
Baughman, John A			
Baughman, Joseph			
Baughman, Josiah		KIA	Chaneysville PA
Beals, George W			
Beals, Nicholas H		Wounded	Cold Harbor
		Wounded	Wilderness
Beard, Daniel		Wounded	Cedar Creek
Beaver, Nicholas			
Beegle, Job M			
Beltz, Adam	165-S	KIA	Cold Harbor
Beltz, George W		Wounded	Cedar Creek
Beltz, John A		Wounded	Cedar Creek
Benner, John			
Biddle, Andrew B	57	Wounded	Wilderness
Binett, Nathan			
Bingham, Linton W			
Bivens, James W	289		
Blackburn, Joseph		Wound.-Died	Monocacy, MD
Blackburn, Levi		Died	
Boor, John A		Died	
Bortz, Martin S	97, 289	Wounded	Cedar Creek
			Petersburg
Briggle, Jacob			
Burge, Joseph			
Burket, David	356	Wounded	Mine Run

Name	Pict. Page	Known Casualties	Battle
Burket, Isaac			
Burket, John N	288		
Calhoun, Christopher P	284		
Carl, Jacob			
Carpenter, Abraham	96, 288	Wounded	Cedar Creek
Carrell, Daniel		MIA	Monocacy, MD
		Wounded	Mine Run
Carrell, Joseph		Wounded	Wilderness
Claar, Jacob C			
Claar, John			
Clark, Samuel M		KIA	Opequon
Claycomb, Conrad	288, 358	Wounded	Wilderness
Cobler, Allen	73	Wounded	Cold Harbor
Cobler, Andrew			
Cobler, Joseph		Wounded	
Conch, Harry			
Conley, Martin L		MIA	Cold Harbor
Cook, David			
Cook, John H		Wound.-POW	Wilderness
Cook, Levi		Wounded	Wilderness
Cook, Reuben W			
Corle, Abraham		Wound.-Died	Wilderness
Corle, Franklin	289, 356		
Corle, William C	288	Wounded	Wilderness
Couch, Harry C			
Craine, David D		KIA	Cedar Creek
Crawford, James S		Wounded	
Croyle, Adam			
Curry, James W			
Defibaugh, Lawrence			
Defibaugh, William H		MIA	Wilderness
Devens, Elisha			
Dicken, John		POW-Died	Mine Run
Dickerhoof, Simon		POW	Sailor's Creek
		Wounded	Brandy Station
Diehl, John			
Dull, Valentine			
Earnest, William		Wounded	Monocacy, MD
Elder, Lewis			
Evans, Harvey			
Fait, John			
Feather, George W		Died	
Feather, Simon	290	Wounded	Spotsylvania
Feight, Abraham		Died	

138th Pennsylvania Infantry Listing

Name	Pict. Page	Known Casualties	Battle	Name	Pict. Page	Known Casualties	Battle
Feight, Henry H				Horton, Oliver	111, 290	Wounded	Cedar Creek
Feight, John W		Wounded	Opequon	Huffman, Josiah		Wounded	Wilderness
Feight, William F		Wounded	Cedar Creek	Hunt, John T			
Ferguson, William		POW	Monocacy, MD	Hunt, Samuel B	238		
Fickus, Cyrus				Ickes, George		Died	
Filler, William T				Ickes, George W	288		
Fisher, Emanuel	290			Imler, Daniel		Died	
Fleegle, George W		Wounded	Cold Harbor	Imler, Ephraim Y			
Foaulke, Carle				Imler, George R	89	Wound.-POW	Monocacy, MD
Foor, Martin T		POW-Died	Monocacy, MD	James, Nathaniel		MIA	Opequon
Foster, William				Kegg, Nathaniel		Wounded	Wilderness
Gardner, Charles		POW	Monocacy, MD	Kellerman, James L	290	Wounded	Cedar Creek
Geller, George		Wounded	Monocacy, MD			Wounded	Monocacy, MD
Geller, John		Wounded	Cedar Creek	Kelly, Henry		MIA	Cold Harbor
Getty, John	289			Kelly, William		Wounded	Cedar Creek
Gilchrist, James A				Kennard, John H		Wounded	Mine Run
Gillam, George		Wounded	Monocacy, MD	King, Harrison H		Wounded	Opequon
Glenn, Josiah						Wounded	Sailor's Creek
Gordon, Isaac	288, 358	POW	Monocacy, MD			Wounded	Wilderness
Gray, George W		Wounded	Petersburg - Final Assault	Kingsley, David			
Gump, John A		Wound.-Died	Cedar Creek	Kinsey, John B	155		
Hammer, John B	290	Wounded	Opequon	Kinton, Allen			
		Wounded	Wilderness	Knisely, David			
Harbaugh, Emanuel		POW	Wilderness	Kurtz, Thomas			
Harden, Calvin				Lape, Jackson			
Hayman, Francis H				Lay, Joseph		MIA	Wilderness
Heckman, James				Layton, John		Wound.-Died	Opequon*
Hellman, Daniel		KIA	Mine Run	Lease, Robert H		Wounded	Sailor's Creek
Hellman, George		Wounded	Wilderness	Leasure, Josiah G		Wounded	Cold Harbor
		Wounded	Petersburg - Final Assault	Leasure, Nathaniel			
Heltzel, Daniel G	135	Died		Lemmon, William			
Heltzel, Simon		Died		Leonard, Henry N			
Heltzel, William				Leonard, John D		Wounded	Wilderness
Henderson, Robert F				Line, William			
Hixon, Erastus J		KIA	Wilderness	Ling, Isaac N	292		
Hochard, John A		Wounded	Mine Run	Ling, William H	358	POW	Monocacy, MD
		Wounded	Wilderness	Long, George			
Hoenstine, Benjamin				Lowery, Emanuel	291	Wounded	Cold Harbor
Hoenstine, David		Died		Lowery, John E		Wound.-Died	Mine Run
Holler, George W	153	Wounded	Opequon	Lowery, William H		Died	Brandy Station
		Wounded	Wilderness	Lowry, Oliver		Wounded	Opequon
Holler, John M	153, 290, 355			Lucas, William		Wounded	Opequon
Hook, Elias				Mauk, John W			
Hook, George				May, Hiram	291	Wounded	Cold Harbor
						Wounded	Opequon

138th Pennsylvania Infantry Listing

Name	Pict.	Known	Battle
May, John W		Wounded	Wilderness
May, Lewis A	57, 284		
May, Marcus	92, 291	Wounded	
May, Samuel M	291		
McCleary, Henry		Wounded	Wilderness
McCoy, Shannon E		MIA	Cold Harbor
McGee, Samuel	174		
McLaughlin, Charles		KIA	Cold Harbor
McVicker, William			
Meloy, Biven D		Wounded	Wilderness
Miller, Abraham		Wounded	Cedar Creek
Miller, Ephraim B	290	Wounded	Cedar Creek
Miller, Henry		Wound.-Died	Wilderness
Miller, Jackson		KIA	Wilderness
Miller, Jesse			
Miller, Thomas	295		
Miller, Thomas J		POW-Died	Wilderness
Miller, Tobias			
Mock, Aaron	47, 293, 358	Wound.-POW	Wilderness
Mock, Emanuel	47, 293, 358	Wounded	Mine Run
		Wounded	Cold Harbor
Mock, Lewis			
Mock, Malachi			
Moore, James			
Moser, Jeremiah		Wounded	Cold Harbor
Mowry, Frederick			
Mullin, John			
Naugle, James			
Neff, Frederick			
Nicodemus, Isaac	114		
Nycum, Bernard			
Nycum, John	177-B	Wound.-Died	Cold Harbor
Oaks, John R	313		
O'Neal, Emanuel	291		
O'Neal, Hezekiah		Wound.-Died	
O'Neal, John E	290, 292	Wounded	Spotsylvania
Over, James E	358	Wounded	Opequon
Owens, Chauncey	89	Wounded	Monocacy, MD
Porter, Philip			
Price, Abraham		Died	
Price, Daniel J	61, 292	Wounded	Spotsylvania
Price, Jacob F			

Name	Pict. Page	Known Casualties	Battle
Price, Joseph J		KIA	Wilderness
Prideaux, Thomas A	292		
Radcliff, James		Died	
Ramsey, William W			
Rea, William H		Died	
Reighard, Matthias			
Reighard, Peter			
Ridenbaugh, Samuel		Wounded	Cedar Creek
Riffle, William		Wounded	Opequon
Risling, Joseph			
Ritchey, Frederick G		Wounded	Cedar Creek
		Wounded	Opequon
Ritchey, Henry C	293		
Ritchey, Jacob K			
Ritchey, John C	293	Wound.-POW	
Robb, George W		KIA	Mine Run
Robb, Samuel			
Robinson, William J		Wounded	Wilderness
Roland, Henry			
Rollins, James			
Rush, David		Wounded	Cold Harbor
Saupp, James			
Scritchfield, Samuel		Wounded	
Sellers, Frederick A	294		
Shaffer, Harvey E	294	Wounded	Monocacy, MD
Shaffer, Thomas		Wounded	Fisher's Hill
Shaffer, Tobias		Wounded	Fisher's Hill
Shroyer, Joseph	159	Wounded	Opequon
Shroyer, Moses		Wounded	Opequon
Slack, Francis M		Wounded	Petersburg - Final Assault
Slick, William S	108	Wounded	Petersburg - Final Assault
Smith, Adam			
Smith, David		Wound.-Died	
Smith, George			
Smith, Jacob F		Wound.-Died	Wilderness
		Wounded	Mine Run
Smith, John W			
Smith, Miles N		Wounded	Wilderness
Smith, Simon R			
Snyder, David F	295	Wounded	
Snyder, Jonathan	96, 145	Wound.-Died	Cedar Creek
		Wounded	Wilderness
Speck, Henry		Wounded	Cedar Creek
			Mine Run

138th Pennsylvania Infantry Listing

Name	Pict. Page	Known Casualties	Battle
Steckman, Francis		Wound.-Died	Cold Harbor
Steckman, John B		Died	
Steckman, Philip H			
Steuby, Conrad G		Wounded	Wilderness
Stevens, Jacob B			
Stiffler, John H		Wound.-Died	Mine Run
Stiffler, Nathaniel		Wounded	
Stineman, William		POW	Monocacy, MD
Stotler, Marion	293		
Stuckey, Elias B	294	Wounded	Petersburg - Final Assault
Stuckey, John S	92, 294	Wounded	Opequon
Stuckey, Simon C		KIA	Mine Run
Stuckey, Wilson H			
Summerville, Charles		MIA	Wilderness
Summerville, John B			
Taylor, Matthew P		Wounded	Mine Run
Thorpe, Jacob			

Name	Pict. Page	Known Casualties	Battle
Thorpe, John W		Died	
Thorpe, Solomon R			
Tipton, Noah	295		
Troutman, George W		Died	
Valentine, John		Wounded	Opequon
Vickroy, James R			
Wagerman, William	147, 289	Died	
Ward, Samuel		POW-Died	Mine Run
Wentz, Philip		KIA	Mine Run
Western, John			
Whip, Jacob		Wound.-Died	Mine Run
Wise, Andrew H			
Witt, Jacob			
Wolford, Daniel	295		
Wolford, Frederick			
Yarnell, Jesse		Wound.-Died	Monocacy, MD
Yarnell, John		Wounded	Cedar Creek

William C Corle, Private

George W Ickes, Private

Isaac Gordon, Private

Abraham Carpenter, Private

John N Burket, Private

Conrad Claycomb, Private

138th Pennsylvania Infantry Listing

Franklin Corle, Private

Private William Wagerman & Captain Martin Bortz

John Getty, 1st Lieutenant

Private James Bivens and his wife

138th Pennsylvania Infantry Listing

Emanuel Fisher, 1st Lieutenant

John M Holler, Private

James L Kellerman, Private

Ephraim B Miller, Private

John E O'Neal, Corporal

Oliver Horton, Captain

Sgt. John B Hammer, wife Leah are seated. Children left to right: Juliann, George and Melissa in 1883.

Simon Feather, Private on the right. Man on the left is unidentified.

138th Pennsylvania Infantry Listing

Private Emanuel and Elizabeth O'Neal married in 1857. Elizabeth was at home with 3 small children while Emanuel was away in the war.

Private Emanuel Lowery and Hannah Burket married in 1864. Emanuel's first wife Louisa passed in 1859, leaving behind 4 young children. Emanuel remarried and Hannah gave birth to a daughter in 1862, the same year Emanuel volunteered.

Five May brothers served in Pennsylvania infantry regiments. Seated left to right are Samuel M, Sgt. - 138th, Daniel H, Corp. - 82nd & Marcus, Corp. - 138th. Standing left to right are Hiram, Pvt. - 138th & John L, Pvt. - 67th.

138th Pennsylvania Infantry Listing

Thomas A Prideaux, 1st Lieutenant

Brothers James R O'Neal, 1st Sergeant - 208th PA on the left & John E O'Neal, Corporal - 138th PA

Daniel J Price, Private

Corporal Isaac N Ling married Delilah Hammer in 1857. Hannah was at home with 2 small children while Isaac was in the war.

138th Pennsylvania Infantry Listing

Henry C Ritchey, Private

Mary Ann & Private John C Ritchey

Margret & Private Aaron Mock

Marion Stotler, Private

138th Pennsylvania Infantry Listing

Harvey E Shaffer, Sergeant

Elias B Stuckey, Corporal

Frederick A Sellers, Private

John S Stuckey, Captain

138th Pennsylvania Infantry Listing

Thomas Miller, Private

David F Snyder, Private

Lavine & Private Noah C. Tipton

The Wolford family (front) Rena, Mirtle, Jennie, (back) Howard, Sarah holding Betty, Corporal Daniel Wolford, Amanda & Alice. Bedford County soldiers Daniel Wolford and John Mauk fired the volley's that killed Confederate General A.P. Hill during the Final Assault on Petersburg.

149th Pennsylvania Infantry Listing

Bedford County Enlistments	53			Known
Known Casualties	25	Year	Major Battles with Casualties	Casualties
Casualty Rate Percentage	47%	1864	Wilderness, May 5-7	7
Killed in Action	2		Spotsylvania, May 8-21	5
Died from Wounds	2		North Anna River, May 23-26	3
Missing in Action	1		Cold Harbor, May 31 - Jun. 12	1
Died	2		Petersburg - Initial Assault, Jun. 15-18	3
Died while POW	4		Globe Tavern, Aug. 18-21	1
Wounded	13			
Injured	1			

Twenty-one percent of the Bedford County soldiers who enlisted in the 149th Pennsylvania Infantry lost their lives during the Civil War. The 149th was organized in Harrisburg in August 1862 and was commonly referred to as the 2nd Bucktail Regiment. The 3 Bucktail regiments were among the most famous Pennsylvania infantry regiments of the Civil War. The 13th PA Reserves, 149th PA Infantry and 150th PA Infantry sported bucktails on their headwear as a trophy of marksmanship. Most Bedford County soldiers joined the 149th in August and September 1863. County soldiers in this regiment took casualties during all the major battles of Grant's Overland Campaign and the Initial Assault on Petersburg in the spring and early summer of 1864. The regiment mustered out on June 24th, 1865 at the conclusion of the war.

Name	Pict. Page	Known Casualties	Battle
Bagley, Henry H	297		
Barnett, Ephraim		POW-Died	North Anna River
Bisel, Noah			
Bousch, Christian		Wound.-Died	North Anna River
Brant, Andrew J		Wounded	Wilderness
Brant, Benjimen F		Wounded	Wilderness
Brechbiel, Daniel			
Carn, Jeremiah		Wounded	Spotsylvania
Clingerman, Harrison	297		
Corley, Benjamin			
Cox, Samuel	196	Injured	Cold Harbor
Croil, John T			
Emerick, Solomon			
Evans, Isaiah		KIA	Spotsylvania
Evans, Lewis			
Fetter, George			
Gates, Jeremiah E		Wounded	Petersburg - Initial Assault
Griffith, Joseph H			
Grim, William		POW-Died	Wilderness
Grubb, George		POW-Died	Globe Tavern
Hardinger, George W		POW-Died	Wilderness
Harr, Christian	76, 358	Wounded	Petersburg - Initial Assault
Huffman, Joseph W			
Hughes, Bailey		Died	
Johnson, Joshua			
Kegg, Levi		Wounded	Wilderness

Name	Pict. Page	Known Casualties	Battle
Kirk, William			
Leasure, Solomon		Died	
Lee, David W	297		
Lingenfelter, David R			
May, Jacob		MIA	Wilderness
McDaniel, Hiram		Wound.-Died	Spotsylvania
McKibbin William L	347		
Miller, George W		Wounded	Spotsylvania
Miller, Levi			
Miller, Michael C			
Nicodemus, Joseph			
Reed, Alexander		Wounded	
Rice, Soloman		Wounded	North Anna River
Ritz, Daniel			
Sanderson, Theodore C	357		
Smith, George			
Steele, Thomas		Wounded	Spotsylvania
Stevens, Denton			
Ward, Henry			
Ward, William S			
Whitfield, John			
Whitfield, William C		Wounded	Petersburg
Wilkins, Josephus			
Williams, Harrison P	79	Wounded	Petersburg - Initial Assault
Williams, S B			
Wolford, Alexander J		Wounded	Weldon RR
Wolford, William		KIA	Wilderness

149th Pennsylvania Infantry Listing

Daughter Martha on left, wife Anna & Private Harrison Clingerman

David W Lee, Private

Sarah & Private Henry Bagley married in 1862.

Unidentified soldiers in the 149th Pennsylvania Regiment during the siege of Petersburg in November 1864.

171st Pennsylvania Infantry Listing

Bedford County Enlistments 116
Known Casualties 15
Casualty Rate Percentage 13%

Died 10
Wounded 5

Bedford County recruits in Company I of the 171st Pennsylvania Infantry received initial training at Camp Curtin in Harrisburg in November 1862. The 171st was sent to North Carolina in January 1863 and took part in several skirmishes. During June 1863, the regiment was part of an army group moved to Fort Monroe in Virginia, to threaten offensive actions on Richmond during the Gettysburg Campaign and took part in pursuing the Confederate Army during their retreat from Gettysburg. The regiment completed their 9-month enlistments on August 8th, 1863. Most of the deaths suffered in the 171st regiment were from disease.

Name	Pict. Page	Known Casualties	Battle
Ball, John		Wounded	
Barns, Joseph			
Barns, Samuel			
Bash, Daniel			
Bayer, Joseph			
Bennett, Abraham			
Bennett, David			
Berkey, Benjamin			
Berkheimer, Samuel	358		
Birkley, David T			
Blattenberger, Daniel	300	Wounded	
Blattenberger, Joseph			
Bowser, Valentine	257		
Bratelbaugh, James B			
Burket, Jacob D	299		
Burns, Oliver P			
Callihan, John			
Callihan, Robert	300		
Carson, Jacob			
Clingerman, Jeremiah		Died	
Clingerman, Joseph			
Conner, David			
Corle, Alexander	301, 356		
Darr, Christian J		Died	
Devore, Michael			
Dick, John C			
Elliott, John K H			
Ellis, Enos			
Emerick, Andrew R	300		
Enos, David			
Feather, Michael	259, 356		
Feathers, Henry			
Fetter, Harrison			
Fetter, Joseph B			
Fetters, Job		Died	
Fisher, Joseph			
Garn, George			
Gordon, James S			
Grove, William A			

Name	Pict. Page	Known Casualties	Battle
Hann, Gaston	301		
Harbaugh, Amos			
Hite, Jacob A			
Holler, Joseph M			
Hook, Elias			
Hook, George			
Hook, James			
Hook, William		Died	
Hoover, Martin		Died	
Howsare, Wesley B			
Huff, James L			
Ickes, Alexander			
Jay, John			
Jay, Thomas	300		
Johnson, Asa			
Johnson, Lewis			
Johnson, William P			
Keel, George			
Keller, John			
Kettering, Jacob T			
King, John			
Knox, James H			
Lape, Abraham		Wounded	New Berne, NC
Lawhead, Thomas			
Leasure, George M			
Leasure, John G			
Leasure, William E			
Linn, Hugh Jr			
Logue, James		Died	
Lunger, Franklin			
McVicker, Jesse	348		
Miller, Bartley			
Miller, Christian	301	Wounded	
Miller, Daniel H			
Miller, David H			
Miller, David P			
Miller, George	358		
Miller, John			

171st Pennsylvania Infantry Listing

Name	Pict. Page	Known Casualties	Battle
Miller, Thomas			
Morris, Israel			
Mowry, Jacob			
Mowry, John	301		
Nelson, Robert	301		
Pennel, Andrew J	254		
Pennington, James F	353		
Pleacher, Andrew		Wounded	
Potts, John A			
Ressler, Harvey M			
Robinette, Amos	299		
Robinette, Jeremiah	299		
Ruby, Henry	332		
Sharp, James		Died	
Shippley, Lorenzo D		Died	
Shrimer, Alex K			
Shull, William			
Slick, Abner W			
Smouse, Abner G			
Snowberger, Elias			
Snowberger, Joseph B			

Name	Pict. Page	Known Casualties	Battle
Spade, Isaac N			
Stirtz, Solomon			
Struckman, Henry			
Stufft, Jacob	299		
Stufft, William S		Died	
Sturtz, Solomon			
Trail, John			
Turner, John			
Walker, Thomas G	181, 254, 358		
Wertz, Talliferro			
Wigfield, Noah			
Wilhelm, Samuel W			
Wilkinson, William	301		
Williams, David F			
Wissinger, Alex			
Witt, John		Died	
Young, Isaac			
Zeller, Michael			

Amos Robinette, Captain & Jeremiah Robinette, Private

Jacob Stufft, Private

Jacob D Burket, Private

171st Pennsylvania Infantry Listing

Daniel Blattenberger, Private

Andrew R Emerick, Private

Thomas Jay, Private

Sergeant Robert Callihan is pictured on the right with a son & grandson

171st Pennsylvania Infantry Listing

William Wilkinson, Private

John Mowry, Private

Alexander Corle, Private

Bottom left to right: Sadie and Private Robert Nelson. Standing: daughter Sara, grandson Cecil Wilber and his wife & son Grover.

Sarah & Private Gaston Hann married in 1854. Sarah tended to the family farm in Monroe Township with 3 small children when Gaston mustered in the army.

Mary Ann & Private Christian Miller were married after the war.

184th Pennsylvania Infantry Listing

Bedford County Enlistments	78
Known Casualties	62
Casualty Rate Percentage	79%
Killed in Action	7
Died from Wounds	4
Died	2
Died while POW	17
Wounded & POW	1
POW	9
Wounded	22

Year	Major Battles with Casualties	Known Casualties
1864	Cold Harbor, May 31 - Jun. 12	21
	Petersburg - Initial Assault, Jun. 15-18	2
	Jerusalem Plank Road, Jun. 21-23	30
	Boydton Plank Road, Oct. 27-28	2
1865	Appomattox, April 9 (at Surrender)	

Thirty-eight percent of the Bedford County soldiers who enlisted in the 184th Pennsylvania Infantry perished during the Civil War, which is the highest percentage of deaths of any Bedford County unit. Most of the soldiers in Company A of the 184th were recruited in Bedford County during May 1864. A lessor number of recruits in this company were from Dauphin County. The regiment was organized at Camp Curtin in Harrisburg before being rushed to the warfront in late May. Three weeks after mustering in the Union Army, Bedford County troops in the 184th took part in one of the most infamous charges of the war at Cold Harbor. Ulysses S Grant ordered an early morning full-frontal assault on June 3rd, on an entrenched Confederate Army. The 184th made 2 desperate charges before the entire attack was called off. Tragically, the Union dead and wounded from the assault remained stranded in the no-man's-land between both armies for several days until a truce was agreed upon and the wounded, who had not already perished, were removed from the battlefield. The 184th suffered the 2nd highest number of Bedford County casualties at Cold Harbor. The following month the 184th PA Infantry was marching through heavily wooded terrain south of Petersburg during an offensive aiming to destroy the Weldon Railroad line. Confederate forces drove through a gap in Union Army lines and enveloped the 184th Regiment, resulting in 30 Bedford County casualties. Four of the 5 county soldiers who were wounded, died during the battle or shortly after, and 25 more were captured. Most of the enlisted men were taken to Andersonville. Tragically 16 of the 25 captured soldiers languished in hellish prisoner of war camps before dying. In less than two months, most Bedford County soldiers in this ill-fated regiment had become a battlefield casualty. In addition to suffering the highest death rate, the 78% casualty rate (total number of deaths, wounded, POW's, etc.) suffered by the 184th was the 2nd highest among all Bedford County units during the war.

Name	Pict. Page	Known Casualties	Battle
Adams, Charles C			
Barnett, David		Wounded	Cold Harbor
		Wounded	Petersburg
Barnett, Samuel		POW-Died	Jerusalem Plank Rd.
Bechtel, Isaac S		KIA	Jerusalem Plank Rd.
Bennett, John E		Wounded	
Berkhimer, Levi		Wounded	Cold Harbor
Blackburn, Henry B			
Blackburn, Martin		POW-Died	Jerusalem Plank Rd.
Bohn, Solomon			
Boston, George W			
Bowers, George		Wound.-Died	Cold Harbor
Bowers, Michael H			
Brown, George D	72	Wounded	Cold Harbor
		Wounded	Petersburg
Brown, Phllip S			
Butts, James B	278		
Carn, Adam B			
Clevenger, Jacob A	185, 303		
Cook, John F		Wounded	Cold Harbor
Croyle, Daniel			
Davidson, Samuel		POW-Died	Jerusalem Plank Rd.
Defibaugh, John			
Defibaugh, John W		POW	Jerusalem Plank Rd.
Devore, Lewis		KIA	Jerusalem Plank Rd.
Dull, John		POW-Died	Boydton Plank Rd.
Earnest, William M			
Ensley, Christopher		POW-Died	Jerusalem Plank Rd.
Evans, Nathan C	303	Wound.-POW	Jerusalem Plank Rd.
Hagan, John		Wounded	Cold Harbor
Imler, Matthias	304		
Irvine, Wilson		POW-Died	Jerusalem Plank Rd.
Johnston, Charles W		Wounded	Cold Harbor
Klahre, Herman T		Wound.-Died	Jerusalem Plank Rd.
Knox, James H		POW-Died	Jerusalem Plank Rd.
Koontz, Charles		Wounded	
Layton, Samuel		POW-Died	Jerusalem Plank Rd.
Lee, John	304	Wounded	Jerusalem Plank Rd.
Lehman, Harry			
Lehman, William H	190		
Leonard, Jacob			
Line, Jacob		KIA	Jerusalem Plank Rd.

184th Pennsylvania Infantry Listing

Name	Pict. Page	Known Casualties	Battle
Lutz, Simon S		Wound.-Died	Cold Harbor
Markey, Joseph	304	Wounded	
Marshall, Henry L		POW-Died	Jerusalem Plank Rd.
Mock, Mathias		KIA	Cold Harbor
Orris, Jacob		POW-Died	Jerusalem Plank Rd.
Otto, Henry S		POW-Died	Jerusalem Plank Rd.
Over, David S		POW	Jerusalem Plank Rd.
Over, Jacob Z		POW	Jerusalem Plank Rd.
Phillips, Daniel W			
Potter, James	304	Wounded	Cold Harbor
Price, John		Died	
Price, Michael H		Wounded	Cold Harbor
Ralston, William H			
Reighard, George		Wounded	Cold Harbor
Rhodes, George		POW	Jerusalem Plank Rd.
Riley, John			
Skillington, Robert	209	Wounded	Cold Harbor
		Wounded	Petersburg - Initial Assault
Sleek, Jesse W		Died	
Smith, Barton C			
Smith, Emanuel			

Name	Pict. Page	Known Casualties	Battle
Smith, Josiah N		Wounded	Cold Harbor
Snowberger, Theodore		Wound.-Died	Petersburg - Initial Assault
		Wounded	Cold Harbor
Snowden, David		KIA	Cold Harbor
Stephens, John G		KIA	Boydton Plank Rd.
		Wounded	Cold Harbor
Stuckey, David H		POW-Died	Jerusalem Plank Rd.
Swoveland, William		POW-Died	Jerusalem Plank Rd.
Teeter, Christian		POW-Died	Jerusalem Plank Rd.
Trout, Sylvester		POW	Jerusalem Plank Rd.
Wagoner, August		KIA	Cold Harbor
Waltman, William H		POW	Petersburg
Watson, Henry S		POW-Died	Jerusalem Plank Rd.
Watson, John R		POW	Jerusalem Plank Rd.
Wilson, James R			
Witman, John		Wounded	Cold Harbor
Wolfhope, John	147	POW-Died	Jerusalem Plank Rd.
Wright, Charles C	80	POW	Jerusalem Plank Rd.
		Wounded	Cold Harbor
Wright, Edmund S		POW-Died	Jerusalem Plank Rd.
Zembower, Josiah A	121	POW	Jerusalem Plank Rd.

Jacob A Clevenger, Private

Nathan C Evans, Captain

184th Pennsylvania Infantry Listing

Markey family - Back: Laura, Alice, Lillian & Frank. Front: John, Corporal Joseph Markey, Mary & Samuel.

James Potter, Private

Corporal John Lee pictured later in life in a uniform.

Matthias Imler, Private

194th Pennsylvania Infantry Listing

Bedford County Enlistments 66
Known Casualties 2
Casualty Rate Percentage 3%
Wounded 2

The 194th Pennsylvania Infantry recruited soldiers in Company I in Bedford and Berks Counties for a 100-day enlistment. The regiment was organized in Harrisburg on July 22nd, 1864, and sent to Baltimore the same day. The 194th remained in the Baltimore area until the regiment was mustered out of the army on November 6th, 1864.

(right) A Mathew Brady photograph of unidentified Soldiers marching in the capitol area during the Civil War.

Name	Pict. Page	Known Casualties	Battle
Abbott, Allison			
Amos, Francis M	306		
Armstrong, David B	225		
Armstrong, Joseph M			
Ashcom, Edward S			
Ashcom, John P			
Baker, Jacob S			
Barndollar, Martin D	278		
Barndollar, William G			
Bayer, Joseph			
Bechtel, John S		Wounded	
Biddle, Jacob S			
Blackburn, Harmon			
Castner, Jacob H			
Chamberlain, Fernando C			
Clark, Alexander			
Clouse, Harmon			
Corbin, George H			
Cramer, William			
Donaldson, Benjamin			
Eichelberger, Jacob A	306		
Garner, Andrew B			
Garner, Thomas G			
Garrett, Levi P			
Grove, Benjamin H			
Gump, Erastus J			
Hall, George W			
Hamer, John C			
Hetrick, Samuel G			
Isett, James M			
Jessner, Joseph			
Kettering, Elijah			
Kettering, Jacob T			
Leonard, William			
Long, William P			
Lucas, Joshua T			
Masters, Frank M			
Masters, William J			
McMahan, William			
Morris, John			
Myers, Henry Sr			
Nicewonger, Andrew			
Peck, Llewellyn H	306		
Penrod, Henry C			
Penrod, John B Jr			
Reed, William B	306		
Replogle, Simon L			
Riffle, Cyrus	150		
Ritchey, Adam S			
Shimer, Isaac			
Shock, Daniel	306		
Skipper, Augustus			
Snowberger, Jacob			
Snyder, William			
Sparks, John	39		
Sparks, John C	283, 359		
Steeley, Jacob E			
Stuby, Jacob E		Wounded	
Swartz, Henry			
Swartz, John W			
Tobias, John B			
Trembath, Samuel D			
Wertz, Thomas			
Williams, John H			
Williams, Richard			
Williams, Samuel D			

194th Pennsylvania Infantry Listing

Jacob A Eichelberger, Corporal

Daniel Shock, Corporal

Francis M Amos, Private

Llewellyn H Peck, Private

William B Reed, Private

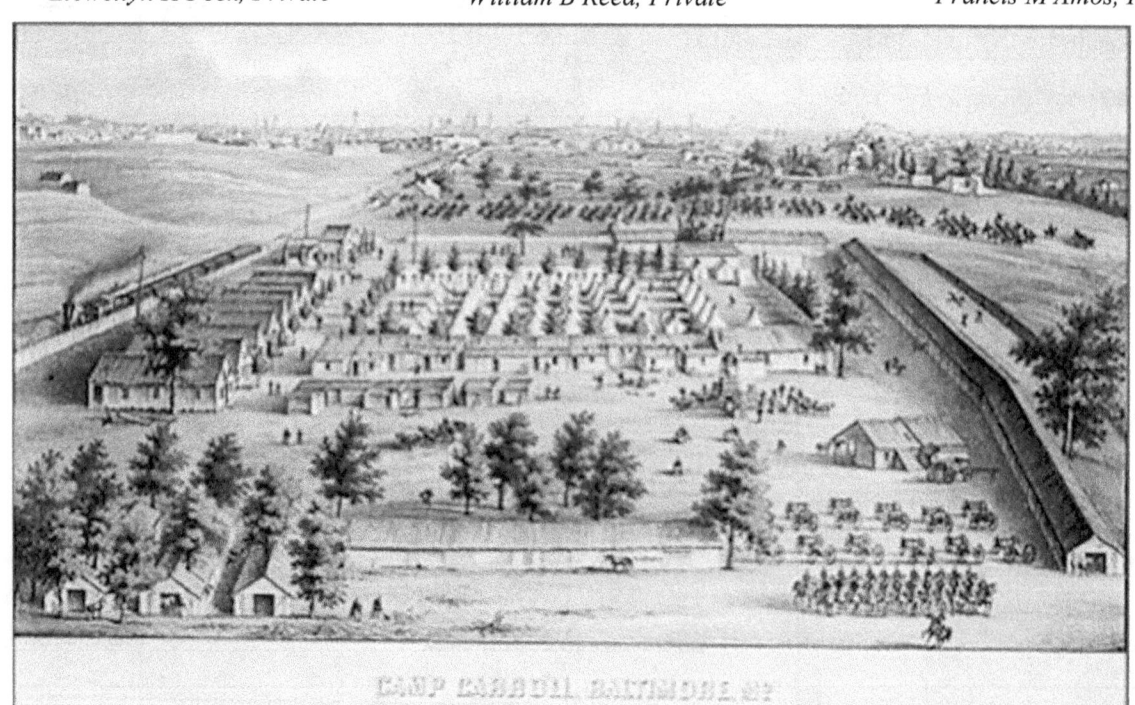

(Left) Camp Carroll near Baltimore is where the 194th PA Infantry was stationed during the Civil War.

205th Pennsylvania Infantry Listing

Bedford County Enlistments	46
Known Casualties	9
Casualty Rate Percentage	20%
Killed in Action	1
Died from Wounds	2
Died	1
Wounded	5

Year	Major Battles with Casualties	Known Casualties
1865	Petersburg - Final Assault, April 2	8

The 205th Pennsylvania Infantry recruited soldiers for some companies in Blair and Huntingdon Counties. Many of the Bedford County soldiers who enlisted in this regiment lived in border areas of these counties. The regiment was organized in Harrisburg on September 2nd, 1864. All known Bedford County casualties in the 205th regiment took place during the Final Assault at Petersburg on April 2nd, 1865. The 205th were among the Union troops who assaulted Ft. Mahone and took part in desperate hand to hand fighting at the Confederate earthwork defensive lines. The regiment mustered out of the army on June 2nd, 1865, after the war ended.

Louis D Speice, Captain

Name	Pict. Page	Known Casualties	Battle
Baker, Alfred	308, 362		
Barr, Reuben			
Bowser, Job		Died	
Burns, Michael			
Cartwright, Franklin J		Wounded	Petersburg - Final Assault
Conrad, Jacob			
Feather, Henry C			
Feather, John A			
Fluke, Oliver B	325		
Fluke, Samuel B	205, 308		
Garretson, Benjamin H	107, 134	Wound.-Died	Petersburg - Final Assault
Grass, Cephas			
Grimes, Jacob R			
Hamm, David K			
Harbaugh, Allen	206		
Harbaugh, John M			
Hart, Isreal			
Hite, John		KIA	Petersburg - Final Assault
Imler, Jonas C	201, 308		
Lingenfelter, George W			
Lingenfelter, Martin			
Long, Samuel L			
Mobley, Ezekiel			

Name	Pict. Page	Known Casualties	Battle
Morgan, William			
Morrow, B Moritimer		Wounded	Petersburg - Final Assault
Moyer, Daniel			
Musselman, George			
Penrose, Joseph	208, 321, 358		
Price, Daniel M		Wounded	Petersburg - Final Assault
Quarry, William C			
Ridenour, Jacob D	308		
Roarabaugh, John			
Rorabaugh, John			
Smeltzer, John B		Wounded	Petersburg - Final Assault
Smith, George			
Smith, Joseph H	308		
Smith, Rufus E			
Snyder, George W Sr.			
Speice, Louis D	307		
Strayer, John			
Strayer, Nicholas		Wound.-Died	Petersburg - Final Assault
Tate, Jacob			
Thomas, John			
Walker, Isaac	308		
Walter, George			
Walter, Moses		Wounded	Petersburg - Final Assault

205th Pennsylvania Infantry Listing

Jacob Ridenour, Private

Alfred Baker, Private

Samuel B Fluke, Musician

Jonas C Imler, Private

Grandson Clifford Mock with Private Isaac Walker

Son - Shannon, wife - Catherine & Private Joseph Smith, circa 1876

208th Pennsylvania Infantry Listing

Bedford County Enlistments	208
Known Casualties	28
Casualty Rate Percentage	13%
Killed in Action	3
Died from Wounds	2
Died	3
Wounded	20

Year	Major Battles with Casualties	Known Casualties
1865	Ft. Stedman, Mar. 25	8
	Petersburg - Final Assault, April 2	13

The 208th Pennsylvania Infantry recruited soldiers for Companies H & K in Bedford County. Most county soldiers mustered in early September 1864 and left Harrisburg for the Bermuda Hundred, northeast of Petersburg, on September 13th. In late 1864, they were moved to the front lines in Petersburg. County Soldiers in the 208th took casualties during the battle of Fort Stedman on March 25th, 1865. Fort Stedman was Robert E Lee's last offensive action in the Civil War. Confederate troops quickly overran Fort Stedman during the desperate attempt to break out of the siege of Petersburg. The 208th Pennsylvania Infantry and other Union regiments near Fort Stedman were successful in pushing back the attacking Confederates and recaptured the Union fort hours after the assault began. Days later, the 208th took part in the Final Assault at Petersburg. Bedford County soldiers took heavy casualties from double cannister cannon shot and rifle fire during an early morning charge on Fort Mahone. Desperate hand to hand fighting took place when enemy trenches were reached. County soldiers were successful in driving the Confederates from their entrenchments during the final battle of the 9-month siege of Petersburg. The 208th PA Infantry took part in the Grand Review victory march in Washington, D.C. on May 23rd before the regiment mustered out of the army on June 1st, 1865.

James H Sparks, Private

Jacob Hammann, Private

208th Pennsylvania Infantry Listing

Name	Pict. Page	Known Casualties	Battle
Akers, Job S	313		
Amick, William	278		
Avey, Joseph			
Baird, George L			
Barkley, Josiah T			
Bartow, Barney	315		
Benkley, Samuel			
Bessor, Philip			
Blake, William B	37	Wounded	
Boemer, John A			
Bookhamer, William			
Bowman, George W			
Brown, Benjamin F		Wounded	Ft. Stedman
Bulger, Daniel B			
Bulger, David B			
Burns, Lafayette W			
Bussard, Emanuel S			
Bussard, Joseph S			
Chamberlain, Eli G		KIA	Ft. Stedman
Chamberlain, Jacob		Wounded	
Chamberlain, Philip			
Clark, Ferdinand		Wounded	Petersburg - Final Assault
Clark, John			
Clark, Simon			
Clark, Willlam W	312		
Colledge, Jacob			
Colledge, Joseph R	314		
College, Simon			
Conner, Adam			
Conner, Emanuel			
Cook, Ezekiel			
Cooper, David A			
Cornell, Daniel			
Cornell, William H			
Coulter, Alexander			
Cramer, Levi			
Crawford, Jacob			
Daugherty, Joseph L			
Davis, John L			
Davis, Porter R			
Davis, William			
Echom, John			
Eichelberger, Winfield S			
Eicholtz, William G	314		
Elwell, John			
Everhart, James H			

Name	Pict. Page	Known Casualties	Battle
Faulkender, William D			
Felton, Peter S		KIA	Petersburg - Final Assault
Fickes, John W			
Fleegle, William W			
Flenner, Stewart			
Fletcher, Jacob			
Fluke, John R			
Foor, Brazella	329		
Foor, Jacob I	314		
Foor, James H	279		
Foor, Peter	313	Wounded	
Foor, Simon P		KIA	Petersburg - Final Assault
Frederick, William			
French, Samuel			
Fry, Joseph			
Funk, John D		Wounded	Petersburg - Final Assault
Furgeson, Thomas			
Gallaher, James W			
Gallbaugh, Henry			
Gaster, Ezekiel W	279		
Gates, Joseph	271		
Gates, William H			
George, Conrad			
Gibson, George G			
Gibson, Henry F		Wounded	Ft. Stedman
Gienger, Andrew J			
Gienger, Jacob			
Giffin, James H			
Gogley, Samuel T			
Hacher, James			
Hammann, Jacob	309		
Hann, Jeremiah W			
Hann, John		Wounded	
Harvey, William			
Heavener, George W		Died	Petersburg
Helmit, Thomas J			
Himes, John	314		
Hollar, Philip V	102	Wounded	Ft. Stedman
Hoopingardner, Joseph			
Householder, James			
Householder, John			
Housenworth, J J			
Hymes, Wiley			
Jackson, Mark J			

208th Pennsylvania Infantry Listing

Name	Pict. Page	Known Casualties	Battle	Name	Pict. Page	Known Casualties	Battle
Johnston, David S				Riley, George W	21		
Karns, Simon				Rinard, David			
Keagy, David F				Rinard, Thomas			
Keagy, George				Ritchey, Daniel S		Wounded	Ft. Stedman
Kelley, David				Ritchey, David		Died	
Kifer, Jacob				Ritchey, George			
King, William B				Ritchey, Jacob			
Kissel, Benjamin				Ritchey, John F			
Kissel, John				Ritchey, John N	314		
Langdon, Samuel		Died		Robinson, Job			
Latta, Abraham				Rohm, Frank			
Lehman, Isaiah	313			Ross, Joseph			
Lehn, Philip		Wounded	Petersburg - Final Assault	Rumel, John E			
Leonard, John B	314	Wounded	Petersburg - Final Assault	Russell, A Sidney			
Long, Joseph C		Wounded	Petersburg - Final Assault	Sams, John W			
Manspeaker, John				Sams, Wilson			
May, Abraham M				Satterfield, John E			
May, Samuel S				Shade, James A			
McDaniel, William				Shaffer, Isaiah A			
McDonald, John	162			Shaffer, Levi M			
Mearkle, Barton	314, 359			Shaffer, William			
Mellott, Jacob L	208			Sheiner, Robert N			
Messersmith, Alexander	156			Shroyer, John			
Messersmith, Joseph S				Skipper, Alexander			
Miller, Bartley H				Slick, John A		Wound.-Died	Petersburg - Final Assault
Miller, Philip S				Smith, William H			
Moser, Martin	275			Smouse, Simon			
Naugle, Jacob				Snider, Augustus			
Nute, William W				Snowberger, Joseph C			
Oaks, John R	313			Snyder, Ferdinand			
O'Neal, James R	292	Wounded	Ft. Stedman	South, James W			
Osborn, Peter		Wound.-Died	Petersburg - Final Assault	Sparks, David G			
Ott, Nicholas		Wounded	Petersburg - Final Assault	Sparks, James		Wounded	Petersburg - Final Assault
Pearson, Thomas K				Sparks, James H	309		
Peck, Jesse				Sparks, Wilson W			
Peck, Simon				Spencer, Israel			
Peightel, James				Spruell, John D			
Ramsey, Oliver C	312, 361			Stailey, George E			
Ramsey, William W				Stailey, Henry C			
Reed, Alexander				Steele, Levi H	357		
Richter, Adam	281			Stephey, Levi			
Riley, Andrew J				Stoner, William			
				Stoudnour, William			

208th Pennsylvania Infantry Listing

Name	Pict. Page	Known Casualties	Battle
Stuckey, Abraham			
Summerville, Robert	314		
Swartz, David H			
Thomas, Joseph			
Thomas, Warner			
Uglow, Nicholas			
Uglow, Samuel			
Walker, William A			
Wall, Lewis			
Way, James H			
Weaverling, Adam	149		
Weaverling, Jacob T	315		
Weaverling, Thomas H	149		
Weimer, David			
Weimert, Stephen			
Wilkins, James B			
Wilkins, Samuel			

Name	Pict. Page	Known Casualties	Battle
Will, John H			
Williams, Alvah R	108, 315	Wounded	Petersburg - Final Assault
Williams, David F			
Williams, Joseph W	315		
Williams, Samuel W		Wounded	Ft. Stedman
Williams, Wilson M			
Wilson, Patrick			
Wilt, Daniel H		Wounded	Ft. Stedman
Wise, Brady B			
Wishart, Harvey S	314, 362		
Witters, George			
Witters, Jacob M			
Woy, Ezekiel C		Wounded	Petersburg - Final Assault
Woy, James H			
Woy, John			
Young, Peter W			

William W Clark, Private

Oliver C Ramsey, Sergeant

208th Pennsylvania Infantry Listing

Isabell & Private Peter Foor

Margaret & Private Isaiah Lehman

Sarah & Private Job S Akers

Mary Anne & Private John R Oaks

208th Pennsylvania Infantry Listing

Harvey S Wishart, Captain

John Himes, Private

William G Eicholtz, 1st Lt.

Joseph R Colledge, Private

Robert Summerville, Private

(left to right) Private John Leonard, Stella Leonard, Russel Weaverling, Sam & Elizabeth Leonard

Barton Mearkle, Private

Jacob I Foor, Private

Catherine & Private John N Ritchey

208th Pennsylvania Infantry Listing

Barney Bartow, Private

Jacob T Weaverling, Private

Mary & Private Joseph W Williams

Susan & Private Alvah R Williams

2nd Pennsylvania Cavalry Listing

Bedford County Enlistments	42
Known Casualties	15
Casualty Rate Percentage	36%
Killed in Action	1
Died from Wounds	2
Died	4
Died while POW	3
POW	2
Wounded	2
Injured	1

Twenty-four percent of county soldiers who enlisted in the 2nd Pennsylvania Cavalry lost their lives during the Civil War. Some soldiers were recruited in Bedford County and the regiment was organized in Philadelphia and Harrisburg from September 1861 to April 1862. The 2nd PA Cavalry fought in many Cavalry battles and skirmishes throughout Virginia and Maryland during the war. Bedford County soldiers in the 2nd PA Cavalry took part in pursuing Lee's army after the Confederate retreat from Petersburg and were eyewitnesses to the surrender at Appomattox on April 9th, 1865. The regiment took part in the Grand Review in Washington D.C. on May 23rd before being merged with 20th Pennsylvania Cavalry on June 17, 1865, forming the 1st Provisional Cavalry.

Name	Pict. Page	Known Casualties	Battle
Anderson, William W		POW	
Andrews, Albert			
Baughman, William	127	Died	
Boor, Jacob B			
Chalfont, William		Wounded	Parker's Store
Defibaugh, Harrison			
Dicken, David			
Dicken, James H		POW-Died	
Ditch, William			
Drenning, Thomas	184		
Elliott, Francis M	316		
Elliott, John		Wound.-Died	
Feight, Frederick	115		
Frederick, Andrew			
Hafer, Frank M			
Hafer, George W			
Hafer, William H		Wounded	
Harlow, Charles E			
Hemming, Augustus		Died	
Hemming, William			
Hite, Perry		Died	

Name	Pict. Page	Known Casualties	Battle
Irvine, Hayes	316		
Leasure, Nathaniel			
Mayers, John		POW-Died	
Miller, Franklin		Injured	
Mock, Harvey			
Morris, David I			
Moyer, John	141	POW-Died	
Nycum, Josiah		Died	
Nycum, Upton	177-B	Wound.-Died	
Raley, Vincent			
Smith, Charles	316		
Smith, Jacob C			
Smith, Nathan		KIA	Rappahannock
Smith, Nathan C			
Snowden, John W		POW	
Spriggs, Asa M			
Suiters, William S			
Swartz, Espy			
Walters, Isaiah			
Watson, William A			
Wilkinson, Emanuel			

Hayes Irvine, Private

Francis M Elliott, Corporal

Charles Smith, Private

18th Pennsylvania Cavalry Listing

Bedford County Enlistments	23
Known Casualties	8
Casualty Rate Percentage	35%
Killed in Action	1
Died	1
Died while POW	3
POW	1
Wounded	2

Year	Major Battles with Casualties	Known Casualties
1863	Gettysburg, Jul. 1-3	1
1864	Wilderness, May 5-7	2

The 18th Pennsylvania Cavalry recruited some soldiers for Company E in Bedford County and was organized in Pittsburgh and Harrisburg from October to December 1862. The 18th PA Cavalry was on the battlefield at Gettysburg. The following is engraved on their monument near Big Round Top, "The Regiment participated in cavalry fights at Hanover on June 30th and Hunterstown on July 2nd, 1863. On July 3rd it occupied this position, and in the afternoon charged with the brigade upon the enemy's infantry behind the stone wall to the north of this point on the outer edge of the woods." The 18th PA Cavalry also took part in the Overland and Shenandoah Valley Campaigns in 1864. County soldiers were participants in the largest cavalry charge of the war at the battle of Opequon in Winchester in September 1864. The 18th PA Cavalry merged with the 22nd Pennsylvania Cavalry to form 3rd Provisional Cavalry in June 1865 and mustered out in October 1865 after the conclusion of the war.

Name	Pict. Page	Known Casualties	Battle
Albright, William	218		
Border, John H			
Bowser, George L			
Davis, Abner O		Wounded	
Earnest, Adam P	318		
Fritz, Daniel E		POW-Died	Wilderness
Green, David M			
Groman, John			
Harclerode, William H			
Hemminger, Abraham O			
Hite, Jacob			
Hoffman, John O	357		

Name	Pict. Page	Known Casualties	Battle
Kegg, Emanuel		Died	
Lowery, William	318	POW	Germanna Ford
Mobley, Denton			
Morgart, William		POW-Died	Wilderness
Moses, Emanuel		POW-Died	Gettysburg
Nelson, John	157, 318	Wounded	Chantilly
Nelson, William N	157		
Oliver, Benjamin F	318		
Shelley, Abraham			
Wareham, Martin			
Wisegarver, William V		KIA	

18th Pennsylvania Cavalry Camp

18th Pennsylvania Cavalry Listing

John Nelson, 1st Lieutenant

Ann & Sergeant William Lowery,
Wedding picture circa 1868

Benjamin F Oliver, Private

A rare Civil War picture of an identified soldier on horseback. John Nelson in 1863, the year his right leg was amputated.

21st Pennsylvania Cavalry Listing

Bedford County Enlistments	104		
Known Casualties	16		
Casualty Rate Percentage	15%		
Killed in Action	1		
Died	3		
POW	1		
Wounded	11		

Year	Major Battles with Casualties	Known Casualties
1864	Cold Harbor, May 31 - Jun. 12	1
	Petersburg - Initial Assault, Jun. 15-18	4
	Boydton Plank Road, Oct. 27-28	3
1865	Five Forks, April 1	1
	Appomattox, April 9 (at Surrender)	

The 21st Pennsylvania Cavalry recruited some soldiers in Bedford County. Many of the soldiers mustered in Company E in July 1863. At the end of the initial 6-month enlistment, some reenlisted and remained in the 21st Cavalry until the war had ended. The 21st PA Cavalry took part in the Overland Campaign during the spring of 1864 and the 9 month siege of Petersburg. County soldiers pursued the fleeing Rebel Army during the Appomattox Campaign from March 28th through April 9th, 1865, and were eyewitnesses to the Confederate surrender at Appomattox. The regiment mustered out after the war ended on July 8th, 1865.

Name	Pict. Page	Known Casualties	Battle
Allison, John H			
Bender, Benjamin F			
Berkeybil, Daniel W			
Blackburn, John M		Wounded	
Bollinger, William H		Died	
Breman, Andrew			
Burkey, Aaron			
Burkhart, Solomon C			
Cargill, James			
Chamberlain, David		Wounded	Five Forks
Cook, Levi	320		
Crissey, John C			
Davis, William H			
Doyle, Martin P		Wounded	Boydton Plank Rd.
			Cold Harbor
Ellis, George N			
Enock, Jennings A			
Enos, David			
Evans, William			
Felix, John			
Fetters, George			
Fickes, Cyrus W	102, 321		
Floyd, John B			
Forgee, Casper			
Gamble, Andrew			
Garretson, Benjamin	107, 134		
Garretson, Edwin			
Gault, Ezekiel			

Name	Pict. Page	Known Casualties	Battle
Glaze, Andrew L			
Gray, John W			
Growden, Thomas			
Guyer, William C			
Harr, Silas			
Helm, Frederick			
Howser, Henry H			
Husband, Johnston			
Jeffries, Howard B	320		
Jenkes, Daniel M			
Jones, Joseph			
Jones, William			
Kell, Ezra			
Keller, Jacob		Wounded	Petersburg - Initial Assault
Lambert, Joseph C	33	Wounded	Amelia Springs
Lambert, Josiah O			
Lester, Omer			
Leter, George D			
Long, Gephart			
Lucas, Robert A			
Manges, Jacob			
Mangis, Levi			
McClincy, William J			
McCormick, James H			
McDonald, William H			
McLaughlin, Edward			
McVicker, James A			
Miller, Aaron J			

21st Pennsylvania Cavalry Listing

Name	Pict.	Known	Battle
Miller, Anthony			
Miller, Armstrong	358		
Miller, George F			
Miller, Jacob H	358		
Miller, Nathan W			
Miller, Samuel W			
Mosser, George W	320		
Nichols, Charles		KIA	Petersburg
		Wounded	Petersburg - Initial Assault
Otto, Emanuel			
Otto, George W			
Pearson, Henry C		Wounded	Boydton Plank Road
Penrose, Joseph	208, 321, 358		
Penrose, William			
Rankin, Frank R			
Rhoades, Peter			
Ripple, Valentine			
Ross, George W			
Rouser, Joseph			
Schellhorn, Albert			
Scritchfield, Hezekiah			
Seaffer, Samuel G			
Shaffer, Daniel L			
Shaffer, Jacob			
Shaffer, William B			

Name	Pict. Page	Known Casualties	Battle
Shank, Joseph			
Shauley, Samuel E			
Shehan, James		POW	Boydton Plank Rd.
Shull, William			
Slick, George W			
Smith, John C			
Smith, Leonard F		Wounded	Rockville, MD
Smith, William F		Wounded	Petersburg - Initial Assault
Snook, Jacob			
Sorber, John			
Stine, Isaac E			
Strunk, George W			
Suters, Emanuel			
Tearnel, William H			
Walker, James			
Walker, William A	181-B	Died	
Watkins, John A			
Watt, David			
Weller, John Q A			
Wertz, Samuel V			
Whitaker, William P			
Whiteneck, George		Wounded	Petersburg - Initial Assault
Williamson, Gideon		Died	
Yarnell, William H			
Yost, Jacob			

Howard B Jeffries, 1st Lieutenant

George W Mosser, Sergeant

Levi Cook, Private

21st Pennsylvania Cavalry Listing

Unidentified 18th PA Cavalry soldiers

Cyrus W Fickes, Private

Penrose family left to right: James and daughter Rhue Gray, Lucretia and Private Joseph Penrose, daughter Mae, Mae's daughter Vera and her husband. The children are likely great grandchildren of Lucretia and Joseph.

22nd Pennsylvania Cavalry Listing

Bedford County Enlistments	163			**Known**
Known Casualties	23	**Year**	**Major Battles with Casualties**	**Casualties**
Casualty Rate Percentage	14%	1864	Opequan, Sep. 19	1
Killed in Action	3		Fisher's Hill, Sep. 21-22	1
Missing in Action	1		Brown's Gap, Sep. 26	3
Died	5			
POW	2			
Wounded	10			
Injured	2			

The 22nd Pennsylvania Cavalry was organized in Chambersburg in February 1864 and merged with the veteran Ringgold Cavalry unit. Some Bedford County soldiers joined the Ringgold Cavalry, one of the most famous Pennsylvania Cavalry units of the Civil War, prior to this consolidation. The 22nd took part in heavy combat during the highly successful Shenandoah Campaign in the fall of 1864. The 22nd were participants in the largest cavalry charge of the Civil War at the Battle of Opequon on September 19, 1864. Over 8000 men on horses thundered across an open plain toward Confederate lines. Colonel James Schoonmaker was awarded the Medal of Honor for leading the 22nd PA, 14th PA and 8th Ohio Cavalry regiments on the charge on Fort Star on the extreme right flank of the Union Army lines. Artillery units at Fort Star were positioned to rain cannon fire on the Union soldiers from an entrenched position on higher ground. Heavy Confederate fire from Fort Star initially halted the Cavalry charge. The 22nd PA and the other two cavalry regiments regrouped and threatened a flanking action, forcing the Confederate troops manning the fort to evacuate. The Medal of Honor citation stated, "At a critical period, Colonel James Schoonmaker gallantly led a cavalry charge against the left of the enemy's line of battle, drove the enemy out of his works, and captured many prisoners." The 22nd PA Cavalry merged with the 18th PA Cavalry on June 24, 1865, to form 3rd Provisional Cavalry. This regiment mustered out of the Union Army on October 31st, 1865.

John C Hawman, Captain

William H Hanks, Sergeant

22nd Pennsylvania Cavalry Listing

Name	Pict. Page	Known Casualties	Battle	Name	Pict. Page	Known Casualties	Battle
Ake, Samuel				Deshong, David D			
Baith, Peter	326			Dicken, James H			
Barkman, Christopher				Dickinson, George			
Barnes, William				Dishong, David D			
Barnett, Joseph E		Wounded		Eichelberger, Alexander		Wounded	
Barton, James A				Eichelberger, David			
Barton, Morgan				Eichelberger, John			
Beeler, James H				Enfield, Americus	43, 186, 202, 355, 359, 362		
Bequeth, Samuel				Ensley, Lewis			
Berkheimer, Daniel B				Ensley, Peter M	326		
Biddle, Jacob S				Felton, John A	91	Injured	Opequon
Blackburn, Cyrus E	325			Felton, Simon P	326		
Blackhart, John W				Filler, William B	93, 202, 326	POW	Browns Gap
Blake, Simon				Fletcher, Henry C		Died	
Blankley, Job				Fletcher, Jacob			
Bollman, George F				Fletcher, Winfield S			
Bonnett, Thomas K				Fluke, John W		Died	
Bookhamer, Thomas				Fluke, Levi			
Border, John	93			Fluke, Oliver B	325		
Bouchman, John H				Foster, Isaiah			
Bowers, Henry H				Garlick, Nicholas			
Bowman, William				Gates, John W			
Brown, James H				Grubb, Harvey	279, 326	Wounded	Mt. Jackson / Forest Hill
Burns, Sylvester							
Burns, William H				Hanks, William H	322	Wounded	Martinsburg
Carpenter, Curtis J	151			Hawman, John C	322		
Castner, Jacob H				Heavner, Michael			
Chamberlain, Jacob		Wounded		Hedding, James E			
Chamberlain, Philip H				Henry, John			
Chambers, William A				Hixon, Perry			
Charleston, John				Hixson, George W			
Chisholm, John P	324			Hollingshead, Oliver S	188, 357		
Clark, Jameson		KIA	Charlestown	Horn, Levi		MIA	
Clevenger, Adam				Houck, George A			
College, Simon				Houck, George W		POW	
Conner, Isaac	168	Injured	Fisher's Hill	Hughes, Scott W			
Conner, Lewis	168, 279			Jackson, Mark J			
Cooper, David M				Jackson, Samuel M			
Cornelius, Peter L				Kagarice, Daniel R			
Cottle, Jacob				Karns, Jacob		Died	
Cramer, John B				Kay, Harry H C			
Croft, George				Kelly, William			
Cullison, John		Wounded		Lamison, James H			
Davis, James P	170			Lasley, Thomas			
Davis, Porter R				Layton, Bartley			

22nd Pennsylvania Cavalry Listing

Name	Pict. Page	Known Casualties	Battle
Leach, Samuel R	324		
Liehty, Christian J			
Linderman, John W			
Livensgood, Charles M			
Long, Amon	326		
Long, Samuel C			
Lucas, Benjamin			
Lyons, Thomas H	239		
Mansberger, Benjamin F			
Martin, Denton O			
Martin, Samuel		KIA	Browns Gap
McDaniel, Lewis		KIA	Berryville
McEldowney, Hezekiah			
McEldowney, James H			
McKee, Alexander H			
Mearkle, Henry		Died	Leetown, MD
Mellott, Hiram			
Mellott, Simon		Wounded	
Messersmith, George	156		
Miller, Jacob B			
Miller, John J			
Miller, Robert C			

Name	Pict. Page	Known Casualties	Battle
Mills, Isaac			
Mortimore, David		Died	
Mowry, Joseph	92		
Nycum, Wilson	177		
Parlett, John C			
Parlett, Thomas W			
Parsons, John E			
Potter, John (1)			
Price, John B			
Ramsey, John			
Ramsey, Wesley A			
Riley, Jacob E			
Riley, James H	280		
Ritchey, James H	224		
Seigle, Simon B			
Sheeder, James T			
Showalter, Simon P			
Sigel, Raphael		Wounded	
Simpson, James			
Sipes, John			
Smith, Emanuel M			
Smith, Seth S			

John P Chisholm, Private

Samuel R Leach, Private

22nd Pennsylvania Cavalry Listing

Name	Pict. Page	Known Casualties	Battle
Smouse, David			
Snider, John W			
Snively, Andrew J			
Snively, John A			
Snyder, Elias J			
Spidle, Bartley			
Spitler, John L			
Sponsler, John W			
Stailey, William A			
Stambaugh, John			
Steele, John W	326		
Stephens, Samuel			
Straley, James			
Swope, George W	326		
Tate, Samuel B			

Name	Pict. Page	Known Casualties	Battle
Wagner, Stephen			
Weaverling, William T			
Weimer, David			
Wertz, George			
Whitaker, Peter			
Whittaker, Jonathan			
Wicks, Isaac D			
Williams, Gideon			
Wilt, Joseph			
Winslow, William			
Woy, John W	182		
Woy, Joseph	325	Wounded	Snicker's Gap - VA
Young, George N			
Young, James H		Wounded	Browns Gap
Young, John			

Joseph Woy, Sergeant

Cyrus E Blackburn, Private

Oliver B Fluke, Sergeant

22nd Pennsylvania Cavalry Listing

Peter M Ensley, Private

Lydia & Private George W Swope

William B Filler, Corporal

Harvey Grubb, Private

Sophia & Private Amon Long

Simon P Felton, Corporal

John Beight and Private Peter Baith. Records show John has a different last name spelling than son Peter.

*Back row: Grandchildren - John & William Steele
Front row: Amanda & Private John Steele*

McKeage's Pennsylvania Militia Listing

On June 12th, 1863, Pennsylvania Governor Curtin issued the following proclamation. "Information has been obtained by the War Department that a large rebel force, composed of cavalry, artillery, and mounted infantry, has been prepared for the purpose of making a raid into Pennsylvania. The importance of immediately raising a sufficient force for the defense of the State cannot be overrated, and the duties of which will be mainly the defense of our own homes, firesides, and property from devastation." Not everyone in Pennsylvania was immediately alarmed. Previous rumors of Rebel Army invasions had failed to materialize. Later in the month, fear spread like wildfire throughout Pennsylvania when it became apparent a Confederate Army invasion was imminent. Two hundred and eighteen Bedford County soldiers responded to Pennsylvania Governor Curtin's urgent call for emergency militia units. Nearly all the county troops in McKeage's Militia mustered on July 2nd, during the battle of Gettysburg. The unit, named after Captain John McKeage, mustered out on August 8th, 1863, after the invading Confederate Army retreated from Pennsylvania. There are no known casualties in McKeage's Militia. Many McKeage's Militia soldiers later enlisted in other regiments.

Name	Pict. Page
Abbott, William	
Adams, John Q	328
Agnew, Hamilton	
Ake, Alexander	
Ake, Castleton	
Alloway, Eli	
Alloway, William	
Amos, Francis M	306
Anderson, Henry	
Anderson, John	
Armstrong, Joseph M	
Arnold, Albert	
Aultz, John K	
Baitzel, Jacob	
Baker, Henry N	
Baker, Samuel N	
Barton, James A	
Beals, Solomon	
Bedell, Edmund	
Berkhimer, Levi	
Blackhart, John W	
Bobb, William	
Bollman, George F	
Bowman, John W	
Brad, Henry	
Bridenstlne, Lemuel E	
Briner, Jacob D	
Brown, James A	
Buck, John	
Burket, Baltzer	
Burns, Joseph	
Bush, Charles	
Carney, Joseph	
Carpenter, Samuel	
Cartwright, Austin	
Chamberlain, Eli G	
Clark, Jameson	
Clark, Simon	

Name	Pict. Page
Clark, Willlam W	312
Clites, Levi	354
Cobler, Francis C	
College, George	
Conner, Lewis	168, 279
Cook, Samuel	
Cooper, Nathan	
Coughenour, James	340
Cremer, George	
Crist, Solomon	
Cruet, A Howard	
Davidson, Samuel	
Davis, John L	
Decker, James M	
Defibaugh, Harrison	
Defibaugh, John W	
Dibert, William	
Diehl, Daniel	
Diehl, Francis M	328
Diehl, Henry	
Diehl, John	
Donley, Walter	
Drenning, William	329
Dull, Lewis	
Ealy, John H	
Edmondson, Samuel	
Eichelberger, Winfield S	
Evans, David V	
Evans, Nathan C	303
Feather, William	
Felton, John A	91
Felton, Peter	
Felton, Simon P	326
Fickes, John W	
Fink, Peter	
Fisher, Andrew	
Fleegle, Simon	
Foor, Brazella	329
Freet, Isaac	

Name	Pict. Page
French, Samuel	
Funk, John D	
Garlick, Nicholas	
Garner, Thomas G	
Gates, Jacob C	
Gates, John W	
Gibson, James	
Gilchrist, David	
Gilchrist, Oliver	
Gordon, Joseph	187, 238, 356
Grass, Cephas	
Gray, Thomas M	
Grove, Henry C	329
Grubb, George	
Hall, John	
Hann, Philip	
Hartagan, Wlliam	
Harvey, William	
Hawman, John C	322
Heffner, Samuel	264
Hicks, Abner	
Hildebrand, Isaac	
Hill, Aaron	340
Himes, William	
Hinish, Thomas P	
Hoffman, John O	357
Hopkins, James	
Horne, William L	
Horton, Jonathan	
Houck, George W	
Hughes, William	
Hull, Jacob	
Hymes, Wiley	
Ickes, William	
Isener, Joseph	
Johnston, William	
Kauffman, David	
Keely, Thomas	
Kelley, David	

McKeage's Pennsylvania Militia Listing

Name	Pict. Page
Lasley, Thomas	
Leader, Henry	
Lehman, Harry	
Leonard, William	
Longston, John A	
Luther, Hiram	
Lysinger, Martin G	
Massey, Isaac D	
May, Jacob	
May, Samuel S	
McCarty, Alvin R	
McConnell, Randolph	
McCulley, James T	
McDaniel, Daniel	281
McDaniel, George W	279
McDaniel, Hiram	
McEnespy, Samuel	
McMullen, Charles	
Mearkle, David S	259
Mearkle, Sansom	328
Mellott, Simon	
Melott, William	328
Messersmith, George	156
Miller, John	
Miller, Nathaniel	
Miller, Samuel	
Mock, Emanuel A	67
Mock, John	
Mock, Tobias	
Moser, Nathaniel	
Myers, Levi	
Naugle, George	
Nigh, Henry	
Nycum, John Q	280, 329
Pearson, Josiah	192
Pittman, William	
Points, Joshua	
Port, George A	
Radebaugh, Daniel W	
Raley, Vincent	
Reed, Andrew J	
Rhodes, George	
Rice, Amos H	
Rice, Samuel	
Ritchey, Daniel S	
Ritchey, Levi	
Shauf, John	199
Sheeden, Aaron	
Shoemaker, Philip T	
Shoop, John	
Showalter, John	
Shuck, John N	
Shultz, David H	
Sigel, Samuel B	
Sloan, Benjamin F	
Smith, Barton C	
Smith, George	
Smith, Josiah	
Smith, William C	
Smith, William P	
Snider, Augustus	
Snowberger, Daniel	
Snyder, Edward	
Snyder, Jacob H	
Sparks, Wilson W	
Steffa, Thomas	
Stombaugh, Joseph	
Stoudenour, John	
Summerville, John B	
Swope, Orlando L	
Trembath, Samuel D	
Trostle, Josiah	
Trout, Alexander	
Trout, Sylvester	
Valentine, Samuel	
VanHorn, John M	
Wagerman, Samuel	
Waltman, William H	
Weaverling, Adam	149
Weimer, George	
Weimert, Jacob F	
Wertz, Jacob A	
Weyandt, Jeremiah	
Whitfield, Benjamin	
Whitfield, Ephraim	
Whitfield, William C	
Wilkins, Harvey	329
Wilkins, Josephus	
Williams, Harrison P	79
Williams, Jonas	
Williams, Samuel	
Williamson, John	
Wilson, John	
Wilt, Joseph	
Wisegarver, Daniel	
Wolf, Jacob	
Wolford, Alexander J	
Wolford, William	
Woodcock, George	
Woodcock, John	
Woodcock, John A	
Yaultz, Franklin	

Sansom Mearkle, Private

John Q Adams, Private

William Melott, Private

Francis Diehl, Private

McKeage's Pennsylvania Militia Listing

Private Harvey Wilkins pictured with grandson James

Jennie & Private Henry Grove

Private Brazella Foor, son Edward and wife Mary

William Drenning, Private

John Q Nycum, 1st Sergeant

1st Maryland PHB Cavalry Listing

Bedford County Enlistments	30
Known Casualties	6
Casualty Rate Percentage	20%
Killed in Action	1
Died	1
POW	3
Wounded	1

The 1st Maryland PHB (Potomac Home Brigade) Cavalry was organized in Frederick, MD during the 2nd half of 1861. During the Antietam Campaign in September 1862, Stonewall Jackson's Rebel troops surrounded the Union garrison at Harper's Ferry. Four companies in the 1st MD PHB Cavalry lead by Major Henry Cole eluded Confederate forces in a daring nighttime escape from Harper's Ferry on September 14th and captured the supply train of General Longstreet the following day. The 1st MD PHB Cavalry took part in many battles, including Gettysburg and the Shenandoah Campaign in 1864. When the regiment mustered out on June 28th, 1865, it was estimated the 1st Maryland PHB Cavalry had traveled over 7000 miles in the saddle.

Name	Pict. Page	Known Casualties	Battle
Bartholow, Samuel			
Bennett, Enos	166	POW	
Boden, John H			
Brant, Grafton			
Cole, Thomas		POW	
Conner, Benjamin F			
Corley, Henry			
Dickens, William			
Gordon, John F			
Hall, Henry H			
Hill, Tolbert A			
Lashley, John W	330		
Leasure, Amos		KIA	Summit Point
Leasure, Riley	330		
Masters, William M			

Name	Pict. Page	Known Casualties	Battle
McKenzie, John			
McLucas, William		Wounded	
Redinger, August			
Riley, Benjamin F			
Riley, John			
Robinett, Mathias			
Robinette, Henry C			
Robinette, Jesse			
Shaffer, Charles H		POW	Hagerstown
Shaffer, Joseph Charles			
Shauf, John J			
Trail, Nathan			
Trail, Nathan B		Died	
Whorrell, Samuel			
Yantz, Peter D			

Riley Leasure, Private

Private John W Lashley pictured with his grandchildren

2nd Maryland PHB Infantry Listing

Bedford County Enlistments	62
Known Casualties	10
Casualty Rate Percentage	16%
Died	2
Died while POW	3
POW	1
Wounded	4

The 2nd Maryland PHB (Potomac Home Brigade) Infantry was organized in Cumberland, Maryland, from August 27th to October 31st, 1861. During much of the war, this regiment was assigned duty guarding the Baltimore and Ohio railroad in Western Maryland and the Shenandoah Valley region. This regiment was in many skirmishes and smaller scale battles with Confederate Cavalry and Infantry units. The 2nd MD PHB infantry mustered out on May 29th, 1865, after the war had ended.

Name	Pict. Page	Known Casualties	Battle
Albright, Jacob			
Beals, George		Died	
Beltz, Oliver			
Beltz, William W	128, 165	POW-Died	Ridgeville, VA
Boyd, William			
Bruner, Israel			
Burket, Albert L			
Burket, John D			
Burley, John		Wounded	
Cessna, John			
Cessna, Samuel S			
Dull, Samuel A		POW-Died	Moorefield Junction
Emerick, Jacob			
Filler, John H			
Fisher, Augustus			
Gonden, John P			

Name	Pict. Page	Known Casualties	Battle
Gordon, George J	332		
Harden, James			
Hillegass, Andrew	331		
Hillegass, Frederick J			
Hite, Benjimen F			
Hite, Joseph		Died	
Holler, Alexander			
Holler, Joseph M			
Hoon, Stacey	332		
Keady, Isaac			
Kegg, Andrew			
Kegg, Levi R			
Kinsey, John B	155		
Kipp, Lewis A	332		
Knee, Philip	217	Wounded	Lynchburg
Leasure, John			

Lydia & Private Andrew Hillegass

Pvt. Benjamin Troutman holding his grandson. Wife Ann is seated beside him. His son and wife are standing.

2nd Maryland PHB Infantry Listing

Name	Pict. Page	Known Casualties	Battle
Leydig, William			
Lybarger, William			
May, Samuel M	291	POW	Moorefield Junction
Metzger, John S			
Murrie, David	192		
Penn, William J			
Perdew, Aaron D			
Rice, Samuel			
Robertson, Hector A	332	Wounded	
Robinett, Eli Z			
Ruby, Henry	322		
Ruby, John			
Scritchfield, Samuel			
Shoup, John N			
Shroyer, Daniel J			
Shroyer, Jacob D	159		
Snyder, John W	179-B	POW-Died	Jonesville
Stoner, Samuel			
Thompson, David Q			
Thorpe, Jacob		Wounded	
Troutman, Benjamin	331		
Troutman, Daniel			
Twigg, Moses			
Walter, James A	211		
Wentling, George			
Whip, William			
Wigfield, Elijah			
Wigfield, Isaac			
Witt, George	322		
Witt, Jonathan			

George J Gordon, Private

Virginia & Sergeant Hector Robertson

Stacey Hoon, Private

Lewis Kipp, Private

Private Henry Ruby with daughter - Carrie

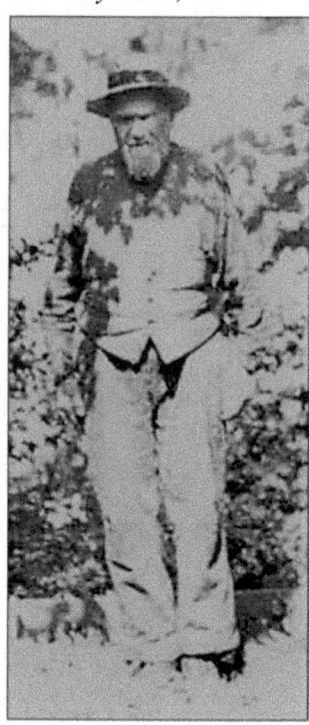

George Witt, Private

3rd Maryland PHB Infantry Listing

Bedford County Enlistments	55
Known Casualties	13
Casualty Rate Percentage	24%
Died	1
POW	9

Year	Major Battles with Casualties	Known Casualties
1862	Harpers Ferry, Sep. 15	7

The 3rd Maryland PHB (Potomac Home Brigade) Infantry was organized in Cumberland, Maryland on October 31st, 1861 and initially assigned duty guarding the railroad lines on the Upper Potomac area in Maryland and Virginia. The entire Harpers Ferry garrison, including the 3rd MD PHB Infantry, surrendered to Stonewall Jackson's troops on September 15th, 1862 during the Antietam Campaign. Many more Bedford County soldiers were likely taken prisoner on that day than the 7 listed soldiers listed, but no additional records have been located. The 3rd MD PHB Infantry also took part in the battle of Monocacy on July 9th, 1864, when heroic efforts of outnumbered Union troops stalled the advance of Confederate forces marching on Washington, D.C. to allow time for reinforcements to reach the capitol. The regiment was disbanded after the war ended on May 29th, 1865.

Prisoner of War Record of Samuel Hendershot of the 3rd Maryland PHB Infantry who was captured at Harpers Ferry, on Sept 14, 1862 and paroled on the following day.

Name	Pict. Page	Known Casualties	Battle
Barney, Dennis A			
Barney, Isaac			
Barney, Jacob	334		
Barney, John H			
Beatty, Henry B			
Bloom, James			
Brooks, William H			
Carson, Robert			
Clevenger, Harrison			
Deneen, Baltzer H			
Deneen, Joseph			
Hebner, Daniel H			
Hebner, Frederick		Died	
Hedding, Ephraim G			
Hedding, James E			
Hedding, Noah			
Hedding, Samuel E			
Hendershot, Charles			
Hendershot, Samuel C		POW	Harpers Ferry
Hess, Jacob			
Hill, Tolbert A			
Karns, Jabez		POW	Harpers Ferry
Leach, Joseph			
Leasure, Amos			
Lee, Henry R			
Lee, William L			
Leighty, George			
Leighty, Jacob			
Leighty, Samuel	334	Wounded	
Linn, Hugh Sr		POW	Harpers Ferry
Linn, Riley		POW	Harpers Ferry
Linn, William		POW	Harpers Ferry
Logsdon, John E			
Lowery, John	25		
Lowery, Joseph			
Lowry, John Jr			
Mellott, Peter			
Metzler, Henry Clay			
Plummer, John T		POW	Moorefield
Richards, David			
Rohm, John S			
Schetrompf, George			
Schetrompf, John F	324		
Schetrompf, Peter C	324		
Shoope, John		POW	Moorefield
Simmons, Thomas H	193		
Smith, Jonathan			
Smith, Joseph B	334		
Smith, Nathan P R		POW	Harpers Ferry
Sponsler, George W	25	POW	Harpers Ferry
		Wounded	
Sponsler, Solomon			
Steckman, John G			
Trail, George T		POW	Moorfield
Truax, Jacob			
Wagerman, Samuel	334		

3rd Maryland PHB Infantry Listing

John F Schetrompf, Private

Peter C Schetrompf, Private

Joseph B Smith, Private

Samuel Leighty, Private

Samuel Wagerman, Private

Jacob Barney, Private

USCT Listing

Bedford County Enlistments	92
Known Casualties	9
Casualty Percentage	10%
Killed in Action	2
Died from Wounds	1
Died	2
Wounded	4

Year	Major Battles with Casualties	Known Casualties
1863	Morris Island	1
1864	2nd Bermuda Hundred, Nov. 28	1
	Petersburg, Jul. 3	1
	Battle of the Crater, Jul. 30	3
1865	Appomattox, April 9 (at Surrender)	

Abraham Lincoln stated in the Emancipation Proclamation on January 1, 1863, African Americans "will be received into the armed service of the United States to garrison forts, positions, stations, and other places, and to man vessels of all sorts in said service." By the end of the war, approximately 185,000 men were serving in 175 USCT regiments, making up approximately one-tenth of the Union Army. The death rate among USCT soldiers was around 20%, a higher death rate than white Union Army soldiers. Sixteen USCT soldiers received the Medal of Honor. It is likely there were many more casualties among the USCT soldiers from Bedford County than are listed on available records. One of the first African American units organized during the Civil War was the Massachusetts 55th Infantry. Joshua Harris of Bedford volunteered in this regiment on June 15th, 1863, and was the 1st known African American from Bedford County to perish in the Civil War. Joshua died of disease on September 8th, 1863 on Folly Island, SC.

It is regrettable more photographs of Bedford County USCT soldiers have not been found, but I consider the picture of DeCharmes Davis on the next page, embracing the American flag, to be among the most memorable photographs in this book. DeCharmes was only 16 years old when he enlisted in the army and was promoted to Corporal during the war. An uncropped version of this picture of DeCharmes standing alongside 3 other Bedford County Civil War Veterans in front of the courthouse is on page 355. Initially, USCT regiments were staffed by white officers. Later in the war, African Americans were being promoted to officers in the Union Army. Photographs of two white officers from Bedford County are included in this book. Prior to enlisting in the 72 USCT, Howard Jeffries volunteered in the 21st PA Cavalry. A photograph of him is on page 320. A photograph of Josiah Slick is on the next page.

Name	Reg't	Pict. Page	Known Casualties	Battle
Allen, Henry T	43rd			
Barks, Alfred	41th			
Barks, John R	32nd			
Barks, Moore	32nd			
Barks, William T	54th			
Barns, Robert	32nd			
Bates, Thomas	24th			
Berry, John W	41th			
Bolden, Elijah	6th			
Boston, James	41th			
Boston, John				
Brice, John	3rd			
Brown, Henry	3rd			
Brown, John	3rd			
Brown, Todd	3rd			
Burk, Cory S	5th			
Burk, Thomas	55th			
Byers, Peter				
Callahan, James	8th			
Carson, John	22nd			
Coleman, George	32nd			
Costler, John				
Costler, Joseph	127th			
Davis, James	24th			
Davis, John	22nd			
	24th			
	8th			

Name	Reg't	Pict. Page	Known Casualties	Battle
Davis, R DeCharmes	32nd	336, 355		
Dean, Andrew				
Dean, Jacob	32nd			
Doogen, Henry				
Fry, Henry	43rd			
Fry, John	43rd		KIA	Petersburg
Ganz, Thomas				
Gates, Reuben	3rd			
Gordon, Daniel	43rd			
Harris, John T	3rd		Wounded	Morris Island, SC
Harris, Joshua	MA 55th		Died	
Hartley, William	2nd			
	34th			
Hollinger, Stephen	43rd		Wound.-Died	Battle of the Crater
Holmes, Philip	45th			
Holsinger, Frank	19th		Wounded	2nd Bermuda Hundred
Jeffries, Howard B	72nd	320		
Johnson, David	1st			
Johnson, Moses	6th			
Johnson, William	3rd			
Jordon, Henry	8th			
Key, James	32nd			
Key, Philip	32nd			
Krausen, E W				

U.S.C.T. Listing

Name	Reg't	Pict. Page	Known Casualties	Battle
Lewis, Bert	32nd		Wounded	
Lewis, Robert	41th			
Lewis, Robert M	8th			
Lisles, George	3rd			
Love, George				
Love, John R	41th			
Luckett, Alexander	32nd			
Lumac, William	43rd			
Lyles, David	3rd			
Lyles, George	38th			
Lyles, James	3rd			
Lyons, George W	41st			
Marshall, Martin	10th			
	75th			
McPherson, Cyrus				
McPherson, John	3rd		Died	
Miller, Charles W	32nd			
Miller, David	32nd			
Parker, James				
Perry, Wythe	3rd			
Plowden, Jacob	3rd			
Reed, Louis	12th			
Slick, Josiah	107th	336		
Smith, Samuel				

Name	Reg't	Pict. Page	Known Casualties	Battle
Stewart, Preston	118th			
Strathers, James	43rd		Wounded	Battle of the Crater
Strathers, Willis	43rd		KIA	Battle of the Crater*
Streets, James	24th			
Streets, Rankin				
Swartz, John	30th			
Tillman, George	41th			
Tillman, Isaac				
Tillman, Jackson	32nd			
Tobias, John B				
Warren, Nimrod	43rd			
Watkins, Hiram	26th			
	55th			
Webster, Daniel				
Willard, Lewis	127th			
Williams, Henry S				
Wilson, Henry W				
Young, Aaron	24th			
Young, Daniel D	24th			
Young, Jacob	127th			
Young, Jacob P	24th			
Young, Peter	127th			

Corporal R DeCharmes Davis, 32nd USCT

Josiah Slick, 2nd Lieut. - 107th USCT

16. Gallery of Bedford County Soldiers

The following pages contain pictures of Bedford County soldiers and their families not already shown in the book. An alphabetical index of all photographs in this book and their page numbers are in the Photograph Index. Please note due to space constraints, Infantry and Cavalry are sometimes abbreviated to Inf or Cav. If a regiment is listed without reference to being Infantry or Cavalry, please assume the regiment listed is an Infantry unit. Some soldiers enlisted in multiple regiments during the war. Due to space limitations in this section of the book, only one regiment is listed. For a more complete service record for each soldier, please refer to the Detailed Alphabetical Listing. Additional information on regiments for most soldiers can be found on the internet including details on battles. The pacivilwar.com & civilwarintheeast.com websites were especially helpful during the research for this book.

(Left) Private John T Crist and Catherine Fetters married on May 27th, 1857. John was 23 and was Catherine 13 years old when they married. John was working on a farm in Union (Pavia) Township when he enlisted in the 12th Pennsylvania Cavalry on March 27th, 1862. Catherine remained home with their two small children while John was away in the war. They were parents to 11 children who survived to adulthood. John passed when he was 68 in 1902. Catherine passed 12 years later at age 69 in 1914. They are buried in the Old Claysburg Cemetery.

(Right) Josiah Cleve Smith was born on April 28, 1823, in Colerain Township. His parents Nancy and Henry Smith moved West when he was a boy. Josiah married Matilda Winn in 1851. He mustered in the 38th Iowa Infantry on November 4th, 1862, when he was 39 years old. Josiah's wife remained on their family farm in Pleasant Valley, Iowa with their 2 small children while he was enlisted. In January 1865, he transferred to the Iowa 34th Infantry and mustered out in August 1865 after the war ended. During the war, his pocket New Testament was stuck by a bullet, likely saving Josiah's life. A photograph of the bible with the bullet clearly visible is below. Josiah passed at age 51 on March 1st, 1875, and is buried in the West Union Cemetery in Fayette County, Iowa.

Josiah C Smith, - 38th Iona Infantry

James F Dicken, MN 1st Cav   *John Bloom, MD 3rd Reg*   *Edward Miller, US 3rd Cav*   *James P Covert, PA 5th Res*

John Gillam, MO 2nd Engineers   *William Garlinger, OH 7th*   *Joseph P Cessna, MO 2nd Cav*   *George & Louvisa Keith, PA 3rd Heavy Art.*

Aaron Hill, 13th PA Inf, wife Alcinda was at home with a toddler during the war   *Reuben Hardinger, MD 1st PHB Inf*   *Reuben A Riley, IN 5th Cavalry*

Gallery of Bedford County Civil War Soldiers 339

Andrew Baker, PA 13th Cav   *John H Karns, Unassigned*   *Jacob H Brown, KS 11th Cav*   *Levi Diehl, US 13th Navy*

Martha & Ephraim Chilcott, IL 8th   *Catherine & Abraham Burket, PA 11th Cav*   *Samuel Livingston, PA 13th Cav*

John J Barclay, PA 11th Cavalry   *Abraham K Arnold, US 5th Cavalry*   *John Welch, MO 7th Cavalry*

Gallery of Bedford County Civil War Soldiers

William J Deremer, MD 5th Inf

Henry Hardinger, WV 17th

Abraham Barnhart, IL 15th

D Alex McKellip, OH 13th

Mary & Simon S Blake, WI 25th Infantry

John A Steckman (center) with unidentified soldiers of the 28th IL Infantry

Joseph Nevitt, IA 35th Infantry

James B Coughenour with wife Pauline & son James M, IA 14th Infantry

(L to R) Fletcher brothers - Baltzer, Samuel - OH 15th & David - PA 211th

Alexander Smith, PA 19th Cav

Gallery of Bedford County Civil War Soldiers

John O Myers, OH 19th Inf   *John Milton Davis, PA 61st*   *Joseph Y Weyandt, PA 13th Cav*   *Hiram Evans, IA 23rd Inf*

Jacob Bridenstine, IA 29th   *Isaac Waters, IL 25th Inf*   *Jacob Hite, IA 28th Inf*   *Nicholas Hite, IA 28th Inf*

Reuben Barley, IN 34th Inf   *Mark Miller, PA 51st Inf*   *Solomon Comp, PA 48th Inf*   *Stiles Jackson, OH 35th*

Jesse W Horton, PA 45th Inf   *Samuel Salkeld, PA 49th Inf*   *David A Hunt, PA 50th Inf*   *Levi Holsinger, IL 34th Inf*

Joseph Wareham pictured with grandson Leonard Masters, IL 45th

Jonathan Wiser pictured with his grandchildren, PA 49th Infantry

David R Potter, IN 47th Infantry

Brice & Mary Twigg pictured with their children, PA 45th Infantry

Louisa & Levi Valentine, PA 50th Inf   *Sarah & Joseph Growden, PA 45th Inf*   *Tobias Boor, PA 50th Infantry*

The Jeremiah Weicht family gathered for this portrait in 1880 on a farm located 3 1/2 miles south of Everett on Milk and Water Road. Seated is Jeremiah Weicht, PA 51st Infantry and his wife Rosie. Daughters: (left to right) are Alice, Lynn and Annie. Sons (left to right) are Frank, William, Simon, George and Robert. Pictured on the lower left is Ellis Weight, the great-grandson of Jeremiah and grandson of William. Ellis was posthumously awarded the Medal of Honor during World War II for his bravery at St. Hippolyte, France, on December 3, 1944.

Hanson Cook, PA 53rd Inf   *Joseph Lybarger, PA 54th Inf*   *John H Akers, PA 56th Inf*   *William H Kelly, PA 54th Inf*

Samuel X Smith, PA 56th Infantry   *Paul McLeary, PA 53rd Inf*   *Henry Mearkle, PA 78th Infantry*

Lemuel Evans, PA 49th Infantry   *Henry College, PA 51st Infantry*   *James A Cook, PA 54th Infantry*

Gallery of Bedford County Civil War Soldiers

 *David R Smith, PA 56th Inf*

 *John McMullen, IL 77th Inf*

 *Samuel Stuckey, Unassigned*

 *Wilson Karns, PA 79th Inf*

 *Andrew McFarland, PA 79th Infantry*

 *David H Fisher, PA 78th Infantry*

 *John L Kellerman, OH 60th Infantry*

 *Caleb Brown, PA 67 Infantry*

 *John R Layton, OH 61st Infantry*

 *Jonathan Dibert, PA 79th Infantry*

Ezra Osborn, IL 93rd Inf

John Wonder, PA 100th

Benson Hanks, PA 87th Inf

David Sparks, IL 93rd

Solomon Sparks, IL 93rd

George Clevenger, PA 97th

George Fockler, PA 93rd Inf

Jacob Rightenour, PA 103rd

Francis P McCoy, PA 81st Inf

Henry Hartman, PA 87th Inf

Jemima & George Truax PA 97th Inf

Gallery of Bedford County Civil War Soldiers 347

Jacob Gettleman, PA 104th *William Watson, PA 105th* *William Stultz, PA 100th* *Israel Grace, PA 102nd*

William McKibbin, PA 130th *Benjamin Blymyer, PA 83rd* *John Fletcher, OH 121st* *Festus Edwards, OH 129th*

John A Miller, PA 100th Inf *Isaac Boose, PA 102nd Inf* *Merrick A Stoner, OH 114th Infantry*

348   Chapter 16

Sam Morningstar, PA 143rd   *Emanuel Holsinger, IL 142nd*   *George Hunt, PA 132nd Inf*   *Francis Reamer, PA 143rd Inf*

George Edwards, OH 174th   *Christopher Knisely, IL 148th*   *John N Salkeld, PA 147th*   *Abraham Sparks, IL 146th Inf*

(seated in the center) Ellen & Jesse McVicker, PA 102nd Infantry are pictured with their Children and Grandchildren. Ellen was at home with 2 small children when Jesse was wounded during the battle of Opequon in 1864. Please note some family members in this picture appear to have superimposed facial photographs. In earlier eras, it was not uncommon for stand-in models to pose in family pictures for missing family members.

Gallery of Bedford County Civil War Soldiers

Sarah & Samuel Murphy, IN 140th Inf. Sarah was at home with 2 children while Sam was away in the war.

Margaret & Christian R Oakes, PA 137th Infantry. Margaret was at home with 7 children while Christian was enlisted.

(seated) Margaret & Josiah Lingenfelter are pictured with their children. Josiah and his brothers Aaron and John were born on a farm near Klahr. Their family moved to Illinois when the three brothers were young. Aaron joined the 55th IL Infantry in 1861 and was wounded twice during the war. Josiah and his 16 year old brother - John joined the 103rd IL Infantry in 1862. John died on March 22nd, 1864. The cause of death is not known. Aaron and Josiah survived the war and both raised families on farms in Fulton County, Illinois.

Dennis B Sipes, PA 158th   Jacob F Salkeld, PA 158th   George G Gordon, PA 158th   Abraham Plessinger, PA 158th

Bartley Hughes, PA 148th Infantry   William H Latta, PA 148th   John E Sparks, IL 151st Infantry

Son Thomas & Benjamin F Lohr, PA 142nd   Charles Ezekiel McMullen, IL 146th Inf. holding grandson Charles E McMullen   Charles Jackson, OH 147th

Wilson S Smith, PA 158th Infantry

Jane & James Barton, PA 158th Infantry. Jane was at home with 3 children while James was enlisted.

(right) Wilson S Smith, pictured during his later years shaking hands with an unidentified man, possibly over the sale of a horse. Wilson was also a minister in Napier Township during the 1880's.

Pictured left to right are Hannah Thomas Himes & husband Samuel Himes, PA 186th Inf, Margaret Himes Thomas & husband Isaac Thomas, PA 186th Inf. Samuel Himes and Isaac Thomas were best friends and married each other's sister. Samuel Himes and Margaret Himes Thomas are also the brother and sister of Andrew John Himes pictured on the front cover. Margaret & Isaac Thomas also raised Andrew's orphaned daughter Eutoka Himes Whitfield.

Bechtel brothers David, PA 183rd on left and Daniel, PA 178th. A third brother - Isaac, PA 184th, was KIA at Petersburg in 1864.   *John Fleegle, PA 191st*   *John Foster, PA 195th*

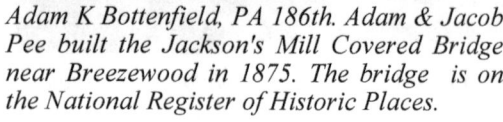

Adam K Bottenfield, PA 186th. Adam & Jacob Pee built the Jackson's Mill Covered Bridge near Breezewood in 1875. The bridge is on the National Register of Historic Places.   *Jeremiah Ramsey, PA 186th*   *Henry Mock, PA 200th*

Gallery of Bedford County Civil War Soldiers

Matthew Kennard, OH 197th Inf with his daughters Ruth and Julia Ann

James Beaston, PA 202nd Infantry

George W Eshelman, PA 186th Infantry

Elizabeth and James F Pennington, PA 188th Inf

Chapter 16

Samuel Z Beam, OH 8th Infantry. Samuel served as a minister at the St. Clairsville German Evangelical Church from 1888—1892

Sarah & Levi Clites, PA 211th Infantry

Solomon J Hankinson, PA 206th Infantry

Susan & Francis T Troutman, PA 211th Inf.

Veterans Group Photographs

Bedford County Veterans gathered at Gettysburg National Cemetery. John Holler—138th PA Infantry is identified on far right.

Bedford County Civil War Veterans gathered in front of the Courthouse prior to a Decoration Day (Memorial Day) Parade on May 30th, 1918. From left to right are Oliver P Fluke - PA 110th Inf, Americus Enfield - 22nd PA Cavalry, DeCharmes Davis - 32nd USCT, Samuel Statler - 55th PA Infantry and Harry C Robison - 49th PA Infantry.

Pavia Veterans Reunion in September 1895. Photograph in enlarged below.

Civil War Veterans – left to right: David Burket - PA 138th Infantry, Joseph Gordon - PA 55th Infantry, Adam Gardner, Adam Corle - Indiana 22nd Infantry, Gabriel Burket - PA 55th Infantry, George G Gordon - PA 158th Infantry, Alexander Corle - PA 171st Infantry, Michael Dively - unknown Civil War regiment, Michael Feather - PA 99th Infantry, Judge Jacob Longenecker - PA 101st Infantry, George W Shaffer and Franklin Corle - PA 138th Infantry.

1912 parade down the hill of present day Rt. 913 in East Saxton.

Civil War Veterans in front of the Eichelberger Store in Saxton following an 1890 parade. Eli Eichelberger was an officer in the Hopewell Rifles Company of the 8th Pennsylvania Reserves and was the proprietor of the store. Pictured left to right are Harry Speece - PA 77th Infantry, Levi Steele - PA 133rd infantry, Coy? (possibly Frank McCoy - PA 81st Infantry), S.K. Sanderson - PA 7th Reserves, Wilson Weaver (possibly John H Weaver - West Virginia 5th Cavalry), T.C. Sanderson - PA 149th Infantry, Adam Maugle - PA 147th Infantry, Christopher Eicher - PA 12th Reserves, John Wonderly - PA 55th Infantry, George Gates - PA 9th Cavalry, John O Hoffman - PA 133rd Infantry, Tom Barr - PA 125th Infantry, Oliver Hollingshead - PA 22nd Cavalry, Dr. M.B. Brenneman - PA 125th Infantry and Captain Eli Eichelberger PA 8th Reserves.

The E.S. Wright GAR Post #333 was located on the third floor on the right-hand side of the above building located on the corner of Main Street and Locust Street in Pleasantville. Access to that level would be by a staircase on the back side of the building. As a tribute, the GAR Post #333 was named after Edmund S. Wright.

1. Possibly Theodore B Potts - 6th PA Heavy Art., 2. Isaac Gordon - 138th PA Inf., 3. Isaac Ling - 55th PA Inf. 4. John H Fleegle - 5th PA Res., 5. Francis W Mock – Spanish - American War Veteran. 6. Peter Nunemaker 99th PA Inf., 7. Aaron Mock - 138th PA Inf., 8. Thomas G Walker—91st PA Inf., 9. Hezekiah Hammer - 55th PA Inf., 10. Joseph Penrose - 21st PA Cav., 11. Conrad Claycomb - 138th PA Inf., 12. Samuel Mock - 91st PA Inf., 13. Christian Harr - 149th PA Inf., 14. George W.Darr - 100th PA Inf., 15. Daniel L. Hetrick - 101st PA Inf., 16. Thomas J Miller - 100th PA Inf., 17. John H Machtley - 82nd PA Inf., 18. Unknown, 19. Jacob H Miller - 21st PA Cav., 20. William A Davis - 84th PA Inf., 21. George Miller - 91st PA Inf., 22. Samuel Berkheimer - 171st PA Inf., 23. George W Adams - Mexican War Veteran & 6th OH Cav., 24. Unknown, 25. William A McGregor - 99th PA Inf., 26. William H Ling - 138th PA Inf., 27. Unknown, 28. James Over - 138th PA Inf., 29. Unknown, 30. Ashel Walker - 84th PA Inf., 31. Armstrong R Miller - 21st PA Cav., 32. Joseph Andrews - 82nd PA Inf., 33. Samuel Hammer - 55th PA Inf., 34. possibly John B Bowser - 192nd PA Inf., 35. George W Rowzer - 107th PA Inf., 36. Samuel H Smith - 99th PA Inf.

Civil War Veterans gathered for a photograph in front of the Bedford County Court House circa 1915. Veterans identified are William Kelly (2 county veterans with the same name) seated 2nd from the left in the 2nd row, Frank McCoy – PA 81st Infantry seated 2nd from the left in the 3rd row, Americus Enfield – PA 22nd Cavalry is the tallest man standing in the middle of the 4th row with the distinctive sideburns and Samuel Statler – PA 55th Infantry is the man against the rail holding the 2nd hat from the left.

Civil War Veterans gathering in 1898 in Tatesville during the celebration of the end of the Spanish American War. Identified are William DeArment Ritchey 3rd from the left, John Sparks 4th from the left and Barton Mearkle 6th from the left.

Verbiage from an undated newspaper article states the picture is Company F, 8th Reserves, 50th Anniversary Reunion at Hopewell, PA held on April 23rd, 1911. Eli Eichelberger and William Whisel, both veterans of the PA 8th Reserves, are identified in the below photograph that appears to have been taken on the same day.

Civil War veterans in a 1911 photograph in Hopewell. Possibly a reunion of 8th Reserve - Hopewell Rifles Company veterans. Identified are Eli Eichelberger standing 2nd from the left in the 1st row and William Whisel 5th from the left in the 1st row. The man holding the flag on the far left of the 1st row is standing on a step below the other men in the picture. David Horton was reportedly one of the tallest men in the 8th Reserve regiment at 6'4" and is quite possibly the soldier holding the flag. Another photograph of David taken during the war is on page 57.

Photograph near Traveler's Rest between Everett and Breezewood in 1897 at the farmhouse of E.C. Weaverling. Regiments are listed for Bedford County Civil War veterans in this picture. Seated from the left are Alexander Davis, Samuel Stailey, Dr. C.N. Hickok, Rev. G.C. Probst - Chaplain and Lewis Piper – PA 8th Reserves. Standing from the left are Oliver Ramsey - PA 208th Infantry, William P Barndollar - PA 76th Infantry, John A Gump, Valentine Steckman, Joseph Weaverling, John Herring - PA 99th Infantry, Simon Nycum - unknown regiment, John P Weaverling, James Piper, Philip Messersmith, Benjamin Mitchell and Christian Wagner - PA 88th Infantry.

Bedford County Civil War Veterans gathered in Gettysburg prior to 1914. Identified are Hendrson Souser - PA 133rd Infantry is 2nd from the left, Isaac Fleegle - PA 55th Infantry is 3rd from the left, Tommy Bridenthal - PA 13th Cavalry is 5th from the left and John Milton Davis (two county Civil War veterans with this name) is 6th from the left.

Bedford County Civil War Veterans pictured prior to 1920. From left to right are Harvey Wishart - PA 208th Infantry, Abel Griffith - PA 8th Reserves, Sam Wishard - PA 20th Cavalry and Al Baker - PA 205th Infantry.

Bedford County Civil War Veterans photograph in 1916. Identified is Henry I Claar - PA 55th Infantry seated in the 2nd row, far left and Americus Enfield - PA 22nd Infantry seated in the 2nd row, 9th from the left.

The Last Surviving Immediate Family Member

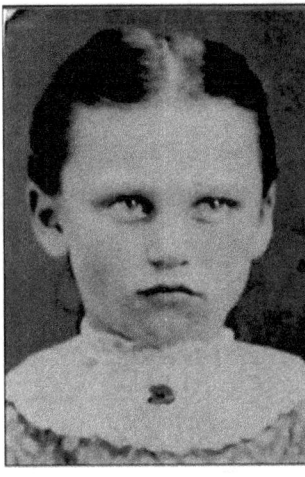

Frances W Pee *Mary Jane Pee Baughman circa 1877* *Sara Karns Pee* *Adam Baughman*

Frances Pee mustered in the 11th Pennsylvania Infantry in September 1861 when he was 21 years old. During his enlistment, Frances suffered a shoulder wound in the Cornfield during the battle of Antietam and a second wound above the elbow on the 1st day of fighting at Gettysburg. Francis survived his wounds and returned to his wife Sara and toddler son in January 1864. The following month, 22-year-old Adam Baughman left a West Providence Township farm to enlist in the 186th Pennsylvania Infantry. The 186th took part in the Overland Campaign in late spring of 1864 and the siege of Petersburg. Adam was hospitalized for Typhoid Fever in November 1864 but survived and returned home on August 15th, 1865. Adam suffered from the effects of contracting Typhoid Fever for the rest of this life. Twenty-three years later, in September 1887, 44-year-old Adam married 15-year-old Mary Jane Pee, the 6th child of Frances and Sara. Twenty-four years later, on July 8th, 1912, Adam and Mary Jane's fourth child, Freda, was born in Everett. Adam passed in 1925 and Mary Jane in 1937. Freda lived until 2008, 143 years after the Civil War had ended. Adam, Mary Jane and Freda are buried at the Everett Cemetery.

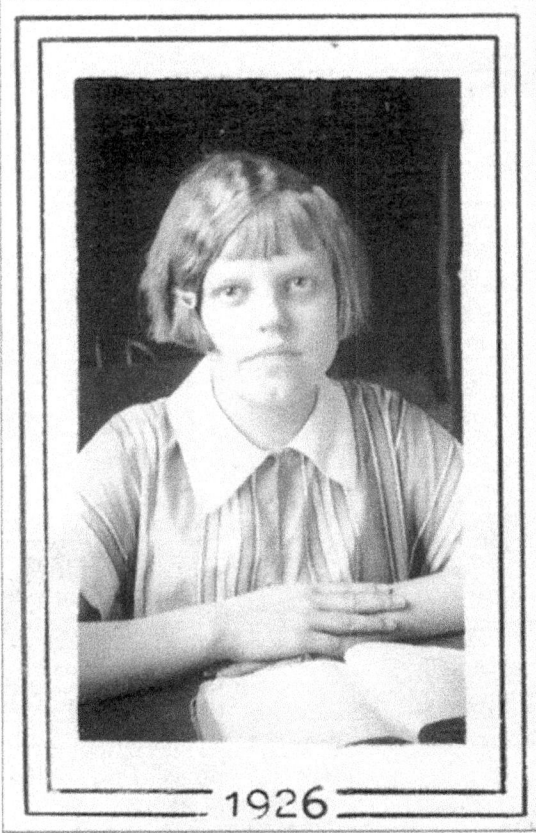

Mary Jane, son Francis and Adam Baughman circa 1900 *Freda Baughman Foor*

17. Unconfirmed Listing of Deaths

The following 170 soldiers were mentioned in various books, listings or records as possibly being from Bedford County or an immediate county border community. After carefully reviewing census records and other available information, it appears some of these fallen soldiers likely never lived in Bedford County. Others on this listing have little to no available documentation to confirm they lived in Bedford County during their lifetimes or perished during or immediately after the Civil War ended. Information on each soldier is contained in the Detailed Alphabetical Listing chapter. If a descendant has verifiable information of a soldier who lived in Bedford County or an immediate border community during their lifetime and their death occurred during the years of 1861 to 1866 due to their military service in the Civil War, please contact the author.

Altman, George W*	Dagenfelt, Joseph*	Kooken, John R*	Quarry, Alfred*
Andrews, Charles (1)*	Davis, Thomas P*	Kurtz, Jacob*	Reed, Samuel*
Andrews, William A*	Deck, Henry*	Lang, James*	Riffle, John*
Anthony, Cyrus*	Demmings, William*	Leippert, Nicholas*	Riley, Edward*
Aulenbach, James*	Deneen, Henry S.*	Lemon, John E*	Robison, Thomas*
Bailey, John W (1)*	Derr, John*	Lenhart, Czar*	Rough, Valentine*
Bailey, John*	Dodds, Mathew*	Lenhart, Peter*	Russell, George D*
Baker, John H*	Doyle, John*	Lewis, William S*	Salkeld, David E*
Bechtel, Daniel*	Elwell, George*	Lindsey, Ephraim N*	Saxon, John C*
Beighel, Burket*	Emeigh, Charles*	Lininger, William P*	Sellers, John F*
Belles, Joseph S*	Epler, Aaron*	Link, Solomon*	Shaffer, Peter*
Bennethum, George*	Evans, Daniel*	Lockhard, John*	Shaw, Zopher P*
Berkheimer, David*	Fair, Henry*	Long, Augustus*	Shawley, Andrew*
Bircamp, Henry*	Ferguson, John W*	Lybarger, Martin*	Shepard, Robert K*
Bisbing, Charles H*	Fidler, Issac M*	Lybarger, Valentine G*	Smith, George W (2)*
Bisbing, Gervase*	Fogle, Nicholas*	Lybarger, William*	Smith, Morris B*
Boose, Moses*	Fritz, Aaron*	Madara, Jacob*	Smith, William R (2)*
Brice, John (1)*	Gates, Jeremiah*	Marshall, George*	Smitman, Henry*
Brightbill, Samuel*	Geiger, Jacob D*	Maxwell, George W*	Snyder, Elias*
Brown, Max*	Gilbert, Daniel*	McCabe, Joseph*	Squint, Henry D*
Brown, William*	Gneill, Conrad*	McCloud, Daniel*	Summers, George*
Bryant, Simon*	Goodman, Frederick*	McConnell, Philip J*	Summers, John*
Burges, Adam*	Gottwalt, Henry*	McCoy, Charles*	Sutters, John*
Burns, James M*	Grace, John*	Mentz, Stephen*	Tannehill, Alfred*
Caldwell, James B*	Greenland, Thomas J*	Meyer, William C*	Tannehill, Eli*
Cameron, Amos M*	Gregor, Solomon*	Miller, John L*	Thompson, John*
Caylor, David H*	Hardinger, Daniel (2)*	Mitchell, John A*	Trott, George*
Chapman, William F*	Harker, Samuel B*	Morris, James*	VanTassel, Russel*
Clay, Henry*	Hartzell, Samuel*	Morse, John*	Vastbinder, Gabriel*
Cochrane, Michael*	Heinbaugh, Jackson*	Moyer, Samuel*	Wallace, John P*
Conrad, Martin*	Heinemyer, Adolph*	Moyer, William*	Warner, Frederick*
Cook, Jacob*	Hobbs, Nelson*	Myers, Daniel*	Wenrick, Daniel*
Cook, James L*	Hockenberry, Samuel*	Newman, John (1)*	Wike, Jacob*
Cook, William A*	Hull, Samuel*	Newman, John R*	Wilkinson, William (2)
Corbett, William*	Ilesmith, Charles*	Noll, William R*	Williams, John*
Corle, Solomon*	James, David*	O'Neil, John*	Wise, Henry*
Cornelius, Peter*	Jayne, Isaac B*	Otto, Abraham*	Womer, Andrew*
Cowan, David*	Johnson, David (2)*	Otto, DeWitt*	Woodcock, Clark*
Craig, George W*	Johnston, Samuel*	Otto, William*	Woodcock, George W (2)*
Crocheron, John F*	Jones, Emanuel*	Packert, Christian*	Younkin, Alfred*
Crosby, Isaac*	Kelly, Oliver F*	Page, Christian*	Younkin, Foster*
Croyle, John*	Kephart, Cyrus*	Peterman, Jacob D*	
Crum, Simon*	Koch, John*	Potter, Levi (2)*	

18. Detailed Alphabetical Listing

This listing contains 4,209 Civil War solders identified as living in Bedford County or an immediate border community during their lifetimes. An additional 571 soldiers are listed whose names were mentioned in various historical listings but appear likely to have never lived in Bedford County or a surrounding area; or lack documentation on their residence. These 571 soldiers are included on this listing with asterisks at the end of their names but are not included in other listings or casualty figures cited in this book.

Approximately half of the identified 4,209 Bedford County soldiers in the Civil War enlisted in regiments organized outside of the county. Conversely, a limited number of soldiers who were residents of other areas of Pennsylvania enlisted in regiments originally organized in Bedford County. A majority of the 571 soldiers listed with asterisks enlisted in the Union Army after 1862. Many of these soldiers were drafted and mustered in as replacement soldiers.

The following are the total numbers of Bedford County soldiers who enlisted in each state. Pennsylvania-3598, Maryland-166, Ohio-82, Illinois-58, Iowa-36, Indiana-24, New York-12, West Virginia-9, Missouri-8, Wisconsin-6, Kansas-4, New Jersey-4, Massachusetts-3, California-3, Kentucky-3, Michigan-3, Minnesota-1, New Hampshire-1 and Vermont-1. Eighty-five soldiers enlisted in USCT, 30 in the US Regular Army and 11 who enlisted in a Confederate state. Regiments of some of the remaining Civil War soldiers could not be determined.

The following is an overview, explanation of terms and abbreviations used in the Detailed Alphabetical Listing. Information on each soldier is listed on two consecutive pages. To assist in readability, the same line items numbers are included on both pages of information on each soldier. If a name is listed on multiple lines, a soldier enlisted in more than one regiment or was a casualty more than once during the war. When a name is listed on the last line of a page, please check the first line of the following page for any additional enlistments or casualties for that soldier.

Some Civil War Soldiers from Bedford County shared the same name. Soldiers who share the exact same name are differentiated by a number within parentheses added to the end of each name, for example: Adams, John (1), Adams, John (2), Adams, John (3). Some soldiers shared the same first and last name but had different middle initials or no middle initial, for example: Allison, John, Allison, John H and Allison, John M are three different soldiers.

Most of the Ages listed on the following pages are the age of a soldier when they first joined the Army during their initial enlistment. Sometimes only a birth-year is stated on available records which necessitated the listing of an approximate Age at muster in. The birthdate for some soldiers is unknown and is left blank. The Rank listed is the highest rank attained by a soldier prior to the Confederate surrender at Appomattox in April 1865. Some soldiers received promotions after hostilities had ceased and just prior to being discharged from the army.

Traditional state abbreviations are used for state Regiments, US is the abbreviation for a US Regular Army Regiment and USCT is the abbreviation for United States Colored Troop Regiments. The following are Regimental unit abbreviations: Inf.-Infantry, Res.-Reserves, Cav.-Cavalry, Art.-Artillery, Mil.-Militia, Lgt. Art.-Light Artillery, VRC-Veterans Reserve Corp and Eng.-Engineering. If no regiment is listed, the regiment for that soldier is unknown. The abbreviation for Company on the header line is Co. If no Muster date is listed, the date is unknown.

The casualty abbreviations are KIA (Killed in Action during a battle), MIA (reported missing during a battle and in most cases assumed to have killed during the battle or died in captivity), Wound.-Died (wounded in battle and died after the battle ended), POW-Died (taken prisoner and died in captivity or shortly after being released), Died (died from a battlefield wound or disease), Wound.-POW (wounded during a battle and taken prisoner), POW (prisoner of war), Wounded (a wound was suffered during a battle), Injured (debilitating injury suffered during a battle).

The Casualty Battle abbreviations are Berm. Hundred (Bermuda Hundred Campaign during May 1864), Culpeper C.H. (Culpeper Court House), Petersburg – Final (Petersburg Final Assault on April 2nd, 1865), Petersburg – Initial (Initial Petersburg Assault during June 15th-18th, 1864), Jerusalem Plank Rd. (Jerusalem Plank Road), Kennesaw Mtn. (Kennesaw Mountain), Moorefield Jct. (Moorefield Junction) and Boydton Plank Rd. (Boydton Plank Road).

Burial locations are listed for each soldier if known. Please note the word Cemetery is sometimes omitted or abbreviated to Cem. because of space limitations. Various other abbreviations of Cemetery names are used. Burial locations outside of Pennsylvania list only the Cemetery name and state abbreviations because of limited space. Most Cemetery information was sourced on the Findagrave.com and the Ancestry.com websites where additional burial location details can be found. Some of the National Cemeteries also list burial plot numbers.

The notes column contains known casualty details including POW camp locations, census records and residential information. Family references (fathers, sons and brothers) are primarily sourced from the Ancestry.com and Findagrave.com websites. The research sources listed in the notes column are detailed in the Authors Notes on Civil War Research at the beginning of the book. The abbreviations B.S.F. History Book-1884 is used for the History of Bedford, Somerset and Fulton Counties, PA book by Waterman. Hist. Data Sys. is the abbreviation for Historical Data Systems. Other abbreviations include Bio.-Biographical, Bros.-Brothers, Cen.-Census, Co.-County, CW-Civil War, Disch. Surg. Cert. or DSC- Discharged on Surgeon's Certificate, Enl.-Enlisted, Rec.-Record, Resid.-Residence, Twp.-Township and Vet.-Veterans. The term "references" is used when residential-type information was found on some records.

Chapter 18

#	Name	Age	Rank	Regiment	Co.	Muster In	Muster Out	Casualty	Casualty Battle
1	Abbott, Allison	19	Pvt.	PA 194th Inf.	I	7/22/64	11/5/64		
2	Abbott, Amos *	18	Pvt.	PA 110th Inf.	C	2/25/64	6/28/65		
3	Abbott, William		Pvt.	PA McKeage's Mil.	G	7/2/63	8/8/63		
4	Ackerman, Chauncey F	20	Pvt.	PA 88th Inf.	A	3/29/65	6/20/65		
5	Adams, Charles C	35	Pvt.	PA 184th Inf.	A	5/12/64	5/30/65		
6	Adams, Daniel	36	Pvt.	PA 8th Res.	F	6/19/61	5/24/64		
7	Adams, David	22	Pvt.	PA 101st Inf.	K	12/14/61	6/25/65	POW	Plymonth, NC
8	Adams, George W (1)	37	Pvt.	OH 6th Cav.	H	6/18/61	6/23/64		
9	Adams, George W (2)	17	Pvt.	PA 55th Inf.	H	2/29/64			
10	Adams, Harry		Pvt.		D				
11	Adams, John (1)	40	Pvt.	PA 55th Inf.	H	9/21/63	8/30/65		
12	Adams, John (2)		Pvt.	PA 20th Cav.	C	1/27/64	6/17/65		
13	Adams, John (3)	29	Pvt.	PA 29th Inf.	C	12/23/64			
14	Adams, John Q	19	Pvt.	PA McKeage's Mil.	D	7/2/63	8/8/63		
15	Adams, Philip	18	Pvt.	PA 55th Inf.	H	10/11/61		POW-Died	Drewry's Bluff
16	Adams, Samuel	33	Pvt.	PA 55th Inf.	H	2/22/64	8/30/65		
17	Adams, Solomon	17	Pvt.	PA 55th Inf.	D	10/12/61		Died	
18	Adams, William (1)	19	Corp.	PA 76th Inf.	E	10/9/61	11/28/64		
19	Adams, William (2)	27	Pvt.	PA 100th Inf.	M	11/14/64	7/24/65		
20	Affierback, George*		Pvt.	PA 110th Inf.	C	3/5/65	6/28/65		
21	Agnew, George H	29	Pvt.	PA 55th Inf.	K	2/19/64		POW-Died	
22	Agnew, Hamilton	19	Corp.	PA McKeage's Mil.	H	7/2/63	8/8/63		
23	Agnew, Levi J	20	Corp.	PA 76th Inf.	E	10/9/61	11/28/64	Wound.-Died	Bermuda Hundred
24	Agnew, William K	28	Pvt.	PA 55th Inf.	H	2/29/64	8/17/65		
25	Ainsworth, James W*		Pvt.	PA 110th Inf.	C	10/24/61	6/28/65		
26	Ake, Alexander	14	Music.	PA McKeage's Mil.	G	7/2/63	8/8/63		
27	Ake, Castleton	41	Sgt.	PA McKeage's Mil.	G	7/2/63	8/8/63		
28	Ake, Castleton		Pvt.	US 5th Light Art.	E				
29	Ake, John G	20	Pvt.	PA 138th Inf.	E	8/29/62	6/10/65	Wounded	Wilderness
30	Ake, Samuel	34	Pvt.	IA 1st Inf.	H	5/14/61	8/20/61		
31	Ake, Samuel		Sgt. QM	PA 22nd Cav.	H	2/27/64	8/14/65		
32	Ake, Thomas	17	Pvt.	PA 11th Cav.	G	9/5/61		Wound.-Died	Carrsville
33	Ake, William H	25	Corp.	PA 138th Inf.	E	8/29/62		KIA	Wilderness
34	Ake, Windfield S	19	Pvt.	OH 72nd Inf.	I	2/18/62	12/10/62		
35	Ake, Windfield S		Sgt.	US 1st Cav.	M	12/10/62	2/12/67		
36	Akers, Abner	35	Pvt.	OH 192nd Inf.	I	3/9/65	9/1/65		
37	Akers, Absalom								
38	Akers, Amos	22	Pvt.	OH 94th Inf.	F	8/24/62		POW-Died	Chickamauga
39	Akers, Job S	39	Pvt.	PA 208th Inf.	H	9/7/64	6/1/65		
40	Akers, John H	42	Pvt.	PA 56th Inf.	F	9/19/64	5/31/65		
41	Akers, John T	16	Pvt.	PA 101st Inf.	D	11/1/61			
42	Albright, George B	27	Pvt.	PA 5th Res.	I	6/11/63			
43	Albright, Jacob	25	Pvt.	MD 2nd PHB Inf.	B	4/4/65	6/28/65		
44	Albright, William	18	Pvt.	VA 114th Mil.	A				
45	Albright, William		Pvt.	PA 18th Cav.	K	11/18/62	10/31/65		
46	Alburgh, George B		Pvt.	PA 188th Inf.	I			Wounded	
47	Alcorn, George	26	Pvt.	PA 76th Inf.	E	2/24/65	7/15/65		
48	Aldstadt, John R	30	Pvt.	OH 63rd Inf.		9/27/64	5/15/65		
49	Alhborn, Augustus	35	Pvt.	PA 101st Inf.	G	12/28/61			
50	Allen, Henry T	18	Pvt.	USCT 43rd Inf.	D	3/26/64	10/20/65		
51	Allen, William (1)	18	Pvt.	PA 55th Inf.	K	11/5/61	11/6/61		
52	Allen, William (1)		Pvt.	PA 53rd Inf.	B	11/6/61			
53	Allen, William (2)	22	Pvt.	PA 110th Inf.	C	10/24/61			
54	Allenbarger, George	32	Pvt.	PA 55th Inf.	K	2/29/64		Wounded	Bermuda Hundred
55	Allison, Andrew	35	Pvt.	PA 99th Inf.	H	2/25/65	5/30/65		
56	Allison, David	20	Pvt.	PA 55th Inf.	K	11/5/61		POW-Died	Drewry's Bluff*
57	Allison, Edward	18	Pvt.	PA 55th Inf.	K	2/10/64		POW-Died	Drewry's Bluff
58	Allison, Jacob W	32	Pvt.	PA 55th Inf.	K	11/4/62		Died	
59	Allison, John	19	Pvt.	PA 55th Inf.	K	11/4/61	8/30/65	Died	
60	Allison, John H	17	Pvt.	PA 21st Cav.	E	7/8/63	7/8/65		
61	Allison, John M	21	Pvt.	PA 55th Inf.	K	9/7/61	2/21/63		
62	Allison, John M		Pvt.	PA 1st Light Art.	M	2/21/63	9/16/64	Wounded	Bermuda Hundred
63	Allison, Joseph D	39	Pvt.	PA 138th Inf.	D	8/29/62		Died	
64	Allison, Nathaniel	18	Pvt.	PA 55th Inf.	K	2/19/64	8/30/65	Wounded	Cold Harbor

	Cemetery	Notes
1		B.S.F. History Book-1884; Historical Data Systems record
2		B.S.F. History Book-1884; Historical Data Systems record; Huntington Co. references
3		1890 Liberty Twp. Veterans Census; B.S.F. History Book-1884
4	St. Clair Cem.-Greensburg	B.S.F. History Book-1884
5		B.S.F. History Book-1884; Born in Bedford Co. in 1837; Discharged on Surgeon's Cert.
6	Hopewell Cem.	B.S.F. History Book-1884; 1860 Broad Top Twp. Census; Historical Data Systems record
7	Anderson Cem.-Cessna	POW-Andersonville 4/20/64; Frank McCoy 1912 Listing; Historical Data Systems record
8	Stone Church Cem.-Fishertown	1890 Napier Twp. Veterans Census; Historical Data Systems record; Mexican War Veteran
9	Wells Valley Presb. Cem.	Deserted 5/31/64; PA Civil War Archives-Bedford Co. residence
10		1890 Brush Creek Twp. Veterans Census
11	Trinity UCC Cem.-Juniata Twp.	PA Civil War Archives-Bedford residence; John (1) & Samuel-brothers
12	Hopewell Cem.	Frank McCoy 1912 Listing; Bed. Inq. 5/22/1908 listing; Historical Data Systems record
13	Everett Cem.	Deserted on way to join regiment; Historical Data Systems record
14	Grandview Cem.-Cambria Co.	B.S.F. History Book-1884; John Q, William, George W (2) and Philip-brothers
15	US Gen. Hosp. Cem.-Annapolis	MIA 5/16/64 in Bedford Inquirer 7/8/64 article; POW until 4/28/65; Died 6/19/65
16	Pleasant Ridge-Buffalo Mills	1890 Juniata Twp. Veterans Census; Historical Data Systems record
17		Died 6/25/62; B.S.F. History Book-1884; Historical Data Systems-Bedford Co. residence
18		B.S.F. History Book-1884; PA Civil War Archives-Bedford residence
19	Cuppett Family-New Paris	1890 Napier Twp. Veterans Census
20		B.S.F. History Book-1884; Bucks Co. references
21		MIA listed in Bedford Inquirer article 6/3/64; POW-Andersonville; Died 11/27/64
22		B.S.F. History Book-1884; possibly George Hamilton
23	Old Presb. Cem.-Bedford	Wounded 5/7/64; Died 3/26/65; Frank McCoy 1912 Listing; 1860 Bedford Census
24	Bedford Cem.	B.S.F. History Book-1884; 1870 Bedford Census; William & Levi-brothers
25		B.S.F. History Book-1884; Huntingdon Co. references
26	West Union Cem.-OH	B.S.F. History Book-1884; PA Civil War Archives-Bedford Co. residence
27	West Union Cem.-OH	B.S.F. History Book-1884; 1860 Union (Pavia) Twp. Census; Alexander-Son
28		Historical Data Systems record; Regiment listed on gravestone
29	Lutheran Cem.-Osterburg	Wounded 5/6/64; PA Civil War Archives-St. Clairsville residence
30	Bedford Cem.	B.S.F. History Book-1884
31		B.S.F. History Book-1884
32	Lutheran Cem.-Osterburg	Wounded 10/15/62; Died 10/16/62; 1860 Union (Pavia) Twp. Census
33	Lutheran Cem.-Osterburg	KIA 5/6/64; B.S.F. History Book-1884; John, Thomas & William-brothers
34		1850 Union Twp. (Pavia) Census
35	Arlington Nat'l ; F 5484-15	1890 Union Twp. (Pavia) Veterans Census
36	Loysburg Cem.	Historical Data Systems record aka Ackers
37	Loysburg Cem.	Bedford Inquirer 5/22/1908 listing
38	Andersonville Cem., 1805	POW-Libby, Danville& Andersonville 9/19/63; Died 6/10/64; Born Akersville-Fulton Co.
39	Shreves Cem.-Robinsonville	1860 Monroe Twp. Cen; B.S.F. History Book-1884; John H & Job-brothers
40	Akersville Cem.-Crystal Springs	1850 Monroe Twp. Census; 1890 Brush Creek Veterans Census
41	Hollidaysburg Presb. Cem.	1860 Colerain Twp. Census; PA Civil War Archives-Bedford Co. residence
42	Fockler Cem.-Saxton	Deserted at Hospital 4/22/64; Liberty Twp. 1890 Veterans Census
43	Hyndman Cem.	1890 Hyndman Veterans Census; B.S.F. History Book-1884
44		1890 Hyndman Veterans Census; Possibly enlisted in 33rd Virginia Infantry
45		Enlisted in Confederate & Union Armies; aka Raley or Railey
46		1890 Liberty Twp. Veterans Census-Right leg broken
47		B.S.F. History Book-1884; Substitute
48	Four Mile House Cem.-OH	1890 Bedford Twp. Veterans Census
49		Deserted 2/15/62; PA Civil War Archives-enlisted in Bedford; Historical Data Systems record
50	Mt. Zion Cem.-Breezewood	1890 Colerain Twp. Veterans Census; 1850 Cumberland Val. Twp. Census
51		B.S.F. History Book-1884; PA Civil War Archives-Bedford Co. residence
52		Historical Data Systems record
53		B.S.F. History Book-1884; PA Civil War Archives-Born in Bedford Co.
54		Wounded 5/20/64; PA Civil War Archives Bedford residence; Historical Data Systems record
55	Mt. Union Cem.-Lovely	1860 Union Twp. (Pavia) Census; Historical Data Systems record
56	Andersonville Cem., 9896	POW-Andersonville; Died 9/27/64; B.S.F. History Book-1884; 1860 Union Twp. Census
57	Andersonville Cem., 2398	POW-Andersonville; Died 6/24/64; PA Civil War Archives-Bedford residence
58	Beaufort Nat'l Cem.	Died 12/14/63; 1850 St. Clair Twp. Census; Edward & Jacob-bros.
59	Friends Cem.-Spring Meadow	Died 1/23/66; in Hospital 6/1/64 - 8/30/65; 1860 Union Twp. Census; Hist. Data Systems Rec.
60		1860 Napier Twp. Cen; Historical Data Systems record; John H & Robert-brothers
61	Mt. Union Cem.-Lovely	PA Civil War Archives-Bedford Co. residence; Historical Data Systems record
62		Wounded May-64-Left Arm Amputated; Historical Data Systems record
63	Culpeper C.H. Nat'l Cem.	Died 1/26/64; 1860 St. Clair Twp. Census; Historical Data Systems record
64	Albrights Cem.-Roaring Springs	Leg wounds 6/3/64-Pension record; Joseph-Father; PA Civil War Archives-Bedford Co. resid.

Chapter 18

	Name	Age	Rank	Regiment	Co.	Muster In	Muster Out	Casualty	Casualty Battle
1	Allison, Noah	19	Pvt.	PA 138th Inf.	D	8/29/62		KIA	Cold Harbor
2	Allison, Robert	28	Pvt.	PA 61st Inf.	I	10/26/64	7/6/65		
3	Alloway, Eli	32	Corp.	PA McKeage's Mil.	H	7/2/63	8/8/63		
4	Alloway, Jonathan	21	Fifer	IA 28th Inf.	E	9/3/62	7/31/65	Wounded	Berwick, LA
5	Alloway, Simon P	28	Pvt.	IA 28th Inf.	E	9/3/62	7/18/63	Died	
6	Alloway, William		Pvt.	PA McKeage's Mil.	H	7/2/63	8/8/63		
7	Almaker, John*	35	Pvt.	PA 110th Inf.	C	7/24/64	6/28/65		
8	Altman, George W*	21	Pvt.	PA 55th Inf.	I	8/28/61		Wound.-Died	Cold Harbor
9	Amick, George W	23	Pvt.	PA 8th Res.	F	6/11/61	5/26/64	POW	Seven Days Battle
10	Amick, George W		Pvt.	PA 99th Inf.	K	3/15/65	7/1/65		
11	Amick, Josiah	16	Pvt.	PA 101st Inf.	D	11/1/61		Died	
12	Amick, Thomas	29	Pvt.	PA 79th Inf.	E	2/23/65	6/14/65		
13	Amick, William	35	Pvt.	PA 133rd Inf.	K	8/15/62	5/26/63		
14	Amick, William		Pvt.	PA 208th Inf.	K	9/7/64	6/1/65		
15	Amick, William B	26	Sgt.	PA 138th Inf.	E	8/29/62		KIA	Wilderness
16	Amick, William M	19	Sgt.	PA 55th Inf.	H	10/11/61		Wound.-Died	Petersburg - Initial
17	Amick, William M		Sgt.	PA 55th Inf.	H			Wounded	Cold Harbor
18	Amos, Francis M	18	Pvt.	PA McKeage's Mil.	G	7/2/63	8/8/63		
19	Amos, Francis M		Pvt.	PA 194th Inf.	I	7/22/64	11/5/64		
20	Amos, John B	28	1st Sgt.	PA 55th Inf.	D	10/12/61	8/30/65	Wounded	Drewry's Bluff
21	Amos, John B		2nd Lt.	PA 55th Inf.	D			Wounded	Cold Harbor
22	Anderson, Daniel R			PA 50th Inf.					
23	Anderson, Henry	19	Pvt.	PA McKeage's Mil.	H	7/2/63	8/8/63		
24	Anderson, Henry		Pvt.	PA 55th Inf.	H	2/19/64	8/30/65	POW	Drewry's Bluff
25	Anderson, James	21	Pvt.	PA 101st Inf.	G	12/28/61	2/5/63	POW-Died	Plymonth, NC
26	Anderson, John (1)	24	Pvt.	IL 55th Inf.	K	22571		Died	
27	Anderson, John (2)	25	Pvt.	PA McKeage's Mil.	G	7/2/63	8/8/63		
28	Anderson, John (2)		Pvt.	PA 63rd Inf.	H	7/11/63		KIA	North Anna River
29	Anderson, Samuel H			PA 19th Cav.					
30	Anderson, William F	27	Pvt.	IL 102nd Inf.	A	9/2/62	3/16/63		
31	Anderson, William W	22	Capt.	PA 2nd Cav.	F	9/14/61	2/18/64	POW	
32	Anderson, William W		Major	PA 20th Cav.	F&S	3/2/64		KIA	
33	Andrews, Albert		Pvt.	PA 2nd Cav.					
34	Andrews, Charles (1)*	22	Pvt.	PA 110th Inf.	C	10/24/61		MIA	Port Republic
35	Andrews, Charles (2)*	22	Sgt.	PA 110th Inf.	C	10/24/61	10/24/64		
36	Andrews, Hiram			PA Inf.				Died	
37	Andrews, Jacob K		Chap.	OH 126th Inf.					
38	Andrews, John		Pvt.	PA 55th Inf.	H	1863	12/30/64		
39	Andrews, Joseph	34	Pvt.	PA 82nd Inf.	A	11/14/64	6/14/65	Wounded	
40	Andrews, William A*		Pvt.	PA 110th Inf.	C	10/24/61		KIA	Wilderness
41	Ansel, Alexander	36	Pvt.	PA 147th Inf.	H	10/6/64	7/15/65		
42	Anthony, Cyrus*	30	Pvt.	PA 55th Inf.	H	9/2/62		Wound.-Died	Bermuda Hundred
43	Antry, Elias	19	Pvt.	PA 20th Cav.	L	7/4/63	8/1/63		
44	Arenze, Charles		Pvt.	PA 11th Inf.	A	11/1/61		Died	
45	Armstrong, Albert	19	Pvt.	PA 138th Inf.	F	5/13/63	6/23/65	Wounded	Wilderness
46	Armstrong, David B	25	1st Sgt.	PA 8th Res.	F	6/19/61	5/26/64	POW	Seven Days Battle
47	Armstrong, David B		2nd Lt.	PA 194th Inf.	I	7/22/64	11/5/64		
48	Armstrong, Henry		Pvt.	OH 49th Inf.	I	8/26/61	8/16/62		
49	Armstrong, Henry		Pvt.	OH 180th Inf.	K	10/6/64	7/12/65		
50	Armstrong, Joseph	34	1st Lt.	PA 11th Inf.	A	9/30/61	7/1/65	Died	
51	Armstrong, Joseph M	21	Pvt.	PA 133rd Inf.	C	8/13/62	4/8/63	Wounded	Fredericksburg
52	Armstrong, Joseph M		1st Sgt.	PA McKeage's Mil.	G	7/2/63	8/8/63		
53	Armstrong, Joseph M		Pvt.	PA 194th Inf.	I	7/22/64	11/5/64		
54	Armstrong, Thomas	35	Pvt.	PA 13th Inf.	G	4/25/61	8/6/61		
55	Armstrong, William H		Pvt.	PA 22nd Cav.	H	2/26/64	6/24/65		
56	Arnold, Abraham Kerns	24	Major	US 5th Cav.		8/3/61			
57	Arnold, Albert		Corp.	PA McKeage's Mil.	H	7/2/63	8/8/63		
58	Arnold, Albin C	21	Pvt.	PA 55th Inf.	K	2/29/64	8/30/65		
59	Arnold, Almon	36	Pvt.	PA 55th Inf.	I	7/20/63	8/30/65		
60	Arnold, Henry H*		Pvt.	PA 55th Inf.	D	9/26/62	8/30/65		
61	Arnold, Humphrey Y	15	Music.	PA 55th Inf.	D	12/30/61	12/30/64		
62	Arnold, Theodore J*		Pvt.	PA 55th Inf.	D	9/2/62	6/11/65		
63	Arnold, William*		Corp.	PA 55th Inf.	D	9/22/62	6/11/65		
64	Ashcom, Edward S	19	Corp.	PA 46th Mil.	I	7/6/63	8/18/63		

	Cemetery	Notes
1	Cold Harbor Nat'l Cem.	KIA 6/5/64; Historical Data Systems-Bedford Co. residence; Noah & John-brothers
2	Fishertown Cem.	1890 W. St. Clair Twp. Veterans Census
3	Broad Top IOOF Cem.	B.S.F. History Book-1884
4	Blair Cem.-NE	Wounded 3/7/64; Jonathan, Eli & Simon-brothers
5	Ridgewood Cem.-IA	Died 8/4/63; Born Broad Top City
6		B.S.F. History Book-1884; enlisted in Bedford
7	Pocono Lake Cem.	B.S.F. History Book-1884
8	Arlington Nat'l Cem., 13-6404	Wounded 6/3/64; Died 6/23/64; B.S.F. History Book-1884; Enlisted in Greenville
9	Clearville Cem.	POW 6/27/62-Castle Thunder & Libby; B.S.F. History Book-1884; 1860 Monroe Twp. Cen.
10		Substitiute; 1890 Monroe Twp. Veterans Census
11	Hampton Nat'l Cem.	Died 11/15/62; aka Eamick; 1860 Monroe Twp. Census; Josiah & George-brothers
12	St. James Luth. Cem.-Bedford	1890 Bedford Twp. Veterans Census; Deserted after the war ended - 6/14/65
13	Rock Hill Cem.-Monroe Twp	B.S.F. History Book-1884
14		Monroe Twp. 1890 Veterans Census; Historical Data Systems record
15		KIA 5/6/64; PA Civil War Archives-Bedford residence; 1860 St. Clair Twp. Census
16	Hampton Nat'l Cem.	Leg Amputated 6/18/64; Died 8/12/64; 1850 Bedford Twp. Census; Hist. Data Systems Rec.
17		Wounded 6/4/63 listed in Bedford Inquirer 7/8/64 article; Historical Data Systems record
18	Bedford Cem.	B.S.F. History Book-1884; 1860 Bedford Twp. Census
19		B.S.F. History Book-1884; Francis & John-brothers
20	Bedford Cem.	Wounded 5/16/64; 1890 Bedford Twp. Veterans Census; B.S.F. History Book-1884
21		Wounded 6/3/64; enlisted as Pvt. & promoted 5 times; B.S.F. History Book-1884
22	Bortz Luth. Cem.-Centerville	Frank McCoy 1912 Listing; 1860 Cumberland Valley Census; B.S.F. History Book-1884
23	Grandview Cem.-Cambria Co.	B.S.F. History Book-1884; Henry, James, John (1) & William F-brothers
24		POW-Libby 5/16/64; Released after 3 months; PA Civil War Archives-Bedford residence
25		POW-Andersonville; 4/20/64; Died 12/1/64 ; 1860 St. Clair Twp. Census
26	Red-Blue Cem.-IL	Died 9/6/62; Ancestry-1929 Family Tree record document
27	Bedford Cem.	B.S.F. History Book-1884; 1860 Bedford residence
28		KIA 5/24/64; PA Civil War Archives-Born in Bedford Co.
29	Duval Cem.-Coaldale	Frank McCoy 1912 Listing; Born 4/25/31
30	Hope Cem.-IL	Ancestry.com-born in Bedford; Discharged on Surgeon's Cert.
31	Bedford Cem.	Historical Data Systems record
32		Thrown from horse 1/17/65 at Harpers Ferry; Historical Data Systems-Bedford Co. residence
33		B.S.F. History Book-1884
34		MIA 6/9/62; B.S.F. History Book-1884; Tyrone references
35		B.S.F. History Book-1884; Tyrone references; 2-Charles Andrews listed in Civil War records
36	Bethal Frame Ch.-Monroe Twp	Died 1/5/65; Bedford Inquirer 5/22/1908 listing; 1860 Monroe Twp. Census
37		1890 Bedford Twp. Veterans Census
38		Drafted; 1850 St. Clair Twp.; B.S.F. History Book-1884; John and Joseph-brothers
39		1890 W. St. Clair Veterans Census
40		KIA 5/7/64; B.S.F. History Book-1884; PA Civil War Archives-Huntingdon residence
41	Jersey Baptist Cem.-Ursina	B.S.F. History Book-1884
42	Hampton Nat'l Cem.	Wounded 5/9/64; Died 7/13/64; Bedford Gazette 2/13/1914 list; Enlisted in Berks County
43		PA Civil War Archives-Bedford Co. residence; Civil Authorities Discharge
44	Alexandria Nat'l Cem, 547	Died 11/29/62; Bedford Gazette 2/13/1914 listing; Historical Data Systems record
45	Everett Cem.	Wounded 5/6/64; B.S.F. History Book-1884; Historical Data Systems record
46	Bedford Cem.	POW 6/27/62-Libby & Belle Island; B.S.F. History Book-1884; David & Joseph-brothers
47		B.S.F. History Book-1884
48	Methodist Cem.-Tatesville	Frank McCoy 1912 Listing; Historical Data Systems record; Discharged on Surgeon's Cert.
49		1890 Hopewell Twp. Veterans Census
50	Everett Cem.	Died 7/12/65; 1860 Hopewell Twp. Census; B.S.F. History Book-1884
51		Wounded 12/13/62; Historical Data Systems-Bedford Co. residence; aka J.M.
52		B.S.F. History Book-1884
53		B.S.F. History Book-1884; 1850 Hopewell Twp. Census
54		B.S.F. History Book-1884; Historical Data Systems-Bedford residence
55		B.S.F. History Book-1884; Historical Data Systems record
56	St. Philip's Church-Garrison, NY	Medal of Honor-Davenport Bridge, VA; born Bedford Co.; Historical Data Systems record
57		B.S.F. History Book-1884; Enlisted in Bedford
58	Burning Bush-Bedford Twp.	1860 Cumberland Valley Twp. Census; 1890 Bedford Twp. Veterans Census
59		B.S.F. History Book-1884; Historical Data Systems record; Sick at Muster Out
60		B.S.F. History Book-1884; Berks Co. references
61	Bedford Cem.	Historical Data Systems record; Humphrey & Abraham K-brothers
62	Lithopolis Cem.-OH	B.S.F. History Book-1884; Berks Co. references
63		B.S.F. History Book-1884; Berks Co. references
64	Everett Cem.	Broad Top 1890 Veterans Census, aka Ashcomb

Chapter 18

	Name	Age	Rank	Regiment	Co.	Muster In	Muster Out	Casualty	Casualty Battle
1	Ashcom, Edward S		Sgt.	PA 194th Inf.	I	7/12/64	11/5/64		
2	Ashcom, George Jr	23	2nd Lt.	PA 133rd Inf.	C	8/13/62	5/26/63	Wounded	Fredericksburg
3	Ashcom, John P	41	Surg.	PA 116th Inf.		7/28/62	3/19/63		
4	Ashcom, John P		Surg.	PA 30th Inf.		6/25/63	7/27/63		
5	Ashcom, John P		Surg.	PA 194th Inf.	I	7/22/64	11/5/64		
6	Ashcom, John P		Surg.	PA 54th Inf.		11/18/64	5/30/65		
7	Atwell, John*		Pvt.	PA 110th Inf.	C	2/19/64	6/27/64	Wounded	1st Deep Bottom
8	Aulenbach, James*		Pvt.	PA 55th Inf.	D	9/22/62		Died	
9	Aultz, John K		Pvt.	PA McKeage's Mil.	G	7/2/63	8/8/63		
10	Auman, Jacob*	19	Pvt.	PA 194th Inf.	I	7/22/64	11/5/64		
11	Avey, Joseph	43	Pvt.	PA 208th Inf.	K	9/7/64	6/1/65		
12	Avey, William H	28	Pvt.	PA 54th Inf.	H	9/30/61	6/26/65	Wounded	Petersburg - Final
13	Ayers, Charles	40	Corp.	PA 55th Inf.	I	8/28/61			
14	Ayres, John	40	Pvt.	PA 84th Inf.	C	9/16/62		Died	
15	Babcock, Francis B*		2nd Lt.	PA 55th Inf.	D	4/30/62	7/31/62		
16	Badgley, George	36	Pvt.	PA 12th Res.	K	6/15/61	2/5/62		
17	Bagley, Henry H		Pvt.	PA 149th Inf.					
18	Bagley, Moses G	25	Pvt.	PA 138th Inf.	E	8/29/62	1/15/65	Wounded	Cold Harbor
19	Bagley, Samuel	30	Pvt.	PA 133rd Inf.	K	8/15/62			
20	Bagley, William		Pvt.	PA 79th Inf.	D	9/2/64	8/18/65		
21	Bailey, James	21	Pvt.	PA 138th Inf.	E	8/29/62			
22	Bailey, John W (1)*	44	Pvt.	PA 184th Inf.	A	5/12/64		KIA	2nd Deep Bottom
23	Bailey, John W (2)	17	Pvt.	PA 138th Inf.	E	8/29/62	6/23/65		
24	Bailey, John*	26	Pvt.	PA 55th Inf.	I	2/15/64		Wound.-Died	Petersburg
25	Bailey, William H	19	Pvt.	PA 138th Inf.	E	8/29/62	6/23/65	POW	Monocacy, MD
26	Bair, Henry A	27	Pvt.	PA 6th Res.	A	4/18/61	5/31/64		
27	Bair, Henry A			PA 191st Inf.		5/31/64			
28	Baird, George L	19	Pvt.	PA 208th Inf.	H	9/5/64	6/1/65		
29	Baird, Joseph*			PA 8th Res.					
30	Baith, Peter	23	Pvt.	PA 22nd Cav.	H	2/26/64	6/24/65		
31	Baitzel, Jacob	18	Pvt.	PA McKeage's Mil.	G	7/2/63	8/8/63		
32	Baker, Abraham	40	Pvt.	PA 76th Inf.	E	10/7/64	7/18/65		
33	Baker, Alfred	20	Pvt.	PA 205th Inf.	D	9/1/64	6/2/65		
34	Baker, Andrew C		Pvt.	PA 46th Inf.	K	7/8/63	8/18/63		
35	Baker, Andrew C	20	Pvt.	PA 13th Cav.	B	2/24/64	7/14/65	Injured	Fredericksburg
36	Baker, David N	18	Pvt.	PA 110th Inf.	D	12/19/61	6/1/62	Wounded	1st Kernstown
37	Baker, David N	20	Pvt.	PA 13th Cav.	C	2/27/64		Died	
38	Baker, Franklin S	18	Pvt.	PA 125th Inf.	E	8/13/62		KIA	Antietam
39	Baker, George W	20	Pvt.	PA 2nd Inf.	B	4/20/61	7/24/61		
40	Baker, George W		Pvt.	PA 133rd Inf.	K	8/15/62			
41	Baker, Henry N		1st Lt.	PA McKeage's Mil.	G	7/2/63	8/8/63		
42	Baker, Jacob S	20	Pvt.	PA 194th Inf.	I	7/22/64	11/5/64		
43	Baker, Jacob S		Pvt.	PA 99th Inf.	D	2/25/65	7/1/65		
44	Baker, John C	23	Sgt.	PA 55th Inf.	I	9/20/61	8/30/65	Wounded	
45	Baker, John H*	32	Pvt.	PA 101st Inf.	C	1/2/64		POW-Died	Plymonth, NC
46	Baker, John T	21	Pvt.	PA 104th Inf.	F	3/9/65	8/25/65		
47	Baker, Samuel N	17	Pvt.	PA McKeage's Mil.	G	7/2/63	8/8/63		
48	Baker, Samuel N		Pvt.	PA 186th Inf.	C	2/10/64	8/15/65		
49	Baker, William	30	Pvt.	PA 55th Inf.	I	10/15/61	10/20/64		
50	Baker, William		Pvt.	PA 99th Inf.	B	2/25/65	7/1/65		
51	Bakerlee, Charles M		Pvt.	PA 7th Cav.	H	2/25/64	11/10/64		
52	Ball, Andrew M	22	Pvt.	PA 87th Inf.	B	6/4/64	5/26/65	Wounded	Opequon
53	Ball, Daniel M	24	Pvt.	PA 138th Inf.	F	9/29/62	5/3/65	Wounded	Monocacy, MD
54	Ball, Daniel M		Pvt.	PA 138th Inf.	F			Wounded	Cedar Creek
55	Ball, John	22	Pvt.	PA 171st Inf.	I	10/11/62	8/8/63	Wounded	
56	Ball, John		Pvt.	PA 82nd Inf.	C	10/14/64	6/14/65	Wounded	
57	Baner, Franklin	17	Pvt.	PA 138th Inf.	F	8/29/62	6/23/65	Wounded	Wilderness
58	Baner, Franklin		Pvt.	PA 138th Inf.	F			Wounded	Cold Harbor
59	Banks, John	18	Pvt.	PA 110th Inf.	C	12/29/63	12/12/64		
60	Banner, Adam	26	Pvt.	PA 103rd Inf.	E	11/20/61	6/25/65	POW	Plymonth, NC
61	Bannon, Joseph J	29	Pvt.	PA 101st Inf.	G	2/18/62	7/23/62		
62	Barber, Alexander R		Pvt.	PA 28th Inf.	O	8/17/61			
63	Barber, James	27	Pvt.	PA 8th Res.	F	7/1/61	5/26/64	POW	Seven Days Battle
64	Barber, John*	26	Corp.	PA 184th Inf.	A	5/12/64	7/14/65		

	Cemetery	Notes
1		Broad Top 1890 Veterans Census
2		Wounded 12/13/62; enlisted in Hopewell; PA Civil War Archives-Bedford Co. residence
3	Everett Cem.	1860 Broad Top Twp. Census; Historical Data Systems record
4		PA Civil War Archives-Hopewell resident; Historical Data Systems record
5		Bedford Inquirer 5/22/1908 listing; Frank McCoy 1912 List; B.S.F. History Book-1884
6		Historical Data Systems record; Promoted to Major
7		B.S.F. History Book-1884; Historical Data Systems record; Tyrone references
8	Point Lookout Cem.-MD	Died 7/20/64; Berks Co. references; Historical Data Systems record
9		B.S.F. History Book-1884
10		B.S.F. History Book-1884; Marklesburg references
11	Everett Cem.	Frank McCoy 1912 Listing; PA Civil War Archives-Bedford Co. residence
12	Everett Cem.	Wounded 4/2/65; lost left forearm; Historical Data Systems record
13		B.S.F. History Book-1884; Deserted 8/30/64 after re-enlisting
14	Sproul Union Cem.-Claysburg	Altoona Tribune-5/3/89 list of war dead; PA Civil War Archives-Claysburg residence
15		B.S.F. History Book-1884; Susquehanna references
16		1860 W. Providence Twp. Census; Historical Data Systems-Bedford Co. resid.; DSC
17	Bedford Cem.	Bedford Cemetery list of Civil War veterans; National Park Service record lists 149th
18	Pleasant Hill Cem.-Shade Gap	Wounded 6/1/64; PA Civil War Archives-Schellsburg residence
19	St. James Luth. Cem.-Bedford	Deserted 9/14/62; 1890 Bedford Twp. Veterans Census
20	St. Thomas Cem.-Bedford	Frank McCoy 1912 Listing; Bedford Inquirer 5/22/1908 listing
21		Deserted 3/7/63, PA Civil War Archives-Schellsburg residence
22	Cypress Hill Cem.-NY	KIA 8/28/64; B.S.F. History Book-1884; Company A recruited in Dauphin & Bedford Co.
23	Helixville Cem.-Napier Twp	Historical Data Systems-Schellsburg residence
24	St. Mary's Cem.-Holidaysburg	Bedford Gazette 2/13/1914 listing; Blair Co. references
25	Schellsburg Cem.	POW 7/9/64; Historical Data Systems-Schellsburg residence
26	Bunker Hill Cem.-Saxton	Historical Data Systemstems record; aka Bear
27		Ancestry.com record
28	Dry Hill Cem.-Woodbury	Historical Data Systems-Bedford Co. residence; Frank McCoy 1912 Listing
29	Claysburg Cem.	Soldiers of Blair Co. Book - 1940
30	New Grenada Cem.-Fulton Co.	1890 Robertsdale Veterans Census
31	Burket Cem.-Hopewell	B.S.F. History Book-1884
32		B.S.F. History Book-1884; Historical Data Systems record
33	Wells Valley Meth. Cem.	1890 Wells Twp. Veterans Census
34		PA Civil War Archives record
35	Green Lawn Cem.-Roaring Spr.	Born Bedford Co.; Broken pelvis-thrown from horse
36	Hickory Bottom Cem.-Woodbury	Wounded 3/23/62; Bedford Inquirer 5/22/1908 listing
37	Hickory Bottom Cem.-Woodbury	Died-Typhoid Fever 8/31/64; Historical Data Systems record; Ancestry.com-Born Woodbury
38	Holsinger Cem.-Bakers Summit	Wound 9/17/62; Died 9/18/62; Bedford Gazette 2/13/1914 listing; Hist. Data Systems record
39		1850 Bedford Census; Historical Data Systems record
40		Historical Data Systems-Deserted 9/21/62-Sharpsburg, MD
41		B.S.F. History Book-1884; enlisted in Hopewell
42	Brumbaugh Cem.-Blair Co.	B.S.F. History Book-1884; Ancestry.com-Born Bedford Co.
43		Historical Data Systems record
44	Albrights Cem.-Roaring Springs	1890 S. Woodbury Twp. Veterans Census; Historical Data Systems record
45	Florence Nat'l Cem.	POW-Andersonville 4/20/64; Died 2/15/65; Bed. Gaz. 2/13/1914 list; Lawrence Co. references
46		Historical Data Systems-Bedford Co. residence
47	Angelus-Rosedale Cem.-CA	B.S.F. History Book-1884; enlisted in Hopewell
48		Findagrave.com record; Historical Data Systems record
49	Holsinger Cem.-Bakers Summit	B.S.F. History Book-1884; Frank McCoy 1912 Listing
50		1890 Bedford Twp. Veterans Census
51		1890 E. St. Clair Twp. Veterans Census
52	Lybarger Cem.-Madley	1890 E. St. Clair Twp. Veterans Census; Wounded 9/19/64-hosp until Feb '65
53	Berlin IOOF Cem.	Wounded 7/9/64; B.S.F. History Book-1884; 1860 Londonderry Twp. Census
54		Wounded 10/19/64; Historical Data Systems record; Daniel, John & Andrew-brothers
55	Ickes Farm Cem.-St. Clair Twp.	1890 E. St. Clair Twp. Veterans Census-Leg Wound
56		Frank McCoy 1912 Listing; 1890 E. St. Clair Twp. Veterans Census
57	Helixville Cem.-Napier Twp.	Wounded 5/6/64; PA Civil War Archives-Bedford Co. residence
58		1890 Juniata Twp. Veterans Census aka Beaner; Wounded 6/1/64
59		B.S.F. History Book-1884; Discharged on Surgeon's Cert.
60	Bedford Cem.	POW-Andersonville & Florence 5/2/64 to 2/26/65; attended 1905 Andersonville Dedication
61		Historical Data Systems-Bedford residence; Discharged on Surgeon's Cert.
62		Deserted 9/16/62; 1890 Broad Top Veterans Census
63		POW 6/27/62; Historical Data Sys.-Bedford Co. residence
64	Christian Union Cem.-IN	B.S.F. History Book-1884; 1860 Licking Creek-Fulton Co. Census

Chapter 18

	Name	Age	Rank	Regiment	Co.	Muster In	Muster Out	Casualty	Casualty Battle
1	Barber, William A	18	1st Lt.	PA 11th Cav.	A	8/18/61		POW-Died	Darbytown Road
2	Barclay, John J	28	1st Lt.	PA 11th Cav.	A	9/21/61	9/28/64	Wound.-POW	Reams Station
3	Barclay, Joseph	28	Sgt.	PA 138th Inf.	F	8/29/62	6/23/65	Wounded	Cedar Creek
4	Barclay, William W	36	Pvt.	CA 1st Cav.	A	2/21/65	5/22/66		
5	Barefoot, George W	30	Pvt.	PA 99th Inf.	H	2/25/65	5/31/65		
6	Barkheimer, John	20	Pvt.	PA 55th Inf.	K	3/8/64	6/20/65	Wounded	
7	Barkley, David T	18	Pvt.	PA 79th Inf.	D	9/1/61	9/24/64		
8	Barkley, George W	21	Corp.	PA 138th Inf.	E	8/29/62	6/13/65		
9	Barkley, Jacob T	27	Pvt.	PA 91st Inf.	B	9/21/64	5/30/65	Wounded	
10	Barkley, Josiah T	34	Pvt.	PA 208th Inf.	H	9/7/64	6/1/65		
11	Barkley, Samuel	23	Corp.	PA 138th Inf.	E	8/29/62	6/23/65		
12	Barkman, Christopher M	19	Pvt.	PA 22nd Cav.	C	7/15/63	6/15/65		
13	Barkman, Daniel	20	Pvt.	PA 101st Inf.	D	11/1/61	11/28/63		
14	Barkman, Daniel		Pvt.	NY 8th Light Art.		11/28/63	6/30/65		
15	Barkman, David	23	Pvt.	PA 138th Inf.	D	8/29/62			
16	Barkman, Hezekiah	33	Corp.	PA 138th Inf.	D	8/29/62	1/12/65	Wounded	Cold Harbor
17	Barkman, Joseph M	28	Corp.	PA 3rd Heavy Art.	D	11/8/62	4/18/64		
18	Barkman, Joseph M		1st Lt.	PA 188th Inf.	G	4/18/64	7/5/65		
19	Barkman, Philip	40	Pvt.	PA 99th Inf.	G	2/24/65	5/31/65		
20	Barkman, Thomas C	19	Pvt.	PA 133rd Inf.	K	8/15/62	5/26/63	Wounded	Fredericksburg
21	Barks, Alfred	30	Sgt.	USCT 41st Inf.	H	9/21/64			
22	Barks, John R	26	Corp.	USCT 32nd Inf.	B	2/12/64	8/22/65		
23	Barks, Moore	20	Pvt.	USCT 32nd Inf.	B	2/7/64	8/22/65		
24	Barks, William T	23	Pvt.	USCT 54th Inf.	D	3/21/63	8/8/65		
25	Barley, Reuben H	21	Music.	IN 34th Inf.		10/12/61	8/12/62		
26	Barmond, John	25	Pvt.	PA 8th Res.	F	6/11/61	4/24/62		
27	Barmond, Nathaniel		Pvt.	PA 8th Res.	F	6/11/61	5/6/64		
28	Barmond, Nathaniel		Pvt.	PA 191st Inf.	H	23512		POW-Died	Globe Tavern
29	Barndollar, Jacob W	21	Pvt.	PA 133rd Inf.	C	8/13/62	5/26/63	Wounded	Fredericksburg
30	Barndollar, Jacob W		Pvt.	PA 186th Inf.	G	8/5/64	8/15/65		
31	Barndollar, James E	18	Pvt.	PA 133rd Inf.	C	8/29/62		KIA	Fredericksburg
32	Barndollar, James J	23	1st Sgt.	PA 133rd Inf.	C	8/29/62	5/26/63		
33	Barndollar, John W	23	Corp.	PA 13th Inf.	G	4/25/61	8/6/61		
34	Barndollar, John W		Pvt.	PA 5th Heavy Art.	C	8/29/64	6/30/65		
35	Barndollar, Martin D	18	Pvt.	PA 133rd Inf.	C	8/8/62	8/26/63		
36	Barndollar, Martin D		Corp.	PA 194th Inf.	I	7/12/64	11/5/64		
37	Barndollar, Martin D		Sgt.	PA 83rd Inf.	K	2/28/65	6/28/65		
38	Barndollar, William G	18	Pvt.	PA 194th Inf.	I	7/21/64	11/5/64		
39	Barndollar, William P	21	2nd Lt.	PA 13th Inf.	G	4/25/61	8/6/61		
40	Barndollar, William P		1st Lt.	PA 76th Inf.	E	10/9/61	3/10/64	POW	
41	Barnes, David		Pvt.	MD 1st PHB Inf.	K	4/23/64	8/5/64	Died	
42	Barnes, Dayton C	29	Pvt.	PA 35th Inf.	C	6/29/63	8/7/63		
43	Barnes, William		Pvt.	PA 22nd Cav.	C	7/7/63			
44	Barnett, David	31	Pvt.	PA 184th Inf.	A	5/12/64	7/14/65	Wounded	Petersburg
45	Barnett, David		Pvt.	PA 184th Inf.	A			Wounded	Cold Harbor
46	Barnett, Ephraim	29	Pvt.	PA 149th Inf.	I	8/14/63		POW-Died	North Anna River
47	Barnett, Henry C	16	Pvt.	PA 9th Res.	G	5/4/61	12/1/62		
48	Barnett, Henry C		Pvt.	PA 46th Inf.	G	7/8/63	8/18/63		
49	Barnett, Joseph E	39	Pvt.	PA 126th Inf.	B	8/12/62	6/20/63		
50	Barnett, Joseph E		Pvt.	PA 22nd Cav.	K	2/25/64	6/24/65	Wounded	
51	Barnett, Samuel (1)	19	Pvt.	PA 184th Inf.	A	5/12/64		POW-Died	Jerusalem Plank Rd.
52	Barnett, Samuel (2)	19	Pvt.	PA 61st Inf.	A	2/25/64	6/28/65		
53	Barney, Dennis A	18	Pvt.	MD 3rd PHB Inf.	K	3/28/64	5/29/65		
54	Barney, Isaac	31	Pvt.	MD 3rd PHB Inf.	C	9/28/61	5/29/65		
55	Barney, Jacob	29	Pvt.	MD 3rd PHB Inf.	H	2/22/64	5/29/65		
56	Barney, John H	22	Pvt.	MD 3rd PHB Inf.	C	11/19/61	5/29/65		
57	Barney, Joseph								
58	Barnhart, Abraham	28	Pvt.	IL 15th Inf.	K	3/9/65	6/30/65		
59	Barnhart, C A*	21	Pvt.	PA 55th Inf.	I	10/6/63	12/30/64	Wounded	Cold Harbor
60	Barnhart, John H	27	2nd Lt.	PA 55th Inf.	D	10/12/61		KIA	Drewry's Bluff
61	Barnitz, Jacob B		1st Lt.	PA 135th Inf.	E	8/12/62	5/24/63		
62	Barnitz, Jacob B		1st Lt.	PA 195th Inf.	E	7/18/64	11/4/64		
63	Barnitz, Jacob B		Capt.	PA 195th Inf.	E	3/16/65	1/31/66		
64	Barns, Joseph	21	Pvt.	PA 171st Inf.	I	11/2/62			

	Cemetery	Notes
1		Wounded & POW-Richmond 10/7/64, Died as POW; PA Civil War Archives-Bedford resid.
2	Bedford Cem.	Wounded & POW 6/29/64 to 9/10/64
3	Trinity UCC Cem.-Juniata Twp.	Wounded 10/19/64; 1890 Juniata Twp. Veterans Census; Historical Data Systems record
4	Presbyterian Cem.-Bedford	B.S.F. History Book-1884; John & William-bros.
5	Hoover Cem.-Alum Bank	1890 W. St. Clair Veterans Census
6	Old Union Cem.-Osterburg	Wounded-Bedford Inquirer 5/27/64 article; Historical Data Sys.-Bedford Co. residence
7	St. John's Cem.-Loysburg	Frank McCoy 1912 Listing; Bedford Inquirer 5/22/1908 listing
8	Everett Cem.	B.S.F. History Book-1884; Obituary-7 brothers in Civil War; 2 may have died before war
9	Lutheran Cem.-Osterburg	1890 King Twp. Veterans Census; 1860 St. Clair Twp. Census
10	Dry Hill Cem.-Woodbury	B.S.F. History Book-1884; Josiah, Jacob, George, Samuel & David brothers
11	St. James Luth. Cem.-Bedford	Historical Data Systems-Bedford residence; 1850 St. Clair Twp. Census
12	Barkman Graveyard	Frank McCoy 1912 Listing; 1860 Southampton Twp. Census; aka Christian
13	Barkman Graveyard	Historical Data Systems-Bedford Co. residence; 1860 Monroe Twp. Census
14		1850 Southampton Census
15	Harlan Cem., IA	Deserted 2/9/63; David, Philip & Hezekiah brothers
16	Barkman Cem.-Monroe Twp.	Wounded 6/1/64; B.S.F. History Book-1884; Discharged on Surgeon's Cert.-Left Hand injury
17	Zion Lutheran Cem.-Clearville	Historical Data Systems record; Co. B 152 Heavy Art. on gravestone
18		1890 Monroe Twp. Veterans Census; 1860 Southampton Twp. Census
19	Barkman Cem.-Monroe Twp.	1890 Monroe Twp. Veterans Census
20		Wounded 12/13/62; B.S.F. History Book-1884; Thomas & Christopher-brothers
21		Frank McCoy 1912 list & Civil Works Admin. 1934 list; Historical Data Systems record
22	Mt. Ross Cem.-Bedford	1860 Cumberland Valley Census; 1890 Bedford Twp. Veterans Census
23	Allegheny Cem.-Pittsburgh	Frank McCoy 1912 & Civil Works Admin. 1934 list; Historical Data Systems record aka Moses
24	Allegheny Cem.-Pittsburgh	Frank McCoy 1912 Listing; William, John, Alfred & Moses-brothers
25	Lutheran Cem.-Osterburg	Frank McCoy 1912 Listing; 1890 St. Clair Twp. Veterans Census
26	Shaffer Cem.-Hyndman	B.S.F. History Book-1884; Historical Data Systems-Bedford Co. residence; Disch. Surg. Cert.
27		B.S.F. History Book-1884; Historical Data Systems-Bedford Co. residence
28		Wounded & POW 8/19/64 & paroled; Died 3/4/65
29	Everett Cem.	Shot through breast 12/13/62; Historical Data Systems record; 1860 W. Providence Twp. Cen.
30		Enlisted in Hopewell; Everett 1890 Veterans Census
31	Fredericksburg Nat'l Cem.	KIA 12/13/62; B.S.F. History Book-1884; 1860 W. Providence Twp. Census
32	Elmwood Cem.-Kansas	B.S.F. History Book-1884; James J, John & Martin-brothers
33	Everett Cem.	Frank McCoy 1912 Listing; Historical Data Systems-Bedford Co. residence
34		Historical Data Systems record
35	Everett Cem.	Historical Data Systems-Bedford Co. residence
36		B.S.F. History Book-1884
37		1890 Everett Veterans Census
38	Wildwood Cem.-Lycoming Co.	B.S.F. History Book-1884; enlisted in Bloody Run
39	Everett Cem.	1890 Everett Veterans Census
40		Frank McCoy 1912 Listing; 1860 W. Providence Twp. Census
41	Antietam Nat'l Cem., 2575	Died 8/5/64; 1860 Southampton Twp. Census; Bedford Inquirer 5/22/1908 listing
42		Historical Data Systems-Bedford Co. residence
43		Deserted 2/15/64; B.S.F. History Book-1884
44	Claysburg Union Cem.	Left Shoulder Wound; B.S.F. History Book-1884; Bedford Gazette 2/13/1914 listing
45		Wounded-Bedford Inquirer 7/1/64 article; B.S.F. History Book-1884
46	Andersonville Cem., 6609	POW-Andersonville 5/23/64; Died 8/23/64; 1860 Woodbury Cen.; Bed. Gazette 2/13/1914 list
47	Evans Cem.-Coaldale	Frank McCoy 1912 Listing; Historical Data Systems record; Discharged on Surgeon's Cert.
48		Died 12/24/66 possibly from effects of Civil War
49	Everett Cem.	Frank McCoy 1912 Listing; 1890 Everett Veterans Census
50		Everett 1890 Veterans Census listed Paralyzed
51	Andersonville Cem.	POW-Andersonville 6/22/64; Died 10/15/64; 1850 M. Woodbury Twp. Cen.
52	Sample Run Cem.-Indiana Co.	Ancestry.com-siblings born in Broad Top; Historical Data Systems record
53	Black Oak Cem.-Fulton Co.	Joseph, Jacob, John, Isaac & Dennis-Brothers
54	Buck Valley Cem.-Fulton Co.	1890 Brush Creek Twp. Veterans Census
55	Fairview Cem.-KS	Findagrave.com- Born in Bedford Co.
56	Clearville Union Cem.	Preacher after the War
57	Clearville Union Cem.	Born 1/19/31; Civil War Grave Marker
58	Bedford Cem.	Historical Data Systems record; 1850 Bedford Twp. Census
59		Wounded 6/3/64; Historical Data Systems record; Mustered-Reading, PA
60	Bedford Cem.	KIA 5/16/64; B.S.F. History Book-1884; 1860 Bedford Twp. Census
61	Dayton Nat'l Cem.-OH	Findagrave.com muster record
62		1869 Bedford Presb. Church Marrige Record-Schellsburg residence
63		PA Civil War Project record
64		Deserted 11/14/62; B.S.F. History Book-1884; enlisted Bedford Co.

	Name	Age	Rank	Regiment	Co.	Muster In	Muster Out	Casualty	Casualty Battle
1	Barns, Robert		Pvt.	USCT 32nd Inf.	G	2/25/64			
2	Barns, Samuel	37	Pvt.	PA 171st Inf.	I	11/2/62			
3	Barr, John	45	Sgt.	PA 3rd Inf.	A	4/20/61	7/29/61		
4	Barr, John*		Pvt.	PA 55th Inf.	I	10/17/63	7/19/65	POW	
5	Barr, Reuben	25	Pvt.	PA 14th Inf.	I	5/2/61	8/7/61		
6	Barr, Reuben		Corp.	PA 137th Inf.	I	8/20/62	6/1/63		
7	Barr, Reuben		Sgt.	PA 205th Inf.	C	8/27/64	6/2/65		
8	Barr, Thomas M	18	Pvt.	PA 3rd Inf.	H	6/20/61	7/30/61		
9	Barr, Thomas M		Corp.	PA 125th Inf.	G	8/13/62	5/18/63		
10	Barron, William H	24	Pvt.	PA 52th Inf.	I	9/1/64	6/1/65		
11	Bartholomew, Borchiel	31	Pvt.	PA 13th Inf.	G	4/25/61	8/6/61		
12	Bartholow, George F	22	Pvt.	PA 45th Inf.	B	7/5/64		Died	
13	Bartholow, Samuel	23	Pvt.	MD 1st PHB Cav.	D	8/13/64	6/28/65		
14	Bartholow, William E	20	Pvt.	PA 99th Inf.	I	3/15/65	7/1/65		
15	Bartlebaugh, John	27	Corp.	PA 55th Inf.	I	9/20/61		MIA	Chaffin's Farm
16	Bartlebaugh, M	23	Pvt.	PA 14th Inf.	I	5/2/61	8/7/61		
17	Bartlebaugh, Philip	23	Pvt.	PA 14th Inf.	H	4/24/61	8/7/61		
18	Bartlebaugh, Philip		Farr.	PA 12th Cav.	L	2/14/62	7/20/65		
19	Bartlebaugh, Silas M	20	Pvt.	PA 14th Inf.	I	5/2/61	8/7/61		
20	Bartlebaugh, Silas M		1st Sgt.	PA 55th Inf.	I	8/28/61	8/30/65	POW	Drewry's Bluff
21	Barton, Asa	20	Pvt.	PA 77th Inf.	F	10/9/61		Died	
22	Barton, James	35	Pvt.	PA 158th Inf.	H	11/4/62	8/12/63		
23	Barton, James A		Pvt.	PA McKeage's Mil.	D	7/3/63	8/8/63		
24	Barton, James A		Pvt.	PA 22nd Cav.	H	2/26/64	6/24/65		
25	Barton, Morgan	21	Pvt.	PA 22nd Cav.	M	2/26/64	6/24/65		
26	Bartow, Barney	29	Pvt.	PA 208th Inf.	H	9/2/64	6/1/65		
27	Bash, Daniel	37	Pvt.	PA 171st Inf.	I	11/10/62	8/8/63		
28	Basore, George	32	Pvt.	PA 76th Inf.	E	9/21/64	6/28/65		
29	Bates, Thomas		Pvt.	USCT 24th Inf.	E	2/22/65	10/1/65		
30	Baughman, Abraham A	27	Pvt.	PA 99th Inf.	A	2/25/65	7/1/65		
31	Baughman, Adam	22	Pvt.	PA 186th Inf.	E	2/24/64	8/15/65		
32	Baughman, George	22	Sgt.	PA 138th Inf.	D	8/29/62		POW-Died	Wilderness
33	Baughman, John A	20	Music.	PA 138th Inf.	E	8/29/62	6/23/65		
34	Baughman, Joseph			PA 138th Inf.					
35	Baughman, Josiah	29	1st Lt.	PA 138th Inf.	D	8/30/62		KIA	Chaneysville PA
36	Baughman, William	24	Pvt.	PA 2nd Cav.	E	11/4/61		Died	
37	Bauman, William*		Pvt.	PA 55th Inf.	K	1863	8/30/65		
38	Bayer, George M	28	Pvt.	PA 133rd Inf.	C	8/13/62	5/26/63		
39	Bayer, Joseph	21	Pvt.	PA 171st Inf.	I	11/2/62	8/8/63		
40	Bayer, Joseph		Pvt.	PA 194th Inf.	I	7/21/64	11/5/64		
41	Bayer, Joseph		Pvt.	PA 16th Cav.	G	2/28/65	8/11/65		
42	Beals, George		Pvt.	MD 2nd PHB Inf.	K	3/25/62		Died	
43	Beals, George W	24	Pvt.	PA 138th Inf.	D	8/29/62	6/23/65		
44	Beals, Nicholas H	22	Pvt.	PA 138th Inf.	D	8/29/62	5/20/65	Wounded	Cold Harbor
45	Beals, Nicholas H		Pvt.	PA 138th Inf.	D			Wounded	Wilderness
46	Beals, Solomon		Pvt.	PA McKeage's Mil.	H	7/2/63	8/8/63		
47	Beam, Daniel	59	Pvt.	PA 101st Inf.	D	11/1/61	6/25/65	POW-Died	Plymonth, NC
48	Beam, Samuel Z	24	Pvt.	OH 8th Inf.	A	5/28/61	12/17/62		
49	Beard, Amos H*	19	Pvt.	PA 194th Inf.	I	7/22/64	11/5/64		
50	Beard, Daniel	25	Sgt.	PA 138th Inf.	E	8/29/62	6/23/65	Wounded	Cedar Creek
51	Beard, George W*		Pvt.	PA 110th Inf.	C	2/25/64	6/10/65	Wounded	1st Deep Bottom
52	Beard, Job M	26	Pvt.	OH 16th Inf.	I	4/27/61	8/18/61		
53	Beard, Job M		Pvt.	OH 102 Inf.	E	9/6/62	6/30/65		
54	Beaston, James		Pvt.	PA 202nd Inf.	G				
55	Beatty, Henry B	43	Pvt.	MD 3rd PHB Inf.	B	10/23/61	10/23/64		
56	Beaver, Nicholas	38	Pvt.	PA 138th Inf.	E	8/29/62	6/23/65		
57	Beaver, Simon J	16	Pvt.	PA 55th Inf.	H	2/19/64	6/15/64	POW	Drewry's Bluff
58	Beaver, William	34	Pvt.	IA 5th Inf.	A	7/15/61		POW-Died	New Madrid, MO
59	Bechtel, Daniel S	31	Pvt.	PA 178th Inf.	G	11/1/62	11/14/62		
60	Bechtel, Daniel*		Pvt.	PA 55th Inf.	D	9/22/62		POW-Died	
61	Bechtel, David S	28	Pvt.	PA 183rd Inf.	D	3/13/62	3/18/65	Wound.-POW	2nd Deep Bottom
62	Bechtel, Isaac S	20	Pvt.	PA 184th Inf.	A	5/12/64		KIA	Jerusalem Plank Rd.
63	Bechtel, John S	27	Pvt.	PA 194th Inf.	I	7/21/64	11/5/64	Wounded	
64	Beck, Henry	21	Pvt.	PA 7th Inf.	D	4/23/61	7/29/61		

	Cemetery	Notes
1		Frank McCoy 1912 list & Civil Works Admin. 1934 list; Hist. Data Systems record aka Barnes
2		Deserted 11/14/62; B.S.F. History Book-1884; enlisted Bedford Co.
3		1890 Saxton Veterans Census
4		B.S.F. History Book-1884; Pottsville references
5		PA Civil War Archives-Bedford Co. residence
6		Historical Data Systems record; 1850 M. Woodbury Twp. Census
7		Historical Data Systems record
8	Fockler Cem.-Saxton	1890 Saxton Veterans Census
9		1890 Saxton Veterans Census
10	Lavansville Luth.-Somerset Co.	B.S.F. History Book-1884
11		B.S.F. History Book-1884; Historical Data Systems-Bedford Co. residence
12	Alexandria Nat'l Cem., B 2983	Died 5/31/65; 1960 Southampton Twp. Census; PA Civil War Project record
13	Chaneysville Meth. Cem.	1860 Southampton Twp. Census; aka Barthlow
14	Union Church Cem.-Clearville	Frank McCoy 1912 Listing; William & Samuel-brothers
15		MIA 9/29/64; PA Civil War Archives-born in Bedford Co.
16		Historical Data Systems-Bedford residence; PA Civil War Project record lists a M and a Silas
17	Holsinger Cem.-Bakers Summit	Frank McCoy 1912 Listing; 1870 Greenfield Twp. Census
18		Philip & Silas-brothers; PA Civil War Archives-Claysburg residence
19	Lutheran Cem.-Newry	PA Civil War Archives-Bedford Co. residence
20		POW 5/16/64 to 12/16/64; PA Civil War Archives-Bedford Co. residence
21	Nashville Nat'l Cem.	Died 10/10/62 in Hospital; 1860 Brush Creek Twp. Census
22	Akersville Cem.-Crystal Springs	1890 Brush Creek Twp. Veterans Census
23		B.S.F. History Book-1884; enlisted in Bloody Run
24		B.S.F. History Book-1884; Historical Data Systems record
25	Akersville Cem.-Crystal Springs	Historical Data Systems record; Asa, James & Morgan-brothers
26	Fairview Cem.-OK	B.S.F. History Book-1884 aka Barton
27	Schellsburg Cem.	1890 Napier Twp. Veterans Census; aka Bosh/Bast
28	Schellsburg Union Cem.	1850 Bedford Twp. Census; B.S.F. History Book-1884
29		Frank McCoy 1912 list & Civil Works Admin. 1934 list; Historical Data Systems record
30	Brumbaugh Cem.-Blair Co	1890 Hopewell Twp. Veterans Census
31	Everett Cem.	Historical Data Systems record; Abraham & Adam-bros.
32	Andersonville Cem., 8653	MIA-Bedford Inquirer 6/24/64; POW-Andersonville; Died 9/13/64; 1850 W. Prov. Twp. Cen.
33		Historical Data Systems record; 1860 Bedford Census; John & Joseph-brothers
34	Fishertown Bretheran Cem.	Frank McCoy 1912 Listing
35	Everett Cem.	Killed by deserter 11/13/62; Historical Data Systems record
36	Baughman Union-W. Prov.	Died 4/26/64; B.S.F. History Book-1884; William, George and Josiah-brothers
37		B.S.F. History Book-1884; Drafted
38	Craig Cem.-NE	B.S.F. History Book-1884; 1870 S. Woodbury Twp. Census
39		B.S.F. History Book-1884; Joseph & George-brothers
40		Historical Data Systems record; 1850 S. Woodbury Twp. Census-aka Boyer
41	Craig Cem.-NE	Historical Data Systems record
42	Burket Cem.-Londonderry Twp.	Died 6/12/62; Bedford Gazette 2/13/1914 listing; Frank McCoy 1912 Listing
43	Workman Cem.-OH	B.S.F. History Book-1884; 1860 Londonderry Twp. Census; George & Nicholas-brothers
44	Palo Alto Cem.-Hyndman	Wounded 6/1/64; 1890 Londonderry Twp. Veterans Census
45		Wounded-Bedford Inquirer 5/6/64
46		B.S.F. History Book-1884; enlisted in Bedford
47	Friends Cove UCC Cem.	POW-Andersonville 4/20/64; Died 8/8/64 (age 61); 1860 Colerain Twp. Census
48		1890 E. St. Clair Twp. Veterans Census; ; Discharged on Surgeon's Cert.
49		B.S.F. History Book-1884; enlisted in Reading
50		Wounded 10/19/64; B.S.F. History Book-1884; 1860 Bedford Census
51		Deserted 7/1/64 & returned 1/18/65; Huntingdon Co. references
52	Old Claysburg Cem.	Historical Data Systems record
53		Historical Data Systems record
54	Mt. Union Cem.-Hunt. Co.	Pension Record, 1880 W. Providence Twp. Census
55	Rio Cem.-WV	1890 Brush Creek Twp. Veterans Census; aka Betty
56	Everett Cem.	1860 Napier Twp. Census; 1890 Harrison Twp. Veterans Census
57	Schellsburg Cem.	POW-Andersonville 5/16/64 to 4/29/65; William is Father
58		Wounded & POW 3/4/62; Died 3/5/62; 1860 Napier Twp. Census
59	Koontz Breth. Cem.-Loysburg	1850 S. Woodbury Twp. Census; Discharged Surg. Cert.; Daniel, David, John & Isaac-bros.
60	Andersonville Cem., 3821	POW-Andersonville; Died 7/23/64; PA Civil War Archives-Berks Co. residence
61	New Enterprise Cem.	Wounded & POW 8/16/64 to 12/20/64; 1850 S. Woodbury Census
62		KIA 6/22/64; B.S.F. History Book-1884; 1860 S. Woodbury Twp. Census
63	New Enterprise Cem.	1890 S. Woodbury Twp. Census
64		Historical Data Systems - Pleasantville Resident; PA Civil War Archives

Chapter 18

	Name	Age	Rank	Regiment	Co.	Muster In	Muster Out	Casualty	Casualty Battle
1	Beck, Henry L*		Sgt. QM	PA 22nd Cav.	C	6/19/63	6/15/65		
2	Bedell, Edmund	31	1st Lt.	PA 55th Inf.	K	11/5/61	5/5/62		
3	Bedell, Edmund		2nd Lt.	PA McKeage's Mil.	G	7/2/63	8/8/63		
4	Beegle, David F	26	1st Lt.	PA 101st Inf.	D	11/1/61	3/15/65	POW	Plymouth, NC
5	Beegle, Job M	25	Corp.	PA 138th Inf.	D	8/18/62	4/27/63		
6	Beegle, John A	22	Pvt.	PA 125th Inf.	E	8/15/62	5/18/63		
7	Beegle, John A		Corp.	PA 110th Inf.	C	2/27/64	6/28/65		
8	Beegle, Solomon	23	Pvt.	PA 207th Inf.	G	9/8/64	5/31/65		
9	Beeler, James H	18	Corp.	PA 22nd Cav.	C	7/11/63	6/15/65		
10	Beeny, Frederick	22	Pvt.	IL 77th Inf.	I	9/2/62	5/15/63	Died	
11	Beichtel, Lewis	20	Pvt.	PA 4th Cav.	E	8/16/61		MIA	
12	Beighel, Burket*		Pvt.	PA McKeage's Mil.	G	7/2/63	8/8/63		
13	Beighel, Burket*		Pvt.	PA 22nd Cav.	I	2/9/64		Died	
14	Beisel, George M	29	Pvt.	PA 55th Inf.	K	3/2/64	10/27/64	Wound.-Died	
15	Beisel, Isaac A	38	Corp.	PA 198th Inf.	F	9/2/64	6/4/65		
16	Bell, James C*	25	Pvt.	PA 3rd Inf.	D	4/20/61	7/29/61		
17	Bell, James C*		1st Sgt.	PA 110th Inf.	C	10/24/61	10/24/64	Wounded	1st Deep Bottom
18	Belles, Joseph S*		Pvt.	PA 55th Inf.	K	9/28/63		KIA	Petersburg - Initial
19	Beltz, Abraham	32	Pvt.	PA 101st Inf.	G	12/2/61		POW-Died	Plymouth, NC
20	Beltz, Adam	24	Pvt.	PA 138th Inf.	E	8/29/62		KIA	Cold Harbor
21	Beltz, Christian		Pvt.	PA 3rd Heavy Art.					
22	Beltz, Daniel E	23	Pvt.	PA 135th Inf.	G	8/16/62	5/24/63		
23	Beltz, Frederick B	20	Pvt.	PA 135th Inf.	G	8/16/62	5/24/63		
24	Beltz, George W		Pvt.	PA 138th Inf.	D	8/29/62	6/23/65	Wounded	Cedar Creek
25	Beltz, John A	33	Pvt.	PA 138th Inf.	D	8/29/62	6/23/65	Wounded	Cedar Creek
26	Beltz, Oliver	21	Corp.	MD 2nd PHB Inf.	C	7/25/64	7/1/65		
27	Beltz, Samuel G	28	Pvt.	PA 88th Inf.	G	3/31/65	7/6/65		
28	Beltz, William H	18	Pvt.	PA 55th Inf.	H	2/19/64	6/11/65		
29	Beltz, William W	26	Pvt.	MD 2nd PHB Inf.	H	9/10/61		POW-Died	Ridgeville, VA
30	Beltz, William Y	25	Pvt.	PA 135th Inf.	G	8/16/62	5/24/63		
31	Bender, Benjamin F	40	Pvt.	PA 21st Cav.	E	7/8/63	2/20/64		
32	Beneigh, John	43	Pvt.	PA 55th Inf.	H	2/29/64	12/30/64		
33	Benkley, Samuel		Pvt.	PA 208th Inf.	H	9/7/64	6/1/65		
34	Benna, Lewis	27	Pvt.	PA 83rd Inf.	H	2/22/65	6/28/65		
35	Bennage, Simon*	33	Pvt.	PA 76th Inf.	E	10/7/64	7/18/65		
36	Benner, John	18	Pvt.	PA 138th Inf.	E	8/29/62	6/23/65		
37	Benner, Samuel B	22	Pvt.	PA 133rd Inf.	C	8/13/62	5/19/63		
38	Bennethum, George*	21	Pvt.	PA 55th Inf.	D	8/25/62		POW-Died	
39	Bennethum, Jonathan*		Pvt.	PA 55th Inf.	D	9/22/62	7/11/65		
40	Bennett, Abraham	33	Pvt.	PA 171st Inf.	I	11/10/62			
41	Bennett, Archer		Pvt.	PA 99th Inf.	F	2/25/65	6/3/65		
42	Bennett, Artemas S	19	Sgt.	PA 76th Inf.	E	10/9/61	11/29/64		
43	Bennett, Daniel	28	Pvt.	PA 55th Inf.	I	10/10/63	3/30/65	Wounded	
44	Bennett, David	27	Pvt.	PA 171st Inf.	I	11/10/62	8/8/63		
45	Bennett, Enos	21	Pvt.	MD 1st PHB Cav.	D	9/7/64	6/28/65	POW	
46	Bennett, George S	31	Pvt.	PA 91st Inf.	F	9/21/64	6/1/65		
47	Bennett, Henry	29	Pvt.	PA 91st Inf.	I	9/21/64	23863		
48	Bennett, Henry R	28	Pvt.	PA 99th Inf.	C	2/28/65	7/1/65		
49	Bennett, Israel M	37	Pvt.	PA 99th Inf.	F	2/25/65	5/31/65		
50	Bennett, Jacob	23	Pvt.	PA 55th Inf.	D	10/12/61		POW-Died	
51	Bennett, John E	19	Pvt.	PA 184th Inf.	A	5/12/64	7/14/65	Wounded	
52	Bennett, Joseph	37	Pvt.	PA 91st Inf.	F	9/21/64	6/1/65		
53	Bennett, Samuel J*			MD					
54	Benseman, Charles*		2nd Lt.	PA 76th Inf.	E	10/9/61	7/18/65		
55	Benton, David H	24	Pvt.	PA 84th Inf.	A	11/24/61			
56	Benton, Emanuel	34	Pvt.	PA 125th Inf.	E	8/13/62	5/18/63		
57	Benton, Emanuel		Pvt.	PA 99th Inf.	B	2/25/65	5/31/65		
58	Bequeath, Frank	21	Pvt.	PA 12th Cav.	F	3/20/61	6/20/65	Wounded	
59	Bequeth, Samuel	37	Pvt.	PA 22nd Cav.	H	2/5/65			
60	Bequeth, William H	17	Pvt.	PA 101st Inf.	D	12/6/61	6/25/65	POW	Plymouth, NC
61	Berchman, Jacob*	43	Pvt.	PA 55th Inf.	K	10/3/63	8/30/65	Wounded	Cold Harbor
62	Berger, George H		Capt.	PA 188th Inf.	G	11/8/62	12/14/65		
63	Berkett, Harvey		Pvt.	PA 55th Inf.	H	1/7/63	8/30/65		
64	Berkey, Benjamin	29	Pvt.	PA 171st Inf.	H	11/1/62	8/8/63		

	Cemetery	Notes
1		Historical Data Systems lists Blair Co. residence; 1860 Warriors Mark Census
2	Evergreen Cem.-IA	B.S.F. History Book-1884; 1860 Union Twp. (Pavia) Cen; aka Edward
3		B.S.F. History Book-1884
4	Oak Ridge Cem.-Altoona	POW-Columbia, SC 4/20/64; Historical Data Systems-Bedford Co. residence
5	Mt. Smith Cem.-Bedford	1860 Colerain Twp. Census; Frank McCoy 1912 Listing; Discharged on Surgeon's Cert.
6	Lutheran Cem.-Osterburg	B.S.F. History Book-1884; 1850 Bedford Twp. Census
7		PA Civil War Archives-born in Bedford Co.
8	Carson Valley-Duncansville	Pension Record, 1860 Bedford Twp. Census
9	Fairview Cem.-Altoona	B.S.F. History Book-1884; 1850 W. Providence Twp. Census
10	Elmwood Twp. Cem.-IL	Died 6/18/64; Hist. Data Systems record; 1850 Napier Twp. Cen.; aka Benny, Binna, Beeney
11	Old Claysburg Cem.	Reinlisted 1/1/64; PA Civil War Project record listed MIA-aka Bechtel
12		B.S.F. History Book-1884
13	Grafton Nat'l Cem.-WV	Died 7/2/65; Huntingdon Co references
14	Arlington Nat'l Cem., 13-11432	Died of Disease 10/27/64; B.S.F. History Book-1884; 1860 St. Clair Twp. Census
15		William Mock record; PA Civil War Project record; Born-St. Clair Twp.
16		Historical Data Systems record; Tyrone references
17		Historical Data Systems record; Tyrone references
18	Poplar Grove Nat'l Cem.	B.S.F. History Book-1884; Drafted; KIA 6/16/64
19	Annapolis Nat'l Cem., B 614	POW-Andersonville 4/20/64; Died 12/3/64 of Starvation; 1850 Schellsburg Census
20	Cold Harbor Nat'l Cem.	KIA 6/5/64; aka Beltze & Belse; Historical Data Systems-Bedford residence
21		1890 Napier Twp. Veterans Census
22	Ligonier Valley Cem.	1850 Harrison Twp. Census; served at Carver Hospital DC
23	Coles Cem.-Derry	1850 Schellsburg Census
24		Wounded 10/19/64; Historical Data Systems-Bedford Co. residence
25	Mt. Olivet Cem.-Manns Choice	Wounded 10/19/64; PA Civil War Archives-Bedford Co. residence
26	Rose Hill Cem.-Cumberland	Enlisted after brother Died as a POW
27	Hyndman Cem.	1890 Londonderry Twp. Veterans Census; Samuel, Oliver, William W & Adam-brothers
28	Homestead Cem.-Munhall	Historical Data Systems-Bedford residence;William H & Daniel-brothers
29	Andersonville Cem., 471	POW-Richmond & Andersonville 1/3/64; Died 4/10/64 - Chronic Diarrhea; Bed. Gaz. 1914 list
30	Maple Grove Cem.-OH	William Y, Frederick, George & Abraham-brothers
31	Daley Cem.-Somerset Co.	Historical Data Systems-Bedford Co. residence
32		Historical Data Systems-Bedford Co. residence; enlisted Schellsburg; aka Benigh
33		B.S.F. History Book-1884; Historical Data Systems-Bedford Co. residence
34	Dry Ridge Cem.-Manns Choice	1860 Harrison Twp. Census; 1890 Juniata Twp. Veterans Census; aka John P. Lewis Benna
35		B.S.F. History Book-1884; Union Co references
36		B.S.F. History Book-1884; 1860 Bedford Twp. Census
37	Elmhurst Cem.-IL	B.S.F. History Book-1884; enlisted in Hopewell
38	Andersonville Cem., 4752	POW-Andersonville; Died 8/5/64, B.S.F. History Book-1884; 1850 Lebanon Co. Census
39		B.S.F. History Book-1884; Schuylkill references
40	Mt. Zion Cem.-Cheneysville	Deserted 11/22/62; Bedford Inquirer 5/22/1908 listing
41		1890 Everett Veterans Census
42		1860 Bedford Twp. Census
43	Bethal Frame Ch.-Monroe Twp.	Wounded-Bedford Inquirer 5/27/64 article; B.S.F. History Book-1884-Bedford residence
44	Mt. Zion Cem.-Chaneysville	1890 Southampton Veterans Census
45	Chaneysville Meth. Cem.	POW listed in Pension record; Enos, Abraham, David, Joseph & Jacob-brothers
46	Bennett Farm-Chaneysville	1890 Southhampton Veterans Census
47	Mt. Union Cem.-Mench	Historical Data Systems record
48		1890 E. Providence Twp. Veterans Census
49	Bennett Cem.-Artemas	1890 Mann Twp. Veterans Census
50	Andersonville Cem., 7477	POW-Andersonville; Died-Scurvy 9/1/64; 1914 McCoy Listing; 1860 Southampton Twp. Cen.
51	Lashley Cem.-Artemas	1890 Mann Twp. Veterans Census Aka Espy S
52	Mt. Zion Cem.-Chaneysville	1890 Southampton Veterans Census
53	Hyndman Cem.	Born 11/18/37; Civil War Marker at Gravesite
54		B.S.F. History Book-1884; Schuylkill Co references
55	Birmingham Cem.-Hunt. Co.	1860 Union Twp. (Pavia) Census; Historical Data Systems-Bedford Co. residence
56	Greenfield Cem.-Queen	Frank McCoy 1912 Listing; 1860 Union Twp. (Pavia) Cen
57		Frank McCoy 1912 Listing; Historical Data Systems record
58	Mt. Pleasant Cem.-Mattie	E. Providence Twp. Honor Roll; Thrown from Horse 1862
59	Mt. Pleasant Cem.-Mattie	1890 Everett Veterans Census
60	Mt. Pleasant Cem.-Mattie	POW-Andersonville 4/20.64; attended 1905 Andersonville Dedication
61		Wounded 6/3/64; B.S.F. History Book-1884; Reading, PA references
62		B.S.F. History Book-1884
63		1890 Junaita Twp. Veterans Census
64	Mt. Zion Cem.-Pavia	aka Berkley; Frank McCoy 1912 Listing; Historical Data Systems record

	Name	Age	Rank	Regiment	Co.	Muster In	Muster Out	Casualty	Casualty Battle
1	Berkey, William*		Pvt.	PA 93rd Inf.	K	9/26/64	6/20/65		
2	Berkeybil, Daniel W		Pvt.	PA 21st Cav.	E	7/2/63	7/8/65		
3	Berkheimer, Daniel B	18	Pvt.	PA 22nd Cav.	M	2/16/64	6/24/65		
4	Berkheimer, David*							Died	
5	Berkheimer, John	18	Pvt.	PA 200th Inf.	C	9/4/64	5/15/65	Died	
6	Berkheimer, Samuel	42	Pvt.	PA 171st Inf.	H	11/1/62	8/8/63		
7	Berkheimer, William S	31	Pvt.	PA 91st Inf.	B	9/21/64	5/30/65		
8	Berkhimer, John		Pvt.	PA 55th Inf.	K	2/16/64			
9	Berkhimer, Levi	17	Pvt.	PA McKeage's Mil.	H	7/2/63	8/8/63		
10	Berkhimer, Levi		Pvt.	PA 184th Inf.	A	5/12/64	11/7/65	Wounded	Cold Harbor
11	Berkhiser, Nicholas*	19	Pvt.	PA 184th Inf.	A	5/12/64	7/14/65		
12	Berkstresser, David S	43	Pvt.	PA 77th Inf.	F	2/27/65	12/13/65		
13	Berkstresser, George	39	Pvt.	PA 188th Inf.	G	2/20/62	7/26/65		
14	Berkstresser, John Y	22	Pvt.	PA 137th Inf.	I	8/20/62	6/1/63		
15	Berkstresser, Levi		Pvt.	PA 20th Cav.	A	6/23/63	1/6/64		
16	Berry, John W	29	Pvt.	USCT 41st Inf.	C	6/4/64	12/10/65		
17	Bessie, Adolph*		Pvt.	PA 55th Inf.	D	3/11/62	7/7/63		
18	Bessie, Adolph*		2nd Lt.	USCT 3rd Inf.		7/7/63			
19	Bessor, John		Corp.	PA 101st Inf.	D	11/6/61		Wounded	Fair Oaks
20	Bessor, John	26	Corp.	PA 101st Inf.	D	11/1/61		POW-Died	Plymouth, NC
21	Bessor, Philip	21	Sgt.	PA 6th Res.	D	8/10/61	6/15/64	Wounded	Fredericksburg
22	Bessor, Philip		Sgt.	PA 6th Res.	D			Wounded	Spotsylvania
23	Bessor, Philip		Sgt.	PA 6th Res.	D			Wounded	Wilderness
24	Bessor, Philip		1st Lt.	PA 208th Inf.	K	9/7/64	6/1/65		
25	Betts, George		Pvt.	PA 14th Cav.	B	11/23/62			
26	Betz, Franklin*		Pvt.	PA 55th Inf.	D	9/26/62	6/11/65		
27	Biddle, Andrew B	16	Pvt.	PA 138th Inf.	E	8/29/62	4/22/65	Wounded	Wilderness
28	Biddle, Jacob S	20	Pvt.	PA 194th Inf.	I	7/22/64	11/5/64		
29	Biddle, Jacob S		Pvt.	PA 22nd Cav.	M	3/14/65	6/24/65		
30	Billman, Adam H*		Pvt.	PA 55th Inf.	D	9/26/62	8/30/65		
31	Biltz, John S		Corp.	NJ 1st Inf.	D	10/1/61	11/22/64		
32	Binett, Nathan		Pvt.	PA 138th Inf.	F	8/3/62			
33	Bingaman, David	26	Pvt.	PA 55th Inf.	H	10/11/61	7/26/62		
34	Bingham, Linton W	32	Pvt.	PA 138th Inf.	F	8/29/62	1/28/65		
35	Bircamp, Henry*			PA 101st Inf.				MIA	
36	Bird, William	20	Pvt.	PA 55th Inf.	I	10/15/61	1/20/63		
37	Bird, William		Pvt.	US 1st Light Art.	M	1/20/63			
38	Birkhimer, Samuel		Pvt.	PA 137th Inf.	I	8/20/62	6/1/63		
39	Birkhimer, Samuel	20	Pvt.	PA 55th Inf.	I	2/20/64	8/30/65		
40	Birkley, David T	21	Pvt.	PA 171st Inf.	I	11/10/62			
41	Bisbing, Charles H*	26	Pvt.	PA 55th Inf.	I	2/15/64		Wound.-Died	Chaffin's Farm
42	Bisbing, Gervase*		Pvt.	PA 76th Inf.	E	10/10/64		Died	
43	Bisel, Noah	34	Pvt.	PA 149th Inf.		7/5/64	5/5/65		
44	Bishop, David P	16	Pvt.	PA 82nd Inf.	C	11/22/64	5/31/65	Wounded	
45	Bivens, James W	25	Pvt.	PA 138th Inf.	D	8/29/62	6/23/65		
46	Bivens, William	19	Pvt.	PA 77th Inf.	F	10/9/61	10/12/65	Wounded	Stones River, TN
47	Black, Andrew J	20	Pvt.	PA 3rd Heavy Art.	G	2/29/64	11/9/65		
48	Black, Erastus	30	Pvt.	PA 77th Inf.	D	2/27/64	12/6/65		
49	Black, Franklin	24	Pvt.	PA 5th Heavy Art.	K	8/31/64	6/30/65		
50	Black, George W Z	19	Capt.	PA 107th Inf.	H	3/5/62	11/22/63	Wounded	Antietam
51	Blackburn, Cyrus E	21	Pvt.	PA 22nd Cav.	I	2/25/64	6/24/65		
52	Blackburn, George	20	Pvt.	PA 200th Inf.	C	8/26/64	5/15/65		
53	Blackburn, Harmon		Pvt.	PA 194th Inf.	G	7/20/64	23656		
54	Blackburn, Harmon	19	Pvt.	PA 77th Inf.	F	2/27/65	12/6/65	Wounded	
55	Blackburn, Henry B	30	Pvt.	PA 184th Inf.	A	5/12/64	2/1/65		
56	Blackburn, John M	45	Sgt. QM	PA 21st Cav.	E	7/8/63	2/20/64	Wounded	
57	Blackburn, Joseph	30	Pvt.	PA 138th Inf.	E	8/17/62		Wound.-Died	Monocacy, MD
58	Blackburn, Levi	26	Pvt.	PA 138th Inf.	E	8/29/62		Died	
59	Blackburn, Martin	40	Pvt.	PA 184th Inf.	A	5/12/64		POW-Died	Jerusalem Plank Rd.
60	Blackburn, Nathan F	18	Pvt.	PA 55th Inf.	H	2/29/64	8/30/65		
61	Blackburn, Samuel R		Pvt.	OH 45th Inf.	G	8/19/62	6/12/65		
62	Blackburn, Samuel S	45	Pvt.	PA 99th Inf.	B	9/21/64	6/31/65		
63	Blackhart, John W	17	Pvt.	PA McKeage's Mil.	D	7/2/63	8/8/63		
64	Blackhart, John W		Pvt.	PA 22nd Cav.	H	2/26/64	6/24/65		

	Cemetery	Notes
1		B.S.F. History Book-1884; Somerset Co. references
2		Historical Data Systems-Bedford Co. residence
3	Salemville Cem.	1890 S. Woodbury Twp. Veterans Census
4	Old Union Cem.-Osterburg	Died 1861-age 21; Civil War Marker at gravesite
5	Lutheran Cem.-Osterburg	Died 9/4/65; Frank McCoy 1912 Listing; Historical Data Systems record
6	Schellsburg Cem.	1890 Colerain Twp. Census; Samuel & John-brothers
7	Trinity UCC Cem.-Osterburg	Civil War Marker at gravesite
8		1890 Greenfield Twp. Veterans Census
9	Rice Family Cem.-New Ent.	B.S.F. History Book-1884
10		Wounded-Bedford Inquirer 7/1/64 article; 1890 S. Woodbury Twp. Veterans Census
11		B.S.F. History Book-1884; enlisted in Harrisburg
12	Hawthorne Cem.-Florida	1870 Liberty Twp. Cen; Born Bedford Co.
13	Fockler Cem.-Saxton	1860 Hopewell Twp. Census; George & David-brothers
14	Fockler Cem.-Saxton	1890 Liberty Twp. Veterans Census; 1860 Woodbury Twp. Census
15	St. Lukes Cem.-Saxton	Frank McCoy 1912 Listing; 1860 Liberty Twp. Cen; Levi & John-brothers
16	Burgess Farm-New Buena Vista	Historical Data Systems record
17		B.S.F. History Book-1884; enlisted in Philadelphia
18		B.S.F. History Book-1884; enlisted in Philadelphia
19		Wounded 5/31/62; B.S.F. History Book-1884; aka Besser
20	Florence Nat'l Cem.	POW-Andersonville* 4/20/64; B.S.F. History Book-1884; 1860 W. Providence Twp. Census
21	Kearney Cem.-NE	Wounded 12/13/62; B.S.F. History Book-1884 list
22		B.S.F. History Book-1884 record; Philip & John-brothers
23		Historical Data Sys.-Bedford Co. residence
24		Historical Data Systems record aka Besser
25	Ursina Cem.-Somerset Co.	Deserted 7/8/63; B.S.F. History Book-1884
26		B.S.F. History Book-1884; Berks Co. references
27	Trinity UCC Cem.-Colerain Twp.	Wounded 5/5/64; B.S.F. History Book-1884; 1860 S. Woodbury Twp. Census
28	Loysburg Hill Cem.	Jacob & Andrew-brothers
29		1890 S. Woodbury Veternas Census
30		B.S.F. History Book-1884; Berks Co. references
31		1890 Hopewell Veterans Census
32		1890 Everett Veterans Census
33	Evergreen Cem.-TX	B.S.F. History Book-1884; Hist. Data Systems-Bedford Co. resid.; Discharged Surgeon's Cert.
34	Carpenter Farm Cem.-Hyndman	PA Civil War Archives-Colerain Twp. residence
35		possibly Boerkamp, Henry; Bed. Gaz. 1914 listed MIA; Trinity UCC Cem. Civil War mem.
36		Historical Data Systems-Bedford Co. residence
37		Historical Data Systems record
38		Historical Data Systems-Born in Bedford Co.
39	Trinity UCC Cem.-Osterburg	PA Civil War Archives-Bedford residence
40		Deserted 11/25/62; B.S.F. History Book-1884
41	Annapolis Gen. Hosp. Cem.	Wounded 9/29/64; Died 10/25/64; Bed. Gazette 2/13/1914 listing; Somerset Co. references
42	Cypress Hill Cem.-NY	Died 4/20/65; B.S.F. History Book-1884; Easton references
43	Stone Church Cem.-Fishertown	1890 Napier Twp. Veterans Census
44	Fairview Cem.-Mercersburg	1890 S. Woodbury Census left shoulder wound
45		1880 W. Providence Twp. Census; Historical Data Systems-Bedford Co. residence
46	Wells Valley Meth. Cem.	1890 Wells Twp. Veterans Census
47	Duval Cem.-Coaldale	1890 Broad Top Veterans Census
48	Broad Top IOOF Cem.	1890 Robertsdale Veterans Census
49	Schellsburg Cem.	1890 Napier Twp. Veterans Census; 1860 Napier Twp. Census
50	Arlington Nat'l Cem., sec. 3	Wounded 9/17/62; Historical Data Systems-Bedford Co. residence
51	Fishertown Cem.	1890 E. St. Clair Veterans Census
52	Fairview Cem.-Altoona	PA Civil War Project record; 1860 St. Clair Twp. Census
53		PA Civil War Project record
54	Sproul Union Cem.-Claysburg	1890 Greenfield Twp. Veterans Census; aka Hirim
55		B.S.F. History Book-1884; 1850 Napier Twp. Census
56	Friends Cem.-Centre Co.	1860 Napier Twp. Census; Cyrus-son
57	Friends Cem.-Spring Meadow	Leg Amputated 7/9/64; Died 9/1/64; B.S.F. History Book-1884
58	Friends Cem.-Spring Meadow	Died of Disease 11/19/62; Historical Data Systems-Schellsburg residence
59	Andersonville Cem., 10,674	POW-Andersonville 6/22/64, Died 10/11/64; 1890 W. St. Clair Twp. Veterans Census (widow)
60	Maple Grove Cem.-KS	Historical Data Sys.-Bedford Co. residence
61	Friends Cem.-Spring Meadow	1890 Napier Twp. Veterans Census
62	Grandview Cem.-Cambria Co.	1890 Napier Twp. Veterans Census
63	Cedar Grove-E. Prov. Twp.	B.S.F. History Book-1884; aka Blachart
64		Frank McCoy 1912 Listing; 1860 E. Providence Twp. Census

Chapter 18

	Name	Age	Rank	Regiment	Co.	Muster In	Muster Out	Casualty	Casualty Battle
1	Blake, Samuel	28	Pvt.	PA 110th Inf.	C	12/19/61			
2	Blake, Simon	32	Pvt.	PA 110th Inf.	C	10/24/61			
3	Blake, Simon		Pvt.	PA 133rd Inf.	K	8/15/62	5/26/63		
4	Blake, Simon		Pvt.	PA 22nd Cav.	B	7/11/63	2/5/64		
5	Blake, Simon S	32	Pvt.	WI 25th Inf.	B	8/20/62	3/20/65	Wounded	Decatur, GA
6	Blake, Thomas	21	Pvt.	PA 110th Inf.	C	10/24/61			
7	Blake, William B	21	Pvt.	PA 125th Inf.	B	8/10/62	5/18/63	Wounded	Chancellorsville
8	Blake, William B		2nd Lt.	PA 208th Inf.	B	8/26/64	6/1/65	Wounded	
9	Blankley, Isaac	27	Pvt.	PA 93rd Inf.	A	11/14/64	6/27/65		
10	Blankley, Job	21	Pvt.	PA 133rd Inf.	C	8/13/62	5/26/63		
11	Blankley, Job		Pvt.	PA 22nd Cav.	I	2/1/64			
12	Blattenberger, Daniel	40	Pvt.	PA 171st Inf.	I	11/10/62	8/8/63	Wounded	
13	Blattenberger, Joseph			PA 171st Inf.					
14	Bloom, David	20	Pvt.	PA 200th Inf.	D	8/25/64	5/27/65		
15	Bloom, Jacob	33	Pvt.	PA 55th Inf.	K	11/5/61	2/21/63		
16	Bloom, Jacob		Pvt.	US 1st Light Art.	M	2/21/63			
17	Bloom, Jacob M	18	Pvt.	PA 74th Inf.	B	3/13/65	6/29/65		
18	Bloom, James	17	Pvt.	MD 3rd PHB Inf.	H	2/22/64	5/29/65		
19	Bloom, John (1)		Pvt.	PA 55th Inf.	K	11/5/61	11/6/61		
20	Bloom, John (1)		Pvt.	PA 76th Inf.	E	11/6/61	11/24/64		
21	Bloom, John (2)	20	Corp.	MD 3rd Reg Inf.	A	10/1/61	7/31/65		
22	Blymyer, Benjamin			PA 122nd Inf.	K				
23	Blymyer, Benjamin	20	Pvt.	PA 83rd Inf.	K	3/9/65	6/28/65		
24	Blymyer, William		Pvt.	PA 20th Cav.	B	1/18/64	7/28/64	POW-Died	
25	Bobb, Alexander*		Capt.	PA 133rd Inf.	C	8/18/62	5/26/63		
26	Bobb, Alexander*		Lt. Col.	PA 208th Inf.		9/7/64	6/1/65		
27	Bobb, William		Corp.	PA McKeage's Mil.	G	7/2/63	8/8/63		
28	Boden, Charles W			MD 3rd Reg Inf.					
29	Boden, John H	18	Pvt.	MD 1st PHB Cav.	K	4/23/64	6/28/65		
30	Boden, William A	28	Sgt.	PA 104th Inf.	F	3/10/65	8/25/65		
31	Boehm, John W	21	Pvt.	PA 13th Inf.	G	4/25/61	8/6/61		
32	Boehm, John W		Corp.	PA 76th Inf.	E	10/9/61	11/28/64		
33	Boemer, John A		Sgt.	PA 208th Inf.	H	9/6/64	6/1/65		
34	Boerkamp, Henry	30	Pvt.	PA 101st Inf.	G	12/28/61	2/5/63	Wounded	Fair Oaks
35	Boher, Samuel	29	Pvt.	PA 79th Inf.	E	2/23/65	5/29/65		
36	Boher, Thomas	21	Pvt.	PA 97th Inf.	A	6/2/64	10/7/65		
37	Bohlman, William H		Corp.	US 19th Inf.	G	4/8/62	4/10/65		
38	Bohn, Solomon	32	Pvt.	PA 184th Inf.	A	5/12/64	7/14/65		
39	Bolden, Elijah		Pvt.	USCT 6th Inf.	E	9/11/63	9/20/65		
40	Bolinger, Alexander	20	Pvt.	PA 76th Inf.	E	9/9/61	11/28/64		
41	Bollinger, David S	23	Pvt.	PA 8th Res.	F	6/2/61	5/15/64	POW	Seven Days Battle
42	Bollinger, David S		Pvt.	PA 191st Inf.	H	5/15/64	6/28/65		
43	Bollinger, Jacob	15		PA 76th Inf.	E	1861	1865		
44	Bollinger, James	26	Pvt.	PA 53rd Inf.	C	10/17/61			
45	Bollinger, James		Pvt.	PA 19th Cav.	C	9/17/63			
46	Bollinger, William H	24	Pvt.	PA 21st Cav.	E	7/8/63		Died	
47	Bollman, David F	43	Pvt.	PA 107th Inf.	K	9/21/64	6/7/65		
48	Bollman, David R	21	Pvt.	PA 55th Inf.	D	2/27/64	8/30/65	Wounded	Drewry's Bluff*
49	Bollman, George F	18	Pvt.	PA McKeage's Mil.	G	7/2/63	8/8/63		
50	Bollman, George F		Pvt.	PA 22nd Cav.	M	2/16/64	6/24/65		
51	Bond, George M	23	Sgt.	PA 12th Cav.	F	1/27/62	7/20/65		
52	Bondebust, John M		Pvt.	PA 82nd Inf.	C	11/23/64	7/13/65		
53	Bonnett, Thomas K		Sgt.	PA 22nd Cav.	B	7/11/63	10/31/65		
54	Bookhamer, John	22	Pvt.	PA 77th Inf.	F	2/28/65	12/6/65		
55	Bookhamer, Thomas		Pvt.	PA 22nd Cav.	D	7/29/63	2/5/64		
56	Bookhamer, Thomas	22	Pvt.	PA 77th Inf.	F	2/28/65	12/6/65		
57	Bookhamer, William	32	Pvt.	PA 208th Inf.	B	8/26/64	7/1/65		
58	Boor, Jacob B			PA 2nd Cav.	E				
59	Boor, John A	21	Pvt.	PA 138th Inf.	F	9/1/62		Died	
60	Boor, Tobias	27	Pvt.	PA 50th Inf.	E	2/24/65	5/9/65		
61	Boor, William A	20	Pvt.	PA 13th Inf.	G	4/25/61	8/6/61		
62	Boor, William A		Sgt.	PA 55th Inf.	D	10/12/61	1865	POW	Drewry's Bluff
63	Boor, William C	30	Pvt.	OH 169th Inf.	H	5/15/64	9/4/64		
64	Boore, Jocob C	43	Pvt.	PA 133rd Inf.	K	8/26/62	5/26/63		

	Cemetery	Notes
1	Fockler Cem.-Saxton	Frank McCoy 1912 Listing; Historical Data Systems-Woodbury residence
2	St. Elizabeths Hosp. East-DC	Historical Data Systems-Woodbury residence; Simon, Thomas & Samuel-brothers
3		Historical Data Systems-Bedford Co. residence; enllisted Bloody Run
4		Civil War Pension Record
5	Oak Ridge Cem.-WI	Wounded 7/22/64; 1830 Woodbury Twp. Census; Historical Data Systems record
6		Historical Data Systems-Woodbury residence; 1860 Woodbury Twp. Census
7	Hopewell Cem.	Wounded 5/3/63; Frank McCoy 1912 Listing; 1890 Broad Top Veterans Census
8		1890 Broad Top Veterans Census
9	Bethal Frame Ch.-Monroe Twp.	1890 Monroe Twp. Veterans Census
10		B.S.F. History Book-1884; Historical Data Systems-Bedford Co. residence
11		B.S.F. History Book-1884
12	Fishertown Bretheran Cem.	1890 E. St. Clair Twp. Veterans Census; Discharged on Surgeon's Cert.
13	Fishertown Bretheran Cem.	Frank McCoy 1912 Listing
14	Trinity UCC Cem.-Osterburg	1890 King Twp. Veterans Census; Deserted & return.
15	Baker Cem.-IA	1850 St. Clair Twp. Census; Historical Data Systems record
16		Historical Data Systems record
17		1890 Napier Twp. Veterans Census; Born in Bedford
18	Shaffer Cem.-Hyndman	1890 Londonderry Twp. Veterans Census
19		Historical Data Systems record; John (1), Jacob & David-brothers
20		Historical Data Systems record
21	Everett Cem.	Everett Cemetery Listing of Civil War Veterans; James & John-Brothers
22		Frank McCoy 1912 Listing; aka Blymire; Benjamin & William-brothers
23	Bedford Cem.	1860 Bedford Census; PA Civil War Archives-Bedford residence
24	Andersonville Cem., 4192	POW-Andersonville; Died 7/28/64; 1860 Bedford Census; aka Blymire; Andersonville Records
25		B.S.F. History Book-1884; Martinsburg references
26		B.S.F. History Book-1884; Martinsburg references
27		B.S.F. History Book-1884
28	Buck Valley Cem.-Fulton Co.	Born 5/17/43; Regiment on gravestone
29	Hyndman Cem.	Findagrave.com record
30		Historical Data Systems-Bedford residence
31	Los Angeles Nat'l Cem.	B.S.F. History Book-1884; PA Civil War Archives-Bedford residence
32		PA Civil War Archives-Bedford residence
33		B.S.F. History Book-1884
34		Wounded 5/31/62; B.S.F. History Book-1884; enlisted Schellsburg
35	Bedford Cem.	1860 Bedford Twp. Census; Samuel & Thomas-brothers
36	Bedford Cem.	1890 Bedford Twp. Veterans Census
37		1890 Snake Spring Valley Veterans Census
38	Lybarger Cem.-Madley	Frank McCoy 1912 Listing; 1860 Juniata Twp. Census
39	Mt. Ross Cem.-Bedford	Frank McCoy 1912 Listing; aka Bouldin
40	Bedford Cem.	Frank McCoy 1912 Listing; 1860 Bedford Census
41	St. Marks Epis.-Chester Co.	POW 6/27/62; B.S.F. History Book-1884
42		B.S.F. History Book-1884
43	Bedford Cem.	Bedford Cemetery Civil War Veterans list; Jacob, William H & Alexander-brothers
44	Yellow Creek Reformed Cem.	1870 Liberty Twp. Census; Findagrave.com information
45		Civil War Pension Record; Absent at Muster Out
46	Cavalry Corps-City Point, VA	Died 11/14/64; Hist. Data Systems-Bedford Co. resid.; 1850 Bedford Census aka Bolinger
47	Yellow Creek Reformed Cem.	Bedford Inquirer 5/22/1908 listing; David F is father of David R & George
48	Ritchey Cem.-Snake Spring V.	Wounded 5/12/64; 1890 Snake Spring Valley Veterans Census
49	Yellow Creek Reformed Cem.	B.S.F. History Book-1884
50		1890 Hopewell Veterans. Census
51	Pleasant Ridge-Buffalo Mills	Findagrave.com & Historical Data Systems record
52		1890 E. St. Clair Veterans Census
53	Vet. Mem. Grove Cem.-CA	1860 Bedford Census; B.S.F. History Book-1884
54	Grandview Cem.-Tyrone	Ancestry.com-Born Bedford Co.
55		1850 M. Woodbury Twp. Census; aka Buckhammer
56	Greenlawn Cem.-Roaring Spring	Historical Data Systems record; William, John & Thomas-brothers
57	Hopewell Cem.	aka Bookhammer; Frank McCoy 1912 Listing
58	Burning Bush-Bedford Twp.	Born 6/5/46; 1860 Cumberland Valley Twp. Census; Jocob and John-brothers
59	Fellowship Cem.-Cumb. Val.	Died-Typhoid Fever 8/17/63; 1860 Cumberland Valley Census; aka Albin
60	Burning Bush-Bedford Twp.	B.S.F. History Book-1884; 1860 Cumberland Valley Twp. Census
61		B.S.F. History Book-1884; 1860 Cumberland Valley Census
62	Bedford Cem.	POW-Andersonville 5/16/64 to 11/14/64; exchanged in Savannah; Hist. Data Systems record
63	Oakwood Cem.-OH	Ancestry.com-Born in Napier Twp.; Civil War Pension Record
64	Shreves Cem.-Monroe Twp.	In Hospital 10/30/62 to 5/19/63; B.S.F. History Book-1884

Chapter 18

	Name	Age	Rank	Regiment	Co.	Muster In	Muster Out	Casualty	Casualty Battle
1	Boose, Isaac	15	Pvt.	PA 102nd Inf.	I	8/18/61	5/16/65	Wounded	Cold Harbor
2	Boose, Moses*		Sgt.	PA 11th Inf.	A	9/30/61		POW-Died	Wilderness
3	Booty, William P	21	Pvt.	PA 101st Inf.	D	12/6/61	1862		
4	Border, Andrew	18	Corp.	PA 110th Inf.	C	10/24/61	6/28/65		
5	Border, David	44		PA Bell's Mil.		5/30/63	8/9/63		
6	Border, Henry	17	Pvt.	PA 133rd Inf.	K	8/15/62	3/27/63	Wounded	Fredericksburg
7	Border, John	20	Pvt.	PA 110th Inf.	C	9/2/61	12/27/62	Wounded	
8	Border, John		Pvt.	PA 22nd Cav.	M	7/11/63	10/31/65		
9	Border, John H	34	Pvt.	PA 18th Cav.	H	2/25/65	6/24/65		
10	Border, John S	21	Pvt.	PA 110th Inf.	C	10/24/61	10/24/64	Wounded	Kelly's Ford
11	Border, John S		Pvt.	PA 110th Inf.	C			Injured	Chancellorsville
12	Boring, Henry J	23	Pvt.	PA 14th Inf.	H	4/24/61	8/7/61		
13	Borland, Zachariah	24	Pvt.	PA 13th Inf.	G	4/25/61	8/6/61		
14	Bortz, Martin S	23	Capt.	PA 138th Inf.	F	8/29/62	6/23/65	Wounded	Cedar Creek
15	Bortz, Martin S		Capt.	PA 138th Inf.	F			Wounded	Petersburg
16	Bose, John*		Pvt.	PA 55th Inf.	D	10/10/63	8/30/65		
17	Boss, John F*		Pvt.	PA 76th Inf.	E	9/23/64	6/28/65		
18	Boston, George W	18	Corp.	PA 184th Inf.	A	5/12/64	7/14/65		
19	Boston, James			USCT 41th Inf.	C				
20	Boston, John			USCT					
21	Bottenfield, Adam K	32	Pvt.	PA 186th Inf.	G	3/24/64	8/15/65		
22	Bottenfield, George H			PA 186th Inf.					
23	Bottenfield, John R	42	Pvt.	PA 67th Inf.	C	11/28/64	7/14/65		
24	Bouchman, John H		Pvt.	PA 22nd Cav.	I	2/1/64	10/31/65		
25	Boughter, David*		Pvt.	PA 55th Inf.	D	11/18/61	12/30/64		
26	Bousch, Christian	25	Pvt.	PA 149th Inf.	H	8/20/63		Wound.-Died	North Anna River
27	Bowen, Alva	24	Pvt.	PA 91st Inf.	C	9/21/64	5/30/65		
28	Bowen, Philip B	33	Pvt.	PA 61st Inf.	A	2/25/64		KIA	Ft. Stevens, DC
29	Bowen, Silas J	37	Pvt.	PA 107th Inf.	D	3/7/62	7/27/63		
30	Bowers, George	19	Pvt.	PA 184th Inf.	A	5/12/64		Wound.-Died	Cold Harbor
31	Bowers, Henry H	27	Pvt.	PA 22nd Cav.	M	2/16/64	6/24/65		
32	Bowers, Jacob	36	Pvt.	PA 99th Inf.	G	2/24/65	7/1/65		
33	Bowers, John	30	Pvt.	PA 13th Inf.	G	4/25/61	8/6/61		
34	Bowers, John C		Pvt.	PA 82nd Inf.	F	1/8/64	1/15/64		
35	Bowers, John C	27	Pvt.	PA 23rd Inf.	C	1/15/64	7/13/65		
36	Bowers, Michael H	31	Pvt.	PA 184th Inf.	A	5/12/64	7/14/65		
37	Bowman, Benjamin F		Pvt.	PA 198th Inf.	F	9/9/64	6/4/65		
38	Bowman, Daniel H	21	Pvt.	PA 110th Inf.	C	9/24/61		Wound.-Died	1st Deep Bottom
39	Bowman, Ephraim	24	Pvt.	PA 143rd Inf.	D	8/17/63	4/15/64		
40	Bowman, George	24	Pvt.	PA 110th Inf.	C	10/24/61		Died	
41	Bowman, George W	53	Pvt.	PA 208th Inf.	K	9/6/64	5/11/65		
42	Bowman, John			PA 55th Inf.					
43	Bowman, John		Pvt.	PA 93rd Inf.	A	9/24/64	6/20/65		
44	Bowman, John W	26	Corp.	PA McKeage's Mil.	D	7/2/63	8/8/63		
45	Bowman, William		Sgt.	PA 13th Inf.	G	4/25/61	8/6/61		
46	Bowman, William	26	Sgt.	PA 55th Inf.	D	9/10/61	10/4/62		
47	Bowman, William		Pvt.	PA 22nd Cav.	I	7/11/63	10/31/65		
48	Bowser, Daniel L	30	Pvt.	PA 55th Inf.	K	3/2/64	8/30/65	Injured	Cold Harbor
49	Bowser, David	22	Pvt.	PA 55th Inf.	K	3/2/64	8/30/65	Wounded	Drewry's Bluff
50	Bowser, Emanuel	18	Pvt.	PA 8th Res.	F	6/19/61		KIA	Fredericksburg
51	Bowser, George L	40	Pvt.	PA 133rd Inf.	A	8/5/62	5/24/63		
52	Bowser, George L		Sgt.	PA 18th Cav.	K	2/29/64	10/31/65		
53	Bowser, Isaac B	34	Pvt.	PA 133rd Inf.	A	8/5/62	5/24/63		
54	Bowser, Isaac B		Sgt.	PA 187th Inf.	K	5/4/64			
55	Bowser, James N*	21	Pvt.	PA 116th Inf.	H	2/15/64		MIA	Spotsylvania
56	Bowser, Job	24	Pvt.	PA 205th Inf.	C	8/27/64		Died	
57	Bowser, John B	15	Pvt.	PA 192nd Inf.	D	2/11/65	8/24/65		
58	Bowser, John J	25	Pvt.	PA 12th Cav.	G	5/18/64	7/20/65		
59	Bowser, Moses	32	Pvt.	PA 49th Inf.	F	8/22/63	11/5/64	Wounded	
60	Bowser, Nicholas	13	Pvt.	PA 55th Inf.	K	10/10/61	6/6/65	Wounded	Drewry's Bluff*
61	Bowser, Valentine	41	Pvt.	PA 171st Inf.	I	11/2/62	7/1/63		
62	Bowser, Valentine		Pvt.	PA 99th Inf.	H	2/25/65	7/1/65		
63	Boyd, William	29	Pvt.	MD 2nd PHB Inf.	D	3/7/65	5/29/65		
64	Boyes, George	41	Pvt.	PA 125th Inf.	F	8/12/62	5/18/63		

	Cemetery	Notes
1	St. Pauls UCC Cem.-Russellville	Wounded 6/3/64; 1890 Hopewell Veterans Census
2	Poplar Grove Nat'l Cem.	Wounded & POW 5/6/64; Died 6/11/64; York County residence record
3	St. Mark's Cem.-Ott Town	Frank McCoy 1912 Listing; Bedford Inquirer 5/22/1908 listing; aka Peter W
4	Bisel Church Cem.-OH	PA Civil War Archives-Yellow Creek residence; Andrew, Henry & John-brothers
5	Schellsburg Cem.	Bedford Inquirer 5/22/1908 listing
6	Bethel Brethren Cem.-Tatesville	Wounded 12/13/62, left arm amputated; 1860 Hopewell Twp. Census; Hist. Data Sys. record
7	Bethel Brethren Cem.-Tatesville	1890 Hopewell Veterans Census lists Hip Wound
8		1890 Hopewell Twp. Veterans Census
9	Homewood Cem.-Pittsburgh	1890 Saxton Veterans Census
10	Potter Creek Cem.-Woodbury	1890 S. Woodbury Twp. Veterans Census-Head Wound
11		Foot crushed by Cannon wheel; B.S.F. History Book-1884
12		Historical Data Systems-Claysburg residence
13		B.S.F. History Book-1884; Historical Data Systems-Bedford residence
14	Union Cem.-Centerville	Wounded 10/19/64; 1890 Cumberland Valley Veterans Census
15		Flesh wound in hip; PA Civil War Archives-Cumberland Valley residence
16		B.S.F. History Book-1884; Drafted; Philadelphia references
17		B.S.F. History Book-1884; Easton references
18		B.S.F. History Book-1884; enlisted in Bedford
19		Frank McCoy 1912 list & Civil Works Admin. 1934 list
20		Frank McCoy 1912 list & Civil Works Admin. 1934 list
21	Union Memorial Cem.-Mench	East Providence Twp. Honor Roll; 1860 E. Providence Twp. Census
22	New Enterprise Cem.	Frank McCoy 1912 Listing
23	Hershberger Cem.-Snake Spr. V.	1890 Everett Veterans Census
24		B.S.F. History Book-1884
25		B.S.F. History Book-1884; Lebonon Co. references
26	Alexandria Nat'l Cem., 2264	Wounded 5/23/64; Died 6/23/64; PA Civil War Archives-Bloody Run residence
27	Everett Cem.	1890 Everett Veterans Census; Alva & Philip-brothers
28	Battleground Nat'l Cem., 14	KIA 7/12/64; born in Bloody Run; PA Civil War Project record; father buried in Everett Cem.
29		Historical Data Systems-Bedford Co. residence; Discharged on Surgeon's Cert.
30	Arlington Nat'l Cem., 13 7991	Wounded-Bedford Inquirer 7/1/64 article; Died 9/9/64; B.S.F. History Book-1884
31	Hopewell Cem.	1890 Broad Top Veterens Census
32	New Paris Comm. Ctr. Cem.	1890 Napier Twp. Veterans Census
33		B.S.F. History Book-1884; PA Civil War Archives-Bedford residence
34		B.S.F. History Book-1884
35	Bedford Cem.	1860 Bedford Census
36	Conrad Mem. Cem.-MT	B.S.F. History Book-1884; Michael & Jacob-brothers
37	IOOF Cem.-Somerset Co.	B.S.F. History Book-1884
38	Potter Creek Cem.-Woodbury	Leg Amputated 7/27/64; Died 10/15/64; Historical Data Systems-Woodbury residence
39	St. John Luth.-Cambria Co.	1890 Hopewell Veterans Census; Discharged on Surgeon's Cert.
40	Potter Creek Cem.-Woodbury	Died of Disease 4/23/62, B.S.F. History Book-1884; George & Daniel-brothers
41		B.S.F. History Book-1884; 1850 Bedford Census; Discharged on Surgeon's Certificate
42	Arnold Cem.-Milligan's Cove	Bedford Inquirer 5/22/1908 listing
43		Frank McCoy 1912 Listing
44		B.S.F. History Book-1884
45		B.S.F. History Book-1884
46	Bedford Cem.	Historical Data Systems-Bedford residence; Discharged on Surgeon's Certificate
47		B.S.F. History Book-1884
48	Greenfield Cem.-Queen	Frank McCoy 1912 Listing; Broken Collar Bone
49	Mt. Hope Cem.-Blue Knob	Shoulder & Lung Wound; 1850 Greenfield Twp. Census; Historical Data Systems record
50		KIA 12/13/62; Historical Data Systems-Bedford Co. residence; 1860 E. Providence Twp. Cen.
51	Benshoff Hill Cem.-Cambria Co.	Historical Data Systems-born in Bedford Co.
52		Historical Data Systems-born in Bedford Co.
53	Benshoff Hill Cem.-Cambria Co.	Isaac, Daniel, David, George, John & Moses-brothers
54		Historical Data Systems record-; Discharged on Surgeon's Cert.; 1850 Greenfield Twp. Census
55		MIA 5/12/64; PA Civil War archives lists born in Bedford Co., Allegheny PA residence
56	City Point Nat'l Cem.-VA	Died-Typhoid Fever 12/12/64; Civil War Pension Record; 1860 Colerain Twp. Census
57	Presbyterian Cem.-Holidaysburg	1860 St. Clair Twp. Census
58		Frank McCoy 1912 Listing; Pension record; PA Civil War Archives; Died in 1866 or 1869?
59	Mt. Hope Cem.-Blue Knob	Wounded-lost use of right arm; PA Civil War Archives record
60	Souls Chapel Cem., MO	Wounded listed in Bedford Inquirer article 6/3/64; 1860 Liberty Twp. Census
61	Mt. Hope Cem.-Blue Knob	1890 Union Twp. (Pavia) Veterans Census
62		PA Civil War Archives; 1860 Union Twp. (Pavia) Census; aka Bouser
63	Everett Cem.	1890 Everett Veterans Census; 1870 Rainsburg Census
64	Fockler Cem.-Saxton	Historical Data Systems record; aka Boyse

Chapter 18

	Name	Age	Rank	Regiment	Co.	Muster In	Muster Out	Casualty	Casualty Battle
1	Boylan, John	22	Pvt.	PA 76th Inf.	B	1/4/65	7/8/65		
2	Boyle, John*		Pvt.	PA 55th Inf.	C	11/5/61	8/30/65		
3	Brad, Henry		Pvt.	PA McKeage's Mil.	G	7/2/63	8/8/63		
4	Bradley, Alexander	27	Pvt.	PA 13th Inf.	G	4/25/61	8/6/61		
5	Bradley, Francis P	24	Pvt.	PA 55th Inf.	H	2/17/64	9/16/64		
6	Bradley, Henry*		Pvt.	PA 55th Inf.	H	1863	12/30/64		
7	Bradley, James A	30	Pvt.	PA 8th Res.	F	4/23/61	2/3/63	Wounded	Antietam
8	Bradley, Thomas	36	Pvt.	PA 125th Inf.	I	8/13/62	5/18/63		
9	Brady, Peter	31	Pvt.	PA 55th Inf.	H	2/15/64	8/1/64		
10	Brallier, Rueben	18	Pvt.	PA 11th Inf.	A	9/30/61		Died	
11	Brallier, William	39	Pvt.	PA 107th Inf.	G	8/21/64	12/30/64	Died	
12	Brant, Andrew J	37	Pvt.	PA 149th Inf.	B	9/23/63	6/24/65	Wounded	Wilderness
13	Brant, Benjimen F	32	Pvt.	PA 149th Inf.	B	8/23/63	6/24/65	Wounded	Wilderness
14	Brant, Grafton	26	Pvt.	MD 1st PHB Cav.	K	4/23/64	6/28/65		
15	Brant, Henry	16	Pvt.	PA 93rd Inf.	H	11/26/64	6/27/65	Wounded	Ft. Fisher
16	Brant, Shannon	19	Pvt.	PA 55th Inf.	D	2/27/64	7/9/65	Wounded	
17	Brant, William	41	Pvt.	PA 107th Inf.	D	9/21/64	5/15/65		
18	Brantner, John H	34	Pvt.	WI 44th Inf.	I	2/4/65	8/4/65		
19	Bratelbaugh, James B	25	Pvt.	PA 171st Inf.	I	11/10/62	8/8/63		
20	Brechbiel, Abraham	18	Pvt.	PA 133rd Inf.	K	8/15/62	5/26/63		
21	Brechbiel, Abraham		Corp.	PA 13th Cav.	D	9/19/63	7/14/65	Wounded	Haw's Shop
22	Brechbiel, Daniel	18	Pvt.	PA 149th Inf.	I	8/26/62	6/30/65		
23	Breman, Andrew		Pvt.	PA 21st Cav.	E	8/8/63	2/20/64		
24	Breneman, Michael B	24	Pvt.	PA 125th Inf.	C	8/11/62	4/6/63	Wounded	Antietam
25	Bretz, Calton L	16	Pvt.	PA 7th Inf.	B	7/4/63	8/11/63		
26	Brice, John (1)*		Pvt.	USCT 6th Inf.	H	9/14/63		KIA	Petersburg
27	Brice, John (2)	23	Pvt.	USCT 3rd Inf.	E	7/8/63	10/31/65		
28	Bridaham, Henry W	29	Pvt.	PA 55th Inf.	H	2/15/64		POW-Died	Drewry's Bluff
29	Bridenstein, John E	26	Pvt.	PA 82nd Inf.	C	11/14/64	5/4/65		
30	Bridenstine, Amos	22	Pvt.	PA	N	4/1/61	6/1/61		
31	Bridenstine, Jacob (1)	34	Pvt.	PA 99th Inf.	B	2/24/65	6/3/65	Wounded	Sailor's Creek
32	Bridenstine, Jacob (2)	33	Sgt.	IA 29th Inf.	E	11/12/62	8/10/65		
33	Bridenstine, Joseph	29	Pvt.	PA 67th Inf.	F	9/29/64	6/20/65		
34	Bridenstlne, Lemuel E	18	Pvt.	PA McKeage's Mil.	H	7/2/63	8/8/63		
35	Bridenthal, David S	43	Pvt.	PA 99th Inf.	F	3/14/65	7/1/65		
36	Bridenthal, Henry B	46	Pvt.	PA 55th Inf.	H	10/11/61	6/26/63		
37	Bridenthal, Thomas	17	Pvt.	PA 13th Cav.	D	9/30/62	7/14/65		
38	Briggle, Jacob	30	Pvt.	PA 138th Inf.	E	8/17/62	12/19/63		
39	Briggle, John M	22	Pvt.	PA 99th Inf.	H	2/25/65	7/1/65		
40	Bright, Lewis*	28	Pvt.	PA 55th Inf.	D	10/14/63	8/30/65		
41	Brightbill, Samuel*		Pvt.	PA 209th Inf.	K	8/31/64		Wound.-Died	Ft. Stedman
42	Brindle, Jonathan	37	Pvt.	PA 78th Inf.	K	2/28/65			
43	Briner, Jacob D	28	Pvt.	PA McKeage's Mil.	H	7/2/63	8/8/63		
44	Brininger, Simon*	41	Pvt.	PA 55th Inf.	I	10/15/61	5/14/62		
45	Brisbin, Ezra D	32	Capt.	PA 110th Inf.	C	10/24/61	6/16/62		
46	Broad, Isaac N	27	Pvt.	PA 55th Inf.	H	9/23/61		Died	
47	Broe, Tobias		Pvt.	PA 51st Inf.	E	2/1/64	5/1/65		
48	Brookins, John	27	Pvt.	PA 55th Inf.	H	8/25/62	5/29/65		
49	Brooks, Jonas W*		Pvt.	PA 110th Inf.	C	12/19/61	6/28/65		
50	Brooks, William H	21	Pvt.	MD 1st Reg Cav.	I	9/3/61	11/1/61		
51	Brooks, William H		Pvt.	MD 3rd PHB Inf.	B	11/1/61	11/1/65		
52	Brower, George		Pvt.	PA 6th Res.	K	4/23/61	5/31/64		
53	Brower, George		Corp.	PA 191st Inf.	E	5/31/64	6/28/65	POW	Globe Tavern
54	Brown, Benjamin F	18	Pvt.	PA 208th Inf.	H	9/5/64	6/1/65	Wounded	Ft. Stedman
55	Brown, Caleb	29	Pvt.	PA 67th Inf.	E	11/28/64	7/14/65		
56	Brown, Charles*		Pvt.	PA 55th Inf.	I	10/10/63			
57	Brown, George	27	Pvt.	PA 8th Res.	F	7/1/61	10/26/62	POW	Seven Days Battle
58	Brown, George			US 6th Cav.		10/26/62			
59	Brown, George D	20	Pvt.	PA 184th Inf.	A	5/12/64	6/9/65	Wounded	Petersburg
60	Brown, George D		Pvt.	PA 184th Inf.	A			Wounded	Cold Harbor
61	Brown, Henry		Pvt.	USCT 3rd Inf.	I	12/6/64	10/31/65		
62	Brown, Jacob	25	Pvt.	PA 14th Inf.	I	5/2/61	8/7/61		
63	Brown, Jacob		Pvt.	PA 125th Inf.	F	8/12/62	5/18/63		
64	Brown, Jacob D	24	Corp.	PA 101st Inf.	D	2/8/62	6/12/65	POW	Plymouth, NC

	Cemetery	Notes
1	Old Union Cem.-Buffalo Mills	1890 Juniata Twp Veterans Census; 1850 Harrison Twp. Census
2		B.S.F. History Book-1884; Johnstown references
3		B.S.F. History Book-1884
4		B.S.F. History Book-1884; Historical Data Systems-Bedford residence
5	St. Patricks Cem.- Newry	B.S.F. History Book-1884; transferred to Veterans Reserve Corp. 9/16/64
6		B.S.F. History Book-1884; Pottsville references
7		Wounded 9/17/62; B.S.F. History Book-1884; Historical Data Systems-Bedford Co. residence
8	Fockler Cem.-Saxton	Frank McCoy 1912 Listing; 1890 Saxton Veterans Census
9		B.S.F. History Book-1884; Frank McCoy 1912 Listing
10	Hinish Family Cem.-Cypher	Died of Disease 2/8/62; 1850 Hopewell Twp. Census; Bedford Gazette 2/13/1914 listing
11	Poplar Grove Nat'l Cem.	Died 12/30/64; PA Civil War Project record; William & Reuben-brothers
12	Zion Park-Cumberland MD	Wounded 5/5/64; 1860 Cumberland Valley Census
13	Centenary Meth.-Cumberland	Wounded 5/5/64; Benjimen & Andrew-brothers
14	Bald Hill Cem.-Inglesmith	1890 Cumberland Valley Veterans Census
15	Trinity UCC Cem.-Juniata Twp.	Wounded 3/25/65; 1890 Junaita Twp. Veterans Census
16	Youngwood-Westmoreland Co.	1890 Juniata Twp. Veterans Census; Shannon & Henry-brothers
17	Trinity UCC Cem.-Juniata Twp.	1890 Harrison Twp. Veterans Census; Shannon & Henry-sons
18	Eastwood IOOF Cem.-OR	1850 Colerain Twp. Census; Historical Data Systems record
19		B.S.F. History Book-1884; Historical Data Sys.-enlisted in Bedford Co.
20	Rose Hill Cem.-Altoona	B.S.F. History Book-1884; 1860 Woodbury Twp. Census; aka Breckbill
21		Wounded 5/28/64
22	Stonerstown Cem.-Liberty Twp.	PA Civil War Project record; Daniel & Abraham-brothers
23		Historical Data Systems-Bedford Co. residence
24	Fockler Cem.-Saxton	Left leg Wound 9/17/62; 1890 Saxton Veterans Census
25	Bedford Cem.	Historical Data Systems record
26		Frank McCoy 1912 list & Civil Works Admin. 1934 list; Armstrong Co references
27		Frank McCoy 1912 list; 1860 Bedford Twp. Census; Historical Data Systems record
28	Andersonville Cem., 7125	MIA listed in Bedford Inquirer 7/8/64 article; POW-Andersonville; Died 7/22/64
29	Wells Valley Chapel Cem.	1890 Wells Twp. Veterans Census
30	Synagogue-Church God-OH	1890 Liberty Twp. Veterans Census
31	Fockler Cem.-Saxton	Wounded 4/6/65, Historical Data Systems record
32	Leib Cem.-KS	Ancestry.com-born in Broad Top City
33	Wells Valley Meth. Cem.	1890 Wells Twp. Veterans Census
34	Walnut Hill Cem.-IA	B.S.F. History Book-1884; Lemuel, Amos & Jacob (1)-brothers
35	Hickory Bottom Cem.-Woodbury	Frank McCoy 1912 Listing; 1860 Woodbury Twp. Census
36	Mt. Smith Cem.-Bedford	B.S.F. History Book-1884; ; Discharged on Surgeon's Cert.; Thomas-Son
37	Schellsburg Cem.	1890 Napier Twp. Veterans Census; Henry-Father
38	Dudenville Cem.-MO	B.S.F. History Book-1884; PA Civil War Archives-St. Clairsville resid.; Discharged Surg. Cert.
39	Greenfield Cem.-Queen	1890 Kimmel Twp. Veterans Census
40		B.S.F. History Book-1884; Reading, PA references
41	Arlington Nat'l Cem., 13 10461	Wounded 3/25/65; Died 4/10/65; Bedford Inquirer 5/22/1908 listing; Lebanon Co. references
42	Reformed Cem.-Marklesburg	1890 Hopewell Veterans Census
43		B.S.F. History Book-1884-enlisted in Bedford
44	Cedar Grove-Huntingdon Co.	B.S.F. History Book-1884; Blair Co. references; Discharged on Surgeon's Cert.
45	Grandview Cem.-Tyrone	Historical Data Systems-Bedford residence; Resigned; Son born Woodbury-1859
46	Cypress Hill Cem.-NY	Died 3/8/64; Historical Data Systems-Bedford Co. residence; enlisted in Schellsburg
47		1890 Cumberland Valley Veterans Census
48		B.S.F. History Book-1884
49		B.S.F. History Book-1884; Huntingdon references
50	Mt. Pleasant Cem.-Mattie	Historical Data Systems record
51		Historical Data Systems record
52	Bedford Cem.	PA Civil War Project record
53		PA Civil War Archives-POW 8/19/64 to 3/12/65
54		Wounded 3/25/65; Historical Data Systems-Bedford Co. residence
55	Everett Cem.	1890 Everett Veterans Census; references as Braun
56		Drafted and Desert 5/ 9/64; B.S.F. History Book-1884
57		POW 6/27/62; B.S.F. History Book-1884; Historical Data Systems-Bedford Co. residence
58		B.S.F. History Book-1884
59	Potter Creek Cem.-Woodbury	Wounded 11/24/64-Arm amputated; George, Jacob, John and Samuel-brothers
60		Wounded-Bedford Inquirer 7/1/64 article
61	Mt. Zion Cem.-Breezewood	1890 Everett Veterans Census
62	Fockler Cem.-Saxton	Historical Data Systems record
63		1890 Robertsdale Veterans Census
64	Potter Creek Cem.-Woodbury	POW-Andersonville; 4/20/64 to 12/6/64; B.S.F. Hist. Book-1884; 1860 Woodbury Twp. Cen.

Chapter 18

	Name	Age	Rank	Regiment	Co.	Muster In	Muster Out	Casualty	Casualty Battle
1	Brown, Jacob H		Pvt.	KS 11th Cav.	E	8/14/63	9/1/65		
2	Brown, James A		1st Lt.	PA McKeage's Mil.	H	7/2/63	8/8/63		
3	Brown, James H		Pvt.	PA 22nd Cav.	I	2/20/64			
4	Brown, James*		2nd Lt.	PA 55th Inf.	I	10/7/63	8/30/65		
5	Brown, Jeremiah	35	Pvt.	PA 13th Inf.	G	4/25/61	8/6/61		
6	Brown, Jeremiah		Pvt.	US 5th Light Art.	F	9/1/61		Died	
7	Brown, John (1)			PA 76th Inf.	E				
8	Brown, John (2)		Pvt.	USCT 3rd Inf.	F	9/7/64	10/31/65		
9	Brown, John D	31	Pvt.	PA 61st Inf.	A	2/25/64	2/28/65		
10	Brown, John W	25	Pvt.	PA 101st Inf.	D	2/8/62	6/20/65	POW	Plymouth, NC
11	Brown, Joseph L	33	Pvt.	PA 101st Inf.	G	12/28/61	6/25/65	Wound.-POW	Plymouth, NC
12	Brown, Max*		Pvt.	PA 55th Inf.	K	9/2/62		KIA	Cold Harbor
13	Brown, Phlip S	24	Pvt.	PA 184th Inf.	A	5/12/64	7/14/65		
14	Brown, Samuel D	19	Pvt.	PA 101st Inf.	D	11/1/61	6/13/65	POW	Plymouth, NC
15	Brown, Todd	22	Pvt.	USCT 3rd Inf.	F	7/10/63	10/31/65		
16	Brown, William P	21	Pvt.	PA 133rd Inf.	C	8/13/62	3/12/63		Fredericksburg*
17	Brown, William*	22	Pvt.	PA 184th Inf.	A	5/12/64		Died	
18	Brumbaugh, David D		Pvt.	PA 50th Inf.	K		7/31/65		
19	Brumbaugh, Francis M		Pvt.	PA 110th Inf.	C	2/24/64	6/28/65	Wound.-Died	Wilderness
20	Brumbaugh, Jacob	23	Pvt.	PA 125th Inf.	B	8/13/62	5/18/63		
21	Brumbaugh, Levi	15	Pvt.	PA 8th Res.	F	6/19/61	5/26/64	POW	Seven Days Battle
22	Bruner, Albert G	17	Pvt.	PA 210th Inf.	C	9/5/64	5/30/65		
23	Bruner, Benjamin	37	Pvt.	PA 27th Inf.	D	6/10/63	7/31/63		
24	Bruner, George	24	Pvt.	PA 126th Inf.	B	8/12/62	5/20/63	Wounded	Chancellorsville
25	Bruner, Israel		Pvt.	MD 2nd PHB Inf.	D	4/8/65	5/29/65		
26	Bruner, Jacob L	22	Pvt.	PA 93rd Inf.	H	8/26/63	6/13/65		
27	Bryant, James	25	Pvt.	PA 125th Inf.	F	8/12/62	5/18/63	Wounded	Antietam
28	Bryant, John	26	Pvt.	PA 107th Inf.	D	2/21/62		Wound.-Died	Culpeper C.H.
29	Bryant, Robert								
30	Bryant, Simon*							Died	2nd Bull Run
31	Bryant, William			MD				POW-Died	
32	Bucher, George H*		Pvt.	PA 55th Inf.	K	9/23/63	8/30/65		
33	Buck, Christian M*		Pvt.	PA 22nd Cav.	I	6/29/63	10/31/65		
34	Buck, John	28	Pvt.	PA 107th Inf.	H	2/24/62	4/2/62		
35	Buck, John		Sgt.	PA McKeage's Mil.	D	7/2/63	8/8/63		
36	Buckenmoyer, Joseph*		Pvt.	PA 76th Inf.	E	9/30/61	3/24/62		
37	Buckland, J H*		Pvt.	PA 76th Inf.	E	9/23/64	3/29/65		
38	Bukherne, William S		Pvt.	PA 71st Inf.	B	9/21/64	6/1/65		
39	Bulger, Andrew	19	Pvt.	PA 110th Inf.	C	10/24/61		Died	
40	Bulger, Daniel B	47	Pvt.	PA 208th Inf.	H	9/8/64	6/1/65		
41	Bulger, David B	35	Pvt.	PA 208th Inf.	H	9/5/64	6/1/65		
42	Bulger, Levi M	20	Corp.	PA 110th Inf.	C	10/24/61	6/28/65		
43	Burch, Thomas H	20	Corp.	PA 133rd Inf.	K	8/18/62	5/26/63		
44	Burge, Joseph		Pvt.	PA 138th Inf.	E	9/19/64	6/23/65		
45	Burger, Joseph S	32	Pvt.	PA 77th Inf.	F	2/28/65	12/6/65		
46	Burges, Adam*		Pvt.	PA 125th Inf.	E	8/13/62		Wound.-Died	Antietam
47	Burget, Isaac	19	Pvt.	PA 133rd Inf.	C	8/8/62	5/26/63		
48	Burk, Cory S			USCT 5th Inf.	A				
49	Burk, Thomas		Pvt.	USCT 55th Inf.	D	6/22/63	9/16/65		
50	Burke, Samuel	36	Pvt.	PA 84th Inf.	A	10/1/61		Wound.-Died	1st Kernstown
51	Burket, Abraham	19	1st Sgt.	PA 11th Cav.	G	9/5/61	8/13/65		
52	Burket, Albert L	15	Pvt.	MD 2nd PHB Inf.	B	2/21/65	6/29/65		
53	Burket, Baltzer	19	Pvt.	PA McKeage's Mil.	G	7/2/63	8/8/63		
54	Burket, Baltzer		Pvt.	PA 55th Inf.	K	3/2/64		KIA	Bermuda Hundred
55	Burket, David (1)	21	Pvt.	PA 138th Inf.	E	8/29/62	6/23/65	Wounded	Mine Run
56	Burket, David (2)	22	Pvt.	PA 77th Inf.	F	3/6/65	12/6/65		
57	Burket, Elias	29	Pvt.	PA 91st Inf.	G	9/21/64	3/30/65		
58	Burket, Elias S	27	Pvt.	PA 84th Inf.	D		1/13/65	Wounded	Chancellorsville
59	Burket, Elias S		Pvt.	PA 57th Inf.	G				
60	Burket, F C	25	Pvt.	PA 63rd Inf.	E	9/9/63			
61	Burket, Francis T	19	Pvt.	PA 210th Inf.	C	9/8/63	6/8/65		
62	Burket, Frederick	33	Pvt.	PA 55th Inf.	K	3/2/64		POW-Died	Drewry's Bluff
63	Burket, Gabriel	26	Pvt.	PA 55th Inf.	K	2/19/64	8/30/65	Wounded	
64	Burket, George S	23	Pvt.	PA 14th Inf.	I	5/2/61	8/7/61		

	Cemetery	Notes
1		Historical Data Systems record; Born Bedford Co.
2		B.S.F. History Book-1884
3	Presby. Cem.-Bedford	B.S.F. History Book-1884
4		B.S.F. History Book-1884; Philadelphia residence
5		B.S.F. History Book-1884; 1860 Bedford Census
6		Graves visited by Vets-Bedford Inquirer 1868 article; Historical Data Systems record
7	Potter Creek Cem.-Woodbury	Morrsions Cove listing; Born 3/1/40
8	Mt. Ross Cem.-Bedford	Frank McCoy 1912 Listing
9	Everett Cem.	Frank McCoy 1912 Listing; PA Civil War Archives-Bedford Co. residence
10	West Union Cem.-OH	POW-Andersonville; 6/20/64 to 2/26/65; 1850 M. Woodbury Twp. Census
11	Schellsburg Cem.	POW 4/20/64 to 2/26/65; Historical Data Systems-Bedford Co. residence
12		KIA 6/3/64; Historical Data Systems-enlisted Philadelphia
13	Greenlawn Cem.-Roaring Spring	B.S.F. History Book-1884; 1860 S. Woodbury Twp. Census
14	Potter Creek Cem.-Woodbury	POW-Andersonville & Florence; 4/20/64 to 12/11/64-Exchanged Charleston, SC
15		Frank McCoy 1912 list; Historical Data Systems record
16		B.S.F. History Book-1884; Historical Data Systems Bedford Co. resid.; DischargedSurg. Cert.
17	Arlington Nat'l Cem., 13 6417	Died 7/19/64; B.S.F. History Book-1884; ; Some in Company A recruited in Harrisburg
18	Spring Hope Cem.-Martinsburg	1860 Woodbury Twp. Census; born 1837
19		Wounded-Died listed in Pension Affidavit-Bed. Hist. Society files; B.S.F. History Book-1884
20	Presb. Cem.-Williamsburg	1860 Woodbury Twp. Census; Jacob & David-brothers
21	Canyon Hill Cem.-ID	POW 6/27/62; B.S.F. History Book-1884; PA Civil War Archives-Bedford Co. residence
22	Mt. Olivet Cem.-Manns Choice	Frank McCoy 1912 Listing; 1860 Cumberland Valley Census
23	Mt. Calvary Cem.-Fayette Co.	Benjimen & Jacob-brothers
24	Bedford Cem.	Wounded 5/3/63; Bedford Inquirer 5/22/1908 listing
25	Centenary Meth.-Cumberland	Frank McCoy 1912 Listing
26	Cooks Mill Cem.-Hyndman	Historical Data Systems record; aka Brunner
27	Stonerstown Cem.-Liberty Twp.	Wounded 9/17/62; Frank McCoy 1912 Listing; James, Simon & John-brothers
28	US Soldiers & Airmans Nat'l	Died 8/2/62; 1890 Robertsdale Veterans Census aka Bryan; 1850 Liberty Township Census
29	Bedford Cem.	Bedford Inquirer 5/22/1908 listing
30		Ancestry.com-Died-Bull Run Aug. 1862; aka Simeon
31		Unknown DOD; POW-Libby; 1890 Bedford Twp. Veterans Census; aka Brant, William
32		B.S.F. History Book-1884; Reading, PA references
33	Tyrone Cem.	B.S.F. History Book-1884; Reading, PA references
34	Green Lawn Cem.-OH	Historical Data Systems-Bedford Co. residence; Discharged on Surgeon's Cert.
35		B.S.F. History Book-1884; 1860 W. Providence Twp. Census
36		B.S.F. History Book-1884; York references; Discharged on Surgeon's Cert.
37		B.S.F. History Book-1884; Wayne Co. references
38		1890 E. St. Clair Veterans Census
39	Dry Hill Cem.-Woodbury	Died of Disease 2/15/63; 1860 Woodbury Twp. Census
40	Dry Hill Cem.-Woodbury	1890 Woodbury Twp. Veterans Census ; aka Bolger
41	Dry Hill Cem.-Woodbury	Frank McCoy 1912 Listing; 1860 Woodbury Twp. Census
42	West Lawn Cem.-OH	Historical Data Systems-Woodbury residence; Levi & Andrew-brothers
43	Forest Cem.-OH	B.S.F. History Book-1884; 1860 S. Woodbury Twp. Census
44		B.S.F. History Book-1884
45	Burger Cem.-Salemville	Historical Data Systems record; aka Berger
46	US Soldiers & Airmans Nat'l	Wounded 9/17/62; Died 9/30/62; Bedford Gazette 2/13/1914 listing; Cambria Co. references
47	Diehl's Crossroads-Henrietta	B.S.F. History Book-1884; aka Burgett
48		1890 Everett Veterans Census
49	Mt. Zion Cem.-Breezewood	Frank McCoy 1912 Listing; Historical Data Systems aka Burke; Discharged on Surgeon's Cert.
50		Wounded 3/23/62; Died 2/22/63; 1860 Union Twp. (Pavia) Census; Hist.Data Systems record
51	Old Claysburg Cem.	1840 Union Twp. (Pavia) Census; Abraham, David & Jacob Stine-half brothers
52	Hyndman Cem.	B.S.F. History Book-1884; 1860 Londonderry Twp. Census
53		B.S.F. History Book-1884; 1860 Greenfield Twp. Census
54		KIA 5/20/64; PA Civil War Archives-Bedford Co. residence; 1860 Greenfield Twp. Census
55	Mt. Zion Cem.-Pavia	Wounded 11/27/63; PA Civil War Archives-St. Clairsville residence
56	Hinish Family Cem.-Cypher	PA Civil War Archives record
57	Mt. Zion Cem.-Pavia	1860 Union Twp. (Pavia) Census; Elias, David (1) & Gabriel-brothers
58	Nicodemus Cem.-Henrietta	1860 Union Twp. (Pavia) Census; Pension Record; aka Burgett
59		Historical Data Systems record
60		Historical Data Systems-Born in Bedford Co.
61	Hyndman Cem.	1890 Londonderry Twp. Veterans Census
62	Andersonville Cem., 11024	Wounded 5/16/64; POW-Andersonville; Died 10/16/64; 1860 Union Twp. (Pavia) Census
63	Mt. Zion Cem.-Pavia	1890 Union Twp. (Pavia) Veterans Cen-Knee Wound; Casualty list Bedford Inquirer - 6/3/64
64		1860 Greenfield Twp. Census; buried in Pavia

	Name	Age	Rank	Regiment	Co.	Muster In	Muster Out	Casualty	Casualty Battle
1	Burket, George S		Pvt.	PA 13th Cav.	D	9/19/63	7/14/65		
2	Burket, Henry	20	Corp.	PA 76th Inf.	E	10/9/61	11/28/64		
3	Burket, Isaac	22	Pvt.	PA 138th Inf.	D	9/2/62	6/28/65		
4	Burket, Jacob	19	Pvt.	PA 55th Inf.	D	10/12/61	8/30/65		
5	Burket, Jacob D	20	Pvt.	PA 171st Inf.	I	11/2/62	8/8/63		
6	Burket, James	20	Pvt.	PA 99th Inf.	H	2/25/65	7/7/65	Died	
7	Burket, John B	21	Pvt.	PA 76th Inf.	E	6/4/64			
8	Burket, John D			MD 2nd PHB Inf.					
9	Burket, John N	22	Pvt.	PA 138th Inf.	D	8/29/62	4/29/65		
10	Burket, Noah	23	Pvt.	PA 12th Cav.	M	2/15/62	6/2/65		
11	Burket, Philip	17	Pvt.	PA 55th Inf.	D	2/27/64		Died	
12	Burket, Samuel	36	Pvt.	PA 107th Inf.	D	9/21/64	6/21/65		
13	Burket, William			PA 99th Inf.				MIA	
14	Burkey, Aaron		Pvt.	PA 21st Cav.	E	7/18/63	2/20/64		
15	Burkhart, Solomon C	22	Pvt.	PA 158th Inf.	B	10/1/62	11/14/62		
16	Burkhart, Solomon C		Pvt.	PA 21st Cav.	L	7/11/64	7/21/65		
17	Burkholder, George		Pvt.	PA 125th Inf.	H	8/14/62	1/13/63	Wounded	Antietam
18	Burley, George W*		1st Lt.	PA 110th Inf.	C	10/24/61	6/16/62		
19	Burley, John	30	Pvt.	MD 2nd PHB Inf.	K	9/25/62	6/11/65	Wounded	
20	Burmingham, Thomas*		Pvt.	PA 55th Inf.	K	7/22/63	8/30/65		
21	Burns, Charles S		1st Sgt.	PA 76th Inf.	E	10/18/64	7/18/65		
22	Burns, Francis P	17	Pvt.	PA 104th Inf.	E	2/18/65	8/25/65		
23	Burns, James	44	Corp.	PA 91st Inf.	B	10/15/61			
24	Burns, James M*	21	Pvt.	PA 55th Inf.	I	8/28/61		Wound.-Died	White Oak Road
25	Burns, John		Pvt.	NY 13th Cav.	L	9/1/63	7/1/65		
26	Burns, Joseph H	27	Pvt.	PA McKeage's Mil.	F	7/3/63	8/8/63		
27	Burns, Joseph H		Pvt.	PA 82nd Inf.	K	9/21/64		KIA	Sailor's Creek
28	Burns, Lafayette W		Pvt.	PA 208th Inf.	H	9/9/64	6/1/65		
29	Burns, Michael	24	Pvt.	PA 205th Inf.	C	8/27/64	6/2/65		
30	Burns, Oliver P		Pvt.	PA 171st Inf.	H	11/1/62			
31	Burns, Sylvester	18	Pvt.	PA 22nd Cav.	K	2/27/64	6/24/65		
32	Burns, William H (1)	24	Sadd.	PA 22nd Cav.	I	2/23/64	10/31/65		
33	Burns, William H (2)		Pvt.	PA 2nd Heavy Art.	D	9/12/62	6/26/65		
34	Burrows, George H	19	1st Sgt.	PA 35th Inf.	C	7/7/63	8/7/63		
35	Burton, Morgan	22	Pvt.	PA 158th Inf.	H	11/4/62			
36	Burton, Morgan		Pvt.	PA 56th Inf.	F	9/19/64	5/31/65		
37	Bush, Charles		Pvt.	PA McKeage's Mil.	H	7/2/63	8/8/63		
38	Bush, Charles	19	Pvt.	PA 55th Inf.	K	3/2/64	5/29/65		
39	Bush, George	17	Pvt.	PA 3rd Heavy Art.	L	2/23/64	11/9/65		
40	Bush, John	21	Pvt.	PA 99th Inf.	B	2/23/65	7/1/65	Wounded	
41	Bussard, Andrew	21	Pvt.	PA 78th Inf.	K	3/7/65	7/18/65		
42	Bussard, Daniel S	25	Pvt.	PA 77th Inf.	F	10/9/61	8/1/63		
43	Bussard, Emanuel S	31	Pvt.	PA 208th Inf.	K	9/7/64	6/26/65		
44	Bussard, John S	34	Pvt.	PA 99th Inf.	G	2/24/65	5/31/65		
45	Bussard, Joseph S	23	Pvt.	PA 133rd Inf.	K	8/7/62	2/14/63		Fredericksburg*
46	Bussard, Joseph S		Corp.	PA 208th Inf.	K	9/7/64	6/26/65		
47	Bussard, Simon S	22	Pvt.	PA 99th Inf.	H	2/25/65	5/30/65		
48	Butler, Andrew	43	Pvt.	PA 55th Inf.	K	11/5/61		Died	
49	Butler, Cyrus*		Pvt.	PA 55th Inf.	K	9/23/63	8/30/65		
50	Butterfield, G W	24	Pvt.	PA 133rd Inf.	K	8/15/62			
51	Butts, James B	24	Music.	PA 133rd Inf.	C	8/13/62	5/26/63		
52	Butts, James B		Music.	PA 184th Inf.	A	5/12/64	7/14/65		
53	Butts, John W	22	Pvt.	PA 2nd Heavy Art.	B	3/25/64	7/13/65		
54	Buxton, George W	21	Pvt.	PA 55th Inf.	D	10/12/61	1/29/63		
55	Byerly, James F	18	Pvt.	PA 55th Inf.	K	11/5/61	11/16/62		
56	Byerly, James F		Pvt.	US 1st Light Art.	M	11/16/62			
57	Byers, Peter			USCT Inf.					
58	Cable, Henry*		Pvt.	PA 55th Inf.	K	11/5/61	11/13/62		
59	Cable, Henry*		Pvt.	US 1st Light Art.	M	11/13/62			
60	Cain, John	43	Pvt.	PA 55th Inf.	D	10/15/63	8/30/65	Wounded	Cold Harbor
61	Caldwell, Charles W*		Pvt.	PA 76th Inf.	E	11/26/61			
62	Caldwell, James B*	21	Pvt.	PA 101st Inf.	G	12/2/61		Died	
63	Calhoun, Christopher P	20	1st Lt.	PA 138th Inf.	F	8/29/62	6/23/65		
64	Calhoun, David C	36	Pvt.	PA 82nd Inf.	C	11/14/64	7/13/65		

	Cemetery	Notes
1		Ancestry.com information
2	New Baltimore Cem.	PA Civil War Archives-Bedford residence; Henry, Albert & Philip-brothers
3	Palo Alto Cem.-Hyndman	Historical Data Systems record; Isaac & Francis-bros.; John B-father
4	Pleasant Hill Cem.-Imlertown	Historical Data Systems record; Frank McCoy 1912 Listing
5	Old Claysburg Cem.	1890 S. Woodbury Twp. Census
6		Died of Disease; Bedford Gazette 2/13/1914 listing
7		1890 Londonderry Twp. Veterans Census; Deserted 3/24/65
8	Burket Cem.-Londonderry Twp.	Frank McCoy 1912 Listing; Bedford Inquirer 5/22/1908 listing
9	Bedford Cem.	Historical Data Systems-Bedford Co. residence
10	Bedford Cem.	1890 Napier Twp. Veterans Census; Discharged on Surgeon's Cert.
11	Mt. Olivet Cem.-Manns Choice	Died of Disease 4/18/64; Historical Data Systems record; Frank McCoy 1912 Listing
12	Mt. Olivet Cem.-Manns Choice	Frank McCoy 1912 Listing; aka Burkhart
13		1890 Kimmel Twp. Veterans Census; Bed. Gazette 2/13/1914 list; 1860 Union Twp. Census
14		Historical Data Systems-Bedford Co. residence
15	Fort Littleton Cem.	Historical Data Systems record; Civil War Pension record; Discharged on Surgeon's Cert.
16		1890 Robertsdale Veterans Census
17	Saxton Cem.	Wounded 9/17/62; Bedford Inquirer 5/22/1908 listing
18		B.S.F. History Book-1884; Resigned; Tyrone references
19	Hyndman Cem.	1890 Londonderry Twp. Veterans Census-leg & arm Wound
20		B.S.F. History Book-1884; Drafted; enlisted Philadelphia
21		B.S.F. History Book-1884
22	St. John Cem.-Altoona	Francis, Michael, Sylvester & William-brothers
23	Hyndman Cem.	Deserted 9/9/62
24		Died 3/30/65; B.S.F. History Book-1884; PA Civil War Archives enlisted in Huntingdon Co.
25		1890 Bedford Twp. Veterans Census
26		B.S.F. History Book-1884
27		KIA 4/6/65; 1860 Londonderry Twp. Census; Bedford Gazette 2/13/1914 listing
28	Dry Hill Cem.-Woodbury	1860 Woodbury Twp. Census; 1890 Woodbury Twp. Veterans Census
29	St. John Cem.-Altoona	1860 Woodbury Twp. Census
30		Deserted 11/22/62; 1850 Napier Twp. Census; Oliver & Joseph-brothers
31	St. John Cem.-Altoona	1890 Woodbury Twp. Veterans Census
32	St. John Cem.-Altoona	B.S.F. History Book-1884; 1860 Woodbury Twp. Census
33		1890 Robertsdale Veterans Census
34		Historical Data Systems-Bedford Co. residence
35	Hendershot Cem.-Fulton Co.	Deserted 12/12/62; Historical Data Systems record
36		1890 Brush Creek Twp. Veterans Census
37		B.S.F. History Book-1884
38	Lybarger Cem.-Madley	PA Civil War Archives-Bedford Co. residence
39	St. Pauls Cem.-Cessna	1850 Union Twp. (Pavia) Census
40	Old Claysburg Cem.	Historical Data Systems record; 1850 Union Twp. (Pavia) Census; John & George-brothers
41	Everett Cem.	Bedford Inquirer 5/22/1908 listing; Frank McCoy 1912 Listing
42	Bethal Frame Ch.-Monroe Twp.	1890 Monroe Twp. Veterans Census; 1860 W. Providence Twp. Census; aka Buzzard
43	Everett Cem.	Historical Data Systems record; Frank McCoy 1912 Listing
44	Bethal Frame Ch.-Monroe Twp.	1890 Monroe Twp. Veterans Census; John, Joseph, Daniel, Emanuel, Simon & Andrew-bros.
45	Everett Cem.	B.S.F. History Book-1884; Buzzard spelling listed; Discharged on Surgeon's Cert.
46		B.S.F. History Book-1884
47	Mt. Union Cem.-Clearville	PA Civil War Project record
48	Beaufort Nat'l Cem.	Died Remittent Fever 10/11/62; Hist. Data Sys.-Bedford Co. resid.; 1850 Greenfield Twp. Cen.
49		B.S.F. History Book-1884; Berks Co. references
50		Deserted 9/14/62; Historical Data Systems-Bedford Co. residence
51	Saint John's Cem.-Loysburg	B.S.F. History Book-1884; 1860 S. Woodbury Twp. Census
52		1890 S. Woodbury Veterans Census
53	Stenger Hills Cem.-Peters Twp.	PA Civil War Archives-Bedford Co. residence; Discharged on Surgeon's Cert.
54	Green Hill Cem.-WV	1850 Southampton Twp. Census; Discharged on Surgeon's Cert.
55		B.S.F. History Book-1884; Historical Data Systems-Bedford Co. residence
56		Historical Data Systems record
57	Union Church-E. Prov. Twp.	Frank McCoy 1912 Listing; Bedford Inquirer 5/22/1908 listing
58		B.S.F. History Book-1884; Somerset Co. references
59		Historical Data Systems record
60		Wounded 6/3/64; B.S.F. History Book-1884
61		Deserted 1/12/65; B.S.F. History Book-1884; Springfield, PA references
62		Died of disease 6/12/62; Bed. Gazette 2/13/1914 list; PA Civil War Archives-Pittsburgh resid.
63	Mt. Union Cem.-Mench	1890 Bedford Twp. Veterans Census; Doctor after war
64	Mt. Union Cem.-Mench	Bedford Inquirer 5/22/1908 listing; Present at Appomatox during surrender

	Name	Age	Rank	Regiment	Co.	Muster In	Muster Out	Casualty	Casualty Battle
1	Callahan, Jacob R (1)	21	Sgt. Maj.	PA 8th Res.	F	6/28/61	5/15/64		
2	Callahan, Jacob R (1)			PA 191st Inf.	F	5/15/64			
3	Callahan, Jacob R (2)	16	Pvt.	PA 54th Inf.	A	10/21/61	5/31/65		
4	Callahan, James		Pvt.	USCT 8th Inf.	D	8/13/63			
5	Callihan, John	26	Corp.	PA 171st Inf.	I	11/10/62	8/8/63		
6	Callihan, Robert	24	Sgt.	PA 171st Inf.	I	11/2/62	8/8/63		
7	Cambell, Robert	19	Pvt.	PA 133rd Inf.	K	8/15/62	5/26/63	Wounded	Fredericksburg
8	Cambell, Samuel L		Sgt.	PA 6th Heavy Art.	K	9/2/64	6/13/65		
9	Cambell, Thomas J		Pvt.	PA 202nd Inf.	B	9/2/64	8/3/65		
10	Cameron, Amos M*		Pvt.	PA 101st Inf.	D	12/6/61		Died	
11	Campbell, Samuel F			PA Cav.					
12	Capstick, James*	21	Pvt.	PA 8th Res.	F	6/3/61	5/26/64	POW	Seven Days Battle
13	Cargill, James		Pvt.	PA 21st Cav.	E	7/8/63	2/20/64		
14	Carl, Jacob	28	Pvt.	PA 138th Inf.	E	8/29/62	6/23/65		
15	Carley, Peter A	23	Pvt.	PA 55th Inf.	H	2/29/64	6/10/65		
16	Carmack, Daniel	25	Pvt.	OH 65th Inf.	G	10/10/61		KIA	Stones River, TN
17	Carmack, John A	35	Pvt.	PA 55th Inf.	I	2/10/64		Wound.-Died	Petersburg
18	Carmack, William		Pvt.	PA 67th Inf.	C	11/22/64	7/15/65		
19	Carmack, Thomas J	15	Corp.	OH 64th Inf.	E	11/10/62	5/22/65	POW	Franklin, TN
20	Carn, Adam B	31	2nd Lt.	PA 184th Inf.	A	5/12/64	10/1/64		
21	Carn, Jeremiah	24	Pvt.	PA 149th Inf.	D	8/27/63	6/24/65	Wounded	Spotsylvania
22	Carnbaugh, Henry		Pvt.	PA 147th Inf.	B	2/2/64	7/15/65		
23	Carnell, David*		Pvt.	PA 55th Inf.	I	2/18/64	8/30/65		
24	Carnell, George W	25	Pvt.	PA 101st Inf.	D	11/1/61	4/5/63		
25	Carnell, James	42	Sgt.	PA 133rd Inf.	C	8/13/62	1/12/63		
26	Carnell, John	20	Pvt.	PA 8th Res.	F	6/19/61	1/7/63	POW	Seven Days Battle
27	Carnell, Samuel	24	Corp.	PA 101st Inf.	D	11/1/61	3/25/65	POW	Plymonth, NC
28	Carney, James M	24	Pvt.	PA 6th Heavy Art.	L	9/3/64	6/13/65		
29	Carney, Joseph	30	Pvt.	PA McKeage's Mil.	H	7/2/63	8/8/63		
30	Carney, Simon S	27	Pvt.	PA 13th Res.	E	1/22/62	5/31/64		
31	Carney, Simon S		Pvt.	PA 190th Inf.		5/31/64			
32	Carothers, William H	22	Pvt.	PA 7th Res.	I	1862	1864		
33	Carothers, William H		Pvt.	PA 190th Inf.	D	1864		POW	Globe Tavern
34	Carpenter, Abraham	20	Corp.	PA 138th Inf.	E	8/29/62	10/19/64	Wounded	Cedar Creek
35	Carpenter, Curtis J	44	Sgt.	PA 22nd Cav.	B	7/11/63	2/5/64		
36	Carpenter, David B	15	Pvt.	PA 110th Inf.	C	11/17/61	10/27/62	POW	
37	Carpenter, David B		Pvt.	US 6th Cav.	L	10/27/62	4/4/65		
38	Carpenter, Samuel		Pvt.	PA McKeage's Mil.	H	7/2/63	8/8/63		
39	Carpenter, Samuel	15	Pvt.	PA 13th Cav.	B	8/27/63	7/14/65		
40	Carpenter, William J	30	Pvt.	PA 50th Inf.	E	2/24/65	5/9/65		
41	Carrell, Daniel	18	Pvt.	PA 138th Inf.	E	8/29/62		Wounded	Mine Run
42	Carrell, Daniel		Pvt.	PA 138th Inf.	E			MIA	Monocacy, MD
43	Carrell, Joseph	20	Pvt.	PA 138th Inf.	E	7/1/63	9/19/65	Wounded	Wilderness
44	Carris, Samuel*		Pvt.	PA 76th Inf.	E	2/14/65	7/18/65		
45	Carson, Daniel	24	Pvt.	PA 133rd Inf.	C	8/13/62	5/26/63		
46	Carson, Jacob	21	Pvt.	PA 171st Inf.	I	11/2/62	8/8/63		
47	Carson, John (1)	16	Pvt.	PA 22nd Inf.	F	9/16/62	9/29/62		
48	Carson, John (2)	48	Pvt.	USCT 22nd Inf.	H	12/29/63	10/16/65		
49	Carson, Robert	33	Pvt.	MD 3rd PHB Inf.	K	8/25/63	5/29/65		
50	Carson, Samuel R	21	Pvt.	PA 55th Inf.	H	10/11/61	2/15/63		
51	Cartwright, Abraham	27	Pvt.	PA 195th Inf.	E	4/7/65	1/31/66		
52	Cartwright, Austin	15	Pvt.	PA McKeage's Mil.	G	7/2/63	8/8/63		
53	Cartwright, Franklin J	16	Pvt.	PA 205th Inf.	C	8/26/64	6/5/65	Wounded	Petersburg - Final
54	Carver, Augustus B	21	Pvt.	PA 83rd Inf.	K	3/9/65	6/28/65		
55	Cashman, Jacob	23	Pvt.	PA 107th Inf.	E	8/27/64		Wound.-Died	White Oak Road
56	Castner, Jacob H	14	Pvt.	PA 133rd Inf.	C	8/13/62	5/26/63		
57	Castner, Jacob H		Pvt.	PA 194th Inf.	I	7/22/64	9/6/64		
58	Castner, Jacob H		Pvt.	PA 22nd Cav.		9/6/64			
59	Castner, John B	42	1st Lt.	PA 133rd Inf.	F&S	8/25/62	2/12/63		
60	Castner, John W	18	Pvt.	PA 110th Inf.	C	10/24/61	10/24/64		
61	Caylor, David H*	25	Pvt.	WI 20th Inf.	I	8/13/62		Died	
62	Cessna, Charles	42	Pvt.	IL 75th Inf.	G	8/25/62	4/4/63	Died	
63	Cessna, George W	23	Pvt.	PA 76th Inf.	E	10/9/61	5/18/63		
64	Cessna, Jacob S	32	Pvt.	PA 67th Inf.	C	11/14/64	7/14/65		

	Cemetery	Notes
1		B.S.F. History Book-1884; Historical Data Systems-Bedford Co. residence
2		Historical Data Systems record
3		Historical Data Systems record; Ancestry.com-born near Bedford Springs
4		Deserted 1/16/64; Frank McCoy 1912 list; Historical Data Systems record
5		B.S.F. History Book-1884; 1850 St. Clair Twp. Census
6	East Lawn Cem.-OH	B.S.F. History Book-1884; Robert & John-brothers
7		Wounded 12/13/62; B.S.F. History Book-1884; aka Camel
8	Saint John's Cem.-Loysburg	Soldiers of Blair Co. Book - 1940; Historical Data Systems record
9		1890 Mann Twp. Veterans Census
10	St. Stephens Epis.-Juniata Co	Died 1/25/62; B.S.F. History Book-1884
11	St. John's Ref. Cem.-Loysburg	Teamster in Cavalry near Hershey; born 3/26/48
12	Fairview Cem.-Altoona	B.S.F. History Book-1884; Blair Co. references
13		Historical Data Systems-Bedford Co. residence
14	St. James Luth. Cem.-Bedford	Frank McCoy 1912 Listing; 1860 Bedford Twp. Census
15		B.S.F. History Book-1884; PA Civil War Archives-Bedford Co. resid.; Discharged Surg. Cert.
16	Stones River Nat'l Cem.-TN	1850 Woodbury Census; Daniel, Thomas & John-brothers
17	Cypress Hill Cem.-NY	Died 10/25/64; B.S.F. History Book-1884; Ancestry.com-born in Osterburg
18		1890 Hopewell Veterans Census
19	West Salem Cem.-OH	POW-Cahaba, AL; Survived Sultana Steamboat Explosion-Memphis, TN
20	Bedford Cem.	1860 Bedford Census; Bedford Inquirer 5/22/1908 listing; Discharged on Surgeon's Cert.
21	Wilmore Cem.-Portage	Wounded 5/8/64 near Laurel Hill; Ancestry.com-Born Union Twp. (Pavia)
22		1890 Hopewell Veterans Census
23		B.S.F. History Book-1884; Washington Co. references
24		PA Civil War Archives-Bedford Co. resid.; Discharged Surg, Cert.; George & Samuel-brothers
25		B.S.F. History Book-1884; PA Civil War Archives-Bedford Co. resid.; Discharged Surg. Cert.
26	Glenwood Cem.-DC	POW 6/27/62; PA Civil War Archives-Bedford Co. residence
27		POW 4/20/64 to 11/20/64; PA Civil War Archives-Bedford Co. residence
28	Bedford Cem.	1850 Bedford Twp. Census
29	Fairmont Cem.-NE	B.S.F. History Book-1884; Joseph, James & Simon-brothers
30	Fairmont Cem.-NE	Historical Data Systems record
31		Historical Data Systems record
32	Elizabeth Cem.-Allegheny Co.	Historical Data Systems record; 1870 Broad Top Twp. Census
33		Historical Data Systems record
34	Highland Cem.-KS	Wounded 10/19/64; Historical Data Systems record; 1860 Londonderry Twp. Census
35	Old Presb. Cem.-Bedford	Frank McCoy 1912 Listing; 1860 Woodbury Census
36	Greenlawn Cem.-Roaring Spring	B.S.F. History Book-1884; David & Samuel-brothers; Curtis-father
37		B.S.F. History Book-1884
38		B.S.F. History Book-1884
39	Saint John's Cem.-Loysburg	Frank McCoy 1912 Listing; Samuel-Father
40	Hyndman Cem.	1890 Londonderry Twp. Veterans Census; William & Abraham-brothers
41		Wounded 11/27/63; 1860 Schellsburg Census; B.S.F. History Book-1884; aka Correl & Crell
42		MIA 7/9/64; Bedford Gazette 2/13/1914 listing; Hist. Data Systems-Schellsburg residence
43		Wounded 5/6/64; B.S.F. History Book-1884; Joseph & Daniel-brothers
44		B.S.F. History Book-1884; Scranton references
45	Salemville Cem.	1890 S. Woodbury Veterans Census; Discharged on Surgeon's Cert.
46	Albrights Cem.-Roaring Springs	1890 Woodbury Veterans Census; Jacob, Daniel & John-brothers
47		1890 Everett Veterans Census
48	Everett Cem.	Frank McCoy 1912 Listing
49	Mckees Cem.-Amaranth	1890 Brush Creek Twp. Veterans Census
50	Maple Grove Cem.-OH	B.S.F. History Book-1884; 1860 Napier Twp. Census; Discharged on Surgeon's Cert.
51	Hopewell Cem.	Frank McCoy 1912 Listing; 1890 Broad Top Veterans Census
52	Hopewell Cem.	Frank McCoy 1912 Listing
53	Hopewell Cem.	Wounded 4/2/65; 1890 Broad Top Veterans Census
54	Bedford Cem.	PA Civil War Project record; 1860 St. Clair Twp. Census; aka Garver
55	Philadelphia Nat'l Cem.	Wounded 3/30/65; Died 5/11/65; Bedford Gazette 2/13/1914 listing; 1860 Woodbury Cen.
56	Shelby-Oakland Cem.-OH	B.S.F. History Book-1884; 1850 M. Woodbury Twp. Census
57		B.S.F. History Book-1884; enlisted in Hopewell
58		B.S.F. History Book-1884
59	Riverside Cem.-OH	1860 Broad Top Twp. Census; Historical Data Systems record; Jacob-son
60		B.S.F. History Book-1884; 1860 Woodbury Twp. Census
61	Springfield Nat'l Cem.-MO	Died-Typhoid Fever 10/28/62; Ancestry.com-born Bedford Co.
62		Died early 1865; 1840 Cumberland Valley Census; Historical Data Systems record
63	Woods Church-Colerain Twp.	B.S.F. History Book-1884; 1860 Colerain Twp. Census; Discharged on Surgeon's Cert.
64	Bethel Meth. Cem.-Centerville	Frank McCoy 1912 Listing; Absent in US Gen Hospital

	Name	Age	Rank	Regiment	Co.	Muster In	Muster Out	Casualty	Casualty Battle
1	Cessna, John	22	Pvt.	MD 2nd PHB Inf.	E	8/27/61	9/29/64		
2	Cessna, Joseph P	35		MI 28th Inf.	F&S	9/6/64	1/27/65		
3	Cessna, Joseph P		1st Lt.	MO 2nd Cav.					
4	Cessna, Martin	24	Pvt.	PA 13th Res.	H	7/22/61	5/31/64	POW	
5	Cessna, Martin		Pvt.	PA 190th Inf.	H	6/1/64		Died	
6	Cessna, Samuel S	18	Pvt.	MD 2nd PHB Inf.	D	4/7/65	5/29/65		
7	Cessna, William	36	Pvt.	PA 55th Inf.	K	2/19/64	6/6/65	Wounded	
8	Chalfont, William	21	Pvt.	PA 2nd Cav.	E	12/3/61	12/4/64	Wounded	Parker's Store
9	Chamberlain, David		Sgt.	PA 21st Cav.	D	7/16/63	7/8/65	Wounded	Five Forks
10	Chamberlain, Eli G	17	Pvt.	PA McKeage's Mil.	G	7/2/63	8/8/63		
11	Chamberlain, Eli G		Pvt.	PA 208th Inf.	K	9/7/64		KIA	Ft. Stedman
12	Chamberlain, Fernando C	16	Pvt.	PA 194th Inf.	I	7/21/64	11/5/64		
13	Chamberlain, Jacob		Corp.	PA 22nd Cav.	B	7/11/63	2/5/64	Wounded	
14	Chamberlain, Jacob	27	Corp.	PA 208th Inf.	K	9/7/64	6/1/65	Wounded	
15	Chamberlain, James*	19	Pvt.	PA 110th Inf.	C	7/24/64			
16	Chamberlain, Joseph	29	Pvt.	PA 133rd Inf.	C	8/13/62	5/26/63		
17	Chamberlain, Joseph		Pvt.	PA 107th Inf.	H	4/7/64		Wounded	Petersburg - Initial
18	Chamberlain, Joseph		Pvt.	PA 107th Inf.	H			MIA	Globe Tavern
19	Chamberlain, Philip	50	Pvt.	PA 208th Inf.	K	9/7/64	6/1/65		
20	Chamberlain, Philip H			PA 22nd Cav.					
21	Chambers, William A		Corp.	PA 22nd Cav.	D	7/17/63			
22	Chaney, Levi	19	Pvt.	PA 107th Inf.	H	2/24/62	2/11/64	Wounded	Antietam
23	Chapman, O W*	39	Pvt.	PA 76th Inf.	E	9/23/64	7/18/65		
24	Chapman, William F*	45	Lt. Col.	IL 38th Inf.	G	9/1/61		Died	
25	Chappell, Charles W	18	Pvt.	PA 202nd Inf.	I	8/30/64	8/3/65		
26	Charleston, John			PA 76th Inf.	E				
27	Charleston, John		Pvt.	PA 22nd Cav.	I	3/5/65	6/24/65		
28	Chester, Edward		Pvt.	PA 76th Inf.	E	10/8/61	12/22/62		
29	Chilcoat, Andrew S	27	Music.	PA 143rd Inf.	G	6/18/63	6/12/65		
30	Chilcoat, Hilany*		Pvt.	PA 110th Inf.	C	2/25/64	6/28/65		
31	Chilcoat, Isaac	22	Pvt.	PA 110th Inf.	C	2/25/64	6/28/65		
32	Chilcott, Ephraim	25	Pvt.	IL 8th Inf.	D	7/25/61	7/30/64		
33	Chilcott, Thomas S	23	Pvt.	PA 13th Cav.	A	2/15/65	7/14/65		
34	Childers, Randall	21	Pvt.	PA 77th Inf.	A	10/9/61	12/6/65		
35	Chisholm, John P	24	Pvt.	PA 22nd Cav.	L	2/27/64	6/24/65		
36	Christ, John	31	Pvt.	PA 107th Inf.	H	9/20/64	6/6/65		
37	Cick, Anthony*		Pvt.	PA 6th Res.	D	4/24/61	6/11/64		
38	Claar, Daniel	42	Pvt.	PA 99th Inf.	H	2/20/65	5/30/65		
39	Claar, Henry I	21	Pvt.	PA 55th Inf.	K	12/4/61	8/30/65	Wounded	
40	Claar, Henry M	36	Pvt.	PA 79th Inf.	E	11/11/64	7/12/65		
41	Claar, Jacob C	27	Pvt.	PA 138th Inf.	E	8/29/62	6/23/65		
42	Claar, John	21	Corp.	PA 138th Inf.	E	8/29/62	6/9/65		
43	Claar, Lewis	23	Pvt.	PA 99th Inf.	H	2/25/65	7/1/65		
44	Claar, Martin			PA 84th Inf.					
45	Claar, Samuel	20	Pvt.	PA 1st Art.	F	7/21/61	12/1/63	Wounded	
46	Claar, Thomas			PA 84th Inf.					
47	Claar, Thomas	24	Pvt.	PA 99th Inf.	H	2/25/65	7/2/65		
48	Clapper, Henry W*		Corp.	PA 147th Inf.	B				
49	Clapper, Jacob H	28	Pvt.	PA 137th Inf.	I	8/20/62	6/1/63		
50	Clapper, Oliver W	19	Pvt.	PA 78th Inf.	K	2/28/65	9/11/65		
51	Clapper, William F	19	Pvt.	PA 19th Cav.	C	9/14/64	6/1/65		
52	Clare, William		Pvt.	PA 1st Cav.	C	6/15/63		Wounded	
53	Clark, Albert B								
54	Clark, Alexander		Pvt.	PA 194th Inf.	I	7/21/64	9/6/64		
55	Clark, Alexander	23	Pvt.	PA 97th Inf.	Ind	9/6/64	7/17/65		
56	Clark, Ferdinand	19	Pvt.	PA 208th Inf.	H	9/8/64	6/1/65	Wounded	Petersburg - Final
57	Clark, George W	40	Capt.	IA 22nd Inf.	K	9/9/62	3/13/65	Wounded	Cedar Creek
58	Clark, Isaiah								
59	Clark, James								
60	Clark, Jameson	19	Pvt.	PA McKeage's Mil.	D	7/2/63	8/8/63		
61	Clark, Jameson		Pvt.	PA 22nd Cav.	H	2/26/64		KIA	Charlestown, WV
62	Clark, Joel	38	Pvt.	IN 142nd Inf.	D	10/13/64		Died	
63	Clark, John (1)	25	Pvt.	PA 208th Inf.	K	9/7/64	6/1/65		
64	Clark, John (2)	20	Pvt.	PA 62nd Inf.	M	8/9/61	7/20/64		

	Cemetery	Notes
1	Bedford Cem.	Historical Data Systems record
2	Christian Cem.-OH	Ancestry.com-Born in Buffalo Mills
3		Unit known as Merrill's Horse Cavalry
4	Bortz Luth. Cem.-Centerville	POW 3/15/64; Frank McCoy 1912 Listing
5		Died 4/11/65; PA Civil War Project record
6	Ellsworth Cem.-OH	Historical Data Systems record; Samuel, Martin & John-brothers
7	Bald Hill Cem.-Everett	Wounded listed in Bed. Inquirer-6/3/64; 1890 Bed. Twp. Vet. Cen.; William & Charles-bros.
8	Everett Cem.	Arm Wound 11/29/63; 1890 Everett Veterans Census
9	Cedar Grove Cem.-Chambersburg	Wounded 4/1/65; 1850 E. Providence Twp. Census; David & Joseph-brothers
10		B.S.F. History Book-1884; Ancestry.com-Born Bedford Co.
11	Rodman Cem.-Roaring Spr.	KIA 3/25/65; Bedford Gazette 2/13/1914 listing; Parents buried Six Mile Run
12	Clearville Union Cem.	B.S.F. History Book-1884; 1860 W. Providence Twp. Census; Philip-father
13		Frank McCoy 1912 Listing; Historical Data Systems
14	Duval Cem.-Coaldale	B.S.F. History Book-1884; Leg Amputated; Jacob & Eli-brothers
15		B.S.F. History Book-1884; Never joined company
16		B.S.F. History Book-1884; Historical Data Systems-Bedford Co. residence
17		Wounded 6/18/64; PA Civil War Project Record; 1860 E. Providence Twp. Census
18		MIA 8/19/64; Hist. Data Systems-Bedford Co. resid.; Pension Record died in Richmond prison
19	Clearville Union Cem.	1890 Monroe Veterans Census-Disabled leg
20	Hopewell Cem.	Frank McCoy 1912 Listing; 110th IL Infantry referenced
21		Deserted 12/15/64; B.S.F. History Book-1884
22	Evans Cem.-Coaldale	Wounded 9/17/62; Transferred VRS 2/11/64; B.S.F. History Book-1884
23		B.S.F. History Book-1884; Easton references
24	Stones River Nat'l Cem.-TN	Fold3-Born Bedford Co.; Died of Disease 11-23-64
25	Mt. Zion Cem.-Pavia	Historical Data Systems record
26	Oakwood Cem.-OH	B.S.F. History Book-1884; 1850 Bedford Census
27		B.S.F. History Book-1884
28		Historical Data Systems-Bedford Co. residence; B.S.F. History Book-1884; Disch. Surg. Cert.
29	Broad Top IOOF Cem.	Historical Data Systems record; Andrew & Isaac-brothers
30		Deserted 7/24/64 & returned 1/15/65; B.S.F. History Book-1884; Huntingdon Co references
31	Coles Valley Cem.-Hunt. Co.	Deserted 7/24/64 & returned 1/15/65 - PA Civil War Project Rec.; 1890 Robertsdale Vet. Cen.
32	Topeka Cem.-KS	Ancestry.com Born in Bedford Co.
33	Hopewell Cem.	1890 Broad Top Twp. Vet. Census; Ephraim & Thomas-bros.; Ancestry.com-born Hopewell
34	Lenoir City Cem.-TN	1890 E. Providence Twp. Veterans Census
35	Union Christian-Crystal Springs	1890 Broad Top Twp. Veterans Census
36		B.S.F. History Book-1884; 1860 Union Twp. (Pavia) Census
37		B.S.F. History Book-1884; aka Zink; Berlin references
38	Lower Claar Cem.-Claysburg	Frank McCoy 1912 Listing; 1860 Union Twp. (Pavia) Census
39	Upper Claar Cem.-Queen	Wounded - summer of 1864; PA State Archives-Bedford residence; Henry & Lewis-brothers
40	Allegheny Cem.-Pittsburgh	B.S.F. History Book-1884; aka Clarr
41	Upper Claar Cem.-Queen	PA Civil War Archives lists St. Clairsville residence
42		PA State Archives-Bedford residence
43	Upper Claar Cem.-Queen	1890 Kimmel Twp. Veterans Census
44	Claar Cem.-Kimmel Twp.	Frank McCoy 1912 Listing
45	Upper Claar Cem.-Queen	Hist. Data Sys. record; Lost use of lower arm; Gunshot wound thru right leg-Pension record
46		Bedford Inquirer 5/22/1908 listing
47	Upper Claar Cem.-Queen	1860 Union Twp. (Pavia) Census
48	Stonerstown Cem.-Liberty Twp.	Findagrave Civil War Marker
49	Geeseytown Cem.-Frankstown	Historical Data Systems record; Jacob & Oliver-brothers
50	St. Pauls UCC Cem.-Russellville	1890 Hopewell Veterans Census
51	Farney Cem.-MO	1850 M. Woodbury Twp. Census
52		Wounded 6/21/64; 1890 Wells Valley Veterans Census
53		1890 Bedford Twp. Veterans Census
54		Historical Data Systems record; Ambulence driver at Gettysburg
55	Union Memorial Cem.-Mench	Historical Data Systems record; Alexander & Philip-Brothers
56		Wounded 4/2/65; B.S.F. History Book-1884
57	Oakland Cem.-IA	Historical Data Systems record; Wounded 10/19/64
58	Fockler Cem.-Saxton	Bedford Inquirer 5/22/1908 listing
59	Fockler Cem.-Saxton	Bedford Inquirer 5/22/1908 listing
60		B.S.F. History Book-1884; 1850 Monroe Twp. census
61	Mt. Pleasant Cem.-Mattie	KIA 8/22/64; Bedford Gazette 2/13/1914 listing; Historical Data Systems record
62	Nashville Nat'l Cem.	Died 6/24/65; Ancestry.com & findagrave.com listed-Born in Bedford
63	Everett Cem.	B.S.F. History Book-1884
64	Sproul Union Cem.-Claysburg	Historical Data Systems record; Soldiers of Blair Co. Book - 1940

Chapter 18

	Name	Age	Rank	Regiment	Co.	Muster In	Muster Out	Casualty	Casualty Battle
1	Clark, John (2)		Pvt.	PA 91st Inf.	K	7/20/64	7/10/65		
2	Clark, Philip	36	Pvt.	PA 91st Inf.	I	9/21/64	5/30/65		
3	Clark, Robert A	23	Pvt.	PA 101st Inf.	D	12/6/61	4/5/63		
4	Clark, Samuel M	18	Pvt.	PA 138th Inf.	E	8/29/62		KIA	Opequon
5	Clark, Simon	22	Pvt.	PA McKeage's Mil.	D	7/2/63	8/8/63		
6	Clark, Simon		Pvt.	PA 208th Inf.	K	9/7/64	6/1/65		
7	Clark, Willlam W		Pvt.	PA McKeage's Mil.	D	7/2/63	8/8/63		
8	Clark, Willlam W	33	Pvt.	PA 208th Inf.	K	9/7/64	6/1/65		
9	Clark, Zachariah	31	Pvt.	PA 107th Inf.	B	10/21/64		Died	
10	Clay, Henry*		Pvt.	PA 184th Inf.	A	5/12/64		POW-Died	Petersburg
11	Claybaugh, James	26	Pvt.	PA 3rd Cav.	G	8/17/61	8/24/64	POW	Spotsylvania
12	Claycomb, Andrew	21	Pvt.	PA 82nd Inf.	C	11/14/64	7/13/65		
13	Claycomb, Conrad	26	Pvt.	PA 138th Inf.	E	8/29/62	6/17/65	Wounded	Wilderness
14	Claycomb, Frederick	40	Pvt.	PA 55th Inf.	K	11/5/61	6/26/63	Wounded	Pocotaligo, SC
15	Claycomb, John	33	Pvt.	PA 55th Inf.	K	11/5/61	3/4/63	Wounded	Pocotaligo, SC
16	Claycomb, Nathaniel	23	Pvt.	PA 1st Light Art.	C	9/14/64	6/21/65		
17	Claycomb, Samuel	26	Pvt.	PA 11th Cav.	G	9/5/61	9/4/64		
18	Claycomb, Thaddeus*								
19	Claycomb, William L	26	Pvt.	PA 99th Inf.	C	2/25/65	5/31/65		
20	Clayes, William		Pvt.	PA 10th Res.	C				
21	Cleaveland, Fredrick W*	18	Pvt.	PA 194th Inf.	I	7/21/64	11/5/64		
22	Cleaver, Irvin B*	19	Pvt.	PA 194th Inf.	I	7/21/64	11/5/64		
23	Cleaver, James	23	2nd Lt.	PA 8th Res.	F	6/11/61	5/26/64	POW	Seven Days Battle
24	Cleaver, James		2nd Lt.	PA 8th Res.	F			Wounded	Spotsylvania
25	Cleaver, James		2nd Lt.	PA 8th Res.	F			Wounded	Fredericksburg
26	Clevenger, Adam	21	Pvt.	PA 126th Inf.	B	8/12/62	5/20/63		
27	Clevenger, Adam			PA 20th Cav.	M	1863	1864		
28	Clevenger, Adam		Pvt.	PA 22nd Cav.	K	2/18/64	6/24/65		
29	Clevenger, Franklin	18	1st Sgt.	MD 3rd Inf.	B	12/1/61	7/26/65	Wounded	Ft. Stedman
30	Clevenger, George W	26	Pvt.	PA 97th Inf.	A	2/21/65	7/23/65		
31	Clevenger, Harrison	19	Pvt.	MD 3rd PHB Inf.	B	2/21/62	5/29/65		
32	Clevenger, Jacob A	15	Pvt.	PA 184th Inf.	A	9/3/64	6/2/65		
33	Clingerman, Harrison	21	Pvt.	PA 149th Inf.	K	8/26/63	6/24/65		
34	Clingerman, Jeremiah	26	Pvt.	PA 171st Inf.	I	11/10/62		Died	
35	Clingerman, Joseph	34	Pvt.	PA 171st Inf.	I	11/5/62	8/8/63		
36	Clingerman, Peter	21	Pvt.	PA 101st Inf.	D	12/6/61	6/25/65	POW	Plymonth, NC
37	Clingerman, Peter		Pvt.	PA 101st Inf.	D			Wounded	Fair Oaks
38	Clites, Levi	18	Pvt.	PA McKeage's Mil.	H	7/2/63	8/8/63		
39	Clites, Levi		Pvt.	PA 211th Inf.	H	9/9/64	6/1/65		
40	Clites, Soloman (2)	36	Pvt.	PA 28th Inf.	H	9/21/64	7/18/65		
41	Clites, Solomon (1)	18	Pvt.	PA 77th Inf.	F	2/27/65	12/6/65		
42	Clouse, Harmon	17	Pvt.	PA 194th Inf.	I	7/21/64	11/5/64		
43	Coble, John*		Pvt.	PA 110th Inf.	C	12/19/61	6/28/65		
44	Cobler, Allen	37	Pvt.	PA 138th Inf.	E	8/29/62	6/23/65	Wounded	Cold Harbor
45	Cobler, Andrew	17	Corp.	PA 138th Inf.	E	8/29/62	6/23/65		
46	Cobler, Francis C	16	Pvt.	PA McKeage's Mil.	G	7/2/63	8/8/63		
47	Cobler, Francis C		Pvt.	PA 55th Inf.	K	3/1/64	8/30/65		
48	Cobler, George W	27	Pvt.	PA 210th Inf.	C	9/5/64	5/30/65		
49	Cobler, John A	18	Pvt.	PA 55th Inf.	K	11/5/61	8/3/65	Wounded	Bermuda Hundred
50	Cobler, Joseph	26	Pvt.	PA 138th Inf.	F	8/29/62	6/23/65	Wounded	
51	Cochrane, Michael*							Died	
52	Coffee, John	35	Pvt.	PA 55th Inf.	K	11/5/61		Died	
53	Cofran, Alphonso*		Pvt.	PA 55th Inf.	I	10/2/63	10/19/64		
54	Cole, John	22	Pvt.	PA 55th Inf.	D	10/15/63			
55	Cole, Samuel	18	Pvt.	PA 55th Inf.	H	2/19/64	8/30/65		
56	Cole, Thomas		Pvt.	MD 1st PHB Cav.	F	4/23/64	6/28/65	POW	
57	Colebaugh, George W	37	Pvt.	PA 210th Inf.	C	9/5/64	5/30/65		
58	Coleman, George	20	Pvt.	USCT 32nd Inf.	G	2/28/64	8/22/65		
59	Colledge, Jacob	29	Pvt.	PA 208th Inf.	K	9/5/64	6/1/65		
60	Colledge, Joseph R	25	Pvt.	PA 208th Inf.	K	9/5/64	6/1/65		
61	College, David	22	Pvt.	PA 110th Inf.	C	10/24/61			
62	College, David		Pvt.	PA 9th Cav.	M	2/4/64	7/18/65		
63	College, George	22	Pvt.	PA McKeage's Mil.	D	7/3/63	8/8/63		
64	College, Henry	31	Pvt.	PA 51st Inf.	A	9/21/64	6/2/65		

	Cemetery	Notes
1		Historical Data Systems record; Soldiers of Blair Co. Book - 1940
2	Union Memorial Cem.-Mench	1890 E. Providence Twp. Veterans Census
3		1850 St. Clair Twp. Census; Hist. Data Systems-Bedford Co. resid.; Discharged Surg. Cert.
4	Mt. Zion Cem.-Pavia	KIA 9/19/64; Bedford Gazette 2/13/1914 listing; Historical Data Systems record
5	Akersville Cem.-Crystal Springs	B.S.F. History Book-1884
6		B.S.F. History Book-1884; 1850 Monroe Twp. Census
7		B.S.F. History Book-1884
8	Union Memorial Cem.-Mench	B.S.F. History Book-1884
9	Alexandria Nat'l Cem.-2807	Died 5/15/65; Historical Data Systems record; 1850 Monroe Twp. Census
10	Andersonville Cem., 7,463	POW-Andersonville; Died 9/1/64; Company A recruited in Dauphin & Bedford Co.
11	Hopewell Cem.	POW 5/15/64; 1890 Broad Top Veterans Census
12	Richland Cem.-Geistown	1890 W. St. Clair Veterans Census
13	Fishertown Cem.	Wounded 5/6/64; 1860 St. Clair Twp. Census
14	Mt. Zion Cem.-Pavia	Wounded 10/22/62; Deb Colledge family record; DSC; PA State Archives
15		Lost Leg 10/22/62; Deb Colledge family record; DSC; Hist. Data Systems-Bedford Co. resid.
16	St. Paul Breth. Cem.-Imler	1860 Union Twp. (Pavia) Census
17	Rose Hill Cem.-Altoona	1850 St. Clair Twp. Census
18	Pleasantville Cem.	Born 12/11/46; Civil War Marker at gravesite
19	Trinity UCC Cem.-Osterburg	Frank McCoy 1912 Listing
20		1890 King Twp. Veterans Census
21		B.S.F. History Book-1884; Reading, PA references
22		B.S.F. History Book-1884; Reading, PA references
23	Bedford Cem.	POW 6/27/62; Bedford Gazette-8th Reserve listing
24		Wounded 5/10/64; B.S.F. History Book-1884
25		Wounded 12/13/62; Frank McCoy 1912 Listing
26	Union Cem.-McConnellsburg	Adam, George, Franklin, Harrison & Jacob-brothers
27		Historical Data Systems record
28		Ancestry.com-family lived in Crystal Springs, Hopewell & W. Providence Twp.
29	Palestine City Cem.-TX	Wounded 3/25/65; aka Charles Franklin & Clevinger
30	Fairview Cem.-Hustontown	Historical Data Systems record
31	Everett Cem.	Frank McCoy 1912 Listing; Ancestry.com-born Bedford Co.
32	Glenwood Cem.-DC	B.S.F. History Book-1884; 1870 W. Providence Township Census
33	Fairview Cem.-Artemas	Frank McCoy 1912 Listing; aka Clingaman
34		Died of Disease 6/29/63; 1860 Southampton Twp. Cen.; Bedford Gazette 2/13/1914 listing
35	Fairview Cem.-Artemas	Frank McCoy 1912 Listing; Joseph, Jeremiah, Harrison & Peter-brothers
36	Fairview Cem.-Artemas	POW-Andersonville & Florence; 4/20/64 to 12/11/64; B.S.F. History Book-1884
37		Wounded 5/31/64-Bedford Histoical Society files
38	Mt. Zion Cem.-Somerset Co.	B.S.F. History Book-1884-aka Clitz
39		Historical Data Systems record
40	Lybarger Cem.-Madley	Frank McCoy 1912 Listing; aka Clitz; Levi is son
41	Farragut Cem.-IA	PA State Archives-Born in Bedford Co.
42	Confluence Baptist-Somerset	B.S.F. History Book-1884; findagrave.com-born Bedford Co.
43		B.S.F. History Book-1884; Huntingdon Co. references
44	Schellsburg Cem.	Wounded 6/1/64; PA Civil War Archives record
45	Pleasant Hill Cem.-Imlertown	Died 12/25/67; B.S.F. History Book-1884; Andrew, John & George-bros.
46		B.S.F. History Book-1884
47		B.S.F. History Book-1884
48	Everett Cem.	1890 Monroe Twp. Veterans Census
49	Trinity UCC Cem.-Osterburg	Wounded 5/20/64-Pension record; PA Civil War Archives record
50	Woods Church-Colerain Twp.	1890 Colerain Twp. Veterans Census-Shot thru arm
51		Bedford Gazette 2/13/1914 listing
52		Died 11/10/62; PA Civil War Archives-Bedford Co. residence
53		B.S.F. History Book-1884; Discharged on Surgeon's Cert.
54	Schellsburg Cem.	Deserted 9/3/64; Historical Data Systems record; 1850 Napier Twp. Census
55		PA Civil War Archives-Bedford residence
56		POW 8/4/64 for 7 months; 1890 Hyndman Veterans Census
57	Lutheran Cem.-Osterburg	1860 Bedford Twp. Census; 1890 King Twp. Veterans Census
58	Mt. Ross Cem.-Bedford	Frank McCoy 1912 Listing; 1890 Bedford Twp. Veterans Census
59	Stonerstown Cem.-Liberty Twp.	B.S.F. History Book-1884; 1860 E. Providence Twp. Census
60	Rosedale Cem.-WV	B.S.F. History Book-1884; Joseph, George & Henry-brothers
61	Duval Cem.-Coaldale	PA Civil War Archives-Yellow Creek residence
62		PA Civil War Archives aka Colledge
63		B.S.F. History Book-1884, Born-E. Providence Twp.; died 1866
64	Mt. Zion Cem.-Breezewood	1890 E. Providence Twp. Veterans Census

Chapter 18

	Name	Age	Rank	Regiment	Co.	Muster In	Muster Out	Casualty	Casualty Battle
1	College, James (1)			PA 133rd Inf.					
2	College, James (2)	28	Pvt.	PA 110th Inf.	C	10/24/61		Wound.-Died	1st Kernstown
3	College, John W	16	Pvt.	PA 110th Inf.	C	10/24/61		Wound.-Died	1st Kernstown
4	College, Simon	27	Pvt.	PA 22nd Cav.	D	7/28/63	2/5/64		
5	College, Simon		Pvt.	PA 208th Inf.	H	9/8/64	6/1/65		
6	Collins, Isaiah	24	Pvt.	PA 91st Inf.	F	9/21/64	6/1/65	Wounded	
7	Collins, John T			PA 99th Inf.				Died	
8	Comp, Adam A	21	Pvt.	PA 82nd Inf.	C	11/14/64	7/15/65		
9	Comp, Solomon	30	Pvt.	PA 48th Inf.	G	9/21/64	6/8/65		
10	Compher, Alexander	41	Capt.	PA 101st Inf.	D	2/13/62	3/12/65	POW	Plymonth, NC
11	Conch, Harry	28	Pvt.	PA 138th Inf.	E	8/29/62	6/23/65		
12	Cone, James*								
13	Conley, Isaiah	31	1st Lt.	PA 101st Inf.	G	2/20/62	6/25/65	POW	Plymonth, NC
14	Conley, Martin L	27	Corp.	PA 138th Inf.	E	8/29/62		MIA	Cold Harbor
15	Conner, Adam	30	Pvt.	PA 208th Inf.	H	9/6/64	6/1/65		
16	Conner, Benjamin F	20	Pvt.	MD 1st PHB Cav.	G	4/23/64	6/28/65		
17	Conner, Benjamin M	22	Corp.	PA 9th Res.	D	5/3/61	5/12/65		
18	Conner, David (1)	21	Pvt.	PA 171st Inf.	I	11/2/62	8/8/63		
19	Conner, David (2)	23	Pvt.	PA 133rd Inf.	K	8/15/62		Died	Fredericksburg*
20	Conner, Emanuel	31	Pvt.	PA 208th Inf.	K	9/7/64	6/1/65		
21	Conner, Isaac	34	Corp.	PA 22nd Cav.	H	2/26/64	6/24/65	Injured	Fisher's Hill
22	Conner, Jonas	19	Pvt.	PA 107th Inf.	H	4/29/64	7/23/65	Wounded	Petersburg - Initial
23	Conner, Lewis	19	Pvt.	PA 133rd Inf.	K	8/15/62	5/26/63		
24	Conner, Lewis		Corp.	PA McKeage's Mil.	D	7/2/63	8/8/63		
25	Conner, Lewis		Sgt.	PA 22nd Cav.	H	2/26/64	6/24/65		
26	Conrad, Jacob	15	Pvt.	PA 205th Inf.	H	9/10/64	6/2/65		
27	Conrad, Martin*		Pvt.	PA 76th Inf.	E	9/2/64	5/5/65	Died	
28	Conrad, Winfield S	16	Pvt.	PA 55th Inf.	I	2/18/64	8/30/65		
29	Cook, David	21	Corp.	PA 138th Inf.	D	9/2/62			
30	Cook, Dennis		Pvt.	IL 51st Inf.	K	10/12/64		MIA	Franklin, TN
31	Cook, Edward H	27	Pvt.	PA 5th Heavy Art.	K	9/1/64	10/5/64	Died	
32	Cook, Ezekiel	22	Pvt.	PA 208th Inf.	K	9/7/64	6/1/65		
33	Cook, Hanson	27	Pvt.	PA 53rd Inf.	B	9/21/64	5/31/65		
34	Cook, Henry		Lt.	PA 46th Inf.	G	7/1/63	8/18/63		
35	Cook, Jacob*							Died	
36	Cook, James A	29	Pvt.	PA 54th Inf.	B	10/10/61	8/18/64	Wounded	Lynchburg
37	Cook, James L*	44	Pvt.	PA 101st Inf.	G	12/2/61		Died	
38	Cook, John	38	Pvt.	PA 99th Inf.	F	9/9/64	5/31/65		
39	Cook, John F	25	Pvt.	PA 13th Inf.	G	4/25/61	8/6/61		
40	Cook, John F		Corp.	PA 184th Inf.	A	5/12/64	7/14/65	Wounded	Cold Harbor
41	Cook, John H	17	Pvt.	PA 138th Inf.	E	6/4/63	6/23/65	Wound.-POW	Wilderness
42	Cook, Joseph M	24	Pvt.	OH 198th Inf.	H	3/27/65	5/8/65		
43	Cook, Joseph S	19	Pvt.	PA 8th Res.	F	6/19/61	5/26/64	POW	Seven Days Battle
44	Cook, Josiah	21	Pvt.	IL 75th Inf.	A	8/9/62	6/12/65		
45	Cook, Levi (1)	18	1st Sgt.	PA 138th Inf.	F	9/29/62	6/23/65	Wounded	Wilderness
46	Cook, Levi (2)	19	Pvt.	PA 21st Cav.	E	7/8/63	2/20/64		
47	Cook, Levi (2)		Pvt.	PA 5th Heavy Art.	K	9/1/64	6/30/65		
48	Cook, Reuben W	18	Capt.	PA 138th Inf.	E	8/29/62	6/23/65		
49	Cook, Samuel	25	Corp.	PA McKeage's Mil.	H	7/2/63	8/8/63		
50	Cook, William A*		Pvt.	IL 106th Inf.	H	9/17/62		Died	
51	Cooper, Barton A	42	Pvt.	PA 107th Inf.	F	9/21/64	6/7/65		
52	Cooper, David A			PA 208th Inf.					
53	Cooper, David M	15	Pvt.	PA 22nd Cav.	C	7/11/63	6/15/65		
54	Cooper, George M		Corp.	PA 77th Inf.	F	10/9/61		Wounded	Liberty Gap, TN
55	Cooper, Jesse V	18	Pvt.	PA 101st Inf.	D	11/1/61		Died	
56	Cooper, Jonathan B	32	Pvt.	PA 158th Inf.	K	11/4/62	8/12/63		
57	Cooper, Joshua H	20	Pvt.	PA 133rd Inf.	C	8/29/62		KIA	Fredericksburg
58	Cooper, Nathan		Pvt.	PA McKeage's Mil.	D	7/2/63	8/8/63		
59	Cooper, Nathan	39	Pvt.	PA 49th Inf.	E	1/4/64	12/14/64	Wounded	Opequon
60	Copelin, Charles*		Capt.	PA 110th Inf.	C	12/19/61	12/17/64	Wounded	1st Deep Bottom
61	Copelin, Isaiah*		Pvt.	PA 110th Inf.	C	1861	10/24/64		
62	Copenhaver, David A		Pvt.	PA 27th Inf.	I	4/26/61	7/1/61		
63	Corbett, William*		Pvt.	PA 76th Inf.	E	11/21/61		MIA	Ft. Wagner, SC
64	Corbin, George H	18	Corp.	PA 194th Inf.	I	7/21/64	11/5/64		

	Cemetery	Notes
1	Stonerstown Cem.-Liberty Twp.	Bedford Gazette 1909 list; Frank McCoy 1912 Listing aka Jacob
2	Yellow Creek Reformed Cem.	Wounded 3/23/62; Died 5/11/62-Typhoid Fever; James (2) and John-brothers
3	Yellow Creek Reformed Cem.	Wounded 3/23/62; Died 3/27/62; B.S.F. History Book-1884
4	Yellow Creek Reformed Cem.	Historical Data Systems record; Simon, John W, David & James (2)-brothers
5		B.S.F. History Book-1884
6	Mt. Zion Cem.-Chaneysville	Died 2/8/67; Frank McCoy 1912 Listing; John & Isiah-brothers
7	Mt. Zion Cem.-Chaneysville	Died 12/15/65; Frank McCoy 1912 Listing; 1860 Southhampton Twp. Census
8	Rose Hill Cem.-Cumberland	1850 Harrison Twp. Census; Adam & Solomon-brothers
9	Trinity UCC Cem.-Juniata Twp.	1890 Harrison Twp. Veterans Census
10	Fairmont Cem.-NE	POW-Columbia 4/20/64; B.S.F. History Book-1884; 1860 Colerain Twp. Census
11		PA Civil War Archives-Bedford residence
12	Yellow Creek Reformed Cem.	Bedford Inquirer 1908 list
13	Schellsburg Cem.	POW-Columbia 4/20/64 to 11/13/64; Isaiah & Martin-brothers
14		MIA 6/1/64; PA Civil War Archives-Schellsburg residence
15	Shreves Cem.-Monroe Twp.	1890 Monroe Twp. Veterans Census
16	Pleasant Union Cem.-Clearville	Benjimen F & David (1)-brothers
17	Allegheny Cem.-Pittsburgh	1870 Napier Twp. Census
18	Rock Hill Cem.-Monroe Twp.	1890 Monroe Twp. Veterans Census; David (1) & Benjamin-brothers
19		Died 1/7/63; Historical Data Systems record aka Conor; 1850 E. Providence Twp. Census
20	Mt. Pleasant Cem.-Mattie	B.S.F. History Book-1884; 1860 W. Providence Twp. Census
21	Mt. Pleasant Cem.-Mattie	Leg injured when Horse was shot out from under him on 8-21-64
22	Mt. Pleasant Cem.-Mattie	Wounded 6/18/64 aka Joseph; Jonas, Adam, David (2), Isaac & Lewis-brothers
23	Rock Hill Cem.-Monroe Twp.	1890 E. Providence Twp. Veterans Census
24		B.S.F. History Book-1884
25		B.S.F. History Book-1884-aka Conor
26		PA Civil War Project record; Jacob & Winfield-bros.
27		Died 5/5/65; B.S.F. History Book-1884; Susquehanna Co. references
28	Riverview Cem.-Huntingdon Co.	B.S.F. History Book-1884; 1860 Hopewell Census aka Conrod; Died 3/11/42
29	Cooks Mill Cem.-Hyndman	Deserted 1/16/63; Historical Data Systems record; 1860 Londonderry Twp. Census
30		MIA 11/30/64; Pension request record; 1860 Londonderry Twp. Census
31	Arlington Nat'l Cem, 13-9104	Died of Disease 10/5/64; Frank McCoy 1912 Listing; Schellsburg Cem. marker
32		B.S.F. History Book-1884; Historical Data Systems-Bedford Co. residence
33	Hyndman Cem.	1890 Hyndman Veterans Census; Hanson & David-brothers
34		1890 Robertsdale Veterans Census
35		Name listed on Trinity UCC Cem. Civil War memorial of soldiers in unknown graves
36	Mt. Union Cem.-Somerset Co.	Arm Amputated-6/18/64 ; James, Levi & Edward-brothers
37	Arlington Nat'l Cem., 13-11695	Died of Disease 8/7/62; Bedford Gazette 2/13/1914 listing; Mckeesport references
38	Bailey Mem. Cem.-IL	1860 W. Providence Twp. Census
39	Old St. Thomas Cem.-Bedford	Historical Data Systems-Bedford residence Aka Franklin
40		Wounded-Bedford Inquirer 6/24/64; 1890 Bedford Twp. Veterans Cen. - Wound above heart
41	Bedford Cem.	Wounded 5/6/64; 4 months in Libby; 1850 E. Providence Twp. Census
42	Mt. Olivet Cem.-Manns Choice	1890 Harrison Twp. Veterans Census
43	Blair Cem.-NE	POW 6/27/62; B.S.F. History Book-1884; 1860 Hopewell Twp. Census
44	Hyndman Cem.	1890 Londonderry Twp. Veterans Census
45	Carpenter Farm Cem.-Hyndman	Wounded 5/6/64; Frank McCoy 1912 Listing
46	Shade Luth. Cem.-Shade Twp.	Historical Data Systems-Bedford Co. residence
47		Historical Data Systems record
48	Everett Cem.	B.S.F. History Book-1884; Reuben, John H-brothers
49	Old St. Thomas Cem.-Bedford	B.S.F. History Book-1884
50		Died-Typhoid Fever 2/15/63; Fold3 record; Born Bedford Co.
51	Clearville Union Cem.	Historical Data Systems record; Pastor before & after war; David M & Joshua-sons
52	Clearville Union Cem.	Bedford Inquirer 5/22/1908 listing; pastor after war
53	Everett Cem.	B.S.F. History Book-1884; David M & Joshua-brothers
54	Chaneysville Meth. Cem.	Wounded 6/25/63; Frank McCoy 1912 Listing
55		Died 7/30/62; PA Civil War archives-Bedford Co. residence
56	Pleasant Grove-Warfordsburg	1870 E. Providence Twp. Census; Jonathan & Barton-bros.
57		KIA 12/13/62; B.S.F. History Book-1884; PA Civil War Archives-Bedford Co. residence
58		B.S.F. History Book-1884
59	Pleasant Union Cem.-Clearville	Wounded 9/19/64-Bedford Historical Society files; Frank McCoy 1912 Listing
60		B.S.F. History Book-1884; Janesville references
61		B.S.F. History Book-1884; Hollidaysburg references
62		1890 Liberty Twp. Veterans Census
63		MIA 7/11/63; B.S.F. History Book-1884; PA Civil War Archives-Cumberland Co. residence
64		1890 Everett Veterans Census

	Name	Age	Rank	Regiment	Co.	Muster In	Muster Out	Casualty	Casualty Battle
1	Corbitt, Henry W*	25	Sgt.	PA 194th Inf.	I	7/21/64	11/5/64		
2	Core, James			PA 49th Inf.	H				
3	Corl, Michael		Pvt.	PA 91st Inf.	D				
4	Corle, Aaron	35	Pvt.	PA 107th Inf.	K	10/6/64	6/7/65		
5	Corle, Abraham	22	Pvt.	PA 138th Inf.	E	8/29/62		Wound.-Died	Wilderness
6	Corle, Adam S	24	Pvt.	IN 22nd Inf.	A	9/20/64	5/8/65		
7	Corle, Alexander	24	Pvt.	PA 171st Inf.	I	11/2/62	8/8/63		
8	Corle, Alexander B	18	Pvt.	PA 55th Inf.	K	2/19/64	5/22/65	POW	Drewry's Bluff
9	Corle, Anthony	19	Pvt.	PA 84th Inf.	A	10/24/61		MIA	Port Republic
10	Corle, Chauncey	24	Corp.	PA 55th Inf.	K	11/5/61		Wound.-Died	Petersburg
11	Corle, Eli	22	Pvt.	PA 55th Inf.	K	11/5/61		Died	
12	Corle, Francis B	37	Pvt.	PA 91st Inf.	G	9/21/64		Wound.-Died	Boydton Plank Rd.
13	Corle, Franklin	18	Pvt.	PA 138th Inf.	E	8/29/62	6/23/65		
14	Corle, Frederick	43	Pvt.	PA 99th Inf.	F	2/25/65	7/1/65		
15	Corle, Henry W	22	Pvt.	PA 46th Inf.	H	7/14/63		Wound.-Died	Resaca, GA
16	Corle, Isaac	26	Pvt.	PA 172nd Inf.	K	10/28/62	8/1/63		
17	Corle, James L	20	Pvt.	PA 55th Inf.	I	8/28/61		KIA	White Oak Road
18	Corle, Jonathan	37	Pvt.	PA 91st Inf.	G	9/21/64		KIA	Boydton Plank Rd.
19	Corle, Leonard	22	Pvt.	PA 11th Inf.	C	9/9/61	4/29/62		
20	Corle, Martin	51	Pvt.	PA 55th Inf.	K	11/4/62	2/16/63		
21	Corle, Michael S	36	Pvt.	PA 55th Inf.	K	11/5/61	4/23/63		
22	Corle, Solomon*							Died	
23	Corle, William C		Music.	PA 93rd Inf.		10/3/61	3/18/62		
24	Corle, William C	21	Pvt.	PA 138th Inf.	D	8/29/62	6/23/65	Wounded	Wilderness
25	Corley, Benjamin	32	Pvt.	PA 149th Inf.	D	8/26/63	6/24/65		
26	Corley, David			PA 82nd Inf.	I				
27	Corley, Henry	40	Pvt.	MD 1st PHB Cav.	B	5/13/63	6/3/65		
28	Corley, Jacob	43	Pvt.	PA 82nd Inf.	C	9/12/64	6/17/65		
29	Corley, John	26	Pvt.	PA 93rd Inf.	D	7/5/64	7/27/65		
30	Corley, Peter A	20	Pvt.	PA 55th Inf.	H	2/29/64	6/10/65	Wounded	
31	Cornelius, Peter L	19	Pvt.	PA 22nd Cav.	K	2/25/64	6/24/65		
32	Cornelius, Peter*	23	Pvt.	PA 107th Inf.	H	2/24/62		MIA	Globe Tavern
33	Cornelius, William M	28	Pvt.	PA 199th Inf.	E	9/26/64	6/7/65		
34	Cornell, Daniel	29	Pvt.	PA 208th Inf.	K	9/7/64	6/1/65		
35	Cornell, William	21	Sgt.	PA 11th Inf.	A	9/30/61	6/14/65	Wounded	Hatcher's Run
36	Cornell, William H	25	Pvt.	PA 208th Inf.	K	9/7/64	6/1/65		
37	Costler, John			USCT Inf.					
38	Costler, Joseph		Pvt.	USCT 127th Inf.	H	9/2/64	10/20/65		
39	Cottle, Jacob	20	Pvt.	PA 133rd Inf.	A	8/5/62	2/16/63		Fredericksburg*
40	Cottle, Jacob		1st Sgt.	PA 22nd Cav.	A	7/16/63	2/2/64		
41	Couch, Harry C	25	Pvt.	PA 138th Inf.	E	8/29/62	6/23/65		
42	Coughenour, James	21	Pvt.	IA 14th Inf.	F	11/2/61		Wounded	Ft. Donelson, TN
43	Coughenour, James		Pvt.	PA McKeage's Mil.	H	7/2/63	8/8/63		
44	Coughenour, Samuel	18	Pvt.	IA 22nd Inf.	H	8/27/62		Wound.-Died	Vicksburg, MS
45	Couiger, Jacob		Pvt.						
46	Coulter, Alexander		Pvt.	PA 208th Inf.	H	9/7/64	6/1/65		
47	Covert, James P	19	Pvt.	PA 5th Res.	F	6/21/61	6/11/64		
48	Covey, Benoni*		Pvt.	PA 76th Inf.	E	10/26/61	6/22/63		
49	Cowan, David*	24	Pvt.	PA 55th Inf.	I	8/28/61		Died	
50	Cox, Dannel		Pvt.	PA 144th Inf.	D	9/1/63	6/1/65		
51	Cox, Samuel	51	Pvt.	PA 149th Inf.	D	10/2/63	6/24/65	Injured	Cold Harbor
52	Cox, Samuel*			PA 49th Inf.	D				
53	Cox, Samuel*			PA 82nd Inf.					
54	Cox, Thomas*		Pvt.	PA 55th Inf.	I	10/5/63			
55	Craig, George W*	33	Pvt.	PA 55th Inf.	I	10/15/61	5/14/62	Wound.-Died	
56	Craine, David D	18	Pvt.	PA 138th Inf.	E	8/29/62		KIA	Cedar Creek
57	Crall, John	34	Pvt.	PA 77th Inf.	A	10/11/61	10/11/64	Wounded	
58	Cramer, Jacob	20	Pvt.	PA 110th Inf.	C	10/24/61		Died	
59	Cramer, John								
60	Cramer, John B	19	Pvt.	PA 22nd Cav.	B	7/11/63	10/31/65		
61	Cramer, Levi	29	Pvt.	PA 208th Inf.	H	9/7/64	6/1/65		
62	Cramer, William	19	Pvt.	PA 194th Inf.	I	7/21/64	9/6/64		
63	Cramer, William		Pvt.	PA 97th Inf.	Ind	9/6/64	7/17/65		
64	Crawford, Jacob	27	Pvt.	PA 208th Inf.	H	9/5/64	6/1/65		

	Cemetery	Notes
1		B.S.F. History Book-1884; Reading, PA references
2	Hopewell Cem.	Bedford Inquirer 5/22/1908 listing; Historical Data Systems record; Born 8/11/24
3	Mt. Zion Cem.-Pavia	Frank McCoy 1912 Listing
4	Grandview Cem.-Cambria Co.	1860 St. Clair Twp. Census; Aaron, Adam & Jonathan-brothers
5	Mt. Zion Cem.-Pavia	Wounded 5/6/64; Died 5/10/64, PA Civil War Archives-St. Clairsville residence
6	Weimer Cem.-IN	Historical Data Systems record; aka Corel
7	Mt. Zion Cem.-Pavia	Frank McCoy 1912 Listing
8	New Paris Comm. Ctr. Cem.	POW-Andersonville 5/16/64 to 11/19/64; Historical Data Systems record; Martin-father
9	Mt. Zion Cem.-Pavia	MIA 6/9/62; PA Civil War Archives record; 1860 Union Twp. (Pavia) Census
10	Cypress Hill Cem.-NY	Died 8/23/64-Ancestry.com listed wounded at Petersburg; Historical Data Systems record
11	Mt. Zion Cem.-Pavia	Died-Cholera 11/21/61; PA Civil War Archives-Bedford Co. residence
12	Mt. Zion Cem.-Pavia	Wounded 10/27/64; Died 10/29/64; 1890 Union Twp. (Pavia) Veterans Census
13	Mt. Zion Cem.-Pavia	PA Civil War Archives-Bedford residence
14	Mt. Zion Cem.-Pavia	Frank McCoy 1912 Listing; 1860 Union Twp. (Pavia) Census
15		Wounded 5/15/64; Died 5/18/64; Bedford Gazette 1914 list; Historical Data Systems record
16		Ancestry.com record; Isaac, William, Chauncey & Eli-bros.
17	Mt. Zion Cem.-Pavia	KIA 3/30/65; PA Civil War Archives-Bedford Co. residence
18	Mt. Zion Cem.-Pavia	KIA 10/27/64; Historical Data Systems record; Jonathan & Francis-cousins
19	Mt. Zion Cem.-Pavia	Bedford Inquirer 5/22/1908 listing; aka Correll; Discharged on Surgeon's Cert.
20	New Paris Comm. Ctr. Cem.	1860 Union Twp. (Pavia) Cen.; Hist. Data Systems record; Disch. Surg. Cert.; Alexander B-son
21	Tidioute Cem.-Warren Co	1860 Union Twp. Census; PA Civil War Archives record; Discharged on Surgeon's Cert.
22	Mt. Zion Cem.-Pavia	Died 8/26/64; Findagrave.com record
23		Historical Data Systems record lists Corl spelling
24	Mt. Union Cem.-Lovely	Wounded 5/6/64 listed in the Bedford Inquirer 6/24/64; PA Civil War Arch.-Bedford Co. resid.
25	Lybarger Cem.-Madley	1890 Juniata Twp. Veterans Census
26	Westphalia City-KS	Benjamin, John & David-brothers
27	Trinity UCC Cem.-Juniata Twp.	Bedford Inquirer 5/22/1908 listing; 1860 Juniata Twp. Census
28	Trinity UCC Cem.-Juniata Twp.	B.S.F. History Book-1884
29	Lybarger Cem.-Madley	1860 Juniata Twp. Census
30	Trinity UCC Cem.-Juniata Twp.	Frank McCoy 1912 Listing; Peter & Henry-brothers
31	Alverton-Westmoreland Co	1870 W. Providence Census; Findagrave.com information; Peter L & William-bros.
32		MIA 8/19/64; B.S.F. History Book-1884; PA Civil War Archives-enlisted in Fulton Co.
33	Everett Cem.	1890 Everett Veterans Census
34	Cornell-Monroe Twp. Cem.	Frank McCoy 1912 Listing; Daniel & William-brothers
35	Union Memorial Cem.-Mench	Head Wound 2/5/65-Penion record; Frank McCoy 1912 Listing; 1850 Monroe Twp. Census
36	Fairview Cem.-Altoona	B.S.F. History Book-1884
37		Frank McCoy 1912 list & Civil Works Admin. 1934 list
38		Frank McCoy 1912 list & Civil Works Admin. 1934 list; Historical Data Systems record
39	Hyndman Cem.	Historical Data Systems record; Discharged on Surgeon's Cert.
40		Historical Data Systems record
41	Riverview Cem.-Huntingdon Co.	PA Civil War Archives-Bedford residence; aka Conch
42	Maple Grove Cem.-IA	Wounded 2/15/62; 1850 Harrison Twp. Census; aka Coughenorn
43		B.S.F. History Book-1884
44	Chalmette Nat'l Cem.	Wounded 5/22/63; Died 10/5/63; 1850 Harrison Twp. Census; Samuel & James-brothers
45		1890 Liberty Twp. Veterans Census
46		B.S.F. History Book-1884
47	Erie Cem.-Erie Co	Historical Data Systems-Bedford Co. residence
48		B.S.F. History Book-1884; Scranton references; Discharged on Surgeon's Cert.
49		Died 7/23/64; Bed. Gaz. 1914 list; B.S.F. Hist. Book-1884; enlisted in Spangs Mill (Roar. Spr.)
50		1890 Union Twp. (Pavia) Veterans Census
51	Mt. Union Cem.-Lovely	Injured leg 6/3/64; PA Civil War Project record; Substitute
52		Frank McCoy 1912 Listing of 49th PA Infantry is an error
53	Mt. Union Cem.-Lovely	Bedford Inquirer 5/22/1908 listing of 82nd PA Infantry is an error
54		Historical Data Systems record
55	Cypress Hill Cem.-NY	Died 6/25/62; B.S.F. History Book-1884; possibly Blair Co. residence
56	Bedford Cem.	KIA 10/19/64; B.S.F. History Book-1884; 1860 Bedford Twp. Census; aka Crane
57		1890 S. Woodbury Twp. Veterans Census
58	Loudon Park Nat'l Cem.	Died-Typhoid 10/25/62; PA Civil War Archives-Woodbury residence; aka Krames
59	Imler St. Paul Breth. Cem.	1890 King Twp. Veterans Census
60	Greenlawn Cem.-Roaring Spring	Historical Data Systems record; John B & William-brothers
61	Dry Hill Cem.-Woodbury	B.S.F. History Book-1884; Bedford Inquirer 5/22/1908 listing
62	Greenlawn Cem.-Roaring Spring	Historical Data Systems record; 1860 S. Woodbury Twp. Census
63		1890 S. Woodbury Twp. Veterans Census
64	Pleasant Union Cem.-Clearville	1890 Monroe Twp. Veterans Census; 1850 Monroe Twp. Census

	Name	Age	Rank	Regiment	Co.	Muster In	Muster Out	Casualty	Casualty Battle
1	Crawford, James S	20	Pvt.	PA 138th Inf.	E	8/17/62	6/27/65	Wounded	
2	Crawford, Joseph	15	Pvt.	WV 15th Inf.	B	2/29/64	1865	POW	Cedar Creek
3	Cremer, George		Pvt.	PA McKeage's Mil.	H	7/2/63	8/8/63		
4	Cresbaugh, John	22	Sgt.	PA 9th Cav.	F	11/16/61	7/18/65	Wounded	Triune, TN
5	Crick, Andrew		Pvt.	PA 76th Inf.	E	10/16/61	11/28/64		
6	Crissey, John C	16	Pvt.	PA 133rd Inf.	D	8/14/62	4/22/63	Wounded	Fredericksburg
7	Crissey, John C		Black.	PA 21st Cav.	E	7/2/63	2/20/64		
8	Crissey, John C		Pvt.	PA 5th Heavy Art.	K	8/26/64	6/30/65		
9	Crissey, Rufus		Pvt.	PA 5th Heavy Art.	D	9/2/64	6/13/65		
10	Crissey, Samuel N	34	Pvt.	PA 53rd Inf.	B	2/9/64	10/16/64	Wounded	North Anna River
11	Crissey, William	20	Pvt.	PA 54th Inf.	B				
12	Crist, David T	19	Pvt.	PA 99th Inf.	A	3/17/65	7/1/65		
13	Crist, Francis T	20	Corp.	PA 55th Inf.	K	3/2/64	8/30/65	Wounded	
14	Crist, Israel	20	Pvt.	PA 54th Inf.	I				
15	Crist, Jacob T	27	Pvt.	IA 14th Inf.	F	11/2/61	11/8/64		
16	Crist, John T (1)	23	Sgt.	PA 55th Inf.	K	11/5/61	8/30/65		
17	Crist, John T (2)	28	Pvt.	PA 12th Cav.	C	3/27/62			
18	Crist, Samuel M		Pvt.	PA 54th Inf.	I	2/23/64	5/31/65		
19	Crist, Solomon		Pvt.	PA McKeage's Mil.	G	7/2/63	8/8/63		
20	Crist, Solomon	49	Pvt.	PA 55th Inf.	K	3/2/64		MIA	Bermuda Hundred
21	Criswell, Joseph	26	Pvt.	PA 110th Inf.	B	2/28/62	6/28/65	Wounded	
22	Critchfield, John J		Pvt.	PA 87th Inf.	D	6/3/64	6/29/65		
23	Critchfield, N B*	25	Chap.	PA 28th Inf.	HQ	5/22/64	7/18/65		
24	Critchfield, N B*		Chap.	PA 171st Inf.		11/28/62	8/8/63		
25	Crocheron, John F*		Pvt.	PA 55th Inf.	K	10/17/63		Died	Petersburg
26	Croft, Alexander	36	Sgt.	PA 110th Inf.	C	10/24/61		Died	
27	Croft, Daniel S	43	Pvt.	PA 101st Inf.	B	3/7/65	6/25/65		
28	Croft, George	57	Pvt.	PA 22nd Cav.	B	7/11/63	2/5/64		
29	Croft, George W	46	Pvt.	PA 51st Inf.	F	7/3/63	9/2/63		
30	Croft, Jeremiah	21	Pvt.	PA 55th Inf.	I	2/15/64			
31	Croft, Levi G	17	Pvt.	OH 11th Inf.	C	9/2/64	6/11/65	Wounded	
32	Croft, Philip P	23	Pvt.	PA 110th Inf.	C	10/24/61		KIA	1st Kernstown
33	Croft, William N	22	Pvt.	OH 72nd Inf.	F	2/18/62	8/7/65	POW	Ripley, MS
34	Croil, John T	16	Pvt.	PA 149th Inf.		8/25/63	5/3/65		
35	Crosby, Isaac*	34	Pvt.	OH 76th Inf.	A	9/2/62		Died	
36	Crosby, John B*	37	Pvt.	IA 24th Inf.	I	9/6/62	7/17/65		
37	Crosby, Joseph A		Pvt.	OH 114th Inf.	C				
38	Crosby, Nelson D*	35	Sgt.	OH 78th Inf.	B	1/8/62	7/11/65		
39	Crosby, William H	27	Pvt.	PA 52th Inf.	E	9/26/64	6/24/65		
40	Crossan, John M		Pvt.	PA 55th Inf.	K	1863	8/30/65		
41	Crouse, Daniel	42	Pvt.	PA 30th Inf.		6/20/63	8/1/63		
42	Crouse, Harry C	21	2nd Lt.	PA 55th Inf.	I	9/20/61		KIA	Petersburg
43	Crouse, Henry	30	Pvt.	PA 55th Inf.	D	10/12/61	1/24/63		
44	Crouse, John H	19	Sgt.	PA 55th Inf.	H	10/11/61	8/30/65		
45	Crouse, Samuel	18	Pvt.	PA 99th Inf.	G	3/15/65	7/1/65		
46	Croyle, Adam	32	Pvt.	PA 138th Inf.	E	8/29/62			
47	Croyle, Archibald	26	Pvt.	PA 54th Inf.	D	10/18/62	7/15/65		
48	Croyle, Bernard*	20	Corp.	PA 55th Inf.	I	2/27/64	8/30/65		
49	Croyle, Daniel	30	Corp.	PA 184th Inf.	A	5/12/64	7/14/65		
50	Croyle, James A	18	Pvt.	PA 133rd Inf.	K	8/4/62	5/26/63		
51	Croyle, James A		Pvt.	PA 55th Inf.	D	2/18/64	8/30/65		
52	Croyle, John*	20	Pvt.	PA 54th Inf.	I	1/17/62		Wound.-Died	Opequon
53	Croyle, Martin		Pvt.	PA 55th Inf.	K	10/21/62	8/30/65		
54	Croyle, William H	18	Pvt.	PA 55th Inf.	H	10/11/61		Wound.-Died	Petersburg - Initial
55	Cruet, A Howard		Sgt.	PA McKeage's Mil.	H	7/2/63	8/8/63		
56	Crum, Simon*		Pvt.	PA 55th Inf.	I	8/28/61		Wound.-Died	Petersburg - Initial
57	Cullison, John	30	Pvt.	PA 22nd Cav.	M	2/27/64		Wounded	
58	Culp, George*	22	Pvt.	PA 55th Inf.	K	10/14/63			
59	Culp, James*		Pvt.	PA 55th Inf.	K	10/14/63	2/21/65		
60	Cumpson, Benjamin	15	Pvt.	PA 99th Inf.	E	3/17/65	7/1/65		
61	Cunard, James		Pvt.	CA 7th Inf.	I	11/17/64	3/31/66		
62	Cunningham, John*		Pvt.	PA 76th Inf.	E	2/24/65	7/18/65		
63	Cunningham, John*		Pvt.	PA 55th Inf.	I	7/27/63			
64	Curry, James W		Chap.	PA 138th Inf.		3/21/63	3/8/64		

	Cemetery	Notes
1	Schellsburg Cem.	1890 Napier Twp. Veterans Census; 1860 Harrison Twp. Census
2	Providence Union Cem.-Everett	POW 10/19/64 to 2/22/65; 1850 Southampton Twp. Census
3		B.S.F. History Book-1884
4		Wounded 6/11/63; 1890 Liberty Twp. Veterans Census aka Gresabauch
5		Historical Data Systems-Bedford Co. residence
6	Mt. Union Cem.-Mench	Frank McCoy 1912 Listing; Shot in leg & Bayonet wound
7		Historical Data Systems record
8		Historical Data Systems record
9	St. Pauls Ref.- Somerset Co.	B.S.F. History Book-1884
10	Mt. Olivet Cem.-Manns Choice	Wounded 5/25/64-Bedford Hist. Soc. Files; 1890 Napier Twp. Veterans Census
11	Messiah Luth.-Dutch Corner	Bedford Inquirer 5/22/1908 listing; William & John-brothers
12	Richland Cem.-Richland	1860 Union Twp. (Pavia) Census
13	Mt. Zion Cem.-Pavia	1890 Union Twp. (Pavia) Veterans Census; Francis & John T (1)-Brothers
14	Mt. Zion Cem.-Pavia	Bedford Inquirer 5/22/1908 listing, aka Isaac
15	Richland Center Cem.-KS	Historical Data Systems record
16	Mt. Zion Cem.-Paint Twp.	PA Civil War Archives-Bedford Co. residence
17	Old Claysburg Cem.	Historical Data Systems record; 1860 Union (Pavia) Twp. Census
18	Mt. Zion Cem.-Paint Twp.	Ancestry.com-born Bedford Co.; Samuel, John & Israel-brothers; 1850 Union Twp. Census
19		B.S.F. History Book-1884
20	Mt. Zion Cem.-Pavia	MIA while on Picket Duty 8/26/64; PA Civil War Archives-Bedford Co. residence
21	Entriken Cem.-Hunt. Co.	1890 Hopewell Veterans Census
22	Carpenter Farm Cem.-Hyndman	Frank McCoy 1912 Listing; 1870 Londonderry Twp. Census
23		B.S.F. History Book-1884; Somerset Co. references
24		B.S.F. History Book-1884; Somerset Co. references
25	City Point Nat'l Cem.-VA	Died 7/18/64; B.S.F. History Book-1884; Muster in-Philadelphia
26	Potter Creek Cem.-Woodbury	Died 2/5/62; PA Civil War Archives-Woodbury residence; Alexander & George-brothers
27	Buck Valley Cem.-Fulton Co.	PA Civil War Project record-born Bedford Co.
28	Potter Creek Cem.-Woodbury	Historical Data Systems record; Enlisted after son Philip had died
29	Dry Hill Cem.-Woodbury	1860 Woodbury Twp. Census
30	Holsinger Cem.-Bakers Summit	Deserted 7/21/64 (estimated); PA Civil War Archives-Bedford Co. residence
31	Dry Hill Cem.-Woodbury	Historical Data Systems record; Levi & William-brothers
32	Potter Creek Cem.-Woodbury	KIA 3/23/62; 1860 Woodbury Twp. Census ; Bedford Gazette 2/13/1914 listing
33	Violet Hill Cem.-IA	1860 Woodbury Twp. Census; Historical Data Systems record
34	Bedford Cem.	1890 Bedford Twp. Veterans Census aka Croyl
35	Vicksburg Nat'l Cem.	Died of Disease 2/17/63; Ancestry.com-Born Bedford Co.
36	Holaday Creek Cem.-IA	Ancestry.com-Born Bedford Co.
37		Ancestry.com-Born Bedford 1-22-33
38		Ancestry.com-Born Bedford Co.; Nelson, Isaac & John-brothers
39	Lebanon Cem.-Somerset Co.	Ancestry.com-Born Schellsburg; William & Joseph-brothers
40		B.S.F. History Book-1884
41	Bedford Cem.	1890 Bedford Twp. Veterans Census
42		KIA 8-20-64; B.S.F. History Book-1884; PA Civil War Archives-Bedford Co. residence
43		Historical Data Systems-Bedford Co. residence; Discharged on Surgeon's Cert.
44		B.S.F. History Book-1884; 1850 Schellsburg Census
45	St. Peters Cem.-Pittsburgh	1890 Napier Twp. Veterans Census; Samuel & John-brothers
46	Imler St. Paul Breth. Cem.	Deserted 2-26-63; PA Civil War Archives-Bedford residence
47	Grandview Cem.-Cambria Co.	Ancestry.com- Osterburg residence; Archibald & Adam-brothers
48		B.S.F. History Book-1884; 1850 Somerset Census
49	Bedford Cem.	Frank McCoy 1912 Listing; 1850 Bedford Twp. Census
50	Dry Hill Cem.-Woodbury	1890 Snake Spring Valley Veterans Census
51		1890 Snake Spring Valley Veterans Census
52		Leg Amputated 9-19-64; Died 12-20-64; PA Civil War Archives-Cambria Co. residence
53		B.S.F. History Book-1884
54	Fishertown Luth. Cem.	Leg Amputated 6-16-64; Died 9-5-64; 1860 St. Clair. Twp. Census
55		B.S.F. History Book-1884
56	Old St. Marys-Hollidaysburg	Wounded 6/18/64; Died 2/4/65; Blair Co. references
57	Greenlawn Cem.-Roaring Spring	1890 S. Woodbury Census-left arm amputated
58		B.S.F. History Book-1884; In hospital from 5/6/64
59		B.S.F. History Book-1884; drafted; Discharged on Surgeon's Cert.
60	Keagy Cem.-Woodbury	1850 M. Woodbury Twp. Census; 1890 Woodbury Twp. Census
61		Ancestry.com-born in Bedford
62		B.S.F. History Book-1884; Substitute
63		B.S.F. History Book-1884; Drafted & Deserted 6/15/64
64		B.S.F. History Book-1884

Chapter 18

	Name	Age	Rank	Regiment	Co.	Muster In	Muster Out	Casualty	Casualty Battle
1	Curtis, Edward*		Pvt.	PA 55th Inf.	I	10/8/63			
2	Cushman, Adam			PA 99th Inf.					
3	Custer, Daniel		Pvt.	PA 56th Inf.	G	8/30/63	7/1/65		
4	Custer, John A	22	Pvt.	PA 54th Inf.	B	11/10/61	8/9/65	Wounded	
5	Custer, John D	22	Pvt.	PA 49th Inf.	E	6/2/64	7/15/65		
6	Custer, Joseph	19	Pvt.	PA 6th Heavy Art.	K	9/3/64	6/15/65		
7	Cutchall, Dutton	33	Pvt.	PA 82nd Inf.	C	11/14/64	7/13/65		
8	Cutler, Jonathan		Pvt.	PA 76th Inf.	E	11/21/61	7/23/63		
9	Cypher, George W	23	Corp.	PA 5th Inf.	D	4/19/61	7/21/61		
10	Cypher, George W		Sgt.	PA 1st Cav.	G	8/28/61	9/9/64		
11	Cypher, Henry S	18	Pvt.	PA 76th Inf.	E	10/9/61	11/28/64		
12	Cypher, Jacob F	19	Corp.	PA 76th Inf.	E	10/9/61	11/28/64		
13	Cypher, Thomas F	27	Pvt.	PA 125th Inf.	F	8/12/62	5/18/63		
14	Cypher, Thomas F		Pvt.	VT		3/10/65	3/29/67		
15	Dagenfelt, Joseph*	25	Pvt.	PA 55th Inf.	D	10/15/63		POW-Died	
16	Dale, Jacob*	24	Pvt.	PA 184th Inf.	A	10/1/64	6/2/65		
17	Dannaker, John	17	Pvt.	PA 55th Inf.	K	11/5/61	2/21/63		
18	Dannaker, John		Pvt.	US 1st Light Art.	M	2/21/63			
19	Dannaker, William A	18	2nd Lt.	PA 55th Inf.	H	10/11/61	8/30/65		
20	Danner, Henry S*	28	Pvt.	PA 55th Inf.	K	9/28/63	8/30/65		
21	Darr, Abraham	20	Sgt.	PA 55th Inf.	H	10/11/61		KIA	Cold Harbor
22	Darr, Charles S	21	Pvt.	PA 12th Inf.	F	12/2/61			
23	Darr, Charles S		Pvt.	PA 11th Cav.	D	3/26/63		Died	
24	Darr, Christian J		Pvt.	PA 171st Inf.	H	11/12/62		Died	
25	Darr, David H	23	Pvt.	PA 55th Inf.	H	12/4/61	8/31/65	Wounded	
26	Darr, George W	29	Pvt.	PA 100th Inf.	K	11/28/64	7/24/65		
27	Darr, Henry H	27	1st Sgt.	PA 55th Inf.	H	10/11/61	7/3/65	Wounded	Petersburg - Initial
28	Darr, Jacob	17	Pvt.	PA 55th Inf.	H	10/11/61	12/30/64		
29	Darr, Joseph A	31	Pvt.	PA 46th Inf.	D	7/1/63	8/19/63		
30	Darr, Michael S			PA 55th Inf.	H				
31	Dasher, Henry	32	Pvt.	PA 78th Inf.	C	2/18/65	9/11/65		
32	Dasher, Jackson	27	Pvt.	WV 2nd Cav.	G	11/1/61			
33	Dasher, James H	23	Pvt.	PA 10th Inf.	G	4/22/61	7/31/61		
34	Dasher, James H		Pvt.	PA 125th Inf.	E	8/13/62	5/18/63		
35	Dasher, James H		Pvt.	PA 55th Inf.	A	2/23/64	5/20/65	POW	Drewry's Bluff
36	Dasher, William H	21	Corp.	PA 8th Res.	F	6/4/61	5/26/64	POW	Seven Days Battle
37	Daugherty, David L	26	Pvt.	PA 55th Inf.	H	8/28/61		Wound.-Died	Petersburg - Initial
38	Daugherty, Jacob W	21	Pvt.	PA 133rd Inf.	C	8/13/62	5/26/63		
39	Daugherty, James	16	Pvt.	PA 110th Inf.	C	10/24/61		KIA	Fredericksburg
40	Daugherty, Joseph L	35	Pvt.	PA 208th Inf.	H	8/15/64	6/8/65		
41	Davidson, Samuel		Pvt.	PA McKeage's Mil.	H	7/2/63	8/8/63		
42	Davidson, Samuel	20	Pvt.	PA 184th Inf.	A	5/12/64		POW-Died	Jerusalem Plank Rd.
43	Davis, Abner O	18	Pvt.	PA 18th Cav.	H	3/19/64	7/14/65	Wounded	
44	Davis, Andrew J	28	Pvt.	PA 107th Inf.	G	9/21/64	6/6/65		
45	Davis, Charles M	24	Pvt.	PA 55th Inf.	H	8/28/61		KIA	Cold Harbor
46	Davis, David	26	Pvt.	PA 49th Inf.	E	8/22/63	8/9/65		
47	Davis, Enos	34	Pvt.	PA 61st Inf.	G	7/5/64	6/28/65		
48	Davis, Ephraim W	27	Pvt.	PA 55th Inf.	H	10/11/61	7/26/62		
49	Davis, Isaiah M	24	Pvt.	PA 8th Res.	F	6/11/61		Died	
50	Davis, James		Pvt.	USCT 24th Inf.	B	3/7/65	10/1/65		
51	Davis, James P	20	Sgt. QM	PA 22nd Cav.	M	7/11/63	6/24/65		
52	Davis, James W	30	Pvt.	PA 13th Inf.	G	4/25/61	8/6/61		
53	Davis, James W		Pvt.	PA 28th Inf.	O	8/17/61	10/28/62	Wounded	Antietam
54	Davis, John		Pvt.	USCT 8th Inf.	D				
55	Davis, John		Pvt.	USCT 22nd Inf.	I				
56	Davis, John		Pvt.	USCT 24th Inf.	E				
57	Davis, John L	19	Pvt.	PA McKeage's Mil.	D	7/2/63	8/8/63		
58	Davis, John L		Pvt.	PA 208th Inf.	K	9/5/64	6/1/65		
59	Davis, John M	20	Pvt.	PA 110th Inf.	C	10/24/61	10/24/64	Wounded	1st Deep Bottom
60	Davis, John Milton	20	Pvt.	PA 61st Inf.	G	10/25/64	6/28/65		
61	Davis, Martin L	34	Pvt.	PA 110th Inf.	C	10/24/61	7/4/64		
62	Davis, Martin L		Pvt.	PA 87th Inf.	C	7/4/64	5/12/65		
63	Davis, Nathan H	29	Corp.	PA 84th Inf.	A	10/24/61	9/26/62		
64	Davis, Porter R	22	Pvt.	PA 110th Inf.	C	12/19/61			

	Cemetery	Notes
1		B.S.F. History Book-1884; NJ references
2	Saint John's Cem.-Loysburg	Bedford Inquirer 5/22/1908 listing
3		1890 W. St. Clair Veterans Census
4	Crum-Oldham-Shade Twp.	1890 Napier Twp. Veterans Census
5		1880 Juniata Twp. Census
6	Schellsburg Cem.	1890 Napier Twp. Veterans Census
7	Hustontown Meth. Cem.	1890 Wells Twp. Veterans Census
8		B.S.F. History Book-1884; Discharged on Surgeon's Cert.
9	Fockler Cem.-Saxton	1890 Saxton Veterans Census
10		Frank McCoy 1912 Listing; 1890 Saxton Veterans Census
11	Wash. Vet. Home Cem.-WA	PA Civil War Archives-Bedford residence
12	Fair Mt. Cem.-Denver	1860 Liberty Twp. Census
13	St. Lukes Cem.-Saxton	Frank McCoy 1912 Listing; 1890 Saxton Veterans Census
14		1890 Saxton Veterans Census; Thomas, Jacob Harry & George-brothers
15	Andersonville Cem., 7053	POW-Andersonville; Died 8/28/64; B.S.F. History Book-1884; Drafted in Reading, PA
16	Forest Hill Cem.-Union Co.	B.S.F. History Book-1884; Harrisburg references
17		1850 Napier Twp. Census; John & William-brothers
18		Historical Data Systems record
19	Mt. View Cem.-CA	Napier Twp. Census; John & William-Brothers
20		B.S.F. History Book-1884; Reading, PA references
21		KIA 6/3/64; B.S.F. History Book-1884; 1860 Schellsburg Census; Abraham and Henry-bros.
22		Deserted 11/12/62
23	Hampton Nat'l Cem.	Died 8/18/63; 1890 Napier Twp. Veterans Census
24	Philadelphia Nat'l Cem.	Died 7/12/63; 1850 Napier Twp. Census; aka C Jackson
25	Schellsburg Cem.	1890 Napier Twp. Veterans Census-wound in right thigh
26	Fishertown Cem.	1890 W. St. Clair Twp. Veterans Census
27	Wooster Cem.-OH	Wounded 6/18/64 listed in Bedford Inquirer 7/8/64 article
28	Lakeview Cem.-SD	1890 Junita Twp. Veterans Census
29	St. John Cem.-Altoona	Joseph, George, Michael, Charles & Christian-brothers
30	Schellsburg Cem.	Born 11/16/37; Pension Record for 55th PA Infantry
31	Yellow Creek Reformed Cem.	Frank McCoy 1912 Listing; 1850 Hopewell Twp. Census; aka Dashin
32	Otterbein Chapel Cem.-OH	Jackson & James-brothers
33	Springville Cem.-Boiling Spgs.	Ancestry.com-born Bedford Co.
34		Historical Data Systems record
35		POW-Andersonville 5/16/64 to 4/29/65
36	Parsons Cem.-WV	POW 6/27/62; 1850 Hopewell Twp. Census; William & Henry-brothers
37	Walker Cem.-Shanksville	Wounded 6/16/64; Died 7/11/64; B.S.F. History Book-1884; 1850 Napier Twp. Census
38	IOOF Riverview Cem.-IN	1850 W. Providence Twp. Census
39	Fredericksburg Nat'l Cem.	KIA 12/13/62; 1860 S. Woodbury Census; Bedford Gazette 1814 list
40	Friends Cem.-Spring Meadow	1890 E. St. Clair Twp. Veterans Census
41		B.S.F. History Book-1884
42	Andersonville Cem., 11628	POW-Andersonville 6/22/64 (est); Died 10/28/64; Bedford Cem. Marker
43	Bedford Cem.	1860 Colerain Twp. Census; Lost sight of Eye
44	Trans Run Cem.-Charlesville	Frank McCoy 1912 Listing; 1860 Colerain Twp. Census
45		KIA 6/5/64; B.S.F. History Book-1884; 1860 Napier Twp. Census
46	Fockler Cem.-Saxton	Deserted 7/27/64 & returned 6/16/65; Frank McCoy 1912 Listing
47	New Paris Comm. Ctr. Cem.	Frank McCoy 1912 Listing; 1860 St. Clair Twp. Census
48	Everett Cem.	1850 Napier Twp. Census; Discharged on Surgeon's Cert.; Ephram & Charles-brothers
49	Everett Cem.	Died of Disease 11/28/61; B.S.F. History Book-1884; 1850 Hopewell Twp. Census
50		Frank McCoy 1912 list & Civil Works Admin. 1934 list; Historical Data Systems record
51	Everett Cem.	Frank McCoy 1912 Listing; 1860 Hopewell Twp. Census
52	Duval Cem.-Coaldale	Frank McCoy 1912 Listing; Historical Data Systems-Bedford Co. residence
53		Leg Amputated 9/17/62; Frank McCoy 1912 Listing
54		Frank McCoy 1912 list & Civil Works Admin. 1934 list
55		Frank McCoy 1912 list & Civil Works Admin. 1934 list
56		Frank McCoy 1912 list & Civil Works Admin. 1934 list
57	Ashbury Meth. Cem.-Graceville	B.S.F. History Book-1884; Bedford Inquirer 5/22/1908 listing
58		1890 E. Providence Twp. Veterans Census; John & William-brothers
59	Everett Cem.	Wounded 7/27/64 listed in Bedford Inquirer 8/12/64; John M (1), James P, Isiah & David-bros.
60	New Paris Comm. Ctr. Cem.	Frank McCoy 1912 Listing; Historical Data Systems record
61	Yellow Creek Reformed Cem.	B.S.F. History Book-1884; Frank McCoy 1912 Listing; Discharged on Surgeon's Cert.
62		Discharged on Surgeon's Cert.
63	Pleasantville Cem.	B.S.F. History Book-1884; 1860 St. Clair Twp. Census
64	Everett Cem.	B.S.F. History Book-1884; 1890 W. Providence Twp. Veterans Census

	Name	Age	Rank	Regiment	Co.	Muster In	Muster Out	Casualty	Casualty Battle
1	Davis, Porter R		Sgt.	PA 22nd Cav.	B	7/11/63	2/5/64		
2	Davis, Porter R		Pvt.	PA 208th Inf.	K	9/7/64			
3	Davis, R DeCharmes	16	Corp.	USCT 32nd Inf.	B	2/12/64	9/2/65		
4	Davis, Richard	31	Pvt.	PA 13th Inf.	G	4/25/61	8/6/61		
5	Davis, Samuel	44	Pvt.	PA 107th Inf.	G	9/21/64	8/8/65		
6	Davis, Thomas P*		Pvt.	PA 55th Inf.	I	8/28/61		KIA	Drewry's Bluff*
7	Davis, William	21	Pvt.	PA 208th Inf.	K	9/7/64	6/1/65		
8	Davis, William A (1)	25	Pvt.	PA 84th Inf.	A	10/24/61		Wounded	1st Kernstown
9	Davis, William A (2)*								
10	Davis, William H	20	Corp.	PA 21st Cav.	G	7/8/63	2/20/64		
11	Davis, Wilson	26	Pvt.	PA 55th Inf.	H	12/4/61	8/30/65		
12	Dawson, Jonathan*		Pvt.	PA 55th Inf.	I	9/29/63	8/30/65		
13	Day, Robert*		Pvt.	PA 55th Inf.	K	10/15/63	8/30/65		
14	Dayton, John C*		Pvt.	PA 55th Inf.	I	1/1/65	8/30/65		
15	Dean, Andrew			USCT Inf.					
16	Dean, Franklin		Pvt.	PA 8th Res.	F	6/11/61	2/7/63	Wounded	Antietam
17	Dean, Jacob	19	Pvt.	USCT 32nd Inf.	H	2/29/64	6/10/65		
18	Deaner, Michael	35	Pvt.	PA 99th Inf.	G	2/24/65	5/29/65		
19	DeArmet, George W	28	Pvt.	PA 29th Mil.	H	6/29/63	8/11/63		
20	DeArmet, George W		Corp.	PA 20th Cav.	E	2/1/64	6/28/65		
21	Deaver, Benjamin A*		Pvt.	PA McKeage's Mil.	G	7/2/63	8/8/63		
22	Deck, Henry*		Pvt.	PA 55th Inf.	H	9/2/62		Died	
23	Deck, John*		Pvt.	PA 55th Inf.	H	9/2/62	4/30/63		
24	Decker, James M	18	Pvt.	PA McKeage's Mil.	H	7/2/63	8/8/63		
25	Decker, James M		Pvt.	PA 53rd Inf.	C	1/4/64		Died	
26	DeCock, Isaac P*		Pvt.	PA 55th Inf.	I	7/22/63			
27	Defebaugh, James C	23	Pvt.	IL 11th Inf.	G	4/3/61	7/30/61		
28	Defebaugh, James C		Sgt.	IL 5th Cav.	L	12/30/61	10/27/64		
29	Deffenbaugh, Harlow W	25	Pvt.	PA 78th Inf.	G	3/2/65	9/11/65		
30	Deffenbaugh, Samuel S	21	Pvt.	PA 5th Inf.	D	4/21/61	7/24/61		
31	Defibaugh, Anderson	33	Pvt.	OH 43rd Inf.	I	12/7/61	7/13/65		
32	Defibaugh, Daniel Jr	18	Pvt.	OH 43rd Inf.	I	2/10/64	7/13/65		
33	Defibaugh, Daniel Sr	61	Sgt.	OH 43rd Inf.	J	7/12/61	7/13/65		
34	Defibaugh, David		Pvt.	PA 110th Inf.	C				
35	Defibaugh, David	24	Pvt.	PA 79th Inf.	E	2/23/65			
36	Defibaugh, Harrison	21	Pvt.	PA 13th Inf.	G	4/25/61	8/6/61		
37	Defibaugh, Harrison		Pvt.	PA McKeage's Mil.	H	7/2/63	8/8/63		
38	Defibaugh, Harrison		Pvt.	PA 2nd Cav.	E				
39	Defibaugh, Jacob F	20	Pvt.	PA 101st Inf.	D	12/6/61		Died	
40	Defibaugh, James C	57	Pvt.	PA 91st Inf.	G	10/8/64	7/10/65		
41	Defibaugh, John (1)	32	Pvt.	PA 184th Inf.	A	5/12/64			
42	Defibaugh, John (2)	34	Pvt.	PA 101st Inf.	G	1861		POW-Died	Plymonth, NC
43	Defibaugh, John W		1st Sgt.	PA McKeage's Mil.	H	7/2/63	8/8/63		
44	Defibaugh, John W	21	Sgt.	PA 184th Inf.	A	5/12/64		POW	Jerusalem Plank Rd.
45	Defibaugh, Lawrence	21	Music.	PA 138th Inf.	E	8/29/62	6/23/65		
46	Defibaugh, William H	18	Pvt.	PA 138th Inf.	E	8/29/62		MIA	Wilderness
47	Dehard, Louis W*		Pvt.	PA 55th Inf.	H	9/26/62	6/11/65		
48	Dehart, Daniel L*		Pvt.	PA 55th Inf.	K	2/20/64	8/30/65		
49	Dell, James	18	Pvt.	PA 12th Cav.	H	3/6/62	7/20/65		
50	Dell, John	23	Pvt.	PA 11th Cav.	G	9/9/61	9/4/64		
51	Dell, Moses	21	Pvt.	PA 1st Light Art.	F	7/8/61	7/11/64	Wounded	Antietam
52	Dell, Peter	20	Pvt.	PA 125th Inf.	E	8/13/62	5/18/63	Wounded	
53	Dell, Samuel	14	Pvt.	PA 76th Inf.	F	2/22/64	7/18/65		
54	Demmings, William*		Pvt.	PA 76th Inf.	E	9/14/63		KIA	Drewry's Bluff
55	Deneen, Baltzer H	28	Pvt.	MD 3rd PHB Inf.	K	1/1/64	5/29/65		
56	Deneen, Henry S.*	28	Pvt.	PA 148th Inf.	B	8/29/63		Died	
57	Deneen, Joseph	22	Pvt.	MD 3rd PHB Inf.	B	6/18/63	5/29/65		
58	Dennis, Adam	17	Pvt.	PA 79th Inf.	G	10/1/61	2/8/64	Wounded	Chickamauga
59	Deppen, Jacob*		Corp.	PA 55th Inf.	D	1/17/65	8/30/65		
60	Deremer, Henry	19	Pvt.	PA 55th Inf.	D	10/12/61			
61	Deremer, Jacob	35	Pvt.	NJ 2nd Cav.	E	9/2/64	7/14/65		
62	Deremer, Joseph C	21	Pvt.	PA 50th Inf.	E	3/13/65	5/9/65		
63	Deremer, William (1)	26	Pvt.	PA 133rd Inf.	D	8/14/62	5/26/63		
64	Deremer, William (2)	19	Pvt.	PA 7th Cav.	M	3/3/64			

	Cemetery	Notes
1		1860 Hopewell Twp. Census; aka David P
2		B.S.F. History Book-1884; Deserted & returned
3	Mt. Ross Cem.-Bedford	1890 Bedford Twp. Veterans Census
4		B.S.F. History Book-1884; Historical Data Systems-Bedford Co. residence
5	Yellow Creek Reformed Cem.	1890 Hopewell Twp. Veterans Census
6	Carson Valley-Duncansville	KIA 5/13/64; B.S.F. History Book-1884
7	Ashbury Meth. Cem.-Graceville	Frank McCoy 1912 Listing; 1890 E. Providence Twp. Veterans Census
8	Friends Cem.-Spring Meadow	Wounded 3-23-62; PA Civil War Archives-Bedford Co. residence
9	Brumbaugh Cem.-Saxton	Civil War Marker at gravesite
10	Pleasantville Cem.	Historical Data Systems-Bedford Co. residence; Ephram & Charles-brothers
11	Mt. Horeb Cem.-OH	Historical Data Systems-Bedford Co. residence; aka John W
12		B.S.F. History Book-1884; Drafted
13		B.S.F. History Book-1884; Drafted; enlisted in Reading, PA
14		B.S.F. History Book-1884
15		Frank McCoy 1912 list & Civil Works Admin. 1934 list; 1860 Bedford Twp. Census
16		Wounded 9/17/62; B.S.F. History Book-1884; Historical Data Systems-Bedford Co. residence
17	Everett Cem.	PA Civil War Project record
18	Hull Baptist Cem.-New Paris	1890 E. St. Clair Veterans Census
19	Prsopect Cem.-Portage	Historical Data Systems record
20		1890 Broad Top Veterans Census aka Dearmont
21		B.S.F. History Book-1884; Fulton Co. references
22		Died 10/30/64; enlisted in Berks Co.
23		B.S.F. History Book-1884; Berks Co. references; Discharged on Surgeon's Cert.
24		B.S.F. History Book-1884; enlisted in Bedford
25		Died 7/22/64; B.S.F. History Book-1884; 1850 Hopewell Twp. Census
26		Deserted 11/16/64; Drafted
27	Freemanton Cem.-IL	Ancestry.com born in Bedford Co.
28		Mother-Elizabeth Isabella S Barndollar
29		PA Civil War Archives-Bedford residence; 1860 St. Clair Twp. Census
30	Bortz Luth. Cem.-Centerville	PA Civil War Archives-Cumberland Valley residence
31	Harvey Cem.-IL	1850 Bedford Twp. Census
32		Historical Data Systems record
33	Harvey Cem.-IL	Anderson, Harrison & Daniel Jr.-sons of Daniel Sr.
34		PA Veterans Burial listing
35	Everett Cem.	Deserted 6/14/65; aka Deffibaugh
36		B.S.F. History Book-1884; Historical Data Systems-Bedford Co. residence
37		B.S.F. History Book-1884; 1890 Bedford Twp. Veterans Census
38		Pension Record
39	Chaneysville Meth. Cem.	Died 1/3/62; Frank McCoy 1912 Listing; Aka Deffibaugh
40	Imler Valley Cem.	Frank McCoy 1912 Listing; Substitute
41	O'Brien Rd. Cem.-Manns Choice	Deserted 5/19/64; B.S.F. History Book-1884; PA Civil War Archives-enlisted in Bedford
42	Andersonville Cem., 4806	Wounded 4/20/64; POW-Andersonville; Died 8/5/64; PA Civil War Archives-Born Bedford Co.
43		B.S.F. History Book-1884; 1860 Bedford Twp. Census
44		POW 6/22/64; B.S.F. History Book-1884
45	Nebraska-unknown gravesite	PA Civil War Archives; 1860 Monroe Twp. Census; Lawrence & David-brothers
46	Fredericksburg Nat'l Cem.	MIA 5/6/64; PA Civil War Archives-Bedford residence; 1860 Bedford Twp. Census
47		B.S.F. History Book-1884; enlisted in Berks Co.
48		B.S.F. History Book-1884; enlisted in Berks Co.
49	St. Patricks Cem.- Newry	1850 Woodbury Twp. Census
50	Fairview Cem.-Altoona	findagrave.com-Born Bedford Co.
51	Holsinger Cem.-Bakers Summit	Wounded 9/17/62; Historical Data Systems record; 1860 Greenfield Twp. Census
52	Carson Valley-Duncansville	Ancestry.com information; Peter, James, John, Samuel & Moses-brothers
53	Oak Ridge Cem.-Altoona	1850 Woodbury Twp. Census
54	North East Cem.-Erie	KIA 5/16/65; B.S.F. History Book-1884; Drafted
55		Ancestry.com-brother born in Londonderry Twp.
56	Culpeper C.H. Nat'l Cem.	Died 3/21/64; B.S.F. History Book-1884; Fulton Co. references
57	Hyndman Cem.	Frank McCoy 1912 Listing; Joseph & Baltzer-brothers
58	Mt. Olivet Cem.-Manns Choice	Wounded 9/19/63, left leg amputated; Historical Data Systems record
59		B.S.F. History Book-1884
60	St. Thomas Cem.-Bedford	Historical Data Systems-Bedford Co. residence; Deserted 10/14/61
61	Mt. Pleasant Cem.-Mattie	Frank McCoy 1912 Listing; Historical Data Systems record
62	Greenwood Cen.-NE	Historical Data Systems record; Joseph & William J-brothers
63	Bortz Luth. Cem.-Centerville	Frank McCoy 1912 Listing; William & Jacob-brothers
64		Deserted 12-11-64; PA Civil War Archives-Born Bedford Co.

	Name	Age	Rank	Regiment	Co.	Muster In	Muster Out	Casualty	Casualty Battle
1	Deremer, William J	20	Pvt.	MD 5th Inf.	C	3/6/65	9/1/65		
2	Derho, Frederick*		Pvt.	PA 76th Inf.	E	10/18/64	7/18/65		
3	Derr, John*		Pvt.	PA 55th Inf.	I	8/19/63		Wound.-Died	Petersburg - Initial
4	Deshong, David D	19	Pvt.	PA 22nd Cav.	M	2/22/64	11/8/65		
5	Detwiler, Jacob	18	Pvt.	PA 143rd Inf.	E	9/22/63	7/12/65		
6	Detwiler, Joseph	43	Pvt.	PA 55th Inf.	K	11/5/61	8/30/65	Wounded	Drewry's Bluff
7	Detwiler, Moses H	22	Corp.	PA 104th Inf.	E	2/21/65	8/25/65		
8	Detwiler, William (1)	21	Pvt.	PA 9th Res.	F	7/28/61	5/13/64		
9	Detwiler, William (2)			PA 110th Inf.					
10	Devens, Elisha	21	Pvt.	PA 138th Inf.	D	8/29/62			
11	Devore, Andrew		Pvt.						
12	Devore, Lewis	18	Pvt.	PA 184th Inf.	B	5/12/64		KIA	Jerusalem Plank Rd.
13	Devore, Michael	32	Pvt.	PA 171st Inf.	I	11/2/62	8/8/63		
14	Devore, Philip		Pvt.	IN 115th Inf.	D	8/5/63	2/25/64		
15	Dibert, Adam	27	Pvt.	PA 79th Inf.	E	2/23/65	5/29/65		
16	Dibert, Andrew E	39	Pvt.	PA 50th Inf.	H	2/23/65	5/11/65		
17	Dibert, David	19	Sgt.	PA 101st Inf.	D	10/9/61	11/12/62	Wounded	
18	Dibert, David		Pvt.	PA 55th Inf.	D	2/27/64	7/14/65	Wounded	
19	Dibert, Isaac	16	Music.	PA 142nd Inf.	D	8/25/62	5/29/65		
20	Dibert, Jacob (1)			IL 135th Inf.	A				
21	Dibert, Jacob (2)	22	Pvt.	PA 82nd Inf.	C	11/22/64	7/13/65		
22	Dibert, Jacob (3)	31	Pvt.	IL 103rd Inf.	K	10/2/62	6/21/65		
23	Dibert, Jacob J	39	Corp.	PA 55th Inf.	K	11/5/61		Died	
24	Dibert, John J	26	3rd Sgt.	PA 55th Inf.	K	8/11/61		KIA	Petersburg - Initial
25	Dibert, John Jackson	29	Pvt.	PA 99th Inf.	H	2/25/65	5/30/65	Died	
26	Dibert, Jonathan	32	Pvt.	PA 79th Inf.	E	2/23/65	5/29/65		
27	Dibert, William	35	2nd Lt.	PA McKeage's Mil.	D	7/2/63	8/8/63		
28	Dick, John C	43	Pvt.	PA 171st Inf.	I	11/10/62	8/8/63		
29	Dicken, David		Pvt.	PA 2nd Cav.	E	3/11/64	6/17/65		
30	Dicken, James	16		PA 2nd Cav.		9/1/61		POW-Died	
31	Dicken, James F	27	Pvt.	MN 1st Cav.	D	10/16/62	11/4/63		
32	Dicken, James H	20	Pvt.	PA 22nd Cav.	D	7/28/63	2/5/64		
33	Dicken, John	21	Pvt.	PA 138th Inf.	D	8/29/62		POW-Died	Mine Run
34	Dicken, Solomon	40	Pvt.	PA 28th Inf.	C	11/28/64	6/9/65		
35	Dicken, Thomas W	19	Pvt.	PA 133rd Inf.	D	8/14/62	5/26/63		
36	Dickens, William			PA 91st Inf.					
37	Dickens, William	29	Pvt.	MD 1st PHB Cav.	A	9/9/64	5/25/65		
38	Dickerhoof, Simon	38	Major	PA 138th Inf.	E	8/30/62	6/23/65	Wounded	Brandy Station
39	Dickerhoof, Simon		Major	PA 138th Inf.	E			POW	Sailor's Creek
40	Dickinson, Enos		Pvt.	PA 6th Cav.	I				
41	Dickinson, George		Pvt.	PA 22nd Cav.	A	2/26/64	10/31/65		
42	Dickinson, Robert	38	Pvt.	PA 4th Cav.	E	8/16/61	8/16/64		
43	Dieffenderfer, Nelson*	30	Pvt.	PA 76th Inf.	E	10/7/64	7/18/65		
44	Dieffenderfer, Paul*	40	Pvt.	PA 76th Inf.	E	10/7/64	7/18/65		
45	Diehl, Adam (1)	26	Pvt.	PA 82nd Inf.	C	11/14/64	7/13/65		
46	Diehl, Adam (2)	20	Pvt.	PA 91st Inf.	F	9/21/64	7/12/65	Died	
47	Diehl, B James	18	Pvt.	PA 55th Inf.	K	3/2/64	8/30/65	Wounded	
48	Diehl, Daniel	18	Pvt.	PA McKeage's Mil.	G	7/2/63	8/8/63		
49	Diehl, Daniel		Pvt.	PA 55th Inf.	D	2/27/64		Died	
50	Diehl, Espy	27	Pvt.	PA 55th Inf.	D	2/27/64		POW-Died	
51	Diehl, Francis M	21	Corp.	PA McKeage's Mil.	G	7/2/63	8/8/63		
52	Diehl, Henry	21	Pvt.	PA McKeage's Mil.	G	7/2/63	8/8/63		
53	Diehl, Henry		Corp.	PA 55th Inf.	D	2/27/64	8/30/65		
54	Diehl, John (1)	23	Pvt.	PA 138th Inf.	F	9/14/62			
55	Diehl, John (2)	20	Pvt.	PA McKeage's Mil.	G	7/2/63	8/8/63		
56	Diehl, John (2)		Pvt.	PA 55th Inf.	D	2/27/64	8/30/65		
57	Diehl, Joshua	20	Sadd.	OH 4th Cav.	H	11/11/61		Died	
58	Diehl, Levi		SMN.	US 13th Navy	3M				
59	Diehl, Michael	38	Pvt.	PA 49th Inf.	F	6/4/64	5/25/65		
60	Diehl, Samuel J	21	Pvt.	PA 55th Inf.	D	10/12/61	8/30/65		
61	Diggins, Jesse	22	Pvt.	PA 77th Inf.	A	2/27/64	12/6/65		
62	Dill, Edward	18	Pvt.	MD 1st Inf.	I	5/27/61		POW	Front Royal
63	Dill, Edward		Capt.	US 5th Cav.	B				
64	Dinges, William J*		Pvt.	PA 110th Inf.	D	8/1/62	5/31/65	Wounded	Jerusalem Plank Rd.

	Cemetery	Notes
1	Utica Cem.-NE	Historical Data Systems record; 1860 Cumberland Valley Census
2		B.S.F. History Book-1884; substitute
3	Hampton, VA	Died 6/18/64; B.S.F. History Book-1884; Mustered in Norristown
4	Sideling Hill CC-Fulton Co.	1890 Hyndman Veterans Census
5	Holsinger Cem.-Bakers Summit	1890 Bloomfield Twp. Veterans Census
6	Fairview Cem.-Altoona	Wounded 5/16/64; B.S.F. History Book-1884
7	Hopewell Cem.	Born in Bedford Co.; Well known Doctor after the war; Historical Data Systems record
8		PA Civil War Archives-Bedford Co. residence
9	Dry Hill Cem.-Woodbury	Frank McCoy 1912 Listing; 1880 Woodbury Census
10		B.S.F. History Book-1884; Deserted 2/6/63; enlisted in Schellsburg
11		1890 Bedford Twp. Veterans Census
12		KIA 6/22/64; 1860 Harrison Twp. Census; Bedford Gazette 2/13/1914 listing
13		B.S.F. History Book-1884; 1860 Londonderry Twp. Census
14		Historical Data Systems record; 1850 Londonderry Twp. Census
15	St. James Luth. Cem.-Bedford	1890 Bedford Twp. Veterans Census; 1850 Bedford Twp. Census
16	Pleasant Hill Cem.-Imlertown	Frank McCoy 1912 Listing; 1860 Bedford Twp. Census
17	Trinity UCC Cem.-Colerain Twp.	1890 Colerain Twp. Veterans Census
18		Wounded listed in Bedford Inquirer article 6/3/64; David, Jacob (3) & Isaac-brothers
19	Friends Cove UCC Cem.	1850 W. Providence Twp. Census
20	Bedford Cem.	Frank McCoy 1912 Listing; Born 4/4/22
21	Pleasant Hill Cem.-Imlertown	1890 Bedford Twp. Veterans Census
22	Greenwood Cem.-IL	Historical Data Systems record; 1850 W. Providence Twp. Census
23	City Point Nat'l Cem.-VA	Died-Chronic Diarrhea 10/26/64; B.S.F. History Book-1884; Mt. Zion Cem.-Pavia marker
24	St. James Luth. Cem.-Bedford	KIA 6/16/64; B.S.F. History Book-1884; John & Adam-brothers
25	Messiah Luth.-Dutch Corner	Died 6/24/65; 1890 Snake Spr. Valley Veterans Census; 1860 Snake Spr. Valley Census
26	Pleasant Hill Cem.-Imlertown	Jonathan & Andrew-brothers
27	Bedford Cem.	B.S.F. History Book-1884; 1850 W. Providence Twp. Census
28	Bunker Hill Cem.-Saxton	1890 Hopewell Twp. Veterans Census
29	Levels Cem.-WV	B.S.F. History Book-1884; 1860 Londonderry Twp. Census
30		B.S.F. History Book-1884-Wounded & Died in enemy hands; 1850 Cumb. Valley Twp. Cen.
31	Kinkead Cem.-MN	1850 Southampton Twp. Census
32		PA Civil War Archives-enlisted in Bedford Co.
33	Annapolis Nat'l Cem.	POW 12/2/63; Died 3/30/64; 1860 Monroe Twp. Census aka Dickin & Deacon
34	Bethel Meth. Cem.-Centerville	1890 Cumberland Valley Veterans Census
35	Floral Cem.-KS	1850 Southampton Twp. Census; Thomas & David-brothers
36	Mt. Zion-Southampton Twp.	Bedford Inquirer 5/22/1908 listing
37	Prosperity Cem.-Clearville	1860 Southampton Twp. Census; 2nd MD Cav referenced
38	Bedford Cem.	Wounded 11/8/62; B.S.F. History Book-1884
39		POW for two hours-Bedford Historical Society files
40	Fockler Cem.-Saxton	1890 Liberty Twp. Veterans Census
41	Fockler Cem.-Saxton	aka Dickson; 1908 Bedford Inquirer list
42	Hopewell Cem.	1890 Broad Top Veterans Census
43		B.S.F. History Book-1884; Northumberland Co. references
44		B.S.F. History Book-1884; Northumberland Co. references
45	Trinity UCC Cem.-Juniata Twp.	Frank McCoy 1912 Listing
46	Friends Cove UCC Cem.	Died 8/2/65; Frank McCoy 1912 Listing; Bedford Inquirer 5/22/1908 listing
47	Lutheran Cem.-Newry	Wounded listed in Bedford Inquirer article 6/3/64; Ancestry.com born Blue Knob
48		1860 Colerain Twp. Census
49	Hampton Nat'l Cem.	Died 7/30/64; B.S.F. History Book-1884; 1860 Colerain Twp. Census
50	Andersonville Cem., 11350	POW-Andersonville; Died 10/23/64; B.S.F. History Book-1884; 1860 Colerain Twp. Census
51	Friends Cove UCC Cem.	B.S.F. History Book-1884; 1860 Colerain Twp. Census
52	Friends Cove UCC Cem.	B.S.F. History Book-1884; 1860 Colerain Twp. Census
53		1890 Colerain Twp. Veterans Census
54		Deserted 10/23/62; 1860 Harrison Twp. Census aka Deihl
55	Bedford Cem.	B.S.F. History Book-1884
56		1890 Bedford Twp. Veterans Census
57		Died of Disease 4/15/62; Ancestry.com-Born in Friends Cove
58	Friends Cove UCC Cem.	Levi & Espy-brothers; born 12/7/43
59	Bethel Frame Ch.-Monroe Twp.	1890 W. Providence Twp. Veterans Census; 1860 Monroe Twp. Cen.; Discharged Surg. Cert.
60	Woodland Union Cem.-OH	1860 Colerain Twp. Census; Civil War Pension Record
61	Broad Top IOOF Cem.	1890 Robertsdale Veterans Census
62	Bedford Cem.	POW-Belle Island 5/23/62 to 9/13/62
63		Historical Data Systems record
64		Wounded 9/22/64; Somerset Co. references

Chapter 18

	Name	Age	Rank	Regiment	Co.	Muster In	Muster Out	Casualty	Casualty Battle
1	Disbrow, Joseph	35	Pvt.	PA 50th Inf.	I	2/23/65	7/5/65		
2	Dishong, David D		Pvt.	PA 22nd Cav.	M	2/23/64	6/24/65		
3	Dishong, David M*	25	Pvt.	PA 158th Inf.	H	11/4/62	8/12/63		
4	Dishong, John R	27	Pvt.	PA 158th Inf.	K	11/4/62			
5	Dishong, John R		Pvt.	IN 88th Inf.	F	1/29/64	6/7/65		
6	Dishong, Morris*	26	Pvt.	IN 34th Inf.	A	10/4/61	12/12/62		
7	Dishong, Morris*		Pvt.	IN 142nd Inf.	C	10/12/64	7/14/65		
8	Dishong, Robert M	39	Pvt.	PA 158th Inf.	H	11/4/62	8/12/63		
9	Ditch, William			PA 2nd Cav.					
10	Ditch, William	24	Pvt.	PA 210th Inf.	L	9/7/64	5/30/65		
11	Dively, Gabriel	20	Pvt.	PA 125th Inf.	E	8/15/62	5/18/63		
12	Dively, Gabriel		Pvt.	PA 12th Cav.	H	3/29/64	7/20/65		
13	Dively, George M	23	Pvt.	PA 77th Inf.	F	2/27/65	12/6/65		
14	Dively, Jacob	35	Pvt.	PA 5th Res.	H	6/21/61	7/11/64		
15	Dively, Jacob O	33	Pvt.	IL 83rd Inf.	C	8/21/62	6/26/65		
16	Dively, James	19	Pvt.	PA 12th Cav.	H	3/19/64	7/20/65		
17	Dively, John	17	Pvt.	PA 110th Inf.	C	2/23/64		POW-Died	
18	Dively, Martin	23	Pvt.	PA 77th Inf.	F	2/28/65	12/6/65		
19	Dively, Michael*								
20	Dively, Morgan	25	Pvt.	PA 77th Inf.	F	2/28/65	12/6/65		
21	Dively, Paul	19	Pvt.	PA 11th Cav.	D	9/10/61	9/10/64		
22	Dixon, George T	22	Pvt.	VA Ind Cav.	B	1/26/63	6/22/65	Wound.-POW	Keyes Switch
23	Dodds, Mathew*		Pvt.	PA 55th Inf.	I	9/23/63		Died	
24	Dodson, Albert	22	Pvt.	PA 19th Cav.	L	9/17/63		Died	
25	Dodson, Andrew J	17	Pvt.	PA 125th Inf.	E	8/15/62	5/18/63		
26	Dodson, Andrew J		Pvt.	PA 19th Cav.	L	9/17/63			
27	Dodson, John (1)	21	Pvt.	PA 97th Inf.	D	9/21/64	6/28/65		
28	Dodson, John (2)	25	Pvt.	PA 1st Light Art.	F	9/1/64	6/9/65		
29	Dodson, Joseph			IN				Died	
30	Dodson, Samuel B	21	Pvt.	PA 19th Cav.	C	9/30/63	5/14/66		
31	Dodson, William	55	Pvt.	PA 13th Cav.	E	2/29/64	7/17/65		
32	Donahoe, Patrick		Pvt.	PA 76th Inf.	E	10/16/61	10/16/64		
33	Donaldson, Amos	29	Pvt.	PA 67th Inf.	C	11/22/64	7/14/65		
34	Donaldson, Benjamin	31	Pvt.	PA 194th Inf.	I	7/21/64	7/22/65		
35	Donley, Walter		Pvt.	PA McKeage's Mil.	G	7/2/63	8/8/63		
36	Doogen, Henry			USCT Inf.					
37	Dorman, Henry		Pvt.	PA 78th Inf.	K	2/28/65	6/6/65		
38	Dorsey, Wlliam C	18	Corp.	PA 55th Inf.	D	2/27/64	8/30/65		
39	Douglass, H C		Surg.						
40	Douglass, Robert	35	Pvt.	PA 29th Inf.	K	12/23/64	7/1/65		
41	Downey, Michael	23	2nd Lt.	PA 133rd Inf.	K	9/6/62	5/26/63		
42	Doyle, John*	37	Pvt.	PA 55th Inf.	I	10/17/63		KIA	Cold Harbor
43	Doyle, Martin P		1st Lt.	PA 21st Cav.	E	7/11/63	1/11/65	Wounded	Cold Harbor
44	Doyle, Martin P		2nd Lt.	PA 21st Cav.	E			Wounded	Boydton Plank Rd.
45	Drake, Adonijah B*		Pvt.	PA 76th Inf.	E	9/23/64	6/28/65		
46	Drenning, Henry G	38	Corp.	PA 55th Inf.	K	11/5/61		KIA	Cold Harbor
47	Drenning, James C		Pvt.	IA 26th Inf.	G	1862	1863		
48	Drenning, Thomas	16	Pvt.	PA 2nd Cav.	E	3/11/64	6/17/65		
49	Drenning, William		Pvt.	PA McKeage's Mil.	D	7/2/63	8/8/63		
50	Drew, Lewis L		Pvt.	IN 8th Cav.	G	9/25/63	7/20/65	Wounded	
51	Drips, Thomas	21	Pvt.	PA 55th Inf.	H	10/11/61	2/22/63		
52	Drips, Thomas		Pvt.	US 1st Light Art.	M	2/22/63			
53	Duffy, James	19	Pvt.	PA 76th Inf.	E	10/9/61	10/31/64	POW	
54	Dull, John	19	Pvt.	PA 184th Inf.	A	5/12/64		POW-Died	Boydton Plank Rd.
55	Dull, Lewis		Pvt.	PA McKeage's Mil.	H	7/2/63	8/8/63		
56	Dull, Lewis	19	Pvt.	PA 55th Inf.	K	2/19/64		Wounded	White Oak Road
57	Dull, Samuel A	19	Pvt.	MD 2nd PHB Inf.	A	4/9/62		POW-Died	Moorefield Jct.
58	Dull, Valentine	45	Pvt.	PA 138th Inf.	E	8/29/62	6/23/65		
59	Duncan, Henry S		Pvt.	PA 148th Inf.	B				
60	Dunkle, David	17	Pvt.	PA 133rd Inf.	K	8/15/62	5/26/63		
61	Dunkle, David		Pvt.	PA 186th Inf.	E	2/20/64	8/15/65		
62	Dunkle, Simon	18	Pvt.	PA 133rd Inf.	K	8/18/62	5/26/63		
63	Dunn, Patrick*	20	Corp.	PA 55th Inf.	I	2/18/64	8/30/65		
64	Durno, William Jr	26	Corp.	PA 133rd Inf.	C	8/13/62	5/26/63		

	Cemetery	Notes
1	Hershberger Cem.-Snake Spr. V.	Frank McCoy 1912 Listing; aka Dishbraw
2		1890 Hyndman Veterans Census; aka Deshong
3	Uniontown Cem.-IN	David, Morris & Robert-brothers
4	Nethken Hill Cem.-WV	Deserted 12/8/62; PA Civil War Archives
5		1890 Londonderry Twp. Veterans Census
6	Lindenwood Cem.-IN	Historical Data Systems record
7		Historical Data Systems record
8	Sideling Hill CC-Fulton Co.	1890 Brush Creek Twp. Veterans Census
9		Frank McCoy 1912 Listing, aka Dietch
10	Everett Cem.	1890 Everett Veterans Census
11	Trinity UCC Cem.-Osterburg	1890 King Twp. Veterans Census
12		1860 Greenfield Twp. Census; Gabriel & John-brothers
13	Greenfield Cem.-Queen	Frank McCoy 1912 Listing; 1860 Greenfield Twp. Census
14	Greenfield Cem.-Queen	Bedford Inquirer 5/22/1908 listing; Divel spelling listed
15	Union Cem.-MI	Jacob O & George M-brothers
16	Old Claysburg Cem.	PA Civil War Project record; James & Paul-brothers
17	Andersonville Cem., 7360	POW-Andersonville; Died 8/31/64; 1860 Greenfield Twp. Census; B.S.F. History Book-1884
18	Upper Claar Cem.-Queen	1850 Greenfield Twp. Census
19	Mt. Zion Cem.-Pavia	Pavia Veterans Group Picture in 1892; Civil War Grave Marker
20	Upper Claar Cem.-Queen	Frank McCoy 1912 Listing, Morgan & Martin-brothers
21	Old Claysburg Cem.	1860 Greenfield Twp. Census
22	Fockler Cem.-Saxton	Side & Leg Wound-Louden Rangers Cav-Union Army
23		B.S.F. History Book-1884; Pottsville references
24	Mound City Nat'l Cem.-IL	Died 4/27/64; Civil War Roll of Honor; 1860 Greenfield Twp. Census
25	Hopewell Cem.	William-father; Bedford Co. sheriff after the war
26		1850 Greenfield Twp. Census
27	Clearville Union Cem.	1890 Monroe Twp. Veterans Census
28	St. Thomas Cem.-Bedford	Frank McCoy 1912 Listing; 1860 E. St. Clair Twp. Census; John (2) & Joseph-brothers
29	St. Thomas Cem.-Bedford	Died 1865 (estimated) - 18-19 years old; Graves visited by Vets-Bedford Inquirer 1868 article
30	Claysburg Union Cem.	Soldiers of Blair Co. Book - 1940; Samuel & Andrew-brothers
31	Claysburg Union Cem.	Soldiers of Blair Co. Book - 1940; Father of Samuel & Andrew; 1860 Greenfield Twp. Cen.
32		B.S.F. History Book-1884
33	Grandview Cem.-Hunt. Co.	1850 Broad Top Twp. Census
34		B.S.F. History Book-1884; Benjamin & Amos-brothers
35		B.S.F. History Book-1884
36		Frank McCoy 1912 list & Civil Works Admin. 1934 list; 1860 Bedford Twp. Census
37	St. Pauls UCC Cem.-Hunt. Co.	1890 Hopewell Twp. Veterans Census
38	Floyd Cem.-IA	1860 Bedford Twp. Census
39		Kernal of Greatness-Bicentennial Book listing
40	Mt. Olivet Cem.-Manns Choice	1890 Napier Twp. Veterans Census
41		Historical Data Systems-Bedford Co. residence
42		KIA 6/3/64; B.S.F. History Book-1884; Muster in Philadelphia
43		Wounded 6/3/64; Historical Data Systems-Bedford Co. residence
44		Wounded 10/27/64; Historical Data Systems-Bedford Co. residence
45		B.S.F. History Book-1884; Pike Co references
46		KIA 6/3/64; Historical Data Systems-Bedford Co. residence
47	Harmony Cem.-IA	1850 Bedford Twp. Census; Civil War Pension Record
48	Rock Hill Cem.-Monroe Twp.	1890 Monroe Twp. Veterans Census
49		B.S.F. History Book-1884
50		1890 Bedford Twp. Veterans Census
51		Historical Data Systems-Bedford Co. residence
52		B.S.F. History Book-1884
53	Leavenworth Nat'l Cem.	POW-Libby & Belle Island; Historical Data Systems-Bedford residence; 1860 Bedford Census
54	Andersonville Cem., 6879	POW-Andersonville; Died 10/28/64; 1860 Napier Twp. Census; aka Dunn
55		B.S.F. History Book-1884
56		Wounded 3/30/65; B.S.F. History Book-1884; Lewis & John-brothers
57		POW-Richmond 1/4/64; Died 2/22/64 Pneumonia; Bed. Gaz 1914 list; 1860 Juniata Twp. Cen.
58	Fishertown Bretheran Cem.	1850 St. Clair Twp. Census; Lewis & John-sons
59		1890 Monroe Twp. Veterans Census
60	Hershberger Cem.-Snake Spr. V.	Historical Data Systems-Bedford Co. residence; David & Simon-brothers
61		1890 Snake Spr.Valley Veterans Census
62	Hershberger Cem.-Snake Spr. V.	B.S.F. History Book-1884; 1850 W. Providence Twp. Census
63		B.S.F. History Book-1884; enlisted in Huntingdon Co
64	Mt. Royal Cem.-Glenshaw	Historical Data Systems-Bedford Co. residence; enlisted in Hopewell

	Name	Age	Rank	Regiment	Co.	Muster In	Muster Out	Casualty	Casualty Battle
1	Duvall, John N	21	Pvt.	PA 97th Inf.	D	9/19/64	6/28/65		
2	Duvall, Thomas	32	Pvt.	PA 77th Inf.	F	2/29/64		Died	
3	Ealy, John C	18	Sgt.	PA 55th Inf.	H	9/2/61	8/31/65	Wounded	Drewry's Bluff
4	Ealy, John H		Pvt.	PA McKeage's Mil.	G	7/2/63	8/8/63		
5	Eamick, Josiah	18	Pvt.	PA 101st Inf.	D	11/1/61		Died	
6	Earnest, Adam P	25	Pvt.	PA 18th Cav.	H	7/5/64			
7	Earnest, Alexander	19	Pvt.	PA 55th Inf.	K	11/5/61	8/30/65		
8	Earnest, Joseph J		Pvt.	OH 115th Inf.	K	9/18/62	6/22/65		
9	Earnest, Joseph W*	18	Pvt.	PA 55th Inf.	H	9/26/62	6/22/65		
10	Earnest, William	32	Pvt.	PA 138th Inf.	F	8/20/62	6/23/65	Wounded	Monocacy, MD
11	Earnest, William M	36	Pvt.	PA 184th Inf.	A	5/12/64	7/14/65		
12	Eastright, Christian		Pvt.	PA 8th Res.	F	7/11/61		POW	Seven Days Battle
13	Eastright, Christian	24	Pvt.	PA 8th Res.	F			Wound.-Died	Spotsylvania
14	Echom, John		Pvt.	PA 208th Inf.	H	9/7/64			
15	Eckels, Francis S	17	Pvt.	PA 76th Inf.	E	2/14/62		POW-Died	
16	Eckels, John F	20	Pvt.	PA 76th Inf.	E	10/9/61		POW-Died	
17	Eckhard, Jacob	18	Pvt.	PA 55th Inf.	I	2/15/64	8/30/65		
18	Edenbo, Daniel H	22	Corp.	PA 55th Inf.	D	9/2/62	6/11/65		
19	Edmondson, Samuel		Pvt.	PA McKeage's Mil.	G	7/2/63	8/8/63		
20	Edmonson, Joseph	17	Pvt.	VA 3rd Cav.	A	8/14/61	8/1/64	Wounded	Boonsboro, MD
21	Edsall, Burton	33	Corp.	NY 1st Cav.	H	8/5/61	6/27/65		
22	Edwards, Allison	27	Pvt.	PA 8th Res.	F	6/11/61	5/15/64	POW	Seven Days Battle
23	Edwards, Allison		Pvt.	PA 8th Res.	F			Wounded	Fredericksburg
24	Edwards, Benjamin		Pvt.	PA 99th Inf.	F	2/23/65			
25	Edwards, Benjamin F		Pvt.	NY 23rd Inf.	E	5/16/61	5/22/63		
26	Edwards, Benjamin F	21	Pvt.	NY 1st Vet Cav.	F	10/10/63	7/20/65		
27	Edwards, Daniel L	15	Pvt.	PA 55th Inf.	K	2/19/64	8/30/65		
28	Edwards, David H	23	Pvt.	PA 78th Inf.	K	2/28/65	8/12/65		
29	Edwards, Festus	20	Pvt.	OH 129th Inf.	G	8/10/63	3/8/64		
30	Edwards, Festus		Pvt.	OH 187th Inf.	B	3/1/65	1/20/66		
31	Edwards, George C	18	Pvt.	OH 174th Inf.	B	8/20/64	6/28/65		
32	Edwards, George W	35	Pvt.	PA 82nd Inf.	C	11/14/64	7/13/65		
33	Edwards, Hiram	21	Pvt.	PA 8th Res.	F	6/11/61		Died	
34	Edwards, John		Pvt.	PA 199th Inf.	H	9/22/64	5/23/65	Wounded	Ft. Bragg
35	Edwards, John R	25	Pvt.	PA 13th Cav.	B	8/1/63	4/6/65	POW	Sulpher Springs
36	Edwards, Jonathan B	32	Corp.	PA 133rd Inf.	C	8/29/62	12/31/62		
37	Edwards, Josiah V	38	Pvt.	PA 55th Inf.	K	3/2/64	3/6/65		
38	Edwardson, Samuel P*	19	Pvt.	PA 55th Inf.	I	2/16/64	5/29/65		
39	Egan, Charles		Pvt.	PA 55th Inf.	K	7/1/63	8/30/65		
40	Eichelberger, Alexander	40	Pvt.	PA 22nd Cav.	M	2/23/64	7/20/65	Wounded	
41	Eichelberger, David	38	Pvt.	PA 22nd Cav.	M	7/11/63	6/24/65		
42	Eichelberger, Eli	24	Capt.	PA 8th Res.	F	6/11/61	5/26/64	POW	Seven Days Battle
43	Eichelberger, Eli		Capt.	PA 8th Res.	F			Wounded	Wilderness
44	Eichelberger, George W	26	Pvt.	PA 49th Inf.	E	6/4/64	7/15/65		
45	Eichelberger, Jacob A	19	Corp.	PA 194th Inf.	I	7/21/64	11/5/64		
46	Eichelberger, John		Pvt.	PA 22nd Cav.	M	2/9/64	6/24/65		
47	Eichelberger, John S	32	Capt.	PA 8th Res.	F	6/11/61	6/10/63	POW	Seven Days Battle
48	Eichelberger, John S		Capt.	PA 8th Res.	F			Wounded	Fredericksburg
49	Eichelberger, John S		Pvt.	PA 99th Inf.	B	2/23/65	7/1/65		
50	Eichelberger, Michael	37	Pvt.	PA 192nd Inf.	B	2/14/65	8/24/65		
51	Eichelberger, William H	21	Pvt.	PA 8th Res.	F	6/11/61	10/31/62	Wounded	
52	Eichelberger, Winfield	15	Pvt.	PA McKeage's Mil.	G	7/3/63	8/8/63		
53	Eichelberger, Winfield		Pvt.	PA 208th Inf.	H	6/8/64	7/1/65		
54	Eicher, Christian C	12	Music.	PA 12th Res.	K	8/10/61	7/11/64		
55	Eicher, Samuel	43	Pvt.	PA 99th Inf.	H	2/23/65			
56	Eicholtz, William G	40	1st Lt.	PA 208th Inf.	H	9/6/64	6/1/65		
57	Eidenbaugh, John	37	Pvt.	PA 107th Inf.	H	1/9/62	2/16/63	Wounded	South Mountain
58	Einerick, Herman								
59	Eipper, John		Pvt.	NY 119th Inf.	B	9/4/62	6/7/65	Wounded	
60	Elbin, Otho	36	Pvt.	PA 91st Inf.	I	9/21/64	5/30/65		
61	Elder, Daniel S	21	Pvt.	PA 133rd Inf.	C	8/13/62	5/26/63		
62	Elder, Lewis	18	Pvt.	PA 138th Inf.	F	8/29/62	2/26/63		
63	Elder, Samuel	34	Pvt.	PA 13th Inf.	G	4/25/61	8/6/61		
64	Eliott, James K		Pvt.	PA 56th Inf.	G	7/5/64	7/1/65		

	Cemetery	Notes
1	Wells Valley Meth. Cem.	1890 Wells Twp. Veterans Census
2	Nashville Nat'l Cem.	Died 8/14/64; 1860 Wells Valley Census
3	Schellsburg Cem.	Wounded 5/13/64 in Bedford Inquirer Article 7/8/64; 1890 Napier Twp. Veterans Census
4		B.S.F. History Book-1884
5		Died 11/15/62; PA Civil War Archives-Bedford Co. residence, enlisted in Rainsburg
6	St. James Luth. Cem.-Bedford	1890 Bedford Twp. Veterans Census; Deserted after war ended 7/12/65
7	Mt. Smith Cem.-Bedford	1890 Bedford Twp. Veterans Census; aka Earnest
8		1850 Bedford Twp. Census; aka Josiah R
9		aka Ernst; B.S.F. History Book-1884; Berks Co. references
10	Bedford Cem.	Wounded 7/9/64; Frank McCoy 1912 Listing
11		B.S.F. History Book-1884; Historical Data Systems-enlisted in Bedford
12		POW 6/27/62; Bedford Gazette-8th Reserve listing
13	Alexandria Nat'l Cem., A 1978	Wounded Corbin's Bridge 5/8/64; Died 5/30/64; 1850 M. Woodbury Census aka Castwright
14		Deserted 9/13/64; B.S.F. History Book-1884
15	Richmond Nat'l Cem.	POW-Richmond; Died 12/22/63; PA Civil War Archives-enlisted in Bedford
16	Salisbury Nat'l Cem.	POW-Salisbury; Died 1/21/64; Historical Data Systems-Bedford Co. residence
17		PA Civil War Archives-Bedford Co. residence; 1870 Union Twp. (Pavia) Census
18		PA Civil War Archives-Snake Spring Valley residence
19		B.S.F. History Book-1884
20	Grandview Cem.-Altoona	Born-Mecklinburg, VA; Right leg Wound on Gettysburg retreat near Boonsboro, MD 7/8/63
21	Bedford Cem.	Frank McCoy 1912 Listing; 1890 Bedford Twp. Veterans Census
22	Evans Cem.-Coaldale	POW 6/27/62; Historical Data Systems-Bedford Co. residence; aka Norris Allieson
23		Wounded 12/13/62; Bedford Gazette-8th Reserve listing
24		1890 Robertsdale Veterans Census
25		Historical Data Systems record
26		Fold3 lists Born in Bedford Co.; Deserted & returned
27	Oak Hill Cem.-IA	B.S.F. History Book-1884; 1860 Union Twp. (Pavia) Census; Josiah-father
28	St. Pauls UCC Cem.-Hunt. Co.	1890 Hopewell Twp. Veterans Census
29	Rockford Cem.-MI	Ancestry.com-Born in Bedford; Historical Data Systems record
30		Festus & George-brothers
31	Oakdale Cem.-OH	Ancestry.com-Born in Bedford; Historical Data Systems record
32	Wells Valley Meth. Cem.	1890 Wells Twp. Veterans Census
33	US Soldiers & Airmans Nat'l	Died of Disease 8/12/61; Historical Data Systems record; 1860 S. Woodbury Census
34		Wounded 4/2/63; 1890 Wells Valley Veterans Census
35	Wooster Cem.-OH	POW 10/12/63, in Andersonville from 3/6/64; 1860 Wells Tannery Census
36	Old Meth. Cem.-Williamsburg	1860 Broad Top Twp. Census
37	Baker Cem.-IA	Historical Data Systems-Bedford Co. residence; Discharged on Surgeon's Cert.; Daniel-son
38		B.S.F. History Book-1884; Born in Centre Co; Discharged on Surgeon's Cert.
39		Deserted 4/28/64; B.S.F. History Book-1884; aka Eagan
40	Yellow Creek Reformed Cem.	1890 Hopewell Veterans Census
41	Claysburg Union Cem.	Soldiers of Blair Co. Book - 1940; 1870 Broad Top Census
42	Everett Cem.	POW-Libby 6/27/62 to 8/5/62; exchanged for Confederate Lt. Thomas J Clay
43		1890 Saxton Veterans Census; Wounded 5/6/64-shot thru thigh
44	Mt. Zion Cem.-Paint Twp.	1890 Juniata Twp. Veterans Census; 1860 Londonderry Twp. Census
45	Hopewell Cem.	1890 Hopewell Veterans Census; Eli & Jacob-brothers
46	Claysburg Union Cem.	Historical Data Systems record
47	Hopewell Cem.	POW-Libby for 42 days; exchanged for Tully Grabill of GA 28th Infantry
48		1890 Broad Top Veterans Census-Shot in Knee 12/13/62
49		Drafted as Pvt.; Historical Data Systems record
50	Hopewell Cem.	1890 Broad Top Veterans Census
51	Hopewell Cem.	1890 Hopewell Veterans Census-shot in arm; William, John & Alexander- brothers
52	Hopewell Cem.	Historical Data Systems record
53		1890 Everett Veterans Census; aka Scott
54	St. Lukes Cem.-Saxton	1850 Liberty Township Census; Historical Data Systems record-aka Eckert
55	Greenfield Cem.-Queen	Deserted 6/15/65; Frank McCoy 1912 Listing
56	Bedford Cem.	Frank McCoy 1912 Listing; 1860 S. Woodbury Twp. Census
57	Everett Cem.	Wounded 9/14/62, 1st MD PHB Cav references; B.S.F. History Book-1884
58		1890 Londonderry Twp. Veterans Census
59		1890 Hopewell Twp. Veterans census; in Hospital for 6 months
60	Fairview Cem.-Artemas	Frank McCoy 1912 Listing; 1860 Southampton Twp. Census
61	St. Lukes Cem.-Saxton	Frank McCoy 1912 Listing; 1860 Woodbury Twp. Census
62	Englevale Cem.-KS	1860 Cumberland Valley Census; Discharged on Surgeon's Cert.
63	Saxton or Valley Cem.	B.S.F. History Book-1884; Historical Data Systems-Bedford residence
64		1890 Cumberland Valley Veterans Census

	Name	Age	Rank	Regiment	Co.	Muster In	Muster Out	Casualty	Casualty Battle
1	Ellenberger, George	32	Pvt.	PA 55th Inf.	K	2/29/64		Wounded	Bermuda Hundred
2	Ellenberger, John	31	Pvt.	PA 99th Inf.	K	3/14/65	7/1/65		
3	Eller, Henry	22	Music.	PA 188th Inf.	A	2/23/64	12/24/65		
4	Elliott, David S		Pvt.	PA 13th Inf.	G	4/25/61	8/6/61		
5	Elliott, David S	19	Music.	PA 76th Inf.	E	10/9/61	11/28/64		
6	Elliott, Francis M	19	Corp.	PA 2nd Cav.	E	6/5/62	5/31/65		
7	Elliott, John	22	Pvt.	PA 2nd Cav.	E	11/4/61		Wound.-Died	
8	Elliott, John K H	27	Pvt.	PA 171st Inf.	I	11/2/62	8/8/63		
9	Ellis, Enos	31	Pvt.	PA 171st Inf.	I	11/10/62	8/8/63		
10	Ellis, Enos		Pvt.	PA 107th Inf.	H	9/21/64	5/15/65		
11	Ellis, George N	34	Sgt.	PA 21st Cav.	E	7/8/63	2/20/64		
12	Ellis, Henry								
13	Elwell, George*	32	Pvt.	PA 85th Inf.	B	11/20/61		Wound.-Died	Fair Oaks
14	Elwell, John	21	Pvt.	PA 208th Inf.	H	9/8/64	6/1/65		
15	Emeigh, Charles		Pvt.	PA 14th Inf.	H	4/24/61	8/7/61		
16	Emeigh, Charles*	20	Pvt.	US 12th Inf.	A			Died	
17	Emeigh, Christopher	34	Corp.	PA 12th Cav.	H	3/29/64	7/14/65		
18	Emeigh, John		Pvt.	US 12th Inf.	A				
19	Emerick, Andrew R	36	Pvt.	PA 171st Inf.	I	11/1/62			
20	Emerick, Jacob	18	Pvt.	MD 2nd PHB Inf.	D	3/13/65	5/29/65		
21	Emerick, Solomon		Pvt.	PA 149th Inf.					
22	Emigh, Abraham*	28	Pvt.	PA 171st Inf.	I	11/10/62	8/8/63		
23	Emigh, Jacob*		Pvt.	PA 22nd Cav.	H	2/24/64			
24	Emme, William		Music.	US 10th Inf.	K	2/11/58	2/11/63		
25	Emme, William		Corp.	US 1st Inf.	G	2/11/64	2/11/66		
26	Enfield, Americus	14		PA Knap's Lgt. Art.	E	6/13/61	2/24/64	Wounded	Gettysburg
27	Enfield, Americus		Surg.	PA 22nd Cav.	G	2/24/64	10/31/65		
28	England, Jacob	21	Pvt.	PA 101st Inf.	D	11/1/61	6/25/65	POW	Plymonth, NC
29	Engle, Barney	17	Pvt.	PA 12th Cav.	G	1/8/62	7/20/65		
30	Engle, Charles	23	Pvt.	PA 55th Inf.	D	10/12/61		Died	
31	Engle, Henry	40	Pvt.	PA 14th Inf.	H	4/24/61	8/7/61		
32	Engle, Henry		Pvt.	PA 121st Inf.	H	8/4/62		POW-Died	Peebles Farm
33	Engle, Joseph	43	Pvt.	PA 12th Cav.	K	3/23/62	7/20/65	POW	
34	Enock, Jennings A		Pvt.	PA 21st Cav.	E	7/8/63	2/20/64		
35	Enos, David	27	Pvt.	PA 21st Cav.	E	7/8/63	2/20/64		
36	Enos, David			PA 171st Inf.					
37	Ensley, Christopher	34	Corp.	PA 184th Inf.	A	5/12/64		POW-Died	Jerusalem Plank Rd.
38	Ensley, Lewis	33	Pvt.	PA 22nd Cav.	L	2/29/64	6/24/65		
39	Ensley, Peter M	36	Pvt.	PA 22nd Cav.	L	2/29/64	6/24/65		
40	Epler, Aaron*		Pvt.	PA 55th Inf.	K	9/23/63		MIA	Drewry's Bluff
41	Ernst, John	41	Pvt.	OH 42nd Inf.	H	11/27/61		Died	
42	Eshelman, Benjamin	28	Pvt.	IL 92nd Inf.	I	9/4/62		Died	
43	Eshelman, George W	21	Pvt.	PA 186th Inf.	G	3/24/64	8/15/65		
44	Eshelman, John W		Pvt.	PA 88th Inf.	A			Died	
45	Estep, Henry C	16	Pvt.	PA 49th Inf.	D	8/30/61	10/31/64		
46	Evans, Daniel*							MIA	
47	Evans, David V	31	Sgt.	PA McKeage's Mil.	G	7/2/63	8/8/63		
48	Evans, Evan S	17	Pvt.	IA 46th Inf.	H	6/10/64	9/23/64		
49	Evans, George W (1)	35	Pvt.	PA 55th Inf.	I	9/20/61	8/30/65		
50	Evans, George W (2)	21	Pvt.	PA 133rd Inf.	C	8/13/62	5/26/63		
51	Evans, Harvey	26	Pvt.	PA 138th Inf.	D	8/29/62			
52	Evans, Hiram	43	Capt.	IA 23rd Inf.	D	9/19/62	1/14/64		
53	Evans, Isaiah	24	Sgt.	PA 101st Inf.	D	11/1/61	1862		
54	Evans, Isaiah		Pvt.	PA 149th Inf.	I	9/29/63	5/8/64	KIA	Spotsylvania
55	Evans, John J	28	Pvt.	PA 84th Inf.	I	11/10/61	8/10/62		
56	Evans, John W	27	Pvt.	WI 14th Inf.	D	10/25/61	10/9/65		
57	Evans, Johnson	22	Pvt.	PA 8th Res.	F	8/2/61	5/26/64	POW	Seven Days Battle
58	Evans, Johnson		Pvt.	PA 8th Res.	F			Wounded	Fredericksburg
59	Evans, Johnson		Pvt.	PA 8th Res.	F			Wounded	Wilderness
60	Evans, Lemuel	39	Pvt.	PA 49th Inf.	E	6/4/64	7/15/65	Wounded	
61	Evans, Lewis	33	Pvt.	PA 149th Inf.	I	8/26/63	6/24/65		
62	Evans, Nathan C	30	2nd Lt.	PA 101st Inf.	D	2/8/62	4/24/63		
63	Evans, Nathan C		Capt.	PA McKeage's Mil.	G	7/2/63	8/8/63		
64	Evans, Nathan C		Capt.	PA 184th Inf.	A	5/13/64	7/14/65	Wound.-POW	Jerusalem Plank Rd.

	Cemetery	Notes
1	Schellsburg Cem.	Wounded 5/20/64; 1850 Napier Twp. Census; Historical Data Systems record
2	Schellsburg Cem.	1860 Juniata Twp. Census; Historical Data Systems record
3	Upper Claar Cem.-Queen	Frank McCoy 1912 Listing; 1860 Greenfield Twp. Census
4		B.S.F. History Book-1884; Historical Data Systems record
5		B.S.F. History Book-1884; Historical Data Systems-Bedford residence
6	Bethel Meth. Cem.-Centerville	1890 Cumberland Valley Veterans Census; Francis & John-brothers
7	US Soldiers & Airmans Nat'l	Leg Amputated 4/1/63; Died 5/5/63; 1860 Cumberland Valley Vet. Cen.; Bed. Gazette 1914 list
8		B.S.F. History Book-1884
9	Messiah Luth.-Dutch Corner	B.S.F. History Book-1884; 1850 Napier Twp. Census
10		1890 Bedford Twp. Veterans Census
11	Ellis Cem.-New Paris	Frank McCoy 1912 Listing; George & Enos-brothers
12	Claar Cem.-Kimmel Twp.	Bedford Inquirer 5/22/1908 listing
13	Yorktown Nat'l Cem.	Wounded 5/21/62, Died 9/24/62; E. Providence Twp. Honor Roll; Fayette Co. references
14	Robinsonville Cem.	B.S.F. History Book-1884; 1870 Monroe Twp. Census; Born in NC
15		Christopher & John-brothers; PA Civil War Archives-Claysburg residence
16	Hampton Nat'l Cem.	Died of Disease 8/31/62; no information to link to the Charles Emeigh enl. in 14th PA Inf.
17	Old Claysburg Cem.	1890 Greenfield Veterans Census
18		1860 Greenfield Twp. Census; born 6/30/41
19	Lybarger Cem.-Madley	Deserted 11/9/62; 1870 Juniata Twp. Census
20	Hyndman Cem.	1890 Londonderry Twp. Veterans Census; 1870 Juniata Twp. Census
21	Shroyer Farm Cem.-Madley	1890 Londonderry Twp. Veterans Census; born 8/14/36
22		B.S.F. History Book-1884; born Cambria Co.
23		B.S.F. History Book-1884; born Cambria Co.
24		Ancestry.com information
25	Everett Cem.	1890 Everett Veterans Census
26	Bedford Cem.	Wounded 7/2/63; GAR Personal War Sketch record-Enlisted under an alias
27		1880 Bedford Census; Well-known Doctor after the war
28	Westfall Cem.-IL	POW-Andersonville & Florence 4/20/64 to 3/27/65; 1860 Colerain Twp. Census
29	Hollidaysburg Presb. Cem.	Ancestry.com information; Henry is Father
30	Beaufort Nat'l Cem.	Died 11/7/62; PA Civil War Archives-Bedford Co. residence
31	Carson Valley-Duncansville	Father on 1830 Woodbury Twp. Census
32		POW-Salisbury 10/1/64; Died 1/16/65; Henry & Joseph-brothers
33		POW-Belle Island for 23 days; Ancestry.com-born Woodbury
34		Historical Data Systems record
35	Trinity Luth. Cem.-Cumberland	1870 Londonderry Twp. Census; Historical Data Systems-Bedford Co. residence
36		Bedford Inquirer 5/22/1908 listing
37	Andersonville Cem., 6889	POW-Andersonville 6/22/64; Died-Dysentery 8/26/64; Ancestry.com born E. Prov. Twp.
38	Mt. Zion Cem.-Breezewood	Lewis, Peter & Christopher-brothers
39	Davenport Cem.-NE	Hist. Data Sys. record; Born-East Providence Twp.-Ancestry; 1860 Brush Creek Twp. Cen.
40		MIA 5/16/64; B.S.F. History Book-1884; Berks Co. residence
41	Jefferson Barracks Cem.-MO	Died 3/22/63; Ancestry.com-Born in Bedford Co., Wife-1830 Hopewell Twp. Census
42	Danville Nat'l Cem.-KY	Died-Typhoid Fever 1/17/63; 1860 E. Providence Twp. Cen.; Bio. Review: Bed. & Som.-1899
43	Everett Cem.	George W, John W & Benjamin-bros.
44	Rays Hill Cem.-Breezewood	Died 6/5/65 after returning home; Frank McCoy 1912 Listing; Pension Record
45	Duval Cem.-Coaldale	1890 Broad Top Veterans Census; Historical Data Systems record
46		Bedford Gazette 2/13/1914 listing; 1850 Broad Top Census
47		B.S.F. History Book-1884
48	Green Bay Cem.-IA	1850 Southampton Twp. Census; Historical Data Systems record
49		B.S.F. History Book-1884
50	Duval Cem.-Coaldale	1860 Broad Top Twp. Census
51		Deserted 12/20/62; Historical Data Systems-Bedford Co. residence
52	Southlawn Cem.-IA	Ancestry.com-born in Clearville
53		B.S.F. History Book-1884; PA Civil War Archives-Bedford Co. residence
54		KIA 5/8/64 at Laurel Hill; 1860 Southampton Cen.; Bedford Gazette 2/13/1914 listing
55	Stone Church Cem.-Fishertown	Frank McCoy 1912 Listing; 1890 E. St. Clair Twp. Veterans Census
56	Southlawn Cem.-IA	Ancestry.com-Married-Bloody Run 1857
57	Hopewell Cem.	POW 6/27/62; 1850 Monroe Twp. Census
58		Wounded 12/13/62; Bedford Gazette-8th Reserve listing
59		Wounded 5/6/64; 1850 Monroe Twp. Census
60	Evans Cem.-Coaldale	Frank McCoy 1912 Listing
61	Fairview Cem.-Artemas	1890 Monroe Twp. Veterans Census; 1850 Monroe Twp. Census
62	Everett Cem.	Nathan, Isaah, Hiram & John W-brothers
63		B.S.F. History Book-1884
64		POW 6/22/64 for 9 months

Chapter 18

	Name	Age	Rank	Regiment	Co.	Muster In	Muster Out	Casualty	Casualty Battle
1	Evans, Septimus		Pvt.	PA 17th Inf.	K	9/17/62	9/28/62		
2	Evans, William	28	Pvt.	PA 133rd Inf.	K	8/15/62	2/14/63		Fredericksburg*
3	Evans, William		Sgt.	PA 21st Cav.	E	7/8/63	2/20/64		
4	Evans, William H	18	Pvt.	PA 101st Inf.	G	12/28/61		Died	
5	Everhart, Alexander	19	Pvt.	PA 11th Res.	F	2/7/62			
6	Everhart, Alexander		Pvt.	PA 99th Inf.	F	2/23/65	7/1/65	Wounded	Petersburg
7	Everhart, David	21	Pvt.	PA 110th Inf.	C	10/24/61		KIA	Fredericksburg
8	Everhart, James H	19	Pvt.	PA 208th Inf.	K	9/7/64	6/1/65		
9	Exline, Jacob	22	Pvt.	PA 55th Inf.	K	10/11/61		POW-Died	Drewry's Bluff
10	Fackler, Samuel O	34	Pvt.	PA 110th Inf.	C	1861	6/28/65		
11	Fagan, Charles A*	19	Sgt.	PA 194th Inf.	I	7/22/64	9/6/64		
12	Fagan, Charles A*		Pvt.	PA 97th Inf.	Ind	9/6/64	2/12/65		
13	Fagans, James	17	Pvt.	PA 125th Inf.	E	8/1/62	5/18/63		
14	Fagans, James		Pvt.	PA 55th Inf.	I	2/15/64	8/30/65		
15	Fahnostock, Walter F*	18	Sgt.	PA 76th Inf.	E	11/27/61	12/3/64		
16	Fair, Henry*	35	Pvt.	PA 53rd Inf.	C	10/17/61		Died	
17	Fait, John	34	Pvt.	PA 138th Inf.	E	8/29/62	6/23/65		
18	Farber, Thomas H	24	Sgt.	PA 55th Inf.	D	10/12/61	8/30/65		
19	Farner, Jeremiah	22	Pvt.	PA 136th Inf.	K	8/20/62	5/29/63		
20	Faulkender, William D	28	Pvt.	PA 208th Inf.	H	9/7/64	6/1/65		
21	Faust, John J		Pvt.	PA 76th Inf.	E	2/17/65	7/18/65		
22	Faust, John R	27	Capt.	PA 173rd Inf.	F	11/1/62	8/16/63		
23	Feaster, John G	20	Pvt.	PA 91st Inf.	B	9/21/64	5/30/65		
24	Feather, David	39	Pvt.	PA 81st Inf.	A	8/25/64	6/29/65		
25	Feather, George W	18	Pvt.	PA 138th Inf.	E	8/29/62		Died	
26	Feather, Henry C	38	Pvt.	PA 205th Inf.	C	8/27/64	6/2/65		
27	Feather, John A	31	Pvt.	PA 205th Inf.	C	8/27/64	6/2/65		
28	Feather, Josiah	39	Pvt.	PA 91st Inf.	G	6/3/64	7/10/65		
29	Feather, Michael	31	Corp.	PA 171st Inf.	I	11/2/62	8/8/63		
30	Feather, Michael		Pvt.	PA 99th Inf.	F	2/25/65	7/1/65		
31	Feather, Simon	17	Pvt.	PA 138th Inf.	E	8/29/62	5/31/65	Wounded	Spotsylvania
32	Feather, William	22	Pvt.	PA McKeage's Mil.	G	7/2/63	8/8/63		
33	Feather, William		Pvt.	PA 55th Inf.	K	3/2/64	8/30/65	Wounded	Drewry's Bluff
34	Feathers, Henry	34	Pvt.	PA 171st Inf.	I	11/10/62	8/8/63		
35	Feathers, Henry		Pvt.	PA 99th Inf.	H	2/25/65	7/1/65		
36	Feese, Edward	18	Pvt.	PA 76th Inf.	E	2/21/65	7/6/65		
37	Feidler, Michael		Pvt.	PA 76th Inf.	E	10/16/61	5/9/63		
38	Feight, Abraham	18	Pvt.	PA 138th Inf.	E	8/29/62		Died	
39	Feight, Frederick	20	Corp.	PA 2nd Cav.	E	11/29/61	6/17/65		
40	Feight, Henry H	16	Pvt.	PA 138th Inf.	E	8/29/62	6/23/65		
41	Feight, John W	25	Capt.	PA 138th Inf.	F	9/30/62	6/23/65	Wounded	Opequon
42	Feight, Joseph	37	Pvt.	OH 43rd Inf.	K	10/8/62	8/18/63		
43	Feight, Levi		Pvt.	PA 57th Inf.	B	9/21/64		Died	
44	Feight, William F	17	Pvt.	PA 138th Inf.	F	8/29/62	6/23/65	Wounded	Cedar Creek
45	Feight, William W	19	Pvt.	PA 55th Inf.	H	10/11/61	8/30/65	Wounded	
46	Felix, John	16	Sgt.	PA 21st Cav.	E	7/8/63	2/20/64		
47	Felton, Christian	42	Pvt.	PA 78th Inf.	K	2/28/65	9/11/65		
48	Felton, Christian K	18	Pvt.	PA 56th Inf.	F	10/3/64	7/1/65		
49	Felton, John A	20	Corp.	PA McKeage's Mil.	D	7/2/63	8/8/63		
50	Felton, John A		Pvt.	PA 22nd Cav.	H	2/26/64	6/24/65	Injured	Opequon
51	Felton, Peter S	18	Pvt.	PA McKeage's Mil.	D	7/2/63	8/8/63		
52	Felton, Peter S		Pvt.	PA 208th Inf.	K	9/7/64		KIA	Petersburg - Final
53	Felton, Simon P	19	Pvt.	PA McKeage's Mil.	D	7/3/63	8/8/63		
54	Felton, Simon P		Corp.	PA 22nd Cav.	H	2/26/64	7/19/65		
55	Feltwell, William V	31	Chap.	NY 5th Inf.	I	7/11/65	8/21/65		
56	Fenton, Henry W*	20	Pvt.	PA 76th Inf.	E	3/8/65	7/18/65		
57	Ferguson, John	19	Pvt.	PA 110th Inf.	C	10/24/61		KIA	1st Kernstown
58	Ferguson, John W*	21	Pvt.	PA 184th Inf.	A	5/12/64		POW-Died	Boydton Plank Rd.
59	Ferguson, William	24	Sgt.	PA 138th Inf.	D	8/29/62	6/23/65	POW	Monocacy, MD
60	Feristh, Daniel					1864	1865		
61	Ferner, Jeremiah	33	Pvt.	PA 136th Inf.	K	8/27/62	5/29/63		
62	Fessler, George H	39	Pvt.	PA 61st Inf.	G	9/19/64	6/20/65		
63	Fessler, John B		Pvt.	PA 151st Inf.	H	11/1/62	7/27/63	Wounded	Gettysburg
64	Fessler, Samuel		Pvt.	PA 107th Inf.	H	1/9/62		KIA	Antietam

	Cemetery	Notes
1		Ancestry.com information; Septimus & George W (2)-brothers
2	Zion Luth. Cem.-Holidaysburg	B.S.F. History Book-1884; Historical Data Systems-Bedford Co. resid.; Discharged Surg. Cert.
3		Historical Data Systems record
4	Yorktown Nat'l Cem.	Died of Disease 6/27/62; B.S.F. History Book-1884; 1860 Broad Top Census
5	Duval Cem.-Coaldale	Deserted 2/18/63; 1890 Coaldale Veterans Census
6		1890 Coaldale Veterans Census-Shot through hip
7		KIA 12/13/62; PA Civil War Archives-Yellow Creek residence; 1860 Hopewell Twp. Census
8	Methodist Cem.-Tatesville	1890 Hopewell Twp. Veterans Census
9	Andersonville Cem., 8066	POW-Andersonville; Died 9/7/64; B.S.F. History Book-1884; 1850 Union Twp. Census
10		PA Civil War Archives-Woodbury residence
11		B.S.F. History Book-1884; Berks Co. references
12		B.S.F. History Book-1884; Berks Co. references
13	Gilmore Cem.-Wash. Co	Ancestry.com information; 1880 Bloomfield Twp. Census
14		B.S.F. History Book-1884; 1870 M. Woodbury Twp. Census
15		B.S.F. History Book-1884; Pittsburgh references
16	Reformed Cem.-Marklesburg	Died of Disease 2/12/62; 1860 Hopewell Twp. (Hunt. Co.) Cen. Buried-Marklesburg
17	Mulligans Cove-Buffalo Mills	1890 Harrison Twp. Veterans Census
18	Bedford Cem.	Frank McCoy 1912 Listing; 1860 St. Clair Twp. Census
19	Hyndman Cem.	1890 Hyndman Veterans Census
20	Waterside Cem.	1890 Woodbury Twp. Veterans Census
21	Mt. Olivet Cem.-Manns Choice	Frank McCoy 1912 Listings 72nd Reg
22	Mt. Olivet Cem.-Manns Choice	1890 Harrison Twp. Veterans Census; Doctor after war
23	Fishertown Cem.	1860 St. Clair Twp. Census
24	Upper Claar Cem.-Queen	Frank McCoy 1912 Listing; 1890 Greenfield Twp. Census; aka Fetters
25		Died 10/25/62; PA Civil War Archives-St. Clairsville residence; B.S.F. History Book-1884
26	Old Claysburg Cem.	Soldiers of Blair Co. Book - 1940; Ancestry.com-born in Pavia
27	Upper Claar Cem.-Queen	Frank McCoy 1912 Listing; 1890 Greenfield Twp. Veterans Census
28	Upper Claar Cem.-Queen	Historical Data Systems record
29	Mt. Zion Cem.-Pavia	1860 Union Twp. (Pavia) Census
30		Historical Data Systems record
31	Mt. Zion Cem.-Pavia	Shot in leg 5/19/64; Michael & Simon-brothers
32	Upper Claar Cem.-Queen	B.S.F. History Book-1884; William & David-brothers
33		Shoulder Wounded 5/15/64; B.S.F. History Book-1884; John & William-brothers
34	Mt. Zion Cem.-Pavia	Frank McCoy 1912 Listing
35		Frank McCoy 1912 Listing; 1860 Union Twp. (Pavia) Census
36		B.S.F. History Book-1884
37		B.S.F. History Book-1884; Discharged on Surgeon's Cert.
38	Schellsburg Cem.	Died 11/13/62; B.S.F. History Book-1884; aka Feicht; PA Civil War Arch.-Schellsburg resid.
39	Mt. Pleasant Meet. House-VA	Historical Data Systems record; 1860 Cumberland Valley Twp. Census
40	Hart Cem.-MI	B.S.F. History Book-1884; 1850 Napier Twp. Census
41	Springhill Cem.-Shippensburg	Wounded 9/19/64; PA Civil War Archives-Rays Hill residence
42	Ohio Vet. Home Cem.-OH	Ancestry.com information; 1860 E. Providence Twp. Census; Joseph & John-brothers
43	Mt. Pleasant Cem.-Mattie	Died 2/21/65; Frank McCoy 1912 Listing; 1860 E. Providence Twp. Census
44	Mt. Pleasant Cem.-Mattie	Wounded 10/19/64, in Hospital at Muster Out; 1860 Cumberland Valley Census
45	Everett Cem.	1850 Harrison Twp. Census; Gunshot Wound in thigh
46	Evergreen Cem.-CO	Historical Data Systems-Bedford Co. residence; 1850 St. Clair Twp. Census
47	Mt. Pleasant Cem.-Mattie	Historical Data Systems record; 1860 E. Providence Twp. Census
48	Topeka Cem.-KS	Ancestry.com information; Christian K & Simon-brothers
49	Mt. Zion Cem.-Breezewood	B.S.F. History Book-1884; 1860 E. Providence Twp. Census
50		B.S.F. History Book-1884; Pension Records
51		B.S.F. History Book-1884
52		KIA 4/2/65; 1860 Monroe Twp. Census; B.S.F. History Book-1884
53	Mt. Pleasant Cem.-Mattie	Frank McCoy 1912 Listing; 1860 E. Providence Twp. Census
54		B.S.F. History Book-1884
55	Bedford Cem.	Historical Data Systems record
56		B.S.F. History Book-1884
57	Yellow Creek Reformed Cem.	KIA 3/23/62; Bedford Inquirer 5/22/1908 listing; 1860 Hopewell Twp. Census; aka Fergusen
58	Andersonville Cem., 11601	POW-Andersonville; Died-Starvation 10/28/64; B.S.F. History Book-1884
59	Brethren Cem.-IA	POW 7/9/64 to 10/18/64; B.S.F. History Book-1884
60		1890 Union Twp. (Pavia) Veterans Census
61	Hyndman Cem.	Findagrave.com information; 1870 Harrison Twp. Census; aka Farner
62	Wells Valley Meth. Cem.	Ancestry.com-born Brush Creek; 1890 Wells Veterans Census
63		Wounded 7/1/63; 1850 Wells Valley Census; John, George & Samuel-brothers
64	Antietam Nat'l Cem, 3736	KIA 9/17/62; 1860 Broad Top Twp. Census; Historical Data Systems record aka Fesler

Chapter 18

	Name	Age	Rank	Regiment	Co.	Muster In	Muster Out	Casualty	Casualty Battle
1	Fetter, George	30	Pvt.	PA 149th Inf.	I	8/26/63	6/24/65		
2	Fetter, Harrison	30	Pvt.	PA 171st Inf.	I	11/10/62	8/8/63		
3	Fetter, John	18	Pvt.	PA 76th Inf.	E	3/18/62	5/12/65		
4	Fetter, Joseph	18	Pvt.	PA 76th Inf.	E	10/9/61		KIA	Ft. Wagner, SC
5	Fetter, Joseph B	24	Pvt.	PA 171st Inf.	I	11/10/62	8/8/63		
6	Fetter, Joseph B		Pvt.	PA 99th Inf.	B	2/25/65	7/1/65		
7	Fetter, Joseph J	20	Pvt.	PA 76th Inf.	E	10/9/61		Died	
8	Fetter, Samuel	18	Pvt.	PA 107th Inf.	H	5/4/64	6/5/65	POW	Dabney's Mills
9	Fetters, George			PA 21st Cav.					
10	Fetters, Job	27	Pvt.	PA 171st Inf.	I	11/2/62		Died	
11	Fetters, John		Pvt.	PA 99th Inf.				KIA	Sailor's Creek
12	Fickes, Cyrus W	19	Pvt.	PA 21st Cav.	E	6/26/63	2/20/64		
13	Fickes, Cyrus W		Pvt.	PA 200th Inf.	C	8/26/64	6/6/65	Wounded	Ft. Stedman
14	Fickes, James M	16	Pvt.	PA 101st Inf.	G	2/18/62	11/8/62	Died	
15	Fickes, John W		Corp.	PA McKeage's Mil.	G	7/2/63	8/8/63		
16	Fickes, John W			PA 208th Inf.					
17	Fickus, Cyrus	21	Pvt.	PA 138th Inf.	E	8/29/62			
18	Fidler, Issac M*		Pvt.	PA 55th Inf.	D	9/2/62		Died	
19	Fidler, Jacob*		Pvt.	PA 55th Inf.	H	9/2/62	3/18/65	Wounded	Bermuda Hundred
20	Fidler, Lewis W*		Pvt.	PA 55th Inf.	D	9/2/62	6/11/65		
21	Figart, David	21	Pvt.	PA 133rd Inf.	C	8/13/62	5/26/63		
22	Figart, Henry J		Pvt.	PA 8th Res.	F	6/11/61		POW	Seven Days Battle
23	Figart, Henry J	21	Pvt.	PA 8th Res.	F			KIA	2nd Bull Run
24	Figart, Levi H	17	Pvt.	PA 107th Inf.	H	1/9/62	2/13/63	Wounded	South Mountain
25	Filebaugh, Rinehart*	37	Pvt.	PA 55th Inf.	I	10/8/63	10/15/64		
26	Filler, John H (1)		Capt.	PA 13th Inf.	G	4/25/61	8/6/61		
27	Filler, John H (1)	32	Lt. Col.	PA 55th Inf.		12/4/61	3/23/65		
28	Filler, John H (2)	23	Sgt.	MD 2nd PHB Inf.	B	9/1/61	5/29/65		
29	Filler, Joseph	49	Lt. Col.	PA 55th Inf.	K	11/5/61	11/13/64	POW	Petersburg
30	Filler, Joseph		Lt. Col.	PA 55th Inf.	K			Wounded	Drewry's Bluff
31	Filler, William B	16	Pvt.	PA 101st Inf.	D	11/1/61	7/11/62		
32	Filler, William B		Corp.	PA 22nd Cav.	C	7/11/63	7/22/65	POW	Browns Gap
33	Filler, William C	17	Pvt.	PA 101st Inf.	D	11/1/61	11/13/62		
34	Filler, William T	19	Pvt.	PA 13th Inf.	G	4/25/61	8/6/61		
35	Filler, William T		Pvt.	PA 138th Inf.	E	8/29/62	6/23/65		
36	Findley, Archibald (1)	33	Pvt.	PA 51st Inf.	A	9/20/64	6/2/65		
37	Findley, Archibald (2)	29	Pvt.	PA 148th Inf.	C	8/20/63	5/15/65	Wounded	
38	Fink, Abraham	29	Pvt.	PA 148th Inf.	C	8/20/63	5/15/65	Wounded	Cold Harbor
39	Fink, John*	40	Pvt.	PA 76th Inf.	E	11/21/61	5/18/63		
40	Fink, Peter		Pvt.	PA McKeage's Mil.	H	7/2/63	8/8/63		
41	Fink, Valentine	26	Pvt.	PA 133rd Inf.	K	8/26/62	6/26/63		
42	Finley, Harris*	22	Corp.	PA 133rd Inf.	K	8/30/62	5/26/63		
43	Finnegan, Daniel	43	Pvt.	PA 55th Inf.	I	9/4/61		Died	
44	Finnegan, Henry	19	Pvt.	PA 55th Inf.	I	2/14/64	10/21/64		
45	Fishel, George W	16	Pvt.	PA 110th Inf.	C	10/24/61	10/24/64		
46	Fisher, Andrew		Pvt.	PA McKeage's Mil.	H	7/2/63	8/8/63		
47	Fisher, Andrew	19	Pvt.	PA 55th Inf.	D	2/27/64	8/30/65		
48	Fisher, Augustus	26	Pvt.	MD 2nd PHB Inf.	E	7/30/61	9/29/64		
49	Fisher, Charles*		Pvt.	PA 46th Inf.	B			Wounded	Antietam
50	Fisher, Charles*	18	Pvt.	PA 46th Inf.	B			Wounded	Cedar Mountain
51	Fisher, Charles*		Pvt.	PA 46th Inf.	B			Wounded	Chancellorsville
52	Fisher, Charles*		Pvt.	PA 46th Inf.	B	9/2/61	9/17/64	Wounded	Winchester
53	Fisher, David H	42	Pvt.	PA 78th Inf.	K	3/3/65	9/11/65		
54	Fisher, Edmund G*	22	Pvt.	PA 55th Inf.	H	9/2/62	6/11/65		
55	Fisher, Emanuel	28	1st Lt.	PA 138th Inf.	D	8/29/62	6/23/65		
56	Fisher, Henry H	25	Pvt.	PA 133rd Inf.	C	8/13/62	5/26/63		
57	Fisher, James N	31	Pvt.	PA 28th Inf.	C	7/5/64			
58	Fisher, John G	25	Pvt.	PA 90th Inf.	I	9/24/61			
59	Fisher, John R	26	2nd Lt.	PA 158th Inf.	H	11/4/62	8/12/63		
60	Fisher, John W	30	Pvt.	PA 133rd Inf.	C	8/13/62	5/26/63		
61	Fisher, John W		Pvt.	PA 186th Inf.	G	3/24/64	8/15/65		
62	Fisher, Joseph	26	Corp.	PA 171st Inf.	I	11/2/62	8/8/63		
63	Fisher, Thomas	36	Pvt.	PA 5th	E	2/24/65	5/15/65		
64	Fitzgerald, Maurice			US 2nd Inf.	C				

	Cemetery	Notes
1	Fountain Park Cem.-IN	1890 Woodbury Twp. Veterans Census; 1870 Saxton Census
2	Dry Hill Cem.-Woodbury	1860 Union Twp. (Pavia) Census
3		B.S.F. History Book-1884
4		KIA 7/11/63; PA Civil War Archives-Bedford residence; Historical Data Systems record
5	Trinity UCC Cem.-Osterburg	B.S.F. History Book-1884; Joseph & Harrison-brothers
6		Pension Record
7	Beaufort Nat'l Cem.	Died 6/9/62; PA Civil War Archives-Bedford residence; 1860 Snake Spring Twp. Census
8		POW 2/6 to 3/3/65; PA Civil War Archives-born in Bedford Co.
9	Bald Hill Cem.-Everett	Frank McCoy 1912 Listing; Tombstone lists "38th US Cav"
10	Casteel Farm-Southampton Twp.	Died 11/27/62; PA Civil War Project record; Frank McCoy 1912 Listing; Job and John-bros.
11		KIA 4/6/65; PA Civil War Archives record; Civil War Pension Record; 1860 Southampton Cen.
12	Fishertown Cem.	1850 E. St. Clair Census; Cyrus & James-brothers
13		Wounded 3/25/65; 1890 E. St. Clair Twp. Veterans Census
14	Fishertown Cem.	Died 12/8/62; Historical Data Systems-Schellsburg residence; 1860 E. St Clair Twp. Cen.; DSC
15		B.S.F. History Book-1884
16	Old Union Cem.-Osterburg	Frank McCoy 1912 Listing
17		Deserted 11/9/62; aka Sierus Fickaus; PA State Civil War Archives.-Bedford resident
18		Died 4/15/64; B.S.F. History Book-1884; enlisted in Berks Co.
19		Wounded-Bedford Inquirer 7/8/64 article; B.S.F. History Book-1884; enlisted in Berks Co.
20		B.S.F. History Book-1884; enlisted in Berks Co.
21	Greenwood Cen.-NE	B.S.F. History Book-1884; 1850 Broad Top Census
22		POW 6/27/62; Bedford Gazette-8th Reserve listing
23	US Soldiers & Airmans Nat'l	Wounded 8/29/62; Died 9/17/62; B.S.F. History Book-1884; 1850 Hopewell Twp. Census
24	Mt. Zion Cem.-Breezewood	Wounded 9/14/62; B.S.F. History Book-1884; Levi & David-brothers
25		B.S.F. History Book-1884; Discharged on Surgeon's Cert.
26		B.S.F. History Book-1884
27		B.S.F. History Book-1884; Historical Data Systems-Bedford Co. residence
28		John H (2) & William B-brothers
29	Allegheny Cem.-Pittsburgh	POW-Macon 7/30/64; Historical Data Systems-Bedford Co. residence
30		Wounded listed in Bedford Inquirer article 12/9/64
31	Woods Church-Colerain Twp.	1850 Colerain Twp. Census; Discharged on Surgeon's Cert.
32		POW-Libby, Belle Island & Salisbury 9/26/64 to 3/8/65; B.S.F. History Book-1884
33	Graceland Cem.-OH	1890 Colerain Twp. Vet. Cen.; Hist. Data Systems-Bedford Co. resid.; Discharged Surg. Cert.
34	Homewood Cem.-Pittsburgh	1860 Bedford Census; Historical Data Systems-Bedford residence
35		B.S.F. History Book-1884; Joseph-father
36	Schellsburg Cem.	Frank McCoy 1912 Listing
37	Replogle Cem.-Woodbury	1890 Woodbury Twp. Veterans Census - shot in left leg
38	Replogle Cem.-Woodbury	Left Leg Wound 6/3/64; 1860 S. Woodbury Twp. Census; PA Civil War Project record
39		B.S.F. History Book-1884; Colombia, MD references; Discharged on Surgeon's Cert.
40	Alms House Grave Yard	Bedford Inquirer 1908; B.S.F. History Book-1884
41	Yellow Creek Reformed Cem.	1890 Hopewell Twp. Veterans Census
42		B.S.F. History Book-1884
43	City Point Nat'l Cem.-VA	Died 10/6/64; B.S.F. History Book-1884; Pension affidavit stated Woodbury residence
44	Greenfield Cem.-Queen	Civil War Records in Ancestry.com
45	Headricks Cem.-Cambria Co.	PA Civil War Archives-Yellow Creek residence
46		B.S.F. History Book-1884
47	St. Columbus Cem.-Johnstown	Deserted 10/2/64 & returned; B.S.F. History Book-1884
48	Hyndman Cem.	1890 Londonderry Twp. Veterans Census
49		Wounded 9/17/62; B.S.F. History Book-1884
50		Wounded 8/9/62; B.S.F. History Book-1884
51		Wounded; in hospital for a year; B.S.F. History Book-1884
52		Bookseller and stationer in Somerset, Pa.
53	Entriken Cem.-Hunt. Co.	1860 Hopewell Census; 1890 Hopewell Veterans Census
54		B.S.F. History Book-1884; references to Berks Co.
55	Norton Cem.-KS	B.S.F. History Book-1884; 1850 Schellsburg Census
56	Saint John's Cem.-Loysburg	Frank McCoy 1912 Listing; 1860 S. Woodbury Twp. Census
57	Bethel Church-Monroe Twp.	1890 Monroe Twp. Veterans Census; Deserted after war ended - 6/19/65
58	Bedford Cem.	PA Civil War Project record
59	Bedford Cem.	1890 Bedford Veterans Census
60	Mt. Zion Cem.-Breezewood	B.S.F. History Book-1884; John W & Henry H-brothers
61		1890 E. Providence Twp. Veterans Census
62		B.S.F. History Book-1884; enlisted in Bedford Co.
63	Bethel Meth. Cem.-Centerville	1890 Cumberland Valley Veterans Census
64	Bedford Cem.	Historical Data Systems record

418   Chapter 18

	Name	Age	Rank	Regiment	Co.	Muster In	Muster Out	Casualty	Casualty Battle
1	Flake, David M								
2	Flake, George			MD					
3	Fleck, Daniel K	27	Pvt.	PA 53rd Inf.	C	10/17/61	7/28/63		
4	Fleck, Frank S	23	Pvt.	PA 102nd Inf.	F	3/16/65	6/28/65		
5	Fleegle, Daniel	18	Pvt.	PA 93rd Inf.	I	3/31/64		KIA	Spotsylvania
6	Fleegle, George W	21	Pvt.	PA 138th Inf.	E	6/10/63	6/23/65	Wounded	Cold Harbor
7	Fleegle, Isaac S	23	Corp.	PA 55th Inf.	K	11/5/61	6/26/63	Wounded	Pocotaligo*
8	Fleegle, Isaac S		Pvt.	PA 5th Res.	B	2/9/64	6/6/64		
9	Fleegle, Isaac S		Pvt.	PA 191st Inf.	C	6/6/64	6/28/65	Wounded	
10	Fleegle, Jacob G	41	Pvt.	PA 76th Inf.	E	10/9/61		Died	
11	Fleegle, John		Pvt.	PA 202nd Inf.	D				
12	Fleegle, John H	18	Pvt.	PA 5th Res.	B	8/8/61	6/6/64	Wounded	
13	Fleegle, John H		Pvt.	PA 191st Inf.		6/6/64	6/28/65		
14	Fleegle, Martin*	18	Pvt.	PA 55th Inf.	I	2/27/64	9/13/65		
15	Fleegle, Simon	16	Pvt.	PA 76th Inf.	E	10/9/61	5/18/63		
16	Fleegle, Simon		Sgt.	PA McKeage's Mil.	H	7/2/63	8/8/63		
17	Fleegle, William		Pvt.	PA 133rd Inf.	K	8/15/62			
18	Fleegle, William	37	Pvt.	PA 208th Inf.	B	8/26/64	6/1/65		
19	Flenner, Stewart	17	Pvt.	PA 208th Inf.	H	9/2/64	6/1/65		
20	Fletcher, David	33	Pvt.	PA 211th Inf.	K	9/5/64	6/2/65		
21	Fletcher, George W*	34	Pvt.	PA 76th Inf.	E	2/14/65	7/18/65		
22	Fletcher, Henry	18	Pvt.	OH 15th Inf.	I	10/16/63	11/21/65		
23	Fletcher, Henry C	22	Pvt.	PA 22nd Cav.	I	2/27/64		Died	
24	Fletcher, Jacob	42	Pvt.	PA 22nd Cav.	C	7/11/63	2/5/64		
25	Fletcher, Jacob		Pvt.	PA 208th Inf.	K	9/7/64	6/1/65		
26	Fletcher, John (1)	25	Pvt.	PA 99th Inf.	G	2/24/65	5/31/65		
27	Fletcher, John (2)	21	Pvt.	OH 121st Inf.	B	9/11/62	6/8/65		
28	Fletcher, John L	42	Sgt.	PA 133rd Inf.	C	8/13/62	5/26/63		
29	Fletcher, Samuel	18	Pvt.	OH 15th Inf.	I	9/7/61	9/18/64	POW	Stones River, TN
30	Fletcher, Winfield S	15	Pvt.	PA 22nd Cav.	C	7/11/63	6/15/65		
31	Flohr, Adam*	36	Pvt.	PA 55th Inf.	K	2/17/64	8/30/65		
32	Flowers, William	26	Pvt.	PA 158th Inf.	C	11/1/62			
33	Floyd, John B		Pvt.	PA 21st Cav.	E	7/8/63	8/3/63		
34	Fluck, Porter	20	Pvt.	PA 133rd Inf.	C	8/13/62	3/13/63		
35	Fluck, Porter		Pvt.	PA 87th Inf.	C	6/4/64		KIA	Petersburg - Final
36	Fluke, Casper	17	Pvt.	PA 195th Inf.	E	4/7/65	8/17/65		
37	Fluke, David	20	Pvt.	MD 13th Reg Inf.	C	4/24/65	5/29/65		
38	Fluke, George	41	Pvt.	PA 99th Inf.	D	2/21/65	5/31/65		
39	Fluke, John R	22	Pvt.	PA 208th Inf.	H	9/8/64	6/1/65		
40	Fluke, John W	16	Pvt.	PA 22nd Cav.	M	2/27/64		Died	
41	Fluke, Levi	22	Sgt.	PA 22nd Cav.	B	7/11/63	2/5/64		
42	Fluke, Levi S	39	Pvt.	PA 28th Inf.	G	9/4/64	7/18/65		
43	Fluke, Oliver B	17	Sgt.	PA 22nd Cav.	B	7/11/63	2/5/64		
44	Fluke, Oliver B		Sgt.	PA 205th Inf.	C	8/27/64	6/2/65		
45	Fluke, Oliver P	26	Pvt.	PA 110th Inf.	C	10/24/61	6/28/62		
46	Fluke, Samuel B	24	Music.	PA 205th Inf.	C	8/26/64	6/2/65		
47	Foaulke, Carle		Pvt.	PA 138th Inf.	E				
48	Focht, George W	20	1st Lt.	PA 107th Inf.	I	1/30/62	10/2/63	POW	2nd Bull Run
49	Fockler, George (1)			PA 107th Inf.					
50	Fockler, George (1)	35	Pvt.	PA 93rd Inf.	F	11/14/64	6/27/65		
51	Fockler, George (2)			PA 11th Res.					
52	Fockler, Jacob L	20	Pvt.	PA 125th Inf.	C	8/11/62	5/18/63		
53	Fockler, Jacob L		Pvt.	PA 101st Inf.	C	3/10/65	6/25/65		
54	Fockler, Lee*			PA 84th Inf.					
55	Fockler, Samuel	31	Corp.	PA 12th Res.	K	8/10/61			
56	Fockler, Samuel		Pvt.	PA 110th Inf.	C	2/27/64	6/28/65		
57	Fogle, Nicholas*	17	Pvt.	PA 101st Inf.	G	12/2/61		Died	
58	Foor, Abraham T	40	Pvt.	PA 107th Inf.	H	1/9/62	11/21/62	Wounded	
59	Foor, Andrew J	31	Pvt.	PA 107th Inf.	H	3/11/62	3/14/63		
60	Foor, Andrew J		Pvt.	PA 199th Inf.	E	10/1/64	6/28/65		
61	Foor, Brazella	20	Pvt.	PA McKeage's Mil.	D	7/2/63	8/8/63		
62	Foor, Brazella		Pvt.	PA 208th Inf.	K	9/7/64	6/1/65		
63	Foor, Daniel V	19	Pvt.	PA 76th Inf.	E	11/21/61		KIA	Ft. Wagner, SC
64	Foor, Francis L	19	Pvt.	PA 101st Inf.	D	11/1/61		POW-Died	Plymonth, NC

	Cemetery	Notes
1	Bedford Cem.	Bedford Cemetery list of Civil War veterans
2	Bedford Cem.	Frank McCoy 1912 Listing; Born 1829
3	Broad Top IOOF Cem.	1890 Robertsdale Veterans Census
4	Duval Cem.-Coaldale	1890 Broad Top Twp. Veterans Census
5	Fredericksburg Nat'l Cem.	KIA 5/12/64; Bedford Gazette 2/13/1914 listing; 1850 Napier Twp. Census
6	Schellsburg Cem.	Wounded 6/1/64; B.S.F. History Book-1884; Simon - Brother
7	Mt. Olivet Cem.-Manns Choice	Right Wrist Wound-1890 E. St. Clair Twp. Veterans Census; Discharged on Surgeons Cert.
8		PA Civil War Project record; Historical Data Systems record
9		PA Civil War Project record; Pension Document-Left Knee Wound-11/2/64 at Weldon RR
10	Beaufort Nat'l Cem.	Died-Typhoid fever 7/9/62; PA Civil War Archives-Bedford residence; Simon & George-sons
11	Rose Hill Cem.-Cumberland	Born 2/28/36; John & Jacob-brothers; NY Cav listed
12	Pleasantville Cem.	1890 W. St. Clair Twp. Veterans Census
13		Historical Data Systems record; John H, Isaac & Daniel-brothers
14		B.S.F. History Book-1884; Somerset references
15	Mt. Olivet Cem.-Manns Choice	1860 Napier Twp. Veterans Census; Discharged on Surgeon's Cert.; Jacob-father
16		B.S.F. History Book-1884
17		Deserted 2/17/63; Historical Data Systems-Bedford Co. residence
18	Helixville Cem.-Napier Twp.	1860 Bedford Twp. Census; Historical Data Systems record
19	Abeline Cem.-KS	B.S.F. History Book-1884; Historical Data Systems-Bedford Co. residence
20	Replogle Cem.-Woodbury	1890 Snake Spring Valley Veterans Census; David, Henry, John (2) & Samuel-brothers
21		B.S.F. History Book-1884; Scranton references
22	Shelby-Oakland Cem.-OH	1850 Cumberland Valley Twp. Census
23	Fletcher Cem.-Clearville	Died 4/20/64; Frank McCoy 1912 Listing; Jacob-father
24	Fletcher Cem.-Clearville	B.S.F. History Book-1884; 1860 Monroe Twp. Census; Father of Henry
25		B.S.F. History Book-1884; Reinlisted after son's death
26	Iowa Veterans Home Cem.-IA	1860 Monroe Twp. Census
27	Shelby-Oakland Cem.-OH	1850 Cumberland Valley Twp. Census
28	Bedford Cem.	1890 Everett Veterans Census; Obituary listed John as Color Bearer
29	Shelby-Oakland Cem.-OH	POW 12/31/62; 1850 Cumberland Valley Twp. Census
30	Everett Cem.	1890 Bedford Twp. Veterans Census; John L / Father
31		B.S.F. History Book-1884
32	Hendershot Cem.-Fulton Co.	Deserted 12/6/62; Born in Frederick Co, VA; Historical Data Systems record
33		Historical Data Systems-Beford Co residence; Discharged on Surgeon's Cert.
34		Historical Data Systems record; 1850 Hopewell Twp. Census; aka Fluke
35		KIA 4/2/65; Civil War Pension Record; 1860 Broad Top Census
36	Lerado Cem.-KS	1860 Broad Top Twp. Census
37	Yellow Creek Reformed Cem.	1890 Coaldale Twp. Veterans Census; 1st MD PHB Infantry listed
38	St. Lukes Cem.-Saxton	Frank McCoy 1912 Listing; 1890 Liberty Twp. Veterans Census
39	Bethel Brethren Cem.-Tatesville	1890 Hopewell Twp. Veterans Census; John R & Porter-brothers
40		Died 7/3/64; 1850 M. Woodbury Twp. Cen aka Fluck
41	Yellow Creek Reformed Cem.	PA Civil War Archives record; 1850 S. Woodbury Twp. Census; Levi, David & Samuel-bros.
42	Riverview Cem.-Huntingdon Co.	1850 S. Woodbury Twp. Census
43	Mt. Royal Cem.-Glenshaw	Historical Data Systems record; Ancestry.com-born in Waterside
44		Historical Data Systems record; 1870 S. Woodbury Twp. Census
45	Yellow Creek Reformed Cem.	Historical Data Systems-Yellow Creek residence
46	St. John's Cem.-Loysburg	1890 Woodbury Twp. Veterans Census; Historical Data Systems record
47		1890 Union Twp. (Pavia) Veterans Census
48		POW 8/30 to 12/19/62; PA Civil War Project record
49	Yellow Creek Reformed Cem.	Frank McCoy 1912 Listing
50		Bedford Inquirer 5/22/1908 listing; Historical Data Systems record
51	Saxton Cem.	Frank McCoy 1912 Listing
52	Fockler Cem.-Saxton	1860 Liberty Twp. Census; Frank McCoy 1912 Listing aka Lee
53		Historical Data Systems record
54		Bedford Inquirer 5/22/1908 listing
55	Washington Co. Cem.-MD	Findagrave.com record; Pension Record for 110th Infantry & 12th Reserves
56		1870 M. Woodbury Twp. Census; aka Fackler
57		Died 7/7/62; Bedford Gazette 2/13/1914 listing; PA Civil War Arch.-Ekin residence
58	Amboy Twp. Cem.-OH	B.S.F. History Book-1884; 1860 E. Providence Twp. Census; Brazella-son
59	Rays Cove CC Cem.	Frank McCoy 1912 Listing; Discharged on Surgeon's Cert.; Andrew, Simon & Peter-brothers
60		PA Civil War Archives record; 1860 E. Providence Twp. Census
61	Amboy Twp. Cem.-OH	1850 E. Providence Twp. Census; Abraham-Father
62		Deserted & returned; Father was Wounded was the reason given
63		KIA 7/11/63; 1860 E. Providence Cen.; PA Civil War Archives-Rays Hill resid.; George-father
64		POW Andersonville & Florence 4/20/64; Died 11/15/64; Hist.Data Sys.-Bedford Co. residence

Chapter 18

	Name	Age	Rank	Regiment	Co.	Muster In	Muster Out	Casualty	Casualty Battle
1	Foor, George W	43	Pvt.	PA 107th Inf.	H	2/10/62		KIA	Antietam
2	Foor, Jacob I	44	Pvt.	PA 208th Inf.	K	9/7/64	6/1/65		
3	Foor, James H	21	Pvt.	PA 133rd Inf.	K	8/4/62	5/26/63	Wounded	
4	Foor, James H		Sgt.	PA 208th Inf.	K	9/7/64	6/1/65		
5	Foor, Jeremiah	23	Pvt.	PA 107th Inf.	H	3/11/62			
6	Foor, Jeremiah		Pvt.	PA 87th Inf.	B	6/3/64	6/9/65		
7	Foor, John T	34	Pvt.	PA 107th Inf.	H	3/24/62			
8	Foor, Jonathan S	22	Pvt.	PA 107th Inf.	H	1/9/62	3/20/65	Wounded	South Mountain
9	Foor, Jonathan S		Pvt.	PA 107th Inf.	H			POW	Gettysburg
10	Foor, Lucius	33	Pvt.	PA 186th Inf.	G	3/24/64	8/15/65		
11	Foor, Mark W	19	Pvt.	PA 8th Res.	F	6/11/61		Died	
12	Foor, Martin T	37	Corp.	PA 138th Inf.	F	8/29/62		POW-Died	Monocacy, MD
13	Foor, Peter	36	Pvt.	PA 208th Inf.	K	9/7/64	6/1/65	Wounded	
14	Foor, Richard T	31	Pvt.	PA 186th Inf.	G	3/24/64	8/15/65		
15	Foor, Samuel H	16	Pvt.	OH 25th Inf.	E	9/5/64	7/13/65		
16	Foor, Samuel S	23	Pvt.	PA 8th Res.	F	6/11/61	5/26/64	Wounded	Seven Days Battle
17	Foor, Samuel S		Pvt.	PA 8th Res.	F			POW	Fredericksburg
18	Foor, Simon P	32	Pvt.	PA 208th Inf.	K	9/7/64		KIA	Petersburg - Final
19	Foor, William H	24	Pvt.	PA 8th Res.	F	6/11/61		KIA	Seven Days Battle
20	Foor, William H H	25	Pvt.	PA 107th Inf.	H	1/9/62	2/11/64	Wounded	
21	Fordan, George G		Pvt.	PA 91st Inf.	G	9/21/64	1/23/65		
22	Fore, James F	31	Pvt.	PA 133rd Inf.	K	8/15/62	5/26/63		
23	Foreman, Samuel		Pvt.	PA 51st Inf.	A	9/12/61	1/29/63		
24	Foreman, Samuel	41	Pvt.	OH 16th Inf.	H	1/30/64	1/1/65		
25	Foreman, William	28	Pvt.	PA 133rd Inf.	C	8/8/62	5/26/63		
26	Foreman, William		Pvt.	PA 2nd Heavy Art.	L	2/29/64		POW-Died	Cold Harbor
27	Forgee, Casper	17	Pvt.	PA 21st Cav.	E	7/8/63	2/20/64		
28	Forquer, Leroy*	20	Pvt.	WV 4th Cav.	I	6/7/63	3/8/64		
29	Foster, Aaron	25	Pvt.	PA 8th Res.	F	6/11/61	5/26/64	POW	Seven Days Battle
30	Foster, Aaron		Pvt.	PA 8th Res.	F			Wounded	Fredericksburg
31	Foster, Aaron		Pvt.	PA 8th Res.	F			Wounded	Spotsylvania
32	Foster, Eli	34	2nd Lt.	IN 30th Inf.	I		9/29/64	POW	Chickamauga
33	Foster, Isaiah	37	Pvt.	PA 22nd Cav.	I	9/3/64	5/24/65		
34	Foster, John W	33	Pvt.	PA 195th Inf.	E	4/7/65	1/31/66		
35	Foster, Joseph E	22	Pvt.	PA 133rd Inf.	C	8/13/62	5/26/63		
36	Foster, William	43	1st Sgt.	PA 138th Inf.	D	9/2/62	6/22/65		
37	Foster, William A	19	Pvt.	PA 55th Inf.	H	10/11/61		Died	
38	Fox, Henry W	28	2nd Lt.	PA 55th Inf.	K	10/11/61	9/17/64		
39	Francis, William	20	Pvt.	PA 55th Inf.	I	10/7/63			
40	Frauenfelter, Adam		Pvt.	PA 55th Inf.	H	9/23/63	5/15/65	Wounded	Cold Harbor
41	Frazey, Frederick L	21	Pvt.	PA 11th Inf.	A	9/30/61	3/18/63	Wounded	Antietam
42	Frazey, Henry P	19	Corp.	PA 11th Inf.	A	9/30/61	7/14/65	Wounded	Dabney's Mills
43	Frazier, William (1)	21	Pvt.	PA 55th Inf.	K	11/5/61		Wound.-Died	
44	Frazier, William (2)*	20	Pvt.	PA 184th Inf.	A	5/12/64	7/14/65		
45	Frederick, Andrew	21	Corp.	PA 2nd Cav.	E	11/5/61	6/17/65		
46	Frederick, William	33	Pvt.	PA 208th Inf.	H	9/7/64	6/1/65		
47	Freeburn, Richard H	20	Pvt.	PA 55th Inf.	K	11/5/61	8/30/65		
48	Freet, Isaac		Pvt.	PA McKeage's Mil.	G	7/2/63	8/8/63		
49	French, Samuel		Pvt.	PA McKeage's Mil.	D	7/2/63	8/8/63		
50	French, Samuel	30	Pvt.	PA 208th Inf.	K	9/5/64	6/2/65		
51	Friend, Isreal	19	Pvt.	PA 104th Inf.	E	2/15/65	8/25/65		
52	Fritz, Aaron*		Pvt.	PA 55th Inf.	K	9/12/62		Died	
53	Fritz, Daniel E	25	Corp.	PA 18th Cav.	K	10/27/62		POW-Died	Wilderness
54	Fry, Henry		Pvt.	USCT 43rd Inf.	G	4/14/64	10/20/65		
55	Fry, John		Pvt.	USCT 43rd Inf.	D	3/26/64		KIA	Petersburg
56	Fry, Joseph		Pvt.	PA 208th Inf.	H	9/8/64	6/1/65		
57	Fry, Michael	38	Pvt.	PA 84th Inf.	A	10/24/61		Died	
58	Fry, Solomon W	31	2nd Lt.	PA 55th Inf.	I	9/20/61	8/30/65	Wounded	Cold Harbor
59	Fryberg, John L		Pvt.	PA 82nd Inf.	C	8/12/61	9/16/64		
60	Fulton, Adam	24	Corp.	PA 133rd Inf.	C	8/13/62	5/26/63		
61	Fulton, William*	19	Pvt.	PA 194th Inf.	I	7/22/64	11/5/64		
62	Fundenberg, George B*	45	Surg.	PA 23rd Inf.		1/18/62	3/26/62		
63	Funk, Enoch	21	Pvt.	PA 11th Inf.	A	9/30/61		KIA	North Anna River
64	Funk, John D		Pvt.	PA McKeage's Mil.	D	7/2/63	8/8/63		

Detailed Alphabetical Listing 421

	Cemetery	Notes
1	Rays Cove CC Cem.	KIA 9/17/62 near Cornfield; Civil War Pension Record; Father of Daniel V
2	Morgart & Morgret-Everett	Frank McCoy 1912 Listing; 1860 W. Providence Twp. Census
3	Rays Cove CC Cem.	1890 E Providence Twp. Veterans Census
4		B.S.F. History Book-1884; E. Providence Honor Roll
5	Trans Run Cem.-Charlesville	Deserted 8/2/62; B.S.F. History Book-1884
6		Historical Data Systems record
7	Lehman Cem.-OH	Deserted 8/2/62; B.S.F. History Book-1884
8	Rays Cove CC Cem.	Wounded 9/14/62; 1890 E. Providence Twp. Veterans Census
9		POW-Andersonville, Charleston & Savannah 7/1/63 to 12/11/64; PA Civil War Project record
10	Rays Cove CC Cem.	Frank McCoy 1912 Listing; Lucius, Abraham, Martin & Richard-brothers
11	Rays Cove CC Cem.	Died 12/4/61; B.S.F. History Book-1884; Mark & William H-brothers
12	Danille Nat'l Cem.-KY	POW 7/9/64; Died 12/6/64; Historical Data Systems-E. Providence Twp. residence
13	Rays Cove CC Cem.	1890 E. Providence Twp. Veterans Census; Peter & Andrew-brothers
14	Mt. Zion Cem.-Breezewood	Frank McCoy 1912 Listing; Richard, Abraham, Lucius & Martin-brothers
15	Amboy Twp. Cem.-OH	Samuel H & Brazella-brothers; Abraham-father
16	Rays Cove CC Cem.	Wounded 6/26/62; Samuel, Francis, Jonathan & William H H-brothers
17		POW 12/13/62; Bedford Gazette-8th Reserve listing; Historical Data Systems record
18	City Point Nat'l Cem.-VA	KIA 4/2/65; findagrave lists Rays Cove CC; 1850 E. Providence Twp. Census
19	Glendale Nat'l Cem.-VA	KIA 6/30/62; B.S.F. History Book-1884; 1860 E. Providence Twp. Census
20	Rays Cove CC Cem.	Historical Data Systems record; 1860 E. Providence Twp. Census
21		1890 Union Twp. (Pavia) Veterans Census
22		B.S.F. History Book-1884; Hist. Data Systems-Bedford Co. residence & enlisted Bloody Run
23		Frank McCoy 1912 Listing
24	Fockler Cem.-Saxton	1850 Liberty Twp. Census; Samuel & William-brothers
25		PA Civil War Archives-enlisted in Hopewell
26	Vicksburg Nat'l Cem.	POW-Andersonville; 6/2/64; Died 4/5/65; 1890 Liberty Twp. Veterans Census
27	Trinity Luth. Cem.-Selinsgrove	Historical Data Systems-Beford Co residence; aka Forgy
28	Jersey Baptist Cem.-Ursina	B.S.F. History Book-1884; Somerset Co. references
29	Duval Cem.-Coaldale	POW 6/27/62; 1860 Broad Top Twp. Census
30		Wounded or Injured shoulder 12/13/62; Bedford Gazette-8th Reserve listing
31		Wounded 5/8/64; PA Civil War Project record.-Bedford Co. residence
32	Oak Woods Cem.-Chicago	Jon Baughman - 10/22/05 Shopper Guide article-Saxton native Eli Foster & the Libby Excape
33	Williamsport Cem.	B.S.F. History Book-1884; Historical Data Systems record
34	Zion Cem.-IA	Ancestry.com-Born in Broad Top; Historical Data Systems record
35		PA Civil War Archives-enlisted inHopewell & Bedford Co. residence
36	Duval Cem.-Coaldale	PA Civil War Archives-enlisted in Schellsburg & Bedford Co. residence
37		Died 8/4/64; Historical Data Systems-enlisted in Schellsburg; 1860 Broad Top Census
38		B.S.F. History Book-1884; Historical Data Systems-Bedford Co. resid.; Discharged Surg. Cert.
39		B.S.F. History Book-1884; Deserted 1/17/65
40		Wounded 6/3/63 listed in Bedford Inquirer 7/8/64 article; Hist. B.S.F. History Book-1884
41	Mt. Hope Cem.-IL	Wounded 9/17/62; E. Providence Honor Roll; 1860 E. Providence Twp. Census
42	Maple Grove Cem.-IA	Wounded 2/6/65; Historical Data Systems record; Henry & Frederick-brothers
43	Arlington Nat'l Cem., 13-12535	Died 6/9/64; B.S.F. History Book-1884; PA Civil War Archives-Bed. Co. residence
44		B.S.F. History Book-1884; enlisted in Harrisburg
45		B.S.F. History Book-1884; Historical Data Systems record
46	Dry Hill Cem.-Woodbury	1890 S. Woodbury Veterans Census
47		B.S.F. History Book-1884; Historical Data Systems-Bedford Co. residence
48		B.S.F. History Book-1884; 1860 Colerain Twp. Census
49		B.S.F. History Book-1884
50	Providence Union Cem.-Everett	B.S.F. History Book-1884; 1860 W. Providence Twp. Census
51	Salemville Cem.	1890 S. Woodbury Twp. Veterans Census; 1850 W. Providence Twp. Census
52		Died 11/4/64; B.S.F. History Book-1884; Bed. Gazette 2/13/1914 list; Muster in Reading, PA
53	Andersonville Cem., 7776	POW 5/5/64; Died-Dysentery 9/4/64; 1860 Juniata Twp. Cen.; PA Civil War Archives record
54		Frank McCoy 1912 list & Civil Works Admin. 1934 list; Historical Data Systems record
55		KIA 7/3/64; 1890 Bedford Twp. Veterans Census lists KIA
56	Uelhing Cem.-NE	B.S.F. History Book-1884; Historical Data Systems-Bedford Co. residence
57	Upper Claar Cem.-Queen	Died 3/16/66; Frank McCoy 1912 Listing; Historical Data Systems-Bedford Co. residence
58	McKeesport & Versailles Cem.	Wounded 6/3/64; Historical Data Systems-Bedford Co. residence
59		1890 Saxton Veterans Census
60		B.S.F. History Book-1884; Historical Data Systems-Bedford Co. residence
61		B.S.F. History Book-1884
62		B.S.F. History Book-1884; Somerset references
63		KIA 5/23/64; Bedford Gazette 1912 list; 1860 E. Providence Twp. Census
64		B.S.F. History Book-1884; John & Enoch-brothers

	Name	Age	Rank	Regiment	Co.	Muster In	Muster Out	Casualty	Casualty Battle
1	Funk, John D		Pvt.	PA 208th Inf.	K	9/7/64	6/1/65	Wounded	Petersburg - Final
2	Furcht, Frederick	41	Pvt.	PA 76th Inf.	E	10/12/64	5/25/65		
3	Furgeson, Thomas	20	Pvt.	PA 208th Inf.	H	9/8/64	6/1/65		
4	Furlong, Edward	36	Pvt.	PA 55th Inf.	D	10/15/63	8/30/65		
5	Furney, Amos	25	Pvt.	PA 67th Inf.	C	11/28/64	7/14/65		
6	Furney, Jacob	28	Pvt.	PA 165th Inf.	G	11/10/62	7/28/63		
7	Furney, Nelson	31	Pvt.	PA 148th Inf.	A	8/26/63	1864	Died	
8	Gabe, Lawrence*	24	Pvt.	PA 76th Inf.	E	11/21/61	7/13/62		
9	Gailey, Joseph*	35	Pvt.	PA 110th Inf.	C	2/27/64	6/28/65		
10	Gallagher, Charles*	27	Pvt.	PA 55th Inf.	I	10/8/63	5/15/65		
11	Gallagher, Edward	32	Pvt.	PA 133rd Inf.	C	8/29/62		KIA	Fredericksburg
12	Gallaher, James W		Pvt.	PA 208th Inf.	H	9/2/64	6/1/65		
13	Gallbaugh, Henry	30	Pvt.	PA 208th Inf.	H	9/5/64	6/1/65		
14	Gamble, Andrew		Pvt.	PA 21st Cav.	E	7/23/63	2/20/64		
15	Gamble, Robert		Pvt.	PA 8th Res.	F	6/11/61	8/15/63	POW	Seven Days Battle
16	Gamble, Robert	35	Pvt.	PA 8th Res.	F			Died	
17	Ganz, Thomas			USCT Inf.					
18	Gardiner, Francis L	37	Pvt.	PA 55th Inf.	K	3/8/64	6/6/65	Wounded	Petersburg - Initial
19	Gardner, Adam	22	Pvt.	PA 55th Inf.	D	2/27/64	5/26/65	Wounded	
20	Gardner, Charles	18	Pvt.	PA 138th Inf.	E	8/29/62	6/24/65	POW	Monocacy, MD
21	Gardner, George	28	Pvt.	PA 14th Cav.	K	11/23/62	5/31/65		
22	Gardner, John	22	Pvt.	PA 55th Inf.	D	10/12/61	8/30/65	Wounded	
23	Gardner, Samuel	37	Corp.	PA 55th Inf.	D	10/12/61	8/30/65	Wounded	
24	Garland, Matthew	40	Pvt.	PA 55th Inf.	I	10/7/63	8/30/65	Wounded	
25	Garlick, Abraham	24	Pvt.	PA 12th Cav.	K	3/14/64	7/20/65	Injured	
26	Garlick, Adam	29	Pvt.	PA 12th Cav.	K	3/16/64	11/24/65	Wounded	
27	Garlick, Christian C	23	Pvt.	PA 8th Res.	F	6/11/61	5/15/64	POW	Seven Days Battle
28	Garlick, Christian C		Pvt.	PA 191st Inf.		5/15/64	7/18/65		
29	Garlick, Nicholas		Pvt.	PA McKeage's Mil.	D	7/2/63	8/8/63		
30	Garlick, Nicholas	21	Pvt.	PA 22nd Cav.	H	2/22/64	10/31/65		
31	Garlinger, Walter E	22	Corp.	PA 55th Inf.	H	10/11/61	8/30/65	POW	Edisto Island, SC
32	Garlinger, Walter E		Corp.	PA 55th Inf.	H			Wounded	Petersburg - Initial
33	Garlinger, William M			OH 7th Inf.	A				
34	Garn, George	24	Pvt.	PA 171st Inf.	I	11/10/62	8/1/63		
35	Garner, Andrew B	18	Pvt.	PA 194th Inf.	I	7/22/64	11/5/64		
36	Garner, Thomas G		Pvt.	PA McKeage's Mil.	H	7/2/63	8/8/63		
37	Garner, Thomas G	19	Pvt.	PA 194th Inf.	I	7/22/64	11/5/64		
38	Garretson, Benjamin H	21	Pvt.	PA 21st Cav.	E	7/8/63	2/20/64		
39	Garretson, Benjamin H		Pvt.	PA 205th Inf.	C	8/27/64		Wound.-Died	Petersburg - Final
40	Garretson, Edwin	24	Corp.	PA 21st Cav.	E	7/8/63	2/20/64		
41	Garretson, George R	17	Music.	PA 101st Inf.	G	12/28/61	11/7/62		
42	Garretson, George R		Pvt.	PA 55th Inf.	H	2/15/64	12/30/64		
43	Garretson, Josiah P	24	Pvt.	PA 55th Inf.	H	2/29/64	5/11/65		
44	Garretson, Moses R	19	Pvt.	PA 55th Inf.	H	10/11/61		Died	
45	Garretson, Thomas	39	Pvt.	PA 84th Inf.	A	1861			
46	Garrett, Albert T	19	Pvt.	PA 110th Inf.	C	3/19/64	6/28/65		
47	Garrett, Alexander A	21	Pvt.	PA 8th Res.	F	6/11/61		KIA	Seven Days Battle
48	Garrett, John C	20	Pvt.	PA 110th Inf.	C	10/24/61	6/9/65	POW	Cold Harbor
49	Garrett, Levi P		Pvt.	PA 194th Inf.	I	7/22/64	11/5/64		
50	Gaster, Ezekiel W	20	Music.	PA 133rd Inf.	K	8/15/62	5/26/63		
51	Gaster, Ezekiel W		Pvt.	PA 208th Inf.	H	9/7/64	6/1/65		
52	Gaster, Isaac R	22	Corp.	KS 11th Cav.	E	9/13/62	8/7/65		
53	Gaster, James H	18	Sgt.	PA 107th Inf.	H	1/9/62		Wound.-Died	Antietam
54	Gates, Andrew G		Pvt.	IL 56th Inf.	C			Wounded	
55	Gates, Charles H		Pvt.	WV 7th Inf.	E	1/26/64	8/1/65		
56	Gates, Charles W			OH				Wounded	
57	Gates, George	20	Pvt.	PA 9th Cav.	M	5/31/64	7/18/65	POW	
58	Gates, George W	18	Pvt.	PA 13th Res.	D	6/26/61	5/31/64	Wounded	Seven Days Battle
59	Gates, George W		Pvt.	PA 13th Res.	D			Wounded	South Mountain
60	Gates, George W		Pvt.	PA 13th Res.	D			Wounded	Spotsylvania
61	Gates, George W		Pvt.	PA 190th Inf.	D	5/31/64	6/28/65	POW	White Oak Road
62	Gates, Jacob C	20	Pvt.	PA 13th Res.	D	6/26/61	9/29/62		
63	Gates, Jacob C		Corp.	PA McKeage's Mil.	H	7/2/63	8/8/63		
64	Gates, Jacob C		Pvt.	PA 13th Cav.	D	12/9/63	7/14/65	POW	Beefsteak Raid

	Cemetery	Notes
1		Wounded 4/2/65; Historical Data Systems-Bedford Co. residence
2		B.S.F. History Book-1884; Substitute; Discharged on Surgeon's Cert.
3	Rock Hill Cem.-Monroe Twp.	1890 Monroe Twp. Veterans Census
4		B.S.F. History Book-1884
5	Mt. Pleasant Cem.-Mattie	Frank McCoy 1912 Listing; Historical Data Systems record
6	Mt. Pleasant Cem.-Mattie	Ancestry.com information; Jacob, David & Nelson-brothers
7	Arlington Nat'l Cem., 13-9346	Died 11/24/64; Historical Data Systems record; 1860 Monroe Twp. Census; aka Fenny
8		B.S.F. History Book-1884; York references; Discharged on Surgeon's Cert.
9		B.S.F. History Book-1884; Blair Co. references
10		B.S.F. History Book-1884
11		KIA 12/13/62; B.S.F. History Book-1884; PA Civil War Archives-Bedford Co. residence
12		PA Civil War Archives-Bedford Co. residence
13		B.S.F. History Book-1884; Historical Data Systems-Bedford Co. residence
14		Historical Data Systems record
15		POW 6/27/62; Bedford Gazette-8th Reserve listing; Historical Data Systems record
16	Alexandria Nat'l Cem., 936	Died 9/2/63; B.S.F. History Book-1884; PA Civil War Archives-Bedford Co. residence
17		Frank McCoy 1912 list & Civil Works Admin. 1934 list
18		Wounded 6/18/64; B.S.F. History Book-1884
19	Mt. Zion Cem.-Pavia	Wounded-Bedford Inquirer 5/27/64 article; Adam, George, John & Samuel-brothers
20		POW 7/9/64 to 2/22/65; Historical Data Systems-Schellsburg residence
21	Hyndman Cem.	1890 Hopewell Twp. Veterans Census; Historical Data Systems record
22		1890 Union Twp. (Pavia) Veterans Census; Historical Data Systems record
23	Hyndman Cem.	1890 Hopewell Twp. Veterans Census
24		Wounded-Bedford Inquirer 5/27/64 article; B.S.F. History Book-1884
25	Mt. Pleasant Cem.-Mattie	Thrown from horse-6/20/63; 1890 Brush Creek Twp. Veterans Census
26	Providence Union Cem.-Everett	Historical Data Systems record; Nicholas-Brother; Obit listed several wounds
27	Rock Hill Cem.-Monroe Twp.	POW 6/27/62; Historical Data Systems record; 1860 Monroe Twp. Census
28		B.S.F. History Book-1884; Historical Data Systems record
29		B.S.F. History Book-1884
30	Providence Union Cem.-Everett	B.S.F. History Book-1884
31	Schellsburg Cem.	POW-Charleston, Columbia & Richmond 3/29/62 for 6 months
32		Wounded 6/18/64 listed in Bedford Inquirer 7/8/64 article; Historical Data Systems record
33	Union Cem.-CA	Ancestry.com-born Schellsburg 2/22/34
34	Smith Cem.-OH	B.S.F. History Book-1884; 1860 Union Twp. (Pavia) Census
35		B.S.F. History Book-1884
36		B.S.F. History Book-1884
37		B.S.F. History Book-1884
38	Friends Cem.-Spring Meadow	Historical Data Systems record; Benjamin & Josiah-brothers
39		Head Wounded 4/2/65; Died 5/27/65; Historical Data Systems-Bedford Co. residence
40	Oak Shade Cem.-IA	PA Civil War Archives-Bed. Co. resident; Historical Data Systems record
41	Harmony Grove Cem.-CA	B.S.F. History Book-1884; Discharged Surg. Cert.; George, Thomas, Edwin & Moses-brothers
42		PA Civil War Archives-enlisted in Schellsburg & Bed. Co. residence
43	Friends Cem.-Spring Meadow	PA Civil War Archives-enlisted in Schellsburg; Historical Data Systems record
44	Dunkard Cem.-IA	Died 10/15/64; Historical Data Systems record; 1860 St. Clairsville Census
45	Dunkard Cem.-IA	1850 St. Clairsville Census; born 12/9/21
46	Prospect Hill Cem.-Newville	Historical Data Systems record; 1850 Hopewell Twp. Census; Albert, John & Alexander-bros.
47	Glendale Nat'l Cem.-VA	KIA 6/30/62; PA Civil War Archives-Bedford Co. residence; aka Absalom
48	Mapleton IOOF-Hunt. Co.	POW 6/1 to 12/10/64; PA Civil War Archives-Yellow Creek residence
49		B.S.F. History Book-1884
50	Hyndman Cem.	Frank McCoy 1912 Listing; Ezekiel, James and Isaac-brothers
51		B.S.F. History Book-1884; 1850 Broad Top Census
52	Woodlawn Cem.-KS	1850 Broad Top Census
53		Wounded 9/17/62; Died 2/6/63; 1850 Broad Top Census
54		B.S.F. History Book-1884; Altoona Tribune 8/8/00 Article
55		B.S.F. History Book-1884
56		Altoona Tribune 8/8/00 article listed OH regiment
57	Fockler Cem.-Saxton	POW-Libby; 1890 Saxton Veterans Census
58	Greenwood Cem.-Altoona	B.S.F. History Book-1884 listed being wounded; 1860 Bedford Twp. Census
59		Wounded 9/17/62 Charging up Turner's Gap in Pension Record; B.S.F. History Book-1884
60		Wounded 5/10/64 in Pension Record; B.S.F. History Book-1884
61		POW 3/31/65; Paroled in April 1865 on Pension Record
62	Clarks Chapel Cem.-OH	Accident with Axe on 9/27/62; B.S.F. History Book-1884; 1860 Bedford Twp. Census
63		B.S.F. History Book-1884; Historical Data Systems record
64		POW-Danville, VA 9/16/64; B.S.F. History Book-1884

	Name	Age	Rank	Regiment	Co.	Muster In	Muster Out	Casualty	Casualty Battle
1	Gates, James		Pvt.	PA 8th Res.	F	6/11/61		POW	Seven Days Battle
2	Gates, James	23	Pvt.	PA 8th Res.	F			Wound.-Died	Antietam
3	Gates, Jeremiah E	17	Pvt.	PA 149th Inf.	I	9/29/63	5/17/65	Wounded	Petersburg - Initial
4	Gates, Jeremiah*	22	Pvt.	PA 84th Inf.	E	1861		KIA	1st Kernstown
5	Gates, John W	21	Pvt.	PA 133rd Inf.	C	8/13/62	5/26/63		
6	Gates, John W		Corp.	PA McKeage's Mil.	D	7/2/63	8/8/63		
7	Gates, John W		Pvt.	PA 22nd Cav.	M	2/27/64	6/24/65		
8	Gates, Joseph	23	Sgt.	PA 110th Inf.	C	12/19/61			
9	Gates, Joseph		Sgt.	PA 208th Inf.	H	9/7/64	6/1/65		
10	Gates, Martin	22	Pvt.	PA 110th Inf.	C	10/24/61	10/24/64	Wounded	Gettysburg
11	Gates, Martin V B	18	Pvt.	PA 14th Inf.	H	4/24/61	8/7/61		
12	Gates, Martin V B		Sgt.	PA 76th Inf.	C	10/17/61	11/28/64	Wounded	Petersburg
13	Gates, Nathaniel		Chap.	PA 13th Res.		1862	1864		
14	Gates, Nathaniel H*	71	Pvt.	PA 1st Light Art.	H	7/30/61	10/16/62		
15	Gates, Reuben	40	Pvt.	USCT 3rd Inf.	C	11/28/64	10/31/65		
16	Gates, Samuel	18	Pvt.	PA 110th Inf.	C	10/24/61		Died	
17	Gates, Samuel K			OH				Wounded	
18	Gates, Theophilus R		Pvt.	PA 13th Inf.	G	4/25/61	8/6/61		
19	Gates, Theophilus R	31	Corp.	PA 55th Inf.	K	2/3/62	8/30/65	Wounded	Cold Harbor
20	Gates, Thomas	15	Pvt.	PA 192nd Inf.	B	2/14/65	8/24/65		
21	Gates, William	38	Pvt.	PA 192nd Inf.	B	2/14/65	8/24/65		
22	Gates, William B*	17	Pvt.	PA 55th Inf.	I	2/15/64	2/15/65	Wounded	Cold Harbor
23	Gates, William H (1)	21	Pvt.	PA 84th Inf.	E	1861			
24	Gates, William H (2)	23	Pvt.	PA 110th Inf.	C	10/24/61	6/28/62		
25	Gates, William H (2)		Sgt.	PA 208th Inf.	K	9/7/64	6/1/65		
26	Gates, William K*	18	Pvt.	PA 55th Inf.	I	2/27/64	8/30/65	POW	Petersburg
27	Gates, William M		Pvt.	PA 210th Inf.	C	9/7/64	5/30/65		
28	Gaudig, Herman	39	Pvt.	MA 15th Inf.	D	7/30/63	7/27/64	Wounded	Wilderness
29	Gaudig, Herman		Pvt.	MA 20th Inf.		7/27/64			
30	Gault, Ezekiel	17	Pvt.	PA 21st Cav.	E	7/22/63	2/20/64		
31	Geiger, Jacob D*	18	Pvt.	PA 55th Inf.	I	8/28/61		KIA	White Oak Road
32	Geisler, Lewis H	37	Pvt.	PA 125th Inf.	E	8/13/62	5/18/63		
33	Geisler, Lewis H			PA 83rd Inf.					
34	Geisler, Lewis H		Sgt.	PA 77th Inf.	F	2/28/65	12/6/65		
35	Geller, George	16	Pvt.	PA 138th Inf.	F	6/15/63	6/23/65	Wounded	Monocacy, MD
36	Geller, Jesse	30	Pvt.	PA 55th Inf.	H	2/29/64	8/30/65	Wounded	Petersburg - Initial
37	Geller, John	34	Sgt.	PA 138th Inf.	F	8/29/62	4/6/65	Wounded	Cedar Creek
38	Geller, Solomon	35	Pvt.	PA 101st Inf.	G	12/28/61		POW-Died	Fair Oaks
39	George, Conrad	40	Pvt.	PA 208th Inf.	K	9/7/64	6/1/65		
40	Gephart, Daniel		Pvt.	PA 49th Inf.	E				
41	Gephart, Jno				B				
42	Gephart, John	21	Pvt.	PA 76th Inf.	E	9/9/61	11/28/64	Wounded	Pocotaligo, SC
43	Gephart, William			PA 76th Inf.				KIA	Mill Springs Gap*
44	German, Ephraim*		Pvt.	PA 55th Inf.	K	10/3/63	6/6/65	Wounded	
45	Gettleman, Jacob	27	Pvt.	PA 104th Inf.	E	2/21/65	8/25/65		
46	Getty, John		1st Lt.	PA 138th Inf.	E	8/30/62	4/13/64		
47	Geyer, John C	20	1st Lt.	PA 55th Inf.	H	10/2/61	7/11/65		
48	Ghast, William*	18	Music.	PA 184th Inf.	A	5/12/64	7/14/65		
49	Gibson, George G	32	Pvt.	PA 208th Inf.	H	9/7/64	6/1/65		
50	Gibson, Henry F	18	Pvt.	PA 133rd Inf.	K	8/15/62	5/26/63		
51	Gibson, Henry F		Corp.	PA 208th Inf.	H	9/5/64	6/1/65	Wounded	Ft. Stedman
52	Gibson, James		Corp.	PA McKeage's Mil.	H	7/2/63	8/8/63		
53	Gibson, James	19	Pvt.	PA 55th Inf.	D	2/27/64	6/3/65		
54	Gibson, John*		Corp.	USCT 2nd Inf.	D	9/1/63	1/1/65		
55	Gibson, John (1)		Pvt.	PA 11th Inf.	C	3/17/64	1/1/65		
56	Gibson, John (2)	39	Pvt.	PA 183rd Inf.	B	3/17/64	9/7/65		
57	Gibson, Joseph	29	Pvt.	PA 11th Inf.	C	9/9/61	6/12/62		
58	Gibson, William Y	34	Pvt.	PA 133rd Inf.	K	8/15/62	5/26/63		
59	Gienger, Andrew J	31		PA 208th Inf.	H	9/6/64	6/1/65		
60	Gienger, Jacob	37	Pvt.	PA 208th Inf.	H	9/6/64	6/1/65		
61	Giffin, James H	22	Pvt.	PA 208th Inf.	K	9/7/64	6/1/65		
62	Giffin, John C	29	Pvt.	PA 158th Inf.	H	11/4/62	8/12/63		
63	Giffin, Peter	22	Pvt.	MD 3rd Reg Inf.	A	8/13/61	7/4/64	Wounded	Chancellorsville
64	Gilbert, Daniel*	35	Pvt.	PA 184th Inf.	A	5/12/64		POW-Died	Jerusalem Plank Rd.

	Cemetery	Notes
1		POW 6/27/62; Bedford Gazette-8th Reserve listing; Historical Data Systems record
2	Yellow Creek Reformed Cem.	Leg Amputated 9/17/62; Died 10/16/62; 1860 Hopewell Twp. Census; Hist. Data Sys. record
3	Bethel Brethren Cem.-Tatesville	Foot Wound 6/18/64 near James River; 1860 Bedford Twp. Census; Hist. Data Systems record
4	Winchester Nat'l Cem.	KIA 3/23/62; Bedford Gazette 2/13/1914 listing; Blair Co. references
5	Maywood Cem.-NE	B.S.F. History Book-1884; Historical Data Systems-Bedford residence
6		B.S.F. History Book-1884
7		B.S.F. History Book-1884; 1850 Hopewell Twp. Census
8	Carson Valley-Duncansville	PA Civil War Archives-Yellow Creek residence
9		B.S.F. History Book-1884
10	Yellow Creek Reformed Cem.	Wounded 7/2/63; B.S.F. History Book-1884; Martin, John W & James-brothers
11	Hopewell Cem.	1890 Hopewell Twp. Veterans Census; Martin V B & George-brothers
12		Wound through right lung 6/30/64-Pension record; Historical Data Systems record
13		B.S.F. Hist. Book-1884; 1860 Bedford Twp. Census; Member of Sanitary Commisssion; DSC
14		Born 1791; 7 sons in Civil War; War of 1812 Veterans
15	Mt. Ross Cem.-Bedford	Frank McCoy 1912 Listing; 1890 Bedford Twp. Veterans Census
16	Yellow Creek Reformed Cem.	Died 3/12/62; PA Civil War Archives-Yellow Creek residence; Samuel & William H (2)-bros.
17		Altoona Tribune 8/8/00 article listed OH regiment
18		Theophilus, Jeremiah, Jacob C, George W, Andrew J, Samuel K & Charles W- brothers
19	Logan Valley Cem.-Bellwood	Wounded 6/3/64; PA Civil War Archives-Bedford residence; 1860 Bedford Twp. Census
20	Duval Cem.-Coaldale	1890 Broad Top Veterans Census
21	Duval Cem.-Coaldale	Ancestry.com information; Thomas-Son
22		Wounded-Pension record; B.S.F. History Book-1884
23	Duval Cem.-Coaldale	Deserted 10/19/62; Frank McCoy 1912 Listing
24	Yellow Creek Reformed Cem.	B.S.F. History Book-1884; 1860 Hopewell Twp. Census; Discharged on Surgeon's Cert.
25		Historical Data Systems-Bedford Co. residence; B.S.F. History Book-1884
26		POW 6/13/64 to 11/28/64; enlisted in Huntingdon Co.; Historical Data Systems record
27		1890 Hopewell Twp. Veterans Census
28	Pleasantville Cem.	1890 W. St. Clair Twp. Census
29		Deserted 1/21/65
30		Historical Data Systems-Bedford Co. residence
31		KIA 3/30/65; PA Civil War Archives-born Somerset Co.; B.S.F. History Book-1884
32	Lutheran Cem.-Osterburg	1880 E. St. Clair Twp. Census
33		Frank McCoy 1912 Listing
34		1890 E. St. Clair Twp. Veterans Census
35	Trinity UCC Cem.-Juniata Twp.	Wounded 7/9/64; B.S.F. History Book-1884; 1850 Harrison Twp. Census; John-father
36	Schellsburg Cem.	Chest Wound 6/18/64 listed in Bedford Inquirer 7/8/64 article; 1890 Juniata Twp. Vet. Cen.
37	Trinity UCC Cem.-Juniata Twp.	Wounded 10/19/64; Right Arm Amputated ; George-son; John & Solomon-brothers
38		Wounded/POW 5/31/62, Died 6/16/62; B.S.F. History Book-1884; 1860 Harrison Twp. Cen.
39	Cedar Grove-E. Prov. Twp.	B.S.F. History Book-1884; 1860 E. Providence Twp. Census
40	Oak Ridge Cem.-Altoona	1890 King Twp. Veterans Census
41	Yellow Creek Reformed Cem.	Co. B PA Vol. Civil War on gravestone
42	Union Cem.-Centerville	Wounded-Thigh 10/22/62; 1890 Cumberland Valley Veterans Census; William & John-brothers
43		Ancestry.com-Died 1864; Bedford Gazette 2/13/1914 listing; 1850 Napier Twp. Census
44		MIA listed in Bedford Inquirer article 6/3/64; B.S.F. History Book-1884
45	Royer Cem.-Blair Co	1860 Woodbury Twp. Census
46		B.S.F. History Book-1884; Historical Data Systems-Bedford Co. resid.; Discharged Surg. Cert.
47	Old Fellows Cem.-MO	1850 Napier Twp. Census; Historical Data Systems-Bedford Co. residence
48		B.S.F. History Book-1884; enlisted in Harrisburg
49	Grandview Cem.-Tyrone	B.S.F. History Book-1884; George, John (2), William & Henry-brothers
50	Bedford Cem.	B.S.F. History Book-1884
51		Leg Wound 3/25/65; 1890 Colerain Twp. Veterans Census
52		B.S.F. History Book-1884
53		B.S.F. History Book-1884; 1850 Bedford Census
54	Everett Cem.	Everett Cemetery Listing of Civil War Veterans; likely VRC and not USCT regiment
55		Frank McCoy 1912 Listing; 1850 W. Providence Twp. Census; transferred to VRC 1/1/65
56	Everett Cem.	Bedford Inquirer 5/22/1908 listing; Substitute; transferred to 43rd VRC in May 1865
57	Everett Cem.	1890 Snake Spring Valley Veterans Census; Discharged on Surgeon's Cert.
58	Hershberger Cem.-Snake Spr. V.	B.S.F. History Book-1884; in Hospital 10/30/62 to discharge
59	Everett Cem.	Everett Cemetery Listing of Civil War Veterans; Andrew & Jacob-brothers; aka Geinger
60	Buck Valley Cem.-Fulton Co.	B.S.F. History Book-1884; 1890 Brush Creek Veterans Census
61	Dudley Meth. Cem.	B.S.F. History Book-1884; 1850 E. Providence Twp. Census
62	Cedar Grove Cem.-Fulton Co.	Historical Data Systems record; 1860 Brush Creek Twp.
63	Hyndman Cem.	Wounded 5/3/63; Frank McCoy 1912 Listing; Peter & James-brothers
64	Andersonville Cem.	POW-Andersonville 6/24/64; Died 12/1/64; Company A recruited in Dauphin & Bedford Co.

	Name	Age	Rank	Regiment	Co.	Muster In	Muster Out	Casualty	Casualty Battle
1	Gilbert, Wiliam A*	26	Sgt.	PA 55th Inf.	H	9/24/63	8/30/65		
2	Gilchrist, David	19	Pvt.	PA McKeage's Mil.	H	7/2/63	8/8/63		
3	Gilchrist, David		Pvt.	PA 54th Inf.	H				
4	Gilchrist, James A	19	Pvt.	PA 138th Inf.	E	8/29/62	5/2/64		
5	Gilchrist, James A		Pvt.	US Signal Corps		5/2/64	6/24/65		
6	Gilchrist, Oliver	17	Pvt.	PA McKeage's Mil.	H	7/2/63	8/8/63		
7	Gillam, George	24	Corp.	PA 138th Inf.	D	8/29/62	6/23/65	Wounded	Monocacy, MD
8	Gillam, John	28	Pvt.	MO 2nd Engineers	A	2/7/63	9/8/63		
9	Gillam, Thomas	34	Pvt.	PA 93rd Inf.	A	11/14/64	5/6/65	Wounded	Petersburg - Final
10	Gille, Charles	29	Pvt.	MD 5th Reg Inf.	K	11/10/64	9/1/65		
11	Gillen, Dominick*	44	Pvt.	PA 76th Inf.	E	11/21/61	3/24/64		
12	Gillet, Anthony		Pvt.	PA 14th Inf.					
13	Gilliam, Michael	21	Pvt.	PA 101st Inf.	D	11/11/61	7/11/62		
14	Gilliam, Michael		Pvt.	PA 55th Inf.	D	2/27/64	8/30/65	Wounded	Cold Harbor
15	Gilliam, Wilson E	19	Pvt.	PA 101st Inf.	D	11/1/61		Died	
16	Gillson, Jackson*	37	Pvt.	PA 110th Inf.	C	9/16/64	5/31/65		
17	Gilson, Walter B	27	Pvt.	PA 56th Inf.	F	9/21/64	5/31/65		
18	Girdan, Isaac		Pvt.	PA 8th Res.	I	9/2/63	1864		
19	Gladwell, George W	28	Pvt.	PA 55th Inf.	D	2/27/64		Died	
20	Glass, George H		Team.	US Inf.					
21	Glass, John G*		Pvt.	PA 55th Inf.	I	2/11/64	7/5/65		
22	Glass, John J*		Pvt.	PA 55th Inf.	I	2/11/64	7/5/65		
23	Glaze, Andrew L		Pvt.	PA 21st Cav.	E	7/22/63	1/6/64		
24	Glaze, Andrew L		Pvt.	PA 117th Inf.	I	1/6/64			
25	Glenn, Josiah	21	Pvt.	PA 138th Inf.	E	8/29/62	1/12/64		
26	Glenn, Josiah			US Signal Corps		1/12/64			
27	Glessner, Philip	34	Pvt.	PA 148th Inf.	G	9/1/63	6/1/65	Wounded	Spotsylvania
28	Glidenell, Thomas	20	Pvt.	PA 76th Inf.	E	2/22/65	7/18/65		
29	Glotfelty, James	19	Pvt.	PA 116th Inf.	E	2/24/64	7/14/65	Wounded	
30	Gneill, Conrad*		Pvt.	PA 55th Inf.	K	10/10/63		Wound.-Died	
31	Gockley, Levi M*		Pvt.	PA 194th Inf.	I	7/22/64	11/5/64		
32	Godfrey, Oscar M*		Corp.	PA 76th Inf.	E	1/7/64	7/18/65		
33	Gogley, Jacob	20	Pvt.	PA 133rd Inf.	C	8/13/62	5/26/63		
34	Gogley, Jacob		Corp.	PA 186th Inf.	C	2/20/64	8/15/65		
35	Gogley, James H	21	Pvt.	PA 133rd Inf.	C	8/29/62	5/26/63	Wounded	Fredericksburg
36	Gogley, James H		Pvt.	PA 186th Inf.	E	2/24/64		Died	
37	Gogley, Samuel T	28	Pvt.	PA 208th Inf.	K	9/7/64	6/1/65		
38	Goheen, James*	18	Music.	PA 55th Inf.	H	9/1/62	6/11/65		
39	Goldsmith, Harvey*	18	Pvt.	PA 76th Inf.	E	2/23/65	7/18/65		
40	Gollipher, Espy	15	Music.	PA 55th Inf.	H	10/11/61	8/30/65		
41	Gollipher, Justice	22	Pvt.	PA 13th Inf.	G	4/25/61	8/6/61		
42	Gollipher, Justice		Pvt.	PA 101st Inf.	G	12/28/61		POW-Died	Plymouth, NC
43	Gollipher, Silas	19	Sgt.	PA 55th Inf.	H	10/11/61	1/15/63	Wound.-POW	Edisto Island, SC
44	Gollipher, Silas		Sgt.	PA 55th Inf.	H			Wounded	Edisto Island, SC
45	Gonden, John P			MD 2nd PHB Inf.	H				
46	Gonden, John W	21	Corp.	PA 55th Inf.	K	11/5/61	6/11/65		
47	Good, George	47	Pvt.	PA 55th Inf.	D	2/9/64	8/30/65		
48	Goodman, Frederick*		Pvt.	PA 55th Inf.	H	9/1/62		POW-Died	Drewry's Bluff
49	Gordon, Daniel		Pvt.	USCT 43rd Inf.	F	3/29/64			
50	Gordon, George G	21	Pvt.	PA 158th Inf.	B	11/4/62	8/12/63		
51	Gordon, George J	27	Pvt.	MD 2nd PHB Inf.	D	3/14/65	5/29/65		
52	Gordon, Isaac	18	Pvt.	PA 138th Inf.	E	8/29/62	6/23/65	POW	Monocacy, MD
53	Gordon, Jacob	34	Pvt.	PA 51st Inf.	E	9/2/61	23285		
54	Gordon, James S	24	Sgt.	PA 171st Inf.	I	11/2/62	8/8/63		
55	Gordon, Jeremiah	50	Pvt.	PA 55th Inf.	D	10/12/61	8/2/62		
56	Gordon, John F	16	Pvt.	MD 1st PHB Cav.	L	4/23/64	6/28/65		
57	Gordon, Joseph		Pvt.	PA McKeage's Mil.	G	7/2/63	8/8/63		
58	Gordon, Joseph	16	Pvt.	PA 55th Inf.	K	2/19/64	7/22/65	Wounded	White Oak Road
59	Gordon, Samuel	41	Pvt.	PA 52th Inf.	C	9/26/64	6/23/65		
60	Gordon, William	22	Pvt.	PA 55th Inf.	K	11/5/61		POW-Died	Drewry's Bluff*
61	Gottwalt, Henry*	27	Corp.	PA 55th Inf.	D	9/2/62		POW-Died	
62	Grace, Israel P	33	Pvt.	PA 102nd Inf.	L	3/15/65	6/30/65	POW	
63	Grace, John*		Pvt.	PA 55th Inf.	I	8/28/61		Died	
64	Gracey, Alfred	18	Sgt.	PA 107th Inf.	H	1/9/62	4/1/65	POW	Gettysburg

	Cemetery	Notes
1		B.S.F. History Book-1884; mustered in Pottsville
2	Bedford Cem.	B.S.F. History Book-1884; David, Oliver & James-brothers
3		1890 Bedford Twp. Veterans Census
4	Oak Grove Cem.-OH	PA Civil War Archives-Bedford residence; B.S.F. History Book-1884
5		1890 Bedford Twp. Veterans Census
6	Old Macksburg Cem.-OH	1860 Bedford Twp. Census; B.S.F. History Book-1884
7	Green Lawn Cem.-MO	B.S.F. History Book-1884; 1850 Monroe Twp. Census
8	Pleasant Grove Cem.-MO	John, Thomas & George-brothers
9	Bethel Frame-Monroe Twp.	Wounded 4/2/65; Frank McCoy 1912 Listing; aka Gillam
10	Burger Cem.-Salemville	Bedford Inquirer 5/22/1908 listing
11		B.S.F. History Book-1884; Discharged on Surgeon's Cert.
12	Burger Cem.-Salemville	Soldiers of Blair County PA book - 1940
13	Union Church Cem.-Clearville	1890 Monroe Twp. Veterans Census; Discharged on Surgeon's Cert.
14		Wounded 6/3/64; 1890 Monroe Twp. Veterans Census
15		Died 5/15/62; PA Civil War Archives-Bedford Co. residence; enlisted in Rainsburg
16		B.S.F. History Book-1884; aka Gilson
17	Bedford Cem.	1860 Bedford Twp. Census; Bedford Cemetery listing of Civil War veterans
18		1890 Everett Veterans Census
19	Prospect Hill Cem.- York	Died 6/20/65; B.S.F. History Book-1884; Draft Registration-Hopewell residence
20	Everett Cem.	1890 W. Providence Twp. Veterans Census
21		B.S.F. History Book-1884
22		B.S.F. History Book-1884; Blair Co. references
23		Historical Data Systems-Bedford Co. residence
24		Historical Data Systems-Bedford Co. residence
25	Allegheny Cem.-Pittsburgh	1850 S. Woodbury Twp. Census; Historical Data Systems record
26		PA Civil War Archives-Bedford residence
27	Grandview Cem.-Cambria Co.	Wounded 5/13/64 while on Picket; Historical Data Systems-Bedford Co. residence
28		B.S.F. History Book-1884; Substitute
29	Hyndman Cem.	1890 Hyndman Veterans Census; aka Gladfelty
30	Hampton Nat'l Cem.	Died 6/29/64; B.S.F. History Book-1884; drafted
31		B.S.F. History Book-1884; Lebanon Co. references
32		B.S.F. History Book-1884; Erie Co. references
33	Everett Cem.	B.S.F. History Book-1884; Historical Data Systems record; Jacob & James-brothers
34		1890 Everett Veterans Census
35	Everett Cem.	Wounded 12/13/62; B.S.F. History Book-1884; Historical Data Systems record
36		Died 1/6/65; Bedford Gazette 2/13/1914 listing; 1860 W. Providence Twp. Census
37	Rose Hill Cem.-Cumberland	B.S.F. History Book-1884; 1870 Bloody Run Census
38		B.S.F. History Book-1884; Berks Co. references
39		B.S.F. History Book-1884; Bradford Co. references
40	Schellsburg Union Cem.	Frank McCoy 1912 Listing; Espy, Justice & Silas-brothers
41		B.S.F. History Book-1884; 1850 Schellsburg Census
42	Florence Nat'l Cem.	POW Andersonville & Florence 4/20/64; Died 10/15/64; PA Civil War Arch.-Schellsburg resid.
43	Schellsburg Union Cem.	Wounded-POW 3/29/62; B.S.F. History Book-1884
44		Wounded 4/17/62; Historical Data Systems record
45	Bayard Cem.-NE	Ancestry.com information; aka John Peter; John P & John W-brothers
46		B.S.F. History Book-1884; 1850 Harrison Twp. Census
47	Mt. Ross Cem.-Bedford	B.S.F. History Book-1884; Mexican War Veteran
48	Andersonville Cem.; 9503	POW-Andersonville; Died 9/22/64; B.S.F. History Book-1884; enlisted in Berks Co.
49		Frank McCoy 1912 & Civil Works Admin. 1934 list; Hist. Data Systems record-Sick at M.O.
50	Mt. Zion Cem.-Pavia	Soldiers Blair Co. Book - 1940; George G, Isaac, Joseph & William-brothers
51	Rock Hill Cem.-Monroe Twp.	1890 Monroe Twp. Veterans Census
52	New Paris Comm. Ctr. Cem.	POW-Danville, VA 7/9/64 to 2/21/65; PA Civil War Archives-St. Clairsville residence
53	Helixville Cem.-Napier Twp.	Findagrave.com record; Jacob & Samuel-brothers
54	Bunker Hill Cem.-Ohio	B.S.F. History Book-1884; Jeremiah-father
55	Fishertown Bretheran Cem.	1890 Napier Twp. Veterans Census; Discharged on Surgeon's Cert.; James-son
56	Perdew Cem.-Clearville	Frank McCoy 1912 Listing; 1860 Southampton Twp. Census
57		B.S.F. History Book-1884
58	Mt. Zion Cem.-Pavia	Wounded 3/30/65; B.S.F. History Book-1884
59	Helixville Cem.-Napier Twp.	1890 Napier Twp. Veterans Census
60		MIA - Bedford Inquirer article 6/3/64; POW-Died 5/22/64 in Richmond; 1860 Union Twp. Cen
61	Andersonville Cem., 2955	POW-Andersonville; Died 7/6/64; B.S.F. History Book-1884; enlisted in Berks Co.
62	Brumbaugh Cem.-Saxton	POW-Libby; 1890 Liberty Twp. Veterans Census
63		Died 10/18/63; B.S.F. History Book-1884; PA Civil War Archives-Muster in Philadelphia
64	Everett Cem.	POW-Andersonville & 4 other prisons 7/1/63 to 2/27/65; attended 1905 Andersonville Ded.

	Name	Age	Rank	Regiment	Co.	Muster In	Muster Out	Casualty	Casualty Battle
1	Gracey, George E	16	Pvt.	PA 107th Inf.	H	2/10/62	7/13/65	POW	
2	Gracey, James A	21	Pvt.	PA 107th Inf.	H			Wound.-POW	2nd Bull Run
3	Gracey, James A		Pvt.	PA 107th Inf.	H	4/4/62	4/29/65	POW	Gettysburg
4	Gracey, William C	42	1st Lt.	PA 107th Inf.	H	2/10/62	3/4/65	Wounded	Gettysburg
5	Grass, Cephas	16	Pvt.	PA McKeage's Mil.	G	7/2/63	8/8/63		
6	Grass, Cephas		Pvt.	PA 205th Inf.	C	8/26/64	6/2/65		
7	Gray, Ellis J		Pvt.	PA 133rd Inf.	K	8/27/62	5/26/63		
8	Gray, George W (1)	35	Pvt.	PA 55th Inf.	I	10/15/61	11/4/64		
9	Gray, George W (2)	41	Sgt.	PA 138th Inf.	E	8/29/62	5/31/65	Wounded	Petersburg - Final
10	Gray, James H	28	Corp.	PA 126th Inf.	H	8/14/62			
11	Gray, John H*	24	Corp.	PA 55th Inf.	I	9/20/61	8/30/65		
12	Gray, John W		Pvt.	PA 21st Cav.	E	7/22/63	2/20/64		
13	Gray, Thomas M	17	Pvt.	PA McKeage's Mil.	H	7/2/63	8/8/63		
14	Grayble, Samuel	29	Sgt. QM	US 15th Inf.	A	3/23/65	3/23/68		
15	Green, Andrew	27	Pvt.	PA 46th Inf.	C	7/1/63	8/19/63		
16	Green, David M	37	Sgt. QM	PA 18th Cav.	L	11/11/62	7/21/65		
17	Green, William B	17		PA 125th Inf.		8/10/62	5/18/63		
18	Green, William R	26	Pvt.	PA 195th Inf.	E	4/13/65	1/31/66		
19	Greene, James								
20	Greenland, Thomas J*	25	Corp.	PA 110th Inf.	C	10/24/61		KIA	Wilderness
21	Greenland, Thomas J*		Corp.	PA 110th Inf.	C	10/24/61		Wounded	Gettysburg
22	Gregor, Solomon*	19	Pvt.	PA 184th Inf.	A	5/12/64		KIA	Jerusalem Plank Rd.
23	Grier, Thomas J*		Capt.	PA 18th Cav.	B	10/25/62	7/21/65		
24	Griffin, Andrew H	24	Pvt.	PA 7th Cav.	B	5/4/61	6/26/65	Wounded	Fredericksburg
25	Griffin, Andrew H		Pvt.	PA 7th Cav.	B			POW	Dallas, GA
26	Griffith, Abel	19	Pvt.	PA 8th Res.	F	3/11/64	5/15/64		
27	Griffith, Abel		Pvt.	PA 191st Inf.	H	5/15/64	6/28/65	POW	Globe Tavern
28	Griffith, Daniel M	44	Pvt.	PA 49th Inf.	E	6/4/64	12/3/64		
29	Griffith, Joseph H	29	Pvt.	PA 149th Inf.	I	2/25/65	5/8/65		
30	Griffith, Michael	46	Pvt.	PA 8th Res.	F	3/25/64	5/15/64		
31	Griffith, Michael		Pvt.	PA 191st Inf.		5/15/64	7/18/65		
32	Grim, William	30	Pvt.	PA 149th Inf.	K	8/26/62		POW-Died	Wilderness
33	Grimes, George W	19	Pvt.	PA 11th Inf.	A	9/30/61	7/1/65		
34	Grimes, Henry W	20	Pvt.	PA 84th Inf.	C	10/21/61		Wounded	1st Kernstown
35	Grimes, Henry W		Pvt.	PA 84th Inf.	C			Wound.-Died	Wilderness
36	Grimes, Jacob R	18	Pvt.	PA 84th Inf.	C	10/12/61	11/6/62		
37	Grimes, Jacob R		Pvt.	PA 205th Inf.	C	8/27/64	6/2/65		
38	Grimes, John C		Corp.	PA 14th Inf.	I	5/2/61	8/7/61		
39	Grimes, John C	49	Pvt.	PA 84th Inf.	C	1861			
40	Groman, John		Pvt.	PA 18th Cav.	H	9/26/64	7/12/65		
41	Groomer, Anthony	37	Pvt.	PA 77th Inf.	F	3/2/65	12/6/65		
42	Grove, Benjamin H	20	Pvt.	PA 194th Inf.	I	7/22/64	11/5/64		
43	Grove, Henry C	18	Pvt.	PA McKeage's Mil.	G	7/2/63	8/8/63		
44	Grove, Henry C		Pvt.	OH 191st Inf.	G	3/8/65	8/27/65		
45	Grove, James A	17	Pvt.	PA 107th Inf.	H	4/26/64		Wounded	Petersburg
46	Grove, Robert C	20	Pvt.	PA 133rd Inf.	K	8/15/62	4/16/63		
47	Grove, William A	34	Sgt.	PA 171st Inf.	I	11/2/62			
48	Grove, William H	16	Pvt.	IN 6th Inf.	E	8/28/62			
49	Growden, John S	50	Pvt.	PA 133rd Inf.	D	8/5/62	5/26/63		
50	Growden, John W	30	Pvt.	PA 56th Inf.	F	9/21/64	6/31/65		
51	Growden, Joseph	31	Pvt.	PA 45th Inf.	C	7/5/64		Wounded	Petersburg - Final
52	Growden, Thomas		Pvt.	PA 21st Cav.	E	7/8/63	2/20/64		
53	Grubb, George	23	Pvt.	PA McKeage's Mil.	D	7/2/63	8/8/63		
54	Grubb, George		Pvt.	PA 149th Inf.	I	8/26/63		POW-Died	Globe Tavern
55	Grubb, Harvey	19	Pvt.	PA 133rd Inf.	K	8/15/62	5/26/63		
56	Grubb, Harvey		Pvt.	PA 22nd Cav.	C	7/11/63	10/31/65	Wounded	Mt. Jackson
57	Grubb, Wilson	25	Pvt.	PA 8th Res.	F	6/5/61	1/5/63	POW	Seven Days Battle
58	Gubernater, Charles	37	Pvt.	LA 3rd Cav.	D	9/1/64			
59	Gump, Erastus J	19	Pvt.	PA 194th Inf.	I	7/21/64	11/5/64		
60	Gump, Frank								
61	Gump, John A	19	1st Lt.	PA 138th Inf.	D	9/12/62		Wound.-Died	Cedar Creek
62	Gump, Simon								
63	Guthridge, Charles N			US Navy					
64	Guthridge, James C*								

	Cemetery	Notes
1		B.S.F. History Book-1884; Alfred, George E and James A-brothers
2	Monument Cem.-NJ	POW 8/30/62 to 9/3/62; Historical Data Systems-Bedford Co. residence
3		POW-Belle Island & Andersonville 7/1/63 to 2/27/65; B.S.F. History Book-1884
4	Mt. Zion Cem.-Breezewood	Neck Wound 7/1/63; B.S.F. History Book-1884; Alfred, George E and James A-Sons
5	Fairview Cem.-Martinsburg	B.S.F. History Book-1884
6		B.S.F. History Book-1884; 1860 Hopewell Twp. Census
7		Promoted to Hospital Steward 8/27/62; Historical Data Systems-Bedford Co. residence
8		B.S.F. History Book-1884; PA Civil War Archives-enlisted in Bloomfield Furnace
9	Milesburg Cem.-Centre Co.	Wounded 4/2/65; B.S.F. History Book-1884; Ancestry.com born in Woodbury
10	Bridgeport Cem.-Fayette Co.	Ancestry.com-Born Bed. Co
11		B.S.F. History Book-1884; Blair Co. references
12		Historical Data Systems-Bedford Co. residence
13		B.S.F. History Book-1884
14	Fockler Cem.-Saxton	1890 Saxton Veterans Census; aka Grable
15	Fairview Cem.-Altoona	1890 Londonderry Twp. Veterans Census
16	Princeton City Cem.-WI	1890 Liberty Twp. Veterans Census
17	Fairview Cem.-Altoona	Ancestry.com-born Bedford Co.; William B & Andrew-brothers
18	Dry Hill Cem.-Woodbury	1890 Woodbury Twp. Veterans Census; Historical Data Systems record
19		1890 Hyndman Veterans census
20		KIA 5/6/64; B.S.F. History Book-1884; Bed.Gazette 2/13/1914 list; Huntingdon Co. references
21		Wounded 7/2/63; Soldiers of Blair County PA book - 1940; Cassville Census
22		KIA 6/22/64; B.S.F. Hist. Book-1884; Company A recruited in Dauphin & Bedford Co.
23		B.S.F. History Book-1884; Somerset Co. references
24	Woods Church-Colerain Twp.	Arm Wound 12/13/62; 1890 Colerain Twp. Veterans Census
25		POW 5/30/64 to 11/26/64; Bedford Historical Society letter
26	Everett Cem.	B.S.F. History Book-1884
27		POW 8/19/64 to 10/7/64; 1890 Wells Valley Veterans Census
28	Mt. Union Cem.-Lovely	1890 Union Twp. (Pavia) Veterans Census
29	Mt. Zion Cem.-Pavia	Frank McCoy 1912 Listing; 1890 Union Twp. (Pavia) Veterans Census
30	Hopewell Cem.	B.S.F. History Book-1884; 1860 Broad Top Twp. Census
31		B.S.F. History Book-1884
32	Florence Nat'l Cem.	POW-Florence 5/5/64; Died 9/20/64; Bed. Gazette 2/13/1914 list; 1860 Southampton Twp. Cen.
33	Hershberger Cem.-Snake Spr. V.	Frank McCoy 1912 Listing; Historical Data Systems record
34		Wounded 3/23/62; PA Civil War Archives-Woodbury residence
35		Wounded 5/6/64; Died 5/16/64
36	Fockler Cem.-Saxton	1890 Liberty Twp. Veterans Census
37		Historical Data Systems record; 1850 M. Woodbury Twp. Census; Jacob & Henry-brothers
38		1890 S. Woodbury Veterans Census
39	Hickory Bottom Cem.-Woodbury	Frank McCoy 1912 Listing; Jacob & Henry-sons
40		1890 Bedford Veterans Can. Deserted after war ended on 7/25/65
41	Hopewell Cem.	1890 Broad Top Veterans Census aka Grumer; 1870 Broad Top Twp. Census
42		B.S.F. History Book-1884
43	Elmwood Cem.-KS	B.S.F. History Book-1884
44		Ancestory.com information; Henry & James-brothers
45	Everett Cem.	Wounded 6/20/64; B.S.F. History Book-1884; 1850 E. Providence Twp. Census
46		Deserted 9/12/62 to 3/10/63; 1860 W. Providence Twp. Census; Discharged on Surgeon's Cert.
47	Yellow Creek Reformed Cem.	Deserted 11/20/62
48	Old Carrollton Cem.-IA	1860 Monroe Twp. Census
49	Alms House Grave Yard	PA Civil War Archives aka Samuel J; Pension record; 1870 Cumberland Valley Census
50	Growden Cem.-Cumb. Valley	1890 Cumberland Valley Veterans Census; 1870 Cumberland Valley Census
51	Bethel Meth. Cem.-Centerville	Wounded 4/2/65; PA Civil War Archives-enlisted in Bedford Co.
52		Historical Data Systems record; 1870 St. Clair Twp. Census
53	Annapolis Nat'l Cem.	B.S.F. History Book-1884; George, Harvey & Wilson-brothers
54		POW Aug. 1864; Died 10/16/64; Bedford Gazette 1914; 1860 W. Providence Twp. Census
55	Bethal Frame Ch.-Monroe Twp.	B.S.F. History Book-1884; PA Civil War Archives-Clearville residence; aka Harry
56		Wounded 9/24/64; Historical Data Systems-Bedford Co. residence
57	Akers Chapel Cem.-IL	POW 6/27/62; B.S.F. History Book-1884; 1850 Monroe Twp. Census
58		Born in Virginia; listed for the first time on the 1870 Bedford Township Census
59		Historical Data Systems record
60		Listed in "From Winchester to Bloody Run" book
61	Mt. Olivet Cem.-Manns Choice	Wounded 10/19/64; Died 10/20/64; Hist. Data Systems-Bedford Co. resid.; John & Simon-bros.
62		1860 Napier Twp. Census; Listed in "From Winchester to Bloody Run" book
63	Brumbaugh Cem.-Saxton	Civil War marker at gravesite; Bedford Gazette 5/21/09 listing aka Nimrod Gutheridge
64	Brumbaugh Cem.-Saxton	Civil War marker at gravesite

Chapter 18

	Name	Age	Rank	Regiment	Co.	Muster In	Muster Out	Casualty	Casualty Battle
1	Guy, Robert		Pvt.	PA 13th Inf.	G	4/25/61	8/6/61		
2	Guyer, William C		Farr.	PA 21st Cav.	E	7/8/63	2/20/64		
3	Guyer, William H	36	Corp.	IL 76th Inf.	G	8/22/62	3/26/63		
4	Hacher, James		Pvt.	PA 208th Inf.	H	9/7/64			
5	Hackler, Samuel								
6	Hafer, Alexander H	21	Pvt.	PA 13th Inf.	G	4/25/61	8/6/61		
7	Hafer, Frank M	21	1st Lt.	PA 2nd Cav.	E	11/5/61	6/17/65		
8	Hafer, George W		Pvt.	PA 2nd Cav.	E	3/16/64	6/17/65		
9	Hafer, William H	21	Pvt.	PA 13th Inf.	G	4/25/61	8/6/61		
10	Hafer, William H		Pvt.	PA 2nd Cav.	E	11/4/61	4/10/65	Wounded	
11	Hafer, Wilson	49	Pvt.	PA 133rd Inf.	D	8/14/62	1/5/63		Fredericksburg*
12	Haffly, David S	27	Pvt.	PA 50th Inf.	I	2/24/65	5/8/65		
13	Haffly, George W		Pvt.	PA 202nd Inf.	K	10/6/64	8/3/65		
14	Haft, Adam*								
15	Hagan, John	18	Pvt.	PA 184th Inf.	A	5/12/64	7/18/65	Wounded	Cold Harbor
16	Hageman, Alexander B		Pvt.	PA 101st Inf.	D	12/6/61	1863		
17	Hagerty, Daniel	19	Corp.	PA 55th Inf.	K	2/19/64	8/30/65		
18	Hahn, Isaac	16	Pvt.	PA 55th Inf.	K	9/18/63	8/30/65		
19	Hainsey, Adam R	15	Corp.	PA 76th Inf.	C	10/17/61	7/18/65		
20	Hainsey, Christopher	27	Pvt.	OH 91st Inf.	K	8/12/62	6/24/65		
21	Hainsey, Frederick	22	Sgt.	PA 55th Inf.	H	9/20/61	8/30/65	Wound.-POW	Drewry's Bluff
22	Hainsey, George	23	Pvt.	PA 76th Inf.	C	10/17/61	11/28/64		
23	Hainsey, Henry	28	Pvt.	OH 181st Inf.	C	10/7/64	7/14/65		
24	Hainsey, John	23	Pvt.	PA 76th Inf.	C	9/4/61	11/28/64		
25	Hainsey, Josiah	28	1st Sgt.	PA 1st Light Art.	A	5/28/61	5/28/64	Wounded	Seven Days Battle
26	Hainsey, Josiah		1st Sgt.	PA 1st Light Art.	A			Wounded	Fredericksburg
27	Hainsey, Josiah		Sgt.	PA 45th Inf.	D	12/15/64	7/17/65		
28	Hainsey, Michael*	60		OH					
29	Hainsey, Valentine	14	Pvt.	PA 55th Inf.	H	8/28/61		Wound.-Died	Cold Harbor
30	Hale, William	27	Pvt.	PA 55th Inf.	I	10/15/61	10/15/64		
31	Haley, Josiah		Music.	PA 55th Inf.	K	10/15/61	11/4/64		
32	Hall, George W	18	Pvt.	PA 194th Inf.	I	7/22/64	11/5/64		
33	Hall, Henry H		Corp.	MD 1st PHB Cav.	K	3/31/64	6/28/65		
34	Hall, John		Pvt.	PA McKeage's Mil.	G	7/2/63	8/8/63		
35	Hall, William M	37	Major	US Inspector		1/1/65	1/1/66		
36	Hallerin, Lawrence		Pvt.	PA 5th Res.	F	6/21/61	6/11/64		
37	Hamer, John C	18	Pvt.	PA 194th Inf.	I	7/22/64	9/6/64		
38	Hamer, John C			PA 97th Inf.		9/6/64			
39	Hamer, Nathaniel		Pvt.	PA 5th Heavy Art.	K	9/2/64	7/1/65		
40	Hamilton, Isaac T*		Lt. Col.	PA 110th Inf.	C	12/5/61	6/28/65	Wounded	Ft. Stedman
41	Hamilton, James C*	19	Capt.	PA 110th Inf.	C	12/19/61	6/28/65		
42	Hamilton, John							Died	
43	Hamilton, John C		Pvt.	PA 110th Inf.	C	10/4/64	6/2/65		
44	Hamilton, Mahlon B*	31	Pvt.	PA 55th Inf.	I	10/15/61	1/15/63		
45	Hamm, David K	30	Pvt.	PA 205th Inf.	C	8/24/64	6/2/65		
46	Hammann, Jacob		Pvt.	PA 158th Inf.	D	11/4/62	7/12/63		
47	Hammann, Jacob		Pvt.	PA 208th Inf.	H	9/6/64	6/1/65		
48	Hammel, Peter S	20	Pvt.	PA 54th Inf.	E	1861		Died	
49	Hammer, Daniel R	29	Pvt.	PA 55th Inf.	H	10/11/61	8/30/65		
50	Hammer, Gordon		Pvt.	OH 13th Cav.	E	5/5/64	7/24/65		
51	Hammer, Hezekiah	21	Capt.	PA 55th Inf.	K	11/5/61	8/30/65	Wounded	Petersburg - Final
52	Hammer, John B	21	Sgt.	PA 138th Inf.	D			Wounded	Wilderness
53	Hammer, John B		Sgt.	PA 138th Inf.	D	8/29/62	6/23/65	Wounded	Opequon
54	Hammer, Joseph D		Pvt.	PA 142nd Inf.	D	8/22/62		Wound.-Died	Gettysburg
55	Hammer, Martin		Pvt.	PA 54th Inf.	H	2/26/64	7/15/65		
56	Hammer, Samuel	22	Pvt.	PA 55th Inf.	H	2/29/64	8/30/65		
57	Hammond, Oliver*	34	Pvt.	PA 55th Inf.	K	10/17/63	8/30/65		
58	Hammond, William*	27	Pvt.	PA 55th Inf.	H	12/31/63	8/30/65		
59	Hand, Henry		Pvt.	PA 55th Inf.	H	10/11/61		POW-Died	Drewry's Bluff
60	Hand, James	26	Pvt.	PA 55th Inf.	I	10/15/61	1/20/63		
61	Hand, James		Pvt.	US 1st Art.	M	1/20/63	10/3/64	Wounded	
62	Hankinson, Solomon J	28	Pvt.	PA 206th Inf.	F	8/26/64	6/26/65	Wounded	
63	Hanks, Albert B	18	Pvt.	PA 133rd Inf.	C	8/13/62	6/26/63		
64	Hanks, Benjamin A	28	Sgt.	PA 101st Inf.	D	11/1/61		POW-Died	Plymonth, NC

	Cemetery	Notes
1		B.S.F. History Book-1884; Historical Data Systems record
2		Historical Data Systems-Bedford Co. residence
3	Woodlawn Cem.-IL	IL Veterans List; Born in Bedford Co.; Discharged on Surgeon's Cert.
4		Deserted 9/13/64; B.S.F. History Book-1884
5		1890 Liberty Twp. Veterans Census
6		B.S.F. History Book-1884; Historical Data Systems-Bedford residence
7		B.S.F. History Book-1884; PA Civil War Archives-Bedford residence
8		B.S.F. History Book-1884; George, Frank & William-brothers
9	Smith Farm Cem.-Cumb. Val.	B.S.F. History Book-1884; 1860 Cumberland Valley Census
10		Leg Amputated 12/31/63; 1890 Cumberland Valley Veterans Census
11	Smith Farm Cem.-Cumb. Val.	Frank McCoy 1912 Listing; Discharged on Surgeon's Cert.; Father of William, George & Frank
12	Diehl's Crossroads-Henrietta	1890 Woodbury Twp. Veterans Census; aka Daniel S Haffley
13		1890 Woodbury Twp. Vet. Cen.; 1850 M. Woodbury Twp. Census; George & David-bros.
14	Chaneysville Meth. Cem.	Civil War marker at gravesite
15		Wounded 6/3/64 in Bedford Inquirer 7/1/64 article; B.S.F. History Book-1884
16		B.S.F. History Book-1884; Discharged on Surgeon's Cert.
17		B.S.F. History Book-1884; PA Civil War Archives-Bedford Co. residence
18	Lutheran Cem.-Osterburg	B.S.F. History Book-1884
19	New Straitsville Joint-OH	Findagrave.com record; 1850 Union (Pavia) Township Census.
20	Keystone Cem.-OH	B.S.F. History Book-1884; Findagrave-Born Bedford Co.;Michael-Father
21	Geeseytown Cem.-Frankstown	POW 5/16 to 8/16/64; B.S.F. History Book-1884; Frederick, Adam & George-brothers
22	Greenlawn Cem.-Roaring Spring	Findagrave: Born Bedford Co.
23	New Straitsville Cem.-OH	Findagrave.com information; Henry, Adam, Valentine, John & Josiah-brothers; aka Hinzy
24	Greenlawn Cem.-Roaring Spring	1840 Greenfield Twp. Census; aka Hinzy
25	Pleasant Hill Meth.-Somerset Co.	Wounded 6/27/62; 1840 Greenfield Twp. Census; aka Hinzy
26		Wounded 12/13/62 listed in Pension reports
27		Ancestry.com lists born in Bedford
28	Keystone Cem.-OH	Findagrave: Born in Bedford Co.; Ohio, U.S., Soldier Grave Registrations
29	Sproul Union Cem.-Claysburg	Wounded 6/3/64; Died 6/17/64; B.S.F. History Book-1884; 1850 Union Twp. Census
30	Albrights Cem.-Roaring Springs	B.S.F. History Book-1884; enlisted Bloomfield Furnace
31		1860 Bedford Census; Historical Data Systems-Bedford Co. residence
32	Maple Grove Cem.-MI	B.S.F. History Book-1884; aka Washington
33		1890 Cumberland Valley Twp. Veterans Census; Historical Data Systems record
34		B.S.F. History Book-1884; 1860 Cumberland Valley Census
35	Bedford Cem.	US inspector Prison Camps; 1890 Bedford Twp. Veterans Census
36		Historical Data Systems-Bedford Co. residence
37		B.S.F. History Book-1884
38		B.S.F. History Book-1884
39		B.S.F. History Book-1884
40		Wounded 3/25/65; B.S.F. History Book-1884; Blair Co. residence
41		B.S.F. History Book-1884; Compropst Mills residence
42	Chaneysville Meth. Cem.	Died 10/8/65; Bedford Gazette 2/13/1914 listing; 1850 Southampton Census
43		B.S.F. History Book-1884
44		B.S.F. History Book-1884; Born in Blair Co.
45	Hopewell Cem.	1890 Broad Top Veterans Census; aka Harn
46		Civil War Veterans Burial Record
47		1890 Brush Creek Veterans Census; Historical Data Systems-Bedford Co. residence
48	Sproul Union Cem.-Claysburg	Died 10/9/61; Civil War Pension Record; PA Civil War Archives; aka Hamel
49		Historical Data Systems record; 1850 Napier Twp. Census
50	Pleasantville Cem.	Frank McCoy 1912 Listing; transferred to 87 OH inf.
51	Pleasantville Cem.	Arm Amputated 4/2/65; B.S.F. History Book-1884; Frank McCoy 1912 Listing
52	Pleasantville Cem.	Wounded 5/6/64; John & Hezekiah-brothers
53		Wounded 9/19/64; B.S.F. History Book-1884
54	Loudon Park Nat'l Cem.	Wounded 7/1/63, Died 9/9/63; Historical Data Systems record; 1850 Napier Twp. Census
55		Ancestry.com information; Martin & Daniel-brothers
56	Schellsburg Cem.	1840 Napier Twp. Census; PA Civil War Archives-Bedford Co. residence
57		B.S.F. History Book-1884; Drafted
58		B.S.F. History Book-1884; Drafted
59	Andersonville Cem., 10538	William H Hand listed MIA 5/16/64 in Bed. Inquirer-7/8/64; POW-Andersonville; Died 9/29/64
60		PA Civil War Archives-enlisted in Bloomfield Furnace; B.S.F. History Book-1884
61		B.S.F. History Book-1884
62	Crum-Oldham-Shade Twp.	1890 Napier Twp. Veterans Census
63	Cedar Grove-Chambersburg	B.S.F. History Book-1884; Historical Data Systems-Bedford Co. residence
64	Andersonville Cem., 9892	POW-Andersonville 4/20/64; Died 9/27/64; B.S.F. Hist. Book-1884; 1860 Monroe Twp. Cen.

	Name	Age	Rank	Regiment	Co.	Muster In	Muster Out	Casualty	Casualty Battle
1	Hanks, Benson	28	Pvt.	PA 87th Inf.	B	6/3/64	6/29/65		
2	Hanks, Caleb	19	Pvt.	PA 101st Inf.	D	11/1/61		POW-Died	Plymonth, NC
3	Hanks, David F	18	Pvt.	PA 101st Inf.	D	11/1/61		POW-Died	Plymonth, NC
4	Hanks, Jacob C	22	Corp.	PA 101st Inf.	D	11/1/61	10/15/62		
5	Hanks, M*								
6	Hanks, Nelson	28	Pvt.	PA 101st Inf.	D	11/1/61		POW-Died	Plymonth, NC
7	Hanks, Thompson	20	Pvt.	PA 101st Inf.	D	11/1/61	10/15/62		
8	Hanks, William H	21	Corp.	PA 133rd Inf.	C	8/13/62	5/26/63		
9	Hanks, William H		Sgt.	PA 22nd Cav.	I	7/11/63	5/12/65	Wounded	Martinsburg, WV
10	Hann, Gaston	34	Pvt.	PA 171st Inf.	I	11/10/62	8/8/63		
11	Hann, Jeremiah W		Pvt.	PA 208th Inf.	H	9/7/64	6/1/65		
12	Hann, John	18	Pvt.	PA 208th Inf.	H	9/7/64	6/1/65	Wounded	
13	Hann, Philip	17	Pvt.	PA 133rd Inf.	K	8/15/62			
14	Hann, Philip		Pvt.	PA McKeage's Mil.	D	7/2/63	8/8/63		
15	Hann, Samuel W*								
16	Hanson, George*	24	Pvt.	PA 55th Inf.	I	10/15/63			
17	Harbaugh, Allen	15	Pvt.	PA 205th Inf.	I	9/1/64	6/2/65		
18	Harbaugh, Amos	24	Corp.	PA 171st Inf.	I	11/2/62	8/8/63		
19	Harbaugh, Eli	16	Pvt.	PA 55th Inf.	K	11/5/61		Died	
20	Harbaugh, Emanuel	19	Pvt.	PA 138th Inf.	D	8/29/62	6/23/65	POW	Wilderness
21	Harbaugh, George W	20	Pvt.	PA 55th Inf.	H	10/11/61		Wound.-Died	Petersburg - Initial
22	Harbaugh, Jason	18	Pvt.	PA 84th Inf.	A	9/9/61	10/24/64		
23	Harbaugh, John	32	Pvt.	PA 55th Inf.	D	10/12/61		Died	
24	Harbaugh, John M	19	Pvt.	PA 205th Inf.	C	8/26/64	6/2/65		
25	Harbaugh, Joseph	16	Pvt.	PA 79th Inf.	I	3/7/65	6/23/65		
26	Harbaugh, Robert	23	Pvt.	PA 55th Inf.	K	2/19/64		KIA	Petersburg - Initial
27	Harbaugh, William H	22	Pvt.	PA 84th Inf.	A	10/24/61		POW	Chancellorsville
28	Harbaugh, Wilson	18	Pvt.	PA 55th Inf.	K	3/2/64		Died	
29	Harbrant, Jacob	22	Pvt.	PA 58th Inf.	C	7/5/64	8/19/65	Wounded	
30	Harclerode, David	20	Pvt.	PA 125th Inf.	E	8/13/62	5/18/63	Wounded	Antietam
31	Harclerode, Solomon	28	Corp.	WI 10th Inf.	F	9/12/61	11/3/64	Wounded	Chickamauga
32	Harclerode, William H	30	Pvt.	PA 18th Cav.	H	2/25/65	7/12/65		
33	Harden, Calvin		Pvt.	PA 138th Inf.	F	9/1/62			
34	Harden, James	31	Pvt.	MD 2nd PHB Inf.	D	2/25/65	5/29/65		
35	Hardinger, Daniel (1)	23	Pvt.	IL 106th Inf.	D	9/17/62	7/12/65		
36	Hardinger, Daniel (2)*	28	Pvt.	OH 100th Inf.	C	9/1/62		Died	Utoy Creek, GA
37	Hardinger, Elias								
38	Hardinger, George W (1)	28	Pvt.	PA 149th Inf.	B	8/26/63		POW-Died	Wilderness
39	Hardinger, George W (2)	28	Pvt.	OH 25th Inf.	D	10/8/64	10/20/65		
40	Hardinger, Henry	25	Pvt.	WV 17th Inf.	K	9/1/64	6/30/65		
41	Hardinger, John	18	Pvt.	PA 7th Cav.	M	3/2/64	8/23/65		
42	Hardinger, Reuben	25	Pvt.	MD 1st PHB Inf.	F	9/4/61	4/8/65		
43	Hare, Henry	16	Pvt.	PA 210th Inf.	K	9/14/64	7/5/65	Wounded	White Oak Road
44	Harker, Jacob	40	Pvt.	PA 143rd Inf.	B	9/18/63	6/12/65		
45	Harker, Samuel B*	33	Pvt.	PA 125th Inf.	I	8/13/62		Wound.-Died	Antietam
46	Harkins, Warner			PA 13th Inf.					
47	Harlow, Charles E	16	Pvt.	PA 2nd Cav.	E	12/17/63	6/28/65		
48	Harp, Henry*		Corp.	PA 55th Inf.	D	9/2/62	6/11/65		
49	Harr, Christian	19	Pvt.	PA 149th Inf.	D	10/2/63	12/13/64	Wounded	Petersburg - Initial
50	Harr, David		Pvt.	PA 3rd Cav.	E	6/14/64	6/6/65		
51	Harr, Silas	23	Sgt.	PA 21st Cav.	E	7/8/63	7/8/65		
52	Harrier, Adam	36	Pvt.	PA 82nd Inf.	K	9/12/64	7/13/65		
53	Harris, John T	27	Sgt.	USCT 3rd Inf.	F	7/10/63	10/31/65	Wounded	Morris Island, SC
54	Harris, Joshua	23	Pvt.	MA 55th Inf.	G	6/15/63		Died	
55	Harris, Samuel O	19	Pvt.	PA 188th Inf.	F	2/27/64	4/1/64		
56	Harris, Samuel O		Pvt.	PA 3rd Art.	E	4/1/64	12/14/65		
57	Hart, Isreal	26	Pvt.	PA 205th Inf.	G	9/1/64	6/2/65		
58	Hart, Thomas*	27	Pvt.	PA 110th Inf.	C	1861	6/28/65		
59	Hartagan, Wlliam	31	Pvt.	PA McKeage's Mil.	G	7/2/63	8/8/63		
60	Hartle, John B	21	Pvt.	PA 99th Inf.	B	2/25/65		Died	
61	Hartley, William	21	Corp.	PA 55th Inf.	D	10/12/61	5/22/63		
62	Hartley, William		1st Lt.	USCT 34th Inf.	F	5/22/63	3/12/64		
63	Hartley, William		Capt.	USCT 2nd Inf.					
64	Hartman, Andrew J	40	Pvt.	PA 210th Inf.	C	9/10/64	5/30/65		

	Cemetery	Notes
1	McKendree Cem.-Crystal Spring	Benson, Thompson & Nelson-brothers
2		POW-Andersonville 4/20/64; Died 10/26/64; Historical Data Systems-Bedford Co. residence
3	Camp Nelson Nat'l Cem.-KY	POW Andersonville & Florence 4/20/64; Died 11/1/64; Hist.Data Systems-Bedford Co. resid.
4	Leavenworth Nat'l Cem.	1860 E. Providence Twp. Census; Discharged on Surgeon's Cert.; Jacob & Caleb-brothers
5	Clearville Union Cem.	Bedford Inquirer 5/22/1908 listing
6	Andersonville Cem., 8804	POW-Andersonville 4/20/64; D. 9/15/64; Hist.l Data Systems record; 1860 Brush Creek Cen.
7	Bethel Meth. Cem.-Centerville	Historical Data Systems-Bedford Co. residence; Discharged on Surgeon's Cert.
8		Historical Data Systems record; William, David & Benjamin-brothers
9	Union Church Cem.-Clearville	Wounded 9/12/64-Pension record; B.S.F. History Book-1884
10	Union Memorial Cem.-Mench	East Providence Honor Roll; 1890 E. Providence Twp. Veterans Census
11		B.S.F. History Book-1884; Historical Data Systems-Bedford Co. residence
12		B.S.F. History Book-1884; Historical Data Systems-Bedford Co. residence
13	Clearville Union Cem.	Deserted 8/19/62; PA Civil War Archives record
14		B.S.F. History Book-1884
15	Pleasantville Cem.	Civil War marker at grave; Born 11/30/21
16		Deserted 3/29/64; B.S.F. History Book-1884
17	Bagdad Cem.-FL	Historical Data Systems record; 1860 Union Twp. (Pavia) Census; Allen, Eli & Emanuel-bros.
18	Pleasantville Cem.	B.S.F. History Book-1884; Amos, Jason & John M-brothers
19	Beaufort Nat'l Cem.	Died-Small Pox 1/27/62; PA Civil War Archives-Bedford Co. residence
20	Mt. Union Cem.-Lovely	POW-Andersonville 5/6/64; Historical Data Systems record; Absent in hospital at Muster Out
21	Hampton Nat'l Cem.	Wounded 6/18/64; Died 7/11/64; PA Civil War Archives record; 1850 St. Clair Twp. Census
22	Pleasantville Cem.	1890 W. St. Clair Twp Veterans Census
23		Died 9/27/63; B.S.F. History Book-1884; Historical Data Systems-Bedford Co. residence
24	Sandyvale Cem.-Johnstown	Ancestry.com information; 1850 St. Clair Twp. Census
25	Mummasburg Men.-Adams Co	Ancestry.com information; Joseph & Wilson-brothers
26		KIA 6/18/64; PA Civil War Archives-Bedford Co. residence; Robert & William-brothers
27	Pleasantville Cem.	POW-Libby - 5/3/63 to 5/19/63; Historical Data Systems record
28	Beaufort Nat'l Cem.	Died 3/28/64; B.S.F. History Book-1884; 1860 St. Clairsville Census
29	St. John Baptist-New Baltimore	1890 Juniata Twp. Veterans Census-wounded in left breast
30	Everett Cem.	Wounded 9/17/62; Frank McCoy 1912 Listing
31	Platteville Cem.-WI	Historical Data Systems record; aka Harkleroad
32	Hershberger Cem.-Snake Spr. V.	1890 Snake Spring Valley Veterans Census; William, Solomon & David-brothers
33		Deserted 2/7/63; B.S.F. History Book-1884; 1860 Londonderry Twp. Census
34	Hyndman Cem.	1890 Hyndman Veterans Census
35	Lake Mound Cem.-IL	1850 Cumberland Valley Cen; IL Civil War Vets list-Born Bedford Co.
36	Marietta Nat'l Cem.-GA	Wounded 8/6/64; Died 9/27/64; Bedford Gazette 2/13/1914 listing
37		
38	Andersonville Cem., 4379	POW-Andersonville; 5/5/64; Died 7/31/64; 1850 Cumb. Val. Cen.; Bed. Gazette 2/13/1914 list
39		Ancestry.com information; 1860 Cumberland Valley Census
40	Bluemont Cem.-WV	Ancestry.com information; 1850 Cumberland Valley Census
41		Ancestry.com information; John, George (2), Daniel (1) & Reuben-brothers
42	Fountain Cem.-OH	Ancestry.com information; 1850 Cumberland Valley Census
43	Holsinger Cem.-Bakers Summit	Wounded 3/31/65; Historical Data Systems record; 13th MD inf. mentioned
44	Newburg Cem.-Hunt. Co.	1890 Broad Top Veterans Census; Jacob & Samuel-brothers
45	Reformed Cem.-Marklesburg	Wounded 9/17/62; Died 11/16/62; 1860 Hopewell Twp. (Hunt. Co.) Cen. Buried-Marklesburg
46	St. Thomas Cem.-Bedford	Frank McCoy 1912 Listing; Bedford Gazette 1909 list
47	Chaneysville Meth. Cem.	Frank McCoy 1912 Listing; B.S.F. History Book-1884
48		B.S.F. History Book-1884; Berks Co. references
49	Mt. Union Cem.-Lovely	Wounded 6/18/64; Historical Data Systems record
50	Hopewell Cem.	Historical Data Systems record
51	Mt. Union Cem.-Lovely	1890 Union Twp. Veterans Census
52	Oak Hill Cem.-IA	Historical Data Systems record; 1860 Juniata Twp. Census
53	Mt. Ross Cem.-Bedford	Face Wound in Aug./Sept. 1863 on Morris Island, SC; 1890 Bedford Veterans Census
54		Died of Disease 9/8/63; Historical Data Systems record; 1860 Bedford Twp. Census
55	Keagy Cem.-Woodbury	Historical Data Systems record
56		Historical Data Systems record
57	Grandview Cem.-Hunt. Co.	1890 Broad Top Veterans Census
58		B.S.F. History Book-1884; PA Civil War Archives-Philadelphia residence
59		B.S.F. History Book-1884; 1860 Colerain Twp. Census
60	Mt. Zion Cem.-Pavia	Died 5/21/65; Frank McCoy 1912 Listing; 1860 Union Twp. (Pavia) Census
61	Bedford Cem.	1890 Bedford Twp. Veterans Census
62		Historical Data Systems record
63		Frank McCoy 1912 Listing; 1860 Snake Spring Valley Census
64	Friends Cove UCC Cem.	1870 S. Woodbury Twp. Census; aka Jackson

	Name	Age	Rank	Regiment	Co.	Muster In	Muster Out	Casualty	Casualty Battle
1	Hartman, Frederick	28		PA 13th Inf.		1861	1861		
2	Hartman, Frederick		Pvt.	PA 133rd Inf.	K	8/15/62	5/26/63		
3	Hartman, George L	21	Corp.	PA 110th Inf.	C	9/24/61		MIA	Chancellorsville
4	Hartman, Henry S	37	Pvt.	PA 87th Inf.	D	7/4/64	7/3/65		
5	Hartman, John P	21	Pvt.	PA 110th Inf.	C	10/24/61	5/30/65	POW	Cold Harbor
6	Hartman, Joseph			MO 13th					
7	Hartsell, Abraham	39	Pvt.	PA 56th Inf.	F	9/19/64	5/31/65		
8	Hartzel, Francis B	22	Pvt.	PA 55th Inf.	D	2/18/64	8/30/65		
9	Hartzell, Samuel*	22	Pvt.	PA 184th Inf.	A	5/12/64		Wound.-Died	Jerusalem Plank Rd.
10	Hartzell, William B	29	Pvt.	PA 82nd Inf.	C	11/14/64	5/30/65	Wounded	
11	Harvey, Alexander*	18	Pvt.	PA 55th Inf.	K	2/13/64	8/30/65		
12	Harvey, William	17	Pvt.	PA McKeage's Mil.	G	7/2/63	8/8/63		
13	Harvey, William		Pvt.	PA 208th Inf.	H	9/5/64	6/1/65		
14	Harwood, Richard	19	Pvt.	PA 110th Inf.	C	10/24/61	6/28/65		
15	Hasenpat, William R	20	Pvt.	PA 7th Inf.	F	9/12/62	9/26/62		
16	Hasenpat, William R		Pvt.	PA 32nd Inf.	F	6/26/63	8/1/63		
17	Hawkins, George C	28	Sgt.	PA 9th Inf.		9/26/62	9/28/62		
18	Hawkins, George C		Pvt.	PA 40th Mil.	E	7/2/63	8/16/63		
19	Hawman, John C	28	1st Lt.	PA 133rd Inf.	C	8/14/62	5/26/63		
20	Hawman, John C		Capt.	PA McKeage's Mil.	D	7/2/63	8/8/63		
21	Hawman, John C		Capt.	PA 22nd Cav.	H	2/26/64	6/8/65		
22	Hayes, William	19	Pvt.	PA 133rd Inf.	K	8/18/62	5/26/63		
23	Hayman, Francis H	23	Pvt.	PA 138th Inf.	F	8/29/62			
24	Hays, Alexander Y	17	Pvt.	PA 110th Inf.	C	10/24/61	4/25/65	Wounded	Gettysburg
25	Hays, Alexander Y		Pvt.	PA 110th Inf.	C	10/24/61	4/25/65	Wounded	Wilderness
26	Hazelett, Moses	19	Pvt.	PA 101st Inf.	B	12/28/61	6/22/65	POW	Plymonth, NC
27	Hazlett, George M	31	Pvt.	PA 76th Inf.	E	9/27/64	6/28/65		
28	Hazlett, Richard	45	Pvt.	PA 56th Inf.	F	10/1/61	5/31/65		
29	Hazzard, David	21	Sgt.	PA 125th Inf.	F	8/12/62	5/18/63		
30	Hazzard, Philip	20	1st Sgt.	PA 76th Inf.	E	10/9/61	11/29/64		
31	Headrick, David	21	Pvt.	PA 8th Res.	F	6/11/61	5/26/64	Wound.-POW	Seven Days Battle
32	Heater, John	30	Pvt.	PA 28th Inf.	A	6/28/61	7/18/65		
33	Heavener, George W	24	Corp.	PA 208th Inf.	K	9/7/64		Died	Petersburg
34	Heavner, Michael	20	Corp.	PA 22nd Cav.	C	7/11/63	2/5/64		
35	Hebner, Daniel H	21	Pvt.	MD 3rd PHB Inf.	B	9/28/61	2/8/65		
36	Hebner, Frederick	23	Pvt.	MD 3rd PHB Inf.	C	11/4/61	11/4/64	Died	
37	Heckman, James	20	Pvt.	PA 138th Inf.	F	8/29/62			
38	Heckman, William	15	Pvt.	PA 107th Inf.	H	2/24/62			
39	Hedding, Ephraim G	42	2nd Lt.	MD 3rd PHB Inf.	H	6/28/61	5/29/65		
40	Hedding, James E	19	Pvt.	PA 22nd Cav.	D	7/16/63	2/5/64		
41	Hedding, James E		Pvt.	MD 3rd PHB Inf.	H	2/11/65	5/29/65		
42	Hedding, Noah	20	Pvt.	MD 3rd PHB Inf.	B	10/12/61	5/29/65		
43	Hedding, Samuel E	18	Pvt.	MD 3rd PHB Inf.	H	2/27/64	5/29/65		
44	Heffleflnger, William*	18	Pvt.	PA 76th Inf.	E	11/21/61	3/29/62		
45	Heffner, Daniel	19	Pvt.	PA 1st Light Art.	D	10/4/64	6/30/65		
46	Heffner, David H	22	Pvt.	PA 167th Inf.	F	11/12/62	8/12/65		
47	Heffner, George		Corp.	PA 8th Res.	F	6/11/61		POW	Seven Days Battle
48	Heffner, George	20	Corp.	PA 8th Res.	F			KIA	2nd Bull Run
49	Heffner, James	29	Pvt.	PA 3rd Heavy Art.	L	9/26/64	10/14/65		
50	Heffner, John	22	Pvt.	PA 125th Inf.	F	8/12/62	5/18/63		
51	Heffner, John F	38	Pvt.	PA 125th Inf.	F	8/8/62		Died	
52	Heffner, Samuel		Pvt.	PA McKeage's Mil.	D	7/2/63	8/8/63		
53	Heffner, Samuel	28	Pvt.	PA 101st Inf.	C	3/10/65	6/25/65		
54	Heinbaugh, Jackson*		Pvt.	PA 212th Inf.				Died	
55	Heinemyer, Adolph*	23	Pvt.	PA 55th Inf.	I	2/27/64		POW-Died	Drewry's Bluff
56	Heinish, James	18	Pvt.	PA 107th Inf.	H	3/12/62			
57	Hellman, Daniel	18	Pvt.	PA 138th Inf.	D	8/29/62		KIA	Mine Run
58	Hellman, George		Pvt.	PA 138th Inf.	D	8/29/62	6/23/65	Wounded	Wilderness
59	Hellman, George	21	Pvt.	PA 138th Inf.	D			Wounded	Petersburg - Final
60	Helm, Edward*		Pvt.	PA 110th Inf.	C	12/19/61	6/28/65		
61	Helm, Frederick		Pvt.	PA 21st Cav.	E	7/2/63	2/20/64		
62	Helm, John B		Sgt.	PA 13th Inf.	G	4/25/61	8/6/61		
63	Helm, John B	27	2nd Lt.	PA 101st Inf.	G	10/8/61	6/25/65	POW	Plymonth, NC
64	Helmit, Thomas J	28	Pvt.	PA 208th Inf.	K	9/7/64	6/1/65		

	Cemetery	Notes
1	Horn Meth. Cem.-Alum Bank	1860 Woodbury Twp. Census; Frank McCoy 1912 Listing
2		B.S.F. History Book-1884; In Hospital 10/30/62 to 5/19/63
3		MIA 5/3/63; PA Civil War Archives-Woodbury residence; George & John-brothers
4	Friends Cove UCC Cem.	Bedford Inquirer 5/22/1908 listing
5	Greenlawn Cem.-Roaring Spring	POW 6/1/64 to 4/28/65; B.S.F. History Book-1884; 1860 Woodbury Twp. Census
6		1890 Everett Veterans Census
7	Pleasant Vale Cumberland-TN	1860 Bedford Twp. Census; aka Hartzel
8	Bedford Cem.	1890 Bedford Twp. Veterans Census; 1850 Colerain Twp. Census
9	Arlington Nat'l Cem., 13-5638	Arm Amputated 6/23/64; Died 6/30/64; B.S.F. History Book-1884; Recruited in Harrisburg
10	Bedford Cem.	B.S.F. History Book-1884; 1890 Bedford Twp. Veterans Census
11		B.S.F. History Book-1884; York Co. references
12	Green Mt. Cem.-Ligonier	B.S.F. History Book-1884
13		B.S.F. History Book-1884; Historical Data Systems record; 1850 Liberty Twp. Census
14	Council Cem.-NE	B.S.F. History Book-1884; 1850 Hopewell Twp. Census
15	Bedford Cem.	Historical Data Systems record
16		Historical Data Systems record
17	Bedford Cem.	Historical Data Systems record
18		1890 Bedford Twp. Veterans list; Historical Data Systems record
19	Everett Cem.	Historical Data Systems-Hopewell residence
20		B.S.F. History Book-1884
21		B.S.F. History Book-1884
22		B.S.F. History Book-1884; Historical Data Systems record; 1860 Snake Spring Twp. Census
23		Deserted 11/6/62; B.S.F. History Book-1884; Historical Data Systems record
24	East Admah Cem.-NE	Wounded 7/2/63; Soldiers of Blair County PA book - 1940; 1850 Hopewell Twp. Census
25		Leg Amputated 5/6/64; Historical Data Systems-Yellow Creek residence
26	St. Pauls Cem.-Cessna	POW 4/20/64 to 2/26/65; Historical Data Systems record; aka Hazlet
27	Mt. Sinai Cem.-Blue Knob	B.S.F. History Book-1884; Historical Data Systems record
28	Ickes Farm-E. St. Clair Twp.	Bedford Inquirer 1908; Mexican War Veteran
29		1890 Liberty Twp. Veterans Census
30	Bedford Cem.	PA Civil War Archives-Bedford residence; B.S.F. History Book-1884
31	Headrick Union-Cambria Co.	POW & Wounded 6/27/62; Historical Data Systems-Bedford Co. residence
32	Replogle Cem.-Woodbury	1890 S. Woodbury Twp. Veterans Census; Historical Data Systems record
33	Arlington Nat'l Cem., 13-10146	Accident 4/1/65; Died 5/9/65, Historical Data Systems record; 1860 Monroe Twp. Census
34	Mt. Zion Cem.-OK	B.S.F. History Book-1884; 1860 Monroe Twp. Census
35	Buck Valley Cem.-Fulton Co.	1890 Brush Creek Twp. Veterans Census
36	Buck Valley Cem.-Fulton Co.	Died 8/6/65; Historical Data Systems record aka Hepner
37		Deserted 9/27/62; B.S.F. History Book-1884
38	Wells Valley Meth. Cem.	Deserted 8/30/62; B.S.F. History Book-1884
39	Everett Cem.	Everett Cemetery Listing of Civil War Veterans; Father of Noah, Samuel & James
40		Historical Data Systems record; aka Jason E
41		Civil War Pension Record; 1880 Everett Census
42		Ancestry.com information; Noah, Samuel & James-brothers
43	Rose Hill Cem.-Altoona	Historical Data Systems record
44		Historical Data Systems-Somerset references; Discharged on Surgeon's Cert.
45	Fockler Cem.-Saxton	Frank McCoy 1912 Listing; 1850 Middle Woodbury Twp. Census
46	Lehigh Zion Cem.	Historical Data Systems record; 1860 Hopewell Twp.(Hunt. Co.) Cen. David & George-bros.
47		POW 6/27/62; Bedford Gazette-8th Reserve listing; Historical Data Systems record
48	Fockler Cem.-Saxton	KIA 8/29/62; Historical Data Systems-Bedford Co. residence
49	Fockler Cem.-Saxton	1890 Hopewell Veterans Census; Discharged on Surgeon's Cert.
50	Fockler Cem.-Saxton	1890 Saxton Veterans Census; John, James, Samuel & Daniel-brothers
51	Arlington Nat'l Cem., 13 11104	Died 3/5/63; 1890 Saxton Veterans Census; 1860 Liberty Twp. Census; reinterred in Arlington
52		B.S.F. History Book-1884
53	Bedford Forge-Yellow Creek	B.S.F. History Book-1884
54		B.S.F. History Book-1884; Somerset Co. references
55	Andersonville Cem., 10903	MIA 5/16/64; POW-Andersonville; B.S.F. History Book-1884; Somerset Co. references
56		Deserted 8/30/62; PA Civil War Archives-Bedford Co. residence
57		KIA 11/27/63; B.S.F. History Book-1884; PA Civil War Archives-Bedford Co. residence
58		Wounded 5/6/64; B.S.F. History Book-1884; Historical Data Systems record
59		Wounded 4/2/65; PA Civil War Archives-Bedford Co. residence
60		B.S.F. History Book-1884; Deserted 3/24/64; Huntingdon Co. references
61		Historical Data Systems-Bedford Co. residence
62		1890 Bedford Twp. Veterans Census; B.S.F. History Book-1884
63	Bedford Cem.	POW-Danville, VA & Macon; 4/20/64; Escaped 2/5/65; B.S.F. History Book-1884
64	Cowan Creek Cem.-OH	B.S.F. History Book-1884; Historical Data Systems-Bedford Co. residence

Chapter 18

	Name	Age	Rank	Regiment	Co.	Muster In	Muster Out	Casualty	Casualty Battle
1	Helsel, Adam	41	Pvt.	OH 91st Inf.	C	9/7/62	6/24/65		
2	Helsel, Edward	35	Pvt.	PA 76th Inf.	C	10/17/61	11/28/64		
3	Helsel, Henry S	35	Pvt.	PA 3rd Inf.	G	4/20/61	7/29/61		
4	Helsel, Henry S		Pvt.	PA 76th Inf.	E	10/18/64	6/23/65		
5	Helsel, John F	20	Pvt.	PA 54th Inf.	A	3/4/65	5/31/65	POW	High Bridge, VA
6	Heltsel, John D	32	Pvt.	PA 82nd Inf.	A	11/14/64	7/13/65		
7	Heltzel, Daniel G	21	Pvt.	PA 138th Inf.	E	8/29/62		Died	
8	Heltzel, David S	21	Pvt.	PA 133rd Inf.	E	8/13/62	5/26/63		
9	Heltzel, Jonathan D	19	Pvt.	PA 110th Inf.	C	10/24/61		KIA	Wilderness
10	Heltzel, Simon		Pvt.	PA 138th Inf.	E	9/29/62		Died	
11	Heltzel, William	18	Pvt.	PA 138th Inf.	E	8/29/62			
12	Helzel, Joseph	23	Pvt.	PA 12th Cav.	G	8/31/64	6/1/65		
13	Heman, George	29	Pvt.	PA 57th Inf.	B	3/10/64	6/29/65		
14	Hemming, Augustus	14	Pvt.	PA 2nd Cav.	E	12/20/61		Died	
15	Hemming, C H								
16	Hemming, Joseph		Pvt.	PA Inf.	E				
17	Hemming, William	38	Pvt.	PA 2nd Cav.	E	11/4/61	1864		
18	Hemminger, Abraham	16	Pvt.	PA 18th Cav.	K	2/25/64	10/31/65		
19	Hendershot, Charles		Pvt.	MD 3rd PHB Inf.	K	7/18/63	5/29/65		
20	Hendershot, Samuel C	27	Pvt.	MD 3rd PHB Inf.	C	10/12/61	5/29/65	POW	Harpers Ferry
21	Henderson, Robert F	30	Pvt.	PA 138th Inf.	F	4/9/63			
22	Henershitz, William*		Pvt.	PA 194th Inf.	I	7/22/64	11/5/64		
23	Henry, Daniel B	18	Corp.	PA 55th Inf.	I	9/20/61	8/30/65		
24	Henry, George A	20	Capt.	IA 4th Inf.	A	8/15/61	7/24/65	Wounded	Pea Ridge, AK
25	Henry, Horner		Pvt.	PA 142nd Inf.	D	8/1/62	5/29/65	POW	
26	Henry, James W	17	Sgt. Maj.	IA 15th Inf.	E	12/1/61	7/24/65	Wounded	Corinth, MS
27	Henry, James W		Sgt. Maj.	IA 15th Inf.	E			Wound.-POW	Atlanta, GA
28	Henry, John	42	Sgt. QM	PA 22nd Cav.	B	7/11/63	2/5/64		
29	Henry, John B	20	Pvt.	IA 2nd Inf.	F	5/27/61	7/12/65		
30	Henry, Porter W	23	Pvt.	IA 1st Cav.	C	7/31/61	11/1/62		
31	Herald, Jacob		Pvt.	PA 77th Inf.	F	3/2/65	12/6/65		
32	Herring, George W	44	Corp.	PA 55th Inf.	K	11/5/61	6/26/63	Died	
33	Herring, John B	44	Pvt.	PA 99th Inf.	F	2/25/65	7/1/65		
34	Hershberger, S Henry			PA 47th Inf.	C				
35	Hershey, James	20	Pvt.	PA 76th Inf.	E	7/2/64	7/18/65		
36	Hess, Benjamin	19	Pvt.	PA 55th Inf.	H	2/19/64	8/30/65		
37	Hess, Daniel A	24	2nd Lt.	PA 55th Inf.	H	10/11/61		Wound.-Died	Sailor's Creek
38	Hess, Daniel A		2nd Lt.	PA 55th Inf.	H			Wounded	Cold Harbor
39	Hess, Jacob	15	Pvt.	MD 3rd PHB Inf.	B	2/12/62	2/12/65		
40	Hess, John	16	Pvt.	IL 50th Inf.	A	9/12/61	7/13/65		
41	Hess, John W	32	Pvt.	PA 127th Inf.	C	8/9/62	5/29/63		
42	Hess, John W		Pvt.	PA 201st Inf.	G	8/26/64	6/21/65		
43	Hess, Samuel B	23	Corp.	IL 50th Inf.	A	9/12/61	7/13/65		
44	Hester, John		Pvt.	PA 198th Inf.	B	1/25/65			
45	Hetrick, Daniel L	16	Pvt.	PA 101st Inf.	D	11/1/61	6/13/65	POW	Plymonth, NC
46	Hetrick, Jacob T	22	Pvt.	PA 99th Inf.	D	2/25/65	7/1/65		
47	Hetrick, Samuel G	20	Pvt.	PA 194th Inf.	I	7/22/64	11/5/64		
48	Hickok, Edwin H	20	2nd Lt.	PA 76th Inf.	E	10/9/61	10/17/63		
49	Hicks, Abner		Pvt.	PA McKeage's Mil.	G	7/2/63	8/8/63		
50	Hicks, Jackson*	33	Pvt.	PA 110th Inf.	C	10/24/61	1/15/65		
51	Hildebrand, Alexander	27	Pvt.	PA 13th Inf.	G	4/25/61	8/6/61		
52	Hildebrand, Isaac	23	Corp.	PA McKeage's Mil.	H	7/2/63	8/8/63		
53	Hileman, Adolphus P	34	Pvt.	PA 13th Cav.	A	2/22/65	7/14/65		
54	Hileman, John	27	Pvt.	PA 55th Inf.	K	11/5/61		Died	
55	Hill, Aaron	22	Pvt.	PA 13th Inf.	G	4/25/61	8/6/61		
56	Hill, Aaron		Corp.	PA McKeage's Mil.	D	7/3/63	8/8/63		
57	Hill, Aaron		Pvt.	PA 6th Heavy Art.	H	9/2/64	3/3/65		
58	Hill, Amos	37	Pvt.	PA 29th Inf.	I	11/28/64	7/17/65		
59	Hill, Henry		Pvt.	PA 136th Inf.	K	8/27/62	1/5/63		
60	Hill, Tolbert A			MD 3rd PHB Inf.					
61	Hill, Tolbert A	17	Pvt.	MD 1st PHB Cav.	G	3/23/64	6/28/65		
62	Hill, William H	44	Pvt.	PA 136th Inf.	K	8/27/62	1/5/63		
63	Hill, William M	31	Pvt.	PA 82nd Inf.	C	9/21/64	6/17/65		
64	Hillebrandt, Henry	21	Corp.	PA 55th Inf.	K	2/19/64	5/3/65		

	Cemetery	Notes
1	Dayton Nat'l Cem.-OH	Historical Data Systems record; Born-Bedford Co.
2	Hopewell Cem.	Bedford Inquirer 5/22/1908 listing
3	Mt. Moriah Cem.-Blue Knob	Historical Data Systems record
4		B.S.F. History Book-1884; Discharged on Surgeon's Certificate
5	St. Patricks Cem.- Newry	POW 4/6/65-Pension record; 1890 Claysburg Veterans Census
6	Pleasant Hill Cem.-Imlertown	Frank McCoy 1912 Listing
7	Camp Relay-MD	Died-Typhoid Fever 11/1/62; PA Civil War Arch.-St. Clairsville resid.; Daniel & William-bros.
8	Oak Hill Cem.-OH	Historical Data Systems record; 1860 Bedford Census
9		KIA 5/6/64; PA Civil War Archives-Woodbury residence; 1860 S. Woodbury Census
10		Died 11/9/62; B.S.F. History Book-1884; PA Civil War Archives-Bedford Co. residence
11		Deserted 6/13/63; PA Civil War Archives-St. Clairsville residence
12	Mt. Moriah Cem.-Blue Knob	PA Civil War Archives record
13		PA Civil War Archives-Bedford Co. residence
14	US Soldiers & Airmans Nat'l	Died 11/20/63; B.S.F. History Book-1884; 1850 Cumberland Valley Census
15	Reformed Cem.-Bedford	Bedford Inquirer 5/22/1908 listing
16		1890 Bedford Twp. Veterans Census
17		B.S.F. History Book-1884
18	Mt. Olivet Cem.-Manns Choice	1890 Harrison Twp. Veterns Census
19	Buck Valley Cem.-Fulton Co.	1890 Brush Creek Twp. Veterans Census; Charles & Samuel-brothers
20	Buck Valley Cem.-Fulton Co.	POW 9/15/62; 1890 Brush Creek Twp. Veterans Census
21		B.S.F. History Book-1884; Deserted 6/29/63
22		B.S.F. History Book-1884; enlisted in Reading
23		B.S.F. History Book-1884; Historical Data Systems-Bedford Co. residence
24	Los Angeles Nat'l Cem.	Severe Wound 10/3/62; Ancestry.com-born Bedford Co.
25		POW-Libby; 1890 Napier Twp. Veterans Census; POW for 3 months
26		Wounded 3/7/62; Ancestry.com-born Bedford Co.
27		Wounded-POW 7/22/64 to 11/10/64; Historical Data Systems record
28	Hetrick Cem.-Woodbury	1890 S. Woodbury Veterans Census
29		John B, George, James & Porter W.-brothers; Ancestry.com : Born in Bedford Co.
30		Ancestry.com: Born-Bedford Co.; Discharged on Surgeon's Cert.
31		1890 Liberty Twp. Veterans Census
32	German Reformed-Bedford	Died 8/25/63 after DSC for Chronic Diarreah; B.S.F. History Book-1884; 1860 Bedford Cen.
33	Everett Cem.	B.S.F. History Book-1884; 1860 W. Providence Twp. Census
34	Everett Cem.	Everett Cemetery Listing of Civil War Veterans
35		B.S.F. History Book-1884
36	Sherbine Cem.-Cambria Co.	B.S.F. History Book-1884; 1860 St. Clair Twp. Census; Benjamin & Daniel-brothers
37	Stone Church Cem.-Fishertown	Wounded 4/6/65 at Rice's Station, Died 4/20/65; Historical Data Systems-Bedford Co. resid.
38		Wounded 6/3/63 listed in Bedford Inquirer 7/8/64 article; B.S.F. History Book-1884
39	Pleasant Grove-Warfordsburg	1890 Brush Creek Twp. Veterans Census; 1850 Monroe Twp. Census
40	Keath Cem.-IL	John & Samuel-brothers; Ancestry.com-born in Bedford Co.
41	Jeannette Mem. Park Cem.	1890 Bedford Twp. Veterans Census
42		B.S.F. History Book-1884; 1880 Manns Choice Census
43	Memory Lane Cem.-OK	Ancestry.com-born in Schellsburg
44		1890 S. Woodbury Veterans Census
45	Pleasantville Cem.	POW-Andersonville & Florence 4/20 to 12/11/64; attended 1905 Andersonville Dedication
46	Ritchey Cem.-Snake Spring V.	1850 S. Woodbury Census; 2nd PA Cavalry Mentioned
47		B.S.F. History Book-1884
48		B.S.F. History Book-1884; Historical Data Systems-Bedford Co. residence
49		B.S.F. History Book-1884
50		B.S.F. History Book-1884; Huntingdon Co. references
51		B.S.F. History Book-1884; Historical Data Systems-Bedford residence
52		B.S.F. History Book-1884
53	Fairview Cem.-Altoona	Ancestry.com-Born & Married in Londonderry Twp.
54		Died 12/1/61; B.S.F. History Book-1884; PA Civil War Archives-Bedford Co. residence
55		B.S.F. History Book-1884; 1850 W. Providence Twp. Census
56		B.S.F. History Book-1884
57		Historical Data Systems record; Discharged on Surgeon's Certificate
58	Oak Lawn Cem.-Baltimore	1890 Brush Creek Twp. Veterans Census
59	Hyndman Cem.	Frank McCoy 1912 Listing
60		Frank McCoy 1912 Listing
61	Fairview Cem.-Artemas	Historical Data Systems lists 1st MD PHB Cavalry
62	Hyndman Cem.	Historical Data Systems record
63	Trinity UCC Cem.-Juniata Twp.	1890 Harrison Twp. Veterans Census
64		B.S.F. History Book-1884; 1860 Juniata Twp. Census; Discharged on Surgeon's Cert.

Chapter 18

	Name	Age	Rank	Regiment	Co.	Muster In	Muster Out	Casualty	Casualty Battle
1	Hillegass, Andrew	23	Pvt.	MD 2nd PHB Inf.	H	4/9/63	5/29/65		
2	Hillegass, Frederick	41		PA 107th Inf.	D	9/21/64	12/27/64		
3	Hillegass, Frederick J	21	Pvt.	MD 2nd PHB Inf.	D	3/3/65	5/29/65		
4	Hillegass, Henry P	30	Pvt.	PA 55th Inf.	H	2/29/64	8/30/65	Wounded	Petersburg - Initial
5	Hillegass, John C	22	Pvt.	PA 55th Inf.	H	2/29/64	12/30/64	Wounded	Petersburg - Initial
6	Hills, John E*	38	Corp.	PA 76th Inf.	E	9/26/64	6/28/65		
7	Hills, Thaddeus*	32	Pvt.	PA 76th Inf.	E	9/26/64	11/23/64		
8	Himes, Adam*	36	Pvt.	PA 76th Inf.	E	2/22/64	7/18/65	Wounded	2nd Deep Bottom
9	Himes, Andrew J			US 5th Art.	A				
10	Himes, Andrew J	29	Pvt.	PA 3rd Heavy Art.	U	3/8/64		Died	
11	Himes, George	37	Pvt.	PA 186th Inf.	G	3/24/64	8/15/65		
12	Himes, John			PA 133rd Inf.					
13	Himes, John	32	Pvt.	PA 208th Inf.	H	9/8/64	6/1/65		
14	Himes, Oliver	19	Pvt.	PA 99th Inf.	G	3/16/65	7/1/65		
15	Himes, Samuel	41	Pvt.	PA 186th Inf.	E	2/24/64	8/15/65		
16	Himes, Thomas	37	Pvt.	PA 91st Inf.	I	9/21/64	5/30/65		
17	Himes, William	54	Pvt.	PA McKeage's Mil.	D	7/2/63	8/8/63		
18	Himes, William		Pvt.	PA 7th Cav.	M	9/15/64	6/23/65		
19	Hiner, John E		Pvt.	PA 67th Inf.	C	1864		Wound.-Died	Cedar Creek
20	Hiner, John M		Pvt.	PA 99th Inf.	F	2/25/65	5/31/65		
21	Hiner, William	47	Pvt.	IA 33rd Inf.	D	9/6/62	5/9/63		
22	Hinish, George W	29	Pvt.	PA 199th Inf.	E	10/1/64	6/28/65		
23	Hinish, Jacob H	22	Pvt.	PA 199th Inf.	E	10/1/64	6/28/65		
24	Hinish, Thomas P	18	Pvt.	PA McKeage's Mil.	G	7/2/63	8/8/63		
25	Hinson, William		Pvt.	MS 10th Inf.		3/1/61	3/27/62		
26	Hinson, William	19	Pvt.	MS 1st Light Art.	I	4/30/62		POW	Vicksburg, MS
27	Hissong, Josiah	22	Pvt.	PA 13th Inf.	G	4/25/61	8/6/61		
28	Hissong, Josiah		2nd Lt.	PA 55th Inf.	H	9/22/61	6/6/65	Wounded	Chaffin's Farm
29	Hissong, Josiah		2nd Lt.	PA 55th Inf.	H			Wounded	Drewry's Bluff
30	Hissong, Josiah		2nd Lt.	PA 55th Inf.	H			Wounded	White Oak Road
31	Hite, Benjimen F	16	Pvt.	MD 2nd PHB Inf.	I	5/18/62	5/1/65		
32	Hite, Daniel	21	Pvt.	PA 200th Inf.	C	8/25/64	7/17/65		
33	Hite, David	15	Pvt.	PA 101st Inf.	G	5/3/64	6/25/65		
34	Hite, David H	21	Pvt.	PA 148th Inf.	C	6/10/63	2/14/64		
35	Hite, George	40	Pvt.	PA 87th Inf.	A	6/18/64	6/29/65		
36	Hite, George W	16	Pvt.	PA 210th Inf.	C	9/8/64	5/30/65		
37	Hite, Jacob (1)	26	Pvt.	IA 28th Inf.	D	9/4/62	7/4/64	Wounded	Champion Hill
38	Hite, Jacob (2)	30	Pvt.	PA 18th Cav.	K	10/27/62	11/1/62		
39	Hite, Jacob A	23	Corp.	PA 171st Inf.	H	11/1/62	8/8/63		
40	Hite, Jacob A		Pvt.	PA 101st Inf.	G	5/3/64	6/25/65		
41	Hite, John	34	Pvt.	PA 205th Inf.	I	8/17/64		KIA	Petersburg - Final
42	Hite, Joseph	19	Pvt.	MD 2nd PHB Inf.	I	10/31/61		Died	
43	Hite, Nicholas		Pvt.	IA 28th Inf.	D	9/4/62	5/8/65		
44	Hite, Perry	17	Pvt.	PA 2nd Cav.	E	6/5/62		Died	
45	Hite, Samuel C	29	Pvt.	PA 55th Inf.	A	2/23/64		POW-Died	
46	Hixon, Akers J	23	Sgt.	PA 101st Inf.	D	11/1/61		Died	
47	Hixon, Amos	33	Pvt.	PA 82nd Inf.	I	11/28/64	7/13/65		
48	Hixon, Erastus J	21	Corp.	PA 138th Inf.	D	8/29/62		KIA	Wilderness
49	Hixon, Henry H	21	Pvt.	PA 11th Inf.	A	9/30/61	7/1/65	Wounded	Fredericksburg
50	Hixon, Henry H		Pvt.	PA 11th Inf.	A			Wound.-POW	Gettysburg
51	Hixon, Henry H		Pvt.	PA 11th Inf.	A			Wounded	Petersburg - Initial
52	Hixon, Jared H	17	Pvt.	PA 22nd Inf.	M	8/9/64	5/24/65		
53	Hixon, Joel B	18	Pvt.	PA 101st Inf.	D	11/1/61		Died	
54	Hixon, Perry	21	Pvt.	PA 22nd Cav.	M	2/26/64	6/24/65		
55	Hixon, William*	54	1st Sgt.	PA 52nd Inf.	I	7/9/63	9/1/63		
56	Hixson, Aquilla	25	Pvt.	PA 158th Inf.	H	11/1/62			
57	Hixson, George W		Pvt.	PA 22nd Cav.	I	7/29/63	6/26/65		
58	Hixson, Lewis B	28	Pvt.	PA 186th Inf.	G	3/3/64	8/15/65		
59	Hobbs, Nelson*	37	Pvt.	PA 76th Inf.	E	2/18/65		Died	
60	Hochard, John A	18	Pvt.	PA 138th Inf.	D	8/29/62	6/23/65	Wounded	Mine Run
61	Hochard, John A		Pvt.	PA 138th Inf.	D			Wounded	Wilderness
62	Hockenbaugh, John*								
63	Hockenberry, John	20	Pvt.	PA 55th Inf.	I	9/20/61	8/30/65		
64	Hockenberry, Jonathan	18	Pvt.	PA 107th Inf.	K	10/6/64		Died	

	Cemetery	Notes
1	Trinity UCC Cem.-Juniata Twp.	1890 Juniata Twp. Veterans Census
2	Schellsburg Cem.	1860 Harrison Twp. Census
3	Schellsburg Cem.	1860 Juniata Twp. Census; Frederick J & Andrew-brothers
4	Hyndman Cem.	Wounded 6/15/64 listed in Bedford Inquirer 7/8/64 article; 1890 Juniata Twp. Veterans Census
5	Schellsburg Cem.	Wounded 6/18/64 listed in Bedford Inquirer 7/8/64 article; 1860 Juniata Twp. Census
6		B.S.F. History Book-1884; Bradford Co. references
7		B.S.F. History Book-1884; Bradford Co. references
8		B.S.F. History Book-1884; Blair Co. references
9		Pension Record
10	Blackheart Farm Cem.-Everett	Died of Dysentery 10/7/64; Frank McCoy 1912 Listing
11	New Enterprise Cem.	Bedford Inquirer 5/22/1908 listing
12		Frank McCoy 1912 Listing; John, Andrew and Samuel-brothers
13	Bunker Hill Cem.-Saxton	1890 Liberty Twp. Veterans Census; Historical Data Systems record
14		1890 Southampton Veterans Census; Historical Data Systems record
15	Mt. Zion Cem.-Breezewood	1890 E. Providence Twp. Veterans Census; 1860 E. Providence Twp. Census
16	Everett Cem.	Frank McCoy 1912 Listing; 1860 Southampton Twp. Census
17		B.S.F. History Book-1884; enlisted in Bloody Run
18		B.S.F. History Book-1884; Ancestry.com born-Woodbury Twp.; aka Hymes
19	Arlington Nat'l Cem., 13 10494	Wounded 10/19/64; Died 4/23/65; Bed. Gazette 1914 list aka Edmund; 1860 Napier Twp. Cen.
20	New Paris Comm. Ctr. Cem.	Frank McCoy 1912 Listing; 1850 Napier Twp. Census; John M & William-brothers
21	Sandyvale Cem.-Johnstown	1850 Napier Twp. Census; Discharged on Surgeon's Cert.
22	Elmwood Cem.-IA	B.S.F. History Book-1884; 1860 E. Providence Twp. Census; George & Jacob-brothers
23	Graceville Luth. Cem.	1890 E. Providence Twp. Veterans Census; B.S.F. History Book-1884
24	Riverside Cem.-OH	B.S.F. History Book-1884; 1860 Hopewell Twp. Census
25		Record provided by Bob Grenke, Hinson/Niley family historian
26	Schellsburg Cem.	Findagrave.com record; POW Ft. Alton, IL; Escaped POW 10/21/63; aka Oliver Niley
27	Schellsburg Union Cem.	B.S.F. History Book-1884
28		Wounded 9/29/64; B.S.F. History Book-1884
29		Wounded 5/16/64
30		Wounded 3/30/65; 1890 Napier Twp. Veterans Census
31	Hyndman Cem.	Historical Data Systems record; 1860 Cumberland Valley Census
32	Greenlawn Cem.-Roaring Spring	1860 Union Twp. (Pavia) Census
33	Grandview Cem.-Cambria Co.	B.S.F. History Book-1884; 1860 Napier Twp. Census
34	Riverside Cem.-NE	David H, John, Samuel & George-brothers
35	Newry Lutheran Cem.	1840 Greenfield Twp. Census
36	Chestnut Hill Cem.-Connellsville	B.S.F. History Book-1884; 1860 Cumberland Valley Census
37	Evergreen Cem.-IA	Wounded 5/16/63; 1840 Union Twp. (Pavia) Census
38		Deserted 11/1/62; PA Civil War Archives-enlisted in Bedford Co.
39	Sandyvale Cem.-Johnstown	PA Civil War Project Record; Jacob A & David-brothers
40		B.S.F. History Book-1884; PA Civil War Archives-born Bedford Co.
41	Meade Station-VA	KIA 4/2/65; Historical Data Systems record; 1840 Greenfield Twp. Census
42		Died 7/7/64; Historical Data Systems record; 1860 Cumberland Valley Census
43	Evergreen Cem.-ID	Nicholas & Jacob (1)-brothers
44	US Soldiers & Airmans Nat'l	Died 11/2/62; Hist. Data Systems record; Bed. Gazette 1914 list; 1860 Cumberland Valley Cen.
45	Andersonville Cem., 3379	POW-Andersonville; Died 7/16/64; Historical Data Systems record; 1840 Greenfield Twp. Cen.
46	Poplar Grove Nat'l Cem.	Died of Disease 7/21/62; 1860 Brush Creek Twp. Census; Akers, Joel & Erastus-brothers
47	McKendree Cem.-Crystal Spring	1890 Brush Creek Twp. Veterans Census
48	Fredericksburg Nat'l Cem.	KIA 5/6/64; 1860 E. Providence Twp. Census; Historical Data Systems record
49	Sparks Family Cem.-Everett	Wounded 12/13/62; 1850 E. Providence Twp. Census
50		Wounded & POW 7/1/63; Historical Data Systems record
51		Wounded 6/18/84
52	Inglewood Park Cem.-CA	Historical Data Systems record aka Jerrard Hixon
53		Died of Disease 10/15/62; 1860 Brush Creek Twp. Census; Historical Data Systems record
54	Lanark City Cem.-IL	Ancestry.com information; Perry, Lewis, Aquilla & Henry-brothers
55		Historical Data Systems record; 1870 Monroe Twp. Census
56	Shiloh Park Cem.-OH	Ancestry.com information; Deserted 6/6/63; aka Hixson
57		B.S.F. History Book-1884
58	Clayton Cem.-KS	Ancestry.com information; 1860 Monroe Twp. Census; aka Hixon
59		Died 4/12/65; B.S.F. History Book-1884; enlisted in Scranton
60	St. John Cem.-New Baltimore	Wounded 11/27/63; Historical Data Systems-Bedford Co. residence
61		Wounded 5/6/64; B.S.F. History Book-1884
62		Bedford Inquirer 5/22/1908 listing; see Hockenberry
63	Hopewell Cem.	B.S.F. History Book-1884
64	Poplar Grove Nat'l Cem.	Died-Fever 12/31/64; Bedford Gazette 2/13/1914 listing; 1860 Juniata Twp. Census

	Name	Age	Rank	Regiment	Co.	Muster In	Muster Out	Casualty	Casualty Battle
1	Hockenberry, Samuel*	23	Pvt.	PA 55th Inf.	I	9/20/61		Died	
2	Hoenstine, Benjamin F	20	Pvt.	PA 138th Inf.	E	8/29/62	6/6/65		
3	Hoenstine, David	18	Pvt.	PA 138th Inf.	E	8/29/62		Died	
4	Hoerkens, Werner	43		IN 35th Inf.	K	1/13/65	9/30/65		
5	Hoffman, Charles A	21	Pvt.	PA 23rd Inf.	B	8/2/61			
6	Hoffman, Jacob	18	Pvt.	PA 76th Inf.	E	11/21/61		Died	
7	Hoffman, John (1)	23	Pvt.	PA 101st Inf.	G	12/28/61		Died	
8	Hoffman, John (2)		Pvt.	MD 3rd Inf.	A	7/15/61	6/23/65		
9	Hoffman, John O	20	Pvt.	PA 133rd Inf.	K	8/15/62	5/26/63		
10	Hoffman, John O		Pvt.	PA McKeage's Mil.	D	7/2/63	8/8/63		
11	Hoffman, John O		Pvt.	PA 18th Cav.	H	3/18/64	6/3/65		
12	Hogan, James	40	Pvt.	PA 55th Inf.	D	11/17/61	2/27/64		
13	Hogan, John	19	Pvt.	PA 55th Inf.	D	10/12/61			
14	Holdcraft, William	38	Pvt.	PA 8th Res.	F	6/11/61	5/26/64	Wound.-POW	Seven Days Battle
15	Holdcraft, William		Pvt.	PA 8th Res.	F			Wounded	Spotsylvania
16	Hollabaugh, George W	34	Pvt.	PA 88th Inf.	D	3/15/64	6/30/65		
17	Hollar, Philip V	40	Pvt.	PA 208th Inf.	K	9/7/64	5/16/65	Wounded	Ft. Stedman
18	Holler, Alexander	31	Pvt.	MD 2nd PHB Inf.	B	2/24/65	5/29/65		
19	Holler, George W	36	Pvt.	PA 138th Inf.	F	8/29/62	6/13/65	Wounded	Wilderness
20	Holler, George W		Pvt.	PA 138th Inf.	F			Wounded	Opequon
21	Holler, James M	18	Pvt.	PA 55th Inf.	K	11/11/61		Died	
22	Holler, John M	15	Pvt.	PA 138th Inf.	F	8/29/62	6/23/65		
23	Holler, Joseph M	34	Pvt.	PA 171st Inf.	I	11/1/62	8/1/63		
24	Holler, Joseph M		Pvt.	MD 2nd PHB Inf.	B	3/21/65	5/29/65		
25	Holler, Samuel	32	Pvt.	PA 49th Inf.	H	9/25/63			
26	Holler, William H	36	Pvt.	PA 82nd Inf.	C	11/14/64	7/13/65		
27	Hollinger, Stephen	17	Pvt.	USCT 43rd Inf.	A	2/27/64	2/20/65	Wound.-Died	Battle of the Crater
28	Hollingshead, Oliver S	16	Pvt.	PA 22nd Cav.	F	2/5/64	6/24/65		
29	Holmes, Philip	19	Pvt.	USCT 45th Inf.	D	7/6/64	11/11/64		
30	Holsinger, Emanuel	26	Pvt.	IL 142nd Inf.	D	6/18/64	10/26/64		
31	Holsinger, Frank	25	Pvt.	PA 8th Res.	F	6/11/61	2/11/64	Wounded	Fredericksburg
32	Holsinger, Frank		Capt.	USCT 19th Inf.	I	3/7/64	1/15/67	Wounded	2nd Berm. Hundred
33	Holsinger, Josiah		Pvt.	PA 110th Inf.	C	10/24/61	5/30/65	POW	Cold Harbor
34	Holsinger, Josiah	19	Pvt.	PA 110th Inf.	C	10/24/61	5/30/65	Wounded	Gettysburg
35	Holsinger, Levi R	21	Corp.	IL 34th Inf.	H	1861	1864	Wounded	Murfreesboro
36	Holsinger, Levi R		Sgt.	IL 78th Inf.	H				
37	Holt, William	24	Pvt.	PA 55th Inf.	D	10/14/63			
38	Holtzmon, David*	18	Pvt.	PA 55th Inf.	H	8/26/62	6/11/65		
39	Homan, James R*		Pvt.	PA 42nd Inf.	D	7/6/63	8/12/63		
40	Homan, James R*	18	Corp.	PA 194th Inf.	I	7/22/64	11/5/64		
41	Homan, William	27	Corp.	PA 125th Inf.	F	8/12/62	5/18/63		
42	Homer, Harry		Pvt.	PA 20th Cav.	E	1863	1864		
43	Homes, William F		1st Sgt.	PA 172nd Inf.	E	11/1/62	8/1/63		
44	Hook, Elias	22	Pvt.	PA 171st Inf.	I	11/10/62	8/8/63		
45	Hook, Elias			PA 138th Inf.					
46	Hook, George	24	Pvt.	PA 171st Inf.	I	11/10/62	8/8/63		
47	Hook, George			PA 138th Inf.					
48	Hook, James	27	Pvt.	PA 171st Inf.	I	11/10/62	8/8/63		
49	Hook, James		Pvt.	PA 79th Inf.	D	9/21/64	7/12/65		
50	Hook, John		Pvt.	PA 56th Inf.	G				
51	Hook, William	23	Pvt.	PA 171st Inf.	I	11/2/62		Died	
52	Hoon, Stacey	30	Pvt.	MD 2nd PHB Inf.	A	3/21/65	6/19/65		
53	Hoopingardner, George	22	Pvt.	PA 82nd Inf.	A	11/14/64	7/13/65		
54	Hoopingardner, Joseph	41	Pvt.	PA 208th Inf.	H	9/6/64	6/1/65		
55	Hoover, Daniel	55	Pvt.	KY 8th Inf.	B	1/30/62			
56	Hoover, Elias		Pvt.	PA 54th Inf.	G				
57	Hoover, John L	21	Pvt.	PA 1st Mil.	K	9/11/62	9/25/62		
58	Hoover, John L		Pvt.	PA 19th Cav.	L	9/9/63	6/30/65		
59	Hoover, Jonathan	23	Pvt.	PA 50th Inf.	I	2/27/65	8/5/65		
60	Hoover, Martin	24	Pvt.	PA 171st Inf.	I	11/2/62		Died	
61	Hoover, Nathaniel	19	Pvt.	PA 55th Inf.	H	2/6/64		Died	
62	Hoover, Rudolph	39	Pvt.	PA 5th Heavy Art.	K	9/1/64	6/30/65		
63	Hoover, Thomas G	20	Pvt.	PA 55th Inf.	H	2/29/64	8/30/65		
64	Hopkins, James		Pvt.	PA McKeage's Mil.	G	7/2/63	8/8/63		

	Cemetery	Notes
1		Died 9/10/64; B.S.F. History Book-1884; enlisted in Blair Co.
2	Cuppett Family-New Paris	In Hospital 9/19/64 to 6/6/65; 1890 Napier Twp. Veterans Census
3		Died 11/4/62; Historical Data Systems-Bedford Co. residence; 1860 Union Twp. (Pavia) Cen.
4	St. Thomas Cem.-Bedford	1860 Bedford Cen aka Herkins; 13th PA Infantry referenced
5	Schellsburg Cem.	Frank McCoy 1912 Listing
6	Beaufort Nat'l Cem.	Died 6/28/62; PA Civil War Archives-Bedford residence; B.S.F. History Book-1884
7	US Soldiers & Airmans Nat'l	Died 5/19/62; PA Civil War Archives-Buffalo Mills residence; B.S.F. History Book-1884
8		1890 Brush Creek Veterans Census
9	Stonerstown Cem.-Liberty Twp.	B.S.F. History Book-1884; In Hospital from 10/30/62 to 2/27/63
10		B.S.F. History Book-1884; Historical Data Systems-Bedford Co. residence
11		Historical Data Systems record
12		B.S.F. History Book-1884; Discharged on Surgeon's Cert.
13		Deserted 2/29/64; Historical Data Systems-Bedford Co. residence
14		Wounded-POW 6/30/62; 1890 Hopewell Twp. Veterans Census
15		Bedford Gazette-8th Reserve listing; Historical Data Systems record
16		1890 Saxton Veterans Census
17	Everett Cem.	Wounded 3/25/65; Historical Data Systems-Bedford Co. residence
18	Hyndman Cem.	Frank McCoy 1912 Listing; William, Joseph, James, George, Samuel & Alexander-brothers
19	Trinity UCC Cem.-Juniata Twp.	Wounded 5/6/64; Historical Data Systems record; John M.-Son
20		Wounded 9/19/64; Historical Data Systems record; 1860 Juniata Twp. Census
21	Beaufort Nat'l Cem.	Died 9/2/62; PA Civil War Archives-Bedford residence; 1860 Juniata Twp. Census
22	Trinity UCC Cem.-Juniata Twp.	Historical Data Systems-Bedford Co. residence; George W.-Father
23	Trinity UCC Cem.-Juniata Twp.	B.S.F. History Book-1884
24		1890 Londonderry Twp. Veterans Census
25	Trinity UCC Cem.-Juniata Twp.	1890 Londonderry Twp. Veterans Census; Deserted 4/30/65 after war ended
26	Mt. Olivet Cem.-Manns Choice	1890 Harrison Twp. Veterans Census
27	Eastern Light Cem.-Altoona	Severe Forearm Wound 7/30/64; Died-Tuberculosis 11-10-66; 1860 Woodbury Twp. Census
28	Fockler Cem.-Saxton	1890 Saxton Veterans Census
29	Hopewell Cem.	PA Civil War Project Record; 1860 Liberty Twp. Census; Discharged on Surgeon's Cert.
30	Soldiers & Sailors Cem.-NE	Historical Data Systems record; 1850 N. Woodbury Twp. Census
31	Forest Hill Cem.-MO	Wounded 12/13/62; Historical Data Systems record; 1860 Woodbury Twp. Census
32		Wounded 11/28/64; B.S.F. History Book-1884; Historical Data Systems record
33		POW-Andersonville 6/1/64-4/28/65; PA Civil War Archives-Woodbury Census; aka Joseph
34	Los Angeles Nat'l Cem.	Wounded 7/2/63; Soldiers of Blair County PA book - 1940; Josiah & Frank-brothers
35	Cedar Hill Cem.-IL	Wounded 1/1/63; Ancestry.com information; Levi & Emanuel-brothers
36		Historical Data Systems record
37		Deserted 4/28/64; B.S.F. History Book-1884; Historical Data Systems record
38		B.S.F. History Book-1884; enlisted in Berks Co.
39		B.S.F. History Book-1884; Reading, PA references
40		B.S.F. History Book-1884; enlisted in Reading, PA
41	Stonerstown Cem.-Liberty Twp.	1890 Liberty Twp. Veterans Census
42		1890 Saxton Veterans Census
43	Mt. Union Cem.-Lovely	Frank McCoy 1912 Listing
44	Union Cem.-Centerville	Frank McCoy 1912 Listing; 1860 Cumberland Valley Census
45		Frank McCoy 1912 Listing; 1860 Cumberland Valley Census
46	Union Cem.-Centerville	B.S.F. History Book-1884; 1850 Cumberland Valley Census
47		Frank McCoy 1912 Listing
48	Fairview Cem.-Artemas	B.S.F. History Book-1884; James, William, George & Elias-brothers
49		Frank McCoy 1912 Listing
50		1890 Cumberland Valley Veterans Census
51	Hampton Nat'l Cem.	Died 12/16/62; B.S.F. History Book-1884; 1860 Cumberland Valley Census
52	Grandview Cem.-Cambria Co.	1890 Napier Twp. Veterans Census; 1870 Juniata Twp. Census
53	Buck Valley Cem.-Fulton Co.	1890 Brush Creek Twp. Veterans Census
54	Buck Valley Cem.-Fulton Co.	B.S.F. History Book-1884; Historical Data Systems-Bedford Co. residence
55		1860 S. Woodbury Twp. Census; Reportedly lied about age to enlist
56	Bunker Hill Cem.-Saxton	Elias, John & Jonathan-brothers
57	New Enterprise Cem.	1860 Liberty Twp. Census
58		1890 Greenfield Twp. Veterans Census
59	Bunker Hill Cem.-Saxton	1890 Liberty Twp. Veterans Census
60	New Bern Nat'l Cem.	Died 5/24/63; B.S.F. History Book-1884; 1860 St. Clairsville Census
61	Beaufort Nat'l Cem.	Died 3/30/64; PA Civil War Archives-Bedford Co. residence; 1860 St. Clairsville Census
62	Replogle Cem.-Woodbury	1860 Woodbury Twp. Census
63	Drain Cem.-OR	B.S.F. History Book-1884; 1850 Napier Twp. Census
64		B.S.F. History Book-1884

Chapter 18

	Name	Age	Rank	Regiment	Co.	Muster In	Muster Out	Casualty	Casualty Battle
1	Horn, Joseph L	21	Pvt.	PA 16th Cav.	H	9/22/62	6/16/65		
2	Horn, Levi		Pvt.	PA 22nd Cav.	D	9/6/62		MIA	
3	Horne, John D	22	1st Lt.	PA 55th Inf.	D	10/12/61	8/30/65		
4	Horne, William L	27	Capt.	PA McKeage's Mil.	H	7/2/63	8/8/63		
5	Horner, Henry			PA 133rd Inf.					
6	Hornig, Frederick	27	Pvt.	PA 13th Inf.	G	4/25/61	8/6/61		
7	Horton, Alexander	36	Pvt.	PA 77th Inf.	F	2/29/64		Died	
8	Horton, David	24	Sgt.	PA 8th Res.	F	6/11/61	5/26/64	POW	Seven Days Battle
9	Horton, David		Sgt.	PA 8th Res.	F			Wounded	Wilderness
10	Horton, George	21	Corp.	PA 8th Res.	F	9/4/61		Wound.-Died	Antietam
11	Horton, George		Corp.	PA 8th Res.	F			Wounded	South Mountain
12	Horton, Jesse W	39	1st Lt.	PA 45th Inf.	C	8/31/61	9/30/62		
13	Horton, Jonathan (1)	22	Corp.	PA 77th Inf.	F	10/9/61		Died	
14	Horton, Jonathan (2)		Corp.	PA McKeage's Mil.	G	7/2/63	8/8/63		
15	Horton, Jonathan A	21	Pvt.	PA 133rd Inf.	C	8/29/62	3/14/63		Fredericksburg*
16	Horton, Milton	22	Corp.	PA 77th Inf.	A	10/9/61	9/2/64		
17	Horton, Oliver	32	Capt.	PA 138th Inf.	D	8/29/62	6/23/65	Wounded	Cedar Creek
18	Horton, Reuben H	20	Pvt.	PA 77th Inf.	F	2/29/64		Wound.-Died	Chattahoochee
19	Horton, Zophar P	19	Pvt.	PA 8th Res.	F	6/11/61	5/15/64		
20	Horton, Zophar P		Pvt.	PA 191st Inf.	H	6/9/64	6/29/65	Injured	Petersburg
21	Hosack, George H*	26	Pvt.	PA 76th Inf.	E	3/28/64	7/18/65		
22	Houck, Amon	21	Corp.	PA 57th Inf.	I	4/7/64		Wounded	2nd Deep Bottom
23	Houck, Ezekiel J	28	Pvt.	PA 53rd Inf.	C	10/17/61			
24	Houck, Ezekiel J		Pvt.	PA 84th Inf.	E	4/8/64		Died	Petersburg - Initial
25	Houck, George A*		Pvt.	PA 125th Inf.	B	8/10/62	5/18/63		
26	Houck, George A*		Pvt.	PA 22nd Cav.	D	7/21/63	6/15/65		
27	Houck, George W		Pvt.	PA McKeage's Mil.	D	7/2/63	8/8/63		
28	Houck, George W		Pvt.	PA 22nd Cav.	H	2/29/64	6/24/65	POW	
29	Houck, McKenzie	18	Pvt.	PA 77th Inf.	F	3/8/62	3/8/65	Wounded	Chickamauga
30	Houp, Arnold	41	Pvt.	PA 79th Inf.	H	2/21/65	4/29/65		
31	Householder, Jacob*	25	Pvt.	PA 110th Inf.	C	2/7/64	5/16/65		
32	Householder, James	44	Pvt.	PA 208th Inf.	K	9/7/64	6/1/65		
33	Householder, John	15	Pvt.	PA 208th Inf.	K	9/7/64	6/1/65		
34	Householder, M C*	18	Pvt.	PA 110th Inf.	C	10/24/61	6/28/65		
35	Housenworth, J J		Music.	PA 208th Inf.	H	9/3/64	6/1/65		
36	Howard, John S*	18	Pvt.	PA 55th Inf.	K	2/1/64	6/25/65		
37	Howard, William	19	Pvt.	PA 101st Inf.	G	12/28/61			
38	Howsare, Thomas	34	Pvt.	PA 158th Inf.	G	11/4/62	8/12/63		
39	Howsare, Wesley B	26	Pvt.	PA 171st Inf.	I	11/10/62			
40	Howser, Henry H		Pvt.	PA 21st Cav.	E	7/8/63	2/20/64		
41	Howser, Jesse			OH 51st Inf.					
42	Howser, Jessie	30	Pvt.	OH 101th Inf.	G	1/3/64	6/1/65		
43	Howser, Lewis	56	Sgt.	PA 172nd Inf.	A	10/28/62			
44	Hoyman, Samuel	22	Pvt.	PA 76th Inf.	D	11/28/64	7/18/65		
45	Hubert, Reuben*		Pvt.	PA 55th Inf.	K	1863	8/30/65	Wounded	
46	Huff, Isaac T	23	Pvt.	PA 6th Res.	D	7/23/61	6/11/64		
47	Huff, James L	21	Pvt.	PA 171st Inf.	I	11/2/62			
48	Huffman, Joseph W	22		PA 149th Inf.	F	8/22/62	6/24/65		
49	Huffman, Josiah	22	Corp.	PA 138th Inf.	D	8/12/62	5/15/65	Wounded	Wilderness
50	Huffman, William B	33	Pvt.	PA 101st Inf.	G	12/28/61	10/5/63	Wounded	Fair Oaks
51	Hughes, Bailey	40	Pvt.	PA 149th Inf.	B	9/22/63		Died	
52	Hughes, Bartley			PA 148th Inf.					
53	Hughes, Edwin	33	Pvt.	PA 55th Inf.	I	10/8/63	2/28/66	Wound.-POW	Chaffin's Farm
54	Hughes, James C*	24	Music.	PA 55th Inf.	K	11/5/61	8/30/65		
55	Hughes, Scott W	22	Sgt.	PA 22nd Cav.	C	7/11/63	6/24/65		
56	Hughes, William	28	Pvt.	PA McKeage's Mil.	D	7/2/63	8/8/63		
57	Hull, Abraham	18	Pvt.	PA 101st Inf.	G	12/28/61	1/21/62		
58	Hull, Daniel	18	Pvt.	NY 88th Inf.	A	4/23/64			
59	Hull, Jacob		Pvt.	PA McKeage's Mil.	H	7/2/63	8/8/63		
60	Hull, Samuel*	23	Pvt.	PA 55th Inf.	D	10/16/63		MIA	Cold Harbor
61	Humbert, Daniel	32	Pvt.	PA 76th Inf.	E	11/21/61	11/28/64		
62	Humbert, Moses	16	Pvt.	PA 5th Heavy Art.	K	9/1/64	6/30/65		
63	Humbert, Wesley C		Corp.	PA 142nd Inf.	C	8/25/62	10/1/64	Wounded	Gettysburg
64	Humbert, Wesley C	19	Corp.	PA 142nd Inf.	C			Wounded	Fredericksburg

Detailed Alphabetical Listing   443

	Cemetery	Notes
1	Zion United Breth.-Franklin Co	Ancestry.com-born in Schellsburg; aka Horne
2		Bedford Gazette 2/13/1914 listing; Historical Data Systems record; aka Horne
3	Bedford Cem.	1890 Bedford Twp. Veterans Census; Historical Data Systems record
4	Bedford Cem.	1890 Bedford Twp. Veterans Census; B.S.F. History Book-1884
5	Fishertown Cem.	Frank McCoy 1912 Listing
6		B.S.F. History Book-1884
7	Wells Valley Presb. Cem.	Died 11/9/64; Historical Data Systems record; Ancestry.com born-Broad Top City
8	Maplewood Cem.-WV	POW 6/27/62; Bedford Gazette-8th Reserve listing; Historical Data Systems record
9		Wounded in the head and knee; 1860 Broad Top Census; David & George-brothers
10	Antietam Nat'l Cem., 3768	Died 9/17/62; B.S.F. History Book-1884; PA Civil War Archives-Bedford Co. residence
11		Wounded 9/14/62; Bedford Gazette-8th Reserve listing; B.S.F. History Book-1884
12	Maple Grove Cem.-Belleville	Alexander & Jesse-brothers
13	Wells Valley-Wells Tannery	Died 12/12/64; B.S.F. History Book-1884; 1850 Broad Top Twp. Census
14		B.S.F. History Book-1884
15	Horton Cem.-NE	B.S.F. History Book-1884; Historical Data Systems-Bedford Co. residence
16	Camp Hill Cem.	Historical Data Sys. record
17	Old Carrollton Cem.-IA	B.S.F. Hist. Book-1884; 1860 Monroe Twp. Cen.; Wound. 10/19/64 by shell that killed 7 men
18	Wells Valley-Wells Tannery	Wounded 7/7/64; Died 8/13/64; 1850 Broad Top Twp. Census
19	Everett Cem.	Historical Data Systems record; Zopher, Jonathan, Milton & Reuben-brothers
20		Leg fracture; B.S.F. History Book-1884
21		B.S.F. History Book-1884; Crawford Co. references
22	Broad Top IOOF Cem.	Arm Amputated 8/16/64; 1890 Robertsdale Veterans Census; Died 1870
23		1850 Hopewell Twp. Census (Huntingdon Co.)
24		Died 6/17/64; Pension request record; Amon & Ezekiel-brothers
25		Historical Data Systems record; Blair County references
26	Fairview Cem.-Altoona	B.S.F. History Book-1884; Historical Data Systems record
27		B.S.F. History Book-1884
28		B.S.F. History Book-1884
29	Wells Valley Meth. Cem.	1890 Wells Twp. Veterans Census
30	Fockler Cem.-Saxton	Frank McCoy 1912 Listing; 1860 Liberty Twp. Census
31		B.S.F. History Book-1884; Blair Co. references
32	Rays Cove CC Cem.	1890 E. Providence Twp. Veterans Census; John-son
33	Rays Cove CC Cem.	Historical Data Systems-Bedford Co. residence; James-father
34		B.S.F. History Book-1884; Green Co. references
35		B.S.F. History Book-1884; Historical Data Systems-Bedford Co. residence
36		B.S.F. History Book-1884; Berks Co. references
37		Deserted 2/5/62; Historical Data Systems-Bedford Co. residence
38	Union Christian-Crystal Springs	East Providence Honor Roll; 1890 E. Providence Twp. Veterans Census
39	Mt. Zion Cem.-Cheneysville	Deserted on 11/18/62; Frank McCoy 1912 Listing; aka Houser
40		Historical Data Systems-Bedford Co. residence
41		Frank McCoy 1912 Listing
42	Bortz Luth. Cem.-Centerville	1890 Cumberland Valley Veteran Census
43	Bethal Frame Ch.-Monroe Twp.	Historical Data Systems record; Civil War grave marker
44	Hyndman Cem.	B.S.F. History Book-1884; In Hospital 3/26/65 to Muster Out
45		Wounded listed in Bedford Inquirer article 6/3/64; B.S.F. History Book-1884; drafted
46	Rays Cove CC Cem.	Frank McCoy 1912 Listing; 1890 E. Providence Twp. Veterans Cen.; Discharged Surg. Cert.
47		Deserted 11/3/62; B.S.F. History Book-1884
48		Pauline Nicodemus-family record on enlistment; 1900 Harrison Township Census
49	Schellsburg Cem.	Shoulder Wound 5/6/64; In Hospital to Muster Out
50	Lybarger Cem.-Madley	3 Wounds suffered on 5/31/62, in Hosp. until Muster Out; Discharged on Surgeon's Cert.
51	US Soldiers & Airmans Nat'l	Died 12/18/63 of Typhoid Fever; 1890 Bedford Veterans Census; 1860 Harrison Twp. Census
52	Everett Cem.	Everett Cemetery Listing of Civil War Vets; 1860 W. Providence Twp. Census
53		Leg Amputated & POW 9/29/64; B.S.F. History Book-1884
54		B.S.F. History Book-1884; Butler Co. references
55	Bedford Cem.	1860 Bedford Twp. Census; B.S.F. History Book-1884
56	Providence Union Cem.-Everett	B.S.F. History Book-1884
57	Hull Farm Cem.-Schellsburg	Historical Data Systems record; 1860 Napier Twp. Census; Discharged on Surgeon's Cert.
58		Fold3-Born Bedford Co.; Deserted 4/23/64
59		B.S.F. History Book-1884; Jacob & Abraham-brothers
60		MIA 6/3/64; B.S.F. History Book-1884; Bed. Gazette 2/13/1914 list; Mustered in Reading
61		B.S.F. History Book-1884
62	St. Johns Breth. Cem.-Somerset	1850 Harrison Twp. Census; Moses & Wesley-brothers
63		Right Arm Wound 7/1/63; transferred to Veterans Reserve Corp. 10/1/64
64	St. Johns Breth. Cem.-Somerset	Leg Wound 12/13/62 on Muster roll record; 1850 Harrison Twp. Census

Chapter 18

	Name	Age	Rank	Regiment	Co.	Muster In	Muster Out	Casualty	Casualty Battle
1	Hummel, William A		Pvt.	PA		8/7/61	12/24/62		
2	Hunt, David A	21	Pvt.	PA 50th Inf.	E	2/23/64	7/30/65		
3	Hunt, David C								
4	Hunt, George E	24	Pvt.	PA 132nd Inf.	A	8/14/62	5/24/63		
5	Hunt, John H	22	Music.	PA 13th Inf.	A	9/12/62	9/26/62		
6	Hunt, John H		Music.	PA 178th Inf.	F	11/5/62	7/27/63		
7	Hunt, John H		Pvt.	PA 187th Inf.	E	1/16/64	8/3/65		
8	Hunt, John T		Pvt.	PA 138th Inf.	F	9/1/62			
9	Hunt, John T	25	Corp.	PA 55th Inf.	K	11/4/62		POW-Died	Drewry's Bluff
10	Hunt, Samuel B		Pvt.	PA 138th Inf.	F	9/1/62			
11	Hunt, Samuel B	28	Pvt.	PA 55th Inf.	K	11/4/62	8/30/65		
12	Hunter, Robert I	30	Sgt.	PA 173rd Inf.	E	11/6/62	8/17/63		
13	Hunter, Robert I		Surg.	IA 2nd Cav.	C		10/3/64		
14	Hurley, Daniel W	18	Sgt.	PA 200th Inf.	G	8/31/64		Died	
15	Hurley, John	18	Pvt.	PA 76th Inf.	F	10/28/61		Died	
16	Hurley, William	32	Pvt.	PA 84th Inf.	E	9/20/62	1/13/65	Wounded	Spotsylvania
17	Hurley, William		Pvt.	PA 57th Inf.	I	1/13/65			
18	Husband, Johnston	19	Sgt.	PA 21st Cav.	E	7/2/63	2/20/64		
19	Husband, Johnston		Sgt.	PA 5th Heavy Art.	K	9/1/64	8/5/65	Wounded	
20	Husler, Thomas J	18	Pvt.	PA 13th Cav.	F	2/11/64	6/13/65		
21	Hutchison, William*	44	Pvt.	PA 76th Inf.	E	11/21/61	8/30/62		
22	Hutton, Henry								
23	Hutton, Jacob H	24	Pvt.	PA 126th Inf.	A	8/11/62	6/20/63		
24	Hyde, Abraham	21	Pvt.	PA 55th Inf.	K	11/5/61		Died	
25	Hyde, John	25	Pvt.	PA 55th Inf.	H	2/29/64	8/30/65		
26	Hyde, Jonathan	20	Pvt.	PA 50th Inf.	D	2/24/65	7/30/65		
27	Hymes, George W		Pvt.	DC 2nd Inf.	A	1/30/62	9/12/65		
28	Hymes, Oliver	19	Pvt.	PA 99th Inf.	G	3/16/65	7/1/65		
29	Hymes, Wiley		Pvt.	PA McKeage's Mil.	D	7/2/63	8/8/63		
30	Hymes, Wiley		Pvt.	PA 208th Inf.	K	9/7/64	6/1/65		
31	Iames, David		Pvt.	PA 99th Inf.	I	2/25/65	5/31/65		
32	Iames, John		Pvt.	PA 99th Inf.	I	9/21/64	7/1/65		
33	Iams, Daniel			PA 101st Inf.					
34	Iams, David			PA 8th Res.					
35	Ibach, James A*	18	Pvt.	PA 194th Inf.	I	7/22/64	9/6/64		
36	Ibach, James A*		Pvt.	PA 97th Inf.		9/6/64	6/17/65		
37	Ickes, Adam (1)		Pvt.	PA 14th Inf.	H	4/24/61	8/7/61		
38	Ickes, Adam (1)	21	Pvt.	US 12th Inf.	A	9/28/61	2/23/67	Wounded	Chancellorsville
39	Ickes, Adam (2)	18	Pvt.	PA 91st Inf.	I	10/7/64	7/6/65		
40	Ickes, Adam H	34	Pvt.	OH 169th Inf.	K	5/15/64	9/4/64		
41	Ickes, Alexander	25	Pvt.	PA 171st Inf.	I	11/2/62	8/8/63		
42	Ickes, Alexander		Pvt.	PA 91st Inf.	B	9/21/64	5/30/65		
43	Ickes, Daniel W	24	Pvt.	PA 91st Inf.	I	9/21/64	5/30/65		
44	Ickes, George	22	Pvt.	PA 138th Inf.	D	9/2/62		Died	
45	Ickes, George W	19	Pvt.	PA 138th Inf.	D	9/2/62	6/23/65		
46	Ickes, Henry	22	Pvt.	PA 55th Inf.	K	3/2/64		Wounded	
47	Ickes, Henry H	20	Pvt.	PA 93rd Inf.	F	11/14/64	6/27/65		
48	Ickes, Jacob	19	Pvt.	US 5th Heavy Art.	E	5/1/62	1/17/65		
49	Ickes, Joseph B		Pvt.						
50	Ickes, Joseph H	24	Pvt.	PA 84th Inf.	A	9/24/61			
51	Ickes, William	23	Pvt.	PA McKeage's Mil.	G	7/2/63	8/8/63		
52	Ickes, William		Pvt.	PA 99th Inf.	B	2/27/65	6/13/65		
53	Ickes, William M	16	Pvt.	PA 91st Inf.	I	10/8/64	7/10/65		
54	Ilesmith, Charles*			PA 1st Light Art.				Died	
55	Imes, Aaron	21	Pvt.	PA 8th Res.	F	6/11/61	12/16/63	Wounded	Fredericksburg
56	Imes, John	22	Pvt.	PA 13th Res.	G	7/4/61		Wound.-Died	Fredericksburg
57	Imler, Adam H (1)	29	Pvt.	PA 91st Inf.	B	10/8/64	7/6/65	Wounded	Weldon Railroad
58	Imler, Adam H (2)	20	Pvt.	PA 133rd Inf.	C	8/13/62	3/15/63		Fredericksburg*
59	Imler, Daniel	22	Pvt.	PA 138th Inf.	E	9/19/64		Died	
60	Imler, Ephraim Y	21	Pvt.	PA 138th Inf.	E	8/29/62	10/1/64		
61	Imler, George R	21	Pvt.	PA 138th Inf.	E	8/29/62	6/23/65	Wound.-POW	Monocacy, MD
62	Imler, Henry W	33	Pvt.	OH 186th Inf.	E	2/27/65	9/18/65		
63	Imler, Isaac M	36	Sgt.	PA 55th Inf.	K	11/5/61		KIA	Petersburg - Initial
64	Imler, John	29	1st Lt.	PA 55th Inf.	K	11/5/61	7/17/65	Wounded	

	Cemetery	Notes
1	St. Pauls UCC Cem.-Hunt. Co.	1890 Liberty Twp. Veterans Census
2		Findagrave.com information; David, John T & Samuel-brothers
3		Ancestry.com-Born Schellsburg
4	Rose Hill Cem.-Altoona	Ancestry.com-Born Schellsburg
5	Fairview Cem.-Danville	Ancestry.com-Born Schellsburg
6		John H, David C & George-brothers
7		Historical Data Systems record
8		B.S.F. History Book-1884; Deserted 9/1/62
9		POW-Libby, Andersonville, Millin & Savanah 5/16/64; Died 10/10/64; PA CW Archives record
10		B.S.F. History Book-1884; Deserted 9/1/62
11	Friends Cove UCC Cem.	1860 Cumberland Valley Census; 1890 Colerain Twp. Veterans Census
12	Wells Valley Presb. Cem.	PA Civil War Project record
13		1890 Wells Twp. Veterans Census; promoted to Hospital Steward
14		Died 1/30/65; PA Civil War Project record; 1850 Harrison Twp. Census
15	Hampton Nat'l Cem.	Died 11/30/61; PA Civil War Project record; 1850 Harrison Twp. Census
16	Greenlawn Cem.-Roaring Spring	Wounded 5/12/64; B.S.F. History Book-1884; 1860 Juniata Twp. Census
17		Ancestry.com information; William, Daniel & John-brothers
18	Evergreen Cem.-IL	Historical Data Systems-Bedford Co. residence
19		Amputated Leg 3/18/65; Historical Data Systems record
20	Fort Littleton Cem.	1890 Wells Twp. Veterans Census
21		B.S.F. History Book-1884; PA Civil War Archives-Phila. residence; Discharged Surgeon's Cert.
22	Bedford Cem.	1870 Bedford Boro Census; Bedford Inquirer 5/22/1908 listing
23	Bedford Cem.	Bedford Cemetery Civil War Veterans list
24	Trinity UCC Cem.-Juniata Twp.	Died 11/30/61; Frank McCoy 1912 Listing; 1850 Colerain Twp. Census
25	Trinity UCC Cem.-Juniata Twp.	1890 Juniata Twp. Veterans Census; John, Jonathan & Abraham-brothers
26	Trinity UCC Cem.-Juniata Twp.	1890 Harrison Twp. Veterans Census
27		1890 Everett Veterans Census; Wash. DC regiment
28	Mt. Hope Cem.-Hewitt	1890 Southhampton Veterans Census, aka Himes
29		B.S.F. History Book-1884
30		B.S.F. History Book-1884; Historical Data Systems-Bedford Co. residence
31		Bedford Inquirer 5/22/1908 listing; aka Iams
32		1860 Southampton Census; 1890 Allegany MD Veterans Census
33	Iams Cem.-Southampton Twp.	Frank McCoy 1912 Listing
34	Mt. Zion-Southampton Twp.	Frank McCoy 1912 Listing; aka Iames & Imes
35		B.S.F. History Book-1884; Historical Data Systems record; enlisted in Harrisburg
36		Historical Data Systems record
37	Mt. Hope Cem.-Claysburg	Historical Data Systems record
38		Wounded 5/3/63-Pension Record
39	Wyuka Cem.-NE	1860 St. Clairsville Twp. Census
40	Riverside Cem.-MI	1850 Union Twp. (Pavia) Census
41	Mt. Zion Cem.-Pavia	B.S.F. History Book-1884
42		Historical Data Systems record; 1860 Union Twp. (Pavia) Census
43	Whittlesey Cem.-OH	Historical Data Systems record; Adam (2) & Daniel brothers
44	Old Ickes Cem.-Pigeon Hills	Died 11/14/62; Frank McCoy 1912 Listing; 1860 St. Clairsville Twp. Census
45	Shelby Cem.-IA	B.S.F. History Book-1884; 1860 Bedford Twp. Census
46		Wounded listed in Bedford Inquirer article 6/3/64; Deserted 8/18/64 at Hospital
47	Mt. Zion Cem.-Pavia	1890 Union Twp. (Pavia) Veterans Census
48	Trinity UCC Cem.-Osterburg	Frank McCoy 1912 Listing; Bedford Inquirer 5/22/1908 listing
49	Highland Cem.-WA	Ancestry.com record
50	Imler St. Paul Breth. Cem.	Deserted 2/23/62 & returned; 1890 King Twp. Veterans Census
51	Mt. Zion Cem.-Pavia	1890 King Twp. Census
52		Historical Data Systems record; Joseph B & William-brothers
53	Albrights Cem.-Roaring Springs	1890 Claysburg Vet Census; Born in St. Clair Twp.
54		Bedford Gazette 2/13/1914 listing; Died of disease
55	Mt. Zion Cem.-Chaneysville	1890 Southampton Twp. Veterans Census.; aka Aaron Iams
56		Wounded 12/13/62; Bed. Gazette 2/13/1914 listing; 1860 W. Providence Census; aka Iames
57	Imler St. Paul Breth. Cem.	1890 King Twp Census; Frank McCoy 1912 Listing; Wounded 1/7/65 listed in obit.
58	Metzgar Cem.-OH	Muster Out in hospital; Discharged on Surgeon's Cert.; Adam H (2), Martin & Jonas-brothers
59	Arlington Nat'l Cem., 13-8070	Died 1/4/65; B.S.F. History Book-1884; PA Civil War Archives-Greenfield Twp. residence
60	Mt. Zion Cem.-Breezewood	PA Civil War Archives-Bedford residence; Veterans Reserve Corp.
61	Lutheran Cem.-Osterburg	1890 Woodbury Twp. Veterans Census
62	Oakwood Cem.-OH	1860 S. Woodbury Twp. Census
63		KIA 6/18/64; PA Civil War Archives-Bedford Co. residence; 1860 Union Twp. (Pavia) Census
64	Mt. Hope Cem.-Claysburg	Wounded listed in Bedford Inquirer article 6/3/64; Historical Data Systems-Bedford Co. resid.

	Name	Age	Rank	Regiment	Co.	Muster In	Muster Out	Casualty	Casualty Battle
1	Imler, John R	34	Pvt.	PA 82nd Inf.	C	10/3/64	7/13/65		
2	Imler, Jonas C	19	Pvt.	PA 205th Inf.	C	8/26/64	6/2/65		
3	Imler, Martin	32	Pvt.	PA 91st Inf.	B	9/21/64	5/30/65		
4	Imler, Matthias	21	Corp.	PA 184th Inf.	A	5/12/64	7/14/65		
5	Imler, William H	21	Pvt.	PA 91st Inf.	B	10/8/64	7/6/65	Wounded	Boydton Plank Rd.
6	Ingoldsby, Lawrence*	22	Pvt.	PA 55th Inf.	D	10/14/63			
7	Inks, Elisha		Pvt.						
8	Irons, John		Pvt.	PA 5th Res.		6/6/61	6/13/64		
9	Irvine, Hayes	24	Pvt.	PA 2nd Cav.	E	11/27/61	11/27/64		
10	Irvine, Wilson	19	Pvt.	PA 184th Inf.	A	5/12/64		POW-Died	Jerusalem Plank Rd.
11	Irwin, James*	21	Pvt.	PA 110th Inf.	C	10/24/61	6/28/65	Wounded	1st Deep Bottom
12	Irwin, Jarrett*	19	Pvt.	PA 110th Inf.	C	3/15/64	6/28/65		
13	Isener, Joseph		Pvt.	PA McKeage's Mil.	G	7/2/63	8/8/63		
14	Isett, James M	19	Pvt.	PA 194th Inf.	I	7/22/64	11/5/64		
15	Jackson, Berdine*	16							
16	Jackson, Charles	27	Pvt.	PA 55th Inf.	H	10/16/63	5/15/65	Wounded	Bermuda Hundred
17	Jackson, Charles		Pvt.	PA 55th Inf.	H			Wounded	Petersburg - Initial
18	Jackson, Charles W	25	Pvt.	OH 147th Inf.	G	5/2/64	8/30/64		
19	Jackson, John*	29	Pvt.	PA 55th Inf.	I	10/8/63			
20	Jackson, Mark J (1)	28	Pvt.	PA 22nd Cav.	M	2/24/64	6/24/65		
21	Jackson, Mark J (2)			PA 208th Inf.					
22	Jackson, Samuel M	25	Corp.	PA 158th Inf.	H	11/4/62	8/12/63		
23	Jackson, Samuel M		Pvt.	PA 22nd Cav.		2/23/64	6/24/65		
24	Jackson, Stiles H	20	Pvt.	OH 35th Inf.	H	9/9/61	9/8/64		
25	Jackson, Stiles H		Pvt.	OH 197th Inf.	H	4/10/65	7/31/65		
26	Jacobs, Thomas*	19	Pvt.	PA 194th Inf.	I	7/22/64	11/5/64		
27	Jacoby, Edward	19	Pvt.	PA 13th Inf.	G	4/25/61	8/6/61		
28	James, David*	40	Pvt.	PA 99th Inf.	K	10/24/61		Died	
29	James, Edward V	19	Pvt.	PA 55th Inf.	K	3/2/64		KIA	Petersburg - Initial
30	James, Jesse T	19	Pvt.	PA 84th Inf.	A	10/24/61		Died	
31	James, John A	14	Pvt.	PA 55th Inf.	K	10/11/61		Wound.-Died	
32	James, John W (1)			PA Militia					
33	James, John W (1)	21	Pvt.	PA 55th Inf.	I	10/22/63	6/5/65	Wound.-POW	Drewry's Bluff
34	James, John W (2)		Pvt.	PA 88th Inf.	B	4/2/64	6/20/65		
35	James, Nathaniel	20	Pvt.	PA 138th Inf.	D	8/29/62		MIA	Opequon
36	Jamison, Benjamin	19	Pvt.	PA 125th Inf.	B	8/13/62	5/18/63	Wounded	Antietam
37	Jamison, Benjamin		Pvt.	PA 110th Inf.	B	2/22/64	5/31/65	POW	Cold Harbor
38	Jay, John	25	Pvt.	PA 171st Inf.	I	11/2/62			
39	Jay, Thomas	22	Pvt.	PA 171st Inf.	I	11/2/62	8/8/63		
40	Jay, Thomas		Pvt.	PA 91st Inf.	F	9/21/64	10/2/65		
41	Jayne, Isaac B*	42	Pvt.	PA 3rd Heavy Art.	K	10/2/62			
42	Jayne, Isaac B*		Pvt.	PA 76th Inf.	E	9/23/64		Died	
43	Jeffries, Howard B		1st Lt.	PA Helm. Mil.		7/6/63	9/7/63		
44	Jeffries, Howard B	20	1st Lt.	PA 21st Cav.	E	7/21/63	2/20/64		
45	Jeffries, Howard B		Capt.	USCT 72nd Inf.	B	10/28/64	10/9/65		
46	Jenkes, Daniel M		Pvt.	PA 21st Cav.	E	7/8/63	2/20/64		
47	Jenkins, David	17	Pvt.	PA 9th Cav.	L	9/25/61	7/18/65		
48	Jenkins, John P	22	Pvt.	PA 16th Cav.	H	2/23/65	8/11/65		
49	Jessner, Joseph	21	Pvt.	PA 194th Inf.	I	7/22/64	9/6/64		
50	Jessner, Joseph		Pvt.	PA 97th Inf.	Ind	9/6/64	7/17/65		
51	Johnson, Abel	40	Pvt.	PA 91st Inf.	A	2/25/65	7/10/65		
52	Johnson, Asa	25	Corp.	PA 171st Inf.	I	11/10/62	8/8/63		
53	Johnson, Asa		Pvt.	PA 91st Inf.	F	9/21/64	6/1/65		
54	Johnson, Charles*		Pvt.	USCT 3rd Inf.	A				
55	Johnson, Charles*		Pvt.	USCT 6th Inf.	I				
56	Johnson, Cromwell O	26	Surg.	PA 5th Res.		3/9/63	9/28/63	Died	
57	Johnson, David (1)	24	Pvt.	USCT 1st Inf.	F	6/21/63	9/27/65		
58	Johnson, David (2)*	37	Pvt.	PA 76th Inf.	E	9/2/63		Wound.-Died	Chester Station
59	Johnson, Edward	26	Pvt.	PA 55th Inf.	D	10/14/63	5/4/64		
60	Johnson, Edward		Pvt.	US 3rd Reg Art.	E	5/4/64			
61	Johnson, Emanuel	37	Pvt.	PA 91st Inf.	F	9/21/64	6/1/65		
62	Johnson, John J	44	Pvt.	PA 91st Inf.	F	9/21/64	6/1/65		
63	Johnson, Joshua	30	Pvt.	PA 149th Inf.	K	8/26/63			
64	Johnson, Lewis	44	Pvt.	PA 171st Inf.	I	11/2/62	8/8/63		

	Cemetery	Notes
1	Lutheran Cem.-Osterburg	1890 King Twp. Veterans Census; George & John R brothers
2	Imler Valley Cem.	1890 King Twp. Veterans Census; 1860 Union Twp. (Pavia) Census
3	Greenfield Cem.-Queen	Frank McCoy 1912 Listing
4	Oak Grove Cem.-OH	B.S.F. History Book-1884; Historical Data Systems record
5	St. Paul Breth. Cem.-Imler	Wounded 10/27/64; 1890 King Twp. Veterans Census; 1860 Union Twp. (Pavia) Census
6		B.S.F. History Book-1884; Deserted 4/27/64; enlisted in Reading, PA
7		1890 Napier Twp. Veterans Census; 1870 Bedford Twp. Census
8		B.S.F. History Book-1884
9	Bedford Cem.	Frank McCoy 1912 Listing; 1850 W. Providence Twp. Census; aka Robert
10	Andersonville Cem., 11560	POW-Andersonville 6/22/64; Died 10/27/64; 1850 W. Prov. Twp. Cen.; aka William & John W
11		B.S.F. History Book-1884; York references
12		B.S.F. History Book-1884; Montgomery Co. residence
13		B.S.F. History Book-1884
14	Pleasant View Cem.-KS	B.S.F. History Book-1884
15		Died 6/30/64; Civil War Grave Marker
16		Wounded 5/13/64 listed in Bedford Inquirer article 7/8/64; B.S.F. History Book-1884
17		Wounded 6/16/64 listed in Bedford Inquirer 7/8/64 article; B.S.F. History Book-1884
18	Polk Grove Cem.-OH	Ancestry.com information; Charles W, Stiles & Berdine-brothers
19		B.S.F. History Book-1884; Deserted 11/25/64
20	Everett Cem.	1890 E. Providence Twp. Veterans Census
21		Frank McCoy 1912 Listing
22	Willow Wild Cem.-TX	Historical Data Systems record; 1850 Brush Creek Twp. Census
23		Historical Data Systems record; 1850 Brush Creek Twp. Census
24	Graceland Cem.-KS	1840 E. Providence Twp. Census
25		Historical Data Systems record; 1870 East Providence Twp. Census
26		B.S.F. History Book-1884; enlisted in Reading, PA
27		B.S.F. History Book-1884; Historical Data Systems-Bedford residence
28	Fockler Cem.-Saxton	Died 1865; Bed. Inquirer 5/22/1908 listing; PA Civil War Project-not on Muster Out rolls
29		KIA 6/18/64; B.S.F. History Book-1884; 1860 Napier Twp. Census
30	Horn Meth. Cem.-Alum Bank	Died 9/28/63; Frank McCoy 1912 List; 1860 St. Clair Twp. Cen.; Edward, Jesse & John-bros.
31	Hampton Nat'l Cem.	Died 6/20/64; B.S.F. History Book-1884; Muster Rolls-enlisted in Bedford
32		Frank McCoy 1912 Listing
33	Bedford Cem.	Wounded 5/16/64; B.S.F. History Book-1884
34	Friends Cove UCC Cem.	1890 Colerain Twp. Veterans Census
35		MIA 9/19/64 ; PA Civil War Archives-Bedford Co. resid.; Muster roll-enlisted in Schellsburg
36	St. John's Cem.-Loysburg	Wound in thigh 9/17/62; Historical Data Systems record; In hospital at Muster Out
37		POW-Andersonville & Florence 6/2/64-3/2/65; 1890 S. Woodbury Veterans Census
38	Mt. Union Cem.-Mench	Deserted 11/26/62; B.S.F. History Book-1884
39	Pine Grove Cem.-Monroe Twp.	B.S.F. History Book-1884; Thomas & John-brothers
40		1850 Southampton Twp. Census
41		Historical Data Systems record; enlisted in Philadelphia
42		Died 4/3/65; B.S.F. History Book-1884; Mustered in Easton
43		Historical Data Systems record
44	Oakside Cem.-FL	Historical Data Systems-Bedford Co. residence
45		Historical Data Systems record
46		Historical Data Systems-Bedford Co. residence
47	Stonerstown Cem.-Liberty Twp.	1890 Liberty Twp. Veterans Census
48	Duval Cem.-Coaldale	1890 Broad Top Veterans Census; Frank McCoy 1912 Listing
49		B.S.F. History Book-1884; enlisted in Hopewell
50		Historical Data Systems record
51	Chaneysville Meth. Cem.	1890 Southhampton Twp Veterans Census
52	Jonathan Creek Cem.-IL	B.S.F. History Book-1884
53		Asa & Abel-brothers; Historical Data Systems record; 1850 Southampton Twp. Census
54		Frank McCoy 1912 list & Civil Works Admin. 1934 list; service unverified on both lists
55		Frank McCoy 1912 list & Civil Works Admin. 1934 list; service unverified on both lists
56	Everett Cem.	Died 12/21/64; B.S.F. History Book-1884; PA Civil War Project record aka O C
57		Ancestry.com.-Born Schellsburg
58	Hampton Nat'l Cem.	Wounded 5/7/64; Died 6/23/64; B.S.F. History Book-1884; mustered in Meadville
59		B.S.F. History Book-1884
60		Historical Data Systems record
61	Mt. Hope Cem.-Hewitt	1890 Southhampton Twp. Census
62	Johnson Cem.-Elbinsville	Frank McCoy 1912 Listing; 1890 Mann Twp. Veterans Census
63	Johnson Cem.-Elbinsville	Frank McCoy 1912 Listing; John J & Joshua-brothers
64	Bethal Frame Ch.-Monroe Twp.	1890 Monroe Twp. Veterans Census; B.S.F. History Book-1884

	Name	Age	Rank	Regiment	Co.	Muster In	Muster Out	Casualty	Casualty Battle
1	Johnson, Moses	23	Pvt.	USCT 6th Inf.	F	9/11/63	9/19/65		
2	Johnson, Samuel J		Pvt.	IN 5th Inf.	K	10/16/62	7/20/63		
3	Johnson, Samuel J		Pvt.	PA 55th Inf.	C	10/26/64	7/23/65		
4	Johnson, William	33	Pvt.	USCT 3rd Inf.	K	7/17/63	10/31/65		
5	Johnson, William P	37	Pvt.	PA 171st Inf.	I	11/2/62	8/8/63		
6	Johnson, William P		Pvt.	PA 91st Inf.	F	9/21/64	6/1/65		
7	Johnston, Charles W	18	Pvt.	PA 184th Inf.	A	5/12/64	7/14/65	Wounded	Cold Harbor
8	Johnston, David S		Pvt.	PA 208th Inf.	H	9/5/64	6/1/65		
9	Johnston, John W	19	Pvt.	PA 133rd Inf.	C	8/31/62	5/26/63	Wounded	Fredericksburg
10	Johnston, Samuel*	19	Pvt.	PA 110th Inf.	C	2/22/64		Wound.-Died	Wilderness
11	Johnston, William		Pvt.	PA McKeage's Mil.	H	7/2/63	8/8/63		
12	Jones, David M	24	Pvt.	PA 28th Inf.	O	8/17/61	10/28/62		
13	Jones, David M		Pvt.	PA 147th Inf.	B	12/29/63	7/15/65		
14	Jones, David W	36	Corp.	PA 133rd Inf.	C	8/13/62	5/26/63		
15	Jones, Emanuel*	18	Pvt.	PA 184th Inf.	A	5/12/64		Died	
16	Jones, Joseph		Pvt.	PA 21st Cav.	E	7/2/63			
17	Jones, Samuel	34	Pvt.	PA 76th Inf.	E	10/9/61	2/18/64		
18	Jones, William		Pvt.	PA 21st Cav.	E	2/10/63	5/3/65		
19	Jones, William F		Pvt.	PA 63rd Inf.	D	8/1/61	9/9/64		
20	Jones, William J	26	Pvt.	PA 13th Cav.	L	11/4/62			
21	Jones, William J		Pvt.	PA 5th Heavy Art.	K	9/1/64	6/30/65		
22	Jordan, Daniel	18	Pvt.	PA 8th Res.	F	6/11/61	12/3/62		
23	Jordan, Daniel		Pvt.	US 5th Light Art.	C	12/3/62			
24	Jordon, Henry		Pvt.	USCT 8th Inf.	H	10/18/64	11/10/65		
25	Juda, George V A		Corp.	PA 8th Res.	F	6/11/61	5/15/64	POW	Seven Days Battle
26	Juda, George V A	20	Corp.	PA 8th Res.	F			Wound.-Died	Spotsylvania
27	Judy, Benjamin		Pvt.	IL 39th Inf.	A	5/1/62	7/1/64	Wounded	
28	Justice, Edward	24	Pvt.	PA 133rd Inf.	C	8/29/62	5/19/63	Wounded	Fredericksburg
29	Justice, Edward		Pvt.	PA 110th Inf.	C	2/1/64		Wound.-Died	Petersburg - Initial
30	Kagarice, Daniel R		Pvt.	IL 69th Inf.	H	6/14/62	9/27/62		
31	Kagarice, Daniel R		2nd Lt.	PA 22nd Cav.	M	7/11/63	6/8/65		
32	Kagarice, Ebenezer	20	Pvt.	PA 76th Inf.	C	7/11/61		KIA	Ft. Wagner, SC
33	Kagarice, John	29	Pvt.	PA 198th Inf.	F	8/31/64	5/17/65	Wound.-Died	White Oak Road
34	Kane, William*	28	Corp.	PA 110th Inf.	C	12/26/63	6/28/65		
35	Karchner, David		Pvt.	PA 13th Inf.	G	4/25/61	8/6/61		
36	Karder, William	34	Pvt.	PA 13th Inf.	G	4/25/61	8/6/61		
37	Karns, Jabez	28	Pvt.	MD 3rd PHB Inf.	C	2/15/62	2/15/65	POW	Harpers Ferry
38	Karns, Jacob	30	Pvt.	PA 22nd Cav.	H	2/26/64		Died	
39	Karns, Jacob Jr	30	Pvt.	PA 209th Inf.	E	9/2/64	6/19/65		
40	Karns, John H	40	Pvt.	PA			1864		
41	Karns, Simon	42	Pvt.	PA 208th Inf.	K	9/7/64	6/1/65		
42	Karns, Wilson	23	Pvt.	PA 79th Inf.	D	2/23/65	6/20/65		
43	Kauffman, David (1)	22	Pvt.	PA 133rd Inf.	C	8/29/62	5/19/63		
44	Kauffman, David (1)		Pvt.	PA McKeage's Mil.	D	7/2/63	8/8/63		
45	Kauffman, David (2)	17	Pvt.	PA 49th Inf.	D	8/19/61	9/10/64		
46	Kauffman, Isaac	15	Pvt.	PA 93rd Inf.	A	11/26/64	6/27/65		
47	Kauffman, John C	25	Pvt.	PA 79th Inf.	H	3/8/65	7/12/65		
48	Kauffman, Samuel B*	19	Pvt.	PA 194th Inf.	I	7/22/64	9/6/64		
49	Kauffman, Samuel B*		Pvt.	PA 97th Inf.	Ind	9/6/64	7/17/65		
50	Kay, Ezra P	31	Pvt.	PA 13th Inf.	G	4/25/61	8/6/61		
51	Kay, Ezra P		Sgt.	PA 110th Inf.	K	12/19/61			
52	Kay, Ezra P		Pvt.	PA 107th Inf.	G	9/21/64	6/6/65		
53	Kay, Harry H C	20	Corp.	PA 13th Inf.	G	4/25/61	8/6/61		
54	Kay, Harry H C		1st Lt.	PA 110th Inf.	C	10/24/61	12/10/62		
55	Kay, Harry H C		Capt.	PA 22nd Cav.	B	7/17/63			
56	Kay, Isaac	33	Surg.	PA 110th Inf.	K	10/24/61	6/20/62		
57	Kay, William H	20	Pvt.	PA 8th Res.	F	6/11/61		Wound.-Died	South Mountain
58	Keady, Isaac	22	Pvt.	MD 2nd PHB Inf.	A	4/6/65	5/29/65		
59	Keagy, David F		2nd Lt.	PA 208th Inf.	H	9/8/64	6/1/65		
60	Keagy, George		Pvt.	PA 208th Inf.	H	9/5/64	6/1/65		
61	Keagy, John T	20	Corp.	PA 101st Inf.	D	11/1/61	5/31/62	Wounded	Fair Oaks
62	Keagy, Samuel	25	Pvt.	PA 133rd Inf.	C	8/13/62	5/26/63		
63	Kean, William C	34	Pvt.	PA 125th Inf.	A	8/10/62	12/24/62		Antietam*
64	Keefe, John	40	Pvt.	CA 3rd Inf.	B	10/19/61	3/31/64		

	Cemetery	Notes
1	Mt. Ross Cem.-Bedford	Frank McCoy 1912 Listing; born in VA
2		1890 Napier Twp. Veterans Census
3		1890 Napier Twp. Veterans Census
4	Mt. Ross Cem.-Bedford	Frank McCoy 1912 Listing; 1890 Bedford Twp. Veterans Census
5	Mt. Zion Cem.-Chaneysville	B.S.F. History Book-1884; 1880 Southampton Twp. Census
6		Frank McCoy 1912 Listing
7		Wounded 6/3/64; B.S.F. History Book-1884; 1860 Woodbury Twp. Census
8		B.S.F. History Book-1884; Historical Data Systems record
9	Claysburg Union Cem.	Wound in forehead 12/13/62; 1890 Greenfield Twp. Veterans Census; John & Charles-brothers
10		Leg Amputated-Died 5/7/64; PA Civil War Archives-Blair Co. residence
11		B.S.F. History Book-1884
12	Fockler Cem.-Saxton	Frank McCoy 1912 Listing; 1860 Broad Top Census
13		Frank McCoy 1912 Listing
14		1860 Broad Top Census; Historical Data Systems-Bedford Co. residence
15	Cypress Hill Cem.-NY	Died 9/17/64; B.S.F. History Book-1884
16		Historical Data Systems-Bedford Co. residence
17		Historical Data Systems-Bedford Co. residence; Discharged on Surgeon's Cert.
18		Historical Data Systems-Bedford Co. residence
19		1890 Brush Creek Twp. Veterans Census
20	Mock Church Cem.-Alum Bank	Frank McCoy 1912 Listing
21		Historical Data Systems record
22		Historical Data Systems-Bedford Co. residence; 1860 S. Woodbury Twp. Census
23		Historical Data Systems record
24		Frank McCoy 1912 list & Civil Works Admin. 1934 list; Historical Data Systems record
25		POW 6/27/62; Bedford Gazette listing on 8th Regiment roster; Historical Data Systems record
26	Arlington Nat'l Cem., 13 6302	Wounded 5/8/64; Died 6/25/64; PA Civil War Archives-Bedford Co. residence
27		1890 Hyndman Veterans Census
28		Wounded 12/13/62; B.S.F. History Book-1884; enlisted in Hopewell
29		Wounded 6/18/64; Died 6/21/64; PA Civil War Archives-Woodbury residence
30		Pension Record
31		B.S.F. History Book-1884; 1860 S. Woodbury Twp. Census
32		KIA 7/11/63; PA CW Project record; Ebenezer & John-brothers; Clear Ridge (Clearville) resid.
33	Arlington Nat'l Cem., 13 9805	Wounded 3/31/65; Died 5/17/65; PA CW Project record; Born in Clear Ridge (Clearville)
34		PA Civil War Archives-born in MD
35		B.S.F. History Book-1884; Historical Data Systems-Bedford residence
36		B.S.F. History Book-1884; Historical Data Systems-Bedford residence
37	Clearville Union Cem.	POW 9/14/62; 1860 Southampton Twp. Census
38	Mt. Pleasant Cem.-Mattie	Died 9/17/64; B.S.F. History Book-1884; 1860 E. Providence Twp. Census
39		1860 Monroe Twp. Census; Died in Wetzel, WV
40	Mt. Pleasant Cem.-Mattie	1890 Monroe Twp. Veterans Census, Unassigned during 51 days in service
41	Baughman Union-W. Prov.	B.S.F. History Book-1884; Frank McCoy 1912 Listing
42	Graceville Luth. Cem.	E. Providence Honor Roll; 1860 W. Providence Census
43	Kauffman Cem.-Curryville	B.S.F. History Book-1884; 1860 Woodbury Twp. Census
44		B.S.F. History Book-1884 lists Bedford Co. residence
45	Union Memorial Cem.-Mench	1860 E. Providence Twp. Census; David (2) & Isaac-brothers
46	Union Memorial Cem.-Mench	1890 E. Providence Twp. Veterans Census; Frank McCoy 1912 Listing
47	Mennonite Cem.-Martinsburg	1890 Woodbury Veterans Census; 1870 M. Woodbury Twp. Census
48		B.S.F. History Book-1884; Muster in Reading, PA
49		B.S.F. History Book-1884
50	Mt. Muncie Cem.-KS	1850 Hopewell Twp. Census
51		Historical Data Systems-Bedford residence; Discharged on Surgeon's Cert.
52		Historical Data Systems-Bedford residence
53	Lone Tree Cem.-CO	1860 Hopewell Twp. Census
54		Resigned; PA Civil War Archives-Bedford Co. residence
55		Historical Data Systems record; Arrested at Muster Out; aka C H Kay
56	Logan Valley Cem.-Bellwood	Ancestry.com information; Isaac, Harry H C Kay & Ezra-brothers
57	Yellow Creek Reformed Cem.	Wounded 9/14/62; Died 9/18/62; Historical Data Systems-Bedford Co. residence
58	Everett Cem.	Everett Cemetery Listing of Civil War Veterans, 1880 Hyndman Census
59	Oak Ridge Cem.-Altoona	B.S.F. History Book-1884; Promoted Pvt to 2nd Lt.; 1860 Woodbury Twp. Census
60	Wellsville Cem.-KS	Historical Data Systems-Bedford Co. residence; George & David-brothers
61	Leavenworth Nat'l Cem.	Wounded 5/31/62; B.S.F. History Book-1884; 1850 M. Woodbury Twp. Census
62	Potter Creek Cem.-Woodbury	1890 Woodbury Twp. Veterans Census
63	Bedford Cem.	1890 Bedford Veterans Census; 1870 Bedford Twp. Census; Discharged on Surgeon's Cert.
64	Bedford Cem.	B.S.F. History Book-1884; 1890 Bedford Twp. Veterans Census; Discharged Surgeon's Cert.

	Name	Age	Rank	Regiment	Co.	Muster In	Muster Out	Casualty	Casualty Battle
1	Keefe, John		2nd Lt.	CA 2nd Inf.	C	5/7/64	5/30/66		
2	Keeffe, Joseph	32	Pvt.	PA 55th Inf.	K	11/5/61	10/18/64		
3	Keel, George	22	Pvt.	PA 171st Inf.	I	11/5/62	8/8/63		
4	Keely, Thomas	17	Pvt.	PA McKeage's Mil.	H	7/2/63	8/8/63		
5	Keely, Thomas		Pvt.	PA 55th Inf.	H	10/16/63			
6	Kegerreis, Peter*								
7	Kegg, Andrew		Pvt.	MD 2nd PHB Inf.	A	9/1/62	5/29/65		
8	Kegg, Emanuel	37	Pvt.	PA 18th Cav.	K	11/14/62		Died	
9	Kegg, Jacob	31	Pvt.	PA 13th Inf.	G	4/25/61	8/6/61		
10	Kegg, Jacob		Pvt.	PA 55th Inf.	D	2/12/62	10/1/64		
11	Kegg, James P	16	Pvt.	PA 55th Inf.	H	2/29/64	8/30/65		
12	Kegg, Joseph	29	Pvt.	PA 77th Inf.	C	2/29/64	12/6/65		
13	Kegg, Levi	40	Corp.	PA 101st Inf.	D	12/1/61	1862		
14	Kegg, Levi		Corp.	PA 149th Inf.	B	9/23/63	11/6/64	Wounded	Wilderness
15	Kegg, Levi R	14	Pvt.	MD 2nd PHB Inf.	A	10/29/61	10/31/64		
16	Kegg, Nathaniel	21	Pvt.	PA 138th Inf.	E	8/29/62	6/23/65	Wounded	Wilderness
17	Kegg, Simon P	23	Pvt.	PA 101st Inf.	D	11/1/61	11/23/64		
18	Keggs, John W	17	Pvt.	OH McL Cav.	B	1/5/61	10/30/65	Wounded	Gladisville, WV
19	Keiser, David O	20	Pvt.	PA 76th Inf.	E	10/9/61	11/28/64		
20	Keith, George J	19	Pvt.	PA 3rd Heavy Art.	C	2/29/64	11/9/65		
21	Keith, Wilson R	21	Pvt.	PA 126th Inf.	B	8/13/62	5/20/63		
22	Kell, Ezra		Pvt.	PA 21st Cav.	E	7/2/63			
23	Keller, Jacob		Sgt.	PA 21st Cav.	E	7/9/63	7/8/65	Wounded	Petersburg - Initial
24	Keller, John	24	Pvt.	PA 171st Inf.	H	11/1/62	8/8/63		
25	Keller, John		Pvt.	PA 50th Inf.	D	2/24/65	7/30/65		
26	Kellerman, James L	28	Pvt.	PA 138th Inf.	F	9/29/62	6/23/65	Wounded	Monocacy, MD
27	Kellerman, James L		Pvt.	PA 138th Inf.	F			Wounded	Cedar Creek
28	Kellerman, John L	26	Pvt.	OH 60th Inf.	H	8/9/62	11/10/62	POW	
29	Kelley, David	18	Pvt.	PA McKeage's Mil.	G	7/2/63	8/8/63		
30	Kelley, David		Pvt.	PA 208th Inf.	H	9/7/64	6/1/65		
31	Kelley, James*	40	Pvt.	PA 76th Inf.	E	9/25/64	6/16/65		
32	Kelley, John		Pvt.	PA 107th Inf.	H	3/14/62	6/19/64	Wounded	Petersburg
33	Kelley, John A	23	Corp.	PA 125th Inf.	D	8/13/62	9/17/62	KIA	Antietam
34	Kellogg, George W*	22	Pvt.	PA 12th Cav.	B	3/15/62			
35	Kellogg, George W*		Corp.	PA 76th Inf.	E	3/24/64	7/18/65		
36	Kelly, David	19	Pvt.	PA 110th Inf.	C	10/24/61	6/2/65		
37	Kelly, George P*	25	Corp.	PA 110th Inf.	C	2/27/64	6/28/65		
38	Kelly, Henry	20	Pvt.	PA 138th Inf.	F	8/29/62		MIA	Cold Harbor
39	Kelly, James*	19	Pvt.	PA 184th Inf.	A	5/12/64	7/14/65		
40	Kelly, John T	36	Pvt.	PA 13th Inf.	G	4/25/61	8/6/61		
41	Kelly, Oliver F*	19	Pvt.	PA 148th Inf.	B	8/31/62		POW-Died	Jerusalem Plank Rd.
42	Kelly, Richard*	24	Pvt.	PA 76th Inf.	E	7/13/63	7/18/65		
43	Kelly, Thomas	18	Pvt.	PA 99th Inf.	B	3/14/65	7/1/65		
44	Kelly, Uriah	37	Pvt.	PA 67th Inf.	D	11/14/64	7/14/65		
45	Kelly, William (1)		Pvt.	PA 22nd Cav.	A	7/16/63	2/5/64		
46	Kelly, William (2)	31	Pvt.	PA 138th Inf.	F	8/29/62	8/16/65	Wounded	Cedar Creek
47	Kelly, William H	17	Pvt.	PA 54th Inf.	G	3/11/64	5/31/65		
48	Kendig, John H*	23	Pvt.	PA 76th Inf.	E	11/21/61	11/28/64		
49	Kennard, John H	21	Pvt.	PA 138th Inf.	D	8/29/62	6/23/65	Wounded	Mine Run
50	Kennard, Matthew P	25	Pvt.	OH 197th Inf.	D	2/13/65	5/29/65		
51	Kennard, Wlliam B	25	Pvt.	PA 101st Inf.	D	11/1/61	7/10/65		
52	Kennedy, James	26	Pvt.	PA 76th Inf.	E	10/7/64	7/18/65		
53	Kennedy, Samuel	20	Corp.	PA 55th Inf.	D	10/12/61		KIA	Drewry's Bluff
54	Kennell, Nathaniel*	13	Pvt.	PA 176th Inf.	G	11/7/62	5/8/63		
55	Kenyon, Matthias*	21	Pvt.	PA 55th Inf.	D	10/14/63			
56	Kephart, Cyrus*		Pvt.	PA 55th Inf.	K	10/11/61		POW-Died	Ft. Harrison
57	Kephart, John	20	Pvt.	PA 13th Cav.	D	9/25/63	7/14/65		
58	Kerns, MacDonald R	36	Pvt.	US 5th Cav.	E	5/1/62			
59	Kerns, Mark C	25	Capt.	PA 1st Light Art.	G	7/26/61		Wounded	Seven Days Battle
60	Kerns, Mark C		Capt.	PA 1st Light Art.	G			KIA	2nd Bull Run
61	Kerns, Robert A	32	Pvt.	PA 158th Inf.	H	11/4/62	8/12/63		
62	Kessler, John	27	Pvt.	PA 55th Inf.	H	9/23/63	8/30/65	Wounded	Bermuda Hundred
63	Kettering, Elijah	19	Pvt.	PA 194th Inf.	I	7/22/64	11/5/64		
64	Kettering, Jacob T	22	1st Lt.	PA 171st Inf.	I	11/11/62	8/8/63		

	Cemetery	Notes
1		1840 Bedford Census; Lt. in Mexican War
2	Bedford Cem.	Pension Record-Hospital Steward; Joseph & John-brothers
3		B.S.F. History Book-1884; 1870 Monroe Twp. Census
4		B.S.F. History Book-1884; 1860 Bedford Twp. Census
5		Deserted 3/20/64; B.S.F. History Book-1884
6	Everett Cem.	Born 1808; Civil War Marker on grave
7	Rose Hill Cem.-Cumberland	Frank McCoy 1912 Listing; Ancestry.com-Andrew & Levi R-brothers
8		Died 6/26/63; PA Civil War Project record; 1860 Colerain Twp. Census
9	Friends Cove UCC Cem.	B.S.F. History Book-1884; 1850 Colerain Twp. Census; Jacob & Emanuel-brothers
10		B.S.F. History Book-1884; Transferred to Veterans Reserve Corp. 10/1/64
11	Grandview Cem.-Cambria Co.	B.S.F. History Book-1884; 1860 Juniata Twp. Census
12	Cedar Hill Cem.-OH	PA Civil War Archives record; 1850 Bedford Twp. Census
13	Friends Cove UCC Cem.	B.S.F. History Book-1884; Historical Data Systems record; 1860 Colerain Twp. Census
14		Wounded 5/5/64; B.S.F. History Book-1884; Transferred to Veterans Reserve Corp.
15	Rose Hill Cem.-Cumberland	Historical Data Systems record; Ancestry.com record
16	Grandview Cem.-Cambria Co.	Wounded 5/6/64; Historical Data Systems-Bedford Co. residence
17	Mansfield Cem.-OH	Historical Data Systems record; 1860 Colerain Twp. Census; Not Captured at Plymouth, NC
18	Pine Hill Cem.-KY	Wounded 7/7/63 in McLaughlins Cavalry unit; Pension Records-born in Bedford Co.
19		B.S.F. History Book-1884; PA Civil War Archives-Bedford residence
20	Pleasant Hill Cem.-NE	1890 Wells Twp. Veterans Census; aka Kieth
21	Wells Valley Chapel Cem.	1890 Wells Twp. Veterans Census; Wilson & George-brothers
22		Historical Data Systems-Bedford Co. residence
23		Wounded 6/18/64; Historical Data Systems-Bedford Co. residence
24	Trinity UCC Cem.-Juniata Twp.	Findagrave.com information; 1860 Juniata Twp. Census
25		Findagrave.com information
26	Cherry Mt. Cem.-KS	Wounded 7/9/64; B.S.F. History Book-1884; Historical Data Systems record
27		Wounded 10/19/64; 1860 Juniata Twp. Census
28	Mapple Grove Cem.-OK	POW-Libby; Findagrave.com information; James & John-brothers
29	Prairie Home Cem.-NE	B.S.F. History Book-1884; 1860 Liberty Twp. Census
30		B.S.F. History Book-1884; Historical Data Systems-Bedford Co. residence
31		B.S.F. History Book-1884
32	Mt. Union Cem.-Mt. Union	Foot shot off- by Cannon ball 6/19/64; 1850 Hopewell Twp. Census
33	Fairview Cem.-Altoona	KIA 9/17/62; Historical Data Systems record; 1850 Broad Top Twp. Census
34		B.S.F. History Book-1884 ; Historical Data Systems record
35		PA Civil War Archives-Crawford Co. residence
36	Hopewell Cem.	B.S.F. History Book-1884; 1860 Liberty Twp. Census
37		B.S.F. History Book-1884; Blair Co. references
38		MIA 6/1/64; PA Civil War Archives-Colerain Twp. residence; Historical Data Systems record
39		B.S.F. History Book-1884
40		B.S.F. History Book-1884; Historical Data Systems-Bedford Co. residence
41	Andersonville Cem., 4895	POW-Andersonville 6/22/64; Died 8/6/64; Bedford Gazette 1914; Somerset Co. residence
42		B.S.F. History Book-1884
43	St. Lukes Cem.-Saxton	Frank McCoy 1912 Listing; 1860 Liberty Twp. Census; Thomas & David Kelley-brothers
44	Schellsburg Cem.	Frank McCoy 1912 Listing; 1860 Harrison Twp. Census
45		B.S.F. History Book-1884; 1890 Liberty Twp. Veterans Census
46		Wounded 10/19/64; PA Civil War Archives-Colerain Twp. residence
47	Lybarger Cem.-Madley	Findagrave.com information; Served under name of William H Breathlin
48		B.S.F. History Book-1884; York references
49	Union Church Cem.-Clearville	Wounded 11/27/63; Historical Data Systems record; 1850 Monroe Twp. Census
50	Asbury Meth. Cem.-OH	Ancestry.com-born Bedford Co.; 1840 W. Providence Twp. Census
51	Union Church Cem.-Clearville	Historical Data Systems record; William & John-brothers
52		B.S.F. History Book-1884
53		KIA 5/16/64; B.S.F. History Book-1884; PA Civil War Archives-Bedford Co. residence
54	Hyndman Cem.	Historial Data Systems record; appears this Nathaniel Kennell not in 176th PA Inf.
55		Deserted 4/27/64; B.S.F. History Book-1884; Drafted in Reading, PA
56	Baptist Cem.-Joanna PA	Leg Amputated 9/24/64, Died 11/18/64; B.S.F. History Book-1884; Berks Co. references
57	Hopewell Cem.	Historical Data Systems record; 1870 Broad Top Twp. Census
58	Presbyterian Cem.-Bedford	Frank McCoy 1912 Listing; MacDonald & Mark-brothers
59		Wounded 6/27/62; Father buried Presb. Cem.-Bedford
60		KIA 8/30/62; PA Civil War Archives; Ancestry.com-born Bedford Boro
61	Black Oak Cem.-Fulton Co.	Ancestry.com-son born-Mann Twp. in 1862
62		Wounded 5/13/64 listed in Bedford Inquirer article 7/8/64; B.S.F. History Book-1884
63	Holsinger Cem.-Bakers Summit	1890 Bloomfield Twp. Veterans Census; aka Ketring & Kettring
64	Highland Cem.-KS	B.S.F. History Book-1884; 1860 S. Woodbury Twp. Census

	Name	Age	Rank	Regiment	Co.	Muster In	Muster Out	Casualty	Casualty Battle
1	Kettering, Jacob T		Sgt.	PA 194th Inf.	I	7/22/64	11/5/64		
2	Key, James	41	Pvt.	USCT 32nd Inf.	G	2/25/64	8/22/65		
3	Key, Philip	41	Pvt.	USCT 32nd Inf.	B	2/12/64	9/22/65		
4	Keyser, Samuel	15	Pvt.	PA 50th Inf.	I	9/28/64	6/2/65		
5	Kichinann, Adam	26	Pvt.	PA 99th Inf.	D	2/25/65	7/1/65		
6	Kiester, Levi		Pvt.	PA 76th Inf.	E	10/7/64	7/15/65		
7	Kifer, Jacob	17	Pvt.	PA 46th Inf.	I	7/1/63	8/19/63		
8	Kifer, Jacob		Pvt.	PA 208th Inf.	B	8/26/64	6/1/65		
9	Kilpatrick, Daniel*	20	Corp.	PA 194th Inf.	I	7/22/64	11/5/64		
10	King, Daniel	45	Pvt.	PA 92nd Inf.	C	1/11/64	6/21/65		
11	King, Erastus	33	Pvt.	PA 148th Inf.	E	5/10/63	6/1/65	Wounded	Po River
12	King, Harrison H	23	Corp.	PA 138th Inf.	E	8/29/62	6/7/65	Wounded	Wilderness
13	King, Harrison H		Corp.	PA 138th Inf.	E			Wounded	Opequon
14	King, Harrison H		Corp.	PA 138th Inf.	E			Wounded	Sailor's Creek
15	King, Hezekiah	23	Pvt.	PA 3rd Inf.	E	4/20/61	7/29/61		
16	King, Hezekiah		Pvt.	PA 16th Cav.	G	2/16/65	8/11/65		
17	King, John	44	Pvt.	PA 171st Inf.	I	11/5/62	8/8/63		
18	King, John T	18	Pvt.	PA 76th Inf.	E	10/9/61	2/18/63		
19	King, John V	50	Major			1863			
20	King, Philip V	22	Sgt.	PA 133rd Inf.	K	8/15/62	5/26/63		
21	King, Philip V		Pvt.	PA 186th Inf.	H	3/14/64	8/15/65		
22	King, Samuel T	15	Pvt.	PA 55th Inf.	H	2/19/64	8/30/65	Wounded	Drewry's Bluff
23	King, Thomas	50	1st Lt.	PA 101st Inf.	G	12/28/61	3/11/65	POW	Plymouth, NC
24	King, Watson	20	Pvt.	PA 76th Inf.	E	10/9/61		Died	
25	King, William B	24	Pvt.	PA 208th Inf.	H	9/7/64	6/1/65		
26	Kingsley, David		Pvt.	PA 138th Inf.	F	9/29/62			
27	Kinley, Jacob	43	Pvt.	PA 55th Inf.	K	11/5/61		Died	
28	Kinley, Samuel*	19	1st Lt.	PA 110th Inf.	C	12/19/61	3/18/65		
29	Kinsey, Benjamin F	33	Pvt.	PA 206th Inf.	K	9/2/64	6/26/65		
30	Kinsey, Dewalt	24	Pvt.	PA 206th Inf.	K	9/2/64	6/26/65		
31	Kinsey, John B	22	Pvt.	PA 138th Inf.	D	8/29/62	4/5/63		
32	Kinsey, John B		Pvt.	MD 2nd PHB Inf.	K	4/5/63	5/29/65		
33	Kinsey, Peter Jr	19	Corp.	PA 55th Inf.	K	11/5/61	8/30/65		
34	Kinsey, Peter Sr	50	Pvt.	PA 55th Inf.	K	11/5/61	4/22/63		
35	Kinton, Allen	39	Corp.	PA 138th Inf.	D	8/29/62	6/23/65		
36	Kinton, David		Pvt.	PA 55th Inf.	K	10/11/61		KIA	Cold Harbor
37	Kinton, John G	34	Pvt.	US 1st Inf.	E	10/28/61	10/1/65		
38	Kipp, Jonas	23	Pvt.	PA 55th Inf.	K	11/5/61	9/4/64		
39	Kipp, Jonas			US 1st Art.					
40	Kipp, Lewis A	19	Pvt.	MD 2nd PHB Inf.	A	3/15/65	5/29/65		
41	Kirk, William	42	Pvt.	PA 149th Inf.		2/25/65	5/5/65		
42	Kissel, Benjamin	21	Pvt.	PA 208th Inf.	K	9/7/64	5/30/65		
43	Kissel, John	19	Pvt.	PA 208th Inf.	K	9/7/64	6/1/65		
44	Klahre, Herman T	20	Pvt.	PA 133rd Inf.	K	8/15/62	5/26/63	Wounded	Fredericksburg
45	Klahre, Herman T		Corp.	PA 184th Inf.	A	5/12/64		Wound.-Died	Jerusalem Plank Rd.
46	Klahre, Theodore M	20	Corp.	PA 76th Inf.	E	10/9/61	11/28/64	Wounded	Pocotaligo, SC
47	Klahre, Theodore M		Corp.	PA 76th Inf.	E			Wounded	Cold Harbor
48	Kline, James S	19	Pvt.	PA 55th Inf.	I	10/11/61		Wound.-Died	Bermuda Hundred
49	Kline, Peter	25	Pvt.	PA 152nd Inf.	F	9/28/64	6/9/65		
50	Klotz, Jacob B	25	Pvt.	NY 59th Inf.	H	10/12/61	12/30/63	Wounded	Fredericksburg
51	Klotz, Joseph B	22	Pvt.	OH 16th Inf.	I	4/27/61	8/18/61		
52	Klotz, Joseph B		Pvt.	IN 30th Inf.	C	9/30/64	6/23/65		
53	Knabb, Albert	20	Sgt.	PA 76th Inf.	E	8/27/63	7/18/65	Wounded	2nd Fair Oaks
54	Knapp, James M		Pvt.	PA 55th Inf.	D	1862	8/30/65		
55	Knee David H	32		VA 14th Mil.					
56	Knee David H		Sgt. QM	VA 18th Cav.	I		4/25/65		
57	Knee, Augustus D	21	Pvt.	IL 53rd Inf.	A	2/18/62		Died	
58	Knee, Philip	36	Pvt.	MD 2nd PHB Inf.	A	9/7/61	5/29/65	Wounded	Lynchburg
59	Knipple, Andrew J	29	Pvt.	PA 101st Inf.	G	2/18/62	1/12/63	POW	
60	Knipple, Andrew J		Pvt.	PA 19th Cav.	L	9/17/63	5/14/66		
61	Knipple, Frederick L	30	Pvt.	PA 13th Cav.	G	2/26/64	7/14/65		
62	Knipple, George W	29	Pvt.	PA 13th Cav.	G	2/26/64	7/14/65		
63	Knipple, Jacob H	17	Pvt.	WV 10th Inf.	D	3/1/62			
64	Knipple, Jacob H		Pvt.	WV 12th Inf.	D	8/30/62			

	Cemetery	Notes
1		B.S.F. History Book-1884; Jacob & Elijah-brothers
2	Mt. Ross Cem.-Bedford	Frank McCoy 1912 Listing; James & Philip-brothers
3	Mt. Ross Cem.-Bedford	Frank McCoy 1912 Listing; 1860 Bedford Twp. Census
4	Hyndman Cem.	Frank McCoy 1912 Listing; Bedford Inquirer 5/22/1908 listing
5		Biographical Review: Bedford & Somerset Co. book-1899; 1880 Hopewell Census
6		B.S.F. History Book-1884
7	Shellytown Fairview-Blair Co.	1860 S. Woodbury Twp. Census
8		Historical Data Systems record
9		B.S.F. History Book-1884; Mustered in Reading, PA
10	Oak Hill Cem.-IL	Ancestry.com information; 1850 S. Woodbury Twp. Census
11	Greenlawn Cem.-Roaring Spring	Knee Wound 5/10/64; Historical Data Systems record; 1850 Schellsburg Census
12		Wounded 5/5/64; PA Civil War Archives-Bedford Co. residence
13		Wounded 9/19/64; Historical Data Systems record
14		Wounded 4/6/65; Historical Data Systems record
15	Rose Hill Cem.-Altoona	1850 Schellsburg Census
16		Historical Data Systems record
17	Mt. Zion Cem.-Pavia	Deserted 11/25/62 & returned 1/14/63; Union Twp. (Pavia) Census
18	Rose Hill Cem.-Altoona	Frank McCoy 1912 Listing; Discharged Surg. Cert.; John T, Erastus, Samuel & Hezekiah-bros.
19		Ancestry.com-Union Commissary General; John & Thomas-brothers
20	Everett Cem.	B.S.F. History Book-1884; 1860 W. Providence Twp. Census
21		Historical Data Systems record
22	Greenwood Cem.-Indiana	Wounded 5/16/64 listed in Bedford Inquirer article 7/8/64; B.S.F. History Book-1884
23	Hartley Family Cem.-Everett	POW 4/20/64 to 3/11/65; Historical Data Systems record; 1850 Hopewell Twp. Census
24	Bedford Cem.	Died-Typhoid Fever 6/18/62; PA Civil War Archives-Bedford residence
25	Asbury Meth.-Harrisonville	B.S.F. History Book-1884; Historical Data Systems-Bedford Co. residence
26		Deserted 10/23/62; Historical Data Systems-Bedford Co. residence
27	Beaufort Nat'l Cem.	Died 10/13/62; B.S.F. History Book-1884; PA Civil War Archives-Bedford Co. residence
28		B.S.F. History Book-1884; Huntington Co. residence
29	Schellsburg Cem.	1890 Napier Twp. Veterans Census
30	Helixville Cem.-Napier Twp.	Historical Data Systems record
31	Schellsburg Cem.	B.S.F. History Book-1884; John, Benjamin, Dewalt, Peter-brothers
32		Historical Data Systems record
33	Grandview Cem.-Cambria Co.	Frank McCoy 1912 Listing; 1860 Juniata Twp. Census
34	Schellsburg Cem.	Disch. Surg. Cert.; B.S.F. Hist. Book-1884; Father of Peter JR, John, Dewalt & Bengamin
35	Mt. Olivet Cem.-Manns Choice	Frank McCoy 1912 Listing; 1890 Napier Twp. Veterans Census
36		KIA 6/3/64; B.S.F. History Book-1884; Muster Rolls-enlisted in Bedford
37	Bedford Cem.	1890 Bedford Twp. Veterans Census; Mexican War Veteran; Discharged on Surgeon's Cert.
38	Dry Ridge Cem.-Manns Choice	1890 Harrison Twp. Veterans Census; Historical Data Systems record; Jonas & Lewis-brothers
39		Historical Data Systems record
40	Hyndman Cem.	1890 Harrison Twp. Veterans Census
41	Fishertown Bretheran Cem.	1890 E. St. Clair Twp. Veterans Census; 1860 St. Clair Census
42	Mt. Union Cem.-Mench	B.S.F. History Book-1884; 1860 W. Providence Twp. Census
43	Mt. Union Cem.-Mench	B.S.F. History Book-1884; John & Benjamin-brothers
44		Wounded 12/13/62; B.S.F. History Book-1884; PA Civil War Archives-Bedford Co. residence
45	Alexandria Nat'l Cem., R 2451	Wounded 6/22/64; Died 7/20/64; B.S.F. History Book-1884; Historical Data Systems record
46	Everett Cem.	Wounded 10/22/63; 1890 Everett Veterans Census; B.S.F. History Book-1884
47		Wounded 6/1/64; 1870 Bloody Run Census
48		Wounded 5/19/64; Died 5/20/64; B.S.F. Hist. Book-1884; PA CW Arch.;born in Bedford Co.
49	Hopewell Cem.	1890 Broad Top Twp. Veterans Census
50	Cosperville Cem.-IN	Wounded 12/13/62; Jacob & Joseph-brothers; transferred to Veterans Reserve Corp.
51	Cosperville Cem.-IN	Ancestry.com-Born in Bedford Co.
52		Historical Data Systems record
53		Wounded 10/27/64; B.S.F. History Book-1884
54		B.S.F. History Book-1884
55	Wardensville Cem.-WV	Civil War record provided by Larry Knee-Knee family historian
56		Historical Data Systems record; Born in Woodbury Twp.; Confederate soldier
57		Historical Data Systems record; Born in Woodbury Twp.
58	Chaneysville Meth. Cem.	Hand Wound 6-18-64; Frank McCoy 1912 Listing; Philip, David & Augustus-brothers
59	Benshoff Hill Cem.-Cambria Co.	B.S.F. History Book-1884 listed POW
60		PA Civil War Archives record
61	Messiah Luth.-Dutch Corner	Findagrave.com.com information; Frederick, Andrew, George, William & John-brothers
62	Greenfield Cem.-Queen	1890 Kimmel Twp. Veterans Census
63	Walker Cem.-Shanksville	Historical Data Systems record; 1850 Napier Twp. Census
64		Historical Data Systems record

	Name	Age	Rank	Regiment	Co.	Muster In	Muster Out	Casualty	Casualty Battle
1	Knipple, John A	21	Pvt.	PA 84th Inf.	A	10/24/61	2/28/63		
2	Knipple, John A		Corp.	PA 3rd Heavy Art.	L	2/23/64	11/9/65		
3	Knipple, William H		Corp.	PA 101st Inf.	G			POW	Plymouth, NC
4	Knipple, William H	18	Corp.	PA 101st Inf.	G	12/28/61	6/25/65	Wounded	Fair Oaks
5	Knisely, Christopher	22	Music.	IL 148th Inf.	E	2/8/65	9/5/65		
6	Knisely, David	21	Pvt.	PA 138th Inf.	F	8/19/62			
7	Knode, Thomas*	31	Pvt.	PA 110th Inf.	C	10/24/61	10/24/64		
8	Knox, James H	24	Sgt.	PA 171st Inf.	I	11/2/62	8/8/63		
9	Knox, James H		1st Sgt.	PA 184th Inf.	A	5/12/64		POW-Died	Jerusalem Plank Rd.
10	Knox, Otho S	16	Pvt.	PA 55th Inf.	D	10/12/61	12/30/64	Wounded	
11	Koch, John*		Pvt.	PA 55th Inf.	K	9/23/63		MIA	Chaffin's Farm
12	Kochendarfer, John Z	22	Pvt.	PA 133rd Inf.	C	8/29/62	5/19/63	Wounded	Fredericksburg
13	Koehler, Martin*	28	Pvt.	PA 76th Inf.	E	2/22/65	7/18/65		
14	Konley, John		Pvt.	PA 93rd Inf.	D	11/14/64	6/27/65		
15	Kooken, John R*		Capt.	PA 110th Inf.	C	6/27/62		KIA	Fredericksburg
16	Koontz, Charles	22	Pvt.	PA 184th Inf.	A	3/3/64	7/14/65	Wounded	
17	Koontz, George	24	Pvt.	PA 55th Inf.	D	10/12/61	8/30/65		
18	Koontz, James	23	Pvt.	IA 5th Cav.	H	11/1/62		KIA	Chehaw, AL
19	Koontz, John	33	Pvt.	IA 5th Cav.	H	11/1/62	6/17/65		
20	Koontz, Peter	26	Pvt.	IA 5th Cav.	H	11/1/62	6/17/65		
21	Koontz, William	38	Pvt.	IA 5th Cav.	H	11/1/62	6/13/65		
22	Krausen, E W		Capt.	USCT Inf.	A	7/1/61	8/1/64		
23	Kreiger, John	31	Pvt.	PA 13th Inf.	G	4/25/61	8/6/61		
24	Kreiger, John		Pvt.	PA 55th Inf.	H	10/11/61	8/30/65		
25	Krielman, Johan A		Pvt.	PA 99th Inf.	D	2/25/65	7/1/65		
26	Kromer, George		Corp.	PA 55th Inf.	H	9/2/62	6/11/65		
27	Krydler, Jacob	36	Pvt.	PA 52th Inf.	D	9/24/63			
28	Kuchman, J Adam		Pvt.	PA 99th Inf.	D	2/25/65	7/1/65		
29	Kuh, William	22	Pvt.	PA 12th Res.	K	6/15/61	6/11/64		
30	Kuh, William		Pvt.	PA 190th Inf.	E	1/23/65	7/3/65		
31	Kuhn, Andrew F*	45	Capt.	PA 93rd Inf.	G	10/26/61	6/27/65		
32	Kuhn, John R*		1st Lt.	PA 93rd Inf.	G	9/10/64	6/27/65	Wounded	Petersburg - Final
33	Kurtz, Jacob*		Pvt.	PA 55th Inf.	K	12/1/63		POW-Died	
34	Kurtz, Thomas	31	Pvt.	PA 138th Inf.	D	9/2/62	6/23/65		
35	Laher, Henry H	25	Sadd.	PA 19th Cav.	C	9/24/64	7/25/65	Wounded	Sugar Creek
36	Lair, Moses	23	Pvt.	PA 55th Inf.	D	11/12/61	6/15/63		
37	Lair, Moses		Pvt.	US 1st Reg Art.		6/15/63			
38	Lake, John	27	Pvt.	PA 82nd Inf.	G	8/26/63	7/13/65		
39	Lamb, Thomas		Pvt.	PA 63rd Inf.	E	9/9/61	9/9/64		
40	Lamberson, Daniel A		Pvt.	PA 126th Inf.	B	8/12/62	5/20/63		
41	Lamberson, Daniel A	41	Pvt.	OH 34th Inf.	E	10/1/64	1/30/65		
42	Lamberson, Daniel A		Pvt.	OH 36th Inf.	E	1/30/65	7/27/65		
43	Lamberson, David	22	Pvt.	PA 133rd Inf.	K	8/15/62	5/26/63		
44	Lambert, Joseph C	18	Corp.	PA 133rd Inf.	D	8/14/62	5/27/63	Wounded	Fredericksburg
45	Lambert, Joseph C		Sgt.	PA 21st Cav.	E	7/9/63	7/8/65	Wounded	Amelia Springs
46	Lambert, Josiah		Pvt.	PA 5th Res.	K	8/1/63	7/1/65		
47	Lambert, Josiah O	20	Pvt.	PA 21st Cav.	E	7/2/63	7/8/65		
48	Lambright, S*				A				
49	Lambright, William	21	Pvt.	PA 84th Inf.	A	10/24/61		POW	Chancellorsville
50	Lamison, David M	22	Pvt.	PA 93rd Inf.	F	11/14/64	6/10/65		
51	Lamison, George W	23	Pvt.	PA 110th Inf.	C	10/24/61		Wound.-Died	Gettysburg
52	Lamison, James H	16	Pvt.	PA 22nd Cav.	B	7/11/63	10/31/65		
53	Lamison, Thomas	22	Pvt.	PA 110th Inf.	C	10/24/61		Died	
54	Lane, David C	18	Sgt.	PA 110th Inf.	C	10/24/61	6/28/65		
55	Lang, James*	32	Pvt.	PA 110th Inf.	C	2/27/64		POW-Died	
56	Langdon, Samuel	36	Sgt.	PA 133rd Inf.	C	8/13/62	5/26/63		
57	Langdon, Samuel		Sgt.	PA 208th Inf.	H	9/7/64	6/1/65	Died	
58	Lanning, James W		Pvt.	OH 159th Inf.	K	1864	1864		
59	Lape, Abraham	48	Pvt.	PA 171st Inf.	H	11/1/62	8/8/63	Wounded	New Berne, NC
60	Lape, Abraham			PA 97th Inf.					
61	Lape, Jackson	37	Pvt.	PA 138th Inf.	D	8/29/62			
62	Larmon, John S*	19	Music.	PA 55th Inf.	I	9/20/61	8/30/65		
63	Larmon, William S*	37	Sgt.	PA 55th Inf.	I	10/10/61	5/14/62		
64	Lashley, Daniel	23	Pvt.	PA 55th Inf.	D	10/12/61	10/26/64		

	Cemetery	Notes
1	Asbury Cem.-Altoona	Historical Data Systems record
2		PA Civil War Archives
3		POW-Andersonville 4/20/64 to 2/24/65; B.S.F. History Book-1884; 1850 Napier Twp. Census
4	Alverton-Westmoreland Co.	Wounded 5/31/62; Historical Data Systems record
5	Ritchey Cem.-Snake Spring V.	1890 Snake Spring Valley Vet. Census; 1860 Hopewell Twp. Cen; 149th PA mentioned
6		Deserted 10/23/62; PA Civil War Archives-Colerain Twp. residence
7		B.S.F. History Book-1884; Huntington Co. references
8		B.S.F. History Book-1884; 1850 Bedford Twp. Census; Otho & James brothers
9	Andersonville Cem., 12695	POW-Andersonville; 6/22/64; Died-Diarrhea 2/23/65; 1860 Bedford Twp. Census
10	Elwood Cem.-IA	Wounded listed in Bedford Inquirer article 6/3/64; PA Civil War Archives-Bedford Co. resid.
11		MIA 9/29/64; B.S.F. History Book-1884; Mustered in Reading, PA
12	New Enterprise Cem.	Hand Wound 12/13/62; 1870 S. Woodbury Twp. Census
13		B.S.F. History Book-1884
14		1890 Londonderry Twp. Veterans Census
15	Fredericksburg Nat'l Cem.	B.S.F. History Book-1884; Died 12/14/62; Historical Data Systems-Blair Co. residence
16	Lashley Cem.-Artemas	1890 Mann Twp. Veterans Census-shot in back
17	Everett Cem.	Historical Data Systems record; 1860 Snake Spring Valley Census
18		KIA 7/18/64; Ancestry.com-born in Cessna
19		Historical Data Systems information aka Koonts; John, James, Peter & William-brothers
20		Historical Data Systems information aka Koonts; Ancestry.com-born in St. Clair Twp.
21		Historical Data Systems information aka Koonts;; Ancestry.com-born in St. Clair Twp.
22		1890 Everett Veterans Census
23		B.S.F. History Book-1884; Historical Data Systems record; 1860 St. Clair Twp. Census
24		Historical Data Systems-Bedford Co. residence
25		1890 Hopewell Veterans Census
26		B.S.F. History Book-1884
27	Mt. Zion Cem.-Pavia	Bedford Inquirer 5/22/1908 listing; 1860 Union Twp. (Pavia) Census; aka Krider or Kreider
28	Saint John's Cem.-Loysburg	Frank McCoy 1912 Listing; 1880 Hopewell Twp. Census
29	Hopewell Cem.	1890 Broad Top Veterans Census; Frank McCoy 1912 Listing
30		1890 Broad Top Veterans Census
31		B.S.F. History Book-1884; Lebanon references
32		B.S.F. History Book-1884; Wounded 4/2/65; Lebanon references
33	Andersonville Cem., 11238	POW-Andersonville; Died 10/21/64; B.S.F. History Book-1884; Mustered in Reading, PA
34	Schellsburg Cem.	B.S.F. History Book-1884; 1860 Napier Twp. Census
35	Everett Cem.	Wounded & Horse Shot-out from underneath him 12/26/64; 1890 Everett Veterans Census
36		B.S.F. History Book-1884; PA Civil War Archives-Bedford Co. residence
37		Historical Data Systems record
38		1890 Hyndman Veterans Census
39	Reformed Church-Hopewell	Bedford Inquirer 5/22/1908 listing
40		1890 Everett Veterans Census
41	Everett Cem.	1890 Everett Veterans Census
42		Historical Data Systems record
43		B.S.F. History Book-1884; In Hospital 10/30/62 to 2/10/63; 1860 Hopewell Twp. Census
44	Grandview Cem.-Cambria Co.	Wounded 12/13/62; Historical Data Systems record
45		Sabre wound of head 4/5/65; Historical Data Systems-Bedford Co. residence
46		B.S.F. History Book-1884
47	McGregor Cem.-Cairnbrook	Historical Data Systems-Bedford Co. residence
48	Mt. Zion Cem.-Pavia	Civil War market at gravesite
49	Mt. Zion Cem.-Pavia	POW 5/3/63; PA Civil War Project record
50	Rosemound Cem.-OK	1870 Hopewell Twp. Census; David, George, Thomas & James- brothers
51	Yellow Creek Reformed Cem.	Wounded 7/2/63; Died 8/3/63; PA Civil War Archives-Yellow Creek residence
52	Osborn Evergreen Cem.-MO	Historical Data Systems record; James, George & Thomas-brothers
53	Yellow Creek Reformed Cem.	Died of Disease 6/23/63; B.S.F. Hist. Book-1884; PA Civil War Archives-Yellow Creek resid.
54		PA Civil War Archives-Snake Spring Valley residence
55	Andersonville Cem.; 10,873	POW-Andersonville; Died 10/13/64; B.S.F. History Book-1884; Blair Co. residence
56	Hopewell Cem.	PA Civil War Archives-Bedford Co. residence
57		Died 6/14/65; B.S.F. History Book-1884; Frank McCoy 1912 Listing
58		1890 Saxton Veterans Census
59	New Paris Comm. Ctr. Cem.	Head Wound 6/1/63; Frank McCoy 1912 Listing; Historical Data Systems record
60		Frank McCoy 1912 Listing
61		Deserted 8/1/64; Historical Data Systems-Bedford Co. residence
62		B.S.F. History Book-1884; Somerset Co. residence
63		B.S.F. History Book-1884; Somerset Co. residence
64	Lashley Cem.-Artemas	PA Civil War Project record; Historical Data Systems record

Chapter 18

	Name	Age	Rank	Regiment	Co.	Muster In	Muster Out	Casualty	Casualty Battle
1	Lashley, Henry C	23	Pvt.	PA 55th Inf.	D	10/12/61	10/27/64	Wounded	
2	Lashley, Henry C		Pvt.	PA 99th Inf.	C	3/16/65	5/15/65		
3	Lashley, Isaac W	21	Pvt.	MD 8th Inf.	C	8/8/62		Wound.-Died	Spotsylvania
4	Lashley, John W	29	Pvt.	MD 1st PHB Cav.	L	4/23/64	6/28/65		
5	Lashley, Lewis H	18	Pvt.	PA 12th Cav.	A	9/3/64	6/1/65	POW	
6	Lashley, Robert	17	Corp.	PA 12th Cav.	A	9/3/64	6/1/65		
7	Lasley, Thomas	19	Pvt.	PA McKeage's Mil.	D	7/2/63	8/8/63		
8	Lasley, Thomas		Pvt.	PA 22nd Cav.		2/25/64			
9	Latta, Abraham	16	Pvt.	PA 208th Inf.	K	9/12/64	6/1/65		
10	Latta, John C	22	Corp.	IL 57th Inf.	K	12/26/61	12/1/62		
11	Latta, John C		Capt.	AL 1st US Cav.	C	12/22/62	10/20/65	Wounded	
12	Latta, William H	22	Pvt.	PA 148th Inf.	B	8/27/63	6/1/65		
13	Laughlin, John*	39	Pvt.	PA 55th Inf.	K	10/2/63	8/30/65		
14	Lauxman, John*	24	Pvt.	PA 110th Inf.	C	3/4/64	6/28/65		
15	Lawhead, Thomas	44	Pvt.	PA 171st Inf.	I	11/2/62	8/8/63		
16	Lay, Joseph		Pvt.	PA 138th Inf.	E	8/29/62		MIA	Wilderness
17	Layton, Bartley	37	Pvt.	PA 22nd Cav.	H	2/26/64			
18	Layton, David	25	Pvt.	PA 101st Inf.	D	11/1/61		Died	
19	Layton, George W	15	Pvt.	PA 99th Inf.	H	2/16/65	7/15/65		
20	Layton, Henry	20	Pvt.	PA 78th Inf.	K	2/28/65		Died	
21	Layton, John (1)	20	Pvt.	PA 101st Inf.	D	2/3/62		KIA	Seven Days Battle
22	Layton, John (2)	37	Pvt.	PA 138th Inf.	D	8/29/62		Wound.-Died	Opequon*
23	Layton, John R	31	Amb.	OH 61st Inf.		1862	1864		
24	Layton, Samuel	32	Pvt.	PA 184th Inf.	A	5/12/64		POW-Died	Jerusalem Plank Rd.
25	Leach, George E	31	1st Sgt.	PA 55th Inf.	K	11/5/61		Died	
26	Leach, Joseph		Pvt.	MD 3rd PHB Inf.	C	11/26/61	5/29/65		
27	Leach, Samuel R	26	Pvt.	PA 22nd Cav.	H	2/26/64	6/24/65		
28	Leader, David F	27	Sgt.	PA 133rd Inf.	K	8/4/62	5/26/63		
29	Leader, David F		Pvt.	PA 186th Inf.	E	2/20/64	8/15/65		
30	Leader, George W	22	Pvt.	PA 8th Res.	F	6/11/61	10/27/62	POW	Seven Days Battle
31	Leader, George W		Pvt.	US 6th Cav.	B	10/27/62			
32	Leader, Henry H		Pvt.	PA McKeage's Mil.	D	7/2/63	8/8/63		
33	Leader, Henry H	20	Pvt.	PA 2nd Heavy Art.	H	2/29/64	1/29/66	POW	Cold Harbor
34	Leader, John	22	Corp.	PA 76th Inf.	E	10/9/61	11/28/64	Wounded	
35	Leader, Simon H	36	Pvt.	PA 99th Inf.	F	2/25/65	7/1/65		
36	Leader, William W	31	Sgt.	MI 12th Inf.	I	12/25/63		KIA	White River, AK
37	Lear, Daniel J	17	Pvt.	PA 55th Inf.	I	9/20/61	8/30/65		
38	Lear, Franklin	15	Pvt.	PA 77th Inf.	F	2/27/65	12/6/65		
39	Lear, John	26	Pvt.	PA 125th Inf.	E	8/13/62	9/17/62	KIA	Antietam
40	Lear, Thomas	27	Pvt.	PA 5th Res.	K	8/25/61	8/25/64		
41	Leary, James M	22	Pvt.	PA 76th Inf.	E	10/9/61	11/28/64	Wounded	
42	Lease, Robert H	21	Pvt.	PA 138th Inf.	D	2/17/64	6/23/65	Wounded	Sailor's Creek
43	Leasure, Amos			MD 3rd PHB Inf.					
44	Leasure, Amos	17	Pvt.	MD 1st PHB Cav.	G	4/23/64		KIA	Summit Point
45	Leasure, George M	27	1st Sgt.	PA 171st Inf.	I	11/2/62	8/8/63		
46	Leasure, George M		Pvt.	PA 100th Inf.	A	9/21/64	6/3/65		
47	Leasure, John		Pvt.	MD 2nd PHB Inf.	B	8/13/62	5/29/65		
48	Leasure, John G	33	Corp.	PA 171st Inf.	I	11/2/62	8/8/63		
49	Leasure, John G		Pvt.	PA 91st Inf.	F	9/21/64	6/1/65		
50	Leasure, Josiah G	20	Pvt.	PA 138th Inf.	D	8/29/62	2/10/65	Wounded	Cold Harbor
51	Leasure, Nathaniel	18	Pvt.	PA 2nd Cav.	E	11/20/61			
52	Leasure, Nathaniel		Pvt.	PA 138th Inf.	D	8/29/62			
53	Leasure, Riley	19	Pvt.	PA 158th Inf.	H	1862			
54	Leasure, Riley		Pvt.	MD 1st PHB Cav.	G	4/23/64	6/28/65		
55	Leasure, Solomon	32	Pvt.	PA 149th Inf.	I	9/29/63		Died	
56	Leasure, William E	21	Pvt.	PA 171st Inf.	I	11/2/62	11/25/62		
57	Leasure, William E			OH 1st Cav.		1862			
58	Leasure, William E		Pvt.	US 1st Eng.	G	10/24/64			
59	Lecrone, William K	23	Pvt.	PA 91st Inf.	C	9/21/64	6/24/65		
60	Lee, Charles*	22	Pvt.	PA 55th Inf.	I	7/22/63			
61	Lee, David W	22	Pvt.	PA 149th Inf.	D	8/26/62	6/24/65		
62	Lee, Henry R	19	Pvt.	MD 3rd PHB Inf.	B	8/16/62	5/29/65		
63	Lee, Henry W	19	Pvt.	PA 133rd Inf.	D	8/14/62	6/26/63		
64	Lee, James	39	Pvt.	PA 55th Inf.	K	10/14/63	8/30/65		

	Cemetery	Notes
1	Chaneysville Meth. Cem.	1890 Everett Veterans Census-shot thru leg
2		Historical Data Systems record; Henry, Daniel, John & Robert-brothers
3	Fredericksburg Nat'l Cem.	Wounded 5/8/64; Died 6/1/64; 1860 Southampton Twp. Census; Hist. Data Systems record
4		Historical Data Systems-1850 Southampton Twp. Census
5	Girard Cem.-KS	POW-Salisbury & Pemberton; PA Civil War Archives record; Lewis & Isaac-brothers
6	Mt. Savage Meth. Cem.-MD	1890 Southhampton Twp. Census
7		B.S.F. History Book-1884; enlisted in Bloody Run
8		B.S.F. History Book-1884; Historical Data Systems record
9	Graceville Luth. Cem.	1890 E. Providence Twp. Veterans Census; Historical Data Systems record
10	Medicine Hill Cem.-SD	Database of Illinois Veterans Index; John, Abraham & William-brothers
11		Arm Wound-March 1865; 1850 E. Providence Township Census; 1st AL Cavalry-Union Reg't
12	Forest Hill Cem.-IL	B.S.F. History Book-1884; 1850 E. Providence Twp. Census
13		B.S.F. History Book-1884; Drafted
14		B.S.F. History Book-1884; Muster in Norristown
15	Mt. Zion Cem.-Cheneysville	B.S.F. History Book-1884; 1860 Southampton Twp. Census
16		MIA 5/6/64; B.S.F. Hist. Book-1884; PA Civil War Archives-Bedford resid.; aka May & Loy
17	McKendree Cem.-Crystal Spring	Deserted after war 7/5/65; B.S.F. History Book-1884
18	New Bern Nat'l Cem.	Died 4/1/63; Historical Data Systems-Bedford Co. residence; 1860 E. Providence Twp. Census
19	Rock Hill Cem.-Monroe Twp.	B.S.F. History Book-1884; George, David, & John (1)-brothers
20	Stevens Chapel-Monroe Twp.	Died of Disease 6/21/65; Historical Data Systems record; 1860 E. Providence Twp. Census
21	Cold Harbor Nat'l Cem.	KIA 6/29/62; Historical Data Systems- enlisted in Clearville; 1860 E. Providence Twp. Census
22	Winchester Nat'l Cem.	Died 10/6/64; B.S.F. History Book-1884; Historical Data Systems-Schellsburg residence
23	Oak Hill Cem.-OH	Ancestry.com-Born Bedford Co.; Ambulance Driver
24	Andersonville Cem.; 3053	POW-Andersonville 6/22/64; Died 8/18/64; B.S.F. Hist. Book-1884; 1860 S. Woodbury Cen.
25		Died 2/26/64; B.S.F. History Book-1884; PA Civil War Archives-Bedford Co. residence
26		1890 Harrison Twp. Veterans Census; Discharged on Surgeon's Cert.
27	Everett Cem.	1890 Harrison Twp. Veterans Census; 1860 Monroe Twp. Census
28	Everett Cem.	1890 Everett Veterans Census
29		1890 Everett Veterans Census; 1860 W. Providence Twp. Census
30	Mt. Union Cem.-Mench	POW 6/27/62; Historical Data Systems record; George W, David & Henry-brothers
31		Historical Data Systems record
32		B.S.F. History Book-1884
33	Mt. Union Cem.-Mench	Bedford Inquirer 5/22/1908 listing; POW 6/2/64 to 5/14/65
34		Wounded-Bed. Inquirer 5/27/64 article; B.S.F. Hist. Book-1884; 1860 Snake Spr. Twp. Cen.
35	Everett Cem.	1890 Everett Veterans Census
36		Bedford Inquirer Obituary 12/16/64; Son of Henry Leader; 1840 Bedford Census
37		B.S.F. History Book-1884; Daniel, Franklin, John & Thomas-brothers
38	Hopewell Cem.	PA Civil War Project record
39	Antietam Nat'l Cem., 3634	KIA 9/17/62; Bedford Gazette 2/13/1914 listing; 1850 M. Woodberry Twp. Census
40	Grays Cem.-Centre Co.	1850 M. Woodbury Twp. Census
41	Bedford Cem.	1890 Bedford Twp. Veterans Census; aka John
42		Wounded 4/6/65; B.S.F. History Book-1884
43		Bedford Gazette 2/13/1914 listing
44		KIA 8/30/64; Historical Data Systems record; 1860 Southampton Twp. Cen.
45		B.S.F. History Book-1884
46		Historical Data Systems record; 1860 Cumberland Valley Census
47		1890 Hopewell Veterans Census
48	Chaneysville Meth. Cem.	B.S.F. History Book-1884
49		Historical Data Systems record; 1860 Southampton Twp. Census
50	Clearville Union Cem.	B.S.F. History Book-1884; Wounded 6/1/64; Josiah & Solomon-brothers
51		Deserted 6/1/62; Historical Data Systems record; 1860 Southampton Twp. Census
52		Deserted 10/22/62; B.S.F. History Book-1884; Nathaniel, Amos & Riley-brothers
53	Shipley Family Cem.-Everett	Fulton Co. Historical Society book on 158th Infantry
54		Findagrave.com-Died 1875 from Civil War; Historical Data Systems record
55	Alexandria Nat'l Cem., 961-2	Died of Disease 1/29/64; Bedford Gazette 2/13/1914 listing; 1860 Southampton Twp. Cen.
56	Pine Grove Cem.-Monroe Twp.	B.S.F. History Book-1884
57	Pine Grove Cem.-Monroe Twp.	Frank McCoy 1912 Listing; 1850 Southampton Twp. Census
58		1890 Monroe Twp. Veterans Census; Engineer
59	Keagy Cem.-Woodbury	1890 Woodbury Twp. Veterans Census; Frank McCoy 1912 Listing
60		Drafted & Deserted 8/23,64; B.S.F. History Book-1884; Philadelphia references
61	Bedford Cem.	1890 Bedford Twp. Veterans Census
62	Buck Valley Zion Cem.	1890 Brush Creek Twp. Veterans Census
63	Bedford Cem.	1890 Bedford Twp. Veterans Census; Henry, Winfield & David-brothers
64	Dry Hill Cem.-Woodbury	B.S.F. History Book-1884; Historical Data Systems record

	Name	Age	Rank	Regiment	Co.	Muster In	Muster Out	Casualty	Casualty Battle
1	Lee, John	18	Corp.	PA 184th Inf.	A	5/12/64	7/14/65	Wounded	Jerusalem Plank Rd.
2	Lee, Thomas P*	27	Corp.	PA 194th Inf.	I	7/22/64	9/6/64		
3	Lee, Thomas P*	27	Sgt.	PA 97th Inf.	Ind	9/6/64	7/17/65		
4	Lee, William L	22	Pvt.	MD 3rd PHB Inf.	C	2/28/62	2/28/65		
5	Lee, Winfield S	18	Pvt.	PA 55th Inf.	K	2/19/64			
6	Lee, Winfield S		Pvt.	OH 197th Inf.	D	4/4/65	7/31/65		
7	Leech, Thomas	35	Corp.	PA 55th Inf.	K	11/5/61	8/30/65		
8	Leech, Wlliam	18	Pvt.	PA 55th Inf.	K	11/5/61		KIA	Pocotaligo, SC
9	Leer, William*	19	Pvt.	PA 110th Inf.	C	2/23/64	6/28/65		
10	Leffingwell, William H		Sgt.	US 4th Art.	G	8/10/50	12/15/62		
11	Lehman, Espy A		Pvt.	PA 93rd Inf.	H	11/26/64	6/27/65		
12	Lehman, Harry		Pvt.	PA 184th Inf.	A	5/12/64			
13	Lehman, Harry	22	Pvt.	PA McKeage's Mil.	H	7/2/63	8/8/63		
14	Lehman, Henry C	21	Pvt.	PA 93rd Inf.	H	9/26/64	6/17/65		
15	Lehman, Isaiah	28	Pvt.	PA 208th Inf.	H	9/6/64	6/1/65		
16	Lehman, Joseph*								
17	Lehman, Josiah M	33	Sgt.	PA 55th Inf.	K	11/5/61	3/28/63		
18	Lehman, William H	16	Pvt.	PA 184th Inf.	A	5/12/64			
19	Lehn, Philip		Pvt.	PA 208th Inf.	H	9/5/64	6/1/65	Wounded	Petersburg - Final
20	Leighty, George (1)	15	Pvt.	MD 3rd PHB Inf.	K	2/26/64	5/29/65		
21	Leighty, George (2)	24	Pvt.	PA 8th Res.	F	2/22/64	5/15/64	Wounded	Wilderness
22	Leighty, George (2)		Pvt.	PA 191st Inf.		5/15/64			
23	Leighty, Jacob	18	Pvt.	MD 3rd PHB Inf.	B	10/21/61	5/29/65		
24	Leighty, John Q	23	Corp.	PA 8th Res.	F	6/11/61		Wound.-Died	Antietam
25	Leighty, Joseph	22	Pvt.	PA 8th Res.	F	6/11/61	5/15/64	Wounded	Spotsylvania
26	Leighty, Joseph		Pvt.	PA 191st Inf.		5/15/64	1/9/65		
27	Leighty, Samuel	22	Pvt.	MD 3rd PHB Inf.	K	2/26/64	5/29/65	Wounded	
28	Leippert, Nicholas*	18	Pvt.	PA 76th Inf.	E	2/15/65		Died	
29	Lemmon, William	19	Pvt.	PA 138th Inf.	E	8/29/62	6/23/65		
30	Lemon, Henry	18	Corp.	PA 55th Inf.	H	10/11/61	4/24/65	Wounded	Drewry's Bluff
31	Lemon, John E*	32	Pvt.	PA 76th Inf.	E	10/7/63		POW-Died	
32	Lenhart, Czar*		Pvt.	IA 2nd Inf.	C	5/28/61		KIA	Ft. Donelson, TN
33	Lenhart, Peter*		Pvt.	PA 29th Inf.	K	12/23/64		Died	
34	Leonard, Adam P	26	Pvt.	PA 107th Inf.	A	9/26/64	6/6/65		
35	Leonard, Benjamin C	21	Corp.	PA 20th Cav.	E	7/1/63	6/17/65	POW	Lynchburg Exped.
36	Leonard, Henry N	17	Pvt.	PA 138th Inf.	E	8/29/62	6/23/65		
37	Leonard, Jacob	28	Pvt.	PA 184th Inf.	A	5/12/64	5/30/65		
38	Leonard, Jerome	25	Sgt.	PA 55th Inf.	D	10/12/61		Wound.-Died	Petersburg - Initial
39	Leonard, John B	27	Pvt.	PA 208th Inf.	K	9/7/64	6/1/65	Wounded	Petersburg - Final
40	Leonard, John D	21	Pvt.	PA 138th Inf.	E	8/29/62	6/23/65	Wounded	Wilderness
41	Leonard, Nathaniel	17	Pvt.	PA 62nd Inf.	M	8/9/61	11/11/65		
42	Leonard, Philip	17	Pvt.	PA 55th Inf.	D	10/12/61	8/30/65		
43	Leonard, William	19	Pvt.	PA McKeage's Mil.	G	7/2/63	8/8/63		
44	Leonard, William		Pvt.	PA 194th Inf.	I	7/22/64	11/5/64		
45	Leopold, John	18	Pvt.	PA 55th Inf.	K	11/5/61	8/30/65		
46	Leppert, Gustavus		Pvt.	PA 55th Inf.	K	10/10/63			
47	Lesh, John A	16	Pvt.	PA 7th Res.	B	2/9/64	5/31/64		
48	Lesh, John A		Pvt.	PA 190th Inf.	I	5/31/64	7/3/65		
49	Lester, Omer		Pvt.	PA 21st Cav.	E	7/21/63	7/8/65		
50	Leter, George D		Pvt.	PA 21st Cav.	E	7/2/63	2/20/64		
51	Levy, Bernard H*	22	Pvt.	PA 55th Inf.	I	8/12/63	4/5/64		
52	Lewis, Bert		Music.	USCT 32nd Inf.	G	2/24/64	6/16/65	Wounded	
53	Lewis, Franklin	22	Pvt.	PA 55th Inf.	D	10/14/63			
54	Lewis, Harrison	21	Corp.	PA 128th Inf.	B	8/14/62	5/19/63		
55	Lewis, James A	26	Pvt.	PA 76th Inf.	E	8/25/63	7/29/65		
56	Lewis, John D		Pvt.	PA 125th Inf.	F	8/12/62		MIA	Chancellorsville
57	Lewis, Robert		Pvt.	USCT 41st Inf.	H	9/21/64	9/30/65		
58	Lewis, Robert M		Pvt.	USCT 8th Inf.	D	1/16/65	11/19/65		
59	Lewis, Simon P	25	Sgt.	PA 133rd Inf.	C	8/13/62	5/26/63		
60	Lewis, William	29	Pvt.	PA 55th Inf.	I	9/24/63	6/5/65		
61	Lewis, William S*	24	Pvt.	PA 76th Inf.	E	7/13/63		Died	
62	Leydig, William	18	Pvt.	MD 2nd PHB Inf.	A	2/25/65	5/29/65		
63	Licher, John S		Pvt.	PA 55th Inf.	H	8/26/62	1/23/64	Wounded	Petersburg - Initial
64	Liehty, Christian J		Pvt.	PA 22nd Cav.	I	3/20/65	6/24/65		

	Cemetery	Notes
1	Woods Church-Colerain Twp.	Wounded 6/22/64; 1890 Southhampton Veterans Census; B.S.F. History Book-1884
2		B.S.F. History Book-1884; Reading, PA references
3		B.S.F. History Book-1884; Reading, PA references
4	Buck Valley Cem.-Fulton Co.	Historical Data Systems record
5		Deserted 6/21/64; PA Civil War Archives-Bedford Co. residence
6	Arlington Nat'l Cem., 20282	Historical Data Systems record; 1860 Cumberland Valley Census
7		B.S.F. History Book-1884; Historical Data Systems-Bedford Co. residence
8		KIA 10/22/62; B.S.F. History Book-1884; PA Civil War Archives-Bedford Co. residence
9		B.S.F. History Book-1884; Blair Co. references
10		1890 Harrison Twp. Veterans Census
11		1850 Harrison Twp. Census
12		1890 Bedford Twp. Veterans Census
13	Bedford Cem.	B.S.F. History Book-1884; 1860 Bedford Twp. Census; aka Harrison
14	Saint Ambrose Catholic-MD	1890 Harrison Twp Veterans Cen.;1860 Juniata Twp. Census; Henry, Espy & William-brothers
15	Buck Valley Cem.-Fulton Co.	B.S.F. History Book-1884; 1850 Harrison Twp. Census
16	Helixville Cem.-Napier Twp.	Bedford Inquirer 5/22/1908 listing; Obituary - didn't serve in Civil War
17	Duval Cem.-Coaldale	Frank McCoy 1912 Listing; 1860 Juniata Twp. Census; Discharged on Surgeon's Cert.
18	Union Dale Cem.-Pittsburgh	B.S.F. History Book-1884; Deserted 3/29/65; PA Civil War Archives-enlisted in Bedford
19		Wounded 4/2/65; B.S.F. History Book-1884; Historical Data Systems-Bedford Co. residence
20	Robinsonville Cem.	1890 Monroe Twp. Veterans Census; Died 1939 at Age 90
21	Dudley Meth. Cem.	Wounded 5/6/65; B.S.F. History Book-1884; George (2), John and Joseph-brothers
22		Historical Data Systems record
23		Jacob, George (1) & Samuel-brothers
24	Antietam Nat'l Cem.	Wounded 9/17/62; Died 9/21/62; B.S.F. History Book-1884; 1850 Hopewell Twp. Census
25	Broad Top IOOF Cem.	Leg Wound 5/12/64; 1890 Robersdale Veterans Census
26		Discharged on Surgeon's Cert.; Hist.Data Systems record; Joseph and John-bros.; aka Leichty
27	Rolla Cem.-KS	1860 Southampton Census
28	Raleigh Nat'l Cem.	Died 7/8/65; B.S.F. History Book-1884; Substitute
29	Metea Baptist Cem.-IN	B.S.F. History Book-1884; PA Civil War Archives-Bedford residence
30		Wounded 5/16/64 listed in Bedford Inquirer on 6/3/64; 1860 Napier Twp. Census
31	Andersonville Cem., 7938	POW-Andersonville; Died 9/5/64; B.S.F. History Book-1884; Mustered-New Brighton
32	Oakdale Mem. Gardens-IA	KIA 2/15/62; aka Cyrus W; B.S.F. History Book-1884
33		B.S.F. History Book-1884; Somerset Co. residence
34	St. Thomas Cem.-Bedford	Frank McCoy 1912 Listing; Adam, Jerome, Henry & John D-brothers
35	Fockler Cem.-Saxton	POW 5/25/64 to 6/5/64; 1890 Broad Top Veterans Census
36	Summit View Cem.-OK	B.S.F. History Book-1884
37		B.S.F. History Book-1884; 1860 Bedford Twp. Census
38	Hampton Nat'l Cem.	Wounded 6/18/64; Died 8/11/64; B.S.F. Hist. Book-1884; Hist. Data Sys.-Bedford Co. resid.
39	Everett Cem.	Wounded 4/2/65; Deserted and returned; Historical Data Systems record
40	Mt. Pleasant-Westmoreland Co.	Wounded 5/6/64; B.S.F. History Book-1884; PA Civil War Archives-Bedford residence
41		PA Civil War Archives-Hopewell residence
42	Bedford Cem.	B.S.F. History Book-1884; Philip & Jacob-brothers
43	Salem Reformed Cem.-Blair Co.	B.S.F. History Book-1884; William, Nathaniel & John B-brothers
44		B.S.F. History Book-1884; 1860 Liberty Twp. Census
45		Historical Data Systems-Bedford Co. residence; Trinity UCC Cem. Civil War Memorial
46		Deserted 6/24/64; B.S.F. History Book-1884
47	Bedford Cem.	enlisted under the name of Calvin R. Harmon
48		1890 Everett Veterans Census; Born in Richmond VA
49		Historical Data Systems-Bedford Co. residence
50		Historical Data Systems-Bedford Co. residence
51		B.S.F. History Book-1884
52	Mt. Ross Cem.-Bedford	1890 Bedford Twp. Veterans Census
53		Deserted 4/27/64; B.S.F. History Book-1884
54	Charles Evens Cem.-Reading	Ancestry.com information; 1880 Saxton Census, Locomotive Engineer
55		B.S.F. History Book-1884; Drafted
56		MIA 5/2/63-Historical Data Systems; 1890 Robersdale Veterans Census
57		1890 Bedford Twp. Veterans Census
58		Frank McCoy 1912 list & Civil Works Admin. 1934 list; Historical Data Systems record
59	Everett Cem.	1890 Everett Veterans Census; Historical Data Systems record
60		B.S.F. History Book-1884; Discharged on Surgeon's Cert.
61	Hampton Nat'l Cem.	Died 6/6/64; B.S.F. History Book-1884
62	Lybarger Cem.-Madley	Frank McCoy 1912 Listing
63		Wounded 6/18/64 in Bed. Inq. 7/8/64 article; B.S.F. History Book-1884; aka Lisher & Lesher
64		B.S.F. History Book-1884

	Name	Age	Rank	Regiment	Co.	Muster In	Muster Out	Casualty	Casualty Battle
1	Light, John H	21	Pvt.	MD 1st Reg Cav.	H	12/3/61	11/1/62	POW	
2	Lightmigator, Ernestus		Pvt.	PA 13th Inf.	G	4/25/61	7/31/61		
3	Lightmigator, Ernestus		Corp.	PA 111th Inf.	G	1/20/64	6/25/64	POW	
4	Lightner, John*	30	Pvt.	PA 110th Inf.	C	10/24/61	10/24/64		
5	Lightner, Peter	28	Pvt.	PA 1st Res.	I	7/17/61	5/31/64		
6	Lightner, Peter*			PA 190th Inf.		5/31/64	6/28/65		
7	Lightningstar, Augustus	25	Pvt.	PA 13th Inf.	G	4/25/61	8/6/61		
8	Lightningstar, Augustus		Pvt.	PA 101st Inf.	G	12/2/61	6/25/65	POW	Plymonth, NC
9	Linderman, Daniel*		Pvt.	PA 194th Inf.	I	7/22/64	11/5/64		
10	Linderman, John W		Pvt.	PA 22nd Cav.	I	2/14/64	10/31/65		
11	Lindsay, Charles B		1st Lt.	PA 76th Inf.	E	9/28/63	7/18/65		
12	Lindsey, Ephraim N*	41	Pvt.	PA 110th Inf.	C	2/22/64		Wound.-Died	
13	Line, Jacob	33	Corp.	PA 184th Inf.	A	5/12/64		KIA	Jerusalem Plank Rd.
14	Line, William	27	Sgt.	PA 138th Inf.	E	8/29/62	6/23/65		
15	Lines, Jacob		Pvt.	PA 8th Res.	F	6/7/61		Died	
16	Ling, David C	20	Sgt.	PA 55th Inf.	K	11/5/61	8/30/65		
17	Ling, Isaac	18	Pvt.	PA 55th Inf.	K	3/2/64	8/30/65	Wounded	Petersburg
18	Ling, Isaac N	29	Corp.	PA 138th Inf.	D	8/29/62	6/23/65		
19	Ling, Thompson	20	Pvt.	PA 5th Heavy Art.	K	9/1/64	6/30/65		
20	Ling, William H	21	Pvt.	PA 138th Inf.	D	8/29/62	6/23/65	POW	Monocacy, MD
21	Lingenfelter, Aaron	19	Pvt.	IL 55th Inf.	A	10/31/61	7/22/65	Wounded	Kennesaw Mtn.
22	Lingenfelter, Aaron			IL 55th Inf.	A			Wounded	
23	Lingenfelter, David	42	Pvt.	PA 55th Inf.	I	2/19/64	6/9/65	POW	Drewry's Bluff
24	Lingenfelter, David R			PA 149th Inf.					
25	Lingenfelter, David R	35	Pvt.	PA 91st Inf.	B	9/21/64	5/30/65	Wounded	Boydton Plank Rd.
26	Lingenfelter, George W	31	Pvt.	PA 205th Inf.	C	8/26/64	6/2/65		
27	Lingenfelter, John	15	Pvt.	IL 103rd Inf.	K	10/2/62		Died	
28	Lingenfelter, Josiah	26	Pvt.	IL 103rd Inf.	G	10/2/62	6/21/65		
29	Lingenfelter, Martin	36	Pvt.	PA 205th Inf.	C	8/26/64	6/2/65		
30	Lingenfelter, Michael	17	Pvt.	PA 19th Cav.	L	10/19/63	6/13/65		
31	Lingenfelter, Thaddeus	16	Pvt.	PA 77th Inf.	F	2/24/65	12/6/65		
32	Lingerfelt, Aaron	16	Pvt.	PA 55th Inf.	A	8/28/61	8/30/65		
33	Lingerfelt, Abram	47	Pvt.	PA 55th Inf.	A	8/28/61	8/30/65		
34	Lingerfelt, Josiah	27	Pvt.	PA 55th Inf.	A	2/25/64	8/30/65		
35	Lininger, Albert*	18	Pvt.	PA 55th Inf.	D	2/23/64	8/30/65		
36	Lininger, William P*		Pvt.	PA 55th Inf.	D	9/22/62		POW-Died	
37	Link, Solomon*	55	Pvt.	PA 101st Inf.	G	12/28/61		Died	
38	Linn, Henry	22	Sgt.	PA 101st Inf.	D	11/1/61	6/25/65	POW	Plymonth, NC
39	Linn, Hugh Jr	23	Pvt.	PA 171st Inf.	I	11/2/62	8/8/63		
40	Linn, Hugh Sr	52	Pvt.	MD 3rd PHB Inf.	B	9/28/61	9/28/64	POW	Harpers Ferry
41	Linn, Jacob B	22	Sgt.	PA 8th Res.	F	6/11/61	5/26/64		
42	Linn, Riley	17	Pvt.	MD 3rd PHB Inf.	B	9/28/61	9/28/64	POW	Harpers Ferry
43	Linn, William	14	Pvt.	MD 3rd PHB Inf.	B	9/28/61	9/28/64	POW	Harpers Ferry
44	Linton, John*	27	Pvt.	PA 55th Inf.	I	10/20/63			
45	Lisles, George	33	Pvt.	USCT 3rd Inf.	E	7/8/63	10/31/65		
46	Little, David	23	Pvt.	PA 55th Inf.	D	9/2/62	6/11/65		
47	Little, Irving	24	Pvt.	PA 55th Inf.	I	10/15/61		Died	
48	Little, Isaac K	29	Capt.	PA 5th Heavy Art.	F	8/30/64	6/30/65		
49	Little, James	27	Pvt.	PA 14th Inf.	I	5/2/61			
50	Little, James		Corp.	PA 55th Inf.	I	10/15/61	2/25/63		
51	Little, James		Corp.	US Inf.		2/25/63	8/30/65		
52	Little, John Pius	34	Pvt.	PA 99th Inf.	B	2/24/64		KIA	Sailor's Creek
53	Little, Silas H	22	Sgt.	IL 93rd Inf.	C	8/13/62	7/23/65		
54	Livensgood, Charles M		Pvt.	PA 22nd Cav.	I	3/20/65	6/24/65		
55	Livingston, John A	22	Capt.	PA 55th Inf.	H	10/11/61	10/11/64	Wounded	Petersburg - Initial
56	Livingston, John A		Capt.	PA 55th Inf.	H			Wounded	Cold Harbor
57	Livingston, Samuel	22	Pvt.	PA 13th Cav.	D	9/30/63	7/14/66	Wounded	Deep Bottom
58	Livingston, Thomas G			PA 11th Inf.					
59	Livingston, Thomas G	24	2nd Lt.	PA 110th Inf.	C	10/24/61	6/28/65	Wounded	Spotsylvania
60	Lockard, Thomas R	19	Pvt.	PA 55th Inf.	H	10/11/61	8/30/65	POW	Edisto Island, SC
61	Lockhard, John*	17	Pvt.	PA 55th Inf.	I	8/28/61		POW-Died	Drewry's Bluff
62	Logsdon, John E	22	Pvt.	MD 3rd PHB Inf.	D	2/23/65	5/29/65		
63	Logsdon, Samuel	25	Pvt.	PA 97th Inf.	G	3/4/65	8/28/65		
64	Logue, James	23	Pvt.	PA 171st Inf.	I	11/2/62	6/26/63	Died	

	Cemetery	Notes
1	Hyndman Cem.	1890 Hyndman Veterans Census-POW 3 months
2		1890 Hopewell Veterans Census
3		POW-Andersonville; 1890 Hopewell Veterans Census
4		B.S.F. History Book-1884; Tyrone references
5		1890 Snake Spring Valley Census
6		PA Civil War Project record
7		B.S.F. History Book-1884
8	Bedford Cem.	POW 4/20/64 to 12/16/64; Historical Data Systems-Schellsburg residence
9		B.S.F. History Book-1884; enlisted in Reading, PA
10		B.S.F. History Book-1884
11		B.S.F. History Book-1884
12	Arlington Nat'l Cem., 27-9	Died 5/19/64; PA Civil War Archives lists Blair Co. residence
13		KIA 6/22/64; B.S.F. History Book-1884; 1860 Bedford Twp. Census; Jacob & William-bros.
14	Bedford Cem.	B.S.F. History Book-1884; 1890 Bedford Twp. Veterans Census
15		Died 5/3/62; B.S.F. History Book-1884
16	Old Union Cem.-Osterburg	B.S.F. History Book-1884; 1850 St. Clair Twp. Census
17	Imler St. Paul Breth. Cem.	B.S.F. History Book-1884; Isaac & David-brothers
18	Mt. Union Cem.-Lovely	Historical Data Systems record; Isaac N & William-brothers
19	Trinity UCC Cem.-Juniata Twp.	1890 Harrison Twp. Veterans Census; B.S.F. History Book-1884
20	Pleasantville Cem.	POW-Libby 7/9/64 to 2/21/65; Historical Data Systems record; Frank McCoy 1912 Listing
21	Bethel Cem.-IL	Wounded 6/27/64-Fulton Co. Illinois Historical Society; John, Aaron & Josiah-brothers
22		Wounded 3/21/65 in South Carolina-Fulton Co. Illinois Historical Society
23	Greenlawn Cem.-Roaring Spring	POW 5/16/64 to 4/29/65; B.S.F. History Book-1884; 1850 Greenfield Twp. Census
24		Bedford Inquirer 5/22/1908 listing; Frank McCoy 1912 Listing
25	Fishertown Luth. Cem.	Wounded 10/27/64; 2nd headstone in Old Claysburg Cem.
26	Greenlawn Cem.-Roaring Spring	1890 Greenfield Twp. Veterans Census
27	High Bridge Cem.-IL	Died 3/22/64; John & Josiah-brothers; Lived in Klahr near Claysburg before moving to Illinois
28	Greenwood Cem.-IL	Ancestry.com born in Klahr
29	Claysburg Union Cem.	Soldiers of Blair Co. Book - 1940; Martin, George & David-brothers
30	Walnut Hills Cem.-OH	1890 Greenfield Twp. Veterans Census
31	Upper Claar Cem.-Queen	Frank McCoy 1912 Listing
32	Calvary Cem.-Altoona	PA Civil War Archives-Bedford Co. residence
33	Cresson Union Cem.	PA Civil War Archives-Bedford Co. residence; Aaron-son
34	Union Cem.-Claysburg	PA Civil War Archives enlisted in Bedford
35	Mt. Lebanon Cemty, Lebanon	B.S.F. History Book-1884; Berks Co. references
36		POW-Richmond; B.S.F. History Book-1884; enlisted in Berks Co.
37	Cypress Hill Cem.-NY	Died 8/24/62; Bedford Gazette 2/13/1914 listing
38		POW 4/20/64 to 2/27/65; Historical Data Systems-Bedford Co. residence
39		B.S.F. History Book-1884; 1890 Robertsdale Veterans Census
40	Robinsonville Cem.	1890 Monroe Twp Veterans Census; 1860 Monroe Twp. Census; Father of William & Riley
41		Historical Data Systems record
42	Bedford Cem.	Monroe Twp. 1890 Veterans census; Marker also at Robinsonville Cem.
43	Robinsonville Cem.	Historical Data Systems record; William & Riley are sons of Hugh (1)
44		Drafted & Deserted 5/6/64; B.S.F. History Book-1884
45	Mt. Ross Cem.-Bedford	Deserted 4/29/65; 1890 Bedford Twp. Veterans Census; PA Civil War Project Record
46	Nichols Cem.-IN	1860 Snake Spring Valley Census; B.S.F. History Book-1884
47	Beaufort Nat'l Cem.	Died 10/12/62; B.S.F. History Book-1884; 1850 Greenfield Twp. Census
48	St. Lukes Cem.-Saxton	1890 Saxton Veterans Census; Superintendant Broad Top & Hunt. RR
49		Civil War PA Archives-Bedford Co. residence
50		B.S.F. History Book-1884; 1850 Greenfield Twp. Census; James & Irving-brothers
51		B.S.F. History Book-1884
52		KIA 4/6/65- wound in mouth; aka Pius; Bed. Gazette 2/13/1914 list; 1850 Bedford Census
53	St. Lukes Cem.-Saxton	1890 Saxton Veterans Census; Silas & Isaac-brothers
54		B.S.F. History Book-1884
55	Oak Hill Cem.-MO	Wounded 6/16/64 listed in Bedford Inquirer 7/8/64 article; 1860 Colerain Twp. Census
56		Wounded 6/3/64 listed in Bedford Inquirer 12/9/64 article; Historical Data Systems record
57	Yellow Creek Reformed Cem.	Historical Data Systems record; 1870 Hopewell Twp. Census
58		Frank McCoy 1912 Listing
59	Everett Cem.	1890 Hopewell Twp. Veterans Census-Hip Wound; Thomas & Samuel-brothers
60	Fairview Cem.-Altoona	PA Civil War Archives-Bedford Co. residence; B.S.F. History Book-1884
61		POW-Richmond 5/16/64; Died 6/4/64; B.S.F. History Book-1884; Centre Co. references
62	Palo Alto Cem.-Hyndman	1890 Hyndman Veterans Census; 1860 Londonderry Twp. Census
63	Palo Alto Cem.-Hyndman	1890 Londonderry Twp. Veterans Census; Samuel & John-brothers
64		Died-Disease 6/26/63; B.S.F. Hist. Book-1884; CW Pension Record; 1850 Harrison Twp. Cen.

Chapter 18

	Name	Age	Rank	Regiment	Co.	Muster In	Muster Out	Casualty	Casualty Battle
1	Logue, James F	21	Pvt.	PA 53rd Inf.	B	9/21/64	1/26/65	Died	Petersburg
2	Lohr, Benjamin F	20	Pvt.	PA 142nd Inf.	D	8/22/62	11/28/63	POW	Gettysburg
3	Lohr, Henry D*		Pvt.	PA 19th Inf.	G	2/22/62	9/12/65	POW	Chickamauga
4	Long, Abraham B	32	Pvt.	PA 99th Inf.	G	2/25/65	5/31/65		
5	Long, Amon	23	Pvt.	PA 22nd Cav.	I	7/16/63	4/15/65		
6	Long, Augustus*	18	Pvt.	PA 55th Inf.	H	9/2/62		POW-Died	Drewry's Bluff
7	Long, George	25	Pvt.	PA 138th Inf.	E	8/29/62	6/23/65		
8	Long, Gephart		Pvt.	PA 21st Cav.	E	7/2/63			
9	Long, John A	19	Corp.	PA 55th Inf.	H	10/11/61			
10	Long, John A		Corp.	PA 79th Inf.	D	2/23/65	7/12/65		
11	Long, Joseph	25	Pvt.	PA 76th Inf.	E	11/21/61	11/28/64		
12	Long, Joseph C	25	1st Sgt.	PA 208th Inf.	H	9/7/64	7/9/65	Wounded	Petersburg - Final
13	Long, Joseph E	26	Pvt.	PA 101st Inf.	C	3/10/65	6/25/65	Died	
14	Long, Levi	30	Pvt.	PA 55th Inf.	D	10/12/61		Died	
15	Long, Samuel C		Buglar	PA 22nd Cav.	F	2/25/64	6/24/65		
16	Long, Samuel L		Pvt.	PA 125th Inf.	A	8/13/62	5/18/63		
17	Long, Samuel L			PA 79th Inf.	A				
18	Long, Samuel L		Corp.	PA 205th Inf.	A	8/31/64	6/2/65		
19	Long, William P	35	Pvt.	IL 39th Inf.	C	8/12/61	7/7/63		
20	Long, William P		Pvt.	PA 194th Inf.	I	7/21/64	9/6/64		
21	Long, William P		Pvt.	PA 97th Inf.	I	9/6/64	7/17/65		
22	Longenecker, George H	20	Pvt.	PA 195th Inf.	C	7/17/64	11/4/64		
23	Longenecker, George H		Pvt.	PA 16th Cav.	G	2/17/65	8/11/65		
24	Longenecker, Jacob H	22	1st Lt.	PA 101st Inf.	D	1/20/62	3/12/65	POW	Plymonth, NC
25	Longenecker, John S	30	Pvt.	PA 133rd Inf.	C	8/13/62	5/26/63		
26	Longenecker, John S		Pvt.	PA 16th Cav.	G	2/17/65	8/11/65		
27	Longston, John A		Sgt.	PA McKeage's Mil.	H	7/2/63	8/8/63		
28	Lorah, Benjamin H*	23	Pvt.	PA 55th Inf.	K	2/23/63	8/30/65		
29	Lorah, Henry H*	28	Pvt.	PA 55th Inf.	K	2/22/64	8/30/65		
30	Lorenze, Charles*	38	Pvt.	PA 55th Inf.	I	10/17/63	7/28/65	POW	Chaffin's Farm
31	Lorow, Franklin*	24	Pvt.	PA 184th Inf.	A	5/12/64		POW	Jerusalem Plank Rd.
32	Love, George			USCT Inf.					
33	Love, John R	53	Pvt.	USCT 41st Inf.	H	9/21/64			
34	Lowery, Daniel	18	Pvt.	IN 2nd Cav.	G	10/7/61		POW-Died	Newnan, GA
35	Lowery, Emanuel	30	Pvt.	PA 138th Inf.	D	8/29/62	6/16/65	Wounded	Cold Harbor
36	Lowery, John E	25	Pvt.	PA 138th Inf.	D	8/29/62		Wound.-Died	Mine Run
37	Lowery, John	46	Pvt.	MD 3rd PHB Inf.	B	9/28/61	5/29/65		
38	Lowery, Joseph		Pvt.	MD 3rd PHB Inf.	B	9/28/61	5/29/65		
39	Lowery, Samuel	34	Pvt.	PA 13th Inf.	G	4/25/61	8/6/61		
40	Lowery, Samuel		Pvt.	PA 28th Inf.	O	8/17/61	10/29/62		
41	Lowery, William	21	Sgt.	PA 18th Cav.	K	10/29/62	4/6/65	POW	Germanna Ford
42	Lowery, William H	21	Corp.	PA 138th Inf.	D	8/29/62		Died	Brandy Station
43	Lowry, John Jr		Pvt.	MD 3rd PHB Inf.	B	8/16/63	5/29/65		
44	Lowry, Oliver	24	Pvt.	PA 138th Inf.	F	9/29/62	6/23/65	Wounded	Opequon
45	Lowry, Samuel		Pvt.	PA 147th Inf.	B	8/17/61	7/15/65		
46	Lowry, William N	28	Pvt.	PA 55th Inf.	I	10/7/63	8/30/65		
47	Loy, Joseph F	36	Capt.	WI 4th Cav.	H	5/21/61	6/13/62		
48	Lucas, Benjamin		Pvt.	PA 22nd Cav.	H	10/3/64	10/4/65		
49	Lucas, Henry		Pvt.	PA 76th Inf.	B	3/11/64	7/18/65		
50	Lucas, Joshua T	19	Pvt.	PA 133rd Inf.	C	8/13/62	5/26/63		
51	Lucas, Joshua T		Pvt.	PA 194th Inf.	I	7/22/64	11/5/64		
52	Lucas, Joshua T		Pvt.	PA 83rd Inf.	K	3/3/65	6/28/65		
53	Lucas, Robert A	18	Pvt.	PA 21st Cav.	E	7/21/63	7/8/65		
54	Lucas, Simeon S		Pvt.	PA 1st Cav.	F	8/16/61	9/18/63		
55	Lucas, William	21	Pvt.	PA 138th Inf.	D	8/29/62	6/16/65	Wounded	Opequon
56	Luckett, Alexander		Pvt.	USCT 32nd Inf.	G	2/24/64	8/22/65		
57	Lumac, William		Pvt.	USCT 43rd Inf.	D	9/10/64	10/20/65		
58	Luman, Aaron	22	Pvt.	PA 99th Inf.	K	3/14/65			
59	Luman, David	15	Pvt.	OH 64th Inf.	A	10/7/64	6/5/65		
60	Luman, John	20	Pvt.	US 16th Inf.	D	1/24/63		Died	
61	Luman, Solomon	21	Pvt.			3/5/65	7/30/65		
62	Luman, William			NJ 2nd Inf.	H				
63	Lunger, Franklin	25	Pvt.	PA 171st Inf.	I	11/2/62			
64	Luther, Frederick H	20	Pvt.	PA 55th Inf.	K	2/19/64	6/28/65		

	Cemetery	Notes
1		Died 1/26/65; 1890 Londonderry Twp. Veterans Census; PA Civil War Project record
2	Grandview Cem.-Cambria Co.	POW 7/1/63; 1890 E. St. Clair Twp. Veterans Census, 1880 E. St. Clair Twp. Census
3		POW Sep. 1863; B.S.F. History Book-1884; Somerset Co. residence
4	Schellsburg Cem.	Frank McCoy 1912 Listing; Abraham & Samuel L-brothers
5	Brumbaugh Cem.-Saxton	Civil War Pension Record
6	Andersonville Cem., 5199	POW-Andersonville; Died 8/10/64; B.S.F. History Book-1884; muster in Womelsdorf, PA
7		B.S.F. History Book-1884; Historical Data Systems-Bedford Co. residence
8		Historical Data Systems-Bedford Co. residence; Discharged on Surgeon's Cert.
9	Duval Cem.-Coaldale	Historical Data Systems record; Discharged on Surgeon's Cert.
10		1890 Broad Top Veterans Census
11		B.S.F. History Book-1884; PA Civil War Archives-Bedford residence
12	Rosehill Cem.-Altoona	Wounded 4/2/65; B.S.F. History Book-1884; 1870 S. Woodbury Twp. Census
13	St. Lukes Cem.-Saxton	Died 8/23/65; Frank McCoy 1912 Listing; PA Civil War Project record
14	Beaufort Nat'l Cem.	Died 7/27/62; Historical Data Systems-Bedford Co. residence; 1860 Colerain Twp. Census
15		1890 Harrison Twp. Veterans Census
16	Duval Cem.-Coaldale	Frank McCoy 1912 Listing
17		Findagrave.com record
18		1890 Broad Top Veterans Census
19	Yellow Creek Reformed Cem.	Findagrave.com record; 1850 Hopewell Twp. Census
20		B.S.F. History Book-1884
21		PA Civil War Archives-Bedford Co. residence
22	Chico Cem.-CA	1860 Woodbury Twp. Census
23		Historical Data Systems record
24	Bedford Cem.	POW at Macon, Savannah, Charleston, Columbia, Charlotte & Salisbury 4/20/64 to 3/2/65
25	Walnut Hill Cem.-KS	Historical Data Systems record; John, Jacob & George-brothers
26		Historical Data Systems record
27		B.S.F. History Book-1884; aka Longsten
28	Mohnsville Cem.-Berks Co	B.S.F. History Book-1884; Berks Co. references
29		B.S.F. History Book-1884; Berks Co. references
30		POW 9/28/64 to 4/12/65; B.S.F. History Book-1884
31		POW 6/22/64; B.S.F. History Book-1884
32		Frank McCoy 1912 list & Civil Works Admin. 1934 list; Historical Data Systems record
33	Eastern Light Cem.-Altoona	Born a Slave in VA in 1811 - Bedford Histoical Society information; 1860 Bedford Twp. Cen.
34	Andersonville Cem., 7162	POW-Andersonville 7/30/64; Died-Diarrhea 8/29/64; Ancestry.com-Born Londonderry Twp.
35	Porter Cem.-Hyndman	Wounded 6/5/64; Historical Data Systems record; 1860 Londonderry Twp. Census
36	Burket Cem.-Queen	Wounded 11/27/63; Died 11/28/63; B.S.F. History Book-1884; 1860 Londonderry Twp. Cen.
37	Everett Cem.	Frank McCoy 1912 Listing; Historical Data Systems record
38	Hyndman Cem.	Frank McCoy 1912 Listing
39	Porter Cem.-Hyndman	B.S.F. History Book-1884; Samuel & Emanuel-brothers
40		Frank McCoy 1912 Listing
41	Green Mt. Cem.-Cumberland	POW-Libby, Danville & Andersonville 11/18/63 to 4/6/65; William & John E-bros.
42	Porter Cem.-Hyndman	Died 4/15/64; Frank McCoy 1912 Listing; B.S.F. History Book-1884
43	Hyndman Cem.	Historical Data Systems record
44	Cochran Cem.-Fayette Co.	Wounded 9/19/64; Historical Data Systems record; 1870 Londonderry Twp. Census
45	Duval Cem.-Coaldale	Historical Data Systems record
46		B.S.F. History Book-1884
47	Woodlawn Cem.-WI	1850 Bedford Census
48		B.S.F. History Book-1884
49		1890 Hopewell Census; Historical Data Systems record
50	Everett Cem.	B.S.F. History Book-1884; Joshua, Simeon & William-brothers
51		B.S.F. History Book-1884; Historical Data Systems-Bedford Co. residence
52		1890 Everett Veterans Census
53		Historical Data Systems-Bedford Co. residence
54	Iowa	Historical Data Systems record; Discharged on Surgeon's Cert.
55	Everett Cem.	Wounded 9/19/64; Frank McCoy 1912 Listing; Historical Data Systems record
56		Frank McCoy 1912 Listing
57		Frank McCoy 1912 list & Civil Works Admin. 1934 list; Historical Data Systems record
58	Lybarger Cem.-Madley	Deserted 6/14/65; Frank McCoy 1912 Listing; Aaron & William-brothers
59	Fairmont Cem.-OH	1850 Cumberland Valley Census; David & John-brothers
60	Nashville Nat'l Cem.	Died-Diarrhea 8/28/63; 1850 Cumberland Valley Census; Born 1844
61	Union Cem.-Centerville	1890 Cumberland Valley Veterans Census
62	Pleasant Ridge-Buffalo Mills	Frank McCoy 1912 Listing; Born 5/12/45
63		Deserted 11/25/62; B.S.F. History Book-1884
64	Rose Hill-Crystal Spring, MD	B.S.F. History Book-1884; 1850 Bedford Boro Census

	Name	Age	Rank	Regiment	Co.	Muster In	Muster Out	Casualty	Casualty Battle
1	Luther, Hiram		Pvt.	PA McKeage's Mil.	H	7/2/63	8/8/63		
2	Lutz, Simon S	24	Pvt.	PA 184th Inf.	A	5/12/64		Wound.-Died	Cold Harbor
3	Lybarger, David	40	Pvt.	PA 49th Inf.	E	6/4/64	7/15/65		
4	Lybarger, George W	33	2nd Lt.	PA 54th Inf.	G	4/14/64	5/31/65		
5	Lybarger, Henry G	22	Pvt.	PA 55th Inf.	D	10/12/61	8/30/65	Wounded	Drewry's Bluff
6	Lybarger, John		Pvt.	PA 12th Cav.	M	3/4/62	7/20/65		
7	Lybarger, Joseph J	30	Pvt.	PA 54th Inf.	G	9/30/61	7/15/65		
8	Lybarger, Martin*	33	Pvt.	PA 101st Inf.	G	12/28/61		POW-Died	Plymonth, NC
9	Lybarger, Valentine G*	16	Pvt.	PA 54th Inf.	G	2/1/62		Died	
10	Lybarger, William		Pvt.	MD 2nd PHB Inf.	G	8/31/61	5/9/63		
11	Lybarger, William*	21	Pvt.	PA 54th Inf.	G	2/23/64		Died	
12	Lyles, David			USCT 3rd Inf.					
13	Lyles, George		Pvt.	USCT 38th Inf.	E	1/23/64			
14	Lyles, James		Pvt.	USCT 3rd Inf.	E	7/8/63	10/31/65		
15	Lyon, Alexander	19	Sgt.	PA 76th Inf.	E	10/9/61	11/12/64		
16	Lyon, Samual	30	1st Lt.	PA 107th Inf.	H	2/10/62	3/11/65		
17	Lyon, William	18	Corp.	PA 76th Inf.	E	9/4/63	7/18/65		
18	Lyon, William M		Music.	PA 76th Inf.	E	9/30/61	11/20/62		
19	Lyons, George W	16	Sgt.	USCT 41st Inf.	F	9/29/64			
20	Lyons, Thomas H	28	Capt.	PA 55th Inf.	D	10/12/61	5/26/63		
21	Lyons, Thomas H		Capt.	PA 22nd Cav.	I	7/14/63	5/29/65		
22	Lysinger, George W	28	1st Sgt.	PA 107th Inf.	H	1/11/62		POW	2nd Bull Run
23	Lysinger, George W		1st Sgt.	PA 107th Inf.	H			POW-Died	Globe Tavern
24	Lysinger, John		Pvt.	PA 107th Inf.	H				
25	Lysinger, John	22	Pvt.	PA 133rd Inf.	C	8/13/62	5/26/63		
26	Lysinger, Joseph H	16	Pvt.	IL 146th Inf.	H	9/8/64	7/8/65		
27	Lysinger, Martin G	18	Corp.	PA McKeage's Mil.	G	7/2/63	8/8/63		
28	MacFarland, Daniel		Pvt.	OH 73rd Inf.	A	6/18/61	7/6/65		
29	Machtley, John	20	Pvt.	PA 82nd Inf.	C	10/3/64	6/24/65		
30	Machtley, William H	24	Pvt.	PA 77th Inf.	A	4/12/65	12/6/65		
31	Mack, Joseph	30	Pvt.	PA 133rd Inf.	F	8/15/62	5/26/63		
32	Mack, Joseph		Pvt.	PA 6th Cav.	F	2/15/65	6/17/65		
33	Madara, David W	21	Capt.	PA 55th Inf.	I	9/20/61	4/20/62		
34	Madara, Jacob*		Pvt.	PA 8th Res.	F	8/23/61	5/15/64		
35	Madara, Jacob*		Pvt.	PA 191st Inf.		5/15/64		POW-Died	Globe Tavern
36	Madara, James	49	Col.	US					
37	Madden, Abisha*	20	Pvt.	PA 55th Inf.	I	9/20/61	8/30/65		
38	Madden, Cyrus	21	Sgt.	PA 133rd Inf.	C	8/13/62	5/26/63		
39	Madden, Daniel*	28	Pvt.	PA 55th Inf.	I	7/27/64	6/11/65		
40	Maheny, William		Pvt.	MD 3rd Inf.	B	7/20/64	5/3/65		
41	Mahoney, John	21	Pvt.	PA 11th Inf.	A	9/30/61	7/1/65		
42	Malcomb, George*	23	Pvt.	PA 55th Inf.	I	10/6/63	8/30/65		
43	Malin, Benjamin F*	22	Pvt.	PA 76th Inf.	E	8/27/62	6/15/65	MIA	2nd Deep Bottom
44	Malone, Charles		Pvt.	PA 8th Res.	F	9/5/62	5/15/64	Wounded	Spotsylvania
45	Malone, Charles		Pvt.	PA 191st Inf.	H	5/15/64		POW-Died	Globe Tavern
46	Malone, Hezekiah*	19	Pvt.	PA 133rd Inf.	K	8/15/62	5/26/63		
47	Malone, John	20	Pvt.	PA 14th Inf.	D	4/22/61	8/7/61		
48	Malone, John		Pvt.	PA 133rd Inf.	C	8/13/62	5/26/63		
49	Malone, John		Pvt.	PA 8th Res.	F	1/28/64	5/15/64		
50	Malone, John		Pvt.	PA 191st Inf.	H	5/15/64	6/26/65	POW	
51	Malone, William	24	Pvt.	PA 8th Res.	F	6/11/61		Wound.-Died	Antietam
52	Maloney, William A	32	Sgt.	PA 55th Inf.	K	3/2/64	8/30/65		
53	Manges, Abraham		Pvt.	PA 14th Inf.	H	4/24/61	8/7/61		
54	Manges, Abraham		Pvt.	US 12th Inf.				Wounded	Seven Days Battle
55	Manges, Abraham	18	Pvt.	US 12th Inf.				POW	
56	Manges, George W	22	Pvt.	PA 55th Inf.	K	2/19/64	6/7/65	Wounded	
57	Manges, Jacob (1)	18	Pvt.	PA 21st Cav.	F	2/17/64	7/8/65		
58	Manges, Jacob (2)	25	Pvt.	PA 61st Inf.	A	9/2/64	6/18/65		
59	Manges, Jacob A	19	Pvt.	PA 12th Cav.	M	3/30/64	7/20/65		
60	Mangis, Levi		Pvt.	PA 21st Cav.	E	7/2/63	2/20/64		
61	Mangis, Levi		Pvt.	PA 5th Heavy Art.	K	8/26/64	7/30/65		
62	Manly, Jonathan*	18	Pvt.	PA 184th Inf.	A	5/12/64	5/27/65	POW	Jerusalem Plank Rd.
63	Manning, John	28	Pvt.	US 2nd Cav.	G	12/11/62	7/7/65	Wounded	
64	Mansberger, Benjamin	18	Pvt.	PA 22nd Cav.	C	6/29/63			

	Cemetery	Notes
1		B.S.F. History Book-1884
2	Bald Hill Cem.-Everett	KIA 6/3/64; B.S.F. History Book-1884; PA Civil War Archives-enlisted in Bedford
3	Lybarger Cem.-Madley	Frank McCoy 1912 Listing; PA Civil War Project record
4	Garrett Union-Somerset Co.	Ancestry.com-brothers born in Bedford Co.
5	Mt. Olivet Cem.-Manns Choice	Wounded 5/15/64; Henry, Joseph, Valentine, William & George-brothers
6		Pauline Nicodemus family record-Londonderry Twp. resident; 1890 E. Conemaugh Vet. Cen.
7	Myersdale Union Cem.	Civil War Pension Record
8		POW-Andersonville; 4/20/64; Died 11/30/64; B.S.F. History Book-1884; Somerset residence
9	White Oak Cem.-Wittenberg	Died of Disease 5/24/63; Somerset Co. references
10		Historical Data Systems record; Ancestry.com-Born Bedford Co.
11		Historical Data Systems record; Somerset Co. references
12		Frank McCoy 1912 list & Civil Works Admin. 1934 list
13	Mt. Ross Cem.-Bedford	Frank McCoy 1912 Listing; Historical Data Systems record
14		Frank McCoy 1912 list & Civil Works Admin. 1934 list; Historical Data Systems record
15		1860 Bedford Boro Census; B.S.F. History Book-1884; Historical Data Systems record
16		B.S.F. History Book-1884; Discharged on Surgeon's Cert.
17		B.S.F. History Book-1884
18		B.S.F. History Book-1884; Discharged on Surgeon's Cert.
19	Union African-Holidaysburg	Historical Data Systems record; 1850 S. Woodbury Twp. Census
20	Old St. Thomas Cem.-Bedford	1890 Bedford Twp. Veterans Census; 1850 Bedford Boro Census; Discharged on Surg. Cert.
21		Served under Rutherford B. Hayes 1/5/64 until mustered out
22	Salisbury Nat'l Cem.	POW 8/30/62 to 12/20/62; Promoted to Sgt. at Gettysburg
23		POW-Salisbury 8/19/64; Died 12/19/64; PA CW Project record; 1860 E. Prov. Twp. Census
24		B.S.F. History Book-1884; John & George-brothers
25	Shelby-Oakland Cem.-OH	B.S.F. History Book-1884; 1850 E. Providence Twp. Census
26	Aurora Cem.-NE	1850 E. Providence Twp. Census; 1890 Aurora, NE Veterans Census
27	Hampton Twp. Cem.-OH	B.S.F. History Book-1884; 1860 Snake Spring Valley Census
28		1890 Snake Spring Valley Census
29	Pleasantville Cem.	Historical Data Systems record; 1860 St. Clair Twp. Census
30	Pleasantville Cem.	Historical Data Systems record; William & John-brothers; aka Mechtley
31	Lower Claar Cem.-Claysburg	Bedford Inquirer 5/22/1908 listing
32		Historical Data Systems record
33	Holsinger Cem.-Bakers Summit	1890 Bloomfield Twp. Veterans Census; 1860 Woodbury Twp. Census
34		B.S.F. History Book-1884; Blair Co. references
35		POW 8/19/64; Died 3/4/65; B.S.F. History Book-1884
36	Holsinger Cem.-Bakers Summit	Soldiers of Blair County PA book - 1940; CW Iron Inspector for US Gov't.; Father of David
37	Yellowspring Cem.-Williamsburg	B.S.F. History Book-1884; Blair Co. references
38	Dayton Nat'l Cem.-OH	Historical Data Systems-Bedford Co. residence; enlisted in Hopewell
39		B.S.F. History Book-1884; Huntington Co residence record
40		1890 Brush Creek Twp. Veterans Census; aka Maloney
41	Everett Cem.	1890 Everett Veterans Census
42		B.S.F. History Book-1884; enlisted in Reading
43	Claremont Cem.-VA	B.S.F. History Book-1884; Elk Co references
44		Wounded 5/8/64; B.S.F. History Book-1884; 1860 Hopewell Twp. Census
45		POW-Salisbury 8/19/64, Died 12/10/64, Historical Data Systems record; aka Maroone
46	Lutheran Cem.-Hollidaysburg	B.S.F. History Book-1884; Blair Co. references
47	Bedford Forge-Yellow Creek	Civil War Pension Record
48		Historical Data Systems record; John, Charles and William-brothers
49		Historical Data Systems-Bedford Co. residence; B.S.F. History Book-1884
50		B.S.F. History Book-1884; Pension affidavit-POW at Andersonville
51	Yellow Creek Reformed Cem.	Wounded 9/17/62; Died 10/24/62; Historical Data Systems record; 1850 Hopewell Twp. Cen.
52		B.S.F. History Book-1884
53		Historical Data Systems record aka Mangus
54		Wounded 6/27/62; Historical Data Systems record
55		POW-Andersonville for 7 mths.; Historical Data Systems record
56	Schellsburg Cem.	Wounded listed in Bedford Inquirer article 6/3/64; 1890 Juniata Twp. Vet. Census-Hip Wound
57	Benshoff Hill Cem.-Cambria Co.	Frank McCoy 1912 Listing; Ancestry.com-born Bedford Co.
58	Mt. Smith Cem.-Bedford	1890 Bedford Twp. Veterans Census; Jacob (2) & Abraham-brothers
59	Daley Cem.-Somerset Co.	1880 Napier Twp. Census
60		Historical Data Systems-Bedford Co. residence
61		Historical Data Systems record
62		POW 6/22/64 to 4/28/65; B.S.F. History Book-1884; No Bedford Co. references
63		1890 Robertsdale Veterans Census
64	Tatesville Methodist Cem.	Deserted 8/12/63; Frank McCoy 1912 Listing

Chapter 18

	Name	Age	Rank	Regiment	Co.	Muster In	Muster Out	Casualty	Casualty Battle
1	Mansberger, Benjamin		Pvt.	PA 143rd Inf.	H	8/10/64	6/12/65		
2	Mansfield, Roy			US 2nd Cav.					
3	Manspeaker, Barclay	23	Pvt.	PA 8th Res.	F	6/19/61		KIA	Fredericksburg
4	Manspeaker, Daniel*	27	Pvt.	PA 77th Inf.	F	10/9/61			
5	Manspeaker, David	24	Pvt.	PA 8th Res.	F	6/11/61		KIA	Spotsylvania
6	Manspeaker, George	18	Pvt.	PA 11th Inf.	A	9/30/61		KIA	Fredericksburg
7	Manspeaker, Jacob	35	Pvt.	PA 56th Inf.	F	9/21/64	5/31/65		
8	Manspeaker, James								
9	Manspeaker, John (1)	34	Pvt.	PA 208th Inf.	K	9/7/64	6/1/65		
10	Manspeaker, John (2)	17	Pvt.	PA 11th Inf.	A	9/30/61	7/24/65	POW	Globe Tavern
11	Manspeaker, Samuel	17	Pvt.	PA 17th Cav.	E	9/2/64	6/16/65		
12	Marbourg, Charles F	21	Pvt.	PA 136th Inf.	K	8/27/62	5/29/63		
13	Marbourg, Jeremiah L	31	Surg.	PA 11th Res.		4/17/62	11/28/62	Wounded	
14	March, William		Pvt.						
15	Mardis, Samuel	21	Pvt.	PA 14th Inf.	I	5/2/61	8/7/61		
16	Mardis, Samuel J	22	Pvt.	PA 4th Cav.	E	2/27/64	7/1/65		
17	Markey, Joseph	30	Corp.	PA 184th Inf.	G	5/17/64	7/14/65	Wounded	
18	Markley, Joseph A	40	Pvt.	PA 195th Inf.	H	3/2/65	1/31/66		
19	Markley, W Harrison	20	Pvt.	PA 20th Cav.	M	7/26/63	10/3/63		
20	Marquart, August*	20	Pvt.	PA 55th Inf.	K	10/16/63	8/30/65		
21	Mars, John		Pvt.	PA 55th Inf.	H	10/11/61		POW	Edisto Island, SC
22	Mars, John	21	Pvt.	PA 55th Inf.	H			POW-Died	Drewry's Bluff
23	Marshall, George*	25	Pvt.	PA 55th Inf.	I	10/3/63		POW-Died	Drewry's Bluff
24	Marshall, Henry	23	Pvt.	PA 8th Res.	F	6/11/61	5/26/64	POW	Spotsylvania
25	Marshall, Henry L	31	Pvt.	PA 55th Inf.	K	11/5/61	12/14/62		
26	Marshall, Henry L		Pvt.	PA 184th Inf.	A	5/12/64		POW-Died	Jerusalem Plank Rd.
27	Marshall, Martin	32	Pvt.	USCT 75th Inf.	E	11/24/62			
28	Marshall, Martin		Pvt.	USCT 10th Inf.	A	5/1/64			
29	Marshall, Moses F	42	Corp.	PA 55th Inf.	K	11/5/61		Died	
30	Martin, Alexander*		Pvt.	PA 55th Inf.	I	9/23/63			
31	Martin, Andrew								
32	Martin, Barney	29	Pvt.	PA 11th Res.	F	6/20/61	6/13/64		
33	Martin, David	22	Pvt.	PA 8th Res.	F	6/11/61		KIA	Seven Days Battle
34	Martin, Denton O	28	Pvt.	PA 22nd Cav.	C	7/11/63	2/5/64		
35	Martin, George W	28	Pvt.	PA 91st Inf.	F	9/21/64	6/1/65		
36	Martin, Henry J*	34	Pvt.	PA 55th Inf.	I	10/7/63	8/30/65		
37	Martin, James P	18	Pvt.	PA 101st Inf.	F	11/1/61	6/3/65	POW	Plymonth, NC
38	Martin, Job H	38	Pvt.	PA 97th Inf.	H	2/21/65	5/28/65		
39	Martin, Josiah	16	Pvt.	PA 107th Inf.	A	10/6/64	6/5/65	Wounded	White Oak Road
40	Martin, Samuel		Pvt.	PA 22nd Cav.	I	7/9/63		KIA	Browns Gap
41	Martin, Thomas	24	Pvt.	PA 76th Inf.	E	10/9/61	11/28/64		
42	Martin, William L	23	Sgt.	PA 55th Inf.	K	11/5/61		KIA	Pocotaligo, SC
43	Massey, Isaac D		Corp.	PA McKeage's Mil.	H	7/2/63	8/8/63		
44	Masters, Frank M	18	Pvt.	PA 194th Inf.	I	7/22/64	11/5/64		
45	Masters, William J	17	Pvt.	PA 194th Inf.	I	7/22/64	11/5/64		
46	Masters, William M	22	Sgt. QM	MD 1st PHB Cav.	L	4/23/64	6/28/65		
47	Mathews, Hiram		Pvt.	PA 55th Inf.	H	9/22/62	12/30/64		
48	Maugle, Adam	18	Pvt.	PA 147th Inf.	B	8/17/61	8/29/64		
49	Maugle, Alexander	21	Pvt.	OH 84th Inf.	K	5/29/62	9/20/62		
50	Maugle, Alexander		Pvt.	OH 14th Inf.	F	2/11/64	7/11/65		
51	Maugle, Joseph	22	Pvt.	PA 8th Res.	F	6/1/61	5/26/64	Wounded	Antietam
52	Maugle, Joseph		Pvt.	PA 8th Res.	F			Wounded	Wilderness
53	Maugle, Solomon	26	Pvt.	PA 133rd Inf.	K	8/26/62	5/26/63		
54	Maugle, Thomas	19	Pvt.	KY 2nd Cav.	E	7/14/61	10/12/64		
55	Maugle, Thomas		Pvt.	OH 19th Cav.	L	11/17/64	7/24/65		
56	Mauk, John W	21	Sgt.	PA 138th Inf.	F	8/29/62	6/23/65		
57	Mauk, Joseph W	25	Sgt.	PA 14th Inf.	H	4/24/61	8/7/61	Died	
58	Maxwell, George W*		Corp.	PA 110th Inf.	C	10/24/61		KIA	1st Deep Bottom
59	Maxwell, Martin M*		2nd Lt.	PA 110th Inf.	C	9/24/61	10/24/64		
60	May, Abraham M	39	Pvt.	PA 208th Inf.	H	9/7/64	6/1/65		
61	May, Daniel H	31	Corp.	PA 82nd Inf.	C	11/14/64	7/13/65		
62	May, Daniel S	21	Pvt.	PA 55th Inf.	D	2/29/64	8/30/65		
63	May, Francis M	36	Pvt.	PA 148th Inf.	G	9/1/63	6/1/65	Wounded	Cold Harbor
64	May, Francis M		Pvt.	PA 148th Inf.	G			Wounded	Watkins House

	Cemetery	Notes
1		Frank McCoy 1912 Listing; aka Frank
2	New Paris Community Cem.	Bedford Inquirer 5/22/1908 listing
3	Fredericksburg Nat'l Cem.	KIA 12/13/62; 1860 E. Providence Twp. Census; PA Civil War Archives-Bedford Co. resid.
4	Mt. Zion Cem.-Breezewood	PA Civil War Archives record - Fulton County residence
5	Fredericksburg Nat'l Cem.	KIA 5/13/64; Civil War Archives-Bedford Co. residence; 1850 E. Providence Twp. Census
6	Fredericksburg Nat'l Cem.	KIA 12/13/62; Bedford Gazette 2/13/1914 listing; George, James, John (2) & Samuel-bros.
7	Ashbury Meth. Cem.-Graceville	Frank McCoy 1912 Listing; 1890 E. Providence Twp. Veterans Census
8	Mont Ida Cem.-KS	East Providence Twp. Honor Roll; No Regiment record found
9	Ashbury Meth. Cem.-Graceville	1890 E. Providence Twp. Veterans Census; John (1), Jacob, David & Daniel*-brothers
10	Mt. View Cem.-SD	POW 8/19/64; PA Civil War Archives-Bedford Co. residence
11	Strange Cem.-MI	1850 E. Providence Twp. Census
12	Schellsburg Cem.	Frank McCoy 1912 Listing; 1870 Harrison Twp. Census
13	Lake View Cem.-WA	1890 Bedford Twp. Veterans Census; B.S.F. History Book-1884
14		1890 W. St. Clair Twp. Veterans Census
15	Bunker Hill Cem.-Saxton	Ancestry.com information; aka Mardus
16	Grandview Cem.-Cambria Co.	Frank McCoy 1912 Listing; aka Mardus
17	St. John's Cem.-Loysburg	Pension reqeust record; 1890 S Woodbury Twp. Vet. Cen.; Loysburg Hotel & Tavern owner
18	Burnt Cabins Presb. Cem.	1850 Wells Tannery Census; Joseph & Harrison-brothers
19	Wells Valley Presb. Cem.	1890 Broad Top Twp. Veterans Census
20		B.S.F. History Book-1884; Drafted
21		POW 3/29/62; B.S.F. History Book-1884; PA Civil War Archives-Bedford Co. residence
22	Andersonville Cem.	MIA 5/16/64 listed in Bedford Inquirer 7/8/64 article; Died 9/30/64; Hist. Data Systems record
23	Millen Cem.-GA	POW-Millen, GA 5/15/62; B.S.F. History Book-1884; enlisted in Philadelphia
24		POW 5/8/64 to 5/12/64; B.S.F. History Book-1884; Historical Data Systems record
25		PA Civil War Archives-Bedford Co. residence; Discharged on Surgeon's Cert.
26	Andersonville Cem.; 11326	POW-Andersonville 6/22/64; Died 10/23/64; B.S.F. Hist.Book-1884; 1860 Harrison Twp. Cen.
27		1890 Bedford Twp. Veterans Census
28		1890 Bedford Twp. Veterans Census
29	Hampton Nat'l Cem.	Died 12/5/61; PA Civil War Archives-Bedford Co. residence; 1860 E. St. Clair Twp. Census
30		Deserted to enemy 8/6/64; Pottsville references
31	Fairview Cem.-Artemas	Bedford Inquirer 1908 Census; Andrew & Barney-brothers; Died 11/8/69
32	Fairview Cem.-Artemas	Frank McCoy 1912 Listing
33		KIA 6/26/62; Historical Data Systems-Bedford Co. residence; 1850 Mann Twp. Census
34	Mays Chapel-Warfordsburg	B.S.F. History Book-1884; Denton & Job-brothers
35	Fairview Cem.-Artemas	Bedford Inquirer 5/22/1908 listing; 1850 Mann Twp. Census; George & David-brothers
36		Drafted; B.S.F. History Book-1884; references in Historical Data Systems record
37	Everett Cem.	POW 4/20/64 to 3/1/65; Historical Data Systems-Bedford Co. residence
38	Milligans Cove-Buffalo Mills	Frank McCoy 1912 Listing; 1890 Harrison Twp. Veterans Census, 1870 Harrison Twp. Census
39	Grandview Cem.-Hunt. Co.	Wounded 3/31/65; 1890 Saxton Veterans Cen; aka Joseph
40	Staunton Nat'l Cem.	KIA 9/26/64; B.S.F. History Book-1884; Bedford Gazette 2/13/1914 listing
41		PA Civil War Archives-Bedford residence; Historical Data Systems record
42		KIA 10/22/62; B.S.F. History Book-1884; PA Civil War Archives-Bedford Co. residence
43		B.S.F. History Book-1884
44	Everett Cem.	B.S.F. History Book-1884; 1890 Everett Veterans Census
45	Historical Cem.-Deadwood, SD	B.S.F. History Book-1884; Frank & William-brothers
46	Hyndman Cem.	Findagrave.com record
47		B.S.F. History Book-1884
48	St. Lukes Cem.-Saxton	1890 Liberty Twp. Veterans Census; 1860 Liberty Twp. Census
49	Farmer Cem.-OH	1850 E. Providence Twp. Census; aka Mangle
50		Historical Data Systems record; aka Mangel
51	Everett Cem.	Wounded 9/17/62; 1890 Liberty Twp. Veterans Census-Thigh Wound
52		Wounded 5/8/64; PA Civil War Archives-Bedford residence; B.S.F. History Book-1884
53	Yellow Creek Reformed Cem.	1860 E. Providence Twp. Census; Solomon & Alexander-brothers
54	Everett Cem.	Frank McCoy 1912 Listing; Everett Cemetery Listing of Civil War Veterans; aka Naugle
55		1850 Liberty Twp. Census; Adam, Joseph & Thomas-brothers
56	Bethel Meth. Cem.-Centerville	Historical Data Systems record; One of two soldiers that killed Confederate Gen. - AP Hill
57	Old Claysburg Cem.	Died 8/3/61; PA Civil War Archives; 1850 Union Twp. Census; aka Mauk
58	Grandview Cem.-Tyrone	KIA 7/27/64; B.S.F. History Book-1884;PA Civil War Archives-Tyrone residence
59		B.S.F. History Book-1884; PA Civil War Archives-Tyrone residence; Discharged Surg. Cert.
60	Mulligans Cove-Buffalo Mills	B.S.F. History Book-1884; 1850 Harrison Twp. Cen; Historical Data Systems record
61	Lybarger Cem.-Madley	1890 Londonderry Twp Veterans Census
62	Mulligans Cove-Buffalo Mills	B.S.F. History Book-1884; 1890 Juniata Twp. Veterans Census
63		Wounded 6/3/64; Historical Data Systems-Bedford Co. residence
64		Historical Data Systems record; Arm Amputated 3/25/65

	Name	Age	Rank	Regiment	Co.	Muster In	Muster Out	Casualty	Casualty Battle
1	May, George			PA 91st Inf.					
2	May, George F	29	Pvt.	PA 82nd Inf.	K	9/21/64	6/17/65		
3	May, Harvey		Pvt.	PA 101st Inf.	G	12/28/61			
4	May, Hiram	17	Pvt.	PA 138th Inf.	F	6/12/63	6/23/65	Wounded	Cold Harbor
5	May, Hiram		Pvt.	PA 138th Inf.	F			Wounded	Opequon
6	May, Jacob	25	Pvt.	PA McKeage's Mil.	D	7/2/63	8/8/63		
7	May, Jacob		Pvt.	PA 149th Inf.	I	8/26/63		MIA	Wilderness
8	May, Jacob L	41	Pvt.	PA 99th Inf.	F	9/20/64		Died	Appomattox
9	May, John L	38	Pvt.	PA 67th Inf.	C	11/28/64	6/28/65	Wounded	Sailor's Creek
10	May, John W	23	Corp.	PA 138th Inf.	F	8/28/62	10/25/64	Wounded	Wilderness
11	May, Joseph	21	Pvt.	PA 55th Inf.	D	10/9/61	10/18/64		
12	May, Joseph C	20	Pvt.	PA 55th Inf.	K	11/5/61	2/21/63		
13	May, Joseph C		Pvt.	US 1st Light Art.	M	2/21/63			
14	May, Josiah	30	Pvt.	PA 99th Inf.	F	3/20/65	7/19/65		
15	May, Lewis A	37	Lt. Col.	PA 138th Inf.	F	9/29/62	6/23/65		
16	May, Marcus		Corp.	PA 13th Inf.	F				
17	May, Marcus	19	Corp.	PA 138th Inf.	F	8/29/62	6/23/65	Wounded	
18	May, P V		Corp.	PA 8th Res.	F	6/8/61	5/15/64		
19	May, Samuel M	22		MD 2nd PHB Inf.		8/31/61	22647		
20	May, Samuel M		Sgt.	PA 138th Inf.	F	8/29/62	3/30/63		
21	May, Samuel M			MD 2nd PHB Inf.		3/30/63	2/27/65	POW	Moorefield Jct.
22	May, Samuel S	17	Pvt.	PA McKeage's Mil.	D	7/2/63	8/8/63		
23	May, Samuel S		Pvt.	PA 208th Inf.	K	9/7/64	6/1/65		
24	May, William A	37	Pvt.	PA 55th Inf.	I	10/2/63	11/4/64		
25	May, William E	22	Pvt.	PA 93rd Inf.	D	7/4/64	6/28/65	Wounded	Ft. Fisher
26	Mayers, John		Pvt.	PA 2nd Cav.	E	10/15/61		POW-Died	
27	McAlwee, Daniel*	20	Pvt.	PA 107th Inf.	H	2/24/62	4/8/63		
28	McBride, Bernard*		Pvt.	PA 76th Inf.	E	11/21/61	3/24/62		
29	McCabe, Joseph*	30	Pvt.	PA 76th Inf.	E	2/18/65		Died	
30	McCarty, Alvin R		Pvt.	PA McKeage's Mil.	G	7/2/63	8/8/63		
31	McChesney, John	26	Corp.	PA 55th Inf.	H	9/20/61	6/18/64	Wounded	Cold Harbor
32	McChesney, John		Corp.	PA 55th Inf.	H			Wounded	Petersburg - Initial
33	McClain, Jesse O	21		PA 3rd Heavy Art.	G	7/29/64	11/7/65		
34	McCleary, George B	20	Pvt.	PA 133rd Inf.	K	8/15/62		Wound.-Died	Fredericksburg
35	McCleary, Henry	22	Sgt.	PA 138th Inf.	D	9/2/62	5/12/65	Wounded	Wilderness
36	McClellan, George W	18	Pvt.	OH 12th Inf.	B	6/20/61			
37	McClellan, John	25	Pvt.	PA 133rd Inf.	K	8/23/62	5/4/63	Wounded	Fredericksburg
38	McClellan, Josiah	22	Pvt.	PA 133rd Inf.	K	8/15/62	5/26/63	Wounded	Fredericksburg
39	McClincy, William J		Pvt.	PA 21st Cav.	E	7/21/63	7/8/65		
40	McCloud, Daniel*	21	Pvt.	PA 55th Inf.	I	10/11/61		Wound.-Died	Petersburg - Initial
41	McClure, Thomas	45	Chap.	PA 151st Inf.		2/3/63	7/27/63		
42	McConnell, Philip J*	25	Pvt.	PA 55th Inf.	I	2/18/64		MIA	Chaffin's Farm
43	McConnell, Randolph		Pvt.	PA McKeage's Mil.	G	7/2/63	8/8/63		
44	McCormick, James H		Pvt.	PA 21st Cav.	E	7/2/63	2/20/64		
45	McCormick, William	41	Corp.	PA 55th Inf.	H	10/19/63	8/30/65		
46	McCoy, Charles*	21	Pvt.	PA 76th Inf.	E	8/27/63		KIA	Drewry's Bluff
47	McCoy, Francis P	15	Pvt.	PA 81st Inf.	E	2/24/64	6/25/65	Wounded	
48	McCoy, Frank*	18	Music.	PA 194th Inf.	I	7/22/64	11/5/64		
49	McCoy, James*	44	Pvt.	PA 110th Inf.	C	2/22/64	6/28/65		
50	McCoy, Robert*	25	Pvt.	PA 55th Inf.	I	10/15/63			
51	McCoy, Shannon E	22	Corp.	PA 138th Inf.	F	8/29/62		MIA	Cold Harbor
52	McCray, Jacob	18	Pvt.	PA 147th Inf.	B	8/7/61	7/15/65		
53	McCray, James	16	Pvt.	PA 77th Inf.	F	3/2/62	7/2/65		
54	McCrossin, John*	31	Pvt.	PA 55th Inf.	D	10/6/63			
55	McCue, William	33	Pvt.	PA 77th Inf.	E	3/8/65	12/6/65		
56	McCulley, James T		Pvt.	PA McKeage's Mil.	E	7/3/63	8/8/63		
57	McCullip, Alexander	22	Pvt.	PA 133rd Inf.	C	8/13/62	5/26/63	Wounded	Fredericksburg
58	McCullip, Alexander		Pvt.	PA 133rd Inf.	C			Wounded	Chancellorsville
59	McDaniel, Daniel	21	Pvt.	PA 133rd Inf.	C	8/13/62	5/26/63		
60	McDaniel, Daniel		Sgt.	PA McKeage's Mil.	D	7/2/63	8/8/63		
61	McDaniel, Daniel		Pvt.	PA 2nd Heavy Art.	K	2/28/64	7/14/65	POW	Cold Harbor
62	McDaniel, George W	18	Pvt.	PA 133rd Inf.	C	8/13/62	5/26/63		
63	McDaniel, George W		Corp.	PA McKeage's Mil.	D	7/2/63	8/8/63		
64	McDaniel, George W		Pvt.	PA 3rd Heavy Art.	L	3/9/64	11/9/65		

	Cemetery	Notes
1	Bethel Graveyard	Frank McCoy 1912 Listing
2	Purcell Field Cem.	1890 Monroe Twp. Veterans Census; 1850 Southampton Twp. Census
3	Fockler Cem.-Saxton	Historical Data Systems record
4	Lybarger Cem.-Madley	Wounded 6/1/64, in Hospital at muster out; B.S.F. History Book-1884
5		Wounded 9/19/64, PA Civil War Project record
6		B.S.F. History Book-1884; enlisted in Bloody Run
7		MIA 5/5/64; Bedford Gazette 2/13/1914 listing; 1863 Draft Reg.-Monroe Twp. residence
8	Mt. Olivet Cem.-Manns Choice	Died 4/12/65; Civil War Pension Record; 1860 Monroe Twp. Census
9	Trinity UCC Cem.-Juniata Twp.	Wounded 4/6/65; Frank McCoy 1912 Listing
10	Bedford Cem.	Wounded 5/6/64; John W, Joseph C & Daniel S-brothers; B.S.F. History Book-1884
11		PA Civil War Archives-Bedford Co. residence; both Joseph & Joseph C May listed in 55th
12	Hyndman Cem.	PA Civil War Archives-Bedford residence; B.S.F. History Book-1884
13		Historical Data Systems record
14	Baughman Union-W. Prov.	Frank McCoy 1912 Listing; Josiah & Jacob-brothers
15	Woods Church-Colerain Twp.	Frank McCoy 1912 Listing; 1860 Cumberland Valley Census
16		Regiment on Headstone
17	Lybarger Cem.-Madley	1890 Veterans Census- Hip Wound; B.S.F. History Book-1884
18		B.S.F. History Book-1884
19	Trinity UCC Cem.-Juniata Twp.	1890 Harrison Twp. Veterans Census
20		PA Civil War Archives record; Samuel M, Marcus, Daniel H, John L & Hiram-brothers
21		POW-Belle Island & Andersonville 1/3/64 to 1/6/65; 1890 Harrison Twp Cen.
22		B.S.F. History Book-1884
23	Mt. Olivet Cem.-Manns Choice	1890 E. Providence Twp. Veterans Census; B.S.F. History Book-1884; Son of Jacob L
24		B.S.F. History Book-1884; Discharged on Surgeon's Cert.
25	Lybarger Cem.-Madley	Wounded 3/25/65; 1890 Londonderry Twp. 1890 Veterans Census; 1860 Juniata Twp. Census
26		POW-Andersonville; Died-Starvation; 1890 Hyndman Veterans Census
27		B.S.F. History Book-1884; Fulton Co. references
28		B.S.F. History Book-1884; PA Civil War Archives-Philadelphia Resid.; Disch. Surg. Cert.
29	Easton-Northampton Co	Died 8/13/65; B.S.F. History Book-1884; buried in Easton, PA
30		B.S.F. History Book-1884
31	Hopewell Cem.	Wounded 6/3/64; PA Civil War Archives-Bedford residence
32		Wounded 6/18/64; Ancestry.com-born in Hopewell
33	New Grenada-Fulton Co.	1890 Robertsdale Veterans Census
34	Yellow Creek Reformed Cem.	Wounded 12/13/62; Died 4/10/63; B.S.F. History Book-1884; 1860 Hopewell Twp. Census
35		Wounded 5/6/64; Historical Data Systems-Bedford Co. residence
36	Fountain Cem.-OH	1870 Colerain Twp. Census; George, John & Josiah-brothers
37	Tonoloway Bapt.-Warfordsburg	Wounded 12/13/62; Historical Data Systems record; 1860 Cumberland Valley Census
38	Woods Church-Colerain Twp.	Wounded 12/13/62; Historical Data Systems-Bedford Co. residence
39		Historical Data Systems-Bedford Co. residence
40	Hampton Nat'l Cem.	Wounded 6/18/64; Died 7/9/64 ; B.S.F. History Book-1884; enlisted in Philadelphia
41	Everett Cem.	Frank McCoy 1912 Listing
42		MIA 9/29/64; B.S.F. History Book-1884; Cambria Co. residence
43		B.S.F. History Book-1884
44		Historical Data Systems-Bedford Co. residence
45		B.S.F. History Book-1884
46		KIA 5/14/64; B.S.F. History Book-1884; No other records located
47	Lybarger Cem.-Madley	1890 Londonderry Twp. Veterans Census
48		B.S.F. History Book-1884; enlisted in Reading, PA
49		PA Civil War Archives-Holidaysburg residence
50		Deserted 5/6/64; B.S.F. History Book-1884; enlisted in Reading, PA
51		MIA 6/1/64; Historical Data Systems-Cumberland Valley residence; 1860 Cumb. Valley Cen.
52	Stonerstown Cem.-Liberty Twp.	Frank McCoy 1912 Listing; Discharged on Surgeon's Cert. 3/15/63; Re-enlisted 2/20/64
53	Stonerstown Cem.-Liberty Twp.	PA Civil War Archives-Bedford Co. residence; James & Jacob-brothers
54		B.S.F. History Book-1884; Drafted & Deserted 4/26/64; Phila references
55	Fockler Cem.-Saxton	Historical Data Systems record
56	Union Cem.-Bellefonte	Ancestry.com-born in Schellsburg; PA 45th Infantry references
57	Congressional Cem.-DC	Wounded 12/13/62; Historical Data Systems-Bedford Co. residence
58		Wounded 5/3/63; Historical Data Systems, enlisted in Hopewell
59	Prairie View Cem.-IA	B.S.F. History Book-1884; Historical Data Systems-Bedford Co. residence
60		B.S.F. History Book-1884
61		POW 6/2/64 to 2/26/65; transferred Andersonville to Macon 9/29/64; Hist. Data Syst. record
62	Baughman Union-W. Prov.	B.S.F. History Book-1884; Historical Data Systems record
63		B.S.F. History Book-1884
64		B.S.F. History Book-1884

Chapter 18

	Name	Age	Rank	Regiment	Co.	Muster In	Muster Out	Casualty	Casualty Battle
1	McDaniel, Henry								
2	McDaniel, Hiram		Pvt.	PA McKeage's Mil.	D	7/2/63	8/8/63		
3	McDaniel, Hiram	19	Pvt.	PA 149th Inf.	I	8/27/63		Wound.-Died	Spotsylvania
4	McDaniel, Jason	25	Pvt.	PA 107th Inf.	F	9/21/64	6/6/65		
5	McDaniel, Lewis	23	Pvt.	PA 133rd Inf.	C	8/13/62	5/26/63		
6	McDaniel, Lewis		Sgt.	PA 22nd Cav.	H	2/27/64		KIA	Berryville
7	McDaniel, Oliver	24	Pvt.	PA 78th Inf.	K	3/7/65	9/11/65		
8	McDaniel, William	20	Pvt.	PA 208th Inf.	K	9/7/64	6/1/65		
9	McDermot, Charles*	26	Pvt.	PA 55th Inf.	I	9/20/63		Wounded	Drewry's Bluff
10	McDonald, James	24	Pvt.	PA 13th Res.	G	5/29/61	12/1/62	Died	
11	McDonald, John	25	Pvt.	PA 208th Inf.	H	9/8/64	6/6/65		
12	McDonald, Samuel	16	Pvt.	PA 110th Inf.	D	12/19/61			
13	McDonald, Samuel		Pvt.	US 6th Cav.	E	10/28/62	8/27/64		
14	McDonald, Samuel		Pvt.	PA 97th Inf.	A	3/3/65	8/28/65	Died	
15	McDonald, William	26	Pvt.	IN 57th Inf.	K	11/14/62	8/15/63		
16	McDonald, William H		Pvt.	PA 21st Cav.	E	7/2/63	2/20/64		
17	McDonald, William, Jr	18	Pvt.	PA 101st Inf.	D	11/1/61		Died	
18	McEldowney, George E	27	Pvt.	PA 101st Inf.	D	11/1/61	6/25/65	POW	Plymonth, NC
19	McEldowney, Hezekiah	27	Corp.	PA 22nd Cav.	C	7/5/63	2/5/64		
20	McEldowney, James H	42	Pvt.	PA 22nd Cav.	M	2/23/64			
21	McEldowney, John*	36	Pvt.	OH 126th Inf.	I	5/12/64	6/25/65		
22	McEldowney, Samuel J	34	1st Sgt.	PA 101st Inf.	D	1/13/62	6/3/65	POW	Plymonth, NC
23	McElroy, John*		Pvt.	PA 55th Inf.	K	10/19/63			
24	McEnespy, James B	16	Pvt.	PA 55th Inf.	D	10/12/61			
25	McEnespy, Samuel	15	Pvt.	PA McKeage's Mil.	H	7/2/63	8/8/63		
26	McEnespy, Samuel		Pvt.	PA 13th Cav.	D	10/5/63		Died	Coggin's Point
27	McFarly, Andrew J	18	Pvt.	PA 79th Inf.	F	3/9/65	7/12/65		
28	McFarland, Daniel M	22	Pvt.	PA 8th Res.	F	6/11/61	5/26/64		
29	McFarland, James		Pvt.	PA 55th Inf.	K	10/13/63			
30	McFarland, Joseph	22	Pvt.	PA 8th Res.	F	9/6/62	5/15/64	Wounded	Spotsylvania
31	McFarland, Joseph		Pvt.	PA 191st Inf.		5/15/64	9/19/64		
32	McFerren, John	49	Pvt.	PA 210th Inf.	I	9/16/64	5/30/65		
33	McFerren, Samuel	27	Pvt.	OH 146th Inf.	C	5/2/64	9/7/64		
34	McGee, David	35	Pvt.	PA 3rd Art.	D	11/8/62	12/3/64		
35	McGee, James	22	Pvt.	PA 55th Inf.	I	10/15/61		POW-Died	Drewry's Bluff
36	McGee, John H	45	Pvt.	PA 147th Inf.	B	11/1/61	5/5/63	Died	
37	McGee, Samuel	37	Pvt.	PA 138th Inf.	H	8/26/62	6/23/65		
38	McGee, William	23	Pvt.	PA 55th Inf.	I	10/15/61		MIA	Chaffin's Farm
39	McGirr, Matthias	25	Pvt.	PA 13th Inf.	G	4/25/61	8/6/61		
40	McGraw, James		Pvt.	PA 45th Inf.	H	9/24/64	6/5/65		
41	McGregor, Elijah	33	Pvt.	PA 91st Inf.	I	9/21/64	5/30/65		
42	McGregor, James	32	Pvt.	PA 5th Heavy Art.	M	9/5/64	6/30/65		
43	McGregor, John	20	Pvt.	PA 55th Inf.	I	9/20/61		Wound.-Died	Cold Harbor
44	McGregor, Robert	47	Pvt.	PA 55th Inf.	I	10/15/61	10/15/64		
45	McGregor, William	51	Pvt.	PA 55th Inf.	I	10/15/61	5/14/62		
46	McGregor, William		Pvt.	PA 13th Cav.	D	9/26/63	6/23/65		
47	McGregor, William A	19	Pvt.	PA 99th Inf.	A	3/17/65	7/1/65		
48	McHugh, William			PA 77th Inf.					
49	McHugh, William	34	Pvt.	PA 11th Res.	F	3/2/65	12/6/65		
50	McIlnay, James	18	Pvt.	PA 110th Inf.	C	10/24/61	6/15/62	Died	
51	McIlnay, John							KIA	Winchester
52	McIlnay, John F	24	Pvt.	PA 14th Inf.	H	4/24/61	8/7/61		
53	McIlnay, John F		Pvt.	PA 137th Inf.	I	8/22/62	6/1/63		
54	McIntyre, J								
55	McKee, Alexander H	25	Pvt.	PA 8th Res.	F	8/23/61	7/15/62		
56	McKee, Alexander H		Pvt.	PA 22nd Cav.	E	8/20/64	5/24/65		
57	McKee, Amos	29	Pvt.	PA 3rd Art.	U	9/28/64	6/9/65		
58	McKee, David	26	Pvt.	PA 14th Inf.	I	5/2/61	8/7/61		
59	McKee, David			PA 55th Inf.					
60	McKellip, D Alexander	28	Pvt.	OH 13th Inf.	B	6/26/61	12/5/65		
61	McKenzie, John		Pvt.	MD 1st PHB Cav.	H	3/23/64	6/10/65		
62	McKibbin William L	24	Pvt.	PA 130th Inf.	A	8/14/62	5/21/63		
63	McKibbin William L		Pvt.	PA 149th Inf.		2/21/65	5/5/65		
64	McLaughlin, Charles P	19	1st Lt.	PA 138th Inf.	F	8/29/62		KIA	Cold Harbor

	Cemetery	Notes
1	Schellsburg Cem.	E. Providence Honor Roll; 1850 E. Providence Twp. Census
2		B.S.F. History Book-1884; Hiram, William & Jason-brothers
3		Head Wound 5/10/64; Died 5/14/64; B.S.F. History Book-1884; 1860 Monroe Twp. Census
4	Mt. Pleasant Cem.-Mattie	1890 E. Providence Twp. Veterans Census; Historical Data Systems record
5		1850 W. Providence Twp. Census; Historical Data Systems record; Lewis & George-brothers
6		KIA 8/21/64; B.S.F. History Book-1884; 1860 W. Providence Twp. Census
7	Mt. Union Cem.-Mench	Frank McCoy 1912 Listing
8	Freeland Cem.-KS	1860 Monroe Twp. Census; B.S.F. History Book-1884
9		Wounded 5/16/64; B.S.F. History Book-1884; enlisted in Philadelphia
10	US Soldiers & Airmans Nat'l	Died 6/25/63-Concussion & bacterial infection; Bed. Gaz. 1914; 1850 S. Woodbury Twp. Cen.
11	Haxtun Cem.-CO	B.S.F. History Book-1884; John, James, Samuel & William-brothers
12	Potter Creek Cem.-Woodbury	Historical Data Systems record; Frank McCoy 1912 Listing; 1850 S. Woodbury Twp. Census
13		Historical Data Systems record; Pension Records
14		Died-Typhoid Fever 9/23/65; Frank McCoy 1912 Listing
15	Milford Cem.-IN	1860 Colerain Twp. Census
16		PA Civil War Archives-Bedford Co. residence; Historical Data Systems record
17	Hampton Nat'l Cem.	Died-Typhoid Fever 7/3/62; PA Civil War Archives-Bedford Co. residence
18	Hope Cem.-OH	POW-Andersonville 4/20/64; 1850 Colerain Twp. Census; B.S.F. History Book-1884
19	Roanoke Twp. Cem.-IL	PA Civil War Archives-Bedford Co. residence; 1860 Colerain Twp. Census
20	Hartford City Cem.-IN	Deserted 3/27/64; PA Civil War Archives; James & John-brothers
21	Ansonia Cem.-OH	Historical Data Systems record; Fulton Co. references
22	Everett Cem.	POW 4/20/64 to 3/1/65; Historical Data Systems record; Samuel, George & Hezekiah-brothers
23		Drafted & Deserted 6/3/64; B.S.F. History Book-1884
24	Dayton Cem.-CA	B.S.F. History Book-1884; 1850 Bedford Boro Census; Discharged on Surgeon's Cert.
25		B.S.F. History Book-1884; Bedford Inquirer 5/22/1908 listing; Samuel & James-brothers
26	Cavalry Corps-City Point, VA	Died 9/16/64 at Beefsteak Raid; Bedford Inquirer 11/11/64 Obituary; 1860 Bedford Boro Cen.
27	Providence Union Cem.-Everett	Frank McCoy 1912 Listing; Bedford Inquirer 5/22/1908 listing; 1860 Monroe Twp. Census
28	Providence Union Cem.-Everett	Historical Data Systems record; Daniel, Andrew & Joseph-brothers
29		Deserted 7/31/65 after war ended; B.S.F. History Book-1884; B.S.F. History Book-1884
30	Providence Union Cem.-Everett	Leg Amputated 5/8/64; Frank McCoy 1912 Listing
31		Historical Data Systems record
32	McFerren Farm Cem.-Bedford	Civil War Pension record; aka McPherson
33	Perry Cem.-OH	Ancestry.com-born in Bedford
34	Holsinger Cem.-Bakers Summit	Frank McCoy 1912 Listing; CW Pension Record; 55th Reg't references; Disch. Surge. Cert.
35	Richmond Nat'l Cem.	POW-Libby 5/16/64; Died 5/27/64; B.S.F. History Book-1884; 1860 Woodbury Twp. Census
36	US Soldiers & Airmans Nat'l	Died 5/5/63; Civil War Pension Record; 1860 Broad Top Twp. Census
37	Bloomfield Furnace Cem.	1850 M. Woodbury Twp. Census; Samuel, William, James & David-brothers
38		MIA 9/29/64; PA Civil War Archives-Bloomfield Furnace resid.; 1860 Woodbury Twp. Cen.
39		B.S.F. History Book-1884; Historical Data Systems-Bedford residence
40	Union Church-E. Prov. Twp.	Frank McCoy 1912 Listing; aka McGrew & McGrew
41	Pleasantville Cem.	1890 W. St. Clair Twp. Veterans Census
42	Whitesburg Cem.-Armstrong Co	Ancestry.com-born in Bedford Co.; Historical Data Systems record
43	Arlington Nat'l Cem., 27-499	Wounded 6/3/64; Died 6/6/64; B.S.F. History Book-1884; 1860 Woodbury Twp. Census
44	Oak Ridge Cem.-Altoona	B.S.F. History Book-1884; 1850 Broad Top Twp. Census; John is son
45	Holsinger Cem.-Bakers Summit	B.S.F. History Book-1884; 1890 Bloomfield Twp. Veterans Census; Discharged on Surg. Cert.
46		Historical Data Systems record
47	Pleasantville Cem.	1890 Pleasantville Veterans Census; PA Civil War Pension Records
48		Frank McCoy 1912 Listing
49	Fockler Cem.-Saxton	1890 Saxton Veterans Census
50	Yellow Creek Reformed Cem.	Died 6/15/62; Civil War Pension Record; Hist. Data Systems record; 1850 Hopewell Twp. Cen.
51	Yellow Creek Reformed Cem.	Died 3/23/63; "Reg PV killed Winchester, Va. GAR" on Headstone; 1850 Hopewell Twp. Cen.
52	Albrights Cem.-Roaring Springs	Ancestry.com-born in Hopewell Twp. in 1837
53		Bedford Inquirer 5/22/1908 listing
54	Duval Cem.-Coaldale	Bedford Inquirer 5/22/1908 listing; James & John McIntyre buried in Duvalls Cemetery
55	Brownsville, PA Cem.	B.S.F. History Book-1884; PA Civil War Archives-Bedford Co. resid.; Discharged Surg. Cert.
56		Historical Data Systems record; PA Civil War Pension document
57	Greenlawn Cem.-Roaring Spring	PA Civil War Archives; Broad Top Twp. Census
58		PA Civil War Archives-Bedford Co. residence
59	Holsinger Cem.-Bakers Summit	Bedford Inquirer 5/22/1908 listing; 1860 Woodbury Twp. Census
60	Viola Cem.-IL	Ancestry.com-born Bedford; Historical Data Systems; aka McKeelip
61	Cooks Mill Cem.-Hyndman	Bedford Inquirer 5/22/1908 listing
62	Buck Valley Cem.-Fulton Co.	1890 Brush Creek Twp. Veterans Census; Doctor after the war
63		Surgeon references
64	Cold Harbor Nat'l Cem.	KIA 6/1/64; B.S.F. History Book-1884; Hist. Data Systems record; 1860 Colerain Twp. Cen.

	Name	Age	Rank	Regiment	Co.	Muster In	Muster Out	Casualty	Casualty Battle
1	McLaughlin, Collin L	19	Pvt.	PA 11th Inf.	C	2/22/62	3/15/64		
2	McLaughlin, Edward		Pvt.	PA 21st Cav.	E	7/21/63	7/8/65		
3	McLaughlin, James A	18	Corp.	PA 102nd Inf.	I	8/16/62	9/3/64	Wounded	Williamsburg
4	McLaughlin, James A		Corp.	PA 102nd Inf.	I			Wounded	Chancellorsville
5	McLaughlin, William H	22	Pvt.	PA 125th Inf.	I	8/13/62	5/18/63		
6	McLaughlin, William H		Pvt.	PA 139th Inf.	B	2/25/64	6/21/65		
7	McLeary, Paul			PA 53rd Inf.					
8	McLucas, William	35	Pvt.	MD 1st PHB Cav.	B	11/27/61	6/28/65	Wounded	
9	McMahan, William	19	Pvt.	PA 194th Inf.	I	7/22/64	11/5/64		
10	McMullen, Absolom	20	Pvt.	IL 146th Inf.	I	9/5/64	6/15/65		
11	McMullen, Charles (1)		Pvt.	PA 55th Inf.	K	9/26/62		Wound.-Died	
12	McMullen, Charles (2)	17	Pvt.	PA McKeage's Mil.	H	7/2/63	8/8/63		
13	McMullen, Charles E	18	Pvt.	IL 146th Inf.	I	9/1/64	6/14/65		
14	McMullen, John H	36	Corp.	IL 77th Inf.	I	9/2/62	7/10/65		
15	McMullen, Samuel	42	Pvt.	IL 47th Inf.	K	3/20/65	1/21/66		
16	McMurtrie, J R		Sgt.	PA 125th Inf.	C	8/11/62	6/18/63		
17	McMurtrie, James E	22	Pvt.	PA 5th Inf.	D	4/21/61	7/25/61		
18	McMurtrie, James E		Pvt.	PA 49th Inf.	D	2/24/62	2/5/63		
19	McPherson, Cyrus			USCT Inf.					
20	McPherson, John		Pvt.	USCT 3rd Inf.	E	7/23/63		Died	
21	McQuillen, Hiram	30	Pvt.	PA 13th Inf.	G	4/25/61	8/6/61		
22	McRean, James	21		NY 91st Inf.		9/21/64			
23	McVicker, J Clay	19	Pvt.	PA 55th Inf.	D	2/27/64		Died	
24	McVicker, James A	45	Sadd.	PA 21st Cav.	E	7/18/63	2/20/64		
25	McVicker, Jesse			PA 171st Inf.					
26	McVicker, Jesse	29	Pvt.	PA 102nd Inf.	K	8/19/64	6/28/65	Wounded	Opequon
27	McVicker, William	30	Pvt.	PA 138th Inf.	D	8/29/62	6/23/65		
28	McWade, John	29	Pvt.	PA 154th Inf.	B	11/11/62	9/29/63		
29	Meade, Haynes P*	17	Pvt.	PA 76th Inf.	C	1/30/65	6/22/65		
30	Means, Edward	23	Pvt.	PA 78th Inf.	K	3/7/65			
31	Mearkle, Barton	20	Pvt.	PA 208th Inf.	K	9/7/64	6/1/65		
32	Mearkle, David S	22	Pvt.	PA McKeage's Mil.	D	7/2/63	8/8/63		
33	Mearkle, David S		Pvt.	PA 99th Inf.	G	2/24/65	5/31/65	Wounded	
34	Mearkle, Henry (1)		Pvt.	PA 22nd Cav.	I	7/23/63		Died	Leetown, MD
35	Mearkle, Henry (2)	43	Pvt.	PA 78th Inf.	K	3/1/65	5/12/65		
36	Mearkle, Sansom	21	Pvt.	PA McKeage's Mil.	D	7/2/63	8/8/63		
37	Mechtley, John E	20	Pvt.	PA 82nd Inf.	C	10/3/64	6/24/65		
38	Medley, William	29	Pvt.	PA 13th Inf.	G	4/25/61	8/6/61		
39	Mellen, Thomas		Pvt.	PA 13th Inf.	G	4/25/61	8/6/61		
40	Mellen, Thomas	35	Pvt.	PA 6th Res.	D	1/1/62		KIA	Fredericksburg
41	Mellen, William S		Pvt.	PA 99th Inf.		3/17/65	7/1/65		
42	Mellott, Caleb	28	Pvt.	PA 82nd Inf.	I	11/28/64		POW-Died	
43	Mellott, Cornelius	20	Pvt.	PA 11th Inf.	A	9/30/61		KIA	2nd Bull Run
44	Mellott, Daniel B	35	Pvt.	PA 82nd Inf.	I	11/28/64	6/15/65	Wounded	
45	Mellott, Frederick	21	Pvt.	PA 12th Res.	K	8/10/61		KIA	South Mountain
46	Mellott, Henry T	30	Pvt.	PA 158th Inf.	H	11/4/62	8/12/63		
47	Mellott, Hiram	18	Pvt.	PA 22nd Cav.	H	2/26/64	2/26/65		
48	Mellott, Jacob B	19	Pvt.	PA 11th Cav.	G	2/27/64	8/13/65		
49	Mellott, Jacob L	19	Pvt.	PA 208th Inf.	K	9/7/64	6/1/65		
50	Mellott, John L	18	Pvt.	PA 11th Inf.	A	9/30/61		Died	
51	Mellott, Peter	16	Pvt.	MD 3rd PHB Inf.	B	11/9/61	5/29/65		
52	Mellott, Simon	20	Pvt.	PA McKeage's Mil.	D	7/2/63	8/8/63		
53	Mellott, Simon		Pvt.	PA 22nd Cav.	H	2/1/64	6/24/65	Wounded	
54	Mellott, Steven	34	Pvt.	PA 12th Res.	A	8/10/61	5/31/64		
55	Mellott, Steven		Pvt.	PA 190th Inf.	B	5/31/64	2/1/65		
56	Mellott, Steven		Pvt.	VRC 9th Inf.		2/1/65	7/15/65		
57	Mellott, Thomas S	16	Pvt.	PA 11th Cav.	G	2/27/64	8/13/65		
58	Melott, William	25	Pvt.	PA McKeage's Mil.	D	7/2/63	8/8/63		
59	Meloy, Biven D	22	Pvt.	PA 138th Inf.	E	8/29/62	6/23/65	Wounded	Wilderness
60	Meloy, John L	32	Pvt.	PA 133rd Inf.	C	8/13/62	5/26/63		
61	Meloy, John L		Pvt.	PA 99th Inf.	D	2/25/65	5/31/65		
62	Mench, John	48	Pvt.	PA 99th Inf.	G	2/25/65	5/29/65		
63	Mentz, Stephen*		Pvt.	PA 76th Inf.	E	10/13/64		Died	
64	Mentzer, Jacob M	31	Pvt.	PA 133rd Inf.	C	8/13/62		KIA	Fredericksburg

	Cemetery	Notes
1	Hollywood Forever-CA	Ancestry.com-born in Bedford
2		Historical Data Systems-Bedford Co. residence
3	Uniondale Cem.-Pittsburgh	Wounded 5/5/62; Ancestry.com-born in Bedford
4		Wounded 5/3/63; Historical Data Systems record; Salem Hights battle
5		PA Civil War Project record; William, James & Collin-brothers
6		PA Civil War Project record; Ancestry.com-born in Bedford
7	Grandview Cem.-Cambria Co.	aka McCreary; 1860 Napier Twp. Census; Pension request
8	McKendree Cem.-Crystal Spring	1890 Everett Veterans Census
9		B.S.F. History Book-1884; 1850 M. Woodbury Twp. Census
10	Oak Ridge Cem.-IL	1850 Schellsburg Census; Illinois Civil War Roster
11		Died 6/20/64; B.S.F. History Book-1884; PA Civil War Muster Record-enlisted in Bedford
12		B.S.F. History Book-1884
13	Billings Union Cem.-OK	Ancestry.com-born in Bedford Co.
14	Elmwood Twp. Cem.-IL	1850 Schellsburg Census; Historical Data Systems record
15	Green Lawn Cem.-MO	John, Absolom & Samuel-brothers; Historical Data Systems record
16	Fockler Cem.-Saxton	Frank McCoy 1912 Listing; aka McMurthie
17	Fockler Cem.-Saxton	Historical Data Systems record
18		1890 Saxton Veterans Census; Discharged on Surgeon's Cert.
19	Mt. Ross Cem.-Bedford	Bedford Inquirer 5/22/1908 listing; aka Macpherson
20		Died-Accidental Wound 8/26/65; Frank McCoy 1912 & Civil Works Admin. 1934 listings
21		B.S.F. History Book-1884; Historical Data Systems-Bedford Co. residence
22		Fold3 Military listing-Born in Bedford Co.
23	Beaufort Nat'l Cem.	Died 4/18/64, 1850 Harrison Twp. Cen; B.S.F. History Book-1884
24	Old Union Cem.-Buffalo Mills	1860 Napier Twp. Census; Historical Data Systems-Bedford Co. residence
25		Frank McCoy 1912 Listing
26	Old Union Cem.-Buffalo Mills	Wounded 9/19/64; Hist. Data Systems record; 1850 Harrison Twp. Cen.; Jesse & James-bros.
27	Miriam Cem.-MI	Historical Data Systems-Bedford Co. residence; William & J Clay-brothers
28		E. Providence Honor Roll; 1880 E. Providence Twp. Census
29		B.S.F. History Book-1884; Scranton residence listed in Ancestry.com
30	Fairview Cem.-Artemas	1890 Brush Creek Twp. Veterans Census
31	Rock Hill Cem.-Monroe Twp.	1890 Monroe Twp. Veterans Census; Historical Data Systems record
32	Rock Hill Cem.-Monroe Twp.	1890 Monroe Twp. Veterans Census; aka Markle
33		1890 Monroe Twp. Census - Leg Wound; David S & Barton-brothers
34	Antietam Nat'l Cem., 26-E-528	Wounded 7/3/64; Died 7/24/64.; B.S.F. History Book-1884; 1850 Monroe Twp. Census
35	Stevens Chapel-Monroe Twp.	1890 Monroe Twp. Veterans Census; in Hospital at Muster Out
36	Rock Hill Cem.-Monroe Twp.	B.S.F. History Book-1884; 1860 Monroe Twp. Census; aka Markle
37	Mechtley Family Cem.	1890 Union Twp. (Pavia) Veterans Census
38		B.S.F. History Book-1884; Historical Data Systems-Bedford Co. residence
39		PA Civil War Archives-Bedford Co. residence; Historical Data Systems record aka Mellan
40		KIA 12/13/62; B.S.F. History Book-1884; PA Civil War Archives aka Mellon
41		B.S.F. History Book-1884
42	Arlington Nat'l Cem., 13 11518	POW-Libby; Died 7/6/65; 1890 Brush Creek Twp. Veterans Census
43	US Soldiers & Airmans Nat'l	KIA 8/30/62; Bedford Gazette 2/13/1914 listing; 1850 E. Providence Twp. Census
44	Buck Valley Cem.-Fulton Co.	Face Wound; 1890 Brush Creek Twp. Veterans Census
45	Antietam Nat'l Cem., 3913	KIA 9/14/62, B.S.F. History Book-1884; PA Civil War Archives-Bedford Co. residence
46	Pleasant Ridge Cem.-Needmore	E. Providence Honor Roll; Historical Data Systems record
47	Hoskins Cem.-IA	B.S.F. History Book-1884; Historical Data Systems record; 1850 E. Providence Twp. Census
48	Sideling Hill Baptist-Fulton Co.	Civil War Pension Record; Jacob & Caleb-brothers
49	Everett Cem.	B.S.F. History Book-1884; Jacob, John, Henry, Frederick Thomas & Cornelius-brothers
50	Mt. Pleasant Cem.-Mattie	Died Remittent Fever 12/31/64; Frank McCoy 1912 Listing; Bedford Gazette 2/13/1914 listing
51	Amaranth Brethren Cem.	1890 Brush Creek Twp. Veterans Census
52	Everett Cem.	B.S.F. History Book-1884; Historical Data Systems record
53		1890 E. Providence Twp. Veterans Census; Historical Data Systems record; Knee wound-1864
54	Piney Plains Meth. Cem.-MD	1860 E. Providence Twp. Census; PA Civil War Project record; Historical Data Systems record
55		PA Civil War Archives; East Providence Honor Roll
56		East Providence Honor Roll
57	Stevens Chapel-Monroe Twp.	1890 E. Providence Twp. Veterans Census; PA Civil War Archives record
58	Hopewell Cem.	B.S.F. History Book-1884; 1870 Hopewell Twp. Census; Historical Data Systems record
59	Everett Cem.	Wounded 5/6/64; 1890 Bedford Twp. Veterans Census; B.S.F. History Book-1884
60	Greenwood Cem.-Altoona	B.S.F. History Book-1884; Historical Data Systems-enlisted in Hopewell
61		Historical Data Systems record; Civil War Pension Record; John & Bevin-brothers
62	Mt. Zion Cem.-Breezewood	Frank McCoy 1912 Listing
63		Died 4/17/65; B.S.F. History Book-1884; Drafted; enlisted in Harrisburg
64		KIA 12/13/62; B.S.F. History Book-1884; 1860 Woodbury Twp. Cen; enlisted in Hopewell

Chapter 18

	Name	Age	Rank	Regiment	Co.	Muster In	Muster Out	Casualty	Casualty Battle
1	Mentzer, Jeremiah D	22	Bugler	PA 13th Cav.	D	9/8/63	7/14/65		
2	Meredith, Charles B	24	Pvt.	PA 76th Inf.	E	11/21/61	11/28/64		
3	Meredith, John	18	Pvt.	PA 78th Inf.	F	10/12/61	11/4/64		
4	Merithew, Horace*	18	Pvt.	PA 76th Inf.	E	10/15/64	6/10/65		
5	Messersmith, Alexander	41	Pvt.	PA 208th Inf.	K	9/7/64	6/1/65		
6	Messersmith, George	17	Pvt.	PA McKeage's Mil.	D	7/2/63	8/8/63		
7	Messersmith, George		Pvt.	PA 22nd Cav.	H	2/26/64	6/24/65		
8	Messersmith, Joseph S	41	Corp.	PA 208th Inf.	K	9/7/64	5/29/65		
9	Metz, Felty	47	Pvt.	PA 13th Inf.	B	9/12/62	9/26/62		
10	Metzger, James	25	Capt.	PA 55th Inf.	D	10/12/61	8/30/65	POW	Drewry's Bluff
11	Metzger, John S	17	Pvt.	MD 2nd PHB Inf.	C	3/28/65	6/28/65		
12	Metzger, Solomon S	23	Capt.	PA 55th Inf.	D	10/12/61	6/28/64	Wounded	Cold Harbor
13	Metzler, Henry Clay	16	Pvt.	MD 3rd PHB Inf.	B	2/17/62	5/29/65		
14	Meyer, William C*	18	Pvt.	PA 148th Inf.	A	8/25/62		Wounded	Po River
15	Meyer, William C*		Pvt.	PA 148th Inf.	A	8/25/62		KIA	2nd Deep Bottom
16	Meyers, Levi	19	Pvt.	PA 55th Inf.	H	3/2/64		POW-Died	Drewry's Bluff
17	Michaels, Henry								
18	Mickel, James			PA 62nd Inf.					
19	Mickel, James		Pvt.	PA 29th Inf.	B	1/18/65	6/20/65		
20	Mickel, William P								
21	Mickey, Rankins	16	Pvt.	PA 55th Inf.	D	10/12/61	8/30/65		
22	Middleton, Edwin	25	Sgt.	PA 15th Cav.	G	10/3/62	6/21/65		
23	Middleton, James M	19	Sgt.	PA 76th Inf.	E	10/9/61	11/29/64		
24	Millburn, William	43	Pvt.	PA 55th Inf.	H	2/29/64	8/30/65		
25	Miller, Aaron J		Pvt.	PA 21st Cav.	M	7/1/63	2/20/64		
26	Miller, Aaron J		Pvt.	PA 5th Heavy Art.	K	2/20/64	6/30/65		
27	Miller, Abraham	18	Pvt.	PA 138th Inf.	F	8/29/62	6/16/65	Wounded	Cedar Creek
28	Miller, Abraham M	30	Pvt.	PA 82nd Inf.	C	11/14/64	7/13/65	Wounded	Ft. Fisher
29	Miller, Andrew (1)		Pvt.	PA 76th Inf.	E	8/24/63	7/18/65		
30	Miller, Andrew (2)	37	Pvt.	PA 110th Inf.	C	10/24/61	10/24/64		
31	Miller, Andrew P	38	Pvt.	PA 107th Inf.	D	9/2/64	7/19/65		
32	Miller, Anthony	33	Pvt.	PA 21st Cav.	E	7/2/63	2/20/64		
33	Miller, Armstrong	20	Pvt.	PA 21st Cav.	E	6/16/63	2/20/64		
34	Miller, Armstrong		Pvt.	PA 200th Inf.	C	8/26/64	5/30/65	Wounded	Ft. Stedman
35	Miller, Bartley	22	Pvt.	PA 171st Inf.	I	11/2/62	8/8/63		
36	Miller, Bartley H	20	Pvt.	PA 208th Inf.	H	9/6/64	6/1/65		
37	Miller, Charles	19	Pvt.	PA 200th Inf.	C	8/26/64	6/4/65		
38	Miller, Charles W	36	Pvt.	USCT 32nd Inf.	G	2/25/64	8/22/65		
39	Miller, Chauncey*		Pvt.	PA 55th Inf.	I	2/29/64			
40	Miller, Christian	29	Pvt.	PA 171st Inf.	I	11/2/62	8/8/63	Wounded	
41	Miller, Clement R		Pvt.	PA 13th Inf.	G	4/25/61	8/6/61		
42	Miller, Clement R	18	Capt.	PA 76th Inf.	E	10/9/61	11/23/64		
43	Miller, Cyrus*		Pvt.	PA 76th Inf.	E	1/18/65	7/18/65		
44	Miller, Daniel H	22	Pvt.	PA 171st Inf.	I	11/10/62	8/8/63		
45	Miller, David (1)	56	Pvt.	PA 55th Inf.	H	3/2/64	6/8/65	Wounded	Petersburg - Initial
46	Miller, David (2)		Pvt.	USCT 32nd Inf.	G	2/22/64			
47	Miller, David H	22	Pvt.	PA 171st Inf.	I	11/10/62	8/8/63		
48	Miller, David P		Pvt.	PA 171st Inf.	H	11/1/62	8/8/63		
49	Miller, David P	33	Pvt.	PA 107th Inf.	K	8/3/64			
50	Miller, David W		Pvt.		B	9/3/64	1865		
51	Miller, Edward J	20	Music.	US 3rd Cav.		5/1/58	3/23/63	Wounded	Valverde, NM
52	Miller, Elijah	27	Pvt.	PA 50th Inf.	D	2/24/65	5/10/65		
53	Miller, Ephraim B	22	Pvt.	PA 138th Inf.	F	8/29/62	6/23/65	Wounded	Cedar Creek
54	Miller, Franklin (1)	18	Pvt.	PA 2nd Cav.	E	11/29/61	11/4/64	Injured	
55	Miller, Franklin (2)	20	Pvt.	PA 54th Inf.	H	3/31/64			
56	Miller, Franklin (3)	26	Pvt.	KS 10th Inf.	K	8/12/61	8/19/64		
57	Miller, George	27	Corp.	PA 171st Inf.	I	11/2/62	8/8/63		
58	Miller, George		Corp.	PA 91st Inf.	I	9/21/64	5/30/65		
59	Miller, George F		Corp.	PA 21st Cav.	E	7/21/63	2/20/64		
60	Miller, George H			PA 11th Cav.	A				
61	Miller, George W (1)	31	Pvt.	PA 149th Inf.	I	8/26/63	6/24/65	Wounded	Spotsylvania
62	Miller, George W (2)	21	Pvt.	IA 11th Inf.	D	10/3/61		Wounded	Corinth, MS
63	Miller, Henry (1)	33	Pvt.	PA 138th Inf.	F	8/29/62		Wound.-Died	Wilderness
64	Miller, Henry (2)*	23	Pvt.	MD 1st Cav.	D	10/17/61	8/8/65		

Detailed Alphabetical Listing 475

	Cemetery	Notes
1	Sinking Valley Cem.-Blair Co	Historical Data Systems record; 1850 W. Providence Twp. Census
2		B.S.F. History Book-1884
3	Oak Hill Cem.-Towanda	1880 Broad Top Census
4		B.S.F. History Book-1884; Substitute
5	Mt. Pleasant Cem.-Mattie	Frank McCoy 1912 Listing; 1890 Monroe Twp. Veterans Census
6	Shreves Cem.-Monroe Twp.	B.S.F. History Book-1884
7	Shreves Cem.-Monroe Twp.	B.S.F. History Book-1884; 1860 Monroe Twp. Census; Alexander-father
8	Everett Cem.	Historical Data Systems record; 1860 Monroe Twp. Census; Frank McCoy 1912 Listing
9	Rinard Family Cem.-Cyper	East Providence Honor Roll; aka Valentine "Fetti"
10		POW-Libby & Belle Island 5/16/64-8 mths.; Historical Data Systems-Bedford Co. residence
11	Frostburg Mem. Park-MD	1850 Harrison Twp. Census; Historical Data Systems record
12	Bedford Cem.	Wounded 6/3/64; B.S.F. History Book-1884; 1850 Schellsburg resid.; Solomon & James-bros.
13	Bedford Cem.	1890 Colerain Twp. Veterans Census
14		Wounded 5/10/64; PA Civil War Archives record
15		KIA 8/14/64; Letters at Bedord Historical Society; Wolf's Store references
16		MIA 5/16/64 listed in Bedford Inquirer 7/8/64 article; POW-Richmond; Died 5/20/64
17		1890 S. Woodbury Veterans Census
18	Friends Cem.-Spring Meadow	Frank McCoy 1912 Listing
19		1890 E. St. Clair Veterans Census
20	Bedford Cem.	Bedford Cemetery Civil War Veterans list; born 1821
21	Dickerson Run-Fayette Co.	Deserted & returned; Historical Data Systems-Bedford Co. residence
22	Bedford Cem.	Bedford Cemetery Civil War Veterans list; Historical Data Systems; aka Middletown
23	Bedford Cem.	B.S.F. History Book-1884; 1860 Bedford Boro Census
24	St. Mark's Cem.-Colerain Twp.	1890 Bedford Twp. Veterans Census; B.S.F. History Book-1884
25		B.S.F. History Book-1884
26		B.S.F. History Book-1884
27		Wounded 10/19/64; Historical Data Systems-Cumberland Valley residence
28	Milligans Cove-Buffalo Mills	1890 Harrison Veterans Census-Scalp Wound; Frank McCoy 1912 Listing
29		B.S.F. History Book-1884
30		B.S.F. History Book-1884; PA Civil War Archives-Woodbury residence
31	St. Paul's Cem.-Cessna	Frank McCoy 1912 Listing; Historical Data Systems record
32		Historical Data Systems-Bedford Co. residence; 1870 Cumberland Valley Census
33	Pleasantville Cem.	1890 Pleasantville Veterans Census
34		Shoulder Wound 3/25/65; 1860 Pleasantville Census
35	Fairview Cem.-Artemas	1890 Mann Twp. Veterans Cen.; B.S.F. Hist. Book-1884; Bartley, Christian & George-bros.
36	Rock Hill Cem.-Monroe Twp.	1890 Monroe Twp. Veterans Census; Historical Data Systems record
37	Fishertown Cem.	1890 E. St. Clair Twp. Veterans Census
38	Mt. Ross Cem.-Bedford	1890 Bedford Veterans Census
39	Grandview Cem.-Cambria Co.	B.S.F. History Book-1884; enlisted in Somerset
40	Fairview Cem.-Artemas	1890 Mann Twp. Veterans Census-Leg Wound; B.S.F. History Book-1884
41		B.S.F. History Book-1884
42	Greenwood Cem.-New Castle	PA Civil War Archives-Bedford residence; B.S.F. History Book-1884
43		B.S.F. History Book-1884; Scranton references
44	Hickory Bottom Cem.-Woodbury	1890 Woodbury Twp. Veterans Census; B.S.F. History Book-1884
45	Schellsburg Cem.	Wounded 6/18/64 listed in Bedford Inquirer article 7/8/64; 1890 Napier Twp. Veterans Census
46	Mt. Ross Cem.-Bedford	Bedford Inquirer 5/22/1908 listing
47	Milligans Cove-Buffalo Mills	1890 Harrison Twp. Veterans Census; David H, John A & Samule W-brothers
48		1890 Napier Twp. Veterans Census
49	Helixville Cem.-Napier Twp.	Frank McCoy 1912 Listing; PA Civil War Project record list; Substitute never joined
50		1890 Cumberland Valley Veterans Census
51	Everett Cem.	Wounded 2/21/62; Frank McCoy 1912 Listing; Doctor in Everett after the war
52	Milligans Cove-Buffalo Mills	Bedford Inquirer 5/22/1908 listing; Elijah, Abraham, Michael & Ephram-brothers
53	Milligans Cove-Buffalo Mills	Arm Wound 10/19/64; Historical Data Systems record; 1890 Harrison Twp. Veterans Census
54	Mt. Olivet Cem.-Manns Choice	Thrown from horse; Justice of Peace after the war
55	Hyndman Cem.	Frank McCoy 1912 Listing
56	Mt. Olivet Baptist Cem.-MO	Ancestry.com-born in Harrison Twp.
57	Fairview Cem.-Artemas	1890 Mann Twp. Veterans Census; B.S.F. History Book-1884
58		Frank McCoy 1912 Listing
59		Historical Data Systems-Bedford Co. residence
60	St. Benedicts Cem.-Carrolltown	Ancestry.com born in Bedford Co.
61	Shreves Cem.-Monroe Twp.	Wounded 5/10/64; 1890 Monroe Twp. Veterans Census
62	Masonic Cem.-IA	Ancestry.com information; Severly wounded 10/4/62
63	Arlington Nat'l Cem., 27-67	Wounded 5/6/64; Died 5/20/64; B.S.F. History Book-1884; 1860 Cumberland Valley Census
64	Duval Cem.-Coaldale	U.S. Nat'l Homes Disabled Vol. Soldiers list

	Name	Age	Rank	Regiment	Co.	Muster In	Muster Out	Casualty	Casualty Battle
1	Miller, Henry S	25	Pvt.	PA 100th Inf.	K	9/21/64	5/30/65		
2	Miller, Henry*		Corp.	PA 55th Inf.	K	1/29/64	8/30/65		
3	Miller, Hezekiah H	18	Pvt.	PA 110th Inf.	C	1861		Died	
4	Miller, Isaac C	31	Pvt.	PA 99th Inf.	B	2/25/65	7/1/65		
5	Miller, Jackson	22	Sgt.	PA 138th Inf.	F	8/29/62		KIA	Wilderness
6	Miller, Jacob (1)		Pvt.	PA 133rd Inf.	C	8/13/62	5/26/63		
7	Miller, Jacob (2)	32	Pvt.	PA 56th Inf.	G	9/21/64	5/31/65		
8	Miller, Jacob (3)	33	Pvt.	WI 12th Inf.	D	2/19/64	7/16/65		
9	Miller, Jacob B	24	Pvt.	PA 133rd Inf.	C	8/13/62	5/19/63		
10	Miller, Jacob B		Pvt.	PA 22nd Cav.	M	2/15/64	6/24/65		
11	Miller, Jacob H	25	Sgt.	PA 21st Cav.	E	6/18/63	2/20/64		
12	Miller, Jacob J		Pvt.	PA 56th Inf.	G	4/6/65	6/6/65		
13	Miller, Jacob Jr	25	Pvt.	MI 6th Inf.	H	8/22/62	7/24/65		
14	Miller, Jacob W	23	Pvt.	PA 133rd Inf.	C	8/13/62	5/26/63	Wounded	Fredericksburg
15	Miller, James	23	Pvt.	OH 79th Inf.	H	8/13/62	2/23/63		
16	Miller, James H	23	1st Lt.	PA 55th Inf.	H	12/4/61	10/11/64	Wounded	Petersburg - Initial
17	Miller, James H		1st Lt.	PA 55th Inf.	H			Wounded	Cold Harbor
18	Miller, James L		Pvt.						
19	Miller, Jeremiah	39	Pvt.	PA 50th Inf.	E	11/14/64			
20	Miller, Jeremiah J	19	1st Sgt.	IA 11th Inf.	D	10/3/61	7/15/65		
21	Miller, Jesse	20	Sgt.	PA 138th Inf.	F	8/29/62	6/23/65		
22	Miller, John		Pvt.	PA McKeage's Mil.	H	7/2/63	8/8/63		
23	Miller, John (1)	33	Pvt.	PA 101st Inf.	G	12/28/61		POW-Died	Plymonth, NC
24	Miller, John (2)	18	Pvt.	PA 55th Inf.	H	10/11/61	8/30/65		
25	Miller, John (3)		Pvt.	PA 171st Inf.	H	10/24/62	8/8/63		
26	Miller, John (4)		Pvt.	MD 3rd Inf.	B	9/12/61	5/29/65		
27	Miller, John A	25	Pvt.	PA 100th Inf.	D	7/19/64	7/24/65	Wounded	
28	Miller, John D	18	Pvt.	PA 55th Inf.	H	9/12/62	6/11/65	Wounded	
29	Miller, John H (1)	21	Pvt.	PA 13th Inf.	G	4/25/61	8/6/61		
30	Miller, John H (2)	37	Pvt.	PA 101st Inf.	G	11/7/61	11/7/62		
31	Miller, John I	14	Pvt.	PA 110th Inf.	C	12/19/61	4/10/64	Wounded	Gettysburg
32	Miller, John J		Sgt.	PA 22nd Cav.	B	7/11/63	4/15/65		
33	Miller, John L*		Pvt.	PA 53rd Inf.	K	1/28/64		POW-Died	
34	Miller, John W	24	Pvt.	PA 55th Inf.	K	11/5/61	12/14/62	Wounded	Pocotaligo, SC
35	Miller, John*		Pvt.	PA 55th Inf.	I	1863			
36	Miller, Joseph	28	Sgt.	PA 55th Inf.	H	1/1/64	8/28/65		
37	Miller, Josiah C	18	Pvt.	PA 142nd Inf.	D	9/10/62	5/14/65	Wounded	Fredericksburg
38	Miller, Levi	28	Pvt.	PA 149th Inf.		1/24/65	5/6/65		
39	Miller, Mark	27	Pvt.	PA 51st Inf.	I	9/27/64	5/17/65		
40	Miller, Martin		Music.	MD 1st Cav.	L	1863	1865		
41	Miller, Martin D	20	Pvt.	PA 101st Inf.	D	2/26/62		KIA	Fair Oaks
42	Miller, Matson J		Pvt.	PA 101st Inf.	D	11/11/61	6/28/62		
43	Miller, Matthew	36	Pvt.	PA 55th Inf.	D	9/26/62		MIA	Drewry's Bluff
44	Miller, Michael C	33	Pvt.	PA 149th Inf.		2/24/65	5/5/65		
45	Miller, Nathan W	23	Pvt.	PA 21st Cav.	E	6/18/63	2/12/64		
46	Miller, Nathaniel		Pvt.	PA McKeage's Mil.	G	7/2/63	8/8/63		
47	Miller, Nathaniel	36	Pvt.	PA 148th Inf.	F	8/22/63	6/1/65	Wounded	
48	Miller, Nelson B	19	Pvt.	PA 55th Inf.	K	11/5/61	11/1/62		
49	Miller, Nelson B		Pvt.	PA 55th Inf.	K	2/27/64	5/31/65		
50	Miller, Peter A	25	Pvt.	PA 74th Inf.	B	3/13/65	8/29/65		
51	Miller, Peter S	29	Pvt.	PA 99th Inf.	B	2/25/65	7/1/65	Died	
52	Miller, Philip S (1)	29	Sgt.	PA 55th Inf.	H	10/11/61		Died	
53	Miller, Philip S (2)	26	Pvt.	PA 208th Inf.	H	9/7/64	6/1/65		
54	Miller, Preston A*	25	Pvt.	PA 76th Inf.	E	8/25/63	5/22/65		
55	Miller, Robert C		Pvt.	PA 22nd Cav.	H	2/24/64	6/24/65		
56	Miller, Samuel		Pvt.	PA McKeage's Mil.	H	7/2/63	8/8/63		
57	Miller, Samuel F	23	Pvt.	PA 74th Inf.	B	3/13/65	8/29/65		
58	Miller, Samuel J		Pvt.	PA 2nd Art.	A	2/5/64	4/30/64		
59	Miller, Samuel W	39	Pvt.	PA 21st Cav.	F	7/14/63	2/20/64		
60	Miller, Solomon	20	Pvt.	PA 74th Inf.	B	3/1/65	8/1/65		
61	Miller, Solomon H	29	Sgt.	PA 55th Inf.	H	10/11/61		POW-Died	Drewry's Bluff
62	Miller, Thomas (1)	21	Pvt.	PA 138th Inf.	D	8/29/62	6/23/65		
63	Miller, Thomas (2)	26	Pvt.	PA 171st Inf.	I	11/2/62			
64	Miller, Thomas J (1)	26	Pvt.	PA 100th Inf.	A	11/14/64	7/24/65		

	Cemetery	Notes
1	Schellsburg Cem.	1890 Napier Twp. Veterans Census
2		B.S.F. History Book-1884; enlisted in Reading, PA
3	Arlington Nat'l Cem., 13-13378	Died-Diphtheria 8/6/64; PA Civil War Archives-Woodbury residence
4	Trinity UCC Cem.-Osterburg	Frank McCoy 1912 Listing; 1860 St. Clair Twp. Census
5	Fredericksburg Nat'l Cem.	KIA 5/6/64; B.S.F. History Book-1884; 1850 Harrison Twp. Census
6		Historical Data Systems-Bedford Co. residence; 3 different Jacob Millers listed in 133rd
7	Bethel Meth. Cem.-Centerville	PA Civil War Project record & Frank McCoy 1912 Listing
8	Woods Nat'l Cem.-WI	Ancestry.com-born in Bedford; Historical Data Systems record
9		1890 Hopewell Twp. Veterans Census
10		Historical Data Systems record
11	Fishertown Cem.	1890 W. St. Clair Veterans Census; Jacob H and Solomon H-brothers
12		1890 Cumberland Valley Veterans Census
13	Millerburg Cem.-MI	MI Volunteers Service Record; 1840 Napier Twp. Census
14		Wounded 12/13/62; B.S.F. History Book-1884; Historical Data Systems record
15	Quitman Cem.-MO	Discharged on Surgeon's Cert.; James, Armstrong, Joseph, Thomas (1) & Charles -brothers
16	Schellsburg Cem.	Wounded 6/18/64 listed in Bedford Inquirer 7/8/64 article; KIA in New Mexico or Utah-1872
17		Wounded 6/3/64 listed in Bedford Inquirer 12/9/64 article; B.S.F. History Book-1884
18		1890 Saxton Veterans Census
19	Union Cem.-Centerville	Frank McCoy 1912 Listing
20	Mt. Hope Cem.-NE	1850 Napier Twp. Census; Historical Data Systems record
21		B.S.F. History Book-1884; Historical Data Systems record; 1860 Cumberland Valley Cen.
22		B.S.F. History Book-1884; unsure of which John Miller served in McKeage's Mil.
23	Andersonville Cem., 5704	POW-Andersonville; Died 8/15/64; PA Civil War Archives-Schellsburg residence
24		B.S.F. History Book-1884; Historical Data Systems-Bedford Co. residence
25		1890 Juniata Twp. Veterans Census
26		1890 Brush Creek Veterans Census
27	Pleasantville Cem.	1890 W. St. Clair Twp. Veterans Census
28		1890 Bedford Twp. Veterans Census
29	Fishertown Luth. Cem.	B.S.F. History Book-1884; Findagrave listed 13th Regiment
30	Schellsburg Cem.	1890 Napier Twp. Veterans Census; Discharged on Surgeon's Cert.
31	Bedford Cem.	Cheek Wound 7/2/63; Frank McCoy 1912 Listing; 1890 Bedford Boro Veterans Census
32		PA Civil War Muster Record-enlisted in Bedford Co.
33	Andersonville Cem., 7119	POW-Andersonville; Died-Fever 8/28/64; unable to confirm
34	Sandyvale Cem.-Johnstown	Wounded 10/22/62; 1890 Bloomfield Twp. Veterans Census
35		B.S.F. History Book-1884; 6/1/64; Reading, PA references
36	Fishertown Cem.	1890 E. St. Clair Twp. Veterans Census
37	Hull Baptist Cem.-New Paris	Foot Wound 12/13/62; 1890 Napier Twp. Veterans Census-gunshot wound-right Foot & Leg
38	Mulligans Cove-Buffalo Mills	Civil War Pension record
39	Pleasantville Cem.	Historical Data Systems record
40		1890 Everett Veterans Census
41		KIA 5/31/62; PA Civil War Archives-Bedford Co. residence, enlisted in Woodbury
42		Historical Data Systems-Bedford Co. residence
43		MIA 5/16/64; B.S.F. History Book-1884; PA Civil War Archives-Harrison Twp. residence
44	Milligans Cove-Buffalo Mills	Frank McCoy 1912 Listing; aka Michael S
45	Fishertown Cem.	1890 E. St. Clair Veterans Census
46		B.S.F. History Book-1884
47	Schellsburg Cem.	Frank McCoy 1912 Listing
48	Parsons Cem.-WV	Historical Data Systems record; Discharged on Surgeon's Cert.; Nelson B & Jackson-brothers
49		1890 Monroe Twp. Veterans Census
50	Helixville Cem.-Napier Twp.	Ancestry.com information; aka Alexander Shafer; Peter, Samuel & Solomon-brothers
51	Schellsburg Cem.	Died 8/24/65; 1890 Napier Twp. Veterans Census; Peter S, John H (1) & Philip S (1)-brothers
52	Beaufort Nat'l Cem.	Died 9/28/62; PA Civil War Archives-Bedford Co. residence, enlisted in Schellsburg
53	Burning Bush-Bedford Twp.	1890 Snake Spring Valley Veterans Census; B.S.F. History Book-1884
54	Rose Hill Cem.-Crawford Co.	B.S.F. History Book-1884; Crawford Co. residence
55		B.S.F. History Book-1884
56		B.S.F. History Book-1884
57	Helixville Cem.-Napier Twp.	Findagrave.com information; aka Franklin P Shaffer
58		1890 Broad Top Twp. Veterans Census; Discharged on Surgeon's Cert.
59	Grandview Cem.-Cambria Co.	PA Civil War Project record; 1850 Napier Twp. Census
60	Helixville Cem.-Napier Twp.	Muster record on Military Burial Record; 1860 Napier Twp. Census
61	Oakwood Cem.-Richmond	Wounded & POW-Libby 5/16/64; Died 6/8/64; PA Civil War Archives Bedford Co. residence
62		B.S.F. History Book-1884; Historical Data Systems-Bedford Co. residence
63		Deserted 11/3/62; B.S.F. History Book-1884
64	Fishertown Cem.	1890 E. St. Clair Twp. Veterans Census

	Name	Age	Rank	Regiment	Co.	Muster In	Muster Out	Casualty	Casualty Battle
1	Miller, Thomas J (2)	23	Pvt.	PA 138th Inf.	D	8/29/62		POW-Died	Wilderness
2	Miller, Tobias	18	Pvt.	PA 138th Inf.	E	8/29/62	5/4/64		
3	Miller, Watson J	20	Pvt.	PA 101st Inf.	D	11/11/61	6/28/62		
4	Miller, Watson J		Pvt.	OH 8th Cav.	D	9/8/64	5/30/65		
5	Miller, William H (1)	26	Pvt.	PA 50th Inf.	E	7/5/64	23863		
6	Miller, William H (2)	19	Pvt.	PA 55th Inf.	H	2/29/64	5/31/65	Wounded	Petersburg
7	Miller, William H (3)	21	Pvt.	PA 84th Inf.	A	10/24/61		Died	
8	Miller, William H (4)		Pvt.	PA 99th Inf.	G	3/17/65	7/1/65		
9	Miller, William H (5)*	18	Pvt.	PA 93rd Inf.	G	9/10/64	6/1/65		
10	Miller, William H (6)	25	Pvt.	PA 61st Inf.	F	10/26/64	5/15/65		
11	Miller, William H (7)	34	Pvt.	PA 100th Inf.	D	12/22/64	7/24/65		
12	Miller, William M*	25	Pvt.	PA 55th Inf.	K	10/14/63	6/21/65	POW	Drewry's Bluff
13	Miller, William O	18	Pvt.	IA 24th Inf.	C	8/29/62		Wound.-Died	Opequon
14	Miller, Wireman	17	Pvt.	IA 11th Inf.	D	8/25/64	6/2/65		
15	Miller, Wyrman S	19	Pvt.	PA 148th Inf.	H	8/16/62		KIA	Chancellorsville
16	Miller, Zachariah			PA 23rd Inf.	H				
17	Millin, William S	17	Pvt.	PA 99th Inf.	D	3/17/65	7/1/65		
18	Mills, Andrew J	21	Pvt.	PA 101st Inf.	D	2/8/62	5/18/65	Wounded	Fair Oaks
19	Mills, Andrew J		Pvt.	PA 101st Inf.	D			POW	Plymonth, NC
20	Mills, David		Pvt.	PA 51st Inf.	K	11/12/61	11/12/64		
21	Mills, Franklin G	18	Music.	PA 101st Inf.	D	12/6/61	6/21/65	Wound.-POW	Plymonth, NC
22	Mills, Isaac	34	Pvt.	PA 22nd Cav.	L	2/22/64	6/24/65		
23	Mills, Jacob	19	Pvt.	PA 133rd Inf.	K	8/18/62	5/26/63		
24	Mills, Jacob H	36	Pvt.	PA 101st Inf.	D	1/23/62	5/24/62		
25	Mills, Samuel	21	Pvt.	PA 76th Inf.	E	8/25/63	7/18/65		
26	Mills, Uriah	34	Sgt.	PA 76th Inf.	E	8/20/63	7/18/65		
27	Mimminger, Jacob*	20	Pvt.	PA 110th Inf.	C	9/4/62	3/25/65	Wounded	1st Deep Bottom
28	Mitchell, James P	18	Pvt.	PA 55th Inf.	H	2/29/64		POW-Died	Drewry's Bluff
29	Mitchell, John A*			PA 55th Inf.				POW-Died	
30	Mittong, John W	19	Corp.	PA 76th Inf.	E	10/9/61	11/28/64		
31	Mixel, Samuel	28	Pvt.	PA 133rd Inf.	K	8/15/62	5/26/63		
32	Mixel, Samuel			US 5th Light Art.	E				
33	Mobley, Denton	42	Sgt.	PA 18th Cav.	K	10/29/62			
34	Mobley, Ezekiel	18	Pvt.	PA 133rd Inf.	B	8/14/62	5/23/63		
35	Mobley, Ezekiel		Pvt.	PA 205th Inf.	C	8/27/64	6/2/65		
36	Mock, Aaron	21	Pvt.	PA 138th Inf.	D	8/29/62	6/20/65	Wounded	Mine Run
37	Mock, Aaron		Pvt.	PA 138th Inf.	D			Wound.-POW	Wilderness
38	Mock, Alexander	32	Pvt.	PA 57th Inf.	H	7/3/63	8/17/63		
39	Mock, Andrew	33	Pvt.	PA 55th Inf.	K	3/2/64		Wounded	Cold Harbor
40	Mock, Andrew	33	Pvt.	PA 55th Inf.	K	3/2/64		Wound.-Died	White Oak Road
41	Mock, Anthony	30	Pvt.	PA 55th Inf.	K	11/5/61	3/4/63		
42	Mock, Daniel	27	Pvt.	PA 61st Inf.	A	2/25/64	6/28/65		
43	Mock, Emanuel	27	Pvt.	PA 138th Inf.	D	8/29/62	2/10/65	Wounded	Cold Harbor
44	Mock, Emanuel A		Pvt.	PA McKeage's Mil.	G	7/2/63	8/8/63		
45	Mock, Emanuel A	22	Pvt.	PA 55th Inf.	K	2/19/64	8/17/64	Wounded	Bermuda Hundred
46	Mock, George (1)	42	Pvt.	PA 115th Inf.	D	7/24/62			
47	Mock, George (1)		Pvt.	PA 110th Inf.	D	6/22/64		Wounded	Boydton Plank Rd.
48	Mock, George (2)	55	Pvt.	IN 129th Inf.	E	3/6/64	4/26/65		
49	Mock, George W	34	Pvt.	PA 125th Inf.	E	8/13/62		Died	
50	Mock, Harrison	19	Pvt.	PA 133rd Inf.	C	8/13/62	5/26/63		
51	Mock, Harrison		Pvt.	PA 50th Inf.	D	2/24/65	7/30/65		
52	Mock, Harvey			PA 2nd Cav.					
53	Mock, Henry	20	Pvt.	PA 200th Inf.	C	8/25/64	5/30/65	Wounded	Ft. Stedman
54	Mock, John (1)	34	Pvt.	PA 55th Inf.	K	2/19/64	4/1/65	Wounded	Petersburg - Initial
55	Mock, John (2)	28	Pvt.	PA McKeage's Mil.	G	7/2/63	8/8/63		
56	Mock, Josiah	38	Pvt.	PA 79th Inf.	E	2/23/65	5/29/65		
57	Mock, Josiah B	26	Sgt.	PA 55th Inf.	K	11/5/61		POW-Died	
58	Mock, Josiah D	28	Pvt.	PA 84th Inf.	A	10/24/61			
59	Mock, Lewis	23	Wagon.	PA 138th Inf.	E	8/29/62	6/23/65		
60	Mock, Malachi	25	Pvt.	PA 138th Inf.	E	8/29/62	6/23/65		
61	Mock, Malachi B	24	Pvt.	PA 55th Inf.	K	11/5/61		Died	
62	Mock, Mathias	24	Pvt.	PA 133rd Inf.	C	8/13/62	5/26/63		
63	Mock, Mathias		Pvt.	PA 184th Inf.	A	5/12/64		KIA	Cold Harbor
64	Mock, Paul S	38	Sgt.	PA 55th Inf.	I	11/10/61	6/3/65	POW	Chaffin's Farm

Detailed Alphabetical Listing

	Cemetery	Notes
1	Andersonville Cem.	POW-Andersonville 5/6/64; Died 9/15/64; PA Civil War Archives-Bedford Co. residence
2	Milligans Cove-Buffalo Mills	1890 Harrison Twp. Veterans Census; Discharged on Surgeon's Cert.; Tobias & Matthew-bros.
3	Pleasant Hill Cem.-Imlertown	B.S.F. History Book-1884; Watson & Bartley H-brothers
4		Historical Data Systems record
5	Bethel Meth. Cem.-Centerville	1890 Cumberland Valley Veterans Census
6		Wounded 7/4/64; B.S.F. History Book-1884
7	Friends Cem.-Spring Meadow	Died 2/24/62; PA Civil War Archives-Bedford Co. residence; Frank McCoy 1912 Listing
8		B.S.F. History Book-1884
9		B.S.F. History Book-1884; Born in Somerset Co.
10	Hyndman Cem.	1890 Hyndman Veterans Census; Discharged on Surgeon's Cert.
11	Grandview Cem.-Cambria Co.	1850 Napier Twp. Census; Regiment on headstone
12		POW 5/16/64 to 4/29/65; B.S.F. History Book-1884
13	Winchester Nat'l Cem.	Wounded 9/19/64; Died 2/13/65; Historical Data Systems record; 1850 Napier Township Cen.
14		Ancestry.com information; Wireman, William O, George W (2) & Jeremiah J-brothers
15	Wilderness burial grounds	KIA 5/3/63; Historical Data Systems record; born in Alum Bank; 1850 St. Clair Twp. Census
16	Schellsburg Cem.	Born 7/29/48; Regiment on gravestone
17	Mt. Pleasant Cem.-Mattie	E. Providence Honor Roll; 1890 E. Providence Twp. Veterans Census
18	Pleasant Union Cem.-Clearville	Wounded 5/31/62; 1850 Monroe Twp. Census
19		POW-Andersonville 4/20/64 to 4/8/65; Historical Data Systems record
20		1890 Greenfield Twp. Veterans Census-aka William D
21		POW 4/20/64 to 11/20/64; B.S.F. History Book-1884; Jacob is Father
22	McKendree Cem.-Crystal Spring	Historical Data Systems-Bedford Co. residence; 1890 Brush Creek Veterans Census
23		Historical Data Systems-Bedford Co. residence, enlisted in Clearville
24		Historical Data Systems-Bedford Co. residence; 1850 Southampton Census
25		B.S.F. History Book-1884
26		B.S.F. History Book-1884
27		B.S.F. History Book-1884; enlisted in Culpville, PA
28	Andersonville Cem., 11081	MIA 5/16/64 in Bed. Inq.-7/8/64; POW-Andersonville; Died 10/17/64; 1860 Napier Twp. Cen.
29		Bedford Gazette 2/13/1914 listing
30	Greenhill Cem.-OK	PA Civil War Archives-Bedford residence; B.S.F. History Book-1884
31	Rock Hill Cem.-Monroe Twp.	B.S.F. History Book-1884 aka Meixel; Historical Data Systems-enlisted in Bloody Run
32		Civil War Pension Record
33	Holsinger Cem.-Bakers Summit	Deserted 10/9/63; Frank McCoy 1912 Listing; 1870 M. Woodbury Twp. Census
34	Carson Valley-Duncansville	Historical Data Systems record; father is Denton
35		1880 Bloomfield Twp. Census; Historical Data Systems record
36	Mt. Union Cem.-Lovely	Wounded 11/27/63; 1890 Union Twp. Veterans Census; Historical Data Systems record
37		Wounded-POW 5/6/64 to 12/16/64
38	Fairview Cem.-Altoona	Historical Data Systems record; Burial Record listed 147th inf.
39	Poplar Grove Nat'l Cem.	Wounded 6/3/64; In hospital until August '64
40	Poplar Grove Nat'l Cem.	Leg Amputated; Died 3/31/65; Civil War Pension Record; 1860 Union Twp. Census
41	Mt. Union Cem.-Lovely	B.S.F. History Book-1884; 1850 Union Twp. (Pavia) Census; Discharged on Surgeon's Cert.
42		PA Civil War Archives-Born in Bedford Co.
43	Ogletown Breth.-Somerset Co.	Wound to left forearm 6/1/64; PA Civil War Archives-enlisted in Schellsurg
44		B.S.F. History Book-1884
45	Mt. Union Cem.-Lovely	1890 Union Twp. Veterans Census-Arm shot off below elbow 5/21/64
46	Bedford Cem.	Deserted 7/15/63 & returned 6/22/64; B.S.F. History Book-1884; Hist. Data Systems record
47		Wounded 10/27/64; Historical Data Systems record; 1860 Bedford Twp. Census
48		Ancestry.com-born in Bedford Co.
49	US Soldiers & Airmans Nat'l	Died-Typhoid Fever 1/23/63; Bed. Gazette 2/13/1914 listing; Ancestry.com-born in Osterburg
50	Barley Luth.-Bakers Summit	PA Civil War Archives-enlisted in Hopewell; Harrison, George W, Josiah & Mathias-brothers
51		Historical Data Systems record; aka Harry
52	Union Cem.-Colerain Twp.	Frank McCoy 1912 Listing
53	Trinity UCC Cem.-Osterburg	Wounded 3/25/65; Frank McCoy 1912 Listing; Henry, Lewis & Malachi-brothers
54		Wounded 6/16/64; B.S.F. History Book-1884; PA Civil War Archives-Born in Bedford
55	Bortz Luth. Cem.-Centerville	B.S.F. History Book-1884
56	Trinity UCC Cem.-Osterburg	Frank McCoy 1912 Listing; aka Mauk
57	Mt. Zion Cem.-Pavia	POW-Salisbury & Libby for 5 months, Died 3/22/65; PA CW Archives-Bedford Co. residence
58	Sixteen Church Cem.-OH	PA Civil War Archives-Bedford Co. residence
59	Coles Cem.-Westmoreland Co.	Promoted to Wagoneer; B.S.F. History Book-1884; 1860 St. Clair Twp. Census
60	Schellsburg Cem.	B.S.F. History Book-1884; Historical Data Systems record; 1860 Union Twp. (Pavia) Census
61	Mt. Zion Cem.-Pavia	Died-Typhoid Fever 11/7/62; Hist. Data Systems record; Malachi B, Josiah B & Tobias-bros.
62		B.S.F. History Book-1884; 1850 Union Twp. (Pavia) Census; enlisted in Hopewell
63		KIA 6/3/64; Civil War Muster Roll Record; Bedford Gazette 2/13/1914 listing; aka Mauk
64	Union Cem.-Claysburg	POW 9/29/64 for 5 months; B.S.F. History Book-1884; 1860 Union Twp. (Pavia) Census

Chapter 18

	Name	Age	Rank	Regiment	Co.	Muster In	Muster Out	Casualty	Casualty Battle
1	Mock, Paul S		Sgt.	PA 55th Inf.	I			Wounded	Bermuda Hundred
2	Mock, Samuel A	20	Pvt.	PA 91st Inf.	I	9/21/64	5/30/65		
3	Mock, Samuel S	36	Pvt.	PA 91st Inf.	B	9/21/64	5/20/65		
4	Mock, Tobias		Pvt.	PA 133rd Inf.	K	8/6/62	5/26/63		
5	Mock, Tobias	22	Pvt.	PA McKeage's Mil.	G	7/2/63	8/8/63		
6	Mock, Tobias B	22	Pvt.	PA 55th Inf.	K	10/11/61	8/7/64	Wound.-Died	Petersburg - Initial
7	Mock, William M	27	Pvt.	PA 91st Inf.	I	9/21/64	5/30/65		
8	Mock, Wlliam A	33	1st Sgt.	PA 55th Inf.	K	11/5/61		KIA	Bermuda Hundred
9	Mohn, Frederick	36	Pvt.	PA 13th Inf.	G	4/25/61	8/6/61		
10	Monihan, James*	30	Pvt.	PA 110th Inf.	C	2/26/64	6/6/65		
11	Montooth, Barnabas*	39	Pvt.	PA 184th Inf.	A	3/17/64	6/21/65		
12	Moore, Abraham	25	Pvt.	PA 74th Inf.	F	3/7/65	6/23/65		
13	Moore, Cyrus B	22	Pvt.	PA 5th Heavy Art.	K	9/8/64	6/30/65		
14	Moore, George	39	Pvt.	PA 14th Cav.	C	8/25/62	5/30/65		
15	Moore, Hiram K*	42	Pvt.	PA 76th Inf.	E	8/27/63	7/18/65		
16	Moore, James		Pvt.	PA 5th Res.	F	7/26/61	6/6/64		
17	Moore, James		Pvt.	PA 138th Inf.	D	9/12/64	6/23/65		
18	Moore, James B		Pvt.	PA 110th Inf.	B	10/24/61	10/23/64		
19	Moore, James E		Sgt.	PA 55th Inf.	D	2/27/64	8/30/65		
20	Moore, John	20	Pvt.	PA 76th Inf.	E	2/29/65	7/18/65		
21	Moore, John B	18	Pvt.	PA 13th Inf.	G	4/25/61	8/6/61		
22	Moore, John B		Sgt.	PA 110th Inf.	C	10/24/61	10/24/64	Wound.-POW	Gettysburg
23	Moore, Joseph	27	Pvt.	PA 125th Inf.	F	8/12/62	5/18/63		
24	Moore, Nelson*		Pvt.	PA 194th Inf.	I	7/22/64	9/6/64		
25	Moore, William G		Capt.	PA 55th Inf.	D	10/1/62	6/11/65		
26	Moore, Wlliam	25	Pvt.	PA 76th Inf.	E	7/9/63	3/11/64		
27	Mophet, Andrew			PA 67th Inf.					
28	Mopps, Edward S	55	1st Lt.	PA 13th Inf.	G	4/25/61	8/6/61		
29	Moran, Thomas	43	Pvt.	PA 55th Inf.	K	11/11/61	6/26/63		
30	Mordus, Samuel	22	Pvt.	PA 14th Inf.	I	5/2/61	8/7/61		
31	Morgan, Dennis	26	Pvt.	PA 110th Inf.	C	10/24/61	10/24/64		
32	Morgan, William		Pvt.	PA 205th Inf.	D	9/1/64	6/2/65		
33	Morgart, Abraham	41	Pvt.	PA 79th Inf.	D	2/23/65		Died	
34	Morgart, William	23	Pvt.	PA 18th Cav.	K	9/29/62		POW-Died	Wilderness
35	Morningstar, Peter	37	Pvt.	PA 84th Inf.	C	10/24/61	1/13/64	POW	2nd Deep Bottom
36	Morningstar, Peter		Pvt.	PA 57th Inf.	H	1/13/65	6/29/65		
37	Morningstar, Samuel	23	Pvt.	PA 143rd Inf.	B	9/8/63	6/12/65		
38	Morris, David I			PA 2nd Cav.					
39	Morris, David I	38	Pvt.	PA 28th Inf.	C	6/14/64	6/21/65		
40	Morris, Edward	34	Pvt.	PA 56th Inf.	K	3/24/62			
41	Morris, Henry	28	Pvt.	PA 76th Inf.	E	8/5/63	7/18/65		
42	Morris, Israel	31	Pvt.	PA 171st Inf.	I	10/4/62	8/8/63		
43	Morris, James*	29	Pvt.	IL 39th Inf.	F	9/9/61		Wound.-Died	Cold Harbor
44	Morris, John	45	Music.	PA 194th Inf.	I	7/22/64	11/5/64		
45	Morrison, John	34	Pvt.	PA 84th Inf.	D	12/24/61			
46	Morrow, B Moritimer	27	Capt.	PA 84th Inf.	C	9/5/61	9/29/62		
47	Morrow, B Moritimer		Major	PA 22nd Inf.	A	7/16/63	2/5/64		
48	Morrow, B Moritimer		Lt. Col.	PA 205th Inf.	F&S	9/2/64	6/3/65	Wounded	Petersburg - Final
49	Morse, David	16	Pvt.	PA 11th Inf.	A	9/30/61		Wound.-Died	Antietam
50	Morse, James	20	Pvt.	PA 99th Inf.	G	2/25/65	6/28/65	Wounded	
51	Morse, Jesse W	22	Pvt.	PA 99th Inf.	F	3/23/65		Died	
52	Morse, John*			PA 11th Inf.				Died	
53	Morse, Joseph S	21	Pvt.	PA 45th Inf.	B	8/26/64	7/18/65		
54	Morse, Morgan	28	Pvt.	PA 133rd Inf.	C	8/13/62	4/27/63		
55	Morse, Samuel	19	Pvt.	PA 101st Inf.	D	2/8/62		Died	
56	Mortimer, John (1)	41	Corp.	PA 76th Inf.	E	8/27/63	7/18/65		
57	Mortimer, John (2)	22	Pvt.	PA 101st Inf.	D	11/1/61			
58	Mortimer, John L	38	Pvt.	PA 67th Inf.	C	11/28/64	6/27/65	Wounded	Ft. Fisher
59	Mortimore, David			PA 101st Inf.					
60	Mortimore, David	21	Pvt.	PA 22nd Cav.	C	7/11/63		Died	
61	Morton, Henry A	18	Pvt.	NY 5th Cav.	F	1/11/64	7/19/65		
62	Mosell, William	24	Corp.	PA 55th Inf.	H	9/20/61		Died	
63	Moser, Abraham	26	Pvt.	OH 3rd Cav.	G	12/11/61	8/4/65		
64	Moser, Jeremiah	21	Corp.	PA 138th Inf.	F	8/29/62	6/23/65	Wounded	Cold Harbor

	Cemetery	Notes
1		Wounded 5/18/64-Pension Record
2	Mt. Union Cem.-Lovely	1890 Union Twp. (Pavia) Veterans Census; Samuel A, Emanuel A & William A-brothers
3	Trinity UCC Cem.-Osterburg	Frank McCoy 1912 Listing; Samuel S, Alexander & George-brothers
4		PA Civil War Archives record
5	Berkey Cem.-Windber	B.S.F. History Book-1884
6	Mt. Zion Cem.-Pavia	Wounded 6/18/64, Died 8/7/64; PA Civil War Archives; 1860 Union Twp. (Pavia) Census
7	Fairfield Cem.-NE	PA Civil War Archives; Civil War Pension Record; William, Henry & Andrew-brothers
8		KIA 5/22/64; PA Civil War Archives-Bedford residence; 1850 Union Twp. (Pavia) Census
9		B.S.F. History Book-1884; Historical Data Systems-Bedford residence
10		B.S.F. History Book-1884; PA Civil War Archives-Huntingdon residence
11	Mt. Zion Cem.-York Co	B.S.F. History Book-1884; York Co. residence
12	Fishertown Cem.	Historical Data Systems record; 1850 St. Clair Twp. Census; aka Moor
13	Union Cem.-Somerset Co.	B.S.F. History Book-1884
14	Bedford Cem.	Historical Data Systems; Bed. Cem. CW Veterans list; pursued Chambersburg Conf. raiders
15		B.S.F. History Book-1884; McKean Co. references
16		B.S.F. History Book-1884
17		B.S.F. History Book-1884; Historical Data Systems record
18	Yellow Creek Reformed Cem.	PA Civil War Archives listed James L
19		B.S.F. History Book-1884; Historical Data Systems record
20		B.S.F. History Book-1884; Historical Data Systems record
21	Yellow Creek Reformed Cem.	B.S.F. History Book-1884
22	St. Paul's Cem.-Cessna	Soldiers of Blair Co. Book - 1940-Arm Wound 7/2/63; 1860 Hopewell Twp. Census
23	St. Pauls UCC Cem.-Russellville	1890 Hopewell Veterans Census
24		B.S.F. History Book-1884; Reading, PA residence
25		B.S.F. History Book-1884; Historical Data Systems record
26		B.S.F. History Book-1884; Historical Data Systems Inf.
27	Fockler Cem.-Saxton	Frank McCoy 1912 Listing; Findagrave.com-regiment record
28		B.S.F. History Book-1884; Historical Data Systems-Bedford residence
29		B.S.F. History Book-1884; Historical Data Systems-Bedford Co. resid.; Discharged Surg. Cert.
30	Bunker Hill Cem.-Saxton	PA Civil War Archives-aka Mordos; Regiment on headstone
31		B.S.F. History Book-1884; Historical Data Systems-Pattonville residence
32		1890 Liberty Twp. Veterans Census
33	Morgart & Morgret Cem.-Everett	Died 6/21/65; Historical Data Systems; Civil War Pension record; aka Morgrett
34	Florence Stockade Cem.	POW-Andersonville 5/5/64; Died 11/1/64; 1860 Cumberland Valley Census; aka Morgan
35	Hickory Bottom Cem.-Woodbury	POW 8/16/64; 1890 Woodbury Twp. Veterans Census; Frank McCoy 1912 Listing
36		Historical Data Systems record
37		1890 Saxton Veterans Census
38		Frank McCoy 1912 Listing
39	Burning Bush-Bedford Twp.	1890 Cumberland Valley Veterans Census
40		PA Civil War Archives-Broad Top residence
41		B.S.F. History Book-1884
42	Rock Hill Cem.-Monroe Twp.	1890 Monroe Twp. Veterans Con.
43	Philadelphia Nat'l Cem.	Ancestry.com questions; Leg Amputated 6/1/64; Died 7/8/64; 1850 Monroe Twp. Census
44		B.S.F. History Book-1884; enlisted in Bloody Run
45	Holsinger Cem.-Bakers Summit	Frank McCoy 1912 Listing
46	Hollidaysburg Presb. Cem.	Ancestry.com-Married in Londonderry Twp. in 1859
47		Historical Data Systems record
48	Holidaysburg Pres. Cem.	Leg Amputated 4/2/65; Died 3/7/67
49	Mt. Olivet Cem.-Frederick	Wounded 9/17/62; Died 1/3/63; Bed. Gazette 2/13/1914 list; 1860 Southampton Twp. Census
50	Fairview Cem.-Artemas	1890 Mann Twp. Veterans Census - Right leg wound
51		Died on 5/15/65; Bedford Gazette 2/13/1914 listing; 1850 Southampton Twp. Census
52		Died of disease; Bedford Gazette 2/13/1914 listing; no records located
53	Fairview Cem.-Artemas	1890 Mann Twp. Veterans Census; Joseph & James-brothers
54	Union Memorial Cem.-Mench	1890 E. Providence Twp. Veterans Census; Morgan, David & Samuel-brothers
55	Bethal Frame Ch.-Monroe Twp.	Died 6/25/62; Frank McCoy 1912 Listing; CW Pension Record; aka Jacob Moss
56		B.S.F. History Book-1884; PA Civil War Project record; David-Son
57		Deserted 4/1/62 in Harrisburg; PA Civil War Archives record; 1860 Colerain Twp. Census
58	Mt. Pleasant-Westmoreland Co	Severe Arm Wound 3/25/65; 1850 W. Providence Twp. Census
59		Bedford Inquirer 5/22/1908 listing; 1850 Monroe Twp. Census
60	Bethal Frame Ch.-Monroe Twp.	Died 9/18/63; Frank McCoy 1912 Listing; Civil War 22nd Reg't on gravestone; aka Mortimer
61		Fold3 record-Born in Bedford Co.
62	Spring Hope Cem.-Martinsburg	Died 4/15/64; B.S.F. History Book-1884; 1860 Woodbury Twp. Census; Pension Record
63	Dimondale Cem.-MI	Ancestry.com information; 1850 Colerain Twp. Census; Twin brother of William
64		Historical Data Systems record; Wounded 6/1/64; Absent from 6/1/64

	Name	Age	Rank	Regiment	Co.	Muster In	Muster Out	Casualty	Casualty Battle
1	Moser, Martin	31	Pvt.	PA 133rd Inf.	K	8/15/62	5/26/63		
2	Moser, Martin		Sgt.	PA 208th Inf.	K	9/7/64	6/1/65		
3	Moser, Nathaniel		Pvt.	PA McKeage's Mil.	H	7/2/63	8/8/63		
4	Moser, Nathaniel	25	Pvt.	PA 29th Inf.	H	9/22/64	7/17/65		
5	Moser, William	26	Pvt.	OH 55th Inf.	B	9/30/61	2/4/63	Wound.-Died	Cross Keys
6	Moser, William S	31	Corp.	PA 55th Inf.	D	2/29/64		Died	
7	Moses, Emanuel	17	Pvt.	PA 18th Cav.	K	10/29/62		POW-Died	Gettysburg
8	Moss, Jacob		Pvt.	PA 101st Inf.	D	2/8/62	3/1/65		
9	Mosser, George W	21	Sgt.	PA 21st Cav.	K	7/9/63	5/15/65		
10	Mountain, George R	30	Pvt.	PA 84th Inf.	E	4/5/64	1/13/65	Wounded	Wilderness
11	Mountain, George R		Pvt.	PA 57th Inf.	I	1/13/65			
12	Mountain, Richard D	16	Pvt.	PA 77th Inf.	F	10/9/61		Wound.-Died	Chickamauga
13	Mower, Abraham C	33	Pvt.	PA 55th Inf.	D	11/2/62	6/15/63		
14	Mower, Alexander C		Sgt.	PA 13th Inf.	G	4/25/61	8/6/61		
15	Mower, Alexander C	27	Music.	PA 55th Inf.	D	10/12/61		Died	
16	Mower, Edward E	18	Music.	PA 55th Inf.	D	10/12/61	8/30/65		
17	Mower, John H	21	Pvt.	PA 101st Inf.	D	11/1/61	5/24/62		
18	Mower, John H		Pvt.	PA 55th Inf.	D	2/27/64	12/28/64		
19	Mowry, Frederick	40	Sgt.	PA 138th Inf.	F	8/29/62	5/2/64		
20	Mowry, Jacob	29	Pvt.	PA 171st Inf.	I	11/10/62	8/1/63		
21	Mowry, John	26	Pvt.	PA 171st Inf.	H	11/27/62	8/6/63		
22	Mowry, Joseph		Pvt.	PA 22nd Cav.	I	7/11/63			
23	Mowry, Richard S	18	Pvt.	PA 55th Inf.	H	2/29/64			
24	Moyer, Alexander	32	Corp.	PA 76th Inf.	E	8/25/63	7/18/65		
25	Moyer, Daniel	22	Pvt.	PA 205th Inf.	D	9/1/64	6/2/65		
26	Moyer, John (1)	30	Pvt.	PA 2nd Cav.	E	11/4/61		POW-Died	
27	Moyer, John (2)	18	Pvt.	PA 55th Inf.	H	10/11/61		Died	
28	Moyer, John (3)	19	Pvt.	PA 5th Res.	G	6/21/61	6/11/64		
29	Moyer, John A*		Pvt.	PA 55th Inf.	H	9/2/62	6/3/65	POW	Chaffin's Farm
30	Moyer, John E*		Sgt.	PA 55th Inf.	H	9/22/62	6/11/65		
31	Moyer, Samuel*		Pvt.	PA 55th Inf.	H	9/2/62	8/30/65	MIA	Cold Harbor
32	Moyer, William M*		Pvt.	PA 55th Inf.	H	9/19/62			
33	Moyer, William*		Pvt.	PA 55th Inf.	H	9/16/62		POW-Died	Drewry's Bluff
34	Mull, William	32	Pvt.	PA 82nd Inf.	K	7/5/64		Died	
35	Mullenix, George	18	Pvt.	PA 107th Inf.	H	1/29/62	2/11/64		
36	Mullin, Alexander S	35	Pvt.	PA 55th Inf.	D	10/12/61	8/30/65		
37	Mullin, David W	33	Pvt.	PA 13th Inf.	G	4/25/61	8/6/61		
38	Mullin, David W		Capt.	PA 101st Inf.	G	2/20/62	5/16/65	POW	Plymonth, NC
39	Mullin, George S	40	Capt.	PA 55th Inf.	H	12/4/61	6/21/62		
40	Mullin, John	47	Pvt.	PA 138th Inf.	D	8/29/62	3/5/64		
41	Mumper, Henry	22	Pvt.	PA 133rd Inf.	K	8/30/62	5/26/63	Wounded	Fredericksburg
42	Munshower, George W	20	Pvt.	PA 13th Inf.	G	4/25/61	8/6/61		
43	Munson, Morrison B*		2nd Lt.	PA 171st Inf.	I	11/11/62	8/8/63		
44	Munson, Morrison B*	34	1st Lt.	PA 184th Inf.	A	5/12/64	7/14/65		
45	Murphy, Dennis*	37	Pvt.	PA 55th Inf.	I	7/20/63	8/30/65		
46	Murphy, Elias	16	Pvt.	PA 55th Inf.	D	9/26/62	1/12/65	Wounded	Chaffin's Farm
47	Murphy, George	29	Pvt.	OH 14th Inf.	E	8/21/61		Died	
48	Murphy, James S	19	Corp.	PA 55th Inf.	D	10/12/61	8/30/65	Wounded	
49	Murphy, James*	22	Pvt.	PA 55th Inf.	I	10/15/63			
50	Murphy, Philip (1)	21	Pvt.	PA 55th Inf.	D	12/20/61		Died	
51	Murphy, Philip (2)	18	Pvt.	OH 189th Inf.	A	3/3/65	9/28/65		
52	Murphy, Samuel	30	Pvt.	IN 140th Inf.	F	10/15/64	7/11/65		
53	Murray, Samuel*	19	Pvt.	PA 110th Inf.	C	7/11/64	6/7/65		
54	Murrie, David	14	Pvt.	MD 2nd PHB Inf.	G	9/20/61	9/28/64		
55	Murrie, William*	40	Pvt.	MD 2nd PHB Inf.	G	9/20/61	9/28/64		
56	Mushbaum, John	35	Pvt.	PA 55th Inf.	K	11/5/61	8/30/65		
57	Musselman, George	16	Pvt.	PA 205th Inf.	C	8/23/64	6/2/65		
58	Musselman, Jacob J	28	Pvt.	PA 13th Cav.	A	2/26/64	7/14/65		
59	Musselman, Simon C	21	Pvt.	PA 55th Inf.	K	3/8/64	8/30/65		
60	Muthardt, Jefferson H*		Pvt.	PA 55th Inf.	K	9/28/63	8/30/65		
61	Myer, John		Pvt.	PA 55th Inf.	K	10/2/63	3/16/64		
62	Myers, Daniel O	28	Pvt.	OH 162nd Inf.	C	5/20/64	9/4/64		
63	Myers, Daniel*	39	Pvt.	PA 110th Inf.	C	2/22/64		Wound.-Died	
64	Myers, Henry Jr								

	Cemetery	Notes
1	Fairmont Cem.-OH	Historical Data Systems-Bedford Co. residence, enlisted in Bloody Run
2		B.S.F. History Book-1884; Martin, William & Abraham-brothers
3		B.S.F. History Book-1884; William S, Nathaniel & Jeremiah-brothers
4	Carpenter Farm Cem.-Hyndman	Frank McCoy 1912 Listing
5	Fairmont Cem.-OH	Wounded 6/8/62; Died 7/25/65 of Wounds; Ancestry.com-born in Colerain Twp.
6	Arlington Nat'l Cem., 27-501	Died-Intermittent Fever 7/15/64; Pension Records; 1860 Londonderry Twp. Census
7		POW-Hagerstown 7/6/63; Died-Pneumonia 11/18/63; 1860 St. Clair Twp. Census
8		B.S.F. History Book-1884; Civil War Muster Roll-enlisted in Rainsburg
9	Old Claysburg Cem.	1890 Greenfield Twp. Veterans Census
10	Riverview Cem.-Huntingdon Co.	Wounded 5/6/64; Frank McCoy 1912 Listing
11		At Hospital at Muster Out; PA Civil War Project Record; 1850 Hopewell Twp. Census
12	Fockler Cem.-Saxton	Wounded 9/19/63; Died 6/23/63; Hist. Data Systems; 1860 Hopewell Twp. Cen. (Hunt. Co.)
13	Woods Church-Colerain Twp.	Frank McCoy 1912 Listing; 1890 Rainsburg Veterans Census; Discharged on Surgeon's Cert.
14		B.S.F. History Book-1884; 1860 Bedford Twp. Census; Alexander, Abraham, John H-brothers
15		Died 1/28/65 in Hospital; B.S.F. History Book-1884; PA Civil War Archives-Bedford Co. resid.
16	McNeeley Cem.-WV	B.S.F. History Book-1884; 1860 Bedford Boro Census
17	Woods Church-Colerain Twp.	Historical Data Systems-Bedford Co. residence
18		B.S.F. History Book-1884; Discharged on Surgeon's Cert.
19	Trinity UCC Cem.-Juniata Twp.	B.S.F. History Book-1884; 1860 Juniata Twp. Census; Discharged on Surgeon's Cert.
20		B.S.F. History Book-1884; Historical Data Systems-enlisted in Bedford Co.
21	Mt. Zion Cem.-Somerset Co.	Findagrave listed died in Bedford Co.
22	Bedford Cem.	Bedford Inquirer 5/22/1908 listing; Deserted 6/18/65 after war ended
23	Schellsburg Cem.	B.S.F. History Book-1884; Deserted 5/31/64
24		B.S.F. History Book-1884
25	Entriken Cem.-Hunt. Co.	1890 Hopewell Veterans Census; Daniel & John (3)-brothers
26	Andersonville Cem., 1006 E	POW-Andersonville; Died 5/10/64; B.S.F. History Book-1884; 1860 Londonderry Twp. Cen.
27	Beaufort Nat'l Cem.	Died 12/11/62; B.S.F. History Book-1884; Civil War Archives-Bedford Co. residence
28	Riverview Cem.-Huntingdon Co.	1860 Hopewell Twp. Census; Historical Data Systems record
29		POW 9/29/64; B.S.F. History Book-1884; enlisted in Berks Co.
30		B.S.F. History Book-1884; enlisted in Berks Co.
31		MIA on 6/3/64; B.S.F. History Book-1884; enlisted in Berks Co.
32		Deserted 8/29/64; B.S.F. History Book-1884; Berks Co. residence
33	Andersonville Cem., 7107	POW-Andersonville; Died 8/28/64, B.S.F. History Book-1884; enlisted in Berks Co.
34	Arlington Nat'l Cem., 13-10946	Died 7/10/65; Bedford Gazette 2/13/1914 listing; 1890 Juniata Twp. Veterans Census
35		B.S.F. History Book-1884; enlisted in Fulton Co.
36	St. Peters Luth.-Clear Spr., MD	B.S.F. History Book-1884; 1850 St. Clair Twp. Census; Mexican War Veteran
37		B.S.F. History Book-1884; George, John & David W-brothers
38	Bedford Cem.	POW-Columbia 4/20/64 to 2/15/65; Historical Data Systems record; 1850 Harrison Twp. Cen.
39	Mt. Olivet Cem.-Manns Choice	B.S.F. History Book-1884; 1890 Hopewell Twp. Veterans Census; Discharged on Surg. Cert.
40	Mt. Olivet Cem.-Manns Choice	B.S.F. History Book-1884; Historical Data Systems-Bedford Co. resid.; Discharged Surg. Cert.
41	Hillcrest Cem.-Clearfield	Wounded 12/13/62; Historical Data Systems-Bedford Co. residence
42		Historical Data Systems-Bedford residence
43		B.S.F. History Book-1884
44		B.S.F. History Book-1884; 1850 Lewisburg Census; Born in Dauphin Co.
45		B.S.F. History Book-1884; Historical Data Systems-enlisted in Philadelphia
46	Lebanon Meth. Cem.-OH	Arm Amputated 9/29/62; B.S.F. History Book-1884; Elias, James & Philip (1)-brothers
47	New Albany Nat'l Cem.-IN	Died-Diarrhea 6/13/64; Ancestry.com-Born Bedford; Sister Born Hopewell Twp.
48	Hampton Nat'l Cem.	1890 Mann Twp. Veterans Census-leg wound
49		Deserted & Executed 1/6/64; B.S.F. History Book-1884; Reading, PA references
50		Died 7/12/62; PA Civil War Archives-Bedford Co. residence; 1860 Southampton Twp. Cen.
51	Maumee Cem.-OH	Ancestry.com-Born in Bedford; Philip (2), Samuel & George-brothers
52	Justus Cem.-IN	Historical Data Systems record
53		B.S.F. History Book-1884
54	Bedford Cem.	1890 Harrison Twp. Veterans Census; Son of William
55	W. Newton-Westmoreland Co.	Historical Data Systems record; Unclear if William ever lived in Bedford Co.
56	Cedar Grove-Chambersburg	B.S.F. History Book-1884; Historical Data Systems-Bedford Co. residence
57	Trinity UCC Cem.-Osterburg	1890 Napier Twp. Veterans Census; Geroge & Simon-brothers
58	Lower Claar Cem.-Claysburg	PA Civil War Project record; Soldiers of Blair Co. Book - 1940
59	Lillydale Cem.-Cambria Co.	B.S.F. History Book-1884; attended 1909 Cold Harbor Dedication
60		B.S.F. History Book-1884; Berks Co. references
61		B.S.F. History Book-1884
62	Union Cem.-OH	Ancestry.com information; 1850 M. Woodbury Twp. Census
63	Soldiers Cem.-Brattleboro, VT	Died 5/19/64; B.S.F. History Book-1884; Blair Co. references
64	Hopewell Cem.	Bedford Inquirer 5/22/1908 listed both Sr. & Jr.; Henry Sr.-father

	Name	Age	Rank	Regiment	Co.	Muster In	Muster Out	Casualty	Casualty Battle
1	Myers, Henry Sr	59	Pvt.	PA 194th Inf.	I	7/22/64	11/5/64		
2	Myers, Jacob*	39	Pvt.	PA 55th Inf.	I	2/6/64	8/30/65		
3	Myers, John O	24	Pvt.	OH 19th Inf.	I	10/8/62	7/22/63		
4	Myers, Joseph		Pvt.	PA 55th Inf.	K	6/17/63	5/15/65	Wounded	
5	Myers, Levi		Pvt.	PA McKeage's Mil.	G	7/2/63	8/8/63		
6	Myers, Martin L	18	Pvt.	PA 99th Inf.	H	3/20/65	7/1/65		
7	Myers, Samuel	18	Pvt.	PA 76th Inf.	E	2/28/65	7/18/65		
8	Myers, William O	20	Pvt.	OH 19th Inf.	I	10/8/62	7/22/63		
9	Myers, William O		Pvt.	OH 104th Inf.	B	3/1/65	6/15/65		
10	Nabona, John			PA 11th Inf.					
11	Naugle, George		Pvt.	PA McKeage's Mil.	D	7/2/63	8/8/63		
12	Naugle, George			PA 2nd Heavy Art.					
13	Naugle, Jacob	26	Pvt.	PA 208th Inf.	K	9/7/64	6/1/65		
14	Naugle, James	39	Pvt.	PA 138th Inf.	D	8/29/62	6/23/65		
15	Neff, Frederick	51	Pvt.	PA 138th Inf.	D	8/29/62	1/15/65		
16	Neff, William L*		Capt.	PA 22nd Cav.	D	7/30/63	2/5/64		
17	Neff, William S*		Pvt.	PA 55th Inf.	D	9/26/62	6/11/65		
18	Negley, David F	24	Pvt.	PA 76th Inf.	E	10/9/61	11/28/64		
19	Nelson, John	41	1st Lt.	PA 18th Cav.	K	12/16/62	5/14/64	Wounded	Chantilly
20	Nelson, Robert	21	Pvt.	PA 171st Inf.	I	11/10/62			
21	Nelson, Robert		Pvt.	IN 74th Inf.	A	3/4/64	6/9/65		
22	Nelson, William N	15	Corp.	PA 18th Cav.	K	10/29/62	11/6/65		
23	Nesbitt, James A	20	Pvt.	PA 20th Cav.	A	6/23/63	1/6/64		
24	Nesbitt, James A		Pvt.	PA 49th Inf.	F	5/30/64	7/15/65		
25	Neville, Henry	15	Pvt.	PA 46th Inf.	B	7/1/63	8/19/63		
26	Neville, Henry		Pvt.	PA 6th Cav.	M	7/18/64	8/25/65		
27	Neville, J Richard	15	Pvt.	VA 23rd Cav.	K	3/14/64			
28	Nevitt, James M	20	Pvt.	PA 133rd Inf.	C	8/13/62	5/26/63	Wounded	Fredericksburg*
29	Nevitt, James M		Pvt.	PA 3rd Heavy Art.	L	3/9/64	11/9/65		
30	Nevitt, Joseph H	26	Pvt.	IA 35th Inf.	F	9/4/62	8/10/65		
31	Nevitt, Thomas								
32	Nevitt, William E	33	Wagon.	OH 67th Inf.	A	11/4/61	12/7/65		
33	Newcomer, Joseph	21	Pvt.	PA 133rd Inf.	K	8/15/62	5/26/63	POW	
34	Newcomer, Joseph		Pvt.	PA 5th Res.	B	2/9/64	6/6/64	Wounded	
35	Newcomer, Joseph		Pvt.	PA 191st Inf.	C	6/6/64	6/16/65	POW	Globe Tavern
36	Newman, John (1)*		Pvt.	PA 55th Inf.	K	10/2/63		KIA	Bermuda Hundred
37	Newman, John (2)		Pvt.	PA 55th Inf.	D	9/12/62	6/11/65		
38	Newman, John R*	34	Pvt.	IL 62nd Inf.	D	4/10/62		Died	
39	Newton, James	17	Pvt.	PA 110th Inf.	C	3/15/62	12/31/64		
40	Nicewonger, Andrew	33	Pvt.	PA 194th Inf.	C	7/20/64	11/6/64		
41	Nichols, Charles	32	Pvt.	PA 21st Cav.	E	7/21/63		Wounded	Petersburg - Initial
42	Nichols, Charles	32	Pvt.	PA 21st Cav.	E	7/21/63		KIA	Petersburg
43	Nicodemus, George	34	Pvt.	IL 15th Inf.	K	3/7/65	7/31/65		
44	Nicodemus, Isaac	29	Pvt.	PA 138th Inf.	E	8/29/62	6/3/65		
45	Nicodemus, Joseph	25	Pvt.	PA 149th Inf.	I	1863	6/24/65		
46	Nicola, Moses		Pvt.	PA 61st Inf.					
47	Nigh, Henry		Pvt.	PA McKeage's Mil.	H	7/2/63	8/8/63		
48	Nine, Harrison H*	21	Pvt.	PA 55th Inf.	H	9/2/62	6/11/65		
49	Nipple, Frederick		Pvt.	PA 36th Inf.	B	7/4/63	8/11/63		
50	Noble, James D	24	Surg.	PA 51st Inf.		9/14/61	7/21/62		
51	Noble, James D		Surg.	US Navy		2/20/65	1/13/66		
52	Noel, Charles	29	Sgt.	PA 103rd Inf.	H	3/3/65	6/25/65		
53	Noffsker, Martin	23	Pvt.	PA 55th Inf.	I	8/28/61		POW	Drewry's Bluff
54	Noffsker, William A*		Pvt.	PA 55th Inf.	I	2/11/64	4/11/65	Wounded	Cold Harbor
55	Noffsker, William H	18	Sgt.	PA 12th Inf.		10/18/61	10/18/64	Wounded	
56	Noland, James*	19	Pvt.	PA 55th Inf.	I	10/15/61	2/25/63		
57	Noland, Thomas*	20	Pvt.	PA 55th Inf.	I	10/15/61	2/25/63		
58	Noll, William R*		Pvt.	PA 55th Inf.	K	9/28/63		KIA	Cold Harbor
59	Norris, Harrison (1)	17	Pvt.	PA 77th Inf.		10/9/61	12/6/65		
60	Norris, Harrison (2)*		Corp.	PA McKeage's Mil.	D	7/2/63	8/8/63		
61	Norton, Franklin G	14	Music.	PA 101st Inf.	G	12/28/61		Died	
62	Norton, James	43	Pvt.	PA 55th Inf.	D	10/12/61	1/17/63		
63	Nottingham, William	31	Pvt.	PA 55th Inf.	D	10/12/61	11/22/64	POW	Drewry's Bluff
64	Null, George W	44	Pvt.	PA 76th Inf.	E	9/21/64	7/18/65		

	Cemetery	Notes
1	Hopewell Cem.	Bedford Inquirer 5/22/1908 listing; Civil War Veterans headstone record aka Meyers
2		B.S.F. History Book-1884; Blair Co. references
3	Union Cem.-OH	Ancestry.com information; John, Daniel & William-brothers
4		Wounded listed in Bedford Inquirer article 6/3/64; B.S.F. History Book-1884
5		B.S.F. History Book-1884
6	Everett Cem.	Died 1/12/42; B.S.F. History Book-1884; 1890 Everett Veterans Census
7		B.S.F. History Book-1884
8	Union Cem.-OH	Findagrave.com born in Bedford; Historical Data Systems record
9		Findagrave.com born in Bedford; Historical Data Systems record
10	Everett Cem.	Bedford Inquirer 5/22/1908 listing
11		B.S.F. History Book-1884
12	Hinish Graveyard-E. Prov. Twp.	B.S.F. History Book-1884; Frank McCoy 1912 Listing
13	Graceville Luth. Cem.	1890 E. Providence Twp. Veterans Census; B.S.F. History Book-1884
14	Mt. Olivet Cem.-Manns Choice	B.S.F. History Book-1884; 1860 Napier Twp. Census
15	Union Cem.-Somerset Co.	B.S.F. History Book-1884
16	Presb. Cem.-Williamsburg	B.S.F. History Book-1884; Ancestry.com Blair Co. residence
17		B.S.F. History Book-1884; PA Civil War Archives-Berks Co. residence
18	Decatur City Cem.-IA	1860 Liberty Twp. Census; PA Civil War Archives-Stonerstown residence
19	Mt. Smith Cem.-Bedford	Leg Amputated 2/26/63, reportedly shot by Col. Mosby (Grey Ghost of Confederacy)
20	Lakeview Cem.-NY	Deserted 11/21/62; B.S.F. History Book-1884
21		Ancestry.com; 1850 Southampton Twp. Census
22	Fairhaven Mem.Park-CA	1890 Saxton Veterans Census; John-Father
23	Grandview Cem.-Cambria Co.	1890 Robertsdale Veterans Census
24		1890 Robertsdale Veterans Census
25	Broad Top IOOF Cem.	1890 Robertsdale Veterans Census
26		Deserted 11/11/64 & returned 12/31/64; B.S.F. History Book-1884
27		Born in Virginia; 1870 Winchester, VA Census
28	Cedar Grove-E. Prov. Twp.	1890 E. Providence Twp. Veterans Census
29		1890 E. Providence Twp. Veterans Census
30	Winfield Scott Twp. Cem.-IA	Ancestry.com information; Joseph, James, William & Thomas-brothers
31	Mt. Pleasant Cem.-Mattie	Bedford Inquirer 5/22/1908 listing
32	Amboy Twp. Cem.-OH	Findagrave.com-born E. Providence Twp.
33	Richland Cem.-Cambria Co.	B.S.F. History Book-1884
34		1890 Napier Twp. Veterans Census-Leg Wound
35	Richland	POW Libby, Belle Island & Andersonville 8/19/64 to 3/1/65; 1890 Napier Twp. Veterans Cen.
36		KIA 5/20/64; B.S.F. History Book-1884; Drafted & Mustered Philadelphia
37		B.S.F. History Book-1884
38	Rose Hill Cem.-IL	Died 11/29/64; 1850 Harrison Twp. Cen; 15 IL Veterans Reserve Corp. referenced
39		Deserted 12/31/64; B.S.F. History Book-1884
40	Potter Creek Cem.-Woodbury	1880 Woodbury Twp. Census
41		Wounded 6/18/64; PA Civil War Project record; Historical Data Systems record
42	Spruce Grove Baptist Cem.	KIA 3/21/65; PA Civil War Archives-Bedford Co. residence
43	Hartley Cem., IA	Historical Data Systems; Pauline Nicodemus family history record-born in Woodbury
44	Schellsburg Cem.	B.S.F. History Book-1884; Historical Data Systems-Schellsburg Co resid.; Disch. Surg. Cert.
45	Oak Shade Cem.-IA	Record on Findagrave-memorial #10196320
46		B.S.F. History Book-1884
47		B.S.F. History Book-1884
48		B.S.F. History Book-1884; Ancestry.com-Schuylkill references
49	Mt. Zion Cem.-Bedford Twp	1860 Union Twp. (Pavia) Census; aka Knipple
50	Saint John's Cem.-Loysburg	Frank McCoy 1912 Listing
51		1870 S. Woodbury Twp. Census; Bedford Inquirer 5/22/1908 listing
52		1890 Broad Top Twp. Veterans Census
53		POW-Millen, GA 5/16/64; B.S.F. History Book-1884; 1870 Broad Top Census; Died 1870
54		Wounded 6/3/64; B.S.F. History Book-1884; Blair Co. references
55	Old Claysburg Cem.	1840 Hopewell Twp. Census; 1890 Greenfield Twp. Veterans Census lists Wounded
56		B.S.F. History Book-1884; Blair Co. references
57	Greenlawn Cem.-Roaring Spring	B.S.F. History Book-1884; Blair Co. references
58		KIA 6/6/64; B.S.F. History Book-1884; Ancestry.com-Berks Co. references
59	Rays Cove CC Cem.	Frank McCoy 1912 Listing; 1890 E. Providence Twp. Veterans Census
60		B.S.F. History Book-1884; Alleghany Co. references
61	New Paris Comm. Ctr. Cem.	Died-Pneumonia 2/23/62; PA Civil War Archives-Bedford Co. residence
62	Horn Farm Cem.-Buffalo Mills	1860 Napier Twp. Cen.; B.S.F. History Book-1884; Disch. Surg. Cert.; Franklin-son
63	Oakwood Cem.-MO	POW 5/16/64 to 11/17/64; 1860 Cumberland Valley Census; B.S.F. History Book-1884
64	Glade Union Cem.-Somerset Co.	Deserted 3/24/65 & returned; B.S.F. Hist. Book-1884; Ancestry.com-Son born in Bedford Co.

	Name	Age	Rank	Regiment	Co.	Muster In	Muster Out	Casualty	Casualty Battle
1	Null, Samuel	39	Pvt.	PA 56th Inf.	G	9/21/64	5/31/65		
2	Nulton, Henry H	22	Pvt.	PA 76th Inf.	E	10/9/61	11/28/64		
3	Nulton, William H	26	Sgt.	PA 13th Inf.	G	4/25/61	8/6/61		
4	Nunemaker, Peter	29	Pvt.	PA 99th Inf.	H	2/25/65	5/31/65		
5	Nute, William W	25	Pvt.	MA 2nd Inf.	B	5/25/61	10/25/62		
6	Nute, William W		Pvt.	US 4th Light Art.	F	10/25/62	5/11/64		
7	Nute, William W		Sgt.	PA 208th Inf.	H	9/8/64	5/19/65		
8	Nycum, Bernard (1)*		Pvt.	PA 133rd Inf.	K	8/8/62	5/25/63		
9	Nycum, Bernard (2)	26	Pvt.	PA 138th Inf.	D	8/29/62	6/23/65		
10	Nycum, George	41	Pvt.	PA 107th Inf.	F	9/21/64	6/6/65		
11	Nycum, John (1)	28	Pvt.	PA 138th Inf.	D	8/29/62		Wound.-Died	Cold Harbor
12	Nycum, John (2)			PA 12th Cav.					
13	Nycum, John Q	21	1st Sgt.	PA McKeage's Mil.	D	7/2/63	8/8/63		
14	Nycum, John Q		Pvt.	PA 186th Inf.	G	3/24/64	8/15/65		
15	Nycum, John W	20	Pvt.	PA 107th Inf.	F	10/6/64	7/13/65		
16	Nycum, Josiah	23	Pvt.	PA 2nd Cav.	E	10/18/61		Died	
17	Nycum, Simon E								
18	Nycum, Upton	22	Sgt.	PA 2nd Cav.	E	11/4/61		Wound.-Died	
19	Nycum, William	28	Pvt.	IN 59th Inf.	F	1/31/65	7/17/65		
20	Nycum, William H	21	Corp.	PA 133rd Inf.	C	8/13/62	5/26/63	Wounded	Fredericksburg
21	Nycum, William H		Sgt.	PA 186th Inf.	G	3/24/64	8/15/65		
22	Nycum, Wilson	20	Pvt.	PA 22nd Cav.	C	7/11/63	2/5/64		
23	Oakes, Christian R	34	Pvt.	PA 137th Inf.	I	8/20/62	6/1/63		
24	Oaks, Jacob R	33	Pvt.	PA 99th Inf.	B	2/24/65	7/1/65		
25	Oaks, John R	24	Pvt.	PA 138th Inf.	D	8/29/62	2/2/64		
26	Oaks, John R		Pvt.	PA 208th Inf.	H	9/5/64	6/1/65		
27	O'Keefe, James*		Pvt.	PA 55th Inf.	K	9/23/63			
28	Oldham, Michael	17	Pvt.	PA 55th Inf.	K	2/15/64		Died	
29	Oldham, Thomas	40	Pvt.	PA 82nd Inf.	C	11/14/64	6/15/65		
30	Olds, A Warren*	29	Pvt.	PA 76th Inf.	E	8/27/63	7/18/65		
31	Oler, James W	20	Pvt.	PA 101st Inf.	D	11/1/61	6/25/65	POW	Plymonth, NC
32	Oler, John W	16	Music.	PA 101st Inf.	D	11/1/61		Died	
33	Olinger, George W*		Pvt.	PA 110th Inf.	C	2/22/64	6/21/65	Wounded	Wilderness
34	Olinger, George W*		Pvt.	PA 110th Inf.	C			Wounded	Petersburg
35	Oliver, Benjamin F	18	Pvt.	PA 18th Cav.	K	10/29/62	10/31/65		
36	Oliver, James M	29	Pvt.	OH 121st Inf.	B	8/18/62		KIA	Kennesaw Mtn.
37	Oliver, Nathaniel H	39	Pvt.	PA 39th Inf.	F	7/1/63	8/2/63		
38	O'Neal, Emanuel	29	Pvt.	PA 138th Inf.	D	8/29/62	6/23/65		
39	O'Neal, Hezekiah	18	Pvt.	PA 138th Inf.	D	8/29/62		Wound.-Died	
40	O'Neal, James R	32	1st Sgt.	PA 208th Inf.	K	9/7/64	5/30/65	Wounded	Ft. Stedman
41	O'Neal, John E	19	Corp.	PA 138th Inf.	D	8/29/62	6/23/65	Wounded	Spotsylvania
42	O'Neal, Samuel	18	Pvt.	PA 195th Inf.	E	4/13/65	1/31/66		
43	O'Neil, John*	30	Capt.	PA 55th Inf.	I	8/28/61		Wound.-Died	Chaffin's Farm
44	Onstadt, Mathias	40	Pvt.	PA 28th Inf.	H	11/14/64	7/18/65		
45	Ormsby, John*	28	Pvt.	PA 55th Inf.	I	10/7/63			
46	Ornst, John		Pvt.	PA 55th Inf.	H	10/9/63	12/30/64	Wounded	Bermuda Hundred
47	Orris, Jacob	35	Pvt.	PA 184th Inf.	A	5/12/64		POW-Died	Jerusalem Plank Rd.
48	Osborn, Ezra	23	Pvt.	IL 93rd Inf.	C	10/13/62	5/27/64	Wounded	
49	Osborn, Peter	37	Pvt.	PA 208th Inf.	K	9/5/64		Wound.-Died	Petersburg - Final
50	Osborn, William	29	Pvt.	PA 133rd Inf.	C	8/29/62	5/26/63		
51	Oster, Samuel C	20	Pvt.	PA 91st Inf.	B	9/21/64	5/9/65	Wounded	Boydton Plank Rd.
52	Ott, Michael	20	Corp.	PA 133rd Inf.	K	8/15/62	5/26/63		
53	Ott, Nicholas	26	Pvt.	PA 158th Inf.	H	11/4/62	8/12/63		
54	Ott, Nicholas		Pvt.	PA 208th Inf.	H	9/9/64	6/30/65	Wounded	Petersburg - Final
55	Otto, Abraham*	20	Pvt.	PA 55th Inf.	I	2/15/64		Wound.-Died	Bermuda Hundred
56	Otto, DeWitt*							Died	
57	Otto, Emanuel		Pvt.	PA 21st Cav.	E	7/8/63	2/20/64		
58	Otto, George W	22	Pvt.	PA 21st Cav.	E	7/8/63	2/20/64		
59	Otto, Henry	22	Pvt.	PA 101st Inf.	G	12/28/61		Died	
60	Otto, Henry S	16	Pvt.	PA 184th Inf.	A	5/12/64		POW-Died	Jerusalem Plank Rd.
61	Otto, Jacob P	34	Pvt.	PA 61st Inf.	F	9/21/64	6/20/65		
62	Otto, John S	29	Pvt.	PA 79th Inf.	F	9/21/64	7/12/65		
63	Otto, William*		Pvt.	OH 5th Cav.	F	2/5/64		KIA	Fayetteville, NC
64	Over, Benjamin	22	Pvt.	PA 133rd Inf.	C	8/13/62	5/26/63	Wounded	Fredericksburg

	Cemetery	Notes
1	Shade Luth. Cem.-Shade Twp.	1890 Bedford Twp. Veterans Census; Samuel & George-brothers
2		B.S.F. History Book-1884; PA Civil War Archives-Bedford residence
3		B.S.F. History Book-1884; PA Civil War Archives-Bedford residence
4	Pleasantville Cem.	1890 W. St. Clair Twp. Veterans Census
5	Cedar Grove-Chambersburg	B.S.F. History Book-1884
6		Historical Data Systems record
7		Historical Data Systems-Bedford Co. residence
8		O'Neils family history book reference; possibly William Nycum
9	Tonoloway Bapt.-Warfordsburg	B.S.F. History Book-1884; 1850 Monroe Twp. Census
10	Shreves Cem.-Monroe Twp.	1890 Monroe Twp. Veterans Census
11	Mt. Pleasant Cem.– Mattie	Wounded 6/1/64; Died 6/28/64; B.S.F. History Book-1884; 1860 Monroe Twp. Census
12	Bethal Frame Ch.-Monroe Twp.	Frank McCoy 1912 Listing
13		B.S.F. History Book-1884
14	Everett Cem.	1890 E. Providence Twp. Veterans Census
15	Mt. Pleasant Cem.-Mattie	Frank McCoy 1912 Listing; John W, George & Josiah-brothers; Nickname-Singing John
16	Shreves Cem.-Monroe Twp.	Died 8/24/64; B.S.F. History Book-1884; 1860 Monroe Twp. Census
17	Everett Cem.	1870 E. Providence Twp. Census; Everett Cemetery Listing of Civil War Veterans
18	Bethal Frame Ch.-Monroe Twp.	Died 12/12/63; B.S.F. Hist. Book-1884; PA Civil War Project record; 1850 Monroe Twp. Cen.
19	Lindenwood Cem.-IN	Ancestry.com lists born Bedford Co.
20	Mt. Zion Cem.-Breezewood	Wounded 12/13/62; William & John Q-brothers
21		Findagrave.com information
22	Lyons Muni. Cem.-KS	B.S.F. Hist. Book-1884; 1860 Monroe Twp. Cen.; Upton, Bernard(2), John(1) & Wilson-bros.
23	Brumbaugh Cem.-Blair Co.	1870 Broad Top Twp. Cen; Ancestry.com born Hopewell
24	Bunker Hill Cem.-Saxton	1890 Liberty Twp. Veterans Census
25		Deserted 2/6/63 and returned; Historical Data Systems record
26	Wooderson Cem.-MO	Historical Data Systems record; 1860 Liberty Twp. Census; John R, Jacob and Christian-bros.
27		B.S.F. History Book-1884; Drafted & Deserted 4/28/64
28	Beaufort Nat'l Cem.	Died 4/6/64; PA Civil War Archives Bedford Co. residence; 1860 Union Twp. (Pavia) Census
29	Mock Church Cem.-Alum Bank	1860 Union Twp. (Pavia) Census; Frank McCoy 1912 Listing; PA 55th references
30	Minden Cem.-NE	B.S.F. History Book-1884; Ancestry.com-McKean Co. references
31	Everett Cem.	POW 4/20/64; Hist. Data Systems record; 1860 Bedford Boro Census; James & John-bros.
32	Everett Cem.	Died 1/2/62; PA Civil War Archives-Bedford Co. residence
33	Mt. Lonestar Cem.-KS	Wounded 5/6/64; Blair Co. references
34		Wounded 4/1/65; Blair Co. references
35	Mechanicsburg Cem.	PA Civil War Archives-Bedford Co. residence
36	Murietta Nat'l Cem.	KIA 6/27/64; Historical Data Systems record; 1850 Cumberland Valley Census
37	Woods Church-Colerain Twp.	Historical Data Systems record; Nathaniel & James-brothers
38		1860 Southampton Twp. Census; B.S.F. History Book-1884
39	Chaneysville Meth. Cem.	Wounded 11/21/63; Died 12/4/63; PA Civil War Archives-Bedford Co. residence
40	Mt. Pleasant Cem.-SD	Wounded 3/25/65; in Hospital at Muster Out; B.S.F. History Book-1884;John & James-bros.
41	Mt. Pleasant Cem.-SD	Wounded-Bedford Inquirer 6/24/64; B.S.F. History Book-1884; 1850 Monroe Twp. Census
42	IOOF Mem. Cem.-Hunt. Co.	1890 Robertsdale Veterans Census
43		Wounded 9/29/64; Died 12/11/64; B.S.F. Hist.Book-1884; PA CW Arch.-Johnstown residence
44	Helixville Cem.-Napier Twp.	Frank McCoy 1912 Listing; 1860 Napier Twp. Census
45		Deserted 5/9/64; B.S.F. History Book-1884
46		Wounded 5/9/64; B.S.F. History Book-1884
47	Bedford Cem.	Wounded & POW 6/22/64; Died 8/1/64; B.S.F. History Book-1884; 1860 Bedford Twp. Cen.
48	Graceland Cem.-IA	1850 Broad Top Census; Historical Data Systems record
49	Everett Cem.	Wounded 4/2/65; Died 4/3/65; B.S.F. Hist. Book-1884; PA CW Archives-Bedford Co. resid.
50	Harvard Cem.-NE	B.S.F. History Book-1884; Peter & William-brothers
51	Harris-Elmore Union Cem.-OH	Wounded 10/27/64; 1860 Union Twp. (Pavia) Census
52	Everett Cem.	B.S.F. History Book-1884; 1860 Colerain Twp. Census
53	Cedar Grove-Chambersburg	Historical Data Systems record; Ancestry.com-Fulton Co. references
54		Arm Amputated 4/2/65; Historical Data Systems-Bedford Co. residence
55		Wounded 5/9/64; Died 5/10/64; Blair Co. references
56		Trinity UCC Cem. Civil War memorial of soldiers who are in unknown graves
57		PA Civil War Archives-Bedford Co. residence; Historical Data Systems record
58		Historical Data Systems-Bedford Co. residence
59	Arlington Nat'l Cem., 11429	Died 9/26/62; B.S.F. History Book-1884; 1850 Napier Twp. Census; aka Ott
60	Andersonville Cem., 9205	POW-Andersonville 6/22/64; Died 9/18/64; B.S.F. Hist. Book-1884; 1860 St. Clair Twp. Cen.
61	Schellsburg Cem.	1890 Napier Twp. Veterans Census
62	New Paris Comm. Ctr. Cem.	1890 Napier Twp. Veterans Census
63		Died 3/10/65; Trinity UCC Cemetery Civil War memorial-soldiers in unknown graves
64		Wounded 12/13/62; B.S.F. History Book-1884; 1860 Woodbury Twp. Census

Chapter 18

	Name	Age	Rank	Regiment	Co.	Muster In	Muster Out	Casualty	Casualty Battle
1	Over, David H	24	Pvt.	PA 99th Inf.	D	3/15/65	7/1/65		
2	Over, David S	18	Pvt.	PA 184th Inf.	A	5/12/64	5/27/65	POW	Jerusalem Plank Rd.
3	Over, Jacob Z	22	Sgt.	PA 101st Inf.	G	2/18/62	4/5/63		
4	Over, Jacob Z		Sgt.	PA 184th Inf.	A	5/12/64	6/17/65	POW	Jerusalem Plank Rd.
5	Over, James E	27	Corp.	PA 138th Inf.	E	8/29/62	6/23/65	Wounded	Opequon
6	Overaker, William	34	Pvt.	PA 186th Inf.	G	3/24/64	8/15/65		
7	Owens, Chauncey	22	Pvt.	PA 138th Inf.	F	8/29/62	6/23/65	Wounded	Monocacy, MD
8	Owens, Richard	37	Pvt.	PA 125th Inf.	F	8/12/62			
9	Oyler, Abraham	36	Pvt.	PA 55th Inf.	D	10/30/62	5/25/65	Wounded	Chaffin's Farm
10	Oyler, William	36	Pvt.	PA 55th Inf.	D	10/12/61	8/30/65		
11	Packert, Christian*		Pvt.	PA 76th Inf.	E	10/25/61		POW-Died	
12	Page, Christian*	21	Pvt.	PA 101st Inf.	D	1/16/62		Died	
13	Pallinger, James		Corp.	PA 19th Cav.	L	9/8/63	10/2/65		
14	Palmer, Casper	22	Corp.	KS 10th Inf.	K	8/12/61		Died	
15	Palmer, Elijah N	22	Pvt.	PA 82nd Inf.	L	11/28/64	7/13/65		
16	Palmer, John	17	Pvt.	PA 55th Inf.	K	2/19/64		POW-Died	
17	Park, Lounzo F		Pvt.	NJ 4th Inf.	G	4/25/61	7/31/61		
18	Park, Lounzo F		Pvt.	PA 99th Inf.	E	2/23/65	5/31/65		
19	Parker, James			USCT Inf.					
20	Parker, William K*		Pvt.	PA 76th Inf.	E	10/16/61	11/23/64		
21	Parlett, John C	22	Pvt.	PA 126th Inf.	B	8/12/62	5/20/63		
22	Parlett, John C		Pvt.	PA 22nd Cav.	M	2/27/64	6/24/65		
23	Parlett, Thomas W	26	Pvt.	PA 22nd Cav.	M	3/17/65	6/24/65		
24	Parsons, George W	20	Pvt.	PA 76th Inf.	E	9/28/64	6/29/65		
25	Parsons, John E		Pvt.	PA 22nd Cav.	H	2/24/64	7/20/65		
26	Parsons, William		Pvt.	PA 55th Inf.	D	9/26/62	6/11/65		
27	Patton, Abraham	25	Corp.	PA 76th Inf.	E	2/9/65	7/18/65		
28	Paul, John	35	Sgt.	PA 8th Res.	F	6/11/61	5/26/64		
29	Peadley, Thomas		Pvt.	PA 7th Res.	I	8/9/62	5/18/63		
30	Peadley, Thomas		Pvt.	PA 103rd Inf.	D	3/4/65	6/25/65		
31	Pearson, Edward P	35	Pvt.	PA 25th Inf.	A	4/18/61	7/26/61		
32	Pearson, Francis	17	Pvt.	PA 110th Inf.	C	10/24/61		Died	
33	Pearson, Henry C		1st Lt.	PA 21st Cav.	H	7/24/63	2/15/65	Wounded	Boydton Plank Rd.
34	Pearson, Josiah	15	Pvt.	PA McKeage's Mil.	H	7/3/63	8/8/63		
35	Pearson, Josiah		Bugler	US 1st Cav.	C	1863	1868		
36	Pearson, Thomas K		Pvt.	PA 208th Inf.	B	8/26/64			
37	Peck, Jacob B	25	Pvt.	PA 13th Inf.	G	4/25/61	8/6/61		
38	Peck, Jacob B		Corp.	PA 55th Inf.	D	10/12/61	10/4/62		
39	Peck, Jacob B		Pvt.	PA 14th Res.		4/15/64	10/20/65		
40	Peck, Jesse	21	Pvt.	PA 133rd Inf.	C	8/13/62	5/26/63		
41	Peck, Jesse		Corp.	PA 208th Inf.	H	9/6/64	6/1/65		
42	Peck, John		Pvt.	PA Ind Light Art.	D	9/19/64	7/13/65		
43	Peck, Llewellyn H	19	Pvt.	PA 194th Inf.	I	7/22/64	11/5/64		
44	Peck, Simon	17	Pvt.	PA 208th Inf.	H	9/7/64	6/1/65		
45	Pee, Frances W	21	Sgt.	PA 11th Inf.	A	9/30/61	1/15/64	Wounded	Antietam
46	Pee, Frances W		Sgt.	PA 11th Inf.	A			Wound.-POW	Gettysburg
47	Peightel, James	17	Pvt.	PA 208th Inf.	H	9/2/64	6/1/65		
48	Penn, William J	18	Pvt.	PA 13th Inf.	G	4/25/61	8/6/61		
49	Penn, William J		Pvt.	MD 2nd PHB Inf.	H	8/31/61	9/29/64		
50	Pennel, Andrew J (1)	22	Pvt.	PA 171st Inf.	I	11/2/62	8/8/63		
51	Pennel, Andrew J (1)		Pvt.	PA 91st Inf.	F	9/21/64	6/1/65	Wounded	Boydton Plank Rd.
52	Pennel, Andrew J (2)	31	Pvt.	PA 45th Inf.	B	9/21/64	6/7/65		
53	Pennell, Henry C	19	Pvt.	PA 76th Inf.	E	10/9/61	6/29/62	Died	
54	Pennington, James F	15	Pvt.	PA 171st Inf.	F	11/4/62	8/7/63		
55	Pennington, James F		Pvt.	PA 188th Inf.	D	2/27/64	12/14/65		
56	Penrod, Amanish*		Pvt.	PA 11th Cav.	G	8/28/61	8/27/64		
57	Penrod, Amanish*		Pvt.	PA 55th Inf.	I	2/27/64	8/30/65		
58	Penrod, George	27	Pvt.	PA 133rd Inf.	B	8/14/62	5/23/63		
59	Penrod, Henry C	18	Pvt.	PA 8th Res.	F	6/11/61	10/27/62		
60	Penrod, Henry C		Pvt.	PA 6th Cav.	K	10/27/62	4/29/64	Wounded	
61	Penrod, Henry C		Pvt.	PA 194th Inf.	I	7/22/64	9/6/64		
62	Penrod, Henry C		Pvt.	PA 97th Inf.	Ind	9/6/64	7/17/65		
63	Penrod, John B Jr	19	Pvt.	PA 8th Res.	F	6/11/61	10/27/62	POW	Seven Days Battle
64	Penrod, John B Jr		Pvt.	PA 6th Cav.	K	10/27/62	4/29/64		

	Cemetery	Notes
1	New Enterprise Cem.	1890 S. Woodbury Twp. Veterans Census aka Ober
2	New Enterprise Cem.	POW 6/22/64 to 4/28/65; Historical Data Systems record
3	McConnellsburg Presb. Cem.	B.S.F. History Book-1884; 1860 Bedford Boro Census; Discharged on Surgeon's Cert.
4		POW Andersonville & Millan, GA; B.S.F. History Book-1884; POW 6/24/64 to 4/24/65
5	Trinity UCC Cem.-Osterburg	Wounded 9/19/64; B.S.F. History Book-1884; 1860 St. Clair Twp. Census
6	St. Matthews Epis. Cem.-NC	1860 E. Providence Twp. Census; Historical Data Systems record
7		Wounded 7/9/64 ; B.S.F. History Book-1884; 1850 Cumberland Valley Census
8	Old Fellows-Clearfield Co	Historical Data Systems-Missing at Chancellorsville; 1870 Coalmont Census
9	Bedford Cem.	Leg Amputated 9/29/64; Frank McCoy 1912 Listing; Abraham & William-brothers
10	Mt. Olivet Cem.-Manns Choice	1890 Harrison Twp. Veterans Census
11		POW-Richmond; Died 1/28/64; B.S.F. History Book-1884; enlisted in Hanover
12	Loudon Park Nat'l Cem.	Died-Typhoid Fever 6/26/62; B.S.F. History Book-1884; Snyder Co. references
13		1890 Liberty Twp. Census
14	Oakwood Cem.-KS	Died 12/30/61; 1850 Harrison Twp. Census; Historical Data Systems record
15	Everett Cem.	1890 Everett Veterans Census
16		POW & Died of Disease; PA Civil War Archives-Bedford Co. resid.; 1860 Union Twp. Cen.
17		1890 Saxton Veterans Census
18		1890 Saxton Veterans Census
19	Mt. Ross Cem.-Bedford	1890 Everett Veterans Census
20		B.S.F. History Book-1884; Hanover references
21	Buck Valley Cem.-Fulton Co.	1890 Brush Creek Veterans Census
22		Historical Data Systems record
23	Great Cacapon Cem.-WV	1890 Brush Creek Veterans Census
24		B.S.F. History Book-1884
25		B.S.F. History Book-1884; Deserted after war
26		B.S.F. History Book-1884
27		B.S.F. History Book-1884
28	Duval Cem.-Coaldale	B.S.F. History Book-1884; Frank McCoy 1912 Listing
29		1890 Saxton Veterans Census
30		1890 Saxton Veterans Census
31		Hist. Data Systems record; 1860 Woodbury Twp. Cen.; Died-3/3/63-cause of death unknown
32	Wagerman Family Cem.-Madley	Died-Pernicious Fever 12/31/62; Civil War Pension Record; 1912 Bedford Gazette listing
33		Wounded 10/27/64; B.S.F. History Book-1884
34	Pleasantville Cem.	Bedford Inquirer 5/22/1908 listing; Josiah, Edward & Francis-brothers
35		Frank McCoy 1912 Listing; 1890 W. St. Clair Twp. Veterans Cen.; US 3rd Cavalry references
36		Deserted 9/4/64; Ancestry.com information; Thomas & Edward-brothers
37	Everett Cem.	B.S.F. History Book-1884; Jacob & Llewellyn-brothers
38		1890 E. St. Clair Twp. Veterans Census; 1860 W. Providence Twp. Census; Disch. Surg. Cert.
39		1890 E. St. Clair Twp. Veterans Census; PA Civil War Project Record
40	Union Memorial Cem.-Mench	1890 E. Providence Twp. Veterans Census; Historical Data Systems record
41		B.S.F. History Book-1884; Historical Data Systems record; Jesse & Simon-brothers
42	Akersville Cem.-Crystal Springs	1890 Brush Creek Twp. Veterans Census
43	Everett Cem.	B.S.F. History Book-1884; 1850 W. Providence Twp. Census
44	Umbria Cem.-Osceola Mills	B.S.F. History Book-1884; 1850 E. Providence Twp. Census
45	Mt. Pleasant Cem.-Mattie	Shoulder Wound 9/17/62; 1850 E. Providence Twp. Census; Frank McCoy 1912 Listing
46		POW & Arm. Wound 7/1/63; Exchanged 7/5/63; transferred to VRC 1/15/64
47		B.S.F. History Book-1884; Historical Data Systems-Bedford Co. residence
48	IOOF Cem.-Berlin	B.S.F. History Book-1884; Historical Data Systems-Bedford Co. residence
49		1890 Berlin Veterans Census
50	Woods Church-Colerain Twp.	B.S.F. History Book-1884; 1870 Colerain Twp. Census
51		Wounded 10/27/64; Historical Data Systems record
52	Pleasant Union Cem.-Clearville	1890 Monroe Twp. Veterans Census
53	Bedford Cem.	Died 6/29/62; B.S.F. History Book-1884; 1860 Bedford Boro Cen. Hist. Data Systems record
54	Everett Cem.	Historical Data Systems record; 1880 Pleasantville Census
55		Frank McCoy 1912 Listing; Pastor after the war
56		Historical Data Systems-Johnstown references
57		B.S.F. History Book-1884; Johnstown references
58	Mt. Hope Cem.-Cambria Co.	Ancestry.com-Son Born in Bedford Co.; Historical Data Systems record
59	Bethel Brethren Cem.-Tatesville	1890 Hopewell Veterans Census; Historical Data Systems record
60		B.S.F. History Book-1884; Historical Data Systems record
61		1890 Hopewell Veterans Census-loss of sight in right eye & right wrist wound
62		B.S.F. History Book-1884; Historical Data Systems record; Henry & John-brothers
63	Rose Hill Cem.-Cumberland	POW 6/27/62; B.S.F. History Book-1884
64		B.S.F. History Book-1884; Historical Data Systems record

Chapter 18

	Name	Age	Rank	Regiment	Co.	Muster In	Muster Out	Casualty	Casualty Battle
1	Penrod, John B Jr		Corp.	PA 194th Inf.	I	7/22/64	9/6/64		
2	Penrod, John B Jr		Pvt.	PA 97th Inf.	Ind	9/6/64	7/17/65		
3	Penrod, John B Sr	47	Pvt.	PA 8th Res.	F	11/12/61	11/26/62		
4	Penrose, Andrew J	18	Sgt.	PA 55th Inf.	D	10/12/61	8/30/65		
5	Penrose, Joseph	18	Pvt.	PA 21st Cav.	E	7/8/63	2/20/64		
6	Penrose, Joseph		Pvt.	PA 205th Inf.	C	8/26/64	6/2/65		
7	Penrose, Mahlon			PA 199th Inf.					
8	Penrose, Mahlon	32	Pvt.	PA 99th Inf.	B	2/25/65	7/1/65		
9	Penrose, William	22	Pvt.	PA 21st Cav.	E	7/8/64	2/15/65		
10	Pensyl, Philip H	23	Surg.	PA 56th Inf.		10/12/64	7/1/65		
11	Perdew, Aaron D	27		MD 2nd PHB Inf.	B	8/30/61	9/28/64		
12	Perdew, Moses	33	Pvt.	PA 206th Inf.	K	9/6/64			
13	Perdew, Nathan	22		PA 16th Cav.	G	9/6/62	12/1/64	Wound.-Died	2nd Deep Bottom
14	Perrin, John	35	Pvt.	PA 133rd Inf.	C	8/13/62		KIA	Fredericksburg
15	Perrin, Jonathan	37	Pvt.	PA 93rd Inf.	A	11/14/64	5/12/65	Wounded	Ft. Fisher
16	Perry, Wythe	51	Pvt.	USCT 3rd Inf.	K	7/23/63	10/31/65		
17	Peterman, Jacob D*	25	Pvt.	PA 76th Inf.	E	8/27/63		Wound.-Died	Spotsylvania
18	Peterson, William A	20	Pvt.	PA 84th Inf.	A	9/6/61	11/15/65	Wound.-POW	Chancellorsville
19	Pfarr, John*		Pvt.	PA 76th Inf.	D	10/16/61	11/28/64		
20	Pfeifier, John M	31	Pvt.	PA 101st Inf.	G	12/28/61	2/26/62		
21	Pfile, William*		Pvt.	PA 55th Inf.	H	9/26/62	6/11/65		
22	Phatic, William								
23	Phillipi, Franklin		1st Lt.	PA 93rd Inf.	E	10/26/61	6/13/65	Wounded	Ft. Fisher
24	Phillips, Daniel W	26	Pvt.	PA 55th Inf.	D	10/12/61	11/20/61		
25	Phillips, Daniel W		Corp.	PA 184th Inf.	A	5/12/64	7/14/65		
26	Phillips, Scott	18	Pvt.	PA 55th Inf.	D	2/27/64	8/30/65		
27	Phillips, William	17	Pvt.	PA 127th Inf.	H	8/12/62	5/29/63		
28	Phillips, William		Pvt.	PA 83rd Inf.	I	2/8/65	6/28/65		
29	Pierrant, Joseph	38	Pvt.	PA 76th Inf.	E	2/16/65	7/18/65		
30	Piles, John	28	Pvt.	PA 28th Inf.	G	1/3/64	7/20/65		
31	Pilkington, James H	17	Pvt.	PA 13th Inf.	G	4/25/61	8/6/61		
32	Pilkington, James H		1st Lt.	PA 133rd Inf.	K	8/15/62	5/26/63		
33	Pilkington, Richard P	20	Pvt.	PA 13th Inf.	G	4/25/61	8/6/61		
34	Pilkington, Richard P		Capt.	PA 76th Inf.	E	10/9/61	7/18/65	Wounded	2nd Deep Bottom
35	Piper, Lewis M	30	Pvt.	PA 8th Res.	F	9/4/61	5/11/63	POW	Seven Days Battle
36	Piper, Lewis M		Pvt.	PA 8th Res.	F			Wounded	Fredericksburg
37	Piper, Luther R		Corp.	PA 8th Res.	F	6/11/61		POW	Seven Days Battle
38	Piper, Luther R	26	Corp.	PA 8th Res.	F			Wound.-Died	Fredericksburg
39	Piper, Thomas	36	Pvt.	PA 50th Inf.	I	3/24/65	5/8/65		
40	Piper, Thompson F	31	Pvt.	PA 84th Inf.	A	10/24/61		Wounded	
41	Pisel, William	20	Pvt.	PA 148th Inf.	B	8/31/63	6/1/65		
42	Pittman, Daniel H	23	Pvt.	PA 101st Inf.		2/8/65	5/3/65		
43	Pittman, John	23	Pvt.	PA 101st Inf.	D	2/8/62	6/21/62		
44	Pittman, William	18	Pvt.	PA McKeage's Mil.	D	7/2/63	8/8/63		
45	Plank, David A	24	Pvt.	PA 7th Res.	H	6/18/61	9/21/61	Wounded	
46	Plants, George*		Pvt.	PA 76th Inf.	E	9/28/64	11/23/64		
47	Plaster, William H*	18	Pvt.	PA 110th Inf.	C	12/19/61	6/28/65		
48	Pleacher, Andrew	34	Pvt.	PA 171st Inf.	I	11/10/62	8/8/63	Wounded	
49	Pleacher, Andrew		Pvt.	PA 55th Inf.	K	3/2/64	6/13/65	POW	
50	Plessinger, Abraham	37	Pvt.	PA 158th Inf.	H	11/4/62	12/7/62		
51	Plowden, Jacob		Pvt.	USCT 3rd Inf.	E	7/6/63	10/31/65		
52	Plummer, John T	21	Pvt.	MD 3rd PHB Inf.	D	3/21/62	5/29/65	POW	Moorefield
53	Plummer, John W*		Sgt.	PA 110th Inf.	C	10/24/61	6/28/65		
54	Points, Joshua	18	Pvt.	PA McKeage's Mil.	H	7/2/63	8/8/63		
55	Points, Joshua		Pvt.	PA 206th Inf.	K	9/7/64			
56	Poleman, William H			US 19th Inf.	G			POW	
57	Polhamus, Theo*		Pvt.	PA 76th Inf.	E	1/30/65	7/18/65		
58	Polta, Augustus	39	Pvt.	PA 55th Inf.	I	9/23/63	2/28/65		
59	Poole, Bushrod V	23	Pvt.	MD 3rd Reg Cav.	I	8/27/63	9/7/65		
60	Poorman, Franklin H	34	Pvt.	PA 13th Inf.	G	4/25/61	8/6/61		
61	Poorman, Franklin H		Pvt.	PA 99th Inf.	G	2/25/65	7/1/65		
62	Port, George A		Pvt.	PA McKeage's Mil.	H	7/2/63	8/8/63		
63	Porter, Andrew J	23	2nd Lt.	PA 55th Inf.	H	12/4/61	10/11/64		
64	Porter, Philip	17	Pvt.	PA 138th Inf.	D	8/29/62			

Detailed Alphabetical Listing

	Cemetery	Notes
1		B.S.F. History Book-1884; Historical Data Systems record
2		B.S.F. History Book-1884; Historical Data Systems record
3	Everett Cem.	B.S.F. Hist. Book-1884; 1860 Hopewell Cen.; Disch. Surg. Cert.; Father of John Jr. & Henry
4		B.S.F. History Book-1884; PA Civil War Archives-Bedford residence
5	Fishertown Cem.	Historical Data Systems-Bedford Co. residence
6		1890 E. St. Clair Twp. Veterans Census
7		Bedford Inquirer 5/22/1908 listing; Mahlon & William-brothers
8	St. Paul's Cem.-Cessna	Frank McCoy 1912 Listing; 1860 St. Clair Twp. Census
9	Hyndman Cem.	1890 Hyndman Veterans Census
10	Everett Cem.	1890 Monroe Twp. Veterans Census
11	Prosperity Cem.-Clearville	1850 Southampton Twp. Census
12	Mt. Hope Cem.-Hewitt	1860 Southampton Twp. Census; Moses & Aaron-brothers
13	Poplar Grove Nat'l Cem.	Wounded 8/16/64; Died 12/1/64; 1860 Southampton Twp. Cen.; Hist. Data Systems record
14	Everett Cem.	KIA 12/13/62; B.S.F. History Book-1884; Historical Data Systems-Bedford Co. residence
15	Everett Cem.	Wounded 3/25/65; Twin Brother with John; 1860 Monroe Twp. Census
16	Mt. Ross Cem.-Bedford	1860 Bedford Twp. Census; Frank McCoy 1912 Listing; Historical Data Systems record
17		Wounded 5/7/64, Died 9/16/64; Armstrong Co. residence
18	Pleasantville Cem.	POW 5/3/63; 1890 W. St. Clair Twp. Veterans Census; suffered 4 gunshot Wounds
19		B.S.F. History Book-1884; Hanover references
20	Dry Ridge Cem.-Manns Choice	B.S.F. History Book-1884; Discharged on Surgeon's Cert.
21		B.S.F. History Book-1884; enlisted in Berks Co.
22	Schellsburg Cem.	Bedford Inquirer 5/22/1908 listing
23		Wounded 3/25/65; B.S.F. History Book-1884
24		Historical Data Systems-Bedford Co. residence; 1860 Bedford Twp. Census; Disch. Surg. Cert.
25	Riverside Cem.-Denver	B.S.F. History Book-1884; Died on 10/11/67; 1860 Bedford Twp. Census
26		B.S.F. History Book-1884; 1860 Bedford Twp. Census
27	Hopewell Cem.	1890 Broad Top Veterans Census; Regiment took part in fighting in town at Fredricksburg
28		1890 Broad Top Veterans Census
29		B.S.F. History Book-1884
30	Bedford Cem.	1890 Bedford Twp. Veterans Census; Deserted after war
31	Evergreen Cem.-IA	B.S.F. History Book-1884
32		B.S.F. History Book-1884; 1850 W. Providence Twp. Census
33	Highmore Cem.-SD	B.S.F. History Book-1884; Richard & James-brothers
34		Wounded 8/16/64 listed in Bedford Inquirer 9/2/64 article; Hist.l Data Systems-Bedford resid.
35	Everett Cem.	POW-Libby & Belle Island 6/27/62 to 8/1/62; B.S.F. History Book-1884
36		Wounded 12/13/62; 1890 W. Providence Twp. Veterans Census
37		POW 6/27/62; Bedford Gazette listing on 8th Regiment; Historical Data Systems record
38	Piper Cem.-Cypher	Leg. Amputated 12/13/63; Died 1/1/63; B.S.F. History Book-1884
39	Old Carrollton Cem.-IA	1860 Hopewell Twp. Census; Historical Data Systems record
40	Bedford Cem.	1890 Bedford Veterans Census-Wound caused blindness
41	Lybarger Cem.-Madley	1890 Londonderry Twp. Veterans Census; PA 53rd references
42	Rock Hill Cem.-Monroe Twp.	Historical Data Systems record; Daniel & John-brothers
43	Rock Hill Cem.-Monroe Twp.	Died in 1866; B.S.F. History Book-1884; 1850 W. Providence Twp. Cen.; Disch. Surg. Cert.
44	Rock Hill Cem.-Monroe Twp.	B.S.F. History Book-1884; 1850 W. Providence Twp. Census
45	Trinity UCC Cem.-Osterburg	1890 E. St. Clair Twp Vet. Cen.; PA CW Project-accidental wound 7/25/61; Doctor after war
46		B.S.F. History Book-1884; Ancestry.com-Bradford references
47		B.S.F. History Book-1884; PA Civil War Archives-Newry residence
48	Mt. Olivet Cem.-Manns Choice	1890 Napier Twp. Veterans Census-Face Wound; B.S.F. History Book-1884
49		POW 6 months; 1890 Napier Twp. Veterans Census; B.S.F. History Book-1884
50	Cedar Grove Cem.-Fulton Co.	1890 Brush Creek Twp. Veterans Census; Discharged on Surgeon's Cert.
51		Frank McCoy 1912 list & Civil Works Admin. 1934 list; Historical Data Systems record
52	Old Coney Cem.-MD	1890 Londonderry Twp. Veterans Census
53		B.S.F. History Book-1884; Tyrone residence
54	Bedford Cem.	B.S.F. History Book-1884
55		1890 Bedford Twp. Veterans Census
56	Everett Cem.	Regiment listed on gravestone; Historical Data Systems record
57		B.S.F. History Book-1884; Scranton references
58		B.S.F. History Book-1884; PA Civil War Archives record
59	Hyndman Cem.	1890 Hyndman Veterans Census; also Hospital Steward
60	New Alex. Union-West. Co	Historical Data Systems-Bedford Co. residence; B.S.F. History Book-1884
61		PA Veterans Burial Card record
62		B.S.F. History Book-1884
63	Fairmont Cem.-NE	PA Civil War Archives-Bedford Co. residence; Enlisted in Schellsburg
64		Deserted 2/9/63; B.S.F. History Book-1884; 1860 Londonderry Twp. Census

	Name	Age	Rank	Regiment	Co.	Muster In	Muster Out	Casualty	Casualty Battle
1	Porter, Philip		Pvt.	IN 1st Heavy Art.	B	12/31/63	1/13/66		
2	Porter, Willam H	15	Pvt.	PA 76th Inf.	E	2/22/65	7/18/65		
3	Pote, Andrew B	28	Pvt.	PA 107th Inf.	E	9/2/64	6/6/65		
4	Pote, Jacob B	30	Pvt.	PA 55th Inf.	I	2/6/64	8/30/65		
5	Pote, Jacob C	41	1st Lt.	OH 131st Inf.	K	5/14/64	8/25/64		
6	Pote, Michael B	41	Pvt.	PA 58th Inf.	G	9/21/64	7/12/65		
7	Potter, David R	22	Corp.	IN 47th Inf.	F	10/16/61	12/6/65		
8	Potter, James	18	Pvt.	PA 184th Inf.	A	5/12/64		Wounded	Cold Harbor
9	Potter, John (1)	49	Pvt.	PA 101st Inf.	D	11/1/61	2/6/62		
10	Potter, John (1)		Pvt.	PA 22nd Cav.	M	2/15/64	6/24/65		
11	Potter, John (2)		Pvt.	PA 133rd Inf.	C	8/13/62	5/26/63		
12	Potter, John W	28	Pvt.	MD 3rd Inf.	B	1/24/62	1/24/65		
13	Potter, Levi (1)	23	Pvt.	PA 56th Inf.	A	1/23/62	7/18/65	Wounded	Gettysburg
14	Potter, Levi (2)*	19	Pvt.	OH 26th Inf.	C	6/15/61		KIA	Kennesaw Mtn.
15	Potter, Martin L	18	Pvt.	PA 101st Inf.	D	11/1/61	6/25/65	POW	Plymonth, NC
16	Potts, Jacob	27	Pvt.	PA 79th Inf.	D	8/26/63	7/12/65		
17	Potts, John A	29	Pvt.	PA 171st Inf.	I	11/10/62	8/8/63		
18	Potts, John A		Pvt.	PA 99th Inf.	I	2/25/65	7/1/65		
19	Potts, Theodore B	21	Pvt.	PA 2nd Light Art.	E	7/23/63	1/23/64		
20	Potts, Theodore B		Pvt.	PA 6th Heavy Art.	B	9/8/64	6/13/65		
21	Powley, Henry*		Pvt.	PA 110th Inf.	C	10/24/61	6/28/65	Wounded	Gettysburg
22	Pressel, David	20	Pvt.	IN 75th Inf.	E	8/1/62	6/8/65		
23	Pressel, Frederick		Pvt.	PA 2nd Heavy Art.	G	11/29/61	11/28/64		
24	Pressel, Jacob W	27	Pvt.	PA 12th Cav.	H	2/24/62	7/20/65		
25	Presser, Philip	27	Pvt.	PA 55th Inf.	K	10/14/63	8/30/65		
26	Price, Abraham	20	Pvt.	PA 138th Inf.	E	8/29/62		Died	
27	Price, Daniel J	21	Pvt.	PA 138th Inf.	E	8/29/62	5/12/64	Wounded	Spotsylvania
28	Price, Daniel J		Pvt.	PA 22nd VRC	B	5/12/64	7/3/65		
29	Price, Daniel M	21	Pvt.	PA 133rd Inf.	C	8/13/62	5/26/63	Wounded	
30	Price, Daniel M		Pvt.	PA 205th Inf.	C	8/27/64	6/2/65	Wounded	Petersburg - Final
31	Price, David J	19	Corp.	PA 110th Inf.	C	10/24/61	10/24/64	Wounded	1st Kernstown
32	Price, George W	16	Pvt.	PA 99th Inf.	D	3/4/65	7/1/65		
33	Price, Jacob F (1)	17	Pvt.	PA 138th Inf.	E	8/29/62	6/23/65		
34	Price, Jacob F (2)	17	Pvt.	PA 56th Inf.	F	9/30/64	5/31/65		
35	Price, John	33	Pvt.	PA 184th Inf.	A	5/12/64		Died	
36	Price, John B	40	Pvt.	PA 22nd Cav.	M	9/26/64	5/24/65		
37	Price, Joseph J	20	Corp.	PA 138th Inf.	D	8/29/62		KIA	Wilderness
38	Price, Michael H	43	Corp.	PA 184th Inf.	A	5/12/64	7/14/65	Wounded	Cold Harbor
39	Price, Richard	19	Pvt.	PA 56th Inf.	F	9/30/64	5/31/65		
40	Price, William	21	Pvt.	PA 3rd Inf.	K	4/20/61	7/29/61		
41	Price, William		Pvt.	PA 54th Inf.	A	2/1/62	7/15/65		
42	Prideaux, Thomas A	22	1st Lt.	PA 138th Inf.	E	9/6/62	6/23/65		
43	Prilles, Joseph	29	Pvt.	PA 76th Inf.	E	2/23/65	7/18/65		
44	Prince, Edwin S	16	Pvt.	OH 162nd Inf.	D	5/20/64	9/4/64		
45	Prince, Edwin S		Pvt.	OH 184th Inf.	F	2/19/65	9/20/65		
46	Probst, George C	34	Chap.						
47	Prosser, Alexander	21	Pvt.	PA 2nd Inf.	B	4/20/61	7/24/61		
48	Prosser, Alexander		1st Lt.	PA 168th Inf.	F	11/20/62	7/23/63		
49	Prosser, David W	15	Corp.	PA 55th Inf.	D	10/12/61	6/15/65	POW	Drewry's Bluff
50	Protheroe, David	42	Pvt.	PA 133rd Inf.	C	8/29/62	5/26/63		
51	Putt, David	29	Pvt.	PA 57th Inf.	E	7/5/64	6/29/65	Wounded	
52	Quarry, Alfred*		Pvt.	PA 191st Inf.	A	2/12/64		POW-Died	Globe Tavern
53	Quarry, Levi	26	Pvt.	PA 2nd Res.	F	12/29/63	6/6/64		
54	Quarry, Levi		Pvt.	PA 191st Inf.	A	6/6/64	6/29/64		
55	Quarry, William C	19	Pvt.	PA 205th Inf.	C	8/27/64	6/2/65		
56	Querry, Matthias	26	Pvt.	PA 53rd Inf.	C	2/18/62	6/30/65	Wounded	Gettysburg
57	Quimby, Henry E*	28	Capt.	PA 194th Inf.	I	7/21/64	11/5/64		
58	Rabe, Frederick W	31	Pvt.	PA 13th Inf.	G	4/25/61	8/6/61		
59	Radcliff, James	18	Pvt.	PA 138th Inf.	D	8/29/62		Died	
60	Radebaugh, Daniel W		Music.	PA McKeage's Mil.	H	7/2/63	8/8/63		
61	Radebaugh, Daniel W	19	Music.	PA 55th Inf.	K	2/19/64	8/30/65		
62	Radebaugh, Jacob L	19	Sgt.	PA 55th Inf.	K	11/5/61	8/30/65	MIA	Drewry's Bluff
63	Rahn, Edwin L	39	Pvt.	PA 55th Inf.	K	9/16/63	8/30/65		
64	Rahn, Jacob M*		Pvt.	PA 194th Inf.	I	7/22/64	11/5/64		

	Cemetery	Notes
1		Historical Data Systems record
2	Moses Cem.-IN	B.S.F. History Book-1884; William & Philip-brothers
3	Holsinger Cem.-Bakers Summit	1890 Bloomfield Twp. Veterans Census; Historical Data Systems record
4	Holsinger Cem.-Bakers Summit	1860 Woodbury Twp. Census; B.S.F. History Book-1884
5	Parish Cem.-OH	Ancestry.com-born in Woodbury; Historical Data Systems record
6	Holsinger Cem.-Bakers Summit	Historical Data Systems record; Michael, Jacob & Andrew-brothers
7	Trinity UCC Cem.-Osterburg	1890 Claysburg Veterans Census; 1850 M. Woodbury Twp. Census
8	Woodland Union Cem.-OH	Wounded 6/3/64; B.S.F. History Book-1884; 1850 M. Woodbury Twp. Census
9		Civil War Archives-Bedford Co. residence, enlisted in Rainsburg; Discharged on Surg. Cert.
10	Potter Creek Cem.-Woodbury	Bedford Inquirer 5/22/1908 listing; Frank McCoy 1912 Listing; Father of David & Martin
11		B.S.F. History Book-1884; Historical Data Systems-Bedford Co. residence
12	Fairview Cem.-Artemas	Historical Data Systems record; John W & Levi (2)-brothers
13	Buck Valley Cem.-Fulton Co.	Wounded 7/1/63; Left Leg Amputated; Hist. Data Systems record; Civil War Headstone record
14	Marietta Nat'l Cem.-GA	KIA 6/27/64; 1860 Woodbury Twp. Cen; Levi(1) & James-bros; unsure if different Levi Potter
15	Elmwood Cem.-MO	POW 4/20/64 to 2/27/65; Historical Data Systems record; Martin L & David-brothers
16	Fairview Cem.-Artemas	1890 Mann Twp. Veterans Census
17		1890 Mann Twp. Veterans Census; 1860 Southampton Census
18	Fairview Cem.-Artemas	Historical Data Systems-Bedford Co. residence
19	Pleasantville Cem.	1890 W. St. Clair Twp. Veterans Census
20		B.S.F. History Book-1884
21		B.S.F. History Book-1884; Blair Co. references
22	Pilgrims Rest Cem.-IN	Historical Data Systems record; Ancestry.com born in Bedford
23		Historical Data Systems record; Frederick & Jacob-brothers
24	Barley Luth.-Bakers Summit	PA Civil War Archives-Claysburg residence; 1870 S. Woodbury Twp. Census
25		B.S.F. History Book-1884
26		Died 10/19/62; PA Civil War Archives-Bedford residence; 1860 Bedford Twp. Census
27	Pleasant Hill Cem.-Imlertown	Wounded 5/12/64; Frank McCoy 1912 Listing; B.S.F. History Book-1884
28		1850 Bedford Twp. Census; 1890 Bedford Twp. Veterans Census
29	Marion Nat'l Cem.-IN	B.S.F. History Book-1884; Hospital 2/14/62 to 2/10/63
30		Wounded 4/2/65; Historical Data Systems record
31	Bedford Cem.	Wounded 3/23/62; B.S.F. History Book-1884; David & Daniel M-brothers
32	Fairview Cem.-Altoona	1860 Woodbury Twp. Census; Historical Data Systems record
33	Arlington Cem.-NE	B.S.F. History Book-1884; Historical Data Systems record
34	Ridgewood Cem.-IA	1860 Bedford Twp. Census; Historical Data Systems record
35	Vaughn Williams (Kegg)-Mench	Died 3/22/65; B.S.F. History Book-1884; Frank McCoy 1912 Listing
36	Greenlawn Cem.-Roaring Spring	1860 S. Woodbury Twp. Census; Veteran Burial record
37	Fredericksburg Nat'l Cem.	KIA 5/6/64; PA Civil War Archives-Bedford Co. residence; 1860 Bedford Twp. Census
38		Wounded-Bedford Inquirer 7/1/64 article; B.S.F. History Book-1884; Michael & John-brothers
39	Bedford Cem.	1890 Bedford Twp. Veterans Census; Jacob F & Richard-brothers
40	Grandview Cem.-Cambria Co.	PA Civil War Project record; William & John B (2); brothers
41		PA Veterans Burial record
42	Lloyd Cem.-Ebensburg	B.S.F. History Book-1884; PA Civil War Archives-Schellsburg residence
43		B.S.F. History Book-1884; Substitute
44	Everett Cem.	1890 Everett Veterans Census aka Edward
45		Historical Data Systems record
46	Mt. Pleasant Cem.-Mattie	1870 E. Providence Twp. Census; Civil War Marker
47	Duval Cem.-Coaldale	1890 Coaldale Veterans Census
48		1890 Coaldale Veterans Census
49	Bedford Cem.	POW-Andersonville 5/16/64 to 4/28/65; attended 1905 Andersonville Dedication
50	Steele Cem.-Yellow Creek	1890 Hopewell Twp. Veterans Census aka Prother & Protherow
51	Grandview Cem.-Liberty Twp.	1890 Hopewell Twp. Veterans Census; Frank McCoy 1912 Listing
52	Salisbury Nat'l Cem.	POW 8/19/64; Died 1/5/65; Historical Data Systems record; Huntingdon Co. references
53	Fockler Cem.-Saxton	Historical Data Systems record; Levi & William-bros.
54		Historical Data Systems record
55	Dry Hill Cem.-Woodbury	1890 Woodbury Veterans Census
56	Entriken Cem.-Hunt. Co.	1890 Hopewell Veterans Census
57		B.S.F. History Book-1884; Berks Co. references
58		B.S.F. History Book-1884; PA Civil War Archives-enlisted in Bedford
59	Cypress Hill Cem.-NY	Died 7/26/64; History PA Civil War Archives-Bedford residence, enlisted in Schellsburg
60		B.S.F. History Book-1884; aka David; Daniel & Jacob-brothers
61	Laurel Hill Cem.-Phila.	B.S.F. History Book-1884; PA Civil War Archives-Bedford Co.residence
62	German Reformed-Bedford	MIA 5/15/64 listed for a Benjamin Radebaugh in Bed. Inquirer-7/8/64; Frank McCoy 1912 List
63		B.S.F. History Book-1884
64		B.S.F. History Book-1884; Berks Co. references

	Name	Age	Rank	Regiment	Co.	Muster In	Muster Out	Casualty	Casualty Battle
1	Raley, Daniel	19	Corp.	PA 133rd Inf.	D	8/1/62	5/26/63		
2	Raley, Daniel		Pvt.	PA 99th Inf.	K	3/16/65			
3	Raley, Vincent	19	Pvt.	PA 2nd Cav.	E	11/4/61			
4	Raley, Vincent		Pvt.	PA McKeage's Mil.	G	7/2/63	8/8/63		
5	Ralston, David E	23	Pvt.	PA 110th Inf.	C	12/19/61		KIA	Chancellorsville
6	Ralston, William H	24	Sgt.	PA 110th Inf.	C	10/24/61	2/11/63		
7	Ralston, William H		1st Sgt.	PA 184th Inf.	A	5/12/64	7/14/65		
8	Ramage, Thomas R	21	Pvt.	PA 76th Inf.	C	10/17/61	11/28/64		
9	Ramsey, Alexander	31	Pvt.	PA 133rd Inf.	C	8/13/62	2/9/63		
10	Ramsey, Eli B	29	Pvt.	PA 13th Inf.	G	4/25/61	8/6/61		
11	Ramsey, Eli B		Pvt.	PA 186th Inf.	E	2/24/64	8/15/65		
12	Ramsey, Jeremiah	33	Pvt.	PA 186th Inf.	E	2/24/64	8/15/65		
13	Ramsey, John	21	Pvt.	PA 22nd Cav.	H	2/29/64	6/24/65		
14	Ramsey, Oliver C	27	Corp.	PA 13th Inf.	G	4/25/61	8/6/61		
15	Ramsey, Oliver C		Sgt.	PA 208th Inf.	K	9/7/64	6/1/65		
16	Ramsey, Wesley A		Pvt.	PA 22nd Cav.	H	2/26/64	6/24/65		
17	Ramsey, William W	36	Pvt.	PA 138th Inf.	D	8/29/62	8/1/63		
18	Ramsey, William W		Pvt.	PA 208th Inf.	K	9/7/64	6/1/65		
19	Randolph, John F*	24	Pvt.	PA 76th Inf.	E	9/23/64	7/1/65		
20	Ranker, Conrad	38	Pvt.	PA 29th Inf.	A	12/23/64	7/6/65	Wounded	
21	Rankin, Frank R	29	1st Sgt.	PA 21st Cav.	E	7/2/63	7/8/65		
22	Rankin, John		Pvt.	WV 4th Inf.	K				
23	Rankin, John	35	Pvt.	PA 50th Inf.	H	9/21/64	12/29/64		
24	Ray, William H		Pvt.	PA 76th Inf.	E	8/29/62	9/7/63		
25	Rea, William H	22	Pvt.	PA 138th Inf.	E	2/1/62	9/7/63	Died	
26	Ream, Isaac	23	Pvt.	PA 55th Inf.	H	10/11/61	8/30/65	POW	Edisto Island, SC
27	Ream, Isaac		Pvt.	PA 55th Inf.	H			Wounded	Petersburg - Initial
28	Reamer, Francis C	40	Surg.	PA 143rd Inf.		9/16/62	2/3/65		
29	Redinger, August	27	Corp.	MD 1st PHB Cav.	D	9/18/61	6/28/65		
30	Redinger, Peter	22	Pvt.	PA 99th Inf.	F	2/25/65	7/11/65		
31	Reed, Alexander (1)	23	Pvt.	PA 149th Inf.	I	8/26/63	6/24/65	Wounded	
32	Reed, Alexander (2)		Pvt.	PA 208th Inf.	H	9/2/64	6/1/65		
33	Reed, Andrew J	17	Pvt.	PA McKeage's Mil.	G	7/2/63	8/8/63		
34	Reed, Andrew J		Pvt.	PA 55th Inf.	D	2/27/64		Died	
35	Reed, Hezekiah C*	32	Pvt.	PA 55th Inf.	I	10/15/61	7/26/62		
36	Reed, Isaac	23	Pvt.	PA 143rd Inf.	E	6/15/63		Wound.-Died	Spotsylvania
37	Reed, James W		Pvt.	PA 6th Heavy Art.	M				
38	Reed, Lewis S (1)	22	Pvt.	PA 84th Inf.	C				
39	Reed, Lewis S (2)		Corp.	PA 188th Inf.	G	2/20/64	4/1/64		
40	Reed, Lewis S (2)		Corp.	PA 3rd Art.		4/1/64	12/14/65		
41	Reed, Louis	22	Pvt.	USCT 12th Inf.	D				
42	Reed, Louis		Pvt.	NJ 12th Inf.	C	3/21/64	7/15/65		
43	Reed, Samuel*	26						Died	
44	Reed, Thomas	22	Pvt.	PA 133rd Inf.	C	8/13/62	5/26/63		
45	Reed, William B			PA 8th Res.					
46	Reed, William B		Pvt.	PA 194th Inf.	I	7/22/64	11/5/64		
47	Reed, William M								
48	Reese, George L		Pvt.	PA 55th Inf.	K	11/11/61			
49	Refley, William	18	Pvt.	PA 133rd Inf.	K	8/18/62			
50	Reigel, Benjamin*		Pvt.	PA 76th Inf.	E	10/16/64	7/18/65		
51	Reighard, George W	17	Pvt.	PA 184th Inf.	A	5/12/64	5/15/65	Wounded	Cold Harbor
52	Reighard, Matthias	22	Pvt.	PA 138th Inf.	E	8/29/62	6/23/65		
53	Reighard, Peter	33	Pvt.	PA 138th Inf.	F	9/12/62			
54	Reiley, Charles*								
55	Reilley, George W		Pvt.	PA 76th Inf.	E	2/17/65	7/18/65		
56	Reily, John M	24	Pvt.	PA 158th Inf.	H	11/4/62	8/12/63		
57	Reily, Michael		Pvt.	PA 55th Inf.	K	10/16/63	4/1/65		
58	Reimund, George*								
59	Renninger, Frederick M	17	Pvt.	PA 84th Inf.	A	10/24/61		Died	
60	Renninger, Josiah	24	Pvt.	PA 100th Inf.	M	11/28/64	7/24/65		
61	Replogle, Simon L	15	Pvt.	PA 194th Inf.	I	7/22/64	11/5/64		
62	Ressler, Abraham	31	Pvt.	PA 101st Inf.	D	2/13/62		Died	
63	Ressler, Harvey M	28	Sgt.	PA 171st Inf.	I	11/2/62	8/8/63		
64	Ressler, William	37	Pvt.	PA 55th Inf.	D	2/27/64	8/30/65		

	Cemetery	Notes
1	Hyndman Cem.	1800 Londonderry Twp. Veterans Census; 1860 Londonderry Twp. Census
2		B.S.F. History Book-1884; Deserted after war
3	Hyndman Cem.	B.S.F. History Book-1884; PA Civil War Project record
4		B.S.F. History Book-1884; Vincent & Daniel-brothers
5	Schuyler Cem.-NE (marker)	KIA 5/3/63; B.S.F. Hist. Book-1884; CWr Pension Record; 1850 M. Woodbury Twp. Cen.
6	Schuyler Cem.-NE	B.S.F. History Book-1884; 1850 S. Woodbury Twp. Census; Discharged on Surgeon's Cert.
7		B.S.F. History Book-1884; William & David-brothers
8	Yellow Creek Reformed Cem.	1890 Hopewell Veterans Census; Frank McCoy 1912 Listing
9	Woodlawn Cem.-MO	1850 W. Providence Twp. Census; Disch. Surg.s Cert.; Alexander, William & Jeremiah-bros.
10		1890 E. Providence Twp. Veterans Census
11		B.S.F. History Book-1884; Historical Data Systems record
12	Bedford Cem.	1890 Bedford Twp. Veterans Census
13	Ramsey Cem.-Ft. Littleton	B.S.F. History Book-1884
14	Mt. Zion Cem.-Breezewood	B.S.F. History Book-1884; 1860 E. Providence Twp. Census
15		B.S.F. History Book-1884
16		B.S.F. History Book-1884; 1870 Monroe Twp. Census
17		B.S.F. History Book-1884; 1860 Bloody Run Census; Discharged on Surgeon's Cert.
18		B.S.F. History Book-1884
19		B.S.F. History Book-1884; 1890 Scranton Veterans Census
20	Hyndman Cem.	1890 Londonderry Twp. Veterans Cenus
21	Buckstown Cem.-Somerset Co.	PA Civil War Archives-Bedford Co. residence; Historical Data Systems record
22		1890 Broad Top Veterans Census
23	Duval Cem.-Coaldale	Frank McCoy 1912 Listing; Discharged on Surgeon's Cert.
24	Shaffer Graveyard	1890 Hyndman Veterans Census; Discharged on Surgeon's Cert.
25	Bedford Cem.	Died 9/23/63; B.S.F. History Book-1884; PA Civil War Archives-Bedford residence
26	High Ridge Cem.-MO	POW 3/29/62; Historical Data Systems-Bedford Co. residence
27		Wounded 6/16/64 listed in Bedford Inquirer 7/8/64 article; Historical Data Systems record
28	Bedford Cem.	B.S.F. History Book-1884; Achieved rank of Major
29	Everett Cem.	1890 Everett Veterans Census
30	Everett Cem.	Frank McCoy 1912 Listing
31	Hopewell Cem.	1890 Liberty Twp. Veterans Census-Shot in ankle
32		B.S.F. History Book-1884; Historical Data Systems-Bedford Co. residence
33		1860 Colerain Twp. Census; B.S.F. History Book-1884
34		Died 11/17/64; B.S.F. Hist. Book-1884; 1860 Colerain Twp. Cen.; PA CW Archives record
35		B.S.F. History Book-1884; 1890 Westmoreland Co Veterans Census; Discharged Surg. Cert.
36	Fredericksburg Nat'l Cem.	Wounded 5/10/64; Died 6/1/64 (estimated), CW Pension Record; 1840 Broad Top Twp. Cen.
37	Oak Ridge Cem.-Altoona	1860 Colerain Twp. Census
38	Odd Fellows-Clearfield Co.	PA Civil War Archives-Saxton residence; Lewis (1), James & Isaac-brothers
39		1890 Liberty Twp. Veterans Census
40		PA Civil War Archives record
41	Mt. Ross Cem.-Bedford	Bedford Inquirer 5/22/1908 listing; Regiment on gravestone
42		Frank McCoy 1912 Listing
43	Woods Church-Colerain Twp.	Died 10/24/61; Civil War Grave Marker
44	Fockler Cem.-Saxton	B.S.F. History Book-1884; Historical Data Systems-Bedford Co. residence
45	Yellow Creek Reformed Cem.	Frank McCoy 1912 Listing
46		1890 Hopewell Veterans Census
47		1890 Liberty Twp. Veterans Census
48		PA Civil War Archives - Deserted 11/21/61
49		Deserted 12/21/62; B.S.F. History Book-1884
50		B.S.F. History Book-1884; Substitute; Berks Co. references
51	Maple Grove Cem.-KS	Wounded-Bedford Inquirer 7/1/64 article; B.S.F. History Book-1884; 1860 Bedford Twp. Cen.
52		PA Civil War Archives-Bedford residence; George & Matthias-brothers
53		Deserted 2/18/63; B.S.F. History Book-1884; Historical Data Systems record
54	Mt. Olivet Cem.-Manns Choice	Civil War marker at gravesite
55		B.S.F. History Book-1884; Substitute
56		Fulton Co. Historical Society book on 158th Infantry; 1850 E. Providence Twp. Census
57		B.S.F. History Book-1884; Historical Data Systems
58	Bedford Cem.	Civil War marker at gravesite
59	Antietam Nat'l Cem., 4184	Died 3/4/62; PA Civil War Archives-Bedford Co. residence; 1860 St. Clair Twp. Census
60	Mock Church Cem.-Alum Bank	1890 W. St. Clair Twp. Veterans Census
61	Fairview Cem.-Altoona	B.S.F. History Book-1884; Historical Data Systems record
62	Yorktown Nat'l Cem.	Died 6/6/62; PA Civil War Archives-Bedford Co. residence; 1860 Colerain Twp. Census
63	Union Cem.-Centerville	1890 Cumberland Valley Veterans Census
64	Woods Church-Colerain Twp.	Frank McCoy 1912 Listing; B.S.F. History Book-1884

	Name	Age	Rank	Regiment	Co.	Muster In	Muster Out	Casualty	Casualty Battle
1	Rhoades, Peter		Pvt.	PA 21st Cav.	E	7/2/63	2/20/64		
2	Rhodes, George		Pvt.	PA McKeage's Mil.	H	7/2/63	8/8/63		
3	Rhodes, George	19	Pvt.	PA 184th Inf.	A	5/12/64	7/11/65	POW	Jerusalem Plank Rd.
4	Rhodes, George M	25	Pvt.	PA 11th Res.	C	6/10/61	8/23/62		
5	Rhodes, Joseph*	46	Pvt.	PA 184th Inf.	A	9/3/64	6/2/65		
6	Rice, Abraham	27	Sgt.	PA 101st Inf.	D	11/1/61	6/22/65	POW	Plymonth, NC
7	Rice, Amos H		Pvt.	PA McKeage's Mil.	H	7/2/63	8/8/63		
8	Rice, Christian		Pvt.	PA 93rd Inf.	C	11/14/61	6/17/65		
9	Rice, Cornelius	28	Pvt.	PA 78th Inf.	K	2/28/65	9/11/65		
10	Rice, Daniel			Conf.					
11	Rice, Fred			Conf.					
12	Rice, George S	32	Pvt.	US 56th Inf.	G	11/28/64	7/1/65		
13	Rice, Henry*	26	Capt.	PA 76th Inf.	E	10/9/61	4/14/64		
14	Rice, Isaac	27	Corp.	PA 101st Inf.	D	11/1/61		POW-Died	Plymonth, NC
15	Rice, Isaiah			PA 50th Inf.					
16	Rice, John	40	Pvt.	PA 82nd Inf.	G	8/24/61			
17	Rice, John H	30	Pvt.	PA 93rd Inf.	C	11/11/64	7/17/65	Wounded	
18	Rice, Jonathan	37	Pvt.	PA 99th Inf.	K	2/24/65	5/29/65		
19	Rice, Samuel		Pvt.	MD 2nd PHB Inf.	I	10/31/61			
20	Rice, Samuel		Pvt.	PA McKeage's Mil.	D	7/2/63	8/8/63		
21	Rice, Soloman	42	Pvt.	PA 149th Inf.	I	8/26/63	5/29/65	Wounded	North Anna River
22	Riceling, William	28	Pvt.	PA 76th Inf.	E	11/21/61		POW-Died	Ft. Wagner, SC
23	Richards, David		Pvt.	MD 3rd PHB Inf.	K	3/28/64	5/29/65		
24	Richards, James J	27	Pvt.	PA 17th Cav.	G	10/2/62	6/14/65		
25	Richards, Jeremiah*		Pvt.	PA 55th Inf.	D	10/3/63			
26	Richards, John B*		Pvt.	PA 194th Inf.	I	7/22/64	11/5/64		
27	Richards, William	30	Pvt.	PA 82nd Inf.	F	8/23/61	9/16/64		
28	Richards, William		Pvt.	PA 87th Inf.	G	3/6/65	6/29/65		
29	Richey, Gideon		Pvt.	PA 79th Inf.	E	2/23/65	8/14/65		
30	Richey, Jacob*		Pvt.	PA 138th Inf.	E	9/19/64	6/23/65		
31	Richter, Adam	43	Pvt.	PA 133rd Inf.	C	8/13/62	5/19/63		
32	Richter, Adam		Pvt.	PA 208th Inf.	H	9/7/64	6/1/65		
33	Ridenbaugh, Samuel	23	Corp.	PA 138th Inf.	E	8/29/62	6/23/65	Wounded	Cedar Creek
34	Ridenour, Jacob D	19	Pvt.	PA 205th Inf.	C	8/27/64	6/2/65		
35	Rider, Adam		Pvt.						
36	Riffle, Albert J	16	Pvt.	PA 55th Inf.	H	2/19/64	8/30/65		
37	Riffle, Cyrus	20	Pvt.	PA 133rd Inf.	C	8/13/62	5/26/63		
38	Riffle, Cyrus		Corp.	PA 194th Inf.	I	7/22/64	11/5/64		
39	Riffle, John*		Corp.	PA 21st Cav.	L	8/4/63	7/8/65	Died	
40	Riffle, Thomas		Pvt.	MO 1st SM Cav.	L				
41	Riffle, Thomas	41	Pvt.	MO 35th Inf.	H	9/1/62		Died	
42	Riffle, William	17	Pvt.	PA 138th Inf.	E	8/29/62	6/23/65	Wounded	Opequon
43	Rightenour, Jacob	18	Pvt.	PA 103rd Inf.	E	3/16/65	7/16/65		
44	Riley, Andrew J	16	Pvt.	PA 107th Inf.	H	2/10/62	5/3/63	Wounded	South Mountain
45	Riley, Andrew J		Pvt.	PA 208th Inf.	K	9/7/64	6/1/65		
46	Riley, Bartley E	20	Pvt.	PA 13th Cav.	I	2/17/65	7/14/65		
47	Riley, Benjamin F	24	Corp.	MD 1st PHB Cav.	G	4/23/64	6/28/65		
48	Riley, Edward*		Pvt.	PA 55th Inf.	D	10/12/61		Died	
49	Riley, George W (1)	41	Sgt.	PA 107th Inf.	H	1/9/62	3/1/63	Wounded	2nd Bull Run
50	Riley, George W (1)		Corp.	PA 208th Inf.	K	9/7/64	6/1/65		
51	Riley, George W (2)	19	Pvt.	PA 107th Inf.	H	1/9/62	7/13/65		
52	Riley, Jacob	41	Pvt.	PA 107th Inf.	H	3/7/62	7/15/63		
53	Riley, Jacob E	20	Pvt.	PA 133rd Inf.	K	8/15/62	5/26/63		
54	Riley, Jacob E		Sgt.	PA 22nd Cav.	H	2/26/64	6/24/65		
55	Riley, James H	22	Pvt.	PA 133rd Inf.	K	8/15/62	5/26/63	Wounded	Fredericksburg
56	Riley, James H		Pvt.	PA 22nd Cav.	H	2/26/64	6/24/65		
57	Riley, John (1)	28	Pvt.	MD 1st PHB Cav.	F	4/23/64			
58	Riley, John (1)		Corp.	PA 184th Inf.	E	6/9/64			
59	Riley, John (2)								
60	Riley, John S		Surg.	TX Waul's Cav.	A			POW	Vicksburg
61	Riley, Reuben A	41	Capt.	IN 8th Inf.	G	4/25/61	8/6/61		
62	Riley, Reuben A		Capt.	IN 5th Cav.	G	10/30/62	12/25/63		
63	Riley, William (1)		Pvt.	PA 11th Inf.	A	9/30/61	4/14/63	Wound.-Died	Antietam
64	Riley, William (2)	53	Pvt.	PA 55th Inf.	D	3/2/64		Wound.-Died	Cold Harbor

	Cemetery	Notes
1		Historical Data Systems record
2		B.S.F. History Book-1884
3	Hickory Grove Cem.-IL	POW 6/22/64 to 4/1/65; B.S.F. History Book-1884; Ancestry.com-siblings born in Bedford Co.
4	Pleasant Hall Cem.-Franklin Co.	1880 Everett Census; Historical Data Systems record; Discharged on Surgeon's Cert.
5		B.S.F. History Book-1884
6	Pleasant Ridge Cem.-KS	POW 4/20/64 to 2/27/65; B.S.F. History Book-1884; Abraham & Isaac-twin brothers
7		B.S.F. History Book-1884; PA Civil War Muster Roll
8	Chaneysville Meth. Cem.	Frank McCoy 1912 Listing
9	Everett Cem.	B.S.F. History Book-1884; In Hospital with Typhoid Fever 8/17/65 to Muster Out
10		Bedford Historical Society record-GAR Voter rolls listing for Kearney in Broad Top Township
11		Bedford Historical Society record-GAR Voter rolls listing for Kearney in Broad Top Township
12	Mt. Pleasant-Allegany Co.	Pension Record references 7th US inf.; 1860 Cumberland Valley Census
13		B.S.F. History Book-1884; Terre Haute, IN residence
14	Beaufort Nat'l Cem.	POW-Andersonville 4/20/64; Died-Chronic Diarrhea 9/21/64; 1850 Monroe Twp. Census
15	Chaneysville Meth. Cem.	Frank McCoy 1912 Listing
16	Clearville Union Cem.	Bedford Inquirer 5/22/1908 listing
17	Robinsonville Cem.	1890 Monroe Twp. Veterans Census-Fractured ribs
18	Shreves Cem.-Monroe Twp.	Jonathan, Soloman, Isaac, Abraham & Cornelius-brothers
19		Deserted on 3/9/62; 1890 Broad Top Veterans Census
20		B.S.F. History Book-1884
21	Shreves Cem.-Monroe Twp.	Wounded 5/23/64; Frank McCoy 1912 Listing
22	Richmond Nat'l Cem.	POW-Richmond 7/11/63; Died 11/2/63, PA Civil War Arch. record; 1860 Bedford Twp. Cen.
23	Deneen Family Cem.-Fulton Co.	1890 Brush Creek Twp. Veterans Census
24	Burning Bush-Bedford Twp.	Deserted 8/24/64 & returned 1/1/65; Frank McCoy 1912 Listing; Hist. Data Systems record
25		Deserted 4/26/64; Executed 3/27/65; B.S.F. History Book-1884; Philadelphia Residence
26		B.S.F. History Book-1884; enlisted in Reading PA
27	Brumbaugh Cem.-Saxton	Frank McCoy 1912 Listing
28		1890 Liberty Twp. Veterans Census
29		1890 Hopewell Twp. Veterans Census
30		B.S.F. History Book-1884; Blair Co. references
31	Waterside Cem.-Woodbury	B.S.F. History Book-1884; Born: Bavaria Germany
32		Bedford Inquirer 5/22/1908 listing; 1870 S. Woodbury Twp. Census
33		Wounded 10/19/64; PA Civil War Archives-Bedford residence; 1860 Colerain Twp. Census
34	Greenlawn Cem.-Roaring Spring	Historical Data Systems record; 1860 Woodbury Twp. Census
35		1890 Broad Top Veterans Census; 1870 Bloody Run Census
36	St. Georges Episcopal Cem.-NJ	B.S.F. History Book-1884; 1850 Harrison Twp. Census; Albert & William-brothers
37	Rippey Cem.-IA	1860 Woodbury Twp. Census; Pension Record
38		B.S.F. History Book-1884; Historical Data Systems record
39		1850 Napier Twp. Census; Bedford Gazette 2/13/1914 listing
40		Ancestry.com information; 1850 Harrison Twp. Census; Father of Albert & William
41	Memphis Nat'l Cem.	Died of Disease 10/6/63; 1850 Harrison Twp.; Historical Data Systems record
42	Calvary Cem.-Pittsburgh	Wounded 9/19/64; B.S.F. History Book-1884; 1860 Bedford Twp. Census
43	Diehl's Crossroads-Henrietta	1890 Robertsdale Veterans Census
44	Rays Cove CC Cem.	Wounded 9/14/62; Died 2/29/68; B.S.F. History Book-1884
45		B.S.F. History Book-1884
46	Riverside Cem.-IA	Historical Data Systems record
47	Bethesda Cem.-MD	Benjamin, John (2) & George W (2)-brothers; 1850 E. Providence Twp. Census
48		Died 9/30/62; B.S.F. History Book-1884; PA Civil War Archives - Frostburg, MD residence
49	Rays Cove CC Cem.	Wounded 8/30/62; B.S.F. History Book-1884; George W (1) & William (1)-brothers
50		B.S.F. History Book-1884
51	Everett Cem.	B.S.F. History Book-1884; Two soldiers share the name George W Riley in 107th
52	Rays Cove CC Cem.	1890 E. Providence Twp. Veterans Census; Discharged on Surgeon's Cert.
53	Avalon Cem.-MO	Hist. Data Systems aka J Emanuel; Civil War Pension records; 1850 E. Providence Twp. Cen.
54		B.S.F. History Book-1884; Jacob, Bartley & James-brothers
55	Avalon Cem.-MO	Wounded 12/13/62; Historical Data Systems- enlisted in Bloody Run; aka Reilly Henry J
56		1860 W. Providence Twp. Cen; B.S.F. History Book-1884
57	Everett Cem.	Deserted and joined 184th Regiment; Historical Data Systems record
58		Deserted; Historical Data Systems record
59	Schellsburg Cem.	Bedford Inquirer 5/22/1908 listing
60		Father listed on 1820 St. Clair Township Cen.; POW-Vicksburg; Excaped captivity in Alton, IL
61	Park Cem.-IN	Ancestry.com-born in Bedford; Historical Data Systems record
62		Reuben & John S-brothers in Union & Confederate Armies; 1820 St. Clair Township Census
63		Wounded 9/17/62; Disch. Surg. Cert. 4/14/63; 1890 Everett Veterans Census lists KIA
64	Rays Cove CC Cem.	Wounded 6/3/64; Died 6/4/64; William (2) is father of Benjamin, George W (2) & John (1)

Chapter 18

	Name	Age	Rank	Regiment	Co.	Muster In	Muster Out	Casualty	Casualty Battle
1	Rimm, Charles*	29	Pvt.	PA 55th Inf.	I	10/9/63	4/25/64		
2	Rinard, David	17	Pvt.	PA 208th Inf.	K	9/7/64	6/1/65		
3	Rinard, Emanuel	22	Corp.	PA 77th Inf.	F	10/9/61		Died	
4	Rinard, George W	24	Pvt.	PA 107th Inf.	G	9/21/64	5/19/65	Wounded	Dabney's Mills
5	Rinard, Jacob	18	Pvt.	PA 84th Inf.	C				
6	Rinard, Samuel	32	Pvt.	PA 5th Inf.	D	4/21/61	7/25/61		
7	Rinard, Samuel		Pvt.	PA 147th Inf.	B	8/17/61	12/24/62		
8	Rinard, Thomas			PA 208th Inf.					
9	Rininger, Eli	21	Pvt.	PA 55th Inf.	H	2/29/64	8/30/65	Wounded	Cold Harbor
10	Rininger, Frederick		Pvt.	PA 84th Inf.	A	10/24/61		Died	
11	Riplett, Joshua*	19	Pvt.	PA 55th Inf.	K	3/2/64	5/23/65	Wounded	
12	Ripple, Valentine		Pvt.	PA 21st Cav.	E	7/2/63	2/20/64		
13	Rislenbatt, August		Pvt.	PA 55th Inf.	H	9/23/63	8/30/65		
14	Risling, John	45	Pvt.	PA 55th Inf.	D	10/12/61	10/29/62		
15	Risling, John H	26	Pvt.	PA 55th Inf.	H	10/11/61	8/30/65	Wounded	Bermuda Hundred
16	Risling, Joseph	18	Pvt.	PA 138th Inf.	D	8/29/62	6/23/65		
17	Risling, William H	19	Pvt.	OH 87th Inf.	I	6/25/62	10/3/62		
18	Ritchey, Adam	17	Pvt.	PA 55th Inf.	D	10/12/61			
19	Ritchey, Adam S	30	Pvt.	PA 133rd Inf.	C	8/13/62	5/26/63		
20	Ritchey, Adam S		Pvt.	PA 194th Inf.	I	7/22/64	9/6/64		
21	Ritchey, Adam S		Pvt.	PA 97th Inf.	Ind	9/6/64	7/17/65		
22	Ritchey, Daniel B	18	2nd Lt.	PA 55th Inf.	K	11/5/61	7/29/65	Wounded	
23	Ritchey, Daniel S	23	Pvt.	PA McKeage's Mil.	D	7/2/63	8/8/63		
24	Ritchey, Daniel S		Pvt.	PA 208th Inf.	K	9/7/64	7/29/65	Wounded	Ft. Stedman
25	Ritchey, David (1)	18	Pvt.	PA 55th Inf.	K	3/2/64	8/30/65	Wounded	Drewry's Bluff*
26	Ritchey, David (2)	28	Pvt.	PA 208th Inf.	K	9/7/64		Died	
27	Ritchey, Ferdinand	44	Corp.	PA 55th Inf.	K	11/5/61	8/30/65		
28	Ritchey, Frederick G	21	Corp.	PA 138th Inf.	F	8/29/62	5/12/65	Wounded	Opequon
29	Ritchey, Frederick G		Corp.	PA 138th Inf.	F			Wounded	Cedar Creek
30	Ritchey, George			PA 208th Inf.					
31	Ritchey, George S	32	Pvt.	PA 77th Inf.	F	2/28/65	7/11/65		
32	Ritchey, Henry C	17	Pvt.	PA 138th Inf.	F	8/29/62	6/23/65		
33	Ritchey, Henry S	21	Corp.	PA 101st Inf.	D	11/1/61		Died	
34	Ritchey, Jacob			PA 208th Inf.					
35	Ritchey, Jacob J	26	Pvt.	PA 93rd Inf.	H	6/4/64	6/27/65		
36	Ritchey, Jacob K	16	Pvt.	PA 138th Inf.	E	9/19/64	6/23/65		
37	Ritchey, Jacob M	37	Pvt.	PA 199th Inf.	E	10/1/64		Died	
38	Ritchey, James H	27	Pvt.	PA 22nd Cav.	M	2/20/64	10/31/65		
39	Ritchey, James T	15	Pvt.	PA 107th Inf.	H	4/26/62	7/13/65		
40	Ritchey, John	20	Pvt.	PA 55th Inf.	K	3/2/64	7/22/65	Wounded	White Oak Road
41	Ritchey, John C	18	Pvt.	PA 138th Inf.	E	8/29/62	6/16/65	Wound.-POW	
42	Ritchey, John F		Pvt.	PA 208th Inf.	K	9/5/64	5/30/65		
43	Ritchey, John N	17	Pvt.	PA 208th Inf.	H	9/8/64	6/1/65		
44	Ritchey, Jonas	18	Pvt.	PA 55th Inf.	K	11/5/61		Wound.-Died	Edisto Island, SC
45	Ritchey, Joseph	22	Pvt.	PA 8th Res.	F	6/11/61	2/26/62		
46	Ritchey, Joseph		Pvt.	PA 186th Inf.	G	3/24/64	8/15/65		
47	Ritchey, Levi	26	Corp.	PA McKeage's Mil.	G	7/2/63	8/8/63		
48	Ritchey, Lewis	22	Pvt.	PA 54th Inf.	G	10/9/62	5/31/65		
49	Ritchey, M L								
50	Ritchey, Samuel Y	26	Pvt.	PA 8th Res.	F	7/29/61	1/1/64	Died	
51	Ritchey, William D	21	Corp.	PA 8th Res.	F	6/11/61	5/15/64	Injured	Seven Days Battle
52	Ritchey, William D		Corp.	PA 8th Res.	F			POW	Spotsylvania
53	Ritchey, William D		Corp.	PA 191st Inf.	H	5/15/64	6/28/65	Wounded	Petersburg
54	Ritz, Daniel	42	Pvt.	PA 149th Inf.	E	8/18/64	9/10/65		
55	Rizer, Levi	31	Corp.	US 19th Inf.	E	12/16/61	2/7/63		
56	Roach, Thomas	43	Pvt.	PA 55th Inf.	I	10/15/61	5/14/62		
57	Roarabaugh, John	34	Pvt.	PA 205th Inf.	C	8/27/64	6/2/65		
58	Robb, Conrad		Pvt.	PA 8th Res.	F	6/11/61	5/15/64	POW	Seven Days Battle
59	Robb, Conrad		Pvt.	PA 191st Inf.		5/15/64			
60	Robb, George W	24	Pvt.	PA 138th Inf.	F	9/4/62		KIA	Mine Run
61	Robb, John M	34	Corp.	PA 55th Inf.	K	11/5/61	8/30/65	Wounded	
62	Robb, Samuel	27	Pvt.	PA 138th Inf.	F	9/4/62	6/23/65		
63	Roberts, John	24	Pvt.	PA 101st Inf.	D	1/20/62	3/1/65		
64	Roberts, William	21	2nd Lt.	PA 110th Inf.	C	10/24/61	12/20/62		

	Cemetery	Notes
1		B.S.F. History Book-1884; Mustered in Reading, PA; Discharged on Surgeon's Cert.
2	Bunker Hill Cem.-Saxton	1890 Liberty Twp. Veterans Census; B.S.F. History Book-1884
3	Murietta Nat'l Cem.	Died 7/30/64; Bedford Gazette 2/13/1914 listing; 1860 Broad Top Twp. Census
4	Brumbaugh Cem.-Saxton	Wounded 2/6/65; 1890 Liberty Twp. Veterans Census; George & David-brothers
5		PA Civil War Archives-Saxton residence
6	Saxton Cem.	Bedford Inquirer 5/22/1908 listing
7		1890 Liberty Twp. Veterans Census; Historical Data Systems record; Discharged on Surg. Cert.
8	Saxton or Valley Cem.	Bedford Inquirer 5/22/1908 listing
9	Greenlawn Cem.-OH	Wounded 6/3/64; B.S.F. History Book-1884; PA Civil War Archives-enlisted in Schellsburg
10	Antietam Nat'l Cem., 4184	Died 3/4/62; Bed. Gaz. 2/13/1914 list; PA CW Project record; 1860 E. St Clair Twp. Cen.
11		Wounded listed in Bedford Inquirer-6/3/64; B.S.F. Hist. Book-1884; Cambria Co. references
12		Historical Data Systems-Bedford Co. residence
13		B.S.F. History Book-1884
14	Methodist Cem.-Bedford	B.S.F. History Book-1884; Discharged on Surgeon's Cert.; John H-son
15	Homewood Cem.-Pittsburgh	Wounded 5/13/64 listed in Bed. Inquirer-7/8/64; B.S.F. Hist. Book-1884; enlisted in Schellsburg
16		B.S.F. History Book-1884; PA Civil War Archives enlisted in Schellsburg
17		1860 Juniata Twp. Census; Civil War Pension Record
18		Deserted 4/21/62; Historical Data Systems-Bedford Co. residence
19	Everett Cem.	B.S.F. History Book-1884; 1850 W. Providence Twp. Census
20		B.S.F. History Book-1884
21		Ancestry.com information; Adam, Daniel, David (2) & Joseph-brothers
22	Graceville Luth. Cem.	B.S.F. History Book-1884; Discharged on Surgeon's Cert.
23	West Union Cem.-OH	B.S.F. History Book-1884
24	Graceville Luth. Cem.	Leg Amputated 3/25/65; Frank McCoy 1912 Listing; Historical Data Systems record
25	Mt. Hope Cem.-Claysburg	1890 Greenfield Twp. Veterans Census-Leg Wound; B.S.F. History Book-1884
26	Richey Farm-W. Prov. Twp.	Died-Malaria Fever 1/21/65; B.S.F. History Book-1884; 1860 W. Providence Twp. Census
27	Mt. Zion Cem.-Pavia	B.S.F. History Book-1884; Historical Data Systems record; Father of John C & Jacob K
28	Fredonia City Cem.-KS	Wounded 9/19/64; Historical Data Systems record; 1860 Juniata Twp. Census
29		Wounded 10/19/64; B.S.F. History Book-1884; Discharged on Surgeon's Cert.
30	Graceville Graveyard	Frank McCoy 1912 Listing
31	Mt. Sinai Cem.-Blue Knob	Historical Data Systems record; Born: Rockingham, VA
32	Center Hill Cem.-KS	B.S.F. History Book-1884; Hist. Data Systems record; Henry, Frederick & Jacob J-brothers
33	US Soldiers & Airmans Nat'l	Died 6/2/62; PA Civil War Archives-Bedford Co. residence; 1860 W. Providence Twp. Census
34	Ritchey-West. Prov. Twp.	Bedford Inquirer 5/22/1908 listing
35	Trinity UCC Cem.-Juniata Twp.	Historical Data Systems record lists G middle initial; PA CW Arch. listed Jacob G Richie
36	Newry Lutheran Cem.	1860 Union Twp. (Pavia) Census; PA Civil War Project record; Jacob K & John C-brothers
37	Ashbury Meth. Cem.-Graceville	Died-Typhoid Fever 12/23/64; Frank McCoy 1912 Listing; 1860 E. Providence Twp. Census
38	Hinish Family Cem.-Cypher	Frank McCoy 1912 Listing; James & William D-brothers; James died in 1868
39	Olivewood Cem.-CA	PA Civil War Archives-born in Bedford Co.; 1860 W. Providence Township Census
40	Mt. Hope Cem.-Claysburg	Wounded 3/30/65; 1890 Greenfield Twp. Veterans Census; B.S.F. History Book-1884
41	Mt. Moriah Cem.-Blue Knob	MIA-Bedford Inquirer 5/20/64; B.S.F. History Book-1884; Attended 1904 Andersonville Ded.
42	Black Oak Cem.-Port Matilda	B.S.F. History Book-1884, Historical Data Systems-Bedford Co. residence
43	St. Johns Cem.-KS	B.S.F. History Book-1884, Historical Data Systems-Bedford Co. residence
44		Wounded on picket duty; Died 3/29/62; PA CW Arch.-Bed. Co. resid.; 1860 Greenfield Cen.
45	Everett Cem.	B.S.F. History Book-1884; Historical Data Systems-Bedford Co. resid.; Discharged Surg. Cert.
46		1890 W. Providence Twp. Veterans Census
47	Mt. Hope Cem.-Blue Knob	B.S.F. History Book-1884; Levi & David (1)-brothers
48	Everett Cem.	Bedford Historical Society-Pension Record; Historical Data Systems record
49		Kernal of Greatness Book by Bedford County Heritage Commission reference
50		Ancestry.com-Died 1864; PA Civil War Archives record; 1850 Napier Twp. Census
51	Bethel Brethren Cem.-Tatesville	Wounded 6/26/62, in hospital until 3/1/63; Findagrave.com information
52		POW 5/8/64 to 5/12/64; Escaped; B.S.F. History Book-1884; Historical Data Systems record
53		1860 Hopewell Twp. Census; Historical Data Systems record
54	Buck Valley Cem.-Fulton Co.	1890 Brush Valley Twp. Veterans Census
55	Hyndman Cem.	1890 Hyndman Veterans Census
56	St. Joseph Cem.-Williamsburg	B.S.F. History Book-1884; Discharged on Surgeon's Cert.
57	Duval Cem.-Coaldale	Historical Data Systems record
58		B.S.F. History Book-1884; Historical Data Systems-Bedford Co. residence
59		B.S.F. History Book-1884
60		KIA 11/27/63; 1860 Juniata Twp. Cen.; B.S.F. History Book-1884; Hist. Data Systems record
61	Old Union Cem.-Buffalo Mills	Wound listed on Pension request; Frank McCoy 1912 Listing; Bedford Inquirer 5/22/1908 list
62	Hyndman Cem.	Historical Data Systems record; 1850 Harrison Twp. Census; Samuel, John & George-brothers
63		PA Civil War Archives-Bedford Co. residence, enlisted in Rainsburg; B.S.F. History Book-1884
64		B.S.F. History Book-1884; 1860 Woodbury Twp. Census

	Name	Age	Rank	Regiment	Co.	Muster In	Muster Out	Casualty	Casualty Battle
1	Robertson, Hector A		Sgt.	MD 2nd PHB Inf.	B	8/25/61	9/8/64	Wounded	
2	Robinett, Eli Z		Pvt.	MD 2nd PHB Inf.	C	2/14/65	5/29/65		
3	Robinett, Mathias	32	Corp.	MD 1st PHB Cav.	B	11/27/61	6/28/65		
4	Robinette, Amos	33	Capt.	PA 171st Inf.	I	11/11/62	8/8/63		
5	Robinette, Henry C	20	Pvt.	MD 1st PHB Cav.	L	4/23/64	6/28/65		
6	Robinette, Jeremiah	21	Pvt.	PA 171st Inf.	I	11/1/62	8/1/63		
7	Robinette, Jesse	27	Pvt.	MD 1st PHB Cav.	L	4/23/64	6/28/65		
8	Robinson, George B*	40	Pvt.	PA 55th Inf.	H	9/2/62	6/11/65		
9	Robinson, James A		1st Lt.	MD 6th Inf.	H	8/21/62	3/17/64		
10	Robinson, Job	30	Pvt.	PA 208th Inf.	H	9/7/64	6/1/65		
11	Robinson, Thomas S		Pvt.	VT 2nd Inf.	H	7/27/63	7/15/65	Wounded	Wilderness
12	Robinson, Tobias		Pvt.	PA 55th Inf.	H	1/1/64	8/30/65		
13	Robinson, William	35	Pvt.	PA 99th Inf.	K	2/24/65		Died	
14	Robinson, William J	23	Pvt.	PA 138th Inf.	E	8/29/62	6/23/65	Wounded	Wilderness
15	Robison, Henry C	17	Pvt.	PA 49th Inf.	K	10/24/61	10/24/64		
16	Robison, Jonas	23	Pvt.	PA 101st Inf.	D	11/1/61		Died	
17	Robison, Philip*		Pvt.	PA 55th Inf.	H	1/1/64	12/30/64		
18	Robison, Thomas*							Died	
19	Roby, George W	30	Pvt.	PA 55th Inf.	D	2/27/64	8/30/65		
20	Rock, George J	32	Pvt.	PA 101st Inf.	G	12/28/61	11/29/62		
21	Rock, George J		Pvt.	PA 67th Inf.	C	7/5/64	7/14/65		
22	Rodgers, Henry C*		Pvt.	PA 76th Inf.	E	11/21/61	11/28/64		
23	Rodgers, James*		Pvt.	PA 55th Inf.	H	9/23/63			
24	Rogan, Patrick*		Pvt.	PA 76th Inf.	E	2/21/65	7/18/65		
25	Rogers, Francis E	18	Corp.	MD 5th PHB Inf.	E	2/23/64	9/1/65		
26	Rohm, David F	16	Pvt.	IL 58th Inf.	E	3/17/65	3/10/66		
27	Rohm, Frank	33	Pvt.	PA 208th Inf.	G	9/5/64	6/1/65		
28	Rohm, John G	21	2nd Lt.	PA 12th Res.	C	6/15/61	6/11/64		
29	Rohm, John S	16	Pvt.	MD 3rd PHB Inf.	K	4/15/64	5/29/65		
30	Rohm, Polk*								
31	Rohm, William H	18	Pvt.	PA 107th Inf.	H	3/4/62			
32	Rohm, William J	20	Pvt.	MD 1st PHB Inf.	M	4/23/64	5/27/64		
33	Roland, Henry	22	Pvt.	PA 138th Inf.	D	8/29/62			
34	Rollins, Andrew	21	Pvt.	PA 55th Inf.	K	11/5/61	9/4/62		
35	Rollins, James	33	Pvt.	PA 138th Inf.	E	8/29/62	7/16/63		
36	Rollins, Thomas J	28	Pvt.	GA 60th Inf.	K	5/10/62	6/8/64	Wounded	2nd Bull Run
37	Rollins, Thomas J		Pvt.	GA 60th Inf.	K				
38	Rollins, Thomas J		Pvt.	GA 60th Inf.	K			POW	Gettysburg
39	Rook, Thomas	28	Pvt.	OH 80th Inf.	A	11/6/61	1/1/64		
40	Rook, Thomas		Pvt.	OH 191st Inf.	A	3/6/65	8/27/65		
41	Rorabaugh, John	34	Pvt.	PA 205th Inf.	C	8/27/64	6/2/65		
42	Rose, Eli E	20	Pvt.	IN 47th Inf.	F	12/3/61	7/26/65	Wounded	Spanish Fort, AL
43	Ross, George W		Corp.	PA 21st Cav.	E	7/2/63	2/20/64		
44	Ross, James*								
45	Ross, Joseph	38	Pvt.	PA 208th Inf.	H	9/8/64	6/1/65		
46	Ross, Oliver P	23	Pvt.	PA 8th Res.	F	6/11/61	5/26/64		
47	Roudabush, Benjamin	16	Pvt.	PA 55th Inf.	H	9/25/63		POW-Died	
48	Roudabush, John M	27	Pvt.	PA 82nd Inf.	C	11/14/64	7/13/65		
49	Rough, Adam*	27	Corp.	PA 125th Inf.	E	8/13/62	5/18/63		
50	Rough, Andrew*	24	1st Lt.	PA 55th Inf.	I	11/20/61	10/4/64		
51	Rough, Benjamin*	40	Capt.	PA 55th Inf.	I	12/4/61	8/17/64		
52	Rough, John	42	Pvt.	PA 76th Inf.	E	2/15/65	7/18/65		
53	Rough, John H*	16	Pvt.	PA 53rd Inf.	C	10/17/61	6/30/65		
54	Rough, Valentine*	35	Pvt.	PA 110th Inf.	H	9/18/62		Died	
55	Rough, William H*	16	Corp.	PA 55th Inf.	I	10/15/61	10/15/64		
56	Rourke, John	29	Pvt.	PA 1st Light Art.	F	6/11/61	7/11/64	Wounded	Bristoe Station
57	Rouse, Elias*								
58	Rouser, Joseph	20	Pvt.	PA 21st Cav.	E	7/8/63	2/20/64		
59	Roush, Ernest	32	Pvt.	PA 76th Inf.	E	11/21/61	7/18/65	POW	Ft. Wagner, SC
60	Roush, Jacob	38	Pvt.	PA 3rd Inf.	G	9/11/62	9/25/62		
61	Roush, Jacob		Pvt.	PA 87th Inf.	B	6/4/64	6/29/65		
62	Roush, James Levi	23	Corp.	PA 6th Res.	D	4/24/61	6/11/64	Wounded	2nd Bull Run
63	Roush, James Levi		Corp.	PA 6th Res.	D	4/24/61	6/11/64		Gettysburg
64	Rowe, Solomon J	35	Pvt.	PA 46th Inf.	C	2/29/64	7/16/65		

	Cemetery	Notes
1		1890 Cumberland Valley Veterans Census
2	Everett Cem.	Historical Data Systems record; Everett Cemetery Listing of Civil War Veterans
3	Everett Cem.	Bedford Inquirer 5/22/1908 listing; Historical Data Systems record
4	Zion Park-Cumberland MD	B.S.F. History Book-1884; Jeremiah & Amos-cousins
5	Hillcrest Mem.-Cumberland	1890 Londonderry Twp. Veterans Census
6	Odd Fellows Cem.-Flintstone	B.S.F. History Book-1884; 1870 Southampton Twp. Census
7	Prosperity Cem.-Clearville	Frank McCoy 1912 Listing; Historical Data Systems record; Jesse & Amos-brothers
8		B.S.F. History Book-1884; Buried in Reading, PA
9		1890 Bedford Twp. Veterans Census
10	Richland Cem.-Allegheny Co	Historical Data Systems record; 1860 Monroe Twp. Cen; Job, Jonas & William-brothers
11		Wounded 5/6/64; B.S.F. History Book-1884
12		B.S.F. History Book-1884
13		Died 5/13/65; 1890 Monroe Twp. Vet. Cen.; PA CW Project record; 1850 Monroe Twp. Cen.
14	Gettysburg Nat'l Cem.	Wounded 5/5/64; PA Civil War Archives-St. Clairsville residence; Hist.Data Systems record
15	Bedford Cem.	PA Civil War Archives record
16	Loudon Park Nat'l Cem.	Died 6/1/62; 1860 Monroe Twp. Census; Hist. Data Systems-Bedford Co. resid.; aka Robinson
17		B.S.F. History Book-1884
18	Bethel Christian Cem.-Clearville	Born 3/15/1790?; Died 2/2/64; Civil War Grave Marker
19	St. Thomas Cem.-Bedford	B.S.F. History Book-1884; 1860 Bedford Twp. Census; Historical Data Systems record
20	Schellsburg Cem.	1890 Napier Twp. Veterans Census; Historical Data Systems record; Discharged on Surg. Cert.
21		1890 Napier Twp. Veterans Census; Discharged on Surgeon's Cert.
22		B.S.F. History Book-1884; Philadelphia references
23		Draft. & Deserted 2/5/64; B.S.F. History Book-1884
24		B.S.F. History Book-1884; Lycoming Co. references
25	Monongahela Cem.-Wash. Co.	1890 Londonderry Twp. Veterans Census
26	Rockville Cem.-IL	Historical Data Systems record; Ancestry.com-born Gapsville
27	Bloomfield Cem.-Perry Co.	E. Providence Twp. Honor Roll; Historical Data Systems record
28		B.S.F. History Book-1884
29	Dickerson Run-Fayette Co.	1890 Cumberland Valley Veterans Census
30	Fockler Cem.-Saxton	Bedford Inquirer 5/22/1908 listing aka Rome
31		Deserted 3/7/62; B.S.F. History Book-1884
32	Woodlawn Cem.-IL	Historical Data Systems record; William & David-brothers
33		Deserted 10/22/62; Historical Data Systems record; killed Josiah Baughman in Chaneysville
34	R.W. Garland-Harrison Twp.	B.S.F. History Book-1884; Frank McCoy 1912 List; Bed. Inquirer 1908 list; Disch. Surg. Cert.
35	Rollins Farm Cem.-Juniata Twp.	1890 Juniata Twp. Veterans Census
36		Wounded 8/28/62; Born Spartanburg, SC; 1850 Spartanburg Census; Findagrave.com info.
37		Deserted 6/6/64 Louisville, KY to Union & Released 6/8/64; Ancestry.com information
38		POW 7/3/63; Historical Data Systems record
39	St. James Luth. Cem.-Bedford	Bedford Inquirer 5/22/1908 listing; Frank McCoy 1912 Listing
40		1890 Bedford Twp. Veterans Census; Historical Data Systems record
41	Duval Cem.-Coaldale	1890 Coaldale Veterans Census; Bedford Inquirer 5/22/1908 listing
42	Belmont Park Cem.-OH	Leg Amputated 3/27/65; 1890 Woodbury Twp. Veterans Census
43		Historical Data Systems-Bedford Co. residence
44	Everett Cem.	Civil War marker at gravesite
45	Hopewell Cem.	1890 Hopewell Veterans Census; Historical Data Systems record
46	Fockler Cem.-Saxton	B.S.F. History Book-1884; 1890 Liberty Twp. Veterans Census; Oliver & Joseph-brothers
47	Andersonville Cem., 10935	POW-Andersonville; Died Diarrhea 9/29/64; Civil War Pension Record; 1860 Union Twp. Cen.
48	Trinity UCC Cem.-Osterburg	1850 Union Twp. (Pavia) Census, Frank McCoy 1912 Listing
49	Greenlawn Cem.-Roaring Spring	Ancestry.com information; Valentine, Adam, Andrew & Benjamin-brothers
50	Montgomery Cem.-Norristown	B.S.F. History Book-1884; 1860 N. Woodbury Twp. Census; Discharged on Surgeon's Cert.
51	Evergreen Cem.-KY	B.S.F. History Book-1884; Father of John H & William; Blair Co. references
52		B.S.F. History Book-1884
53	Fairview Cem.-Altoona	Ancestry.com information; 1850 N. Woodbury Twp. Census
54	Grandview Cem.-Tyrone	Died 2/9/64; Ancestry.com information; Blair Co. references
55	Evergreen Cem.-KY	B.S.F. History Book-1884; William & John H-brothers; Blair Co. references
56	Ave Maria Cem.-Dudley	Wounded 10/13/63; at Mansion Hospital-Lee's Alexandria home; Ancestry.com information
57	Mt. Ross Cem.-Bedford	Civil War marker at gravesite
58	Hull Baptist Cem.-New Paris	1890 Napier Twp. Veterans Census; Historical Data Systems-aka Rowzer
59		POW 7/11/63; PA Civil War Archives-Bedford residence; Historical Data Systems record
60	Rays Cove CC Cem.	Historical Data Systems record
61		1890 E. Providence Twp. Veterans Census; Frank McCoy 1912 Listing
62	St. Patricks Cem.- Newry	Face wound at 2nd Bull Run-Fold3; PA Civil War Archives-Sarah Furnace residence
63		Medal of Honor recipient-Gettysburg during 2nd day of battle near Little Round Top
64	Bedford Cem.	Bedford Inquirer 5/22/1908 listing

Chapter 18

	Name	Age	Rank	Regiment	Co.	Muster In	Muster Out	Casualty	Casualty Battle
1	Rowland, John	34	Pvt.	PA 88th Inf.	G	9/24/64	6/31/65	Died	
2	Rowser, Philip	20	Pvt.	PA 55th Inf.	H	10/11/61		Died	
3	Rowzer, George	22	Pvt.	PA 107th Inf.	I	9/21/64	6/2/65	Wounded	Dabney's Mills
4	Rowzer, John S	25	Pvt.	PA 55th Inf.	H	10/11/61	2/22/63		
5	Rowzer, John S		Pvt.	US 1st Light Art.	M	2/22/63			
6	Rowzer, Joseph O	25	Pvt.	PA 99th Inf.	B	2/25/65	7/1/65		
7	Roy, James	24	Pvt.	PA 133rd Inf.	C	8/13/62	5/26/63	Wounded	Fredericksburg
8	Royal, Clark		Pvt.	PA 76th Inf.	E	11/21/61	2/19/62		
9	Ruben, Joseph		Pvt.	PA 55th Inf.	I	10/8/63			
10	Ruby, Henry	38	Corp.	PA 171st Inf.	I	11/2/62			
11	Ruby, Henry		Pvt.	MD 2nd PHB Inf.	D	3/14/65	5/29/65		
12	Ruby, John (1)	21	Pvt.	PA 101st Inf.	D	11/1/61		Died	
13	Ruby, John (2)	23	Pvt.	PA 55th Inf.	D	2/27/64		Died	
14	Ruby, John (3)			MD 2nd PHB Inf.					
15	Ruggles, Albert*		Pvt.	PA McKeage's Mil.	B	6/3/63	8/8/63		
16	Ruggles, Albert*	26	Corp.	PA 55th Inf.	I	2/11/64	8/30/65		
17	Rumel, John E	22	Pvt.	PA 208th Inf.	H	9/7/64	6/1/65		
18	Rush, David	29	Pvt.	PA 138th Inf.	F	9/12/62	5/15/65	Wounded	Cold Harbor
19	Rush, Jacob	32	Pvt.	PA 46th Inf.	F	7/5/64	7/16/65		
20	Rush, James L	31	Major	MO 16th Inf.	F&S	9/11/64	7/1/65		
21	Rush, William T*								
22	Russell, A Sidney	40	QM	PA 208th Inf.	K	9/7/64	6/1/65		
23	Russell, Abraham	23	Pvt.	PA 91st Inf.	D	3/15/65	6/29/65		
24	Russell, David	32	Pvt.	PA 91st Inf.	D	3/15/65	7/10/65		
25	Russell, George D*	41	Pvt.	PA 78th Inf.	K	3/7/65	6/7/65	Died	
26	Russell, Isaac		Pvt.	PA 91st Inf.	D	3/15/65	7/10/65		
27	Russell, John		Pvt.	PA 91st Inf.	D	3/15/65	7/10/65		
28	Salkeld, Bernard	47	Pvt.	PA 158th Inf.	H	11/4/62	8/12/63		
29	Salkeld, David E*	17						KIA	
30	Salkeld, Jacob F	25	Pvt.	PA 158th Inf.	H	11/4/62	8/12/63		
31	Salkeld, James W	18	Pvt.	PA 13th Cav.	D	10/3/63	6/19/65	POW	Coggin's Point
32	Salkeld, John F	15	Music.	PA 107th Inf.	H	1/20/62	7/13/65		
33	Salkeld, John N	21	Pvt.	PA 28th Inf.	O	8/17/61			
34	Salkeld, John N		Pvt.	PA 147th Inf.	B	4/1/62	4/3/65		
35	Salkeld, Samuel W	23	Pvt.	PA 126th Inf.	B	8/12/62	5/20/63	Wounded	Chancellorsville
36	Salkeld, Samuel W		Pvt.	PA 49th Inf.	E	6/4/64	7/15/65		
37	Salkeld, Thomas L	31	Pvt.	PA 107th Inf.	H	1/20/62	2/11/64		
38	Sampsel, Napoleon*	31	Pvt.	PA 184th Inf.	A	10/1/64	6/2/65		
39	Sams, John W	43	Corp.	PA 208th Inf.	K	9/7/64	6/1/65		
40	Sams, Wilson	45	Pvt.	PA 208th Inf.	H	9/7/64	6/1/65		
41	Sanderson, Samuel K	16	Pvt.	PA 7th Res.	I	7/4/63	8/11/63		
42	Sanderson, Samuel K		Pvt.	PA 188th Inf.	K	2/24/64	12/14/65		
43	Sanderson, Theodore C	15	Pvt.	PA 149th Inf.	A	2/25/65	6/26/65		
44	Sanno, Frederick	43	Corp.	PA 55th Inf.	K	11/5/61		Died	
45	Sansom, Samuel	25	1st Sgt.	IL 76th Inf.	B	8/1/62	7/22/65		
46	Satler, Frederick*		Pvt.	PA 55th Inf.	H	12/16/63	8/30/65		
47	Satterfield, John E	41	Pvt.	PA 208th Inf.	K	9/7/64	6/1/65		
48	Saupp, Frank D	22	1st Lt.	PA 55th Inf.	K	11/5/61	10/16/64	Wounded	
49	Saupp, James	20	Pvt.	PA 13th Inf.	G	4/25/61	8/6/61		
50	Saupp, James		Pvt.	PA 138th Inf.	E	8/29/62	6/23/65		
51	Saupp, John	17	Pvt.	PA 55th Inf.	K	11/5/61		KIA	Edisto Island, SC
52	Savits, Henry	37	Pvt.	PA 99th Inf.	B	2/24/65	7/1/65		
53	Saxon, John C*	18	Pvt.	PA 55th Inf.	I	2/11/64		KIA	White Oak Road
54	Saxton, Henry C	20	Pvt.	PA 125th Inf.	F	8/12/62	5/18/63		
55	Saylor, Andrew J*	16	Pvt.	PA 133rd Inf.	E	8/14/62	5/26/63	Wounded	Fredericksburg
56	Saylor, Andrew J*		Pvt.	PA 22nd Cav.	I	2/15/64	6/24/65		
57	Saylor, William	37	Pvt.	PA 82nd Inf.	K	11/18/64		Died	
58	Schaffer, Michael*	28	Pvt.	PA 55th Inf.	H	9/22/62	6/11/65		
59	Schellhorn, Albert		Pvt.	PA 21st Cav.	E	7/2/63	2/20/64		
60	Schetrompf, George	21	Pvt.	MD 3rd PHB Inf.	C	1/8/62	1/8/65		
61	Schetrompf, John F	17	Pvt.	MD 3rd PHB Inf.	B	10/28/61	5/29/65		
62	Schetrompf, Peter C	17	Pvt.	MD 3rd PHB Inf.	B	10/7/63	5/29/65		
63	Schinich, William H		Pvt.	PA 82nd Inf.	C	9/14/64	7/13/65		
64	Schlag, Paul B*	21	Corp.	PA 52th Inf.	C	9/26/64	6/24/65		

	Cemetery	Notes
1	Bennett Farm-Chaneysville	Died 8/4/65 of Disease contracted in army; 1890 Southampton Veterans Census
2	Hull Baptist Cem.-New Paris	Died 12/30/61; B.S.F. History Book-1884; 1850 Napier Twp. Census
3	Pleasantville Cem.	Wounded 2/6/65; 1890 W. St. Clair Twp. Veterans Census
4	Hull Baptist Cem.-New Paris	B.S.F. History Book-1884; aka Rouser & Rowser
5		Historical Data Systems record
6	Mock Church Cem.-Alum Bank	PA Civil War Archives record aka Rouser
7		Wounded 12/13/62; Historical Data Systems-Bedford Co. residence
8	Dayton Nat'l Cem.-OH	B.S.F. History Book-1884; Discharged on Surgeon's Cert.
9		Deserted 6/11/64; B.S.F. History Book-1884
10	Chaneysville Meth. Cem.	Deserted 11/20/62; B.S.F. History Book-1884
11		1890 Monroe Twp. Veterans Census
12	Woods Church-Colerain Twp.	Died 12/30/61; Frank McCoy 1912 Listing; PA Civil War Archives-enlisted in Rainsburg
13		Died 4/29/64; PA Civil War Archives-enlisted in Bedford; B.S.F. History Book-1884
14	Ruby Cem.-Beans Cove	Civil War Pension Record; 1850 Southampton Twp. Census
15		Historical Data Systems record
16		B.S.F. History Book-1884; Blair Co. references
17	Union Cem.-McConnellsburg	B.S.F. History Book-1884; Historical Data Systems-Bedford Co. residence
18		Wounded 6/1/64; B.S.F. History Book-1884; Historical Data Systems-Bedford Co. residence
19	Lybarger Cem.-Madley	1890 Londonderry Twp. Veterans Census
20	Marshfield Cem.-MO	Historical Data Systems record; 1850 Bedford Borough Census
21	Old St. Thomas Cem.-Bedford	Civil War market at gravesite
22	Bedford Cem.	B.S.F. History Book-1884; Historical Data Systems-Bedford Co. residence
23	Grandview Cem.-Hunt. Co.	1890 Hopewell Veterans Census; Abraham, Isaac & David-brothers
24	St. Pauls UCC Cem.-Hunt. Co.	1890 Hopewell Veterans Census
25	St. Pauls UCC Cem.-Russellville	Died 10/30/65; 1890 Hopewell Veterans Census; 1860 Hopewell Twp. (Huntingdon Co.) Cen.
26		1890 Hopewell Veterans Census; 1850 Hopewell Census
27		1890 Hopewell Veterans Census
28	South Fork Cem.-Cambria Co.	1890 Brush Creek Twp. Vet.Cen.; 1850 Colerain Twp. Cen.; Father of David, John F & James
29		Ancestry.com-KIA 1865; Findagrave.com-David H Salkeld in Illinois Regiment died 5/2/65
30	Center Cem.-Waterfall	Ancestry.com-Waterfall residence; Historical Data Systems record
31	Dunmyer Cem.-Cambria Co.	POW 9/16/64; 1850 Colerain Twp. Census; Historical Data Systems record
32	S. Fork Cem.-Cambria Co.	Historical Data Systems record; B.S.F. History Book-1884; David, John F & James-brothers
33	Congressional Cem.-DC	Historical Data Systems record; John N, Bernard, Jacob F, Samuel & Thomas-brothers
34		Ancestry.com-Waterfall residence; Historical Data Systems record
35		Wounded 5/ 3/63; Civil War Pension Records; 1890 Broadtop Twp. Veterans Census
36	Duval Cem.-Coaldale	1880 Broad Top Twp. Census; Historical Data Systems record
37	Arlington Nat'l Cem.	Ancestry.com-married in East Providence Twp.; B.S.F. History Book-1884
38		B.S.F. History Book-1884; Union & Snyder Co. references
39	Cedar Grove-E. Prov. Twp.	1890 E. Providence Twp. Veterans Census; B.S.F. History Book-1884
40	Mansfield Cem.-OH	1860 E. Providence Twp. Census; B.S.F. History Book-1884; Wilson & John-brothers
41	Fockler Cem.-Saxton	1890 Saxton Veterans Census; Historical Data Systems record
42		1890 Saxton Veterans Census; Frank McCoy 1912 Listing
43	East Harrisburg Cem.	1890 Saxton Veterans Census; Theodore & Samuel-brothers
44	Beaufort Nat'l Cem.	Died 3/6/63; B.S.F. History Book-1884; PA Civil War Archives-Bedford Co. residence
45	Rockville Cem.-MO	Historical Data Systems record; 1860 Bedford Twp. Census
46		B.S.F. History Book-1884; Mustered Reading, PA
47	Duval Cem.-Coaldale	1890 Broad Top Veterans Census; 1860 S. Woodbury Twp. Cen.; B.S.F. History Book-1884
48	Calvary Cem.-Pittsburgh	Wounded in Bedford Inquirer article 6/3/64; 1850 Bedford Boro Cen.; Frank & John-bros.
49	Fair Cem.-IN	B.S.F. History Book-1884
50		1860 Colerain Twp. Census; Historical Data Systems-Bedford residence
51	St. Thomas Cem.-Bedford	KIA on Picket 3/29/62; possibly 1st Bedford Co. resident KIA; Historical Data Systems record
52	Fockler Cem.-Saxton	1890 Liberty Twp. Veterans Census
53		KIA 3/30/65; B.S.F. Hist. Book-1884; PA CW Archives-born Centre Co.; enlisted in Hunt. Co.
54	Riverview Cem.-Huntingdon Co.	1890 Liberty Twp. Veterans Census
55	Pleasant Hill Meth.-Somerset Co.	B.S.F. History Book-1884; Somerset residence; Historical Data Systems record
56		B.S.F. History Book-1884
57	Poplar Grove Nat'l Cem.	Died-Typhoid 5/31/65; Bed. Gaz. 2/13/1914 list; Ancestry-Daughter born in Lovely in 1855
58		B.S.F. History Book-1884; enlisted in Berks Co.
59		Historical Data Systems-Bedford Co. residence
60	Buck Valley Cem.-Fulton Co.	Findagrave.com information; George, John & Peter-brothers
61	Seward Cem.-OK	Ancestry.com-Buck Valley references
62	Everett Cem.	Historical Data Systems record aka Shetrompf; Everett Cemetery Listing of Civil War Veterans
63		1890 Wells Twp. Veterans Census
64	IOOF Mem. Cem.-Stoystown	B.S.F. History Book-1884; Somerset Co. references

Chapter 18

	Name	Age	Rank	Regiment	Co.	Muster In	Muster Out	Casualty	Casualty Battle
1	Schlottman, John	30	1st Sgt.	US 5th Cav.	E				
2	Schmidt, George			MO					
3	Schmittle, George*	16	Pvt.	PA 110th Inf.	C	1861	10/24/64		
4	Schoener, John F*	28	1st Lt.	PA 55th Inf.	D	11/25/62	6/25/63		
5	Schroder, Charles*		Music.	PA 110th Inf.	C	10/24/61	6/28/65		
6	Schultz, Jacob*		Pvt.	PA 55th Inf.	I	10/7/63			
7	Schwartz, Samuel B	19	Sgt.	PA 110th Inf.	C	10/24/61	6/28/65		
8	Sclotheim, Edmund*		Pvt.	PA 55th Inf.	H	10/16/63		MIA	Drewry's Bluff
9	Scott, Cornelius W	17	Pvt.	PA 46th Inf.	I	7/1/63	8/19/63		
10	Scott, Cornelius W		Pvt.	PA 110th Inf.	B	2/22/64	6/28/65	Wounded	
11	Scoville, Marshall L	18	Pvt.	OH 7th Inf.	F	3/14/64	7/13/64	Wounded	
12	Scritchfield, Hezekiah	17	Pvt.	PA 133rd Inf.	A	8/5/62	5/24/63		
13	Scritchfield, Hezekiah		Corp.	PA 21st Cav.	F	8/10/63	2/20/64		
14	Scritchfield, Hezekiah		Sgt.	PA Lamb. Cav.	Ind	8/12/64	11/25/64		
15	Scritchfield, Samuel	18	Pvt.	PA 138th Inf.	E	8/29/62	4/5/63		
16	Scritchfield, Samuel		Pvt.	MD 2nd PHB Inf.	K	9/7/61	10/31/64		
17	Scutchall, David	22	Pvt.	PA 8th Res.	F	6/11/61		Died	
18	Scutchall, John	20	Pvt.	PA 133rd Inf.	C	8/13/62	5/26/63		
19	Scutchall, Samuel	18	Pvt.	PA 133rd Inf.	C	8/13/62		KIA	Fredericksburg
20	Seabrooks, George	29	Pvt.	PA 110th Inf.	C	10/24/61			
21	Seaffer, Samuel G		Pvt.	PA 21st Cav.	E	7/18/63	2/20/64		
22	Seigle, Simon B		Pvt.	PA 22nd Cav.	H	2/25/64	6/24/65		
23	Sellers, Frederick A	29	Pvt.	PA 138th Inf.	D	8/29/62	6/23/65		
24	Sellers, John F*	22	Pvt.	PA 2nd Cav.	E	11/4/61	6/28/65	POW-Died	
25	Semler, Reuben J	18	Pvt.	PA 55th Inf.	D	12/20/61		Wound.-Died	Cold Harbor
26	Sexton, William S		Pvt.	PA 82nd Inf.	K	7/5/64	7/13/65		
27	Shade, James A	25	Music.	PA 133rd Inf.	C	8/13/62	5/26/63		
28	Shade, James A		Music.	PA 208th Inf.	H	9/7/64	6/1/65		
29	Shaeffer, Sebastian		Pvt.	PA 55th Inf.	K	11/5/61	8/30/65		
30	Shafer, Alexander		Pvt.	PA 74th Inf.	B	3/13/65	8/29/65		
31	Shafer, Josiah		Pvt.	PA 147th Inf.	B	2/15/64	7/15/65	Wounded	
32	Shaffer, Abraham	31	Corp.	PA 133rd Inf.	K	8/15/62	5/26/63		
33	Shaffer, Abraham		Corp.	PA 186th Inf.	G	3/24/64	8/15/65		
34	Shaffer, Charles H			MD 6th PHB Inf.					
35	Shaffer, Charles H	18	Pvt.	MD 1st PHB Cav.	A	9/11/62	5/12/65	POW	Hagerstown
36	Shaffer, Daniel L	24	Corp.	PA 21st Cav.	E	7/2/63	7/8/65		
37	Shaffer, George W	15	Pvt.	PA 55th Inf.	K	2/19/64	7/7/65	Wounded	Drewry's Bluff
38	Shaffer, Harvey E	21	Sgt.	PA 138th Inf.	F	8/29/62	6/23/65	Wounded	Monocacy, MD
39	Shaffer, Isaiah A	40	Pvt.	PA 208th Inf.	K	9/7/64	6/1/65		
40	Shaffer, Jacob		Pvt.	PA 21st Cav.	E	7/4/63	2/20/64		
41	Shaffer, Jacob J	19	Pvt.	PA 55th Inf.	H	2/19/64		POW-Died	Drewry's Bluff
42	Shaffer, Joseph C		Pvt.	MD 1st PHB Cav.	A	8/10/62	1865		
43	Shaffer, Joseph H		Pvt.	MD 6th Inf.	D	8/20/62	6/20/65	POW	Wilderness
44	Shaffer, Levi M	35	Pvt.	PA 208th Inf.	K	9/7/64	6/1/65		
45	Shaffer, Peter*	25	Pvt.	PA 55th Inf.	I	8/28/61		POW-Died	Drewry's Bluff
46	Shaffer, Samuel	29	Pvt.	PA 133rd Inf.	K	8/15/62	5/26/63		
47	Shaffer, Thomas	29	Pvt.	PA 138th Inf.	F	8/29/62	6/12/65	Wounded	Fisher's Hill
48	Shaffer, Tobias	23	Pvt.	PA 138th Inf.	F	8/29/62	6/23/65	Wounded	Fisher's Hill
49	Shaffer, William	38	Pvt.	PA 208th Inf.	K	9/7/64	6/1/65		
50	Shaffer, William B		Pvt.	PA 21st Cav.	E	7/8/63	2/20/64		
51	Shaffer, Zachariah A	38	Pvt.	PA 91st Inf.	F	9/21/64	5/15/65		
52	Shank, Joseph		Corp.	PA 21st Cav.	E	7/8/63	7/8/65		
53	Shank, Joshua*	33	Pvt.	PA 55th Inf.	I	2/15/64	8/30/65		
54	Sharp, George	39	Corp.	PA 61st Inf.	G	9/26/64		Died	
55	Sharp, James	28	Pvt.	PA 171st Inf.	I	11/10/62		Died	
56	Shauf, Cornelius	18	Pvt.	PA 8th Res.	F	6/11/61		Died	
57	Shauf, John	52	Pvt.	PA 107th Inf.	H	2/10/62	5/8/62		
58	Shauf, John		Pvt.	PA McKeage's Mil.	D	7/3/63	8/8/63		
59	Shauf, John J	17	Pvt.	MD 1st PHB Cav.	M	3/19/64	6/28/65		
60	Shauley, Samuel E		Pvt.	PA 21st Cav.	E	7/8/63	2/20/64		
61	Shaw, Matthew P		Pvt.	PA 8th Res.	F	6/19/61		POW	Seven Days Battle
62	Shaw, Matthew P	22	Pvt.	PA 8th Res.	F			KIA	South Mountain
63	Shaw, Zopher P*	24	Pvt.	PA 133rd Inf.	K	8/15/62		KIA	Fredericksburg
64	Shawley, Andrew*			PA 21st Cav.				Died	

	Cemetery	Notes
1		Historical Data Systems-Bedford Co. residence
2	Union Cem.-Colerain Twp.	Frank McCoy 1912 Listing
3	Rose Hill Cem.-Altoona	B.S.F. History Book-1884; Blair Co. references
4	Union Cem.-Berks Co.	B.S.F. History Book-1884; Berks Co. references
5		B.S.F. History Book-1884; PA Civil War Archives-Philadelphia residence
6		Deserted 5/6/64; B.S.F. History Book-1884
7		B.S.F. History Book-1884; PA Civil War Archives-Woodbury residence
8		MIA 5/16/64 Bedford Inquirer 7/8/64 article; B.S.F. Hist. Book-1884; Mustered Reading, PA
9	Broad Top IOOF Cem.	Historical Data Systems record
10		1890 Robertsdale Veterans Census-Chest Wound
11	St. Bartholomew-Cambria Co.	Arm shot off 5/25/64; 1890 Union Twp. (Pavia) Vet Census
12	Grandview Cem.-Cambria Co.	Historical Data Systems record; Born in Bedford Co.; Hezekiah & Samuel-brothers
13		1850 Napier Twp. Census; Historical Data Systems record
14		Ancestry.com information; Regiment on gravestone; aka Screachfield
15	Milligans Cove-Buffalo Mills	B.S.F. History Book-1884; Transferred back to 2nd MD PHB Infantry on 4/4/63
16		PA Civil War Project record; Historical Data Systems record; possible unauthorized absence
17		Died 1/5/63; Historical Data Systems-Bedford Co. residence; 1850 Broad Top Twp. Census
18	Ashbury Meth. Cem.-Graceville	Frank McCoy 1912 Listing; Historical Data Systems record; John, David & Samuel-brothers
19		KIA 12/13/62; B.S.F. History Book-1884; 1860 Broad Top Twp. Census
20		Deserted 12/28/61; PA Civil War Archives-Woodbury Census
21		Historical Data Systems-Bedford Co. residence
22		B.S.F. History Book-1884
23	Mt. Olivet Cem.-Manns Choice	1890 Napier Twp. Veterans Census; B.S.F. History Book-1884
24		POW-Andersonville; B.S.F. History Book-1884; Philadelphia references
25	Cold Harbor Nat'l Cem.	Wounded 6/3/64; Died 6/9/64; B.S.F. History Book-1884; 1860 Bedford Twp. Census
26		1890 Napier Twp. Veterans Census
27	Potter Creek Cem.-Woodbury	B.S.F. History Book-1884; Historical Data Systems-Bedford Co. residence
28		B.S.F. History Book-1884; Historical Data Systems record
29		B.S.F. History Book-1884; Historical Data Systems-Bedford Co. residence
30		1890 Napier Twp. Veterans Census
31		1890 Hopewell Veterans Census
32	Providence Union Cem.-Everett	B.S.F. History Book-1884; Historical Data Systems record
33		B.S.F. History Book-1884; Abraham & Samuel-brothers
34		Frank McCoy 1912 Listing
35	Hyndman Cem.	POW 7/30/64; Historical Data Systems record
36		Historical Data Systems-Bedford Co. residence; 1860 Napier Twp. Census
37	Mt. Zion Cem.-Pavia	Wounded 5/13/64-Pension record; 1890 Union Twp. (Pavia) Veterans Census
38	Woods Church-Colerain Twp.	Left Ankle Wound 7/9/64; PA Civil War Arch.-Colerain Twp. Cen.; Harvey & Thomas-bros.
39	Rock Hill Cem.-Monroe Twp.	1890 Monroe Twp. Veterans Census; Historical Data Systems record
40		Historical Data Systems-Bedford Co. residence
41	Andersonville Cem.	MIA 5/16/64 in Bed. Inq.-7/8/64; POW-Andersonville; Died 9/30/64; 1850 Napier Twp. Cen.
42		1890 Hyndman Veterans Census
43		POW 5/5/64-10 mths.; 1890 Hyndman Veterans Census
44	Graceville Luth. Cem.	1890 E. Providence Twp. Veterans Census
45		POW 5/16/64; Died 6/4/64; B.S.F. Hist. Book-1884; PA CW Archives-Born in Huntingdon Co.
46	Providence Union Cem.-Everett	B.S.F. History Book-1884; aka Shafer; 1850 W. Providence Twp. Census
47		Leg Amputated 9/21/64; Historical Data Systems-Bedford Co. residence
48	Newton Union Cem.-IA	Wounded 9/22/64, Hospital at Must. Out; Historical Data Systems record
49	Rays Cove CC Cem.	1860 E. Providence Twp. Census; B.S.F. History Book-1884; William & Levi-bros.
50		Historical Data Systems-Bedford Co. residence
51	Mt. Zion-Southampton Twp.	Frank McCoy 1912 Listing; Discharged on Surgeon's Cert.; Zachariah & Isiah-brothers
52		Historical Data Systems-Bedford Co. residence
53		B.S.F. History Book-1884
54		Died 2/19/65; Bedford Gazette 2/13/1914 listing; 1860 S. Woodbury Twp. Census
55		Died 7/3/63; B.S.F. History Book-1884; PA Civil War Archives enlisted in Bedford Co.
56	Rock Hill Cem.-Monroe Twp.	Died of Disease 12/17/61; Hist. Data Systems record; Cornelius & John-brothers; aka Shoaff
57	Mt. Zion Cem.-Breezewood	1850 Colerain Twp. Census; Frank McCoy 1912 Listing; Discharged on Surgeon's Cert.
58		Historical Data Systems record; John Shoaff is father of John J Shauf & Cornelius Shoaff
59	Mt. Zion Cem.-Breezewood	1890 Broad Top Veterans Census
60		Historical Data Systems-Bedford Co. residence
61		POW 6/27/62; Bedford Gazette-8th Reserve List; Historical Data Systems record
62	Antietam Nat'l Cem.	KIA 9/14/62; B.S.F. Hist. Book-1884; Hist. Data Systems record; 1860 Monroe Twp. Cen.
63		KIA 12/13/62; PA Civil War Archives-enlisted in Fulton Co.; Taylor Twp. references
64		Died of Disease; Bedford Gazette 2/13/1914 listing

Chapter 18

	Name	Age	Rank	Regiment	Co.	Muster In	Muster Out	Casualty	Casualty Battle
1	Shay, Charles*		Pvt.	PA 76th Inf.	E	2/21/65	7/18/65		
2	Sheaffer, Anthony	23	Pvt.	PA 101st Inf.	D	11/1/61	3/5/63		
3	Shearer, Daniel*	25							
4	Sheeden, Aaron		Pvt.	PA McKeage's Mil.	H	7/2/63	8/8/63		
5	Sheeder, Henry F	25	Pvt.	PA 53rd Inf.	C	10/17/61	6/30/65	Wounded	Fredericksburg
6	Sheeder, James T	17	Pvt.	PA 22nd Cav.	M	2/17/64	6/24/65		
7	Shehan, James		Pvt.	PA 21st Cav.	E	7/8/63	7/8/65	POW	Boydton Plank Rd.
8	Sheiner, Robert N		Corp.	PA 208th Inf.	H	9/6/64	6/1/65		
9	Shellar, William	19	Pvt.	PA 13th Inf.	G	4/25/61	8/6/61		
10	Shelley, Abraham	33	Pvt.	PA 18th Cav.	H	2/24/64	10/31/65		
11	Shepard, Robert K*		Pvt.	PA 55th Inf.	I	8/28/61		POW-Died	Drewry's Bluff
12	Shields, James		Pvt.	PA 8th Res.	F	6/11/61	2/22/62		
13	Shields, Michael*		Pvt.	PA 55th Inf.	K	7/20/63			
14	Shimer, Isaac	17	Pvt.	PA 194th Inf.	G	7/20/64	11/6/64		
15	Shimer, John	14	Pvt.	PA 200th Inf.	C	8/25/64	4/21/65		
16	Shimer, William H H	20	Sgt.	PA 110th Inf.	C	10/24/61	6/28/65		
17	Shine, James		Pvt.	PA 55th Inf.	D	7/20/63			
18	Shinn, Job	24	Pvt.	PA 53rd Inf.	I	10/10/61	11/6/64	Wounded	Fair Oaks
19	Shinn, Job		Pvt.	PA 53rd Inf.	I			Wounded	Reams Station
20	Shinn, William		Pvt.	PA 28th Inf.	O	8/17/61	12/28/62		
21	Shinn, William		Pvt.	PA 147th Inf.	B	12/29/62	7/15/65		
22	Shippley, Lorenzo D	40	Pvt.	PA 171st Inf.	I	11/2/62		Died	
23	Shock, Daniel		Corp.	PA 194th Inf.	G	7/20/64	11/6/64		
24	Shock, Daniel	41	2nd Lt.	PA 77th Inf.	F	3/7/65	12/6/65		
25	Shoemaker, Austin	20	Pvt.	PA 110th Inf.	C	10/24/61	6/28/65	POW	Jerusalem Plank Rd.
26	Shoemaker, Benjamin	18	Sgt.	PA 110th Inf.	C	10/24/61	6/28/65		
27	Shoemaker, George F	19	Corp.	PA 101st Inf.	D	11/1/61	6/3/65	POW	Plymonth, NC
28	Shoemaker, Isaac F	26	Corp.	PA 101st Inf.	D	12/6/61		Died	
29	Shoemaker, Jacob W		Pvt.	PA 101st Inf.	D	11/1/61			
30	Shoemaker, Jacob W		Pvt.	OH 86th Inf.					
31	Shoemaker, John	40	Corp.	PA 28th Inf.	O	8/17/61	10/28/62		
32	Shoemaker, John		Corp.	PA 147th Inf.	I	11/3/62	5/31/65		
33	Shoemaker, Philip T		Pvt.	PA McKeage's Mil.	G	7/2/63	8/8/63		
34	Shoeman, David	24	Pvt.	PA 14th Inf.	I	5/2/61	8/7/61		
35	Shoeman, Peter	30	Pvt.	PA 49th Inf.	H	9/26/63		Wound.-Died	Spotsylvania
36	Shoenfelt, Henry	24	Pvt.	PA 55th Inf.	D	10/12/61	8/30/65		
37	Shoenfelt, Jacob*	17	Corp.	PA 55th Inf.	D	2/16/64	8/30/65		
38	Sholl, Isaac	20	Pvt.	PA 55th Inf.	H	9/2/62	6/11/65	Wounded	Drewry's Bluff
39	Sholl, Isaac		Pvt.	PA 55th Inf.	H			Wounded	Petersburg - Initial
40	Shontz, George W	31	Pvt.	PA 78th Inf.	K	2/28/65	9/11/65		
41	Shontz, Philip*	25	Corp.	PA 76th Inf.	E	8/27/63	7/18/65		
42	Shoop, John	16	Pvt.	PA McKeage's Mil.	G	7/2/63	8/8/63		
43	Shoop, Joseph L	22	Pvt.	PA 55th Inf.	I	10/11/61		Wound.-Died	Cold Harbor
44	Shoop, William	25	Pvt.	PA 1st Light Art.	F	7/13/61	7/20/64	Wounded	
45	Shoope, John		Pvt.	MD 3rd PHB Inf.	A	9/12/61	9/12/64	POW	Moorefield
46	Shoup, David	38	Pvt.	PA 78th Inf.	K	2/28/65	9/11/65		
47	Shoup, John N	19	Pvt.	MD 2nd PHB Inf.	A	8/27/61	5/29/65		
48	Showalter, Absolom	17	Pvt.	PA 48th Inf.	B	1/20/65	7/17/65		
49	Showalter, Jacob	25	Pvt.	PA 12th Cav.	K	3/16/64	7/20/65		
50	Showalter, John		Pvt.	PA McKeage's Mil.	H	7/2/63	8/8/63		
51	Showalter, Simon P	15	Pvt.	PA 8th Res.	F	4/23/61	10/27/62		
52	Showalter, Simon P		Pvt.	US 6th Cav.	K	10/27/62			
53	Showalter, Simon P		Pvt.	PA 22nd Cav.	H	2/26/64	6/24/65		
54	Showalters, Henry		Pvt.	PA 8th Res.	F	6/11/61	5/15/64	Wounded	Seven Days Battle
55	Showalters, Henry		Pvt.	PA 8th Res.	F			Wound.-POW	2nd Bull Run
56	Showalters, Henry		Pvt.	PA 191st Inf.		5/15/64	10/17/64		
57	Showman, William	47	Pvt.	PA 101st Inf.	G	12/2/61		Died	
58	Shrader, Auterbine	18	Pvt.	PA 55th Inf.	H	2/19/64		POW-Died	
59	Shrader, William O	18	Pvt.	PA 55th Inf.	H	2/29/64		POW-Died	Drewry's Bluff
60	Shrimer, Alex K	21	Pvt.	PA 171st Inf.	I	11/10/62	8/8/63		
61	Shroyer, Andrew G	20	Pvt.	PA 133rd Inf.	K	8/15/62	5/26/63	Wounded	Fredericksburg
62	Shroyer, Andrew G		Pvt.	PA 186th Inf.	E	2/24/64	8/15/65		
63	Shroyer, Daniel J	22	Pvt.	MD 2nd PHB Inf.	H	8/31/61	9/29/64		
64	Shroyer, Jacob D	37	Pvt.	PA 19th Cav.	F	8/11/63			

	Cemetery	Notes
1		Historical Data Systems; B.S.F. History Book-1884; Substitute
2		B.S.F. History Book-1884; Historical Data Systems enlisted in Rainsburg; Disch. Surg. Cert.
3	Clearville Union Cem.	Civil War marker at gravesite
4		B.S.F. History Book-1884
5	Everett Cem.	1890 Everett Veterans Census; B.S.F. History Book-1884
6	Everett Cem.	Civil War Pension Record; Everett Cemetery Listing of Civil War Veterans
7		POW 10/27/64; Historical Data Systems-Bedford Co. residence
8		B.S.F. History Book-1884; Historical Data Systems record-Bedford Co. residence
9		B.S.F. History Book-1884; 1860 Bedford Boro Census
10	Wooster Cem.-OH	B.S.F. History Book-1884; 1860 Hopewell Twp. Census
11		POW-Richmond 5/16/64; Died 6/29/64; B.S.F. History Book-1884; Mustered in West Chester
12		B.S.F. History Book-1884; Historical Data Systems record; General Court Marshall
13		Deserted 7/20/63; B.S.F. History Book-1884
14	Fairview Cem.-Altoona	1850 Union Twp. (Pavia) Census, Frank McCoy 1912 Listing; Isaac & John-brothers
15	Greenlawn Cem.-Roaring Spring	1890 Bedford Twp. Veterans Census; 51st PA Infantry references
16	Greenlawn Cem.-Roaring Spring	B.S.F. History Book-1884; 1860 Bedford Twp. Census; Civil War Pension records
17		Deset. 5/3/64; B.S.F. History Book-1884
18	Everett Cem.	Wounded May 31-Jun 1, 1862; 1890 Everett Veterans Census
19		Wounded 8/29/64; Everett Cemetery Listing of Civil War Veterans
20		1890 W. Providence Twp. Veterans Census
21		1890 W. Providence Twp. Veterans Census
22	US Soldiers & Airmans Nat'l	Died of Disease 7/21/63; B.S.F. History Book-1884; Bed. Gazette 2/13/1914 list; aka L. Dow
23		Historical Data Systems record; 1860 Greenfield Twp. Cen.; Ancestry.com Born-Bedford Co.
24	Claysburg Union Cem.	County Commissioner and State Legislator
25	Dry Hill Cem.-Woodbury	POW-Andersonville 6/23/64 to 4/28/65; attended 1905 Andersonville Dedication
26	Holsinger Cem.-Bakers Summit	PA Civil War Archives-Woodbury residence; Austin & Benjamin-brothers
27	St. Marks Cem.-King	POW-Andersonville 4/20/64 to 3/1/65, attended 1905 Andersonville Dedication
28	New Bern Nat'l Cem.	Died 11/10/64; B.S.F. History Book-1884; 1860 Colerain Twp. Census
29	Fountain Cem.-OH	Deserted 9/15/62; Historical Data Systems-Bedford Co. residence
30		Historical Data Systems record; 1850 Colerain Twp. Census
31	Bedford Cem.	Frank McCoy 1912 Listing; Historical Data Systems record; aka Shumaker
32	Stonerstown Cem.-Liberty Twp.	Historical Data Systems record; Bedford Inquirer 5/22/1908 listing; aka Shumaker
33		B.S.F. History Book-1884
34	Fairview Cem.-Martinsburg	Historical Data Systems record; 1850 M. Woodbury Twp. Census; David & Peter-brothers
35	Arlington Nat'l Cem., 27 200	Wounded 5/10/64; Died 5/18/64; Hist. Data Systems record; 1850 M. Woodbury Twp. Cen.
36	Bedford Cem.	1890 Bedford Twp. Veterans Census; Frank McCoy 1912 Listing
37		B.S.F. History Book-1884; aka Shoenfeld; Westmoreland Co. references
38		Wounded 5/16/64 listed in Bedford Inquirer article 7/8/64; 1850 Union Twp. (Pavia) Census
39		Wounded 6/18/64 listed in Bedford Inquirer 7/8/64 article; B.S.F. History Book-1884
40	Entriken Cem.-Hunt. Co.	1890 Hopewell Veterans Census
41		B.S.F. History Book-1884; Drafted
42		1890 Hyndman Veterans Census also lists MD regiment; B.S.F. History Book-1884
43	Arlington Nat'l Cem., 13-7641	Wounded 6/3/64; Died 8/7/64; B.S.F. History Book-1884; 1860 Greenfield Twp. Census
44	St. John's Cem.-Loysburg	1890 S. Woodbury Twp. Census
45		POW 6/29/62; 1890 Hyndman Twp. Census; aka Shoe
46	Old Stone Ch.-Marklesburg	1840 Hopewell Twp. Census; Historical Data Systems record
47	Hyndman Cem.	Historical Data Systems record aka Shupe; Civil War Pension Record
48	Union Cem.-Crystal Springs	Findagrave.com information; Substitute; Absolom, Jacob and Simon-brothers
49	Stevens Chapel-Monroe Twp.	Civil War Pension Record; Bedford Inquirer 5/22/1908 listing
50		B.S.F. History Book-1884
51	Providence Union Cem.-Everett	1860 W. Providence Twp. Census; PA Civil War Archives-Bedford Co. residence
52		Historical Data Systems record
53		Historical Data Systems record
54		POW 6/27/62; B.S.F. History Book-1884; Historical Data Systems-Bedford Co. residence
55		Wounded-POW; Bedford Gazette-8th Reserve List; Historical Data Systems record
56		B.S.F. History Book-1884; Historical Data Systems-Bedford Co. resid.; Disch. Surgeon's Cert.
57	Cypress Hill Cem.-NY	Died of Typhoid 7/9/62, B.S.F. History Book-1884; 1860 Juniata Twp. Census; aka Sherman
58	Andersonville Cem.	POW-Andersonville; Died 9/29/64; B.S.F. Hist. Book-1884; PA CW Arch.-enlist. Schellsburg
59	Andersonville Cem.	MIA on 5/16/64 listed in Bedford Inquirer on 7/8/64; POW-Andersonville; Died 8/27/64
60		B.S.F. History Book-1884
61	Oak Ridge Cem.-Altoona	Wounded 12/13/62; B.S.F. History Book-1884; 1850 Londonderry Twp. Census
62		PA Civil War Project record
63	Rose Hill Cem.-Cumberland	Civil War Pension Record; Daniel, John & Joseph-brothers
64	Comp Cem.-Somerset Co.	Civil War burial record; aka Schreyer; Father of Joseph

Chapter 18

	Name	Age	Rank	Regiment	Co.	Muster In	Muster Out	Casualty	Casualty Battle
1	Shroyer, Jacob D		Pvt.	MD 2nd PHB Inf.	D	3/4/65	5/29/65		
2	Shroyer, John	16	Pvt.	PA 208th Inf.	A	8/18/64	6/1/65		
3	Shroyer, Joseph	19	Pvt.	PA 138th Inf.	F	8/29/62	6/23/65	Wounded	Opequon
4	Shroyer, Moses	17	Pvt.	PA 138th Inf.	D	8/29/62	6/23/65	Wounded	Opequon
5	Shuck, Daniel J*								
6	Shuck, James H								
7	Shuck, John N	20	Pvt.	PA McKeage's Mil.	D	7/2/63	8/8/63		
8	Shull, Henry R	27	Pvt.	PA 55th Inf.	K	11/5/61	2/21/63		
9	Shull, Henry R		Pvt.	US 1st Light Art.	M	2/21/63			
10	Shull, Isaac		Pvt.	IN 84th Inf.	E	8/11/62		KIA	Chickamauga
11	Shull, Jacob	17	Pvt.	PA Ind Light Art.	C	2/25/64	6/30/65		
12	Shull, John	19	Pvt.	PA 5th Heavy Art.	K	9/1/64	6/30/65		
13	Shull, Joseph	21	Corp.	IN 84th Inf.	E	7/26/62	6/14/65		
14	Shull, William (1)		Pvt.	PA 21st Cav.	E	6/23/63	2/20/64		
15	Shull, William (1)	26	Pvt.	PA 5th Heavy Art.	K	9/1/64	6/30/65		
16	Shull, William (2)	22	Pvt.	PA 171st Inf.	I	11/2/62	8/8/63		
17	Shultz, Anthony B	38	Pvt.	PA 78th Inf.	K	3/3/65	9/11/65		
18	Shultz, David H		Pvt.	PA McKeage's Mil.	H	7/2/63	8/8/63		
19	Shultz, Frederick	24	Pvt.	PA 55th Inf.	E	10/2/63			
20	Shultz, Henry	42	Pvt.	PA 78th Inf.	K	3/3/65	9/11/65		
21	Shultz, Jacob		Pvt.	PA 58th Inf.	E	11/1/64	5/30/65		
22	Shultz, Washington	32	Pvt.	PA 102nd Inf.	I	3/28/65	6/28/65		
23	Shunk, Jacob	33	Pvt.	PA 76th Inf.	E	9/26/64	6/28/65		
24	Shur, William*	24	Pvt.	PA 55th Inf.	K	9/28/63	8/30/65		
25	Shuss, Adam	35	Pvt.	PA 56th Inf.	F	9/21/64	5/31/65		
26	Sigel, Raphael	33	Pvt.	PA 11th Res.	A	9/30/61	3/5/62	Wounded	
27	Sigel, Raphael		Pvt.	PA 22nd Cav.	I	2/26/64	8/11/65	Wounded	
28	Sigel, Samuel B		Music.	PA McKeage's Mil.	D	7/2/63	8/8/63		
29	Sigel, Stephen	36	Pvt.	PA 11th Inf.	A	9/30/61	1/8/63	Wounded	2nd Bull Run
30	Sigel, Stephen		Pvt.	PA 11th Inf.	A			Wounded	Antietam
31	Siler, James P	20	Pvt.	PA 101st Inf.	D	12/6/61	6/25/65	POW	Plymonth, NC
32	Simmons, Thomas H	14	Sgt.	MD 3rd PHB Inf.	B	11/11/61	6/29/65		
33	Simmons, William	26	Corp.	PA 17th Cav.	G	10/2/62			
34	Simmons, William		Pvt.	PA 91st Inf.	K	3/21/65	7/25/65		
35	Simpson, James	28	Pvt.	PA 22nd Cav.	M	2/26/64	8/19/65		
36	Singleton, Samuel	22	Pvt.	PA 133rd Inf.	D	8/5/62	5/24/63	Wounded	Fredericksburg*
37	Singleton, Samuel		2nd Lt.	PA 209th Inf.	C	9/1/64	5/31/65		
38	Sipes, Charles W	24	Pvt.	IA 28th Inf.	F	9/15/62		POW	Opequon
39	Sipes, Dennis B	36	Pvt.	PA 158th Inf.	H	11/4/62	8/12/63		
40	Sipes, John		Pvt.	PA 22nd Cav.	I	2/11/64	6/24/65		
41	Skillington, John	24	Pvt.	PA 49th Inf.	F	6/14/64		Died	
42	Skillington, Robert M	21	Pvt.	PA 133rd Inf.	C	8/13/62	5/26/63		
43	Skillington, Robert M	23	Pvt.	PA 184th Inf.	A	5/12/64	7/14/65	Wounded	Cold Harbor
44	Skillington, Robert M		Pvt.	PA 184th Inf.	A			Wounded	Petersburg - Initial
45	Skipper, Alexander	31	Sgt.	PA 20th Cav.	M	7/28/63	10/3/63		
46	Skipper, Alexander		Sgt. QM	PA 208th Inf.	H	9/8/64	6/1/65		
47	Skipper, Augustus	18	Pvt.	PA 194th Inf.	I	7/22/64	9/6/64		
48	Slack, Francis M	23	Pvt.	PA 13th Inf.	G	4/25/61	8/6/61		
49	Slack, Francis M		1st Sgt.	PA 138th Inf.	E	8/29/62	6/13/65	Wounded	Petersburg - Final
50	Slack, George	32	Pvt.	PA 99th Inf.	G	2/25/65	7/1/65		
51	Slaughter, George W		Pvt.	MD 3rd Cav.	C	7/28/63	3/17/65		
52	Slayman, William	21	Corp.	MD 1st Reg Cav.	H	12/3/61	8/8/65		
53	Sleek, Andrew J	25	Pvt.	PA 55th Inf.	K	11/5/61			
54	Sleek, Jesse W		Pvt.	PA 184th Inf.	C	5/15/64	10/29/64	Died	
55	Sleek, Jesse W		Pvt.	VRC 33rd Inf.		10/29/64	9/4/65		
56	Sleek, Thomas J	39	Pvt.	PA 97th Inf.	C	3/4/65	7/15/65		
57	Slick, Abner W	29	Pvt.	PA 171st Inf.	I	11/2/62	11/10/62		
58	Slick, Allen	48	Pvt.	PA 55th Inf.	H	2/19/64		KIA	Cold Harbor
59	Slick, George W	18	Pvt.	PA 21st Cav.	E	7/8/63	2/20/64		
60	Slick, George W		Pvt.	PA 5th Heavy Art.	K	9/1/64	7/30/65		
61	Slick, Henry	21	Pvt.	IL 112th Inf.	A	9/20/62	6/20/65		
62	Slick, Hezekiah B	22	Pvt.	PA 55th Inf.	H	10/11/61	1/13/63		
63	Slick, Hezekiah B		Pvt.	PA 55th Inf.	H	2/19/64		POW-Died	
64	Slick, James		Pvt.	IL 112th Inf.	A				

	Cemetery	Notes
1		Ancestry.com information; Father of John & Joseph
2	Salisbury IOOF-Somerset Co.	1850 Harrison Twp. Census; B.S.F. History Book-1884
3	Lybarger Cem.-Madley	Wounded 9/19/64; 1890 Londonderry Twp. Veterans Census; Historical Data Systems record
4	Cooks Mill Cem.-Hyndman	Wounded 9/19/64; Died 1868; Frank McCoy 1912 Listing; Andrew & Moses-brothers
5	Bedford Cem.	Findagrave lists as a Civil War Veteran
6	Bedford Cem.	Bedford Cemetery Civil War Veterans list; Civil War Burial Record
7	Oak Ridge Cem.-Altoona	B.S.F. History Book-1884; 1850 Bedford Boro Census
8	West Union Cem.-OH	B.S.F. History Book-1884; Historical Data Systems-Bedford Co. residence
9		Ancestry.com-born Bedford Co.; Henry, Isaac, Joseph & Jacob-brothers
10		KIA 9/20/63; Ancestry.com-born Bedford Co.; 1850 Union Twp. Census
11	Saratoga Cem.-IN	Historical Data Systems record
12	Oak Grove Cem.-OK	Ancestry.com information; 1850 Cumberland Valley Census; John & William (1)-brothers
13	Newport Cem.-WA	Ancestry.com-born Bedford Co.
14		Civil War Pension Records; Historical Data Systems-Bedford Co. residence
15	Schellsburg Cem.	1890 Napier Twp. Veterans Census
16		B.S.F. History Book-1884
17	Reformed Cem.-Marklesburg	Ancestry.com information; 1850 Hopewell Census; Anthony, Washington & Henry-brothers
18		B.S.F. History Book-1884
19		B.S.F. History Book-1884; Historical Data Systems record; Discharged on Surgeon's Cert.
20	Reformed Cem.-Marklesburg	Historical Data Systems record; Absent-Sick at Muster Out
21		1890 Brush Creek Twp. Veterans Census
22	Reformed Cem.-Marklesburg	1890 Hopewell Veterans Census; aka Sheets
23	Pleasant Hill Cem.-Imlertown	B.S.F. History Book-1884; 1860 Bedford Twp. Census
24		B.S.F. History Book-1884; Buried in Berks County
25	Mt. Union Cem.-Mench	1890 E. Providence Twp. Veterans Census; Frank McCoy 1912 Listing
26	Mt. Pleasant Cem.-Mattie	1890 E. Providence Twp. Veterans Census; Historical Data Systems record; Disch. Surg. Cert.
27		1860 E. Providence Twp. Census; Foot Wound-Sep. 1864 near Mt. Verrow Forge, VA
28		B.S.F. History Book-1884
29	Clearville Union Cem.	Wounded 8/28/62; aka Seigel; B.S.F. History Book-1884; Historical Data Systems record
30		Wounded 9/17/62; Pension Document record
31		POW 4/20/64; B.S.F. History Book-1884
32	Providence Union Cem.-Everett	1890 Broad Top Twp. Veterans. Census
33	Providence Union Cem.-Everett	Deserted 7/7/63; Regiment at Gettysburg; Bedford Inquirer 5/22/1908 listing
34		Findagrave.com information; 1870 W. Providence Twp. Census; William & Thomas-brothers
35	Akersville Cem.-Crystal Springs	1890 E. Providence Twp. Veterans Census
36		1890 Hopewell Veterans Census-Thigh wound
37		Historical Data Systems record
38	Cook Center Cem.-IA	POW 9/19/64; Ancestry.com information; 1850 Colerain Twp. Census
39	Wells Valley Meth. Cem.	1890 Wells Twp. Veterans Census
40		B.S.F. History Book-1884
41	Alexandria Nat'l Cem., 3363	Died 9/14/64; 1850 Bedford Boro Census; Pension Record
42	Mt. Zion Cem.-Breezewood	1890 E. Providence Twp. Veterans Census
43		Wounded 6/3/64; 1860 Snake Spring Twp. Census; John & Robert-brothers
44		B.S.F. History Book-1884; Exploding Shell Wound 6/16/64; in Hospital at Muster Out
45	Green Hill Presb.-Fulton Co.	Historical Data Systems-Bedford Co. residence
46		Historical Data Systems record; 1850 Wells Twp. Census (Fulton Co.)
47		B.S.F. History Book-1884; enlisted in Hopewell
48		B.S.F. History Book-1884
49		Wounded 4/2/65; Historical Data Systems-Bedford residence; B.S.F. History Book-1884
50	Schellsburg Cem.	1890 Napier Twp. Veterans Census
51		1890 Hyndman Veterans Census
52	Everett Cem.	Historical Data Systems record; Everett Cemetery Listing of Civil War Veterans
53		Deserted 11/13/61; B.S.F. History Book-1884; Historical Data Systems record
54		1860 St. Clair Twp. Census; 1/15/66 Pension Record; Historical Data Systems record
55		Pension Record-DOD 1/15/66; PA Civil War Project record
56	St. Pauls Cem.-Cessna	1890 E. St. Clair Twp. Veterans Census; Deserted After war
57	Pleasantville Cem.	Findagrave.com info.; Disch. Surgeon's Cert.; Abner, Thomas W, Samuel & Josiah-brothers
58		KIA 6/3/64; B.S.F. History Book-1884; PA Civil War Archives record; 1850 Napier Twp. Cen.
59	Arlington Nat'l Cem., 21325	Historical Data Systems-Bedford Co. residence
60		Historical Data Systems record
61	Slick Cem.-Somerset Co.	Ancestry.com-born in Bedford
62		Historical Data Systems record; Discharged on Surgeon's Cert.
63		POW-Salisbury; Died 2/6/65; PA CW Archives-Bedford Co. resid.; 1850 Napier Twp. Cen.
64	Fairview Cem.-IA	Ancestry.com Born in St. Clairsville; James, Henry & Philip-brothers

	Name	Age	Rank	Regiment	Co.	Muster In	Muster Out	Casualty	Casualty Battle
1	Slick, John A	33	Pvt.	PA 208th Inf.	H	9/8/64		Wound.-Died	Petersburg - Final
2	Slick, Josiah	15	Corp.	PA 55th Inf.	H	10/11/61	12/6/64	Wounded	
3	Slick, Josiah		2nd Lt.	USCT 107th Inf.		12/6/64			
4	Slick, Nicholas	28	Pvt.	PA 55th Inf.	D	2/18/64	8/30/65		
5	Slick, P Abraham	25	Pvt.	OH 87th Inf.	E	6/10/62	10/3/62		
6	Slick, P Abraham		Pvt.	OH 156th Inf.	H	5/2/64	9/1/64		
7	Slick, Philip	18	Pvt.	IL 9th Cav.	C	9/19/61		POW-Died	Tupelo, Miss.
8	Slick, Samuel K	20	Pvt.	PA 101st Inf.	G	12/28/61		Wound.-Died	
9	Slick, Thomas	26	Pvt.	OH 156th Inf.	H	5/15/64	9/1/64		
10	Slick, Thomas W	23	Pvt.	PA 101st Inf.	G	12/28/61	6/25/65	POW	Plymouth, NC
11	Slick, William B	24	Pvt.	IA 27th Inf.	A	9/3/62	8/8/65		
12	Slick, William S	18	Sgt.	PA 138th Inf.	D	9/2/62	6/23/65	Wounded	Petersburg - Final
13	Slick, William W (1)	45	Pvt.	PA 101st Inf.	G	2/18/62	8/26/62		
14	Slick, William W (2)	18	Pvt.	PA 55th Inf.	H	2/19/64	6/25/65	Wounded	
15	Sloan Samuel B	21	Corp.	IA 12th Inf.	H	10/23/61	12/13/63		
16	Sloan, Andrew J	26	Pvt.	IA 12th Inf.	H	10/23/61	1/20/66		
17	Sloan, Benjamin F	18	Pvt.	PA McKeage's Mil.	H	7/2/63	8/8/63		
18	Sloan, Benjamin F		Pvt.	PA 13th Cav.	B	3/21/64	7/14/65		
19	Slonaker, John G	35	Pvt.	PA 99th Inf.	H	2/25/65	5/31/65		
20	Slott, Samuel		Sgt.	PA 76th Inf.	E	8/20/63	7/18/65		
21	Smeltzer, John B	18	Pvt.	PA 205th Inf.	C	8/27/64	6/2/65	Wounded	Petersburg - Final
22	Smith, Adam	29	Pvt.	PA 138th Inf.	F	8/29/62	10/1/63		
23	Smith, Albert	21	Pvt.	PA 133rd Inf.	D	8/14/62	5/26/63		
24	Smith, Albert		Pvt.	NJ 2nd Cav.	E	9/2/64	6/29/65		
25	Smith, Alexander C	25	Pvt.	PA 12th Cav.	H	3/4/62			
26	Smith, Alexander C		Pvt.	PA 19th Cav.	L	9/17/63		Wounded	
27	Smith, Amos	18	Corp.	PA 101st Inf.	D	1/13/62	6/3/65	POW	Plymouth, NC
28	Smith, Andrew J	23	Pvt.	PA 101st Inf.	D	12/6/61	8/14/62		
29	Smith, Aquilla	21	Pvt.	PA 45th Inf.	B	9/21/64	6/7/65		
30	Smith, Bartley	25	Pvt.	PA 91st Inf.	I	9/21/64	5/30/65		
31	Smith, Barton C		Corp.	PA McKeage's Mil.	G	7/2/63	8/8/63		
32	Smith, Barton C	18	Sgt. QM	PA 184th Inf.	A	5/12/64	7/14/65		
33	Smith, Benjamin S	22	Pvt.	PA 55th Inf.	D	9/26/62	6/11/65		
34	Smith, Charles	35	Pvt.	PA 2nd Cav.	E	11/20/61	6/17/65		
35	Smith, Charles S	18	Pvt.	PA 8th Res.	F	6/11/61	5/6/64		
36	Smith, Charles S		Pvt.	PA 191st Inf.	D	5/2/64	6/28/65	Wounded	Jerusalem Plank Rd.
37	Smith, Daniel (1)	34	Pvt.	PA 55th Inf.	H	2/29/64	8/30/65		
38	Smith, Daniel (2)	33	Pvt.	PA 158th Inf.	K	11/4/62	8/12/63		
39	Smith, Daniel (2)		Pvt.	PA 56th Inf.	F	9/19/64	5/31/65		
40	Smith, Daniel A		Pvt.	OH 18th Inf.	E	9/23/61	1/30/64		
41	Smith, David (1)	19	Pvt.	PA 138th Inf.	F	8/29/62		Wound.-Died	
42	Smith, David (2)	29	Pvt.	PA 50th Inf.	D	2/24/65	5/9/65		
43	Smith, David (3)	37	Pvt.	PA 45th Inf.	E	6/4/64	6/15/65		
44	Smith, David H		Pvt.	PA 48th Inf.	F	7/2/61	11/26/63		
45	Smith, David L	20	Pvt.	PA 67th Inf.	E	9/10/62	6/20/65		
46	Smith, David R	24	Corp.	PA 56th Inf.	B	10/1/64	5/31/65		
47	Smith, David S	22	Pvt.	PA 110th Inf.	C	12/19/61	10/24/64		
48	Smith, Emanuel	27	Pvt.	PA 184th Inf.	A	3/3/64	6/21/65		
49	Smith, Emanuel M	43	Pvt.	PA 22nd Cav.	C	7/15/63	2/5/64		
50	Smith, George (1)		Pvt.	PA 138th Inf.	F	8/29/62	6/23/65		
51	Smith, George (2)		Pvt.	PA McKeage's Mil.	H	7/2/63	8/8/63		
52	Smith, George (2)	20	Pvt.	PA 149th Inf.	I	8/17/63			
53	Smith, George (3)	35	Pvt.	PA 205th Inf.	C	9/3/64	6/2/65		
54	Smith, George E	19	Pvt.	PA 79th Inf.	C	3/9/65	7/12/65		
55	Smith, George G	17	Pvt.	PA 49th Inf.	C	2/8/64	7/15/65		
56	Smith, George L*								
57	Smith, George W (1)	21	Pvt.	PA 101st Inf.	D	2/8/62		Died	
58	Smith, George W (2)*	18	Pvt.	PA 110th Inf.	C	8/16/62		KIA	Petersburg - Initial
59	Smith, George*		Pvt.	PA 55th Inf.	I	7/20/63			
60	Smith, Gideon	25	Pvt.	PA 45th Inf.	B	6/1/64	5/30/65	Wounded	
61	Smith, Henry								
62	Smith, Henry D*		Pvt.	PA 55th Inf.	D	9/2/62	6/11/65		
63	Smith, Isaac B*		Pvt.	PA 55th Inf.	D	9/2/62	6/11/65		
64	Smith, Isaac M	34	Pvt.	PA 99th Inf.	H	7/16/63	9/19/64		

	Cemetery	Notes
1	City Point Nat'l Cem.-VA	Wounded 4/2/65; Died 4/5/65; 1860 St. Clair Twp. Census; Pension Record
2	Riverside Cem.-NE	B.S.F. History Book-1884; 1850 St. Clair Twp. Census
3		Family bible stated "Died as result of Wounds in Civil War" in 1901
4	Grandview Cem.-Cambria Co.	1890 Harrison Twp. Veterans Census; B.S.F. History Book-1884
5	Pleasant Valley Baptist-OH	1850 E. St. Clair Twp. Census
6		Historical Data Systems record
7	Andersonville Cem., 10663	POW-Andersonville 7/5/64; Died 10/11/64; Ancestry-Born Bed.Co.; 1840 St. Clair Twp. Cen.
8	Fishertown Luth. Cem.	Died of Wounds 12/19/62; PA Civil War Archives-Schellsburg residence
9	Fairview Cem.-OH	Ancestry.com information; Thomas & P Abraham-brothers
10	Old Union Cem.-Osterburg	POW 4/20/64 to 2/26/65; B.S.F. History Book-1884; 1860 S. Clair Twp. Census
11	Orchard Beach Cem.-WI	1850 St. Clair Twp. Census; Historical Data Systems record; PA Veteran Pension data
12	Pleasantville Cem.	Wounded 4/2/65 listed in Bed. Inq. 4/14/65; Frank McCoy 1912 List; 1850 St. Clair Twp. Cen.
13	Grandview Cem.-Cambria Co.	PA Civil War Archives-Bedford residence; Discharged on Surgeon's Cert.
14	Mt. Smith Cem.-Bedford	1890 Hopewell Twp. Veterans Census
15	Grant View Cem.-IA	Hist. Data Systems record; transferred to Vet. Res. Corp. 12/13/63; Samuel & Andrew-bros.
16	Platt Cem.-IA	Awarded Medal of Honor-Nashville 12/16/64; 1850 St. Clair Twp. Census
17	Newbury Cem.-Somerset Co.	B.S.F. History Book-1884
18		Historical Data Systems record
19	Mt. Union Cem.-Lovely	1890 W. St. Clair Twp. Veterans Census
20		B.S.F. History Book-1884
21	Yellow Creek Reformed Cem.	Wounded 4/2/65; 1890 Broad Top Veterans Census
22	Smith Cem.-Londonderry Twp.	Frank McCoy 1912 Listing; 1860 Londonderry Census; Disch. Surg. Cert.; Adam & Jesse-bros.
23	Bedford Cem.	Historical Data Systems record
24		1890 Cumberland Valley Veterans Census
25	Claysburg Union Cem.	PA Civil War Archives record
26		1890 Greenfield Twp. Veterans Census
27		POW 4/20/64 to 3/1/65; B.S.F. History Book-1884
28	Pleasant Union Cem.-Clearville	1890 Monroe Twp. Veterans Census
29	Fairview Cem.-Artemas	1890 Monroe Twp. Veterans Census; Aquilla, Philip, Gideon and James-brothers
30	Fairview Cem.-Artemas	Frank McCoy 1912 Listing; Bartley, Nathan P R; William H-brothers
31		B.S.F. History Book-1884
32	Forest Lawn Mem. Park-NE	B.S.F. History Book-1884; 1860 Colerain Twp. Census
33		B.S.F. History Book-1884; 1860 Southampton Twp. Census
34	Bedford Cem.	B.S.F. History Book-1884; Frank McCoy 1912 Listing
35	Elbridge Township Cem.-MI	B.S.F. History Book-1884; Ancestry.com-born S. Woodbury Twp.
36		Wounded 6/23/64; Ancestry.com information; Charles S & David S-brothers
37	Schellsburg Cem.	PA Civil War Archives-Bedford Co. residence; Frank McCoy 1912 Listing
38	Buck Valley Cem.-Fulton Co.	1890 Brush Creek Twp. Veterans Census
39		Findagrave.com record
40	Nordhoff Cem.-CA	1850 Southampton Twp. Census; Findagrave record
41	Loudon Park Nat'l Cem.	Accidental Wounded 9/20/62; Died 10/20/62; PA Civil War Archives-Colerain Twp. residence
42	Trinity UCC Cem.-Juniata Twp.	1890 Harrison Twp. Veterans Census
43	Bennett Cem.-Artemas	Frank McCoy 1912 Listing; David (3) & Levi-brothers
44		1890 Harrison Twp. Census
45	Unity Cem.-Latrobe	Ancestry.com-Born in Bedford Co.
46	Mt. Smith Cem.-Bedford	1890 Bedford Twp. Veterans Census
47		PA Civil War Archives-Woodbury residence; B.S.F. History Book-1884
48	Evergreen Cem.-Adams Co.	B.S.F. History Book-1884; Sick in Hospital almost entire enlistment
49	Bedford Cem.	1890 Bedford Twp. Veterans Census
50	Smith Cem.-Londonderry Twp.	1890 Londonderry Twp. Veterans Census
51		B.S.F. History Book-1884
52	Mt.Hope-Southampton Twp.	Frank McCoy 1912 Listing; Sick in Washington, DC hospital at Muster Out
53	Yellow Creek Reformed Cem.	1890 Hopewell Twp. Census
54	Everett Cem.	1890 Everett Veterans Census
55	Everett Cem.	Ancestry.com information; George G, Aquilla, Gideon, Philip E & James S-brothers
56	Woods Church-Colerain Twp.	Civil War marker at gravesite
57		Died of Disease; PA Civil War Archives-Bedford Co. residence, enlisted in Clearville
58		KIA 6/18/64; B.S.F. History Book-1884; Historical Data Systems-Blair Co. residence
59		Drafted & Deserted 4/25/64; B.S.F. History Book-1884; Cincinnati references
60	Fairview Cem.-Artemas	1890 Mann Twp. Veterans Census; Frank McCoy 1912 Listing
61	Stonerstown Cem.-Liberty Twp.	Bedford Inquirer 5/22/1908 listing
62		B.S.F. History Book-1884; Berks Co. references
63		B.S.F. History Book-1884; Berks Co. references
64	Mt. Union Cem.-Lovely	Frank McCoy 1912 Listing; Historical Data Systems record

Chapter 18

	Name	Age	Rank	Regiment	Co.	Muster In	Muster Out	Casualty	Casualty Battle
1	Smith, Isaac W (1)	35	Pvt.	PA 91st Inf.	F	9/26/64	6/1/65		
2	Smith, Isaac W (2)		Pvt.	WV 6th Inf.	H	2/6/65	6/10/65		
3	Smith, Jacob (1)	32	Pvt.	PA 28th Inf.	H	9/21/64		Died	
4	Smith, Jacob (2)	43	Sgt.	PA 133rd Inf.	K	8/15/62	3/21/63		Fredericksburg*
5	Smith, Jacob (3)			PA 185th Inf.					
6	Smith, Jacob (4)	28	Pvt.	PA 99th Inf.	K	2/24/65	7/1/65		
7	Smith, Jacob C	22	Pvt.	PA 2nd Cav.	E	10/31/61	7/13/65		
8	Smith, Jacob F	32	Pvt.	PA 138th Inf.	F	8/29/62	6/23/65	Wounded	Mine Run
9	Smith, Jacob F		Pvt.	PA 138th Inf.	F			Wounded	Wilderness
10	Smith, Jacob N	22	Pvt.	PA 133rd Inf.	C	8/13/62	5/26/63		
11	Smith, Jacob W	23	Pvt.	PA 53rd Inf.	C	10/10/61	7/1/63		
12	Smith, James		Pvt.	PA 13th Cav.	K	2/27/64	7/14/65		
13	Smith, James M								
14	Smith, James S	35	Pvt.	PA 45th Inf.	G	8/4/64	7/17/65	Wounded	
15	Smith, Jasper W	31	Pvt.	PA 55th Inf.	D	9/26/62		POW-Died	
16	Smith, Jeremiah	39	Pvt.	PA 55th Inf.	K	11/5/61	7/3/63		
17	Smith, Jesse	40	Pvt.	PA 55th Inf.	D	2/17/64		Wound.-Died	Drewry's Bluff
18	Smith, John		Pvt.	PA 107th Inf.	K	8/3/64			
19	Smith, John (1)*	24	Pvt.	PA 55th Inf.	I	10/7/63			
20	Smith, John B	21	Pvt.	PA 84th Inf.	A	10/24/61	10/27/64		
21	Smith, John C		Pvt.	PA 21st Cav.	E	7/18/63	2/20/64		
22	Smith, John F	37	Pvt.	PA 107th Inf.	D	9/21/64	6/6/65		
23	Smith, John P			PA 133rd Inf.					
24	Smith, John P	21	Pvt.	PA 50th Inf.	D	2/24/65	7/30/65		
25	Smith, John W (1)	22	Corp.	PA 110th Inf.	C	10/24/61	10/24/64	Wound.-POW	Port Republic
26	Smith, John W (1)		Corp.	PA 110th Inf.	C			POW	Chancellorsville
27	Smith, John W (1)		Corp.	PA 110th Inf.	C			Wounded	Wilderness
28	Smith, John W (2)	19	Pvt.	PA 138th Inf.	F	9/29/62	6/23/65		
29	Smith, Jonathan	20	Pvt.	MD 3rd PHB Inf.	K	2/26/64	7/31/65		
30	Smith, Joseph	24	Pvt.	PA 28th Inf.	A	11/4/64	7/14/65	Died	
31	Smith, Joseph A	19	Pvt.	PA 101st Inf.	D	11/1/61	6/3/65	POW	Plymonth, NC
32	Smith, Joseph B	21	Pvt.	MD 3rd PHB Inf.	B	2/8/62	5/29/65		
33	Smith, Joseph H		Pvt.	PA 205th Inf.	G	9/1/64	6/2/65		
34	Smith, Joseph L	19	Pvt.	PA 101st Inf.	G	12/2/61	4/5/63		
35	Smith, Joseph*			PA 138th Inf.					
36	Smith, Josiah		Pvt.	PA McKeage's Mil.	H	7/2/63	8/8/63		
37	Smith, Josiah C	39	Pvt.	IA 38th Inf.	F	11/4/62	1/1/65		
38	Smith, Josiah C		Pvt.	IA 34th Inf.	K	1/1/65	8/15/65		
39	Smith, Josiah N	22	Sgt.	PA 184th Inf.	A	5/12/64	7/14/65	Wounded	Cold Harbor
40	Smith, Leonard F	25	Pvt.	PA 21st Cav.	E	7/8/63	7/20/65	Wounded	Rockville, MD
41	Smith, Leonard W	19	Pvt.	PA 147th Inf.	B	2/2/64	7/15/65	Wounded	
42	Smith, Levi	24	2nd Lt.	PA 76th Inf.	E	10/9/61	11/28/64	Wounded	2nd Deep Bottom
43	Smith, Lewis M	31	Pvt.	PA 91st Inf.	I	9/21/64	5/30/65		
44	Smith, Mathias	31	Pvt.	PA 56th Inf.	B	9/21/64	7/18/65		
45	Smith, Miles N	18	Pvt.	PA 53rd Inf.	A	9/18/61			
46	Smith, Miles N		Pvt.	PA 138th Inf.	E	8/29/62	5/11/65	Wounded	Wilderness
47	Smith, Morris B*		Pvt.	PA 76th Inf.	E	8/27/63		POW-Died	Drewry's Bluff
48	Smith, Nathan	21	Pvt.	PA 2nd Cav.	E	11/4/61		KIA	Rappahannock
49	Smith, Nathan C	42	Pvt.	PA 2nd Cav.	M	10/17/62	6/17/65		
50	Smith, Nathan P R	18	Pvt.	MD 3rd PHB Inf.	B	10/21/61	6/29/65	POW	Harpers Ferry
51	Smith, Peter	21	Pvt.	PA 11th Cav.	A	12/30/63	6/6/65	Wounded	Five Forks
52	Smith, Philip	24	Pvt.	PA 158th Inf.	H	11/4/62	8/12/63		
53	Smith, Philip E	23	Pvt.	PA 55th Inf.	D	10/12/61	11/23/64	Wounded	
54	Smith, Robert C	17	Sgt.	PA 55th Inf.	H	12/4/61	8/30/65	Wounded	Chaffin's Farm
55	Smith, Robert*		Pvt.	PA 55th Inf.	H	9/23/63			
56	Smith, Rufus E	19	Corp.	PA 205th Inf.	C	8/27/64	6/2/65		
57	Smith, Samuel (1)		Pvt.	PA 91st Inf.	D	12/31/64	7/10/65		
58	Smith, Samuel (2)			USCT Inf.					
59	Smith, Samuel (3)		Pvt.	PA 55th Inf.	I	9/20/63		KIA	Drewry's Bluff
60	Smith, Samuel B	28	Corp.	PA 49th Inf.	H	8/2/63	7/15/65		
61	Smith, Samuel H (1)		Pvt.	PA 110th Inf.	C	2/25/64	6/15/65	Wounded	1st Deep Bottom
62	Smith, Samuel H (2)	17	Pvt.	PA 99th Inf.	A	2/26/64	7/1/65		
63	Smith, Samuel X			PA 56th Inf.					
64	Smith, Seth S	20	Pvt.	PA 76th Inf.	E	10/9/61	11/28/64	Wounded	

	Cemetery	Notes
1	Mt. Union Cem.-Lovely	1890 Mann Twp. Veterans Census; Bedford Inquirer 5/22/1908 listing
2	Union Grove Cem.-MD	Ancestry.com information; 1860 Southampton Census; Isaac W (2) & Jasper-brothers
3		Died-Typhoid Fever 5/19/65; Bed.Gazette 2/13/1914 list; 1860 Londonderry Twp. Census
4	Everett Cem.	B.S.F. History Book-1884; Frank McCoy 1912 Listing; Discharged on Surgeon's Cert.
5	Everett Cem.	Bedford Inquirer 5/22/1908 listing
6	Bethel Christian Cem.-Clearville	Historical Data Systems record; Jacob (4) & Joseph A-brothers
7	Schellsburg Cem.	1890 Napier Twp. Veterans Census
8	Kammerer Cem.-Somerset Co.	Wounded 11/27/63; Historical Data Systems record; 1860 Londonderry Twp. Census
9		Wounded 5/6/64 in Bed. Inquirer 5/27/64 article; PA CW Archives-Londonderry Twp. resid.
10		B.S.F. History Book-1884; Historical Data Systems record
11	Newry Lutheran Cem.	Ancestry.com born Greenfield Twp.; Jacob & Samuel (3)-brothers
12	Dodsons Cem.-Claysburg	Historical Data Systems record
13	Rinard Family Cem.-Cyper	East Providence Twp. Civil War Honor Roll
14	Fairview Cem.-Artemas	1890 Mann Twp. Veterans Census; Frank McCoy 1912 Listing
15		POW-Richmond; Died 5/20/64; PA CW Archives record; 1860 Southampton Twp. Cen.
16	Friends Cove UCC Cem.	1890 Colerain Twp. Veterans Census; Discharged on Surgeon's Cert.
17	Hampton Nat'l Cem.	Wounded 5/16/64; Died 5/27/64; B.S.F. History Book-1884; 1860 Londonderry Twp. Census
18		Bates-History of Pennsylvania Volunteers book; Never joined Company
19		B.S.F. History Book-1884; Drafted & Deserted 5/6/64
20	Pleasantville Cem.	PA Civil War Archives-Bedford Co. residence
21		Historical Data Systems-Bedford Co. residence
22	Shroyer Farm Cem.-Madley	1890 Londonderry Twp. Veterans Census; John F & Jacob F-brothers
23	Schellsburg Cem.	Frank McCoy 1912 Listing
24		1890 Harrison Twp. Veterans Census
25	Yellow Creek Reformed Cem.	Frank McCoy 1912 Listing; Hip Wounded & POW 6/9/62 for short time
26		POW-Libby & Belle Island for 1 month - May 1863; Bio. Review: Bed. & Somerset Co. -1899
27		Elbow Wound 5/6/64; Biographical Review: Bedford & Somerset Co. Book-1899
28		PA Civil War Archives-Colerain Twp. residence; Historical Data Systems record
29	Fairview Cem.-Artemas	1890 Mann Twp. Veterans Census; Frank McCoy 1912 Listing
30	Smith Cem.-Londonderry Twp.	Ancestry.com-Died 1865; Frank McCoy 1912 Listing; 1860 Londonderry Twp. Census
31	Fairview Cem.-Artemas	POW 4/20/64 to 3/1/65; Frank McCoy 1912 Listing; Historical Data Systems record
32	Fairview Cem.-Artemas	1850 Southampton Census; Historical Data Systems record
33		Ancestry.com record; PA Civil War Project record
34		Historical Data Systems-Bedford Co. resid., enlisted in Rainsburg; Disch. on Surgeon's Cert.
35		Bedford Inquirer 5/22/1908 listing
36		B.S.F. History Book-1884
37	West Union Cem.-IA	Ancestry.com-Born Colerain Twp.
38		Historical Data Systems record
39	Presb. Cem.-Williamsburg	Leg, Neck & Hand Wound; 1890 Woodbury Twp. Veterans Census; B.S.F. Hist. Book-1884
40	Winona Cem.-OR	Wounded 7/13/64; Hist. Data Systems record; 1870 Monroe Twp. Cen.; 6th Cavalry references
41	Fockler Cem.-Saxton	1890 Hopewell Veterans Census
42		Wounded 8/16/64 in Bedford Inquirer 9/2/64 article; 1890 Bedford Twp. Veterans Census
43	Fairview Cem.-Artemas	1890 Mann Twp. Veterans Census
44	Mt. Smith Cem.-Bedford	Frank McCoy 1912 Listing; Historical Data Systems record
45	Grandview Cem.-Cambria Co.	B.S.F. History Book-1884; Miles & David R-brothers
46		Wounded 5/5/64; PA Civil War Archives-Schellsburg resid.; Historical Data Systems record
47	Richmond Nat'l Cem.	Wounded & POW-Richmond 5/16/64; Died 5/28/64; B.S.F. Hist. Book-1884; Maine reference
48	Arlington Nat'l Cem., 7581	KIA 11/13/63; B.S.F. History Book-1884; 1860 Napier Twp. Census
49		1890 Napier Veterans Census
50	Fairview Cem.-Artemas	POW 9/15/62; Frank McCoy 1912 Listing; 1870 Southampton Twp. Census
51	Mt. Olivet Cem.-Manns Choice	Wounded 4/1/65, loss of arm; Findagrave.com information
52	Buck Valley Cem.-Fulton Co.	1890 Brush Creek Twp. Veterans Census
53	Fairview Cem.-Artemas	1890 Mann Twp. Veterans Census; Historical Data Systems record
54	Fishertown Cem.	Historical Data Systems record; Robert & Nathan C-brothers
55		B.S.F. History Book-1884; Drafted & Deserted 4/4/64
56	Yellow Creek Reformed Cem.	1890 Hopewell Veterans Census; Rufas & John W (1)-brothers
57		1890 Union Twp. (Pavia) Veterans Census
58		1890 Bedford Twp. Veterans Census
59		Died 5/16/64; B.S.F. History Book-1884; Ancestry.com-born Greenfield Twp.
60	Mt. Union Cem.-Lovely	1890 W. St. Clair Twp. Veterans Census; Samuel B & Isaac-brothers
61		Wounded 7/27/64 listed in Bedford Inquirer 8/7/64 article; B.S.F. History Book-1884
62	Pleasantville Cem.	1860 St. Clair Twp. Census; Samuel H & John B-brothers
63	Mt. Smith Cem.-Bedford	Frank McCoy 1912 Listing; Civil War Veterans Burial Record; Born 1827
64	Bedford Cem.	1890 Bedford Twp. Veterans Census; Historical Data Systems record

Chapter 18

	Name	Age	Rank	Regiment	Co.	Muster In	Muster Out	Casualty	Casualty Battle
1	Smith, Seth S		Pvt.	PA 22nd Cav.	I	4/5/65	6/24/65		
2	Smith, Simon R	19	Pvt.	PA 138th Inf.	F	5/31/64	6/23/65		
3	Smith, Soloman	20	Pvt.	PA 28th Inf.	A	9/21/64	7/18/65		
4	Smith, William C		Pvt.	PA McKeage's Mil.	H	7/2/63	8/8/63		
5	Smith, William F (1)	19	Pvt.	PA 21st Cav.	K	2/12/64	7/8/65	Wounded	Petersburg - Initial
6	Smith, William F (2)*		Capt.						
7	Smith, William H (1)	19	Pvt.	PA 208th Inf.	K	9/7/64	6/1/65		
8	Smith, William H (2)		SMN.	US Navy		1/25/65			
9	Smith, William H (3)	18	Pvt.	PA 99th Inf.	I	3/15/65	7/17/65	Wounded	Sailor's Creek
10	Smith, William L	27	Pvt.	PA 186th Inf.	H	4/1/64	8/15/65		
11	Smith, William P	25	Corp.	PA McKeage's Mil.	A	7/3/63	8/8/63		
12	Smith, William R (1)	19	Pvt.	PA 32nd Inf.	F	6/26/63			
13	Smith, William R (2)*		Pvt.	PA 138th Inf.	E	12/7/63		Wound.-Died	Wilderness
14	Smith, William*		Pvt.	PA 55th Inf.	K	10/10/63	8/30/65		
15	Smith, Wilson S	17	Pvt.	PA 158th Inf.	C	11/1/62	8/12/63		
16	Smith, Wilson S		Pvt.	PA 187th Inf.	D	2/19/64	8/3/65		
17	Smitman, Henry*	42	Pvt.	PA 55th Inf.	I	10/14/63		Died	
18	Smouse, Abner G		Music.	PA 171st Inf.	I	11/10/62	8/8/63		
19	Smouse, Abner G	29	Pvt.	PA 91st Inf.	C	9/21/64	5/30/65	Wounded	
20	Smouse, David	18	Pvt.	PA 22nd Cav.	L	2/22/64	6/24/65		
21	Smouse, Samuel			PA 91st Inf.					
22	Smouse, Simon	21	Music.	PA 208th Inf.	K	9/7/64	6/1/65		
23	Snare, Calvin L*	19	Pvt.	PA 194th Inf.	I	7/22/64	11/5/64		
24	Snave, Joseph W		Pvt.	PA 76th Inf.	E	10/9/61	11/28/64		
25	Snellrider, David		Pvt.	PA 55th Inf.	K	9/23/63	8/30/65		
26	Snider, Augustus	20	Pvt.	PA McKeage's Mil.	D	7/3/63	8/8/63		
27	Snider, Augustus		Pvt.	PA 208th Inf.	K	9/7/64	6/1/65		
28	Snider, John W	22	Pvt.	PA 22nd Cav.	C	7/15/63	2/5/64		
29	Snider, Leonard N	37	Pvt.	MO 1st SM Cav.	D	4/29/62	3/16/63	Wounded	
30	Snively, Andrew J	55	Pvt.	PA 22nd Cav.	C	7/11/63	2/5/64		
31	Snively, John A		Pvt.	PA 22nd Cav.	C	7/11/63	2/5/64		
32	Snook, Emanuel	21	Corp.	PA 55th Inf.	H	12/4/61		Wound.-Died	Petersburg
33	Snook, Jacob	26	Corp.	PA 21st Cav.	E	6/8/63	2/20/64		
34	Snow, William H	19	Pvt.	PA 207th Inf.	C	8/24/64	6/16/65	Wounded	Ft. Stedman
35	Snow, William H		Pvt.	PA 207th Inf.	C			Wounded	Petersburg - Final
36	Snow, William J	22	Pvt.	PA 77th Inf.	F	10/9/61	9/26/64		
37	Snowberger, Daniel	33	Pvt.	PA McKeage's Mil.	G	7/2/63	8/8/63		
38	Snowberger, Daniel		Pvt.	PA 107th Inf.	G	10/21/64	2/1/65		
39	Snowberger, David	31	Pvt.	PA 55th Inf.	D	10/12/61		Died	
40	Snowberger, Elias	23	Pvt.	PA 171st Inf.	I	11/2/62	8/8/63		
41	Snowberger, Elias		Pvt.	PA 29th Inf.	K	12/23/64	7/17/65		
42	Snowberger, Jacob		Pvt.	PA 194th Inf.	G	7/20/64	11/6/64		
43	Snowberger, Joseph B	28	Corp.	PA 171st Inf.	I	11/10/62	8/8/63		
44	Snowberger, Joseph C		Pvt.	PA 208th Inf.	H				
45	Snowberger, Theodore	20	Pvt.	PA 184th Inf.	A	5/12/64		Wound.-Died	Petersburg - Initial
46	Snowberger, Theodore		Pvt.	PA 184th Inf.	A			Wounded	Cold Harbor
47	Snowden, David	24	Pvt.	PA 184th Inf.	A	5/12/64		KIA	Cold Harbor
48	Snowden, John W	19	Pvt.	PA 2nd Cav.	E	10/3/61		POW	
49	Snowden, Joseph B	34	Pvt.	PA 53rd Inf.	B	9/21/64	5/24/65		
50	Snyder, Arthur	23	Pvt.	NY 102nd Inf.	E	12/10/61	1/3/63	Wounded	
51	Snyder, Arthur			US 1st Art.	K	1/3/63			
52	Snyder, Christopher	35	Black.	PA 13th Cav.	D	3/21/64			
53	Snyder, David F	22	Pvt.	PA 138th Inf.	D	8/29/62	6/23/65	Wounded	
54	Snyder, Edward		Pvt.	PA McKeage's Mil.	H	7/2/63	8/8/63		
55	Snyder, Elias J	23	Pvt.	PA 22nd Cav.	I	7/22/63	10/31/65		
56	Snyder, Elias*							Died	
57	Snyder, Ferdinand		Pvt.	PA 208th Inf.	K	9/7/64	6/1/65		
58	Snyder, George W Jr.		Pvt.	PA 125th Inf.	E	8/13/62	5/18/63		
59	Snyder, George W Jr.	23	Pvt.	PA 55th Inf.	I	2/19/64	8/30/65		
60	Snyder, George W Sr.	50	Pvt.	PA 125th Inf.	E	8/13/62	4/2/63		
61	Snyder, George W Sr.		Pvt.	PA 205th Inf.	C	8/17/64	7/2/65		
62	Snyder, Henry			NH 10th Inf.					
63	Snyder, Jacob H	15	Pvt.	PA McKeage's Mil.	G	7/2/63	8/8/63		
64	Snyder, James	23	Music.	PA 76th Inf.	C	10/17/61	7/18/65		

	Cemetery	Notes
1		B.S.F. History Book-1884; 1860 Bedford Boro Census
2	Kammerer Cem.-Somerset Co.	B.S.F. History Book-1884; 1860 Londonderry Twp. Census
3	Lybarger Cem.-Madley	Frank McCoy 1912 Listing; 1890 Londonderry Twp. Census
4		B.S.F. History Book-1884; 1890 Everett Veterans Census
5	Bedford Cem.	Wounded 6/19/64; 1890 Bedford Twp. Veterans Census
6		Obituary-Confederate Veteran; Born in Whips Cove in Fulton Co.; No Civil War record located
7	Indian Springs-Monroe Twp.	B.S.F. History Book-1884
8		Deserted 6/9/65; 1890 Saxton Veterans Census
9	Rock Hill Cem.-Monroe Twp.	Wounded 4/6/65; Ancestry.com information; 1860 Southampton Twp. Census
10	Greenlawn Cem.-OH	1870 Bloody Run Census; PA Civil War Project record
11		Ancestry.com information; 1850 Southampton Twp. Census
12	Dry Hill Cem.-Woodbury	Historical Data Systems record; 134th IL Infantry referenced
13	Philadelphia	Wounded 5/6/64; Died 2/11/65; B.S.F. Hist. Book-1884; PA CW Archives-Franklin Co. resid.
14		B.S.F. History Book-1884; Drafted
15	New Prov. Ch. God-Lancaster	Ancestry.com information; Substitute; 1880 Napier Twp. Census
16		Minister at Napier Twp. Church in 1880s
17		Died 1/8/65; B.S.F. History Book-1884; Mustered in Reading, PA
18		1890 Liberty Twp. Veterans Census- Shot in right side
19	St. Mark's Cem.-Ott Town	Died of Wounds-1875; Bedford Inquirer 5/22/1908 listing; 1860 Colerain Twp. Census
20	Bald Hill Cem.-Everett	Bedford Inquirer 5/22/1908 listing
21	St. Mark's Cem.-Ott Town	Frank McCoy 1912 Listing
22	Everett Cem.	Historical Data Systems record; 1860 W. Providence Twp. Census
23		B.S.F. History Book-1884; Huntingdon Co references
24		Historical Data Systems-Bedford Co. residence
25		B.S.F. History Book-1884
26	Carson Valley-Duncansville	PA Civil War Archives record
27		Findagrave.com information; 1880 W. Providence Twp. Census
28	Everett Cem.	B.S.F. History Book-1884; 1850 Monroe Twp. Census; John W (1) & Leonard W (2)-bros.
29	Bethal Frame Ch.-Monroe Twp.	Frank McCoy 1912 Listing; in Hospital 10/30/62 to 4/1/63
30	Schellsburg Cem.	Historical Data Systems record; Father of John
31	Schellsburg Cem.	B.S.F. History Book-1884; 1870 Schellsburg Census
32	Schellsburg Cem.	Died 7/6/65 at home from Wounds; Frank McCoy 1912 Listing
33	Horn Meth. Cem.-Alum Bank	1890 E. St. Clair Twp. Veterans Census; Historical Data Systems record
34	Bedford Cem.	Wounded 3/25/65; Historical Data Systems record
35		Wounded 4/2/65; Historical Data Systems record
36	Wells Valley Meth. Cem.	1890 Wells Twp. Veterans Census
37	Diehl's Crossroads-Henrietta	B.S.F. History Book-1884
38		Historical Data Systems record; 1860 Woodbury Twp. Census
39	Cypress Hill Cem.-NY	Died 1/25/64; Historical Data Systems record; 1860 Snake Spring Twp. Census
40	Fishertown Bretheran Cem.	B.S.F. History Book-1884; aka Harry Elias
41		Historical Data Systems record; 1860 St. Clair Twp. Census; Elias & Jacob-brothers
42	Claysburg Union Cem.	1890 Claysburg Veterans Census; Historical Data Systems record
43	New Enterprise Cem.	B.S.F. History Book-1884; 1890 Hopewell Twp. Veterans Census
44	Eshelman Cem.-Woodbury	Soldiers of Blair Co. Book - 1940
45	Arlington Nat'l Cem., 13-7403	Wounded 6/18/64; Died 9/1/64; B.S.F. History Book-1884; 1860 S. Woodbury Twp. Census
46		Wounded-Bedford Inquirer 7/1/64 article; B.S.F. History Book-1884
47	Cold Harbor Nat'l Cem.	KIA 6/3/64; B.S.F. History Book-1884; 1860 Napier Twp. Census
48		POW-Andersonville; B.S.F. Hist. Book-1884 listed-Died; Hist. Data Sys. enl.-1st Cav. 6-17-65
49	Dry Hill Cem.-Woodbury	Historical Data Systems record
50	Hyndman Cem.	1890 Londonderry Twp. Veterans Census
51		Findagrave.com information
52	Nicodemus Cem.-Henrietta	PA Civil War Project record; 1890 Roaring Springs Veterans Census; Father of William B
53	Clearville Union Cem.	1890 Monroe Twp. Veterans Census; Historical Data Systems record
54		B.S.F. History Book-1884
55	Oakwood Cem.-KS	B.S.F. History Book-1884
56		Bedford Gazette 2/13/1914 listing; an Elias Snyder - Hunt. Co resid. enlisted in 12th PA Res.
57		B.S.F. History Book-1884; 1850 Harrison Twp. Census
58	Greenlawn Cem.-Roaring Spring	Ancestry.com information; George W Jr & James-brothers
59		B.S.F. History Book-1884; PA Civil War Archives-Bedford Co. residence
60	Greenlawn Cem.-Roaring Spring	Ancestry.com information; 1850 Harrison Twp. Census; Discharged on Surgeon's Cert.
61		Ancestry.com information; Father of George W Jr & James
62	Mt. Olivet Cem.-Manns Choice	Frank McCoy 1912 Listing; Bedford Inquirer 5/22/1908 listing
63	Shreves Cem.-Monroe Twp.	B.S.F. History Book-1884
64	Greenlawn Cem.-Roaring Spring	Ancestry.com information; 1850 Harrison Twp. Census

Chapter 18

	Name	Age	Rank	Regiment	Co.	Muster In	Muster Out	Casualty	Casualty Battle
1	Snyder, John W (1)	19	Pvt.	MD 2nd PHB Inf.	H	9/11/61		POW-Died	Jonesville
2	Snyder, John W (2)	19	Pvt.	PA 49th Inf.	B	8/17/61	7/15/65	Wounded	Petersburg - Final
3	Snyder, John*		Pvt.	PA 55th Inf.	H	10/16/63			
4	Snyder, Jonathan		1st Sgt.	PA 138th Inf.	D	8/29/62		Wounded	Wilderness
5	Snyder, Jonathan	30	1st Sgt.	PA 138th Inf.	D	8/29/62		Wound.-Died	Cedar Creek
6	Snyder, Joseph		Pvt.	PA 91st Inf.	F	9/21/64	6/1/65		
7	Snyder, Joseph N		Pvt.	PA 4th Cav.	A				
8	Snyder, Leonard N	21	Pvt.	IL 26th Inf.	D	8/17/61	7/20/65		
9	Snyder, William	17	Pvt.	PA 133rd Inf.	K	8/29/62	5/26/63		
10	Snyder, William		Music.	PA 194th Inf.	I	7/22/64	11/5/64		
11	Snyder, William H	16	Pvt.	PA 125th Inf.	D	8/12/62	5/18/63		
12	Sohn, Calvin		Pvt.	PA 76th Inf.	E	10/9/61	7/18/65		
13	Soistman, Theodore*	47	Pvt.	PA 76th Inf.	E	10/18/64	7/18/65		
14	Songster, Calvin A	23	Sgt.	IL 55th Inf.	G	10/31/61	8/14/65		
15	Songster, John A			PA 142nd Inf.					
16	Songster, John A		Sgt.	PA 206th Inf.	K		1/26/65		
17	Sorber, John		Pvt.	PA 21st Cav.	E	7/2/63	2/20/64		
18	Sorber, Martin V	23	Capt.	PA 55th Inf.	H	9/29/61	8/30/65		
19	Souser, Henderson	21	Corp.	PA 133rd Inf.	K	8/15/62	5/26/63		
20	South, James W	24	Pvt.	PA 208th Inf.	K	9/7/64	6/1/65		
21	Spade, Isaac N	19	Pvt.	PA 171st Inf.	I	11/2/62	8/8/63		
22	Spang, David R P	22	Pvt.	PA 19th Cav.	L	11/17/63	6/6/65		
23	Spangle, Daniel*	28	Pvt.	PA 76th Inf.	E	8/27/63	7/9/65		
24	Spangler, Jeremiah	24	Pvt.	PA 125th Inf.	H	8/8/62	5/18/63		
25	Spangler, William H		Pvt.	PA 99th Inf.	F	2/24/65	7/1/65		
26	Spark, Nelson*		Pvt.	PA 208th Inf.	K	8/18/64	9/10/64		
27	Sparks, Abraham J		Capt.	IL 146th Inf.	H	9/17/64	7/8/65		
28	Sparks, David G	25	Pvt.	PA 208th Inf.	K	9/7/64	6/1/65		
29	Sparks, David W	27	Pvt.	IL 12th Inf.	H	5/2/61	8/1/61		
30	Sparks, David W		1st Lt.	IL 93rd Inf.	F&S	10/13/62	11/15/62	Injured	
31	Sparks, George W	18	Pvt.	OH 163rd Inf.	C	5/12/64	9/19/64		
32	Sparks, Henry H	20	Pvt.	OH 64th Inf.	A	10/2/61	12/3/65		
33	Sparks, Jacob	21	Pvt.	PA 133rd Inf.	K	8/29/62		Died	
34	Sparks, James	33	Pvt.	PA 208th Inf.	K	9/7/64	6/1/65	Wounded	Petersburg - Final
35	Sparks, James H	21	Pvt.	PA 133rd Inf.	K	8/29/62	5/26/63		
36	Sparks, James H		Pvt.	PA 208th Inf.	K	9/7/64	6/1/65		
37	Sparks, John	15	Pvt.	PA 194th Inf.	I	7/22/64	11/5/64		
38	Sparks, John C	18	Pvt.	PA 133rd Inf.	K	8/29/62	5/26/63		
39	Sparks, John C		Pvt.	PA 194th Inf.	I	7/22/64	11/5/64		
40	Sparks, John C		Pvt.	PA 82nd Inf.	C	11/14/64	7/13/65		
41	Sparks, John E	22	Music.	IL 57th Inf.	E	12/26/61	3/20/62		
42	Sparks, John E		1st Lt.	IL 151st Inf.	F	2/21/65	1/24/66		
43	Sparks, John M	18	Pvt.	OH 25th Inf.	D	6/8/61	3/17/62		
44	Sparks, John M		Pvt.	OH 12th Light Art.		3/17/62	6/25/64		
45	Sparks, Joseph H	21	Music.	PA 133rd Inf.	K	8/29/62	5/26/63		
46	Sparks, Joseph H		Pvt.	US					
47	Sparks, Joseph R			IL 151st Inf.					
48	Sparks, Mahlon	42	1st Sgt.	OH 64th Inf.	A	10/2/61	5/21/65		
49	Sparks, Silas H	22	Pvt.	PA 133rd Inf.	K	8/29/62	5/26/63		
50	Sparks, Silas H		Pvt.	PA 186th Inf.	G	3/24/64	8/15/65		
51	Sparks, Solomon C	41	Wagon.	IL 93rd Inf.	C	10/13/62	6/23/65		
52	Sparks, Uriah	24	Sgt.	PA 107th Inf.	H	3/12/62	7/13/65	Wounded	Gettysburg
53	Sparks, Wesley W	24	Sgt.	OH 163rd Inf.	C	5/12/64	9/10/64		
54	Sparks, William	15	Pvt.	PA 101st Inf.	D	11/1/61	6/25/65	POW	Plymonth, NC
55	Sparks, Wilson W		Sgt.	PA McKeage's Mil.	D	7/2/63	8/8/63		
56	Sparks, Wilson W	23	2nd Lt.	PA 208th Inf.	K	9/7/64	6/1/65		
57	Speck, Henry	39	Pvt.	PA 138th Inf.	E	8/29/62	6/23/65	Wounded	Mine Run
58	Speck, Henry		Pvt.	PA 138th Inf.	E			Wounded	Cedar Creek
59	Speece, Harry	24	Pvt.	PA 77th Inf.	F	3/2/65	12/6/65		
60	Speer, William H	18	Pvt.	PA 110th Inf.	C	3/19/64	6/28/65	Wounded	Spotsylvania
61	Speice, Louis D	23	1st Sgt.	PA 133rd Inf.	K	8/15/62	5/26/63		
62	Speice, Louis D		Capt.	PA 205th Inf.	C	8/31/64	6/2/65		
63	Spencer, Israel	40	Pvt.	PA 208th Inf.	K	9/7/64	6/1/65		
64	Spidel, Matthew	18	Pvt.	PA 76th Inf.	E	3/3/62	3/13/65		

Detailed Alphabetical Listing

	Cemetery	Notes
1	Andersonville Cem., 718	POW 1/3/64; Died-Chronic Diarrhea 4/24/64; 1850 Monroe Twp. Cen.; Hist. Data Systems
2	Duval Cem.-Coaldale	Wounded 4/2/65; Died 6/5/71; Union Army Veterans Headstone
3		Deserted 4/4/64; B.S.F. History Book-1884; enlisted in Reading, PA
4		Wounded 5/6/64; Historical Data Systems record; 1850 Monroe Twp. Census; aka Snider
5	Ash-Snider Cem.-Monroe Twp.	Wounded 10/19/64; Died 10/22/64; B.S.F. History Book-1884; 1850 Monroe Twp. Census
6		1890 Bedford Twp. Veterans Census
7	Okanogan City Cem.-WA	1850 Monroe Twp. Census; Nat'l Park Sys. Record; aka Snider
8	Cripple Creek Cem.-CO	Ancestry.com information; Joseph, John W (2), Leonard & Jonathan-brothers
9		Historical Data Systems-Bedford Co. residence
10		B.S.F. History Book-1884
11	Duval Cem.-Coaldale	1890 Broad Top Veterans Census
12		B.S.F. History Book-1884; Historical Data Systems-Bedford Co. residence
13		B.S.F. History Book-1884; Cambria Co. references
14	Exeter Cem.-NE	Ancestry.com-Born in Bedford; Calvin & John-brothers
15		Bedford Cemetery Civil War Veterans list; 142nd & 206th Regiments on gravestone
16	Bedford Cem.	1890 Broadtop Veterans Cen.; 1870 Schellsburg Cen.; aka Sorgeter; Frank McCoy 1912 List
17		Histitorial Data Sysetms-Bedford Co. residence
18	IOOF Cem.-Somerset Co.	Historical Data Systems-Bedford Co. residence; Mustered in as Corporal
19	Mt. Olivet Cem.-Manns Choice	1890 Napier Twp. Veterans Census; B.S.F. History Book-1884
20	Rose Hill Cem.-SD	B.S.F. History Book-1884; 1870 W. Providence Twp. Census
21	Maple Hill Cem.-OH	B.S.F. History Book-1884; Born in VA; 1860 Hampshire Co., VA Census
22	St. Lukes Cem.-Saxton	Historical Data Systems record
23	Allemansville Cem.-Clearfield	B.S.F. History Book-1884; Drafted; Clearfield Co. references
24	Moore Cem.-WV	1890 Harrison Twp. Census
25		1890 Wells Twp. Veterans Census
26		Historical Data Systems record
27		Historical Data Systems record
28	Sparks Family Cem.-Everett	B.S.F. History Book-1884; David, Jacob & James H-brothers
29	Greenwood Memory Cem.-AZ	Ancestry.com born in Ray's Hill
30		Thrown from Horse; Historical Data Systems-Born Bedford Co.
31	Odd Fellows Cem.-MI	Ancestry.com-born in Bedford Co.
32	Lakeview Cem.-MI	Historical Data Systems record
33	Sparks Family Cem.-Everett	Died 3/16/63; B.S.F. History Book-1884; Frank McCoy 1912 Listing
34	Sparks Family Cem.-Everett	Arm Wound 4/2/65; B.S.F. History Book-1884
35	Sparks Family Cem.-Everett	Historical Data Systems record; 1860 W. Providence Twp. Cen.; James H & Jacob-twin bros.
36		B.S.F. History Book-1884; Historical Data Systems-Bedford Co. residence
37	Providence Union Cem.-Everett	B.S.F. History Book-1884; John, Uriah & William-brothers
38	Everett Cem.	Historical Data Systems record; 1850 W. Providence Twp. Census; John C & Joseph-brothers
39		Historical Data Systems-Mustered in Bloody Run; B.S.F. History Book-1884
40		Historical Data Systems-Mustered in Bedford
41	Hilltop Cem.-CO	Ancestry.com information; 1850 E. Providence Twp. Census
42		Ancestry.com information; John E & David W, Joseph R, Abraham & Solomon-brothers
43	Avalon Cem.-MI	Ancestry.com-born in Bedford Co.
44		Ancestry.com information; John M, Henry, George & Wesley-brothers
45	Union Church Cem.-Clearville	Historical Data Systems record; In Hospital at Sharpsburg from 10/30/62
46		1890 Monroe Twp. Veterans Census
47	Forest Hill Cem.-IL	Jack Sparks family history documentation
48	Avalon Cem.-MI	Ancestry.com information; 1840 W. Prov. Twp. Cen.; Father of Wesley, John M & Henry
49	Union-Graham Cem.-KS	B.S.F. History Book-1884; Historical Data Systems record
50		Historical Data Systems record; 1860 W. Providence Twp. Census
51	Montrose Cem.-MI	1850 E. Providence Twp. Census; Historical Data Systems record
52	Providence Union Cem.-Everett	Wounded 7/1/63; PA Civil War Archives-Born Bedford Co.; Uriah, John & William-brothers
53	Clarinda Cem.-IA	Historical Data Systems record
54	Wilmore Breth.-Cambria Co.	POW-Andersonville 4/20/64 to 3/3/65; attended 1905 Andersonville Dedication
55		B.S.F. History Book-1884
56	Everett Cem.	B.S.F. History Book-1884; 1860 W. Providence Twp. Census
57	Sixteen Church Cem.-OH	Wounded 11/27/63; PA Civil War Archives-Bedford residence; Historical Data Systems record
58		Wounded 10/19/64; B.S.F. History Book-1884; Historical Data Systems record
59	Fockler Cem.-Saxton	1890 Saxton Veterans Census; Bedford Inquirer 5/22/1908 listing; aka Henry
60	Rose Hill Cem.-Altoona	Wounded 5/12/64-Pension record; 1890 Coaldale Twp. Veterans Census; Born in Bedford Co.
61	W. Laurel Hill-Montgomery Co	Historical Data Systems-Bedford Co. residence
62		Historical Data Systems record
63	Rays Cove CC Cem.	1890 E. Providence Twp. Veterans Census; B.S.F. History Book-1884
64	Bedford Cem.	1890 Bedford Twp. Veterans Census; Matthew, Samuel R & Wilson-brothers

Chapter 18

	Name	Age	Rank	Regiment	Co.	Muster In	Muster Out	Casualty	Casualty Battle
1	Spidel, Samuel R		Pvt.	PA 165th Inf.	B	11/4/62	7/28/63		
2	Spidle, Bartley	33	Pvt.	PA 22nd Cav.	I	3/4/65	6/24/65		
3	Spidle, Samuel		Pvt.	IN 3rd Cav.	M	12/11/61	4/15/65		
4	Spidle, Wilson	22	Pvt.	PA 55th Inf.	D	10/12/61		Wound.-Died	Petersburg - Initial
5	Spielman, Martin V	22	Pvt.	PA 133rd Inf.	K	8/15/62	1/16/63		Fredericksburg*
6	Spitler, John L	18	Corp.	PA 22nd Cav.	C	7/15/63	2/5/64		
7	Sponsler, George W	18	Pvt.	MD 3rd PHB Inf.	C	11/5/61	5/29/65	POW	Harpers Ferry
8	Sponsler, George W		Pvt.	MD 3rd PHB Inf.	C			Wounded	
9	Sponsler, John W	17	Pvt.	PA 22nd Cav.	C	7/22/63	6/15/65		
10	Sponsler, Solomon	16	Pvt.	MD 3rd PHB Inf.	B	11/5/61	5/19/65		
11	Sponsler, William	16	Pvt.	PA 100th Inf.	B	3/4/65	7/24/65	Wounded	Ft. Stedman
12	Spriggs, Asa M	27	Pvt.	PA 2nd Cav.	E	11/4/61	2/17/65		
13	Spriggs, John*			USCT Inf.					
14	Sproat, George R			US 2nd Cav.					
15	Sproat, James R	26	Pvt.	PA 88th Inf.	D	3/11/65	6/20/65		
16	Sproat, Joseph R	26	Corp.	PA 133rd Inf.	C	8/13/62	5/26/63		
17	Sproat, William A	23	Pvt.	PA 6th Cav.		3/2/65	6/16/65		
18	Sprowl, Jacob R*	23	Pvt.	PA 21st Cav.	E	7/2/63	2/20/64		
19	Spruell, John D		Pvt.	PA 208th Inf.	H	9/7/64			
20	Squint, Henry D*		Pvt.	PA 55th Inf.	D	9/2/62		POW-Died	
21	Stahley, Henry		Pvt.	PA 55th Inf.	D	9/23/63	8/30/65		
22	Stailey, George E	21	Pvt.	PA 133rd Inf.	C	8/13/62	2/7/63	Wounded	Fredericksburg
23	Stailey, George E		Corp.	PA 208th Inf.	K	9/7/64	6/1/65		
24	Stailey, Henry C	21	Music.	PA 208th Inf.	K	9/7/64	6/1/65		
25	Stailey, Thomas	17		IN 147th Inf.		1863	1864		
26	Stailey, Thomas			US 18th Inf.		1864	9/8/68		
27	Stailey, William A	26	Corp.	PA 22nd Cav.	H	2/26/64	6/24/65		
28	Staley, Adam S	33	Pvt.	PA 78th Inf.	K	3/2/65	5/12/65		
29	Stambaugh, David	20	Pvt.	PA 148th Inf.	A	4/26/63	6/30/65		
30	Stambaugh, John			PA 22nd Cav.					
31	Stambaugh, Joseph	29	Pvt.	PA 55th Inf.	K	3/2/64	4/4/65	Wounded	
32	Stanbaugh, Samuel	23	Pvt.	PA 136th Inf.	K	8/27/62	5/29/63		
33	Stanehfield, Ivory N*	21	Pvt.	PA 76th Inf.	E	8/27/63	5/16/64		
34	Stathers, James								
35	Statler, Samuel F	13	Pvt.	PA 55th Inf.	H	10/10/61	8/8/65	Wounded	Bermuda Hundred
36	Statler, Samuel F		Pvt.	PA 55th Inf.	H			Wounded	Petersburg
37	Statler, William	20	Pvt.	PA 13th Inf.	G	4/25/61	8/6/61		
38	Staub, George G		Pvt.	MD 3rd Reg Cav.	I	9/7/63	7/25/64		
39	Steckler, Charles	22	Pvt.	PA 55th Inf.	H	7/22/63		POW-Died	
40	Steckman, Daniel H	19	Pvt.	PA 76th Inf.	E	10/9/61		KIA	Ft. Wagner, SC
41	Steckman, David B	38	Pvt.	PA 91st Inf.	I	9/21/64	5/30/65		
42	Steckman, Francis	18	Corp.	PA 138th Inf.	E	8/29/62		Wound.-Died	Cold Harbor
43	Steckman, John			PA 9th Res.					
44	Steckman, John A	36	Pvt.	IL 28th Inf.	I	3/12/64	3/4/65		
45	Steckman, John B	18	Corp.	PA 138th Inf.	F	8/29/62		Died	
46	Steckman, John G	26	Pvt.	MD 3rd PHB Inf.	A	1/5/64	7/31/65		
47	Steckman, Levi	19	Pvt.	PA 55th Inf.	D	2/18/64	7/12/65	Wounded	Drewry's Bluff
48	Steckman, Peter	36	Pvt.	PA 61st Inf.	G	6/4/64	6/28/65		
49	Steckman, Philip H	20	Pvt.	PA 138th Inf.	D	8/29/62	6/23/65		
50	Steel, David S	23	Pvt.	PA 13th Inf.	G	4/25/61	8/6/61		
51	Steele, David F	21	Pvt.	PA 133rd Inf.	K	8/15/62		KIA	Fredericksburg
52	Steele, Edward	16	Pvt.	PA 133rd Inf.	K	8/18/62	10/9/63		
53	Steele, John W	27	Pvt.	PA 22nd Cav.	L	2/26/64	6/24/65		
54	Steele, Levi H	18	Pvt.	PA 133rd Inf.	K	8/15/62	1/31/63		Fredericksburg*
55	Steele, Levi H		Pvt.	PA 208th Inf.	K	9/7/64	6/1/65		
56	Steele, Thomas	23	Pvt.	PA 149th Inf.	K	8/26/63	6/24/65	Wounded	Spotsylvania
57	Steeley, Jacob E		Pvt.	PA 194th Inf.	I	7/22/64	9/6/64		
58	Steeley, Jacob E			PA 97th Inf.	Ind	9/6/64	7/17/65		
59	Steffa, Thomas		Pvt.	PA McKeage's Mil.	G	7/2/63	8/8/63		
60	Steinman, Mathew C	23	Pvt.	PA 15th Inf.	H	4/23/61	8/8/61		
61	Steinman, Mathew C		Pvt.	PA 62nd Inf.	M	8/9/61	9/8/64	Wounded	Gettysburg
62	Steinman, Mathew C		Pvt.	PA 16th Inf.	D	1/30/65	5/29/65		
63	Stengal, Jacob	28	Pvt.	PA 55th Inf.	K	11/5/61	6/26/63		
64	Stephens, Jacob	51	Pvt.	PA 11th Cav.	A	12/24/63		POW-Died	Reams Station

	Cemetery	Notes
1	Bedford Cem.	Mustered Out at Gettysburg; Historical Data Systems record
2	Bedford Cem.	1890 Bedford Twp. Veterans Census; aka Spidie & Spidel
3	Bedford Cem.	Bedford Inquirer 5/22/1908 listing
4	Hampton Nat'l Cem.	Wounded 6/18/84; Died 7/11/64; B.S.F. History Book-1884; 1860 Bedford Boro Census
5	Yellow Creek Reformed Cem.	1890 Hopewell Twp. Veterans Census; in York hospital at MO; Discharged on Surgeon's Cert.
6	Philipsburg Cem.	B.S.F. History Book-1884
7	Everett Cem.	POW 9/15/62-Pension Record; Frank McCoy 1912 Listing; 1860 Monroe Twp. Census
8		Pension record of accidental gunshot wound at Camp Parole-5/1/63
9	Everett Cem.	Historical Data Systems record; George W, Solomon, William & John W-brothers
10	Everett Cem.	1890 Monroe Twp. Veterans Census
11	Rock Hill Cem.-Monroe Twp.	Wounded 3/25/65; Leg Amputated; Hist. Data Systems record; Barb Sponsler Miller record
12	Bedford Cem.	1890 Bedford Twp. Veterans Census; Frank McCoy 1912 Listing
13		Frank McCoy 1912 list & Civil Works Admin. 1934 list
14	Los Angeles Nat'l Cem.	Regiment on gravestone; Born 12/13/30
15	Masonic Cem.-IA	1890 Brush Creek Veterans Census
16	Mt. Zion Cem.-Breezewood	Findagrave.com information; Joseph, James William & George-brothers
17	Topeka Cem.-KS	Historical Data Systems record
18	Dunmyer Cem.-Cambria Co.	Historical Data Systems-Bedford Co. residence; Cambria Co. references
19		Deserted 9/13/64; B.S.F. History Book-1884
20		POW-Richmond; Died 5/18/64, B.S.F. History Book-1884; Berks Co. references
21		B.S.F. History Book-1884; Bedford Inquirer listed KIA on 6/3/64
22	Westminster-Montgomery Co.	Wounded 12/13/62; discharged at DC Hospital; Historical Data Systems record
23		1870 E. Providence Twp. Census; 1890 Everett Veterans Census
24	Everett Cem.	1890 Everett Veterans Census
25	South Hill Cem.-IL	Ancestry.com record
26		Ancestry.com record; Thomas, George E, Henry & William-brothers
27	Mt. Zion Cem.-Breezewood	1890 E. Providence Twp. Veterans Census
28	W. Kishacoquillas Presb. Cem.	1890 Saxton Veterans Census; discharged at Nashville Hospital
29	Osterburg Community Cem.	1890 E. St. Clair Twp. Census
30	Mock Church Cem.-Alum Bank	Frank McCoy 1912 Listing
31	Mt. Moriah Cem.-Blue Knob	Wound listed in Bedford Inquirer article 6/3/64; Disch. Surg. Cert. at Davis Island, NY
32	St. Paul Breth. Cem.-Imler	PA Civil War Project record; aka Stambaugh
33	Evergreen Cem.-ME	B.S.F. History Book-1884; Maine references; Discharged on Surgeon's Cert.
34	Mt. Ross Cem.-Bedford	Bedford Inquirer 5/22/1908 listing
35	Everett Cem.	Wounded 5/19/64 listed in Bedford Inquirer 7/8/64 article; 1890 Bedford Twp. Veterans Cen.
36		Wound listed on GAR Personal War Sketch
37		B.S.F. History Book-1884; Historical Data Systems-Bedford Co. residence
38	Hyndman Cem.	Frank McCoy 1912 Listing; Bedford Inquirer 5/22/1908 listing
39	Andersonville Cem.	POW-Andersonville; Died-Diarrhea 9/29/64; 1860 Napier Twp. Census aka Stickler
40		Listed as KIA 7/11/63 and Died 7/18/64; 1860 Bedford Boro Cen.; Hist. Data Systems record
41	Bethal Frame Ch.-Monroe Twp.	Bedford Inquirer 5/22/1908 listing; 1860 W. Providence Twp. Census; David & John B-bros.
42		Wound. 6/1/64; Died 6/5/64; B.S.F. Hist. Book-1884; 1850 Bed. Cen.; Francis & Daniel-bros.
43	Black Valley Graveyard	Frank McCoy 1912 Listing
44	Andalusia Cem.-IL	Ancestry.com born in Bedford
45	Steckman Farm-Monroe Twp.	Died of Typhoid Fever 12/23/62; B.S.F. History Book-1884; 1860 Monroe Twp. Census
46	Hyndman Cem.	1850 Harrison Twp. Census; Historical Data Systems record
47	Pleasant Union Cem.-Clearville	Forearm Amputated 5/16/64; 1890 Monroe Twp. Veterans Census; B.S.F. History Book-1884
48	Bedford Cem.	Frank McCoy 1912 Listing; Historical Data Systems record
49	Brandon Cem.-IA	B.S.F. History Book-1884; 1860 Monroe Twp. Census
50	Steele Cem.-Yellow Creek	B.S.F. History Book-1884; 1860 Hopewell Twp. Census
51		KIA 12/13/62; B.S.F. History Book-1884; 1850 Hopewell Twp. Census
52		Historical Data Systems record; Edward, David F, Levi, John & Thomas-brothers
53	Yellow Creek Reformed Cem.	1890 Hopewell Veterans Census
54	Grandview Cem.-Hunt. Co.	1890 Liberty Twp. Veterans Census; in Hospital at Muster Out; Discharged on Surgeon's Cert.
55		B.S.F. History Book-1884; 1860 Broad Top Census
56	Bethel Brethren Cem.-Tatesville	Wounded 5/8/64 in Laurel Hill during Spottsylvania Campaign; 1890 Hopewell Twp. Vet. Cen.
57		B.S.F. History Book-1884
58		Historical Data Systems record
59		B.S.F. History Book-1884
60	Grandview Cem.-Tyrone	1890 Saxton Veterans Census
61		1890 Saxton Veterans Census; Leg. Wound 7/2/63; US Home of Disabled Vets record
62		1890 Saxton Veterans Census
63	Mt. Sinai Cem.-Blue Knob	Historical Data Systems record; aka Stingle; Discharged on Surgeon's Cert.
64	Andersonville Cem., 11069	POW-Andersonville 6/29/64; Died-Scorbutus 10/17/64; 1860 S. Woodbury Twp. Census

	Name	Age	Rank	Regiment	Co.	Muster In	Muster Out	Casualty	Casualty Battle
1	Stephens, John G	19	Pvt.	PA 184th Inf.	A	5/12/64		KIA	Boydton Plank Rd.
2	Stephens, John G		Pvt.	PA 184th Inf.	A			Wounded	Cold Harbor
3	Stephens, Samuel		Pvt.	PA 22nd Cav.	I	2/4/64			
4	Stephenson, Cyrus*		Pvt.	PA 55th Inf.	I	2/18/64	8/30/65		
5	Stephey, Levi	28	Pvt.	PA 208th Inf.	H	9/5/64	6/1/65		
6	Steuby, Conrad G	37	Pvt.	PA 138th Inf.	F	8/29/62	3/6/65	Wounded	Wilderness
7	Stevens, Denton	57	Pvt.	PA 149th Inf.	I	9/29/63	4/21/64		
8	Stevens, Jacob B	21	Pvt.	PA 138th Inf.	E	8/29/62	6/23/65		
9	Stevens, Nicholas	18	Corp.	PA 9th Cav.	M	10/24/61	7/18/65		
10	Stewart, John		SMN.	US Navy					
11	Stewart, Preston		Pvt.	USCT 118th Inf.	C	8/15/64			
12	Stewart, William A*		Pvt.	PA McKeage's Mil.	G	7/2/63	8/8/63		
13	Stickler, Henry	38	Corp.	PA 122nd Inf.	E	8/11/62	5/15/63		
14	Stickler, Samuel	23	Pvt.	PA 55th Inf.	D	10/12/61		POW-Died	Drewry's Bluff
15	Stickman, John G	23	Pvt.	MD 3rd Reg Inf.	A	9/15/61	7/31/65		
16	Stiffler, George C	36	Pvt.	PA 55th Inf.	H	2/19/64	8/30/65		
17	Stiffler, John H	18	Pvt.	PA 138th Inf.	E	8/29/62	7/16/63	Wound.-Died	Mine Run
18	Stiffler, Nathaniel	21	Pvt.	PA 138th Inf.	E	8/29/62	6/23/65	Wounded	
19	Stiffler, Thomas H	17	Pvt.	PA 99th Inf.	H	3/20/65	7/1/65		
20	Stine, David	30	Pvt.	PA 11th Cav.	G	9/5/61		POW-Died	Suffolk, VA
21	Stine, Isaac E		Pvt.	PA 21st Cav.	E	7/22/63	2/20/64		
22	Stine, Jacob	39	Corp.	PA 77th Inf.	F	2/27/65	9/14/65		
23	Stine, Thomas	31	Pvt.	PA 54th Inf.	E	9/15/62	7/15/65		
24	Stineman, Daniel		Pvt.	PA 55th Inf.	I	9/23/63	1/18/65		
25	Stineman, John	26	Pvt.	PA 91st Inf.	B	9/21/64		KIA	Boydton Plank Rd.
26	Stineman, Thomas B	23	Pvt.	PA 91st Inf.	B	9/21/64	5/23/65	Wounded	Boydton Plank Rd.
27	Stineman, William	18	Pvt.	PA 138th Inf.	E	8/29/62	6/23/65	POW	Monocacy, MD
28	Stiner, Samuel*		Sgt.	PA 55th Inf.	I	9/20/61			
29	Stingle, Jacob		Pvt.	PA 55th Inf.	K	11/5/61	6/26/63		
30	Stirtz, Solomon	36	Pvt.	PA 171st Inf.	I	11/10/62	8/8/63		
31	Stoler, Jacob	22	Pvt.	IN 100th Inf.	F	8/18/62	6/8/65		
32	Stombaugh, Joseph	41	Pvt.	PA McKeage's Mil.	G	7/2/63	8/8/63		
33	Stone, Reuben M	26	Sgt.	PA 101st Inf.	D	11/1/61	3/4/65	POW	Plymonth, NC
34	Stoner, Daniel F	44	Pvt.	PA 91st Inf.	B	9/21/64	5/30/65	Wounded	Boydton Plank Rd.
35	Stoner, David E	18	Pvt.	PA 77th Inf.	A	10/8/62	10/7/65	Wounded	Franklin, TN
36	Stoner, J Daniel	19	Sgt.	PA Ind Light Art.	B	10/11/61	10/12/65		
37	Stoner, John R	21	Pvt.	PA 165th Inf.	A	11/4/62	7/28/63		
38	Stoner, Joshua	21	Pvt.	PA 133rd Inf.	C	8/13/62	5/26/63		
39	Stoner, Merrick A	23	Pvt.	OH 120th Inf.	H	8/21/62	11/27/64		
40	Stoner, Merrick A		Pvt.	OH 114th Inf.	K	11/27/64	7/27/65		
41	Stoner, Samuel		Pvt.	MD 2nd PHB Inf.	H	10/15/61	10/15/64		
42	Stoner, William	36	Pvt.	PA 208th Inf.	H	9/9/64	6/1/65		
43	Stonerook, Aaron B	21	Pvt.	PA 110th Inf.	C	2/27/64	11/29/64		
44	Stonerook, Aaron B		Pvt.	PA 18th VRC	I	11/29/64	8/1/65		
45	Stonerook, Jacob	19	Pvt.	PA 50th Inf.	I	3/15/65	7/31/65		
46	Stonerook, Simon B	20	Sgt.	PA 110th Inf.	C	10/24/61	8/5/65	Wounded	Spotsylvania
47	Stotler, Marion	17	Pvt.	PA 138th Inf.	F	9/29/62	6/23/65		
48	Stotzer, Paul*	35	Pvt.	PA 55th Inf.	I	10/15/63			
49	Stoudenour, John (1)	21	Music.	PA 76th Inf.	E	10/9/61	7/30/63		Ft. Wagner*
50	Stoudenour, John (2)		Sgt.	PA McKeage's Mil.	H	7/2/63	8/8/63		
51	Stoudnour, Jacob	16	Pvt.	PA 76th Inf.	E	10/9/61	11/28/64		
52	Stoudnour, William	28	Pvt.	PA 133rd Inf.	C	8/13/62	5/26/63		
53	Stoudnour, William		Pvt.	PA 208th Inf.	K	9/7/64	6/1/65		
54	Stouffer, George		Pvt.	PA 13th Cav.	D	10/14/63	7/14/65		
55	Stout, Richard F		Pvt.	PA 110th Inf.	C	9/3/64	6/9/65		
56	Stoutnour, James H	20	Music.	PA 55th Inf.	D	10/12/61	8/30/65		
57	Stoutnour, Samuel R	20	Pvt.	PA 133rd Inf.	K	8/18/62	5/26/63		
58	Stoutnour, Samuel R		Pvt.	PA 11th Inf.	A	3/21/64	1/1/65	Wounded	North Anna River
59	Strahan, Robert V*		Pvt.	PA 76th Inf.	E	10/7/64	7/18/65		
60	Straight, Abraham		Pvt.	PA 5th Res.					
61	Straley, James	23	Pvt.	PA 110th Inf.	C	12/19/61	6/25/62		
62	Straley, James		Pvt.	PA 22nd Cav.	B	7/11/63			
63	Straney, Edward	25	Pvt.	PA 55th Inf.	D	10/12/61	8/30/65		
64	Strathers, James	27	Pvt.	USCT 43rd Inf.	B	3/14/64	10/20/65	Wounded	Battle of the Crater

Detailed Alphabetical Listing   521

	Cemetery	Notes
1		KIA 10/27/64; B.S.F. History Book-1884; 1860 S. Woodbury Twp. Census; Son of Jacob
2		Wounded listed in Bedford Inquirer 7/1/64 article; B.S.F. History Book-1884
3		B.S.F. History Book-1884
4		B.S.F. History Book-1884; Historical Data Systems-Blair Co. residence
5	Dry Hill Cem.-Woodbury	B.S.F. History Book-1884
6	Tacoma Cem.-WA	Bedford Inquirer 5/27/64 serious leg wound on 5/6/64; B.S.F. History Book-1884
7	Mt. Zion Cem.-Chaneysville	Frank McCoy 1912 Listing; Bedford Inquirer 5/22/1908 listing; Discharged on Surgeon's Cert.
8	Norton Cem.-KS	B.S.F. History Book-1884; PA Civil War Archives-Bedford residence
9	Hopewell Cem.	1890 Broad Top Veterans Census
10	Bedford Cem.	Bedford Inquirer 5/22/1908 listing; On U.S. ship named Pensacola
11		1890 Bedford Twp. Veterans Census
12		B.S.F. History Book-1884; Blair Co. references
13	Grandview Cem.-Cambria Co.	1890 Napier Twp. Veterans Census aka Strichler
14		MIA 5/16/64 in Bedford Inquirer on 7/8/64; POW-Richmond; Died 6/18/64; PA CW Archives
15	Hyndman Cem.	1890 Londonderry Twp. Veterans Census
16	Bedford Cem.	1890 Bedford Twp. Veterans Census; B.S.F. History Book-1884
17	Stiefler Cem.-Blue Knob	Wounded 11/27/63; Died 1/3/64; PA CW Arch.-St. Clairsville resid.; 1850 Colerain Twp. Cen.
18	Salemville Cem.	1890 S. Woodbury Twp. Census
19	Stiefler Cem.-Blue Knob	1890 Kimmel Twp. Veterans Census; Thomas & John-brothers
20	Andersonville Cem.	POW 11/10/63; Died in Andersonville 10/8/64; 1890 Greenfield Twp. Veterans Census
21		Historical Data Systems-Bedford Co. residence
22	Old Claysburg Cem.	Soldiers of Blair Co. Book - 1940; Jacob & David are half brothers with Abraham Burket
23	Bedford Cem.	1890 Bedford Veterans Census
24		B.S.F. History Book-1884; Transferred to Veterans Reserve Corp.
25		KIA 10/27/64; Bedford Gazette 2/13/1914 listing; 1860 Union Twp. (Pavia) Census
26	Graceland Cem.-IA	Wounded 10/27/65; Ancestry.com information; Thomas, William & John-brothers
27		B.S.F. History Book-1884; PA Civil War Archives-St. Clairsville residence
28		Deserted 10/22/64; PA Civil War Archives-Huntingdon residence
29	Mt. Sinai Cem.-Blue Knob	B.S.F. History Book-1884; aka Stengal; Discharged on Surgeon's Cert.
30		B.S.F. History Book-1884
31	Lakeview Cem.-IN	Ancestry.com information; Abraham-father in 1840 Hopewell Twp. Census
32	Mt. Union Cem.-Lovely	B.S.F. History Book-1884
33	Greenlawn Cem.-OH	POW 4/20/64 to 2/27/65; Historical Data Systems-Bedford Co. residence
34	Albrights Cem.-Roaring Springs	Wounded 10/27/64; Ancestry.com information; 1850 S. Woodbury Twp. Census
35	Old Covenenter Cem.-Fayetteville	Ancestry.com-Born Bedford Co.
36	Mt. Union Meth.-Franklin Co.	Findagrave.com-Born Bedford Co.
37	Mentzers Cem.-Fayetteville	Ancestry.com information; John, J Daniel & David-brothers
38		B.S.F. History Book-1884; Joshua & David-brothers
39	Bedford Cem.	1890 Bedford Twp. Veterans Census
40		Historical Data Systems record; Coroner and cabinet maker in Bedford after the war
41	Buck Valley Cem.-Fulton Co.	1890 Brush Creek Twp. Veterans Census
42	McConnellsburg Luth. Cem.	B.S.F. History Book-1884; William & Merrick-brothers
43	Horton Cem.-NE	B.S.F. History Book-1884; Ancestry.com-born Woodbury
44		B.S.F. History Book-1884
45	Spring Hope Cem.-Martinsburg	Pension Record; Jacob, Aaron & Simon-brothers
46	Masonic Cem.-IA	Wounded in May 1864; B.S.F. History Book-1884
47	Greenwood Cem.-TX	B.S.F. History Book-1884; Historical Data Systems-Bed. Co residence; aka Statler
48		B.S.F. History Book-1884; enlisted in Norristown
49	Mt. View Cem. & Maus.-CA	B.S.F. History Book-1884; PA CWArchives-Bedford resid.; aka Stoudnour; Disch. Surg. Cert.
50		B.S.F. History Book-1884
51	Bedford Cem.	1890 Bedford Twp. Veterans Census; Historical Data Systems record
52	Bedford Cem.	Frank McCoy 1912 List; Hist. Data Systems record; 1850 Bedford Twp. Cen.; aka Stoudenour
53		B.S.F. History Book-1884; William, Jacob & John- brothers
54		1890 Everett Veterans Census
55		B.S.F. History Book-1884
56	Everett Cem.	1890 Everett Veterans Census
57	Everett Cem.	B.S.F. Hist. Book-1884; in Sharpsburg Hospital at Muster Out; Historical Data Systems record
58		Wounded 5/25/64-Pension record; Frank McCoy 1912 List; transferred-Vet. Res. Corp. 1/1/65
59		B.S.F. History Book-1884; Drafted; Union Co. references
60		B.S.F. History Book-1884
61	Eshelman Cem.-Woodbury	PA Civil War Archives-Woodbury residence; Historical Data Systems record
62		Hist. Data Systems record; 1860 S. Woodbury Twp. Cen.; Died-1875; Discharged Surg. Cert.
63		B.S.F. History Book-1884; Historical Data Systems-Bedford Co. residence
64	Mt. Ross Cem.-Bedford	Neck Wound 7/30/64; Frank McCoy 1912 Listing; aka Strawthers; James & Willis-brothers

	Name	Age	Rank	Regiment	Co.	Muster In	Muster Out	Casualty	Casualty Battle
1	Strathers, Willis	24	Pvt.	USCT 43rd Inf.	C	3/19/64		KIA	Battle of the Crater*
2	Stratton, Jeremiah	19	Pvt.	PA 55th Inf.	K	2/19/64	8/30/65	Wounded	
3	Strayer, Jacob M R	26	Pvt.	PA 148th Inf.	E	10/30/63	6/1/65	Wounded	Deep Bottom Run
4	Strayer, John	24	Pvt.	PA 205th Inf.	C	8/27/64	6/2/65		
5	Strayer, Nicholas	30	Pvt.	PA 205th Inf.	C	8/27/64		Wound.-Died	Petersburg - Final
6	Streets, James		Music.	USCT 24th Inf.	A	1/10/65	9/1/65		
7	Streets, Rankin		Pvt.	USCT Inf.					
8	Streight, John	33	Pvt.	PA 11th Inf.	A	9/30/61	9/30/64		
9	Streightif, Samuel	30	Pvt.	PA 107th Inf.	I	9/21/64	5/3/65		
10	Strellie, John*		Pvt.	PA 76th Inf.	E	2/23/65	7/18/65		
11	Strong, William	19	Pvt.	PA 101st Inf.	D	11/1/61	1863		
12	Stroud, George W	16	Pvt.	NY 13th Light Art.		9/15/64	8/30/65		
13	Struble, George A		Pvt.	PA 198th Inf.	I	9/13/64			
14	Struckman, Augustus	37	Pvt.	PA 50th Inf.	D	11/14/64	7/30/65		
15	Struckman, Charles	20	Pvt.	PA 55th Inf.	H	10/11/61	10/11/64	Wounded	
16	Struckman, Henry	37	Pvt.	PA 171st Inf.	I	11/2/62	8/8/65		
17	Strunk, George W		Pvt.	PA 21st Cav.	E	7/8/63	2/20/64		
18	Stuby, Jacob E	36	Pvt.	PA 194th Inf.	I	7/21/64	11/5/64	Wounded	
19	Stuckey, Abraham	33	Pvt.	PA 208th Inf.	K	9/7/64	6/1/65		
20	Stuckey, David H	28	Corp.	PA 184th Inf.	A	5/12/64		POW-Died	Jerusalem Plank Rd.
21	Stuckey, Elias B	19	Corp.	PA 138th Inf.	D	9/2/62	6/23/65	Wounded	Petersburg - Final
22	Stuckey, George W	31	Pvt.	PA 107th Inf.	F	9/21/64	6/6/65		
23	Stuckey, John S	27	Capt.	PA 138th Inf.	D	9/2/62	2/8/65	Wounded	Opequon
24	Stuckey, Samuel								
25	Stuckey, Simon C	22	1st Sgt.	PA 138th Inf.	D	9/2/62		KIA	Mine Run
26	Stuckey, Ted*		Music.						
27	Stuckey, William H	20	Corp.	PA 101st Inf.	D	11/1/61	10/12/64		
28	Stuckey, Wilson H	36	Pvt.	PA 138th Inf.	D	8/29/62	6/23/65		
29	Stufft, Jacob	35	Pvt.	PA 171st Inf.	I	11/10/62	8/8/63		
30	Stufft, Michael	41	Pvt.	PA 91st Inf.	G	9/21/64		Died	
31	Stufft, William S	32	Pvt.	PA 171st Inf.	I	11/2/62		Died	
32	Stull, William (1)	21	Pvt.	PA 17th Cav.	G	9/26/62	6/16/65	Wounded	
33	Stull, William (2)			PA 3rd Heavy Art.					
34	Stull, William (2)	27	Pvt.	PA 188th Inf.	A	2/26/64	12/14/65	Wounded	Cold Harbor
35	Stultz, William	34	Pvt.	PA 100th Inf.	A	9/21/64	7/24/65	Wounded	
36	Sturtz, Jesse	21	Pvt.	OH 80th Inf.	H	2/23/64	8/13/65		
37	Sturtz, Solomon	36	Pvt.	PA 171st Inf.	I	11/1/62	8/8/63		
38	Sturtz, Solomon		Pvt.	PA 53rd Inf.	B	9/21/64	5/31/65		
39	Sturtz, Solomon A	18	Pvt.	OH 9th Cav.	K	12/15/63	7/20/65		
40	Sturtzman, James E	32	Pvt.	PA 49th Inf.	D	1/25/62	1/25/65		
41	Sturtzman, James E		Pvt.	PA 192nd Inf.	B	2/13/65			
42	Suder, Charles H	23	Music.	PA 125th Inf.	E	8/13/62	5/19/63		
43	Suder, Charles H		Sgt.	PA 13th Cav.	D	10/6/63			
44	Suiters, William S	20	Pvt.	PA 2nd Cav.	E	11/20/61	1864		
45	Sullivan, John O*	25	Pvt.	PA 55th Inf.	H	10/16/63	12/30/64		
46	Sullivan, John*	21	Pvt.	PA 55th Inf.	H	10/16/63	12/30/64		
47	Summerfield, Alexander*		Pvt.	PA 55th Inf.	I	7/20/63			
48	Summerland, John	24	Pvt.	PA 55th Inf.	I	9/20/61			
49	Summerland, Peter J	18	Pvt.	PA 55th Inf.	I	2/15/64	8/30/65	Wounded	Petersburg - Initial
50	Summerlatt, Charles	27	Music.	MD 2nd Inf.		9/24/61	8/16/62		
51	Summers, George*	27	Pvt.	PA 55th Inf.	H	9/25/63		KIA	Petersburg - Initial
52	Summers, John*	26	Pvt.	PA 55th Inf.	I	10/15/63		MIA	Drewry's Bluff
53	Summerville, Abner	35	Pvt.	PA 55th Inf.	D	10/26/62		POW-Died	
54	Summerville, Charles	21	Pvt.	PA 138th Inf.	D	8/29/62		MIA	Wilderness
55	Summerville, John B		Pvt.	PA McKeage's Mil.	D	7/2/63	8/8/63		
56	Summerville, John B	35	Pvt.	PA 138th Inf.	D	2/24/64	6/23/65		
57	Summerville, Robert	27	Pvt.	PA 208th Inf.	K	9/7/64	6/1/65		
58	Summerville, Sylvanus B	22	Pvt.	PA 55th Inf.	D	10/12/61		Wound.-Died	Chaffin's Farm
59	Suter, William S	36	Pvt.	PA 82nd Inf.	K	7/5/64	7/13/65		
60	Suters, Emanuel	18	Pvt.	PA 21st Cav.	E	7/8/63	2/20/64		
61	Sutters, John*		Pvt.	PA 55th Inf.	I	8/28/61		POW-Died	Drewry's Bluff
62	Sutton, Jonathan A	31	Pvt.	PA 110th Inf.	C	2/27/64	6/28/65	Wounded	1st Deep Bottom
63	Sutton, Joseph		Pvt.	PA 76th Inf.	E	10/9/61	2/9/65		
64	Swaney, David R P	24	Pvt.	PA 133rd Inf.	C	8/13/62	5/26/63		

	Cemetery	Notes
1		KIA 7/30/64*; Frank McCoy 1912 list & Civil Works Admin. 1934 list
2		Wounded listed in Bedford Inquirer article 6/3/64; 1860 Liberty Twp. Census
3	Albrights Cem.-Roaring Springs	Thigh Wound 8/28/64; Historical Data Systems record
4	Albrights Cem.-Roaring Springs	1890 Bloomfield Twp. Veterans Census; John, Jacob & Nicholas-brothers
5	Arlington Nat'l Cem., 13 10559	Wounded 4/2/65; Died 5/12/65; Historical Data Systems record; 1860 Woodbury Twp. Census
6	Allegheny Cem.-Pittsburgh	1890 Bedford Twp. Veterans Census
7		Frank McCoy 1912 list & Civil Works Admin. 1934 list; 1850 Bedford Twp. Census
8	Union Memorial Cem.-Mench	Deserted 10/19/62 & returned 4/21/63; 1890 E. Providence Twp. Veterans Census
9	Robinsonville Cem.	1890 Brush Creek Twp. Veterans Census; Discharged on Surgeon's Cert.
10		B.S.F. History Book-1884; Mustered in Easton
11	St. Mark's Cem.-Ott Town	Frank McCoy 1912 Listing; Historical Data Systems record; Discharged at Plymouth NC
12	Tioga Point Cem.-Bradford Co	Historical Data Systems-enlisted in Bedford Co.
13		Historical Data Systems record-buried Broad Top
14	Benna Cem.-Manns Choice	1890 Juniata Twp. Veterans Census; Historical Data Systems aka Strupman
15	Schellsburg Cem.	1890 Juniata Veterans Census-Shoulder wound
16	Schellsburg Cem.	1890 Napier Twp. Veterans Census
17		Historical Data Systems-Bedford Co. residence
18	Yellow Creek Reformed Cem.	1890 Hopewell Twp. Veterans Census; Frank McCoy 1912 Listing
19	Everett Cem.	B.S.F. History Book-1884; Monroe Twp. Census
20	Andersonville Cem., 12081	POW-Andersonville 6/22/64; Died-Scorbutus 11/18/64 1860 Colerain Twp. Census
21	Chico Cem.-CA	Wounded 4/2/65; B.S.F. History Book-1884; 1850 M. Woodbury Twp. Census
22	Friends Cove UCC Cem.	1890 Everett Veterans Cen.; PA Civil War Archives record; George, William & David-bros.
23	Evergreen Cem.-NE	Leg Amputated 9/19/64; Historical Data Systems record; 1860 Napier Township Census
24		Listed in the "From Winchester to Bloody Run" book
25	Mt. Olivet Cem.-Manns Choice	KIA 11/27/63; B.S.F. History Book-1884; Hist.Data Systems record; 1850 Napier Twp. Cen.
26		Listed in the "From Winchester to Bloody Run" book
27	Everett Cem.	1860 Colerain Twp. Census; 1890 Everett Veterans Census
28	New Enterprise Cem.	1890 S. Woodbury Twp. Census; B.S.F. History Book-1884
29	Pleasantville Cem.	1890 W. St. Clair Twp. Census; Frank McCoy 1912 Listing; Jacob, Michael & William S-bros.
30	Mt. Zion Cem.-Pavia	Died 6/22/65; Frank McCoy 1912 Listing; 1860 Union Twp. Census
31	New Bern Nat'l Cem.	Died 6/13/63; B.S.F. History Book-1884; 1860 Union Twp. Census; 1890 King Twp. Vet. Cen.
32	Holsinger Cem.-Bakers Summit	1890 Woodbury Twp. Veterans Census
33		Civil War Pension index record; Pension request record
34	Hopewell Cem.	Wounded 6/1/64; 1890 Broad Top Veterans Census
35	Schellsburg Cem.	1890 Napier Twp. Veterans Census-Back Wound
36	Palo Alto Cem.-Hyndman	1890 Londonderry Twp. Veterans Census
37	Lybarger Cem.-Madley	1890 Londonderry Twp. Veterans Census
38		Bedford Inquirer 5/22/1908 listing
39	Palo Alto Cem.-Hyndman	1890 Londonderry Twp. Veterans Census; Solomon A & Jesse-brothers
40	Broad Top IOOF Cem.	1890 Robertsdale Veterans Census
41		Historical Data Systems record
42	Walters Cem.-Duncansville	PA Civil War Project record; aka Suter
43		1890 S. Woodbury Twp. Veterans Census; Promoted to Sgt.
44		B.S.F. History Book-1884
45		B.S.F. History Book-1884; Drafted
46		B.S.F. History Book-1884; Drafted
47		Drafted & Deserted 8/31/64; B.S.F. History Book-1884; born Chester Co
48		Deserted 9/30/64; B.S.F. History Book-1884; 1840 S. Woodbury Twp. Census
49	Loudon Park Nat'l Cem.	Wounded 6/18/65; B.S.F. History Book-1884; Peter & John-brothers
50	St. Lukes Luth.-Cumberland	1890 Southampton Twp. Veterans Census; Historical Data Systems record
51		KIA 6/16/64; B.S.F. History Book-1884; Muster in Pottsville
52		MIA 5/16/64; B.S.F. History Book-1884; Muster In Reading, PA
53	Lawton National Cemetery	POW-Andersonville; Died 10/31/64 in Millin, GA; PA Civil War Archives-Bedford Co. resid.
54		MIA 5/6/64; B.S.F. History Book-1884; 1850 Southampton Twp. Census
55		B.S.F. History Book-1884; Charles, John, Robert, Sylvanus & Abner-bros.; aka Summerfield
56	Indian Springs-Monroe Twp.	B.S.F. History Book-1884; Frank McCoy 1912 Listing
57	Circle Hill Cem.-Punxsutawny	1890 Monroe Twp. Veterans Census; B.S.F. History Book-1884
58	Annapolis Nat'l Cem.	Wounded 9/29/64; Died 10/10/64; B.S.F. History Book-1884; 1860 Southampton Twp. Cen.
59	Mt. Olivet Cem.-Manns Choice	Frank McCoy 1912 Listing; Historical Data Systems record; 1860 Napier Twp. Census
60		Historical Data Systems-Bedford Co. residence; enlisted in Bedford
61	Jefferson Barracks Cem.-MO	POW 5/16/64; Died disease 5/7/65; B.S.F. Hist. Book-1884; PA CW Arch.-Somerset Co. resid.
62		Wounded 7/27/64-Bedford Inquirer on 8/7/64; PA Civil War Archives-Woodbury residence
63		B.S.F. History Book-1884; Historical Data Systems-Bedford Co. residence
64	Hickory Bottom Cem.-Woodbury	B.S.F. History Book-1884; Historical Data Systems-Bedford Co. residence

	Name	Age	Rank	Regiment	Co.	Muster In	Muster Out	Casualty	Casualty Battle
1	Swaney, David R P		Corp.	PA 110th Inf.	C	2/27/64	6/28/65		
2	Swaney, Samuel J	16	Pvt.	PA 110th Inf.	C	2/28/64	6/28/65		
3	Swaney, William S	20	Pvt.	PA 110th Inf.	C	3/5/64	6/2/65	Wounded	Wilderness
4	Swaney, William S		Pvt.	PA 110th Inf.	C			Wounded	Spotsylvania
5	Swaney, William S		Pvt.	PA 14th VRC	B	2/1/65	6/2/65		
6	Swank, George W	24	Pvt.	PA 133rd Inf.	C	8/13/62	5/26/63	Wounded	Fredericksburg
7	Swank, George W		Pvt.	PA 210th Inf.	I	9/18/64	5/30/65		
8	Swartz, Abraham	17	Pvt.	PA 11th Inf.	A	9/8/61	7/1/65	Wounded	2nd Bull Run
9	Swartz, David H	19	Pvt.	PA 208th Inf.	H	9/7/64	6/1/65		
10	Swartz, Espy	15	Pvt.	PA 2nd Cav.	F	12/10/61	12/13/64		
11	Swartz, Francis	18	Pvt.	PA 55th Inf.	D	10/12/61	6/15/63	Wounded	
12	Swartz, Francis		Pvt.	US 1st Art.		6/15/63			
13	Swartz, Henry	16	Pvt.	PA 194th Inf.	I	7/22/64	11/5/64		
14	Swartz, John	25	Sgt.	PA 55th Inf.	D	10/12/61	12/30/64		
15	Swartz, John		2nd Lt.	USCT 30th Inf.	A	1/1/65	12/10/65		
16	Swartz, John W	17	Pvt.	PA 194th Inf.	I	7/22/64	9/6/64		
17	Swartz, John W		Pvt.	PA 97th Inf.	Ind	9/6/64	7/17/65		
18	Swartz, Josiah	31	Pvt.	PA 58th Inf.	E	11/4/64	11/9/65		
19	Swartz, William B	44	Pvt.	PA 107th Inf.	I	3/8/62	12/10/62	Wounded	2nd Bull Run
20	Sweitzer, Daniel F	21	Pvt.	PA 101st Inf.	D	2/13/62	4/5/65		
21	Swisher, Daniel		Pvt.	PA 125th Inf.	E	8/13/62	5/18/63		
22	Swope, Allen			US Navy		1864	8/11/65		
23	Swope, George W	20	Pvt.	PA 22nd Cav.	M	2/6/64	6/24/65		
24	Swope, Orlando L		Pvt.	PA McKeage's Mil.	H	7/2/63	8/8/63		
25	Swope, Thomas J		Corp.	PA 76th Inf.	E	8/25/63	7/18/65		
26	Swoveland, William	31	Pvt.	PA 184th Inf.	A	5/12/64		POW-Died	Jerusalem Plank Rd.
27	Tallifaerro, Lawrence		Major	US		5/14/57	8/27/63		
28	Tannehill, Alfred*			IA 14th Inf.				POW-Died	
29	Tannehill, Eli*	41	Pvt.	PA 2nd Heavy Art.	K	11/11/62		KIA	Petersburg
30	Tasker, George	22	Pvt.	PA 110th Inf.	C	10/24/61	10/24/64		
31	Taslinger, Walter E		Pvt.	PA 55th Inf.	H	9/12/61	8/30/65		
32	Tate, George H	18	Pvt.	PA 101st Inf.	D	2/8/62		Died	
33	Tate, Jacob	20	Pvt.	PA 46th Inf.	G	7/9/63	8/19/63		
34	Tate, Jacob		Pvt.	PA 205th Inf.	I	8/27/64	6/2/65		
35	Tate, Samuel B	25	Pvt.	PA 13th Inf.	G	4/25/61	8/6/61		
36	Tate, Samuel B		Capt.	PA 133rd Inf.	K	8/15/62	1/27/63		
37	Tate, Samuel B		2nd Lt.	PA 22nd Cav.	L	2/22/64	8/8/65		
38	Taylor, Ambrose K	19	Sgt.	PA 110th Inf.	C	10/24/61		KIA	1st Deep Bottom
39	Taylor, George W	24	Pvt.	PA 99th Inf.	H	2/25/65	7/1/65		
40	Taylor, Henry C*		Pvt.	PA 76th Inf.	E	2/17/65	11/23/65		
41	Taylor, James	34	Pvt.	PA 76th Inf.	E	10/9/61	5/18/63		
42	Taylor, John W	21	Pvt.	PA 133rd Inf.	C	8/13/62	5/26/63		
43	Taylor, John W		Corp.	PA 185th Inf.	H	4/1/64	6/15/65		
44	Taylor, Matthew P	31	Pvt.	PA 138th Inf.	D	8/29/62	4/17/64	Wounded	Mine Run
45	Taylor, Thomas (1)*	39	Pvt.	PA 4th Cav.	D	2/24/64		Died	
46	Taylor, Thomas (2)*		Pvt.	PA 55th Inf.	H	9/24/63			
47	Taylor, Thomas A	20	Pvt.	PA 8th Res.	F	7/29/61	5/24/64	POW	Seven Days Battle
48	Taylor, William Y	14	Sgt.	PA 110th Inf.	D	6/28/62	5/31/65		
49	Tearnel, William H		Pvt.	PA 21st Cav.	F	2/21/65			
50	Teeter, Christian	22	Pvt.	PA 184th Inf.	A	5/12/64		POW-Died	Jerusalem Plank Rd.
51	Teeter, Samuel	23	Pvt.	PA 99th Inf.	D	2/25/65	7/1/65		
52	Tetwiler, Henry		Pvt.	IA 7th Cav.	G	5/23/63	5/17/66		
53	Tetwiler, Jacob D	23	Sgt.	PA 110th Inf.	C	12/19/61	10/24/64	Wounded	
54	Tetwiler, Peter	24	Pvt.	PA 53rd Inf.	C	10/17/61		Died	
55	Tetwiler, William	28	Pvt.	PA 110th Inf.	C	10/24/61	10/24/64		
56	Tewell, Daniel	34	Corp.	PA 4th Cav.	D	9/16/61		POW-Died	Sulpher Springs
57	Tewell, George	33	Pvt.	PA 91st Inf.	F	9/21/64	7/20/65		
58	Tewell, Joseph	19	Corp.	PA 55th Inf.	K	11/5/61		Wound.-Died	Cold Harbor
59	Tewell, Moses	14	Pvt.	PA 4th Cav.	D	9/16/61	9/18/64		
60	Thatcher, Bartholomew*		Pvt.	PA 76th Inf.	E	11/21/61	11/28/64		
61	Thomas, Augustus	22	Pvt.	PA 28th Inf.	C	11/14/64			
62	Thomas, Frank								
63	Thomas, George W	34	Corp.	PA 21st Cav.	E	7/2/63	7/8/65		
64	Thomas, Isaac	40	Pvt.	PA 186th Inf.	G	3/1/64	8/1/65		

	Cemetery	Notes
1		1860 Woodbury Twp. Census; PA Civil War Archives-Woodbury residence
2	Oak Ridge Cem.-Altoona	PA Civil War Archives-Woodbury residence; Historical Data Systems record
3	Seneca City Cem.-KS	PA Civil War Archives-Bedford residence; William Samuel & David-brothers
4		Obituary stated 7 wounds suffered during the battles of the Wilderness & Spottsylvania
5		Historical Data Systems record; Civil War Pension Record; Discharged on Surgeon's Cert.
6	Eckis Cem.-OH	Wounded 12/13/62; Historical Data Systems-Bedford Co. residence
7		Historical Data Systems record
8	Bedford Cem.	Wounded 8/30/62; Frank McCoy 1912 Listing; Abram, David & Henry-brothers
9	Rimersburg Cem.-Clarion Co.	B.S.F. History Book-1884; 1860 E. Providence Twp. Census; aka Swarts
10	Webb City Cem.-MO	Ancestry.com information; Espy & Francis-brothers
11	Riverview Cem.-NE	B.S.F. History Book-1884; Historical Data Systems record; 1860 Bedford Twp. Census
12		PA Civil War Project record; Historical Data Systems record
13	Richey Farm-W. Prov. Twp.	B.S.F. History Book-1884; 1860 Hopewell Twp. Census
14		Historical Data Systems-Bedford Co. residence
15		Historical Data Systems-Bedford Co. residence
16	Fairview Cem.-Altoona	B.S.F. History Book-1884
17		Historical Data Systems record
18	Everett Cem.	1890 Everett Veterans Census; substitute
19	Dry Hill Cem.-Woodbury	Wounded 8/28/62; Soldiers of Blair Co. Book - 1940; Father of John W
20	Everett Cem.	Frank McCoy 1912 Listing; 1870 Broad Top Cen.; aka Frederick; Discharged on Surg. Cert.
21	Hopewell Cem.	1890 Broad Top Veterans Census
22		1890 Robersdale Veterans Census; USS Antona
23	Wells Valley Meth. Cem.	1890 Wells Twp. Veterans Census
24		B.S.F. History Book-1884
25		B.S.F. History Book-1884
26	Andersonville Cem.; 5215	POW-Andersonville 6/22/64; Died 8/10/64; B.S.F. Hist. Book-1884; children born-Bedford Co.
27	Bedford Cem.	Bedford Inquirer 5/22/1908 listing; Frank McCoy 1912 Listing
28	Jefferson Barracks Cem.-MO	POW-Libby; Died 1/10/65; B.S.F. History Book-1884; 27th IA listed; Somerset Co. references
29	Poplar Grove Nat'l Cem.	KIA 8/25/64, B.S.F. History Book-1884; Somerset Co. references
30		B.S.F. History Book-1884; Historical Data Systems record
31		1890 Napier Twp. Veterans Census
32	New Bern Nat'l Cem.	Died of Disease 7/24/63; PA Civil War Archives-Bedford Co. resid.; enlisted in Bloody Run
33	Hopewell Cem.	1890 Broad Top Veterans Census
34		1890 Broad Top Veterans Census
35	Everett Cem.	B.S.F. History Book-1884
36		B.S.F. History Book-1884; Frank McCoy 1912 Listing; Bedford Inquirer 5/22/1908 listing
37		B.S.F. History Book-1884; Historical Data Systems-Bedford Co. residence
38		KIA 7/27/64; PA Civil War Archives-Yellow Creek residence; 1860 Hopewell Twp. Census
39	Hyndman Cem.	1890 Hyndman Veterans Census
40		B.S.F. History Book-1884; Susquehanna references
41		B.S.F. History Book-1884; PA Civil War Archives-Bedford resid.; Discharged on Surg. Cert.
42	Curtis Cem.-NE	B.S.F. History Book-1884; Historical Data Systems-Bedford Co. residence
43		Ancestry.com information; John, William & Thomas A-brothers
44	Togus Nat'l Cem.-ME	Arm Amputated 11/27/63; Historical Data Systems-Bedford Co. residence
45	Arlington Nat'l Cem., 13-5565	Died 10/23/64; Unsure this Thomas Taylor was actually from Bedford County
46		Deserted 4/4/64; PA Civil War Archives-Pottsville residence
47	Grand Island Cem.-NE	POW 6/27/62; B.S.F. History Book-1884; 1850 Hopewell Twp. Census
48	Marion Nat'l Cem.-IN	B.S.F. History Book-1884; 1860 Hopewell Twp. Census; 115th PA references
49		1890 Union Twp. (Pavia) Veterans Census
50	Andersonville Cem.; 8619	POW-Andersonville; 6/22/64; Died 9/13/64; B.S.F. Hist. Book-1884; 1860 S. Woodbury Cen.
51	Dry Hill Cem.-Woodbury	1890 Woodbury Veterans Census; Historical Data Systems record
52	Nevada Municipal Cem.-IA	Ancestry.com information; Henry & Jacob-brothers
53	Waterside Cem.-Woodbury	Loss of right arm; 1890 S. Woodbury Twp. Veterans Census
54	Pote Cem.-Bakers Summit	Died 4/13/65; Frank McCoy 1912 Listing; Bedford Inquirer 5/22/1908 listing
55	Dry Hill Cem.-Woodbury	1890 S. Woodbury Twp. Veterans Census; B.S.F. History Book-1884
56	Andersonville Cem., 1145	POW-Andersonville 10/12/63; Died-Pneumonia 5/16/64; PA CW Archives-Bedford Co. resid.
57	Mt. Zion Cem.-Pavia	Frank McCoy 1912 Listing; Historical Data Systems record
58	Mt. Zion Cem.-Chaneysville	Leg Amputated 6/2/64; Died 6/17/64; Frank McCoy 1912 List; 1860 Southampton Twp. Cen.
59	Waco Cem.-NE	PA Civil War Archives-Bedford Co. residence
60		B.S.F. History Book-1884; Huntington, NJ references
61	Hyndman Cem.	1890 Harrison Twp. Veterans Census
62	Hopewell Cem.	Bedford Inquirer 5/22/1908 listing
63		Historical Data Systems-Bedford Co. residence
64	Mt. Zion Cem.-Breezewood	1890 E. Providence Twp. Veterans Census; B.S.F. History Book-1884

Chapter 18

	Name	Age	Rank	Regiment	Co.	Muster In	Muster Out	Casualty	Casualty Battle
1	Thomas, John	22	Pvt.	PA 205th Inf.	D	8/29/64	6/2/65		
2	Thomas, Joseph	43	Pvt.	PA 208th Inf.	K	9/7/64	6/1/65		
3	Thomas, Joseph A	19	Pvt.	PA 13th Inf.	C	4/25/61	8/6/61		
4	Thomas, Joseph A		Pvt.	PA 12th Res.	I	3/29/62		POW-Died	Globe Tavern
5	Thomas, Samuel	40		MO Home Brigade	F			Died	
6	Thomas, Warner	38	Pvt.	PA 208th Inf.	K	9/7/64	6/1/65		
7	Thompson, Charles H		Pvt.	PA 7th Res.	G	3/14/63	8/13/65		
8	Thompson, David	42	Pvt.	PA 110th Inf.	C	10/24/61		POW-Died	Petersburg
9	Thompson, David Q		Pvt.	MD 2nd PHB Inf.	B	4/7/62	5/29/65		
10	Thompson, Jeremiah H	32	Pvt.	PA 55th Inf.	D	10/12/61	1/18/64	POW	Drewry's Bluff
11	Thompson, John*		Pvt.	PA 55th Inf.	D	10/14/63		MIA	Cold Harbor
12	Thompson, William	30	Pvt.	PA 133rd Inf.	K	8/15/62	5/26/63		
13	Thorpe, David W*		Pvt.	PA 62nd Inf.	K				
14	Thorpe, Jacob	27	Pvt.	MD 2nd PHB Inf.	K	9/7/61	10/31/64	Wounded	
15	Thorpe, Jacob		Pvt.	PA 138th Inf.	D	8/29/62	1/17/63		
16	Thorpe, John W	21	Music.	PA 138th Inf.	D	8/29/62		Died	
17	Thorpe, Solomon R	19	Music.	PA 138th Inf.	D	8/29/62	6/23/65		
18	Tiday, Henry	28	Pvt.	OH 1st Inf.	A	8/9/62	6/13/65		
19	Tiday, Josiah	35	Corp.	PA 49th Inf.	E	7/19/64	7/15/65		
20	Tillman, George	39	Pvt.	USCT 41th Inf.	H	10/3/64			
21	Tillman, Isaac			USCT Inf.					
22	Tillman, Jackson	40	Pvt.	USCT 32nd Inf.	G	2/24/64	8/22/65		
23	Tipton, Levi	21	Pvt.	PA 77th Inf.	F	2/27/65	12/6/65		
24	Tipton, Noah (1)	37	Pvt.	PA 61st Inf.	G	9/26/64	6/20/65	Died	
25	Tipton, Noah C	21	Pvt.	PA 138th Inf.	F	9/29/62			
26	Tobias, Calvin	24	Pvt.	PA 5th Inf.	D	4/21/61	7/25/61		
27	Tobias, Calvin		Pvt.	PA 28th Inf.	O	8/17/61	10/28/62	POW	
28	Tobias, Calvin		Pvt.	PA 147th Inf.	B	12/29/63	7/15/65		
29	Tobias, John			PA 110th Inf.					
30	Tobias, John B	20	Corp.	PA 8th Res.	F	6/19/61	5/26/64	POW	Seven Days Battle
31	Tobias, John B		Corp.	PA 8th Res.	F			Wounded	Fredericksburg
32	Tobias, John B		1st Sgt.	PA 194th Inf.	I	7/21/64	11/5/64		
33	Tobias, John B		1st Sgt.	USCT Inf.		11/5/64	1865		
34	Tobias, Samuel H G	19	Pvt.	PA 13th Inf.	G	4/25/61	8/6/61		
35	Tobias, Samuel H G		1st Sgt.	PA 110th Inf.	C	10/24/61		Wounded	Port Republic
36	Tobias, Samuel H G		1st Sgt.	PA 110th Inf.	C			KIA	Gettysburg
37	Trail, George T	16		MD 3rd PHB Inf.	D	12/31/61	9/30/62	POW	Moorefield
38	Trail, George T		Corp.	PA 199th Inf.	A	9/29/64	6/28/65		
39	Trail, Hugh	15	Pvt.	MD 1st PHB Inf.	K	2/13/65	6/28/65		
40	Trail, John	42	Pvt.	PA 171st Inf.	I	11/2/62	8/8/63		
41	Trail, John		Pvt.	PA 93rd Inf.	A	11/14/64	6/27/65		
42	Trail, Nathan	17	Pvt.	MD 1st PHB Cav.	G	9/11/64	10/23/64		
43	Trail, Nathan B	26	Pvt.	MD 1st PHB Cav.	G	2/27/64		Died	
44	Travis, John L		Pvt.	PA 55th Inf.	H	10/16/63			
45	Treese, Adie Bell	17	Pvt.	PA 55th Inf.	I	2/15/64	6/7/65	Wounded	Cold Harbor
46	Treese, Francis	19	Pvt.	PA 78th Inf.	K	2/28/65	9/9/65		
47	Treese, Willaim		Pvt.	PA 91st Inf.	B	3/15/65	7/10/65		
48	Trembath, Samuel D		Sgt.	PA McKeage's Mil.	G	7/2/63	8/8/63		
49	Trembath, Samuel D	19	Corp.	PA 194th Inf.	I	7/22/64	11/5/64		
50	Tricker, George	41	Pvt.	PA 8th Res.	F	6/11/61	1/31/63		
51	Trostle, Josiah	20	Pvt.	PA McKeage's Mil.	H	7/2/63	8/8/63		
52	Trott, Benjamin	36	Pvt.	PA 55th Inf.	H	3/2/64		Died	
53	Trott, Benjamin		Pvt.	PA 55th Inf.	H			Wounded	Bermuda Hundred
54	Trott, George*	40	Pvt.	PA 158th Inf.	H	11/4/62		Died	
55	Trout, Alexander	38	Pvt.	PA 125th Inf.	B	8/10/62	5/18/63		
56	Trout, Alexander		Pvt.	PA McKeage's Mil.	G	7/2/63	8/8/63		
57	Trout, John C	29	Pvt.	PA 79th Inf.	D	6/4/64			
58	Trout, Sylvester		Pvt.	PA McKeage's Mil.	H	7/2/63	8/8/63		
59	Trout, Sylvester	19	Pvt.	PA 184th Inf.	A	5/12/64		POW	Jerusalem Plank Rd.
60	Troutman, Benjamin	25	Pvt.	MD 2nd PHB Inf.	D	3/11/65	3/29/65		
61	Troutman, Daniel	38	Pvt.	MD 2nd PHB Inf.	C	8/27/61	5/29/65		
62	Troutman, Francis T	23	Pvt.	PA 211th Inf.	I	9/8/64	6/16/65	Wounded	Petersburg
63	Troutman, George W	16	Pvt.	PA 138th Inf.	F	8/29/62		Died	
64	Troutman, James W	31	Pvt.	PA 91st Inf.	I	9/21/64	5/30/65	Wounded	

	Cemetery	Notes
1	Evans Cem.-Coaldale	1890 Broad Top Veterans Census
2	Union Cem.-McConnelsburg	Hist. Data Systems record; 1860 E. Providence Twp. Cen.; Joseph (2) Warner & Isaac-bros.
3		Findagrave.com record
4	Duval Cem.-Coaldale	POW- Danville, VA 8/19/64; Died 4/24/65; Historical Data Systems record
5	Montrose Cem.-IL	Died of Disease 8/18/61; Findagrave.com-Woodbury residence
6	Saltillo Cem.-Hunt. Co.	B.S.F. History Book-1884; enlisted in Bloody Run
7		1890 Snake Spring Valley Veterans Census
8	Poplar Grove Nat'l Cem.	POW-Lynchburg 7/12/64; Died 7/23/64; PA Civil War Archives-Woodbury residence
9		1890 Hyndman Veterans Census
10	Everett Cem.	POW 5/16/64 to 11/19/64; B.S.F. History Book-1884
11		MIA 6/3/64; B.S.F. History Book-1884; Drafted & Mustered in Philadelphia
12		Historical Data Systems-Bedford Co. residence; enlisted in Bloody Run
13	Hyndman Cem.	National Park Service listing
14	Hyndman Cem.	1890 Hyndman Veterans Census aka Tharp
15		B.S.F. History Book-1884; Jacob, John & Solomon-brothers
16		Died 8/2/64; PA Civil War Archives-Bedford Co. residence, enlisted in Schellsburg
17	Porter Cem.-Hyndman	B.S.F. History Book-1884; PA Civil War Arch.-Bedford Co. residence
18	Ash Valley Cem.-KS	Ancestry.com information; 1850 Napier Twp. Census; Henry & Josiah-brothers
19	Mt. Olivet Cem.-Manns Choice	1890 Harrison Twp. Veterans Census
20	Mt. Ross Cem.-Bedford	Historical Data Systems record; Born in VA; George & Jackson-brothers
21		Frank McCoy 1912 list & Civil Works Admin. 1934 list; 1850 Bedford Boro Census
22	Mt. Ross Cem.-Bedford	Frank McCoy 1912 Listing; 1860 Bedford Census; Born in VA; 43rd USCT references
23	Old Claysburg Cem.	1890 Greenfield Veterans Census; PA Civil War Archives record
24	Burkhart-Somerset Co.	Died-Malaria 10/65; B.S.F. Hist. Book-1884; 1850 Harrison Twp. Cen.; PA 165th references
25	Hyndman Cem.	Deserted 3/18/63; PA Civil War Archives-Londonderry Twp. residence
26	Hopewell Cem.	1890 Everett Veterans Census; Calvin & John B-brothers
27		POW-Libby & Andersonville; 1860 Hopewell Twp. Census; Frank McCoy 1912 Listing
28		Frank McCoy 1912 Listing; 1890 Everett Veterans Census; Shermans March to Sea Campaign
29	Hickory Bottom Cem.-Woodbury	Bedford Inquirer 5/22/1908 listing
30	Everett Cem.	POW 6/27/62; B.S.F. History Book-1884; Historical Data Systems-Bedford Co. residence
31		Wounded 12/13/62; Bedford Gazette-8th Reserve List; B.S.F. History Book-1884
32		1890 Everett Veterans Census; B.S.F. History Book-1884
33		B.S.F. History Book-1884
34	Hickory Bottom Cem.-Woodbury	B.S.F. History Book-1884
35	Hickory Bottom Cem.-Woodbury	Wounded 6/9/62; Historical Data Systems-Bedford residence
36		KIA 7/2/63; B.S.F. History Book-1884; 1860 Bedford Boro Census
37	Mt. Zion Cem.-Chaneysville	POW 6/29/62-Pension Record; B.S.F. History Book-1884
38		1890 Southampton Veterans Census; references to 99th PA Infantry
39	Piney Plains Meth. Cem.-MD	1850 Monroe Twp. Census; Historical Data Systems record
40	Mt. Zion Cem.-Chaneysville	Deserted 11/25/62 & returned 1/13/63; B.S.F. History Book-1884
41		Historical Data Systems record; Civil War Pension Record; Father of George T & Nathan
42	Trail Family Cem.-Little Orleans	Historical Data Systems record; 1850 Monroe Twp. Census; Hugh & Nathan-brothers
43	Rock Hill Cem.-Monroe Twp.	Died of Disease 4/9/64; Frank McCoy 1912 Listing; 1850 Southampton Twp. Census
44		B.S.F. History Book-1884; Drafted & Deserted 4/4/64
45	Unity Cem.-Latrobe	Wounded 6/3/64; B.S.F. History Book-1884
46	Russellville Meth. Cem.	1890 Broad Top Veterans Census; Historical Data Systems record
47		1890 Broad Top Veterans Census; Historical Data Systems record
48		B.S.F. History Book-1884
49		B.S.F. History Book-1884; enlisted in Hopewell
50		B.S.F. History Book-1884; Historical Data Systems-Bedford Co. resid.; Disch. Surgeon's Cert.
51	Union Cem.-Greensburg	B.S.F. History Book-1884
52	Schellsburg Cem.	Died 6/28/65; PA Civil War Archives-Bedford Co. residence; 1860 Napier Twp. Census
53		Wounded 5/9/64 in Bedford Inquirer 7/8/64 article; B.S.F. History Book-1884
54	Beaufort Co. Cem.-NC	Died-Typhoid Fever 6/28/63, George & Benjamin-brothers; Fulton Co. references
55	Riverview Cem.-IL	Historical Data Systems record; Alexander & John C-brothers
56		B.S.F. History Book-1884
57	Bedford Cem.	Deserted 6/23/65; 1890 Bedford Twp. Veterans Census
58		B.S.F. History Book-1884
59	Morris Hill Cem.-ID	POW-Andersonville 6/22/64 to 4/1/65; 1860 Bedford Boro Census; Bed. Gazette 2/13/1914 list
60	Comp Cem.-Somerset Co.	1910 Bedford Twp. Census; Hist Data Systems record
61	Highland Cem.-KS	Ancestry.com information; 1850 Londonderry Twp. Census; Daniel & George-brothers
62	Green Mt. Cem.-Cumberland	Wounded 4/1/65; 1890 Flintstone Veterans Census; Francis & James-brothers
63	Trinity UCC Cem.-Juniata Twp.	Died 12/20/62; Frank McCoy 1912 Listing; Bedford Inquirer 5/22/1908 listing
64	Clearville Union Cem.	1890 Southampton Twp. Veterans Census-Shoulder Wound

Chapter 18

	Name	Age	Rank	Regiment	Co.	Muster In	Muster Out	Casualty	Casualty Battle
1	Truax, George M	19	Pvt.	PA 101st Inf.	D	11/1/61		Died	
2	Truax, Jacob			MD 3rd PHB Inf.					
3	Trueax, George M	29	Pvt.	PA 97th Inf.	A	2/21/65	5/28/65		
4	Tucker, Benjamin F	39	Pvt.	PA 199th Inf.	G	9/26/64	6/28/65		
5	Turner, Andrew J	20	Corp.	PA 55th Inf.	K	11/5/61		Died	
6	Turner, John	29	Pvt.	PA 171st Inf.	I	11/11/62	8/8/63		
7	Turner, Thomas*		Pvt.	PA 184th Inf.	A	5/12/64	5/27/65	POW	Jerusalem Plank Rd.
8	Twigg, Brice	33	Pvt.	PA 45th Inf.	G	11/14/64	7/17/65	Wounded	
9	Twigg, Moses	20	Pvt.	MD 2nd PHB Inf.	B	10/16/62	5/29/65		
10	Tyson, Samuel H	14	Music.	PA 110th Inf.	C	10/24/61			
11	Uglow, Nicholas	25	Pvt.	PA 208th Inf.	H	9/7/64	6/1/65		
12	Uglow, Samuel	21	Corp.	PA 208th Inf.	H	9/7/64	6/6/65		
13	Ullery, John E	17	Pvt.	PA 57th Inf.	I	9/1/64	5/31/65		
14	Uperaft, John		Pvt.	PA 61st Inf.	F	8/1/61	12/18/64	Wounded	Fair Oaks
15	Valentine, John	42	Pvt.	PA 138th Inf.	F	8/29/62	6/23/65	Wounded	Opequon
16	Valentine, Levi	30	Pvt.	PA 50th Inf.	E	9/21/64	8/16/65		
17	Valentine, Samuel		Pvt.	PA McKeage's Mil.	H	7/2/63	8/8/63		
18	Valentine, Samuel		Pvt.	PA 210th Inf.	C	9/5/64	5/30/65		
19	VanHorn, James F	18	Sgt.	PA 55th Inf.	D	10/12/61	8/30/65		
20	VanHorn, John M	17	Pvt.	PA 133rd Inf.	C	8/13/62	5/26/63		
21	VanHorn, John M		Corp.	PA McKeage's Mil.	D	7/2/63	8/8/63		
22	VanHorn, John M		Pvt.	PA 186th Inf.	G	8/5/64	8/15/65		
23	VanOrman, John W	16	Pvt.	PA 79th Inf.	A	1/14/65	8/1/65		
24	VanOrmer, William W	20	Capt.	PA 53rd Inf.	I	10/10/61	6/30/65	Wounded	Antietam
25	VanOrmer, William W		Capt.	PA 53rd Inf.	I			Wounded	Gettysburg
26	VanOrmer, William W		Capt.	PA 53rd Inf.	I			Wounded	Spotsylvania
27	VanTassel, Russel*		Pvt.	PA 76th Inf.	E	8/26/63	3/12/64		
28	VanTassel, Russel*	32	Pvt.	PA 57th Inf.	B	7/3/63	8/17/63		
29	VanTassel, Russel*			US 18th Inf.	E	3/12/64		Died	New Hope Church
30	Vastbinder, Gabriel*		Pvt.	PA 76th Inf.	E	8/26/63		KIA	Drewry's Bluff
31	Vastbinder, Gabriel*	24	Pvt.	PA 105th Inf.	D	8/28/61	11/11/62		
32	Vaughan, Aaron C	25	Surg.	PA 105th Inf.		4/29/63	9/3/64	Wounded	
33	Vaughan, Ephraim	18	Music.	PA 101st Inf.	D	11/1/61	6/26/65	POW	Plymonth, NC
34	Vaughan, John W	20	Music.	PA 101st Inf.	D	11/1/61	6/28/62		
35	Veach, David C	31	Pvt.	PA 186th Inf.	G	3/24/64	8/15/65		
36	Veatch, John G	22	Pvt.	PA 11th Inf.	A	12/6/61	11/3/64		
37	Veatch, Samuel	26	Pvt.	PA 101st Inf.	D	12/6/61		Wound.-Died	
38	Vickroy, James R	52	Pvt.	PA 138th Inf.	F	8/30/62	11/28/63		
39	Vickroy, Orrin G	26	Sgt.	PA 55th Inf.	D	10/12/61	10/26/64		
40	Wade, John		Corp.	IA 14th Inf.	C	6/16/63	8/8/65		
41	Wagerman, Jacob	31	Pvt.	IL 146th Inf.	A	8/31/64	7/8/65		
42	Wagerman, Samuel (1)	20	Pvt.	MD 3rd PHB Inf.	E	3/10/62	5/29/65		
43	Wagerman, Samuel (2)		Pvt.	PA McKeage's Mil.	H	7/2/63	8/8/63		
44	Wagerman, William	24	Pvt.	PA 138th Inf.	F	9/29/62		Died	
45	Wagner, Christian	27	Pvt.	PA 88th Inf.	A	2/25/65			
46	Wagner, Joseph H	31	Pvt.	PA 131st Inf.	D	8/8/62	5/23/65	Wounded	Fredericksburg
47	Wagner, Stephen	41	Pvt.	PA 22nd Cav.	H	2/26/64			
48	Wagoner, August		Pvt.	PA 184th Inf.	B	5/12/64		KIA	Cold Harbor
49	Walker, Asahel	22	Pvt.	PA 84th Inf.	A	10/24/61	10/24/64		
50	Walker, Benjamin H	30	Corp.	PA 84th Inf.	A	10/24/61	1/13/65		
51	Walker, Benjamin H		Corp.	PA 57th Inf.	G	1/13/65	6/29/65		
52	Walker, Charles W		Pvt.	PA 76th Inf.	E	9/26/64	6/28/65		
53	Walker, Isaac	19	Pvt.	PA 205th Inf.	C	8/26/64	6/2/65		
54	Walker, James		Pvt.	PA 21st Cav.	E	7/8/63	2/20/64		
55	Walker, Morris	25	Pvt.	PA 84th Inf.	A	10/24/61			
56	Walker, Thomas G	31	Pvt.	PA 171st Inf.	I	11/2/62	8/8/63		
57	Walker, Thomas G		Pvt.	PA 91st Inf.	I	9/21/64	5/30/65		
58	Walker, William A (1)	24	Pvt.	PA 21st Cav.	E	7/8/63		Died	
59	Walker, William A (2)		Pvt.	PA 208th Inf.	H	8/29/64	6/1/65		
60	Walker, William M	22	Sgt.	PA 55th Inf.	H	10/11/61	8/30/65		
61	Wall, Albert		Pvt.	PA 76th Inf.	E	2/15/65	7/18/65		
62	Wall, Lewis		Pvt.	PA 208th Inf.	H	9/7/64			
63	Wallace, John P*		Pvt.	PA 55th Inf.	H	4/4/62		Died	
64	Wallace, Samuel G	34	Pvt.	PA 110th Inf.	C	12/18/63	11/25/64		

	Cemetery	Notes
1	Hampton Nat'l Cem.	Died of Disease 11/30/62; PA Civil War Archives-Bedford Co. resid., enlisted in Rainsburg
2	Rock Hill Cem.-Monroe Twp.	Frank McCoy 1912 Listing; Jacob & George-brothers
3	Sideling Hill Baptist-Fulton Co.	1890 Brush Creek Twp. Veterans Census
4	Schellsburg Cem.	Frank McCoy 1912 Listing
5	Schellsburg Cem.	Died 12/3/61; PA Civil War Project record; B.S.F. Hist. Book-1884; 1860 Harrison Twp. Cen.
6	Trinity UCC Cem.-Juniata Twp.	Frank McCoy 1912 Listing; 1890 Harrison Twp. Veterans Census
7		POW 6/22/64-4/28/65; B.S.F. Hist. Book-1884; Comp. A recruited in Dauphin & Bedford Co.
8	Centenary Meth.-Cumberland	1890 Cumberland Valley Veterans Census-Leg Wound
9	Hyndman Cem.	1890 Londonderry Twp. Veterans Census
10		B.S.F. History Book-1884; Mustered in Bedford
11	Leavenworth Nat'l Cem.-KS	B.S.F. History Book-1884; Nicholas & Samuel-brothers
12	Forest Home Cem.-IL	B.S.F. History Book-1884; Historical Data Systems-Bedford Co. residence
13	Lillydale Cem.-Cambria Co.	1890 Broad Top Veterans Census
14		Shoulder Wound 5/31/62; 1890 Harrison Twp. Veterans Census
15	Union Cem.-Centerville	Frank McCoy 1912 Listing; 1890 Cumberland Valley Veterans Census
16	Woods Church-Colerain Twp.	Frank McCoy 1912 Listing; Bedford Inquirer 5/22/1908 listing
17		B.S.F. History Book-1884
18		B.S.F. History Book-1884
19		B.S.F. History Book-1884; Historical Data Systems-Bedford Co. residence
20	Mt. Union Cem.-Mench	B.S.F. History Book-1884; 1850 Colerain Twp. Census; Historical Data Systems record
21		B.S.F. History Book-1884
22		1870 Bloody Run Census; Historical Data Systems record
23	Jacobs Church Cem.-Hesston	1890 Hopewell Veterans Census
24	Schellsburg Union Cem.	Wounded 9/17/62; 1890 Napier Twp. Veterans Census
25		Wounded 7/2/63; B.S.F. History Book-1884
26		B.S.F. History Book-1884
27		B.S.F. History Book-1884
28	Moore Cem.-Jefferson Co	Historical Data Systems record; Jefferson Co. references
29		Died 5/31/64; Ancestry.com record; PA Civil War Archives record
30		KIA 5/14/64; B.S.F. History Book-1884
31		Historical Data Systems record; Jefferson Co. references; Discharged on Surgeon's Cert.
32	Riverside Cem.-OH	1890 Veterans Census-Leg Wound; 1860 Colerain Twp. Census
33		POW-Andersonville; B.S.F. History Book-1884; Ephram, Aaron & John-brothers
34		B.S.F. History Book-1884; Historical Data Systems record; Discharged on Surgeon's Cert.
35	Oak Ridge Cem.-Altoona	David, Samuel and John; 1860 W. Providence Twp. Census
36	Mt. Union Cem.-Mench	Frank McCoy 1912 Listing; Guarded Gen. McClelland's HQ-Findagrave
37	Point Lookout Cem.-MD	Died 8/1/62; Historical Data Sys.-Bedford Co. residence, enlisted in Rainsburg
38	Bethel Cem.-Centerville	1860 Cumberland Valley Veterans Census; Frank McCoy 1912 Listing
39	Forest Lawn Mem. Park-NE	Historical Data Systems record;1860 Cumberland Valley Veterans Cen.; Disch. Surgeon's Cert.
40	Mt. Zion Cem.-Breezewood	Frank McCoy 1912 Listing
41		Ancestry.com record; Jacob, Samuel & William-brothers
42	Shroyer Farm Cem.-Madley	1890 Londonderry Twp. Veterans Census listed PA 138th Infantry
43		B.S.F. History Book-1884 aka Waugherman
44	Wagerman Family Cem.-Madley	Died-Typhoid Fever; 12/8/63; 1850 Londonderry Twp. Census; Frank McCoy 1912 Listing
45	Everett Cem.	Frank McCoy 1912 Listing; 1890 Everett Veterans Census; Discharged on Surgeon's Cert.
46	Dry Hill Cem.-Woodbury	Wounded 12/13/62; Frank McCoy 1912 Listing
47	Stevens Chapel-Monroe Twp.	1890 Monroe Twp. Veterans Census; Frank McCoy 1912 Listing
48		KIA 6/4/64; Bedford Gazette 2/13/1914 listing; 1860 Bedford Twp. Census
49	Pleasantville Cem.	1890 W. St. Clair Twp. Veterans Census
50	Union Cem.-Beaver Co.	PA Civil War Project record
51		Historical Data Systems record
52		B.S.F. History Book-1884
53	Pleasantville Cem.	PA Civil War Project record; William Mock family record
54		Historical Data Systems-Bedford Co. residence
55	Pleasantville Cem.	PA Civil War Archives-Beford residence; 1860 E. St. Clair Twp. Census
56	Pleasantville Cem.	B.S.F. History Book-1884; Thomas, Asahel, Benjamin, Issac, Morris & William A (1)-bros.
57		Historical Data Systems record
58	Friends Cem.-Spring Meadow	Died-Typhoid Fever 1/4/64 after returning home; Frank McCoy 1912 Listing
59		B.S.F. History Book-1884; Historical Data Systems-Bedford Co. residence
60		B.S.F. History Book-1884; PA Civil War Archives-Bedford Co. residence
61		B.S.F. History Book-1884; Substitute
62		Deserted 9/13/64; B.S.F. History Book-1884
63	Beaufort Nat'l Cem.	Died 11/10/62; Historical Data Systems record; Huntingdon Co. references
64		B.S.F. History Book-1884; Discharged on Surgeon's Cert.

Chapter 18

	Name	Age	Rank	Regiment	Co.	Muster In	Muster Out	Casualty	Casualty Battle
1	Walter, George		Pvt.	PA 205th Inf.	C	9/2/64	6/2/65		
2	Walter, George I	27	Pvt.	PA 13th Cav.	A	2/26/64	7/14/65		
3	Walter, Herman	37	Pvt.	PA 28th Inf.	A	9/21/64			
4	Walter, Jacob		Pvt.	PA 77th Inf.	F	2/28/65	12/6/65		
5	Walter, James A	18	Pvt.	MD 2nd PHB Inf.	A	10/17/63	5/19/65		
6	Walter, John H	33	Pvt.	PA 84th Inf.	A	10/24/61	12/10/65	Wounded	
7	Walter, Joseph H	33	Pvt.	PA 19th Cav.	L	9/17/63	9/16/64	Died	
8	Walter, Michael H	23	Pvt.	MD 1st Reg Cav.	L	9/4/61	9/4/64	Wounded	Deep Run
9	Walter, Moses	17	Pvt.	PA 205th Inf.	C	8/26/64	8/8/65	Wounded	Petersburg - Final
10	Walter, Samuel								
11	Walter, Samuel H	20	Pvt.	PA 19th Cav.	L	9/17/63	5/14/66	Wounded	
12	Walters, David		Pvt.	PA 55th Inf.	D	2/29/64	8/30/65		
13	Walters, Isaiah	22	Pvt.	PA 2nd Cav.	E	11/4/61			
14	Walters, Jacob D	33	Pvt.	PA 77th Inf.	F	2/27/65	12/6/65		
15	Walters, Jacob W		Pvt.	PA 77th Inf.	F	2/28/65	12/6/65		
16	Walters, William H	20	Pvt.	PA 126th Inf.	G	8/11/62	5/20/63		
17	Walters, William H		Pvt.	PA 22nd Inf.	L	2/27/64	6/24/65		
18	Waltman, William H	20	Pvt.	PA McKeage's Mil.	G	7/3/63	8/8/63		
19	Waltman, William H		Pvt.	PA 184th Inf.	A	5/12/64	2/1/65	POW	Petersburg
20	Waltz, Lewis B	37	Capt.	PA 8th Res.	F	6/11/61	5/26/64	POW	Seven Days Battle
21	Wambaugh, Lewis	28	Pvt.	PA 76th Inf.	D	7/5/64	7/18/65		
22	Ward, Henry	34	Pvt.	PA 149th Inf.	I	9/26/63	12/30/63		
23	Ward, Henry		Pvt.	PA 28th Inf.	H	1/7/65			
24	Ward, James*								
25	Ward, Jeremiah	46	Pvt.	PA 107th Inf.	E	10/21/64	3/26/65		
26	Ward, Samuel	32	Pvt.	PA 138th Inf.	E	8/29/62		POW-Died	Mine Run
27	Ward, William S	40	Pvt.	PA 149th Inf.	B	8/16/63	6/24/65		
28	Wareham, John	42	Pvt.	PA 107th Inf.	E	9/21/64	6/6/65	Wounded	
29	Wareham, Joseph	25	Pvt.	IL 45th Inf.	E	11/20/61	4/24/62		
30	Wareham, Martin		Pvt.	PA 18th Cav.	H	2/24/65	7/20/65		
31	Warner, Frederick*		Pvt.	PA 55th Inf.	I	10/11/61		POW-Died	Drewry's Bluff
32	Warner, Pius	26	Pvt.	PA 55th Inf.	D	10/12/61	8/30/65		
33	Warner, Simon*		Corp.	PA 76th Inf.	E	8/27/63	4/24/65		
34	Warren, Nimrod	22	Pvt.	USCT 43rd Inf.	C	3/19/64	10/20/65		
35	Warsing, Alexander		Pvt.	PA 8th Res.	F	6/19/61	3/4/63	POW	Seven Days Battle
36	Warsing, Alexander	33	Pvt.	PA 8th Res.	F			Died	
37	Washabaugh, Daniel	19	Pvt.	PA 126th Inf.	D	8/12/62	2/7/63		
38	Washabaugh, William H	25	Pvt.	PA 13th Inf.	G	4/25/61	8/6/61		
39	Washabaugh, William H		Pvt.	PA 76th Inf.	E	10/9/61		KIA	Ft. Wagner, SC
40	Waters, David	30	Pvt.	PA 55th Inf.	D	2/22/63	7/25/65	Wounded	
41	Waters, David Y	31	Sgt.	IL 20th Inf.	A	6/13/61		Died	
42	Waters, Isaac O	24	Corp.	IL 25th Inf.	K	8/9/61	8/30/64		
43	Waters, James B	20	Pvt.	IL 76th Inf.	B	11/17/63	6/13/65		
44	Waters, John		Pvt.	PA 195th Inf.	C	7/16/64	11/1/64		
45	Waters, John F		Sgt.	IL 20th Inf.	A	6/13/61	11/1/61		
46	Waters, John F		Pvt.	IL 2nd Cav.	I	4/28/64	11/22/65		
47	Waters, Josiah	34	Pvt.	PA 93rd Inf.	G	9/10/64			
48	Waters, William	16	Pvt.	IL 25th Inf.	K	8/9/61	9/5/64		
49	Watkins, Hiram			USCT 55th Inf.					
50	Watkins, Hiram	28	Pvt.	USCT 26th Inf.	K	2/27/64	9/1/65		
51	Watkins, Jesse	33	Pvt.	PA 55th Inf.	I	10/15/61	3/2/63	Died	
52	Watkins, John A	18	Pvt.	PA 21st Cav.	E	7/6/63	7/8/65		
53	Watson, Henry S	18	Pvt.	PA 184th Inf.	A	5/12/64		POW-Died	Jerusalem Plank Rd.
54	Watson, John N			PA 99th Inf.					
55	Watson, John R	45	Pvt.	PA 184th Inf.	E	5/12/64		POW	Jerusalem Plank Rd.
56	Watson, William A			PA 2nd Cav.					
57	Watson, William H	25	Surg.	PA 105th Inf.		9/16/62	5/27/65		
58	Watt, David		Pvt.	PA 21st Cav.	E	7/18/63	2/20/64		
59	Watters, Samuel*			PA 2nd Inf.	E				
60	Way, James H	39	Corp.	PA 208th Inf.	H	9/7/64	6/1/65		
61	Weaver, Geo W*	24	Pvt.	PA 55th Inf.	I	9/20/61	8/30/65		
62	Weaver, John H	23	Corp.	WV 5th Cav.	I	1/26/64			
63	Weaver, Samuel	22	Pvt.	PA 91st Inf.	D	3/15/65	7/10/65		
64	Weaverling, Adam	42	2nd Lt.	PA 11th Inf.	A	9/30/61	1/20/63	Wounded	2nd Bull Run

	Cemetery	Notes
1		1890 S. Woodbury Twp. Census
2	St. John's Cem.-Loysburg	1890 S. Woodbury Twp. Census; George I, Michael, Samuel & Joseph-brothers
3	Hyndman Cem.	1890 Hopewell Twp. Veterans Cen.; Frank McCoy 1912 Listing; Deserted after the war ended
4		1890 Claysburg Veterans Census
5	St. Jacobs Cem.-OH	Ancestry.com information; 1870 Juniata Twp. Census; Son of Herman
6	Upper Claar Cem.-Queen	1890 Bedford Twp. Veterans Census; John & Moses-brothers
7	Memphis Nat'l Cem.	Died 9/3/64; Hist. Data Systems record; 1850 Greenfield Twp. Cen.; Born Southampton Twp.
8	New Paris Comm. Ctr. Cem.	Arm & Thigh Wound 8/16/64-Pension record; 1850 Greenfield Twp. Census
9	Rose Hill Cem.-Altoona	Wounded 4/2/65; Ancestry.com information; 1860 Greenfield Twp. Census
10	Mt. Zion Cem.-Pavia	PA Veterans Burial Record; Findagrave.com marker
11	Koontz Breth. Cem.-Loysburg	1890 S. Woodbury Twp. Veterans Census; Wounded-Eye Loss
12		B.S.F. History Book-1884
13		Deserted 6/1/62; B.S.F. History Book-1884; 1860 Southampton Census
14	Claysburg Union Cem.	1890 Claysburg Veterans Census; PA Civil War Project Record
15	Old Claysburg Cem.	1890 Greenfield Twp. Veterans Census; PA Civil War Project Record
16	Akersville Cem.-Crystal Springs	1890 Brush Creek Twp. Veterans Census
17		1890 Brush Creek Twp. Veterans Census
18	St. Mark's Cem.-Ott Town	B.S.F. History Book-1884; Historical Data Systems record
19		POW-Andersonville-5 months; 1890 Colerain Twp. Veterans Census; transferred-VRC 2/1/65
20	St. Lukes Cem.-Saxton	POW 6/27/62; Exchanged 8/27/62 for Lewis M Slaughter 17th VA; Hist.Data Systems record
21	St. John Baptist-New Baltimore	Historical Data Systmes record; 1860 Juniata Twp. Census
22	Robinsonville Cem.	1890 Monroe Twp. Veterans Census
23		Deserted 3/15/65 (estimated); Frank McCoy 1912 Listing
24	Robinsonville Cem.	Civil War Marker at gravestone
25	Rock Hill Cem.-Monroe Twp.	Findagrave.com information; Jeremiah, James & Henry-brothers
26	Andersonville Cem., 4338	POW-Andersonville 11/27/63; Died 7/30/64; PA Civil War Archives-Bedford residence
27	St. Pauls UCC Cem.-Hunt. Co.	1890 Hopewell Veterans Census; 1870 W. Providence Twp. Census
28	Ritchey Cem.-Snake Spring V.	1890 W. Providence Twp. Vet. Cen.- lost his leg during the war; John, Martin & Joseph-bros.
29	Cedar Hill Cem.-IL	Historical Data Systems record; 1850 S. Woodbury Twp. Census
30		Ancestry.com information
31		POW-Richmond 5/16/64; Died 7/15/64; B.S.F. History Book-1884; Drafted; Mustered Phila.
32		B.S.F. History Book-1884; Historical Data Systems-Bedford Co. residence
33		B.S.F. History Book-1884; Elk Co. references; Discharged on Surgeon's Cert.
34	Mt. Ross Cem.-Bedford	1890 Bedford Twp. Veterans Census
35		POW 6/27/62; Bedford Gazette-8th Reserve List; Historical Data Systems record
36	Rinard Family Cem.-Cyper	Died 8/14/64; Discharged on Surg. Cert.; PA Civil War Archives-Bedford Co. residence
37	Bedford Cem.	Historical Data Systems record
38		B.S.F. History Book-1884; Historical Data Systems-Bedford residence
39	Bedford Cem.	KIA 7/11/63; B.S.F. History Book-1884; 1860 Bedford Twp. Census
40	Mt. Olivet Cem.-Manns Choice	Wounded listed in Bed. Inquirer-6/3/64; 1890 Harrison Twp. Vet. Cen.; David & Josiah-bros.
41	New Bern Nat'l Cem.	Died 3/8/65; Ancestry.com-Born in Bedford; 1840 Bedford Census
42	Jones Cem.-MO	1837 Christening record at Trinity Lutheran Church in Bedford
43	Danville Nat'l Cem.-IL	1850 Bedford Twp. Census; Historical Data Systems record
44	Everett Cem.	Frank McCoy 1912 Listing; Bedford Inquirer 5/22/1908 listing
45		Disch. Surgeon's Cert.; Ancestry.com info; John F, David Y, Isaac, James & William-bros.
46		Illinois Veterans index; Historical Data Systems record
47	Grandview Cem.-Cambria Co.	Ancestry.com-born in Harrison Twp.
48		1850 Bedford Twp. Census; Historical Data Systems record
49		Frank McCoy 1912 list & Civil Works Admin. 1934 list; Regiment on Headstone
50	Everett Cem.	1890 Everett Veterans Census; MA 55th references
51	Beaufort Nat'l Cem.	Disc. Surg. Cert. & Died 4/26/63; B.S.F. History Book-1884; 1860 Woodbury Twp. Census
52	Pleasantville Cem.	PA Civil War Pension Record; Married in Alum Bank in 1872
53	Andersonville Cem.; 12380	POW-Andersonville 6/22/64; Died 1/2/65; PA Civil War Archives-Bedford Co. residence
54	Pleasant Union Cem.-Clearville	Frank McCoy 1912 Listing
55	Dry Hill Cem.-Woodbury	POW-Andersonville 6/22/64; 1870 M. Woodbury Twp. Cen.; father is Henry
56	Bedford Cem.	Frank McCoy 1912 Listing
57	Bedford Cem.	Medical Director 3rd Army Corp.; B.S.F. History Book-1884; 1860 Bedford Census
58		Historical Data Systems-Bedford Co. residence; enlisted in Bedford
59	Bedford Cem.	Died 8/25/76; Mexican War Vet.; No record-Civil War Vet.; Bedford Cem. Civil War Vet. list
60	Friends Cem.-Spring Meadow	B.S.F. History Book-1884; 1860 St. Clair Twp. Census
61		B.S.F. History Book-1884; PA Civil War Archives-born in Blair Co.
62	Duval Cem.-Coaldale	1890 Broad Top Veterans Census; Merged 6 WV Cav 12/14/64
63	Fockler Cem.-Saxton	1890 Hopewell Veterans Census
64	Everett Cem.	Knee Wound 8/30/62; B.S.F. History Book-1884

Chapter 18

	Name	Age	Rank	Regiment	Co.	Muster In	Muster Out	Casualty	Casualty Battle
1	Weaverling, Adam		1st Lt.	PA McKeage's Mil.	D	7/2/63	8/8/63		
2	Weaverling, Adam		Capt.	PA 208th Inf.	K	9/10/64	6/1/65		
3	Weaverling, David	22	2nd Lt.	PA 11th Inf.	A	9/30/61	6/19/65	Wounded	Antietam
4	Weaverling, David		2nd Lt.	PA 11th Inf.	A			Wounded	North Anna River
5	Weaverling, Jacob P	33	Pvt.	PA 11th Inf.	A	3/21/62		KIA	Antietam
6	Weaverling, Jacob T	29	Pvt.	PA 208th Inf.	K	9/5/64	6/1/65		
7	Weaverling, James T	21	Corp.	PA 133rd Inf.	K	8/15/62	3/31/63	Wounded	Fredericksburg
8	Weaverling, James T		Corp.	PA 186th Inf.	H	4/1/64	6/15/65		
9	Weaverling, Stephen	37	Pvt.	PA 186th Inf.	E	2/24/64	8/15/65		
10	Weaverling, Thomas H	16	Pvt.	PA 208th Inf.	K	9/7/64	6/1/65		
11	Weaverling, William (1)	16	Corp.	PA 133rd Inf.	K	8/15/62	5/26/63		
12	Weaverling, William (1)		Sgt.	PA 22nd Cav.	L	2/22/64	10/31/65		
13	Weaverling, William (2)	26	Sgt.	IL 45th Inf.	H	12/24/61	7/12/65		
14	Webster, Daniel			USCT Inf.					
15	Weicht, Jeremiah	44	Pvt.	PA 51st Inf.	C	9/21/64	12/23/64		
16	Weimer, David	24	Sgt.	PA 22nd Cav.	C	7/11/63	2/5/64		
17	Weimer, David		Pvt.	PA 208th Inf.	K	9/7/64	6/1/65		
18	Weimer, George	33	Pvt.	PA McKeage's Mil.	D	7/2/63	8/8/63		
19	Weimer, Henry H*	27	Pvt.	PA 95th Inf.	H	3/11/65	7/17/65		
20	Weimer, John S	20	Pvt.	PA 99th Inf.	F	3/17/65	7/1/65		
21	Weimer, Joseph S	23	Pvt.	PA 186th Inf.	E	2/24/64	8/15/65		
22	Weimer, William			PA 12th Inf.	D				
23	Weimert, Jacob F	17	Pvt.	PA McKeage's Mil.	D	7/2/63	8/8/63		
24	Weimert, Jacob F		Pvt.	PA 20th Cav.	C	1/27/64	7/13/65		
25	Weimert, Stephen	26	Pvt.	PA 208th Inf.	H	9/8/64	6/1/65		
26	Weisel, Jacob								
27	Weisel, John H	32	Pvt.	PA 84th Inf.	A	12/5/61	1/13/65		
28	Weisel, John H		Pvt.	PA 57th Inf.	G	1/13/65	6/29/65		
29	Weisel, William W	22	Pvt.	PA 55th Inf.	D	2/27/64	8/30/65		
30	Welch, John	27	Pvt.	MO 7th Cav.	D	10/2/61	9/1/62	POW	Independence, MO
31	Weller, John Q A	25	Capt.	PA 21st Cav.	E	7/21/63	2/20/64		
32	Welsh, John F	25	Pvt.	PA 55th Inf.	K	2/19/64	8/30/65		
33	Welsh, William F	34	Corp.	PA 133rd Inf.	K	8/15/62	5/26/63		
34	Wenling, A J		Pvt.	PA 210th Inf.	C	9/8/64	6/27/65		
35	Wenrick, Daniel*	22	Pvt.	PA 55th Inf.	D	9/23/63		POW-Died	
36	Wentling, George		Corp.	PA 13th Inf.	G	4/25/61	8/6/61		
37	Wentling, George	27	Sgt.	MD 2nd PHB Inf.	B	10/31/61	5/29/65		
38	Wentling, George*			PA 133rd Inf.					
39	Wentling, Samuel J	18	Pvt.	PA 210th Inf.	C	9/5/64	5/30/65	Wounded	
40	Wentting, Adam G*		Pvt.	PA 13th Inf.	G	1861			
41	Wentting, Adam G*		Pvt.	MD 2nd		1864	6/1/65		
42	Wentz, Adam	21	Pvt.	PA 55th Inf.	K	11/5/61	2/21/63		
43	Wentz, Adam		Pvt.	US 1st Light Art.	M	2/21/63			
44	Wentz, Henry	26	Pvt.	PA 55th Inf.	K	11/5/61	1865	Wounded	Petersburg - Initial
45	Wentz, Isaac	21	Pvt.	PA 55th Inf.	K	11/5/61	8/30/65	Wounded	Drewry's Bluff
46	Wentz, John	32	Pvt.	PA 55th Inf.	K	11/5/61	6/3/62	Died	
47	Wentz, Philip	25	Pvt.	PA 138th Inf.	D	8/29/62		KIA	Mine Run
48	Werning, John	39	Pvt.	PA 55th Inf.	H	10/11/61	8/30/65	POW	Edisto Island, SC
49	Wertz, George		Pvt.	PA 22nd Cav.	H	2/26/64	6/24/65		
50	Wertz, George F	35	Pvt.	PA 56th Inf.	F	9/19/64	5/31/65		
51	Wertz, Henry	21	Pvt.	PA 133rd Inf.	C	8/13/62	5/26/63		
52	Wertz, Hugh*								
53	Wertz, Jacob A	31	Pvt.	PA McKeage's Mil.	F	7/3/63	8/8/63		
54	Wertz, Jacob A		Pvt.	PA 76th Inf.	A				
55	Wertz, John A	37	Pvt.	PA 82nd Inf.	K	11/28/64	6/3/65		
56	Wertz, John S	25	Sgt.	IA 14th Inf.	K	11/6/61	11/16/64	Wounded	Shiloh
57	Wertz, John W	25	Pvt.	PA 6th Cav.	B	3/1/65	6/17/65		
58	Wertz, Samuel V			PA 21st Cav.					
59	Wertz, Talliferro	24	Pvt.	PA 171st Inf.	I	11/2/62	8/1/63		
60	Wertz, Thomas	18	Pvt.	PA 194th Inf.	I	7/22/64	11/5/64		
61	Wertz, William								
62	Western, John		Pvt.	PA 138th Inf.	F	1/30/65			
63	Weyandt, Jacob		Pvt.	PA 84th Inf.	A	9/24/61	1/13/65		
64	Weyandt, Jacob D	18	Pvt.	PA 3rd Heavy Art.	L	2/23/64	11/9/65		

	Cemetery	Notes
1		B.S.F. History Book-1884
2		Historical Data Systems record; 1860 W. Providence Twp. Census
3	Everett Cem.	Wounded 9/17/62; Frank McCoy 1912 Listing; 1860 W. Providence Twp. Census
4		Wounded 5/23/64; Frank McCoy 1912 Listing; Historical Data Systems record
5	Antietam Nat'l Cem.	KIA 9/17/62; Bedford Gazette 2/13/1914 listing; 1850 W. Providence Twp. Census
6	Everett Cem.	Historical Data Systems record; Jacob T, James & David-brothers
7	Irvine Cem.-Beaver Co.	B.S.F. History Book-1884; Wounded 12/13/62; Disch. Surg. Cert. & in Hospital at Muster Out
8		Historical Data Systems record
9	Cedar Grove-E. Prov. Twp.	Frank McCoy 1912 Listing; Stephen, Jacob P & William T (2)-brothers
10	Bedford Cem.	B.S.F. History Book-1884; his father Adam enlisted in same regimental 3 days after Thomas
11	Everett Cem.	Historical Data Systems record; Promoted to Corp. 8/16/62 at 16 years old one day after muster
12		1860 W. Providence Twp. Census; 1890 Everett Veterans Census
13	Plainville Cem.-KS	Historical Data Systems record; 1860 W. Providence Twp. Census
14		1890 Bedford Twp. Veterans Census
15	Everett Cem.	Frank McCoy 1912 Listing; Discharged on Surgeon's Certificate
16	Wilson Weimer Cem.-Clearville	B.S.F. History Book-1884; David & George-brothers
17		1890 Monroe Twp. Veterans Census; B.S.F. History Book-1884
18	Everett Cem.	B.S.F. History Book-1884
19	Walter Cem.-Somerset	B.S.F. History Book-1884; Historical Data Systems record; Somerset Co. references
20	Pleasant Union Cem.-Clearville	1890 Monroe Twp. Veterans Census; John & Joseph-brothers
21	Clearville Union Cem.	1890 Monroe Twp. Veterans Census
22	Hopewell Cem.	Frank McCoy 1912 Listing; Bedford Inquirer 5/22/1908 listing
23	Yellow Creek Reformed Cem.	B.S.F. History Book-1884; Jacob & Stephen-brothers
24		Frank McCoy 1912 Listing; aka Weimer
25	Yellow Creek Reformed Cem.	B.S.F. History Book-1884
26	Lutheran Cem.-Osterburg	Bedford Inquirer 5/22/1908
27	St. Thomas Cem.-Cambria Co.	Deserted 11/28/62 & returned 10/8/64; PA Civil War Archives-Bedford Co. residence
28		1860 St. Clair Twp. Census ; PA Civil War Archives record
29	Bedford Cem.	1890 Bedford Twp. Veterans Census; B.S.F. History Book-1884
30	Oquawka Cem.-IL	Obituary listed taken POW, Paroled in 1862 & Discharged; Born Ray's Hill
31	Arlington Nat'l Cem., 6 8548	Historical Data Systems-Bedford Co. residence
32	Everett Cem.	Frank McCoy 1912 Listing; PA Civil War Archives; MIA listed in Bedford Inquirer-6/3/64
33	Unmarked Grave-OK	B.S.F. History Book-1884; 1860 Bedford Boro Census; William & John F-brothers
34		B.S.F. History Book-1884
35	Salisbury Nat'l Cem.	POW-Salisbury; Died 12/8/64; B.S.F. History Book-1884; Drafted & Mustered in Pottsville
36		Historical Data Systems-Bedford Co. residence; B.S.F. History Book-1884
37	Samuel Wentling-Cumb. Val.	Frank McCoy 1912 Listing; 1850 Cumberland Valley Twp. Census; George & Samuel-bros.
38		Frank McCoy 1912 Listing
39	Bethel Meth. Cem.-Centerville	1890 Cumberland Valley Veterans Census-Left side Wound
40		1890 Cumberland Valley Veterans Census; aka George Wentling?
41		1890 Cumberland Valley Veterans Census; aka George Wentling?
42		B.S.F. History Book-1884; Historical Data Systems-Bedford Co. residence
43		1850 Union Twp. (Pavia) Census; Historical Data Systems record
44	Mt. Zion Cem.-Pavia	Arm & Side Wounded 6/18/64; Histo. Data Systems record; Henry, Adam, Isaac & John-bros.
45	Mt. Zion Cem.-Pavia	Thumb Shot-off 5/16/64; 1890 Union Twp. (Pavia) Veterans Census
46	Burket Cem.-Queen	Died 10/15/65; Historical Data Systems record-Bedford Co. residence, enlisted in Bedford
47		KIA 11/27/63; B.S.F. History Book-1884; 1850 Union Twp. (Pavia) Census
48	Lancaster Cem.	POW 3/29/62; History Data Sys.-enlisted in Schellsburg; Historical Data Systems record
49		B.S.F. History Book-1884
50	Bedford Cem.	1890 Cumberland Valley Veterans Census
51		Historical Data Systems-Bedford Co. residence; enlisted in Hopewell
52	Milligans Cove-Buffalo Mills	Civil War marker at gravesite
53	Bortz Luth. Cem.-Centerville	Bedford Inquirer 5/22/1908 listing
54	Bortz Luth. Cem.-Centerville	Frank McCoy 1912 Listing; Regiment on gravestone
55	Bedford Cem.	1890 Bedford Twp. Veterans Census
56	Aspen Grove Cem. IA	1850 Harrison Twp. Census; Historical Data Systems record
57	Utahville Cem.-Clearfield Co	PA Civil War Project record; 1850 Greenfield Twp. Census
58	Hopewell Cem.	Frank McCoy 1912 Listing
59		B.S.F. History Book-1884; PA Civil War Archives-enlisted in Bedford Co.
60	Oak Hill Cem.-MO	B.S.F. History Book-1884; PA Civil War Archives-enlisted in Hopewell
61	Bortz Luth. Cem.-Centerville	Bedford Inquirer 5/22/1908 listing; William & Jacob-brothers
62		B.S.F. History Book-1884
63		1890 King Twp. Veterans Census
64	Old Claysburg Cem.	PA Civil War Archives record; Jacob D & Samuel-brothers

Chapter 18

	Name	Age	Rank	Regiment	Co.	Muster In	Muster Out	Casualty	Casualty Battle
1	Weyandt, James	27	Pvt.	PA 99th Inf.	H	2/25/65	7/1/65	Died	
2	Weyandt, Jeremiah	16	Pvt.	PA McKeage's Mil.	G	7/2/63	8/8/63		
3	Weyandt, Joseph Y	33	Pvt.	PA 13th Cav.	A	2/26/64		Wounded	
4	Weyandt, Samual S	18	Pvt.	PA 125th Inf.	E	8/13/62	5/18/63		
5	Weyandt, Samual S		Pvt.	PA 3rd Heavy Art.	L	2/23/64	11/9/65		
6	Weyant, Alexander	34		PA 3rd Heavy Art.		2/25/64	4/1/64		
7	Weyant, Alexander		Pvt.	PA 188th Inf.	A	4/1/64	12/14/65	Wounded	
8	Weyant, Isaac	28	Sgt.	US 5th Art.	K	8/19/61	2/9/67		
9	Weyant, Joseph	17	Pvt.	PA 84th Inf.	A	10/24/61	1/13/65		
10	Weyant, Joseph		Pvt.	PA 57th Inf.	G	1/13/65	6/29/65		
11	Weyant, Lafayette	15	Pvt.	PA 13th Cav.	A	2/26/63	8/14/65	POW	Winchester
12	Wharton, John W		Pvt.						
13	Wheeler, Daniel A	18	Music.	PA 55th Inf.	I	2/6/64	8/30/65		
14	Whip, Jacob	20	Sgt.	PA 138th Inf.	F	9/29/62		Wound.-Died	Mine Run
15	Whip, William	19	Pvt.	MD 2nd PHB Inf.	D	4/8/65	5/29/65		
16	Whisel, William H	20	Pvt.	PA 8th Res.	F	6/19/61	1/24/63	POW	Seven Days Battle
17	Whisel, William H		Pvt.	PA 8th Res.	F			Wounded	Fredericksburg
18	Whitaker, Christian	20	Pvt.	PA 55th Inf.	H	10/11/61	2/22/63	POW	Edisto Island, SC
19	Whitaker, Christian		Pvt.	US 1st Art.	M	2/22/63	8/30/65	Died	
20	Whitaker, John H	16	Pvt.	PA 91st Inf.	I	10/7/64	6/19/65	Wounded	Five Forks
21	Whitaker, Joseph	33	Pvt.	PA 91st Inf.	F	9/21/64	5/30/65		
22	Whitaker, Peter	19	Pvt.	PA 22nd Cav.	C	7/15/63	5/15/65		
23	Whitaker, William P	18	Pvt.	PA 21st Cav.	E	6/18/63	2/20/64		
24	Whitaker, William P		Pvt.	PA 99th Inf.	A	2/26/65	7/1/65		
25	White, Andrew J		Pvt.	PA 55th Inf.	H	8/10/62	11/10/64	Wounded	
26	White, Dexter	17	Sgt.	PA 122nd Inf.	K	8/11/62	5/15/63		
27	White, Edmund H	25	Corp.	PA 8th Res.	F	6/11/61	1/14/63	POW	Seven Days Battle
28	White, Edmund H		Corp.	PA 8th Res.	F			Wounded	Antietam
29	White, James S	41	Pvt.	PA 55th Inf.	D	9/2/62		Wound.-Died	Drewry's Bluff
30	White, Silas	29	Pvt.	PA 84th Inf.	C	11/1/61			
31	White, William A								
32	Whited, David L	23	Pvt.	PA 133rd Inf.	C	8/13/62	5/26/63		
33	Whited, David L		Pvt.	PA 99th Inf.	H	2/23/65	5/30/65		
34	Whitehead, Charles R*		Pvt.	PA 194th Inf.	I	7/22/64	11/5/64		
35	Whiteneck, George W		Pvt.	PA 21st Cav.	I	7/21/63	7/8/65	Wounded	Petersburg - Initial
36	Whitfield, Benjamin	23	Pvt.	PA McKeage's Mil.	D	7/3/63	8/8/63		
37	Whitfield, Benjamin		Pvt.	PA 49th Inf.	E	6/4/64	7/15/65		
38	Whitfield, Ephraim	17	Pvt.	PA McKeage's Mil.	D	7/3/63	8/8/63		
39	Whitfield, Ephraim		Pvt.	PA 49th Inf.	E	6/4/64	7/15/65		
40	Whitfield, John	31	Pvt.	PA 149th Inf.	B	8/26/63	9/7/63		
41	Whitfield, William C		Pvt.	PA McKeage's Mil.	D	7/2/63	8/8/63		
42	Whitfield, William C	35	Pvt.	PA 149th Inf.	I	8/26/63	1/9/65	Wounded	Petersburg
43	Whitman, Benjamin F*	20	Pvt.	PA 194th Inf.	I	7/21/64	11/5/64		
44	Whittaker, Jonathan	18	Pvt.	PA 133rd Inf.	C	8/13/62	5/26/63		
45	Whittaker, Jonathan		Pvt.	PA 22nd Cav.	I	7/11/63			
46	Whorrell, Samuel		Pvt.	MD 2nd PHB Cav.	M				
47	Whorrell, Samuel		Pvt.	MD 1st PHB Cav.	D	9/6/64	6/28/65		
48	Whysong, John			PA 55th Inf.					
49	Whysong, Josiah*								
50	Whysong, Samuel	29	Pvt.	PA 55th Inf.	K	11/5/61	8/3/65	Wounded	Drewry's Bluff
51	Wicks, Isaac D		Corp.	PA 22nd Cav.	I	2/23/64	10/31/65		
52	Widner, Jacob	24	Pvt.	PA 137th Inf.	I	8/20/62	6/1/63		
53	Widner, Jacob		Pvt.	PA 9th Cav.	M	8/26/64	5/29/65		
54	Wigaman, Henry*		Sgt.	PA 55th Inf.	D	9/2/62	6/11/65		
55	Wigfield, Elijah	25	Pvt.	MD 2nd PHB Inf.	C	8/27/61	5/29/65		
56	Wigfield, Ephraim A		Pvt.	PA 49th Inf.	G	6/4/64	8/30/65		
57	Wigfield, Isaac	24	Sgt.	MD 2nd PHB Inf.	A	8/27/61	5/29/65		
58	Wigfield, James W	26	Pvt.	PA 57th Inf.	E	7/5/64	6/29/65		
59	Wigfield, Jonathan	21	Pvt.	WV 1st Light Art.	E	2/29/64	6/26/65		
60	Wigfield, Moses	24	Pvt.	PA 45th Inf.	E	7/5/64	7/17/65		
61	Wigfield, Noah	29	Pvt.	PA 171st Inf.	I	11/2/62	8/8/63		
62	Wigfield, Wesley	21	Pvt.	PA 99th Inf.	I	2/25/65	7/1/65		
63	Wike, Jacob*	38	Pvt.	PA 101st Inf.	G	12/2/61		Wound.-Died	Fair Oaks
64	Wilcox, John								

	Cemetery	Notes
1	Upper Claar Cem.-Queen	Died 1865; Civil War Pension record; 1860 Union Twp. (Pavia) Census
2		B.S.F. History Book-1884; 1860 Union Twp. (Pavia) Census
3	Upper Claar Cem.-Queen	1890 Greenfield Twp. Vet. Cen.-Torso Wound; Transferred to VRC.; Joseph & James-bros.
4	Old Claysburg Cem.	1890 Greenfield Twp. Veterans Census; Historical Data Systems record
5		Historical Data Systems record
6	Diehl's Crossroads-Henrietta	1860 Woodbury Twp. Census; PA Civil War Project record
7		1890 Woodbury Veterans Census-Leg Wound
8	Rose Hill Cem.-Altoona	Ancestry.com information; 1850 Greenfield Twp. Census; aka Wyant & Weyandt
9	Trinity UCC Cem.-Osterburg	1890 King Twp. Veterans Census; Joseph & Isaac-brothers
10		Historical Data Systems record; PA Civil War Project Record
11	Grandview Cem.-Cambria Co.	POW 6/13/63 to 7/22/63-Historical Data Systems; 1860 Union Township Census
12		1890 Londonderry Twp. Veterans Census
13		B.S.F. History Book-1884
14	Nave Farm Cem.-Cumb. Val.	Died 11/27/63; B.S.F. History Book-1884; 1860 Cumberland Valley Census
15	Bortz Luth. Cem.-Centerville	Historical Data Systems record; Jacob & William-brothers
16	Everett Cem.	POW-Castle Thunder & Belle Island - 6/27/62 to 8/6/62; Historical Data Systems record
17		Arm Amputated 12/13/62; Historical Data Systems record; Everett Postmaster
18		Findagrave.com information; 1850 E. St. Clair Twp. Census
19	Whitaker Farm-Somerset Co.	Died 1865; 1850 Union Twp. Census; Historical Data Systems-Bedford Co. residence
20	Pleasantville Cem.	Leg Wound 3/31/65; 1890 Napier Twp. Veterans Census
21	Rose Hill Cem.-Altoona	Civil War Pension record; 1870 Union Twp. (Pavia) Census
22	Trinity UCC Cem.-Osterburg	Findagrave.com information; 1860 E. St. Clair Twp. Census; Peter Christian & Joseph-bros.
23	Pleasantville Cem.	1890 W. St. Clair Twp. Veterans Census; William & John-brothers
24		1890 W. St. Clair Twp. Veterans Census
25		1890 Harrison Twp. Veterans Census - Left leg shot off
26	Bedford Cem.	1890 Bedford Veterans Census
27	Fockler Cem.-Saxton	POW 6/27/62; Bedford Gazette-8th Reserve List; B.S.F. History Book-1884
28		Wounded 9/17/62; B.S.F. History Book-1884; 1890 Hopewell Veterans Census
29	Hampton Nat'l Cem.	Leg Amputated 5/14/64; Died 5/29/64; B.S.F. History Book-1884; 1850 Broad Top Census
30	Fockler Cem.-Saxton	1890 Saxton Veterans Census
31	St. Thomas Cem.-Bedford	1890 Cumberland Valley Veterans Census
32	Hopeland Cem.-IL	1860 W. Providence Twp. Census; B.S.F. History Book-1884
33		Historical Data Systems record
34		B.S.F. History Book-1884; Blair Co. references
35		Wounded 6/18/1864; Historical Data Systems-Bedford Co. residence
36	Union Memorial Cem.-Mench	E. Providence Twp. Honor Roll; Benjamin & William-brothers
37		1890 E. Providence Twp. Veterans Census
38	Buck Valley Cem.-Fulton Co.	E. Providence Twp. Honor Roll
39		B.S.F. History Book-1884; 1890 Mann Twp. Veterans Census
40	Rock Hill Cem.-Monroe Twp.	Frank McCoy 1912 Listing; Rejected by Medical Board
41		B.S.F. History Book-1884
42	Union Memorial Cem.-Mench	Forearm Amputated 6/19/64; East Providence Honor Roll
43		B.S.F. History Book-1884; Berks Co. references
44	Bly Cem.-MI	1850 Monroe Twp. Census; PA Civil War Archives-Bedford Co. residence
45		B.S.F. History Book-1884; Deserted 12/3/64 after re-inlisting
46		1890 Hopewell Twp. Veterans Census; Aka Whomal & Whirl
47		Historical Data Systems record
48	Mt. Zion Cem.-Pavia	Bedford Inquirer 5/22/1908 listing
49	Mt. Zion Cem.-Pavia	Civil War Marker at gravesite
50	Mt. Zion Cem.-Pavia	MIA listed in Bedford Inquirer article 6/3/64; 1890 Union Twp. (Pavia) Veterans Census
51	Mt. Union Cem.-Hunt. Co.	B.S.F. History Book-1884 aka Weiks; PA Civil War Project record
52	Reformed Cem.-Marklesburg	1890 Hopewell Veterans Census; Historical Data Systems record
53		1890 Hopewell Veterans Census aka Weidner
54		B.S.F. History Book-1884; Berks Co. references
55	Dayton Nat'l Cem.-OH	Ancestry.com-born in Bedford Co.
56		1890 Mann Twp. Veterans Census
57	John Growden Farm-Cumb. Val.	Frank McCoy 1912 Listing; Isaac & Elijah-brothers
58	Fairview Cem.-Artemas	1890 Mann Twp. Veterans Census
59	Twigg Cem. #2-Flintstone	1890 Allegany MD Veterans Census; Jonathan & Moses-brothers
60	Odd Fellows Cem.-Flintstone	1890 Southampton Twp. Veterans Census
61	Fairview Cem.-Artemas	Frank McCoy 1912 Listing; Noah & Wesley-brothers
62	Fairview Cem.-Artemas	Frank McCoy 1912 Listing; 1890 Mann Twp. Veterans Census
63	Hampton Nat'l Cem.	Wounded 5/31/62; Died 6/11/62; Bedford Gazette 2/13/1914 listing; Allegheny Co references
64	Hopewell Cem.	Bedford Inquirer 5/22/1908 listing

Chapter 18

	Name	Age	Rank	Regiment	Co.	Muster In	Muster Out	Casualty	Casualty Battle
1	Wilds, George*	25	Corp.	PA 22nd Cav.	H	2/29/64	7/19/65		
2	Wilds, Sylvester*		Pvt.	PA 22nd Cav.	H	2/29/64	6/24/65		
3	Wilds, William C*	24	Corp.	PA 158th Inf.	H	11/4/62	8/12/63		
4	Wilds, William C*		Sgt.	PA 22nd Cav.	H	2/29/64	10/24/65		
5	Wilhelm, Samuel W	24	Pvt.	PA 171st Inf.	I	11/2/62	8/8/63		
6	Wilkins, Ephraim	33	Pvt.	PA 99th Inf.	F	2/25/65	7/1/65		
7	Wilkins, Harvey	33	Pvt.	PA McKeage's Mil.	D	7/2/63	8/8/63		
8	Wilkins, James B		Pvt.	PA 208th Inf.	K	9/7/64	6/1/65		
9	Wilkins, John H	16	Pvt.	PA 186th Inf.	E	2/24/64	7/24/65		
10	Wilkins, Josephus	21	Pvt.	PA 11th Inf.	A	9/30/61			
11	Wilkins, Josephus		Pvt.	PA McKeage's Mil.	D	7/3/63	8/8/63		
12	Wilkins, Josephus		Pvt.	PA 149th Inf.	I	8/26/63	7/24/65		
13	Wilkins, Samuel		Pvt.	PA 208th Inf.	K	9/7/64	6/1/65		
14	Wilkins, William	19	Pvt.	PA 11th Inf.	A	9/30/61		Died	
15	Wilkinson, Emanuel	22	Pvt.	PA 2nd Cav.	E	11/4/61	1864		
16	Wilkinson, Philip	18	Pvt.	PA 79th Inf.	D	3/7/65	7/12/65	Wounded	
17	Wilkinson, William (1)	19	Pvt.	PA 133rd Inf.	C	8/29/62	5/26/63		
18	Wilkinson, William (2)	30	Pvt.	PA 171st Inf.	I	11/2/62	8/8/63		
19	Will, John H	18	Corp.	PA 208th Inf.	H	9/7/64	6/1/65		
20	Willard, Lewis		Pvt.	USCT 127th Inf.	A	8/22/64	10/20/65		
21	Willett, William	20	Pvt.	PA 77th Inf.	F	10/9/61	4/12/65		
22	Willetts, Nathaniel*	24	Pvt.	PA 55th Inf.	H	9/23/63	8/30/65		
23	Williams, Alvah R	33	Pvt.	PA 208th Inf.	K	9/7/64	6/8/65	Wounded	Petersburg - Final
24	Williams, Charles	27	Pvt.	PA 133rd Inf.	C	8/29/62	5/19/63		
25	Williams, David F	35	Pvt.	PA 171st Inf.	I	11/2/62	8/8/63		
26	Williams, David F			PA 208th Inf.					
27	Williams, Gideon	29	Pvt.	PA 22nd Cav.	H	2/26/64	10/31/65		
28	Williams, Harrison P	26	Sgt.	PA McKeage's Mil.	D	7/2/63	8/8/63		
29	Williams, Harrison P		Pvt.	PA 149th Inf.	I	8/26/63	6/2/65	Wounded	Petersburg - Initial
30	Williams, Henry S			USCT Inf.					
31	Williams, Jacob	26		PA 133rd Inf.	C				
32	Williams, John H	24	Sgt.	PA 8th Res.	F	6/11/61	5/26/64		
33	Williams, John H		1st Lt.	PA 194th Inf.	I	7/21/64	11/5/64		
34	Williams, John P	26	Pvt.	PA 8th Res.	F	9/5/62	5/15/64	Wounded	Spotsylvania
35	Williams, John P		Pvt.	PA 191st Inf.	H	5/15/64	6/1/65		
36	Williams, John*		Pvt.	PA 55th Inf.	I	9/23/63		Wound.-Died	Drewry's Bluff
37	Williams, Jonas	23	Music.	PA McKeage's Mil.	D	7/2/63	8/8/63		
38	Williams, Joseph W	35	Pvt.	PA 208th Inf.	K	9/7/64	6/1/65		
39	Williams, Richard (1)	45	Pvt.	PA 194th Inf.	I	7/18/64	11/6/64		
40	Williams, Richard (2)	30	Pvt.	PA 13th Inf.	G	4/25/61	8/6/61		
41	Williams, S B			PA 149th Inf.					
42	Williams, Samuel		Pvt.	PA McKeage's Mil.	D	7/2/63	8/8/63		
43	Williams, Samuel D	22	2nd Lt.	PA 133rd Inf.	C	8/29/62	11/26/62		
44	Williams, Samuel D		2nd Lt.	PA 194th Inf.	I	7/21/64	11/5/64		
45	Williams, Samuel W		Corp.	PA 208th Inf.	K	9/7/64	5/23/65	Wounded	Ft. Stedman
46	Williams, William C*		2nd Lt.	PA 55th Inf.	I	9/20/61	2/2/63		
47	Williams, Wilson M	31	Sgt.	PA 208th Inf.	K	9/7/64	6/1/65		
48	Williamson, Gideon	39	Pvt.	PA 110th Inf.	H	9/5/62	12/15/62		
49	Williamson, Gideon		Pvt.	PA 21st Cav.	E	7/21/63		Died	
50	Williamson, John	18	Pvt.	PA McKeage's Mil.	H	7/2/63	8/8/63		
51	Willis, Harman		Pvt.	PA 16th Cav.	H	10/13/62			
52	Wills, Samuel W*								
53	Wills, William*		Pvt.	PA 138th Inf.	D	8/20/62	6/12/65	Wounded	Cold Harbor
54	Wilson, David B*	25	Pvt.	PA 11th Cav.	G	10/30/62	8/13/65		
55	Wilson, George W	50	Pvt.	PA 101st Inf.	D	3/7/62		POW-Died	Plymonth, NC
56	Wilson, Henry W		Pvt.	USCT Inf.	L				
57	Wilson, Hugh	17	Pvt.	PA 13th Inf.	G	4/25/61	8/6/61		
58	Wilson, James A		Pvt.	PA 110th Inf.	C	1/14/65	6/28/65		
59	Wilson, James R	18	Pvt.	PA 184th Inf.	A	5/12/64	7/14/65		
60	Wilson, John (1)	40	Pvt.	PA 55th Inf.	K	10/11/61		Wound.-Died	
61	Wilson, John (2)		Music.	PA McKeage's Mil.	H	7/2/63	8/8/63		
62	Wilson, Joseph B		Pvt.	PA 11th Inf.	C	10/22/62	6/7/65		
63	Wilson, Patrick		Pvt.	PA 208th Inf.	H	9/7/64	6/1/65		
64	Wilson, William	31	Pvt.	PA 13th Inf.	G	4/25/61	8/6/61		

	Cemetery	Notes
1	Fort Littleton Cem.	B.S.F. History Book-1884; Fulton Co. references
2		B.S.F. History Book-1884; Fulton Co. references
3		Historical Data Systems record; Fulton Co. references
4		B.S.F. History Book-1884; Fulton Co. references
5	Hyndman Cem.	B.S.F. History Book-1884; 1860 Londonderry Twp. Census
6	Union Memorial Cem.-Mench	1890 E. Providence Twp. Veterans Census
7	Union Memorial Cem.-Mench	B.S.F. History Book-1884; Josephus, Harvey and Ephraim were brothers
8	Union Cem.-McConnellsburg	1890 Robertsdale Veterans Census; James & John-brothers
9	Puttstown Cem.-Hunt. Co.	Ancestry.com information; 1850 W. Providence Twp. Census; 22nd PA Cavalry references
10	Union Memorial Cem.-Mench	Deserted 11/12/62 from Hospital; Frank McCoy 1912 Listing
11		Historical Data Systems record
12		Drafted; Historical Data Systems record
13		B.S.F. History Book-1884; Historical Data Systems-Bedford Co. residence
14	Alexandria Nat'l Cem.	Died 7/12/62; Bedford Gazette 2/13/1914 listing; 1850 W. Providence Twp. Census
15	Chaneysville Meth. Cem.	B.S.F. History Book-1884; Emanuel, Philip & William-brothers
16	Mt. Zion Cem.-Breezewood	1890 Snake Spring Valley Veterans Census
17		B.S.F. History Book-1884; Historical Data Systems-Bedford Co. residence
18		B.S.F. Hist. Book-1884; 1850 Southampton Twp. Cen.; Couldn't confirm William died in 1864
19	Glade Union Cem.-Somerset Co.	B.S.F. History Book-1884; Historical Data Systems-Bedford Co. residence
20	Snowberger-Bloomfield Twp.	Frank McCoy 1912 Listing; Civil War Pension record
21	Graceland Memorial Park-IN	1890 Wells Twp. Veterans Census
22		B.S.F. History Book-1884; Dauphin Co. references
23	Rock Hill Cem.-Monroe Twp.	Arm Amputated 4/2/65; 1890 Monroe Twp. Veterans Census
24		B.S.F. History Book-1884; Historical Data Systems-Bedford Co. residence
25	Rock Hill Cem.-Monroe Twp.	B.S.F. History Book-1884; David, Alvah, Jonas, Harrison & Wilson-brothers
26		Bedford Inquirer 5/22/1908 listing
27	Lawnridge Cem.-IL	B.S.F. History Book-1884; Gideon & Joseph-brothers
28	Rock Hill Cem.-Monroe Twp.	B.S.F. History Book-1884
29		Wounded 6/18/64; 1890 Monroe Twp. Veterans Census; 1850 Monroe Twp. Census
30		Frank McCoy 1912 list & Civil Works Admin. 1934 list
31	Everett Cem.	Everett Cemetery Listing of Civil War Veterans; Jacob & Samuel D-brothers
32	Salemville Cem.	B.S.F. History Book-1884; Historical Data Systems-Bedford Co. residence
33		B.S.F. History Book-1884; Historical Data Systems-Bedford Co. residence
34	Salemville Cem.	Wounded 5/8/64; B.S.F. History Book-1884
35		Ancestry.com information
36		Wounded 5/16/64; Died 5/25/64; B.S.F. History Book-1884; Drafted & Mustered-Reading, PA
37		B.S.F. History Book-1884
38	Mt. Union Cem.-Mench	B.S.F. History Book-1884; 1860 E. Providence Twp. Census
39		B.S.F. History Book-1884
40		B.S.F. History Book-1884; Historical Data Systems-Bedford Co. residence
41	Everett Cem.	Frank McCoy 1912 Listing; Bedford Inquirer 5/22/1908 listing
42		B.S.F. History Book-1884
43	Everett Cem.	Obituary-Mustered out for sickness; Discharged on Surg. Cert.; 1890 Everett Veterans Cen.
44		B.S.F. History Book-1884; Historical Data Systems record; 1860 W. Providence Twp. Census
45		Wounded 3/25/65; in Hospital at Muster Out; B.S.F. History Book-1884
46		B.S.F. History Book-1884; Blair Co. references
47	Mt. Pleasant Cem.-Mattie	1890 Monroe Twp. Veterans Census; B.S.F. History Book-1884
48		Historical Data Systems-Bedford Co. residence
49	Arlington Nat'l Cem.*	Died 12/31/64; Historical Data Systems-Bedford Co. residence; PA Civil War Project record
50	Mt. Smith Cem.-Bedford	B.S.F. History Book-1884
51		Deserted 1/4/63; Historical Data Systems record
52	Bedford Cem.	Civil War Marker at gravesite
53		Wounded 6/1/64; PA Civil War Project Record; Mongomery Co. references
54	Grandview Cem.-Cambria Co.	B.S.F. History Book-1884; Cambria Co. references
55	Florence Nat'l Cem.	POW-Andersonville 4/20/64; Died 10/26/64; PA Civil War Archives-enlisted in Rainsburg
56		1890 Bedford Twp. Veterans Census
57	Chaneysville Meth. Cem.	B.S.F. History Book-1884; 1860 Southampton Twp. Census
58		B.S.F. History Book-1884
59		B.S.F. History Book-1884; 1890 Huntingdon Co. Veterans Census; enlisted in Bedford
60	Arlington Nat'l Cem., 13 5636	Died 6/24/64; PA Civil War Archives-Born Bedford Co., enlisted in Bedford
61		B.S.F. History Book-1884
62		1890 Robertsdale Veterans Census
63		Historical Data Systems-Bedford Co. residence
64		Historical Data Systems-Bedford Co. residence

	Name	Age	Rank	Regiment	Co.	Muster In	Muster Out	Casualty	Casualty Battle
1	Wilt, Daniel H	42	Pvt.	PA 208th Inf.	K	9/7/64	6/1/65	Wounded	Ft. Stedman
2	Wilt, Henry*								
3	Wilt, Joseph	17	Pvt.	PA McKeage's Mil.	D	7/2/63	8/8/63		
4	Wilt, Joseph		Pvt.	PA 22nd Cav.	H	2/15/64	6/24/65		
5	Wilt, Silas D*		Pvt.	PA 110th Inf.	C	1861	10/24/64		
6	Wiltner, James		Pvt.	PA 76th Inf.	E	10/9/61			
7	Winslow, William		Pvt.	PA 22nd Cav.	I	3/4/64			
8	Winters, John J	21	Pvt.	PA 133rd Inf.	K	8/15/62			
9	Wise, Andrew H	40	Pvt.	PA 138th Inf.	E	8/29/62	6/23/65		
10	Wise, Brady B		Pvt.	PA 208th Inf.	H	9/7/64	6/1/65		
11	Wise, Henry H*	18	Pvt.	PA 76th Inf.	E	2/21/65	7/18/65		
12	Wise, Henry*		Pvt.	PA 11th Inf.	A	11/15/61		Died	
13	Wisegarver, Daniel	19	Pvt.	PA McKeage's Mil.	H	7/3/63	8/8/63		
14	Wisegarver, Daniel		Pvt.	PA 13th Cav.	D	9/30/63		Died	
15	Wisegarver, David	21	Corp.	PA 55th Inf.	H	10/11/61	8/30/65		
16	Wisegarver, George W	39	Pvt.	PA 78th Inf.	K	3/1/65			
17	Wisegarver, William H	17	Pvt.	IL 107th Inf.	B	9/4/62	7/12/65		
18	Wisegarver, William S	23	Pvt.	IA 12th Inf.	H	10/23/61	11/30/64		
19	Wisegarver, William V	21	Pvt.	PA 18th Cav.	K	2/29/64		KIA	
20	Wisel, George C	18	Pvt.	PA 55th Inf.	H	2/10/64	8/20/65		
21	Wiser, Jonathan	34	Pvt.	PA 49th Inf.	E	11/4/63	7/6/65	Wounded	
22	Wishart, Harvey S	24	1st Sgt.	PA 126th Inf.	B	8/12/62	5/20/63		
23	Wishart, Harvey S		Capt.	PA 208th Inf.	H	9/11/64	6/1/65		
24	Wishart, Henry	29	Capt.	PA 77th Inf.	F	10/9/61	2/2/63		
25	Wishart, James	25	2nd Lt.	PA 77th Inf.	F	10/9/61	4/24/62		
26	Wishart, Samuel	28	2nd Lt.	PA 20th Cav.	M	7/26/63	10/3/63		
27	Wissinger, Alex	29	Pvt.	PA 171st Inf.	I	11/2/62	8/8/63		
28	Witman, John	30	Corp.	PA 184th Inf.	A	5/12/64	7/14/65	Wounded	Cold Harbor
29	Witt, Dennis	30	Pvt.	PA 61st Inf.	F	9/26/64	6/20/65		
30	Witt, George	25	Pvt.	MD 2nd PHB Inf.	K	10/31/61	1/31/62		
31	Witt, George		Pvt.	MD 5th Inf.	C	11/10/64	9/1/65		
32	Witt, Jacob	21	Pvt.	PA 138th Inf.	D	9/2/62	1/5/65		
33	Witt, John	22	Pvt.	PA 171st Inf.	H	11/1/62		Died	
34	Witt, Jonathan	20	Pvt.	MD 2nd PHB Inf.	A	3/30/65	5/29/65		
35	Witters, George	19	Pvt.	PA 208th Inf.	H	9/7/64	6/1/65		
36	Witters, Jacob M	20	Pvt.	PA 84th Inf.	E	1861			
37	Witters, Jacob M		Corp.	PA 208th Inf.	H	9/7/64	6/1/65		
38	Witters, William H	35	Pvt.	PA 87th Inf.	C	6/4/64	6/29/65		
39	Wogan, James P	24	Sgt.	PA 55th Inf.	H	12/4/61	8/30/65		
40	Woldkill, Thomas		Pvt.	PA 6th Inf.	I	3/22/61	7/29/61		
41	Wolf, Edmund	15	Pvt.	PA 55th Inf.	H	2/19/64	8/30/65		
42	Wolf, Jacob		2nd Lt.	PA McKeage's Mil.	H	7/2/63	8/8/63		
43	Wolf, John D	20	Pvt.	PA 77th Inf.	A	10/11/61	10/26/64	POW	Chickamauga
44	Wolf, Richard	18	Pvt.	PA 55th Inf.	H	2/29/64	8/30/65	Wounded	Cold Harbor
45	Wolf, Samuel Y*	17	Music.	PA 76th Inf.	E	10/9/61		POW	
46	Wolf, William H*	23	Corp.	PA 25th Inf.	G	4/18/61	7/26/61		
47	Wolf, William H*		Pvt.	NJ 5th Inf.	H	7/17/64	5/7/65		
48	Wolfhope, John	25	Pvt.	PA 184th Inf.	A	5/12/64		POW-Died	Jerusalem Plank Rd.
49	Wolford, Alexander J		Corp.	PA McKeage's Mil.	G	7/2/63	8/8/63		
50	Wolford, Alexander J	24	Pvt.	PA 149th Inf.	B	9/23/63	6/24/65	Wounded	Weldon Railroad
51	Wolford, Daniel	21	Corp.	PA 138th Inf.	F	9/29/62	6/23/65		
52	Wolford, Frederick	19	Pvt.	PA 138th Inf.	F	8/29/62	6/23/65		
53	Wolford, George W	18	Pvt.	PA 101st Inf.	D	12/6/61	6/25/65		
54	Wolford, Samuel	50		PA 49th Inf.	E			POW	
55	Wolford, William	30	Pvt.	PA McKeage's Mil.	H	7/2/63	8/8/63		
56	Wolford, William		Pvt.	PA 149th Inf.	B	9/24/63		KIA	Wilderness
57	Womer, Andrew*	25	Corp.	PA 125th Inf.	A	8/10/62		KIA	Antietam
58	Wonder, John S	36	Pvt.	PA 100th Inf.	A	1/4/65	7/24/65		
59	Wonderly, John B	17	Pvt.	PA 55th Inf.	I	2/22/64	10/24/64		
60	Wonderly, William H	19	Pvt.	PA 55th Inf.	I	2/20/64	8/30/65	POW	Drewry's Bluff
61	Wonders, Daniel M	21	Sgt. QM	PA 55th Inf.	H	10/2/62	8/30/65		
62	Wonders, Henry	24	Pvt.	PA 55th Inf.	I	2/19/64	1/28/65		
63	Wonders, Jacob	24	Pvt.	PA 100th Inf.	K	6/4/64	7/24/65		
64	Wonech, Michael	22	Pvt.	PA 13th Inf.	G	4/25/61	8/6/61		

	Cemetery	Notes
1	Mt. Zion Cem.-Breezewood	Wounded 3/25/65; Frank McCoy 1912 Listing; Historical Data Systems record
2	Mt. Zion Cem.-Breezewood	Civil War Marker at gravesite
3	St. Francis River Cem.-MN	B.S.F. History Book-1884; 1860 E. Providence Twp. Census
4		B.S.F. History Book-1884; aka Whilt
5		B.S.F. History Book-1884; Blair Co. references
6		Deserted 6/12/64; Historical Data Systems-Bedford Co. residence
7		Deserted 4/21/65; B.S.F. History Book-1884
8		Deserted 9/21/62; B.S.F. History Book-1884-Bedford Co. residence
9	Bedford Cem.	Absent sick from 1/18/64; PA Civil War Archives-St. Clairsville resid.; Discharged Surg. Cert.
10		Historical Data Systems-Bedford Co. residence; B.S.F. History Book-1884
11		B.S.F. History Book-1884
12	US Soldiers & Airmans Nat'l	Died disease 11/28/61; Bedford Gazette 2/13/1914 listing; Historical Data Systems record
13		B.S.F. History Book-1884; Daniel, Cavid & William S-brothers
14	Arlington Nat'l Cem., 13 10351	Died 2/29/64; 1860 St. Clair Twp. Census; PA Civil War Project record
15	Drain Cem.-OR	B.S.F. History Book-1884; 1860 St. Clair Twp. Census
16	Fairview Cem.-Altoona	PA Civil War Archives record; Ancestry.com-born in St. Clairsville
17		Historical Data Systems record; 1850 St. Clair Twp. Census
18	Fairview Cem.-OK	1860 St. Clair Twp. Census; Historical Data Systems record
19	Sandyvale Cem.-Johnstown	Accidental gunshot 4/1/65; PA Civil War Arch.-Bedford Co. resid.; St. Clair Twp. references
20	Fishertown Cem.	Frank McCoy 1912 Listing; George & John-brothers
21	Broad Top IOOF Cem.	1890 Hopewell Veterans Census-Head Wound
22	Wells Valley Presb. Cem.	Historical Data Systems record; 1860 Wells Tannery Census
23		B.S.F. History Book-1884; Doctor after war
24	Grandview Cem.-Cambria Co.	Historical Data Systems record; 1860 Wells Tannery Census
25	Zion Cem.-IA	Historical Data Systems record; 1870 Wells Tannery Census
26	Everett Cem.	Historical Data Systems record; Samuel, Henry, James & Harvey-brothers
27		B.S.F. History Book-1884; PA Civil War Archives-enlisted in Bedford Co.
28		Wounded listed in Bedford Inquirer 7/1/64 article; B.S.F. History Book-1884
29	Sparks Cem.-Fayette Co.	PA Civil War Project record
30	St. George's Episcopal-MD	Historical Data Systems record; 1850 Londonderry Twp. Census
31		Civil War Pension Record; George, Jonathan, Dennis, Jacob, & John-brothers
32	Palo Alto Cem.-Hyndman	1890 Londonderry Twp. Veterans Census
33		Died 2/17/63; Historical Data Systems record; 1860 Londonderry Twp. Census
34	Palo Alto Cem.-Hyndman	1890 Londonderry Twp. Veterans Census
35	Garden Plain Cem.-NE	B.S.F. History Book-1884; Historical Data Systems-Bedford Co. residence
36	Riverside Cem.-NE	Civil War Pension Record; Jacob & George-brothers
37		B.S.F. History Book-1884; Historical Data Systems-Bedford Co. residence
38	Byers Cem.-Woodbury	PA Civil War Project record; 1860 Woodbury Twp. Census
39	Wilmington & Brandywine-DE	B.S.F. History Book-1884; Historical Data Systems record; 1860 St. Clair Twp. Census
40	First Meth. Cem.-Lewistown	1890 Bedford Twp. Veterans Census
41	Webb City Cem.-MO	B.S.F. History Book-1884; Edmund & Richard-brothers
42		B.S.F. History Book-1884; Historical Data Systems record
43	Salemville Cem.	POW-Belle Island; Frank McCoy 1912 Listing; POW 10/19/63 to 5/11/64
44	Fishertown Cem.	Wounded-Top of left Ear clipped off-6/1/64; 1890 E. St. Clair Twp. Veterans Census
45		POW-Andersonville B.S.F. History Book-1884; Cumberland Co. references
46	Union-West End-Allentown	Civil War Soldier Records and Profiles
47		1890 Robertsdale Veterans Census
48	Andersonville Cem.; 10315	POW-Andersonville 6/22/64; Died-Dysentery 10/4/64; Ancestry.com-New Baltimore references
49		B.S.F. History Book-1884; 1890 Colerain Twp. Veterans Census
50	Woods Church-Colerain Twp.	Wounded 9/20/64 while on picket; 1890 Colerain Twp Vet. Census; Alexander & William-bros.
51	Lybarger Cem.-Madley	PA Civil War Archives-Londonderry Twp. residence; one of the soldiers that shot AP Hill
52	Lybarger Cem.-Madley	1890 Londonderry Twp. Census; Daniel & Frederick-brothers
53	Harrisburg Cem.	B.S.F. History Book-1884; Historical Data Systems record; 1850 Southampton Twp. Census
54	Mt. Olivet Cem.-Manns Choice	POW Feb. 1865-Pension record; Frank McCoy 1912 List; Born 1812; likely in 50s at muster
55		B.S.F. History Book-1884
56		KIA 5/5/64; Historical Data Systems record; 1850 Cumberland Valley Census
57	Antietam Nat'l Cem., 3669	KIA 9/17/62; Bedford Gazette 2/13/1914 listing; Blair Co. references
58	Helixville Cem.-Napier Twp.	1890 Napier Twp. Veterans Census
59	Little Valley Cem.-Hunt. Co.	B.S.F. History Book-1884; 1860 Hopewell Twp. (Huntingdon Co.) Census
60	Benshoff Hill Cem.-Cambria Co.	POW 5/16/64 to 11/19/64; B.S.F. History Book-1884
61	Schellsburg Union Cem.	1890 Napier Twp. Veterans Census
62		B.S.F. History Book-1884
63	Carpenter Cem.-New Germany	Bedford Inquirer 5/22/1908 listing; 1850 Napier Twp. Census; Jacob & John-brothers
64		B.S.F. History Book-1884; Historical Data Systems-Bedford Co. residence

	Name	Age	Rank	Regiment	Co.	Muster In	Muster Out	Casualty	Casualty Battle
1	Woodcock, Clark*	29	Pvt.	PA 110th Inf.	C	2/27/64		KIA	Sailor's Creek
2	Woodcock, George		Pvt.	PA McKeage's Mil.	H	7/2/63	8/8/63		
3	Woodcock, George W (1)	18	Pvt.	PA 53rd Inf.	C	1/29/64		Wound.-Died	Spotsylvania
4	Woodcock, George W (2)*	30	Pvt.	OH 76th Inf.	H	1/11/62		Died	
5	Woodcock, John		Pvt.	PA McKeage's Mil.	H	7/2/63	8/8/63		
6	Woodcock, John A		Sgt.	PA McKeage's Mil.	G	7/2/63	8/8/63		
7	Woodcock, John A	23	1st Sgt.	PA 202nd Inf.	K	9/5/64	8/3/65		
8	Woodcock, Oliver E	30	Pvt.	PA 11th Inf.	E	7/5/64	6/11/65		
9	Woodcock, Walter W	21	Pvt.	PA 126th Inf.	B	8/12/62	5/20/63		
10	Woodcock, Walter W		Corp.	PA 107th Inf.	I	9/19/64	6/7/65		
11	Woodcock, William L		Lt.	PA 77th Inf.	F	10/9/61	8/1/63		
12	Woods, James C			Conf.					
13	Woodward, George	25	Pvt.	PA 55th Inf.	I	7/15/63	8/30/65		
14	Woodward, James A		Pvt.	PA 110th Inf.	C	1/21/65			
15	Woolett, Sylvester B	23	Pvt.	PA 110th Inf.	C	12/19/61	10/24/64		
16	Woomer, Henry		Pvt.	PA 20th Inf.	E				
17	Woy, David M	21	Pvt.	PA 12th Cav.	F	2/14/62		Died	
18	Woy, Ezekiel C	16	Pvt.	PA 208th Inf.	K	9/7/64	6/6/65	Wounded	Petersburg - Final
19	Woy, George	22	Pvt.	PA 77th Inf.	F	11/22/61		KIA	Liberty Gap, TN
20	Woy, James H	19	Pvt.	PA 126th Inf.	B	8/12/62	5/20/63	Wounded	Fredericksburg
21	Woy, James H		Pvt.	PA 126th Inf.	B			Wounded	Chancellorsville
22	Woy, James H		Corp.	PA 208th Inf.	H	9/5/64	6/1/65		
23	Woy, John		Pvt.	PA 208th Inf.	H	8/18/64	6/1/65		
24	Woy, John W	19	Pvt.	PA 22nd Cav.	H	2/22/64	6/24/65		
25	Woy, Joseph	25	Pvt.	PA 22nd Cav.	L	2/29/64	6/24/65	Wounded	Snicker's Gap - VA
26	Wray, William H	37	Pvt.	PA 76th Inf.	E	2/7/62	5/18/63		
27	Wright, Charles C	22	Pvt.	PA 184th Inf.	A	5/12/64	5/20/65	Wounded	Cold Harbor
28	Wright, Charles C		Pvt.	PA 184th Inf.	A			POW	Jerusalem Plank Rd.
29	Wright, Darwin P	20	Pvt.	PA 200th Inf.	C	8/26/64	5/30/65		
30	Wright, Edmund S	20	Pvt.	PA 184th Inf.	A	5/12/64		POW-Died	Jerusalem Plank Rd.
31	Wright, Edwin V	30	Pvt.	PA 107th Inf.	F	9/21/64	6/21/65	Wounded	
32	Wright, Isaac	35	Pvt.	PA 23rd Inf.	D	8/6/61	9/8/64		
33	Wright, Isaac		Pvt.	PA 48th Inf.	G	1/2/65	8/9/65		
34	Wright, Jacob	31	Pvt.	PA 13th Cav.	K	2/26/64	7/14/65		
35	Wright, Jacob*		Pvt.	PA 55th Inf.	K				
36	Wright, James (1)	21	Pvt.	PA 99th Inf.	H	2/25/65		Died	
37	Wright, James (2)*	42	Pvt.	PA 55th Inf.	I	10/8/63	5/29/65		
38	Wright, Paul	21	Pvt.	PA 126th Inf.	B	8/5/62	5/20/63		
39	Wright, Paul		Pvt.	PA 56th Inf.	H	9/19/64	5/31/65		
40	Wright, Thomas	20	Pvt.	PA 125th Inf.	E	8/13/62	5/18/63		
41	Wright, Thomas		Pvt.	PA 13th Cav.	E	2/26/64	6/26/65	POW	2nd Deep Bottom
42	Wyant, John			PA 74th Inf.	K				
43	Yantz, Henry	36	Pvt.	PA 67th Inf.	C	11/28/64		Died	
44	Yantz, Peter D	27	Pvt.	MD 1st PHB Cav.	M	4/23/64	6/28/65		
45	Yarnell, Jesse	43	Pvt.	PA 138th Inf.	D	8/29/62		Wound.-Died	Monocacy, MD
46	Yarnell, John	17	Pvt.	PA 138th Inf.	D	8/29/62	6/23/65	Wounded	Cedar Creek
47	Yarnell, William H	23	Pvt.	PA 21st Cav.	F	2/28/65	7/8/65		
48	Yaultz, Franklin		Pvt.	PA McKeage's Mil.	H	7/2/63	8/8/63		
49	Yeader, William*	16	Pvt.	PA 184th Inf.	A	5/12/64	11/7/64	Wounded	Cold Harbor
50	Yeagle, Simon B	19	Pvt.	PA 133rd Inf.	K	8/15/62	5/26/63		
51	Yeck, Frederick		Pvt.	PA 55th Inf.	I	10/20/63	6/7/65		
52	Yingling, Martin M	21	Pvt.	NY 16th Cav.	A	9/26/62	23863		
53	Yost, Francis F		Pvt.	PA 55th Inf.	D	9/26/62	6/11/65		
54	Yost, Jacob		Pvt.	PA 21st Cav.	E	7/2/63	2/20/64		
55	Young, Aaron	21	Corp.	USCT 24th Inf.	D	2/16/65	10/1/65		
56	Young, Alexander	19	Pvt.	PA 8th Res.	F	6/11/61	5/15/64		
57	Young, Alexander		Pvt.	PA 191st Inf.	H	5/15/64	6/28/65	Wounded	Five Forks
58	Young, Daniel		Pvt.	KY 14th Inf.	I	6/30/64	10/31/65		
59	Young, Daniel D	29	Sgt.	USCT 24th Inf.	D	2/15/65	10/1/65		
60	Young, David*	28	Pvt.	MA 30th Inf.	C	12/4/61	7/25/63		
61	Young, Edward H	25	Corp.	PA 119th Inf.	D	8/19/62	7/18/65	Wounded	Sailor's Creek
62	Young, Edwin	25	Pvt.	PA 110th Inf.	C	10/24/61			
63	Young, George N	20	Pvt.	PA 110th Inf.	C	10/24/61			
64	Young, George N		2nd Lt.	PA 22nd Cav.	M	7/11/63	6/24/65		

Detailed Alphabetical Listing 541

	Cemetery	Notes
1	Rose Hill Cem.-Altoona	KIA 4/6/65; B.S.F. History Book-1884; Blair Co. references
2		B.S.F. History Book-1884
3	Arlington Nat'l Cem., 27 311	Severe Wound: Forearm, Leg & Neck - 5/12/64; Died 5/23/64; 1860 Wells Twp. Census
4	Vicksburg Nat'l Cem.	Died of Disease 6/23/63; Unsure if he was ever a Bedford Co. or Wells Valley resident
5		B.S.F. History Book-1884
6		B.S.F. History Book-1884
7	Union Cem.-Bellefonte	1860 Wells Township Census; Historical Data Systems record
8	Wells Valley Meth. Cem.	1890 Wells Twp. Veterans Census
9	Lake View Cem.-WA	1860 Wells Twp. Census; Historical Data Systems record
10		Civil War Pension Record; Walter & Oliver-brothers
11		1860 Wells Twp. Census; William L & George-brothers
12		Listed in 1890 Broad Top Veterans Census as a Confederate living in Broad Top Township.
13		B.S.F. History Book-1884
14		B.S.F. History Book-1884; Historical Data Systems record
15	Union Cem.-McConnellsburg	B.S.F. History Book-1884; PA Civil War Archives-Woodbury residence
16	Broad Top IOOF Cem.	Bedford Inquirer 5/22/1908 listing; aka Womer
17	Mt. Zion Cem.-Breezewood	Died 2/1/63; Frank McCoy 1912 Listing; 1860 W. Providence Twp. Census
18	Everett Cem.	Wounded 4/2/65; B.S.F. History Book-1884; Ezekiel, David, James & John-brothers
19	Stones River Nat'l Cem.-TN	KIA 6/25/63; 1850 E. Providence Twp. Census; PA Civil War Project record-aka Way
20	Hyndman Cem.	Wounded-bayonet in foot 12/13/62; Frank McCoy 1912 Listing; Historical Data Systems record
21		Severe Arm Wound 5/3/63; Bedford Inquirer 5/22/1908 listing; B.S.F. History Book-1884
22		208th PA Infantry Muster Record
23		1890 W. Providence Twp. Veterans Census
24	Ashbury Meth. Cem.-Graceville	B.S.F. History Book-1884
25	Forest Hills Cem.-TN	Wounded 7/17/64; Ancestry.com information; Joseph & George-brothers
26		PA Civil War Archives-Bedford residence; Discharged on Surgeon's Cert.
27	Pleasantville Cem.	Chest Wound 6/3/64; B.S.F. History Book-1884; Charles & Edmund-brothers
28		POW-Andersonville 6/22/64 to 4/29/65; attended 1905 Andersonville Dedication
29	Pleasantville Cem.	1890 W. St. Clair Twp. Veterans Census; Darwin & Edwin-brothers
30	Wilmington Nat'l Cem.	POW-Andersonville; 6/22/64; Died 3/2/65; 1860 St. Clair Twp. Cen.; B.S.F. Hist. Book-1884
31	New Paris Comm. Ctr. Cem.	1890 Napier Twp. Veterans Census
32	Chaneysville Meth. Cem.	Civil War Pension Record
33		PA Civil War Archives-Bedford Co. residence; Frank McCoy 1912 Listing
34	Upper Claar Cem.-Queen	1890 Union Twp. (Pavia) Veterans Census
35		Frank McCoy 1912 Listing
36	Arlington Nat'l Cem., 13-12700	Died 5/27/65; Bedford Gazette 2/13/1914 listing; 1860 Union (Pavia) Twp. Census
37		B.S.F. History Book-1884; Drafted; Muster In-Philadelphia
38	Bedford Cem.	1890 Bedford Twp. Veterans Census
39		Frank McCoy 1912 Listing
40	Greenfield Cem.-Queen	PA Civil War Archives-Bedford Co. residence; Jacob, James & Thomas-brothers
41		POW-Andersonville 8/18/64 to 3/3/65; 1890 Kimmel Twp. Veterans Census
42		Bedford Historical Society record on file
43	Arlington Nat'l Cem., 13-11944	Died-Typhoid Fever 6/15/65; Bedford Gazette 2/13/1914 list; Registered in Londonderry Twp.
44	Bald Hill Cem.-Inglesmith	1880 Cumberland Valley Census; Historical Data Systems record
45	Philadelphia Nat'l Cem.	Wound 7/9/64; Died 7/22/64; PA CW Arch.-enlisted in Schellsburg; Father of John & William
46	Bradner Cem.-Ohio	Wounded 10/19/64 in Bed. Inquirer on 11/4/64; Obituary-Standing by father when he was shot
47	Pleasantville Cem.	1860 St. Clair Twp. Census; Frank McCoy 1912 Listing; William & John-brothers
48		B.S.F. History Book-1884
49		Shoulder Wound 5/31/64; B.S.F. Hist. Book-1884; Comp. A recruited-Dauphin & Bedford Co.
50		In Hospital 10/10/62 to 2/16/63; Historical Data Systems-Bedford Co. residence
51		B.S.F. History Book-1884; Drafted
52	Claysburg Union Cem.	1890 Claysburg Veterans Census
53		B.S.F. History Book-1884
54		Historical Data Systems-Bedford Co. residence
55	Mt. Ross Cem.-Bedford	1890 Bedford Twp. Veterans Census; Frank McCoy 1912 Listing
56	Coon Rapids Cem.-IA	B.S.F. History Book-1884; Alexander & Peter W-brothers
57		Wounded 4/1/65; B.S.F. History Book-1884
58		1890 Bedford Twp. Veterans Census
59	Mt. Ross Cem.-Bedford	1890 Bedford Twp. Veterans Census; Frank McCoy 1912 Listing
60	Tatesville Methodist Cem.	1860 Broad Top Twp. Census; 25th US Infantry references; Discharged on Surgeon's Cert.
61	Hopewell Cem.	Wounded 4/6/65; Bedford Gazette 1909 article listing
62	Young Farm Cem.-Hopewell	Frank McCoy 1912 Listing aka Edward; Historical Data Systems record
63	Old Carrollton Cem.-IA	PA Civil War Archives-Yellow Creek residence; B.S.F. History Book-1884
64		B.S.F. History Book-1884

	Name	Age	Rank	Regiment	Co.	Muster In	Muster Out	Casualty	Casualty Battle
1	Young, Isaac	27	Pvt.	PA 171st Inf.	H	11/1/62	8/8/63		
2	Young, Jacob P			USCT 127th Inf.					
3	Young, Jacob P	33	Pvt.	USCT 24th Inf.	G	2/28/65	10/1/65		
4	Young, James H	20	Pvt.	PA 22nd Cav.	H	2/26/64	6/6/65	Wounded	Browns Gap
5	Young, Joel T	28	Pvt.	PA 8th Res.	F	6/19/61	8/7/62		
6	Young, John		Pvt.	PA 22nd Cav.	H	2/8/64	6/24/65		
7	Young, Peter	28	Pvt.	USCT 127th Inf.	F	8/27/64	10/20/65		
8	Young, Peter W	21	Pvt.	PA 208th Inf.	K	9/7/64	6/1/65		
9	Young, Thomas J		Pvt.	PA 76th Inf.	E	9/27/64	6/29/65		
10	Younkin, Alfred*		Pvt.	KS 11th Cav.	L	3/31/64		Died	
11	Younkin, Foster*	23	Pvt.	PA 1st Art.	H	6/17/63	1/1/64	Died	
12	Younkin, Frederick J*	31	Pvt.	PA 52nd Inf.	I	9/26/64	6/2/65		
13	Zeller, John M*								
14	Zeller, Michael	28	Pvt.	PA 171st Inf.	I	11/2/62	8/8/63		
15	Zembower, Josiah	35	Pvt.	PA 186th Inf.	G	7/26/64	8/15/65		
16	Zembower, Josiah A	21	Pvt.	PA 184th Inf.	G	5/17/64	6/27/65	POW	Jerusalem Plank Rd.
17	Zeth, George W	20	Pvt.	PA 19th Cav.	C	9/9/63	5/14/66		
18	Zimmerman, Jacob J*	25	Pvt.	PA 93rd Inf.	G	9/17/64	6/1/65		
19	Zimmerman, John H*		Pvt.	PA 93rd Inf.	G	9/17/64	6/1/65		
20	Zimmerman, Samuel	24	Pvt.	PA 77th Inf.	F	2/27/65	12/6/65		
21	Zimmerman, William	17	Pvt.	PA 77th Inf.	F	2/27/65	12/6/65	Wounded	
22	Zinn, John	27	Pvt.	PA 17th Cav.	F	12/30/63	6/27/65	POW	Berryville

	Cemetery	Notes
1	Oldham Cem.-Shade Twp.	1890 Napier Twp. Veterans Census
2	Mt. Ross Cem.-Bedford	Frank McCoy 1912 Listing
3	Mt. Ross Cem.-Bedford	Frank McCoy 1912 & Civil Works Admin. 1934 list;; Jacob P, Aaron, Peter & Daniel D-bros.
4	Graceville Luth. Cem.	Wounded 6/26/64; 1890 E. Providence Twp. Veterans Census; B.S.F. History Book-1884
5		B.S.F. History Book-1884; 1850 Hopewell Twp. Census; Discharged on Surgeon's Cert.
6		B.S.F. History Book-1884
7	Mt. Ross Cem.-Bedford	Frank McCoy 1912 Listing; 1860 Bedford Census
8	Carroll City Cem.-IA	Historical Data Systems-Bedford Co. residence; B.S.F. History Book-1884
9		B.S.F. History Book-1884
10		Died 9/18/65; B.S.F. History Book-1884; Somerset Co. references
11	Jersey Baptist Cem.-Ursina	Died 9/30/64; B.S.F. History Book-1884; Somerset Co. references
12	Delilah Younkin-Somerset Co.	B.S.F. History Book-1884; Somerset Co. references
13		1890 Juniata Twp. Veterans Census
14		B.S.F. History Book-1884
15	Mt. Union Cem.-Mench	PA Civil War Project record
16	Centenary Meth.-Cumberland	POW-Andersonville 6/22/64 to 4/28/65; 1890 Cumberland Valley Veterans Census
17	Old Claysburg Cem.	1890 Greenfield Twp. Veterans Census; Historical Data Systems record
18	Horner-Mt.Tabor-Somerset Co.	B.S.F. History Book-1884; Somerset Co. references
19	Stoystown IOOF-Somerset Co.	B.S.F. History Book-1884; Somerset Co. references
20	Old Claysburg Cem.	Ancestry.com information
21	Old Claysburg Cem.	1890 Greenfield Twp. Veterans Census
22	Lutheran Cem.-Osterburg	POW-Richmond & Salisbury 9/24/64 to 2/28/65; PA Civil War Project record

Photograph Index

This index includes photographs of soldiers and their families. Battlefield photographs contained in the book can be found by referencing the pages listed for each Battle in the Contents on page vii. Gold Star family member photographs are designated with a dash after the page number and a family member abbreviation: -B(Brother), -D(Daughter), -M (Mother), -P(Parents), -S(Sister), -So(Son) and -W(Wife). Only the names of the Civil War veterans are included in this index.

Name	Page(s)	Name	Page(s)	Name	Page(s)
Adams, George W (1)	358	Bennett, Henry	255	Callahan, Jacob R (2)	184
Adams, John Q	328	Bennett, Jacob	166-B	Callihan, Robert	300
Adams, Philip	163-M	Berkheimer, John	129	Carpenter, Abraham	96, 288
Adams, Samuel	234	Berkheimer, Samuel	358	Carpenter, Curtis J	151
Akers, Job S	313	Berkheimer, William S	253	Carpenter, David B	151
Akers, John H	344	Biddle, Andrew B	57	Cessna, George W	243
Albright, William	218	Bivens, James W	289	Cessna, Joseph P	338
Amick, George W	225	Black, Erastus	247	Cessna, William	235
Amick, William	278	Blackburn, Cyrus E	325	Chilcott, Ephraim	339
Amick, William M	126	Blake, Simon S	340	Chisholm, John P	324
Amos, Francis M	306	Blake, William B	37	Claar, Henry I	114, 362
Amos, John B	65, 234	Blattenberger, Daniel	300	Claar, Thomas	257
Anderson, James	164-M,S	Bloom, John (2)	338	Clark, George W	97
Anderson, John (1)	164-M,S	Blymyer, Benjamin	347	Clark, Joel	130
Andrews, Joseph	358	Bond, George M	204	Clark, Philip	255
Armstrong, David B	225	Boor, Tobias	343	Clark, Willlam W	312
Armstrong, Joseph	126	Boor, William A	65	Claycomb, Conrad	288, 358
Arnold, Abraham Kerns	214, 339	Boose, Isaac	347	Claycomb, John	13
Arnold, Albin C	235	Border, John	93	Cleaver, James	5
Bagley, Henry H	297	Border, John S	272	Clevenger, George W	346
Baith, Peter	326	Bortz, Martin S	97, 289	Clevenger, Jacob A	185, 303
Baker, Alfred	308, 362	Bottenfield, Adam K	352	Clingerman, Harrison	297
Baker, Andrew C	339	Bowser, Isaac B	167	Clites, Levi	354
Barclay, John J	339	Bowser, John B	358	Cobler, Allen	73
Barley, Reuben H	341	Bowser, John J	167-B	Colledge, Joseph R	314
Barndollar, Jacob W	33	Bowser, Moses	167	College, Henry	344
Barndollar, James J	278	Bowser, Valentine	257	Collins, Isaiah	87
Barndollar, Martin D	278	Brant, Henry	204	Comp, Solomon	341
Barndollar, William P	45, 361	Brant, William	267	Compher, Alexander	52, 264
Barney, Jacob	334	Breneman, Michael B	28, 357	Conley, Isaiah	263
Barnhart, Abraham	340	Bridenstine, Jacob (2)	341	Conner, David (2)	168-B
Barr, Thomas M	274, 357	Bridenthal, Thomas	361	Conner, Isaac	168
Barton, James	351	Brisbin, Ezra D	11, 271	Conner, Lewis	168, 279
Bartow, Barney	315	Brown, Caleb	345	Conrad, Winfield S	203
Baughman, Abraham A	258	Brown, George D	72	Cook, Hanson	344
Baughman, Adam	363	Brown, Jacob H	339	Cook, James A	344
Baughman, George	127	Burket, Abraham	339	Cook, John	259
Baughman, William	127	Burket, David (1)	356	Cook, Levi (2)	320
Beam, Samuel Z	354	Burket, Elias S	37	Cooper, Barton A	267
Beaston, James	353	Burket, Frederick	129	Corle, Adam S	356
Beaver, William	128	Burket, Gabriel	73, 356	Corle, Alexander	301, 356
Bechtel, Daniel S	352	Burket, Jacob	235	Corle, Chauncey	130, 169-P
Bechtel, David S	352	Burket, Jacob D	299	Corle, Eli	169-P
Beegle, David F	53	Burket, John N	288	Corle, Francis B	131
Beltz, Adam	165-S	Butts, James B	278	Corle, Franklin	289, 356
Beltz, William W	128, 165	Calhoun, Christopher P	284	Corle, Frederick	258
Bennett, Enos	166	Callahan, Jacob R (1)	225	Corle, William C	288

Photograph Index

Name	Page(s)
Corley, David	249
Corley, Jacob (1)	249
Coughenour, James	340
Covert, James P	338
Cox, Samuel	196
Crist, David T	205, 258
Crist, John T (2)	337
Darr, Abraham	131
Darr, David H	235
Darr, George W	358
Davis, Isaiah M	170-B
Davis, James P	170
Davis, John Milton	341, 361
Davis, R DeCharmes	336, 355
Davis, William A (1)	358
Defibaugh, David	272
Defibaugh, John (2)	132, 263
Deremer, William J	340
Dibert, Jacob J	132, 236
Dibert, Jonathan	345
Dicken, James F	338
Dicken, Thomas W	205
Diehl, B James	236
Diehl, Francis M	328
Diehl, Levi	339
Diehl, Samuel J	236
Dively, George M	247
Dively, Martin	247
Dively, Michael*	356
Dodson, Andrew J	274
Drenning, Thomas	184
Drenning, William	329
Dunkle, David	283
Durno, William Jr	278
Edwards, Daniel L	152, 185
Edwards, Festus	347
Edwards, George C	348
Edwards, Jonathan B	279
Edwards, Josiah V	152, 236
Eichelberger, Eli	19, 222, 225, 357, 360
Eichelberger, Jacob A	306
Eicher, Christian C	357
Eicholtz, William G	314
Elliott, Francis M	316
Emerick, Andrew R	300
Enfield, Americus	43, 186, 202, 355, 359, 362
England, Jacob	119

Name	Page(s)
Ensley, Peter M	326
Eshelman, Benjamin	133
Eshelman, George W	353
Estep, Henry C	186
Evans, Hiram	341
Evans, Lemuel	344
Evans, Nathan C	303
Farber, Thomas H	115
Feaster, John G	87, 253
Feather, Michael	259, 356
Feather, Simon	290
Feight, Frederick	115
Felton, John A	91
Felton, Simon P	326
Fickes, Cyrus W	102, 321
Fickes, James M	133
Filler, William B	93, 202, 326
Fink, Valentine	279
Fisher, David H	345
Fisher, Emanuel	290
Fisher, Henry H	279
Fisher, John W	275
Fleegle, Isaac S	237, 361
Fleegle, John H	352, 358
Fletcher, David	340
Fletcher, John (2)	347
Fletcher, Samuel	340
Fluke, George	258
Fluke, Oliver B	325
Fluke, Oliver P	355
Fluke, Samuel B	205, 308
Fockler, George (1)	346
Fockler, Samuel	272
Foor, Andrew J	22
Foor, Brazella	329
Foor, Daniel	150-B
Foor, Francis L	134
Foor, George	150-So
Foor, Jacob I	314
Foor, James H	279
Foor, Jeremiah	268
Foor, Jonathan S	40
Foor, Peter	313
Foster, Aaron	60
Foster, John W	352
Frazey, Frederick L	29
Gardner, Adam	238, 356
Garlick, Christian C	19
Garlinger, William M	338

Name	Page(s)
Garretson, Benjamin H	107, 134
Garretson, Josiah P	107, 237
Garretson, Moses R	171-M
Gaster, Ezekiel W	279
Gates, George	357
Gates, Joseph	271
Gates, Theophilus R	195, 237
Gephart, John	245
Gettleman, Jacob	347
Getty, John	289
Gillam, John	338
Gollipher, Silas	13
Gordon, George G	350
Gordon, George J	332
Gordon, Isaac	288, 358
Gordon, Joseph	187, 238, 356
Gordon, William	172-P
Grace, Israel P	347
Gracey, Alfred	40
Gracey, George E	187
Griffith, Abel	85, 362
Grove, Henry C	329
Grove, James A	85, 265
Growden, Joseph	343
Grubb, Harvey	279, 326
Hainsey, Adam R	245
Hainsey, Valentine	173-P
Haley, Josiah	238
Hammann, Jacob	309
Hammer, Hezekiah	109, 237, 358
Hammer, John B	290
Hammer, Samuel	358
Hankinson, Solomon J	354
Hanks, Benson	346
Hanks, William H	322
Hann, Gaston	301
Harbaugh, Allen	206
Harbaugh, Jason	251
Harclerode, David	29, 273
Hardinger, Henry	340
Hardinger, Reuben	338
Harr, Christian	76, 358
Hartman, Henry S	346
Hasenpat, William R	206
Hawman, John C	322
Heffner, Samuel	264
Helm, John B	263
Heltzel, Daniel G	135
Henry, George A	206

545

Photograph Index

Name	Page(s)
Herring, John B	361
Hetrick, Daniel L	188, 358
Hill, Aaron	338
Hillegass, Andrew	331
Himes, Andrew J	i-P&W, 135
Himes, John	314
Himes, Samuel	352
Himes, Thomas	255
Hinson, William	219
Hissong, Josiah	105, 238
Hite, Jacob (1)	341
Hite, Nicholas	341
Hoffman, John O	357
Hollar, Philip V	102
Holler, George W	153
Holler, John M	153, 290, 355
Hollingshead, Oliver S	188, 357
Holsinger, Emanuel	348
Holsinger, Levi R	341
Hoon, Stacey	332
Horton, David	57, 360
Horton, Jesse W	341
Horton, Jonathan A	206
Horton, Oliver	111, 290
Horton, Zophar P	79
Hughes, Bartley	350
Humbert, Moses	189
Hunt, David A	341
Hunt, George E	348
Hunt, Samuel B	238
Ickes, Adam (2)	253
Ickes, George W	288
Ickes, William M	253
Imler, George R	89
Imler, Isaac M	137
Imler, Jonas C	201, 308
Imler, Matthias	304
Irvine, Hayes	316
Jackson, Charles W	350
Jackson, Stiles H	341
Jamison, Benjamin	28, 272
Jay, Thomas	300
Jeffries, Howard B	320
Johnston, John W	283
Juda, George V A	137
Karns, John H	339
Karns, Wilson	345
Kean, William C	273
Kegg, James P	189, 239
Kegg, Simon P	15
Keith, George J	338
Kellerman, James L	290
Kellerman, John L	345
Kelly, William H	344
Kennard, Matthew P	353
King, Samuel T	207
Kinsey, Benjamin F	155
Kinsey, Dewalt	155
Kinsey, John B	155
Kinsey, Peter Jr	155, 239
Kinsey, Peter Sr	155
Kipp, Lewis A	332
Klahre, Theodore M	13
Knee David H	217
Knee, Augustus D	217-B
Knee, Philip	217
Knipple, John A	251
Knipple, William H	251
Knisely, Christopher	348
Lambert, Joseph C	33
Lashley, John W	330
Latta, John C	219
Latta, William H	350
Layton, Henry	139
Layton, John R	345
Leach, Samuel R	324
Leasure, Riley	330
Lee, David W	297
Lee, John	304
Lehman, Isaiah	313
Lehman, William H	190
Leighty, Samuel	334
Leonard, Adam P	266
Leonard, John B	314
Leonard, Philip	239
Ling, Isaac	235, 358
Ling, Isaac N	292
Ling, William H	358
Lingenfelter, Josiah	349
Lingenfelter, Thaddeus	190
Livingston, John A	76, 228, 240
Livingston, Samuel	339
Livingston, Thomas G	11, 61
Lohr, Benjamin F	40, 350
Long, Amon	326
Longenecker, Jacob H	53, 264, 356
Lowery, Emanuel	291
Lowery, John	25
Lowery, William	318
Lybarger, Henry G	240
Lybarger, Joseph J	344
Lyons, Thomas H	239
Lysinger, Joseph H	191
Machtley, John	358
Madara, James	197
Manges, George W	240
Manges, Jacob A	207
Markey, Joseph	304
Maugle, Adam	357
May, Daniel H	249, 291
May, Hiram	291
May, John L	291
May, Joseph C	241
May, Lewis A	57, 284
May, Marcus	92, 291
May, Samuel M	291
McChesney, John	78
McClellan, Josiah	281
McCoy, Francis P	191, 346, 357, 359
McDaniel, Daniel	281
McDaniel, George W	279
McDonald, James	161-M
McDonald, John	162
McDonald, Samuel	162-S
McDonald, William, Jr	162-W
McEldowney, Samuel J	117
McFarland, Andrew J	345
McFerren, John	198
McGee, James	174-B
McGee, Samuel	174
McGee, William	174-B
McGregor, William A	208, 358
McKellip, D Alexander	340
McKibbin William L	347
McLaughlin, Collin L	227
McLeary, Paul	344
McMullen, Charles E	350
McMullen, John H	345
McVicker, Jesse	348
Mearkle, Barton	314, 359
Mearkle, David S	259
Mearkle, Henry (2)	344
Mearkle, Sansom	328
Mellott, Cornelius	175-P
Mellott, Daniel B	249
Mellott, Frederick	175-P
Mellott, Jacob L	208
Mellott, John L	175-P

Photograph Index

Name	Page(s)
Melott, William	328
Messersmith, Alexander	156
Messersmith, George	156
Miller, Armstrong	358
Miller, Christian	301
Miller, Edward J	338
Miller, Ephraim B	290
Miller, George	358
Miller, Jacob H	358
Miller, James H	71
Miller, Jeremiah J	176
Miller, John A	347
Miller, John I	43
Miller, Joseph	240
Miller, Mark	341
Miller, Thomas (1)	295
Miller, Thomas J (1)	358
Miller, William O	176-P
Miller, Wireman	176
Millin, William S	260
Mills, Andrew J	264
Mock, Aaron	47, 293, 358
Mock, Emanuel A	67
Mock, Henry	352
Mock, Malachi B	139
Mock, Samuel A	358
Morgart, Abraham	141
Morningstar, Samuel	348
Moser, Martin	275
Mosser, George W	320
Mower, Edward E	241
Mowry, John	301
Mowry, Joseph	92
Moyer, John (1)	141
Murphy, Samuel	349
Murrie, David	192
Myers, John O	341
Myers, Martin L	203
Nelson, John	157, 318(2)
Nelson, Robert	301
Nelson, William N	157
Nevitt, James M	280
Nevitt, Joseph H	340
Nicodemus, Isaac	114
Nunemaker, Peter	358
Nycum, John (1)	177-B
Nycum, John Q	280, 329
Nycum, John W	267
Nycum, Simon E	361
Nycum, Upton	177-B

Name	Page(s)
Nycum, William H	280
Nycum, Wilson	177
Oakes, Christian R	349
Oaks, John R	313
Oliver, Benjamin F	318
O'Neal, Emanuel	291
O'Neal, James R	292
O'Neal, John E	290, 292
Osborn, Ezra	346
Oster, Samuel C	255
Over, Benjamin	282
Over, James E	358
Owens, Chauncey	89
Pearson, Josiah	192
Peck, Jacob B	241
Peck, Llewellyn H	306
Pee, Frances W	40, 227, 363
Pennel, Andrew J (1)	254
Pennington, James F	353
Penrose, Joseph	208, 321, 358
Pilkington, Richard P	243
Piper, Lewis M	361
Piper, Luther R	125
Plessinger, Abraham	350
Porter, Andrew J	239
Potter, David R	342
Potter, James	304
Potts, Theodore B	358
Price, Daniel J	61, 292
Prideaux, Thomas A	292
Probst, George C	361
Ramsey, Alexander	282
Ramsey, Jeremiah	352
Ramsey, Oliver C	312, 361
Reamer, Francis C	348
Reed, William B	306
Rice, Abraham	143
Rice, Cornelius	178
Rice, Isaac	143
Richter, Adam	281
Ridenour, Jacob D	308
Riffle, Cyrus	150, So,D
Rightenour, Jacob	346
Riley, George W (1)	21
Riley, James H	280
Riley, John S	217
Riley, Reuben A	217, 338
Ritchey, Henry C	293
Ritchey, Jacob M	144

Name	Page(s)
Ritchey, James H	224
Ritchey, James T	193
Ritchey, John C	293
Ritchey, John N	314
Ritchey, William D	224, 359
Robertson, Hector A	332
Robinette, Amos	299
Robinette, Jeremiah	299
Robison, Henry C	355
Rock, George J	264
Roudabush, John M	249
Roush, James Levi	41, 213
Rowzer, George	209, 358
Ruby, Henry	332
Salkeld, Bernard	158
Salkeld, Jacob F	350
Salkeld, James W	158
Salkeld, John Nelson	348
Salkeld, Samuel W	341
Salkeld, Thomas L	22, 267
Sanderson, Samuel K	357
Sanderson, Theodore C	357
Saupp, Frank D	242
Saylor, William	144
Schetrompf, John F	334
Schetrompf, Peter C	334
Sellers, Frederick A	294
Shaffer, Abraham	283
Shaffer, George W	356
Shaffer, Harvey E	294
Shauf, Cornelius	199-P
Shauf, John	199
Shock, Daniel	306
Shoemaker, Benjamin	272
Shoemaker, George F	121
Showalter, Absolom	209
Shroyer, Jacob D	159
Shroyer, Joseph	159
Simmons, Thomas H	193
Simmons, William	254
Sipes, Dennis B	350
Skillington, Robert M	209
Slack, George	260
Slick, Josiah	336
Slick, Thomas W	119
Slick, William S	108
Sloan Samuel B	215
Sloan, Andrew J	215-W,M
Smith, Alexander C	340
Smith, Charles	316

Photograph Index

Name	Page(s)
Smith, David R	345
Smith, John W (1)	37
Smith, Joseph B	334
Smith, Joseph H	308
Smith, Josiah C	337
Smith, Robert C	241
Smith, Samuel H (2)	358
Smith, Samuel X	344
Smith, William H (3)	111, 260
Smith, Wilson S	351 (2)
Snyder, Christopher	160
Snyder, David F	295
Snyder, James	245
Snyder, John W (1)	179-B
Snyder, John W (2)	108
Snyder, Jonathan	96, 145
Snyder, Joseph N	179
Snyder, Leonard N	179
Snyder, William H	160
Souser, Henderson	282, 361
Spangler, William H	259
Sparks, Abraham J	348
Sparks, David W	346
Sparks, George W	210
Sparks, James H	309
Sparks, John	39
Sparks, John C	283, 359
Sparks, John E	350
Sparks, Silas H	283
Sparks, Solomon C	346
Sparks, Uriah	39
Sparks, William	53
Speece, Harry	357
Speice, Louis D	307
Sponsler, George W	25
Statler, Samuel F	67, 183, 202, 355, 359
Steckman, John A	340
Steele, David F	180-P
Steele, John W	326
Steele, Levi H	357
Stine, David	145
Stoner, Merrick A	347
Stotler, Marion	293
Stoutnour, James H	242
Streightif, Samuel	268
Stuckey, Elias B	294
Stuckey, John S	92, 294
Stuckey, Samuel	345
Stuckey, William H	15, 263

Name	Page(s)
Stufft, Jacob	299
Stultz, William	347
Summerville, Abner	146
Summerville, Robert	314
Swartz, Francis	242
Swope, George W	326
Thomas, Isaac	352
Tipton, Noah (1)	146
Tipton, Noah C	295
Troutman, Benjamin	331
Troutman, Francis T	354
Trueax, George M	346
Twigg, Brice	342
Valentine, Levi	343
VanOrmer, William W	43
Veatch, John G	227
Wagerman, Samuel (1)	334
Wagerman, William	147, 289
Wagner, Christian	361
Walker, Asahel	358
Walker, Isaac	308
Walker, Thomas G	181, 254, 358
Walker, William A (1)	181-B
Walter, James A	211
Waltz, Lewis B	18, 225
Wareham, John	265
Wareham, Joseph	342
Waters, Isaac O	341
Watson, William H	347
Weaver, John H	357
Weaverling, Adam	149
Weaverling, Jacob T	315
Weaverling, Thomas H	149
Weicht, Jeremiah	343
Weisel, William W	242
Welch, John	339
Wentz, Isaac	234
Weyandt, Joseph Y	341
Weyandt, Samual S	210
Weyant, Joseph	251
Whisel, William H	225, 360
Whitaker, John H	105
Whitaker, Joseph	254
Whited, David L	282
Whysong, Samuel	242
Wilkins, Ephraim	260
Wilkins, Harvey	329
Wilkinson, William (2)	301
Williams, Alvah R	108, 315

Name	Page(s)
Williams, Harrison P	79
Williams, Jacob	282
Williams, Joseph W	315
Wiser, Jonathan	342
Wishart, Harvey S	314, 362
Wishart, James	247
Wishart, Samuel	362
Witt, George	332
Wolf, Edmund	235
Wolf, Richard	73
Wolfhope, John	147
Wolford, Daniel	295
Wonder, John S	346
Wonderly, John B	357
Wonders, Daniel M	228
Woodcock, William L	211
Woy, David M	182-B
Woy, George	148
Woy, John W	182
Woy, Joseph	325
Wright, Charles C	80
Zembower, Josiah A	121

Acknowledgments

I found research for this book to be analogous to becoming a parent for the first time. You really have no idea what you are getting yourself into, and you are more than appreciative when someone provides a helping hand. A debt of gratitude is owed to many who made this book possible.

Large numbers of people were contacted through the Ancestry and Find-A-Grave websites. An overwhelming number of people responded to requests for pictures and information on their ancestors. To all who responded and contributed to this book, your enthusiastic support was a source of inspiration throughout this project.

Much time and many efforts are given by volunteers at our local historical societies and museums, to safeguard irreplaceable treasures and catalog innumerable historical documents. The enthusiastic support of Gillian Leach and everyone at the Bedford Historical Society was a crucial part of this project. Much information and many photographs were gained while researching their archives. I also want to acknowledge Gillian's efforts to acquire additional photographs and documents in possession of historical society members and others in our community.

The Bloody Run Historical Society provided significant numbers of photographs and information on individual soldiers. I would be remiss if I did not acknowledge a great local historian, Barbara Sponsler Miller, for her efforts and many contributions to this book.

There are limited numbers of pictures and letters in existence of Bedford County soldiers who died during or in the immediate aftermath of the Civil War. The Fort Bedford Museum provided photographs, correspondence, and information on some of these soldiers. Their support of my requests is appreciated.

I extend my gratitude for the photographs and information provided by Jeff Whetstone and Steve Hollingshead. Their book, "From Winchester to Bloody Run" set the bar on what a good local history book on the Civil War should be.

William Roy Mock provided information and photographs on an impressive listing of Civil War ancestors. William's work in Veterans organizations and his many good books on local history deserve to be recognized.

I am grateful to John Crider for sharing his knowledge of the Civil War. His efforts in supporting local Historical Societies, the preservation of GAR documents and support of Veterans causes, including arranging for replacement gravestones for local Civil War soldiers, are commendable.

Debra Topinka graciously provided feedback on some battlefield details in this book. It is certainly helpful to reach out to someone with such an impressive knowledge of the Civil War.

Last but not least, I want to thank my wife, Valinda, for walking with me on this journey for the last 4 years and reviewing the materials in this book.

Bibliography and Photograph Courtesy Listing

P. 1	Union soldiers along the west bank of the Rappahannock River at Fredericksburg, Virginia in 1863. Photograph courtesy of Library of Congress
	"Civil War Casualties." *American Battlefield Trust. (n.d)*. https://www.battlefields.org/learn/articles/civil-war-casualties
	"Civil War Deadlier Than Previously Thought?" *History.com,* A&E Television Networks, 31 August 2018, https://www.history.com/news/civil-war-deadlier-than-previously-thought
	"Civil War Weapons, Firearms, and Small Arms." *American Civil War,* Thomas Legion, (n.d.), http://www.thomaslegion.net/americancivilwar/civilwarweaponsfirearmssmallarms.html
	"The Guns, Rifles, Pistols, Muskets, Shotguns, and Weapons of the Union and Confederate Armies." *American Civil War,* Thomas Legion, (n.d.), http://www.thomaslegion.net/civil_war_small_arms_and_firearms.html
P. 2	Shoop, Isaac. "Small but Deadly: The Minié Ball." *The Gettysburg Compiler,* Civil War Institute, 20 April 2019, https://gettysburgcompiler.org/2019/04/30/small-but-deadly-the-minie-ball/
	"Minie Ball." *History.com,* A&E Television Networks, 21 Aug. 2018, https://www.history.com/topics/american-civil-war/minie-ball
	"Civil War Diseases." *Civil War Academy.com,* (n.d.), https://www.civilwaracademy.com/civil-war-diseases
	"American Civil War Disease Facts." *American Civil War Facts,* (n.d.), http://www.civil-war-facts.com/Interesting-Civil-War-Facts/American-Civil-War-Diseases-Facts.html
	Goellnitz, Jenny. "Civil War Medicine: An Overview of Medicine." *Ohio State University Department of History,* (n.d.), https://ehistory.osu.edu/exhibitions/cwsurgeon/cwsurgeon/introduction
	"Civil War Technology." *History.com,* A&E Television Networks, 21 Aug. 2018, https://www.history.com/topics/american-civil-war/civil-war-technology
	55th Regiment Reunion in 1913. Photograph courtesy of Mary Pearl Hunt, *Bedford Gazette Old Picture Album* (undated)
P. 3	Maps of Central and Eastern US. *D-maps.com. Modified by Kevin Mearkle*
	"77th Pennsylvania Regiment." *Pennsylvania Volunteers of the Civil War,* Pennsylvania Civil War Volunteers, (n.d.), http://www.pacivilwar.com/regiment/77th.html
P. 4	Boatner, Mark Mayo. *The Civil War Dictionary.* New York, Random House, 1959. pp. 610.
	"Civil War Index - PA Union Regiments" *Civil War Index,* (n.d.), https://civilwarindex.com/pennsylvania-regiments.html
P. 5	Map of Bedford County, Pennsylvania, United States with township and municipal boundaries. 11 Apr 2006. *US Census website*
	James Cleaver photograph courtesy of the Bedford County Courthouse; biographical information sourced on Ancestry.com
	Rev. Charles Cleaver information referenced in the Minutes of the 97th Session of the Baltimore Annual Conference of the Methodist Episcopal Church, 9-15 Mar. 1881.
P. 6	Over, D. "Patriotic Meetings." *Bedford Inquirer,* 26 Apr 1861, pp. 2. https://www.newspapers.com/image/329774674/?terms=bedford%20inquirer&match=1 . Accessed 23 Jan 2021
P. 7	Map of PA, MD, VA. *D-maps.com.* Modified by Kevin Mearkle
	"Two soldiers kneeling along Bull Run" photograph courtesy of Library of Congress
	"First Battle of Bull Run." *Britannica,* (n.d.), https://www.britannica.com/event/First-Battle-of-Bull-Run-1861
P. 9	Map of PA, MD, VA, SC. *D-maps.com. Modified by Kevin Mearkle*
	"George McClellan." *History.com,* A&E Television Networks, 10 Jun. 2019. https://www.history.com/topics/american-civil-war/george-b-mcclellan
P. 10	Map of PA, MD, VA. *D-maps.com.* Modified by Kevin Mearkle
	Drawing by Frank Leslie. *Famous Leaders and Battle Scenes of the Civil War*, New York, NY, Mrs. Frank Leslie, 1896.
	"First Battle of Kernstown - March 23, 1862." *National Park Service,* US Department of the Interior, 22 Aug. 2020, https://www.nps.gov/cebe/learn/historyculture/first-battle-of-kernstown.htm
	"Battle of Kernstown: Stonewall Jackson's Only Defeat." *American Battlefield Trust,* (n.d.), https://www.battlefields.org/learn/articles/battle-kernstown-stonewall-jacksons-only-defeat
P. 11	Thomas Livingston photograph courtesy of Barbara Sponsler Miller; biographical information sourced on Ancestry family trees
	Ezra Brisbin photograph courtesy of Robin Cucinotta collection; biographical information sourced on Ancestry family trees
P. 12	Map of SC. *D-maps.com.* Modified by Kevin Mearkle
	"Recently freed slaves at the Edisto Island, SC plantation of James Hopkinson in 1862." photograph courtesy of Library of Congress
	"Battle of Battle of Pocotaligo 21st, 22nd and 23rd. October 1862." *The Battle of Pocotaligo,* (n.d.), https://www.battleofpocotaligo.com/history.html
	Dyer, F. H. *A compendium of the War of the Rebellion Volume 1.* Des Moines, IA, Dyer Publishing Company, 1908. pp. 831.
	Moore, Frank and Everett, Edward. *The Rebellion Record: A Diary of American Events Vol. 4.* New York, GP Putnam, 1864. pp. 506-507.
	History of Bedford, Somerset and Fulton Counties, Pennsylvania. Chicago, Waterman, Watkins & Co. 1884, pp. 134-135.
P. 13	Theodore Klahre photograph courtesy of Ponder170 in Ancestry; biographical information sourced on Ancestry family trees
	Silas Gollipher photograph courtesy of Bedford Historical Society; biographical information sourced on Ancestry family trees
	John Claycomb photograph courtesy of Deb College; biographical information sourced on Ancestry family trees
P. 14	Map of MD, VA. *D-maps.com.* Modified by Kevin Mearkle
	"Union Troops at Fair Oaks minutes before a Confederate attack" photograph courtesy of Library of Congress
	"Peninsula Campaign." *History.com,* A&E Television Networks, 21 Aug. 2018, https://www.history.com/topics/american-civil-war/peninsula-campaign
	"Seven Pines Fair Oaks." *American Battlefield Trust,* (n.d.), https://www.battlefields.org/learn/civil-war/battles/seven-pines

Bibliography and Photograph Courtesy Listing 551

P. 14	*The Union army; a history of military affairs in the loyal states, 1861-65 - records of the regiments in the Union army.* Madison, WI, Federal Publishing Company, 1908. pp. 384-390.
	History of Bedford, Somerset and Fulton Counties, Pennsylvania. Chicago, Waterman, Watkins & Co. 1884, pp.146.
P. 15	Simon P Kegg photograph courtesy of Ronn Palm; biographical information sourced on Ancestry.com
	William H Stuckey photograph courtesy of Jaycet Pittman Collection, Jeff Whetstone, Steve Hollingshead; biographical information sourced on Ancestry family trees
P. 16	Map of VA. *D-maps.com.* Modified by Kevin Mearkle
	"Union field hospital during the 7 Days Battle" photograph courtesy of Library of Congress
P. 16	"Seven Days Battles Around Richmond." *American Civil War,* Thomas Legion, (n.d.), http://www.thomaslegion.net/seven_days_battles_around_richmond_civil_war.html
	History of Bedford, Somerset and Fulton Counties, Pennsylvania. Chicago, Waterman, Watkins & Co. 1884. pp.119-120.
P. 18	"Photograph of Ellerson's Mill" courtesy of Library of Congress
	Lewis B Waltz photograph courtesy of Ronn Palm; biographical information sourced on Ancestry family trees
P. 19	Eli Eichelberger photograph courtesy of Linda Bunch; biographical information sourced on Ancestry family trees
	Christian C Garlic photograph courtesy of Ronn Palm; biographical information sourced on Ancestry family trees
P. 20	Map of PA, MD, VA. D-maps.com. Modified by Kevin Mearkle
	"Children playing across from Calvary Troops at Sudley's Ford on the Bull Run battlefield" photograph courtesy of U. S. Army Military History Institute, Carlisle Barracks
	"Second Battle of Bull Run." *History.com,* A&E Television Networks, 11 Dec. 2019, https://www.history.com/topics/american-civil-war/second-battle-of-bull-run
	Leepson, Mark. "Second Battle of Bull Run." *Britannica,* (n.d.), https://www.britannica.com/event/Second-Battle-of-Bull-Run-1862
P. 21	George & Susan Foor Riley photograph courtesy of Bloody Run Historical Society; Bayart, Jsuriano88 in Ancestry; biographical information sourced on Ancestry family trees
P. 22	Map of PA, MA VA. *D-maps.com.* Modified by Kevin Mearkle
	Andrew J Foor photograph courtesy of ConnerFamilykkkk in Ancestry; biographical information sourced on Ancestry family trees
	Thomas L Salkeld photograph courtesy of Bonnie Salkeld Beglin; biographical information sourced on Ancestry family trees
	Hoptak, John D. "Battlefields And Beyond: Battle Of South Mountain." *Hisotrynet,* (n.d.), https://www.historynet.com/battle-of-south-mountain.htm
	Hartwig, Scott. "The Maryland Campaign of 1862." *American Battlefield Trust,* (n.d.), https://www.battlefields.org/learn/articles/maryland-campaign-1862
P. 23	Drawing of Union Troops charging Confederates holding the higher ground at Turners Gap during the Battle of South Mountain by H. Charles McBarron in 1862
P. 24	Map of PA, MD, VA. D-maps.com. Modified by Kevin Mearkle
	"Harper's Ferry in 1862" photograph courtesy of Library of Congress
	Hartwig, Scott. "The Maryland Campaign of 1862." *American Battlefield Trust,* (n.d.), https://www.battlefields.org/learn/articles/maryland-campaign-1862
P. 25	John Lowery photograph courtesy of Barbara Sponsler Miller, Carolyn Miller Carroll; biographical information provided by Barbara and Carolyn.
	George W Sponsler photograph courtesy of Barbara Sponsler Miller, Carolyn Miller Carroll; biographical information sourced on Ancestry family trees
P. 26	Map of PA, MD, VA. D-maps.com. Modified by Kevin Mearkle
	"Confederate dead lying near the Dunker Church in one of the most iconic images of the Civil War" photograph courtesy of Library of Congress
	Hartwig, Scott. "The Maryland Campaign of 1862." *American Battlefield Trust,* (n.d.), https://www.battlefields.org/learn/articles/maryland-campaign-1862
	"Antietam." *American Battlefield Trust,* (n.d.), https://www.battlefields.org/learn/articles/antietam
	"Farming the Cornfield: D. R. Miller's 1862 Harvest of Death." *Antietam's Cornfield,* 18 Oct. 2017, https://antietamscornfield.com/2017/10/18/farming-the-cornfield-d-r-millers-1862-harvest-of-death/
P. 27	Hartwig, Scott. "The Maryland Campaign of 1862." *American Battlefield Trust,* (n.d.), https://www.battlefields.org/learn/articles/maryland-campaign-1862
	"Antietam." *American Battlefield Trust,* (n.d.), https://www.battlefields.org/learn/articles/antietam
	"Farming the Cornfield: D. R. Miller's 1862 Harvest of Death." *Antietam's Cornfield,* 18 Oct. 2017, https://antietamscornfield.com/2017/10/18/farming-the-cornfield-d-r-millers-1862-harvest-of-death/
P. 28	"Knap's Pennsylvania Light Artillery Battery immediately after the battle of Antietam" photograph courtesy of Library of Congress
	Benjamin F Jamison photograph courtesy of Gaylord W Little; biographical information sourced on Ancestry family trees
	Michael B Breneman photograph and information courtesy of "Biographical review : containing life sketches of leading citizens of Bedford and Somerset Counties, Pennsylvania (1899)"
P. 29	Close up view of "Knap's Pennsylvania Light Artillery Battery immediately after the battle of Antietam" photograph courtesy of Library of Congress
	Frederick L Frazey photograph courtesy of jeagle00 in Ancestry; biographical information sourced on Ancestry family trees
	David Harclerode photograph courtesy of Thomas McGuire; biographical information sourced on Ancestry family trees
P. 30	Map of PA, MD, VA. D-maps.com. Modified by Kevin Mearkle
	"Marye's Heights" photograph courtesy of Library of Congress

P. 30	"Confederate dead at stone wall" photograph courtesy of Library of Congress
P. 32	Cangelosi, Glen C. "Eyewitness Account of Firing Upon the Enemy." *The Washington Artillery at Mayre's Heights, Confederate Hall Museum,* (n.d.), http://www.washingtonartillery.com/Killing%20a%20Man%20page.htm
	"Battle of Fredericksburg." *History.com,* A&E Television Networks, 11 Dec. 2019, https://www.history.com/topics/american-civil-war/battle-of-fredericksburg
	"Battle of Fredericksburg." *History.com,* A&E Television Networks, 11 Dec. 2019, https://www.history.com/this-day-in-history/battle-of-fredericksburg
	"Fredericksburg". *American Battlefield Trust,* (n.d.). https://www.battlefields.org/learn/articles/fredericksburg
P. 33	Jacob W Barndollar photograph courtesy of Ronn Palm; biographical information sourced on Ancestry family trees
	Snowberger, Ella M. "Civil War Days Recalled", *Recollections of by-gone days in the Cove,* Morrisons Cove Herald, 1933. pp.75-77.
	Joseph C Lambert photograph courtesy of RettaSell in Ancestry; biographical information sourced on Ancestry family trees
P. 35	Map of PA, MD, VA, SC. D-maps.com. Modified by Kevin Mearkle
	"Civil War Casualties." *American Battlefield Trust,* (n.d.), https://www.battlefields.org/learn/articles/civil-war-casualties
	"Gettysburg: High Watermark of the Confederacy." *U.S. History - Pre-Columbian to the New Millennius,* ushistory.org, (n.d.), https://www.ushistory.org/us/33g.asp
P. 35	"The war in 1863." *Britannica,* (n.d.), https://www.britannica.com/event/American-Civil-War/The-war-in-1863#ref1202563
P. 36	Map of PA, MD, VA. *D-maps.com.* Modified by Kevin Mearkle
	"110th Pennsylvania Infantry is pictured on April 24th, 1863 in Falmouth, VA" photograph courtesy of Library of Congress
	"Chancellorsville." *American Battlefield Trust,* (n.d.), https://www.battlefields.org/learn/civil-war/battles/chancellorsville
	"Battle of Chancellorsville." *History.com,* A&E Television Networks, 11 Dec. 2019, https://www.history.com/topics/american-civil-war/battle-of-chancellorsville
	"Robert E. Lee's Right Arm." *Warfare History Network, Sovereign Media,* 3 Feb. 2019, https://warfarehistorynetwork.com/2019/02/03/robert-e-lees-right-arm/
P. 37	Elias S Burket photograph courtesy of Jeanette - bamaburket in Ancestry; biographical information sourced on Ancestry family trees
	William B Blake photograph courtesy of Jeff Whetstone; biographical information sourced on Ancestry family trees
	John W Smith photograph courtesy of ladycca in Ancestry; biographical information sourced on Ancestry family trees
P. 38	Map of PA. D-maps.com. Modified by Kevin Mearkle
	"Union soldiers near McPherson's Woods on the 1st Day of Gettysburg " photograph courtesy of Library of Congress
	"Gettysburg Campaign." *American Battlefield Trust,* (n.d.), https://www.battlefields.org/learn/civil-war/gettysburg-campaign
	"1863-06-13 Winchester II." *Civil War Times,* 2016. http://www.civilwartroops.org/1863-06-13-winchester-ii/
	"Battle of Gettysburg." *Britannica, (n.d.),* https://www.britannica.com/event/Battle-of-Gettysburg
	"Pickett's Charge at Gettysburg." *American Civil War,* Thomas Legion, (n.d.), http://www.thomaslegion.net/americancivilwar/gettysburgpickettschargebattle.html
P. 39	Uriah & John Sparks photograph courtesy of Jack and Carolyn Sparks; biographical information sourced on Ancestry family trees
P. 40	Jonathan S Foor photograph courtesy of Foor Family Military History Book; biographical information sourced on Ancestry family trees
	Alfred Gracey photograph courtesy of Barbara Sponsler Miller; biographical information sourced on Ancestry family trees
	Frances W Pee photograph courtesy of Niels Witkamp; biographical information sourced on Ancestry family trees
	Benjamin Lohr photograph courtesy of Bedford Historical Society; biographical information sourced on Ancestry family trees
P. 41	James Levi Roush photograph courtesy of Bedford Historical Society; biographical information sourced on Ancestry family trees
	Beyer, Walter F. "Deeds of Valor - how America's heroes won the Medal of Honor". Detroit, Perrien-Keydel Company, 1901. pp.244.
	"Little Round Top is pictured in the middle background" photograph courtesy of Library of Congress
P. 42	"110th Pennsylvania Infantry, Company C in early 1863" photograph courtesy of Library of Congress
	"Confederate dead gathered for burial at the edge of the Rose Woods, July 5, 1863" photograph courtesy of Library of Congress
P. 43	John I Miller photograph courtesy of Bedford Historical Society; biographical information sourced on Ancestry family trees
	William W VanOrmer photograph courtesy of Ronn Palm; biographical information sourced on Ancestry family trees
	Americus Enfield photograph courtesy of Bedford Historical Society; biographical information sourced on Ancestry family trees
P. 44	Map of SC. D-maps.com. Modified by Kevin Mearkle
	"View of Ft. Wagner after being evacuated by the Confederates on September 7, 1863" photograph courtesy of Library of Congress
	"The Defenses of Fort Wagner." *American Battlefield Trust, (n.d.),* https://www.battlefields.org/learn/articles/defenses-fort-wagner
	Hickman, Kennedy. "American Civil War: Battles of Fort Wagner." ThoughtCo, Dotdash Publishing, 29 Aug. 2020, https://www.thoughtco.com/battles-of-fort-wagner-2360930
	"Assault on Fort Wagner - July 18, 1863." *American Battlefield Trust,* (n.d.), https://www.battlefields.org/learn/maps/assault-fort-wagner-july-18-1863
P. 45	William P Barndollar photograph courtesy of Barbara Sponsler Miller & Bloody Run Historical Society; biographical information sourced on Ancestry family trees
	"The Storming of Fort Wagner" was illustrated in an 1890 lithograph by Kurz and Allison
P. 46	Map of PA, MD, VA. D-maps.com. Modified by Kevin Mearkle
	"U.S. Signal Corps on a hill above the Rapidan River overlooking the Mine Run Battlefield" photograph courtesy of Library of Congress

Bibliography and Photograph Courtesy Listing 553

P. 46	Singel, Kati. "Mine Run Campaign." *Encycolpedia Virginia,* Virginia Humanities, (n.d.), https://www.encyclopediavirginia.org/Mine_Run_Campaign#start_entry
P. 47	Aaron Mock photograph courtesy of Charles Fetters; biographical information sourced on Ancestry family trees
	Aaron Mock Pension record compiled by Jim Huttinger, member of No. 333 G.A.R. Wright Post in 1988
P. 49	Map of PA, MD, VA. D-maps.com. Modified by Kevin Mearkle
	"I Will Send a Barrel of This Wonderful Whiskey to Every General in the Army." *Quote Investigator,* WordPress, (n.d.), https://quoteinvestigator.com/2013/02/18/barrel-of-whiskey/
	"Ulysses S. Grant's Overland Campaign: Six Bloody Weeks." *History.com,* A&E Television Networks, 21 Apr. 2020, https://www.history.com/news/grants-overland-campaign-civil-war
	"Civil War Casualties." *American Battlefield Trust,* (n.d.), https://www.battlefields.org/learn/articles/civil-war-casualties
	"Sheridan's 1864 Shenandoah Campaign." *Shenandoah Valley Battlefields,* 9 Jun. 2015, https://www.shenandoahatwar.org/history/sheridans-shenandoah-campaign-1864/
P. 50	Map of NC. D-maps.com. Modified by Kevin Mearkle
	"Photograph of the ironclad ship, the CSS Albemarle taken from the waterfront in Plymouth" photograph courtesy of Library of Congress
	"The Battle of Plymouth." *Washington County Historical Society,* (n.d.), http://portoplymouthmuseum.org/about-port-o-plymouth-museum/the-battle-of-plymouth/
	Martin, Jonathan. "Battle of Plymouth." *North Carolina History Project,* John Locke Foundation, (n.d.), https://northcarolinahistory.org/encyclopedia/battle-of-plymouth-1864/
P. 52	Alexander Compher photograph courtesy of Amber Cannon; biographical information sourced on Ancestry family trees
	"Field of Honor" photograph and information courtesy of Civil War Plymouth Pilgrims Descendants Society
P. 53	Jacob H Longenecker photograph courtesy of Chuck Fogtman; biographical information sourced on Ancestry family trees
	William Sparks photograph courtesy of Jack & Carolyn Sparks; biographical information sourced on Ancestry family trees
	David F Beegle photograph courtesy of Civil War Plymouth Pilgrims Descendant Society; biographical information sourced on Ancestry family trees
P. 54	Map of PA, MD, VA. D-maps.com. Modified by Kevin Mearkle
	"Battle of the Wilderness." *History.com, A&E Television Networks,* 21 Aug. 2018, https://www.history.com/topics/american-civil-war/battle-of-the-wilderness
	Adams, Simon. "Battle of the Wilderness." *Britannica,* (n.d.), https://www.britannica.com/event/Battle-of-the-Wilderness
	"The Wilderness." *American Battlefield Trust,* (n.d.), https://www.battlefields.org/learn/civil-war/battles/wilderness
	Morris, Roy Jr. "Battle of the Wilderness." *Historynet,* Apr 1997, https://www.historynet.com/battle-of-the-wilderness
P. 56	"Fire in the Wilderness" illustration courtesy of Library of Congress
	"Skulls and bones of unburied soldiers on the Wilderness Battlefield in 1865" photograph courtesy of Library of Congress
P. 57	Lewis A May photograph courtesy of Ronn Palm; biographical information sourced on Ancestry family trees
	David Horton photograph courtesy of Ronn Palm; biographical information sourced on Ancestry family trees
	Andrew B Biddle photograph courtesy of Ronn Palm; biographical information sourced on Ancestry family trees
P. 58	Map of PA, MD, VA. D-maps.com. Modified by Kevin Mearkle
	"Confederate entrenchments at the Muleshoe Salient" photograph courtesy of Library of Congress
P. 59	Horn, Joshua, "Battle of Spotsylvania – The Bloody Angle." *The Civil War 150th Blog,* 12 May, 2014, https://civilwar150th.blogspot.com/2014/05/battle-of-spotsylvania-bloody-angle.html
	"Battle of Spotsylvania Court House." *History.com,* A&E Television Networks, 21 May 2020, https://www.history.com/topics/american-civil-war/battle-of-spotsylvania-court-house
P. 60	Aaron Foster photograph courtesy of Shoppers Guide Weekend Edition, May 26 & 27, 2012. A Moment in History.
	Aaron Foster information courtesy of Shoppers Guide Weekend Edition, May 26 & 27, 2012. A Moment in History by Jon Baughman
	Aaron Foster additional biographical information sourced on Ancestry family trees
	"Attack at the Bloody Angle" painting by Thure De Thulstrup" courtesy of Library of Congress
P. 61	Thomas G Livingston photograph courtesy of Bedford Hist. Society; biographical information sourced on Ancestry family trees
	Daniel J Price photograph courtesy of Susan Recchia; biographical information sourced on Ancestry family trees
P. 62	Map of VA. D-maps.com. Modified by Kevin Mearkle
	"Confederate Fort Darling on Drewry's Bluff" photograph courtesy of Library of Congress
	Chick, Sean Michael. "Drewry's Bluff: Victory Without Satisfaction." *Emerging Civil War,* 18 Mar. 2020, https://emergingcivilwar.com/2020/03/18/drewrys-bluff-victory-without-satisfaction/
	"Drewry's Bluff." *National Park Service,* US Department of the Interior, 1 Feb. 2018, https://www.nps.gov/rich/learn/historyculture/drewrys-bluff.htm
	Coffey, Walter. "The Second Battle of Drewry's Bluff." *The Civil War Months,* 16 May, 2019, https://civilwarmonths.com/2019/05/16/the-second-battle-of-drewrys-bluff/
P. 63	"View from within Fort Darling of the James River taken during the Civil War" photograph courtesy of Library of Congress
P. 65	William A Boor photograph courtesy of adamegg in Ancestry; biographical information sourced on Ancestry family trees
	John B Amos photograph courtesy of Ann Sinton & Rachel Shultzaberger; biographical information sourced on Ancestry family trees
P. 66	"Union earthworks at the Bermuda Hundred near the Point of Rocks battlefield" photograph courtesy of Library of Congress

P. 67	Emanuel A Mock photograph courtesy of Bedford Historical Society - *Bedford Gazette Old Picture Album* (undated); biographical information sourced on Ancestry family trees
	Samuel F Statler photograph courtesy of Cliffhouse49 in Ancestry; biographical information sourced on Ancestry family trees
	"Bermuda Hundred Landing, Virginia" photograph courtesy of Library of Congress
P. 68	Map of PA, MD, VA. D-maps.com. Modified by Kevin Mearkle
P. 70	"Extreme left of Confederate lines at Cold Harbor" photograph courtesy of Library of Congress
P. 71	Grant, Ulysses S. "Personal Memoirs of U.S. Grant", New York, C.L. Webster Publishing, 1885–86. pp.264-278.
	Thompson, Robert N. "Battle of Cold Harbor." *HistoryNet*, Nov. 2006, https://www.historynet.com/cold-harbor
	James H Miller photograph courtesy of Ronn Palm; biographical information sourced on Ancestry family trees
P. 72	George D Brown photograph courtesy of Jacquie1649 in Ancestry; biographical information sourced on Ancestry family trees
	"The killing fields of the Cold Harbor battlefield after the Civil War had ended in 1865" photograph courtesy of Library of Congress
P. 73	Richard Wolf photograph courtesy of Edward Schoenberger; biographical information sourced on Ancestry family trees
	Allen Cobler photograph courtesy of Ronn Palm; biographical information sourced on Ancestry family trees
	Gabriel Burket photograph courtesy of Barbara Sponsler Miller; biographical information sourced on Ancestry family trees
P. 74	Map of PA, MD, VA. D-maps.com. Modified by Kevin Mearkle
	"Pontoon bridge across the James River " photograph courtesy of Library of Congress
	Calkins, Chris. "The Wearing Down of Lee's Army." *American Battlefield Trust,* (n.d.), https://www.battlefields.org/learn/articles/petersburg-wearing-down-lees-army#:~:text=%22We%20must%20destroy%20this%20Army,fear%20is%20becoming%20a%20reality.
P. 75	"The initial Assaults." *Petersburg Siege.org,* (n.d.), http://www.petersburgsiege.org/assaults.htm
	Laidig, Scott. "U.S. Grant and Operations." *Ohio State University eHistory,* Department of History, (n.d.), https://ehistory.osu.edu/articles/us-grant-and-operations-0
P. 76	Petersburg Map by Kevin Mearkle
	John A Livingston photograph courtesy of Ronn Palm; biographical information sourced on Ancestry family trees
	Christian Harr photograph courtesy of Bedford Historical Society - *Bedford Gazette Old Picture Album (undated);* biographical information sourced on Ancestry family trees
P. 78	John and Mary Elizabeth McChesney photograph courtesy of A—kwm16637 in Ancestry; biographical information sourced on Ancestry family trees
	"Union troops standing on the captured ground of outer Confederate lines in front of Petersburg during the 1st Assault" photograph courtesy of Library of Congress
P. 79	Zophar P and Mary Horton photograph courtesy of Deloresdetwiler4 in Ancestry; biographical information sourced on Ancestry family trees
	Harrison P Williams family photograph courtesy of irish_Michael in Ancestry; biographical information sourced on Ancestry family trees
P. 80	Petersburg Map by Kevin Mearkle
	Clews, C. "1864-06-21 Jerusalem Plank Road." *Civil War Times, (n.d.),* http://www.civilwartroops.org/1864-06-21-jerusalem-plank-road/
	"The Battle of Jerusalem Plank Road: June 21-24, 1864." *The Siege of Petersburg Online,* (n.d.), http://www.beyondthecrater.com/resources/bat-sum/petersburg-siege-sum/second-offensive-summaries/the-battle-of-jerusalem-plank-road-summary/
	"The Battle of the Jerusalem Plank Road." *Petersburg Siege.org,* (n.d.), https://www.petersburgsiege.org/jplank.htm
	Charles C Wright photograph courtesy of Bedford Historical Society; biographical information sourced on Ancestry family trees
P. 82	Petersburg Map by Kevin Mearkle
	"The Crater" photograph courtesy of Library of Congress
	"The Crater." *American Battlefield Trust,* (n.d.), https://www.battlefields.org/learn/civil-war/battles/crater
	"The Battle of the Crater." *Historic Petersburg Foundation,* (n.d.), http://www.historicpetersburg.org/the-battle-of-the-crater/
	"Calamity in the Crater." *American Battlefield Trust,* (n.d.), https://www.battlefields.org/learn/articles/calamity-crater
P. 83	Mickley, Jeremiah Marion. *The Forty-Third Regiment United States Colored Troops,* Gettysburg, J. E. Wible-Printer, 1866.
	"USCT Troops at Petersburg" photograph courtesy of Library of Congress
P. 84	Petersburg Map by Kevin Mearkle
	"Blick House" photograph courtesy of National Park Service
	Green, Wilson A. "The Fight for the Weldon Railroad." *American Battlefield Trust,* (n.d.), https://www.battlefields.org/learn/articles/fight-weldon-railroad
	Hickman, Kennedy. "Battle of Globe Tavern." *ThoughtCo, Dotdash Publishing,* 6 Mar. 2017, https://www.thoughtco.com/battle-of-globe-tavern-2360928#:~:text=The%20Battle%20of%20Globe%20Tavern%20was%20fought%20August,the%20Battle%20of%20Globe%20Tavern%20they%20were%20successful.
P. 85	Abel Griffith photograph courtesy of Barbara Sponsler Miller; biographical information sourced on Ancestry family trees
	James A Grove photograph courtesy of PeggyKelleyMack50 in Ancestry; biographical information sourced on Ancestry family trees
P. 86	Petersburg Area Map by Kevin Mearkle
	"A Union soldier on picket duty in front of Petersburg" photograph courtesy of National Archives
	OR XLII P1 #229: Report of Maj. Gen. Gregg. (4 Nov. 1864). *Official Army Report*

P. 86	OR XLII P1 #127: Report of Brig. Gen. Griffin (29 Oct. 1864). *Official Army Report*
	"The Battle of Boydton Plank Road: October 27-28, 1864." *The Siege of Petersburg Online,* (n.d.), http://www.beyondthecrater.com/resources/bat-sum/petersburg-siege-sum/sixth-offensive-summaries/the-battle-of-boydton-plank-road-october-27-28-1864/
P. 87	Isaiah Collins photograph courtesy of Huguette23 in Ancestry; biographical information sourced on Ancestry family trees
	John G Feaster photograph courtesy of Donna Miller; biographical information sourced on Ancestry family trees
P. 88	Map of PA, MD, VA. D-maps.com. Modified by Kevin Mearkle
	"Monocacy Junction" photograph courtesy of Library of Congress
	Quint, Ryan. "A General Redeemed: Lew Wallace and the Battle of Monocacy." *Emerging Civil War,* 23 Sep. 2013, https://emergingcivilwar.com/2013/09/23/a-general-redeemed-lew-wallace-and-the-battle-of-monocacy/
	"Battle of Monocacy Junction. *Exploring Off the Beaten Path,* (n.d.), http://exploringoffthebeatenpath.com/Battlefields/Monocacy/
	"Monocacy National Battlefield. *National Park Service,* US Department of the Interior, (n.d.), https://unidescription.org/account/project/export/620
P. 89	George R Imler photograph courtesy of Bedford Historical Society; biographical information sourced on Ancestry family trees
	Chauncey Owens photograph courtesy of Ronn Palm; biographical information sourced on Ancestry family trees
P. 90	Map of PA, MD, VA. D-maps.com. Modified by Kevin Mearkle
	"War's Largest Cavalry Charge" 1886 Chromolithograph by Louis Prang & Company courtesy of Library of Congress
	Hickman, Kennedy. "American Civil War: Third Battle of Winchester (Opequon)." *ThoughtCo,* Dotdash Publishing, 26 Aug. 2020, https://www.thoughtco.com/third-battle-of-winchester-opequon-2360265
	"Greatest Charges of the Civil War." *Amercian Battlefield Trust,* (n.d.), https://www.battlefields.org/learn/articles/greatest-charges-civil-war
P. 91	John A Felton photograph courtesy of Silas Felton; biographical information sourced on Ancestry family trees
P. 92	Marcus May photograph courtesy of facebook@138thPVI; biographical information sourced on Ancestry family trees
	John S Stuckey photograph courtesy of Bedford Historical Society Stuckey and Huffman Cousins History book; biographical information sourced on Ancestry family trees
	Joseph Mowry photograph courtesy of Ronn Palm; biographical information sourced on Ancestry family trees
P. 93	John Border photograph courtesy of Ronn Palm; biographical information sourced on Ancestry family trees
	William B Filler photograph courtesy of Ronn Palm; biographical information sourced on Ancestry family trees
P. 94	Map of PA, MD, VA. D-maps.com. Modified by Kevin Mearkle
	"Cedar Creek Belle Grove." *American Battlefield Trust, (n.d.),* https://www.battlefields.org/learn/civil-war/battles/cedar-creek
	"Overview of the Battle of Cedar Creek." *National Park Service,* US Department of the Interior, 11 Nov. 2017, https://www.nps.gov/cebe/learn/overview-of-the-battle-of-cedar-creek.htm
	Hickman, Kennedy. "Battle of Cedar Creek." *ThoughtCo,* Dotdash Publishing, 26 Aug. 2020, https://www.thoughtco.com/battle-of-cedar-creek-2360937
	"Sheridan's Ride" Thure de Thulstrup painting courtesy of Library of Congress
P. 95	"Members of the Sheridan's Veterans Association" photograph courtesy of Library of Congress
P. 96	Abraham L Carpenter photograph courtesy of Edward Schoenberger; biographical information sourced on Ancestry family trees
	Isaiah Collins photograph courtesy of Mason323 in Ancestry; biographical information sourced on Ancestry family trees
P. 97	George Clark photograph courtesy of Jeffry Burden; biographical information sourced on Ancestry family trees
	Martin S Bortz photograph courtesy of Carol Waugerman Stout; biographical information sourced on Ancestry family trees
	Battlefield photograph courtesy of U.S. Army Military History Institute, Carlisle
P. 99	Map of Petersburg to Appomattox by Kevin Mearkle
	Laidig, Scott. "U.S. Grant and Operations." *Ohio State University eHistory,* History Department, (n.d.), https://ehistory.osu.edu/articles/us-grant-and-operations-0
	Young, Neely. "Sherman's March To The Sea." *GeorgiaTrend,* 1 Dec. 2014, https://www.georgiatrend.com/2014/12/01/shermans-march-to-the-sea/
	"Earthwork fortifications created by both the Union and Confederate armies at Petersburg" photograph courtesy of Library of Congress
P. 100	Map of Petersburg by Kevin Mearkle
	"Fort Steadman picket lines" photograph courtesy of Library of Congress
	"Today in the Petersburg Campaign: March 25, 1865." *The Siege of Petersburg Online,* (n.d.), https://www.beyondthecrater.com/resources/tipc/march-1865/march-25-1865/
	Ray, Fred L. "Pre-Dawn Assault on Fort Stedman." *American Battlefield Trust, (n.d.),* https://www.battlefields.org/learn/articles/americas-civil-war-pre-dawn-assault-fort-stedman
	"Fort Stedman." *PetersburgSeige,* (n.d.), http://www.petersburgsiege.org/stedman.htm
P. 101	"View from outside of Ft. Steadman" photograph courtesy of National Archives
P. 102	Philip V Hollar photograph courtesy of Rachael Hall; biographical information sourced on Ancestry family trees
	Cyrus W Fickes photograph courtesy of Maj. Frank E. Thompson; biographical information sourced on Ancestry family trees
P. 103	"Two views from inside Fort Steadman" photographs courtesy of Library of Congress
P. 104	Map of Petersburg area by Kevin Mearkle
	"Captured Confederate Soldiers at Five Forks" photograph courtesy of Library of Congress

P. 104	White, Kristopher D. "The Downfall of a Federal Corps Commander: Warren-Sheridan and the Five Forks Controversy: Part Three" *Emerging Civil War*, 1 Apr. 2015, https://emergingcivilwar.com/2015/04/01/the-downfall-of-a-federal-corps-commander-warren-sheridan-and-the-five-forks-controversy-part-three/
	"The Battle of Lewis's Farm: March 29, 1865." *The Siege of Petersburg Online*, (n.d.), http://www.beyondthecrater.com/resources/batsum/petersburg-siege-sum/ninth-offensive-summaries/the-battle-of-lewiss-farm-march-29-1865/
P. 105	John H Whitaker photograph courtesy of Bedford Historical Society; biographical information sourced on Ancestry family trees
	Josiah Hissong photograph courtesy of Jeff Hissong, Tom & Christina Turner and ruthelliott155 in Ancestry; biographical information sourced on Ancestry family trees
P. 106	Map of Petersburg by Kevin Mearkle
	"The Third Battle of Petersburg: April 2, 1865." *The Siege of Petersburg Online*, (n.d.), http://www.beyondthecrater.com/resources/batsum/petersburg-siege-sum/ninth-offensive-summaries/the-third-battle-of-petersburg-april-2-1865/
	"The Final Assaults and the Fall of Petersburg." *Internet Archive Wayback Machine*, 19 Oct. 2020, http://web.archive.org/web/20060620044748/http://members.aol.com/siege1864/final.html
	Guttman, John. "The Man Who Shot A.P. Hill." *HistoryNet*, American's Civil War, Jan. 2010, https://www.historynet.com/man-shot-p-hill.htm
	Alexander, Edward S. "Mapping the Attack on Fort Mahone, April 2, 1865." *Emerging Civil War*, 13 Oct. 2017, https://emergingcivilwar.com/2017/10/13/mapping-the-attack-on-fort-mahone-april-2-1865/
P. 107	Josiah & Benjamin Garretson photographs courtesy of Charles Garretson & Becky Garretson Perigo; biographical information sourced on Ancestry family trees
	Becky Garretson Perigo: Garretson brothers letters on the events of 2 Apr. 1865.
P. 108	"Dead Confederate soldiers at Ft. Mahone" photographs courtesy of Library of Congress
	William S Slick photograph courtesy of Danielervinusa in Ancestry; biographical information sourced on Ancestry family trees
	Alva R Williams photograph courtesy of Dan Layton; biographical information sourced on Ancestry family trees
P. 109	"Ft. Gregg the Confederate Alamo" photograph courtesy of Library of Congress
	Fox III, John J. "Confederate Alamo: Outnumbered rebels at Petersburg." *HistoryNet*, Feb. 2017, https://www.historynet.com/confederate-alamo-outnumbered-rebels-petersburg.htm
	Alexander, Edward S. "Shoot and Be Damned" *Emerging Civil War*, 12 Nov. 2017, https://emergingcivilwar.com/2017/11/16/shoot-and-be-damned-lawrence-berry-at-fort-gregg/
	O.R. XLII P3 #089 - *Hezekiah Hammer Promotion - Official Army Report*
	Hezekiah Hammer photograph courtesy of Kelly Allison; biographical information sourced on Ancestry family trees
P. 110	Map of Petersburg to Appomattox by Kevin Mearkle
	"Sailor's Creek Battlefield" photograph courtesy of Library of Congress
	Hickman, Kennedy. " Battle of Sayler's Creek." *ThoughtCo*, Dotdash Publishing, 26 Aug. 2020, https://www.thoughtco.com/battle-of-saylers-creek-2360935
	"10 Facts: Sailor's Creek." *American Battlefield Trust*, (n.d.), https://www.battlefields.org/learn/articles/10-facts-sailors-creek
	Calkins, Chris M. "Battle of Sailor's Creek." *HistoryNet* , Civil War Times Magazine, Jan. 2006, https://www.historynet.com/battle-of-sailors-creek.htm
P. 111	William H Smith photograph courtesy of patsmith1869 in Ancestry; biographical information sourced on Ancestry family trees
	Oliver Horton photograph courtesy of Cyndi Young; biographical information sourced on Ancestry family trees
P. 112	Map of Petersburg to Appomattox by Kevin Mearkle
	"McLean's House at Appomattox in 1865" photograph courtesy of Library of Congress
	Hundon, Miles. "Battle of Appomattox Court House." *Britannica*, (n.d.), https://www.britannica.com/event/Battle-of-Appomattox-Court-House
	"Appomattox Campaign." *HistoryNet*, (n.d.), https://www.historynet.com/appomattox-campaign
	"Robert E. Lee surrenders." *History.com*, A&E Television Networks, 7 Apr. 2020, https://www.history.com/this-day-in-history/robert-e-lee-surrenders
P. 113	"Unidentified Union Soldiers during the Confederate Surrender at Appomattox" photograph courtesy of Library of Congress
P. 114	Isaac Nicodemus photograph courtesy of facebook@138thPVI; biographical information sourced on Ancestry family trees
	Henry I Claar photograph courtesy of Ronn Palm; biographical information sourced on Ancestry family trees
P. 115	Thomas H Faber photograph courtesy of Fort Bedford Museum; biographical information sourced on Ancestry family trees
	Frederick Feight photograph courtesy of Sheilaw142 in Ancestry; biographical information sourced on Ancestry family trees
P. 116	Map of GA. D-maps.com. Modified by Kevin Mearkle
	"The 1864 Andersonville Prison Camp Atrocity." *Spartacus Educational*, (n.d.). https://spartacus-educational.com/USACWandersonville.htm
	"Horrors in Georgia." *History Collection*, (n.d.), https://historycollection.com/horrors-andersonville-prison-civil-wars-worst-pow-camp/2/
	Skoch, George. "Andersonville Prison Camp." *HistoryNet*, Civil War Times Magazine, Oct. 2007, https://www.historynet.com/andersonville-prison-camp
	"Unidentified emaciated Andersonville POW after his release in 1865" photograph courtesy of Library of Congress

P. 117	Samuel J McEldowney photograph courtesy of Bayart in Ancestry; biographical information sourced on Ancestry family trees
P. 118	"Rations being issued in Andersonville Prison on 17 Aug. 1864" photograph courtesy of Library of Congress
P. 119	Thomas W Slick photograph courtesy of Civil War Plymouth Pilgrims Descendants Society; biographical information sourced on Ancestry family trees
	Jacob England photograph courtesy of DaBost in Ancestry; biographical information sourced on Ancestry family trees
P. 120	"Bird's eye view photograph of Andersonville Prison taken on 17 Aug. 1864" photograph courtesy of Library of Congress
P. 121	Josiah A Zembower photograph courtesy of Mspsgt1 & EdwardAskey in Ancestry and John Zembower; biographical information sourced on Ancestry family trees
	George F Shoemaker photograph courtesy of Michael Mattix & Garry Barney; biographical information sourced on Ancestry family trees
P. 122	"Andersonville Prison Camp in 1864" photograph courtesy of Library of Congress
P. 123	"Photograph of the Burial of Prisoners at Andersonville taken on 17 Aug. 1864" photograph courtesy of U. S. Army Military History Institute, Carlisle Barracks
P. 124	Warren G Harding address at the Tomb of Unknown Soldier, 11 Nov. 1921
P. 125	Luther R Piper photograph courtesy of Ronn Palm; biographical information sourced on Ancestry family trees
	"A Creature of Its Time: The Pension Bureau." *Clara Barton Museum, 6 Apr. 2017,* https://www.clarabartonmuseum.org/pension/
	Ephrat Livni and Dan Kopf, "The decline of the large US family, in charts." *Quartz Media,* 11 Oct. 2017, US Census Data. https://qz.com/1099800/average-size-of-a-us-family-from-1850-to-the-present/
P. 126	William M Amick photograph courtesy of Ronn Palm; biographical information sourced on Ancestry family trees
	Joseph Armstrong photograph courtesy of Fort Bedford Museum; biographical information sourced on Ancestry family trees
P. 127	William J Baughman photograph courtesy of Barbara Sponsler Miller; biographical information sourced on Ancestry family trees and Barbara Sponsler Miller
	George Baughman photograph courtesy of Barbara Sponsler Miller; biographical information sourced on Ancestry family trees and Barbara Sponsler Miller
P. 128	William Beaver photograph courtesy of JeanneDiehl42 in Ancestry; biographical information sourced on Ancestry family trees
	William W Beltz photograph courtesy of srobertsond in Ancestry; biographical information sourced on Ancestry family trees
P. 129	John Berkheimer photograph courtesy of Fort Bedford Museum; biographical information sourced on Ancestry family trees
	Frederick Burket photograph courtesy of Richard & Sandy Allison; biographical information sourced on Ancestry family trees
P. 130	Joel Clark photograph courtesy of Thomas E VanHorn; biographical information sourced on Ancestry family trees
	Chauncey Corle photograph courtesy of Darlene Boggs; biographical information sourced on Ancestry family trees
P. 131	Francis Corle photograph courtesy of Eric Berkhimer; biographical information sourced on Ancestry family trees
	Abraham Darr photograph courtesy of Ronn Palm; biographical information sourced on Ancestry family trees
P. 132	John Defibaugh photograph courtesy of Eve Smith; biographical information sourced on Ancestry family trees
	Jacob J Dibert photograph courtesy of Bedford Historical Society; biographical information sourced on Ancestry family trees
P. 133	Benjamin Eshelman photograph courtesy of Deborah Snow Pascoe; biographical information sourced on Ancestry family trees
	James M Fickes photograph courtesy of Jeff Hissong; biographical information sourced on Ancestry family trees
P. 134	Francis L Foor photograph courtesy of Bedford Historical Society; biographical information sourced on Ancestry family trees
	Benjamin Garretson photographs courtesy of Charles Garretson & Becky Garretson Perigo; biographical information sourced from Charles Garretson & Becky Garretson Perigo
P. 135	Daniel G Heltzel photograph courtesy of Michael Dudley Greer; biographical information sourced on Ancestry family trees
	Andrew J Himes photograph courtesy of the late Mildred Himes Lawson, Hope Creighton & Shaun Creighton; biographical information sourced on Ancestry family trees
P. 137	Isaac M Imler photograph courtesy of Bedford Historical Society; biographical information sourced on Ancestry family trees
	George V A Juda photograph courtesy of Ronn Palm; biographical information sourced on Ancestry family trees
P. 139	Henry Layton photograph courtesy of Gillian K Leach; biographical information sourced on Ancestry family trees & Kevin Mearkle
	Malachi B Mock photograph courtesy of Bedford Historical Society & *Bedford Gazette Old Picture Album (undated);* biographical information sourced on Ancestry family trees
P. 141	Abraham Morgart photograph courtesy of Historical Data Systems; biographical information sourced on Ancestry family trees
	John Moyer photograph courtesy of Charles Robert Alcaraz Moyer; biographical information sourced on Ancestry family trees
P. 143	Isaac & Abraham Rice photograph courtesy of Kevin Mearkle; biographical information sourced on Ancestry family trees & Kevin Mearkle
P. 144	Jacob M Ritchey photograph courtesy of Joanne Fesler; biographical information sourced on Ancestry family trees
	William Saylor photograph courtesy of Carol Saylor Scott; biographical information sourced on Ancestry family trees
P. 145	Jonathan Snyder photograph courtesy of Mason323 in Ancestry; biographical information sourced on Ancestry family trees
	David Stine photograph courtesy of Jkgamoyer in Ancestry; biographical information sourced on Ancestry family trees
P. 146	Abner Summerville photograph courtesy of NancySomer in Ancestry; biographical information sourced on Ancestry family trees
	Noah Tipton photograph courtesy of Collectornuts in Ancestry; biographical information sourced on Ancestry family trees
P. 147	William Wagerman photograph courtesy of Carol Waugerman Stout; biographical information sourced on Ancestry family trees
	John Wolfhope photograph courtesy of John Halliday; biographical information sourced on Ancestry family trees
P. 148	George Woy photograph courtesy of Jeanette - bammaburkett in Ancestry; biographical information sourced on Ancestry family trees
P. 149	Thomas & Adam photographs courtesy of Barbara Sponsler Miller; biographical information sourced on Ancestry family trees and Barbara Sponsler Miller
P. 150	Leonard J Foor photograph courtesy of frmartin77 in Ancestry; biographical information sourced on Ancestry family trees

P. 150	Cyrus Riffle, Eve Stevens Custer and Jacob Stevens Jr. photograph courtesy of juditha196 in Ancestry; biographical information sourced on Ancestry family trees
P. 151	David Carpenter photograph courtesy of Bedford Historical Society; biographical information sourced on Ancestry family trees
	Curtis and Delilah photograph courtesy of Robert Peck; biographical information sourced on Ancestry family trees
P. 152	Edwards photographs courtesy of Gari Jensen; biographical information sourced on Ancestry family trees
P. 153	Holler photographs courtesy of Aimee Stout Benitez; biographical information sourced on Ancestry family trees
P. 155	Dewalt, Peter Sr, Peter Jr and John Dwalt photographs courtesy of Bedford Historical Society and the "A History of Jacob Kinsey book"; biographical information sourced on Ancestry family trees
	Benjamin Kinsey photograph courtesy of BCSDMD4 in Ancestry; biographical information sourced on Ancestry family trees
P. 156	Messersmith photographs courtesy of Sandra Messersmith Millin; biographical information sourced on Ancestry family trees
P. 157	Nelson photographs courtesy of Stephanie Perry; biographical information sourced on Ancestry family trees
P. 158	Salkeld photographs courtesy of Bonnie Salkeld Beglin; biographical information sourced on Ancestry family trees
P. 159	Joseph Shroyer photograph courtesy of Bud Evans; biographical information sourced on Ancestry family trees
	Jacob Shroyer photograph courtesy of John Lewis; biographical information sourced on Ancestry family trees
P. 160	Christopher & Catherine Snyder photographs courtesy of Jim Snyder Jr; biographical information sourced on Ancestry family trees
	William Snyder photograph courtesy of marciecallan & capitalman in Ancestry; biographical information sourced on Ancestry family trees
P. 161	Civil War Pension Records of William, James and Samuel McDonald
	Ann Croft McDonald photograph courtesy of Rebekkah Jackson; biographical information sourced on Ancestry family trees
P. 162	John, Sarah McDonald & children photographs courtesy of Rebekkah Jackson; biographical information sourced on Ancestry family trees
	Margaret McDonald Shimer photograph courtesy of Rebekkah Jackson
	Clara Annie memoir provided by Rebekkah Jackson
	Hannah Spotts McDonald photograph courtesy of Sallys32 in Ancestry.com
P. 163	Nancy Anna Schrader photograph courtesy of Craig B Adams; biographical information sourced on Ancestry family trees
P. 164	Elizabeth and Susan Anderson photograph courtesy of Reif Hammond; biographical information sourced on Ancestry family trees
P. 165	Anna Beltz Kipp photograph courtesy of Kenneth McClellan; biographical information sourced on Ancestry family trees
	William W Beltz photograph courtesy of srobertsond in Ancestry.com; biographical information sourced on Ancestry family trees
P. 166	Enos Bennett photograph courtesy of Bedford Historical Society; biographical information sourced on Ancestry family trees
P. 167	Isaac Bowser photograph courtesy of Wmreichart in Ancestry.com; biographical information sourced on Ancestry family trees
	Moses Bowser photograph courtesy of Mary Jo Martin; biographical information sourced on Ancestry family trees
P. 168	Lewis and Isaac Conner photographs courtesy of Ronn Palm; biographical information sourced on Ancestry family trees
P. 169	Joseph and Mary Crist Corle photographs courtesy of S_Murphy in Ancestry.com; biographical information sourced on Ancestry family trees
P. 170	James P Davis photograph courtesy of Ronn Palm; biographical information sourced on Ancestry family trees
P. 171	Hannah Miller Garretson photograph courtesy of Karen Clites Shampine; biographical information sourced on Ancestry family trees
P. 172	Rachel and Uriah Gordon photographs courtesy of Darlene Boggs; biographical information sourced on Ancestry family trees
P. 173	Hainsey brothers pension documentation provided by Jim Snyder Jr
	Elizabeth and Adam Hainsey photographs courtesy of Jim Snyder Jr; biographical information sourced on Ancestry family trees
P. 174	Samuel and Margarette McGee photographs courtesy of Gpatricia24 in Ancestry.com; biographical information sourced on Ancestry family trees
P. 175	John and Mary Anne Mellott photograph courtesy of Mary Lou McElhaney; biographical information sourced on Ancestry family trees
P. 176	Miller photographs courtesy of cavaleto in Ancestry.com; biographical information sourced on Ancestry family trees
P. 177	Wilson Nycum photograph courtesy of Bedford Historical Society and Nycum Family History Book; biographical information sourced on Ancestry family trees
P. 178	Jane and Cornelius Rice photograph courtesy of Kevin Mearkle; biographical information sourced on Ancestry family trees & Kevin Mearkle
P. 179	Snyder photographs courtesy of mason323 in Ancestry.com; biographical information sourced on Ancestry family trees
P. 180	Fannie & Solomon Steele photograph courtesy of Slhinish in Ancestry.com; biographical information sourced on Ancestry family trees
P. 181	Thomas G Walker photograph courtesy of Ronn Palm; biographical information sourced on Ancestry family trees
P. 182	John W Woy photograph courtesy of Photosbysan in Ancestry.com; biographical information sourced on Ancestry family trees
P. 183	Samuel Statler photographs courtesy of Fort Bedford Museum and Cliffhouse49 in Ancestry.com; biographical information sourced on Ancestry family trees
	Samuel Statler Obituary, *Bedford Gazette - 20 Jun. 1936*
	"Children in the Civil War." *Digital History*, (n.d.), http://www.digitalhistory.uh.edu/active_learning/explorations/children_civilwar/child_soldiers.cfm
	Davis, Burk. *The Civil War: Strange & Fascinating Facts.* Random House, Dec. 1988. pp.65.
	"Child soldiers in the American Civil War." *Wikipedia*, 25 Apr. 2021, https://en.wikipedia.org/wiki/Child_soldiers_in_the_American_Civil_War
	Grove, Greg. "The Youngest and Oldest Soldiers of the Civil War." *gregsegroves.blogspot.com*, 13 Jul. 2013, https://gregsegroves.blogspot.com/2013/07/the-youngest-boy-wounded-in-civil-war.html
P. 184	Jacob R Callihan photograph courtesy of Pittsburgh Post-Gazette 13 Jun 1915 obituary; biographical information sourced on Ancestry family trees

P. 184	Thomas Drenning photograph courtesy of melboys33 in Ancestry.com; biographical information sourced on Ancestry family trees
P. 185	Jacob Clevenger photograph courtesy of Asperjt in Ancestry.com; biographical information sourced on Ancestry family trees
	Daniel Edwards photograph courtesy of GariJensen in Ancestry.com; biographical information sourced on Ancestry family trees
P. 186	Americus Enfield photograph courtesy of JNicholson in Findagrave.com; biographical information sourced on Ancestry family trees
	Henry Estep photograph courtesy of Vicki4801 in Ancestry.com; biographical information sourced on Ancestry family trees
P. 187	George Gracey photograph courtesy of Stacychuck1 in Ancestry.com; biographical information sourced on Ancestry family trees
	Joseph Gordon photograph courtesy of Roy C Wilson; biographical information sourced on Ancestry family trees
P. 188	Daniel Hetrick photograph courtesy of Bedford Historical Society; biographical information sourced on Ancestry family trees
	Oliver Hollingshead photograph courtesy of Steve Hollingshead; biographical information sourced on Ancestry family trees
P. 189	Moses Humbert photograph courtesy of Ldevl8 in Ancestry.com; biographical information sourced on Ancestry family trees
	James and Elizabeth Knupp Kegg photograph courtesy of Mary Lynne Kelley; biographical information sourced on Ancestry family trees
P. 190	William Lehman photograph courtesy of Jlehman1597 in Ancestry.com; biographical information sourced on Ancestry family trees
	Thaddeus Lingenfelter photograph courtesy of jkgamoyer in Ancestry.com; biographical information sourced on Ancestry family trees
P. 191	Joseph H Lysinger photograph courtesy of "History of Hamilton and Clay Counties Nebraska" compiled by Dale P Stough - Chicago ; biographical information sourced on Ancestry family trees
	Francis McCoy photograph courtesy of Bedford Historical Society; biographical information sourced on Ancestry family trees
P. 192	David Murrie photograph courtesy of 4michaelsimon in Ancestry.com; biographical information sourced on Ancestry family trees
	Josiah Pearson photograph courtesy of Neal Jones; biographical information sourced on Ancestry family trees
P. 193	James Ritchey photograph courtesy of Gina Girmes; biographical information sourced on Ancestry family trees
	Thomas Simmons photograph courtesy of Nora Calhoun; biographical information sourced on Ancestry family trees
P. 195	Theophilus/Nathaniel Gates photograph courtesy of Kyler Gates; biographical information sourced on Ancestry family trees
	"Survivor of Famous Bucktail Regiment is Still Active at 93." *Altoona Tribune*, 2 Jun. 1923, pp.13
	"A Patriotic Family Nathaniel Gates." *Altoona Tribune*, 8 Aug. 1900, pp.7
	"The Blind Preacher Nathaniel R. Gates." *Altoona Tribune*, 3 Oct. 1878, pp.2
	Reimer, Terry. "Recruiting Exams and Disqualifications for Military Service." National Museum of Civil War Medicine, 9 Nov. 2002, https://www.civilwarmed.org/surgeons-call/exams/
	"Life expectancy (from birth) in the US from 1860 to 2020." *Statista*, (n.d.), https://www.statista.com/statistics/1040079/life-expectancy-united-states-all-time/
	Downie, James, "How Much Did A Civil War Soldier Carry?" *The New Republic*, 12 Apr. 2011, https://newrepublic.com/article/86592/fort-sumter-civil-war-soldiers-march-weight
	Commanger, Henry, "How The Civil War Soldiers Marched." *CivilWarHome*, (n.d.), https://www.civilwarhome.com/soldiersmarch.html
P. 196	Samuel and Susanna Cox photograph courtesy of Dolores Myers; biographical information sourced on Ancestry family trees
	Freer, Edward K. "The Lost Children of the Alleghenies." *Bedford Gazette, 1975*
P. 197	James Madara photograph courtesy of SLJ in Findagrave.com; biographical information sourced on Ancestry family trees, Altoona Tribune Obituary, 8 May 1879; History of Bedford, Somerset and Fulton Counties, Pennsylvania. Chicago, Waterman, Watkins & Co. 1884, pp. 350.
P. 198	Susan and John McFerren photographs courtesy of Barbara Jones Simmons; biographical information sourced on Ancestry family trees
P. 199	Mary and John Shauf photograph courtesy of altshauf in Ancestry.com; biographical information sourced on Ancestry family trees
P. 201	Jonas and Elizabeth Imler photograph courtesy of Bedford Historical Society, *Bedford Gazette Old Photo Album (undated)*, originally submitted by Pauline Crist ; biographical information sourced on Ancestry family trees
	"Jonas Imler Obituary." Altoona Tribune, 9 Sep. 1938
	Biographical review : containing life sketches of leading citizens of Bedford and Somerset Counties, Pennsylvania, Boston, Biographical Review Pub. Co, 1899, pp.62
	"Drummer Boy Albert Henry Woolson." *Geni,* (n.d.), https://www.geni.com/people/Drummer-Boy-Albert-Woolson/6000000021851668584
	Bellis, Mary. "Photography Timeline." *ThoughtCo*, Dotdash Publishing, 23 Jan. 2020, https://www.thoughtco.com/photography-timeline-1992306
	"U.S. Railroad History: A Timeline." *American Rails,* (n.d.),https://www.american-rails.com/history.html
	Mabee, Carleton. "Samuel F.B. Morse American artist and inventor." *Britannica*, 23 Apr. 2021, https://www.britannica.com/biography/Samuel-F-B-Morse
	Bellis, Mary. "The Most Important Inventions of the 19th Century." *ThoughtCo*, Dotdash Publishing, 26 Jul. 2019, https://www.thoughtco.com/inventions-nineteenth-century-4144740
P. 202	Samuel Statler photograph courtesy of Cliffhouse49 in Ancestry.com; biographical information sourced on Ancestry family trees
	Americus Enfield and William B Filler photograph courtesy of Bedford Historical Society, *Bedford Gazette Old Photo Album (undated)*, biographical information sourced on Ancestry family trees
P. 203	Martin L Myers photograph courtesy of Barbara Sponsler Miller; biographical information sourced on Ancestry family trees
	Winfield Scott Conrad photograph courtesy of mrpadrat_1 in Ancestry.com; biographical information sourced on Ancestry family trees
P. 204	George Bond photograph courtesy of SUEBEE in Findagrave.com; biographical information sourced on Ancestry family trees
	Henry and Hester Brant photograph courtesy of Bedford Historical Society, *Bedford Gazette Old Photo Album (undated);* biographical information sourced on Ancestry family trees

Bibliography and Photograph Courtesy Listing

P. 205	David Crist photograph courtesy of Lindamiller382 in Ancestry.com; biographical information sourced on Ancestry family trees
	Samuel Fluke photograph courtesy of Randy Gladstone; biographical information sourced on Ancestry family trees
	Thomas Dicken family photograph courtesy of DebHarrington43 in Ancestry.com; biographical information sourced on Ancestry family trees
P. 206	Allen Harbaugh photograph courtesy of Bedford Historical Society; biographical information sourced on Ancestry family trees
	William R Hasenpat photograph courtesy of John Risdon; biographical information sourced on Ancestry family trees
	George Henry photograph courtesy of Carol Gandrud; biographical information sourced on Ancestry family trees
	Jonathan Horton photograph courtesy of Jen Styskal Faust; biographical information sourced on Ancestry family trees
P. 207	Samuel King photograph courtesy of CHQ21 in Ancestry.com; biographical information sourced on Ancestry family trees
	Jacob Manges photograph courtesy of Richard Ringler; biographical information sourced on Ancestry family trees
P. 208	McGregor family photograph courtesy of *Bedford Gazette Old Picture Album (undated)*; biographical information sourced on Ancestry family trees
	Jacob Mellott photograph courtesy of Bedford Historical Society; biographical information sourced on Ancestry family trees
	Joseph Penrose photograph courtesy of Carole Carlson; biographical information sourced on Ancestry family trees & Recollections of the Underground Railroad by Joseph Penrose, Fishertown, Bedford County, Pennsylvania, written in 1904
P. 209	Robert M Skillington photograph courtesy of Jill Hocking; biographical information sourced on Ancestry family trees
	George Rowzer photograph courtesy of Bedford Historical Society, *Bedford Gazette Old Photo Album (undated)*; biographical information sourced on Ancestry family trees
	Absolom Showalter photograph courtesy of Philip Brambley; biographical information sourced on Ancestry family trees
P. 210	George Sparks photograph courtesy of Eugene Reynolds; biographical information sourced on Ancestry family trees
	Samuel and Mary Weyandt photograph courtesy of Sharon Snyderburn; biographical information sourced on Ancestry family trees
P. 211	James Walter family photograph courtesy of Mlehman4128 in Ancestry.com; biographical information sourced on Ancestry family trees
	William Woodcock photograph courtesy of Altoona Tribune 9/18/35 Obituary; biographical information sourced on Ancestry family trees
P. 213	James Levi Roush photograph courtesy of Congressional Medal of Honor Society
	Pennsylvania Reserves charging toward the Wheatfield Illistration by Franklin Briscoe
P. 213	Sterner, Doug. "Hometown Heroes of the Keystone State." *Pennsylvania Medal of Honor Recipients*. 2008, http://www.homeofheroes.com/moh/states/pa.html
	Doughty, Heather, and Mary M. Geis. "Medal of Honor Recipients." *Commonwealth of PA*, 10 Nov. 1994
	Williams, Hershel. "Take Out the Enemy: Reflections on the Mears Party's Valor", *Hallowed Ground Magazine, 1 Mar. 2019*
	Beyer, Walter F. *Deeds of Valor: From Records in the Archives of the United States, Volume 1*, Perrien-Keydel Co., 1907, pp.244-245
P. 214	Abraham Kerns Arnold photograph courtesy of Asinton in Ancestry.com; biographical information sourced on Ancestry family tree and Historical Data Systems
	"United States Army Medal of Honor", 11 Sep. 2020, https://www.army.mil/medalofhonor/citations1.html
	Kurz & Allison lithograph Battle of Todd's Tavern, 1886, Library of Congress
	"The Battle of Todd's Tavern", *National Park Service*, Department of the Interior, 31, Mar. 2012, https://www.nps.gov/frsp/planyourvisit/ttav.htm
P. 215	Photograph of Shy's Hill courtesy of Library of Congress
	Samuel Sloan photograph courtesy of Pikespeak1012 in Ancestry.com; biographical information sourced on Ancestry family trees
	Mary Ann "Polly" Cuppett Sloan picture is courtesy of OctApril in Ancestry.com; biographical information sourced on Ancestry family trees
	Martha Sterling photograph courtesy of Roger Handke; biographical information sourced on Ancestry family trees & Roger Handke
	"Roster and Records of Iowa Soldiers, War of the Rebellion." *Iowa In the Civil War*, (n.d.), http://iagenweb.org/civilwar/books/logan/mil404.htm
	Roger Handke, *Rescued From Obscurity: The Story Of A Genuine American Hero, unpublished research paper*, 12 Feb. 2015, https://www.ancestry.com/mediaui-viewer/collection/1030/tree/7381795/person/24888276886/media/2c21ca49-2f45-460a-8640-8f6cefe1b23d?_phsrc=Ota937&usePUBJs=true
P. 217	Reuben A Riley photograph courtesy of saylesworth1& NJMcM in Ancestry.com; biographical information sourced on Ancestry family trees & Find A Grave
	John S Riley photograph courtesy of *The Daily Ardmoreite*, Ardmore, OK, 11 Aug. 1907; biographical information sourced on Ancestry family trees and The Daily Ardmoreite
	David H Knee photograph courtesy of Larry Knee and Jack Yocum; biographical information sourced on Ancestry family trees
	Knee, Larry. "Knees in the Civil War." Larry Knee Publisher, 2008.
	Philip Knee photograph courtesy of Jack Yocum
P. 218	William Raley Albright photograph courtesy of Bedford Historical Society; biographical information sourced from Jeff Whetstone and Ancestry family trees
	"The Manassas Campaign, Virginia, 16-22 July 1861 - 114th Virginia Militia." *Firstbullrun.co.uk*, Jonathan Soffe, (n.d.), http://firstbullrun.co.uk/Shenandoah/virginia-militia.html

Bibliography and Photograph Courtesy Listing 561

P. 218	"Confederate Units-Militia." *West Virginia - The Other History*, (n.d.), https://sites.google.com/site/wvotherhistory/confederate-units-militia
	Hawks, Steve A. "The Battle of Gettysburg.", *Stone Sentinels,* (n.d.), https://gettysburg.stonesentinels.com/union-monuments/pennsylvania/pennsylvania-cavalry/18th-pennsylvania-cavalry/
P. 219	William Hinson photograph courtesy of Bob Grenke; biographical information sourced from Ancestry.com
	Glessner, Rusty. "Exploring the 1806 Old Log Church in Bedford County", *PA Bucket List,* (n.d.), https://pabucketlist.com/exploring-the-1806-old-log-church-in-bedford-county/
	William Hinson - Civil War information on marker at gravesite
	"Galvanized Yankees." *Civil War Soldier Search,* 19 Feb. 2015, http://civilwarsoldiersearch.com/galvanized-yankees.html
	"The Galvanized Yankees." *National Park Service,* U.S. Department of the Interior, Jul. 1992, https://www.nps.gov/jeff/learn/historyculture/upload/galvanized_yankees.pdf
	"6 Southern Unionist Strongholds During the Civil War." *History.com,* A&E Television Networks, 19 Feb. 2015, http://www.history.com/news/history-lists/6-unionist-strongholds-in-the-south-during-the-civil-war
	"1st Alabama Cavalry." (n.d.), http://www.1stalabamacavalryusv.com/Default.aspx
	O'Brein, Sean Michael. *Mountain Partisans: Guerrilla Warfare in the Southern Appalachians 1861-1865,* Praeger Publishing, Sep 1999, pp.92.
	John C Latta photograph courtesy of Jack and Carolyn Sparks; biographical information sourced on Ancestry family trees
P. 220	GAR voter roll listing at Bedford Historical Society
P. 221	Soldier holding flag photograph courtesy of the Libruary of Congress
	"America's War's." *Department of Veterans Affairs,* Nov. 2020, https://www.va.gov/opa/publications/factsheets/fs_americas_wars.pdf
P. 222	Eichelberger Store in Saxton photograph courtesy of Bedford Historical Society
	History of Bedford, Somerset and Fulton Counties, Pennsylvania. Chicago, Waterman, Watkins & Co. 1884, pp.120-122.
	Photograph and Article courtesty of the *Bedford Gazette (undated)*
P. 224	Ritchey brothers photograph courtesy of Charles Coulter; biographical information sourced on Ancestry family trees
P. 225	Eli Eichelberger photograph courtesy of Bedford Historical Society; biographical information sourced on Ancestry family trees
	David B Armstrong photograph courtesy of Historical sketches of Morrisons Cove 1933 by C. W Karns ; biographical information sourced on Ancestry family trees
	Jacob R Callahan photograph courtesy of Ronn Palm; biographical information sourced on Ancestry family trees
	George W Amick photograph courtesy of Bedford Historical Society; biographical information sourced on Ancestry family trees
	Lewis Waltz photograph courtesy of Historical Data Systems; biographical information sourced on Ancestry family trees
	William H Whisel photograph courtesy of JoLaughlin82 in Ancestry.com; biographical information sourced on Ancestry family trees
P. 226	"Fredericksburg - Slaughter Pen Fighting." *American Battlefield Trust,* (n.d.), https://www.battlefields.org/learn/maps/fredericksburg-slaughter-pen-fighting-december-13-1862
	11th Pennsylvania Regiment." *Pennsylvania Civil War Volunteers,* (n.d.), http://www.pacivilwar.com/regiment/11th.html
	"Antietam - The West Woods." *American Battlefield Trust,* (n.d.), https://www.battlefields.org/learn/maps/antietam-west-woods-september-17-1862-6am-7am
P. 227	"Gettysburg Sculptures." (n.d.), http://gettysburgsculptures.com/11th_pennsylania_infantry_monument/e_a_kretschman_sculptor_of_the_11th_pa_infantry_monument
	11th PA Infantry Photograph courtesy of gettysburg.stonesentinels.com
	Colin McLaughlin photograph courtesy of Wcfancyfab in Ancestry.com; biographical information sourced on Ancestry family trees
	Frances & Sarah Pee photograph courtesy of Benmertz in Ancestry.com; biographical information sourced on Ancestry family trees
	John & Adaline Veach photograph courtesy of Reedja77414 in Ancestry.com; biographical information sourced on Ancestry family trees
P. 228	"55th Pennsylvania Regiment." *Pennsylvania Civil War Volunteers,* (n.d.), http://www.pacivilwar.com/cwpa55history.html
	John A Livingston photograph courtesy of Ronn Palm
	Daniel M Wonders photograph courtesy of Ronn Palm
P. 234	Samuel Adams photograph courtesy of Craig B. Adams
	John B Amos photograph courtesy of Barbara Sponsler Miller
	Isaac Wentz photograph courtesy of Bedford Historical Society
P. 235	Albin C. Arnold photograph courtesy of Midas2b in Ancestry.com
	Jacob Burket photograph courtesy of Ronn Palm
	Edmund Wolf photograph courtesy of amandadawn1986 of Ancestry.com
	David H Darr photograph courtesy of Craig B. Adams
	William Cessna photograph courtesy of Ronn Palm
	Isaac Ling photograph courtesy of Bedford Historical Society
P. 236	Josiah Edwards photograph courtesy of GariJensen in Ancestry.com
	James B Diehl photograph courtesy of Marilyn Larson
	Samuel J Diehl photograph courtesy of BoweK52 in Ancestry.com

P. 236	Jacob J Dibert photograph courtesy of Rick Darby
P. 237	Isaac S Fleegle photograph courtesy of Darlam59 in Ancestry.com
	Theophilus R Gates photograph courtesy of Bedford Historical Society - undated Gazette Old Picture Album picture
	Hezekiah Hammer photograph courtesy of Johann Frantz Hammer Palantine Pioneer book by Lowell Varner Hammer
	Josiah P Garretson photograph courtesy of Charles Garretson & Becky Garretson Perigo
P. 238	Joseph U Gordon photograph courtesy of Bedford Historical Society
	Adam Gardner photograph courtesy of Bedford Historical Society
	Samuel B Hunt photograph courtesy of Bedford Historical Society
	Josiah Haley photograph courtesy of Jeff Hissong
	Josiah & Elizabeth Hissong photograph courtesy of Ruthelliott155 in Ancestory.com; biographical information sourced on Ancestry family trees
P. 239	Peter Kinsey Jr photograph courtesy of Scaramuccia - Ancestry.com
	James Polk Kegg photograph courtesy of Mary Lynne Kelley
	Thomas H Lyons drawing courtesy of *Bedford Gazette Obituary - 24 Jan. 1913*
	Andrew & Matilda Porter photograph courtesy of quiltinchicken in Ancestry.com; biographical information sourced on Ancestry family trees
	Philip Leonard photograph courtesy of Jeff Hissong
P. 240	Henry G Lybarger photograph courtesy of Pauline Nicodemus
	Philip Leonard photograph courtesy of Kim Bollinger
	John and Mary Livingston photograph courtesy of Pcoley50 in Ancestry.com; biographical information sourced on Ancestry family trees
	George and Sarah Manges photograph courtesy of John Williams; biographical information sourced on Ancestry family trees
P. 241	Joseph C May photograph courtesy of Jeff Hissong
	Edward E Mower photograph courtesy of Jeff Hissong
	Jacob B Peck photograph courtesy of vivkifretz in Ancestry.com
	Robert C Smith photograph courtesy of Melissa Moorhead Flynn
P. 242	Frank D Saupp photograph courtesy of Pittsburgh Sunday Post Obituary - Dec. 30th, 1906
	Frances Swartz photograph courtesy of barbarapfeifferswartz62 & stufferdarwin in Ancestry.com
	James H Stoutnour photograph courtesy of Bloody Run Historical Society & Barbara Sponsler Miller
	Samuel Whysong photograph courtesy of Bedford Historical Society
	William W Weisel photograph courtesy of Jeff Hissong
P. 243	Richard P Pilkington photograph courtesy of Ronn Palm
	George W Cessna photograph courtesy of Tjohnson610 in Ancestry.com
	Hawks, Steve A. "76th PA Vol. Infantry Reg't." *The Civil War in the East,* (n.d.), https://civilwarintheeast.com/us-regiments-batteries/pennsylvania/76th-pennsylvania-infantry/
P. 245	John Gephart photograph courtesy of Gephart in Ancestry.com
	Adam R Hainsey photograph courtesy of Jim Snyder Jr
	James Snyder photograph courtesy of Jim Snyder Jr
P. 246	Kelly, Martin. "10 Deadliest U.S. Civil War Battles." *thoughtco.com,* Dotdash Publishing, https://www.thoughtco.com/ten-bloodiest-civil-war-battles-104527
	"77th Pennsylvania Regiment." *Pennsylvania Civil War Volunteers,* (n.d.), http://www.pacivilwar.com/regiment/77th.html
P. 247	James Wishart photograph courtesy of Michael Schwartz
	George M Dively photograph courtesy of Mdiveley1954 in Ancestry.com
	Erastus Black photograph courtesy of Dgrenade10181 in Ancestry.com+E618
	Martin Dively photograph courtesy of Vincent Dively
P. 248	"82nd Pennsylvania Regiment." *Pennsylvania Civil War Volunteers,* (n.d.), http://www.pacivilwar.com/regiment/82nd.html
P. 249	John & Elizabeth Roudabush photograph courtesy of Richard Stewart; biographical information sourced on Ancestry family trees
	Daniel, Mary & son Riley Mellott photograph courtesy of Donna Morris; biographical information sourced on Ancestry family trees
	Jacob Corley photograph courtesy of Eland63 in Ancestry.com
	Daniel H May photograph courtesy of www.minerd.com/annualreview2018.htm
	David Corley photograph courtesy of Garner, KS Historical Society
P. 250	"84th Pennsylvania Regiment." *Pennsylvania Civil War Volunteers,* (n.d.), http://www.pacivilwar.com/regiment/84th.html
P. 251	Jason Harbaugh photograph courtesy of Bedford Historical Society
	Knipple brothers photograph courtesy of K. Ann Cornwall
	Joseph Weyant photograph courtesy of Jeff Whetstone
P. 252	"91st Pennsylvania Regiment." *Pennsylvania Civil War Volunteers,* (n.d.), http://www.pacivilwar.com/regiment/91st.html
P. 253	John G Feaster photograph courtesy of Donna Miller
	William S Berkheimer photograph courtesy of Bedford Historical Society
	William M Ickes photograph courtesy of Bedford Historical Society
	Adam Ickes photograph courtesy of Nathan D. Linn
P. 254	William Simmons photograph courtesy of Kerry Maddox
	Thomas G Walker photograph courtesy of Bedford Historical Society
	Joseph Whitaker photograph courtesy of Glassfused93 in Ancestry.com
	Andrew J Pennel photograph courtesy of Bedford Historical Society & *Bedford Gazette Old Photograph Album (undated)*

P. 255	Samuel C Oster photograph courtesy of Terry Sautter
	Philip & Sarah Clark photograph courtesy of Steve Hollingshead & Kristina Covalt Aldred; biographical information sourced on Ancestry family trees
	Henry Bennett and Ann Perdew Bennett photograph courtesy of *Bedford Gazette Old Photograph Album - 1 Oct. 1973*; biographical information sourced on Ancestry family trees
	Thomas Himes photograph courtesy of Ashley McQuillen
P. 256	"99th Pennsylvania Regiment." *Pennsylvania Civil War Volunteers*, (n.d.), http://www.pacivilwar.com/regiment/99th.html
P. 257	Valentine Bowser photograph courtesy of GLENDAKUELPER60 in Ancestry.com
	Thomas Claar photograph courtesy of Rich & Sandy Allison
P. 258	George Fluke photograph courtesy of Slhinish in Ancestry.com
	Fred & Sophia Corle photograph courtesy of Deb College; biographical information sourced on Ancestry family trees
	David T Crist drawing courtesy of Lindamiller382 in Ancestry.com & Jeff Hissong
	Abraham A Baughman photograph courtesy of Benmertz in Ancestry.com
P. 259	David S Mearkle photograph courtesy of Kevin Mearkle
	Peter Nunemaker photograph courtesy of jonathanpardew in Ancestry.com
	William H Spangler photograph courtesy of Laloba27 in Ancestry.com
	John Cook photograph courtesy of Matthew W. Cumberledge
	Michael Feather photograph courtesy of Ronn Palm
P. 260	George Slack photograph courtesy of Bedford Historical Society
	William & Maria Smith photograph courtesy of Patsmith1869 in Ancestry.com
	William S Millin photograph courtesy of Sandra Messersmith Millin
	Ephraim Wilkins photograph courtesy of Sandra Messersmith Millin
P. 261	"101st Pennsylvania Veteran Volunteer Infantry." (n.d.), https://www.101stpa.com/index.php/history/
	"101st Pennsylvania Regiment." *Pennsylvania Civil War Volunteers*, (n.d.), http://www.pacivilwar.com/regiment/101st.html
P. 263	John Defibaugh photograph courtesy of Bedford Historical Society
	William H Stuckey photograph courtesy of Ronn Palm
	Officers of 101st photograph courtesy of Civil War Plymouth Pilgrims Descendants Society
P. 264	Jacob H Longenecker photograph courtesy of History of the 101st regiment, Pennsylvania veteran volunteer infantry 1861-1865 book
	George J Rock photograph courtesy of Bedford Historical Society
	Andrew J Mills photograph courtesy of Bedford Historical Society
	Alexander & Barbara Compher photograph courtesy of Amber Cannon; biographical information sourced on Ancestry family trees
	Samuel Heffner photograph courtesy of Ronn Palm
P. 265	*History of Bedford, Somerset and Fulton Counties, Pennsylvania*. Chicago, Waterman, Watkins & Co. 1884, pp.149-150.
	"107th Pennsylvania Regiment." *Pennsylvania Civil War Volunteers*, (n.d.), http://www.pacivilwar.com/regiment/107th.html
	Hawks, Steve A. "107th PA Vol. Infantry Reg't." *The Civil War in the East*, (n.d.), https://civilwarintheeast.com/us-regiments-batteries/pennsylvania/107th-pennsylvania-infantry/
	John Wareham photograph courtesy of Historical Data Systems
	James A Grove photograph courtesy of Bedford Historical Society
P. 267	Thomas L Salkeld photograph courtesy of Bonnie Beglin
	John W Nycum photograph courtesy of Nycum Family History Book - Bedford Historical Society
	William Brant photograph courtesy of Ggeiselman in Ancestory.com
	Barton A Cooper photograph courtesy of the Everett Press in 1898 Obituary
P. 268	Adam & Emma Leonard photograph courtesy of Karenswegleholt in Ancestry.com; biographical information sourced on Ancestry family trees
	Samuel Streightif photograph courtesy of Theresa in Findagrave.com
	Jeremiah Foor family picture courtesy of Guy Calhoun
P. 269	110th Gettysburg Stature courtesy of Michael Mazaika
	Pennsylvania State Memorial at Gettysburg photographs courtesy of gettysburg.stonesentinels.com
	"Battle of Kernstown: Stonewall Jackson's Only Defeat." *American Battlefield Trust*, (n.d.), https://www.battlefields.org/learn/articles/battle-kernstown-stonewall-jacksons-only-defeat
	"110th Pennsylvania Regiment." *Pennsylvania Civil War Volunteers*, (n.d.), http://www.pacivilwar.com/regiment/110th.html
P. 271	Joseph Gates picture courtesy of Mike Ford
	Ezra D Brisbin picture courtesy of Robin Cucinotta Collection
P. 272	John S Border photograph courtesy of Squitchel in Ancestry.com
	Benjamin Shoemaker photograph courtesy of SandraKBreighne in Ancestry.com
	David Defibaugh photograph courtesy of Collectornuts in Ancestry.com
	Samuel Fockler photograph courtesy of NLM in Ancestry.com
	Benjamin Jamison family photograph courtesy of Gaylord W. Little
P. 273	Gettysburg Photograph courtesy of history culture images
	Hawks, Steve A. "125th PA Vol. Infantry Reg't." *The Civil War in the East*, (n.d.), https://civilwarintheeast.com/us-regiments-batteries/pennsylvania/125th-pennsylvania-infantry/
	Hawks, Steve A. "125th PA Vol. Infantry Reg't." *Stone Sentinels*, (n.d.), https://antietam.stonesentinels.com/monuments/pennsylvania/125th-pennsylvania/
	"Col Jacob Higgins' Official Report." *Antietam on the Web*, (n.d.), https://antietam.aotw.org/exhibit.php?exhibit_id=343
	William C Kean photograph courtesy of Irened153 in Ancestry.com

P. 273	David Harclerode photograph courtesy of Jil in Findagrave.com
P. 274	Andrew J Dodson photograph courtesy of Vava59 in Ancestry.com
	Thomas M Barr photograph courtesy of 125th Regimental History Book in Ancestry.com
P. 275	History of Bedford, Somerset and Fulton Counties, Pennsylvania. Chicago, Waterman, Watkins & Co. 1884, pp.152-153.
	"133rd Pennsylvania Regiment." *Pennsylvania Civil War Volunteers*, (n.d.), http://www.pacivilwar.com/regiment/133rd.html
	Martin Moser picture courtesy of Sheila May Wolf
	John W Fisher picture courtesy of Gaylord Little
P. 278	James J Barndollar picture courtesy of Barbara Sponsler Miller
	Martin D Barndollar picture courtesy of Barbara Sponsler Miller
	William Durno photograph courtesy of Bess Jr in Ancestry.com
	William & Sophia Amick photograph courtesy of Donna Miller; biographical information sourced on Ancestry family trees
	James B Butts photograph courtesy of janekilgore7 in Ancestry.com
P. 279	Ezekiel W Gaster photograph courtesy of *Bedford Gazette Obituary - 20 Apr. 1900*
	Harvey Grubb photograph courtesy of Kristina Covalt
	Jonathan B Edwards photograph courtesy of Kwm16637 in Ancestry.com
	Valentine Fink photograph courtesy of Blair County Genealogical Society
	George W McDaniel photograph courtesy of Bedford Historical Society - Gazette Old Picture Album (undated)
	James H Foor photograph courtesy of Hshelley1958 in Ancestry.com
	Henry & Catherine Fisher photograph courtesy of Gaylord W. Little; biographical information sourced on Ancestry family trees
	Lewis Conner photograph courtesy of Larry Conner
P. 280	Nycum family photograph courtesy of aubstrap in Ancestry.com; biographical information sourced on Ancestry family trees
	James M Nevitt photograph courtesy of Bedford Historical Society
	James & Sarah Riley photograph courtesy of Loisbillco in Ancestry.com
P. 281	Josiah McClellan photograph courtesy of Eltootall in Ancestry.com
	Adam Richter photograph courtesy of Squitchel in Ancestry.com
	McDaniel family photograph courtesy of Janina Hawley; biographical information sourced on Ancestry family trees
P. 282	David & Rebecca Whited photograph courtesy of Roman2007 in Ancestry.com
	Jacob Williams photograph courtesy of Harde5046 in Ancestry.com
	Alexander Ramsey photograph courtesy of Birdwatcher20 in Ancestry.com
	Henderson Souser photograph courtesy of Bedford Historical Society - Gazette Old Picture Album (undated)
	Benjamin Over photograph courtesy of Bedford Historical Society - Gazette Old Picture Album (undated)
P. 283	John W Johnston photograph courtesy of Rich Allison
	Lucy and David Dunkle photograph courtesy of Deborah Dunkle-Jones
	Humphrey's Division photograph courtesy of National Park Service And Rob Shenk
	Silas H Sparks photograph courtesy of Jack Sparks
	Abraham Shaffer courtesy of Kathy Williams
	John C Sparks photograph of likeness displayed on his gravestone
P. 284	"Battle of the Wilderness - May 6th." *American Battlefield Trust*, (n.d.), https://www.battlefields.org/learn/maps/battle-wilderness-may-6-1864
	History of Bedford, Somerset and Fulton Counties, Pennsylvania. Chicago, Waterman, Watkins & Co. 1884, pp.154-156.
	Christopher P Calhoun photograph courtesy of Ronn Palm
	Lewis A May photograph courtesy of Terry Alwine
P. 288	William C Corle photograph courtesy of Glen Gordon Collection
	George W Ickes photograph courtesy of Bedford Historical Society
	Isaac Gordon photograph courtesy of Darlene Boggs
	Abraham Carpenter photograph courtesy of Russ Middleton
	John N Burket photograph courtesy of Rob Hall
	Conrad Claycomb photograph courtesy of Bedford Historical Society
P. 289	Franklin Corle photograph courtesy of Bedford Historical Society
	William Wagerman & Captain Martin Bortz photograph courtesy of Carol Waugerman Stout
	John Getty photograph courtesy of Shari Mock
	James & Louisa Bivens photograph courtesy of RogerG77 in Ancestry.com
P. 290	Emanuel Fisher photograph courtesy of Oldergraver in Findagrave.com
	John M Holler photograph courtesy of Aimee Stout Benitez
	James L Kellerman photograph courtesy of Shari Mock
	Ephraim B Miller photograph courtesy of Terry Diehl
	John E O'Neal photograph courtesy of Deborah Stilley
	Oliver Horton photograph courtesy of Cyndi Young
	John B Hammer photograph courtesy of Jo Ann Linck
	Simon Feather photograph courtesy of Shari Mock
P. 291	Elizabeth & Emanuel O'Neal photograph courtesy of Ann Sinton; biographical information sourced on Ancestry family trees
	James & Louisa Bivens photograph courtesy of Snorris129 in Ancestry.com; biographical information sourced on Ancestry family trees
	May brothers photograph courtesy of Minerd family annual review
P. 292	O'Neal brothers photograph courtesy of Deborah Stilley

P. 292	Thomas A Prideaux photograph courtesy of Shari Mock
	Delilah & Isaac Ling photograph courtesy of Bedford Historical Society; biographical information sourced on Ancestry family trees
	Daniel J Price photograph courtesy of Susan Recchia
P. 293	Henry C Ritchey photograph courtesy of 138th Pennsylvania Infantry Volunteer – Facebook page
	Mary Ann & John C Ritchey photograph courtesy of Mary Lou Pease
	Margret & Aaron Mock photograph courtesy of William Roy Mock
	Marion Stotler photograph courtesy of Holly Badman
P. 294	Harvey E Shaffer photograph courtesy of Jma1965 in Ancestry.com
	Elias B Stuckey photograph courtesy of David Yunt
	Frederick A Sellers photograph courtesy of Bedford Historical Society
	John S Stuckey photograph courtesy of Ronn Palm
P. 295	Thomas Miller photograph courtesy of Bedford Historical Society
	David F Snyder photograph courtesy of Bedford Historical Society
	Lavine & Noah Tipton photograph courtesy of Collectornuts in Ancestry.com
	Wolford family photograph courtesy of Bedford Historical Society and Larry Yantz
P. 296	"149th Pennsylvania Regiment." *Pennsylvania Civil War Volunteers*, (n.d.), http://www.pacivilwar.com/regiment/149th.html
P. 297	Harrison Clingerman family photograph courtesy of Sally Weed
	David W Lee photograph courtesy of Walteroldham1 in Ancestry.com
	Sarah & Henry Bagley photograph courtesy of Bart Sugden
	149th Pennsylvania Regiment photograph courtesy of Dolores Myers & Library of Congress
P. 298	*History of Bedford, Somerset and Fulton Counties, Pennsylvania*. Chicago, Waterman, Watkins & Co. 1884, pp.160.
	"171st Pennsylvania Regiment." *Pennsylvania Civil War Volunteers*, (n.d.), http://www.pacivilwar.com/regiment/171st.html
P. 299	Robinette photograph courtesy of Jeff Whetstone
	Jacob Stufft photograph courtesy of Edward Reimer
	Jacob D Burket photograph courtesy of Margaret Bucklen
P. 300	Andrew R Emerick photograph courtesy of jamesirc in Ancestry.com
	Daniel Blattenberger photograph courtesy of OmaraBlatt in Ancestry.com
	Robert Callihan family photograph courtesy of Thomas Gregg
	Thomas Jay photograph courtesy of Michael S. Caldwell
P. 301	William Wilkinson photograph courtesy of Kaie17 in Ancestry.com
	John Mowry photograph courtesy of EauGallie1972 in Ancestry.com
	Alexander Corle photograph courtesy of Deb Colledge
	Sarah & Gaston Hann photograph courtesy of Daniel Pyle; biographical information sourced on Ancestry family trees
	Robert Nelson family photograph courtesy of Narmerding in Ancestry.com
	Mary Ann & Christian Miller photograph courtesy of 1_judyinva in Ancestry.com; biographical information sourced on Ancestry family trees
P. 302	*History of Bedford, Somerset and Fulton Counties, Pennsylvania*. Chicago, Waterman, Watkins & Co. 1884, pp.162-163.
	Clews, C. "1864-06-21 Jerusalem Plank Road." *Civil War Times*, (n.d.), http://www.civilwartroops.org/1864-06-21-jerusalem-plank-road/
	"The Battle of Jerusalem Plank Road: June 21-24, 1864." *The Siege of Petersburg Online*, (n.d.), http://www.beyondthecrater.com/resources/bat-sum/petersburg-siege-sum/second-offensive-summaries/the-battle-of-jerusalem-plank-road-summary/
	"The Battle of the Jerusalem Plank Road." *Petersburg Siege.org*, (n.d.), https://www.petersburgsiege.org/jplank.htm
P. 303	Jacob A Clevenger photograph courtesy of Asperjt in Ancestry.com
	Nathan C Evans photograph courtesy of *Bedford Gazette Obituary - 14 Jul. 1899*
P. 304	Markey family photograph courtesy of Nancy Long
	James Potter photograph courtesy of Leroy Higley
	John Lee photograph courtesy of Sammie Lee
	Matthias Imler photograph courtesy of mssewnso in Ancestry.com
P. 305	"194th Pennsylvania Regiment." *Pennsylvania Civil War Volunteers*, (n.d.), http://www.pacivilwar.com/regiment/194th.html
	Photograph of unidentified Soldiers marching in the capitol courtesy of Library of Congress
P. 306	Jacob A Eichelberger photograph courtesy of Bedford Historical Society
	Daniel Shock etching courtesy of the History of Blair County, Pennsylvania book, page 118.
	Llewellyn H Peck photograph courtesy of vivkifretz in Ancestry.com
	Jacob A Clevenger photograph courtesy of Asperjt in Ancestry.com
	William B Reed photograph courtesy of Bedford Historical Society
	Francis M Amos photograph courtesy of *Bedford Gazette - 23 Oct. 1908, pp.1*
	Camp Carroll drawing courtesy of wikimedia.org
P. 307	Alexander, Edward S. "Mapping the Attack on Fort Mahone.", *Emerging Civil War*, 13 Oct. 2017, https://emergingcivilwar.com/2017/10/13/mapping-the-attack-on-fort-mahone-april-2-1865/
	"205th Pennsylvania Regiment." *Pennsylvania Civil War Volunteers*, (n.d.), http://www.pacivilwar.com/regiment/205th.html
	Louis D Speice photograph courtesy of 4dandg in Ancestry.com
P. 308	Jacob D Ridenour photograph courtesy of Bedford Historical Society
	Alfred Baker photograph courtesy of Barbara Sponsler Miller; biographical information sourced on Ancestry family trees
	Samuel B Fluke photograph courtesy of Randy Gladstone
	Jonas C Imler photograph courtesy of Bedford Historical Society

P. 308	Clifford Mock and Isaac Walker photograph courtesy of William Roy Mock
	Shannon, Catherine & Joseph Smith photograph courtesy of Kim Bollinger
P. 309	"208th Pennsylvania Regiment." *Pennsylvania Civil War Volunteers*, (n.d.), http://www.pacivilwar.com/regiment/208th.html
	Alexander, Edward S. "Mapping the Attack on Fort Mahone.", *Emerging Civil War*, 13 Oct. 2017, https://emergingcivilwar.com/2017/10/13/mapping-the-attack-on-fort-mahone-april-2-1865/
	James H Sparks photograph courtesy of "Jaycet Pittman Collection" shown in the "From Winchester to Bloody Run" book
	Jacob Hammann photograph courtesy of Barbara Brown
P. 312	Willlam W Clark photograph courtesy of Bloody Run Historical Society & Barb Sponsler Miller
	Oliver C Ramsey photograph courtesy of Bedford Historical Society
P. 313	Isabell & Peter Foor photograph courtesy of Hshelley1958 in Ancestry.com
	Margaret & Isaiah Lehman photograph courtesy of Deborah Sasser
	Sarah & Job S Akers photograph courtesy of Bedford Historical Society
	Mary Anne & John R Oaks photograph courtesy of Jan Davis
P. 314	Harvey S Wishart photograph courtesy of Barb Sponsler Miller
	Robert Summerville photograph courtesy of Melboys33 in Ancestry.com
	Barton Mearkle photograph courtesy of Kevin Mearkle
	Jacob I Foor photograph courtesy of Rockhold in Ancestry.com
	John Himes photograph courtesy of Shaun Creighton
	William G Eicholtz photograph courtesy of MarthaW52 in Ancestry.com
	Joseph R Colledge photograph courtesy of Deb Colledge
	John Leonard family photograph courtesy of Jeff Whetstone
	Catherine & John Ritchey photograph courtesy of Marcialj in Ancestry.com
P. 315	Barney Bartow photograph courtesy of Sps16 in Ancestry.com
	Jacob T Weaverling photograph courtesy of Barb Sponsler Miller
	Mary & Joseph W Williams photograph courtesy of Harde5046 in Ancestry.com
	Alvah R & Susan Williams photograph courtesy of Dan Layton
P. 316	*History of Bedford, Somerset and Fulton Counties, Pennsylvania.* Chicago, Waterman, Watkins & Co. 1884. pp.168.
	"2nd Pennsylvania Cavalry." *Pennsylvania Civil War Volunteers*, (n.d.), http://www.pacivilwar.com/regiment/59th.html
	Hayes Irvine photograph courtesy of Dlwartell in Ancestry.com
	Francis M Elliott photograph courtesy of Matt Quinn
	Charles Smith photograph courtesy of *Bedford Gazette obituary -11 Mar. 1910*
P. 317	18th PA Cavalry photograph courtesy of the Library of Congress
	"18th Pennsylvania Cavalry." *Pennsylvania Civil War Volunteers*, (n.d.), http://www.pacivilwar.com/regiment/163rd.html
	Hawks, Steve A. "The Battle of Gettysburg.", *Stone Sentinels*, (n.d.), https://gettysburg.stonesentinels.com/union-monuments/pennsylvania/pennsylvania-cavalry/18th-pennsylvania-cavalry/
P. 318	John Nelson photograph courtesy of Stephanie Perry
	Ann & William Lowery photograph courtesy of Tony Klingensmith
	Benjamin F Oliver photograph courtesy of Harrisburg Telegraph Obituary 20 Jun. 1928
	John Nelson photograph courtesy of Stephanie Perry
P. 319	"21st Pennsylvania Cavalry." *Pennsylvania Civil War Volunteers*, (n.d.), http://www.pacivilwar.com/regiment/182nd.html
P. 320	Howard B Jeffries photograph courtesy of Madonna Jervis Wise
	George W Mosser photograph courtesy of D McNelis in Findagrave.com
	Levi Cook photograph courtesy of Brian J. Ensley
P. 321	Unidentified 21st PA Cavalry soldiers photograph courtesy of Yvonne Whetstone
	Cyrus W Fickes photograph courtesy of Jeff Hissong
	Joseph Penrose family photograph courtesy of Carole Carlson; biographical information sourced from Ancestry.com
P. 322	"Medal of Honor at The Star Fort." *22nd Pennsylvania Cavalry*, (n.d.), https://22ndpacavalry.com/3rd-winchester
	"James M. Schoonmaker." *Congressional Medal of Honor Society*, (n.d.), https://www.cmohs.org/recipients/james-m-schoonmaker
	"22nd Pennsylvania Cavalry." *Pennsylvania Civil War Volunteers*, (n.d.), http://www.pacivilwar.com/regiment/185th.html
	John C Hawman photograph courtesy of Ronn Palm
	William H Hanks photograph courtesy of Ronn Palm
P. 324	John P Chisholm photograph courtesy of Dennis Blubaugh
	Samuel R Leach photograph courtesy of Gillian K Leach
P. 325	Joseph Woy photograph courtesy of Coleary310 in Ancestry.com
	Cyrus E Blackburn photograph courtesy of Aflorimonte1 in Findagrave.com
	Oliver B Fluke photograph courtesy of Laura L Burgess
P. 326	Peter M Ensley photograph courtesy of Brian Ensley
	Lydia & George W Swope photograph courtesy of Rachael Hall and Mm3854 in Ancestry.com
	William B Filler photograph courtesy of Bedford Historical Society
	Harvey Grubb photograph courtesy of Kristina Covalt Aldred
	Sophia & Amon Long photograph courtesy of MichelleCapwellKloske in Ancestry.com
	Simon P Felton photograph courtesy of pesj551 in Ancestry.com
	Peter Baith photograph courtesy of Fred Wadley
	Steele family photograph courtesy of D57h1 in Ancestry.com
P. 327	"Proclamation by Governor Curtin of PA, 6/12/63, *House Divided*, Dickinson College, 21 Apr. 20113, http://hd.housedivided.dickinson.edu/node/40019

P. 327	Gayley, Alice J. "Militia Troops of 1863." *Pennsylvania in the Civil War*, 2015, https://www.pa-roots.com/pacw/1863militia/index.html
P. 328	Sansom Mearkle photograph courtesy of Rebecca in Ancestry.com
	John Q Adams photograph courtesy of Craig B. Adams
	William Melott photograph courtesy of Clayton Betts
	Francis Diehl photograph courtesy of Dv1t042 in Ancestry.com
P. 329	Harvey Wilkins photograph courtesy of jbrubaker in Findagrave.com
	Jennie & Henry Grove photograph courtesy of Rmcneal103 in Ancestry.com
	Brazella Foor photograph courtesy of Birgit Irmgard Condon
	William Drenning photograph courtesy of Ronn Palm
	John Q Nycum photograph courtesy of Margaret Bucklen and aubstrap in Ancestry.com
P. 330	Hawks, Steve A. "1st MD PHB Cavalry" *The Civil War in the East*, (n.d.), https://civilwarintheeast.com/us-regiments-batteries/maryland/1st-maryland-potomac-home-brigade-cavalry/
	The 1st Reg't Potomac Home Brigade, *The American Civil War*, USWars.com, http://www.americancivilwar101.com/units/usa-md/md-cav-reg-01-potomac.html
	Riley Leasure photograph courtesy of Beth Leasure-Hudson
	John W Lashley photograph courtesy of Denise Hedberg
P. 331	Hawks, Steve A. "2nd MD PHB Infantry." *The Civil War in the East*, (n.d.), https://civilwarintheeast.com/us-regiments-batteries/maryland/2nd-maryland-potomac-home-brigade/
	Lydia & Andrew Hillegass photograph courtesy of an undated Gazette Old Photo Album Photograph submitted by Randy Hillegass
	Benjamin Troutman family photograph courtesy of the Bedford Historical Society
P. 332	George J Gordon photograph courtesy of skm150 in Ancestry.com
	Virginia & Hector Robertson photograph courtesy of 81rsboyd in Ancestry.com
	Stacey Hoon photograph courtesy of the Bedford Historical Society
	Lewis Kipp photograph courtesy of Kenneth McClellan
	Henry Ruby and daughter photograph courtesy of Frederick Neff Wilson
	George Witt photograph courtesy of Karen Gaumer & Rezmer1149 in Ancestry.com
P. 333	Hawks, Steve A. "3rd MD PHB Infantry." *The Civil War in the East*, (n.d.), https://civilwarintheeast.com/us-regiments-batteries/maryland/3rd-maryland-potomac-home-brigade/
	Samuel Hendershot POW document courtesy of rhendershot119 in Ancestry.com
P. 334	John F Schetrompf photograph courtesy of Barbara Sponsler Miller
	Peter C Schetrompf photograph courtesy of Barbara Sponsler Miller
	Joseph B Smith photograph courtesy of Debbie Crousern
	Samuel Leighty photograph courtesy of an undated Gazette Old Photo Album Photograph
	Samuel Wagerman photograph courtesy of Aimee Stout Benitez
	Jacob T Weaverling photograph courtesy of Edward Schoenberger
P. 335	"Black Soldiers in the Civil War." *National Archives*, 19 Mar. 2019, https://www.archives.gov/education/lessons/blacks-civil-war/article.html
	"United States Colored Troops.", *Wikipedia*, (n.d.), https://en.wikipedia.org/wiki/United_States_Colored_Troops
P. 336	R DeCharmes Davis photograph courtesy of an undated Gazette Old Photo Album Photograph sourced at Bedford Historical Society
	Josiah Slick photograph courtesy of David Johnston
P. 337	John T Crist and Catherine Fetters photograph courtesy *Bedford Gazette Old Photo Album (undated) at Bedford Historical Society*; biographical information sourced from Ancestry.com
	Josiah C Smith photograph courtesy of Steven Showers; biographical information sourced from Ancestry.com
P. 338	James F Dicken photograph courtesy of Rose in Findagrave.com
	John Bloom photograph courtesy of Jen in Findagrave.com
	Edward J Miller photograph courtesy of Everett Republican Obituary - 6 Oct. 1911
	James P Covert photograph courtesy of Tom Jones
	Reuben Hardinger photograph courtesy of Karen Gaumer & Rezmer1149 in Ancestry.com
	George Witt photograph courtesy of Bssatksc in Ancestry.com
	Reuben A Riley photograph courtesy of saylesworth1 & NJMcM in Ancestry.com
	William M Garlinger photograph courtesy of David Garlinger
	Aaron Hill photograph courtesy of Brenda Johnson; biographical information sourced from Ancestry.com
	John Gillam photograph courtesy of Christine Gillam Austin
	Joseph P Cessna photograph courtesy of the Grand Rapids Library Collection
	George & Louvisa Keith photograph courtesy of Ccruff41 in Ancestry.com
P. 339	Andrew Baker photograph courtesy of Paula Mcdevitt
	John Welch photograph courtesy of Lynnwont in Ancestry.com
	Jacob H Brown photograph courtesy of Laura Belle Stoddard
	Levi Diehl photograph courtesy of Bedford Historical Society
	Martha & Ephraim Chilcott photograph courtesy of Mary Lou Rogers
	Catherine & Abraham Burket photograph courtesy of Bedford Historical Society
	Samuel Livingston photograph courtesy of Barbara Sponsler Miller
	John J Barclay photograph courtesy of field_iii in Findagrave.com
	Abraham K Arnold photograph courtesy of v8m8i in Ancestry.com
	John H Karns photograph courtesy of Charles Clark

P. 340	William Deremer photograph courtesy of Susan Clausen
	Henry Hardinger photograph courtesy of Davewith in Ancestry.com
	Abraham Barnhart photograph courtesy of ggramly in Ancestry.com
	D Alex McKellip photograph courtesy of Michael N Meyers
	Mary & Simon S Blake photograph courtesy of Vava59 in Ancestry.com
	John A Steckman photograph courtesy of Emma Simmons
	Joseph Nevitt photograph courtesy of vkentw in Ancestry.com
	James B Coughenour family photograph courtesy of criss5 in Ancestry.com
	Fletcher brothers photograph courtesy of Barbara Jones Simmons
	Alexander Smith photograph courtesy of Rich Allison & Bill Cox
P. 341	John O Myers photograph courtesy of Toldot386 in Ancestry.com
	John Milton Davis photograph courtesy of Bob Way
	Joseph Y Weyandt photograph courtesy of David Gay
	Hiram Evans photograph courtesy of 3336bob in Ancestry.com
	Jacob Bridenstine photograph courtesy of Dlwhitmarsh in Ancestry.com
	Isaac Waters photograph courtesy of Marleneinlv in Ancestry.com
	Jacob Hite photograph courtesy of Kandrevans in Ancestry.com
	Nicholas Hite photograph courtesy of Kandrevans in Ancestry.com
	Reuben Barley photograph courtesy of Amanda Smith
	Mark Miller photograph courtesy of Richard Barnes
	Solomon Comp photograph courtesy of Rich & Sandy Allison
	Stiles Jackson photograph courtesy of Lmn0352 in Ancestry.com
	Jesse W Horton photograph courtesy of Jack R. Box
	Samuel Salkeld photograph courtesy of Bonnie Beglin
	David A Hunt photograph courtesy of Todd Williams
	Levi Holsinger photograph courtesy of DJGodden in Ancestry.com
P. 342	Joseph Wareham photograph courtesy of Csm1963 in Ancestry.com
	Jonathan Wiser photograph courtesy of Sprowlseeker in Ancestry.com
	David R Potter photograph courtesy of David Hensarling
	Brice & Mary Twigg courtesy of Jetjr1970 & Skyblu222us in Ancestry.com
P. 343	Louisa & Levi Valentine courtesy of Scott Bumbaugh
	Sarah & Joseph Growden courtesy of Trenda Roch von Rochsburg
	Tobias Boor courtesy of Jeff Whetstone
	Jeremiah Weicht family courtesy of Jimmie Turner; biographical information sourced from Ancestry.com
P. 344	Hanson Cook photograph courtesy of Bud Evans
	Joseph Lybarger photograph courtesy of Linda Marker
	John H Akers photograph courtesy of Connie Chrisman Hatch
	William H Kelly photograph courtesy of Bud Evans
	Samuel X Smith photograph courtesy of Llmaxey1 in Ancestry.com
	Paul McLeary photograph courtesy of Ms7987 & dmsmith130 in Ancestry.com
	Henry Mearkle photograph courtesy of Gillian K Leach
	Lemuel Evans photograph courtesy of Xflygrl in Ancestry.com
	Henry College photograph courtesy of Deb College
	James A Cook photograph courtesy of Clark Brocht
P. 345	David R Smith photograph courtesy of Bedford Historical Society of *Bedford Gazette Old Picture Album (undated)*
	John McMullen photograph courtesy of AliciaMay606 in Ancestry.com
	Samuel Stuckey photograph courtesy of Jeff Whetstone
	Wilson Karns photograph courtesy of Kim Karns Bowser
	Andrew McFarland photograph courtesy of Karla Banninger
	David H Fisher photograph courtesy of Bedford Historical Society undated Gazette Old Picture Album picture
	Caleb Brown photograph courtesy of Jil in Findagrave.com
	John R Layton photograph courtesy of Anne Patricia Brown
	Jonathan Dibert photograph courtesy of Aimee Stout Benitez
P. 346	Ezra Osborn photograph courtesy of Carole Steele
	John Wonder photograph courtesy of Ms7987 in Ancestry.com
	Benson Hanks photograph courtesy of Glen Hanks & TRADITIONS71 in Ancestry.com
	David Sparks photograph courtesy of Amy Howell
	Solomon Sparks photograph courtesy of Jack Sparks
	George Clevenger photograph courtesy of asperjtc in Findagrave.com
	George Fockler photograph courtesy of Don Fockler
	Jacob Rightenour photograph courtesy of Ksm185 in Ancestry.com
	Francis P McCoy photograph courtesy of Gfk111 in Ancestry.com
	Henry Hartman photograph courtesy of Troy Hartman
	Jemima & George Truax photograph courtesy of Krysten Betts
P. 347	Jacob Gettleman photograph courtesy of Cnwkoch in Ancestry.com

P. 347	William Watson photograph courtesy of Letters of a Civil War Surgeon by Paul Fatout
	William Stultz photograph courtesy of Bedford Historical Society
	Israel Grace photograph courtesy of Joseph Lankard
	William McKibbin photograph courtesy of Socialworkersteph in Ancestry.com
	Benjamin Blymyer photograph courtesy of Don Shires
	John Fletcher photograph courtesy of Bssatksc in Ancestory.com
	Festus Edwards photograph courtesy of Dennis L Farley
	John A Miller photograph courtesy of Vtrimble04 in Ancestry.com
	Isaac Boose photograph courtesy of Brenda Tyree Holder
	Merrick A Stoner photograph courtesy of Gwencott & WVaGal in Ancestry.com
P. 348	Samuel Morningstar photograph courtesy of Shelly Foster
	Emanuel Holsinger photograph courtesy of Gail Davis
	George Hunt photograph courtesy of GHASE1 in Ancestry.com
	Francis Reamer photograph courtesy of History of Bedford, Somerset, and Fulton Counties book
	Christopher Knisely photo courtesy of A History of the Church of the Brethren in the Middle District of Pennsylvania: 1781-1925
	George Edwards photograph courtesy of Jeandpetty in Ancestry.com
	John N Salkeld photograph courtesy of Bonnie Beglin
	Abraham Sparks photograph courtesy of Historical Data Systems
	Ellen & Jesse McVicker photograph courtesy of John Carolus; biographical information sourced from Ancestry.com
P. 349	Sarah & Samuel Murphy photograph courtesy of Mrobison161 in Ancestry.com; biographical information sourced from Ancestry.com
	Margaret & Christian R Oakes photograph courtesy of Sam Dutrow; biographical information sourced from Ancestry.com
	Margaret & Josiah Lingenfelter photograph courtesy of Heather Alexander; biographical information sourced from Ancestry.com
P. 350	Abraham Plessinger photograph courtesy of Hope Creighton
	Dennis B Sipes photograph courtesy of Laloba27 in Ancestry.com
	Jacob F Salkeld photograph courtesy of Dnblubaugh in Ancestry.com
	George G Gordon photograph courtesy of Darlene Boggs
	Bartley Hughes photograph courtesy of Barbara Sponsler Miller
	William H Latta photograph courtesy of Barbara Sponsler Miller
	John E Sparks photograph courtesy of Dawn Kelley
	Thomas & Benjamin F Lohr photograph courtesy of Bedford Historical Society
	Charles Ezekiel McMullen photograph courtesy of Abshaw1 in Ancestry.com
	Charles Jackson photograph courtesy of Nicole Bouas
P. 351	Wilson S Smith photograph courtesy of Paul & Deborah Oesch
	Jane & James Barton photograph courtesy of Joanne Fesler; biographical information sourced from Ancestry.com
	Wilson S Smith photograph courtesy of Dbkso in Ancestry.com
P. 352	Himes and Thomas photographs courtesy of the late Mildred Himes Lawson, Hope Creighton & Shaun Creighton & Ancestry.com
	Bechtel brothers photograph courtesy of Judith Hinkle
	John Fleegle photograph courtesy of Bedford Historical Society
	John Foster photograph courtesy of Jim Childers
	Adam K Bottenfield photograph courtesy of ERVetTeck2 in Ancestry.com; biographical information sourced from Ancestry.com
	Jeremiah Ramsey photograph courtesy of Birdwatcher20 in Ancestry.com
	Henry Mock photograph courtesy of Todd Ligas
P. 353	Matthew Kennard photograph courtesy of Emily Kennard
	James Beaston photograph courtesy of Michael Beaston
	George W Eshelman photograph courtesy of Ronn Palm
	Elizabeth and James F Pennington photograph courtesy of Dorcus Cromwell
P. 354	Samuel Z Beam photograph courtesy of Kimberly Maher
	Sarah & Levi Clites photograph courtesy of Tracy Venice Avins
	Solomon J Hankinson photograph courtesy of Sherry lynn Rickard
	Susan & Francis T Troutman photograph courtesy of Dawncripham in Ancestry.com
P. 355	John Holler photograph courtesy of Aimee Stout Benitez
	DeCharmes Davis and Veterans photograph courtesy of Bedford Historical Society & *Bedford Gazette (undated)*
P. 356	Pavia Veterans photograph courtesy of Bedford Historical Society
P. 357	Saxton Parade photograph courtesy of heysaxton.com
	Saxton Veterans photograph courtesy of Bedford Historical Society
P. 358	Pleasantville Veterans photograph courtesy of Bedford Historical Society & Hubert Whysong featured in the 2014 Spring Pioneer
P. 359	Courthouse Veterans photograph courtesy of Bedford Historical Society & *Bedford Gazette - 11 Jun. 1984*
	Tatesville Veterans photograph courtesy of Barbara Sponsler Miller
P. 360	Hopewell Veterans photograph courtesy of Bedford Historical Society & *Bedford Gazette (undated). Photo submitted by Cindy Diehl*
	Hopewell Veterans photograph courtesy of Barbara Sponsler Miller
P. 361	Travelers Rest photograph courtesy of Barbara Sponsler Miller from a Gazette Old Picture Album picture submitted by Herb Gump
	Veterans photograph courtesy of Bedford Historical Society in an undated Gazette Old Picture Album
P. 362	Four Veterans photograph courtesy of Barbara Sponsler Miller
	Veterans photo courtesy of Bedford Historical Society & *Bedford Gazette Old Photo Album (undated) submitted by Chet Musselman*
P. 363	Baughman & Pee family photographs courtesy of Niels Witkamp; Biographical information sourced from Ancestry.com

www.ingramcontent.com/pod-product-compliance
Lightning Source LLC
Chambersburg PA
CBHW080633170426
43209CB00008B/1559